Sustainable Use of Feed Additives in Livestock

Georgios Arsenos • Ilias Giannenas
Editors

Sustainable Use of Feed Additives in Livestock

Novel Ways for Animal Production

Springer

Editors
Georgios Arsenos
School of Veterinary Medicine
Aristotle University of Thessaloniki
Thessaloniki, Greece

Ilias Giannenas
School of Veterinary Medicine
Aristotle University of Thessaloniki
Thessaloniki, Greece

ISBN 978-3-031-42854-8 ISBN 978-3-031-42855-5 (eBook)
https://doi.org/10.1007/978-3-031-42855-5

This Springer imprint is published by the registered company Springer Nature Switzerland AG
The registered company address is: Gewerbestrasse 11, 6330 Cham, Switzerland

Paper in this product is recyclable.

Prof Georgios Arsenos
To my family

Dr Ilias Giannenas
I would like to express my gratitude to my cherished wife Alcmene for her support and guidance in every step of my academic career and her participation in exploratory visits in numerous countries or places in Greece and abroad. Also, I dedicate this book to my beloved children, Olga and Marios-Alexandros.

Preface

Feed additives are products, such as microorganisms or non-nutritive substances, which are added to animal diets to improve the quality of feed and animal products. Feed additives are expected to result directly or indirectly in feed digestion and nutrient utilization, as well as animal performance and health. Feed additives is a vast subject. Therefore, an effort was made to limit the topics presented in this book to those principles and concepts that are pertinent to the practice of both ruminant and monogastric animals nutrition. The idea was that a new book was needed to cover all aspects of feed additives considering up-to-date information of novel products and their applicable options in practice. Hence, this book was created.

This book provides a collection of the most recent knowledge regarding natural feed additives, both in vitro and in vivo. It can be used by researchers, nutrition specialists, animal science and veterinary students, livestock consultants and all other experts and stakeholders in different livestock sectors. The topics covered in this book span over 33 chapters delivered by experts in the fields of animal and poultry nutrition as well as animal and veterinary science. All contributors take a comprehensive approach to show the principles governing production and uses of both natural and industrialized feed additives focusing on animal performance and health issues. All major monogastric and ruminant animal species are covered in the book. Some chapters were created specifically to deal with aqua culture, rabbit and horse nutrition. There are also chapters covering insect production and its role in animal nutrition either as feed additive or as novel feedstuff. Other chapters highlight the value of herbal plants, their extracts and derivatives, cold-pressed and essential oils, by-products of fruits, immunomodulators, organic acids, probiotics and nanoparticles that are used as alternatives to antibiotic growth promoters in animal nutrition. Across chapters, the unique characteristics of feed additives and their use towards environmental enrichment and sustainability are explored.

As editors of this book, we are very grateful to all authors who contributed their time and expert knowledge towards the common goal of delivering a valuable book. We are also indebted to all those experts who reviewed chapters and offered guidance and constructive advice; Thank you.

Thessaloniki, Greece

Georgios Arsenos
Ilias Giannenas

Contents

About the Editors

Georgios Arsenos has over 25 years of experience in Animal Science research and education. He is an academic with well-established international links in research and extension services to industry. His scientific expertise includes veterinary medicine, animal breeding and genetics, data analysis and sustainability of livestock production systems. He held research and academic positions in the UK and has been leader or partner in several national, industry and EU-funded research projects (104 research projects; coordinator in one EU funded: www.isage.eu). Highlights of service outside the university include numerous seminars for farmers, regular farm visits to individual farms, farm cooperatives and dairies that have a contract for extension services with the Laboratory of Animal Husbandry. He is author of one book and co-author in 2 books and several technical papers as well as more than 136 papers in refereed journals (1745 citations, h-index: 23). He participated in numerous national and international conferences either as invited speaker or as lecturer with over 300 presentations. He acts as reviewer in several scientific journals.

Ilias Giannenas DVM, PhD, Dip ECPVS, Director of Laboratory of Nutrition, is a veterinarian and Associate Professor at the School of Veterinary Medicine, Faculty of Health Sciences, Aristotle University of Thessaloniki. Formerly he served as Lecturer in the Veterinary Faculty of University of Thessaly. He did sabbatical studies in Newcastle University to investigate the interaction between dietary fibre and coccidiosis in broiler chickens. He is particularly interested in issues of applied poultry nutrition, the use of aromatic plants, plant extracts and essential oils as alternative growth promoters or natural anticoccidials. He is co-author of 3 books, 12 book chapters and over 100 papers in refereed journals. He regularly serves as invited speaker in national or international congresses and presented more than 100 conference papers. He has been scientifically responsible or member of the research group in more than 30 projects and established collaboration with the feed and feed additive industry. He served as reviewer and editorial board member for several journals and evaluated national and international projects. His work has more than 3000 citations and an h-index of 33.

Part I

Introductory Chapters and Historical Abstract

Feed Additives to Ensure Sustainability

Georgios Arsenos and Sotiria Vouraki

Abstract

Farm animals (and animals in general) require many nutrients, in various amounts and proportions, to fulfil their needs for maintenance, growth, and reproduction. The main objective of this chapter is to discuss how feed additives contribute to balancing the diet of farm animals and hence ensure sustainability in livestock production systems. The latter is crucial for the future of such systems and hence balanced nutrition is a key component of success. The notion is that the dietary choices and food intake of animals are the outcome of the animal's internal state and knowledge of its feeding environment. Balanced feed formulation including feed additives provides farm animals with necessary nutrients while minimizing nutrient excretion and greenhouse gas emissions. The underlying hypothesis of using feed additives in animal nutrition is that they can improve animal efficiency and profitability as well as environmental sustainability. Feed additives can support animal health and particularly in ruminants can also contribute to the reduction of enteric methane production. There is a plethora of experimental studies dealing with many components that are considered feed additives. However, the role of feed additives in the sustainability of livestock systems is a question that merits further investigation. The issue of sustainability is certainly much more complicated than efficiency and profitability in livestock production system that is related with the nutritional content of food alone. A large body of experimental evidence generally acknowledges that feed additives involve a broad category of substances that can affect to a large extent animal productivity as well as animal health and welfare, and hence food security and consumer safety.

G. Arsenos (✉) · S. Vouraki
Laboratory of Animal Husbandry, Department of Veterinary Medicine, School of Health Sciences, Aristotle University, Thessaloniki, Greece
e-mail: arsenosg@vet.auth.gr

G. Arsenos, I. Giannenas (eds.), *Sustainable Use of Feed Additives in Livestock*,
https://doi.org/10.1007/978-3-031-42855-5_1

Keywords

Farm animals · Feed additives · Diet · Sustainability

1 Introduction

The use of feed additives in animal diets spans thousands of years. The addition of different substances to animal feeds for technological or organoleptic purposes is a practice traced back over millennia. The use of additives in animal feeds increased considerably during the end of the last century because of intense research and increased knowledge regarding their technological and organoleptic properties. The notion is that feed additives are substances that affect or are expected to affect directly or indirectly the characteristics of feeds used in animal nutrition. Farm animals (and animals in general) require many nutrients, in various amounts and proportions, to fulfil their needs for maintenance, growth, and reproduction. Of the many materials, which serve as food for animals, few can supply all the required nutrients. Most consist of several, often numerous, different chemical compounds, the relative levels of which can vary substantially even among different items of the same food type. Similarly, many materials are toxic or harmful if eaten, or eaten in large quantities. If the animal is to meet its nutrient requirements, and avoid toxicity, it must make a choice or choices from the available foods. The nutritional choices available to farm animals relate to one of the following two questions: (i) how much to eat, and (ii) what to eat over a particular time period. The first question applies mainly to the simple feeding situation where animals are given free and continuous access to a single homogeneous food. In this system the only way available to the animal to regulate its diet is to eat more or less of the food. The second question applies to situations where animals are given access to two or more foods as a choice or in extensive systems. Irrespective of whether animals have a limited degree of choice or unlimited choices, their dietary choice is then a description of what an animal does when it encounters food (Kyriazakis 1997; Arsenos 2001).

The difficulties in describing adequately the feeding requirements of animals have led animal feed industries to direct efforts toward additives with intrinsic sustainability properties. The first hypothesis stems from the idea of "responsible consumption and production" set forward by the United Nations. It implies that the feed industry should seek additives making the most out of available feed sources while improving feed conversion, maintaining feed quality, and using alternative ingredients. A common paradigm is that ingredients used as additives are those preserving the environment by reducing ammonia and methane emissions and fulfilling energy requirements, as well as those reducing antibiotics usage to maintain their efficacy for future generations. A second hypothesis assumes that feed additives optimize food conversion ratio of available feeds consumed and hence are a sustainable option considering the outputs produced, such as growth, milk, meat, or eggs. The notion is that additives that improve feed conversion can directly save resources subject to gut health of animals in question but also reduce ammonia production by improved protein digestion in animals. Hence, additives that improve feed conversion also indirectly improve gut health. Furthermore, another idea is that feed additives save resources by preventing downgrading of feed quality and feed spoilage (Johnson and Johnson 1995; Pearlin et al. 2020; Ku-Vera et al. 2020).

The idea behind this chapter is that proper user of feed additives in animal nutrition can contribute to sustainability by enhancing animal's ability to balance its diet. While large amount of research effort over many years has been aimed at identifying the nutrient requirements of farm animals and investigating whether they are able to select a diet that meets their requirements, there has been less effort put into understanding how farm animals reach decisions related to their feeding behavior both in terms of food intake and diet selection. Studies using various feed additives in different production systems and animal species concentrate on describing the outcomes in terms of performance, efficiency, and health.

2 Feed Additives to Increase Environmental Sustainability

The agricultural sector is currently considered responsible for one-third of global greenhouse gas emissions (GHGs). Specifically, the sector accounts for 82% of nitrous oxide, 44% of methane, and 13% of carbon dioxide produced from anthropogenic activities (IPCC 2019). Among agricultural GHG emissions, ca. 15% is attributed to livestock farming, 44% accounts for enteric methane, and 29% for nitrous oxide and ammonia (Ahmed et al. 2020). Moreover, the livestock sector contributes significantly to water pollution through animal waste, especially nitrogen and phosphorus excretions (Wang et al. 2020).

European Commission is working toward tackling the above environment-related challenges. According to the European Green Deal, all actors across the food chain, including livestock farmers, will need to adjust to new standards under the "Farm to Fork" strategy with special emphasis on reducing GHG emissions and improving water quality (European Commission 2019). Toward this end, measures related to animal nutritional management are warranted.

Among nutritional strategies, feed additives have been widely studied during the last decade and considered a viable solution for improving farm environmental performance. Certain feed additives and their compounds (Scheme 1) can improve digestion and modulate microbial populations, thus reducing methane and ammonia emissions as well as phosphorus and nitrogen excretions (Lewis et al. 2015). This section of the present chapter discusses the state-of-the-art knowledge on the use of feed additives in animal nutrition toward improved environmental sustainability.

2.1 Plant Extracts

2.1.1 Plant Oils: Lipid Supplementation

In ruminants, plant oils have been reported to effectively mitigate methane emissions given that lipids reduce the activity of rumen methanogenic and protozoan

Methane emissions	Ammonia emissions	Nitrogen and phosphorus excretions
• Plant oils • Essential oils • Plant secondary metabolites • Algae • Probiotics • Ionophores • Organic acids • Exogenous enzymes • Synthetic anti-methanogenic compounds • Biochar	• Plant oils • Plant secondary metabolites • Organic acids • Bentonite • Biochar	• Essential oils • Ionophores • Exogenous enzymes

Scheme 1 Feed additives with the potential to mitigate methane and ammonia emissions, and nitrogen and phosphorus excretions from livestock

populations (McCauley et al. 2020; Prathap et al. 2021). In Swiss Brown Hill cattle, diet supplementation with coconut oil at 3.5% and 7% of dry matter reduced methane emissions by 28% and 73%, respectively (Machmüller and Kreuzer 1999). In Brahman cattle, the use of cottonseeds, sunflower seeds, and coconut kernels resulted in methane reduction by up to 50% (Chuntrakort et al. 2014). Additionally, methane mitigation by 18% was reported in Hereford × Friesian cattle that grazed on pasture sprayed with canola oil (Pinares-Patiño et al. 2016). Similar effects of lipid supplementation on reduction of methane emissions have been also indicated in sheep. Machmüller et al. (2000) investigated the inclusion of coconut oil, sunflower seed, rapeseed, and linseed in the diet of Swiss White Hill lambs that resulted in methane reduction by 26%, 27%, 19%, and 10%, respectively. Moreover, Mao et al. (2010) found that Huzhou lambs supplemented with soybean oil at 3% of dry matter produced ca. 14% less methane compared to those fed a control diet. It should be noted that in the latter study, soybean oil supplementation also significantly reduced ammonia emissions.

2.1.2 Essential Oils

Essential oils and their compounds (isoprenes, terpenes, diterpenes, etc.) can inhibit rumen protozoa and bacteria and the deamination of amino acids (McCauley et al. 2020; Prathap et al. 2021). These properties could help toward reducing GHG emissions, especially of methane and ammonia. Several studies have investigated the potential of essential oils to reduce such emissions from ruminants and promising results have been obtained.

In dairy cows, supplementation with a commercial blend of coriander seed oil, eugenol, and geranyl acetate (Agolin Ruminant; 1 g/head/day) resulted in significantly less methane and ammonia intensity (Hart et al. 2019; Carrazco et al. 2020). Moreover, in the in vitro study of Rossi et al. (2022), total gas and methane emissions were significantly reduced in dairy cows supplemented with a blend of the aforementioned essential oils combined with bioflavonoids and tannins. In Hu sheep, Wu et al. (2018) suggested that intermittent supplementation with citrus essential oils could help mitigate methane emissions by reducing microbial adaptation. Additionally, in the study of Soltan et al. (2018), supplementation of Santa Inês sheep with a microencapsulated blend of cinnamaldehyde, carvacrol, capsicum oleoresin, and eugenol (400 mg/kg of dry matter) reduced methane emissions and nitrogen excretion by ca. 28% and 10%, respectively; however, no significant effects on ammonia emissions were observed. Lewis et al. (2015) also reported the potential of thymol and eucalyptus oils to mitigate both methane and ammonia emissions in sheep (Lewis et al. 2015). Finally, controversial results have been obtained regarding oregano and garlic oil supplementation in ruminants with some studies (reviewed by Honan et al. 2021) reporting reduced methane emissions and others observing no significant differences compared to control diets.

2.2 Plant Secondary Metabolites

Plant secondary metabolites, especially tannins and saponins, modulate rumen microbial populations and can effectively mitigate GHG emissions. In beef cattle, condensed tannins from *Acacia mimosa* extract were found to significantly reduce methane emissions (Fagundes et al. 2020). Moreover, in dairy cattle, a blend of tannins with essential oils and bioflavonoids reduced methane emissions and overall GHG emissions (Rossi et al. 2022). In Mandya sheep, Baruah et al. (2019) reported lower methane production following diet supplementation with tanniferous leaves (*Syzygium cumini* and *Machilus bombycina*). Likewise, Santa Inês sheep supplemented with tanniferous legumes (*Leucaena leucocephala*) had decreased methane emissions (Moreira et al. 2013). Moreover, in the study of Liu et al. (2011), Rideau Arcott sheep supplemented with chestnut tannins (30 g/kg of dry matter) produced less methane and ammonia (ca. 27% in both cases) compared to those fed a conventional diet. Furthermore, tannins from *Castanea sativa* wood extract at a level of 2% in lambs' diet reduced ammonia emissions by 56% (Śliwiński et al. 2002).

The potential of saponins to mitigate GHG emissions has been also widely studied. Saponins from *Pennisetum purpureum* at a dose of 0.30–0.45 kg/day reduced methane emissions from Pelibuey lambs (Albores-Moreno et al. 2017). *Yucca schidigera* extracts, which are also rich in saponins, have been found to mitigate methane and ammonia production when studied in vitro (Narvaez et al. 2013; Guyader et al. 2017). Similar results were obtained from some in vivo studies involving beef cattle (Hristov et al. 1999) and lambs (Śliwiński et al. 2002). However, other similar literature in ruminants reported no significant effects (Zúñiga-Serrano et al. 2022). Finally, *Yucca schidigera* extracts are considered a successful strategy for mitigating ammonia emissions in aquaculture ponds. Specifically, ammonia concentration in the water was reduced in studies with Nile tilapia, striped catfish, mirror carp, and seabass juveniles (Paray et al. 2021).

2.3 Algae

Macroalgae contain compounds such as haloforms and dihalomethanes that have anti-methanogenic properties (Prathap et al. 2021; Ahmed et al. 2022). Although the effect of several macroalgae species on methane mitigation has been investigated in ruminants, the most promising results have been obtained for *Asparagopsis* spp. (McCauley et al. 2020). In dairy cows, inclusion of *Asparagopsis armata* at a 0.5% level in the total mixed ratio reduced methane production, yield, and intensity by ca. 26%, 20%, and 27%, respectively. At an inclusion level of 1%, the latter methane metrics were reduced by ca. 67%, 42%, and 60%, respectively (Roque et al. 2019). In Merino-cross sheep, methane production was reduced following diet supplementation with *Asparagopsis taxiformis* at 0.5%, 1%, 2%, and 3% levels; the highest reduction (ca. 80%) was observed at the 3% inclusion level (Li et al. 2016). Favorable effects of *Asparagopsis taxiformis* on methane mitigation have also been reported in vitro at an inclusion level $>2\%$ of organic dry matter (Kinley et al. 2016). Finally, several microalgae species have been investigated regarding their potential to reduce methane emissions in ruminants; however, results were less promising compared to macroalgae (McCauley et al. 2020).

2.4 Probiotics

Several studies have suggested the potential of probiotics to mitigate methane emissions from ruminants. This is attributed to their ability to influence microbial activity and stabilize rumen pH (McCauley et al. 2020). Specifically, methane production was found to be significantly reduced (ca. 6%) in rams and lambs supplemented with *Bacillus licheniformis* and *Ruminococcus flavefaciens*, respectively (Deng et al. 2018; Hassan et al. 2020). Moreover, in Simmental cattle, *Saccharomyces cerevisiae* and gambier leaves waste supplementation decreased methane production by 57% (Ningrat et al. 2020). The latter probiotic and its combination with *Bacillus amyloliquefaciens* have also been reported to mitigate methane emissions in Bali cattle due to competition with methanogenic bacteria in the rumen (Prathap et al. 2021). Likewise, Latham et al. (2019) reported reduced methane production from Holstein steers supplemented with *Paenibacillus* 79R4. Moreover, an in vitro study of rumen fluid in dairy cattle suggested the potential of propionic acid bacteria, especially *Propionibacterium thoenii*, to reduce methane production by ca. 20% (Chen et al. 2020). Although relevant research is limited, probiotics have also been suggested to help reduce ammonia emissions in swine (Wang et al. 2020).

2.5 Ionophores

Ionophores promote the growth of Gram-positive bacteria in the rumen, which increases propionic acid compared to acetic and butyric acid. This change in the ratio of fermentation acids inhibits methane production (McCauley et al. 2020; Honan et al. 2021). Specifically, in Holstein cows (Perna Junior et al. 2020) and Angus steers (Guan et al. 2006), diet supplementation with monensin resulted in ca. 26% and 30% reduction in methane production, respectively. Similar favorable effects of monensin on methane mitigation have been reported in dairy goats (Li et al. 2018) and buffaloes (Gupta et al. 2019). Besides methane mitigation, inclusion of monensin in the diet of ruminants has been associated with lower ammonia, carbon dioxide, and nitrogen excretions (Lewis et al. 2015). Reduced ammonia emissions have also been reported following supplementation with salinomycin (Lewis et al. 2015).

2.6 Organic Acids

In ruminants, organic acids have been found to inhibit methanogenesis by promoting propionate production. Supplementation of malic acid (7.5% of dry matter) in Charolais cattle (Foley et al. 2009) and allicin (2 g/head/day) in Dorper × thin-tailed Han crossbred ewes (Ma et al. 2016) reduced methane emissions by 16% and 8%, respectively. In Holstein steers, supplementation with fumaric acid (2% of dry matter) resulted in reduced methane production by 23% (Bayaru et al. 2001). Moreover, in Welsh Mule cross lambs, methane production was reduced by 75% following the use of encapsulated fumaric acid (Wallace et al. 2006).

In monogastrics, organic acids could help to mitigate ammonia emissions and phosphorus excretions. Specifically, in pigs, supplementation with adipic, benzoic, and fumaric acids resulted in ca. 25%, 23%, and 12% reduction in ammonia production, respectively (Lewis et al. 2015). Moreover, inclusion of citric acid in the diet of poultry reduced phosphorus excretion by ca. 16% (Lewis et al. 2015).

2.7 Exogenous Enzymes

In ruminants, exogenous enzymes are known to improve fiber digestibility, which could help toward reducing methane production (Beauchemin et al. 2008; Prathap et al. 2021). However, studies regarding the association of exogenous enzymes with methane mitigation are limited. In vitro evidence suggests the possibility of reducing methane production by including cellulase and xylanase in ruminant nutrition (Zhao et al. 2015). Moreover, Arriola et al. (2011) reported that Holstein cows fed low- and high-concentrate diets supplemented with fibrolytic enzymes produced less methane compared to the ones with the respective conventional diets. However, in the study of Oh et al. (2019), where dairy cows were supplemented with a commercial enzyme extract from *Aspergillus oryzae* and *Aspergillus niger*, no significant effects on methane production were reported. Likewise, addition of xylanase or cellulase in the diets of goats did not affect rumen fermentation, digestibility, and methane emissions (Lu et al. 2015). Based on the above, further research is needed to evaluate the suitability of exogenous enzymes to mitigate methane emissions in ruminants (Prathap et al. 2021).

In swine, dietary inclusion of phytase and protease has been associated with reduced environmental pollution from animal waste (Lei et al. 2013; Wang et al. 2020). Phytase improves digestion of phytate and, hence, reduces phosphorus excretion. Specifically, in the study of Lei et al. (2013) a decrease in fecal phosphorus up to 50% was reported in pigs supplemented with phytase. Moreover, a global meta-analysis in swine showed that protease could help minimize nitrogen excretion by ca. 18% (Wang et al. 2020). Although relevant literature in poultry and fish is scarce, similar benefits could be expected by phytase and protease supplementation.

2.8 Synthetic Anti-Methanogenic Compounds

Although the use of chemical anti-methanogenic compounds is prohibited in some countries due to anti-nutritional effects, it has been reported to drastically reduce methane emissions from ruminants. Among the most widely studied synthetic anti-methanogenic compounds is bromochloromethane, which has been found to reduce methane production by ca. 95% and 90% in steers and rams, respectively (Matsui et al. 2020; Prathap et al. 2021). Another such synthetic compound is 3-nitrooxypropanol (3-NOP). The inclusion of 3-NOP in the diet of Angus cattle at 4.5 mg/kg of body weight and 2 g/day reduced methane production by 33% and 60%, respectively (Romero-Perez et al. 2015). Similar results were reported in dairy cows following supplementation with 3-NOP at 60 mg/kg dry matter (Lopes et al. 2016). Finally, in Segureña ewes, 3-NOP and ethyl-3-nitrooxy propionate reduced methane production by 25% and 14%, respectively (Martínez-Fernández et al. 2014).

2.9 Other Feed Additives

Bentonite, an additive currently used for its anticoagulant properties, has been suggested to reduce ammonia emissions by ca. 41% in poultry (Lewis et al. 2015). In the latter species, favorable effects on ammonia mitigation (by up to 17%) have also been reported by supplementation with biochar (2–6%), the solid residue of biomass following decomposition with pyrolysis (Kalus et al. 2020).

In ruminants, however, biochar has been studied for its potential to reduce methane emissions. The latter is thought to be linked to its ability to act as an electron acceptor in the rumen (Schmidt et al. 2019). In vitro studies have indicated that inclusion of biochar at 0.5%, 1%, and 2% could reduce methane production by up to 10%, 13%, and 22%, respectively (Leng et al. 2012a; Saleem et al. 2018). Moreover, in the study of Leng et al. (2012b), cattle supplemented with biochar at 0.6% of dry matter produced 20% less methane compared to those fed a conventional diet. A lower methane reduction of ca. 7% was reported in the study of Phanthavong et al. (2015) following the inclusion of biochar in cattle diet at 1% of dry matter. According to Schmidt et al. (2019), differences in obtained results could be related to pyrolysis feedstock and temperature as well as to post-pyrolytic treatments. Hence, future research under different conditions could help to investigate the potential of biochar to mitigate GHG emissions.

3 Conclusion

Overall, literature suggests that a plethora of feed additives could be included in the diet of livestock to increase the sustainability of the sector. Farm environmental performance could be substantially improved with the use of certain plant extracts, algae, probiotics, organic acids, exogenous enzymes, and synthetic compounds. Such feed additives reduce methane and ammonia emissions as well as phosphorus and nitrogen excretions by improving digestion and modulating microbial populations. Future research on the economic viability of diet supplementation with different feed additives could help to increase their adoption by livestock farmers.

References

Ahmed M, Ahmad S, Waldrip HM, Ramin M, Raza MA (2020) Whole farm modeling: a systems approach to understanding and managing livestock for greenhouse gas mitigation, economic viability and environmental quality. In: Animal manure: production, characteristics, environmental concerns, and management, vol 67, pp 345–371

Ahmed E, Batbekh B, Fukuma N, Hanada M, Nishida T (2022) Evaluation of different Brown seaweeds as feed and feed additives regarding rumen fermentation and methane mitigation. Fermentation 8(10):504

Albores-Moreno S, Alayón-Gamboa JA, Ayala-Burgos AJ, Solorio-Sánchez FJ, Aguilar-Pérez CF, Olivera-Castillo L, Ku-Vera JC (2017) Effects of feeding ground pods of Enterolobium cyclocarpum Jacq. Griseb on dry matter intake, rumen fermentation, and enteric methane production by Pelibuey sheep fed tropical grass. Tropl Anim Health Prod 49(4):857–866

Arriola KG, Kim SC, Staples CR, Adesogan AT (2011) Effect of fibrolytic enzyme application to low-and high-concentrate diets on the performance of lactating dairy cattle. J Dairy Sci 94(2): 832–841

Arsenos G (2001) The role of learning in the selection for dietary protein by sheep. PhD thesis, The University of Edinburgh, UK

Baruah L, Malik PK, Kolte AP, Goyal P, Dhali A, Bhatta R (2019) Rumen methane amelioration in sheep using two selected tanniferous phyto-leaves. Carbon Manag 10(3):299–308

Bayaru E, Kanda S, Kamada T, Itabashi H, Andoh S, Nishida T, Ishida M, Itoh T, Nagara K, Isobe Y (2001) Effect of fumaric acid on methane production, rumen fermentation and digestibility of cattle fed roughage alone. Nihon Chikusan Gakkaiho 72(2):139–146

Beauchemin KA, Kreuzer M, O'mara F, McAllister TA (2008) Nutritional management for enteric methane abatement: a review. Aust J Exp Agric 48(2):21–27

Carrazco AV, Peterson CB, Zhao Y, Pan Y, McGlone JJ, DePeters EJ, Mitloehner FM (2020) The impact of essential oil feed supplementation on enteric gas emissions and production parameters from dairy cattle. Sustainability 12(24):10347

Chen J, Harstad OM, McAllister T, Dörsch P, Holo H (2020) Propionic acid bacteria enhance ruminal feed degradation and reduce methane production in vitro. Acta Agric Scand A Anim Sci 69(3):169–175

Chuntrakort P, Otsuka M, Hayashi K, Takenaka A, Udchachon S, Sommart K (2014) The effect of dietary coconut kernels, whole cottonseeds and sunflower seeds on the intake, digestibility and enteric methane emissions of Zebu beef cattle fed rice straw based diets. Livest Sci 161:80–89

Deng KD, Xiao Y, Ma T, Tu Y, Diao QY, Chen YH, Jiang JJ (2018) Ruminal fermentation, nutrient metabolism, and methane emissions of sheep in response to dietary supplementation with Bacillus licheniformis. Anim Feed Sci Technol 241:38–44

European Commission (2019) Communication from the Commission to the European Parliament, the European Council, the Council, the European Economic and Social Committee and the Committee of the Regions. The European Green Deal. https://ec.europa.eu/info/sites/info/files/european-green-dealcommunication_en.pdf

Fagundes GM, Benetel G, Welter KC, Melo FA, Muir JP, Carriero MM, Souza RLM, Meo-Filho P, Frighetto RTS, Berndt A, Bueno IC (2020) Tannin as a natural rumen modifier to control methanogenesis in beef cattle in tropical systems: friend or foe to biogas energy production? Res Vet Sci 132:88–96

Foley PA, Kenny DA, Callan JJ, Boland TM, O'Mara FP (2009) Effect of DL-malic acid supplementation on feed intake, methane emission, and rumen fermentation in beef cattle. J Anim Sci 87(3):1048–1057

Guan H, Wittenberg KM, Ominski KH, Krause DO (2006) Efficacy of ionophores in cattle diets for mitigation of enteric methane. J Anim Sci 84(7):1896–1906

Gupta S, Mohini M, Malla BA, Mondal G, Pandita S (2019) Effects of monensin feeding on performance, nutrient utilisation and enteric methane production in growing buffalo heifers. Tropl Anim Health Prod 51(4):859–866

Guyader J, Eugène M, Doreau M, Morgavi DP, Gérard C, Martin C (2017) Tea saponin reduced methanogenesis in vitro but increased methane yield in lactating dairy cows. J Dairy Sci 100(3): 1845–1855

Hart KJ, Jones HG, Waddams KE, Worgan HJ, Zweifel B, Newbold CJ (2019) An essential oil blend decreases methane emissions and increases milk yield in dairy cows. Open J Anim Sci 9(03):259

Hassan A, Gado H, Anele UY, Berasain MA, Salem AZ (2020) Influence of dietary probiotic inclusion on growth performance, nutrient utilization, ruminal fermentation activities and methane production in growing lambs. Anim Biotechnol 31(4):365–372

Honan M, Feng X, Tricarico JM, Kebreab E (2021) Feed additives as a strategic approach to reduce enteric methane production in cattle: modes of action, effectiveness and safety. Anim Prod Sci

Hristov AN, McAllister TA, Van Herk FH, Cheng KJ, Newbold CJ, Cheeke PR (1999) Effect of Yucca schidigera on ruminal fermentation and nutrient digestion in heifers. J Anim Sci 77(9): 2554–2563

IPCC (2019) Climate change and land, an Ipcc special report on climate change, desertification, land degradation, sustainable land management, food security, and greenhouse gas fluxes in terrestrial ecosystems. Inter Governmental Panel on Climate Change, Geneva, Switzerland, p 41

Johnson KA, Johnson DE (1995) Methane emissions from cattle. J Anim Sci 73(8):2483–2492. https://doi.org/10.2527/1995.7382483x

Kalus K, Konkol D, Korczyński M, Koziel JA, Opaliński S (2020) Effect of biochar diet supplementation on chicken broilers performance, NH3 and odor emissions and meat consumer acceptance. Animals 10(9):1539

Kinley RD, de Nys R, Vucko MJ, Machado L, Tomkins NW (2016) The red macroalgae Asparagopsis taxiformis is a potent natural antimethanogenic that reduces methane production during in vitro fermentation with rumen fluid. Anim Prod Sci 56(3):282–289

Ku-Vera JC, Jiménez-Ocampo R, Valencia-Salazar SS, Montoya-Flores MD, Molina-Botero IC, Arango J, Gómez-Bravo CA, Aguilar-Pérez CF, Solorio-Sánchez FJ (2020) Role of secondary plant metabolites on enteric methane mitigation in ruminants. Front Vet Sci 7. https://doi.org/10.3389/fvets.2020.00584

Kyriazakis I (1997) The nutritional choices of farm animals: to eat or what to eat? In: Forbes JM, Lawrence TLJ, Rodway RG, Varley MA (eds) Animals choices. Occasional Publication, British Society of Animal Science, Edinburgh, pp 55–65

Latham EA, Pinchak WE, Trachsel J, Allen HK, Callaway TR, Nisbet DJ, Anderson RC (2019) Paenibacillus 79R4, a potential rumen probiotic to enhance nitrite detoxification and methane mitigation in nitrate-treated ruminants. Sci Total Environ 671:324–328

Lei XG, Weaver JD, Mullaney E, Ullah AH, Azain MJ (2013) Phytase, a new life for an "Old" enyzme. Annu Rev Anim Biosci 1:283–309

Leng RA, Inthapanya S, Preston TR (2012a) Biochar lowers net methane production from rumen fluid in vitro. Livest Res Rural Dev 24(6):103

Leng RA, Preston TR, Inthapanya S (2012b) Biochar reduces enteric methane and improves growth and feed conversion in local "Yellow" cattle fed cassava root chips and fresh cassava foliage. Livest Res Rural Dev 24(11):1–7

Lewis KA, Tzilivakis J, Green A, Warner DJ (2015) Potential of feed additives to improve the environmental impact of European livestock farming: a multi-issue analysis. Int J Agric Sustain 13(1):55–68

Li X, Norman HC, Kinley RD, Laurence M, Wilmot M, Bender H, de Nys R, Tomkins N (2016) Asparagopsis taxiformis decreases enteric methane production from sheep. Anim Prod Sci 58(4):681–688

Li ZJ, Ren H, Liu SM, Cai CJ, Han JT, Li F, Yao JH (2018) Dynamics of methanogenesis, ruminal fermentation, and alfalfa degradation during adaptation to monensin supplementation in goats. J Dairy Sci 101(2):1048–1059

Liu H, Vaddella V, Zhou D (2011) Effects of chestnut tannins and coconut oil on growth performance, methane emission, ruminal fermentation, and microbial populations in sheep. J Dairy Sci 94(12):6069–6077

Lopes JC, De Matos LF, Harper MT, Giallongo F, Oh J, Gruen D, Ono S, Kindrmann M, Duval S, Hristov AN (2016) Effect of 3-nitrooxypropanol on methane and hydrogen emissions, methane isotopic signature, and ruminal fermentation in dairy cows. J Dairy Sci 99(7):5335–5344

Lu Q, Jiao J, Tang S, He Z, Zhou C, Han X, Wang M, Kang J, Odongo NE, Tan Z (2015) Effects of dietary cellulase and xylanase addition on digestion, rumen fermentation and methane emission in growing goats. Arch Anim Nutr 69(4):251–266

Ma T, Chen D, Tu Y, Zhang N, Si B, Deng K, Diao Q (2016) Effect of supplementation of allicin on methanogenesis and ruminal microbial flora in Dorper crossbred ewes. J Anim Sci Biotechnol 7(1):1–7

Machmüller A, Kreuzer MCJAS (1999) Methane suppression by coconut oil and associated effects on nutrient and energy balance in sheep. Can J Anim Sci 79(1):65–72

Machmüller A, Ossowski DA, Kreuzer M (2000) Comparative evaluation of the effects of coconut oil, oilseeds and crystalline fat on methane release, digestion and energy balance in lambs. Anim Feed Sci Technol 85(1–2):41–60

Mao HL, Wang JK, Zhou YY, Liu JX (2010) Effects of addition of tea saponins and soybean oil on methane production, fermentation and microbial population in the rumen of growing lambs. Livest Sci 129(1–3):56–62

Martínez-Fernández G, Abecia L, Arco A, Cantalapiedra-Hijar G, Martín-García AI, Molina-Alcaide E, Kindermann M, Duval S, Yáñez-Ruiz DR (2014) Effects of ethyl-3-nitrooxy propionate and 3-nitrooxypropanol on ruminal fermentation, microbial abundance, and methane emissions in sheep. J Dairy Sci 97(6):3790–3799

Matsui H, Tajima K, Itabashi H (2020) Diversity of prokaryotes in the rumen of steers fed a diet supplemented with or without Bromochloromethane, an anti-methanogenic compound. Jpn Agric Res Q 54(2):179–183

McCauley JI, Labeeuw L, Jaramillo-Madrid AC, Nguyen LN, Nghiem LD, Chaves AV, Ralph PJ (2020) Management of enteric methanogenesis in ruminants by algal-derived feed additives. Curr Pollut Rep 6(3):188–205

Moreira GD, Lima PDMT, Borges BO, Primavesi O, Longo C, McManus C, Abdalla A, Louvandini H (2013) Tropical tanniniferous legumes used as an option to mitigate sheep enteric methane emission. Tropl Anim Health Prod 45(3):879–882

Narvaez N, Wang Y, McAllister T (2013) Effects of extracts of Humulus lupulus (hops) and Yucca schidigera applied alone or in combination with monensin on rumen fermentation and microbial populations in vitro. J Sci Food Agric 93(10):2517–2522

Ningrat RWS, Zain M, Elihasridas MM, Putri EM, Sari YC (2020) Effect of dietary supplementation based on ammoniated palm frond with saccharomyces cerevisiae and Gambier leaves waste on nutrient intake and digestibility, daily gain and methane production of simmental cattle. Adv Anim Vet Sci 8(12):1325–1332

Oh J, Harper M, Melgar A, Compart DP, Hristov AN (2019) Effects of Saccharomyces cerevisiae-based direct-fed microbial and exogenous enzyme products on enteric methane emission and productivity in lactating dairy cows. J Dairy Sci 102(7):6065–6075

Paray BA, El-Basuini MF, Alagawany M, Albeshr MF, Farah MA, Dawood MA (2021) Yucca schidigera usage for healthy aquatic animals: potential roles for sustainability. Animals 11(1):93

Pearlin BV, Muthuvel S, Govidasamy P, Villavan M, Alagawany M, Ragab Farag M, Dhama K, Gopi M (2020) Role of acidifiers in livestock nutrition and health: a review. J Anim Physiol Anim Nutr 104(2):558–569. https://doi.org/10.1111/jpn.13282

Perna Junior F, Vásquez DCZ, Gardinal R, Meyer PM, Berndt A, Friguetto RTS, de Abreu Demarchi JJA, Rodrigues PHM (2020) Short-term use of monensin and tannins as feed additives on digestibility and methanogenesis in cattle. Rev Bras Zootec 49

Phanthavong V, Viengsakoun N, Sangkhom I, Preston TR (2015) Effect of biochar and leaves from sweet or bitter cassava on gas and methane production in an in vitro rumen incubation using cassava root pulp as source of energy. Livest Res Rural Dev 27. Article #72

Pinares-Patiño CS, Franco FE, Molano G, Kjestrup H, Sandoval E, MacLean S, Battistotti M, Koolaard J, Laubach J (2016) Feed intake and methane emissions from cattle grazing pasture sprayed with canola oil. Livest Sci 184:7–12

Prathap P, Chauhan SS, Leury BJ, Cottrell JJ, Dunshea FR (2021) Towards sustainable livestock production: estimation of methane emissions and dietary interventions for mitigation. Sustainability 13(11):6081

Romero-Perez A, Okine EK, McGinn SM, Guan LL, Oba M, Duval SM, Kindrmann M, Beauchemin KA (2015) Sustained reduction in methane production from long-term addition of 3-nitrooxypropanol to a beef cattle diet. J Anim Sci 93(4):1780–1791

Roque BM, Salwen JK, Kinley R, Kebreab E (2019) Inclusion of Asparagopsis armata in lactating dairy cows' diet reduces enteric methane emission by over 50 percent. J Clean Prod 234:132–138

Rossi CAS, Grossi S, Dell'Anno M, Compiani R, Rossi L (2022) Effect of a blend of essential oils, bioflavonoids and tannins on in vitro methane production and in vivo production efficiency in dairy cows. Animals 12(6):728

Saleem AM, Ribeiro Jr GO, Yang WZ, Ran T, Beauchemin KA, McGeough E J, Ominski KH, Okine EK, McAllister TA (2018) Effect of engineered biocarbon on rumen fermentation, microbial protein synthesis, and methane production in an artificial rumen (RUSITEC) fed a high forage diet. J Anim Sci 96(8):3121–3130

Schmidt HP, Hagemann N, Draper K, Kammann C (2019) The use of biochar in animal feeding. PeerJ 7:e7373

Śliwiński BJ, Kreuzer M, Wettstein HR, Machmüller A (2002) Rumen fermentation and nitrogen balance of lambs fed diets containing plant extracts rich in tannins and saponins, and associated emissions of nitrogen and methane. Arch Anim Nutr 56(6):379–392

Soltan YA, Natel AS, Araujo RC, Morsy AS, Abdalla AL (2018) Progressive adaptation of sheep to a microencapsulated blend of essential oils: ruminal fermentation, methane emission, nutrient digestibility, and microbial protein synthesis. Anim Feed Sci Technol 237:8–18

Wallace RJ, Wood TA, Rowe A, Price J, Yanez DR, Williams SP, Newbold CJ (2006, July) Encapsulated fumaric acid as a means of decreasing ruminal methane emissions. In: International congress series, vol 1293. Elsevier, pp 148–151

Wang H, Long W, Chadwick D, Velthof GL, Oenema O, Ma W, Wang J, Qin W, Hou Y, Zhang F (2020) Can dietary manipulations improve the productivity of pigs with lower environmental and economic cost? A global meta analysis. Agric Ecosyst Environ 289:106748

Wu P, Liu ZB, He WF, Yu SB, Gao G, Wang JK (2018) Intermittent feeding of citrus essential oils as a potential strategy to decrease methane production by reducing microbial adaptation. J Clean Prod 194:704–713

Zhao L, Peng Y, Wang J, Liu J (2015) Effects of exogenous fibrolytic enzyme on in vitro ruminal fiber digestion and methane production of corn Stover and corn Stover based mixed diets. Life Sci J 12(2s):1–9

Zúñiga-Serrano A, Barrios-García HB, Anderson RC, Hume ME, Ruiz-Albarrán M, Bautista-Martínez Y, Sánchez-Guerra NA, Vázquez-Villanueva J, Infante-Rodríguez F, Salinas-Chavira J (2022) Antimicrobial and digestive effects of Yucca schidigera extracts related to production and environment implications of ruminant and non-ruminant animals: a review. Agriculture 12(8):1198

Feed Additives, Their Role, and Technological Properties

Olivera Djuragic, Ivana Čabarkapa, Marija Milašinović Šeremešić, Slađana Rakita, and Zorica Tomičić

Abstract

The main objectives of precision feeding are maximum production results and maintaining animal health. It is of great importance for large production systems that depend on high investments. Nowadays, the feed industry uses significant number of components of different origins to meet all requirements. Feed additives are substances used in animal nutrition to improve the quality of feed and food of animal origin. They are intended to improve animal performance and health (e.g., improving feed digestibility). To produce high-quality and safe feed, it is necessary to take into account the physical and chemical characteristics of the ingredients and additives. Mode of action and effect is the main criterion for component selection, and it serves as the foundation for feed formulation and nutrient value achievement. The choice of additives in relation to their market price is vital to comprehend the similarities, differences, and restrictions as well as the specifics of the technological procedure. Environmentally acceptable feed additives that can replace currently employed chemotherapeutic drugs or the previously added antibiotics to feed that have been discontinued are being sought. Probiotics, prebiotics, symbiotics, organic acids, herbs, and plant extracts are among these supplements that have the highest impact.

Keywords

Feed · Additives · Probiotics · Prebiotics · Organic acids · Phytogenic additives · Enzymes

O. Djuragic (✉) · I. Čabarkapa · M. M. Šeremešić · S. Rakita · Z. Tomičić
Institute of Food Technology, University of Novi Sad, Novi Sad, Serbia
e-mail: olivera.djuragic@fins.uns.ac.rs; ivana.cabarkapa@fins.uns.ac.rs; marija.milasinovic@fins.uns.ac.rs; sladjana.rakita@fins.uns.ac.rs; zorica.tomicic@fins.uns.ac.rs

G. Arsenos, I. Giannenas (eds.), *Sustainable Use of Feed Additives in Livestock*,
https://doi.org/10.1007/978-3-031-42855-5_2

1 Types of Additives

Maximizing production and preserving animal health are the main goals of precision feeding. It is crucial for large production systems that rely on significant financial investments. To meet these requirements, the feed business now uses a variety of components from different origins. Animal nutrition uses feed additives to enhance the quality of the feed and foods of animal origin. They are intended to enhance the health and performance of animals (e.g., improving feed digestibility). Usually incorporated in feed, feed additives may influence feed composition, feed effectiveness, feed intake, and animal products (milk, meat, egg, wool) (Innocenti 2017). To reduce feed losses and environmental emissions, it may be possible to improve feed preservation during handling and storage (FAO Guidelines 2019).

According to the European Union (EU) systematization (https://food.ec.europa.eu/safety/animal-feed/feed-additives/legislation-feed-additives_en), there are a few categories of additives. The regulation covers the following feed additive categories (Fig. 1):

1. **Technological additives**—e.g., preservatives, antioxidants, emulsifiers, acidity regulators, silage additives. They do not have a direct impact on the feed's nutritional value, but may have an indirect effect by making feed easier to handle or more hygienic. An example of such an addition is an organic acid used to preserve feed.
2. **Sensory additives**—e.g., flavors, colorants. It increases the palatability of a diet by provoking appetite. For instance, vanilla extract might help piglets consume their food better.
3. **Nutritional additives**—offers the exact nutrient(s) that the animal needs for best growth. Examples include additives like vitamins, amino acids (AAs), or trace minerals necessary for the animal growth.
4. **Zootechnical additives**—e.g., digestibility enhancers. It enhances the animal's performance causing more effective utilization of the nutrients already present in the feed rather than supplying specific nutrients. An example would be enzymes or microorganisms, which improve the circumstances of the digestive tract and

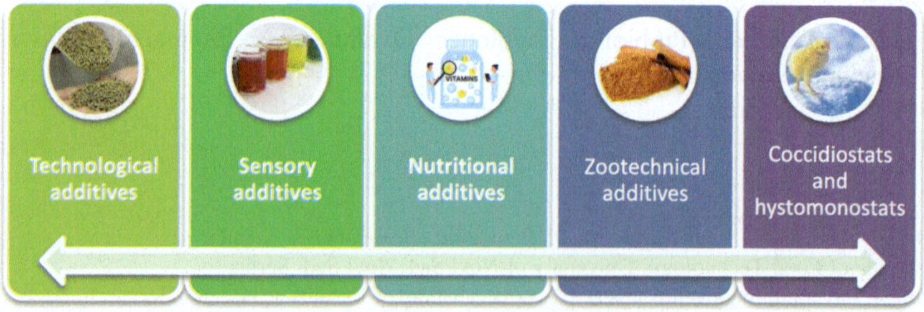

Fig. 1 Categories of additives

Table 1 Types of additives according to action

Name of additive group	Type	Function	Reference source
Probiotics	Different types of microorganisms	Improvement of digestion; stimulator of growth	Brzóska and Stecka (2007), Janocha et al. (2022)
Prebiotics	Different types of oligosaccharides	Improvement of digestion; stimulator of growth	Brzóska and Stecka (2007), Janocha et al. (2022)
Phytogenic additives	Herbs and essential oils	Improvement of digestion; antimicrobial effects; gut health	Puvača et al. (2018), Janocha et al. (2022)
Organic acids	Carboxyl acids (citric, propionic, ascorbic)	Increasing feed efficiency; improvement of digestion; feed preservation	Brzóska and Stecka (2007), Lević et al. (2007)
Enzymes and amino acids	Plant enzymes and essential amino acids	Antibacterial; improvement of digestion; antioxidative	Velázquez-De et al. (2021)
Functional elements and nanoparticles	Different metal oxides; different clays and minerals	Improvement of digestion; antimicrobial effects	Bąkowski et al. (2018), Yusof et al. (2019)

make it possible for nutrients to assimilate more efficiently. They are frequently pro-nutrients or probiotics.

5. **Coccidiostats and histomonostats**—the compounds known as coccidiostats and histomonostats can be made synthetically or by microorganisms. The role is to prevent or eliminate the protozoan parasites that infect animals and cause coccidiosis and histomoniasis.

According to the active ingredients and mode of the action, additives could be categorized as presented in Table 1:

To produce high-quality and safe feed, it is necessary to take into account the physical and chemical characteristics of the ingredients and additives. Mode of action and effect is the main criterion for component selection, and it serves as the foundation for feed formulation and nutrient value achievement. The choice of additives in relation to their market price is vital to comprehend the similarities, differences, and restrictions as well as the specifics of the technological procedure.

Some of the physical characteristics of powder feed additives that are frequently utilized in the production of feed, such as bulk density, mean particle diameter, angle of repose, and dusting potential, are very important (Đuragić et al. 2007; Mubeen et al. 2021).

Using recognized techniques, dusting potential should be evaluated. The amount of the active substance in the dust should be measured when an occupational exposure limit is established or when there is a known or suspected toxicity following inhalation exposure (Regulation [EC] No 1831/2003).

If the nature of the additive allows the possibility of the presence of nanoparticles, initially a particle size analysis of the additive by laser diffraction should be made. If the particle size analysis of the additive indicates that more than 1% of particles below 1 μm are present, this fraction should be further characterized by scanning electronic microscopy (wet method). Results should be expressed as a percentage of the total number of particles. It should be indicated if the product is a nanomaterial as defined by European legislation (Regulation [EC] No 1831/2003).

Data on vapor pressure, viscosity, specific weight, and (pH dependent) solubility or dispersibility should be included for liquid additives if the additive is meant to be used in water (Đuragić et al. 2007; Regulation [EC] No 1831/2003). The label must contain the name of the animal food additive made of nanomaterials as well as all relevant details, such as its chemical identity and composition; physical, chemical, and biological characteristics; as well as specifications defining the minimum concentrations of the desired component(s) and identifying and limiting reaction byproducts and other impurities (Rychen et al. 2017).

2 Probiotics

Increasing interest in probiotics has been accompanied by many research studies and the use of different definitions. The working group experts of the Food and Agriculture Organization of the United Nations and the World Health Organization define probiotics as "live microorganisms which when administered in adequate amounts confer a health benefit on the host" (FAO/WHO 2002). In addition to a set of major safety and functionality criteria in the selection process for defining probiotic species (Hill et al. 2014; Markowiak and Slizewska 2017), the probiotic activity undergoes numerous clinical trials against gastric or enterovirulent bacterial pathogens, which constitute a major problem for human and animal health. The majority of probiotics in use today include species of lactic acid bacteria of genus *Lactobacillus* and *Bifidobacterium*, the most used and tested probiotic bacteria, as well as *Escherichia coli*, bacilli, and the best known probiotic yeast *Saccharomyces boulardii* (Tiago et al. 2012; Tomičić et al. 2016a). For many decades, antibiotics have been widely used as growth promoters in the poultry industry. However, overuse of antibiotics for veterinary purposes leads to harmful consequences, including the emergence of antibiotic-resistant bacteria and the development of microbiota disorders—such infections are responsible for productivity loss and increased mortality—as well as the presence of antibiotic residues in poultry products that pose the health hazard to the consumer, which is why they are now prohibited in many countries (Al-Fatah 2020; Wu et al. 2019a, b). Therefore, controlling pathogenic bacteria without antibiotics has become a big challenge. For this reason, there is an urgent demand for new antibiotic alternatives in livestock production. The scientific and clinical evidence indicates that the use of probiotics shows therapeutic potential in the maintenance of health treatment of various intestinal disorders by enhancing the beneficial components of the intestinal microbiota (O'Hara and Shanahan 2007). They improve the health status, production performances, feed conversion rate, and immune response of poultry and farm animals (Getachew 2016).

2.1 Mode of Action

The primary action of probiotics is toward pathogenic bacteria relating to their abilities to: produce antimicrobial substances that inhibit their development as bacteriocins, compete with pathogens for adhesion to the epithelium and nutrients, modulate the host immune system, and inhibit bacterial toxin production (Markowiak and Slizewska 2017; Tomičić et al. 2016b). The beneficial effects of probiotics are lowering the pH of the intestine through the production of primary metabolites such as organic acids and hydrogen peroxide, and later found volatile fatty acids that suppress the colonization of pathogens in the digestive tract (Ayalew et al. 2022; Khan and Naz 2013). Various primary and secondary metabolites lower intestinal pH to inhibit the growth of pathogens such as *E. coli* and *Salmonella* (Marteau et al. 2004).

Probiotics adhere to and colonize the epithelial surface of the gut and compete with the pathogen to the adhesion site forming the enterocyte's complexity. A previous study reported that spent culture supernatants of *Bifidobacterium* strains can antagonize the growth and the adhesion of *Clostridium difficile* to cultured enterocytes (Al-Fatah 2020; Trejo et al. 2006). Probiotics have positive effects on health and growth and have beneficial nutritional effects; they improve health and growth of animals by inducing immune responses and possess the ability to limit inflammation and infection in the gastrointestinal tract, increasing fiber digestion and enzyme activity to be efficient in the use of nutrients in feed. Furthermore, they help in utilization of nutrients such as digestible protein, vitamins, minerals, and enzymes, and have important role in proper growth and metabolism (Al-Fatah 2020). It has been proven that probiotic yeast *S. boulardii* is effective in the context of many infectious and inflammatory gastrointestinal disorders. It possesses a wide range of beneficial effects against bacterial infections, such as modulating the immune system, degrading *C. difficile* toxins A and B and their respective receptors on colonic mucosa, inhibiting cholera toxin action, modulating the transduction pathway when activated by enteropathogenic bacteria, downregulating production of inflammatory molecules (such as interleukin [IL]-8 and tumor necrosis factor [TNF]-α), or increasing the levels of secretory IgA (Douradinha et al. 2014; Tiago et al. 2012). Physicochemical properties of meat are also important because they largely determine the possibilities for its storage or further processing. The color of the meat is determined by measuring its lightness, redness, and yellowness and is important for the consumer's perception of the freshness and quality of the meat (Droval et al. 2012; Prommachart et al. 2021). Probiotics exhibit good potential for improving the quality of poultry meat by affecting the protein and fat contents (Popova 2017). Higher protein content and lower fat content were reported in the treatment with probiotic *Bacillus subtilis* (Hossain et al. 2012; Král et al. 2013). The use of probiotics in water broilers supplemented with *Lactobacillus fermentum* significantly increased redness in the breast (Hačik et al. 2015), while improvement in color characteristics in breast meat in terms of lightness and yellowness was found in a study of Abdurrahman et al. (2016). In addition, pH and meat color are important indicators of meat quality, which should be used especially for further storage (Abdulla et al. 2015).

Probiotics have beneficial nutritional effects, and recent research indicated that the use of probiotics as feed supplements shows potential as a natural way to improve animal health and growth, and in that context they are a good alternative to antibiotics. Moreover, supplementation of livestock diets with probiotics significantly appears to improve characteristics and meat quality.

3 Prebiotics

Prebiotics are beneficial non-digestible feed ingredients that affect the host animal by promoting the growth and/or activity of bacteria within a colon (Gaggìa et al. 2010).

Prebiotics can be found in a variety of foods, including oats, barley, dandelion greens, chicory, chia seeds, flax seeds, onion, garlic, almonds, and artichoke (Davani-Davari et al. 2019). Containing water-soluble sulfated polysaccharides, Green algae (Chlorophyta) are also regarded as prebiotics because they modulate the gut microbiota and the immune system as well as possess antioxidant, antibacterial, anti-hyperlipidemia, and anti-diabetic properties.

Carbohydrates and oligosaccharides with different molecular structures make up the majority of prebiotics that have been identified. Dietary carbohydrates like fibers are a candidate for both human and animal diets. However, non-digestible oligosaccharides (NDOs) show the greatest potential as prebiotics. The recent development of the prebiotic industry may benefit from the use of novel carbohydrates like resistant starch from a variety of starch sources (Milašinović et al. 2010).

Currently, in the prebiotic category, two major groups are dominant: (1) Fructans (fructooligosaccharides [FOS] and inulin), and (2) Galactans (galactooligosaccharides or GOS).

Their influence is manifested through the enrichment of *Lactobacillus* and/or *Bifidobacterium* spp. ("healthier" microbiota) activity resulting in the increased growth rate and improved feed efficiency.

The following potential prebiotics can be used in feed formulations: fructan, oligofructose, inulin, fructooligosaccharides, galactan, galactooligosaccharides, xylooligosaccharides (XOS), pectin, fiber components, and milk oligosaccharides. β-glucan, D-mannose, and other mannanoligosaccharides (MOSs), which are derived from *Saccharomyces cerevisiae*'s cell wall, are examples of refined functional carbohydrates (RFCs) that are easily accessible as prebiotics for use by animals (Dallies et al. 1998). Among these, fructooligosaccharides and mannanoligosaccharides are the most widely used commercial feed ingredients in the production of poultry feed.

Prebiotics generally alter gut microorganisms, allowing for an increase in their population, increased digestibility, a decrease in pathogenic bacteria, an increase in the absorbability of minerals and vitamins, the maintenance of optimal intestinal pH, and the maximum utilization of nutrients (Kulshreshtha and Bhatnagar 2014; Murate et al. 2014).

3.1 Mode of Action

Among other effects that can have an impact on host health, prebiotics have the potential to modify microbial metabolism, produce metabolites like lactic acid, and improve the integrity of epithelium cells (Yaqoob et al. 2022). Prebiotics are used to manage the interactions between the host and the gastrointestinal ecosystem, which includes altering the intestinal microbiota, supporting the immune system, enhancing the epithelium, and regulating the host–intestinal microbiota interaction. There are many references in the literature that pointed out prebiotics as a selective substrate for beneficial bacteria, changing the colon's microflora in the direction of a more wholesome gastrointestinal environment (Gibson and Fuller 2000).

As a result of the competitive exclusion of pathogenic microbes and the selective colonization by beneficial microbes, they act as a substrate for endogenous beneficial bacteria (Biggs et al. 2007). The prebiotic feed supplements have been shown to alter the pH of the intestine, which makes them unfavorable to pathogenic organisms (Mazanko et al. 2018).

By increasing the abundance of *Lactobacilli* and *Bifidobacteria* and decreasing the titers of *Coliform*, it establishes a healthy microbial community in the intestine of broilers (Po Yun and Woo 2018). Manase enzymes are found in *Bifidobacterium* and *Lactobacillus*; they selectively bind mannan oligosaccharides only for harmful bacteria, which typically lack this enzyme. Broiler weight gain per day increases by 4–8% as a result of mannan oligosaccharides (Sinovec and Markovic 2005).

Prebiotics are an exceptional tool to promote the growth of good gut microbes that ferment them, producing short-chain fatty acids (SCFAs), vitamins, and other fragmented molecules as well as some antibacterial substances like bacteriocin against pathogenic microorganisms (Bogusławska-Tryk et al. 2012). This mechanism is attributed to the fact that prebiotics are sustained in acidic environments and remain resistant to specific digestive enzymes in the small intestine. The fermented products derived from the use of prebiotics and the act of beneficial microbes also strengthen intestinal epithelial cells, which increases nutrient absorption and boosts animal growth performance (Po Yun and Woo 2018).

Some research indicate oligosaccharides have immunomodulatory beneficial effects on the gut, including altering the effectiveness with which pathogenic bacteria are cleared, activating T-cell-dependent immune responses, and suppressing pro-inflammatory cytokines (Troy and Kasper 2010). Pathogen-associated molecular patterns are generated by pathogenic microorganisms and can be reduced by preventing pathogen colonization with prebiotics (Tizard 2013).

4 Phytogenic Feed Additives (PFAs)

Phytogenic feed additives (PFAs) are plant-derived products used in animal feeding to improve the feed quality, performance, and health of agricultural livestock (Ayalew et al. 2022; Windisch et al. 2008). PFAs are becoming attractive by the ban on antibiotics as growth promoters in animal feed (Windisch et al. 2008). In addition, the ban on antibiotics as growth promoters has significantly increased the incidence of infections caused by pathogens and consequently negatively impacted livestock performance. Therefore, the use of alternative feed additives such as PFAs has become an important strategy to reduce the risk of infection and maintain animal health. The name "phytogenic compound" (PC) refers to plant materials such as seeds, fruits, vegetables, roots, and leaves (Čabarkapa et al. 2020; Steiner and Syed 2015). PCs can be used in solid, dried, and ground forms or as extracts (crude or concentrated). According to the process used to derive the active ingredients, PCs can be classified as essential oils (EOs; volatile lipophilic substances obtained by cold extraction or by steam or alcohol distillation) and oleoresins (extracts derived by non-aqueous solvents) (Gadde et al. 2017). Biologically active PCs have been divided into three main groups: (1) terpenoids (plant volatiles, sterols, carotenoids, saponins, and glycosides); (2) phenolic compounds (flavonoids, phenolic acids, lignin, lignans, coumarins, stilbenes, and tannins); and (3) nitrogen-containing substances (alkaloids, glucosinolates, and cyanogenic glycosides) (Li et al. 2020). The composition and concentration of bioactive compounds in plants vary and depend on plant species, parts of the plant, geographical origin, harvesting season, environmental factors, storage conditions, and processing technique. These variations can greatly influence the bioactive potential of plant products, and this needs to be taken into account when selecting plants for therapeutic purposes (Gadde et al. 2017; Čabarkapa et al. 2019; Wu et al. 2021). According to European Union legislation (EC 1831/2003), PFAs are categorized as sensory and flavoring compounds derived from herbs or spices as antibiotic alternatives and recognized as "generally recognized as safe" (GRAS).

4.1 Mode of Action

A growing body of evidence has shown that supplementation of PCs in the form of PFAs in animal diets improves the flavor and palatability of feed, thus enhancing production performance (Windisch et al. 2008), increases digestibility (Syed et al. 2021), reduces inflammation (Olfati and Hosseini 2022; Chang et al. 2022), modulates intestinal microbiota (Oetting et al. 2006; Yang et al. 2020), and improves antioxidative activities (Puvača et al. 2019). Since the feed industry has recognized the potential benefits of using plant-based substances for various animal species, PCs are being utilized more frequently, particularly in feeding programs for pigs (Liotta et al. 2015; Chang et al. 2022), poultry (Windisch et al. 2008; Puvača et al. 2019; Wang et al. 2021), ruminants (Piran Filho et al. 2021), and fish (Bulfon et al. 2015). The modes of action of PCs are associated with their biological activities. Based on their properties, PCs could be classified as antimicrobials, intestinal function modulators, antioxidants, anti-inflammatory agents, and growth promoters (Wang et al. 2021). Taking into account a large number of aromatic plants, herbs, and single compounds, the mode of their action is distinguished by diversity and will depend on their chemical composition, concentration, and particular combination. One of the remarkable properties of PCs is their antimicrobial activity. Antimicrobial properties of PCs derived from herbs and spices have been extensively researched (Burt 2004; Burt et al. 2005; Lee et al. 2011; Čabarkapa et al. 2019). The majority of published studies revealed that the largest antimicrobial effect was to exhibit phenols, which have been present in the highest percentage in the EOs (Čabarkapa et al. 2016, 2019), followed by alcohols, aldehydes, ketones, ethers, and hydrocarbons (Burt 2004). Studies conducted until now related to the antimicrobial activity of PCs proposed several target positions and their modes of antimicrobial action: degradation of the cell wall, alteration of membrane proteins, alteration of the permeability cytoplasmic membrane, leakage of cell contents, coagulation of the cytoplasm, and depletion of the proton motive force (Ultee et al. 2000; Burt 2004). Other mechanisms of action may be related to the inhibition of nutrient absorption, enzymatic inhibition, and synthesis of DNA, RNA, and proteins by bacterial cells (Lambert et al. 2001). Considering that PCs have well-documented antimicrobial activity in vitro, it is expected that the dietary application of PCs in animal feed influences the composition of the gastrointestinal microbiota. Du et al. (2016) have shown that PCs can effectively combat harmful microorganisms such as *Escherichia coli* and *Clostridium perfringens*, indicating a decreased risk of developing colibacillosis and necrotic enteritis in broiler chickens. Published results are, however, inconsistent. While most conducted studies showed that dietary PFA supplementation was effective on the selected intestinal microbial populations (Oetting et al. 2006; Manafi et al. 2016; Yang et al. 2020), some studies reported no impact, or a negative impact on intestinal microbiota (Vukic-Vrajnes et al. 2013; Ahsan et al. 2018). Similar conclusions were drawn by Ahsan et al. (2018). These authors investigated the influence of different levels of PFAs on growth performance, cecal microbiota, and intestinal morphology in broilers. According to these authors, the main reason for the reduced effect of PFAs on intestinal microbiota was the low levels of applied essential oils in the phytogenic product used in this study. However, in addition to the applied concentration, interactions between different classes of compounds incorporated in PFA should be taken into account, considering that certain mixtures can exhibit antagonistic effects (Čabarkapa et al. 2020). Antioxidant activity is another biological characteristic of phytogenic compounds, which may play an important role in preventing some diseases caused by free radicals, such as cancer and heart diseases (Gheisar and Kim 2018). Antioxidative properties of herbs and plants are based mainly on phenolic compounds (Čabarkapa et al. 2020), flavonoids, or anthocyanins (Dias et al. 2021). The primary role of antioxidants is to neutralize harmful free radical activity, helping protect the body from diseases caused by oxidative stress. Among the various plants containing antioxidants, the EOs from the Lamiaceae family have been attracting the greatest interest, especially products from oregano and thyme (Ultee et al. 2002; Marinelli et al. 2018). Their antioxidative activity arises from phenolic terpenes, such as monoterpenes thymol and carvacrol (Čabarkapa et al. 2019; Gholami-Ahangaran et al. 2022). Other Lamiaceae species with significant antioxidative properties are rosemary and sage. The most abundant components with antioxidant activity in these plants primarily represent diterpenes such as carnosic acid and carnosol. Their content depends on the species, variety, plant growth conditions, sample treatment, and type of extract preparation (Tounekti and Munné-Bosch 2012), and their content increases with aging (Abreu et al. 2008). Herbs and bioactive components of herbs added to chicken diets can improve the quality of broiler chicken meat by altering the content of fatty acids. Previous studies indicate that dietary supplementation with sage, oregano, and rosemary positively affected the fatty acid profile, improved meat tenderness, and reduced lipid oxidation in broiler breast and thigh meat (Pisarski et al. 2007). Lobo et al. (2010) have proposed two modes of action for antioxidants. The first is a chain-breaking mechanism in which the primary antioxidant donates an electron to the free radical present in the system. The second mechanism involves the removal of reactive oxygen species (ROS)/reactive nitrogen species initiators (secondary antioxidants) by quenching a chain-initiating catalyst. The antioxidants act as part of an organism's defense system by preventing, radical scavenging, repairing, and de novo, the fourth line of defense, by adaptation (Lobo et al. 2010). The anti-inflammatory activity of PCs is well documented. As a result, many researchers have been drawn to PCs in order to learn more about their roles in various disease processes such as inflammatory diseases, tumorigenesis, and neurodegeneration. These studies have shown that the health-promoting activities of PCs are attributed to their ability to improve host defense against microbial infection (Lillehoj et al. 2018; Chang et al. 2022). PCs play a significant role in human and animal health, in the prevention and treatment of inflammatory conditions. Widely is known that the inflammation process involves a cascade of dynamic responses, including cellular and vascular actions with specific humoral secretions (Abdulkhaleq et al. 2018). Inflammation processes are commonly followed by the overproduction of pro-inflammatory cytokines and other inflammatory mediators (IL-1, IL-6, TNF-α, nitric oxide [NO], and prostaglandin E2) (Kim et al. 2020). The attention of most researchers was directed to nuclear factor-κB (NF-κB) as a central transcription factor that regulates pro-inflammatory genes' expression during inflammation and has become an interesting target for the development of new anti-inflammatory drugs (Kim et al. 2020). Next to the antimicrobial, antioxidant, and anti-inflammatory properties of PCs, comprehensively reviewed reports indicated that PCs are distinguished by proven palatability, a stimulating effect on digestive activity, an overall improvement of feed intake, and growth performance (Windisch et al. 2008; Ahmed et al. 2013; Gadde et al. 2017; Gholami-Ahangaran et al. 2022). The addition of PCs in animals' diets improves the palatability of the diet through enhanced flavor and odor, which results in increased feed intake and consequently leads to improved growth performance (Ahmed et al. 2013; Steiner and Syed 2015). However, improving feed palatability cannot be applied to poultry due to their insensitivity to odor (Steiner and Syed 2015). It has been hypothesized that the antioxidant properties of PCs are linked to improved feed palatability owing to their preservative function and ability to prevent the release of undesirable odors (Franz et al. 2010). Moreover, the addition of PCs improves feed digestibility, which leads to better nutrient absorption (Ahmed et al. 2013). In addition, the results of some studies have demonstrated that phytogenic products may be able to stimulate intestinal mucus production, which may further contribute to the relief from pathogen pressure by inhibiting adherence to the mucosa (Gheisar and Kim 2018). PCs exhibited effects on egg and meat quality (Liotta et al. 2015; Kikusato 2021).

5 Organic Acids

Organic acids have been used in the animal feed industry for many years as an effective alternative to antibiotics because of their ability to prevent undesirable bacterial and fungal growth in feed, and improve gastrointestinal health and nutrient utilization that may further contribute to enhanced animal performance (Ayalew et al. 2022). Use of organic acids has been permitted in the diets of all the livestock species and according to EU legislations on the feed additives, they belong to the group of preservatives (Adil et al. 2010). Various organic acids used in animal diets are: acetic, butyric, citric, formic, propionic, malic, tartaric, lactic, benzoic, fumaric, and sorbic acids. Organic acids are weak acids containing a carboxylic acid group (R-COOH) (Costa et al. 2013; Ayalew et al. 2022). Commercial organic acids can be usually found in the form of their salts (sodium, potassium, calcium salts), because they are solid, odorless, less volatile, less corrosive, and more soluble in water and therefore more convenient for the use in feed manufacturing than free acids (Huyghebaert et al. 2011; Vukmirović et al. 2017). The application of organic acids in animal feed varies among the countries due to the different regulations and other factors. In some countries they are added before any thermal treatment, while in others acids are added directly to drinking water or produced feed (Costa et al. 2013). Organic acids can be effectively used in combination with other feed additives (Suiryanrayna and Ramana 2015). Despite that some hypotheses have been suggested, the mechanisms of action of organic acids have been still unclear. The effectiveness of organic acids to inhibit the growth of a microorganism depends on the concentration, the type of acid, time of exposure, pKa, type of microorganism control, the composition of the feed, and the age of the animal (Costa et al. 2013).

5.1 Mode of Action

The addition of organic acids inhibits the growth of pathogenic bacteria through the reduction of pH and buffering capacity of the diet and reduction of pH in the gastrointestinal tract of animals. Acidic conditions limit the survival of undesirable microorganisms in the stomach and hence prevent passage to the intestines. Only undissociated form of organic acid delivers antimicrobial effect as these molecules can diffuse freely through the bacterial cell membrane and reach the cytoplasm (Pearlin et al. 2020). Inside the bacterial cell, the acid tends to dissociate into ions and protons, decrease cytoplasmic pH, and disrupt the normal physiology of the bacteria by suppressing enzymatic reactions and protein and DNA denaturation (Mani-Lopez et al. 2012). That results in the inhibition of bacterial growth and the death of harmful pH-sensitive bacteria including *Salmonella*, *Clostridia*, *E. coli*, *Listeria*, *Campylobacter*, and some coliforms that cause gastrointestinal distress. These bacteria are not able to tolerate the variation between internal pH in the bacteria and pH in the stomach (Suiryanrayna and Ramana 2015; Vukmirović et al. 2017; Kumar et al. 2022).

The effectiveness of organic acids to suppress the growth of microorganisms depends on its pKa value, which is the pH value at which the acid is half dissociated. Organic acids with higher pKa value (the higher proportion of undissociated form) are weaker acids and thus are more effective in preventing the growth of undesirable microorganisms. On the other hand, organic acids with lower pKa have greater effect on the reduction of pH by acidifying the feed and the stomach, but have less antimicrobial efficiency in the digestive tract (Costa et al. 2013; Suiryanrayna and Ramana 2015). Organic acids are more effective toward the inhibition of Gram-positive bacteria. This is associated to the structural differences between Gram-negative and Gram-positive bacteria. Namely, Gram-negative bacteria have the additional outer layer surrounding the cytoplasmic membrane, which consists of lipopolysaccharides. This layer provides the bacteria with immanent antimicrobial resistance (Suiryanrayna and Ramana 2015). Gómez-García (2019) observed that formic acid followed by propionic acid was the most effective against *E. coli* and *Salmonella* spp. in pigs. Hassan et al. (2010) observed a decrease in *E. coli* and *Salmonella* spp. in the intestinal microflora of broilers fed diet supplemented with combination of organic acids or salts.

Organic acids are able to improve gut health by promoting the beneficial bacterial growth in the intestines. It was found that acidic conditions in the stomach are suitable environment for the growth of *Lactobacilli*, which prevent the colonization and proliferation of pathogens by obstructing the adhesion site or by producing lactic acid that reduces gastric pH and suppresses the pathogen growth (Partanen and Mroz 1999; Suiryanrayna and Ramana 2015). Akyurek et al. (2011) reported that broiler chickens fed diet supplemented with blends of organic acids had greater number of *Lactobacilli* and less pathogenic bacteria such as *Clostridia* and *E. coli* in ileum than chickens fed diet with antibiotic growth promoters.

The inclusion of organic acids in animal diet has been shown to enhance the diet digestibility and nutrient retention (Costa et al. 2013). In general, organic acid acidify the gut environment and enhance gastric retention time and the activity of proteolytic enzymes. Because of the lower gastric emptying rate, acidic digesta is retained longer in gastrointestinal tract, and proteins are hydrolyzed more efficiently, which stimulate apparent ileal digestibilities of proteins and amino acid (Partanen and Mroz 1999; Suiryanrayna and Ramana 2015). Some authors reported that dietary administration of organic acids supports the protein digestion in the gastro-intestinal tract by converting inactive pepsinogen into active pepsin (Suiryanrayna and Ramana 2015). By providing acidic gastric conditions, organic acids improve pepsin activity and digestibility of minerals (Christian and Mellor 2011;

Letlhogonolo et al. 2020). Generally, dietary supplementation of organic acid in animal diets showed to have beneficial effect on the apparent digestibility of protein, amino acids, fat, and nitrogen in gastrointestinal tract and consequently increasing the energy digestibility (Partanen and Mroz 1999, Costa et al. 2013).

Organic acids can improve digestion of diets with poor protein quality. Diets with poor protein quality cause more indigestible proteins in gastrointestinal tract, intense protein fermentation, and production of volatile fatty acids, ammonia, and other gases, which may cause discomfort in the animal and thus adversely affect growth rate (Ikker et al. 2007; Diether and Willing 2019). Organic acids can form complexes with phosphorus, calcium, magnesium, and zinc and enhance digestion of these minerals, which results in the lower excretion of minerals and nitrogen and lower environmental pollution (Partanen and Mroz 1999; Suiryanrayna and Ramana 2015). It was also documented that organic acids have stimulative effect on exocrine pancreatic secretion and biliary secretion (Partanen and Mroz 1999; Costa et al. 2013).

Organic acids are a powerful tool to improve overall production performance by reducing pathogenic bacterial growth and bacterial competition with the animal for nutrients, while promoting beneficial bacteria and nutrient availability (Suiryanrayna and Ramana 2015). This leads to maximized nutrient utilization efficiency, enhanced feed conversion rate and body weight gain (Bafundo et al. 2002; Hassan et al. 2010; Qaisrani et al. 2015; Lan et al. 2005). Further, the organic acids prevent diarrhea and enhance growth performance in piglets (Sutton et al. 1991). However, later it was found that organic acids also have beneficial effect on the performance of fattening pigs and sows. Partanen and Mroz (1999) reported that dietary supplementation of organic acids can improve growth performance and modify intestinal microbiota in pigs. Furthermore, Hassan et al. (2010) demonstrated that mixtures of organic acids supplemented to the broiler diet were more efficient than antibiotic growth promoter in improving broiler growth, feed conversion rate, and nutrient utilization.

It has been documented that feeding ruminants with organic acids can ameliorate methanogenesis in the rumen. Organic acids act as an alternative sink for metabolic hydrogen in the rumen, thereby decreasing the amount of hydrogen used in methane production (Lila et al. 2004; Newbold et al. 2005). Dicarboxylic organic acids (fumarate and malate) stimulate production of propionic acid in the rumen, which improves hydrogen utilization (Song et al. 2011). Lopez et al. (1999) concluded that sodium fumarate is convenient feed additive as it redirects hydrogen from methane formation and increases the number of cellulolytic bacteria and stimulates fiber breakdown. An increasing dietary inclusion of malic acid in beef cattle diets resulted in a decrease of total daily emission of methane (Foley et al. 2009). It has been proposed that supplementation of fumarate and malate can diminish the risk of lactic acidosis through an enhanced utilization of lactic acid in the rumen and prevention of large decline in pH (Carro and Ungerfeld 2015).

6 Enzymes

Structurally, enzymes are mostly proteins capable of catalyzation of specific biochemical reactions. Each enzyme disintegrates very specific substrates for very specific reaction sites. The stability of enzymes during high-temperature feed production and transit through the gastrointestinal tract is significantly impacted by their protein nature. In order to achieve the most advantages from enzyme addition, it is important to make sure the enzymes are selected based on the substrates that are part of the specific composition of feed formulations. Furthermore, for optimal enzyme activity there must be a number of reaction conditions such as moisture content, temperature, pH, and enzyme and substrate concentration (Bedford and Schulze 1998).

In the feed industry, enzymes are the most important and useful additives. They can be obtained from various plants, animals, and microorganisms (Ojha et al. 2019). Enzymes are produced by the fermentation of fungi and bacteria and used as feed additives in the broiler industry to increase feed conversion efficiency (FCE) (Letlhogonolo et al. 2020). Exogenous enzymes are added to feed to improve animal performance despite the fact that animals naturally produce endogenous enzymes involved in digestion; they are unable to fully break down and utilize all of the nutritional components of feed (Ravindran 2013; Ojha et al. 2019). The most abundant exogenous enzymes in feed formulations are pectinases, amylases, cellulases, galactosidases, β-glucanases, xylanases, associated enzyme phytases, proteases, and lipases. Mostly, monogastric animals like pigs and chickens use these exogenous enzymes (Velázquez-De et al. 2021).

In broilers, xylanase has been shown to increase crude protein (CP) digestibility, feed intake, nitrogen and fiber absorption, and weight gain (Babalola et al. 2006). Phytases improve the utilization of phytate phosphorus in feeds (Khattak et al. 2006). Avizyme, a multi-enzyme complex made up of xylanases, proteases, and amylases, is used to increase net energy, increase body weight, decrease mortality, and improve nutritional quality (Café et al. 2002). Furthermore, it enhances animal intestinal health (Ohimain and Ofongo 2014).

According to numerous studies, exogenous feed enzymes are increasingly used in poultry diets to counteract the negative effects of anti-nutritional factors, enhance dietary component digestion, and enhance bird performance.

6.1 Mode of Action

Varieties of feed enzymes have different pathways or mechanisms. The following mechanisms are thought to be involved in at least some of the observed benefits (Bedford and Schulze 1998; Adeola and Cowieson 2011):

- Decomposition of specific bonds in the feed that typically do not hydrolyze by digestive enzymes existing in the body.

- Decomposition of anti-nutritional factors that directly restrict nutrient digestion and/or raise indirect viscosity of intestinal digestion.
- Breaking endosperm integrity resulting in the release of nutrients bound or contained by the cell wall.
- Changing digestion that is more effective.
- Decrease in endogenous secretions and protein losses from the gut.
- Decrease in the intestinal tract weight and modification of the intestinal morphology.
- Alterations in the microflora profile of the small intestine. Utilization of enzymes directly affects the gut having a significant impact on the bacteria that comprise communities of microflora.
- Enhance the internal digestive enzymes that are either not enough or lacking in the bird, leading to an improvement in digestion. This will be particularly accurate for young chickens with immature digestive systems.

Increased digesta viscosity and subsequent growth performance depression are caused by broiler diets high in non-starch polysaccharide (NSP) (Jia et al. 2009). To solve this problem carbohydrase enzymes are added to broiler diets, which enhances nutrient uptake and boosts their productivity. For instance, encapsulated nutrients are released when non-starch polysaccharides (NSPs) are hydrolyzed by carbohydrase into smaller oligosaccharides.

The advantages of prebiotics may also be present in the small oligosaccharides that are produced during NSP hydrolysis (Courtin et al. 2008). Prebiotic-like properties are produced when using enzymes to partially hydrolyze the hemicellulose found in agro-industrial byproducts, particularly Palm kernel expeller (PKE) and oligosaccharides (DP-6).

The enzymes start biological reactions and regulate how quickly substrates are transformed into useful products. Beneficial bacteria like *Bifidobacterium* and *Lactobacilli* spp. ferment the NSP hydrolysis products resulting in generation of short-chain fatty acids (SCFAs). Increased SCFA concentration is frequently correlated with an increase in the population of beneficial bacteria and a decrease in the population of pathogenic bacteria. Some SCFAs are also utilized by the host as a growth-supportive energy source (Lee et al. 2017).

Supplementing broilers with glucose oxidase (GOD) has been shown to increase daily body weight gain, enhance meat quality, and improve their ability to digest feed as evidenced by the nutrients' apparent digestibility and digestive enzyme levels. Principally, dietary GOD supplements may boost broiler growth performance in two ways: (1) by enhancing gut digestion, as evidenced by increased nutrient apparent digestibility and digestive enzyme, and (2) by boosting the population of good bacteria like *Faecalibacterium prausnitzii*, Ruminococcaceae, and Firmicutes (Wu et al. 2019a, b).

7 Amino Acids

Dietary amino acids (AAs) are essential for animal nutrition. They are crucial for building protein and represent ingredients necessary for desirable animal performances and profitability. Depending on whether an amino acid is synthesized in the body, it is traditionally classified as nutritionally essential amino acids (EAAs) or non-essential amino acids (NEAAs) (Han and Lee 2000).

Due to its favorable amino acid (AA) profile and high digestibility, soybean meal is the most commonly used protein source in the diets of monogastric farm animals like pigs and poultry (Cromwell 2000). Additionally, a significant part of the protein consumed by monogastric farm animals comes from cereals (e.g., maize, wheat, and barley) (Milašinović-Šeremešić et al. 2019; Milašinović-Šeremešić et al. 2021; Han and Lee 2000; Zilic et al. 2011). To meet the AA requirement, however, the AA composition of cereals and other plant protein sources like rapeseed meal is not as good as the AA composition of soybean meal (Milašinović-Šeremešić et al. 2021). Besides, the decrease in crude protein (CP) can lead to use of free AA (e.g., Lys, Met, Thr, Trp, and Val) to guarantee a sufficient supply of AA while making the animal diet more cost-effective and simple to prepare. L-lysine and DL-methionine are two of the crystalline amino acids that are produced in large quantities and are frequently used to replace or supplement natural protein sources that are deficient in lysine and methionine (Han and Lee 2000).

Dietary AA can be given as intact proteins, peptides, or free AA to farm animals with monogastric gastrointestinal systems. Free amino acids and small peptides can be absorbed by the gut directly, but intact proteins and larger peptides must first be hydrolyzed through digestion in the gastrointestinal tract (Wu 1998; Krehbiel and Matthews 2003). As a result, different dietary forms of AA are absorbed at various rates, which may have an impact on how AA is metabolized after eating (postprandial metabolism of AA).

Synthetic amino acid additions to animal diets are crucial not only from a nutritional and economic standpoint, but also from an environmental standpoint. Utilizing a highly utilizable source for animal feeds to reduce the nitrogen excretion in animal manure is the primary solution to the ecological problem (Garcia-Launay et al. 2014). Synthetic amino acids, as is well known, increase the digestibility of amino acids (nitrogen) and encourage high lean growth. It is just one of the crucial functions that synthetic amino acids play in animal nutrition. Animals' production of antibodies is also impacted by amino acids.

Commercial animal diets frequently include free AA in the form of pure, crystalline L-lysine, DL-methionine, L-threonine, L-tryptophan, and L-valine.

7.1 Mode of Action

Several studies have shown that certain AAs (e.g., Asn, Thr, Ser, and Val) have effects on many metabolic parameters; they modulate metabolic pathways (Wu 2010, 2013), regulate protein synthesis and turnover, maintain carbohydrate and AA metabolism (Dai et al. 2013; Wu 2010), and manage feed intake (Trevisi et al. 2018). Further, other AAs (e.g., Arg, Gln, Glu, Gly, and Trp) contribute to signaling, immunity, and antioxidative responses (Wu 2010). Gln supplementation in the diet of young broiler chickens has shown beneficial effects on the intestinal health with longer villi, higher intestinal IgG and IgA concentrations, and improved growth performance (Chalvon-Demersay et al. 2021; Bartell and Batal 2007). Some authors found that the diet supplementation with Ala, Ser, and Pro resulted in the decrease in stress hormones and stress-related behavior in young chicks (Kurauchi et al. 2009).

Recently, the research on the roles of AA in immune function has gotten more attention. Methionine acts as a methyl donor for vital synthesis of polyamines, DNA methylation, and is a precursor to form Cys, which is required to generate glutathione as a crucial antioxidant (Ruan et al. 2017). Tryptophan participates in numerous metabolic pathways including immune response and the synthesis of serotonin, a neurotransmitter involved in controlling appetite (Li et al. 2007). Further, threonine can be incorporated into immunoglobulins and preferentially utilized by the gut in pigs for the formation of mucosal proteins (Tang et al. 2021). Le Floc'h et al. (2018) found the health of growing pigs can be supported by the utilization of AAs such as Trp, Gln, Leu, and Ala as feed ingredients.

There is little data on the long-term effects on metabolism and physiology of dietary forms of AA because most recent studies focus on their short-term effects.

8 Functional Elements and Nanoparticles

Nanotechnology is obtaining nano-sized materials for medicinal, antimicrobial, medication, electronics, cosmetic, food packaging, and encapsulation purposes with increased efficiency. It uses organic and inorganic matter in trace amounts, typically between 1 and 100 nm (Gangadoo et al. 2016). Nanoparticles are increasingly being used in animal nutrition, particularly as feed additives, health boosters, and antimicrobials for animals. A small item known as a nanoparticle is one that functions as a single unit throughout transit and has a high level of bioavailability and biodegradability. Nano-feed additives can increase feed output and quality while lowering feed costs and boosting feed efficiency (El Sabry et al. 2018). Nano-biosensors could be used to find toxin-producing insects or fungi in bulk grain storage silos, which would help to increase feed quality and safety (Mathivanan 2021). They serve a variety of technological (binding, fluidifying, and anti-caking), nutritional (increasing nutrient digestibility, slowing transit speed), and hygienic (gastric and intestinal) purposes in animal feed. According to the available scientific evidence, materials that have undergone nano-scale manipulation through the application of nanotechnology may end up with novel physicochemical properties. The biological activity of a substance is the result of its physical properties, such as surface area, volume ratio, shape, surface features, and charge. Furthermore, biodistribution, biocompatibility, or toxicity of a material could be changed as a result of nanotechnology manipulation. For instance, altering a substance's particle size may have an impact on how it is absorbed and transported inside the body. The bioavailability of an animal food ingredient in a nanomaterial range may differ from larger-scaled material with the same chemical composition (Swain et al. 2015). For instance, while some tea polyphenols act as antioxidants in their bulk form, they act as pro-oxidants at the nano scale (Alotaibi et al. 2013). It may change the minimum amount of the animal food ingredient required to achieve the intended technical or nutritional effect (CVM GFI #220 Use of Nanomaterials in Food for Animals (2018)).

8.1 Mode of Action

Additionally, nanoparticles developed to safeguard fats in ruminant diets, to reduce rumen fermentation disruption, to protect amino acids, and increase their availability for the host animal, may be useful in the future (El Sabry et al. 2018).

There are successful and encouraging examples of application of various nanomaterials in poultry and cattle nutrition (Moyosore et al. 2019).

The principal application of nanotechnology in animal nutrition is the creation of nano-minerals, particularly trace minerals with limited bioavailability (Hill and Li 2017).

Minerals in nanoparticle form also lessen intestinal mineral antagonism, which lowers excretion and minimizes environmental contamination. According to studies, feeding livestock and poultry nanoparticles enhanced their performance by strengthening their immune and digestive system (Janocha et al. 2022; Gopi et al. 2017). Nano-forms of micro and macro elements most frequently increase the body weight, and daily weight gain, and improve the feed conversion ratio (FCR) (Bąkowski et al. 2018; Yusof et al. 2019). For instance, silver nanoparticles have a potent antibacterial action (Fondevila et al. 2009). However, use of nano-minerals, such as nano-selenium, nano-chromium, or nano-zinc, may improve animal production, health, and quality of animal products. Research has shown that nanomaterials can be used more effectively than the chelates and inorganic salts of these elements, widely used in the feed industry (Konkol and Wojnarowski 2018).

For example, nano-Zn can be a good substitute for traditional Zn sources in cattle feeding. Nano-Zn has shown several positive benefits, including encouraging growth and acting as an antimicrobial and immunomodulator. It supports all physiological processes and fulfils the same functions as traditional Zn sources. As a result, nano-Zn can be utilized in livestock feed at lower dosages than conventional Zn sources to produce superior results, indirectly preventing environmental damage (Partha et al. 2016). According to Prokhorova et al. (2016), feed additives containing microelements have shown a positive impact on the body of broiler chickens. Some surface-functionalized nano-components and nano-minerals can bind and eliminate viruses and poisons. For many years, as structural elements of soil, clays have been used for different purposes. Kaolinites, smectites, and sepiolites are the most frequent, due to their high surface area, cation exchange capability, and physical and chemical stability (Papaioannou et al. 2005). Among other qualities, clays have a high metal sorption capacity (Abad-Álvaro et al. 2019; Kumari and Mohan 2021). In recent years, mycotoxins have been frequent contaminant of animal feed and cereal grains (Kos et al. 2014, 2018; Čolović et al. 2019). The use of nutritionally inert adsorbents in the animal diet that bind mycotoxins is one of the most recent solutions of this global problem. It reduces intestinal absorption preventing toxic effects on cattle and the transfer of toxin compounds to animal products (Čolović et al. 2019).

9 Conclusions

Feed additives are utilized on a global scale for a variety of purposes. Due to the feed manufacturer's increased emphasis on extending the feed's shelf life and satisfying all of the nutritional needs of animals, the need for feed additives permanently increases. They may have a favorable impact on technological capabilities and product quality. Nevertheless, the health of animals with high growth rates is a key consideration when choosing feed additives.

It should be mentioned that there are numerous additives on the market today, all of varied quality and cost. Additionally, a lot of products could not share the same manufacturing and processing characteristics. In terms of business, it is crucial that the feed additive genuinely helps with enhancing production parameters or resolving particular issues. Today, the market for feed additives includes two types of products: dry and liquid. Because they are simple to mix with feed, easy to store, and simple to handle, the dry form is more favorable among feed producers. Customers now have more alternatives for mixing because of its availability in pellet and mash forms, which should assist the segment's growth. However, high reactivity of PFAs (EOs), interruption of their mode of action, degradation, oxidation, mixing with the basic feed constituents, and organoleptic properties are the main limits on the application and incorporation of PFAs into feed. Temperature, light, metals, and the availability of water and oxygen in the production processes can affect the stability and bioactivity of EOs. To overcome these constraints, cutting-edge technologies such as encapsulation and microencapsulation seem to be the best approaches to preserving the biological activities of EOs and minimizing their effects on feed sensory qualities.

Additionally, production costs should be in relation to the social responsibility of feed processing, which may include energy consumption, carbon footprint, water resource use, and life cycle assessment. Feed technology is a crucial component of animal production system that produces high-quality, nutrient-rich feed by processing components and creating animal feeds. Finally, the main goal is to improve nutrient use of compound feeds and conversion of poor ingredients into higher-value feed components. Therefore, animal feed producers have a social responsibility to support more environmentally friendly food production methods.

References

Abad-Álvaro I, Trujillo C, Bolea E, Laborda F, Fondevila M, Latorre MA, Castillo JR (2019) Silver nanoparticles-clays nanocomposites as feed additives: characterization of silver species released during in vitro digestions. Effects on silver retention in pigs. Microchem J 149:104040. https://doi.org/10.1016/j.microc.2019.104040

Abdulkhaleq LA, Assi MA, Abdullah R, Zamri-Saad M, Taufiq Yap YH, Hezmee MNM (2018) The crucial roles of inflammatory mediators in inflammation: a review. Veterinary World 11 (5):627–635. https://doi.org/10.14202/vetworld.2018.627-635

Abdulla NR, Zamri ANM, Sabow AB, Kareem KY, Nurhazirah S, Ling FH, Sazili AQ, Loh TC (2015) Physico-chemical properties of breast muscle in broiler chickens fed probiotics, antibiotics or antibiotic-probiotic mix. J Appl Anim Res 45(1):64–70

Abdurrahman ZH, Pramono YB, Suthama N (2016) Meat characteristic of crossbred local chicken fed inulin of Dahlia Tuber and Lactobacillus sp. Med Pet 39:112–118

Abreu ME, Müller M, Alegre L, Munné-Bosch S (2008) Phenolic diterpene and α-tocopherol contents in leaf extracts of 60 Salvia species. J Sci Food Agric 88:2648–2653

Adeola O, Cowieson AJ (2011) Opportunities and challenges in using exogenous enzymes to improve nonruminant animal production. J Anim Sci 89:3189–3218

Adil S, Tufail B, Gulam AB, Masood S, Manzoor R (2010) Effect of dietary supplementation of organic acids on performance, intestinal histomorphology, and serum biochemistry of broiler chicken. Vet Med Int:1–7. https://doi.org/10.4061/2010/479485

Ahmed ST, Hossain ME, Kim GM, Hwang JA, Ji H, Yang CJ (2013) Effects of resveratrol and essential oils on growth performance, immunity, digestibility and fecal microbial shedding in challenged piglets. Asian-Australas J Anim Sci 26(5):683–690

Akyurek H, Ozduven ML, Okur AA, Koc F, Samli HE (2011) The effect of supplementing an organic acid blend and/or microbial phytase to a corn-soybean based diet fed to broiler chickens. Afr J Agric Res 6:642–649. https://doi.org/10.5897/AJAR10.469

Al-Fatah MA (2020) Probiotic modes of action and its effect on biochemical parameters and growth performance in poultry. Iran J Appl Anim Sci 10(1):9–15

Alotaibi A, Bhatnagar P, Najafzadeh M, Gupta KC, Anderson D (2013) Tea phenols in bulk and nanoparticle form modify DNA damage in human lymphocytes from colon cancer patients and healthy individuals treated in vitro with platinum-based chemotherapeutic drugs. Nanomedicine (Lond) 8(3):389–401. https://doi.org/10.2217/nnm.12.126

Ahsan E, Kuter I, Raza BH, Köksal Ö, Cengiz M, Yıldız PK, Kızanlık M, Kaya O, Tatlı Ö, Sevim (2018) Dietary supplementation of different levels of phytogenic feed additive in broiler diets: the dynamics of growth performance caecal microbiota and intestinal morphometry brazilian. J Poult Sci 20(4):737–746. https://doi.org/10.1590/1806-9061-2017-0698

Ayalew H, Zhang H, Wang J, Wu S, Qiu K, Qi G, Tekeste A, Wassie T, Chanie D (2022) Potential feed additives as antibiotic alternatives in broiler production. Front Vet Sci 9:916473. https://doi.org/10.3389/fvets.2022.916473

Babalola TO, Apata DF, Atteh JO (2006) Effect of b-xylanase supplementation of boiled castor seed meal-based diets on the performance, nutrient absorbability and some blood constituents of pullet chicks. Trop Sci 46:216–223. https://doi.org/10.1002/ts.181

Bafundo KW, Cox LA, Bywater R (2002) Review lends perspective to recent scientific findings on virginiamycin, antibiotic resistance debate. Feed Stuffs 75:26–27

Bąkowski M, Kiczorowska B, Samolińska W, Klebaniuk R, Lipiec A (2018) Silver and zinc nanoparticles in animal nutrition—a review. Ann Anim Sci 18(4):879–898. https://doi.org/10.2478/aoas-2018-0029

Bartell S, Batal A (2007) The effect of supplemental glutamine on growth performance, development of the gastrointestinal tract, and humoral immune response of broilers. Poult Sci 86(9): 1940–1947. https://doi.org/10.1093/ps/86.9.1940

Bedford MR, Schulze H (1998) Exogenous enzymes in pigs and poultry. Nutr Res Rev 11:91–114

Biggs P, Parsons C, Fahey G (2007) The effects of several oligosaccharides on growth performance nutrient digestibilities and cecal microbial populations in young chicks. Poult Sci 86(11):2327–2336. https://doi.org/10.3382/ps.2007-00427

Bogusławska-Tryk M, Piotrowska A, Burlikowska K (2012) Dietary fructans and their potential beneficial influence on health and performance parametrs. In broiler chickens. J Cent Eur Agric [Internet] 13(2). https://hrcak.srce.hr/83284

Brzóska F, Stecka K (2007) Effect of probiotic, prebiotic and acidifier on the body weight of broiler chickens, feed conversion, and carcass and meat composition. Ann Anim Sci 7:279–288

Bulfon C, Volpatti D, Galeotti M (2015) Current research on the use of plant-derived products in farmed fish. Aquacult Res 46:513–551. https://doi.org/10.1111/are.12238

Burt S (2004) Essential oils: their antibacterial properties and potential applications in foods—a review. Int J Food Microbiol 94(3):223–253. https://doi.org/10.1016/j.ijfoodmicro.2004.03.022

Burt SA, Vlielander R, Haagsman HP, Veldhuizen EJ (2005) Increase in activity of essential oil components carvacrol and thymol against Escherichia coli O157:H7 by addition of food stabilizers. J Food Prot 68(5):919–926. https://doi.org/10.4315/0362-028x-68.5.919

Čabarkapa I, Đuragić O, Lj K (2016) Essential oils: mode of antimicrobial activity and potential application in food systems. Agro Food Ind Hi Tech 27:61–64

Čabarkapa I, Čolović R, Đuragić O, Popović S, Kokić B, Milanov D, Pezo L (2019) Antibiofilm activities of essential oils rich in carvacrol and thymol against Salmonella Enteritidis. Biofouling 35(3):361–375

Čabarkapa I, Puvača N, Popović S, Čolović D, Kostadinović LJ, Tatham E, Lević J (2020) Aromatic plants and their extracts pharmacokinetics and in vitro/in vivo mechanisms of action. In: Florou-Paneri P et al (eds) Aromatic plants and herbs in animal nutrition and health. Elsevier, pp 75–85

Café MB, Borges CA, Fritts CA, Waldroup PW (2002) Avizyme improves performance of broilers fed corn-soybean meal-based diets. J Appl Poultry Res 11:29–33. https://doi.org/10.1093/japr/11.1.29

Carro MD, Ungerfeld EM (2015) Utilization of organic acids to manipulate ruminal fermentation and improve ruminant productivity. In: Puniya A, Singh R, Kamra D (eds) Rumen microbiology: from evolution to revolution. Springer, New Delhi. https://doi.org/10.1007/978-81-322-2401-3_13

Chalvon-Demersay T, Luise D, Le-Floc'h N, Tesseraud S, Lambert W, Bosi P, Trevisi P, Beaumont M, Corrent E (2021) Functional amino acids in pigs and chickens: implication for gut health. Front Vet Sci 8:496. https://doi.org/10.3389/fvets.2021.663727

Chang SY, Song MH, Lee JH et al (2022) Phytogenic feed additives alleviate pathogenic Escherichia coli-induced intestinal damage through improving barrier integrity and inhibiting inflammation in weaned pigs. J Anim Sci Biotechnol 13:107. https://doi.org/10.1186/s40104-022-00750-y

Christian L, Mellor S (2011) The use of organic acids in animal nutrition, with special focus on dietary potassium deformity under European and Austral-Asian conditions. Recent Adv Anim Nutr Aus 4:123–130

Čolović R, Puvača N, Cheli F, Avantaggiato G, Greco A, Đuragić O, Kos J, Pinotti L (2019) Decontamination of mycotoxin-contaminated feedstuffs and compound feed. Toxins 11(11):617. https://doi.org/10.3390/toxins11110617

Costa LB, Luciano FB, Miyada VS, Gois FD (2013) Herbal extracts and organic acids as natural feed additives in pig diets. S Afr J Anim Sci 43(2):181–193. https://doi.org/10.4314/sajas.v43i2.9

Courtin C, Broekaert W, Swenen K, Lescroart O, Onagbesan O, Buyse J et al (2008) Dietary inclusion of wheat bran ararbinoxylooligosacchairdes induces beneficial nutritional effects in chickens. Cereal Chem 85:607–613. https://doi.org/10.1094/CCHEM-85-5-0607

Cromwell GL (2000) Utilization of soy products in swine diets. In: Drackley JK (ed) Soy in animal nutrition. Federal Animal Science Society, Savoy, pp 258–282

CVM GFI #220 Use of Nanomaterials in Food for Animals (2018)

Dai Z, Wu Z, Yang Y, Wang J, Satterfield MC, Meininger CJ, Bazer FW, Wu G (2013) Nitric oxide and energy metabolism in mammals. Biofactors 39(4):383–391. https://doi.org/10.1002/biof.1099

Dallies N, François J, Paquet V (1998) A new method for quantitative determination of polysaccharides in the yeast cell wall application to the cell wall defective mutants of Saccharomyces cerevisiae. Yeast 14:1297–1306. https://doi.org/10.1002/1097-0061(1998100)14:143.0.CO;2-L

Davani-Davari D, Negahdaripour M, Karimzadeh I, Seifan M, Mohkam M, Masoumi SJ, Berenjian A, Ghasemi Y (2019) Prebiotics: definition types sources mechanisms and clinical applications foods 8(3):92. https://doi.org/10.3390/foods8030092

Dias MC, Pinto DCGA, Silva AMS (2021) Plant flavonoids: chemical characteristics and biological activity. Molecules 26(17):5377. https://doi.org/10.3390/molecules26175377

Diether NE, Willing BP (2019) Microbial fermentation of dietary protein: an important factor in diet-microbe-host interaction. Microorganisms 7:19. https://doi.org/10.3390/microorganisms7010019

Douradinha B, Reis V, Rogers M, Torres F, Evans J, Marques E (2014) Novel insights in genetic transformation of the probiotic yeast Saccharomyces boulardii. Bioengineered 5(1):21–29

Droval AA, Benassi VT, Rossa A, Prudencio SH, Paiao FG, Shimokomaki M (2012) Consumer attitudes and preferences regarding pale: soft, and exudative broiler breast meat. J Appl Poultry Res 21:502–507

Du E, Wang W, Gan L, Li Z, Guo S, Guo Y (2016) Effects of thymol and carvacrol supplementation on intestinal integrity and immune responses of broiler chickens challenged with Clostridium perfringens. J Anim Sci Biotechnol 7:19. https://doi.org/10.1186/s40104-016-0079-7

Đuragić O, Sredanović S, Lević J (2007) Physical properties of feedstuffs. Contemp Agric 56(3–4): 157–162. https://agris.fao.org/agris-search/search.do?recordID=RS2007001186

EC (European Commission) (2003) Regulation 1831 of the European Parliament and of the Council of 22 September 2003 on additives for use in animal nutrition. Off J Eur Union L 268:29 https://eur-lex.europa.eu/legal-content/EN/TXT/?uri=CELEX%3A32003R1831

El Sabry M, McMillin K, Sabliov C (2018) Nanotechnology considerations for poultry and livestock production systems—a review. Ann Anim Sci 18(2):319–334. https://doi.org/10.1515/aoas-2017-0047

FAO (2019) Environmental performance of feed additives in livestock supply chains—guidelines for assessment (draft for public review). Livestock environmental assessment and performance partnership (LEAP). FAO, Rome, Italy https://www.fao.org/3/ca5562en/ca5562en.pdf

FAO/WHO (2002) Guidelines for the evaluation of probiotics in food. Food and Agriculture Organization of the United Nations and World Health Organization Working Group Report. ftp://ftp.fao.org/es/esn/food/wgreport2.pdf

Foley PA, Kenny DA, Callan JJ, Boland TM, O'Mara FP (2009) Effect of DL-malic acid supplementation on feed intake, methane emission, and rumen fermentation in beef cattle. J Anim Sci 87:1048–1057. https://doi.org/10.2527/jas.2008-1026

Fondevila M, Herrer R, Casallas MC, Abecia L, Ducha JJ (2009) Silver nanoparticles as a potential antimicrobial additive for weaned pigs. Anim Feed Sci Technol 150(3–4):259–269. https://doi.org/10.1016/j.anifeedsci.2008.09.003

Franz C, Baser KHC, Windisch W (2010) Essential oils and aromatic plants in animal feeding—an European perspective: a review. Flavour Fragr J 25:327–340. https://doi.org/10.1002/ffj.1967

Gadde U, Kim WH, Oh ST, Lillehoj HS (2017) Alternatives to antibiotics for maximizing growth performance and feed efficiency in poultry: a review. Anim Health Res Rev 18(1):26–45. https://doi.org/10.1017/S1466252316000207

Gaggìa F, Mattarelli P, Biavati B (2010) Probiotics and prebiotics in animal feeding for safe food production. Int J Food Microbiol 141:S15–S28. https://doi.org/10.1016/j.ijfoodmicro.2010.02.031

Gangadoo S, Stanley D, Hughes JR, Moore JR, Chapman J (2016) Nanoparticles in feed: progress and prospects in poultry research. Trends Food Sci Technol 58:115–126. https://doi.org/10.1016/j.tifs.2016.10.013

Garcia-Launay F, van der Werf H, Nguyen TTH, Le Tutour L, Dourmad JY (2014) Evaluation of the environmental implications of the incorporation of feed-use amino acids in pig production using life cycle assessment. Livest Sci 161:158–175. https://doi.org/10.1016/j.livsci.2013.11.027

Getachew T (2016) A review on effects of probiotic supplementation in poultry performance and cholesterol levels of egg and meat. J World's Poult Res 6:31–36

Gheisar MM, Kim IH (2018) Phytobiotics in poultry and swine nutrition—a review. Ital J Anim Sci 17(1):92–99. https://doi.org/10.1080/1828051X.2017.1350120

Gholami-Ahangaran M, Ahmadi-Dastgerdi A, Azizi S, Basiratpour A, Zokaei M, Derakhshan M (2022) Thymol and carvacrol supplementation in poultry health and performance. Vet Med Sci 8(1):267–288. https://doi.org/10.1002/vms3.663

Gibson G, Fuller R (2000) Aspects of in vitro and in vivo research approaches directed toward identifying probiotics and prebiotics for human use. J Nutr 130(2):391S–395S. https://doi.org/10.1093/jn/130.2.391S

Gómez-García M, Sol C, de Nova PJG et al (2019) Antimicrobial activity of a selection of organic acids, their salts and essential oils against swine enteropathogenic bacteria. Porc Health Manag 5:32. https://doi.org/10.1186/s40813-019-0139-4

Gopi M, Pearlin B, Kumar RD, Shanmathy M, Prabakar G (2017) Role of nanoparticles in animal and poultry nutrition: modes of action and applications in formulating feed additives and food processing. Int J Pharm 13:724–731. https://doi.org/10.3923/ijp.2017.724.731

Hačik P, Trembecka L, Bobko M, Čubon J, Bučko O, Tkačova J (2015) Evaluation of meat quality after application of different feed additives in diet of broiler chickens. Potravinarstvo 9:174–182

Han I, Lee J (2000) The role of synthetic amino acids in monogastric animal production—review. Anim Biosci 13(4):543–560. https://doi.org/10.5713/ajas.2000.543

Hassan HM, Mohamed MA, Youssef AW, Hassan ER (2010) Effect of using organic acids to substitute antibiotic growth promoters on performance and intestinal microflora of broilers. Asian-Australas J Anim Sci 20(23):1348–1353

Hill EK, Li J (2017) Current and future prospects for nanotechnology in animal production. J Anim Sci Biotechnol 8:26. https://doi.org/10.1186/s40104-017-0157-5

Hill C, Guarner F, Reid G, Gibson GR, Merenstein DJ, Pot B, Morelli L, Canani RB, Flint HJ, Salminen S, Calder PC, Sanders ME (2014) Expert consensus document: the international scientific association for probiotics and prebiotics consensus statement on the scope and appropriate use of the term probiotic. Nat Rev Gastroenterol Hepatol 11:506–514

Hossain ME, Ko SY, Kim GM, Firman JD, Yang CJ (2012) Evaluation of probiotic strains for development of fermented Alisma canaliculatum and their effects on broiler chickens. Poult Sci 91:3121–3131. https://doi.org/10.3382/ps.2012-02333

Huyghebaert G, Ducatelle R, Van Immerseel F (2011) An update on alternatives to antimicrobial growth promoters for broilers. Vet J 187:182–188. https://doi.org/10.1016/j.tvjl.2010.03.003

Ikker P, Dirkzwager A, Fledderus J, Trevisi P, Le Huërou-Luron I, Lallès J, Awati A (2007) Dietary protein and fermentable carbohydrates contents influence growth performance and intestinal characteristics in newly weaned pigs. Livest Sci 108:194–197. https://doi.org/10.1016/j.livsci.2007.01.057

Innocenti ML (2017) Guidance on the identity, characterisation and conditions of use of feed additives. EFSA J 15(10):5023. https://doi.org/10.2903/j.efsa.2017.5023

Janocha A, Milczarek A, Kosmalski M, Gajownik-Mućka P, Radzikowski D (2022) Effect of feed additives supplementation on the growth performance, gastrointestinal tract characteristics, and carcass composition in Turkey Hens. Animals 12(24):3464. https://doi.org/10.3390/ani12243464

Jia W, Slominski BA, Bruce HL, Blank G, Crow G, Jones O (2009) Effects of diet type and enzyme addition on growth performance and gut health of broiler chickens during sub clinical clostridium perfingens challenge. Poult Sci 88:132–140. https://doi.org/10.3382/ps.2008-00204

Khan RU, Naz S (2013) The applications of probiotics in poultry production. World's Poult Sci 69: 621–631

Khattak FM, Pasha TN, Hayat Z, Mahmud A (2006) Enzymes in poultry nutrition. J Anim Plant Sci 16:1–2

Kikusato M (2021) Phytobiotics to improve health and production of broiler chickens: functions beyond the antioxidant activity. Anim Biosci 34(3):345–353. https://doi.org/10.5713/ab.20.0842

Kim T, Song B, Cho KS, Lee IS (2020) Therapeutic potential of volatile terpenes and terpenoids from forests for inflammatory diseases. Int J Mol Sci 21(6):2187. https://doi.org/10.3390/ijms21062187

Konkol D, Wojnarowski K (2018) The use of nanominerals in animal nutrition as a way to improve the composition and quality of animal products. J Chem. 5927058, 7 p. https://doi.org/10.1155/2018/5927058

Kos J, Lević J, Đuragić O, Kokić B, Miladinović I (2014) Occurrence and estimation of aflatoxin M1 exposure in milk in Serbia. Food Control 38:41–46. https://doi.org/10.1016/j.foodcont.2013.09.060

Kos J, Janic Hajnal E, Šaric B, Jovanov P, Mandic A, Đuragic O, Kokic B (2018) Aflatoxins in maize harvested in the Republic of Serbia over the period 2012–2016. Food Addit Contam B 11: 246–255. https://doi.org/10.1080/19393210.2018.1499675

Král M, Angelovičova M, Alfaig E, Walczycka M (2013) Meat quality of broiler chickens fed diets with Bacillus subtilis and malic acid additives. Sci Paper Anim Sci Biotechnol 46:375–378

Krehbiel C, Matthews J (2003) Absorption of amino acids and peptides. In: D'Mello J (ed) Amino acids in animal nutrition, 2nd edn. CABI International, Wallingford, pp 41–70. https://doi.org/10.1079/9780851996547.0041

Kulshreshtha Nupur S, Bhatnagar P (2014) Mushroom as a product and their role in mycoremediation AMB. Express 4(1). https://doi.org/10.1186/s13568-014-0029-8

Kumar A, Toghyani M, Kheravii SK, Pineda L, Han Y, Swick RA, Wu SB (2022) Organic acid blends improve intestinal integrity, modulate short-chain fatty acids profiles and alter microbiota of broilers under necrotic enteritis challenge. Anim Nutr 8(1):82–90. https://doi.org/10.1016/j.aninu.2021.04.003

Kumari N, Mohan C (2021) In: Nascimento G (ed) Basics of clay minerals and their characteristic properties in clay and clay minerals. IntechOpen, London. https://doi.org/10.5772/intechopen.97672

Kurauchi I, Yamane H, Tsuneyoshi Y, Denbow DM, Furuse M (2009) Central l-alanine reduces energy expenditure with a hypnotic effect under an acute stressful condition in neonatal chicks. Amino Acids 36(1):131–135. https://doi.org/10.1007/s00726-008-0042-9

Lambert RJ, Skandamis PN, Coote PJ, Nychas GJ (2001) A study of the minimum inhibitory concentration and mode of action of oregano essential oil, thymol and carvacrol. J Appl Microbiol 91(3):453–462. https://doi.org/10.1046/j.1365-2672.2001.01428.x

Lan Y, Verstegen MWA, Tamminga S, Williams BA (2005) The role of the commensal gut microbial community in broiler chickens. Worlds Poult Sci J 61:95–104. https://doi.org/10.1079/WPS200445

Le Floc'h N, Wessels A, Corrent E, Wu G, Bosi P (2018) The relevance of functional amino acids to support the health of growing pigs. Anim Feed Sci Technol 245:104–116. https://doi.org/10.1016/j.anifeedsci.2018.09.007

Lee SH, Lillehoj HS, Jang SI, Lee KW, Park MS, Bravo D, Lillehoj EP (2011) Cinnamaldehyde enhances in vitro parameters of immunity and reduces in vivo infection against avian coccidiosis. Br J Nutr 106(6):862–869. https://doi.org/10.1017/S0007114511001073

Lee S, Apajalahti K, Vienola G, Gonzalez-Ortiz C, Bedford M (2017) Age and dietary xylanase supplementation affects ileal sugar residues and short chain fatty acid concentration in the ileum and caecum of broiler chickens. Anim Feed Sci Technol 234:29–42. https://doi.org/10.1016/j.anifeedsci.2017.07.017

Letlhogonolo AS, Zahra MH, Tlou GM, Monnye M (2020) The current status of the alternative use to antibiotics in poultry production: an African perspective. Antibiotics 9:1–18. https://doi.org/10.3390/antibiotics9090594

Lević J, Sredanović S, Đuragić O, Jakić D, Lević LJ, Pavkov S (2007) New feed additives based on phytogenics and acidifiers in animal nutrition. Biotechnol Anim Husb 23(5-6-1):527–534. https://doi.org/10.2298/BAH0701527L

Li P, Yin Y, Li D, Woo Kim S, Wu G (2007) Amino acids and immune function. Br J Nutr 98(2): 237–252. https://doi.org/10.1017/s000711450769936x

Li Y, Kong D, Fu Y, Sussman MR, Wu H (2020) The effect of developmental and environmental factors on secondary metabolites in medicinal plants. Plant Physiol Biochem 148:80–89

Lila ZA, Mohammed N, Yasui T, Kurokawa Y, Kanda S, Itabashi H (2004) Effects of a twin strain of *Saccharomyces cerevisiae* live cells on mixed ruminal microorganism fermentation in vitro. J Anim Sci 82:1847–1854. https://doi.org/10.2527/2004.8261847x

Lillehoj H, Liu Y, Calsamiglia S, Miyakawa MEF, Chi F, Craven RL, Oh S, Gay CG (2018) Phytochemicals as antibiotic alternatives to promote growth and enhance host health. Vet Res 49:1–18

Liotta L, Chiofalo V, D'Alessandro E, Lo Presti V, Chiofalo B (2015) Supplementation of Rosemary extract in the diet of Nero Siciliano pigs: evaluation of the antioxidant properties on meat quality. Animal 9(6):1065–1072. https://doi.org/10.1017/S1751731115000348

Lobo V, Patil A, Phatak A, Chandra N (2010) Free radicals, antioxidants and functional foods: impact on human health. Pharmacogn Rev 4(8):118–126. https://doi.org/10.4103/0973-7847.70902

Lopez S, Valdez C, Newbold CJ, Wallace RJ (1999) Influence of sodium fumarate on rumen fermentation in vitro. Br J Nutr 81:59–64. https://doi.org/10.1017/S000711459900015X

Manafi M, Hedayati M, Khalaji S (2016) Effectiveness of phytogenic feed additive as alternative to bacitracin methylene disalicylate on haematological parameters,intestinal histomorphology and microbial population and production performance of japanese quails. Asian-Australas J Anim Sci 29:1300–1308

Mani-Lopez E, García HS, López-Malo A (2012) Organic acids as antimicrobials to control Salmonella in meat and poultry products. Food Res Int 45:713–721. https://doi.org/10.1016/j.foodres.2011.04.043

Marinelli L, Di Stefano A, Cacciatore I (2018) Carvacrol and its derivatives as antibacterial agents. Phytochem Rev 17:903–921. https://doi.org/10.1007/s11101-018-9569-x

Markowiak P, Slizewska K (2017) Effects of probiotics, prebiotics, and synbiotics on human health. Nutrients 9:1–30

Marteau P, Seksik P, Lepage P, Dore J (2004) Cellular and physiological effects of probiotics and prebiotics. Mini Rev Med Chem 4(8):889–896

Mathivanan S (2021) Perspectives of nano-materials and nanobiosensors in food safety and agriculture. Novel nanomaterials. IntechOpen. https://doi.org/10.5772/intechopen.95345

Mazanko M, Gorlov I, Prazdnova E, Makarenko M, Usatov A, Bren A, Chistyakov V, Tutelyan A, Komarova Z, Mosolova N, Pilipenko D, Krotova O, Struk A, Lin A, Chikindas M (2018) Bacillus probiotic supplementations improve laying performance egg quality hatching of laying hens and sperm quality of roosters probiotics and antimicrobial. Proteins 10(2):367–373. https://doi.org/10.1007/s12602-017-9369-4

Milašinović M, Radosavljević M, Lj D (2010) Effects of autoclaving and pullulanase debranching on the resistant starch yield of normal maize starch. J Serb Chem Soc 75(4):449–458. https://doi.org/10.2298/JSC090904027M

Milašinović-Šeremešić M, Radosavljević M, Srdić J, Tomičić Z, Đuragić O (2019) Physical traits and nutritional quality of selected Serbian maize genotypes differing in kernel hardness and colour. Food Feed Res 46(1):51–59

Milašinović-Šeremešić M, Đuragić O, Radosavljević M, Dokić LJ (2021) Chapter 2, Maize (Zea mays L.) suitability for wet milling and animal nutrition in relation to physical and chemical quality parameters. In: Dunn S (ed) Zea mays L.: cultivation, and uses. Nova Science Publishers, New York, pp 51–81. ISBN: 978-1-53619-181-3

Moyosore A, Elghandour M, Barbabosa-Pliego A, Cedillo Monroy J, Mellado M, Ravi Kanth Reddy P, Salem A (2019) Nanoparticles in equine nutrition: mechanism of action and application as feed additives. J Equine Vet 78:29–37. https://doi.org/10.1016/j.jevs.2019.04.001

Mubeen B, Ansar AN, Rasool R, Ullah I, Imam SS, Alshehri S, Ghoneim MM, Alzarea SI, Nadeem MS, Kazmi I (2021) Nanotechnology as a novel approach in combating microbes providing an alternative to antibiotics. Antibiotics (Basel) 10(12):1473. https://doi.org/10.3390/antibiotics10121473

Murate L, Paião F, De Almeida A, Berchieri A Jr, Shimokomaki M (2014) Efficacy of prebiotics, probiotics, and Synbiotics on laying hens and broilers challenged with salmonella Enteritidis. J Poult Sci 52(1):52–56. https://doi.org/10.2141/jpsa.0130211

Newbold CJ, Lopez S, Nelson N, Ouda JO, Wallace RJ, Moss AR (2005) Propionate precursors and other metabolic intermediates as possible alternative electron acceptors to methanogenesis in ruminal fermentation in vitro. Br J Nutr 94:27–35. https://doi.org/10.1079/BJN20051445

O'Hara AM, Shanahan F (2007) Mechanisms of action of probiotics in intestinal diseases. Sci World J 7:31–46

Oetting LL, Utiyama CE, Giani PA, Ruiz US, Miyada VS (2006) Effects of antimicrobials and herbal extracts on intestinal microbiology and diahrrea incidence in weanling pigs. Rev Bras De Zootec 35:2013–2017

Ohimain EI, Ofongo R (2014) Enzyme supplemented poultry diets: benefits so far—a review. Int J Adv Biotechnol Res 3:31–39

Ojha BK, Singh PK, Shrivastava N (2019) Enzymes in the animal feed industry. In: Mohammed K (ed) Enzymes in food biotechnology. Academic Press, Cambridge, MA, pp 93–109. https://doi.org/10.1016/B978-0-12-813280-7.00007-4

Olfati A, Hosseini S (2022) The effects of dietary supplementation of encapsulated thyme essential oil on growth pro-inflammatory cytokines and serum amino acid profiles of broiler chicks challenged with salmonella typhimurium abstract annals. Anim Sci 22(1):189–200. https://doi.org/10.2478/aoas-2021-0029

Papaioannou D, Katsoulos PD, Panousis N, Karatzias H (2005) The role of natural and synthetic zeolites as feed additives on the prevention and/or the treatment of certain farm animal diseases: a review. Microporous Mesoporous Mater 84(1):161–170. https://doi.org/10.1016/j.micromeso.2005.05.030

Partanen KH, Mroz Z (1999) Organic acids for performance enhancement in pig diets. Nutr Res Rew 12(1):117–145. https://doi.org/10.1079/095442299108728884

Partha S, Somu R, Rajendran D, Dominic G, Selvaraju S (2016) Nano zinc, an alternative to conventional zinc as animal feed supplement: a review. Anim Nutr 2:3. https://doi.org/10.1016/j.aninu.2016.06.003

Pearlin BV, Muthuvel S, Govidasamy P, Villavan M, Alagawany M, Ragab Farag M, Dhama K, Gopi M (2020) Role of acidifiers in livestock nutrition and health: a review. J Anim Physiol Anim Nutr 104(2):558–569

Piran Filho FA, Turner TD, Mueller I, Daniel JLP (2021) Influence of phytogenic feed additive on performance of feedlot cattle. Front Anim Sci. https://doi.org/10.3389/fanim.2021.767034

Pisarski RK, Szkucik K, Zieba M (2007) The chemical composition of muscles and sensory traits of meat in broiler chickens fed herbs substituted for antibiotic growth promoter. Herba Polonica 53:350–354

Popova T (2017) Effect of probiotics in poultry for improving meat quality. Curr Opin Food Sci 14: 72–77

Po Yun T, Woo K (2018) Review: roles of prebiotics in intestinal ecosystem of broilers frontiers. Vet Sci 5. https://doi.org/10.3389/fvets.2018.00245

Prokhorova YV, Gavrikov AV, Yoshchik VV (2016) The significance of trace elements for poultry. Poultry Farming 6:32–35

Prommachart R, Cherdthong A, Navanukraw C, Pongdontri P, Taron W, Uriyapongson J, Uriyapongson S (2021) Effect of dietary anthocyanin-extracted residue on meat oxidation and fatty acid profile of male dairy cattle. Animals (Basel) 11(2):322. https://doi.org/10.3390/ani11020322

Puvača N, Đuragić O, Popović S (2018) Aromatic and medicinal herbs in the diet of broiler chickens. Monographyc Publication, University of Novi Sad, Institute of Food Technology, pp 1–123. ISBN: 978-86-7993-053-7

Puvača N, Čabarkapa I, Petrović A, Bursić V, Prodanović R, Soleša D, Lević J (2019) Tea tree (Melaleuca alternifolia) and its essential oil: antimicrobial, antioxidant and acaricidal effects in poultry production. Worlds Poult Sci J 75(2):235–246

Qaisrani S, Van Krimpen M, Kwakkel R, Verstegen M, Hendriks W (2015) Diet structure, butyric acid, and fermentable carbohydrates influence growth performance, gut morphology, and cecal fermentation characteristics in broilers. Poult Sci 94:2152–2164. https://doi.org/10.3382/ps/pev003

Ravindran V (2013) Feed enzymes: the science, practice, and metabolic realities. J Appl Poultry Res 22:628–636. https://doi.org/10.3382/japr.2013-00739

Ruan T, Li L, Peng X, Wu B (2017) Effects of methionine on the immune function in animals. Health 9:857–869. https://doi.org/10.4236/health.2017.95061

Rychen G, Aquilina G, Azimonti G, Bampidis V, Bastos ML, Borie G, Chesson A, Cocconcelli PS, Flachowsky G, Gropp J, Kolar B, Kouba M, López-Alonso M, López Puente S, Mantovani A, Mayo B, Ramos F, Saarela M, Villa RE, Wallace RJ, Wester P, Anguita M, Galobart J, Innocenti ML (2017) Guidance on the identity, characterisation and conditions of use of feed additives. EFSA J 15(10):5023. https://www.efsa.europa.eu/en/efsajournal/pub/5023#:~:text=https%3A//. https://doi.org/10.2903/j.efsa.2017.5023

Sinovec Z, Markovic R (2005) Use of pre-biotics in poultry nutrition Biotechnology in Animal Husbandry Biotehnologija u stocarstvu 21(5–6):235–239. https://doi.org/10.2298/BAH0506235S

Song MK, Li XZ, Oh YK, Lee CK, Hyun Y (2011) Control of methane emission in ruminants and industrial application of biogas from livestock manure in Korea. Asian-Aust J Anim Sci 24(1): 130–136. https://doi.org/10.5713/ajas.2011.r.02

Steiner T, Syed B (2015) Phytogenic feed additives in animal nutrition. In: Máthé Á (ed) Medicinal and aromatic plants of the world. Medicinal and aromatic plants of the world, vol 1. Springer, Dordrecht. https://doi.org/10.1007/978-94-017-9810-5_20

Suiryanrayna MV, Ramana JV (2015) A review of the effects of dietary organic acids fed to swine. J Animal Sci Biotechnol 6(45). https://doi.org/10.1186/s40104-015-0042-z

Sutton AL, Mathew AG, Scheidt AB, Patterson JA, Kelly DT (1991) Effects of carbohydrate sources and organic acids on in-testinal microflora and performance of the weanling pig. Paper presented at 5th international symposium on digestive physiology in the pigs, Wageningen, The Netherlands, 24–26 Apr 1991

Swain PS, Rajendran D, Rao SB, Dominic G (2015) Preparation and effects of nano mineral particle feeding in livestock: a review. Vet World:888–891. https://doi.org/10.14202/vetworld.2015.888-891

Syed B, Kesselring J, Sánchez J, Gracia M (2021) Growth performance and nutrient digestibility in broiler chickens fed with an encapsulated blend of a phytogenic feed additive. J World Poult Res 11(3):278–285. https://doi.org/10.36380/jwpr.2021.33

Tang Q, Tan P, Ma N, Ma X (2021) Physiological functions of threonine in animals: beyond nutrition metabolism. Nutrients 13(8):2592. https://doi.org/10.3390/nu13082592

Tiago FCP, Martins FS, Souza ELS, Pimenta PFP, Araujo HRC, Castro IM, Brandão RL, Nicoli JR (2012) Adhesion to the yeast cell surface as a mechanism for trapping pathogenic bacteria by Saccharomyces probiotics. J Med Microbiol 61(9):1194–1207

Tizard I (2013) Veterinary immunology - E-book, ninth Edition, Elsevier/Saunders, St. Louis, Mo. ISBN: 9781455703630

Tomičić Z, Čolović R, Čabarkapa I, Vukmirović Đ, Đuragić O, Tomičić R (2016a) Beneficial properties of probiotic yeast Saccharomyces boulardii. Food Feed Res 43(2):103–110

Tomičić Z, Zupan J, Matos T, Raspor P (2016b) Probiotic yeast Saccharomyces boulardii (nom. nud.) modulates adhesive properties of Candida glabrata. Med Mycol 54(8):835–845

Tounekti T, Munné-Bosch S (2012) Enhanced phenolic diterpenes antioxidant levels through non-transgenic approaches. Crit Rev Plant Sci 31:505–519. https://doi.org/10.1080/07352689.2012.696457

Trejo FM, Minnaard J, Perez PF, De Antoni GL (2006) Inhibition of Clostridium difficile growth and adhesion to enterocytes by Bifidobacterium supernatants. Anaerobe 12(4):186–193

Trevisi P, Priori D, Jansman AJ, Luise D, Koopmans SJ, Hynönen U, Palva A, Van Der Meulen J, Bosi P (2018) Molecular networks affected by neonatal microbial colonization in porcine jejunum, luminally perfused with enterotoxigenic Escherichia coli, f4ac fimbria or Lactobacillus amylovorus. PloS One 13(8):e0202160. https://doi.org/10.1371/journal.pone.0202160

Troy E and Kasper D (2010) Beneficial effects of bacteroides fragilis polysaccharides on the immune system Frontiers. Biosci 15(1):25. https://doi.org/10.2741/3603

Ultee A, Slump RA, Steging G, Smid EJ (2000) Antimicrobial activity of carvacrol toward Bacillus cereus on rice. J Food Prot 63(5):620–624. https://doi.org/10.4315/0362-028x-63.5.620

Ultee A, Bennik MH, Moezelaar R (2002) The phenolic hydroxyl group of carvacrol is essential for action against the food-borne pathogen Bacillus cereus. Appl Environ Microbiol 68(4): 1561–1568. https://doi.org/10.1128/AEM.68.4.1561-1568.2002

Velázquez-De L, Brianda S, Hernández-Domínguez EM, Villa-García M, Gerardo Díaz-Godínez G, Mandujano-Gonzalez V, Mendoza-Mendoza B, Álvarez-Cervantes J (2021) Exogenous enzymes as zootechnical additives in animal feed: a review. Catalysts 11(7):851. https://doi.org/10.3390/catal11070851

Vukic-Vrajnes M, Tolimir N, Vukmirovic D, Colovic R, Stanacev V, Ikonic P, Pavkov S (2013) Effect of phytogenic additives on performance, morphology and caecal microflora of broiler chickens. Biotechnol Anim Husb 29:311–319

Vukmirović ĐM, Rakita SM, Spasevski NJ, Kokić BM, Banjac VV, Čabarkapa IS (2017) A review of possibilities for control of Salmonella and other pathogenic bacteria in pig feed. Food Feed Res 44(2):151–162

Wang J, Su S, Pender C, Murugesan R, Syed B, Kim WK (2021) Effect of a phytogenic feed additive on growth performance, nutrient digestion, and immune response in broiler-fed diets with two different levels of crude protein. Animals (Basel) 11(3):775. https://doi.org/10.3390/ani11030775

Windisch W, Schedle K, Plitzner C, Kroismayr A (2008) Use of phytogenic products as feed additives for swine and poultry. J Anim Sci 86(14 Suppl):E140–E148. https://doi.org/10.2527/jas.2007-0459

Wu G (1998) Intestinal mucosal amino acid catabolism. J Nutr 128(8):1249–1252. https://doi.org/10.1093/jn/128.8.1249

Wu G (2010) Functional amino acids in growth, reproduction, and health. Adv Nutr 1(1):31–37. https://doi.org/10.3945/an.110.1008

Wu G (2013) Functional amino acids in nutrition and health. Amino Acids 45(3):407–411. https://doi.org/10.1007/s00726-013-1500-6

Wu S, Li T, Niu H, Zhu Y, Liu Y, Duan Y, Sun Q, Yang X (2019a) Effects of glucose oxidase on growth performance, gut function, and cecal microbiota of broiler chickens. Poult Sci 98(2): 828–841. https://doi.org/10.3382/ps/pey393

Wu Y, Wang B, Zeng Z, Liu R, Tang L, Gong L, Weifen L (2019b) Effects of probiotics Lactobacillus plantarum 16 and Paenibacillus polymyxa 10 on intestinal barrier function, antioxidative capacity, apoptosis, immune response, and biochemical parameters in broilers. Poult Sci 98(10):5028–5039

Wu S, Pang Y, He Y, Zhang X, Peng L, Guo J et al (2021) A comprehensive review of natural products against atopic dermatitis: flavonoids, alkaloids, terpenes, glycosides and other compounds. Biomed Pharmacother 140:111741. https://doi.org/10.1016/j.biopha.2021.111741

Yaqoob M, Wang G, Wang M (2022) An updated review on probiotics as an alternative of antibiotics in poultry—a review. Anim Biosci 35(8):1109–1120. https://doi.org/10.5713/ab.21.0485

Yang C, Kennes YM, Lepp D, Yin X, Wang Q, Yu H, Yang C, Gong J, Diarra MS (2020) Effects of encapsulated cinnamaldehyde and citral on the performance and cecal microbiota of broilers vaccinated or not vaccinated against coccidiosis. Poult Sci 99:936–948. https://doi.org/10.1016/j.psj.2019.10.036

Yusof HM, Mohamad R, Zaidan UH, Rahman NAA (2019) Microbial synthesis of zinc oxide nanoparticles and their potential application as an antimicrobial agent and a feed supplement in animal industry: a review. J Anim Sci Biotechnol 10(57). https://doi.org/10.1186/s40104-019-0368-z

Zilic S, Milasinovic M, Terzic D, Barac M, Ignjatovic-Micic D (2011) Grain characteristics and composition of maize specialty hybrids. Span J Agric Res 9(1):230–241. https://doi.org/10.5424/sjar/20110901-053-10

Food Additives as Functional Ingredients in Food Products

Sonia A. Socaci, Liana Salanţă, Anca C. Fărcaş, Silvia Amalia Nemeş, and Mădălina Medeleanu

Abstract

In recent years, there has been a notable advancement in the field of food science and technology, with a growing emphasis on adopting sustainable practices in food production. This includes a focus on the selection and sourcing of food additives. Food additives are widely acknowledged for their significant contribution to enhancing the overall quality and stability of food products. These additives serve various purposes, such as imparting color, flavor, and texture, as well as preserving the food and enhancing its resistance to transportation and handling. The utilization of additives is deemed acceptable solely when they contribute to enhancing the overall quality and organoleptic characteristics of food while posing no threats to humans' health. The physicochemical and sensory properties of food products may undergo changes due to various deterioration processes, including microbiological, enzymatic, physical, and chemical factors. These processes can lead to a decline in both the nutritional quality and food safety of the products. In order to safeguard the well-being of consumers, preservatives (antimicrobials, and anti-browning agents), antioxidants from

S. A. Socaci (✉) · A. C. Fărcaş · M. Medeleanu
Faculty of Food Science and Technology, University of Agricultural Sciences and Veterinary Medicine Cluj-Napoca, Cluj-Napoca, Romania

Institute of Life Sciences, University of Agricultural Sciences and Veterinary Medicine Cluj-Napoca, Cluj-Napoca, Romania
e-mail: sonia.socaci@usamvcluj.ro

L. Salanţă
Faculty of Food Science and Technology, University of Agricultural Sciences and Veterinary Medicine Cluj-Napoca, Cluj-Napoca, Romania

S. A. Nemeş
Institute of Life Sciences, University of Agricultural Sciences and Veterinary Medicine Cluj-Napoca, Cluj-Napoca, Romania

G. Arsenos, I. Giannenas (eds.), *Sustainable Use of Feed Additives in Livestock*,
https://doi.org/10.1007/978-3-031-42855-5_3

natural sources, natural pigments as food colorants (anthocyanins, carotenoids, betalains, and chlorophyll), and hydrocolloids (carboxymethylcellulose, xanthan, β-glucans) are included in the food. Unfortunately, depending on the dosage, there is a fine line between safe and hazardous. However, regulatory agencies and law enforcement organizations are responsible for overseeing the development of the food industry and have implemented rigorous laws to regulate the licensing and supervision of food additives.

Keywords

Functional foods · Antioxidants · Pigments · Food colorants · Hydrocolloids

1 Introduction

The global food system is kept under control by the national and international authorities responsible for food quality, safety, and security (those from the USA and Europe being the most important) (Wu et al. 2022b). One of the most significant roles of European legislation is laying down the rules on food additives used in the food industry, ensuring a high level of consumer protection, including human health. Even though there are some inconsistencies regarding the approval and use of certain food additives between the European Union (EU) and the USA (e.g., additives permitted in some countries are banned in others such as sodium sorbate, fast green, and fluorescein, which are permitted in the USA but prohibited in the EU, or amaranth and carmoisine dyes, which are prohibited in the USA but allowed in the EU [M'Arcio Carocho et al. 2014]), the responsible authorities have a common principle: to ensure a clean and safe alimentation for consumers worldwide. The European Food Safety Authority (EFSA) and the European Commission, based on EFSA assessment and regulations (Cox et al. 2021), define a food additive as a "substance not normally consumed as a food in itself and not normally used as a characteristic ingredient of food, whether or not it has nutritive value, the intentional addition of which to food for a technological purpose in the manufacture, processing, preparation, treatment, packaging, transport, or storage of such food results, or may be reasonably expected to result, in it or its by-products becoming directly or indirectly a component of such foods" (https://eur-lex.europa.eu/legal-content/EN/TXT/?uri=celex%3A32008R1333 (EU), C. R. Regulation (EC) No 1333/2008 of the European Parliament and of the Council of 16 December 2008 on Food Additives).

The use of food additives has had ancient roots in history; in the time of Homer, Greeks used a mixture of salt and sodium nitrite to preserve meat, while, 3000 years ago, Egyptians used sulfur dioxide for wine preservation (Saltmarsh 2013). For preservation and to improve the look of meals, the Romans utilized potassium nitrate, spices, and colors (EUFIC 2021). Due to social, scientific, and technological evolution, nowadays food additives gained interest in society and the population is increasingly concerned about the composition of the food. Currently, in Europe over 330 authorized additives are used to increase the food's quality and shelf life (40 are colors, 19 are sweeteners, and 275 are other than colors and sweeteners) (Eloi Chazelas et al. 2020).

Food additives are recognized for their role in enhancing quality and stability, being used to color, flavor, preserve, or improve the texture of food, or conferring resistance to transportation and handling. Their use is justified only when they bring a benefit to the quality of the food and/or its organoleptic properties without presenting risks to public health.

Firstly, an additive can be considered suitable to be placed on the market and used in food only if it passes the approval process for new additives (Wu et al. 2022b). EFSA Panel on Food Additives and Nutrient Sources added to Food (ANS) from March 7, 2021, refers to the application for the authorization or re-authorization of a new additive, and of an already authorized additive, respectively. This document is built on four sections (European Food Safety Authority [EFSA] 2021):

(A) Chemistry and specifications: the section aims to identify the food additive, and potential hazards from its manufacturing, and to define the material tested through specifications.
(B) Existing authorizations and evaluation: the purpose of this section is to provide an overview of previous risk assessments on the additive and their conclusions.
(C) Proposed uses and exposure assessment: the section aims to estimate dietary exposure based on the proposed uses and use levels, as well as the consumption of the proposed foods by various age groups in the EU population.
(D) Toxicological studies: the purpose of this section is to describe the methods that can be used to identify and characterize hazards.

Notable is the fact that food additives permitted prior to January 20, 2009, are undergoing a risk assessment re-evaluation by the European Food Safety Authority (EFSA). This procedure was executed in three distinct phases, each with its own completion date. The first group, colorants and additives, must be evaluated by 2015; the second group, texturizing agents, must be evaluated by 2018; and, finally, the compliance of sweeteners must be evaluated by 2020 (EU; Claudia Pașca and Socaci 2018).

Depending on the dosage of additives used, there is a fine line between safe and hazardous, with the minimum dose required to achieve the desired effect being used the majority of the time (GA Blekas 2016). The "Acceptable Daily Intake" (ADI) represents the amount that can be safely consumed every day, throughout the entire life, without putting the consumer's health at risk. The ADI and acceptance rate of the additive are adjusted by EFSA's and World Health Organization-Food and Agriculture Organization (WHO–FAO) Joint FAO/WHO Expert Committee on Food Additives (JECFA) scientific panels (Saltmarsh 2013). This procedure can be applied to determine the maximum authorized amount of a specific additive or a chemically related group of additives. These concentrations are usually expressed in milligrams of additive per kilogram of body weight; ADI can also be unspecified at *quantum satis*, which is a Latin word that can be associated with the use of the additive in a suitable concentration to achieve the desired effect, but according to the Good Manufacturing Practice (Authority).

According to their properties, the approved food additives are included in Annexes to Regulation EC No. 1333/2008. As such, Annex I contains classifications

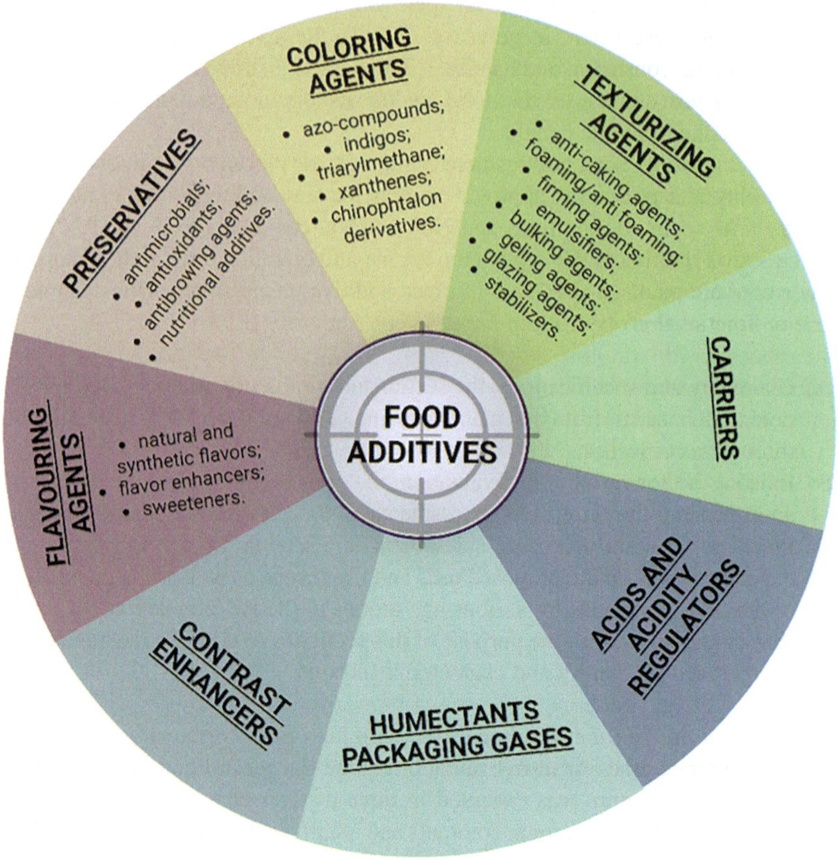

Fig. 1 Food additives classification based on their functionality

of food additives based on their functionality (27 functional classes): sweeteners, colors, preservatives, antioxidants, carriers, acids, acidity regulators, anti-caking agents, anti-foaming agents, bulking agents, emulsifiers, emulsifying salts, firming agents, flavor enhancers, foaming agents, gelling agents, glazing agents, humectants, stabilizers, and so on (Fig. 1). Also, the same regulation stipulates the Union list of food additives approved for use in foods and conditions of use (Annex II; EU). Additionally, food additives can be divided according to their origin as natural (as a result of plant or animal substrate purifying) and synthetic additives (as a result of chemical synthesis) (Wu et al. 2022b). Natural additives are an important future tool for food preservation due to their health benefits and synergistic properties (Márcio Carocho and Ferreira 2015).

In Europe, a scheme in which each additive receives a unique number called an "E-number" is used for the individualization and easiest classification of food additives. The label of each commercialized food product must include the

E-number or the name of the food additive and a specific reference for food use (e.g., acidulants and preservatives). Nowadays, numbering and coding additives is a key step in gaining consumers' trust, but this has not always been so. There was a campaign against "E-numbers" in the 1980s due to their irresponsible use; foods containing additives were considered dangerous, which is why they were as much to be avoided as foods containing genetically modified ingredients. The public's view of the relationship between "food additives" and "chemicals" has compounded matters. The focus on additives in the 1980s was prompted by a 1986 change in labeling laws that mandated the listing of each additive in the ingredient list of the majority of pre-packaged products. Prior to that, the use of additives was denoted by generic functional groups, such as "preservatives," "antioxidants," and "colors." As a consequence of the new labeling requirements, some food labels now contain lengthy inventories of additives, including lengthy chemical names. Some products appeared to be nothing more than a few basic constituents held together by a chemical dictionary. The "E" number system, which was intended as a short code for some of the longer chemical names and to indicate common European safety approval, became the focal point of the criticism against the use of additives, and consumers voted with their feet by abandoning products with lengthy "E" number lists. The anti-additives campaign and subsequent consumer pressure to eliminate or reduce the use of additives resulted in inevitable changes to manufacturing and marketing practices. It is therefore time to re-evaluate the role and application of additives in the food supply, keeping in mind that they will always be necessary for food preparation, quality, and preservation (M'Arcio Carocho et al. 2014). According to the EFSA's scientific forum, the introduction of government bodies that examine potential consumer dangers has caused these issues to dissipate and people's trust to rise (M'Arcio Carocho et al. 2014).

2 Preservatives

Food products' physicochemical and sensory properties can alter as a result of deterioration processes (microbiological, enzymatic, physical, and chemical), which also reduces the nutritional quality and food safety. To ensure the safety of food products for consumers, several methods (biological, physical, and chemical) have been developed to extend the shelf life of food products without changing any of the sensory properties. One of the most frequent conservation methods that sustains the quality of food items and decreases the incidence of foodborne diseases is the incorporation of preservatives into the technological processes of food production (Novais et al. 2022). Preservatives are one of the most important families of additives, with E values ranging from E200 to E399. The preservatives are classified into three functional groups: antioxidants, antimicrobials, and anti-browning agents (Carocho et al. 2018). Antimicrobials are substances that preserve food for a longer period of time by preventing natural spoilage and inhibiting spoilage caused by fungi, bacteria, and other microorganisms (Carocho et al. 2018; Wu et al. 2022a). Enzymatic and non-enzymatic food browning is often prevented by using

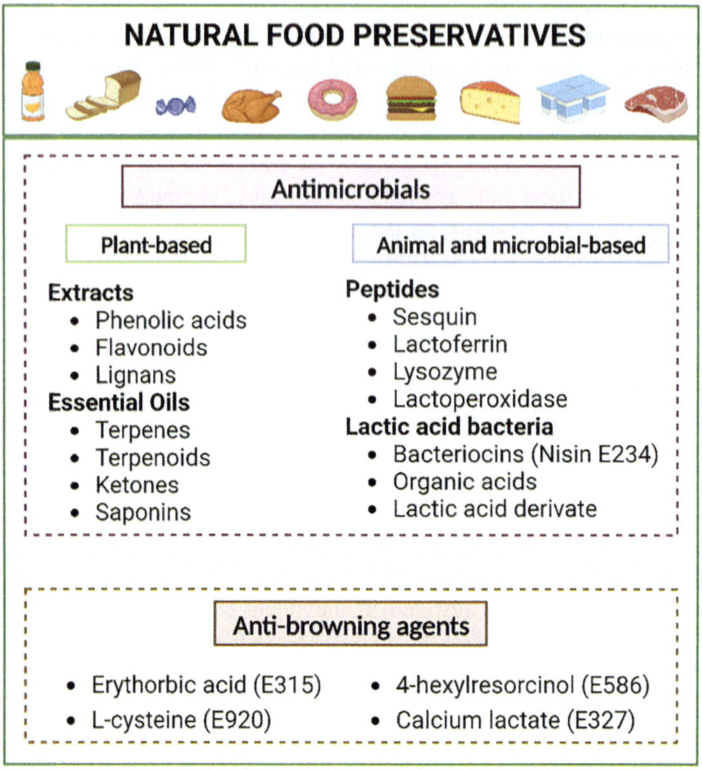

Fig. 2 Classification of natural food preservatives

anti-browning chemicals when handling, processing, and storing food products (Carocho et al. 2018). Compounds with an antioxidant action are a type of preservative used to extend the shelf life of food products by preventing oxidative processes such as rancidity, color and texture deterioration, and odor change (Carocho et al. 2018). In this book chapter, a special section is devoted to discuss the preservatives with antioxidant properties in more detail.

Based on their origin, food preservatives are divided into two groups: artificial preservatives and natural preservatives. Food preservatives are most commonly found in synthetic forms in food products, such as sorbates, nitrates, and sulfites (Novais et al. 2022). However, the impact of artificial preservatives on the human body can produce an intestinal microbiota imbalance that may have consequences for human health (Ruiz-Rico et al. 2023). Improving food quality and safety is critical for human well-being, therefore, previous research articles have focused on finding alternatives to conventional food preservatives, such as plant-based and animal- or microbial-based antimicrobial compounds (Ruiz-Rico et al. 2023; Gokoglu 2019). A classification of natural preservatives is illustrated in Fig. 2.

Various plant parts, including the leaves, stems, flowers, fruits, and roots, are used to produce plant extracts that are rich in compounds with preserved functions. Herbs and spices extracts are significant examples of plant-based preservatives; depending on the species, habitat, and concentration, they may be used as natural preservatives (Gokoglu 2019). Polyphenols, flavonoids, tannins, alkaloids, terpenoids, isothiocyanates, lectins, and polypeptides are the main plant-based phytochemicals utilized as natural preservatives alongside essential oils (EOs) (Gokoglu 2019). Polyphenols represent a large class of natural compounds abundantly distributed in many plant sources, with more than 8000 phenolic structures known up to now. The primary industrial application of polyphenolic compounds is to prevent oxidative processes in food products, although studies demonstrate that polyphenolic compounds can also have an antibacterial effect on foods. The antimicrobial effect's mechanism of action is based on polyphenols' ability to change the permeability of the cell membrane, modify intracellular functions via links formed between phenolic compounds and enzymes, or degrade the cell wall via interactions between polyphenols and the cell membrane (Olszewska et al. 2020). The antibacterial potential of pomegranate extract, vanillin, and geraniol, which were utilized as food preservatives in strawberry juice, was examined by Tomadoni and colleagues (2016). Each antimicrobial preservative was used in two concentrations: pomegranate extract at 180 and 360 μg/mL; vanillin at 2.5 and 5 mg/mL; and geraniol at 0.6 and 1.2 μL/mL. The native microflora on strawberry juice was significantly reduced by more than three log cycles by geraniol and vanillin at both concentrations tested, increasing the product's microbiological shelf life. Additionally, both antimicrobials increased the security of the product by lowering inoculated *Escherichia coli* O157:H7. The highest concentration of pomegranate extract lowered the concentrations of mesophilic and psychrophilic bacteria, whereas it had no effect on yeasts, molds, and the inoculated *E. coli* bacterium (Tomadoni et al. 2016). Moreover, the leaf extract of *Basilicum polystachyon* rich in phenolic compounds (gallic acid, trans-cinnamic acid, ellagic acid, quercetin, vanillic acid, caffeic acid, *p*-coumaric acid, and rosmarinic acid) expressed antimicrobial activities against *Bacillus subtilis*, *Staphylococcus aureus*, *Mycobacterium smegmatis*, *E. coli*, and *Candida albicans* investigated through disk diffusion assay and minimum inhibitory concentration assay (Das et al. 2022). Furthermore, solvents such as methanol, ethanol, and acetone were utilized to extract bioactive chemicals having biological activity (antioxidant, antibacterial, and antimutagenic) from brewers' spent grain. All of the phenolic extracts have been found to have antibacterial properties on bacterial and fungal strains, particularly the *Candida albicans* fungus (Socaci et al. 2018). Natural food preservatives, on the other hand, such as phenolic extracts with antibacterial effects, may have a functional role in the human body due to their potential to reduce cellular damage or cellular death, lowering the incidence risk of chronic disease (Nemes et al. 2022). However, the bioavailability and bioaccessibility of phenolic compounds determine their efficiency in reaching target areas and performing any protective functionality on the human body (Nemes et al. 2022). Ferulic acid, one of the most abundant phenolic compounds in plant-based sources, is a high-value bioactive compound due to its preservation effects on food products and functional potential in the human body, such as anti-hyperlipidemic, anti-oxidative, and anti-inflammatory activities (Nemes et al. 2022; Bumrungpert et al. 2018). A recent study examined the impact of ferulic acid supplementation on lipid profiles, oxidative stress, and inflammation in hyperlipidemia patients, and found that it has the potential to lower cardiovascular diseases risk factors like total cholesterol, low-density lipoprotein (LDL)-cholesterol, triglycerides, oxidative stress biomarkers, and inflammatory markers (Bumrungpert et al. 2018).

Essential oils (EOs) are volatile oils with strong aromatic properties that provide a characteristic aroma and odor to aromatic plants (Pavela 2015). EOs are among the plant-based components that have been utilized since ancient times in fields such as medicine, flavoring agents, and food preservatives (Tiwari and Dubey 2022). More than 17,500 plant species, primarily from the angiospermic families Myrtaceae, Lamiaceae, Asteraceae, Rutaceae, and Zingiberaceae, have been identified to produce EO metabolites (Pavela 2015). Because of their antioxidant, antifungal, and antibacterial activities, EOs and their bioactive components are now widely used as novel green preservatives in the food industry. Aromatic phytoproducts are synthesized from many plant parts, including seeds, flowers, bark, rhizomes, roots, buds, fruits, and leaves (Tiwari and Dubey 2022). EOs contain a complex mix of bioactive chemical components such as terpenes, terpenoids, and phenolic compounds (Falleh et al. 2020). Terpenes, represented by pinene, myrcene, limonene, terpinene, or *p*-cymene, are hydrocarbons with a simple structure. Terpenoids are hydrocarbons that contain oxygen molecules and are derived from the structure of terpenes, by binding functional groups and methyl oxylated groups in various positions (Masyita et al. 2022). Mustard essential oil with allyl isothiocyanate (98.4%), thyme with linalool (14.6%), and Mexican oregano with carvacrol and *p*-cymene (26.9% and 20.0%, respectively) were found to have effective antibacterial properties against *Salmonella typhimurium* with minimum inhibitory concentration values ranging from 0.025 to >5 μg/mL (Meenu et al. 2023). Avocado leaf EOs indicated antibacterial activity against *Staphylococcus epidermidis* and *Staphylococcus aureus* (Nasri et al. 2022). Moreover, the bioactive components in EOs provide beneficial functional effects on the human body, including anti-inflammatory and antinociceptive effects (Huang et al. 2019), antibacterial and immunomodulatory activities (Valdivieso-Ugarte et al. 2021), and antioxidative and mucolytic effects on respiratory diseases (Li et al. 2023). Table 1 summarizes the plant-based, and animal and microbial-based food preservatives that provide a functional role in human health.

Another class of compounds with biological activity that can act as natural preservatives for food products are those derived from plants and microbial sources. Antimicrobial peptides, which are effective against bacteria, fungi, protozoa, and some viruses, are one of the animal-based natural preservatives (Novais et al. 2022). Specific bioactive peptides, such as Sesquin, are already in use in the food additives sector due to their ability to inhibit unwanted microorganisms' growth while preserving product quality and nutritional benefits. *Botrytis cinerea* and *Fusarium oxysporum*, which have the capacity to affect common products such as wheat and grapes, are the most critical microorganisms for which scientists are looking for

Table 1 Plant-based and animal- and microbial-based food preservatives with functional effects on the human body

Source		Bioactive constituent	Food application	Health benefits	References
Plant-based extracts	Leguminosae/ Fabaceae family	Prenylated isoflavonoids: Glabridin 6,8-Diprenylgenistein	Antimicrobial activity against *Listeria monocytogenes* in vitro tested on fresh-cut cantaloupe	Antidiabetic activity Obesity prevention Cardioprotection Immunomodulation Neuroprotection Osteoprotection	Bombelli et al. (2023), Chang et al. (2021)
	Asparagus Bean seeds (*Vigna sesquipedalis*)	Sesquin peptide	Food biopreservation Synergistic antimicrobial action	Hormone-like beneficial activities	Ramos-Martín et al. (2022), Hayes and Bleakley (2018)
	Lemon Basil leaf Mint leaf Lemongrass	Phenolic compounds	Sugarcane juice with a storage stability of 26 days at 4°C, preserved with natural preservatives Health/therapeutic drink	Prevention of diseases, like obesity, diabetes, and cardiovascular diseases	Bag et al. (2022), González-Molina et al. (2010)
	Kenaf seed	Peptide mixture	Antifungal effects on tomato puree Increased shelf life of tomato puree for up to 23 days at 4°C	Nutritional and physiological benefits	Arulrajah et al. (2021)
	Polyphenolic extracts	Eugenol Vanillin Ferulic acid	Food biopreservation	Free eugenol increased the Lachnospiraceae and Akkermansiaceae families Immobilized phenolics enhanced the Bacteroides and reduced the ratio of Firmicutes to Bacteroidetes	Ruiz-Rico et al. (2023)
	Sisymbrium officinale extract	Phenolic and flavonoids content	Strong antimicrobial activity against *Escherichia coli* and *Streptococcus* bacteria	Anticancer activity against breast cancer cell line	Khalid et al. (2022)

(continued)

Table 1 (continued)

Source		Bioactive constituent	Food application	Health benefits	References
	Henna (*Lawsonia inermis*) extract	Catechin acid Methyl gallate Ellagic acid Coumaric acid	High antibacterial efficacy against *Bacillus cereus*, *E. coli*, and *Pseudomonas aeruginosa* on yoghurt during 15 days of storage	Maintains beneficial probiotic concentrations at optimal levels	Ghazy et al. (2023)
Essential oils (EOs)	Cinnamon EOs Clove EOs White thyme EOs	Cinnamic aldehyde Thymol Eugenol	Extended food products' shelf life	Anti-inflammatory properties Immunomodulatory activities	Valdivieso-Ugarte et al. (2021)
	Anise (*Pimpinella anisum*) EOs	Anethole Estragole Fenchone	Green preservative in food and agricultural industries In situ minerals and macronutrient preservation of rice seeds Fungitoxic, aflatoxin inhibitory, and antioxidant potency	Ayurvedic treatments in intestinal parasitic infections Carminative Diuretic Stomachic Antispasmodic Expectorant	Das et al. (2021)
	Juniperus communis L. EOs	β-myrcene Monoterpene hydrocarbons	Antimicrobial activity against *Escherichia coli*, *Listeria monocytogenes*, *Salmonella* spp., and sulfite-reducing clostridia after 225 days of sausages fermentation Alternative for sodium nitrite in dry-fermented sausages	It reduced the risk of the formation of carcinogenic *N*-nitroso-compounds	Tomović et al. (2020)
	Acorus calamus EOs *Allium sativum* EOs *Mucuna pruriens* EOs	Hydroxylamine 1,2,4-Trimethoxy-5-1-propenyl Dodecanoic acid 1,2,3-Propanetriyl ester 2-Diisoprophylphosphinnoethane	In vitro antibacterial and antifungal properties against plum fruit spoilage microbes such as *Aspergillus niger*, *Aspergillus flavus*, and *Rhizopus microsporus*	Antiviral, antioxidant, and antimicrobial effects	Arasu et al. (2019)

	Sesamum indicum L. EOs				
	Sweet orange (*Citrus sinensis*) EOs Lemon (*Citrus limonum*) EOs	Limonene β-myrcene β-pinene α-pinene Citral Z Citral E Linalool	Preserving fresh-cut apples Weight loss, color, texture, and microbial growth of fresh-cut Jazz apples were reduced	Low toxicity Health benefits Pharmacological applications	Sumonsiri et al. (2020), Rasool et al. (2022)
Microbial-based sources	Antimicrobial enzymes	Lactoperoxidase Thiocyanate Hydrogen peroxide	The shelf life of trout fillets was extended by 4 days The growth of *Shewanella putrefaciens*, *Pseudomonas fluorescens*, and psychrotrophic and mesophilic bacteria was significantly reduced	A key component of the nonspecific immune response involved in oral health care	Jasour et al. (2015)
	Lactobacillus sakei subsp. *sakei* 2a	Bacteriocins	The growth of *Listeria monocytogenes serotypes* 4b and 1/2a was inhibited in three cheese spread trials	Potential probiotic effects	Martinez et al. (2015)
	Lactic acid bacteria *Lactococcus lactis* C15	*Lactococcin B* gene	Nisin-like bacteriocin Prevent food spoilage and inhibit foodborne pathogens The growth of *Escherichia coli* in Ultra Heat Treatment (UHT) milk was reduced	Improve intestinal diseases Increase immunity	Lei et al. (2022)
Animal-based sources	White shrimp (*Litopenaeus vannamei*)	C-type lysozyme	Antibacterial activities against *Escherichia coli*, *Vibrio splendidus*, *Micrococcus luteus*, *Vibrio parahaemolyticus*, and *Staphylococcus aureus*	Increased immune response against invading pathogens	Hu et al. (2022)

(continued)

Table 1 (continued)

Source		Bioactive constituent	Food application	Health benefits	References
	Chicken egg white	Lysozyme	*L. monocytogenes*, total aerobic microbial, yeasts, and molds in smoked salmon samples were inactivated	–	Min et al. (2005)

bio-preservative alternatives. Fungi's capacity to affect every part of the plant, such as grapes, tomatoes, and strawberries, at any stage of development limits the use of synthetic fungicides (Ramos-Martín et al. 2022). Brisha Arulrajah and colleagues created a microbial-based preservative with antifungal activity by fermenting kenaf seeds to produce a mixture of antifungal peptides (Arulrajah et al. 2021). The peptide mixture produced during lacto-fermentation with *Lactobacillus pentosus* RK3 presented fungicidal effects against *Aspergillus niger* and *Fusarium* sp. when tested on tomato puree (Arulrajah et al. 2021).

3 Antioxidants from Natural Sources

Plants or plant extracts rich in polyphenols can be used as antimicrobial and antioxidant food additives, as flavoring agents, or as natural sources of antioxidants with health benefits. With the ability to interact with free radicals and antimicrobial effect, phenolic compounds have antibacterial, antioxidant, anti-hyperlipidemic, antitumoral, antidiabetic, cardioprotective, and neuroprotective properties (Zeb 2020). Antioxidants of natural origin are "generally recognized as safe" (GRAS) by regulatory bodies, but they need to meet some criteria: low-concentration efficacy, maintaining stability during food preparation and storage, compatibility with food, and consumer safety regarding mutagenicity, carcinogenicity, teratogenicity, and toxicity due to higher amounts used than synthetic antioxidants (Lourenço et al. 2019). Table 2 presents examples of natural sources of antioxidants with applications in food products, along with their biological activities.

A method to assure stability and bioavailability of phenolic compounds in functional food products (bread, butter, yogurt, cake, biscuits) or beverages (milk, juice) is nano-formulation using various matrices and carriers that are non-toxic for consumers (Jampilek et al. 2019). Phenolic compounds with antioxidant properties like phenolic acids, flavonoids, stilbenes, coumarins, lignans, and tannins may be found in all parts of the plant: fruits, roots, leaves, seeds, and barks (Shahidi and Ambigaipalan 2015). Alasalvar et al. (2021) reviewed eight specialty seeds—black cumin, chia, hemp, flax, perilla, pumpkin, quinoa, and sesame—to assess nutrients, bioactive compounds, health benefits, and consumer safety. Rich in essential nutrients like amino acids, minerals, and vitamins, specialty seeds contain bioactive ingredients such as tocopherols, carotenoids, phytosterols, and polyphenols (flavonoids, phenolic acids, lignans, and isoflavones) that have anti-inflammatory, hypoglycemic, blood pressure and lipid metabolism regulation, antioxidant, and antimicrobial activities with multiple health benefits (Alasalvar et al. 2021).

Functional foods enriched with plant antioxidants may contribute to dementia and Alzheimer's disease treatment as well as maintaining a proper renal, hepatic, cardiovascular, and digestive status in older patients (Wilson et al. 2017). Red raspberries possess anthocyanins and ellagitannins, polyphenol compounds that can reduce the risk of metabolism or oxidative-related diseases, increasing cardiovascular and brain health (Burton-Freeman et al. 2016). Green coffee and tea antioxidants (chlorogenic and caffeic acids, caffeine, and trigonelline) can be incorporated in various

Table 2 Examples of natural sources of antioxidants, biological activities, and applications in food products

Antioxidant	Source	Food application	Effects/benefits for human health	Studies (in vitro/ in vivo)	Functional properties	References
Polyphenols	Green coffee bean	Bread	Improved the serum levels of fasting blood sugar, insulin, triglycerides, and high-density lipoprotein (HDL)-cholesterol	In vivo	↑ Phenolic content and antioxidant properties	Zain et al. (2018), Morvaridi et al. (2020)
Anthocyanins, procyanidins, flavan-3-ol	Blueberry	Biscuits	Antioxidant, anti-inflammatory, antihypertensive, and antidiabetic activities with cardiovascular effects and gut-microbiome modulation with prebiotic activity and decrease in pro-inflammatory cytokines	In vitro and in vivo	↑ Phenolic content	Aksoylu et al. (2015), Pap et al. (2021)
Carnosic acid, carnosol, rosmarinic acid, diterpenoids, flavonoids	Rosemary	Yoghurt	Decrease post-prandial glycemia	In vivo	↑ Antioxidant properties and α-amylase activity	Shori (2020)
Caffeoylquinic acid derivatives	Sweet potato	Bread	Hepato- and cardio-protective, anti-inflammatory, antihypertensive properties; modulate lipid metabolism and glucose	In vitro	↑ Phenolic content, antioxidant properties, and dietary fibers	Mau et al. (2020), Sun et al. (2014)
Isoflavones	Soy	Bread	↓ Pro-inflammatory cytokines in patients with chronic pancreatitis and modulate gut microbiota,	In vivo on humans	↑ Isoflavones content with the anti-inflammatory role	Ahn-Jarvis et al. (2020)

			gut immune function, and gut integrity			
Phenolic acids (ferulic acid, *p*-hydroxybenzoic acid) and anthocyanins (luteolinidin and apigeninidin)	Sorghum	Pasta	Prevention of diabetes type 2 and improved intestinal health; increased antioxidant and superoxide dismutase (SOD) activity and decreased oxidative stress markers	In vivo on humans	Enhanced antioxidant potential	Khan et al. (2015), Khan et al. (2013)
Phenolic compounds, condensed tannins (proanthocyanidins), flavonoids (luteolinidin, 5-methoxyluteolinidine)	Sorghum	Cereals	Improved inflammation and oxidative stress in individuals with chronic kidney disease on hemodialysis	In vivo on humans	↑ Phenolic content and antioxidant properties	Lopes et al. (2018)
Phenolic compounds (secoiridoid derivatives—hydroxytyrosol, tyrosol, and verbascoside)	Olive oil by-products	Refined oils	Inhibition of lipid peroxidation on LDL molecules; antiproliferative effects; scavenging activity	In vitro	Preserving α-tocopherol content and ↓ negative volatile compounds during frying	Esposto et al. (2015), Araújo et al. (2015)
Polyphenols, flavonoids	Pollen	Kombucha beverage	Cytotoxic and antitumoral activities; antimicrobial, antimutagenic, antioxidant properties	In vitro	Enhancement of pollen phytonutrients bioavailability due to Kombucha fermentation	Uțoiu et al. (2018), Pascoal et al. (2014)
Furanocoumarins, quercetin 3-*O*- or 7-*O*-glucosides	Pollen	Sheep, goat, and cow milk yoghurt	Antioxidant	In vitro	Increased TPC and antioxidant capacity	Karabagias et al. (2018)
Phenolic compounds	Bamboo shoots	Crackers	Antioxidant; modulate the composition of the human gut microbiome	In vivo	Increased TPC, vitamin C and E, and phytosterols	Santosh et al. (2021), Fraga et al. (2019)
	Walnut green husk	Sausages	Increased paraoxonase (PON-1), catalase (CAT),	In vivo		Salejda et al. (2016),

(continued)

Table 2 (continued)

Antioxidant	Source	Food application	Effects/benefits for human health	Studies (in vitro/ in vivo)	Functional properties	References
Rosmarinic acid, chlorogenic acid, quercetin derivatives			and superoxide dismutase (SOD) enzymatic activities and γ-tocopherol levels and decreased lipoperoxides in patients with high cardiovascular risk		Increased polyphenol content; increased sensory acceptability	Sánchez-Muniz et al. (2012)
Hydroxycinnamic acids (danshensu, caftaric acid, fertaric acid, salviaflaside, and rosmarinic acid), Flavonoids (quercetin dihexoside, kaempferol dihexoside, quercetin hexoside)	Chia	Cookies	Antioxidant properties; prebiotic capacity	In vitro	Increased antioxidant capacity and polyphenol content	Lucini Mas et al. (2020)
Low-molecular-weight peptides	Flaxseed	Bakery products	Lower total and LDL-cholesterol; antioxidant properties	In vivo, patients with cardiovascular disease	Increased antioxidant capacity	Edel et al. (2015), Wu et al. (2019)

TPC total phenolic content

matrices—muffins, bread, and donuts—offering increased antioxidant levels and lower levels of acrylamide formed during the frying process of donuts. Sensory acceptability and polyphenol bioavailability are increased when the extracts are micro- or nano-encapsulated (Aguiar et al. 2016). Epicatechin derivatives found in green tea may increase bone mineral density due to the inhibition of bone resorption and apoptosis of the osteoclasts, and it can represent a good source of antioxidants for osteoporosis prevention (Arnold et al. 2021).

Several plants used as traditional herbal medicines can be included in functional foods with human health benefits. *Angelica dahurica* root contains coumarins and furanocoumarins with in vitro antioxidant and antiproliferative activities (Bai et al. 2016). *Moringa oleifera* leaves contain ascorbic acid, flavonoids, phenolics, carotenoids with antioxidant properties, and calcium, iron, copper, potassium, and folate with good bioavailability (Peñalver et al. 2022). Powdered fruits and vegetables like mango, apple, carrot, pumpkin, jaboticaba, guava, pomegranate, blueberry, grape, orange, mushroom, and grapefruit can be added to biscuits as functional ingredients, increasing mineral and fiber content but also carotenoids and polyphenols levels, with human health benefits (Salehi 2020). Antioxidant additives also contribute to an efficient production process. The addition of pollen to beverages or dairy products as a fermentation activator, natural antioxidant, or fermentation feedstock improved the final product, with an increase in polyphenols and flavonoids content, an improvement in sensory and textural or rheological properties, and increased alcohol formation (Kostić et al. 2020). The incorporation of plant antioxidants into meat products increased shelf life and inhibited the formation of chemical toxins produced during preparation (Jiang and Xiong 2016).

4 Natural Pigments as Food Colorants

Colorants are used in the food industry to increase consumer acceptability (Solymosi et al. 2015). Plants, especially fruits, are good sources of natural colorants. The main natural pigments in the plant kingdom are anthocyanins, carotenoids, betalains, and chlorophyll, and their use in the food industry is challenged by low stability, which can be increased by micro- and nano-encapsulation (Rodriguez-Amaya 2019). Anthocyanins are a family of natural pigments responsible for the purple, red, blue, and orange colors of fruits, vegetables, and flowers (Zhang et al. 2014). The most common anthocyanidins are pelargonidin, cyanidin, peonidin, delphinidin, petunidin, and malvidin.

Fortified food with anthocyanins has multiple health benefits: anticancer, anti-inflammatory, neuroprotective, cardioprotective, anti-obesity, and antidiabetic activities (Li et al. 2017). The selection of a specific colorant is made considering its solubility and the matrix that needs to incorporate it: anthocyanins, which offer the red-blue-purple color, and betalains, which offer the red color, are water soluble, while carotenoids, which offer the yellow-orange-red color, and chlorophylls, which offer the green color, are lipid soluble (Sharma et al. 2021). Most natural origin colorants also have antioxidant effects. García-Cruz et al. (2017) compared two different species of pitaya fruit, one with white pulp and one with red pulp, and found different amounts of betalains and phenolic compounds, with similar antioxidant capacities. Carotenoids are also colorants with antioxidant properties and can be found in mango, pumpkin, carrots (β-carotene), tomatoes (lycopene), green leafy vegetables, or algae (lutein and zeaxanthin) (Xu et al. 2017). Flavonoids and anthocyanins are added to bakery products, biscuits, dairy products, or beverages for increasing antioxidant properties and stability, or improving the final product color (like yellow or orange-colored cheeses) (Neri-Numa et al. 2020). In the process of making naturally colored foods, other beneficial compounds are made. Red yeast rice receives its red color after rice fermentation by the *Monascus purpureus* fungus.

During the fermentation, a series of molecules that inhibit cholesterol (monacolin A and KA) are formed, and their effect is identical to that of lovastatin, a drug that inhibits 3-hydroxy-3-methyl-glutaryl-coenzyme A (HMG-CoA)-CoA reductase and lowers serum cholesterol levels, especially LDL-cholesterol (Poli et al. 2018). Bioactive compounds may also be found in by-products, and their use is encouraged to contribute to sustainable production. Veneziani et al. (2017) reviewed the application of recovered bioactive compounds in food products with a focus on olive oil production wastes. Olive mill wastewater is rich in phenolic compounds such as phenolic acids, comselogoside, secoiridoids, flavonoids, and verbascoside and can be used as a fortifier in oil, milk beverages, and meat products with antioxidant and antimicrobial activities (Veneziani et al. 2017). Anthocyanins, carotenoids, anthoxanthins, and chlorophyll can be extracted from wine pomace, rice bran, tomato by-products, berries, potatoes, citrus peels, or green leafy vegetables, and used as natural-source colorants in the food industry (Faustino et al. 2019). Lombardelli et al. (2021) efficiently extracted betalains from unsold red beets, which can be further used as a food colorant in desserts and confectioneries, dry mixes, and dairy and meat products (Lombardelli et al. 2021). Bagasse, skin, and seeds of kiwi are wasted, which are rich in bioactive compounds such as flavonoids, tocopherols, phenolic compounds, and anthocyanins, with antioxidant, antitumoral, and anti-inflammatory activities and gastrointestinal benefits (Chamorro et al. 2022). Ooi et al. (2021) analyzed the carotenoids and phenolic compounds content of the skin of various sweet potato species. The orange peel had the highest carotenoid content while the purple peel had the highest phenolic compounds content, both with high antioxidant capacity, being a sustainable option for food colorant or antioxidant (Ooi et al. 2021). Table 3 presents examples of natural sources of colorants with applications in food products, along with their biological activities.

5 Hydrocolloids' Functionality in Foods

Hydrocolloids are high-molecular-weight, long-chain hydrophilic polymers having amino and carbonyl functional groups. They are used in the food industry to thicken, gel, replace fat, and produce films (Zhang et al. 2021). Starch, xanthan gum, agar, pectin, gellan gum, alginate, inulin, carob bean gum, gum Arabic, and carrageenan are some of the most well-known hydrocolloids (Zhang et al. 2021; Pirsa and Hafezi

Table 3 Examples of natural sources of colorants, biological activities, and applications in food products

Colorant	Source	Food application	Effects/benefits for human health	Studies (in vitro/ in vivo)	Functional properties	References
Anthocyanins (cyanidin 3-rutinoside, cyanidin 3-glucoside, peonidin 3-rutinoside, peonidin 3-glucoside, and pelargonidin 3-rutinoside)	Sweet cherry skins	Yoghurt; marshmallows	Prebiotic effect for yoghurt; antioxidant activity	In vitro	↑ Antioxidant activity; anthocyanins levels ↓ after 2 days of storage of yoghurt while ↑ for marshmallows	Milea et al. (2019), Kumar and Kumar (2016)
Carotenoids (β-carotene)	Carrot pomace	Biscuits; pasta	Contribute to eye, skin, and mucosal membrane health; antioxidant; maintain immunity	In vitro + in vivo	↑ Carotenoid retention with good sensory acceptability for 8% pomace powder addition in biscuits ↑ Fiber content; ↑ nutritional value of pasta; good sensory acceptability of orange color given by carotenoids	Bellur Nagarajaiah and Prakash (2015), Gull et al. (2015), Ahmad et al. (2019)
Anthocyanins (cyanidin-based, cyanidin-3-xylosyl-sinapoyl-glucosyl-galactoside being predominant)	Black carrot pomace	Cake	Antioxidant, cardioprotective, antidiabetic, anticancer effects; contribute to visual health	In vitro + in vivo	↑ Phenolic content and antioxidant capacity	Kamiloglu et al. (2017), Khoo et al. (2017)
Anthocyanins (glycosylated derivative of delphinidin)	Eggplant (fruit, pulp, epicarp)	Pastry cream	Antioxidant, antimicrobial, cytotoxic, and	In vitro	Stable purple-colored cream with antioxidant activity and ↑ phenolic	Pantuzza Silva et al. (2021), Horincar et al. (2020)

(continued)

Table 3 (continued)

Colorant	Source	Food application	Effects/benefits for human health	Studies (in vitro/ in vivo)	Functional properties	References
			hepatoprotective effects		content, with good rheological behavior	
Anthocyanins (cyanidin 3-rutinoside, peonidin 3-rutinoside)	Fig peels and blackthorn fruits	Beijinho—Brazilian traditional pastry and donuts icings	Antioxidant; antimicrobial; cytotoxic	In vitro	↑ Firmness and consistency of doughnuts ↑ Softness and chewiness of beijinho Stable pink color for fig extract	Backes et al. (2020)
Anthocyanins (delphinidin-3-*O*-glucoside, petunidin-3-*O*-glucoside, malvidin-3-*O*-glucoside)	Black beans; blue maize; chard	Snacks	Reactive oxygen species (ROS) inhibition; decreased glucose uptake	In vitro	Dark red color snacks with good retention of health beneficial compounds due to extrusion cooking and microwave heating	Mojica et al. (2017), Neder-Suárez et al. (2021)
Anthocyanins (cyanidin-3-*O*-glucoside)	Passion fruit epicarp	Cake	Antioxidant, antimicrobial, and cytotoxic activities	In vitro	Good sensory acceptability with increase of dietary fibers	Ghada et al. (2020), Oliveira et al. (2016)
Carotenoids (β-carotene, α-carotene, β-cryptoxanthin, zeaxanthin, lycopene)	Mandarin epicarp	Bakery products (cake and bread)	Anticancer (by anti-inflammatory and antioxidant mechanisms)	In vitro	Stable orange color of the final product; may be used instead of tartrazine in bakery products	Ordóñez-Santos et al. (2021), Saini et al. (2022)
Betaxanthines	Pitayas	Gummies; beverages	Antioxidant	In vitro	Match the yellow color of commercially available products (synthetically colored); increased stability of color in	Rodríguez-Sánchez et al. (2017)

					gummies than in beverages	
Chlorophyll a	*Spirulina* sp. and *Chlorella* sp.	Breadsticks	Antioxidant; anti-inflammatory; gut microbiota modulator	In vitro	Spirulina pigments have lower stability than *Chlorella* pigments for breadsticks' final green color	Igual et al. (2022), Zhou et al. (2023)
Anthocyanins (cyanidin-3-*O*-sambubioside)	Black elder flowers and fruits	Jellies	Scavenging activities; antioxidant	In vitro	A combination of fruit and flower dye had the best sensory acceptability with good antioxidant capacity (due to phenolic acids from flowers) and a pleasant purple color	Salejda et al. (2016)
Betalains (betacyanins, betaxanthins); anthocyanins (pelargonidin mainly)	Red beetroot, opuntia; Hibiscus, red radish	Soy-based yoghurt alternative	Beneficial in oxidative stress-, inflammation-, and dyslipidemia-related diseases	In vitro and in vivo	Non-encapsulated red radish extract offered an appealing pink color while encapsulated opuntia extract had increased health benefits, maintaining the pink color	Dias et al. (2020), Rahimi et al. (2019)
Carotenoids; chlorophyll	Pumpkin pulp and peel	Biscuits	Antioxidant; anti-fatigue; antibacterial	In vitro and in vivo *(mice)*	Sensory acceptability was the best at 3% pumpkin pulp flower addition to the biscuits	Abdulaali and George (2020), Hussain et al. (2022)
Anthocyanins (delphinidin)	*Clitoria ternatea* flowers	Muffins	Antioxidant, antimicrobial, anticancer, and antidiabetic activities	In vitro and in vivo	Encapsulated *C. ternatea* with maltodextrin assured blue color stability and improved shelf life of the muffins	Ab Rashid et al. (2021), Jeyaraj et al. (2021)

(continued)

Table 3 (continued)

Colorant	Source	Food application	Effects/benefits for human health	*S*tudies (in vitro/ in vivo)	Functional properties	References
Anthocyanins (delphinidin-3-*O*-glucoside, cyanidin-3-*O*-glucoside)	*Jabuticaba* epicarp	Macarons	Anticancer activity; hepatic protection; effects on metabolic syndrome	In vitro and in vivo	Stable color of the final product	Albuquerque et al. (2020), Fernandes et al. (2022)

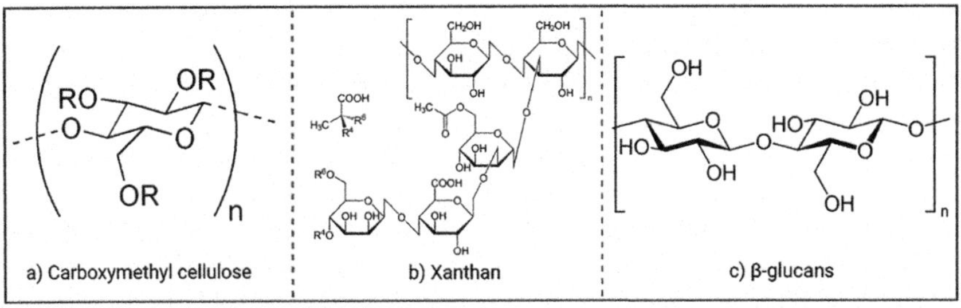

Fig. 3 Structure of functional hydrocolloids

2023). Furthermore, because of their preservation properties, several hydrocolloids are used in the food additives industry.

Carboxymethylcellulose (Fig. 3a) is nowadays one of the most commonly used food additives in foods such as ice cream, sweets, biscuits, cakes, cookies, candies, juices, liquid beverages, dairy and meat products, frozen foods, instant pasta, and fruit compotes (Pirsa and Hafezi 2023). Carboxymethylcellulose can perform numerous roles in food products, including thickening, emulsification, water retention, and stabilization. Carboxymethylcellulose also has rheological properties, which makes it a viable alternative for gelatins. Therefore, carboxymethylcellulose is economical, improves the sensory properties of foods, and increases their shelf life (Pirsa and Hafezi 2023).

Xanthan (Fig. 3b) is another potential additive hydrocolloid. Xanthan is an anionic bacterial heteropolysaccharide derived from *Xanthomonas campestris* fermentation. It has a linear cellulose backbone consisting of β-*d*-glucose substituted on every two units with a pendant trisaccharidic side chain containing a β-*d*-glucuronic acid between an inner α-*d*-mannose and a terminal β-*d*-mannose (Abou Dib et al. 2023). The antimicrobial properties of xanthan gum-based edible coatings on fresh-cut lotus root were recently analyzed (Lara et al. 2020). The findings of the experiments indicated that the spray-coating treatments with xanthan gum solutions successfully reduced the enzymatic browning of fresh-cut lotus root during storage, and potentially improved its market shelf life. Throughout the first 24 hours of incubation, the xanthan gum-based spray-coating significantly inhibited the growth of *Bacillus subtilis* (Lara et al. 2020).

β-glucans (Fig. 3c) are polymers of glucose that are found in numerous bacteria, fungi, algae, and higher plants. One of the largest amounts of β-glucans is found in *Saccharomyces* sp., where the cell walls contain 55–65% of β-glucan (Caruso et al. 2022). The beer industry is one of the richest sources of yeasts. Moreover, brewer's spent yeast is one of the main by-products of the technological process of beer manufacturing, accounting for around 1.5–2.5% of total beer produced annually (1.82 billion hL in 2020) (Caruso et al. 2022). Regarding their functional role, the European Food Safety Authority has recognized β-glucans as safe food additives since 2011. Also, the Food and Drug Administration (FDA) classifies β-glucans as nutritional supplements and food additives in the USA, and there is precedent in

Argentina for the inclusion of the first beer by-product in the Argentinian Codex Alimentarius CAA (Caruso et al. 2022). Therefore, many research works are underlining the use of β-glucans as a thickener or stabilizer in food products, such as salad dressing, soups, sauces, and dairy products (Kayanna et al. 2022). In a recent study performed by Shuya Xu and colleagues, various concentrations (0%, 1%, 3%, and 5%) of oat β-glucan were added to Chinese steamed bread as food additives, and the effect on water mobility, starch retrogradation characteristics, and product quality was investigated (Xu et al. 2021). The properties of β-glucan slow the staling processes of bread by inhibiting the migration of water and the retrogradation of starch, sustaining its use as a natural food additive (Xu et al. 2021). Another study investigated the effect of carboxymethylcellulose, locust bean gum, and psyllium husk powder on wheat dough rheological behavior and bread quality (Sim et al. 2015). A 0.2% concentration of psyllium husk powder ensured the strength and extensibility of the dough, creating a balanced ratio between bread stretch and volume. The study's findings highlighted that the addition of non-starch polysaccharides, such as hydrocolloids, in the dough production process, improves the quality of the resulting bread, slows the aging process, and extends the shelf life (Sim et al. 2015).

In addition to the role of natural food additives, hydrocolloids have many beneficial effects on the human body, such as lowering blood cholesterol, regulating blood lipids and blood sugar (Xu et al. 2021), and prebiotic, immunomodulatory (Nemes et al. 2022), antitumoral, anti-inflammatory, and antioxidant effects (Caruso et al. 2022). A meta-analysis conducted in the USA, Canada, and Europe included 15 randomized controlled studies on human subjects that attempted to measure the effect of consuming β-glucan-rich oatmeal in individuals in the early stages of diabetes. Following the analysis, it was revealed that consuming 3 mg of oat β-glucans for at least eight weeks can considerably reduce insulin, blood sugar, and glycosylated hemoglobin (Bao et al. 2014). The health benefits of beta-glucans, including their ability to reduce serum LDL-cholesterol in humans, were also highlighted in a randomized clinical trial conducted by Wolever Thomas and colleagues. The physicochemical properties of β-glucans influence their effects on lowering cholesterol. An extruded breakfast cereal containing 3 g oat-glucan, consumed daily, with a high molecular weight (2,210,000 g/mol) or a medium molecular weight (530,000 g/mol), similarly reduced LDL-cholesterol (Wolever et al. 2010). However, hydrocolloids represent a valuable class of bioactive compounds that can act as both natural food additives and functional compounds with a beneficial role for the human body.

6 Conclusions

Over the last decades, food science and technology have progressed toward greener and environmentally friendly ways of producing foods, thus including the types of food additives and their sources of origin. The current trend is focused on finding new sources of natural compounds that can be used in foods and feed due to their bioactive properties. Besides their protective role in products, they also have beneficial effects on human health. The large number of chemicals in nature, the biological activities of natural extracts, and the synergies with other compounds provide limitless sources of novel compounds with potential uses in foods and feeds. Nonetheless, the development in the food industry is being overseen by regulatory agencies and law enforcement organizations that have developed stringent laws controlling the licensing and supervision of all food additives.

Acknowledgment This work was supported by grants of the Romanian Ministry of Education and Research, CNCS-UEFISCDI, project number PN-III-P4-ID-PCE-2020-1847 and project number PN-III-P4-ID-PCE-2020-2306, within PNCDI III.

References

Codex Stan 192-1995. General Standard for Food Additives

Food Additives and Why They Are Used. Royal Society of Chemistry, Essential Guide to Food Additives

(EU), C. R. Regulation (EC) No 1333/2008 of the European Parliament and of the Council of 16 December 2008 on Food Additives

(EUFIC), T. E. F. I. C (2021) What are food additives and how are they regulated in the Eu?

Ab Rashid S, Tong WY, Leong CR, Abdul Ghazali NM, Taher MA, Ahmad N, Tan WN, Teo SH (2021) Anthocyanin microcapsule from Clitoria Ternatea: potential bio-preservative and blue colorant for baked food products. Arab J Sci Eng 46:65–72

Abdulaali S, George S (2020) Preparation of pumpkin pulp and Peel flour and study their impact in the biscuit industry. J Biol Agric Healthcare

Abou Dib M, Gore E, Grisel M (2023) Intrinsic and rheological properties of hydrophobically modified xanthan synthesized under green conditions. Food Hydrocoll 138:108461

Aguiar J, Estevinho BN, Santos L (2016) Microencapsulation of natural antioxidants for food application—the specific case of coffee antioxidants—a review. Trends Food Sci Technol 58: 21–39

Ahmad T, Cawood M, Iqbal Q, Ariño A, Batool A, Sabir Tariq RM, Azam M, Akhtar S (2019) Phytochemicals in Daucus Carota and their health benefits—review article. Foods 8:1–22

Ahn-Jarvis J, Lombardo E, Cruz-Monserrate Z, Badi N, Crowe O, Kaul S, Komar H, Krishna SG, Lesinski GB, Mace TA, Ramsey ML, Roberts K, Stinehart K, Traczek M, Conwell DL, Vodovotz Y, Hart PA (2020) Reduction of inflammation in chronic pancreatitis using a soy bread intervention: a feasibility study. Pancreatology 20:852–859

Aksoylu Z, Çağindi Ö, Köse E (2015) Effects of blueberry, grape seed powder and poppy seed incorporation on physicochemical and sensory properties of biscuit. J Food Qual 38:164–174

Alasalvar C, Chang SK, Bolling B, Oh WY, Shahidi F (2021) Specialty seeds: nutrients, bioactives, bioavailability, and health benefits: a comprehensive review. Compr Rev Food Sci Food Saf 20: 2382–2427

Albuquerque BR, Pinela J, Barros L, Oliveira MBPP, Ferreira ICFR (2020) Anthocyanin-rich extract of Jabuticaba Epicarp as a natural colorant: optimization of heat- and ultrasound-assisted extractions and application in a bakery product. Food Chem 316:126364

Arasu MV, Viayaraghavan P, Ilavenil S, Al-Dhabi NA, Choi KC (2019) Essential oil of four medicinal plants and protective properties in plum fruits against the spoilage bacteria and fungi. Ind Crop Prod 133:54–62

Araújo M, Pimentel FB, Alves RC, Oliveira MBPP (2015) Phenolic compounds from olive mill wastes: health effects, analytical approach and application as food antioxidants. Trends Food Sci Technol 45:200–211

Arnold M, Rajagukguk YV, Gramza-Michałowska A (2021) Functional food for elderly high in antioxidant and chicken eggshell calcium to reduce the risk of osteoporosis—a narrative review. Foods 10

Arulrajah B, Muhialdin BJ, Qoms MS, Zarei M, Hussin ASM, Hasan H, Saari N (2021) Production of cationic antifungal peptides from Kenaf seed protein as natural bio preservatives to prolong the shelf-life of tomato puree. Int J Food Microbiol 359:109418

Authority EFS Food Addit. https://www.Efsa.Europa.Eu/En/Topics/Topic/Food-Additives

Backes E, Leichtweis MG, Pereira C, Carocho M, Barreira JCM, Kamal Genena A, José Baraldi I, Filomena Barreiro M, Barros L, Ferreira ICFR (2020) Ficus Carica L and Prunus Spinosa L extracts as new anthocyanin-based food colorants: a thorough study in confectionery products. Food Chem 333:127457

Bag BB, Panigrahi C, Gupta S, Mishra HN (2022) Efficacy of plant-based natural preservatives in extending shelf life of sugarcane juice: formulation optimization by Moga and correlation study by principal component analysis. Appl Food Res 2:100164

Bai Y, Li D, Zhou T, Qin N, Li Z, Yu Z, Hua H (2016) Coumarins from the roots of Angelica Dahurica with antioxidant and antiproliferative activities. J Funct Foods 20:453–462

Bao L, Cai X, Xu M, Li Y (2014) Effect of oat intake on glycaemic control and insulin sensitivity: a meta-analysis of randomised controlled trials. Br J Nutr 112:457–466

Bellur Nagarajaiah S, Prakash J (2015) Nutritional composition, acceptability, and shelf stability of carrot pomace-incorporated cookies with special reference to total and Β-carotene retention. Cogent Food Agric 1

Bombelli A, Araya-Cloutier C, Vincken J-P, Abee T, Den Besten HMW (2023) Impact of food-relevant conditions and food matrix on the efficacy of prenylated isoflavonoids Glabridin and 6,8-diprenylgenistein as potential natural preservatives against listeria monocytogenes. Int J Food Microbiol 390:110109

Bumrungpert A, Lilitchan S, Tuntipopipat S, Tirawanchai N, Komindr S (2018) Ferulic acid supplementation improves lipid profiles, oxidative stress, and inflammatory status in hyperlipidemic subjects: a randomized, double-blind, placebo-controlled clinical trial. Nutrients 10:713

Burton-Freeman BM, Sandhu AK, Edirisinghe I (2016) Red raspberries and their bioactive polyphenols: cardiometabolic and neuronal health links. Adv Nutr 7:44–65

Carocho M, Morales P, Ferreira ICFR (2018) Antioxidants: reviewing the chemistry, food applications, legislation and role as preservatives. Trends Food Sci Technol 71:107–120

Caruso MA, Piermaria JA, Abraham AG, Medrano M (2022) Β-glucans obtained from beer spent yeasts as functional food grade additive: focus on biological activity. Food Hydrocoll 133: 107963

Chamorro F, Carpena M, Fraga-Corral M, Echave J, Riaz Rajoka MS, Barba FJ, Cao H, Xiao J, Prieto MA, Simal-Gandara J (2022) Valorization of kiwi agricultural waste and industry by-products by recovering bioactive compounds and applications as food additives: a circular economy model. Food Chem 370:131315

Chang SK, Jiang Y, Yang B (2021) An update of prenylated phenolics: food sources, chemistry and health benefits. Trends Food Sci Technol 108:197–213

Claudia Paşca AC, Socaci S (2018) Risks and benefits of food additives—review. Bull Uasvm Anim Sci Biotechnol

Cox S, Sandall A, Smith L, Rossi M, Whelan K (2021) Food additive emulsifiers: a review of their role in foods, legislation and classifications, presence in food supply, dietary exposure, and safety assessment. Nutr Rev 79:726–741

Das S, Kumar Singh V, Kumar Dwivedy A, Kumar Chaudhari A, Deepika, Kishore Dubey N (2021) Nanostructured Pimpinella Anisum essential oil as novel green food preservative against fungal infestation, aflatoxin B1 contamination and deterioration of nutritional qualities. Food Chem 344:128574

Das S, Sultana KW, Chandra I (2022) Characterization of polyphenols by Rp-Hplc in Basilicum Polystachyon (L.) Moench with their antioxidant and antimicrobial properties. S Afr J Bot 151: 926–940

Dias S, Castanheira EMS, Fortes AG, Pereira DM, Gonçalves MST (2020) Coloring soy-based yogurt alternative. Foods 9:1–13

Edel AL, Rodriguez-Leyva D, Maddaford TG, Caligiuri SPB, Alejandro Austria J, Weighell W, Guzman R, Aliani M, Pierce GN (2015) Dietary flaxseed independently lowers circulating cholesterol and lowers it beyond the effects of cholesterol-lowering medications alone in patients with peripheral artery disease. J Nutr 145:749–757

Eloi Chazelas MD, Srour B, Kesse-Guyot E, Julia C, Alles B, Druesne-Pecollo N, Galan P, Hercberg S, Latino-Martel P, Esseddik Y, Szabo F, Slamich P, Gigandet S, Touvier M (2020) Food additives: distribution and co-occurrence in 126,000 food products of the French market. Sci Rep

Esposto S, Taticchi A, Di Maio I, Urbani S, Veneziani G, Selvaggini R, Sordini B, Servili M (2015) Effect of an olive phenolic extract on the quality of vegetable oils during frying. Food Chem 176:184–192

European Food Safety Authority (EFSA) (2021) Guidance for submission for food additive evaluations. EFSA J

Falleh H, Ben Jemaa M, Saada M, Ksouri R (2020) Essential oils: a promising eco-friendly food preservative. Food Chem 330:127268

Faustino M, Veiga M, Sousa P, Costa EM, Silva S, Pintado M (2019) Agro-food byproducts as a new source of natural food additives. Molecules 24:1–23

Fernandes IDAA, Maciel GM, Maroldi WV, Bortolini DG, Pedro AC, Haminiuk CWI (2022) Bioactive compounds, health-promotion properties and technological applications of Jabuticaba: a literature overview. Measurement Food 8:100057

Fraga CG, Croft KD, Kennedy DO, Tomás-Barberán FA (2019) The effects of polyphenols and other bioactives on human health. Food Funct 10:514–528

Blekas G (2016) Food additives: classification, uses and regulation. Encyclopedia Food Health

García-Cruz L, Dueñas M, Santos-Buelgas C, Valle-Guadarrama S, Salinas-Moreno Y (2017) Betalains and phenolic compounds profiling and antioxidant capacity of pitaya (Stenocereus Spp.) fruit from two species (S. Pruinosus and S. Stellatus). Food Chem 234:111–118

Ghada B, Pereira E, Pinela J, Prieto MA, Pereira C, Calhelha RC, Stojkovic D, Sokóvic M, Zaghdoudi K, Barros L, Ferreira ICFR (2020) Recovery of anthocyanins from passion fruit Epicarp for food colorants: extraction process optimization and evaluation of bioactive properties. Molecules 25

Ghazy OA, Fouad MT, Morsy TA, Kholif AE (2023) Nanoemulsion formulation of Lawsonia Inermis extract and its potential antimicrobial and preservative efficacy against foodborne pathogens. Food Control 145:109458

Gokoglu N (2019) Novel natural food preservatives and applications in seafood preservation: a review. J Sci Food Agric 99:2068–2077

González-Molina E, Domínguez-Perles R, Moreno DA, García-Viguera C (2010) Natural bioactive compounds of citrus Limon for food and health. J Pharm Biomed Anal 51:327–345

Gull A, Prasad K, Kumar P (2015) Effect of millet flours and carrot pomace on cooking qualities, color and texture of developed pasta. Lwt 63:470–474

Hayes M, Bleakley S (2018) 21 Peptides from plants and their applications. In: Koutsopoulos S (ed) Peptide applications in biomedicine, biotechnology and bioengineering. Woodhead Publishing

Horincar G, Enachi E, Barbu V, Andronoiu DG, Râpeanu G, Stănciuc N, Aprodu I (2020) Value-added pastry cream enriched with microencapsulated bioactive compounds from eggplant (Solanum Melongena L) Peel. Antioxidants:9

Hu F, Wang Y, Hu J, Bao Z, Wang M (2022) A novel C-type lysozyme from Litopenaeus Vannamei exhibits potent antimicrobial activity. Fish Shellfish Immunol 131:729–735

Huang X-L, Li X-J, Qin Q-F, Li Y-S, Zhang WK, Tang H-B (2019) Anti-inflammatory and antinociceptive effects of active ingredients in the essential oils from Gynura Procumbens, a traditional medicine and a new and popular food material. J Ethnopharmacol 239:111916

Hussain A, Kausar T, Sehar S, Sarwar A, Ashraf AH, Jamil MA, Noreen S, Rafique A, Iftikhar K, Quddoos MY, Aslam J, Majeed MA (2022) A comprehensive review of functional ingredients, especially bioactive compounds present in pumpkin Peel, flesh and seeds, and their health benefits. Food Chem Adv 1:100067

Igual M, Uribe-Wandurraga ZN, García-Segovia P, Martínez-Monzó J (2022) Microalgae-enriched breadsticks: analysis for vitamin C, carotenoids, and chlorophyll A. Food Sci Technol Int 28: 26–31

Jampilek J, Kos J, Kralova K (2019) Potential of nanomaterial applications in dietary supplements and foods for special medical purposes. Nanomaterials 9

Jasour MS, Ehsani A, Mehryar L, Naghibi SS (2015) Chitosan coating incorporated with the Lactoperoxidase system: an active edible coating for fish preservation. J Sci Food Agric 95: 1373–1378

Jeyaraj EJ, Lim YY, Choo WS (2021) Extraction methods of butterfly pea (Clitoria Ternatea) flower and biological activities of its phytochemicals. J Food Sci Technol 58:2054–2067

Jiang J, Xiong YL (2016) Natural antioxidants as food and feed additives to promote health benefits and quality of meat products: a review. Meat Sci 120:107–117

Kamiloglu S, Ozkan G, Isik H, Horoz O, Van Camp J, Capanoglu E (2017) Black carrot pomace as a source of polyphenols for enhancing the nutritional value of cake: an in vitro digestion study with a standardized static model. Lwt 77:475–481

Karabagias IK, Karabagias VK, Gatzias I, Riganakos KA (2018) Bio-functional properties of bee pollen: the case of "Bee Pollen Yoghurt". Coatings 8

Kayanna N, Suppavorasatit I, Bankeeree W, Lotrakul P, Punnapayak H, Prasongsuk S (2022) Production of prebiotic Aubasidan-like Β-glucan from Aureobasidium Thailandense Nrrl 58543 and its potential as a functional food additive in gummy jelly. Lwt 163:113617

Khalid M, Amayreh M, Sanduka S, Salah Z, Al-Rimawi F, Al-Mazaideh GM, Alanezi AA, Wedian F, Alasmari F, Faris Shalayel MH (2022) Assessment of antioxidant, antimicrobial, and anticancer activities of Sisymbrium Officinale plant extract. Heliyon 8:E10477

Khan I, Yousif A, Johnson SK, Gamlath S (2013) Effect of sorghum flour addition on resistant starch content, phenolic profile and antioxidant capacity of durum wheat pasta. Food Res Int 54: 578–586

Khan I, Yousif AM, Johnson SK, Gamlath S (2015) Acute effect of sorghum flour-containing pasta on plasma total polyphenols, antioxidant capacity and oxidative stress markers in healthy subjects: a randomised controlled trial. Clin Nutr 34:415–421

Khoo HE, Azlan A, Tang ST, Lim SM (2017) Anthocyanidins and anthocyanins: colored pigments as food, pharmaceutical ingredients, and the potential health benefits. Food Nutr Res:61

Kostić A, Milinčić DD, Barać MB, Shariati MA, Tešić ŽL, Pešić MB (2020) The application of pollen as a functional food and feed ingredient—the present and perspectives. Biomol Ther 10

Kumar A, Kumar D (2016) Development of antioxidant rich fruit supplemented probiotic yogurts using free and microencapsulated lactobacillus Rhamnosus culture. J Food Sci Technol 53:667–675

Lara G, Yakoubi S, Villacorta CM, Uemura K, Kobayashi I, Takahashi C, Nakajima M, Neves MA (2020) Spray technology applications of xanthan gum-based edible coatings for fresh-cut lotus root (Nelumbo Nucifera). Food Res Int 137:109723

Lei W, Hao L, You S, Yao H, Liu C, Zhou H (2022) Partial purification and application of a Bacteriocin produced by probiotic Lactococcus Lactis C15 isolated from raw milk. Lwt 169: 113917

Li D, Wang P, Luo Y, Zhao M, Chen F (2017) Health benefits of anthocyanins and molecular mechanisms: update from recent decade. Crit Rev Food Sci Nutr 57:1729–1741

Li J, Chen W, Liu H, Liu H, Xiang S, You F, Jiang Y, Lin J, Zhang D, Zheng C (2023) Pharmacologic effects approach of essential oils and their components on respiratory diseases. J Ethnopharmacol 304:115962

Lombardelli C, Benucci I, Mazzocchi C, Esti M (2021) A novel process for the recovery of Betalains from unsold red beets by low-temperature enzyme-assisted extraction. Foods 10

Lopes RDCSO, De Lima SLS, Da Silva BP, Toledo RCL, Moreira MEDC, Anunciação PC, Walter EHM, Carvalho CWP, Queiroz VAV, Ribeiro AQ, Martino HSD (2018) Evaluation of the health benefits of consumption of extruded tannin sorghum with unfermented probiotic milk in individuals with chronic kidney disease. Food Res Int 107:629–638

Lourenço SC, Mold M, Alves VD (2019) Antioxidants of natural plant origins : from sources to food industry applications. Molecules:14–16

Lucini Mas A, Brigante FI, Salvucci E, Pigni NB, Martinez ML, Ribotta P, Wunderlin DA, Baroni MV (2020) Defatted chia flour as functional ingredient in sweet cookies. How do processing, simulated gastrointestinal digestion and colonic fermentation affect its antioxidant properties? Food Chem 316:126279

M'Arcio Carocho MFB, Morales P, Ferreira ICFR (2014) Adding molecules to food, pros and cons: a review on synthetic and natural food additives. Compr Rev Food Sci Food Saf

Márcio Carocho PM, Ferreira ICFR (2015) Natural food additives: quo vadis? Trends Food Sci Technol

Martinez RCR, Staliano CD, Vieira ADS, Villarreal MLM, Todorov SD, Saad SMI, Franco BDGDM (2015) Bacteriocin production and inhibition of listeria monocytogenes by lactobacillus Sakei Subsp. Sakei 2a in a potentially synbiotic cheese spread. Food Microbiol 48:143–152

Masyita A, Mustika Sari R, Dwi Astuti A, Yasir B, Rahma Rumata N, Emran TB, Nainu F, Simal-Gandara J (2022) Terpenes and terpenoids as main bioactive compounds of essential oils, their roles in human health and potential application as natural food preservatives. Food Chem X 13: 100217

Mau JL, Lee CC, Yang CW, Chen RW, Zhang QF, Lin SD (2020) Physicochemical, antioxidant and sensory characteristics of bread partially substituted with aerial parts of sweet potato. Lwt 117:108602

Meenu M, Padhan B, Patel M, Patel R, Xu B (2023) Antibacterial activity of essential oils from different parts of plants against Salmonella and Listeria Spp. Food Chem 404:134723

Milea AS, Vasile AM, Cîrciumaru A, Dumitrascu L, Barbu V, Râpeanu G, Bahrim GE, Stanciuc N (2019) Valorizations of sweet cherries skins phytochemicals by extraction, microencapsulation and development of value-added food products. Foods 8:1–12

Min S, Harris LJ, Han JH, Krochta JM (2005) Listeria monocytogenes inhibition by whey protein films and coatings incorporating lysozyme. J Food Prot 68:2317–2325

Mojica L, Berhow M, Gonzalez De Mejia E (2017) Black bean anthocyanin-rich extracts as food colorants: physicochemical stability and antidiabetes potential. Food Chem 229:628–639

Morvaridi M, Rayyani E, Jaafari M, Khiabani A, Rahimlou M (2020) The effect of green coffee extract supplementation on cardio metabolic risk factors: a systematic review and meta-analysis of randomized controlled trials. J Diabetes Metab Disord 19:645–660

Nasri C, Halabi Y, Aghzaf S, Nounah I, Brunel M, Oubihi A, El-Guorrami O, Harhar H, Costa J, Tabyaoui M (2022) Seven Persea Americana varieties essential oils comparison: chemical composition, toxicity, antibacterial, and antioxidant activities. Biocatal Agric Biotechnol 44: 102468

Neder-Suárez D, Lardizabal-Gutiérrez D, Zazueta-Morales JDJ, Meléndez-Pizarro CO, Delgado-Nieblas CI, Wong BR, Gutiérrez-Méndez N, Hernández-Ochoa LR, Quintero-Ramos A (2021) Anthocyanins and functional compounds change in a third-generation snacks prepared using extruded blue maize, Black Bean, and chard: an optimization. Antioxidants:10

Nemes SA, Călinoiu LF, Dulf FV, Fărcas AC, Vodnar DC (2022) Integrated technology for cereal bran valorization: perspectives for a sustainable industrial approach. Antioxidants 11:2159

Neri-Numa IA, Arruda HS, Geraldi MV, Maróstica Júnior MR, Pastore GM (2020) Natural prebiotic carbohydrates, carotenoids and flavonoids as ingredients in food systems. Curr Opin Food Sci 33:98–107

Novais C, Molina AK, Abreu RMV, Santo-Buelga C, Ferreira ICFR, Pereira C, Barros L (2022) Natural food colorants and preservatives: a review, a demand, and a challenge. J Agric Food Chem 70:2789–2805

Oliveira DVR, Preto LT, De Oliveira Schmidt H, Komeroski M, Silva DVL, De Oliveira Rios A (2016) Physicochemical and sensory evaluation of cakes made with passion fruit and orange residues. J Culin Sci Technol 14:166–175

Olszewska MA, Gędas A, Simões M (2020) Antimicrobial polyphenol-rich extracts: applications and limitations in the food industry. Food Res Int 134:109214

Ooi SF, Sukri SAM, Zakaria NNA, Harith ZT (2021) Carotenoids, phenolics and antioxidant properties of different sweet potatoes (Ipomoea Batatas) varieties. Iop Conf Ser Earth Environ Sci:756

Ordóñez-Santos LE, Esparza-Estrada J, Vanegas-Mahecha P (2021) Ultrasound-assisted extraction of Total carotenoids from mandarin Epicarp and application as natural colorant in bakery products. Lwt 139:110598

Pantuzza Silva GF, Pereira E, Melgar B, Stojković D, Sokovic M, Calhelha RC, Pereira C, Abreu RMV, Ferreira ICFR, Barros L (2021) Eggplant fruit (Solanum Melongena L.) and bio-residues as a source of nutrients, bioactive compounds, and food colorants, using innovative food technologies. Appl Sci 11:1–24

Pap N, Fidelis M, Azevedo L, Do Carmo MAV, Wang D, Mocan A, Pereira EPR, Xavier-Santos D, Sant'ana AS, Yang B, Granato D (2021) Berry polyphenols and human health: evidence of antioxidant, anti-inflammatory, microbiota modulation, and cell-protecting effects. Curr Opin Food Sci 42:167–186

Pascoal A, Rodrigues S, Teixeira A, Feás X, Estevinho LM (2014) Biological activities of commercial bee pollens: antimicrobial, antimutagenic, antioxidant and anti-inflammatory. Food Chem Toxicol 63:233–239

Pavela R (2015) Essential oils for the development of eco-friendly mosquito Larvicides: a review. Ind Crop Prod 76:174–187

Peñalver R, Martínez-Zamora L, Lorenzo JM, Ros G, Nieto G (2022) Nutritional and antioxidant properties of Moringa Oleifera leaves in functional foods. Foods 11:1–13

Pirsa S, Hafezi K (2023) Hydrocolloids: structure, preparation method, and application in food industry. Food Chem 399:133967

Poli A, Barbagallo CM, Cicero AFG, Corsini A, Manzato E, Trimarco B, Bernini F, Visioli F, Bianchi A, Canzone G, Crescini C, De Kreutzenberg S, Ferrara N, Gambacciani M, Ghiselli A, Lubrano C, Marelli G, Marrocco W, Montemurro V, Parretti D, Pedretti R, Perticone F, Stella R, Marangoni F (2018) Nutraceuticals and functional foods for the control of plasma cholesterol levels. An intersociety position paper. Pharmacol Res 134:51–60

Rahimi P, Abedimanesh S, Mesbah-Namin SA, Ostadrahimi A (2019) Betalains, the nature-inspired pigments, in health and diseases. Crit Rev Food Sci Nutr 59:2949–2978

Ramos-Martín F, Herrera-León C, Antonietti V, Sonnet P, Sarazin C, D'amelio N (2022) The potential of antifungal peptide Sesquin as natural food preservative. Biochimie 203:51–64

Rasool N, Saeed Z, Pervaiz M, Ali F, Younas U, Bashir R, Bukhari SM, Mahmood Khan RR, Jelani S, Sikandar R (2022) Evaluation of essential oil extracted from ginger, cinnamon and lemon for therapeutic and biological activities. Biocatal Agric Biotechnol 44:102470

Rodriguez-Amaya DB (2019) Update on natural food pigments—a mini-review on carotenoids, anthocyanins, and Betalains. Food Res Int 124:200–205

Rodríguez-Sánchez JA, Victoria CY, M. T. & Barragán-Huerta, B. E. (2017) Betaxanthins and antioxidant capacity in Stenocereus Pruinosus: stability and use in food. Food Res Int 91:63–71

Ruiz-Rico M, Renwick S, Vancuren SJ, Robinson AV, Gianetto-Hill C, Allen-Vercoe E, Barat JM (2023) Impact of food preservatives based on immobilized phenolic compounds on an in vitro model of human gut microbiota. Food Chem 403:134363

Saini RK, Ranjit A, Sharma K, Prasad P, Shang X, Gowda KGM, Keum YS (2022) Bioactive compounds of citrus fruits: a review of composition and health benefits of carotenoids, flavonoids, limonoids, and terpenes. Antioxidants:11
Salehi F (2020) Recent applications of powdered fruits and vegetables as novel ingredients in biscuits: a review. Nutrire 45
Salejda AM, Janiewicz U, Korzeniowska M, Kolniak-Ostek J, Krasnowska G (2016) Effect of walnut green husk addition on some quality properties of cooked sausages. Lwt 65:751–757
Saltmarsh M (2013) Essential guide to food additives
Sánchez-Muniz FJ, Canales A, Nus M, Bastida S, Guillén M, Corella D, Olmedilla-Alonso B, Granado-Lorencio F, Benedí J (2012) The antioxidant status response to low-fat and walnut paste–enriched meat differs in volunteers at high cardiovascular risk carrying different Pon-1 polymorphisms. J Am Coll Nutr 31:194–205
Santosh O, Bajwa HK, Bisht MS, Chongtham N (2021) Antioxidant activity and sensory evaluation of crispy salted snacks fortified with bamboo shoot rich in bioactive compounds. Appl Food Res 1:100018
Shahidi F, Ambigaipalan P (2015) Phenolics and polyphenolics in foods, beverages and spices: antioxidant activity and health effects—a review. J Funct Foods 18:820–897
Sharma M, Usmani Z, Gupta VK, Bhat R (2021) Valorization of fruits and vegetable wastes and by-products to produce natural pigments. Crit Rev Biotechnol 41:535–563
Shori AB (2020) Inclusion of phenolic compounds from different medicinal plants to increase α-amylase inhibition activity and antioxidants in Yogurt. J Taibah Univ Sci 14:1000–1008
Sim SY, Noor Aziah AA, Cheng LH (2015) Quality and functionality of Chinese steamed bread and dough added with selected non-starch polysaccharides. J Food Sci Technol 52:303–310
Socaci SA, Fărcaş AC, Diaconeasa ZM, Vodnar DC, Rusu B, Tofană M (2018) Influence of the extraction solvent on phenolic content, antioxidant, antimicrobial and antimutagenic activities of brewers' spent grain. J Cereal Sci 80:180–187
Solymosi K, Latruffe N, Morant-Manceau A, Schoefs B (2015) Food colour additives of natural origin
Sumonsiri N, Danpongprasert W, Thaidech K (2020) Comparison of sweet orange (citrus Sinensis) and lemon (Citrus Limonum) essential oils on qualities of fresh-cut apples during storage. Sci Stud Res Chem Chem Eng Biotechnol Food Ind 21:47–57
Sun H, Mu T, Xi L, Zhang M, Chen J (2014) Sweet potato (Ipomoea Batatas L.) leaves as nutritional and functional foods. Food Chem 156:380–389
Tiwari S, Dubey NK (2022) Nanoencapsulated essential oils as novel green preservatives against fungal and mycotoxin contamination of food commodities. Curr Opin Food Sci 45:100831
Tomadoni B, Viacava G, Cassani L, Moreira MR, Ponce A (2016) Novel biopreservatives to enhance the safety and quality of strawberry juice. J Food Sci Technol 53:281–292
Tomović V, Šojić B, Savanović J, Kocić-Tanackov S, Pavlić B, Jokanović M, Đorđević V, Parunović N, Martinović A, Vujadinović D (2020) New formulation towards healthier meat products: Juniperus Communis L essential oil as alternative for sodium nitrite in dry fermented sausages. Foods 9:1066
Uțoiu E, Matei F, Toma A, Diguță CF, Ştefan LM, Mănoiu S, Vrăjmaşu VV, Moraru I, Oancea A, Israel-Roming F, Cornea CP, Constantinescu-Aruxandei D, Moraru A, Oancea F (2018) Bee collected pollen with enhanced health benefits, produced by fermentation with a Kombucha consortium. Nutrients 10:1–24
Valdivieso-Ugarte M, Plaza-Diaz J, Gomez-Llorente C, Lucas Gómez E, Sabés-Alsina M, Gil Á (2021) In vitro examination of antibacterial and immunomodulatory activities of cinnamon, white thyme, and clove essential oils. J Funct Foods 81:104436
Veneziani G, Novelli E, Esposto S, Taticchi A, Servili M (2017) Applications of recovered bioactive compounds in food products
Wilson DW, Nash P, Singh H, Griffiths K, Singh R, De Meester F, Horiuchi R, Takahashi T (2017) The role of food antioxidants, benefits of functional foods, and influence of feeding habits on the health of the older person: an overview. Antioxidants 6:1–20

Wolever TMS, Tosh SM, Gibbs AL, Brand-Miller J, Duncan AM, Hart V, Lamarche B, Thomson BA, Duss R, Wood PJ (2010) Physicochemical properties of oat B-glucan influence its ability to reduce serum Ldl cholesterol in humans: a randomized clinical trial. Am J Clin Nutr 92:723–732

Wu L, Zhang C, Long Y, Chen Q, Zhang W, Liu G (2022a) Food additives: from functions to analytical methods. Crit Rev Food Sci Nutr 62:8497–8517

Wu L, Zhang CH, Long YX, Chen Q, Zhang WM, Liu GZ (2022b) Food additives: from functions to analytical methods. Crit Rev Food Sci Nutr 62:8497–8517

Wu S, Wang X, Qi W, Guo Q (2019) Bioactive protein/peptides of flaxseed: a review. Trends Food Sci Technol 92:184–193

Xu DP, Li Y, Meng X, Zhou T, Zhou Y, Zheng J, Zhang JJ, Li BH (2017) Natural antioxidants in foods and medicinal plants: extraction, assessment and resources. Int J Mol Sci 18:20–31

Xu S, Gong Y, Rafique H, He T, Hu X (2021) Effect of oat B-glucan addition on the staling properties of wheat-oat blended flour Chinese steamed bread. Bioact Carbohydr Diet Fibre 26: 100285

Zain MZM, Baba AS, Shori AB (2018) Effect of polyphenols enriched from green coffee bean on antioxidant activity and sensory evaluation of bread. J King Saud Univ Sci 30:278–282

Zeb A (2020) Concept, mechanism, and applications of phenolic antioxidants in foods. J Food Biochem 44:1–22

Zhang N, Zhou Q, Fan D, Xiao J, Zhao Y, Cheng K-W, Wang M (2021) Novel roles of hydrocolloids in foods: inhibition of toxic Maillard reaction products formation and attenuation of their harmful effects. Trends Food Sci Technol 111:706–715

Zhang Y, Butelli E, Martin C (2014) Engineering anthocyanin biosynthesis in plants. Curr Opin Plant Biol 19:81–90

Zhou J, Wang M, Bäuerl C, Cortés-Macías E, Calvo-Lerma J, Carmen Collado M, Barba FJ (2023) The impact of liquid-pressurized extracts of spirulina, chlorella and Phaedactylum Tricornutum on in vitro antioxidant, antiinflammatory and bacterial growth effects and gut microbiota modulation. Food Chem:401

History of Feed Additives

Giannenas Ilias, Panitsidis Ioannis, Iliadi Charmana Vasiliki, Dokou Stella, Vasilopoulou Konstantina, Chantzi Paraskevi, and Kumar Praffula

Abstract

Feed additives have been used in animal nutrition for centuries, with the earliest recorded use dating back to ancient civilizations. Initially, farmers and nutritionists relied on natural products such as plants, minerals, and waste products to supplement animal diets. Over time, with advancements in technology and scientific understanding, the use of feed additives evolved to include synthesized vitamins, minerals, and growth promoters. The twentieth century saw significant growth in feed additive industry, driven by increased demand for animal protein as a source of food for a rapidly growing human population. In the latter part of the twentieth century, the industry continued to expand, with the introduction of new products such as antioxidants, antibiotics, and enzymes. The use of antibiotics as feed additives was especially popular, with farmers and nutritionists relying on them to prevent and treat bacterial infections in livestock and poultry. In recent decades, there has been increasing concern about the use of antibiotics as feed additives, with evidence suggesting that the widespread use of antibiotics has contributed to the development of antibiotic-resistant bacteria. This has led to stricter regulations, with many countries now limiting or banning, i.e., the EU, the use of antibiotics as feed additives. In response, the industry has shifted toward the use of alternative feed additives such as probiotics, prebiotics, and essential oils, which seem to have a positive impact on animal health and performance.

G. Ilias (✉) · P. Ioannis · I. C. Vasiliki · D. Stella · V. Konstantina · C. Paraskevi
Laboratory of Nutrition, Faculty of Veterinary Medicine, Aristotle University of Thessaloniki, Thessaloniki, Greece
e-mail: igiannenas@vet.auth.gr

K. Praffula
Vets Plus, Inc., Menomonie, WI, USA

G. Arsenos, I. Giannenas (eds.), *Sustainable Use of Feed Additives in Livestock*,
https://doi.org/10.1007/978-3-031-42855-5_4

Keywords

Basic feed additives · Technological additives · Sensory additives · Nutritional additives · Zootechnical additives

1 Introduction

Feed additives are substances added to animal feeds to improve the nutritional value and confer health benefits to animals and further improve the quality of produced products of animal origin. The notion is that feed additives can enhance growth, improve feed efficiency, prevent disease, and improve overall animal performance. Marketing of feed additives is highly regulated subject to rigorous scientific evidence regarding their impact on animal health and the environment as well as the health of consumers of animal products. Hence, authorizations are for designated use in feed of certain animal species or groups, as well as under specified circumstances. Due to the large variety of additives, they are divided into the following groups:

- Technological additives (preservatives, antioxidants, emulsifiers, stabilizing agents, acidity regulators, silage additives)
- Sensory additives (flavorings, colorants)
- Nutritional additives (vitamins, minerals, amino acids, trace elements)
- Zootechnical additives (digestibility enhancers, gut flora stabilizers)
- Coccidiostats and histomonostats

These categories are further subdivided into functional groupings according to additive's primary functions. Antibiotics other than coccidiostats and histomonostats are not permitted as feed additives in Europe. Authorization of feed additives has become an important issue, especially in current century. In the previous era, and soon after the Second World War, growth rate of animal production was continuous and constantly positive. Simultaneously, development and usage of different feed additives was easy and easily applied. However, since the establishment of resistant strains of bacteria to antibiotics, the idea of using antimicrobial feed additives has been under thorough scientific scrutiny. Today, there are two kinds of licensing permitting the use of specific feed additives:

- Licenses issued to a holder of authorization. Those are granted for additives characterized as "zootechnical additives" and "coccidiostats and histomonostats," as well as those that consist of, contain, or are produced from genetically modified organisms (GMOs).
- Licenses not issued to a holder of authorization for substances characterized as "technological additives" and "sensory additives." The terms "additional ingredients" and "nutritional additives" are used interchangeably.

Both types of licensing authorizations are valid for ten years throughout the EU and the European Economic Area (EEA). Those authorizations are renewable for ten-year period. Generally, the Feed Additives Regulation (Regulation [EC] No 1831/2003 of September 2003) establishes a common procedure for authorizing feed additives and lays down rules for their marketing, labeling, and use. The regulation applies to all feed additives and premixtures except the following:

- Processing aids
- Veterinary medicinal products (VMPs) that are regulated by "Regulation (EU) 2019/6 on veterinary medicinal products that repeals Directive 2001/82/EC" (Regulation EU 2019, 2018)

The European Commission is now reassessing existing regulations on Feed Additives. The aim is to assess if the current legislation framework catered to the needs of citizens, businesses, and public authorities as expected and also to identify any gaps. The whole process is being carried out according to *Better Regulation Guidelines* (2021).

As stated above, there is a large variety of feed additives that are commonly used in animal nutrition. The major categories are the following (Fig. 1):

1. Antibiotics: Antibiotics are used to prevent and treat bacterial infections in animals. They are classified either as growth promoters, or as therapeutic agents to treat specific diseases. Their use in animal feeds has been a controversial issue, as there are concerns about the development of antibiotic-resistant bacteria.
2. Probiotics and Prebiotics: Probiotics are live microorganisms that are added to animal feed to improve the health of the digestive system and boost immune status. They can also reduce the risk of infection and improve feed efficiency. Some common probiotics used in animal nutrition include lactic acid bacteria, yeast, and bifidobacteria. Prebiotics on the other hand are compounds that promote the well-being of microflora of the gut. Common prebiotics are fructooligosaccharides, galactooligosaccharides, and xylooligosaccharides.
3. Enzymes: Enzymes are proteins contributing to breakdown of complex molecules into simpler components and hence facilitating their absorbance in the digestive tract. Their addition to animal feed improves digestibility, nutrient utilization, and feed efficiency.
4. Vitamins and Minerals: Vitamins and minerals are essential nutrients that are necessary for optimal animal health and performance. They are added to animal feed to ensure a balanced and complete diet, especially in cases where feedstuffs are deficient in certain nutrients.
5. Acidifiers: Acidifiers can lower pH of the digestive tract and hence reduce the risk of digestive problems and improve feed efficiency. Some common acidifiers used in animal nutrition include organic acids such as propionic acid and formic acid, as well as inorganic acids such as hydrochloric acid.
6. Antioxidants: Antioxidants protect against oxidative damage, which can occur because of normal metabolic processes. They improve overall health and

Fig. 1 Types of feed additives

performance of animals by reducing the risk of disease and improving feed efficiency. Some common antioxidants are vitamin E, selenium, and beta-carotene.

7. Essential Oils: Essential oils are highly concentrated liquids extracted from plants that contain volatile aromatic compounds. These compounds have various therapeutic and nutritional benefits, including antimicrobial and antioxidant properties. In animal nutrition, essential oils are often used as feed additives to enhance gut health, improve feed efficiency and growth performance, and promote overall health and wellness in livestock and poultry.
8. Amino Acids: Amino acids are the building blocks of proteins and are essential for growth and maintenance in animals. Their addition ensures that animals receive a balanced and complete diet, especially in cases where feedstuffs are deficient in certain amino acids (Kim and Kim 2017; D'Mello 2015).

The use of feed additives in animal nutrition can have a significant impact on animal performance and overall health. For example, the addition of antibiotics to animal feed has been shown to improve growth rates, feed efficiency, and overall health, while the addition of probiotics reduces the risk of digestive problems and improves gut health. However, it is important to use feed additives responsibly and only as directed by a veterinarian, as the overuse of certain additives can lead to the development of resistance, which can reduce their effectiveness over time (D'Mello 2015).

The history of feed additives in animal nutrition can be traced back to the early twentieth century, when the discovery of vitamins and minerals revolutionized the field of animal nutrition. Prior to this time, farmers relied on natural feed sources, such as grass and grain, to provide the necessary nutrients for their animals. However, the lack of knowledge about the specific nutritional requirements of different species of animals, as well as the variability in the quality of natural feed sources, led to poor performance and health outcomes for many animals (Grimble 1998). After the discovery of vitamins and minerals, farmers were able to supplement the diets of their animals with specific nutrients to improve their health and performance. This marked the beginning of the use of feed additives in animal nutrition. Over the following decades, advances in the field of animal nutrition, such as the discovery of antibiotics, probiotics, and enzymes, led to the development of a wide range of feed additives that could be used to improve the health and performance of animals.

One of the earliest uses of feed additives in animal nutrition was the addition of antibiotics to animal feed. Antibiotics were first introduced into animal feed in the 1950s, and quickly became a popular tool for improving growth and preventing disease in livestock. Antibiotics are still widely used in animal nutrition today, although their use has become controversial due to concerns about the development of antibiotic-resistant bacteria (Hornitzky and Rawnsley 2015). Another important development in the history of feed additives was the introduction of probiotics. Probiotics are live microorganisms that can improve digestive health and boost the immune system in animals. They were first introduced into animal feed in the 1980s and have since become a popular tool for improving gut health and preventing digestive problems in livestock (Gaur and Singh 2020). The use of enzymes in animal nutrition also has a long history, with the first commercial enzyme products being introduced into animal feed in the 1970s. Enzymes are proteins that help to break down complex molecules into simpler components that can be more easily absorbed by the animal, and they can be added to animal feed to improve digestibility and nutrient utilization (Zhou et al. 2021). The use of vitamins and minerals in animal nutrition has a similarly long history, with farmers having been supplementing the diets of their animals with these essential nutrients for over a century. Vitamins and minerals are still widely used in animal feed today, as they play a crucial role in maintaining the health and performance of animals (Bai et al. 2021).

Despite the long history of feed additives in animal nutrition, their use remains a subject of debate, with some experts arguing that the overuse of certain additives can lead to the development of resistance, reducing their effectiveness over time. However, there is also a growing body of research demonstrating the benefits of feed additives, such as improved growth, feed efficiency, and overall health in animals, leading many experts to conclude that their use is important for the sustainability of animal production.

2 History of Antibiotics as Feed Additives

Back in the 1930s, antibiotics' use, not only in human medicine but also in food and animal production, began to get established with the ultimate goal of effective disease prevention and rapid enhancement of animal growth. The first antibiotic for animal use was none other than synthetic sulfonamides that present antibacterial activity against Gram-positive and certain Gram-negative bacteria (White and Cooper 2005). The first sulfonamide drug was inaugurated by Bayer AG pharmaceutical company under the brand name Prontosil in 1935.

Thenceforth, because of World War II, drug manufacturing in Europe deteriorated in comparison to US pharmaceutical companies such as Merck and Pfizer that proceeded toward production of many synthetic and biological antibiotics. Meanwhile, researchers trialed mass-medication of entire herds and flocks. In 1948 Merck entered the animal health market with sulfaquinoxaline, the first antibiotic officially licensed to be used as a food or water additive in poultry. Sulfaquinoxaline in final concentration of 0.0125% in feed produces anticoccidiotic effects (Campbell 2008). After this, the use of antibiotics expanded to other animal species such as beehives, farmed fish, and cattle.

Vitamin B_{12} was then known as the main growth promoter in farm animal, but later studies revealed that dried mycelia of fungi caused better growth in chicks (Stokstad and Jukes 1949). Given the fact that antibiotics were growth-promoting substance in fungal mycelia, the idea of using antibiotics in animals' diets as growth-promoting factors prevailed, as, among others, they were also inexpensive in comparison to B_{12}. Moore et al. (1946) first proved that addition of antibiotics in chickens' diet contributed to weight gain. A few years later, many researchers found similar effects on pigs and chicks (Groschke and Evans 1950; Stokstad and Jukes 1951; Oleson et al. 1950).

The use of antibiotics in low doses as animal feed additives seemed very promising for both farmers and pharmaceutical companies. The benefits of prophylactic protection against bacterial diseases, growth promotion along with faster paces in weight gain, and therapeutic results of antibiotics were profound. Thereafter, in Midwest antibiotic growth promoters (AGP) feeds were officially licensed in 1951. After the prevalence of AGPs in the United States, European countries one after another began to adopt the idea and license AGPs without veterinary prescription in early 1950s (Thoms 2012; Kirchhelle 2018a, b). Penicillin, oxytetracycline, and chlortetracycline growth promoters were among the most commonly used AGPs in Europe during this decade. Contrary to US market situation, European farmers were only allowed to purchase premixed antibiotic solution and feeds (Jones 2003; Smith-

Howard 2017), thus avoiding the unrestrained use of antibiotics. According to researchers, some interesting facts are that in 1958 up to 50% of British pigs were fed antibiotics (Smith 1958) and by 1966 West Germany's Ministry of Agriculture's estimation was that 80% of mixed feeds for young pigs, veal calves, and poultry contained antibiotic additives (Kirchhelle 2016). Japan also licensed its own antimicrobial additives since 1953 (Morita 1997), both in livestock and fish production. In a need of modernization of their agriculture methods, Africa, South America, and Southeast Asia imported antibiotic additives for their livestock from the United States.

In the scenario of population increase and emerging economic development, there was intense need for a fast-paced agricultural production that resulted in a global outspread of the use of antibiotics as feed additives. However, concerns by experts about antimicrobial resistance (AMR) did not take long to arise. Thus, in late 1960s, national regulatory organizations and governments started to establish monitoring programs for antibiotics use and set certain limitations on their usage.

Scandinavian countries were the first to take precautions concerning AGPs. Sweden banned all AGPs' use in 1986. Norway started to adopt vaccinations instead of antibiotics in aquaculture since 1987. In Denmark, due to the association between the occurrence of resistance to vancomycin *Enterococcus faecium* (VREF) in healthy pigs and poultry (Bager et al. 1997) and use of avoparcin, the latter antibiotic was banned in 1995 and general AGP consumption decreased by 103,503 kg when producers decided to voluntarily give up on AGPs (Aarestrup and Engberg 2001; Kahn 2016).

Following Scandinavian countries' measures, EU's Scientific Committee on Animal Nutrition (SCAN) member states banned four AGPs and established the European Resistance Surveillance System (EARSS) in 1998 (Kirchhelle 2018a, b). Japan banned avoparcin and orienticin as feed additives in 1997 and South Korea restricted 45 different antimicrobial additives to only be administered after veterinary prescription.

However despite the partial restrictions, the use of AGPs still remains even in higher levels in most high-income countries than in 1960s (Kirchhelle 2018a, b). For example, there is significant difference in consumption of antibiotics between countries of Southwest Europe like Italy and Spain and Northern countries that have limited their use. In 2017, several Indian pharmaceutical businesses promoted colistin growth enhancers (Davies and Walsh 2018). By 2010, China surpassed the United States as the world's largest consumer of agricultural antimicrobials (about 23% of worldwide consumption) (Van Boeckel et al. 2015).

3 History of Probiotics

The use of probiotics in animal nutrition can be traced back to the early twentieth century, when scientists first began to study the role of microorganisms in the digestive system of animals. In the late nineteenth and early twentieth centuries, researchers such as Elie Metchnikoff began to study the role of microorganisms in the gut and their impact on health. Metchnikoff was particularly interested in the role of lactic acid bacteria, which he believed were responsible for the longevity of Bulgarian peasants. This early work laid the foundation for the development of probiotics as a tool for improving animal health and performance (Bengmark 1998). In the mid-twentieth century, scientists made significant advances in the production and formulation of probiotics. In 1958, the first probiotic product, *Lactobacillus acidophilus*, was commercialized by Yakult, a Japanese company (Fuller 1989). Over the following decades, the use of probiotics in animal nutrition grew rapidly, as researchers and farmers alike recognized the potential benefits of these beneficial bacteria. One of the first studies to demonstrate the efficacy of probiotics in animal nutrition was conducted by Fullers and Bradley in the 1980s. The study found that supplementing the diet of chickens with a mixture of *Lactobacillus* and *Bifidobacterium* species improved growth performance, feed efficiency, and overall health (Fuller and Bradley 1985). This work helped to establish the role of probiotics in animal nutrition and sparked a wave of research into the use of probiotics in livestock and aquaculture.

The use of probiotics in animal nutrition has continued to evolve and expand over the past few decades. Today, probiotics are widely used in the production of livestock and aquaculture species, including chickens, pigs, cows, fish, and shellfish (Bari and Roy 2016). The benefits of probiotics in animal nutrition are well-established, and include improved growth performance, feed efficiency, disease resistance, and gut health (Rao et al. 2018). In recent years, the use of probiotics in animal nutrition has become increasingly sophisticated, with the development of novel probiotic strains and delivery systems. For example, researchers have developed probiotic products that are specifically formulated for different species of animals, based on their unique digestive physiology (Kim et al. 2019). Additionally, advances in genomics and microbial ecology have allowed scientists to better understand the complex relationships between probiotics, the gut microbiota, and animal health (Rao et al. 2020).

In conclusion, the use of probiotics in animal nutrition has a long and rich history, dating back to the early twentieth century. Today, probiotics are widely used in the production of livestock and aquaculture species, and their benefits in animal health and performance are well-established. With continued advancements in our understanding of probiotics and their mechanisms of action, the use of probiotics in animal nutrition is poised to play an increasingly important role in sustainable and responsible animal production in the coming years.

4 History of Prebiotics

The concept of prebiotics is relatively new and has only been established in the last few decades. However, the history of prebiotics can be traced back to the early twentieth century when a Russian scientist named Élie Metchnikoff first proposed the idea that consuming fermented foods could promote longevity and improve health (Mackowiak 2013).

The term "prebiotics" was first used in the late 1990s and early 2000s to describe non-digestible food ingredients that selectively stimulate the growth and activity of beneficial bacteria in the gut (G. R. Gibson and Roberfroid 1995). This definition may correlate with the definition of dietary fiber (De Vrese 2001). Later on, prebiotics were defined as "selectively fermented ingredients that allow specific changes, both in composition and/or activity in the GI microflora that confer benefits upon host wellbeing and health" (Gibson et al. 2004, p. 269). Finally, Roberfroid et al. (2011) updated the definition of prebiotics based on International Life Sciences Institute (ILSI) Europe task force.

It is likely that prebiotics have been consumed in some form since prehistoric times, as they are naturally occurring in many foods. Archaeological evidence from dry cave deposits in the northern Chihuahuan Desert shows that desert plants high in inulin were extensively used (Slavin 2013). While the concept of prebiotics and their specific definition as non-digestible food ingredients is a relatively recent development, it is possible that our ancestors consumed prebiotic-rich foods as part of their diets. For example, fermented foods, such as yogurt, kefir, and sauerkraut, contain both prebiotics and probiotics, and have been consumed for thousands of years in various cultures (Jew et al. 2009). It is important to note, however, that the human gut microbiome has changed significantly over the course of human evolution, and our diets today are quite different from those of our prehistoric ancestors. Despite this, consuming prebiotic-rich foods is still considered important for maintaining gut health and promoting overall well-being (Moran et al. 2019).

Prebiotics have received increased scientific attention, and several studies have been conducted to determine the potential health benefits of consuming prebiotics. Today, prebiotics are considered an important component of a healthy diet and are often used in functional foods and dietary supplements to promote gut health (Roberfroid 2000). While research in this field is ongoing, there is growing evidence to suggest that consuming prebiotics can support the growth of beneficial bacteria in the gut, which can have a positive impact on overall health and well-being (Moran et al. 2019).

Prebiotics are commonly used as feed additives in human and animal nutrition. The use of prebiotics as feed additives is based on the idea that they can improve gut health by promoting the growth of beneficial bacteria in the gut and reducing the growth of harmful bacteria (Dahiya and Nigam 2022). The use of prebiotics as feed additives in animal nutrition is a relatively recent development, with the first commercial products appearing in the market in the late 1990s and early 2000s (Davani-Davari et al. 2019). Since then, the use of prebiotics in animal feed has grown rapidly, and they are now widely used in various livestock and poultry species (Alloui et al. 2013). Figure 2 shows a timeline of the use of prebiotics as feed additives. It is important to note that this timeline is a rough approximation and that the specific timeline of the use of prebiotics as feed additives may vary in different regions and countries. However, the overall trend has been one of steady growth in the use of prebiotics as feed additives in animal nutrition.

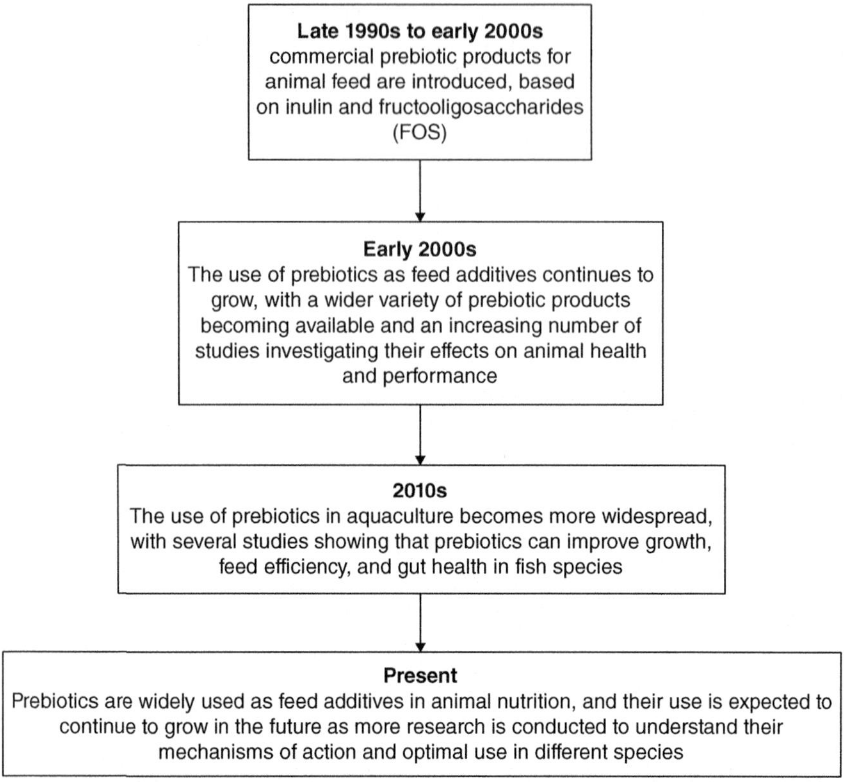

Fig. 2 The use of prebiotics in time

5 History of Essential Oils

The use of essential oils as feed additives in animal nutrition has a long history dating back to the early 1900s. Essential oils composed of a complex mixture of volatile compounds, including terpenoids, phenols, and aldehydes, which give essential oils their distinctive odor and flavor (Alfermann and Njardvik 1998). One of the earliest uses of essential oils in animal nutrition was for their antimicrobial properties, which were thought to improve animal health and growth by controlling gut pathogens. For example, the use of oregano and thyme essential oils as feed additives was shown to have potent antimicrobial effects against various bacteria and parasites, such as *Escherichia coli*, *Salmonella*, and *Ascaris suum* (Burt 2004). These findings paved the way for the widespread use of essential oils as feed additives in animal nutrition, especially in poultry and swine production. Over time, the use of essential oils as feed additives has expanded to include a wide range of applications, including improved growth performance, feed efficiency, and gut health. For example, the addition of peppermint oil to pig diets was shown to improve weight gain and feed efficiency, while also reducing the incidence of diarrhea (Wang et al. 2010). Similarly, the use of rosemary and eucalyptus essential oils in poultry diets was shown to improve growth performance, feed efficiency, and gut health by modulating the gut microbiota (Abdelqader et al. 2013). More recently, the use of essential oils as feed additives has expanded to include their potential as natural growth promoters, immunomodulators, and antioxidants. For example, the addition of clove and cinnamon essential oils to broiler chicken diets was shown to improve growth performance, feed efficiency, and immune function (Hossain et al. 2016). Additionally, the use of lemon and ginger essential oils was found to have potent antioxidant effects in pig diets, which could improve animal health and prevent oxidative stress (Sari et al. 2019).

Despite the long history and growing body of research on the use of essential oils in animal nutrition, there is still much to be learned about the mechanisms of action and optimal usage of these feed additives. For example, the complex composition of essential oils, including their volatile and highly reactive compounds, poses challenges for their safe and effective use in animal nutrition (Alfermann and Njardvik 1998). Additionally, there is a need for more research to determine the optimal dose and form of essential oils for different species and production systems.

In conclusion, the use of essential oils as feed additives in animal nutrition has a long history and a growing body of research to support its benefits for animal health and performance. While there are still challenges to overcome and much to be learned about their optimal usage, essential oils have the potential to be a valuable tool for improving animal nutrition and sustainability (Giannenas et al. 2013).

6 History of Minerals and Vitamins

The incorporation of supplemented minerals and vitamins in animal diets was a milestone for the increment in their productivity. Although they make up a small part of an animal's diet, vitamins and minerals are vital to maintaining their health, comfort, and performance. These nutrients perform crucial roles in metabolic process. The scarcity of vitamins and minerals in animal nutrition can negatively affect growth and performance yields (McDowell 1996).

The mineral content in animal tissues and feeds varies widely. Before the mid-nineteenth century, there was limited understanding of the mineral constituents in plant and animal tissues and their role. The first evidence of the importance of minerals in nutrition came from Fordyce (1791), who indicated the need of a supplement "calcareous earth" for canaries to stay healthy and produce eggs. References on the importance of salt in cattle nutrition were made in 1850 (Boussingault 1847) along with the consideration of iodine as significant for maintaining a healthy organism (Chatin 1854). Later on, during the second quarter of the twentieth century, several researchers focused on mineral investigation regarding deficiencies, imbalances, or toxicities, and basic studies with laboratory animals fed specialized diets were performed. Sodium, potassium, calcium, magnesium, phosphorus, chlorine, sulfur, iron, coper, cobalt, iodine, and zing were recognized as essential minerals and efforts took place to meet their requirements when formulating diets (Haag 1951). Great knowledge regarding mineral metabolism was provided by Underwood and Suttle (1999) in all livestock species. Changes in animal husbandry and immense intensification of livestock sector have led to the substantial need of balancing trace elements to maintain both production and health status. Currently, mineral supplementation in concentrate feed is an indispensable practice. New technological and biomedical applications, for example providing minerals in the form of nanoparticles, are adopted to increase bioavailability and absorption (Bhagat and Singh 2022).

Vitamins are a group of organic compounds, essential for normal growth and production processes. Dietary supplementation provides the small quantities of vitamins required, due to their inability of being directly synthesized through animal metabolic pathways. The term vitamin was first introduced by the biochemist Casimur Funk in 1912, originating from the Latin word vita (life) and amine, based on the original perception that vitamins contained an amino acid (Halver 2003). In the early 1900s, vitamins research began to rise in the effort to treat several disorders identified in humans, such as scurvy, night blindness, beriberi, and pellagra. Experiments with animals including rats, mice, chickens, pigeons, guinea pigs, and dogs contributed significantly to the advances made in vitamin research (McDowell 2006).

Monogastric animals depend solely on external sources to meet their vitamin requirements, while in the rumen symbiotic processes lead to the production of vitamin B complex (McDonald et al. 2011). Livestock animals' optimal performance is the ultimate goal under commercial conditions. Their requirements cannot be met only by providing raw materials such as forage and cereal grains, hence additional supplementation is needed (McDowell 2006).

7 History of Enzymes

Proteins called enzymes operate as biological catalysts by quickening chemical reactions. Enzyme catalysts are required for the majority of metabolic processes in the cell to proceed at speeds quick enough to maintain life (Stryer et al. 2002). Enzymes serve a plethora of an organism's functions including signal transduction, cell regulation, transportation of cargos, and of course metabolic activity.

Enzymes that are aftereffect of the metabolic activity of animal's system or of the microorganisms that constitute their natural microflora are so called endogenous enzymes. The main activity of endogenous enzymes is to promote the digestion of animal feed and therefore the weight gain and animal growth. However, despite their beneficial results, endogenous enzymes do not sufficiently contribute to their goal, thus rendering necessary their supplementation in animal's feed with exogenous enzymes (Velázquez-De Lucio et al. 2021).

The most significant and practical additives in the animal feed sector are enzymes (Ojha et al. 2019). Although its usage in the livestock feed business was modest a few years ago (Walsh et al. 1993), the poultry sector was the first to express interest in it in the 1980s, and, over time, their use in animal feed increased significantly, with an estimated commercial value of $1280 million in 2019 (Brufau 2014).

Exogenous enzymes are zootechnical additions that enhance the uniformity and nutritional value of animal feed, increase digestibility, especially of carbohydrates, and animal performance, decrease the effect of antinutrients, maintain intestinal health, and reduce the viscosity that would otherwise have a negative impact on energy intake by generating unstirred layers on the epithelium's surface (Rainbird et al. 1984). Celluloses, β-glucanases, xylanases, and associated enzymes phytases, proteases, lipases, and galactosidases are the most widely utilized exogenous enzymes in the animal feed business. Exogenous enzymes are largely used in monogastrics in the animal feed sector, like poultry and swine, to counteract the negative effects of cereals' viscous, non-starchy polysaccharides (NSPs). In the livestock feed industry, other enzyme products are being tested. These include proteases to enhance protein digestion, lipases to enhance fat digestion, galactosidases to mitigate certain antinutritional elements in non-cereal feedstuffs, and amylases to facilitate starch digestion in young animals (Adeola and Cowieson 2011).

In poultry supplementation, the most commonly used enzymes are xylanases, glucanases, pectinases, cellulases, proteases, amylases, phytases, and galactosidases. The use of enzymes has grown to the point where the majority of intensively raised poultry diets now include carbohydrases to improve nutrient bioavailability and assimilation and to alleviate digestive issues caused by viscosity issues (Ojha et al. 2019). Because phytases aid in the absorption of metals like Ca, Mg, and Zn as well as amino acids during the past few decades, they are now widely employed as broiler additives in Europe and the United States (Acosta 2006). Exogenous enzymes are utilized in swine feeding to shorten the maturation and fattening stage of piglets. Pigs need phytases because their endogenous counterparts have almost minimal action, which results in the complete excretion of phosphorus. Amylolytic enzymes are a necessary component of high-starch forages for ruminants, and using phytases has been shown to minimize phosphorus in feces from the perspective of the environment (Pariza and Cook 2010). Due to their capacity to hydrolyze the polysaccharides found in plant diet, phytases and carbohydrases are the enzymes most frequently used in aquaculture. Concerning the use of exogenous enzymes in companionship animals, it is far well known about the administration of amylases, lipases, and proteases in dogs with exocrine pancreatic insufficiency (EPI). Furthermore, some vets propose the use of exogenous enzymes in all pets for digestibility promotion and enhancement of immune system.

Traditionally, the enzymes were isolated from animals, plants, and microorganisms. Recombinant DNA technology facilitated enzyme production for industrial use with expression of single enzyme genes of certain microorganisms by bacterial expression systems in large volumes and with low cost. The search for alternate sources of enzymes, such as ruminal fluid from slaughterhouses, which is rich in cellulases and xylanases but also contains toxic ammonia and phosphorus, has also been prompted by the current environmental crisis (Sarteshnizi et al. 2018a, b). The design of thermostable enzymes with better, particular features and their incorporation in granulated meals are future issues regarding research and development in new enzyme systems (Ravindran 2013). This will enable the manufacturing of high-value products at cheaper costs.

8 History of Antioxidants

The use of antioxidants in animal nutrition, as the most of feed additives, dates back to the early twentieth century, when scientists first discovered the role of vitamins in maintaining animal health. Since then, the field of animal nutrition has made tremendous progress in understanding the specific nutritional requirements of different species of animals, and the role that antioxidants play in maintaining animal health.

One of the earliest uses of antioxidants in animal nutrition was the addition of vitamin E to animal feed. Vitamin E is a fat-soluble vitamin that plays an important role in maintaining the integrity of cell membranes and protecting against oxidative damage. In the 1950s, it was discovered that supplementing animal diets with vitamin E could improve growth, health, and fertility in livestock (Bogden et al. 1987). In addition to vitamin E, other antioxidants, such as vitamin C, carotenoids, and selenium, have also been shown to play important roles in animal nutrition. Carotenoids, for example, are pigments that are found in plant-based diets and have been shown to have antioxidant and immune-boosting properties. Selenium is an essential trace mineral that is required for proper immune function and has been shown to play a role in reducing oxidative stress in animals (Goff and Horst 1997).

In recent years, the use of antioxidants in animal nutrition has become increasingly important as the global demand for animal products continues to rise. The increased demand for animal products has led to increased production pressures and has resulted in animals being exposed to a variety of stressors, such as high-density housing, transportation, and disease challenges. These stressors can lead to oxidative stress and reduced performance and health outcomes for animals (Wang et al. 2018). In response to these challenges, the use of antioxidants in animal nutrition has become a major area of research and development. Scientists are working to develop new and innovative methods for delivering antioxidants to animals, such as encapsulated antioxidants, which can protect the antioxidants from degradation and improve their efficacy in the gut (Niu et al. 2017). Lately, extracts and essential oils of aromatic and so called medicinal plants have been largely exploited as excellent sources of food or diet antioxidants, among which oregano can be regarded as one of the most efficient (Giannenas et al. 2018).

In conclusion, the history of antioxidants in animal nutrition dates back to the early twentieth century, and the use of antioxidants has become increasingly important as the global demand for animal products continues to rise. Antioxidants play a critical role in maintaining animal health and protecting against oxidative stress, and the field of animal nutrition continues to make important advancements in the development of new and innovative methods for delivering antioxidants to animals.

9 History of Amino Acids

Amino acids are essential nutrients for animals, and their importance in animal nutrition has long been recognized. They are the building blocks of proteins and play a crucial role in many physiological processes, including growth, reproduction, and immune function. The use of amino acids as feed additives in animal nutrition has therefore become an important area of research and development, with scientists seeking to optimize their use to improve the health, performance, and efficiency of livestock and poultry.

The amino acids as feed additives in animal nutrition have a hundred-year history, when scientists first began to understand the role of amino acids in animal growth and development. Amino acids are the building blocks of proteins, and are essential for many physiological processes, including muscle growth, immune function, and tissue repair. The discovery of individual amino acids and their specific functions, combined with advances in feed technology, led to the development of amino-acid-based feed additives for animals.

One of the earliest uses of amino acids in animal nutrition was the addition of methionine to poultry diets. Methionine is an essential amino acid that is required for the synthesis of proteins but is often limiting in poultry diets due to its low availability in plant-based feed ingredients. The addition of methionine to poultry diets has been shown to improve growth performance, egg production, and feed efficiency (Wang et al. 2016). In addition to methionine, other essential and non-essential amino acids, such as lysine, threonine, and tryptophan, have been used as feed additives in animal nutrition. These amino acids are added to animal diets to address specific nutritional deficiencies, and to improve the balance of amino acids in the diet. For example, lysine supplementation has been shown to improve growth performance, feed efficiency, and meat quality in pigs (Yao et al. 2015). More recently, advances in animal nutrition research have led to the development of new and innovative amino-acid-based feed additives, such as peptides and nucleotides. Peptides are short chains of amino acids that are rapidly absorbed and utilized by animals, and have been shown to improve growth performance, immune function, and gut health in various species of animals (Geng et al. 2018). Nucleotides are the building blocks of DNA and RNA, and have been shown to improve immune function, tissue repair, and gut health in animals (Xie et al. 2019).

In conclusion, the use of amino acids as feed additives in animal nutrition has a long and rich history. Today, advances in animal nutrition research have led to the development of new and innovative amino-acid-based feed additives, such as peptides and nucleotides, which offer new opportunities for improving animal performance and health.

10 History of Acidifiers

Acidifiers have been used as feed additives in animal nutrition for several decades. Their primary function is to regulate the pH of the digestive system, which can have a profound impact on the health and performance of animals. Acidifiers work by lowering the pH in the digestive tract, creating a more acidic environment that is less hospitable to harmful bacteria and other pathogens. This can help to reduce the risk of disease, improve nutrient absorption, and enhance overall health and performance.

One of the earliest uses of acidifiers in animal nutrition was the addition of organic acids, such as formic and propionic acid, to animal feed (Kirsch et al. 2006). These organic acids have been shown to have a number of benefits for animal health and performance, including improved gut health, enhanced feed utilization, and increased weight gain. In addition to organic acids, other types of acidifiers have been developed and used in animal nutrition, such as mineral acids and feed-grade enzymes. Mineral acids, such as sulfuric and hydrochloric acids, have been shown to be effective in reducing the pH level of the digestive tract, which can help to control harmful bacteria and improve feed utilization (Flesner et al. 2011). Feed-grade enzymes, such as phytases, can also help to improve the efficiency of feed utilization by breaking down plant fibers and releasing the nutrients for absorption (Shan et al. 2016).

Despite the long history and widespread use of acidifiers in animal nutrition, their use is not without controversy. Some research has suggested that excessive use of acidifiers may lead to nutrient imbalances, digestive disorders, and reduced feed utilization (Kirsch et al. 2006). Therefore, it is important to use acidifiers in animal nutrition with caution and only under the guidance of a veterinarian or nutritionist.

In conclusion, acidifiers have played a significant role in the development of modern animal nutrition, providing a means to improve the health and performance of livestock through the manipulation of the digestive environment. However, the use of acidifiers should be approached with caution and only used under the guidance of a professional in order to ensure that their benefits are realized without any adverse effects (Fig. 3).

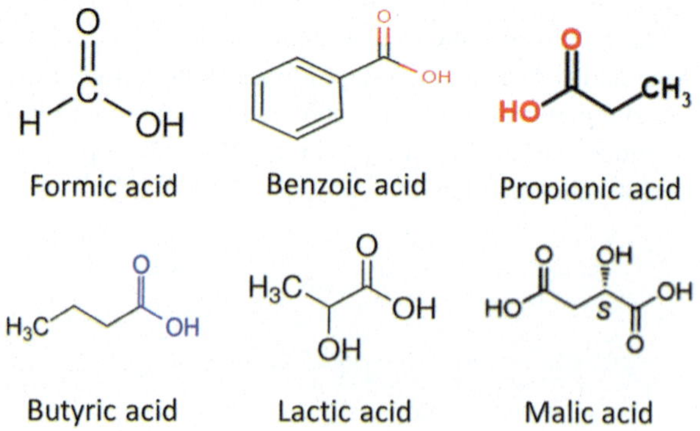

Fig. 3 Common type of acidifiers used in animal nutrition

References

Aarestrup FM, Engberg J (2001) Antimicrobial resistance of thermophilic Campylobacter. Vet Res 32(3–4):311–321. https://doi.org/10.1051/vetres:2001127

Abdelqader M, Alsheikh A, Al-Ahmad A, Al-Mufarrej SI, Al-Ahmadi NJ (2013) Rosemary and eucalyptus essential oils supplementation to broiler diets: effects on growth performance, digestive enzyme activities and oxidative status. Livest Sci 154(1–3):167–173

Acosta A (2006) Cárdenas, M. Enzimas en la alimentación de las aves. Fitasas Rev Cuba Cienc Agríc 40:377–387

Adeola O, Cowieson AJ (2011) Board-invited review: opportunities and challenges in using exogenous enzymes to improve non ruminant animal production. J Anim Sci 89:3189–3218. https://doi.org/10.2527/jas.2010-3715

Alfermann D, Njardvik U (1998) Essential oils and their volatile components in feedstuffs for livestock. Anim Feed Sci Technol 74(1–2):199–216

Alloui MN, Szczurek W, Świątkiewicz S (2013) The usefulness of prebiotics and probiotics in modern poultry nutrition: a review. Ann Anim Sci 13(1):17–32. https://doi.org/10.2478/v10220-012-0055-x

Bager F, Madsen M, Christensen J, Aarestrup FM (1997) Avoparcin used as a growth promoter is associated with the occurrence of vancomycin-resistant enterococcus faecium on Danish poultry and pig farms. Prev Vet Med 31(1–2):95–112. https://doi.org/10.1016/s0167-5877(96)01119-1

Bai X, Li J, Xia C (2021) Nutritional and functional roles of vitamins and minerals in animal nutrition and production. Anim Nutr 7(2):438–448. https://doi.org/10.1016/j.aninu.2020.11.010

Bari ML, Roy P (2016) Probiotics in animal nutrition: a review. J Anim Sci Adv 6(1):17–23. https://doi.org/10.5455/jasa.20151210150129

Bengmark S (1998) Probiotics: the role of gut microbiota. Curr Opin Gastroenterol 14(6):753–759

Better Regulation Guidelines (2021). https://commission.europa.eu/law/law-making-process/planning-and-proposing-law/better-regulation/better-regulation-guidelines-and-toolbox_en

Bhagat S, Singh S (2022) Nanominerals in nutrition: recent developments, present burning issues and future perspectives. Food Res Int:111703

Bogden JD, L'pez MC, Whanger PD (1987) The biological functions of selenium in animals. In: The selenium-tellurium cycle. Springer, Boston, MA, pp 97–142

Boussingault JB (1847) Ann Chem Phys 3rd Ser. 19, 117; 22, 116; Comptes Rendus des Séances de l'Académie des Sciences 25, 72Y. Cited by McCollum, E.V. (1957) A history of nutrition. Houghton Mifflin, Boston

Brufau J (2014) Introducción al uso de enzimas en la alimentación animal un proceso de innovación nutriNews, 1:17–21

Burt S (2004) Essential oils: their antibacterial properties and potential applications in foods—a review. Int J Food Microbiol 94(3):223–253

Campbell WC (2008) History of the discovery of sulfaquinoxaline as a coccidiostat. J Parasitol 94(4):934–945. https://doi.org/10.1645/GE-1413.1

Chatin GA (1854) Recherche de l'iode dans l'air, les eaux, le sol et les produits alimentaires du Jura, du Valais, de la Lombardie, de l'Allemagne et de la Belgique. Impr. Mallet-Bachelier

Dahiya D, Nigam PS (2022) Probiotics, prebiotics, synbiotics, and fermented foods as potential biotics in nutrition improving health via microbiome-gut-brain axis. Fermentation 8(7). https://doi.org/10.3390/fermentation8070303

Davani-Davari D, Negahdaripour M, Karimzadeh I, Seifan M, Mohkam M, Masoumi SJ, Berenjian A, Ghasemi Y (2019) Prebiotics: definition, types, sources, mechanisms, and clinical applications. Foods 8(3):1–27. https://doi.org/10.3390/foods8030092

Davies M, Walsh TR (2018) A colistin crisis in India. Lancet Infect Dis 18(3):256–257. https://doi.org/10.1016/S1473-3099(18)30072-0

De Vrese M (2001) Probiotics, prebiotics, and synbiotics—approaching a definition 1–3 73(14)

D'Mello JPF (2015) Feed additives for non-ruminants: modes of action and effects on gut microflora. In: Gut health and disease in farm animals. CABI, pp 259–287

Flesner M, Pedersen BK, Jensen BB (2011) Effect of sulfuric acid on broiler performance, gut morphology, and gut microflora. Poult Sci 90(1):216–222

Fordyce G (1791) Treatise on the digestion of food. 2nd edn. London. Cited by McCollum, E.V. (1957) A history of nutrition. Houghton Mifflin, Boston

Fuller R (1989) Probiotics in man and animals. J Appl Bacteriol 66(5):365–378

Fuller R, Bradley R (1985) The effect of Lactobacillus supplementation on feed conversion in broiler chickens. Br Poultry Sci 26(1):83–87

Gaur R, Singh H (2020) Probiotics in animal nutrition: a review. J Appl Anim Res 47(1):79–89. https://doi.org/10.1080/09712119.2019.1658201

Geng X, Liu F, Zhang Q, Li H (2018) Effects of different peptides supplementation on growth performance, nutrient digestibility, and gut microbiota of broilers. Poult Sci 97(3):862–870

Giannenas I, Bonos E, Christaki E, Florou-Paneri P (2013) Essential oils and their applications in animal nutrition. Med Aromat Plants 2. https://doi.org/10.4172/2167-0412.1000140

Giannenas I, Bonos E, Christaki E, Florou-Paneri P (2018) Oregano: a feed additive with functional properties. In: Holban AM, Grumezescu AM (eds) Therapeutic foods, handbook of food engineering, vol 8. Elsevier Academic Press, London. ISBN: 978-0-12-811517-6, pp 179–208

Gibson GR, Roberfroid MB (1995) Dietary modulation of the human colonic microbiota: introducing the concept of prebiotics. J Nutr 125(6):1401–1412. https://doi.org/10.1093/jn/125.6.1401

Gibson GR, Probert HM, Van Loo J, Rastall RA, Roberfroid MB (2004) Dietary modulation of the human colonic microbiota: updating the concept of prebiotics. Nutr Res Rev 17(2):259–275. https://doi.org/10.1079/nrr200479

Goff JP, Horst RL (1997) Vitamins in animal nutrition: comparisons of requirements and dietary adequacy. J Nutr 127(2):373S–379S

Grimble RF (1998) The use of feed additives in animal nutrition: a review. Nutr Res Rev 11(1): 149–168

Groschke AC, Evans RJ (1950) Effect of antibiotics, synthetic vitamins, vitamin B12 and an APF supplement on chick growth. Poult Sci 29(4):616–618

Haag JR (1951) Minerals for livestock

Halver JE (2003) The vitamins. In: Fish nutrition. Academic Press, pp 61–141

Hornitzky MA, Rawnsley H (2015) Antibiotic resistance in animal production: an overview. Aust J Agr Res 66(7):791–798. https://doi.org/10.1071/AR14063

Hossain MA, Islam MR, Kim IH (2016) Effects of clove and cinnamon essential oils on growth performance, immune function, and antioxidant capacity of broiler chickens. Poult Sci 95(12): 2901–2907

Jew S, Abumweis SS, Jones PJH (2009) Evolution of the human diet: linking our ancestral diet to modern functional foods as a means of chronic disease prevention. J Med Food 12(5):925–934. https://doi.org/10.1089/jmf.2008.0268

Jones SD (2003) Valuing animals. Veterinarians and their patients in modern, America. Johns Hopkins University Press, London

Kahn LH (2016) One health and the politics of antimicrobial resistance. Johns Hopkins University Press, Baltimore

Kim SS, Kim IH (2017) Review: functions and safety of feed additives in animal production. Anim Sci J 88(1):1–13

Kim YJ, Lee J, Park JH, Lee JS, Kim W (2019) Probiotics for poultry: current status and future prospects. Front Microbiol 10:754

Kirchhelle C (2016) Toxic confusion: the dilemma of antibiotic regulation in west German food production (1951-1990). Endeavour 40(2):114–127. https://doi.org/10.1016/j.endeavour.2016.03.005. Epub 2016 Apr 13

Kirchhelle C (2018a) Pharming animals: a global history of antibiotics in food production (1935–2017). Palgrave Commun 4:96. https://doi.org/10.1057/s41599-018-0152-2

Kirchhelle C (2018b) Swann song: antibiotic regulation in British livestock production (1953-2006). Bull Hist Med 92(2):317–350. https://doi.org/10.1353/bhm.2018.0029

Kirsch P, Rodehutscord M, Deuling B (2006) Performance, digestion, and microbial activity in broiler chickens fed diets containing different levels of formic acid. Br Poultry Sci 47(5): 557–564

Mackowiak PA (2013) Recycling Metchnikoff: probiotics, the intestinal microbiome and the quest for long life. Front Public Health 1:1–3. https://doi.org/10.3389/fpubh.2013.00052

McDonald P, Edwards R, Greenhalgh JFD, Morgan CA, Sinclair LA (2011) Animal nutrition, 7th edn. Pearson

McDowell LR (1996) Feeding minerals to cattle on pasture. Anim Feed Sci Technol 60(3–4): 247–271

McDowell LR (2006) Vitamin nutrition of livestock animals: overview from vitamin discovery to today. Can J Anim Sci 86(2):171–179

Moore PR, Evension A, Luckey TD, McCoy E, Elvehjam CA, Hart EB (1946) Use of sulfasuxidine, streptothricin and streptomycin in nutritional studies with the chick. J Biol Chem 165:437–441. https://doi.org/10.1016/S0021-9258(17)41154-9

Moran NA, Ochman H, Hammer TJ (2019) Evolutionary and ecological consequences of gut microbial communities. Annu Rev Ecol Evol Syst 50:451–475. https://doi.org/10.1146/annurev-ecolsys-110617-062453

Morita K (1997) Policies for use of anti-infective drugs in food animal production in Japan. Working papers—medical impact of antimicrobial use in food animals. WHO, Geneva

Niu M, Cai Y, Zhang H, Li X (2017) Antioxidants in animal nutrition: a review. Anim Feed Sci Technol 230:1–14

Ojha BK, Singh PK, Shrivastava N (2019) Enzymes in the animal feed industry. In: Mohammed K (ed) Enzymes in food biotechnology. Academic Press, Cambridge, pp 93–109. https://doi.org/10.1016/B978-0-12-813280-7.00007-4

Oleson JJ, Hutchings BL, Whttehtll AR (1950) The effect of feeding aureomycin on the vitamin B12 requirement of the chick. Arch Biochem 29:334–338

Pariza MW, Cook M (2010) Determining the safety of enzymes used in animal feed. Regul Toxicol Pharmacol 56(3):332–342. https://doi.org/10.1016/j.yrtph.2009.10.005

Rainbird AL, Low AG, Zebrowska T (1984) Effect of guar gum on glucose and water absorption from isolated loops of jejunum in conscious growing pigs. Br J Nutr 52:489–498. https://doi.org/10.1079/bjn19840116

Rao S, Huang X, Yang Y, Peng Y (2018) Probiotics in animal nutrition: a review. Front Microbiol 9:2690

Rao S, Huang X, Zhang Y, Gao X, Peng Y (2020) The gut microbiome and its interactions with host and diet in poultry. Front Microbiol 11:589

Ravindran V (2013) Feed enzymes: the science, practice, and metabolic realities. J Appl Poultry Res 22(3):628–636. https://doi.org/10.3382/japr.2013-00739

Regulation (EC) No 1831/2003. https://eur-lex.europa.eu/legal-content/EN/TXT/PDF/?uri=CELEX:02003R1831-20210327&from=EN

Regulation (EU) (2018) 2019 of the European Parliament and of the Council of 11 December 2018 on veterinary medicinal products and repealing Directive 2001/82/EC

Roberfroid MB (2000) Prebiotics and probiotics: are they functional foods? Am J Clin Nutr 71(6): 1682–1687. https://doi.org/10.1093/ajcn/71.6.1682s

Roberfroid M, Gobson GR, Hoyles L, McCartney AL, Rastall R, Rowland I, Wolvers D, Watzl B, Szajewska H, Stahl B (2011) Prebiotic effects: metabolic and health benefits. Br J Nutr 101(104):1–63. https://doi.org/10.1017/S0007114509990511

Sari M, Kahraman R, Ay G, Gürbüz I (2019) Antioxidant and anti-inflammatory effects of lemon and ginger essential oils in pig diets. Livest Sci 223:166–172

Sarteshnizi FR, Benemar HA, Seifdavati J, Greiner R, Salem AZ, Behroozyar HK (2018a) Production of an environmentally friendly enzymatic feed additive for agriculture animals by spray drying abattoir's rumen fluid in the presence of different hydrocolloids. J Clean Prod 197: 870–874. https://doi.org/10.1016/j.jclepro.2018.06.220

Sarteshnizi FR, Seifdavati J, Abdi-Benemar H, Salem AZ, Sharifi RS, Mlambo V (2018b) The potential of rumen fluid waste from slaughterhouses as an environmentally friendly source of enzyme additives for ruminant feedstuffs. J Clean Prod 195:1026–1031
Shan X, Ren X, Liu Z, Chen Y, Li H (2016) Influence of different levels of dietary phytase supplementation on growth performance, apparent total tract digestibility, and nutrient utilization of weaned pigs. Anim Sci J 87(5):719–725
Slavin J (2013) Fiber and prebiotics: mechanisms and health benefits. Nutrients 5(4):1417–1435. https://doi.org/10.3390/nu5041417
Smith HW (1958) Antibiotics in animal husbandry. Br Med J 1(5078):1062
Smith-Howard K (2017) Healing animals in an antibiotic age: veterinary drugs and the professionalism crisis, 1945-1970. Technol Cult 58(3):722–748. https://doi.org/10.1353/tech.2017.0079
Stokstad EL, Jukes TH (1949) The multiple nature of the animal protein factor. J Biol Chem 180 (2):647–654
Stokstad ELR, Jukes TH (1951) Effect of various levels of vitamin B12 upon growth response produced by aureomycin in chicks. Proc Soc Exp Biol Med 76(1):73–76
Stryer L, Berg JM, Tymoczko JL (2002) Biochemistry, 5th edn. W.H. Freeman, San Francisco
Thoms U (2012) Zwischen Verrheißung und Bedrohung. Antibiotika in Nahrungsmitteln in Westdeutschland 1950-1980 [Between promise and threat : antibiotics in foods in west Germany 1950-1980]. NTM 20(3):181–214. German. https://doi.org/10.1007/s00048-012-0073-x
Underwood EJ, Suttle NF (1999) The mineral nutrition of livestock
Van Boeckel TP, Brower C, Gilbert M, Grenfell BT, Levin SA, Robinson TP, Teillant A, Laxminarayan R (2015) Global trends in antimicrobial use in food animals. Proc Natl Acad Sci 112(18):5649–5654. https://doi.org/10.1073/pnas.1503141112
Velázquez-De Lucio BS, Hernández-Domínguez EM, Villa-García M, Díaz-Godínez G, Mandujano-Gonzalez V, Mendoza-Mendoza B, Álvarez-Cervantes J (2021) Exogenous enzymes as zootechnical additives in animal feed: a review. Catalysts 11:851. https://doi.org/10.3390/catal11070851
Walsh GA, Ronan FP, Denis RH (1993) Enzymes in the animal-feed industry. Trends Biotechnol 11:424–430. https://doi.org/10.1016/0167-7799(93)90006-u
Wang X, Smith J, Johnson A (2010) Enhancing pig performance: peppermint oil as a promising additive for improved feed efficiency, and diarrhea reduction in swine diets. J Anim Sci 28 (3):245–256
Wang Y, Zhou Y, Zeng J, Li H, Fan H, Du Q (2016) Effects of supplemental methionine on growth performance, egg production, and serum biochemical characteristics in laying hens. Poult Sci 95(4):870–876
Wang X, Sun L, Qi J, Zhang L, Wu X, Wang Y, Yang J (2018) Antioxidant capacity of plasma and liver in response to different feed levels of dried red pepper and allicin in weaned piglets. Anim Sci J 89(5):753–759
White R, Cooper R (2005) Silver sulphadiazine: a review of the evidence. Wounds UK. 1
Xie M, Zhang R, Liu Z, Liu X, Zhang J, Li Y (2019) The effect of nucleotides on the growth performance, gut morphology, and immune function of broilers. Poult Sci 98(7):3145–3151
Yao J, Cai Y, Li D (2015) The effects of dietary lysine supplementation on growth performance, carcass characteristics, and meat quality of growing-finishing pigs. J Anim Sci Biotechnol 6(1): 1–8
Zhou X, Huang Y, Li Y (2021) Enzymes in animal nutrition: an overview. Animals 11(4):744. https://doi.org/10.3390/ani11040744

Part II

Sources and Types of Feed Additives

Enzymes as Feed Additives

Georgios A. Papadopoulos and Styliani Lioliopoulou

Abstract

Enzyme dietary supplementation in monogastric diets has been implemented widely in the last decade. Starting from phytase, enzymes have been used to assist in the utilization of essential minerals such as calcium (Ca) and phosphorus (P). Nevertheless, new opportunities have arisen by introduction of up-to-date enzymatic technology. As most diets for poultry and swine contain plant grains as basic ingredients, the fundamental issue to combat was the efficient breakdown of cell walls and the liberation of entrapped nutrients. Therefore, supplementation of specific enzymes has led to improved utilization of nutrients, while it had generated positive effects on gastrointestinal environment. Apparently, the evolution of feed enzyme technology has led to production of single- or multi-enzymes, which are now considered as indispensable in feed formulation for monogastric species. Yet, majority of the related studies have focused either on improving digestibility of major nutrients or on enhancing performance indexes. However, few studies have addressed the effects of enzyme supplementation on improving the properties of meat and eggs. The chapter covers the main effects of single enzymes, with emphasis on the use of xylanase enzymes, on the efficient breakdown of polysaccharides and their potential effects on derived products' quality.

Keywords

Enzymes · Poultry · Swine · Monogastric · Broilers · Layers

G. A. Papadopoulos (✉) · S. Lioliopoulou
Laboratory of Animal Husbandry, Faculty of Veterinary Medicine, School of Health Sciences, Aristotle University, Thessaloniki, Greece
e-mail: geopaps@vet.auth.gr

G. Arsenos, I. Giannenas (eds.), *Sustainable Use of Feed Additives in Livestock*,
https://doi.org/10.1007/978-3-031-42855-5_5

1 Introduction

Efficient breakdown of non-starch polysaccharides (NSPs) has been under debate in monogastric diets, because the enzymes needed for their digestion are not among those secreted by the organism. Therefore, it was hypothesized that addition of enzymes could be beneficial in multiple ways. Firstly by reducing viscosity in the gut, as undigested NSPs accumulating in the gut can provoke such conditions. When present in excess, soluble NSP causes increased digesta viscosity due to its strong capacity to store water. Secondly, by improving nutrient digestibility. This effect could be mediated either by the reduction of viscosity or by liberation of nutrients that were encircled by vegetable cell walls. Thirdly, by exerting the so-called prebiotic effect, mainly by the products of polysaccharide breakdown, such as the oligosaccharides or the xylooligosaccharides. On the other hand, the beneficial microbiota may decompose a large part of soluble NSP, which leads to the synthesis of short chain fatty acids, enhanced gut microflora, and less conflict between the host and microbiome for important nutrients (Bedford 2000). Meanwhile, substantial quantities of soluble NSP also guarantee that the digesta passage rate is not excessively rapid, allowing for the absorption of nutrients (Mateos et al. 2012). Especially in broilers, it could preserve motility and digesta transit in the gastrointestinal system by water retention, while it can increase the activity of the gizzard and proventriculus, boosting feed grinding and peptic digestion in the gizzard (Yokhana et al. 2016). Mostly, the use of enzymes such as xylanase has been formulated to target specific dietary ingredients with high NSP content such as wheat and barley. However, in soybean, arabinogalactans, galactans, and galactomannans occur primarily over arabinoxylans and glucans, which are the primary NSPs in cereal grains like maize (Slominski 2011). In this direction, multi-enzyme compounds with various NSP breakdown abilities have received much interest in an effort to boost enzyme efficiency (Waititu et al. 2018). Interestingly, recent studies have shown improvement in egg quality characteristics, which indicate a further potential use for the enzymes to improve quality of the end product.

2 Enzyme Supplementation in Monogastric Diets: An Unavoidable Necessity

Exogenous enzymes have been widely utilized in poultry diets to increase productivity and decrease feed costs. This is one of the most significant nutritional developments of the last 50 years. Due to the complexity of plant cell walls, monogastric animals, such as pigs and poultry, lack the required enzymes to break down certain plant chemicals prevalent in animal diets (Khattak et al. 2006; Adeola and Cowieson 2011). Additionally, poultry diets may contain antinutritional substances that negatively impact poultry performance. Antinutritional substances, such as non-starch polysaccharides (NSP), are present in varying concentrations in cereal grains, which are the primary source of energy for poultry (Raza et al. 2019). Enzymes are utilized to raise the digestibility of foods, diminish the effect of antinutritional factors, and enhance nutrient availability (Acamovic 2001; Cowieson et al. 2006; Costa et al. 2008), either by substituting enzymes that are not released in the gastrointestinal system at all, or by quantitatively supporting enzymes that are produced in modest amounts.

Protein molecules called enzymes catalyze certain reactions by functioning on a single or a small number of chemicals, also referred to as substrates (Khattak et al. 2006). Every organism produces them; however the majority of the enzymes utilized in the food business are made by microbes like bacteria, fungi, and yeast (Khattak et al. 2006). The following modes of action may be responsible for the favorable effects of enzymes on the utilization of poultry feed: (a) phytate hydrolysis releases phosphorus (P); (b) nutrients contained in elaborate structures are released; (c) NSP deterioration; and (d) hydrolysis of certain protein–carbohydrate bonds. Plants store P in phytate, also referred to as phytic acid (Naghshbandi and Moghimi 2020). Because poultry species lack the necessary enzymes to break down phytate, P bioavailability is decreased and P excretion is increased. These effects lead to environmental problems including eutrophication. Moreover, it is known that phytate chelates with essential minerals, and therefore reduces the bioavailability of important cations (Kumar et al. 2019; Zhu et al. 2019).

Plant byproducts often include bioactive substances that could improve the health and performance of birds, but they also frequently contain antinutritional elements that restrict their use. Enzymes could therefore be added to improve the nutritional content of alternative feed byproducts. It is well established that NSPs decrease nutrient digestibility, feed effectiveness, and ultimately growth performance by making intestinal digesta more viscous and promoting small intestine fermentation (Bedford and Classen 1991; Choct and Annison 1992a, b; Friesen et al. 1992; Bhat 2000; Nian et al. 2011; Van Hoeck et al. 2021). Additionally, sticky droppings are a result of the high water-holding ability of glucans and arabinoxylans (Slominski 2011). There are no endogenous carbohydrases produced by poultry and swine that can hydrolyze NSPs. By removing the nutrient-encapsulating effect and lowering digesta viscosity, the enzymes that hydrolyze NSPs increase the digestibility and utilization of nutrients (Chesson 2001; Slominski 2011; Van Hoeck et al. 2021).

Finally, Slominski (2011) asserts that hydrolysis of particular carbohydrate–protein connections enhances amino acid bioavailability.

The type of the enzymes and their inclusion level, the type of the microflora present, the type of the birds' physiology, whether or not the diets have been processed, the species and age of the birds, etc. will all affect how much benefit using enzymes in poultry diets will have (Acamovic 2001; Khattak et al. 2006). In the poultry feed industry, the most often employed enzymes include glucanases, xylanases, phytases, proteases, lipases, and galactosidases (Khattak et al. 2006). In order to be more efficient across a variety of substrates, enzyme combinations or preparations are frequently utilized in the poultry business. This is because the majority of feed ingredients contain a number of chemicals that can have negative impacts on bird performance. For instance, wheat mostly contains arabinoxylans, barley primarily contains β-glucans, and corn has lectins, phytate, and resistant starch (Acamovic 2001). Additionally, potential synergistic interactions between

Table 1 Enzymes used in poultry feeds and substrates that are more effective

Enzymes	Substrates
β-Glucanases	Barley, oats
Xylanases	Wheat, rye, triticale, rice bran
β-Galactosidases	Grain legumes, lupins
Phytases	Plant feedstuffs
Proteases	Proteins
Lipases	Lipids
Amylases	Starch

Adapted/Modified from Khattak et al. (2006)

enzymes in combinations can further increase their performance. For instance, supplementing phytase with NSP-degrading enzymes increases its potency (Slominski 2011). The main substrates in which each category of enzymes is more effective is shown in Table 1.

Altogether, the use of enzymes in poultry diets enhances the availability of nutrients and lessens the impact of antinutritional factors utilizing the previously stated processes, which improves the performance of the birds. Economic benefits of using enzymes in poultry include lower feeding costs, the ability to develop alternative diets, and improved performance, health, and litter quality (Costa et al. 2008; Zakaria et al. 2010). Minimizing feeding costs is one of the key issues facing the poultry business, particularly in developing nations. According to Acamovic (2001), Alagawany et al. (2018), and Raza et al. (2019), feeding costs account for roughly 70–75% of the total production costs in the poultry sector. The cost of conventional feed ingredients is rising, which has generated interest in less expensive options as fruit and vegetable waste. As has already been mentioned, the use of enzymes in these processes is crucial for both the economy and environmental protection since without them, plant byproducts would be left in the environment to decay. Additionally, according to Alagawany et al. (2018), enzyme supplementation reduces the excretion of nitrogen (N) and P in poultry manure, which is crucial for environmental conservation.

3 Use of Enzymes in Broilers

Compared to other animal species, broilers have had more comprehensive enzyme research since it is more affordable and yields findings more quickly (Dosković et al. 2013). This in turn accelerates the production of additional enzyme products. Furthermore, broilers are susceptible to digestive problems, which can be avoided by using particular feed enzymes (Annison 1997). In general, broiler studies have found that adding enzymes exerts positive impacts on digestible energy, energy use efficacy, body weight increase, and Feed Conversion Ratio (FCR) (Zanella et al. 1999; Douglas et al. 2000; Kocher et al. 2001; Brufau et al. 2002; Olukosi et al. 2007; Zhou et al. 2009). However, other research studies did not demonstrate any beneficial effects of dietary enzyme supplementation in broilers (Iji et al. 2003; Günal et al. 2004; Sayyazadeh et al. 2006).

One of the key methods for enhancing broiler growth and nutrient digestion is the addition of NSP-degrading enzymes to cereal-based diets. Increasing levels of β-glucanase in broilers offered barley-based diets resulted in higher feed intake and live weight, enhanced feed efficiency, and decreased excreta dry matter content, based on a study by Hesselman et al. (1982). According to Shirmohammad and Mehri (2011), using a multi-enzyme preparation that contained β-glucanase and protease decreased broiler belly fat and increased the nutritional content of a corn–soybean diet. Santos et al. (2013) exhibited improved growth outcomes in broilers fed rye-based diets when xylanase was added as a supplement in their initial growth stages. Elsewhere, broilers' ability to metabolize hemicelluloses, neutral detergent fibers, and crude proteins was improved by an enzyme complex composed of endo-1,4-β-glucanase, endo-1,4-β-xylanase, and endo-1,3(4)-β-glucanase (Glamočić et al. 2011). Diets with less metabolizable energy showed these effects more overtly. In the study by Gao et al. (2008), adding xylanase to broilers fed a wheat-based regimen increased production performance, and changed the digesta's viscosity and pH after 21 days of age. In the study by Esmaeilipour et al. (2012), xylanase supplementation in broilers' wheat-based diet increased nutrient availability and enhanced performance. This resulted in a reduction in the viscosity of the digestive tract. In their investigation, Van Hoeck et al. (2021) presented data that were similar. In the latter study, a novel intrinsically thermostable xylanase altered the viscosity of the intestinal tract in broilers, positively impacted their performance and carcass characteristics, and also modified intestinal pH. In broilers fed diets including various types of wheat, xylanase caused also a decrease in pH in the duodenum, jejunum, and cecum, as demonstrated by Engberg et al. (2004). Changes in intestinal microbiota may be responsible for the effect of xylanase on intestinal pH. More particular, it was demonstrated that xylanase supplementation enhanced the number of *Lactobacillus* spp. and *Ruminococcaceae* in the colon of broilers (Vahjen et al. 1998; Singh et al. 2021). Intestinal pH is decreased by lactic acid, which is produced by *Lactobacillus* spp. and other helpful gut bacteria. An increase in pancreatic production of digestive enzymes and the segregation of bile acid for lipid emulsification are afterward thought to result from a decrease in intestinal pH. These results enhance nutrient digestion and absorption even more (Panda et al. 2009; Martínez et al. 2021). When broilers were infected with *Salmonella typhimurium*, the negative effects could be reduced by mixing xylanase and a probiotic called *Lactobacillus plantarum* (Vandeplas et al. 2009). Supplementing broilers with xylanase also lessened the negative effects of *Clostridium perfringens* on the intestinal mucosal barrier (Liu et al. 2012). The Van Hoeck et al. (2021) study highlighted the prebiotic effect of xylanase. More specifically, xylanase supplementation at the dose of 30,000 U/kg (10 g/t) increased *Lactobacilli* development in the intestinal tract in addition to enhancing broiler performance, nutritional digestibility, and gut health. The xylanase-mediated variations in the cecal temperature of broilers suggested a potential modification in the distal gastrointestinal tract's fermentative activity (Cowieson and Masey O'Neill 2013).

For optimum growth, broilers need elevated dietary protein levels. In general, for the stages of bird development, not only the appropriate amino acid profile but also an appropriate supply of total proteins is required (Dosković et al. 2013). Soybean meal, one of the most popular ingredients in poultry feed, has a comparatively high protein level and a well-balanced amino acid profile (Dosković et al. 2013). However, because it contains antinutritional elements including trypsin inhibitors, lectins, and NSP, it may not be totally absorbed by poultry (Marsman et al. 1997; Pack and Bedford 1997). Traditional corn–soybean diets for broilers include a variety of protein molecules that are difficult for young broilers to digest because they lack the necessary enzymes in the early stages of life and the feed passes through the digestive system quickly (Uni et al. 1999). Amylase and protease supplementation had no appreciable impact on broiler development and feed intake during the course of the trial, according to Mahagna et al. (1995), but it had an impact on FCR from the first to the seventh day of life. When protease is added to broiler diets, protein hydrolysis can be improved when there are antinutritional factors present (Huo et al. 1993; Ghazi et al. 2002). Additionally, protease supplementation decreased the requirement for the meal to contain both amino acids and energy (Dosković et al. 2013). Due to the comparatively high cost of protein and amino acids, such solutions could minimize both nitrogen excretion in the environment and feed costs (Aletor et al. 2000; Bregendahl et al. 2002). Supplementing with protease increased body weights and Apparent Metabolizable Energy (AME) in broilers fed diets including soybean meal (Ghazi et al. 1997). A commercial carbohydrase and protease preparation increased the energy value of feed in broilers fed a corn/soybean diet in the Douglas et al. (2000) trial. Numerous studies on broilers demonstrated the advantages of including protease in multi-enzyme complexes for the animals (Ranade and Rajmane 1992; Morgan and Bedford 1995; Simbaya et al. 1996). Irrespective of the nutritional value of the diets, an enzyme complex combining protease, amylase, β-glucanase, xylanase, pectinase, cellulose, and phytase increased growth performance and FCR (Perić et al. 2008). By providing broilers a mixture of xylanase, amylase, and protease, researchers were able to increase the ileal digestibility of N, amino acids, and energy (Zanella et al. 1999; Douglas et al. 2000). The addition of combined protease and amylase to broiler diets has a positive impact on growing performance (Café et al. 2002; Odetallah et al. 2005). Another study using broilers found that the combination of xylanase, amylase, and protease had a dose-dependent impact (Cowieson and Ravindran 2008).

4 Use of Enzymes in Laying Hens

Research about the use of enzymes in laying hens is not so extensive in comparison with the research in broilers. The mechanism of action of the main enzymes used in poultry has already been discussed in detail previously, so this section will focus mainly on their effects on birds' health, performance, and product quality. The NSP-degrading enzymes in laying hens increase the efficiency of the hens overall by releasing nutrients that have been encased and decreasing the viscosity of digesta

Fig. 1 Changes in egg yolk coloration due to xylanase supplementation at increasing levels in laying hens fed wheat-based diets. (Modified from Papadopoulos et al. 2022)

and levels of cecal ammonia (Lima et al. 2019; Chesson 2001). Compared to older birds, the impact of enzymes is more on younger ones as xylanase is more pronounced during the early phases of growth. It has been proposed that the benefits of xylanase diminish with age in hens because layers have larger and more developed digestive systems (Annison et al. 1968; Salih et al. 1991; Campbell and Bedford 1992). In the study by Hahn-Didde and Purdum (2014), laying hens fed diets with moderate and low AME levels received a supplement of an enzyme complex composed of xylanase, amylase, and protease. The findings revealed greater AME retention values for the initial phase of the production cycle as well as higher retention values for calcium (Ca), P, and Crude Protein (CP) in later phases. In a recent study, laying hens were fed a wheat–corn diet with the addition of rye, and xylanase supplementation generated positive effects such as enhanced eggshell calcium content and thickness (Muszyński et al. 2022). Regardless of the dietary addition of the type of rye, xylanase supplementation was also observed to increase bone strength. Xylanase improved the nutritional value of wheat–rye-based diets, as evidenced by the increased weight of the birds and the color of their yolks in laying hens (Pirgozliev et al. 2010). The beneficial effect of dietary xylanase on egg yolk coloration was also presented in the recent studies of Nguyen et al. (2021), Papadopoulos et al. (2022), and Lioliopoulou et al. (2023) (Fig. 1). A possible explanation for these findings is that NSP-degrading enzymes enhance the release of encapsulated nutrients such as carotenoids, which are the main pigments that define egg yolk coloration. It is conceivable that any enhancement in xylanase-mediated fatty acid digestion and absorption potentially indirectly impacted the absorption of carotenoids (Papadopoulos et al. 2022). Yolk carotenoids exist as lipid-dissolved compounds and may be linked to big membrane lipoproteins

(Papadopoulos et al. 2022). The latter results might be explained by the fact that xylanase enzymes improved the availability of lipids and other components by reducing intestinal viscosity. Additionally, by minimizing bile salt deconjugation, xylanase improves fat digestion. These findings are of importance, particularly for the animal feed industry, since alternative feed ingredients may be used to replace partially or fully conventional feed ingredients for laying hens such as corn, without a detrimental effect on the properties of the egg yolk coloration.

Besides improved nutrient digestion and absorption, the changes in digesta viscosity caused by NSP-degrading enzymes also can affect birds' droppings, resulting in a reduction in the numbers of dirty eggs (Chesson 2001). Similar findings have been demonstrated in an early study in laying hens, where high doses of a commercial enzyme complex containing β-glucanases, hemicellulases, and pectinases improved excreta quality (Francesch et al. 1995). Egg weight was also increased in this study by the high enzyme dose of 1 g/t. However, this effect on excreta quality is not consistent. For instance, in the study conducted by Pirgozliev et al. (2010), the addition of xylanase had no effect on the quantity of dirty eggs. In the study by Bobeck et al. (2014), xylanase supplementation boosted egg production, egg mass, and feed efficiency in first-cycle laying hens fed corn–soybean meal-dried distiller's grain. Increased body weight of hens was observed during the entire 12-week experiment in second-cycle laying hens supplemented with an enzyme mixture that included xylanases, β-glucanases, mannanases, pectinases, and proteases (Gunawardana et al. 2009). The performance of laying hens in the second production cycle can be enhanced by a multi-carbohydrase complex containing phytase, which can offset the detrimental effects of diets with less nutrients (Lima et al. 2019). Abreu et al. (2018) noted higher egg weight in laying hens fed nutritionally reduced diets supplemented with 100 g/t of xylanase, β-glucanase, and phytase and found similar results in laying hen performance. In laying hens fed corn–oat–wheat middling-based diets, Jaroni et al. (1999) found that supplementing with both xylanase and protease enhanced egg weight and improved feed conversion. Xylanase enhanced AME and N retention when added to the "wheat distiller's dried grains with solubles" diets of laying hens in the study by Whiting et al. (2019). According to studies by Lei et al. (2018), Sun and Kim (2019), Roberts and Choct (2006), and Lioliopoulou et al. (2023), eggshell color was lightened by xylanase supplementation alone or in combination with other enzymes. Last but not least, investigations by Jia et al. (2008), Westbrook and Cherian (2019), and Papadopoulos et al. (2022) have demonstrated that enzyme supplementation can modify the lipid profile of egg yolk. Lipid profile of egg yolk has been indispensably associated with the egg functional properties and further research is needed toward improvement of egg yolk profile concurrently with xylanase supplementation. The addition of protease to laying hen diets can optimize nutrient absorption, enhance feed conversion efficiency, increase eggshell thickness, and reduce production costs (Dosković et al. 2013). It was suggested that including protease in the diets of laying hens could lower the amount of dietary crude protein by 1% (Yadav and Sah 2006). The addition of protease to a multi-enzyme complex increased laying hens' egg yield (Adams 1989). There is very little information regarding protease in laying hen diets when used alone; in the majority of research, protease has been combined with other enzymes.

5 Use of Enzymes in Swine: Need for Research Beyond the Thresholds

Unexpected health benefits and discrepancies in performance indicators urge for a deeper comprehension of xylanase's in vivo mode of action (Petry and Patience 2020). All across the world, pig diets frequently include corn as an energy source; when financially feasible, industrial corn byproducts are also used. Starch makes up the majority of the energy in corn, with smaller amounts coming from protein, fat, and non-starch polysaccharides (NSPs) (Petry and Patience 2020). The majority of NSP in corn and its coproducts is made up of arabinoxylans (Petry and Patience 2020). In spite of diet composition, xylanase enhanced the digestibility of dry matter, gross energy, and crude protein, according to a meta-analysis of 67 trials that included a diversity of ingredients in growing finishing pigs by Torres-Pitarch et al. (2019). However, growth performance responses were less commonly noticed. Pigs fed diets based primarily on corn still exhibited less digestibility reactions (Torres-Pitarch et al. 2019). It is frequently assumed that corn Arabinoxylan (AX) is resistant to enzymatic breakdown because of its complicated makeup and resultant insolubility when xylanase is unsuccessful (Petry and Patience 2020). However, the quantity of studies reporting increased fiber digestibility as well as the size of those improvements undoubtedly suggest that xylanase is hydrolyzing fiber either directly or indirectly through microbiota modification (Petry and Patience 2020). Recent research indicates that xylanase likely affects gut health through pathways that enhance gut barrier integrity, ameliorate oxidative stress, and modify immune response (Li et al. 2018; Duarte et al. 2019; Petry et al. 2020). The interaction between xylanase and the gut flora is most likely what caused these benefits in the pig's overall and digestive health (Petry and Patience 2020).

It has been suggested that supplementing swine diets with ingredients such as emulsifiers and enzymes is another approach that might be used to increase nutrient usage by enhancing nutrient bioavailability (Papadopoulos et al. 2014). In particular, the weaned pig's largely milk-oriented enzymatic system suggests that adding extra enzymes to the feed may help young pigs develop in ways that are favorable to them (Kyriazakis and Whittemore 2006). Li et al. (1996) found that adding β-glucanase to pig diets increased the amounts of nutrients and energy that were digested and absorbed in the small intestine and decreased the amount of bacteria that fermented them in the large intestine. The small intestine's ability to digest fat specifically increased. The efficiency of energy use increased as a result of this change in the energy's disappearance from the large to the small intestine. Additionally, a multi-carbohydrase supplement improved the digestibility of non-starch polysaccharides and ileal dry matter (Kiarie et al. 2007). Moreover, adding enzymes to weaned piglet diet dramatically increased the digestibility of total amino acids in the intestine (Vahjen et al. 2007; Ayoade et al. 2012). Elsewhere, it has been suggested that the

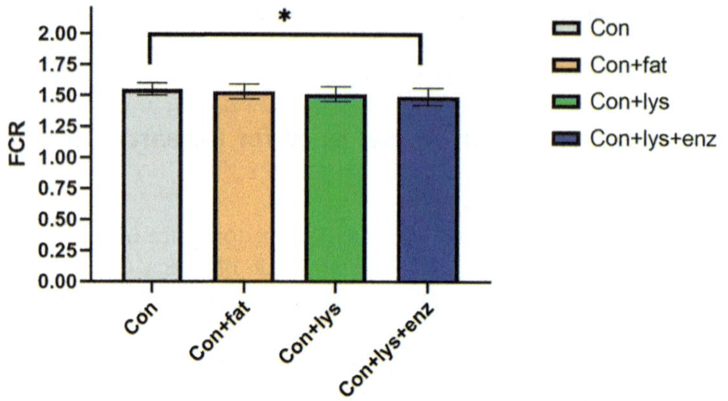

Fig. 2 Effects on weaned piglets' performance after supplementation of weaned piglet diets (Con) with extra fat content at 7.0 g/kg (Con+fat), lysolecithin at 0.5 g/kg (Con+lys), or both with lysolecithin and an enzyme mixture at 0.5 g/kg (Con+lys+enz). (Adapted from Papadopoulos et al. [2014]; *mean values are significantly different at $p < 0.05$)

use of emulsifiers, which can assist with nutrient digestion and absorption from the small intestine, could be an additional strategy for increasing nutritional use in weaned pigs. This concept is based on previous findings that the NSP level of the diets may have an effect on fat digestion (Freire et al. 1998). In particular, the digestibility of fat and energy was decreased when the NSP concentration was increased (Freire et al. 1998). Based on this concept, it was shown that the addition of lysolecithin in combination with a multi-NSP-enzyme in barley–wheat-based diets can improve feed efficiency in weaned piglets (Fig. 2; Papadopoulos et al. 2014).

In order to employ enzymes in swine, it would appear that more research is required. It is necessary to create clear recommendations on how to use enzymes in pigs and distinguish between piglets and sows. Less fibrous diets are offered to piglets than to sows. In the meanwhile, piglets need to use their fat and energy efficiently in the first few days after being weaned. Given the contradictions in the available data and the predominance of corn in weaned piglet diets, it may initially appear that enzyme supplementation is of no added value. Despite being less than other viscous cereals, corn nevertheless has a significant amount of NSP. Co-supplementation of commercially available absorption enhancers could also generate benefits in this crucial growth phase. On the other hand, sow diets are richer in fibrous content, and its efficient utilization has been under debate in the recent years. The modern sow has increased considerably in size and weight, gives birth to more than 15 live-born piglets, and has a stressful and prolonged farrowing duration, while its metabolic needs to support lactation are high. Therefore, it is now crucial to use dietary fiber effectively in order to maximize available energy and prevent constipation-related problems. According to this perspective, recent research demonstrated that supplementing with xylanase enhanced breastfeeding sow performance and altered the microbiome's composition in their feces (Vermeulen et al. 2022). Especially noteworthy, the xylanase-supplemented sows lost less backfat throughout lactation than the control sows, showing an enhanced nutrient utilization irrespective of the diet's fiber level (Vermeulen et al. 2022). Overall, it is anticipated that the use of enzymes in swine will receive major attention in the upcoming years, and that they will play a key role in the adoption of feed sources other than maize.

References

Abreu MT, Fassani EJ, Silveira MMBM, Viveiros MP (2018) Enzymatic complex based on xylanase, beta-glucanase and phytase on feeds for light commercial layers at peak production. Bol Ind Anim 75(1):17–24. https://doi.org/10.17523/bia.v75n1p17

Acamovic T (2001) Commercial application of enzyme technology for poultry production. Worlds Poult Sci J 57(3):225–242. https://doi.org/10.1079/WPS20010016

Adams CA (1989) Kemzyme and animal feed digestion. Krmiva: Časopis o hranidbi životinja, proizvodnji i tehnologiji krme 31(7–8):139–144

Adeola O, Cowieson AJ (2011) Board-invited review: opportunities and challenges in using exogenous enzymes to improve nonruminant animal production. J Anim Sci 89(10): 3189–3218. https://doi.org/10.2527/jas.2010-3715

Alagawany M, Elnesr SS, Farag MR (2018) The role of exogenous enzymes in promoting growth and improving nutrient digestibility in poultry. Iran J Vet Res 19(3):157

Aletor VA, Hamid II, Niess E, Pfeffer E (2000) Low-protein amino acid-supplemented diets in broiler chickens: effects on performance, carcass characteristics, whole-body composition and efficiencies of nutrient utilization. J Sci Food Agric 80(5):547–554. https://doi.org/10.1002/(SICI)1097-0010(200004)80:5<547::AID-JSFA531>3.0.CO;2-C

Annison G (1997) The use of exogenous enzymes in ruminant diets. Recent Adv Anim Nutr Australia 97:8–16

Annison EF, Hill KJ, Kenworthy R (1968) Volatile fatty acids in the digestive tract of the fowl. Br J Nutr 22(2):207–216. https://doi.org/10.1079/BJN19680026

Ayoade DI, Kiarie E, Woyengo TA, Slominski BA, Nyachoti CM (2012) Effect of a carbohydrase mixture on ileal amino acid digestibility in extruded full fat soybeans fed to finishing pigs. J Anim Sci 90(11):3842–3847. https://doi.org/10.2527/jas.2011-4761

Bedford MR (2000) Exogenous enzymes in monogastric nutrition their current value and future benefits. Anim Feed Sci Technol 86:1–13. https://doi.org/10.1016/S0377-8401(00)00155-3

Bedford MR, Classen HL (1991) Influence of dietary xylanase on intestinal viscosity and molecular weight distribution of carbohydrates in rye-fed broiler chicks. Prog Biotechnol 7:361–370

Bhat MK (2000) Cellulases and related enzymes in biotechnology. Biotechnol Adv 18(5):355–383. https://doi.org/10.1016/S0734-9750(00)00041-0

Bobeck EA, Nachtrieb NA, Batal AB, Persia ME (2014) Effects of xylanase supplementation of corn-soybean meal-dried distiller's grain diets on performance, metabolizable energy, and body composition when fed to first-cycle laying hens. J Appl Poultry Res 23(2):174–180. https://doi.org/10.3382/japr.2013-00841

Bregendahl K, Sell JL, Zimmerman DR (2002) Effect of low-protein diets on growth performance and body composition of broiler chicks. Poult Sci 81(8):1156–1167. https://doi.org/10.1093/ps/81.8.1156

Brufau J, Francesch M, Pérez-Vendrell AM (2002) Exogenous enzymes in poultry feeding. Recent developments. Paper presented at the 11th European poultry conferences, Bremen, Germany, 6–10 Sept 2002

Café MB, Borges CA, Fritts CA, Waldroup PW (2002) Avizyme improves performance of broilers fed corn-soybean meal-based diets. J Appl Poultry Res 11(1):29–33. https://doi.org/10.1093/japr/11.1.29

Campbell GL, Bedford MR (1992) Enzyme applications for monogastric feeds: a review. Can J Anim Sci 72(3):449–466. https://doi.org/10.4141/cjas92-058

Chesson A (2001) Non-starch polysaccharide degrading enzymes in poultry diets: influence of ingredients on the selection of activities. Worlds Poult Sci J 57(3):251–263. https://doi.org/10.1079/WPS20010018

Choct M, Annison G (1992a) The inhibition of nutrient digestion by wheat pentosans. Br J Nutr 67(1):123–132. https://doi.org/10.1079/BJN19920014

Choct M, Annison G (1992b) Anti-nutritive effect of wheat pentosans in broiler chickens: roles of viscosity and gut microflora. Br Poultry Sci 33(4):821–834. https://doi.org/10.1080/00071669208417524

Costa FGP, Goulart CC, Figueiredo DF, Oliveira CFS, Silva JHV (2008) Economic and environmental impact of using exogenous enzymes on poultry feeding. Int J Poult Sci 7(4):311–314

Cowieson AJ, Masey O'Neill HV (2013) Effects of exogenous xylanase on performance, nutrient digestibility and caecal thermal profiles of broilers given wheat-based diets. Br Poultry Sci 54(3):346–354. https://doi.org/10.1080/00071668.2013.780200

Cowieson AJ, Ravindran V (2008) Effect of exogenous enzymes in maize-based diets varying in nutrient density for young broilers: growth performance and digestibility of energy, minerals and amino acids. Br Poultry Sci 49(1):37–44. https://doi.org/10.1080/00071660701812989

Cowieson AJ, Hruby M, Pierson EM (2006) Evolving enzyme technology: impact on commercial poultry nutrition. Nutr Res Rev 19(1):90–103. https://doi.org/10.1079/NRR2006121

Dosković V, Bogosavljević-Bosković S, Pavlovski Z, Milošević B, Škrbić Z, Rakonjac S, Petričević V (2013) Enzymes in broiler diets with special reference to protease. Worlds Poult Sci J 69(2):343–360. https://doi.org/10.1017/S0043933913000342

Douglas MW, Parsons CM, Bedford MR (2000) Effect of various soybean meal sources and Avizyme on chick growth performance and ileal digestible energy. J Appl Poultry Res 9(1):74–80. https://doi.org/10.1093/japr/9.1.74

Duarte ME, Zhou FX, Dutra WM, Kim SW (2019) Dietary supplementation of xylanase and protease on growth performance, digesta viscosity, nutrient digestibility, immune and oxidative stress status, and gut health of newly weaned pigs. Anim Nutr 4:351–358. https://doi.org/10.1016/j.aninu.2019.04.005

Engberg RM, Hedemann MS, Steenfeldt S, Jensen BB (2004) Influence of whole wheat and xylanase on broiler performance and microbial composition and activity in the digestive tract. Poult Sci 83(6):925–938. https://doi.org/10.1093/ps/83.6.925

Esmaeilipour O, Moravej H, Shivazad M, Rezaian M, Aminzadeh S, Van Krimpen MM (2012) Effects of diet acidification and xylanase supplementation on performance, nutrient digestibility, duodenal histology and gut microflora of broilers fed wheat based diet. Br Poultry Sci 53(2):235–244. https://doi.org/10.1080/00071668.2012.681771

Francesch M, Perez-Verdrell AM, Esteve-Garcia E, Brufau J (1995) Enzyme supplementation of a barley and sunflower-based diet on laying hen performance. J Appl Poultry Res 4(1):32–40. https://doi.org/10.1093/japr/4.1.32

Freire JPB, Peiniau J, Cunha LF, Almeida AA, Aumaitre A (1998) Comparative effects of dietary fat and fibre in Alentejano and large white piglets: digestibility, digestive enzymes and metabolic data. Livest Prod Sci 53(1):37–47. https://doi.org/10.1016/S0301-6226(97)00142-5

Friesen OD, Guenter W, Marquardt RR, Rotter BA (1992) The effect of enzyme supplementation on the apparent metabolizable energy and nutrient digestibilities of wheat, barley, oats, and rye for the young broiler chick. Poult Sci 71(10):1710–1721. https://doi.org/10.3382/ps.0711710

Gao F, Jiang Y, Zhou GH, Han ZK (2008) The effects of xylanase supplementation on performance, characteristics of the gastrointestinal tract, blood parameters and gut microflora in broilers fed on wheat-based diets. Anim Feed Sci Technol 142(1–2):173–184. https://doi.org/10.1016/j.anifeedsci.2007.07.008

Ghazi S, Rooke JA, Galbraith H, Morgan A (1997) Effect of adding protease and alpha-galactosidase enzymes to soyabean meal on nitrogen retention and true metabolisable energy in broilers. Br Poultry Sci 38:28

Ghazi S, Rooke JA, Galbraith H, Bedford MR (2002) The potential for the improvement of the nutritive value of soya-bean meal by different proteases in broiler chicks and broiler cockerels. Br Poultry Sci 43(1):70–77. https://doi.org/10.1080/00071660120109935

Glamočić D, Polovinski-Horvatović M, Ivković M, Beuković D, Bjedov S (2011) Effects of enzymes supplementation on digestibility and energy utilizations of broilers diets with different metabolizable energy level. Biotechnol Anim Husb 27(3):583–590. https://doi.org/10.2298/BAH1103583G

Günal M, Yaşar S, Forbes JM (2004) Performance and some digesta parameters of broiler chickens given low or high viscosity wheat-based diets with or without enzyme supplementation. Turkish J Vet Anim Sci 28(2):323–327

Gunawardana P, Roland DA Sr, Bryant MM (2009) Effect of dietary energy, protein, and a versatile enzyme on hen performance, egg solids, egg composition, and egg quality of Hy-Line W-36 hens during second cycle, phase two. J Appl Poultry Res 18(1):43–53. https://doi.org/10.3382/japr.2008-00047

Hahn-Didde D, Purdum SE (2014) The effects of an enzyme complex in moderate and low nutrient-dense diets with dried distillers grains with solubles in laying hens. J Appl Poultry Res 23(1):23–33. https://doi.org/10.3382/japr.2013-00764

Hesselman K, Elwinger K, Thomke S (1982) Influence of increasing levels of β-glucanase on the productive value of barley diets for broiler chickens. Anim Feed Sci Technol 7(4):351–358. https://doi.org/10.1016/0377-8401(82)90004-9

Huo GC, Fowler VR, Inborr J, Bedford M (1993) The use of enzymes to denature antinutritive factors in soyabean. In: Poel AFB, Huisman J, Saini HS (eds) Proceedings, 2nd international workshop on antinutritional factors in legume seed. Wageningen, The Netherlands, pp 517–521

Iji PA, Khumalo K, Slippers S, Gous RM (2003) Intestinal function and body growth of broiler chickens on diets based on maize dried at different temperatures and supplemented with a microbial enzyme. Reprod Nutr Dev 43(1):77–90. https://doi.org/10.1051/rnd:2003007

Jaroni D, Scheideler SE, Beck M, Wyatt C (1999) The effect of dietary wheat middlings and enzyme supplementation. 1. Late egg production efficiency, egg yields, and egg composition in two strains of Leghorn hens. Poult Sci 78(6):841–847. https://doi.org/10.1093/ps/78.12.1664

Jia W, Slominski BA, Guenter W, Humphreys A, Jones O (2008) The effect of enzyme supplementation on egg production parameters and omega-3 fatty acid deposition in laying hens fed flaxseed and canola seed. Poult Sci 87(10):2005–2014. https://doi.org/10.3382/ps.2007-00474

Khattak FM, Pasha TN, Hayat Z, Mahmud A (2006) Enzymes in poultry nutrition. J Anim Pl Sci 16(1–2):1–7

Kiarie E, Nyachoti CM, Slominski BA, Blank G (2007) Growth performance, gastrointestinal microbial activity, and nutrient digestibility in early-weaned swine fed diets containing flaxseed and carbohydrase enzyme. J Anim Sci 85(11):2982–2993. https://doi.org/10.2527/jas.2006-481

Kocher A, Choct M, Morrisroe L, Broz J (2001) Effects of enzyme supplementation on the replacement value of canola meal for soybean meal in broiler diets. Aust J Agr Res 52(4):447–452. https://doi.org/10.1071/AR00072

Kumar A, Kumar V, Krishnan V, Hada A, Marathe A, Jolly M, Sachdev A (2019) Seed targeted RNAi-mediated silencing of GmMIPS1 limits phytate accumulation and improves mineral bioavailability in soybean. Sci Rep 9(1):1–13. https://doi.org/10.1038/s41598-019-44255-7

Kyriazakis I, Whittemore C (2006) Appetite and voluntary feed intake. In: Kyriazakis I, Whittemore CT (eds) Whittemore's science and practice of pig production, 3rd edn. Blackwell Publishing, Oxford, pp 417–436

Lei XJ, Lee KY, Kim IH (2018) Performance, egg quality, nutrient digestibility, and excreta microbiota shedding in laying hens fed corn-soybean-meal-wheat-based diets supplemented with xylanase. Poult Sci 97(6):2071–2077. https://doi.org/10.3382/ps/pey041

Li S, Sauer WC, Huang SX, Gabert VM (1996) Effect of β-glucanase supplementation to the hulless barley- or wheat-soybean meal diets on the digestibilities of energy, protein, β-glucans, and amino acids. J Anim Sci 74:1649–1656. https://doi.org/10.2527/1996.7471649x

Li Q, Gabler NK, Loving CL, Gould SA, Patience JF (2018) A dietary carbohydrase blend improved intestinal barrier function and growth rate in weaned pigs fed higher fiber diets. J Anim Sci 96:5233–5243. https://doi.org/10.1093/jas/sky383

Lima MR, Costa FGP, Vieira DV, Cardoso AS, Lima GS, Cavalcante et al (2019) Xylanase, glucanase, and phytase in the diet of light laying hens. J Appl Poultry Res 28(4):1150–1155

Liolìopoulou S, Papadopoulos GA, Giannenas I, Vasilopoulou K, Squires C, Fortomaris P, Mantzouridou FT (2023) Effects of dietary supplementation of pomegranate Peel with xylanase on egg quality and antioxidant parameters in laying hens. Antioxidants 12(1):208. https://doi.org/10.3390/antiox12010208

Liu D, Guo S, Guo Y (2012) Xylanase supplementation to a wheat-based diet alleviated the intestinal mucosal barrier impairment of broiler chickens challenged by Clostridium perfringens. Avian Pathol 41(3):291–298. https://doi.org/10.1080/03079457.2012.684089

Mahagna M, Nir I, Larbier M, Nitsan Z (1995) Effect of age and exogenous amylase and protease on development of the digestive tract, pancreatic enzyme activities and digestibility of nutrients in young meat-type chicks. Reprod Nutr Dev 35(2):201–212

Marsman GJ, Gruppen H, Van der Poel AF, Kwakkel RP, Verstegen MW, Voragen AG (1997) The effect of thermal processing and enzyme treatments of soybean meal on growth performance, ileal nutrient digestibilities, and chyme characteristics in broiler chicks. Poult Sci 76(6): 864–872. https://doi.org/10.1093/ps/76.6.864

Martínez Y, Gonzalez A, Botello A, Perez K (2021) Effect of a combination of propionic-acetic acid on body weight, relative weight of some organs, lactic acid bacteria and intestinal pH of neonatal broilers. Braz J Poult Sci 23: eRBCA-2020. https://doi.org/10.1590/1806-9061-2020-1252

Mateos GG, Jimenez-Moreno E, Serrano MP, Lazaro RP (2012) Poultry response to high levels of dietary fiber sources varying in physical and chemical characteristics. J Appl Poult Sci 21:156–174. https://doi.org/10.3382/japr.2011-00477

Morgan AT, Bedford MR (1995) Wheat specific feed enzymes. Feed Compounder 15:25–25

Muszyński S, Kasperek K, Świątkiewicz S, Arczewska-Włosek A, Wiącek D et al (2022) Assessing bone health status and eggshell quality of laying hens at the end of a production cycle in response to inclusion of a hybrid Rye to a wheat–corn diet. Vet Sci 9(12):683. https://doi.org/10.3390/vetsci9120683

Naghshbandi MP, Moghimi H (2020) Stabilization of phytase on multi-walled carbon nanotubes via covalent immobilization. Methods Enzymol 630:431–451. https://doi.org/10.1016/bs.mie.2019.10.013

Nguyen XH, Nguyen HT, Morgan NK (2021) Dietary soluble non-starch polysaccharide level and xylanase supplementation influence performance, egg quality and nutrient utilization in laying hens fed wheat-based diets. Anim Nutr 7(2):512–520. https://doi.org/10.1016/j.aninu.2020.05.012

Nian F, Guo YM, Ru YJ, Li FD, Peron A (2011) Effect of exogenous xylanase supplementation on the performance, net energy and gut microflora of broiler chickens fed wheat-based diets. Asian Australas J Anim Sci 24(3):400–406. https://doi.org/10.5713/ajas.2011.10273

Odetallah NH, Wang JJ, Garlich JD, Shih JCH (2005) Versazyme supplementation of broiler diets improves market growth performance. Poult Sci 84(6):858–864. https://doi.org/10.1093/ps/84.6.858

Olukosi OA, Cowieson AJ, Adeola O (2007) Age-related influence of a cocktail of xylanase, amylase, and protease or phytase individually or in combination in broilers. Poult Sci 86(1): 77–86. https://doi.org/10.1093/ps/86.1.77

Pack M, Bedford M (1997) Feed enzymes for corn-soybean broiler diets: a new concept to improve nutritional value and economics. AFMA Matrix (South Africa) 6(4):18–21

Panda AK, Rao SV, Raju MVLN, Sunder GS (2009) Effect of butyric acid on performance, gastrointestinal tract health and carcass characteristics in broiler chickens. Asian Australas J Anim Sci 22(7):1026–1031. https://doi.org/10.5713/ajas.2009.80298

Papadopoulos GA, Müller K, Schertling D, Di Benedetto M (2014) Supplementation of lysolecithin in combination with a multi-non-starch polysaccharides enzyme improves the feed efficiency during the post-weaning period in piglets. Acta Agric Scand A Anim Sci 64(2):130–136. https://doi.org/10.1080/09064702.2014.978360

Papadopoulos GA, Lioliopoulou S, Ordoudi SA, Giannenas I, Van Hoeck V, Morisset D et al (2022) Xylanase supplementation in wheat-based diets of laying hens affects the egg yolk color, carotenoid and fatty acid profiles. Foods 11(15):2209. https://doi.org/10.3390/foods11152209

Perić L, Milošević N, Đukić-Stojčić M, Bjedov S, Rodić V (2008) Effect of enzymes on performances of broiler chickens. Biotechnol Anim Husb 24(5–6):45–51

Petry AL, Patience JF (2020) Xylanase supplementation in corn-based swine diets: a review with emphasis on potential mechanisms of action. J Anim Sci 98(11):skaa318. https://doi.org/10.1093/jas/skaa318

Petry AL, Huntley NF, Bedford MR, Patience JF (2020) Xylanase increased the energetic contribution of fiber and improved the oxidative status, gut barrier integrity, and growth performance of growing pigs fed insoluble cornbased fiber. J Anim Sci 98(7):1–11. https://doi.org/10.1093/jas/skaa233

Pirgozliev V, Bedford MR, Acamovic T (2010) Effect of dietary xylanase on energy, amino acid and mineral metabolism, and egg production and quality in laying hens. Br Poultry Sci 51(5):639–647. https://doi.org/10.1080/00071668.2010.514325

Ranade AS, Rajmane BV (1992) Effect of enzyme feed supplement on commercial broilers. In: Proceedings of the 19th world's poultry congress, vol 2, Amsterdam, The Netherlands, pp 485–487

Raza A, Bashir S, Tabassum R (2019) An update on carbohydrases: growth performance and intestinal health of poultry. Heliyon 5(4):e01437. https://doi.org/10.1016/j.heliyon.2019.e01437

Roberts JR, Choct M (2006) Effects of commercial enzyme preparations on egg and eggshell quality in laying hens. Br Poultry Sci 47(4):501–510

Salih ME, Classen HL, Campbell GL (1991) Response of chickens fed on hull-less barley to dietary β-glucanase at different ages. Anim Feed Sci Technol 33(1–2):139–149. https://doi.org/10.1016/0377-8401(91)90052-T

Santos CI, Ribeiro T, Ponte PIP, Fernandes VO, Falcão L, Freire JP et al (2013) The effects of restricting enzyme supplementation in rye-based diets for broilers. Anim Feed Sci Technol 186(3–4):214–217. https://doi.org/10.1016/j.anifeedsci.2013.10.011

Sayyazadeh H, Rahimi G, Rezaei M (2006) Influence of enzyme supplementation of maize, wheat and barley-based diets on the performance of broiler chickens. Pak J Biol Sci 9(4):616–621

Shirmohammad F, Mehri M (2011) Effects of dietary supplementation of multi-enzyme complex on the energy utilization in rooster and performance of broiler chicks. Afr J Biotechnol 10(38):7541–7547

Simbaya J, Slominski BA, Guenter W, Morgan A, Campbell LD (1996) The effects of protease and carbohydrase supplementation on the nutritive value of canola meal for poultry: in vitro and in vivo studies. Anim Feed Sci Technol 61(1–4):219–234. https://doi.org/10.1016/0377-8401(95)00939-6

Singh AK, Mishra B, Bedford MR, Jha R (2021) Effects of supplemental xylanase and xylooligosaccharides on production performance and gut health variables of broiler chickens. J Anim Sci Biotechnol 12(1):1–15. https://doi.org/10.1186/s40104-021-00617-8

Slominski BA (2011) Recent advances in research on enzymes for poultry diets. Poult Sci 90(9):2013–2023. https://doi.org/10.3382/ps.2011-01372

Sun HY, Kim IH (2019) Effects of multi-enzyme on production performance, egg quality, nutrient digestibility, and excreta noxious gas emission of early phase Hy-line brown hens. Poult Sci 98(10):4889–4895

Torres-Pitarch A, Manzanillaac EG, Gardiner GE, O'Doherty JV, Lawlor PG (2019) Systematic review and meta-analysis of the effect of feed enzymes on growth and nutrient digestibility in grow-finisher pigs: effect of enzyme type and cereal source. Anim Feed Sci Technol 251:153–165. https://doi.org/10.1016/j.anifeedsci.2018.12.007

Uni Z, Noy Y, Sklan D (1999) Posthatch development of small intestinal function in the poult. Poult Sci 78(2):215–222. https://doi.org/10.1093/ps/78.2.215

Vahjen W, Gläser K, Schäfer K, Simon O (1998) Influence of xylanase-supplemented feed on the development of selected bacterial groups in the intestinal tract of broiler chicks. J Agric Sci 130(4):489–500. https://doi.org/10.1017/S0021859698005498

Vahjen W, Osswald T, Schäfer K, Simon O (2007) Comparison of a xylanase and a complex of non-starch polysaccharide-degrading enzymes with regard to performance and bacterial metabolism in weaned piglets. Arch Anim Nutr 61(2):90–102. https://doi.org/10.1080/17450390701203881

Van Hoeck V, Papadopoulos GA, Giannenas I, Lioliopoulou S, Tsiouris V, Mantzios T et al (2021) New intrinsically thermostable xylanase improves broilers' growth performance, organ weights, and affects intestinal viscosity and pH. Agriculture 11(12):1235. https://doi.org/10.3390/agriculture11121235

Vandeplas S, Dauphin RD, Thiry C, Beckers Y, Welling GW, Thonart P, Thewis A (2009) Efficiency of a Lactobacillus plantarum-xylanase combination on growth performances, microflora populations, and nutrient digestibilities of broilers infected with Salmonella Typhimurium. Poult Sci 88(8):1643–1654. https://doi.org/10.3382/ps.2008-00479

Vermeulen K, Papadopoulos GA, Wealleans A, Van Hoeck V, Giannenas I, Lioliopoulou S, Tassis P, Papageorgiou K, Fotomaris P (2022) Xylanase supplementation improves performance and alters the microbiome of lactating sows fed low or high fibre diets. Anim Sci Proc 13(2):183

Waititu SM, Sanjayan N, Hossain MM, Leterme P, Nyachoti CM (2018) Improvement of the nutritional value of highprotein sunflower meal for broiler chickens using multi-enzyme mixtures. Poult Sci 97:1245–1252. https://doi.org/10.3382/ps/pex418

Westbrook LA, Cherian G (2019) Egg quality, fatty-acid composition and gastrointestinal morphology of layer hens fed whole flaxseed with enzyme supplementation. Br Poultry Sci 60(2): 146–153. https://doi.org/10.1080/00071668.2018.1556783

Whiting IM, Rose SP, Mackenzie AM, Amerah AM, Pirgozliev VR (2019) Effect of wheat distillers dried grains with solubles and exogenous xylanase on laying hen performance and egg quality. Poult Sci 98(9):3756–3762. https://doi.org/10.3382/ps/pez063

Yadav JL, Sah RA (2006) Supplementation of corn-soybean based layers diets with different levels of acid protease. J Inst Agric Anim Sci 27:93–102. https://doi.org/10.3126/jiaas.v27i0.701

Yokhana J, Parkinson G, Frankel T (2016) Effects of insoluble fiber supplementation applied at different ages on digestive organ weight and digestive enzymes of layer-strain poultry. Poult Sci 95:550–559. https://doi.org/10.3382/ps/pev336

Zakaria HA, Jalal MA, Ishmais MAA (2010) The influence of supplemental multi-enzyme feed additive on the performance, carcass characteristics and meat quality traits of broiler chickens. Int J Poult Sci 9(2):126–133. https://doi.org/10.3923/ijps.2010.126.133

Zanella I, Sakomura NK, Silversides FG, Fiqueirdo A, Pack M (1999) Effect of enzyme supplementation of broiler diets based on corn and soybeans. Poult Sci 78(4):561–568. https://doi.org/10.1093/ps/78.4.561

Zhou Y, Jiang Z, Lv D, Wang T (2009) Improved energy-utilizing efficiency by enzyme preparation supplement in broiler diets with different metabolizable energy levels. Poult Sci 88(2): 316–322. https://doi.org/10.3382/ps.2008-00231

Zhu A, Tan H, Cao L (2019) Isolation of phytase-producing yeasts from rice seedlings for prospective probiotic applications. 3 Biotech 9(6):1–8. https://doi.org/10.1007/s13205-019-1746-0

Medicinal and Aromatic Plants as a Source of Potential Feed and Food Additives

Katerina Grigoriadou, Antigoni Cheilari, Eva Dina, Stella Alexandri, and Nektarios Aligiannis

Abstract

In the last decades, the use of medicinal and aromatic plants (MAPs) shows an increase in both animal and human nutrition and demonstrates a plethora of beneficial properties attributed to the bioactive compounds that have been investigated in their extracts and essential oils, which are considered as a great source of antioxidant and antimicrobial constituents. Specifically, the antioxidant capacity of MAPs and their inhibitory effects against pathogens have been evaluated for their use as natural food and feed additives. In feeds, they are used for improving their flavor, as colorants, as nutritional additives and for other livestock production parameters. The processing of MAPs from the food and feed industries has as derivatives several by-products that are not commercially accepted. These by-products are rich in bioactive natural compounds with antioxidant and antimicrobial properties. A number of plants have been evaluated for their antioxidant activity and their phenolic and flavonoid content so as for their extracts to be used in the development of products with health benefits. Moreover, MAPs are used in food products for the enhancement of color, taste, odor, and texture. Due to their antioxidant properties, aromatic plants and their derivatives are used instead of synthetic preservatives and antioxidants.

Keywords

Extracts · Essential oils · By-products · Bioactive compounds · Mediterranean flora

K. Grigoriadou (✉) · S. Alexandri
Institute of Plant Breeding and Genetic Resources, Hellenic Agricultural Organization – DIMITRA, Thessaloniki, Greece

A. Cheilari · E. Dina · N. Aligiannis
Department of Pharmacognosy and Natural Products Chemistry, Faculty of Pharmacy, National and Kapodistrian University of Athens, Athens, Greece

G. Arsenos, I. Giannenas (eds.), *Sustainable Use of Feed Additives in Livestock*,
https://doi.org/10.1007/978-3-031-42855-5_6

1 Introduction

Medicinal aromatic plants (MAPs) have many properties and their use is becoming more extensive and widespread. Especially, the Mediterranean basin is rich in many indigenous aromatic species, known for their medicinal properties that derive from secondary metabolites produced in the plants. Terpenes (or isoprenoids), phenolic compounds (phenylpropanoids and flavonoids), and nitrogen-containing compounds (alkaloids, cyanogenic glycosides, and glucosinolates) are the main categories of plant secondary metabolites (PSMs), which are important for the adaptation of plants to their environment, as they demonstrate multiple roles (Bourgaud et al. 2001). They have defensive and protective roles to abiotic stresses that are related to changes in environmental parameters (temperature, light levels, water status, etc.) (Wink and Schimmer 1999; Khan et al. 2011) and they also act as plant growth regulators or modulators in signal transduction (Kaufman et al. 1999; Zhao et al. 2005).

Several studies have been made for the investigation of the content of their compounds, in dry extracts and essential oils. Rosemary (*Rosmarinus officinalis*) contains various compounds such as carnosol, carnosic acid, rosmanol, camphor, and α-pinene, valuable as feed supplements as well. The essential oil of *Origanum vulgare* shows antimicrobial, antifungal, and antiparasitic activities (Adam et al. 1998; Karagouni et al. 2005; Viuda-Martos et al. 2007). The majority of the substances of the essential oil of *O. vulgare* subsp. *hirtum* are phenolic antioxidants, containing carvacrol, p-cymene, and γ-terpinene (Anastasiou et al. 2019; Tsitlakidou et al. 2022). Another plant with various properties is *Crocus sativus*, which has antidepressant, hypolipidemic, anti-inflammatory, nervo-stimulation, antitumor, antistress, and free radical scavenging activities. Different parts of *C. sativus* contain various compounds, such as flavonoids, vitamins, volatile compounds, and fatty acids (Assimopoulou et al. 2005; Moshiri et al. 2015; Mzabri et al. 2019). The genus *Thymus* is, also, well known for their biological and pharmacological properties, as the various species are rich in thymol and carvacrol that provide antifungal, antibacterial, and antioxidant activities (Miguel et al. 2004; Ebrahimi et al. 2008; Kontogiorgis et al. 2016).

Essential oils are extracted from parts of the plants, such as seeds, flowers, leaves, buds, fruit, bark, twigs, and roots, and are volatile natural complex compounds with strong odor usually obtained by steam or hydrodistillation (Bakkali et al. 2008; Seow et al. 2014) (Fig. 1a). Each oil can be composed from a few substances to a complex of more than 100 single constituents (Blitzke 2009). In addition, the chemical composition may vary in the raw material as it is influenced by plant health, habitat, growth stage, and harvest time (Figueiredo et al. 2008; Schmidt 2010). Essential oils include antioxidants, such as terpenoid and phenolic components, and some oils present differences of cytotoxicity due to various PSM compounds, contents, and phenolic substances (Bakkali et al. 2005). Their composition is mainly investigated through chromatographic methods coupled to spectroscopic techniques—gas chromatography-mass spectrometry (GC-MS), high-performance liquid

Fig. 1 Distillation (**a**), essential oil (**b**), and by-products (**c**) after distillation of *Origanum vulgare* subsp. *hirtum*, (**d**) broiler chicken feed with *Tenebrio molitor*, (**e**) larvas grown in substrate with postdistillation residues, (**f**) lavender fields in Provence, France, (**g**) distillation of *Rosa damascena* in Bulgaria, and (**h**) by-products after distillation of *R. damascena* in Bulgaria

chromatography (HPLC), and liquid chromatography-mass spectrometry (LC-MS) (Turek and Stintzing 2013).

As the essential oils of MAPs contain many different secondary metabolites with medicinal properties, their application is known in medical microbiology, pharmaceutical botany, and in food industry, as food flavors and preservatives (Svoboda et al. 2006; Oroojalian et al. 2010). For this reason, MAPs play a significant role as food and feed additives. Their use is widespread as the demand for natural preservatives, as safer alternatives, is becoming more and more increased. Also, natural products have better biodegradability and wide availability. Essential oils are, also, used in livestock feeds for animals (Burt 2004). The complex mixture of components can interact with multiple molecular sites, and for this reason essential oils, in many cases, show better beneficial effects than a chemically synthesized compound (Svoboda et al. 2006), but some substances of essential oils may affect the bioavailability of other components (Seow et al. 2014). Thus, their uses can be extended to many sectors of food and feed industry for the production of novel products with environmentally friendly substances.

2 Use of MAPs as Feed Additives

In the last decades, due to regulations related to human health, many antibiotics and hormones used in animals are banned, as since 2006 the use of antimicrobial growth promoters has been banned in European Union due to the developed antibiotic resistance (Giannenas et al. 2016a). As a result, there is need for herbal products to replace antibiotics in animal feed. Some probiotics, organic acids, enzymes, symbiotics, and plant essential oils, which are nonantibiotic additives, have been tested for their use instead of conventional antibiotics. Essential oils and extracts from MAPs are used to replace synthetic chemicals in animals' antibiotics and medicines (Franz et al. 2010). In general, as feed additives, aromatic and medicinal plants, their essential oils, and their extracts are used. Especially, essential oils are widely spread for their use but also extracts and by-products of MAPs play an important role in animal feed.

Different methods of extraction and solvents determine the composition of extracts, as certain molecules are extracted with certain solvents. Plant extracts comprise a plethora of molecules with more than one active substance. Often, extracts are processed to be enriched in a selected category of substances or bioactive molecules are purified and used otherwise (Franklc et al. 2009). Feeds are supplemented with herbs and their extracts in order to activate appetite in animals and to accelerate the time of digestion (Platel and Srinivasan 2001; Suresh and Srinivasan 2007; Franklc et al. 2009). A study on rats with the use of extracts of curcuma, red and black pepper, cumin, nutmeg, and ginger showed anti-inflammatory effects (Srinivasan 2005; Manjunatha and Srinivasan 2006). Lippens et al. (2005) studied the effect of supplementation of chicken feed with plant extracts (cinnamon, oregano, thyme, cayenne, and citrus). The supplementation enhanced the body weight of chickens as a result of increased feed consumption. Cardozo et al.

(2005) found that among different plant extracts (garlic, cinnamon, anise, yucca, oregano, and capsicum extract) and three secondary plant metabolites (cinnamaldehyde, eugenol, and anethole), and garlic, capsicum, yucca, and cinnamaldehyde altered ruminal fermentation in favor of propionate, which provides more energy.

Essential oils contain antimicrobial, antioxidative, anticoccidial, and antifungal properties and stimulate the immune response (Vidanarachchi et al. 2005; Wang et al. 2007; Giannenas et al. 2016a). They also promote nitrogen absorption (Gill 2001) and inhibition of ammonia control (Varel 2002). All these elements are important for the optimization of animal diet and the improvement of some feed characteristics. Antioxidant compounds from essential oils can control lipid quality deterioration and, therefore, the meat industry can take advantage of such possibilities so as microbiological and sensory properties to be evaluated. Research has been made for the use of MAPs in the diet of chickens and piglets.

The benefits of essential oils have been reported for their use in broiler diet. Essential oils from MAPs improve the oxidative stability of chicken meat (Jiménez-Colmenero et al. 2001; Zhang et al. 2010; Giannenas et al. 2013). The supplementation of oregano essential oil to their feed or water showed an increase in broilers' body weight and feed conversion ratio (Bassett 2000; Alçiçek et al. 2003). Giannenas et al. (2016a) refer that broiler chickens fed with oregano or a mixture of oregano and laurel showed better body weight and feed to gain ratio compared to those of control. Also, groups of broilers supplemented with oregano presented higher antioxidant capacity of breast and thigh meat, but no changes were noted in mineral content between the two groups. A study, where the essential oils of *O. vulgare* subsp. *hirtum* and *Salvia fruticosa* were used as dietary supplements in chicken diet, showed that the feed mixture was well accepted by broiler chickens and the oxidative status was improved compared to control. In cooked breast and thigh meat samples, a similar profile was found (Giannenas et al. 2016b).

Most studies, in poultry, show enhanced growth and improved feed conversion rate by essential oil additives, but no significant change in feed intake. The enrichment of feed with oregano herb in 2–20 g/kg feed resulted in better performance (Halle et al. 2004). The use of carvacrol (50 mg/kg), p-cymene, and γ-terpinene (25 mg/kg) showed no significant effects (Westendarp et al. 2006). Dried oregano leaves (1.25–3.75 g/kg) added to turkeys' feed, showed an improvement in feed conversion rate (Bampidis et al. 2005). Also, the addition of 60 ppm thyme oil rich in carvacrol resulted in higher body weight, better feed efficiency, and lower abdominal fat weight (Denli et al. 2004). Papageorgiou et al. (2003) studied the diet of turkeys supplemented with certain quantities of α-tocopherol or oregano oil or oregano oil plus α-tocopherol and their effect on iron-induced lipid oxidation of turkey breast, thigh, liver, and heart tissues. The diet with oregano oil delayed the lipid oxidation better than with α-tocopherol, followed by the combination of α-tocopherol and oregano oil. Thigh tissue was more susceptible to oxidation than breast tissue, and lipid oxidation in heart was high. The results present that tissue α-tocopherol is an important factor that influences the level of lipid oxidation. Breast, thigh, and heart tissues from oregano groups showed higher levels of α-tocopherol compared to control, and the increase is positive correlated with the supplementation level.

There are also studies for piglets, where different essential oils such as thyme and oregano were used as feed additives containing bioactive compounds. A rise in feed intake is observed with growth-promoting additives and reflects the higher consumption capacity of larger animals, but not always an enhancement of voluntary feed consumption due to improved palatability. The results for each case are influenced by the type and the origin of the essential oil, the quantity added to the feed, and the environmental conditions (Franz et al. 2010). As in large-scale piggeries, feed improvements with 250–500 mg oregano oil/kg on zootechnical parameters of up to 20% and a decrease in mortality of weaners and piglets could be obtained (Tsinas et al. 1998). In an investigation of acceptance of thyme and oregano herbs, animals freely selected between control feed, two concentrations of herbs (0.12 and 1.2%, respectively) or mixtures of herbs (0.06 and 0.6%, respectively), corresponding to 20 and 200 mg essential oil/kg feed, respectively. Piglets preferred more the control feed, and between treatments with herbs, the preference was higher for thyme compared to oregano, maybe because of the stronger flavor and taste of oregano. In general, no better palatability was achieved with the addition of herbs.

Shell eggs are stable against oxidation, but processed eggs can be oxidized during refrigerated storage. Synthetic antioxidants are used for increasing their oxidative stability, although natural antioxidants that can replace the synthetics could be more attractive to consumers and the poultry industry would satisfy the demands of consumers for products without harmful substances (Botsoglou et al. 2010). Studies have suggested feed supplementation with herbs, such as rosemary, oregano, and thyme, as an effective way for the improvement of oxidative stability of egg yolk and meat (Radwan Nadia et al. 2008; Botsoglou et al. 2010). *Crocus sativus* has been researched among other spices, as crocins and crocetins exhibit antioxidant activity (Pham et al. 2000; Martinez-Tome et al. 2001). An inhibitory factor for its use as a feed additive could be its cost, as it is the most expensive cultivated spice. At least 30% of the saffron samples do not have adequate specifications and are considered as waste. These by-products have the same composition as the rest of the spice and are discarded for aesthetic reasons. For this reason, saffron can be investigated as a promising feed additive for its antioxidant and coloring properties (Botsoglou et al. 2010).

The references are very limited about the use of saffron as a feed additive and are limited to layers and broiler chickens. Botsoglou et al. (2005a) investigated the effect of feed supplementation with red stigmas of Greek saffron on the oxidative stability of shell eggs and yolks during refrigerated storage. Laying hens had a supplemented diet with 10 or 20 mg saffron or 200 mg α-tocopherol acetate/kg feed. The two treatments differed significantly concerning the extent of lipid oxidation in shell eggs. The enrichment of eggs with antioxidant constituents of saffron may inhibit the chain reaction in oxidation of the consumed lipids, decreasing oxidation products into yolk. Botsoglou et al. (2005b) studied the effect of rosemary, oregano, saffron, or α-tocopherol acetate included in feed on egg quality. The extent of lipid oxidation in shell eggs differed among the treatments but did not change with storage time.

Yolk color was improved in the group of saffron and the oregano group presented higher oxidation rate than the saffron group. Another study about crocins, which are the main active constituents of saffron, showed that they can prevent the oxidation of linoleic acid due to their antioxidative activity (Pham et al. 2000).

During the procedure of MAPs, a great number of by-products are produced, such as solid residues from the essential oil process, and for some plant species the production of by-products can be more than 60% (Grigoriadou et al. 2018) (Fig. 1c). These products were treated as waste, but they can be used as they contain bioactive compounds with biological properties and can add value to the final products (Saha and Basak 2020).

The distillation of MAPs produces a quantity of by-products that is often discharged (Fig. 1h). These by-products can be exploited and can be used as feed additives, and this use has been studied in the substrate on which insect species are developed (Fig. 1). Insects are consumed in many countries of Asia, Africa, and Central and South America, as they provide energy and protein and are a good source of micronutrients (Yen 2009; Chiet al. 2019). They exhibit high feed conversion efficiency, a fact that implies that their body temperature depends on the environment rather than the feed energy (Dobermann et al. 2017; Van Huis and Oonincx 2017). Thus, insects can grow on different types of substrates (organic, industrial, and agricultural waste streams) and they convert them into protein of high quality (Van Huis and Oonincx 2017).

Tenebrio molitor, one of the most commercial edible insect species high in protein content (Adámková et al. 2020), has been researched for its nutritional value after its growth on different substrates and the supplementation of the substrate with postdistillation residues of MAPs (lavender, Greek oregano, rosemary, and olive; 1:1:1:1 ratio) (Fig. 1d, e). The results showed that MAPs had a positive effect on the total phenolic content and antioxidant activity of each substrate, with the substrate with rice bran to be the highest in content and capacity. This study indicates the use of alternative substrate and their enrichment with natural phenolics influence *T. molitor* growth (Andreadis et al. 2021).

3 Use of MAPs as Food Additives

MAPs and their by-products, as a source of several bioactive compounds, such as terpenes, phenolic compounds, and flavonoids, can have an extend use as food antimicrobial and antioxidant additives for the production of high added-value products, such as aromatic edible oils, herbal teas and instant beverages. By-products of four different Greek cultivated species (*Matricaria recutita*, *Origanum vulgare* subsp. *hirtum*, *Sideritis scardica*, and *Thymus vulgaris*) were determined for their content in bioactive compounds through high-performance thin layer chromatography (HPTLC) and liquid chromatography-mass spectrometry (LC-MS) techniques. It was proved that the residues of the process, after the analyses carried out, contained a similar chemical profile and a high phenolic content mainly in thyme, oregano, and Greek mountain tea. Also, GC-MS analyses of essential oils of oregano and thyme by-products revealed the existence of carvacrol, thymol, γ -terpinene, and p-cymene (Dina et al. 2022).

Essential oils are used as flavoring agents in the food industry, and they are preferred for their antimicrobial properties against foodborne pathogens and as a source of natural preservatives. Interactions between essential oils and food characteristics may affect their antimicrobial properties. The antimicrobial effect of essential oils of oregano and garlic against *Salmonella* spp. and *Listeria monocytogenes* was studied on fat, protein, pH, $a_{w,}$ and food additives. Protein, pH, a_W, presence of beef extract, sodium lactate, and nitrates did not influence their antimicrobial effect. Pork fat had a negative effect against essential oils of both species associated with their dilution of the lipid content and also food phosphates had a negative influence probably because of their emulsification properties (García-Díez et al. 2017).

MAPs are known for their nutritional role and the benefits to health from long-term use as foods (Franz et al. 2010). Foodborne illnesses constitute a major problem all over the world and food safety is of great importance. There are many bacteria that develop antibiotic resistance because of the extensive use of antibiotics. Food industries are looking for safer alternatives and natural preservatives as they exhibit wide availability and better biodegradability. For this reason, in recent years, the use of natural antimicrobials in food systems is of great interest (Seow et al. 2014).

The activity of antimicrobial essential oils has been researched in natural and synthetic food matrices or packaging in order to eliminate food spoilage pathogens in food manufacturing. Antimicrobial activities of essential oils were studied in apple juice and were increased with juice clarity, incubation temperature, and time and were dependent on the miscibility of antimicrobial components in the juice (Friedman et al. 2004). Wang et al. (2007), after the coverage of strawberries with eugenol, menthol, and thymol, successfully delayed the rate of berry decay and promoted decay resistance through an increase in phenolic compounds, anthocyanins, flavonoids, and antioxidant capabilities.

Essential oils have been used for the reduction of microbial contents into food additives. In an instant soup containing broccoli and white flour, the addition of essential oils of *Pelargonium* sp. was used to improve shelf-life against *Staphylococcus aureus* (Lis-Balchin et al. 2003), while Gutierrez et al. (2009) studied the effect of lemon balm, marjoram, oregano, and thyme essential oils using model systems of meat, milk, and lettuce against *Listeria* spp. Both studies revealed that several factors could affect antimicrobial activities of PSMs presented and their actions are dependent on the existence of simple sugars, high protein concentrations, and the pH of the media. It was observed also that certain essential oils' combination showed synergistic activity and they reduced undesirable side effects while they improved organoleptic characteristics (Gutierrez et al. 2009). Bakery products have, also, been investigated for the improvement of their nutrient profile. Supplementation with turmeric (*Curcuma longa*), cilantro, fennel, ginger (*Zingiber officinale*), oregano, and black and green tea (*Camellia* spp.) extracts leads to a better antioxidant activity (Dziki et al. 2014). Si et al. (2006) studied the antimicrobial activity of 66 essential oils and compounds against *Salmonella typhimurium* DT104 and

Escherichia coli O157:H. Most of the oils showed high efficacy against the two bacterial pathogens and *E. coli* with K88 pili, with a simultaneous tolerance to the low pH.

Apart from the use of essential oils into food, there is also an increasing interest for their use into food packaging. This practice aims to extend shelf-life and improve consumer safety of products by decreasing pathogens in packed foods and packaging materials (Seydim and Sarikus 2006). The advantage of essential oils for their use in food packaging is their highly volatile and antimicrobial nature, as oils do not require direct contact with food, reducing possible sensory changes and provide an alternative way to undesirable synthetic additives (Becerril et al. 2007). In general, the literature is limited on active food packaging with essential oils. Packaging with oregano and cinnamon essential oils affected the action of *E. coli* and *S. aureus*, and their death time curves showed shorter optimum time of exposure against bacteria than that of pure essential oils in vapor phase (Becerril et al. 2007). Beef slices and fish were preserved with active packaging using protein-based and gelatin–chitosan films incorporated with essential oils. Milk protein- based film with a mix of essential oils successfully reduced *E. coli* O157:H7 and *Pseudomonas* spp. microbial load after seven days of storage at 4 °C (Oussalah et al. 2004; Gómez-Estaca et al. 2010).

The simultaneous and synergistic effect between essential oils, food preservatives, and antibiotics has also been extensively studied. Synergistic action occurs when two or more factors can give an improved result greater than the sum of each separately. However, combined use should be done carefully and after research on the concentrations, as essential oils when used together can cause unpleasant changes in organoleptic properties (Seow et al. 2014). The synergistic action of essential oils has been used in different combinations and concentrations, as in the case of the inactivation of *Listeria monocytogenes* in salads, which was achieved using a mixture of thyme verbena, thyme red, Spanish oregano, ajowan, tea tree, clove, sage, and rosemary oils (Molinos et al. 2009). Also, an additive effect occurred when a mixture of carvacrol and thymol was used in *Pseudomonas aeruginosa* and *S. aureus* (Lambert et al. 2001). Specific compounds contained in the essential oil of thyme (linalool, thymol, or carvacrol) showed synergistic antimicrobial activity against six stains (Iten et al. 2009) while total inhibition was achieved when carvacrol and thymol mixed at certain concentrations (Lambert et al. 2001; Iten et al. 2009).

4 MAPs By-Products

The processing of MAPs generates a significant amount of by-products that are typically considered waste and discarded. These by-products come not only from food and feed industries, but also from post-harvest materials, such as branches and low-quality leaves, as well as hydrolates and solid residues from the essential oil (EO) process. These materials are not commercially accepted, even though they can make up more than 60% of production, and are often burned or discarded in the fields, polluting the environment. Despite being considered waste, these by-products are rich in bioactive secondary metabolites with antioxidant and antimicrobial properties and are increasingly being studied as valuable sources of phenolic compounds (Dina et al. 2022). Research studies show evidence that these residual by-products are potential sources of bioactive compounds, as they contain the same ingredients and properties as the final product (Navarrete et al. 2011; Redondo et al. 2017). Five plant species that are considered among the most important MAPs of the Mediterranean region are cultivated for commercial use for culinary purposes or for the production of their EOs.

Rose oil is produced on industrial scale from the petals of *Rosa damascena* Mill (Rosaceae) by hydrodistillation and is often referred to as "liquid gold." Rose petals are rich in flavonoids such as quercetin, kaempferol, and respective glycoside derivatives. In addition, their extracts contain anthocyanins, gallic acid derivatives, and other phenolic compounds (Velioglu and Mazza 1991). In parallel, the wastewater (i.e., aqueous extract of petals) produced from the hydrodistillation and the solid petal residue remain unexploited. It is worth noting that rose water, which contains less than 1% essential oil, was also considered waste in the past. The yield of rose hydrodistillation is very low, as it takes about 3000 kg of rose petals to produce 1 kg of rose oil, resulting in tons of waste (Zhu et al. 2012). In Bulgaria, *R. damascena* is an important commercial MAP, and the country is responsible for about 50% of the world's production of rose oil (about 870–2000 kg annually) (Fig. 1g). The chemical composition of the wastewater as well as the aqueous and hydroalcoholic extracts from the rose petals remaining after hydrodistillation has been studied, and they have been found to be rich in phenolic compounds including flavonoids (Kumar et al. 2008; Rusanov et al. 2014; Schieber et al. 2005). Moreover, these extracts have been studied for their tyrosinase inhibition activity due to the presence of quercetin, kaempferol and ellagic acid suggesting their potential use for bleaching purpose (Solimine et al. 2016). Furthermore, it has been suggested that the presence of these compounds may make the extracts useful for treating acute or chronic inflammation (Wedler et al. 2016a; Wedler et al. 2016b). In addition, enriched polyphenols extracts after processing with adsorption resins have been found to have antioxidant activity and to activate proteostatic modules, exhibiting potential antiaging properties (Dina et al. 2021).

The stigmas of *Crocus sativus*, commonly known as saffron, are often used in the production of food and medicine, but the remaining parts of the plant, including the styles, the tepals, and the petals, are typically discarded as waste. The cultivation of the plant is found principally in Iran, India, and Morocco but also in the eponymous region Crocos (Kozani) of central Macedonia, Greece. During collection, only a small part of the plant is used (stigmas), while the rest is discarded and hence is considered one of the most expensive crops of medicinal plant products in the world, earning it the nickname "red gold" (Abdullaev and Espinosa-Aguirre 2004; Mzabri et al. 2021). Due to the considerable annual production of tepals and petals, the utilization of by-products for the preparation of extracts with significant antioxidant and possibly anti-aging properties is an attractive alternative for the saffron industry. Tepals are rich in flavonoids (kaempferol, rutin, quercetin, luteolin, etc.), carotenoids, phenolic acids, tannins, and anthocyanins and have shown significant radical scavenging activity (Mykhailenko et al. 2019). Pharmacological properties of *C. sativus* products have been reviewed extensively, and aqueous extracts have shown very high content of phenolic compounds, antioxidant, and antibacterial activity, as well as good activity in in vitro and ex vivo models of inflammation and oxidative stress. Moreover, antitumor properties as well as antiproliferative activity of hydroalcoholic extracts have been exhibited (Chichiriccò et al. 2019; Mykhailenko et al. 2019). Beside the potential applications in the cosmetic and pharmaceutical industries, by-products of *C. sativus* contain carotenoids, which can be used as natural colorants in food products (Rubio Moraga et al. 2013).

The lavender (*Lavandula angustifolia* Mill) essential oil has a great commercial interest due to its widespread use in perfumery, aromatherapy, and pharmaceutical products (Fig. 1f). It is produced on an industrial scale from aerial parts of the plant by steam distillation and is rich in monoterpenes, sesquiterpenes, and their oxidized derivatives. The essential oil of *L. angustifolia* has shown remarkable bacteriostatic and bactericidal properties against several strains of *Staphylococcus* and *Proteus* species as well as the antifungal properties of the essential oil have been tested against yeasts, dermatophytes, and mucosal strains. Given that the biological activity of essential oils is often attributed to their main constituents, the bacteriostatic and antifungal activity of pure linalool appears to be responsible for the observed result, while the synergistic action of other components of the oil is noteworthy. Leaf extracts have a noteworthy content of secondary metabolites, including flavonoids (flavones, anthocyanins, and flavonoid glycosides), phenolic acids, hydroxybenzoic acids and derivatives, hydroxycinnamic acids and derivatives, coumarins, benzofuran derivatives, and others with significant pharmacological properties (Héral et al. 2021). Although there are few data on the composition of lavender by-products from distillation, research results show that they contain compounds with significant antimicrobial, anti-inflammatory, and antioxidant properties (Ciocarlan et al. 2021; Méndez-Tovar et al. 2015; Moon et al. 2006; Turrini et al. 2021). The main applications of lavender by-products in the pharmaceutical and cosmetic industry are as raw materials for the synthesis of derivatives or as active ingredients in final products. However, their potential use as natural pesticides in agriculture and their biological activity against pests and diseases have not yet been fully exploited.

Thymus vulgaris is a native plant of southern Europe and has a widespread distribution in all Mediterranean countries as well as in the temperate zones of Asia and America. The essential oil of thyme, whose main constituent is thymol, is used in cosmetology and the food industry for its antimicrobial, antifungal, and antioxidant properties (Patil et al. 2021). Commercially, it is used as a spice or with the form of paste as a condiment, while the main components of essential oil such as thymol and carvacrol have been incorporated, beside cosmetics, and in many dental products. There have been numerous studies supporting the potential future use of both essential oil and extracts in the food industry as natural preservatives, replacing synthetic ones (Pesavento et al. 2015). *T. vulgaris* has a notable content of secondary metabolites such as flavonoids (rutin and caffeic acid), terpenoids (thymol and carvacrol), flavones (apigenin and luteolin), sterols, alkaloids, and saponins (Patil et al. 2021). The extracts of thyme have shown antispasmodic, antiparasitic, and anti-inflammatory activities (Begrow et al. 2010; Braga et al. 2006). By-products of thyme's EO distillation have been studied and can be potentially exploited in the polymers industry as they can provide stability to thermo-oxidative processes. Studies have been conducted for the incorporation of thyme by-products in food, showing that they can significantly improve fatty acid profile of lamb meat, retarding lipid oxidation in raw and cooked meat of turkey thighs, and are useful as natural antioxidants and antimicrobial agents in pork meat processing (Mielnik et al. 2008; Nieto 2013; Šojić et al. 2020). Finally, the presence of bioactive secondary metabolites in thyme by-products underlies their potent use not only as food antimicrobial and antioxidant additives, but also in the preparation of high added-value products, such as enriched aromatic edible oils or herbal preparations such as instant beverages (Dina et al. 2022).

Oregano (*Origanum vulgare*) is a perennial aromatic plant, widely distributed as a wild plant and one of the most cultivated herbs in the Mediterranean basin. Its EO contains significant amounts of the antimicrobial substances thymol and carvacrol, which are transferred to the hydrosol in a percentage, while the water extract contains flavonoids and other antioxidants. In particular, the major constituents include terpenoids (monoterpenes, sesquiterpenes, diterpenes, and triterpenes), phenols, phenolic acids, and flavonoids, and as minors can be found alkaloids and saponins (Fig. 1b). Due to the presence of these metabolites, oregano extracts exhibit a broad spectrum of pharmacological properties including anticancer activities in adrenocortical tumor, hepatocellular carcinoma, HeLa, human breast adenocarcinoma, and human colon adenocarcinoma cell lines (Begnini et al. 2014; Elshafie et al. 2017; Rubin et al. 2018). In addition, several studies show the anti-inflammatory properties of extracts and EOs (Bukovská et al. 2007; Ocaña-Fuentes et al. 2010; Paur et al. 2010). Oregano EO has been tested against several microbial strains, exhibiting antimicrobial activity against pathogens including *Malassezia furfur*, *Trichophyton rubrum*, *Trichosporon beigelii,* and *Helicobacter pylori* (Adam et al. 1998; Stamatis et al. 2003). The oregano oil finds applications in the food industry and is also an ingredient in cosmetic products. Specifically, it is used in alcoholic beverages, in baking, in meat products, as a seasoning, in dairy products, in processed vegetables, and in butter and oils. Research studies on oregano by-products prove their antioxidant and antimicrobial properties and indicate their use as food and feed additives as well as for the production of high added-value products (Bouloumpasi et al. 2021; Dina et al. 2022).

To conclude, despite the potential value of these by-products, their utilization is still limited, and most of them are discarded or burned. However, some research has been done on the potential uses of these materials, and several studies have shown that they can be used as ingredients in functional foods, nutraceuticals, and natural cosmetics. For example, solid residues (petals, leaves, and stems) of MAPs, which are commonly used in the production of EOs, can be used to extract antioxidants and other bioactive compounds, which can be incorporated into food products or dietary supplements. Similarly, wastewaters, which are rich in polyphenols, can be appropriately processing (e.g., adsorption resins) and used as ingredients in functional foods. Finally, by valorizing these by-products, it is possible to reduce waste and pollution, and to create new opportunities for the food and feed industries.

References

Abdullaev FI, Espinosa-Aguirre JJ (2004) Biomedical properties of saffron and its potential use in cancer therapy and chemoprevention trials. Cancer Detect Prev 28(6):426–432. https://doi.org/10.1016/J.CDP.2004.09.002

Adam K, Sivropoulou A, Kokkini S, Lanaras T, Arsenakis M (1998) Antifungal activities of *Origanum vulgare* subsp. *hirtum, Mentha spicata, Lavandula angustifolia,* and *Salvia fruticosa* essential oils against human pathogenic fungi. J Agric Food Chem 46(5):1739–1745

Adámková A, Mlcek J, Adámek M, Borkovcová M, Bednárová M, Hlobilová V, Knížková I, Juríková T (2020) *Tenebrio molitor* (Coleoptera: Tenebrionidae)—optimization of rearing conditions to obtain desired nutritional values. J Insect Sci 20:24

Alçiçek A, Bozkurt M, Çabuk M (2003) The effect of an essential oil combination derived from selected herbs growing wild in Turkey on broiler performance. S Afr J Animal Science 33:89–94

Anastasiou TI, Mandalakis M, Krigas N, Vézignol T, Lazari D, Katharios P, Dailianis T, Antonopoulou E (2019) Comparative evaluation of essential oils from medicinal-aromatic plants of Greece: chemical composition, antioxidant capacity and antimicrobial activity against bacterial fish pathogens. Molecules 25(1):148

Andreadis SS, Panteli N, Mastoraki M, Rizou E, Stefanou V, Tzentilasvili S, Sarrou E, Chatzifotis S, Krigas N, Antonopoulou E (2021) Towards functional insect feeds: Agri-food by-products enriched with post-distillation residues of medicinal aromatic plants in *Tenebrio molitor* (Coleoptera: Tenebrionidae) breeding. Antioxidants 2022 11:68. https://doi.org/10.3390/antiox11010068

Assimopoulou AN, Sinakos Z, Papageorgiou V (2005) Radical scavenging activity of *Crocus sativus* L. extract and its bioactive constituents. Phytother Res: An International Journal Devoted to Pharmacological and Toxicological Evaluation of Natural Product Derivatives 19(11):997–1000

Bakkali F, Averbeck S, Averbeck D, Zhiri A, Idaomar M (2005) Cytotoxicity and gene induction by some essential oils in the yeast Saccharomyces cerevisiae. Mutat Res Genet Toxicol Environ Mutagen 585(1–2):1–13

Bakkali F, Averbeck S, Averbeck D, Idaomar M (2008) Biological effects of essential oils–a review. Food Chem Toxicol 46(2):446–475

Bampidis VA, Christodoulou V, Florou-Paneri P, Christaki E, Chatzopoulou PS, Tsiligianni T, Spais AB (2005) Effect of dietary dried oregano leaves on growth performance, carcase characteristics and serum cholesterol of female early maturing turkeys. Br Poult Sci 46(5):595–601

Bassett R (2000) Oregano's positive impact on poultry production. World Poultry-Elsevier 16:31–34

Becerril R, Gómez-Lus R, Goni P, López P, Nerín C (2007) Combination of analytical and microbiological techniques to study the antimicrobial activity of a new active food packaging containing cinnamon or oregano against *E. coli* and *S. aureus*. Anal Bioanal Chem 388(5):1003–1011

Begnini KR, Nedel F, Lund RG, Carvalho PHDA, Rodrigues MRA, Beira FTA, Del-Pino FAB (2014) Composition and Antiproliferative effect of essential oil of *Origanum vulgare* against tumor cell lines. J Med Food 17(10):1129–1133. https://doi.org/10.1089/JMF.2013.0063

Begrow F, Engelbertz J, Feistel B, Lehnfeld R, Bauer K, Verspohl EJ (2010) Impact of thymol in thyme extracts on their antispasmodic action and ciliary clearance. Planta Med 76(4):311–318. https://doi.org/10.1055/S-0029-1186179/ID/1

Blitzke T (2009) Ätherische Öle und Aromaextrakte in der Kosmetik, der Aromatherapie und im Lebensmittelbereich. In: Hoppe B, Hoppe K, Junghanns W, Kabelitz L, Pank F, Schliephake E (eds) Handbuch des Arznei- und Gewürzpflanzenbaus. Band 1 Grundlagen des Arznei- und Gewürzpflanzenbaus I. Verein für Arznei- und Gewürzpflanzen SALUPLANTA, Bernburg, pp 383–394

Botsoglou NA, Florou-Paneri P, Nikolakakis I, Giannenas I, Dotas V, Botsoglou EN, Aggelopoulos S (2005a) Effect of dietary saffron (*Crocus sativus* L.) on the oxidative stability of egg yolk. Br Poult Sci 46(6):701–707

Botsoglou N, Florou-Paneri P, Botsoglou E, Dotas V, Giannenas I, Koidis A, Mitrakos P (2005b) The effect of feeding rosemary, oregano, saffron and α-tocopheryl acetate on hen performance and oxidative stability of eggs. S Afr J Anim Sci 35(3):143–151

Botsoglou E, Govaris A, Giannenas I, Botsoglou N (2010) Use of saffron (*Crocus sativus* L.) as a feed additive for improving growth and meat or egg quality in poultry. Saffron Global Science Books, UK/Japan, pp 98–107

Bouloumpasi E, Hatzikamari M, Lazaridou A, Chatzopoulou P, Biliaderis CG, Irakli M (2021) Antibacterial and antioxidant properties of oregano and rosemary essential oil distillation by-products. Biol Life Sci Forum 6(1):47. https://doi.org/10.3390/FOODS2021-11020

Bourgaud F, Gravot A, Milesi S, Gontier E (2001) Production of plant secondary metabolites: a historical perspective. Plant Sci 161(5):839–851

Braga PC, Dal Sasso M, Culici M, Bianchi T, Bordoni L, Marabini L (2006) Anti-inflammatory activity of thymol: inhibitory effect on the release of human neutrophil elastase. Pharmacology 77(3):130–136. https://doi.org/10.1159/000093790

Bukovská A, Čikoš Š, Juhás Š, Il'ková G, Rehák P, Koppel J (2007) Effects of a combination of thyme and oregano essential oils on TNBS-induced colitis in mice. Mediat Inflamm 2007: 23296. https://doi.org/10.1155/2007/23296

Burt S (2004) Essential oils: their antibacterial properties and potential applications in foods – a review. Int J Food Microbiol 94(3):223–225

Cardozo PW, Calsamiglia S, Ferret A, Kamel C (2005) Screening for the effects of natural plant extracts at different pH on *in vitro* rumen microbial fermentation of a high-concentrate diet for beef cattle. J Anim Sci 83(11):2572–2579

Chichiriccò G, Ferrante C, Menghini L, Recinella L, Leone S, Chiavaroli A, Brunetti L, di Simone S, Ronci M, Piccone P, Lanza B, Cesa S, Poma A, Vecchiotti G, Orlando G (2019) *Crocus sativus* by-products as sources of bioactive extracts: pharmacological and toxicological focus on anthers. Food Chem Toxicol 126:7–14. https://doi.org/10.1016/J.FCT.2019.01.040

Ciocarlan A, Lupascu L, Aricu A, Dragalin I, Popescu V, Geana EI, Ionete RE, Vornicu N, Duliu OG, Hristozova G, Zinicovscaia I (2021) Chemical composition and assessment of antimicrobial activity of lavender essential oil and some by-products. Plants 10(9):1829. https://doi.org/10.3390/PLANTS10091829

Denli M, Okan F, Uluocak AN (2004) Effect of dietary supplementation of herb essential oils on the growth performance, carcass and intestinal characteristics of quail (Coturnix coturnix japonica). South African J Anim Sci 34(3):174–179

Dina E, Sklirou AD, Chatzigeorgiou S, Manola MS, Cheilari A, Louka XP, Argyropoulou A, Xynos N, Skaltsounis AL, Aligiannis N, Trougakos IP (2021) An enriched polyphenolic extract obtained from the by-product of *Rosa damascena* hydrodistillation activates antioxidant and proteostatic modules. Phytomedicine 93:153757. https://doi.org/10.1016/J.PHYMED.2021.153757

Dina E, Vontzalidou A, Cheilari A, Bagatzounis P, Agapidou E, Giannenas I, Grigoriadou K, Aligiannis N (2022) Sustainable use of Greek herbs' by-products, as an alternative source of biologically active ingredients for innovative products. Front Nutr 9:867666. https://doi.org/10.3389/FNUT.2022.867666/BIBTEX

Dobermann D, Swift JA, Field LM (2017) Opportunities and hurdles of edible insects for food and feed. Nutr Bull 42(4):293–308

Dziki D, Rózyło R, Gawlik-Dziki U, Swieca M (2014) Current trends in the enhancement of antioxidant activity of wheat bread by the addition of plant materials rich in phenolic compounds. Trends Food Sci Technol 40(1):48–61. https://doi.org/10.1016/j.tifs.2014.07.010

Ebrahimi SN, Hadian J, Mirjalili MH, Sonboli A, Yousefzadi M (2008) Essential oil composition and antibacterial activity of *Thymus caramanicus* at different phenological stages. Food Chem 110(4):927–931

Elshafie HS, Armentano MF, Carmosino M, Bufo SA, de Feo V, Camele I (2017) Cytotoxic activity of *Origanum vulgare* L. on hepatocellular carcinoma cell line HepG2 and evaluation of its biological activity. *Molecules 22*(9):1435. https://doi.org/10.3390/MOLECULES22091435

Figueiredo AC, Barroso JG, Pedro LG, Scheffer JJC (2008) Factors affecting secondary metabolite production in plants: volatile components and essential oils. Flavour Fragr J 23:213–226

FrankIc T, Voljč M, Salobir J, Rezar V (2009) Use of herbs and spices and their extracts in animal nutrition. Acta Agric Slov 94(2):95–102

Franz C, Baser KHC, Ungerhofeh W (2010) Essential oils and aromatic plants in animal feeding–a European perspective. A review. Flavour Frag J 25(5):327–340

Friedman M, Henika PR, Levin CE, Mandrell RE (2004) Antibacterial activities of plant essential oils and their components against *Escherichia coli* O157: H7 and *salmonella enterica* in apple juice. J Agric Food Chem 52(19):6042–6048

García-Díez J, Alheiro J, Pinto AL, Soares L, Falco V, Fraqueza MJ, Patarata L (2017) Influence of food characteristics and food additives on the antimicrobial effect of garlic and oregano essential oils. Foods 6(6):44

Giannenas I, Bonos E, Christaki E, Florou-Paneri P (2013) Essential oils and their applications in animal nutrition. Med Aromat Plants 2(140):2167–0412

Giannenas I, Tsinas A, Bartzanas T, Skoufos I, Tzora A, Karamoutsios A (2016a) Oregano and sage essential oils improve antioxidant status of raw and cooked breast and thigh chicken meat: *in vivo* investigation of antioxidant constituents of oregano and sage on chicken breast and thigh meat. Agro Food Industry Hi-Tech, 27(IKEEART-2017-1228), 24–27

Giannenas I, Tzora A, Sarakatsianos I, Karamoutsios A, Skoufos S, Papaioannou N, Anastasiou F, Skoufos I (2016b) The effectiveness of the use of oregano and laurel essential oils in chicken feeding. Ann Anim Sci 16(3):779–796

Gill C (2001) Safe and sustainable feed ingredients. Feed Int 22:40–45

Gómez-Estaca J, De Lacey AL, López-Caballero ME, Gómez-Guillén MC, Montero P (2010) Biodegradable gelatin–chitosan films incorporated with essential oils as antimicrobial agents for fish preservation. Food Microbiol 27(7):889–896

Grigoriadou K, Giannenas I, Bagatzounis P, Agapidou E, Aligiannis N, Christaki E, Florou-Paneri P (2018) An holistic approach to sustainable use of Greek native medicinal/aromatic plants' by-products to produce innovative feeds for domestic animals. 30th International Symposium on the Chemistry of Natural Products and the 10th International Congress on Biodiversity. 25–29/11, Athens, Greece

Gutierrez J, Barry-Ryan C, Bourke P (2009) Antimicrobial activity of plant essential oils using food model media: efficacy, synergistic potential and interactions with food components. Food Microbiol 26(2):142–150

Halle I, Thomann R, Bauermann U, Henning M, Köhler P (2004) Effects of a graded supplementation of herbs and essential oils in broiler feed on growth and carcass traits. Landbauforshung Volkenrode 54.219–229

Héral B, Stierlin É, Fernandez X, Michel T (2021) Phytochemicals from the genus *Lavandula*: a review. Phytochem Rev 20(4):751–771. https://doi.org/10.1007/S11101-020-09719-Z/TABLES/3

Iten F, Saller R, Abel G, Reichling J (2009) Additive antimicrobial effects of the active components of the essential oil of *Thymus vulgaris*–chemotype carvacrol. Planta Med 75(09):PJ100

Jiménez-Colmenero F, Carballo J, Cofrades S (2001) Healthier meat and meat products: their role as functional foods. Meat Sci 59(1):5–13

Karagouni E, Athanassopoulou F, Lytra A, Komis C, Dotsika E (2005) Antiparasitic and immunomodulatory effect of innovative treatments against *Myxobolus* sp. infection in Diplodus puntazzo. Vet Parasitol 134(3–4):215–228

Kaufman PB, Cseke LJ, Warber S, Duke JA, Brielmann HL (1999) Natural products from plants. CRC Press, Boca Raton, FL

Khan TA, Mazid M, Mohammad F (2011) Status of secondary plant products under abiotic stress: an overview. J Stress Physiol Biochem 7(2):75–98

Kontogiorgis C, Ntella M, Mpompou L, Karallaki F, Athanasios P, Hadjipavlou-Litina D, Lazari D (2016) Study of the antioxidant activity of *Thymus sibthorpii* Bentham (Lamiaceae). J Enzyme Inhib Med Chem 31(sup4):154–159

Kumar N, Bhandari P, Singh B, Gupta AP, Kaul VK (2008) Reversed phase-HPLC for rapid determination of polyphenols in flowers of rose species. J Sep Sci 31(2):262–267. https://doi.org/10.1002/JSSC.200700372

Lambert RJW, Skandamis PN, Coote PJ, Nychas GJ (2001) A study of the minimum inhibitory concentration and mode of action of oregano essential oil, thymol and carvacrol. J Appl Microbiol 91(3):453–462

Lippens MG, Huyghebaert E, Cerchiari E (2005) Effect of the use of coated plant extracts and organic acids as alternatives for antimicrobial growth promoters on the performance of broiler chickens. Arch Geflugelkd 69:261–266

Lis-Balchin M, Steyrl H, Krenn E (2003) The comparative effect of novel *Pelargonium* essential oils and their corresponding hydrosols as antimicrobial agents in a model food system. Phytother Res: An International Journal Devoted to Pharmacological and Toxicological Evaluation of Natural Product Derivatives 17(1):60–65

Manjunatha H, Srinivasan K (2006) Protective effect of dietary curcumin and capsaicin on induced oxidation of low-density lipoprotein, iron-induced hepatotoxicity and carrageenan-induced inflammation in experimental rats. FEBS J 273(19):4528–4537

Martinez-Tome M, Jimenez AM, Ruggieri S, Frega N, Strabbioli R, Murcia MA (2001) Antioxidant properties of Mediterranean spices compared with common food additives. J Food Prot 64(9):1412–1419

Méndez-Tovar I, Herrero B, Pérez-Magariño S, Pereira JA, Asensio-S-Manzanera MC (2015) By-product of *Lavandula latifolia* essential oil distillation as source of antioxidants. J Food Drug Anal 23(2):225. https://doi.org/10.1016/J.JFDA.2014.07.003

Mielnik MB, Sem S, Egelandsdal B, Skrede G (2008) By-products from herbs essential oil production as ingredient in marinade for Turkey thighs. LWT Food Sci Technol 41(1): 93–100. https://doi.org/10.1016/J.LWT.2007.01.014

Miguel G, Simoes M, Figueiredo AC, Barroso JG, Pedro LG, Carvalho L (2004) Composition and antioxidant activities of the essential oils of *Thymus caespititius*, *Thymus camphoratus* and *Thymus mastichina*. Food Chem 86(2):183–188

Molinos AC, Abriouel H, López RL, Omar NB, Valdivia E, Gálvez A (2009) Enhanced bactericidal activity of enterocin AS-48 in combination with essential oils, natural bioactive compounds and chemical preservatives against *Listeria monocytogenes* in ready-to-eat salad. Food Chem Toxicol 47(9):2216–2223

Moon T, Wilkinson JM, Cavanagh HMA (2006) Antibacterial activity of essential oils, hydrosols and plant extracts from Australian grown *Lavandula* spp. Int J Aromather 16(1):9–14. https://doi.org/10.1016/J.IJAT.2006.01.007

Moraga A, Ahrazem O, Rambla JL, Granell A, Gómez Gómez L (2013) Crocins with high levels of sugar conjugation contribute to the yellow Colours of early-spring flowering crocus Tepals. PLoS One 8(9):e71946. https://doi.org/10.1371/JOURNAL.PONE.0071946

Moshiri M, Vahabzadeh M, Hosseinzadeh H (2015) Clinical applications of saffron (*Crocus sativus*) and its constituents: a review. Drug Res 65(06):287–295

Mykhailenko O, Kovalyov V, Goryacha O, Ivanauskas L, Georgiyants V (2019) Biologically active compounds and pharmacological activities of species of the genus Crocus: a review. Phytochemistry 162:56–89. https://doi.org/10.1016/J.PHYTOCHEM.2019.02.004

Mzabri I, Addi M, Berrichi A (2019) Traditional and modern uses of saffron (Crocus sativus). Cosmetics 6(4):63

Mzabri I, Charif K, Rimani M, Kouddane N, Boukroute A, Berrichi A (2021) History, biology, and culture of *Crocus sativus*: overview and perspectives. Arab J Chem Environ Res 8:1–28. www.mocedes.org

Navarrete A, Herrero M, Martín A, Cocero MJ, Ibáñez E (2011) Valorization of solid wastes from essential oil industry. J Food Eng 104(2):196–201. https://doi.org/10.1016/J.JFOODENG.2010.10.033

Nieto MG (2013) Incorporation of by-products of rosemary and thyme in the diet of ewes: effect on the fatty acid profile of lamb. Eur Food Res Technol 236(2):379–389. https://doi.org/10.1007/S00217-012-1892-7/TABLES/7

Ocaña-Fuentes A, Arranz-Gutiérrez E, Señorans FJ, Reglero G (2010) Supercritical fluid extraction of oregano (*Origanum vulgare*) essentials oils: anti-inflammatory properties based on cytokine response on THP-1 macrophages. Food Chemical Toxicol: An International Journal Published for the British Industrial Biological Research Association 48(6):1568–1575. https://doi.org/10.1016/J.FCT.2010.03.026

Oroojalian F, Kasra-Kermanshahi R, Azizi M, Bassami MR (2010) Phytochemical composition of the essential oils from three Apiaceae species and their antibacterial effects on food-borne pathogens. Food Chem 120(3):765–770

Oussalah M, Caillet S, Salmiéri S, Saucier L, Lacroix M (2004) Antimicrobial and antioxidant effects of milk protein-based film containing essential oils for the preservation of whole beef muscle. J Agric Food Chem 52(18):5598–5605

Papageorgiou G, Botsoglou N, Govaris A, Giannenas I, Iliadis S, Botsoglou E (2003) Effect of dietary oregano oil and α-tocopheryl acetate supplementation on iron-induced lipid oxidation of Turkey breast, thigh, liver and heart tissues. J Anim Physiol Anim Nutr 87(9–10):324–335

Patil SM, Ramu R, Shirahatti PS, Shivamallu C, Amachawadi RG (2021) A systematic review on ethnopharmacology, phytochemistry and pharmacological aspects of *Thymus vulgaris* Linn. Heliyon 7(5):e07054. https://doi.org/10.1016/J.HELIYON.2021.E07054

Paur I, Balstad TR, Kolberg M, Pedersen MK, Austenaa LM, Jacobs DR, Blomhoff R (2010) Extract of oregano, coffee, thyme, clove, and walnuts inhibits NF-κB in monocytes and in transgenic reporter mice. Cancer Prev Res 3(5):653–663. https://doi.org/10.1158/1940-6207.CAPR-09-0089

Pesavento G, Calonico C, Bilia AR, Barnabei M, Calesini F, Addona R, Mencarelli L, Carmagnini L, di Martino MC, Lo Nostro A (2015) Antibacterial activity of Oregano, Rosmarinus and Thymus essential oils against *Staphylococcus aureus* and *Listeria monocytogenes* in beef meatballs. Food Control 54:188–199. https://doi.org/10.1016/J.FOODCONT.2015.01.045

Pham TQ, Cormier F, Farnworth E, Tong VH, Van Calsteren MR (2000) Antioxidant properties of crocin from *Gardenia jasminoides* Ellis and study of the reactions of crocin with linoleic acid and crocin with oxygen. J Agric Food Chem 48(5):1455–1461

Platel K, Srinivasan K (2001) Studies on the influence of dietary spices on food transit time in experimental rats. Nutr Res 21(9):1309–1314

Radwan Nadia L, Hassan RA, Qota EM, Fayek HM (2008) Effect of natural antioxidant on oxidative stability of eggs and productive and reproductive performance of laying hens. Int J Poult Sci 7(2):134–150

Redondo D, Arias E, Oria R, Venturini ME (2017) Thinned stone fruits are a source of polyphenols and antioxidant compounds. J Sci Food Agric 97(3):902–910. https://doi.org/10.1002/JSFA.7813

Rubin B, Manso J, Monticelli H, Bertazza L, Redaelli M, Sensi F, Zorzan M, Scaroni C, Mian C, Iacobone M, Armanini D, Bertolini C, Barollo S, Boscaro M, Pezzani R (2018) Crude extract of

Origanum vulgare L. induced cell death and suppressed MAPK and PI3/Akt signaling pathways in SW13 and H295R cell lines. Nat Prod Res 33(11):1646–1649. https://doi.org/10.1080/14786419.2018.1425846
Rusanov K, Garo E, Rusanova M, Fertig O, Hamburger M, Atanassov I, Butterweck V (2014) Recovery of polyphenols from rose oil distillation wastewater using adsorption resins - a pilot study. Planta Med 80(17):1657–1664. https://doi.org/10.1055/S-0034-1383145
Saha A, Basak BB (2020) Scope of value addition and utilization of residual biomass from medicinal and aromatic plants. Ind Crop Prod 145:111979
Schieber A, Carle R, Berardini N, Mihalev K, Mollov P (2005) Flavonol glycosides from distilled petals of *Rosa damascena* Mill. Zeitschrift Fur Naturforschung - Section C J Biosci 60(5–6): 379–384. https://doi.org/10.1515/ZNC-2005-5-602/MACHINEREADABLECITATION/RIS
Schmidt E (2010) Production of essential oils. In: Başer KH, Buchbauer G (eds) Handbook of essential oils. In: science, technology, and applications. CRC Press, Boca Raton, Fla, pp 83–119
Seow YX, Yeo CR, Chung HL, Yuk HG (2014) Plant essential oils as active antimicrobial agents. Crit Rev Food Sci Nutr 54(5):625–644
Seydim AC, Sarikus G (2006) Antimicrobial activity of whey protein based edible films incorporated with oregano, rosemary and garlic essential oils. Food Res Int 39(5):639–644
Si W, Gong J, Tsao R, Zhou T, Yu H, Poppe C, Johnson R, Du Z (2006) Antimicrobial activity of essential oils and structurally related synthetic food additives towards selected pathogenic and beneficial gut bacteria. J Appl Microbiol 100(2):296–305
Šojić B, Tomović V, Kocić-Tanackov S, Kovačević DB, Putnik P, Mrkonjić Ž, Đurović S, Jokanović M, Ivić M, Škaljac S, Pavlić B (2020) Supercritical extracts of wild thyme (*Thymus serpyllum* L.) by-product as natural antioxidants in ground pork patties. LWT 130:109661. https://doi.org/10.1016/J.LWT.2020.109661
Solimine J, Garo E, Wedler J, Rusanov K, Fertig O, Hamburger M, Atanassov I, Butterweck V (2016) Tyrosinase inhibitory constituents from a polyphenol enriched fraction of rose oil distillation wastewater. Fitoterapia 108:13–19. https://doi.org/10.1016/J.FITOTE.2015.11.012
Srinivasan K (2005) Spices as influencers of body metabolism: an overview of three decades of research. Food Res Int 38(1):77–86
Stamatis G, Kyriazopoulos P, Golegou S, Basayiannis A, Skaltsas S, Skaltsa H (2003) *In vitro* anti-Helicobacter pylori activity of Greek herbal medicines. J Ethnopharmacol 88(2–3):175–179. https://doi.org/10.1016/S0378-8741(03)00217-4
Suresh D, Srinivasan K (2007) Studies on the *in vitro* absorption of spice principles–curcumin, capsaicin and piperine in rat intestines. Food Chem Toxicol 45(8):1437–1442
Svoboda K, Brooker JD, Zrustova J (2006) Antibacterial and antioxidant properties of essential oils: their potential applications in the food industries. Acta Hort (ISHS) 709:35–44
Tsinas AC, Giannakopoulos CG, Papasteriades A, Alexopoulos C, Mavromatis J, Kyriakis S C (1998) Use of Origanum essential oils as growth promoter in pigs. In: Proceedings of the 15th IPVS Congress (Vol. 221)
Tsitlakidou P, Papachristoforou A, Tasopoulos N, Matzara A, Hatzikamari M, Karamanoli K, Mourtzinos I (2022) Sensory analysis, volatile profiles and antimicrobial properties of *Origanum vulgare* L. essential oils. Flavour Fragr J 37(1):43–51
Turek C, Stintzing FC (2013) Stability of essential oils: a review. Compr Rev Food Sci Food Saf 12(1):40–53
Turrini F, Beruto M, Mela L, Curir P, Triglia G, Boggia R, Zunin P, Monroy F (2021) Ultrasound-assisted extraction of lavender (*Lavandula angustifolia* miller, cultivar rosa) solid by-products remaining after the distillation of the essential oil. Appl Sci (Switzerland) 11(12):5495. https://doi.org/10.3390/APP11125495/S1
Van Huis A, Oonincx DG (2017) The environmental sustainability of insects as food and feed. A review. Agron Sustain Dev 37(5):1–14
Varel VH (2002) Carvacrol and thymol reduce swine waste odor and pathogens: stability of oils. Curr Microbiol 44:38–43

Velioglu YS, Mazza G (1991) Characterization of flavonoids in petals of *Rosa damascena* by HPLC and spectral analysis. J Agric Food Chem 39:467

Vidanarachchi J, Mikkelsen L, Sims I, Iji P, Choct M (2005) Phytobiotics: alternatives to antibiotic growth promoters in monogastric animal feeds. Rec Adv Anim Nutr Austr 15:131–144

Viuda-Martos M, Ruiz-Navajas Y, Fernández-López J, Pérez-Álvarez JA (2007) Antifungal activities of thyme, clove and oregano essential oils. J Food Saf 27(1):91–101

Wang CY, Wang SY, Yin JJ, Parry J, Yu LL (2007) Enhancing antioxidant, antiproliferation, and free radical scavenging activities in strawberries with essential oils. J Agric Food Chem 55(16): 6527–6532

Wedler J, Rusanov K, Atanassov I, Butterweck V (2016a) A polyphenol-enriched fraction of rose oil distillation wastewater inhibits cell proliferation, migration and TNF- α -induced VEGF secretion in human immortalized keratinocytes. Planta Med 82(11–12):1000–1008. https://doi.org/10.1055/s-0042-105158

Wedler J, Weston A, Rausenberger J, Butterweck V (2016b) *In vitro* modulation of inflammatory target gene expression by a polyphenol-enriched fraction of rose oil distillation waste water. Fitoterapia 114:56–62. https://doi.org/10.1016/j.fitote.2016.08.019

Westendarp H, Klaus P, Halle I, Mörlein D, Henning M, Köhler P (2006) Aus dem Institut für Tierernährung und dem Institut für Tierzucht Mariensee Landbauforsch. Völkenrode 3/4(56): 149

Wink M, Schimmer O (1999) Modes of action of defensive secondary metabolites. In: Wink M (ed) Functions of plant secondary metabolites and their exploitation in biotechnology. CRC Press, Boca Raton, FL, pp 17–112

Yen AL (2009) Edible insects: traditional knowledge or western phobia? Entomol Res 39(5): 289–298

Zhang W, Xiao S, Samaraweera H, Lee EJ, Ahn DU (2010) Improving functional value of meat products. Meat Sci 86(1):15–31

Zhao J, Davis LC, Verpoorte R (2005) Elicitor signal transduction leading to production of plant secondary metabolites. Biotechnol Adv 23(4):283–333

Zhu G, Xiao Z, Zhou R, Niu Y, Yi F, Zhu J (2012) The utilization of aromatic plant waste resource. Adv Mater Res 518–523:3561–3565. https://doi.org/10.4028/WWW.SCIENTIFIC.NET/AMR.518-523.3561

The Role of Insects in Novel Sustainable Animal Production Systems

T. Veldkamp, I. Belghit, S. Chatzfiotis, M. Mastoraki, A. J. M. Jansman, G. Radhakrishnan, A. Schiavone, S. Smetana, and L. Gasco

Abstract

Insects are a natural component of animal diets. They contain a high amount of digestible protein and fat and are also rich in micronutrients such as copper, iron, magnesium, selenium and zinc, as well as riboflavin, pantothenic acid and biotin. In addition, insects contain bioactive and immunostimulatory constituents such as lauric acid, antimicrobial peptides and chitin. These nutritional and functional properties make insects a promising feed ingredient to replace conventional feed ingredients and as such sustainability in animal production may be improved. In the European Union (EU), since 2017 eight insect species are authorized for

T. Veldkamp (✉) · A. J. M. Jansman
Wageningen Livestock Research, AH, Wageningen, The Netherlands
e-mail: teun.veldkamp@wur.nl; alfons.jansman@wur.nl

I. Belghit · G. Radhakrishnan
Feed and Nutrition group, Institute of Marine Research, Bergen, Norway
e-mail: Ikram.Belghit@hi.no; Gopika.Radhakrishnan@hi.no

S. Chatzfiotis · M. Mastoraki
Institute of Marine Biology, Biotechnology and Aquaculture, Hellenic Centre for Marine Research, Crete, Greece
e-mail: stavros@hcmr.gr; mmastora@bio.auth.gr

A. Schiavone
Department of Veterinary Sciences, Università degli Studi di Torino, Grugliasco, Italy
e-mail: achille.schiavone@unito.it

S. Smetana
German Institute of Food Technologies (DIL e.V.), Quakenbrück, Germany
e-mail: s.smetana@dil-ev.de

L. Gasco
Department of Agricultural, Forest, and Food Sciences, Università degli Studi di Torino, Grugliasco, Italy
e-mail: laura.gasco@unito.it

G. Arsenos, I. Giannenas (eds.), *Sustainable Use of Feed Additives in Livestock*,
https://doi.org/10.1007/978-3-031-42855-5_7

aquafeed. Recent relaxation of the EU feed ban rules and animal by-products legislation on September 7, 2021, also allowed the use of insect proteins in poultry and pig diets. Among the authorized insect species, some are more promising for feed purposes as they can be theoretically mass produced. Moreover, these species apply the circular economy concept by bioconverting organic substrates, which find minor applications for other purposes. Advances in the development of the European insect industry are often associated with more favourable sustainability potentials of insects, compared to traditional protein sources. For the EU there is a significant overlap in ingredient use in diets for pigs and poultry. Despite multiple studies on economic feasibility, social acceptance and environmental impact, many open questions are left for the industry to deal with. Life Cycle Assessment (LCA), relying on a modular modelling approach to cover the complete spectrum of insect production and processing parameters, is a methodology which can provide viable answers and recommendations on environmental sustainability performance. For different livestock animal species, the current animal production systems, volumes of feed and composition of conventional diets are presented. The digestibility of insect meals and effects of their use on growth performance, product quality and health at different dietary inclusion levels are reviewed. Finally, the contribution of dietary inclusion of insect protein in animal production systems to sustainability is discussed.

Keywords

Insects · Aquaculture · Poultry · Pigs · Sustainability

1 Introduction

The Food and Agriculture Organization (FAO) calculated that increasing global food production by 60% from 2005–2007 to 2050 would be necessary to feed the world's estimated 9.1 billion people (Alexandratos and Bruinsma 2012). Such significant changes can only be effectively accomplished in a circular economy that takes resource constraints into account (Raworth 2017). The European Commission (EC) has adopted an ambitious new Circular Economy Package to assist European businesses and consumers in making the transition to a stronger and more circular economy, in which resources are utilized in a more sustainable way (EC 2018). With long-term goals to decrease landfilling and boost recycling and reuse, this collection of legislative measures on waste (reduction) comprises those. The package also contains an action plan to promote the circular economy at every stage of the value chain, from production to consumption, repair and manufacturing, waste management and feedback of secondary raw materials, in order to close loops in product lifecycles. The by-products of food production can be greatly improved by insects in a productive way. The very effective use of insects in the food chain urgently has to be developed and expanded in order to contribute to the circular economy anticipated by the EC and to create more sustainable, resilient supply chains with high consumer acceptability and appealing business potential. The most potential alternative protein sources for the feed and food markets in Europe are insects. Insects are the "missing link" that can close the loops in the product lifecycle; therefore, they help ensure the sustainability of the food chain (Van Huis et al. 2013). Insects can effectively transform low-grade biomass into important feed and food ingredients, whereas human food production necessitates the direct utilization of high-grade resources. Therefore, insects are increasingly being explored as a new or alternative source for producing protein, either directly for human consumption or indirectly via reconstituted foods (including extracted protein from insects), as well as a good source of protein for feed. Edible insect production offers unique advantages such as insects can be grown on organic side streams, and its conversion is highly efficient (Van Huis et al. 2013), insects emit less greenhouse gases and ammonia than conventional livestock (Halloran et al. 2017; Oonincx et al. 2010; Van Huis and Oonincx 2017), and edible insects contain high-quality protein, amino acids and vitamins for animal and human health (Nowak et al. 2016; Rumpold and Schlüter 2013).

Since the effect of dietary insects protein may differ between animal species, different levels of insect meal inclusion in commercial feed are performed to assess the animal performance, health and product quality in feeding trials with rainbow trout, sea bass, salmon, laying hens, broilers and piglets. Effects of insect products on growth performance will be largely affected by the way of exchange of ingredients in diets and differences in nutritional value for which either or not has been properly corrected for. So the response to a diet is mostly not towards a specific feed ingredient but towards the entire feed. For different livestock animal species, the current animal production systems, volumes of feed and composition of conventional diets are presented for the respective animal species. The digestibility of insect meals and effects of their use as feed ingredient on the growth performance, product quality and health at different dietary inclusion levels are reviewed as well as the contribution of dietary inclusion of insect protein in animal production systems to sustainability is discussed.

As a basis for the animal species targeted in the next paragraphs, the chemical nutritional composition of (common) insect-based feed ingredients is presented in Table 1.

2 Fish Species

2.1 Rainbow Trout

The rainbow trout (*Oncorhynchus mykiss*) is a freshwater carnivorous fish belonging to the Salmonidae family. It originated from North America and is now introduced worldwide (Stanković et al. 2015). The first record of its presence in Europe dates at the end of the nineteenth century (Stanković et al. 2015).

Reproduction naturally occurs during the winter period (from November to March), but in captivity and to obtain eggs all over the year, the reproductive

Table 1 Chemical composition of common insect-based feed ingredients (adapted from Shah et al. (2022))

Insect species	Percentage (% DM)								
	DM	CP	CF	Ash	Ca	P	Mg	K	Na
Black soldier fly larvae[a]	27.40	56.10	23.20	9.85	2.14	1.15	0.39	1.35	0.13
Mealworm larvae[b]	38.10	55.83	25.19	4.84	0.21	1.06	0.30	1.12	0.21
Housefly larvae[c]	29.00	62.19	19.20	7.73	0.49	1.09	0.23	1.27	0.54

DM, dry matter; CP, crude protein; CF, crude fat; Ash; Ca, calcium; P, phosphorus; Mg, magnesium; K, potassium; Na, sodium.

[a] (Cullere et al. 2016; De Marco et al. 2015; Jayanegara et al. 2017; St-Hilaire et al. 2007)

[b] (Gugołek et al. 2020; Marono et al. 2015)

[c] (Hussein et al. 2017; Khan 2018; Makkar et al. 2014)

behaviour can be altered through photoperiod manipulation. To reach sexual maturity, normally 2 and 3 years is needed for males and females, respectively. Each mature female produces about 2000 eggs per kg of somatic mass and fertilization is external. In captivity, artificial reproduction is performed collecting gametes (ova and sperm) by stripping (abdominal massage) (Beirão et al. 2019). Eggs have a diameter of about 0.5 cm. The incubation time is temperature dependent and, at a constant temperature of 10 °C, eggs hatch in about 30 days.

Rainbow trout is commercially farmed in over 90 countries, with a global production of more than 848,000 tonnes. The top producers are Iran, Turkey, Chile, Norway and Peru. The first two account for 42% of the global production (Kamalam et al. 2020; Seafish 2022b). In 2019, the European Union (EU) member states produced 192,450 tonnes of trout (mostly rainbow trout) for a market value of more than €650 million. More than half of the production takes place in France (35,097 tonnes), Italy (34,460 tonnes) and Denmark (30,904 tonnes) (EUFOMA 2021). Considering a feed efficiency of 0.8–1 for the whole cycle, the global feed produced for rainbow trout can be estimated at 0.8–1.06 million tonnes (Kamalam et al. 2020). Applying the same estimation, in the EU, the feed used for this species accounts for 0.19–0.24 million tonnes.

Rainbow trout is a carnivorous fish with a protein requirement, varying from 40 to 50% crude protein. When formulating diets, particular attention is taken not only to reach the recommended dietary levels for essential amino acids (EAAs) but also on the optimal ratio between EAAs and non-EAAs, which is reported to be 57:43 (Kamalam et al. 2020). Covering the requirements and considering the whole EAA contribution of ingredients, a reduction in the dietary crude protein content enables to optimize protein deposition and therefore to avoid nitrogen losses and optimize feed costs.

Extrusion technology is widely used to prepare feeds, and in the past, fishmeal (FM) and fish oil (FO) were the main ingredients of a trout diet. However, to decrease reliance on such finite unsustainable resources, nowadays feeds contain large quantities of plant ingredients (oilseed meals, plant protein concentrates and cereal by-products) and processed animal proteins (PAPs; poultry by-products, meat and bone meals and blood meal) (Gaylord et al. 2017). Paying attention to properly supply of limiting EAAs, vitamins and minerals, nowadays diets without FM or FO are used (Kamalam et al. 2020). However, in particular when using high levels of plant proteins, gastrointestinal tract inflammations are often reported mainly because of unbalanced nutrient profile or the presence of antinutritional factors (Colombo 2020; Venold et al. 2012).

Since some years, there is a growing concern about the food-feed competition (Breewood and Garnett 2020). Moreover, in the EU, restrictions on the use of PAPs apply as a consequence of the bovine spongiform encephalopathy issue (EC 2001, 2022). Innovative and more sustainable ingredients are therefore tested to develop alternative aquafeeds. Among them, the most promising are insects-derived products.

Insect-derived proteins, mainly from black soldier fly (*Hermetia illucens,* HI) and yellow mealworm (*Tenebrio molitor,* TM), have been proven to be efficient in partial or total substitution of FM in rainbow trout feeding. Inclusion levels up to 32% resulted in an increase or no impact on performances parameters (Biasato et al. 2022b; Caimi et al. 2021; Chemello et al. 2021; Józefiak et al. 2019; Kumar et al. 2021; Melenchón et al. 2022; Rema et al. 2019). However, in some cases, reduced performances are reported (Cardinaletti et al. 2019; Dumas et al. 2018). Impaired performances are ascribed to insect meal form (full-fat, defat), nature of the production process or related nutrient imbalances in diets as well as related to fish age (Hua 2021; Weththasinghe et al. 2022).

As for other animal species, the formulation of diets based on digestible values is fundamental to avoid nutritional unbalances that could lead to a poor growth and welfare of fish, as well as environmental issues. So far, only one paper deals with the digestibility of different insect meals as test ingredients in rainbow trout (Gasco et al. 2022). In this paper, authors investigated the apparent digestibility coefficient (ADC) of dry matter, crude protein, ether extract, gross energy, amino acids, and the main fatty acids of one *Hermetia illucens*, one *Alphitobius diaperinus*, and two *Tenebrio molitor* defatted meals. Differences in ADCs of dry matter (from 66 to 89%), crude protein (from 75 to 91%) and gross energy (from 75 to 86%) were reported, with *Alphitobius diaperinus* showing the lowest values. No differences appeared for ether extract digestibility that was above 96% in all treatments. *Alphitobius diaperinus* meal also reported the lowest ADC values for methionine, cysteine and tyrosine. The ADCs of the main fatty acids (C12:0, C14:0, C16:0, C18:1 c9, C18:2 n-6 and C18:3 n-3) were higher than 85% and did not differ significantly in insect meals. Another recent paper investigated the digestibility of black soldier fly meals with different chitin contents (Eggink et al. 2022). Authors reported a decrease in the ADC of dry matter, crude protein, nitrogen-free extract and chitin with the increase in the level of chitin supporting the hypothesis that chitin might act as an antinutrient. In the proximal and distal intestine, an increase in exochitinase (β-*N*-acetylglucosaminidase) activity was reported, with increasing inclusion of chitin suggesting that exochitinase production can be upregulated depending on the dietary chitin intake (Eggink et al. 2022).

When feeds containing insect meals are concerned, digestibility is often reported as high and supports the performance parameters (Biasato et al. 2022b; Caimi et al. 2021; Chemello et al. 2021). However, due to indigestible chitin, in general high levels of inclusion reduce ADC values (Hua 2021; Weththasinghe et al. 2022).

Concerning product quality, and the impact of insect meals on rainbow trout fillet, some trials describe small differences in crude protein or lipid content among treatments while others do not report differences (Borgogno et al. 2017; Dumas et al. 2018; Melenchón et al. 2022; Melenchón et al. 2021).

An important effect to consider is the impact that insect meal inclusion in feeds can have on sensory properties (Borgogno et al. 2017; Sealey et al. 2011). This parameter is fundamental, together with psychological aspects for consumers, in the acceptance of fish fed insect meals (Baldi et al. 2021; Mancuso et al. 2016). In this context, insects perceived as protein sources to produce more sustainable food products are of major importance.

Insect-derived products also receive attention because of bioactive compounds they contain, which seems able to exert positive health effects (Aragão et al. 2022; Gasco et al. 2021; Palomba et al. 2022). Recently, Kumar et al. (2021) reported how inclusion of 8 or 16% of black soldier fly larvae (BSFL) is able to prevent intestinal enteritis caused by soybean meal in a low-FM diet. A positive effect on the stimulation of the immune system based on the gene expression and some innate immunity parameters was also reported when using BSFL oil (Kumar et al. 2021). Moreover, a positive impact on gut microbiota, i.e. in terms of selection of short-chain fatty acids-producing bacteria and reduction of foodborne disease-causing pathogens (Biasato et al. 2022b; Rimoldi et al. 2021; Rimoldi et al. 2019; Terova et al. 2021), and a small increase in villi and enterocyte height (Melenchón et al. 2022) have been shown.

Considering the importance of using new feedstuff to contribute to a sustainable growth of the aquaculture sector supporting not only the fish growth but also welfare, health and high-quality products, further research on the effects of these alternative sources is recommended.

2.2 Sea Bass

European sea bass (*Dicentrarchus labrax*) is a coastal, marine, carnivorous fish, which belongs to the Moronidae family. Its distribution spreads from the Black Sea and the Mediterranean to the north-eastern Atlantic Ocean (Vandeputte et al. 2019). Wild sea bass spawns naturally in the winter, between November and March. However, in aquaculture conditions, the manipulation of the photoperiod and water temperature allows the reproduction to take place throughout the year (Kissil et al. 2000). Sea bass is gonochoristic and can reach sexual maturity at one or three years for males and females, respectively (Kissil et al. 2000). Females can produce 200,000 eggs/kg, the fertilization is external, and eggs hatch in three to five days (Vandeputte et al. 2019).

European sea bass is of major importance for the Mediterranean aquaculture, and 88% of the global 2020 production was produced in Turkey, Greece, Egypt and Spain. Turkey and Greece produced about 67% of the global production (FAO 2022). In the EU, the total production of sea bass in 2020 was 81,404 tonnes valued for €552 million (FAO 2022). Greece produced 50% of the total EU production (40,763 tonnes), followed by Spain with 25% (22,764 tonnes) (FAO 2022). The reported average feed conversion ratio for the whole production cycle is 2:1 (Pulina et al. 2018a; Seafish 2022a). Using this feed conversion ratio, the 2020 production of feed for European sea bass can be estimated to 0.55 million tonnes and 0.16 million tonnes at a global level and the EU, respectively.

European sea bass is a carnivorous fish and has a protein requirement of about 33–60% depending on the life stage. A ratio 50/50 between essential amino acids to nonessential amino acids has been reported for enhanced growth performance and 60/40 for efficient protein and energy utilization (Kousoulaki et al. 2015). Levels of fishmeal up to 70% are used in diet formulations, which use fishmeal as the only protein source (Gasco et al. 2016; Pulina et al. 2018b). However, due to the global initiative to reduce the use of finite resources, the use of fishmeal in the commercial diets has dropped to 15–20% over the last 10 years (Seafish 2022a).

To replace the widely used fishmeal, alternative ingredients must be used and plant proteins are most commonly included in diets of European sea bass (wheat and corn gluten, soybean meal and oilseed meals (Kousoulaki et al. 2015)). However, plant ingredients possess dietary fibre, antinutritional factors, an imbalanced amino acid profile and low phosphorus availability (Delgado and Reyes-Jaquez 2018; Hajra et al. 2013; Jannathulla et al. 2019; Jobling 2016). In European sea bass the replacement of fishmeal by plant proteins has been associated with the disruption of several metabolic and physiological processes (Dias et al. 2005; Geay et al. 2011). A shift to more novel ingredients is therefore needed and insect meals are great candidates for the substitution of fishmeal.

Insect meals that have mostly been studied for the incorporation in diets of European sea bass are the ones originated from *Tenebrio molitor* and *Hermetia illucens*. Inclusion of these insect meals at a level up to 25% did not adversely affect the growth performance of sea bass (Abdel-Tawwab et al. 2020; Gasco et al. 2016; Magalhães et al. 2017; Mastoraki et al. 2020). Nevertheless, a reduced growth performance and/or feed efficiency was observed at higher inclusion levels (Gasco et al. 2016; Reyes et al. 2020). In diets devoid of fishmeal, inclusion of 15% *Hermetia illucens* meal, as a plant protein substitute, acted beneficially and improved the growth performance as well as feed efficiency and nutrient digestibility (Pérez-Pascual et al. 2020).

The documented results on the digestibility of insect meals in European sea bass are very encouraging. Inclusion of 19.5% *Tenebrio molitor*, *Musca domestica* (Mastoraki et al. 2022) or *Hermetia illucens* (Magalhães et al. 2017) meals did not have any effect on nutrient digestibility coefficients. Moreover, dietary incorporation of 25% *Tenebrio molitor* meal or 19.5% *Zophobas morio* meal even improved protein digestibility (Gasco et al. 2016; Mastoraki et al. 2022). Chitin, the *N*-acetylglucosamine polymer of the insect exoskeleton, appears to act as an antinutrient reducing the insect meal digestibility in many reported cases. Reduced digestibility coefficients, observed when fishmeal was substituted by *Hermetia illucens* or *Alphitobius diaperinus* meals (19.5% inclusion), were attributed to the dietary presence of chitin, indigestible fibre and ash (Mastoraki et al. 2022). Additionally, when comparing different insect meals, better apparent digestibility coefficients (ADCs) were reported in *Tenebrio molitor* meal (79% dry matter ADC, 92% protein ADC and 85% energy ADC), followed by *Hermetia illucens* meal (71% dry matter ADC, 89% protein ADC and 83% energy ADC) and lastly locust meal (70% dry matter ADC, 91% protein ADC and 80% energy ADC) due to a higher presence of fibre, chitin and ash in the last two insect meals (Basto et al. 2020).

As regards product quality, the fillet composition of European sea bass was not affected by dietary insect meals (Moutinho et al. 2020; Reyes et al. 2020). In addition, Moutinho et al. (2020) did not report any differences in the skin or fillet colour following a *Hermetia illucens* dietary incorporation. They even observed a reduced fillet pH after three days of refrigeration, which indicates prolonged shelf-life of the product (Moutinho et al. 2020). A consequence that may occur from dietary insect meals is the alteration of fillet fatty acid profile. When full-fat insect meals are used, fish lose some amount of their essential fatty acids (omega-3 polyunsaturated fatty acids, eicosapentaenoic acid-EPA and docosahexaenoic acid-DHA) due to the lower fish oil dietary inclusion (Gasco et al. 2016; Mastoraki et al. 2020). Hence, the use of defatted *Hermetia illucens* did not reduce the omega-3, EPA and DHA content of the fillet or the whole body (Mastoraki et al. 2020; Moutinho et al. 2021).

Despite adverse effects of the chitin content on nutrient digestibility, the major advantage of chitin and its derivative, chitosan, is their antioxidant, anti-inflammatory and bacteriostatic activity (Abdel-Ghany and Salem 2020). Indeed, even a low level of inclusion (7.25%) of *Hermetia illucens* meal in diets of European sea bass strengthened the phagocytic capacity of leukocytes, enhanced the immune response and increased the survival after a *Vibrio alginolyticus* challenge (Abdel-Latif et al. 2021). In addition, *Tenebrio molitor* meal (24.8% inclusion) acted as an immunostimulant, resulted in increased serum lysozyme and trypsin-inhibition activities and strengthened the antiparasitic defence of European sea bass (Henry et al. 2018). Regarding the antioxidant properties of insect meals, incorporation of *Hermetia illucens* (up to 19.5%) enhanced the activity of antioxidant enzymes (Abdel-Latif et al. 2021) and reduced lipid peroxidation (Moutinho et al. 2020). However, presumably only small doses of insect meals or insect chitin can act beneficially because it has been found that higher levels of insect meals (50% *Tenebrio molitor*) can lead to cellular damage due to oxidative stress, which triggers hepatic autophagy (Mente et al. 2022). The intolerance of European sea bass to high levels of insect meals is also evident in the gut microbiome, in which incorporation of 19.5% of insect meals from *Tenebrio molitor*, *Hermetia illucens* or *Musca domestica* laid the foundation for beneficial bacterial species to grow (Panteli et al. 2021), but higher inclusions (50% *Tenebrio molitor*) resulted in an abundance of

Firmicutes and Proteobacteria in the gut bacterial communities (Antonopoulou et al. 2019).

A partial fishmeal substitution will allow the aquaculture sector to continue to grow more sustainably. In addition, more interest in insect meals will drive the insect farming sector to achieve industrialization and large-scale production to cover the needs for aquaculture. Further research is needed to assess the effects of insect meals on fish growth as well as fish welfare and overall health.

2.3 Atlantic Salmon

Atlantic salmon (*Salmo salar*) are anadromous fish, found throughout the North Atlantic Ocean, belonging to family Salmonidae. The whole production of Atlantic salmon lasts for about 3 years. In autumn, selected broodstocks are stripped for eggs and fertilized in freshwater facilities. A mature female can lay up to 1500–1800 eggs per kg of body weight. After 10–16 months, smolt fish are grown to approximately 80–120 g and are then transferred to open seawater cages, where they grow till a harvestable size of 4–5 kg over a period of 12–24 months.

Atlantic salmon are typically fed on high-energy diets, containing large amounts of protein and lipid (~40 and ~ 20% of the diet, respectively), and low levels of carbohydrates (NRC 2011). Salmon are among the most efficient animals since the feed conversion ratio is low and ranges from 1.15 to 1.5 kg of feed per 1 kg salmon body weight. Thus, to cover 2.6 metric tons of global salmon production, the industry needs around 3–4 metric tons of feed per year. Traditionally, major sources of protein and lipid in salmon feeds have been fishmeal and fish oil obtained from wild-caught fish. The use of these marine ingredients in aquafeed has been gradually reduced and replaced with plant-based ingredients such as soy, wheat, corn, beans, peas, sunflower and rapeseed oil. For example, in 2020, the Norwegian farmed salmon, which dominates the global production, produced a total of 1.5 metric tons using 2 metric tons of ingredients (wet weight), consisting of 12.1% fishmeal, 10.3% fish oil, 40.5% vegetable protein sources, 20.1% vegetable oils, 12.5% carbohydrate sources and 4.1% microingredients (Aas et al. 2022). However, the choice of ingredients and formulation of fish diets are critical due to the rapid expansion of aquaculture. Therefore, there are continuous improvements in this sector to develop alternative ingredients since the replacement of marine ingredients with plant-based ingredients shifted resource demand from oceans to lands, adding pressure to the environment and food production systems destined to human consumption.

Despite the approval of eight insect species in aquafeed (EC 2017), black soldier fly larvae (BSFL) meal is the only insect meal used in diets of Atlantic salmon so far, probably due to upscaling production and cost. Dietary inclusion of processed BSFL meal and/or oil up to 60% in salmon diets has shown improved (Weththasinghe et al. 2021) or similar growth performance as the control group (Belghit et al. 2019a; Belghit et al. 2018; Leeper et al. 2022; Weththasinghe et al. 2021). However, two feeding trials showed a decrease in growth rate (25–55% of relative specific growth) compared to the control group with an inclusion between 25 and 30% in salmon diets

(Fisher et al. 2020; Lock et al. 2016). Presumably some of this difference is due to the source and/or processing of BSFL meal (Hua 2021; Liland et al. 2021; Weththasinghe et al. 2022).

Until today, there are limited data on nutrient apparent digestibility coefficients (ADCs) of BSFL meal in Atlantic salmon. Two studies only reported the ADC of BSFL meal, which varied between 90–100 and 96–100% for protein and lipid digestibility, respectively (Fisher et al. 2020; Radhakrishnan et al. 2022). Furthermore, digestibility values of diets containing black soldier fly (BSF) larvae meal were generally above 80%, ranging from 82 to 98% and 84 to 99% for protein and lipid digestibility, respectively (Belghit et al. 2019a; Belghit et al. 2018; Belghit et al. 2019b; Weththasinghe et al. 2022; Weththasinghe et al. 2021). Also, higher ADCs were reported for amino acids, such as arginine (91%), suggesting that BSF larvae meal is a valuable source of this essential amino acid (AA) compared to plant protein ingredients, where arginine is a limiting AA (Belghit et al. 2018). Similarly, ADCs and fatty acids were highly digestible and were efficiently utilised by Atlantic salmon (ADC >95%) (Belghit et al. 2019b).

Dietary inclusion of BSFL meal at 15% did not impair the physicochemical properties of the fillet from Atlantic salmon post-smolt (Bruni et al. 2020). The results demonstrated that moisture, ash and total lipids were not affected, but the crude protein was found to be higher when 10% of BSFL meal was included in the diet of salmon. Furthermore, Atlantic salmon fed BSFL meal-based diet did not alter the flavour, odour or texture in the fillet (Lock et al. 2018). However, in the study conducted by (Belghit et al. 2019a), it was found that the baked fillet produced rancid flavour when fishmeal was completely replaced with BSFL meal. Also, the intensity of the colour of the cooked fillet with 10% of BSFL meal was lower in post-smolt salmon.

Insects contain bioactive substances such as chitin, lauric acid and antimicrobial peptides. These compounds improve the health of Atlantic salmon pre-smolts when fed with insect-based diets. It was shown that inclusion of 20% full-fat BSFL meal in Atlantic salmon pre-smolts positively modulates the gut microbiota enhancing better response against invading pathogens, and improves the overall immune response (Weththasinghe et al. 2022). Also Leeper et al. (2022) in their study revealed that inclusion of 10% BSF larvae meal in salmon diets had a potentially prebiotic impact on the gut microbiome community alpha diversity compared with conventional protein sources alone. In the study conducted by Li et al. (2020), it was found that 15% of defatted BSFL meal reduced enterocyte steatosis in the proximal intestine of seawater-phase Atlantic salmon. Moreover, partial or complete replacement of fishmeal with BSFL meal downregulated the expression of oxidative stress genes from leukocytes of Atlantic salmon head kidney under a bacterial challenge (Stenberg et al. 2019).

The aquaculture feed industry needs new sources of feed ingredients, and insects are on their way to become one of these novel sources of proteins and fats. When overcoming the obstacle of production quantities, both insect protein and lipid should have a natural place in aquafeeds determined by the balance between price and nutritional value. In general, dietary inclusion of BSFL meal shows promising effects on the growth performance, nutrient digestibility and general health in Atlantic salmon. However, most of these studies were performed as small-scale feeding trials with mostly juvenile salmon. This was mainly due to limitations in cost and availability in producing large quantities of insect meal. Therefore, further research is needed on a large scale, to study inclusion of insect-derived proteins in salmon diets under semicommercial conditions.

3 Poultry

The animals that fall in this category are domesticated fowls and are used for the production of both meat and eggs. The term "poultry" regroups different species such as chicken, turkey, duck, goose, ostrich, pheasant, quail, guinea fowl and peafowl. Chickens are the largest part of this group of animals worldwide and represent the major source of protein (meat and egg) for humans due to the absence of economic, social and religious obstacles. Over the last century, intensive poultry farming selected two different chicken strains: the laying hen, specialized in egg production, and the broiler chicken (or broiler), characterized by fast growth and specialized in meat production.

The laying hen and broiler chicken (*Gallus gallus domesticus*) originate from the red junglefowl (*Gallus gallus*). Domestication probably occurred 7000–10,000 years ago in Southeast Asia and Oceania. Chickens (*Gallus domesticus*) have a long history and exhibit differences when one traces their ancestry. It was used throughout the Roman Empire, which is associated with the development of breeds, particularly for egg production (Elson et al. 2011). In the first part of the nineteenth century, new varieties of chicken were created for exhibitions across Europe (Yamada 1988). Some breeds have contributed to those used in current chicken production. By carefully combining the most productive strains of pure-bred layers, hybrid chickens have been created in the latter part of the twentieth century. Hybrid laying hens are specifically designed to be docile, not go broody, and to produce a lot of eggs in their first year. Some traits of the original parent stock may still be present in some hybrid hens. The best of them can produce an egg almost every day. Due to improved health and disease prevention, nutrition, genetics, and flock management, each laying hen produces almost 300 eggs per year (Mavromati et al. 2017). The egg weight varies between 50 and 70 g.

At the same time, the poultry industry started an intense selection activity in order to obtain fast-growing chickens, suitable for intensive breeding. Broiler chicken differs considerably from the production of the so-called traditional chicken, both from a morphophysiological and production point of view. Broilers have been genetically selected to gain weight exponentially, to have high consumption efficiency, an optimal growth rate and considerable breast and thigh yields (Wang et al. 2012). Separate rearing of broiler breeder males and females is the preferred method to control the growth of the skeletal structure and body weight, as it allows animals to grow at specific and separate body weights throughout life (Garmon and Hogan 2010). Broilers have become heavier than ever before, and this necessity is dictated by the need to shorten breeding time to reach a certain commercial weight. Nowadays, a broiler is slaughtered at the age of 42 days at an average weight of 2.650 kg. In 1976, the weight reached at the same age was around 1.050 kg (Zuidhof et al. 2014).

3.1 Poultry Feeds: Production Volume, Ingredients and Nutrient Composition

In the next decades, a continuous increase in poultry product (meat and eggs) demand for human consumption is expected (OECD 2022). Over 90% of poultry production comes from specialized broiler or laying hen systems where diets are mainly constituted by cereal grains and oilseed cakes, having a high environmental impact (in terms of land, energy and water uses) (Mottet and Tempio 2017).

Over the past decade, global egg production has grown impressively. According to data from the FAO (2020), global hen egg production reached 77 million tonnes in 2018, a 50% increase from the 2000 level. Asia is by far the main producing region, accounting for 60% of the global production in 2018, followed by the Americas (21%), Europe (14%), Africa (4%) and Oceania (0.4%). European egg production (in billion eggs) according to Statista (2021) is dominated by six countries: France (14.4), Italy (13.4), Germany (13.0), Spain (12.3), the United Kingdom (11.3) and the Netherlands (10.5).

In 2018, the world's meat production was close to 350 million tons and represented by pigs (35%), chickens (33%), cattle (20%) and others (12%). The United States was the first region in the world for chicken meat production, followed by China and Brazil (FAO 2020). Growth in global consumption of meat proteins over the next decade is projected to increase by 14% by 2030 compared to the base period average of 2018–2020, driven largely by income and population growth. Protein availability from beef, pork, poultry, and sheep meat is projected to grow 5.9%, 13.1%, 17.8% and 15.7%, respectively, by 2030. Meat consumption has been shifting towards poultry. In lower-income developing countries this reflects the lower price of poultry as compared to other meats, while in high-income countries this indicates an increased preference for white meats, which are more convenient to prepare and perceived as a healthier food choice. Globally, poultry meat is expected to represent 41% of all the protein from meat sources in 2030 (OECD 2022).

The global feed production for broilers according to Alltech (2022) was 350,921 million metric tons and the feed production in the EU for broilers was 54,727 million metric tons in 2021. The global feed production for layers according to Alltech (2022) was 158,789 million metric tons in 2021 and the feed production for laying hens in the EU was estimated at 22,200 million metric tons in 2021.

Feed for both broilers and laying hens is formulated based on least cost formulation matching the nutritional requirements. On average the crude protein concentration in layer feed is about 16.0–16.5%, while in broiler chickens this is about 20.0–23.0%. Poultry feed, however, is most formulated based on standardized ileal digestible (SID) amino acids to reach maximum production results (egg production and feed conversion). Requirements are based on a laying hen with an egg production rate of 95%, producing an egg with a weight of 60 g. Based on a daily feed intake of 115 and 128 g per laying hen, corresponding with body weights of 1.5 and 2.0 kg, respectively, the estimated requirements for SID amino acids (in g/kg) in laying hens are: lysine (6.9–6.2), methionine (3.8–3.4), methionine + cysteine (6.1–5.5), threonine (4.8–4.3), tryptophan (1.5–1.3), valine (5.4–4.8) and isoleucine (5.5–4.9) (CVB 2018). In broilers, on average, the daily feed intake varies from 10–20 g at day 1 to 150–160 g at day 30 and 230–240 g at day 55. The estimated requirements for SID amino acids (in g/kg) in broilers vary depending on the age, being the highest for the starter period and the lowest for the finishing period. The lysine requirement decreases from 12.0–12.3 g/kg during the growing period to 9.7–10.3 g/kg for the finishing period (CVB 2018). Similarly, other essential amino acids follow the same pattern over age, as methionine (from 5.3 to 4.3 g/kg), methionine + cysteine (from 9.6 to 7.8 g/kg), threonine (from 8.4 to 6.6 g/kg), tryptophan (from 2.0 to 1.6 g/kg), valine (from 9.6 to 7.6 g/kg) and isoleucine (from 8.4 to 6.8 g/kg) (Aviagen 2022).

The demand for eggs and poultry meat is anticipated to grow quickly on a global scale. As a result, there will be a rise in the demand for vegetable proteins for use in poultry feed, such as soybean meal (Alexandratos and Bruinsma 2012). In Europe, self-sufficiency for soybean meal is only 5% (EU 2017). Due to its high dependence on this crucial protein source for poultry diets and its low level of self-sufficiency, the EU livestock sector is susceptible to fluctuations in price and trade distortions. In addition, there are concerns on deforestation of tropical rainforest to provide sufficient arable land for soybean cultivation (Van Gelder and Kuepper 2012; WNF 2011). Biodiversity loss (the proportion of extinct species) grows as a result of the conversion of natural ecosystems into agricultural, even if the current situation has already surpassed the planned boundary by more than ten times (Rockström et al. 2009). Large-scale soybean cultivation may also worsen water and soil contamination, put small farmers out of business, and displace native populations (WNF 2011). This is the driving force to discover substitutes for imported soybean meal in poultry diets. Additionally, the environmental impact of poultry production is influenced by the emission of nutrients like nitrogen and phosphorus through poultry manure (Kumari et al. 2016).

A lower inclusion of soybean protein in both laying hen and broiler diets is one way to reduce the use and import of soybeans and soybean meal. This can be done by partially replacing soybean products with other protein commodities such as peas, rapeseed/rapeseed meal, sunflower meal, corn gluten meal and/or potato protein or novel proteins such as insect protein. In this context, the use of insects as substitutes for traditional protein sources in feed is recognized as one of the potential solutions, helping in the achievement of a more sustainable production system. Recently, the European Union has authorized seven insect species for poultry feed (Commission Regulation (EU) 2021/1372 of 17 August 2021): *Hermetia illucens* (black soldier fly), *Musca domestica* (common housefly), *Tenebrio molitor* (yellow mealworm), *Alphitobius diaperinus* (lesser mealworm), *Acheta domesticus* (house cricket), *Gryllodes sigillatus* (banded cricket), and *Gryllus assimilis* (field cricket). In poultry nutrition the black soldier fly and yellow mealworm larvae have been mostly studied. Dried larvae meal of these species can be provided as is or it can be further processed to obtain oil-extracted meal (partially or totally defatted) (Dörper et al. 2021). In the last decades, a large number of studies have been published regarding the effects of inclusion of insect meal in poultry diets, which were recently reviewed by Dörper et al. (2021) and Elahi et al. (2022) and focused mainly on black soldier fly, mealworm and housefly larva meal, considering their effects on growth performance and product (meat and egg) quality. Studies showed variable and sometimes conflicting results but, as a whole, the inclusion level of up to 15% for BSF meal and housefly meal is considered to be applicable.

3.2 Use of Insects in Broiler Chicken Diets

Only few studies have been performed on the digestibility of insects-derived products in poultry species, being broiler chicken, the most animal model studied. The available studies focused on either full-fat or partially defatted insect meal, being BSFL meal the most studied ingredient, followed by mealworm meal. The dry matter digestibility for BSFL meal has been reported to be between 59% (De Marco et al. 2015) and 63% (Schiavone et al. 2017b). The crude protein digestibility for BSFL meal has been reported to be between 51% (De Marco et al. 2015) and 62% (Schiavone et al. 2017b), being lysine and methionine digestibility around 80% (Schiavone et al. 2017a). The BSFL meal lipid digestibility is always more than 90% (De Marco et al. 2015; Schiavone et al. 2017b). The apparent metabolizable energy (AME) and nitrogen-corrected apparent metabolizable energy (AME_n) values vary depending on the amount of BSFL meal lipid content, being 17.4 and 16.6 MJ/kg for full-fat meal (De Marco et al. 2015), and 16.2 and 9.9 MJ/kg for partially defatted meal (Schiavone et al. 2017b), respectively. De Marco et al. (2015) reported the nutritional value of full-fat mealworm meal including the following data: dry matter digestibility (60%), crude protein digestibility (60%), lysine and methionine digestibility (85% and 80%, respectively), lipid digestibility (88%), AME and AME_n (16.9 MJ/kg and 16.0 MJ/kg, respectively). Matin et al. (2021) described the AME_n in a rooster assay, being 14.9 and 21.8 MJ/kg for full-fat mealworm and partially defatted BSLF meal, respectively. When mealworm meal or BSFL meal is included in complete diets, the digestibility of the whole diet may be negatively affected, even if at a low extent. Some authors reported reduced crude protein digestibility in the starter period of Muscovy duck (Gariglio et al. 2019), while this negative effect was not observed in quail (Cullere et al. 2016) and adult Muscovy duck (Gariglio et al. 2019).

A substantial number of studies showed that black soldier fly meal (BSFM) and mealworm are valuable insect protein sources to improve growth performance, carcass composition, and meat quality in broiler chickens, and those aspects have been recently reviewed by Elahi et al. (2022). Briefly, insect meal can replace conventional protein sources, like soybean meal, without affecting broiler performance, or, in some cases, with slight negative effect when included in the diet over

10–15%. The meat fatty acid composition is linearly influenced by the fatty acid composition of insect meal (Elahi et al. 2022) or insect fat (Schiavone et al. 2017a; Schiavone et al. 2019), being the saturated fatty acid the most represented group.

The broiler welfare cannot be preserved without considering their gut health, which is a complex, multifactorial concept that takes into account several intestinal features (such as the histology traits, the microbiota, and the mucin dynamics). The gut health of broilers may be influenced by both intrinsic (i.e. age, sex, breed) and extrinsic (i.e. diet, environment) factors, thus, in turn, influencing the growth performance of birds. Dietary insect meal inclusion in broiler diets produced conflicting results. In the majority of the studies, the histology traits, in terms of villus height and crypt depth, have been preserved as well as the proximodistal gradient from duodenum to ileum. The caecal microbiota was positively influenced in slow-growing free-range chickens fed a diet in which *Tenebrio molitor* was included at 7.5% (Biasato et al. 2018). However *Tenebrio molitor* meal inclusion in broiler chicken produced positive effects when included at low levels (5%), while negative effects were displayed at higher-level inclusion (10% and 15%) in terms of partial alteration of the physiological microbial population and reduction of the potential beneficial bacteria (Biasato et al. 2019, 2020b). A similar trend was observed for BSFL meal supplementation in broiler diets. In fact, dietary BSFL meal utilization at low inclusion levels (i.e. 5%) positively influenced either the caecal microbiota or the gut mucin dynamics in terms of selection of potentially beneficial bacteria and increase in villi mucins. However, high inclusion levels (in particular 15%) exerted a negative influence in terms of partial reduction of microbial complexity, reduction of potentially beneficial bacteria, selection of bacteria with mucolytic activity, and decrease in villi mucins (Biasato et al. 2020a).

The use of whole insect's larvae (both live and dried) in poultry has been recently proposed as a dietary supplementation to promote bird's welfare. Indeed, live insects can be considered as an environmental enrichment, as they are able to stimulate the curiosity of poultry as a result of their motility (Star et al. 2020). Biasato et al. (2022a) have recently reported an improvement in the welfare of broilers fed diets supplemented with 5% yellow mealworm (*Tenebrio molitor*) or 5% live black soldier fly larvae (*Hermetia illucens*) (BSFL), as well as an increase in physical activity. Furthermore, insects are a natural feed source for poultry, and are capable of promoting the foraging behaviour of birds, as they capture their attention and reduce the aggressiveness of chickens towards conspecifics (Clara et al. 2009; Star et al. 2020). Nevertheless, the provision of live larvae to broilers has been poorly investigated. The few published papers about the use of live insect larvae in poultry nutrition have mainly focused on the provision of live black soldier fly larvae to turkeys (Veldkamp and Van Niekerk 2019), broiler chickens (Bellezza Oddon et al. 2021) and laying hens (Ipema et al. 2020; Tahamtani et al. 2021), and have evaluated the effects on the growth, health status, and slaughtering performance of birds. Veldkamp and Van Niekerk (2019) revealed a higher daily feed intake, and body weight gain in turkeys fed 10% live BSFL (on an as-fed basis) than control groups. The physical exertion of broilers, as well as their hock burn score, can be improved by the administration of live larvae (supplemented as feeds of 5% or 10% live BSFL, twice or four times/day), as reported by Ipema et al. (2020). Dietary supplementation with live yellow mealworm (at 5% in the feed) further improved the feed conversion ratio of broilers in a trial conducted by Bellezza Oddon et al. (2021). In addition to these parameters, the recording of the larva consumption time could be useful, and this method has already been experimented by Veldkamp and Van Niekerk (2019) and Bellezza Oddon et al. (2021) to evaluate the appreciation of larvae by birds, as well as to monitor their level of confidence in larva consumption over time.

3.3 Use of Insects in Laying Hen Diets

Literature on nutrient digestibility of insect-based products in laying hen diets is scarce. It was found that chitin can hinder the digestion of protein by the presence of chitin-protein matrixes in insect-based products. Dry matter, organic matter and crude protein digestibility decreased, as the percentage of BSFL meal in hen diet increased, due to chitin's negative effect, which was more remarkable in the diet, which was 50% of the protein content replaced by BSFL meal (inclusion level 14.6%) (Bovera et al. 2018). In another study also a reduced nutrient digestibility was found by inclusion of BSFM. Apparent ileal digestibility of protein, but also organic matter digestibility, was reduced due to inclusion of 17% defatted BSFL meal (Cutrignelli et al. 2018).

Variable outcomes in production performance were noted in studies that were conducted on the effects of black soldier fly larvae (BSFL) meal in laying hens. The growth performance, apparent digestibility of crude protein and crude fat were improved and immunoglobulin A and glutathione peroxidase in blood plasma of Hy-Line Brown laying hens fed on 3% BSFL meal were higher than in control diets without insect meal (Chu et al. 2020). Egg weight, albumin weight, egg shell thickness, albumin height and plasma calcium were all enhanced in laying hens (Julia; 168 days old) fed a diet containing 10% BSFL meal, and additionally, a diet containing 10% BSFL meal significantly increased the egg yolk colour score (Kawasaki et al. 2019). Egg weight, egg mass, nitrogen and metabolizable energy were all increased in Lohman brown classic laying hens fed a diet comprising 15% defatted BSFL meal and fat (Heuel et al. 2021). Xuefeng black-bone laying hens fed a diet with 3% BSFL meal increased egg weight, Haugh unit, egg shell weight, yolk C14:00, C17:00, C20:2 fatty acids, yolk amino acids (glutamic acid, methionine, phenylalanine and leucine), plasma total superoxide dismutase and plasma avian influenza virus antibody, but decreased egg shell thickness and plasma interleukin-2 (Liu et al. 2021). In general terms, this means that most product quality and immunity parameters were improved but also some quality and immunity parameters were adversely affected by inclusion of BSFL meal, which needs to be studied further. While 5% BSFL meal inclusion in the diet increased body weight (23 weeks) and shank breaking strength but decreased hen day egg production, egg weight, egg mass and feed intake, 7.5% BSFL meal inclusion in the diet increased body weight (27 weeks), yolk colour score and shell thickness in Shaver white leghorn hens (Mwaniki et al. 2018). Lohman Brown Classic laying hens fed a diet containing 17%

BSFL meal, to completely replace soybean meal, resulted in poor growth and laying rate percentage, and decreased blood lipids, blood chloride and blood creatine. However, percentage of small, medium and extra-large size eggs, blood globulin and blood calcium were increased by the dietary inclusion of BSFL meal (Marono et al. 2017). These results were most related to the lower feed intake of hens fed diets with BSFL meal. In comparison to a control diet, the performance of laying hens fed diets containing 12 or 24% BSFL meal to replace 50 or 100% soybean meal was unaffected (egg production, feed consumption). Faecal dry matter was increased in the diet in which soybean meal was substituted by 24% BSFL meal (Maurer et al. 2016). Similar productive performance with no negative effects on the nutritional and physical quality of eggs was observed when laying hens were fed microwave-dried BSFL meal at two different substitution levels (2% and 4%) of soybean meal (Park et al. 2021). Feed intake, weight gain, Haugh unit and hatchability were not affected by dietary treatments at 1 and 5% inclusion of BSFL meal in Arabic strain laying hens. However, there was a significant improvement in hen day egg production and hen house egg production due to dietary treatments of BSFL meal. In addition, a significant improvement was observed in appearance, texture, taste and acceptance of eggs of hens fed 5% BSFL meal. The odour was not affected by dietary treatments (Al-Qazzaz et al. 2016). Egg shell percentage and thickness were significantly reduced in eggs laid by Hy-line Brown hens fed a diet with 50% substitution of vegetable proteins by BSFL meal (Secci et al. 2020). Commercial Single Comb White Leghorn hens were fed isocaloric, isonitrogenous diets with three treatment levels of BSFL meal (8, 16 and 24%). Dietary amino acid balance at the highest level of inclusion (24% BSFL meal) indicated that arginine and tryptophan were limiting and average daily feed intake, body weight and egg production were reduced. However, 8 and 16% BSFL meal levels had no negative impact on performance and were not significantly different from controls. Yolk colour was again higher among BSFL meal treatments compared to the control. The authors concluded that BSFL meal can be used as dietary energy, protein and amino acids for hen maintenance, egg production and yolk coloration, although there may be upper limits of dietary inclusion (Patterson et al. 2021).

There is little information on the effects of mealworm meal (*Tenebrio molitor*; TM) on the relative organ weights, caecum microbiota, ileum morphology and digesta viscosity in laying hens. All of these factors may have an impact on an animal's performance, including egg production and quality. The use of 2 to 5% of TM meal in a hen's diet may be used as a sustainable and applicable protein feed (Stastnik et al. 2021). Hy-Line Brown laying hens were fed diets without TM meal and three inclusion levels of 1, 2, and 3% TM meal. Dietary TM meal supplementation induced no improvement in egg production but positively affected the egg quality by improving the yolk colour and increasing the yolk content of linolenic acid and unsaturated fatty acid (Ko et al. 2020).

Next to BSFL meal, black soldier fly larvae have also been fed as dried larvae or as live larvae to reduce production costs or to improve the welfare of laying hens. ISA brown laying hens were fed ad libitum indoors with a wheat-soy-based diet formulated according to breed requirements. Black soldier fly-fed hens were offered dried BSF larvae ad libitum on the range. Black soldier fly-fed hens consumed on average 15 ± 1.7 g BSF larvae/hen per day. There were no differences between BSF and control hens for any of the performance parameters obtained. Egg weight, shell weight and shell thickness of eggs from BSF hens were significantly lower than those of eggs from control hens. The authors conclude that feed formulation should be adjusted for the intake of the choice fed source (Ruhnke et al. 2018). Replacing soya with live black soldier fly larvae and local protein sources had no adverse effect on production performance and egg quality. Additionally, random and steady provision of larvae to older laying hens with intact beaks had a positive effect on feather condition (Star et al. 2020). Voluntary intake of live BSF larvae was studied in Bovan White laying hens provided with 0, 10 and 20% or ad libitum portions of live larvae from 18 to 30 weeks of age. Ad libitum-fed hens, consumed 163 ± 41 g larvae/hen/day, consumed less compound feed and gained more weight than all other treatments. They also had an overall higher consumption of protein, fat and energy. There was no effect of larvae consumption on egg production, egg weight, shell thickness, shell breaking strength or Haugh unit (Tahamtani et al. 2021).

In general, it can be stated that at high inclusion levels (>10%) of insect-based products, the nutrient digestibility of a diet can be adversely affected by chitin present in the exoskeleton of insects. However, partial replacement of soybean meal by insect-based products in feed with inclusion levels between 10 and 15% seems to be applicable for poultry and therefore these products can be used as a valuable replacement for conventional protein sources. Research on providing live larvae to laying hens as welfare enhancement is currently ongoing.

4 Pigs

Pig production in many parts of the world is an important contributor to the total volume of animal sourced food (ASF) produced. China, the EU and the United States are the top three producers. The world volume, currently being 112 million metric tons per year, is expected to grow in the coming decade (Shahbandeh 2022). For feeding pigs the sector relies on diets containing a variety of ingredients ranging from cereals, cereal by-products, oil seeds and their extracted meals, legume seeds and fishmeal to a variety of mainly other by-products and rest products from the food processing industry. EU compound feed production (EU27) for farmed animals in 2021 was estimated at 150 million metric tons, of which one third was pig feed (FEFAC 2022). To decrease the feed-to-food competition and increase the application of more circular feeding concepts, increasing amounts of rest and by-products are considered feed ingredients for pigs. To cover the animal's requirements for adequate and efficient growth, the focus is on matching nutrient requirements, including requirements for protein and essential amino acids, of different categories of pigs (piglets, growing finishing pigs and gestating and lactating sows) by using precision feeding concepts. In search for novel and more sustainable protein sources as feed ingredients for pigs, interest in using insects as protein and energy sources has grown with particular focus on mealworm (*Tenebrio molitor* L.) and black soldier fly larvae (*Hermetia illucens*), either or not after processing to separate the fat and protein fraction for specific applications. These sources have a crude protein content of 46–54 and 34–42%, respectively, and a fat content of 25–36 and 25–58% both on dry matter basis, respectively (Hawkey et al. 2021). Partly defatting by chemical or mechanical extraction increases the protein content and reduces the fat and energy content. As protein and amino acid requirements are relatively high in young piglets, most focus in research has been given to using insects in diets for this category of pigs.

Various studies evaluated the effects of inclusion of insect products in diets for piglets on the growth performance. In such studies insect products are included at the expense of other protein and energy sources using a 1:1 exchange approach or via an exchange in which diets are fully nutritionally balanced on contents of energy and ileal digestible essential amino acids. In such studies, assumed rather than in vivo determined nutritional values for the insect-derived ingredients are used. The former might imply a risk for using diets, which are not entirely nutritionally balanced, which could affect the performance response of animals and conclusions drawn from such studies. Protein and energy digestibility of insects-based products could vary rather largely depending on the species, nature of the substrate on which insects were grown and the processing (grinding, defatting, heat and/or enzyme treatment) and fineness of grinding applied. Hawkey et al. (2021) and Hong and Kim (2022) summarized studies with piglets in which the growth performance of pigs fed diets with insect-based ingredients was evaluated. Full soybean replacement with either full-fat or partially defatted black soldier fly prepupae showed no differences in body weight gain and feed intake (Spranghers et al. 2018). Inclusion of partially defatted black soldier fly larvae meal, up to 10% of the diet, did not affect growth performance in weaned piglets (Biasato et al. 2019) while partial replacement (50%) of dried plasma in the diet (5% during phase 1 and 2.5% during phase 2), with black soldier fly prepupae improved performance in early-weaned pigs in nonadditionally amino acid-supplemented diets but tended to reduce performance when diets were balanced for essential amino acids and at 100% replacement (Newton et al. 2005). Similarly, in weaned pigs, partial replacement of soybean meal with mealworms was found to increase feed intake and body weight gain (Jin et al. 2016). Generally, it can be concluded that the growth performance is maintained in piglets when insect-based protein-rich products are included at the expense of other protein sources such as soybean meal, fishmeal or plasma proteins when the nutritional value, based on digestible nutrients, is adequately taken into account. Limited studies in growing pigs come to the same conclusion (Yu et al. 2019).

Only a few studies have been performed on the digestibility of insect products in pigs. Crosbie et al. (2020) determined the digestibility of full-fat and partially defatted BSFL meal in growing pigs. Standardized ileal digestibility (SID) of protein was determined at 80 and 81%, respectively, and corresponding values for essential amino acids ranged from 80 to 95% and 79 to 96%, for the two respective ingredients. A lower digestibility value for protein, relative to amino acids, could be related to the presence of chitin and the low digestible nitrogen in *N*-acetylglucosamine subunits of chitin and to protein encapsulated by chitin structures.

Tan et al. (2020) evaluated the SID of essential amino acids of nondefatted housefly and BSFL meal in growing pigs on average at 92% and 81%, respectively. These data show that protein and amino acid digestibility of insect-derived ingredients in pigs seems relatively high. However, more information on nutrient digestibility in pigs is required for a wider range of insect-based ingredients of different origins and on the effects of processing of such ingredients.

Information on the effects of feeding insect-based ingredients on carcass and meat quality is scarce. If insect oil or high-fat insect products are included, fatty acid composition in carcass fat can be affected related to fatty acid profile of insect oil. Yu et al. (2019) found that inclusion of 4% BSFL meal in the diet improved carcass traits and muscle chemical composition of finishing pigs and affected the meat quality via upregulating the expression of genes related to retention of lipid in the body.

Håkenåsen et al. (2021) indicated that insect-based products, besides as a protein and energy source, are also considered as potential functional ingredients in relation to gut health of pigs and other animal species, including post-weaning piglets. If not largely defatted, BSF and other insect species are high in fat, with lauric acid (C12:0) being high in such triglycerides. Such medium-chain fatty acids (MCFAs) have antimicrobial properties, especially against Gram-positive bacteria (Zentek et al. 2011). Spranghers et al. (2018) found that extracted fat of BSF larvae at 0.6 g C12:0 per 100 mL in vitro suppressed growth of lactobacilli and D-streptococci. Insects and insect larvae also contain chitin, a polysaccharide that can function as a prebiotic and an immunostimulant (Elieh Ali Komi et al. 2018). Also antimicrobial peptides, e.g. cecropins, defensins and drosomycins (Brady et al. 2019; Jozefiak and Engberg 2017), which are part of the insect immune system, can be identified in insect-based products as potential antimicrobial agents with low risk of development of bacteria resistance (Lewies et al. 2019). For the former reasons, over the last decade attention has also been given to identify the functional properties of insect-based ingredients in piglets. Håkenåsen et al. (2021) fed diets with full-fat BSFL meal at 5, 10, and 20% to post-weaning piglets. Feed intake and body weight gain tended to be reduced in the 5% inclusion compared to the control group, while feed efficiency was not affected by BSF inclusion. Intestinal morphometry, fatty acid composition in colon digesta and colonic microbiome composition were not affected, except for a lower abundance of the *Lactobacillus* genus when BSF was included in the diet. Overall, gut health and function were not affected in this study. Meyer et al. (2020) concluded that *Tenebrio molitor* meal can be used as a dietary source of protein at a level of 10% in weaned pigs without causing adverse effects on intermediary metabolism when considering the plasma metabolome and muscle transcriptome of these animals. Biasato et al. (2019) studied haematological, biochemical, morphometric and histopathological parameters in weaned piglets fed 0, 5 or 10% partially defatted BSFL meal at the expense of soybean meal. They neither observed effects of feeding diets containing BSFL meal as protein sources on the growth performance nor on any of the gut and systemic health-related parameters. In contrast, Kar et al. (2021) reported that dietary BSFL meal included at the expense of soybean meal influenced intestinal microbiome composition, increased microbial diversity in the jejunum and ileum and changed the amine metabolite profile in blood. Sarcosine and methionine sulfoxide were increased in blood of BSFL meal-fed piglets, and α-aminobutyric acid, a precursor of glutathione, and taurine, an amino acid derivative involved in regulating immune responses and restoring tight junctions in the gut mucosa, were also higher. Feeding BSFL meal did not cause systemic inflammatory responses in healthy animals based on analysis of pro- and anti-inflammatory cytokines and chemokines in blood. The former provides some indications that BSFL could have functional effects in relation to gut and animal health.

A study with growing pigs showed that inclusion of partially defatted BSFL meal at the expense of soybean meal in adult pig feeds had no impact on meat quality but the concentration of lauric acid levels in backfat was increased, reflecting the high concentration of lauric acid in insect oils.

It can be concluded that, in line with observations of (Veldkamp and Vernooij 2021), insects-based ingredients are suitable protein, and depending on their fat content, energy sources in diets for piglets and growing pigs. They can be used for at least partly replacing more common and traditionally used dietary protein sources. As the origin, conditions and substrates used for culture of insects vary rather largely and processing techniques applied to defat or purify the protein fraction differ as well, more attention should be given to further characterization of the nutritional value of a variety of insect-based ingredients for pigs. Although functional effects of insect-based ingredients in relation to gut health and function in pigs have been suggested, their nature and options for application should be further explored.

5 Contribution of Animal Production Systems to Sustainability by Dietary Inclusion of Insect Protein

Animal production is often related to having a large share of environmental impacts (e.g. livestock emits 65 Tg N yr.$^{-1}$, equivalent to one-third of current human-induced N emissions) (Uwizeye et al. 2020). Moreover, it is expected that by 2050 there will be a need to nourish 10 billion people. Such an increase in food demand should fulfil additional calorific energy supply in 15 exajoules (increase from 30 to 45 exajoules) (Bodirsky et al. 2020). It will trigger an additional pressure on farming systems to supply animals and people with required crops for feed and food. Currently, the farm stage is responsible for 61–81% of greenhouse gas emissions (GHGEs), 79% of acidification and 95% of eutrophication of food-related impacts (Poore and Nemecek 2018).

In order to deal with the issues of increased environmental impact of food systems and specifically animal production chains, specific measures are proposed. One of the strategies aims for the elimination or reduction by at least 50% of meat and animal-derived products in diets; however, such an approach will trigger the increase in plant food consumption at least by 100% (Willet 2019). Complete elimination of animal-derived products is also associated with negative social and economic consequences. Meat products play an important cultural and social role for societies and deliver valuable nutrients such as proteins, essential amino acids and microelements (Bohrer 2017). Animal-derived products supply around 17% of global food and around 40–58% of proteins (González et al. 2020). According to market analysis data, the value of meat industry on a global market is expected to grow 20% more till 2025. Other strategies of sustainability improvement of animal-derived products are associated with the potential to rely on underutilized or recycled nutrients, often derived from waste or underutilized side streams.

Direct utilization of food wastes and organic side streams is not feasible from many perspectives. Direct use of wastes for food or livestock feed is associated with high sanitary risks and has clear legislation barriers (Gasco et al. 2020; Varelas 2019). Transformation of valuable nutrients, available in wastes or underutilized biomass, on the other hand, is possible if different transforming agents are used. Bacteria, fungi, insects and microalgae are among such successful agents (Pleissner and Smetana 2020; Salomone et al. 2017; Smetana et al. 2022). Their efficiency, performance and suitability are quite different if considered in the scope of various animal production chains. They should be thoroughly analysed to define the relevance of their application.

Inclusion of insects in the animal feeding from a sustainability perspective will depend on a number of factors. It is quite obvious that different animal species react differently to the provision of insect-based feeds (Table 2). Moreover, not only the type of insect species, but also a type of processing (fresh, frozen, meal and fat) influences the growth performance and other potential effects in the target animal. The analysis of the available studies indicates that feeding animals completely on insects is not feasible and would result in negative consequences on performance, health and survival. Partial substitution of more common feed ingredients mostly results in a growth performance equivalent to the control treatment. At the same time, sustainable benefits are maximized only if high-impacting feed ingredients are substituted. The impact in this case can refer to the biodiversity, emissions, economic or other impacts. The highest sustainability benefits can be highlighted for the cases of fishmeal substitution in fish feeding, when inclusion of a small amount of insect biomass or insect fractions results in considerable improvements in terms of growth performance or health. It is also indicated in a few studies that substitution of a larger part of the feed with insect meal or fat results in negative effects on growth performance and health and thus also negatively impacts sustainability indicators.

It should be kept in mind that there are several sustainability trade-offs within the insect production chains, when higher impacts upstream (higher-quality feed associated with higher environmental impacts) offset the impacts associated with insect production (e.g. shorter production time, higher survival rates and more beneficial feed conversion rates) (Spykman et al. 2021). Sustainability trade-offs, indicated for insect production chains, transfer into more complex trade-offs, when the feeding and farming type of insects used for feed affects the development and performance of animal production chains.

Despite the complexity of the feed-insects-animals chain trade-offs, it is possible to draw some connections. First of all, the impact of biomass available and fed to insects would, to a large extent, determine the environmental performance and

Table 2 Sustainability relevance of insect inclusion in the diet of fish, poultry and pigs

Animal species	Insect species	Inclusion rates	Observed effects	Sustainability relevance	Sources
Rainbow trout	Black soldier fly (HI), yellow mealworm (TM)	Up to 32%	Increased performance or no change.	Potential reduced reliance on the global protein chains, potential for the environmental impact reduction.	Multiple, see in text.
Rainbow trout	Defatted meals HI, TM, and *Alphitobius diaperinus*	30%	*Alphitobius diaperinus* shows the lowest apparent digestibility coefficient. The digestibility is reduced with higher chitin content.	Separation of fat from the meal increases the content of chitin in the meal, which partially diminished the benefits of high protein content.	Eggink et al. (2022), Gasco et al. (2022)
Rainbow trout	HI	8–16% larvae or oil	Prevention of intestinal enteritis, stimulation of the immune system, positive impact on gut microbiota.	Inclusion of low levels of insect biomass or derived fractions could improve the animal well-being and reduce the environmental impact.	Multiple, see in text.
Sea bass	HI, TM	Up to 25% insect meal	No negative effect on growth performance. Higher inclusions rates decreased the growth rate.	Sustainability performance depends on the differences of impacts between fishmeal and insect meal. Currently no negative sustainable effects are observed.	Multiple, see in text.
Sea bass	HI	15% meal	Positive effect and improved growth in diets	Minor improvements in sustainability	Pérez-Pascual et al. (2020)

(continued)

Table 2 (continued)

Animal species	Insect species	Inclusion rates	Observed effects	Sustainability relevance	Sources
			limiting fishmeal.	indicators can be assumed.	
Sea bass	TM, MD, HI, ZM	19.5–25% meals	Digestibility is not affected; a slight improvement of protein digestibility might be observed. High amount of chitin—anti-nutrient and not improving the digestibility.	Minor improvements in sustainability indicators can be assumed.	Gasco et al. (2016), Magalhães et al. (2017), Mastoraki et al. (2020), Mastoraki et al. (2022)
Sea bass	HI, TM	7.25–50% meal	Low inclusion levels improve immune and antioxidant properties of feed. High doses (around 50%) cause nutritional stress and imbalances of gut microbiome.	Moderate improvements in sustainability indicators can be assumed, connected with low inclusion rates and observed beneficial effects.	Multiple, see in text.
Atlantic salmon	HI	Up to 60% meal or oil	Improved or similar growth rates. In some cases, a decrease is observed. Positive health effects are observed.	Sustainable benefits or impacts may vary depending on the growth performance.	Multiple, see in text.
Broiler chickens	HI or TM	5–15% larvae meal	Improved or similar growth rates. In some cases, a decrease is observed.	Potential reduced reliance on the global protein chains, potential for environmental impact reduction.	Multiple, see in text.

(continued)

Table 2 (continued)

Animal species	Insect species	Inclusion rates	Observed effects	Sustainability relevance	Sources
Broiler chickens	HI or TM	Up to 10% live larvae	Improved or similar growth rates. Beneficial effects on bird's welfare.	Potential reduced reliance on the global protein chains, potential for environmental impact reduction.	Multiple, see in text.
Laying hens	HI or TM	5–15% larvae meal	Improved or similar production performance. At high inclusion levels sometimes a decrease is observed.	Reduced reliance on global protein chains.	Multiple, see in text.
Laying hens	HI or TM	Up to 10% live larvae	Improved or similar production performance. Beneficial effect on bird's welfare.	Reduced reliance on global protein chains.	Multiple, see in text.
Pig	HI or TM	Up to 10% larvae meal	Improved or similar production performance.	Reduced reliance on global protein chains.	Multiple, see in text.

sustainability of insect biomass. More processed and more fractionated insect biomass would have a higher impact than living, fresh or frozen whole insects. From sustainability point of view, it is most feasible to include relatively low levels of insect biomass (especially nonprocessed or minimally processed) into the feed of animals, considering that the ingredients positively influence animal health or performance. In this case insect feed acts as a valuable feed ingredient. However, other ways of application of insect meals and fat are possible, but it requires a holistic assessment in each specific case.

References

Aas TS, Ytrestøyl T, Åsgård T (2022) Utilization of feed resources in Norwegian farming of Atlantic salmon and rainbow trout in 2020. Professional report., Tromso, Norway

Abdel-Ghany HM, Salem MES (2020) Effects of dietary chitosan supplementation on farmed fish; a review. Rev Aquac 12:438–452. https://doi.org/10.1111/raq.12326

Abdel-Latif HMR, Abdel-Tawwab M, Khalil RH, Metwally AA, Shakweer MS, Ghetas HA, Khallaf MA (2021) Black soldier fly (*Hermetia illucens*) larvae meal in diets of European seabass: effects on antioxidative capacity, non-specific immunity, transcriptomic responses, and resistance to the challenge with *vibrio alginolyticus*. Fish Shellfish Immunol 111:111–118. https://doi.org/10.1016/j.fsi.2021.01.013

Abdel-Tawwab M, Khalil RH, Metwally AA, Shakweer MS, Khallaf MA, Abdel-Latif HMR (2020) Effects of black soldier fly (*Hermetia illucens* L.) larvae meal on growth performance, organs-somatic indices, body composition, and hemato-biochemical variables of European sea bass, *Dicentrarchus labrax*. Aquaculture 522:735136. https://doi.org/10.1016/j.aquaculture.2020.735136

Alexandratos N, Bruinsma J (2012) World agriculture towards 2030/2050: the 2012 revision

Alltech (2022) Agri-Food Outlook 2022. Available at: https://www.alltech.com/agri-food-outlook

Al-Qazzaz MFA, Ismail D, Akit H, Idris LH (2016) Effect of using insect larvae meal as a complete protein source on quality and productivity characteristics of laying hens. Rev Bras Zootec 45: 518–523. https://doi.org/10.1590/S1806-92902016000900003

Antonopoulou E, Nikouli E, Piccolo G, Gasco L, Gai F, Chatzifotis S, Mente E, Kormas KA (2019) Reshaping gut bacterial communities after dietary *Tenebrio molitor* larvae meal supplementation in three fish species. Aquaculture 503:628–635. https://doi.org/10.1016/j.aquaculture.2018.12.013

Aragão C, Gonçalves AT, Costas B, Azeredo R, Xavier MJ, Engrola S (2022) Alternative proteins for fish diets: implications beyond growth. Animals 12:1211. https://doi.org/10.3390/ani12091211

Aviagen (2022) Broiler performance objectives. Available at: http://eu.aviagen.com/tech-center/

Baldi L, Mancuso T, Peri M, Gasco L, Trentinaglia M (2021) Consumer attitude and acceptance toward fish fed with insects: a focus on the new generations. J Insects Food Feed. https://doi.org/10.3920/JIFF2021.0109

Basto A, Matos E, Valente LMP (2020) Nutritional value of different insect larvae meals as protein sources for European sea bass (*Dicentrarchus labrax*) juveniles. Aquaculture 521:735085. https://doi.org/10.1016/j.aquaculture.2020.735085

Beirão J, Boulais M, Gallego V, O'Brien JK, Peixoto S, Robeck TR, Cabrita E (2019) Sperm handling in aquatic animals for artificial reproduction. Theriogenology 133:161–178. https://doi.org/10.1016/j.theriogenology.2019.05.004

Belghit I, Liland NS, Waagbø R, Biancarosa I, Pelusio N, Li Y, Krogdahl Å, Lock E-J (2018) Potential of insect-based diets for Atlantic salmon (*Salmo salar*). Aquaculture 491:72–81. https://doi.org/10.1016/j.aquaculture.2018.03.016

Belghit I, Liland NS, Gjesdal P, Biancarosa I, Menchetti E, Li Y, Waagbø R, Krogdahl Å, Lock E-J (2019a) Black soldier fly larvae meal can replace fish meal in diets of sea-water phase Atlantic salmon (*Salmo salar*). Aquaculture 503:609–619. https://doi.org/10.1016/j.aquaculture.2018.12.032

Belghit I, Waagbø R, Lock E-J, Liland NS (2019b) Insect-based diets high in lauric acid reduce liver lipids in freshwater Atlantic salmon. Aquac Nutr 25:343–357. https://doi.org/10.1111/anu.12860

Bellezza Oddon S, Biasato I, Imarisio A, Pipan M, Dekleva D, Colombino E, Capucchio MT, Meneguz M, Bergagna S, Barbero R (2021) Black soldier Fly and yellow mealworm live larvae for broiler chickens: effects on bird performance and health status. J Anim Physiol Anim Nutr 105:10–18. https://doi.org/10.1111/jpn.13567

Biasato I, Ferrocino I, Biasibetti E, Grego E, Dabbou S, Sereno A, Gai F, Gasco L, Schiavone A, Cocolin L (2018) Modulation of intestinal microbiota, morphology and mucin composition by dietary insect meal inclusion in free-range chickens. BMC Vet Res 14:1–15. https://doi.org/10.1186/s12917-018-1690-y

Biasato I, Ferrocino I, Grego E, Dabbou S, Gai F, Gasco L, Cocolin L, Capucchio MT, Schiavone A (2019) Gut microbiota and mucin composition in female broiler chickens fed diets including yellow mealworm (*Tenebrio molitor*, L.). Animals 9:213. https://doi.org/10.3390/ani9050213

Biasato I, Ferrocino I, Dabbou S, Evangelista R, Gai F, Gasco L, Cocolin L, Capucchio MT, Schiavone A (2020a) Black soldier fly and gut health in broiler chickens: insights into the relationship between cecal microbiota and intestinal mucin composition. J Anim Sci Biotechnol 11:1–12. https://doi.org/10.1186/s40104-019-0413-y

Biasato I, Ferrocino I, Grego E, Dabbou S, Gai F, Gasco L, Cocolin L, Capucchio MT, Schiavone A (2020b) Yellow mealworm inclusion in diets for heavy-size broiler chickens: implications for intestinal microbiota and mucin dynamics. Animals 10:1909. https://doi.org/10.3390/ani10101909

Biasato I, Bellezza Oddon S, Chemello G, Gariglio M, Fiorilla E, Dabbou S, Pipan M, Dekleva D, Macchi E, Gasco L (2022a) Welfare implications for broiler chickens reared in an insect larvae-enriched environment: focus on bird behaviour, plumage status, leg health, and excreta corticosterone. Front Physiol 13:930158. https://doi.org/10.3389/fphys.2022.930158

Biasato I, Chemello G, Oddon SB, Ferrocino I, Corvaglia M, Caimi C, Resconi A, Paul A, van Spankeren M, Capucchio M (2022b) *Hermetia illucens* meal inclusion in low-fishmeal diets for rainbow trout (*Oncorhynchus mykiss*): effects on the growth performance, nutrient digestibility coefficients, selected gut health traits, and health status indices. Anim Feed Sci Technol 290: 115341. https://doi.org/10.1016/j.anifeedsci.2022.115341

Bodirsky BL, Dietrich JP, Martinelli E, Stenstad A, Pradhan P, Gabrysch S, Mishra A, Weindl I, Le Mouël C, Rolinski S, Baumstark L, Wang X, Waid JL, Lotze-Campen H, Popp A (2020) The ongoing nutrition transition thwarts long-term targets for food security, public health and environmental protection. Sci Rep 10:19778. https://doi.org/10.1038/s41598-020-75213-3

Bohrer BM (2017) Review: nutrient density and nutritional value of meat products and non-meat foods high in protein. Trends Food Sci Technol 65:103–112. https://doi.org/10.1016/j.tifs.2017.04.016

Borgogno M, Dinnella C, Iaconisi V, Fusi R, Scarpaleggia C, Schiavone A, Monteleone E, Gasco L, Parisi G (2017) Inclusion of *Hermetia illucens* larvae meal on rainbow trout (*Oncorhynchus mykiss*) feed: effect on sensory profile according to static and dynamic evaluations. J Sci Food Agric 97:3402–3411. https://doi.org/10.1002/jsfa.8191

Bovera F, Loponte R, Pero ME, Cutrignelli MI, Calabrò S, Musco N, Vassalotti G, Panettieri V, Lombardi P, Piccolo G (2018) Laying performance, blood profiles, nutrient digestibility and inner organs traits of hens fed an insect meal from *Hermetia illucens* larvae. Res Vet Sci 120:86–93. https://doi.org/10.1016/j.rvsc.2018.09.006

Brady D, Grapputo A, Romoli O, Sandrelli F (2019) Insect cecropins, antimicrobial peptides with potential therapeutic applications. Int J Mol Sci 20:5862. https://doi.org/10.3390/ijms20235862

Breewood H, Garnett T (2020) What is feed-food competition. In: Foodsource: building blocks. Food Climate Research Network, University of Oxford, Oxford

Bruni L, Belghit I, Lock EJ, Secci G, Taiti C, Parisi G (2020) Total replacement of dietary fish meal with black soldier fly (*Hermetia illucens*) larvae does not impair physical, chemical or volatile composition of farmed Atlantic salmon (*Salmo salar* L.). J Sci Food Agric 100:1038–1047. https://doi.org/10.1002/jsfa.10108

Caimi C, Biasato I, Chemello G, Oddon SB, Lussiana C, Malfatto VM, Capucchio MT, Colombino E, Schiavone A, Gai F (2021) Dietary inclusion of a partially defatted black soldier fly (*Hermetia illucens*) larva meal in low fishmeal-based diets for rainbow trout (*Oncorhynchus mykiss*). J Anim Sci Biotechnol 12:1–15. https://doi.org/10.1186/s40104-021-00575-1

Cardinaletti G, Randazzo B, Messina M, Zarantoniello M, Giorgini E, Zimbelli A, Bruni L, Parisi G, Olivotto I, Tulli F (2019) Effects of graded dietary inclusion level of full-fat *Hermetia illucens* Prepupae meal in practical diets for rainbow trout (*Oncorhynchus mykiss*). Animals 9: 251. https://doi.org/10.3390/ani9050251

Chemello G, Biasato I, Gai F, Capucchio MT, Colombino E, Schiavone A, Gasco L, Pauciullo A (2021) Effects of *Tenebrio molitor* larvae meal inclusion in rainbow trout feed: myogenesisrelated gene expression and histomorphological features. Ital J Anim Sci 20:1211–1221. https://doi.org/10.1080/1828051X.2021.1945959

Chu X, Li M, Wang G, Wang K, Shang R, Wang Z, Li L (2020) Evaluation of the low inclusion of full-fatted *Hermetia illucens* larvae meal for layer chickens: growth performance, nutrient digestibility, and gut health. Front Vet Sci 7:585843. https://doi.org/10.3389/fvets.2020.585843

Clara E, Regolin L, Vallortigara G, Rogers LJ (2009) Chicks prefer to peck at insect-like elongated stimuli moving in a direction orthogonal to their longer axis. Anim Cogn 12:755–765. https://doi.org/10.1007/s10071-009-0235-y

Colombo SM (2020) Chapter 2 - Physiological considerations in shifting carnivorous fishes to plant-based diets. In: Benfey TJ, Farrell AP, Brauner CJ (eds) Fish physiology. Academic Press, pp 53–82. https://doi.org/10.1016/bs.fp.2020.09.002

Crosbie M, Zhu C, Shoveller AK, Huber L-A (2020) Standardized ileal digestible amino acids and net energy contents in full fat and defatted black soldier fly larvae meals (*Hermetia illucens*) fed to growing pigs. Translational. Anim Sci 4:txaa104. https://doi.org/10.1093/tas/txaa104

Cullere M, Tasoniero G, Giaccone V, Miotti-Scapin R, Claeys E, De Smet S, Dalle Zotte A (2016) Black soldier fly as dietary protein source for broiler quails: apparent digestibility, excreta microbial load, feed choice, performance, carcass and meat traits. Animal 10:1923–1930. https://doi.org/10.1017/S1751731116001270

Cutrignelli MI, Messina M, Tulli F, Randazzo B, Olivotto I, Gasco L, Loponte R, Bovera F (2018) Evaluation of an insect meal of the black soldier Fly (*Hermetia illucens*) as soybean substitute: intestinal morphometry, enzymatic and microbial activity in laying hens. Res Vet Sci 117:209–215. https://doi.org/10.1016/j.rvsc.2017.12.020

CVB (2018) CVB Table Booklet Feeding of Poultry 2018. CVB-series no 61. Federatie Nederlandse Diervoederketen

De Marco M, Martínez S, Hernandez F, Madrid J, Gai F, Rotolo L, Belforti M, Bergero D, Katz H, Dabbou S, Kovitvadhi A, Zoccarato I, Gasco L, Schiavone A (2015) Nutritional value of two insect larval meals (*Tenebrio molitor* and *Hermetia illucens*) for broiler chickens: apparent nutrient digestibility, apparent ileal amino acid digestibility and apparent metabolizable energy. Anim Feed Sci Technol 209:211–218. https://doi.org/10.1016/j.anifeedsci.2015.08.006

Delgado E, Reyes-Jaquez D (2018) Extruded aquaculture feed: a review. In: Qamar SZ (ed) Extrusion of metals, polymers and food products. InTech, Croatia, pp 145–163

Dias J, Alvarez MJ, Arzel J, Corraze G, Diez A, Bautista JM, Kaushik SJ (2005) Dietary protein source affects lipid metabolism in the European seabass (*Dicentrarchus labrax*). Comp Biochem Physiol A Mol Integr Physiol 142:19–31. https://doi.org/10.1016/j.cbpb.2005.07.005

Dörper A, Veldkamp T, Dicke M (2021) Use of black soldier fly and house fly in feed to promote sustainable poultry production. J Insects Food Feed 7:761–780. https://doi.org/10.3920/JIFF2020.0064

Dumas A, Raggi T, Barkhouse J, Lewis E, Weltzien E (2018) The oil fraction and partially defatted meal of black soldier fly larvae (*Hermetia illucens*) affect differently growth performance, feed efficiency, nutrient deposition, blood glucose and lipid digestibility of rainbow trout (*Oncorhynchus mykiss*). Aquaculture 492:24–34. https://doi.org/10.1016/j.aquaculture.2018.03.038

EC (2001) Commission Regulation (EU) No. 999/2001 of the European Parliament and of the Council of 22 May 2001 laying down rules for the prevention, control and eradication of certain transmissible spongiform encephalopathies. Official Journal L 147, 1–40, Brussels, Belgium

EC (2017) Commission regulation (EU) 2017/893 of 24 may 2017 amending annexes I and IV to regulation (EC) no 999/2001 of the European Parliament and of the council and annexes X, XIV and XV to commission regulation (EU) no 142/2011 as regards the provisions on processed animal protein. Off J Eur Union 138: 92–116

EC (2018) 2018 circular economy package. In: E. Commission (Ed.), Brussels, Belgium

EC (2022) Commission regulation (EU) no 2022/1104 of 1 July 2022 amending regulation (EU) no 68/2013 on the catalogue of feed materials. Official journal L177: 1–71, Brussels, Belgium

Eggink KM, Pedersen PB, Lund I, Dalsgaard J (2022) Chitin digestibility and intestinal exochitinase activity in Nile tilapia and rainbow trout fed different black soldier fly larvae meal size fractions. Aquac Res 53:5536–5546. https://doi.org/10.1111/are.16035
Elahi U, Xu C-c, Wang J, Lin J, Wu S-g, Zhang H-j, Qi G-H (2022) Insect meal as a feed ingredient for poultry. Anim Biosci 35:332. https://doi.org/10.5713/ab.21.0435
Elieh Ali Komi D, Sharma L, Dela Cruz CS (2018) Chitin and its effects on inflammatory and immune responses. Clin Rev Allergy Immunol 54:213–223. https://doi.org/10.1007/s12016-017-8600-0
Elson H, Gleadthorpe A, Vale M, Mansfield U (2011) Housing and husbandry of laying hens: past, present and future. Lohmann Infor 46:16–24
EU (2017). Available at: https://ec.europa.eu/agriculture/market-observatory/crops/oilseeds-protein-crops/balancesheets_en
EUFOMA (2021) European market observatory for fisheries and aquaculture products - The Eu Fish Market - 2021 edition. Available at: https://www.eumofa.eu/
FAO (2020) World food and agriculture—statistical yearbook 2020. World Food and Agriculture-Statistical Yearbook
FAO (2022) FishstatJ. Fisheries and aquaculture software. Software for fishery and aquaculture statistical time series bibliographic citation [online] Rome Updated August 5, 2022
FEFAC (2022) EU Compound feed production in 2021/Market Outlook 2022. Available at: https://fefac.eu/newsroom/news/eu-compound-feed-production-in-2021-market-outlook-2022/
Fisher HJ, Collins SA, Hanson C, Mason B, Colombo SM, Anderson DM (2020) Black soldier fly larvae meal as a protein source in low fish meal diets for Atlantic salmon (*Salmo salar*). Aquaculture 521:734978. https://doi.org/10.1016/j.aquaculture.2020.734978
Gariglio M, Dabbou S, Biasato I, Capucchio MT, Colombino E, Hernández F, Madrid J, Martínez S, Gai F, Caimi C (2019) Nutritional effects of the dietary inclusion of partially defatted *Hermetia illucens* larva meal in Muscovy duck. J Anim Sci Biotechnol 10:1–10. https://doi.org/10.1186/s40104-019-0344-7
Garmon J, Hogan M (2010) Managing the modern broiler male. Available at: www.aviagen.com
Gasco L, Henry M, Piccolo G, Marono S, Gai F, Renna M, Lussiana C, Antonopoulou E, Mola P, Chatzifotis S (2016) *Tenebrio molitor* meal in diets for European sea bass (*Dicentrarchus labrax* L.) juveniles: growth performance, whole body composition and in vivo apparent digestibility. Anim Feed Sci Technol 220:34–45. https://doi.org/10.1016/j.anifeedsci.2016.07.003
Gasco L, Biancarosa I, Liland NS (2020) From waste to feed: a review of recent knowledge on insects as producers of protein and fat for animal feeds. Curr Opin Green and Sustain Chem 23: 67–79. https://doi.org/10.1016/j.cogsc.2020.03.003
Gasco L, Józefiak A, Henry M (2021) Beyond the protein concept: health aspects of using edible insects on animals. J Insects Food Feed 7:715–741. https://doi.org/10.3920/JIFF2020.0077
Gasco L, Caimi C, Trocino A, Lussiana C, Oddon SB, Malfatto V, Anedda R, Serra G, Biasato I, Schiavone A (2022) Digestibility of defatted insect meals for rainbow trout aquafeeds. J Insects Food Feed 8(11):1385–1399. https://doi.org/10.3920/JIFF2021.0160
Gaylord TG, Sealey WM, Barrows FT, Myrick CA, Fornshell G (2017) Evaluation of ingredient combinations from differing origins (fishmeal, terrestrial animal and plants) and two different formulated nutrient targets on rainbow trout growth and production efficiency. Aquac Nutr 23: 1319–1328. https://doi.org/10.1111/anu.12507
Geay F, Ferraresso S, Zambonino-Infante JL, Bargelloni L, Quentel C, Vandeputte M, Kaushik S, Cahu CL, Mazurais D (2011) Effects of the total replacement of fish-based diet with plant-based diet on the hepatic transcriptome of two European sea bass (*Dicentrarchus labrax*) half-sibfamilies showing different growth rates with the plant-based diet. BMC Genomics 12:1–18. https://doi.org/10.1186/1471-2164-12-522
González N, Marquès M, Nadal M, Domingo JL (2020) Meat consumption: which are the current global risks? A review of recent (2010–2020) evidences. Food Res Int 137:109341. https://doi.org/10.1016/j.foodres.2020.109341

Gugołek A, Strychalski J, Juśkiewicz J, Żary-Sikorska E (2020) The effect of fish and mealworm larvae meals as alternative dietary protein sources on nutrient digestibility and gastrointestinal function in *Chinchilla lanigera*. Exp Anim 69:70–79. https://doi.org/10.1538/expanim.19-0072

Hajra A, Mazumder A, Verma A, Ganguly DP, Mohanty BP, Sharma AP (2013) Antinutritional factors in plant origin fish feed ingredients: the problems and probable remedies. Adv Fish Res 5:193–202

Håkenåsen IM, Grepperud GH, Hansen JØ, Øverland M, Ånestad RM, Mydland LT (2021) Full-fat insect meal in pelleted diets for weaned piglets: effects on growth performance, nutrient digestibility, gastrointestinal function, and microbiota. Anim Feed Sci Technol 281:115086. https://doi.org/10.1016/j.anifeedsci.2021.115086

Halloran A, Hanboonsong Y, Roos N, Bruun S (2017) Life cycle assessment of cricket farming in North-Eastern Thailand. J Clean Prod 156:83–94. https://doi.org/10.1016/j.jclepro.2017.04.017

Hawkey KJ, Lopez-Viso C, Brameld JM, Parr T, Salter AM (2021) Insects: a potential source of protein and other nutrients for feed and food. Annu Rev Anim Biosci 9:333–354. https://doi.org/10.1146/annurev-animal-021419083930

Henry M, Gasco L, Chatzifotis S, Piccolo G (2018) Does dietary insect meal affect the fish immune system? The case of mealworm, *Tenebrio molitor* on European sea bass, *Dicentrarchus labrax*. Dev Comp Immunol 81:204–209. https://doi.org/10.1016/j.dci.2017.12.002

Heuel M, Sandrock C, Leiber F, Mathys A, Gold M, Zurbrügg C, Gangnat IDM, Kreuzer M, Terranova M (2021) Black soldier fly larvae meal and fat can completely replace soybean cake and oil in diets for laying hens. Poult Sci 100:101034. https://doi.org/10.1016/j.psj.2021.101034

Hong J, Kim YY (2022) Insect as feed ingredients for pigs. Anim Biosci 35:347. https://doi.org/10.5713/ab.21.0475

Hua K (2021) A meta-analysis of the effects of replacing fish meals with insect meals on growth performance of fish. Aquaculture 530:735732. https://doi.org/10.1016/j.aquaculture.2020.735732

Hussein M, Pillai VV, Goddard JM, Park HG, Kothapalli KS, Ross DA, Ketterings QM, Brenna JT, Milstein MB, Marquis H, Johnson PA, Nyrop JP, Selvaraj V (2017) Sustainable production of housefly (*Musca domestica*) larvae as a protein-rich feed ingredient by utilizing cattle manure. PLoS One 12:e0171708. https://doi.org/10.1371/journal.pone.0171708

Ipema AF, Gerrits WJ, Bokkers EA, Kemp B, Bolhuis JE (2020) Provisioning of live black soldier fly larvae (*Hermetia illucens*) benefits broiler activity and leg health in a frequency-and dose-dependent manner. Appl Anim Behav Sci 230:105082. https://doi.org/10.1016/j.applanim.2020.105082

Jannathulla R, Rajaram V, Kalanjiam R, Ambasankar K, Muralidhar M, Dayal JS (2019) Fishmeal availability in the scenarios of climate change: inevitability of fishmeal replacement in aquafeeds and approaches for the utilization of plant protein sources. Aquac Res 50:3493–3506. https://doi.org/10.1111/are.14324

Jayanegara A, Yantina N, Novandri B, Laconi E, Nahrowi N, Ridla M (2017) Evaluation of some insects as potential feed ingredients for ruminants: chemical composition, in vitro rumen fermentation and methane emissions. Journal of Indonesian tropical animal. Agriculture 42: 247–254. https://doi.org/10.14710/jitaa.42.4.247-254

Jin X, Heo P, Hong J, Kim N, Kim Y (2016) Supplementation of dried mealworm (*Tenebrio molitor* larva) on growth performance, nutrient digestibility and blood profiles in weaning pigs. Asian Australas J Anim Sci 29:979. https://doi.org/10.5713/ajas.15.0535

Jobling M (2016) Fish nutrition research: past, present and future. Aquac Int 24:767–786. https://doi.org/10.1007/s10499-014-9875-2

Jozefiak A, Engberg RM (2017) Insect proteins as a potential source of antimicrobial peptides in livestock production. A review. J Anim Feed Sci 26:87–99. https://doi.org/10.22358/jafs/69998/2017

Józefiak A, Nogales-Mérida S, Mikołajczak Z, Rawski M, Kierończyk B, Mazurkiewicz J (2019) The utilization of full-fat insect meal in rainbow trout nutrition: the effects on growth performance, intestinal microbiota and gastrointestinal tract Histomorphology. Ann Anim Sci 19:747–765. https://doi.org/10.2478/aoas-2019-0020

Kamalam BS, Rajesh M, Kaushik S (2020) Nutrition and feeding of rainbow trout (*Oncorhynchus mykiss*), fish nutrition and its relevance to human health. CRC Press, Boca Raton, pp 299–332

Kar SK, Schokker D, Harms AC, Kruijt L, Smits MA, Jansman AJ (2021) Local intestinal microbiota response and systemic effects of feeding black soldier fly larvae to replace soybean meal in growing pigs. Sci Rep 11:1–16. https://doi.org/10.1038/s41598-021-94604-8

Kawasaki K, Hashimoto Y, Hori A, Kawasaki T, Hirayasu H, Iwase S-i, Hashizume A, Ido A, Miura C, Miura T, Nakamura S, Seyama T, Matsumoto Y, Kasai K, Fujitani Y (2019) Evaluation of black soldier Fly (*Hermetia illucens*) larvae and pre-pupae raised on household organic waste, as potential ingredients for poultry feed. Animals 9:98. https://doi.org/10.3390/ani9030098

Khan SH (2018) Recent advances in role of insects as alternative protein source in poultry nutrition. J Appl Anim Res 46:1144–1157. https://doi.org/10.1080/09712119.2018.1474743

Kissil GK, Tandler A, Colorni A, Elizur A (2000) Sea bass culture. In: Stickney RR (ed) Encyclopedia of aquaculture. John Wiley & Sons, Inc., Toronto

Ko H, Choi Y, Khun S, Cho E, Kim Y, Pi J, Park K, Kim J, Lee S, Kim J (2020) Laying performance, egg quality, haematological traits, and faecal noxious gas emission of laying hens fed with Tenebrio molitor meal. Eur Poult Sci 2020:84

Kousoulaki K, Sæther BS, Albrektsen S, Noble C (2015) Review on European sea bass (*Dicentrarchus labrax*, Linnaeus, 1758) nutrition and feed management: a practical guide for optimizing feed formulation and farming protocols. Aquac Nutr 21:129–151. https://doi.org/10.1111/anu.12233

Kumar V, Fawole FJ, Romano N, Hossain MS, Labh SN, Overturf K, Small BC (2021) Insect (black soldier fly, *Hermetia illucens*) meal supplementation prevents the soybean meal-induced intestinal enteritis in rainbow trout and health benefits of using insect oil. Fish Shellfish Immunol 109:116–124. https://doi.org/10.1016/j.fsi.2020.12.008

Kumari K, Chandra K, Kumar D (2016) Effect of protein levels in diet on production and emissions in WHL layer. Adv Anim Vet sci 4:357–362

Leeper A, Benhaïm D, Smárason BÖ, Knobloch S, Òmarsson KL, Bonnafoux T, Pipan M, Koppe W, Björnsdóttir R, Øverland M (2022) Feeding black soldier fly larvae (*Hermetia illucens*) reared on organic rest streams alters gut characteristics of Atlantic salmon (*Salmo salar*). J Insects Food Feed 8:1355–1372. https://doi.org/10.3920/jiff2021.0105

Lewies A, Du Plessis LH, Wentzel JF (2019) Antimicrobial peptides: the Achilles' heel of antibiotic resistance? Probiotics Antimicrob Proteins 11:370–381. https://doi.org/10.1007/s12602-018-9465-0

Li Y, Kortner TM, Chikwati EM, Belghit I, Lock E-J, Krogdahl Å (2020) Total replacement of fish meal with black soldier fly (*Hermetia illucens*) larvae meal does not compromise the gut health of Atlantic salmon (*Salmo salar*). Aquaculture 520:734967. https://doi.org/10.1016/j.aquaculture.2020.734967

Liland NS, Araujo P, Xu XX, Lock E-J, Radhakrishnan G, Prabhu AJP, Belghit I (2021) A meta-analysis on the nutritional value of insects in aquafeeds. J Insects Food Feed 7:743–759. https://doi.org/10.3920/jiff2020.0147

Liu X, Liu X, Yao Y, Qu X, Chen J, Xie K, Wang X, Qi Y, Xiao B, He C (2021) Effects of different levels of *Hermetia illucens* larvae meal on performance, egg quality, yolk fatty acid composition and oxidative status of laying hens. Ital J Anim Sci 20:256–266. https://doi.org/10.1080/1828051X.2021.1878946

Lock ER, Arsiwalla T, Waagbø R (2016) Insect larvae meal as an alternative source of nutrients in the diet of Atlantic salmon (*Salmo salar*) postsmolt. Aquac Nutr 22:1202–1213. https://doi.org/10.1111/anu.12343

Lock E-J, Biancarosa I, Gasco L (2018) Insects as raw materials in compound feed for aquaculture, edible insects in sustainable food systems. Springer, Cham, pp 263–276. https://doi.org/10.1007/978-3-319-74011-9_16

Magalhães R, Sánchez-López A, Leal RS, Martínez-Llorens S, Oliva-Teles A, Peres H (2017) Black soldier fly *(Hermetia illucens)* pre-pupae meal as a fish meal replacement in diets for European seabass (*Dicentrarchus labrax*). Aquaculture 476:79–85. https://doi.org/10.1016/j.aquaculture.2017.04.021

Makkar HPS, Tran G, Heuzé V, Ankers P (2014) State-of-the-art on use of insects as animal feed. Anim Feed Sci Technol 197:1–33. https://doi.org/10.1016/j.anifeedsci.2014.07.008

Mancuso T, Baldi L, Gasco L (2016) An empirical study on consumer acceptance of farmed fish fed on insect meals: the Italian case. Aquac Int 24:1489–1507. https://doi.org/10.1007/s10499-016-0007-z

Marono S, Piccolo G, Loponte R, Di Meo C, Attia YA, Nizza A, Bovera F (2015) *In vitro* crude protein digestibility of *Tenebrio Molitor* and *Hermetia Illucens* insect meals and its correlation with chemical composition traits. Ital J Anim Sci 14:3889. https://doi.org/10.4081/ijas.2015.3889

Marono S, Loponte R, Lombardi P, Vassalotti G, Pero M, Russo F, Gasco L, Parisi G, Piccolo G, Nizza S (2017) Productive performance and blood profiles of laying hens fed Hermetia illucens larvae meal as total replacement of soybean meal from 24 to 45 weeks of age. Poult Sci 96: 1783–1790

Mastoraki M, Mollá Ferrándiz P, Vardali SC, Kontodimas DC, Kotzamanis YP, Gasco L, Chatzifotis S, Antonopoulou E (2020) A comparative study on the effect of fish meal substitution with three different insect meals on growth, body composition and metabolism of European sea bass (*Dicentrarchus labrax* L.). Aquaculture 528:735511. https://doi.org/10.1016/j.aquaculture.2020.735511

Mastoraki M, Panteli N, Kotzamanis YP, Gasco L, Antonopoulou E, Chatzifotis S (2022) Nutrient digestibility of diets containing five different insect meals in gilthead sea bream (*Sparus aurata*) and European sea bass (*Dicentrarchus labrax*). Anim Feed Sci Technol 292:115425. https://doi.org/10.1016/j.anifeedsci.2022.115425

Matin N, Utterback P, Parsons CM (2021) True metabolizable energy and amino acid digestibility in black soldier fly larvae meals, cricket meal, and mealworms using a precision-fed rooster assay. Poult Sci 100:101146. https://doi.org/10.1016/j.psj.2021.101146

Maurer V, Holinger M, Amsler Z, Früh B, Wohlfahrt J, Stamer A, Leiber F (2016) Replacement of soybean cake by *Hermetia illucens* meal in diets for layers. J Insects Food Feed 2:83–90

Mavromati E, Mavromati J, Sena L (2017) Comparative evaluation of the egg production performance indicators of Hy-line hybrid kept in traditional cage system versus the enriched cages one comparative evaluation of the egg production performance indicators of Hy-line hybrid kept in traditional cage system versus the enriched cages one. Eur Acad Res V:1474

Melenchón F, Larrán AM, de Mercado E, Hidalgo MC, Cardenete G, Barroso FG, Fabrikov D, Lourenço HM, Pessoa MF, Tomás-Almenar C (2021) Potential use of black soldier fly (*Hermetia illucens*) and mealworm (*Tenebrio molitor*) insect meals in diets for rainbow trout (*Oncorhynchus mykiss*). Aquac Nutr 27:491–505. https://doi.org/10.1111/anu.13201

Melenchón F, de Mercado E, Pula HJ, Cardenete G, Barroso FG, Fabrikov D, Lourenço HM, Pessoa M-F, Lagos L, Weththasinghe P, Cortés M, Tomás-Almenar C (2022) Fishmeal dietary replacement up to 50%: a comparative study of two insect meals for rainbow trout (*Oncorhynchus mykiss*). Animals 12:179. https://doi.org/10.3390/ani12020179

Mente E, Bousdras T, Feidantsis K, Panteli N, Mastoraki M, Kormas KA, Chatzifotis S, Piccolo G, Gasco L, Gai F, Martin SAM, Antonopoulou E (2022) Tenebrio molitor larvae meal inclusion affects hepatic proteome and apoptosis and/or autophagy of three farmed fish species. Sci Rep 7: 121. https://doi.org/10.1038/s41598-021-03306-8

Meyer S, Gessner DK, Braune MS, Friedhoff T, Most E, Höring M, Liebisch G, Zorn H, Eder K, Ringseis R (2020) Comprehensive evaluation of the metabolic effects of insect meal from *Tenebrio molitor* L. in growing pigs by transcriptomics, metabolomics and lipidomics. J Anim Sci Biotechnol 11:20. https://doi.org/10.1186/s40104-020-0425-7

Mottet A, Tempio G (2017) Global poultry production: current state and future outlook and challenges. Worlds Poult Sci J 73:245–256. https://doi.org/10.1017/S0043933917000071

Moutinho S, Pedrosa R, Magalhães R, Oliva-Teles A, Parisi G, Peres H (2020) Black soldier fly (*Hermetia illucens*) pre-pupae larvae meal in diets for European seabass (*Dicentrarchus labrax*) juveniles: effects on liver oxidative status and fillet quality traits during shelf-life. Aquaculture 533:736080. https://doi.org/10.1016/j.aquaculture.2020.736080

Moutinho S, Pedrosa R, Magalhães R, Oliva-Teles A, Parisi G, Peres H (2021) Black soldier fly (*Hermetia illucens*) pre-pupae larvae meal in diets for European seabass (*Dicentrarchus labrax*) juveniles: effects on liver oxidative status and fillet quality traits during shelf-life. Aquaculture 533:736080. https://doi.org/10.1016/j.aquaculture.2020.736080

Mwaniki Z, Neijat M, Kiarie E (2018) Egg production and quality responses of adding up to 7.5% defatted black soldier fly larvae meal in a corn–soybean meal diet fed to shaver white leghorns from wk 19 to 27 of age. Poult Sci 97:2829–2835

Newton L, Sheppard C, Watson DW, Burtle G, Dove R (2005) Using the black soldier fly, *Hermetia illucens*, as a value-added tool for the management of swine manure. Animal and Poultry Waste Management Centre, North Carolina State University, Raleigh, NC, USA

Nowak V, Persijn D, Rittenschober D, Charrondiere UR (2016) Review of food composition data for edible insects. Food Chem 193:39–46. https://doi.org/10.1016/j.foodchem.2014.10.114

NRC (2011) Nutrient requirements of fish and shrimp. Available at: https://doi.org/10.17226/13039

OECD (2022) OECD-FAO Agricultural Outlook 2021–2030. Available at: https://www.oecd-ilibrary.org/agriculture-and-food/oecd-fao-agricultural-outlook-2021-2030_19428846-en

Oonincx DG, Van Itterbeeck J, Heetkamp MJ, Van Den Brand H, Van Loon JJ, Van Huis A (2010) An exploration on greenhouse gas and ammonia production by insect species suitable for animal or human consumption. PLoS One 5:e14445. https://doi.org/10.1371/journal.pone.0014445

Palomba A, Melis R, Biosa G, Braca A, Pisanu S, Ghisaura S, Caimi C, Biasato I, Oddon SB, Gasco L, Terova G, Moroni F, Antonini M, Pagnozzi D, Anedda R (2022) On the compatibility of fish meal replacements in Aquafeeds for rainbow trout. A combined Metabolomic, proteomic and histological study. Front Physiol 13:920289. https://doi.org/10.3389/fphys.2022.920289

Panteli N, Mastoraki M, Lazarina M, Chatzifotis S, Mente E, Kormas KA, Antonopoulou E (2021) Configuration of gut microbiota structure and potential functionality in two teleosts under the influence of dietary insect meals. Microorganisms 9:699. https://doi.org/10.3390/microorganisms9040699

Park S-H, Kim H-R, Baek Y-C, Ryu C-H, Ji S-Y, Jeong J-Y, Kim M, Jung H, Kim B (2021) Effects of dietary inclusion level of microwave-dried and press-defatted black soldier Fly (*Hermetia illucens*) larvae meal on productive performance, Cecal volatile fatty acid profile, and egg quality in laying hens. Animals 11:1486. https://doi.org/10.3390/ani11061486

Patterson PH, Acar N, Ferguson AD, Trimble LD, Sciubba HB, Koutsos EA (2021) The impact of dietary black soldier Fly larvae oil and meal on laying hen performance and egg quality. Poult Sci 100:101272. https://doi.org/10.1016/j.psj.2021.101272

Pérez-Pascual D, Estellé J, Dutto G, Rodde C, Bernardet J-F, Marchand Y, Duchaud E, Przybyla C, Ghigo J-M (2020) Growth performance and adaptability of European Sea bass (*Dicentrarchus labrax*) gut microbiota to alternative diets free of fish products. Microorganisms 8:1346. https://doi.org/10.3390/microorganisms8091346

Pleissner D, Smetana S (2020) Estimation of the economy of heterotrophic microalgae- and insect-based food waste utilization processes. Waste Manag 102:198–203. https://doi.org/10.1016/j.wasman.2019.10.031

Poore J, Nemecek T (2018) Reducing food's environmental impacts through producers and consumers. Science 360:987–992. https://doi.org/10.1126/science.aaq0216

Pulina P, Arru B, Madau F, Furesi R, Gasco L (2018a) Insect meal in the fish diet and feeding cost: first economic simulations on European Sea bass farming by a case study in Italy. https://doi.org/10.22004/ag.econ.275929

Pulina P, Arru B, Madau F, Furesi R, Gasco L (2018b) Insect meal in the fish diet and feeding cost: first economic simulations on European sea bass farming by a case study in Italy. In 30th International Conference of Agricultural Economists. International Association of Agricultural Economists (IAAE), Vancouver, British Columbia, Canada, pp 1–17

Radhakrishnan DK, Kumar S, Ali IA (2022) Different animal feeds and their role in aquaculture, aquaculture science and engineering. Springer, Berlin, pp 99–129

Raworth K (2017) Doughnut economics: seven ways to think like a 21st-century economist. Chelsea Green Publishing, Vermont

Rema P, Saravanan S, Armenjon B, Motte C, Dias J (2019) Graded incorporation of defatted yellow mealworm (*Tenebrio molitor*) in rainbow trout (*Oncorhynchus mykiss*) diet improves growth performance and nutrient retention. Animals 9:187. https://doi.org/10.3390/ani9040187

Reyes M, Rodríguez M, Montes J, Barroso FG, Fabrikov D, Morote E, Sánchez-Muros MJ (2020) Nutritional and growth effect of insect meal inclusion on seabass (*Dicentrarchuss labrax*) feeds. Aust Fish 5:16. https://doi.org/10.3390/fishes5020016

Rimoldi S, Gini E, Iannini F, Gasco L, Terova G (2019) The effects of dietary insect meal from *Hermetia illucens* Prepupae on autochthonous gut microbiota of rainbow trout (*Oncorhynchus mykiss*). Animals 9:143. https://doi.org/10.3390/ani9040143

Rimoldi S, Antonini M, Gasco L, Moroni F, Terova G (2021) Intestinal microbial communities of rainbow trout (*Oncorhynchus mykiss*) may be improved by feeding a *Hermetia illucens* meal/low-fishmeal diet. Fish Physiol Biochem 47:365–380. https://doi.org/10.1007/s10695-020-00918-1

Rockström J, Steffen W, Noone K, Persson Å, Chapin FS, Lambin EF, Lenton TM, Scheffer M, Folke C, Schellnhuber HJ (2009) A safe operating space for humanity. Nature 461:472–475

Ruhnke I, Normant C, Campbell DLM, Iqbal Z, Lee C, Hinch GN, Roberts J (2018) Impact of on-range choice feeding with black soldier fly larvae (*Hermetia illucens*) on flock performance, egg quality, and range use of free-range laying hens. Anim Nutr 4:452–460. https://doi.org/10.1016/j.aninu.2018.03.005

Rumpold BA, Schlüter OK (2013) Potential and challenges of insects as an innovative source for food and feed production. Innovative Food Sci Emerg Technol 17:1–11. https://doi.org/10.1016/j.ifset.2012.11.005

Salomone R, Saija G, Mondello G, Giannetto A, Fasulo S, Savastano D (2017) Environmental impact of food waste bioconversion by insects: application of life cycle assessment to process using *Hermetia illucens*. J Clean Prod 140:890–905. https://doi.org/10.1016/j.jclepro.2016.06.154

Schiavone A, Cullere M, De Marco M, Meneguz M, Biasato I, Bergagna S, Dezzutto D, Gai F, Dabbou S, Gasco L, Dalle Zotte A (2017a) Partial or total replacement of soybean oil by black soldier fly larvae (*Hermetia illucens* L.) fat in broiler diets: effect on growth performances, feed-choice, blood traits, carcass characteristics and meat quality. Ital J Anim Sci 16:93–100. https://doi.org/10.1080/1828051X.2016.1249968

Schiavone A, De Marco M, Martínez S, Dabbou S, Renna M, Madrid J, Hernandez F, Rotolo L, Costa P, Gai F, Gasco L (2017b) Nutritional value of a partially defatted and a highly defatted black soldier fly larvae (*Hermetia illucens* L.) meal for broiler chickens: apparent nutrient digestibility, apparent metabolizable energy and apparent ileal amino acid digestibility. J Anim Sci Biotechnol 8:51. https://doi.org/10.1186/s40104-017-0181-5

Schiavone A, Dabbou S, Petracci M, Zampiga M, Sirri F, Biasato I, Gai F, Gasco L (2019) Black soldier fly defatted meal as a dietary protein source for broiler chickens: effects on carcass traits, breast meat quality and safety. Animal 13:2397–2405. https://doi.org/10.1017/S1751731119000685

Seafish (2022a) European sea bass *Dicentrachus labrax* – Feed. Available at: https://www.seafish.org/responsible-sourcing/aquaculture-farming-seafood/species-farmed-in-aquaculture/aquaculture-profiles/european-sea-bass/feed/

Seafish (2022b) Rainbow trout *Oncorhynchus mykiss* - Sources, Quantities and Cultivation Methods. Available at: https://www.seafish.org/responsible-sourcing/aquaculture-farming-seafood/species-farmed-in-aquaculture/aquaculture-profiles/rainbow-trout/sources-quantities-and-cultivation-methods/

Sealey WM, Gaylord TG, Barrows FT, Tomberlin JK, McGuire MA, Ross C, St-Hilaire S (2011) Sensory analysis of rainbow trout, *Oncorhynchus mykiss*, fed enriched black soldier Fly

Prepupae, *Hermetia illucens*. J World Aquacult Soc 42:34–45. https://doi.org/10.1111/j.1749-7345.2010.00441.x

Secci G, Bovera F, Parisi G, Moniello G (2020) Quality of eggs and albumen technological properties as affected by *Hermetia Illucens* larvae meal in Hens' diet and hen age. Animals 10. https://doi.org/10.3390/ani10010081

Shah AA, Totakul P, Matra M, Cherdthong A, Hanboonsong Y, Wanapat M (2022) Nutritional composition of various insects and potential uses as alternative protein sources in animal diets. Anim Biosci 35:317–331. https://doi.org/10.5713/ab.21.0447

Shahbandeh M (2022) Global pork production 2021 & 2022, by country. Available at: https://www.statista.com/statistics/273232/net-pork-production-worldwide-by-country/

Smetana S, Zeidler VZ, Pleissner D (2022) Introduction: waste to food – returning nutrients to the food chain, waste to food. Wageningen Academic Publishers, Wageningen, pp 9–12. https://doi.org/10.3920/978-90-8686-929-9_i

Spranghers T, Michiels J, Vrancx J, Ovyn A, Eeckhout M, De Clercq P, De Smet S (2018) Gut antimicrobial effects and nutritional value of black soldier fly (*Hermetia illucens* L.) prepupae for weaned piglets. Anim Feed Sci Technol 235:33–42. https://doi.org/10.1016/j.anifeedsci.2017.08.012

Spykman R, Hossaini SM, Peguero DA, Green A, Heinz V, Smetana S (2021) A modular environmental and economic assessment applied to the production of *Hermetia illucens* larvae as a protein source for food and feed. Int J Life Cycle Assess 26:1959–1976. https://doi.org/10.1007/s11367-021-01986-y

Stanković D, Crivelli AJ, Snoj A (2015) Rainbow trout in Europe: introduction, naturalization, and impacts. Rev Fish Sci Aquac 23:39–71. https://doi.org/10.1080/23308249.2015.1024825

Star L, Arsiwalla T, Molist F, Leushuis R, Dalim M, Paul A (2020) Gradual provision of live black soldier Fly (*Hermetia illucens*) larvae to older laying hens: effect on production performance, egg quality, feather condition and behavior. Animals 10:216. https://doi.org/10.3390/ani10020216

Stastnik O, Novotny J, Roztocilova A, Kouril P, Kumbar V, Cernik J, Kalhotka L, Pavlata L, Lacina L, Mrkvicova E (2021) Safety of mealworm meal in layer diets and their influence on gut morphology. Animals 11:1439. https://doi.org/10.3390/ani11051439

Statista (2021). Available at: https://www.statista.com/statistics/768350/egg-production-in-western-europe/

Stenberg OK, Holen E, Piemontese L, Liland NS, Lock E-J, Espe M, Belghit I (2019) Effect of dietary replacement of fish meal with insect meal on *in vitro* bacterial and viral induced gene response in Atlantic salmon (*Salmo salar*) head kidney leukocytes. Fish Shellfish Immunol 91:223–232. https://doi.org/10.1016/j.fsi.2019.05.042

St-Hilaire S, Sheppard C, Tomberlin JK, Irving S, Newton L, McGuire MA, Mosley EE, Hardy RW, Sealey W (2007) Fly Prepupae as a feedstuff for rainbow trout, *Oncorhynchus mykiss*. J World Aquacult Soc 38:59–67. https://doi.org/10.1111/j.1749-7345.2006.00073.x

Tahamtani FM, Ivarsson E, Wiklicky V, Lalander C, Wall H, Rodenburg TB, Tuyttens FAM, Hernandez CE (2021) Feeding live black soldier Fly larvae (*Hermetia illucens*) to laying hens: effects on feed consumption, hen health, hen behavior, and egg quality. Poult Sci 100:101400. https://doi.org/10.1016/j.psj.2021.101400

Tan X, Yang HS, Wang M, Yi ZF, Ji FJ, Li JZ, Yin YL (2020) Amino acid digestibility in housefly and black soldier fly prepupae by growing pigs. Anim Feed Sci Technol 263:114446. https://doi.org/10.1016/j.anifeedsci.2020.114446

Terova G, Gini E, Gasco L, Moroni F, Antonini M, Rimoldi S (2021) Effects of full replacement of dietary fishmeal with insect meal from *Tenebrio molitor* on rainbow trout gut and skin microbiota. J Anim Sci Biotechnol 12:30. https://doi.org/10.1186/s40104-021-00551-9

Uwizeye A, de Boer IJ, Opio CI, Schulte RP, Falcucci A, Tempio G, Teillard F, Casu F, Rulli M, Galloway JN (2020) Nitrogen emissions along global livestock supply chains. Nature Food 1:437–446. https://doi.org/10.1038/s43016-020-0113-y

Van Gelder J, Kuepper B (2012) Verdeling van de economische waarde van de mondiale sojateelt. Profundo, Amsterdam, p 44

Van Huis A, Oonincx DG (2017) The environmental sustainability of insects as food and feed. A review. Agron Sustain Dev 37:1–14. https://doi.org/10.1007/s13593-017-0452-8

Van Huis A, Van Itterbeeck J, Klunder H, Mertens E, Halloran A, Muir G, Vantomme P (2013) Edible insects: future prospects for food and feed security. Food and agriculture organization of the United Nations, Rome

Vandeputte M, Gagnaire PA, Allal F (2019) The European sea bass: a key marine fish model in the wild and in aquaculture. Anim Genet 50:195–206. https://doi.org/10.1111/age.12779

Varelas V (2019) Food wastes as a potential new source for edible insect mass production for food and feed: a review. Fermentation 5:81. https://doi.org/10.3390/fermentation5030081

Veldkamp T, Van Niekerk T (2019) Live black soldier fly larvae (*Hermetia illucens*) for Turkey poults. J Insects Food Feed 5:301–311. https://doi.org/10.3920/JIFF2018.0031

Veldkamp T, Vernooij AG (2021) Use of insect products in pig diets. J Insects Food Feed 7:781–793. https://doi.org/10.3920/jiff2020.0091

Venold FF, Penn MH, Krogdahl Å, Overturf K (2012) Severity of soybean meal induced distal intestinal inflammation, enterocyte proliferation rate, and fatty acid binding protein (Fabp2) level differ between strains of rainbow trout (*Oncorhynchus mykiss*). Aquaculture 364-365:281–292. https://doi.org/10.1016/j.aquaculture.2012.08.035

Wang S, Hu X, Wang Z, Li X, Wang Q, Wang Y, Tang Z, Li H (2012) Quantitative trait loci associated with body weight and abdominal fat traits on chicken chromosomes 3, 5 and 7. Genet Mol Res 11:956–965

Weththasinghe P, Hansen JØ, Nøkland D, Lagos L, Rawski M, Øverland M (2021) Full-fat black soldier fly larvae (*Hermetia illucens*) meal and paste in extruded diets for Atlantic salmon (*Salmo salar*): effect on physical pellet quality, nutrient digestibility, nutrient utilization and growth performances. Aquaculture 530:735785. https://doi.org/10.1016/j.aquaculture.2020.735785

Weththasinghe P, Hansen JØ, Mydland LT, Øverland M (2022) A systematic meta-analysis based review on black soldier fly (*Hermetia illucens*) as a novel protein source for salmonids. Rev Aquac 14:938–956. https://doi.org/10.1111/raq.12635

Willet J (2019) INCUBATOR lab: where artists collaborate with life. Publica 30:46–55. https://doi.org/10.1386/public.30.59.46_1

WNF (2011) Ontbossing. Available at: https://www.wwf.nl/wat-we-doen/focus/bossen/ontbossing

Yamada Y (1988) The contribution of poultry science to society1. Worlds Poult Sci J 44:172–178

Yu M, Li Z, Chen W, Rong T, Wang G, Li J, Ma X (2019) Use of *Hermetia illucens* larvae as a dietary protein source: effects on growth performance, carcass traits, and meat quality in finishing pigs. Meat Sci 158:107837. https://doi.org/10.1016/j.meatsci.2019.05.008

Zentek J, Buchheit-Renko S, Ferrara F, Vahjen W, Van Kessel AG, Pieper R (2011) Nutritional and physiological role of medium-chain triglycerides and medium-chain fatty acids in piglets. Anim Health Res Rev 12:83–93. https://doi.org/10.1017/S1466252311000089

Zuidhof MJ, Schneider BL, Carney VL, Korver DR, Robinson FE (2014) Growth, efficiency, and yield of commercial broilers from 1957, 1978, and 2005. Poult Sci 93:2970–2982. https://doi.org/10.3382/ps.2014-04291

Probiotics, Prebiotics, Paraprobiotics, Postbiotics

Ioannis Panitsidis, Florence Barbe, Eric Chevaux, Ilias Giannenas, and Vanessa Demey

Abstract

Antibiotics have long been used to preserve animal health, improve growth, and increase efficiency in animal industries, but the overuse of antibiotics as growth promoters led to the development of resistant bacteria. As a result of this predicament, researchers investigated safer alternative biotechnological breakthroughs, which led notably to the use of probiotics (bacteria and yeast) and prebiotics as feed additives in livestock. Since then, the focus of research teams has shifted to more varied fields of application of probiotics and prebiotics, in animal nutrition. Nowadays, these additives are commonly included in feeds for various species of production animals (swine, poultry, ruminants, and aquaculture). Recently, paraprobiotics and postbiotics are also being studied. The reason for the inclusion of these "biotics" in animal feeds is wide: promotes animal health and productivity by enhancing gut health, nutrient utilization, as well as boosting immune system functionality and reducing foodborne pathogen carriage. In more recent years, an increased concern of the consumer on the sustainability of animal production has been recorded. This chapter aims at reviewing research findings on the potential application of "biotics" into an integrated approach of sustainable farming practices, which is not only limited in terms of environmental concerns but also about of combining profitability with an increase in animal welfare.

I. Panitsidis · I. Giannenas (✉)
Laboratory of Nutrition, Faculty of Veterinary Medicine, School of Health Sciences, Thessaloniki, Greece
e-mail: igiannenas@vet.auth.gr

F. Barbe · E. Chevaux · V. Demey
Lallemand SAS, Blagnac, France

G. Arsenos, I. Giannenas (eds.), *Sustainable Use of Feed Additives in Livestock*,
https://doi.org/10.1007/978-3-031-42855-5_8

Keywords

Probiotics · Prebiotics · Yeasts · Yeast extracts · Immunity

1 Introduction

Since their discovery, antibiotics have been used to increase growth in a variety of animal species by adding them to their meals, to cure infectious diseases, or prevent them when given in sub-therapeutic levels. However, besides killing the vulnerable bacteria, the administration of antibiotics results in the development of some resistant strains. These resistant strains grow in number over time and may spread resistant genes to other bacteria. These resistant bacteria may spread from one host to another, either directly or indirectly, reduce the potency of medications, and lead to antibiotic resistance (Palma et al. 2020). Such gene transfer can occur horizontally between bacteria or vertically during reproduction of the bacteria, from one ecosystem to another (animal to human, environment to human but also human to animal, e.g., (Beukers et al. 2018) in intensively reared cattle, (Hedman et al. 2020) in poultry farming systems with the risk of sharing at once the multiple resistance genes hosted by one microorganism to another bacterial species (Sun et al. 2019; Lerminiaux and Cameron 2019). As a result, a global movement to reduce the use of antibiotics in animals has been applied, and many countries have outlawed their usage (Castanon 2007; FDA 2012).

Several microbes have developed resistance as a result of improper antibiotic treatment: *Pseudomonas aeruginosa* has developed resistance to carbapenems, quinolones, aminoglycosides, cephalosporins, penicillins, and monobactam (Odoi 2016); Resistance to β-lactams has evolved in *Staphylococcus* species (Mamza et al. 2010); Ampicillin, tetracycline, trimethoprim, ciprofloxacin, and sulfamethoxazole are no longer effective against *Salmonella* spp. and *Escherichia coli* (Medeiros et al. 2011; Van den Bogaard and Stobberingh 2000). More bacteria have already emerged or may do so in the near future if the current epidemic of antibiotic resistance does not end.

In order to prevent the reduction in animal production performance and financial losses, alternatives are more in demand as a result of the limited use or ban of antibiotics in feed. Since the late 1990s up until the present, nutritionists and pharmacists have been working on producing various replacements to preserve or enhance animal health and performance. The use of probiotics and prebiotics, which have demonstrated their effectiveness for both humans and animals, is one alternative with high expectations. Additionally, during the past 10 years, studies on paraprobiotics and postbiotics have produced encouraging results that justify their use.

In terms of definition, probiotics are living bacteria that benefit their host when given in sufficient quantities (Arora and Baldi 2015). Regarding prebiotics, the sixth Meeting of the International Scientific Association of Probiotics and Prebiotics (ISAPP) defined them as "a selectively fermented ingredient that results in specific changes in the composition and/or activity of the gastrointestinal microbiota, thus conferring benefit(s) upon host health" (Gibson et al. 2010). Paraprobiotics are characterized as non-viable, inactivated microbial cells that provide the host with health benefits (Siciliano et al. 2021). Last but not least, postbiotics are the complex mixture of metabolic products secreted by probiotics in cell-free supernatants such as enzymes, secreted proteins, short chain fatty acids, vitamins, secreted biosurfactants, amino acids, peptides, organic acids, etc. (Nataraj et al. 2020).

Probiotics and prebiotics have the potential to reduce the environmental impact of meat, milk, egg, and fish production. By improving feed efficiency, probiotics can contribute to a better utilization of the feed, hereby reducing the carbon footprint per unit of food produced. Their potential effect on digestibility allows the farmers to have more flexibility on the feed stuffs they provide to the animals. This can also allow for more locally grown raw materials to be included in feed rations. In addition, better utilization of the feed by the animal should result in lower losses on N and P into the environment via the animal excretions. The positive effects probiotics and prebiotics can have on animals' immunity and disease resistance, can contribute to an improved animal welfare. In addition to probiotics, postbiotics have also been gaining attention as a potential solution for improving animal health and productivity. Postbiotics refer to the metabolic by-products of probiotics, such as organic acids and peptides, that have been shown to have health-promoting effects. Unlike probiotics, which are live microorganisms, postbiotics are non-viable and therefore do not carry the risk of pathogen contamination. The use of postbiotics in animal nutrition has the potential to have all the benefits of probiotics without their side-effects, making it a promising area of research. However, the research is still in its early stages, and there are many new things to learn for all the type of "biotics" and their potential applications. This review will attempt to explain all the recent advances in the field of probiotics, prebiotics, paraprobiotics, and postbiotics, along with what the most recent research has revealed about how they function in various organs and how they improve animal health.

2 Probiotics

History The phrase "probiotic" is contrasted to the term "antibiotic" and is derived from the Greek words "pro" and "bios," that together mean "for life." Elie Metchnikoff, a researcher at the Pasteur Institute in Paris, was the pioneer in the field of probiotics. In 1907, as he was working in Bulgaria, he discovered that Bulgarian peasants who consumed a lot of spoiled milk lived longer than average. This confirmed Metchnikoff's hypothesis that bacteria from the sour milk would have an impact on the lower gut and general health. From there, Metchinkoff made the decision to continue his research and his findings encouraged other researchers from around the world to start their own research on probiotics. Later in 1965 Probiotics were redefined by Lilley and Stillwell. They defined probiotics as microbes that would promote the growth of other beneficial microbes in the intestines (Lilly and Stillwell 1965). Probiotics were more clearly described by

Fuller in 1992 as "a live microbial feed additive that benefits the host animal by enhancing its gut microbial balance" (McFarland 2015). Since the discovery of *Lactobacillus acidophilus* in 1890 and Henry Tissier's discovery of the species of *Bifidobacteria* later that same decade, scientists have been able to isolate numerous strains of good bacteria for the digestive tract. After that, Henri Boulard discovered the probiotic yeast *Saccharomyces boulardii* in 1923 and Minoru Shirota identified a novel strain of *Lactobacillus casei* in the 1930s. Since then, other helpful bacteria have been discovered, and there will undoubtedly be many more in the future.

The most frequently used probiotics in animal feed are the genera *Bifidobacterium* (*thermophilum, pseudolongum, longum*) (Mattarelli and Biavati 2018), *Bacillus* (*subtilis*) (Cutting 2011), *Lactobacillus* (*acidophilus, bulgaricus, plantarum*), *Lactococcus* (*lactis*), *Enterococcus* (*faecalis, faecium*), and *Saccharomyces* (*cerevisiae, boulardii*) (Muzaffar et al. 2021). Probiotics can be given in feed or water as a single strain, a multi-strain mixture, or as an addition to other feed additives. It is generally agreed upon that probiotics containing multiple strains are more beneficial than those containing a single strain due to their synergistic effects (Bhogoju and Nahashon 2022).

For their use to be optimized, probiotics must first be evaluated for their safety, and this is a challenging process. Probiotic organisms that adhere to the digestive tract must endure harsh conditions and must have a positive impact on the stability and security of the intestinal environment. Additionally, they should affect the way digestion, metabolism, and the immune system respond (Patel et al. 2015). The selection of novel probiotic organisms entails choosing the strains and even genera of microbes that have the most advantageous or targeted effects. The evaluation mainly focuses on security and the benefit-to-risk ratio connected with the use of a specific probiotic strain. Furthermore, probiotic strains added to feed should be resistant to pelleting temperatures and pressures, humidity, and the effects of harmful compounds during feed handling and storage, such as heavy metals or mycotoxins. The probiotics' peak duration of activity in feed and premixes cannot be less than 4 months. Encapsulating formulas allows for a longer survival time for strains, extending that time frame (Hollister et al. 1989) (Fig. 1).

Mechanism of Action The mechanisms of action of probiotics are still not entirely understood because they are numerous, diversified, and strain-specific. Some known mechanisms include pathogen exclusion and bacteriocin production, modulation of fecal enzymatic activities, production of short-chain and branched-chain fatty acids, cell adhesion and mucin production, immune system modulation, and interaction with the brain-gut axis via regulation of endocrine and neurologic functions.

Competitive exclusion occurs when one species of bacteria outcompetes other species for receptor sites in the intestinal tract (Bermudez-Brito et al. 2012). These events can occur as a result of a decrease in the pH of the surrounding environment, competition for nutritional sources, and the production of bacteriocin or bacteriocin-like substances (Collado et al. 2010). Moreover, probiotics can change the metabolism of bile acids in the gut lumen, which in turn affects how much cholesterol is absorbed. Most known probiotics and bacterial species from numerous genera

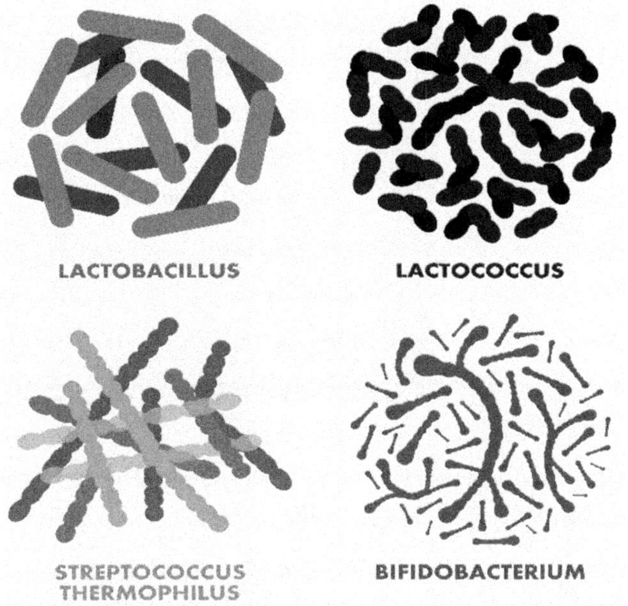

Fig. 1 Shape of different probiotic bacteria

associated with the digestive tract generate bile salt hydrolase, an enzyme that may take part in the initial reaction of the deconjugation of biliary salts (Pavlovic et al. 2012). The immune system is modulated by the gut microbiota through the synthesis of chemicals that have anti-inflammatory and immunomodulatory properties and can activate immune cells. Probiotic bacteria interact with epithelial cells and DCs, as well as macrophages and lymphocytes, to provide these immunomodulatory effects (D'Amelio and Sassi 2017). The imbalance of the gut microbiota can influence the brain-gut axis and may be directly related to stress brought on by either physical or environmental reasons (Dinan et al. 2006). Overall, the neurological, endocrine, and immunologic mechanisms underlying the impacts of the gut intestinal microbiota on the central nervous system are complex, although these effects are thought to primarily occur via the creation of bacterial metabolites (Ong et al. 2018). Gut bacteria produce a wide range of neuroactive substances, including dopamine, γ-aminobutyric acid, histamine, acetylcholine, and tryptophan, which is a precursor in the production of serotonin (Fig. 2).

Probiotics on Performance Although more research is needed to determine the exact mechanism by which probiotics improve growth performance, research findings have shown that they are growth promoters, they decrease the impact of stress and mitigate the severity of disease by altering the gut environment and enhancing gut barrier function through the strengthening of useful gut microflora, competitive exclusion of pathogens, and immune system stimulation.

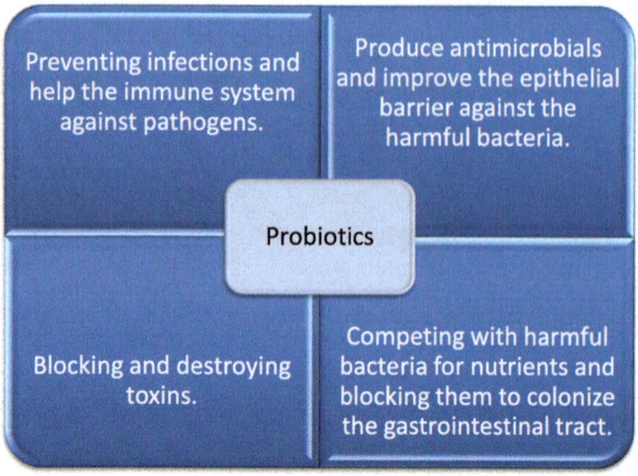

Fig. 2 Mechanism of action

Lactobacillus has demonstrated to be quite powerful in broiler development performance, according to numerous studies (Gao et al. 2017). When *Lactobacillus* was added to the diet of layer chickens, the egg weight increased (Pambuka et al. 2014). The same results were observed in terms of growth, performance, and survival rate when fish were supplemented with *Lactobacillus* (Selvaraj and Bogar 2019; Rahman et al. 2019). In ruminants, *Lactobacillus* is used in young calves, where it has been observed to improve weight gain (Al-Saiady 2010). *Lactobacillus* increased the rate of weight gain in pigs weaned at 21 days because the synthesis of digestive enzymes improved meal conversion (Tian et al. 2020). A substantial portion of the animal microbiota is made up of the genus *Lactobacillus*, which is composed of Gram-positive, anaerobe, non-spore producing bacteria. There are 44 species of *Lactobacillus* presently (Zheng et al. 2020).

Bifidobacterium greatly boosts production effectiveness, feed conversion, and weight gain in poultry (Palamidi et al. 2016). Similar outcomes *Bifidobacterium* generates in the fish affecting the overall health and the growth performance (Khalafalla et al. 2020). The genus *Bifidobacterium* is a Gram-positive, non-motile, and frequently branched anaerobic bacteria. One of the most important bacterial genera that contributes to the microbiota of the gastrointestinal tract in mammals is the *Bifidobacteria*.

Another probiotic strain that has been used as a growth promoter is *Bacillus* sp. which have been shown to improve broiler overall body weight gain (Wang et al. 2016a). Moreover, *Bacillus* has been shown to boost egg production, egg production, eggshell calcium content, thickness, and weight in laying hens (Wang et al. 2021; Li et al. 2006). A more recent study found that *Bacillus pumilus* increased final weight and weight gain, and improved feed conversion ratio when administered in broilers. It also increased counts of *Lactobacillus* and *Bacillus* spp. in the ileum and the ceca (Bonos et al. 2021). In fish, *Bacillus* improved the growth rate, survival, and weight gain (Sadat Hoseini Madani et al. 2018) and when administered in rainbow trout improved the growth rate and the feed efficiency (Giannenas et al. 2014). In Holstein calves the intake of dry matter and average daily gain were both enhanced by *Bacillus* (Zhang et al. 2016). Finally, in swine production inclusion of *Bacillus* complex to the diet increased daily average gains and feed consumption (Ahmed et al. 2014). Additionally, Cai et al. (2015) found that adding dietary probiotics based on the *Bacillus* strain enhanced average daily gain during days 0–14 following weaning and had a beneficial impact on raising gain/feed ratios between days 0 and 42.

Lastly, *Saccharomyces cerevisiae*, a probiotic yeast that is frequently employed, has been demonstrated in studies to dramatically boost the body weight gain of broilers feed consumption and feed conversion efficiency (Shareef and Al-Dabbagh 2009). Additionally, a study revealed that layers treated with *Saccharomyces cerevisiae* produced noticeably more eggs (Hassanein and Soliman 2010). The same outcomes were observed in terms of improved feed consumption and growth performance when *Saccharomyces* were administered to fish (Jahan et al. 2021). Moreover, supplementing horses with yeast can enhance dietary digestion and assimilation of food nutrients which results in better equine performance (Julliand et al. 2006). *Saccharomyces cerevisiae* has been also shown to enhance performance in dairy ruminants, with the most recurrent benefits being an increase in dry matter intake and milk output (Poppy et al. 2012). Furthermore, it has been demonstrated that probiotic yeast increases some beef cattle productivity indicators which includes the ability to digest their food and improved growth performance (Batista et al. 2022). In swine industry supplementing sows with *S. cerevisiae* increased the number of piglets born and the percentage of live births. Another study found no difference in growth performance between pigs fed diets supplemented with yeast and pigs fed antibiotic growth boosters (Agazzi et al. 2015; Shen et al. 2009).

Probiotics on Immune Response An organism must be healthy and have a strong immune system in order to thrive and prosper. Because of this, antibiotics have been used for a long time to encourage growth or act as a prophylactic against potentially hazardous microorganisms. However, due to excessive antibiotic use, the emergence of novel pathogens that are resistant to antibiotics is something that veterinarians frequently discover in livestock. For this reason, probiotics are currently regarded as the best antibiotic alternative that secures health and encourages growth.

It has been demonstrated that *Lactobacillus* affected the chemokine gene expression and cytokine production in the intestinal epithelium of broiler chickens. A further way that *Lactobacillus* affects the immune system is through boosting β-lymphocytes production of antibodies, increased the blood CD4+ lymphocyte count as well as interferon and tumor necrosis factor expression in the ileum (Haghighi et al. 2008). According to another study, vaccination and *Lactobacillus* together with probiotic mix may have worked synergistically to reduce the amount of *Salmonella enteritidis* in broilers (Praharaj et al. 2015). The same outcomes against salmonella were reported in laying hens following vaccination and *Lactobacillus* supplementation (Groves et al. 2021). Except *Salmonella*, *Lactobacillus* has shown good results against *Eimeria* sp. by lowering the quantity of oocysts in the feces and raising the levels of CD3, CD4, and CD8 lymphocytes as well as IL-2 interleukin in the intestines of poultry with *Eimeria acervulina* infection (Dalloul et al. 2003). The same results against *Eimeria* were reported by Giannenas et al. (2012) who administered a combination of *Lactobacillus*, *Bacillus*, and *Bifidobacterium* in broiler chickens challenged with *Eimeria* spp. In fish, *Lactobacillus* boosted lysozyme and IgM, enhanced immunological responses, and increased disease resistance (Hassani et al. 2020). Additionally, *Lactobacillus* spp. led to a significant improvement in immune responses as well as a reduction in the overall mortality rate after challenged with *Vibrio harveyi* infection (Geng et al. 2012]). A similar type of animal response on reduced mortality was reported by Castex et al. (2008, 2010) in Vibrio challenged shrimps supplemented with a strain of *Pediococcus acidilactici*.

In horses, *Lactobacillus* in combination with *Enterococcus* lowered the frequency of diarrhea in newborn foals and reduced the incidence of *Salmonella* shedding by about 65% in probiotic-treated horses (Ward et al. 2004). Furthermore, in another study, strains of *Lactobacillus* reduced pathological score in inflammatory bowel disease, increased colon length, and controlled the cytokine release (Qin et al. 2022). In ruminants, the administration of a mixture consisting of various *Lactobacillus* species to dairy calves resulted in an improvement of the animals' overall health. The mortality and morbidity rates tended to be lower (Maldonado et al. 2017). In young calves, *Lactobacillus* bacteria are most frequently employed because they lower the number of Gram-positive bacteria present. Early *Lactobacillus* administration to calves is particularly advantageous as it enhances immunocompetence, lowers the occurrence of diarrhea cases by more than 70%, and speeds up T-cell transformation (Zhang et al. 2016; Al-Saiady 2010). *Lactobacillus* raises insulin and blood sugar levels in dairy cows and stimulates macrophages to generate cytokines (Matsuguchi et al. 2003; Oetzel et al. 2007). In pregnant cows, *Lactobacillus* reduced the incidence of metritis, uterine infections, and laminitis (Deng et al. 2015). Finally, *Lactobacillus* altered the gut microbiota of goats by decreasing *Salmonella*/*Shigella*-like enterobacteria and by increasing other lactic bacteria (Apas et al. 2010).

Supplementing neonatal pigs with *Lactobacilli* aids in the early establishment of a stable gut flora, boosts the immune system, and stops diarrhea. In neonatal pigs, oral treatment of *Lactobacillus* enhanced intestinal health, increased *Lactobacilli* and *Bifidobacterium*, and decreased the number of possible entero-pathogens such as *E. coli* and *Clostridia* (Liu et al. 2014; Chiang et al. 2015). Similar outcomes were observed earlier when *Lactobacillus* was administered to piglets that had been exposed to *E. coli*; *Lactobacillus* improved T-cell differentiation and triggered cytokine production in the ileum (Wang et al. 2009). *Lactobacillus* exhibited similar advantages when administered to grower-finisher pigs (Giang et al. 2011). In companion animals, *Lactobacillus* administration to puppies prevented gastrointestinal infection (Fernández et al. 2019). When administered to dogs with gastrointestinal disorders, lactic acid bacteria populations increased, clostridia populations declined, and some Gram-negative bacterial genera were also reduced. Blood samples also showed improvement in total protein, cholesterol, and ALT levels (Strompfová et al. 2017). Similar results were shown in cats, where *Lactobacillus* increased the beneficial bacteria and decreased the harmful bacteria *Clostridium* spp.

and *Enterococcus faecalis*. Additionally, it reduced plasma endotoxin levels, causing systemic and immunomodulatory modifications in the treated cats (Marshall-Jones et al. 2006).

Despite not being as popular as *Lactobacillus*, using *Lactococcus* has demonstrated some positive effects on the regulation of the immune system. In broilers, *Lactococcus* and inulin administration affected the development of central and peripheral lymphatic organs and improved immune responses related to stimulation of Peyer's patch and cecal tonsil colonization by T-cells (Madej et al. 2015). The administration of *Lactococcus* in fish increased lysosomal activities and production of IL-12 and IFN-g. Additionally, it demonstrated potent antibacterial properties against the dangerous pathogens *Streptococcus* and *Enterococcus*. Only 5.7% of the control group survived at 20 days after being challenged with *Streptococcus iniae*, compared to 65.7% of those who consumed *Lactococcus*, demonstrating a considerable level of protection (Kim et al. 2013). Similar findings were made in a different study that suggested *Lactococcus* regulated innate immune parameters and provided protection against *Streptococcosis* (Hasan et al. 2018). According to results of another study, *Lactococcus* can be employed as a vaccine carrier that may provide protection against multiple Vibrio species and can enhance T-cell numbers, cytokine production, and innate immune responses (Lee et al. 2021). In ruminants, *Lactococcus* effects have been studied extensively as a therapy for mastitis in dairy cows. All studies have concluded that it is a promising candidate strain for either mastitis prevention or therapy (Bouchard et al. 2015; Malvisi et al. 2016). In swine, *Lactococcus* is frequently administered to weaning piglets to lessen the likelihood of diarrhea and to regulate intestinal immunity. According to a study, *Lactococcus* increased the expression of the genes for the anti-inflammatory cytokines while suppressing the expression of interleukin-g in the jejunum and interleukin-22 in the ileum (Yu et al. 2021).

Bifidobacterium is another useful probiotic for the immune system. *Bifidobacterium*, in particular, it has typically been viewed as an anti-inflammatory, it has been shown to interact with immune cells, and modify immunological pathways. *Bifidobacterium* treatment improved IFN levels in chickens with *Clostridium* infections, had an anti-*Campylobacter* effect, and increased the production of antibodies against the Newcastle disease virus in layers (Kim et al. 2007; Santini et al. 2010; Hatab et al. 2016). Additionally, in broilers, *Bifidobacterium* and lactic acid bacteria increased systemic immunity and local antibody production (Haghighi et al. 2006). Finally, it has been shown that *Bifidobacterium* probiotic strains raised the levels of immunoglobulin (Ig) A, G, and M in broilers and turkeys, which improves the disease resistance (Abdel-Moneim et al. 2019). The same outcomes were observed in fish, where *Bifidobacteria* were used to stimulate the creation of immunostimulants in the fish intestines that prevented the growth of several pathogenic bacteria (Mussatto and Mancilha 2007). These conclusions are supported by two further studies, where *Bifidobacterium* use led to an enhanced overall fish health (Hassani et al. 2020; Khalafalla et al. 2020).

In ruminants, *Bifidobacterium* usage is more prevalent in young calves. Young calves given *Bifidobacterium* and *Lactobacillus* throughout the early stages of life gained more weight and experienced fewer diarrheal episodes. Likewise, giving probiotic *Bifidobacteria* to calves reduced the levels of *E. coli* and total coliforms while raising the relative abundance of *Lactobacilli* (Shehta et al. 2019; Geigerova et al. 2016). Another study found that the administration of *Bifidobacterium* was able to raise interferon (IFN) levels. This could have a major impact on intestinal viral defense because of IFN's ability to boost the expression of antiviral proteins that can reduce or inhibit viral replication (Kobayashi et al. 2017). In weaned pigs, *Bifidobacterium* improved gut health and immunity, and decreased the pathogen number after the *Salmonella* challenge (Herfel et al. 2013; Barba-Vidal et al. 2017). Similar to this, a recent study discovered that taking *Bifidobacterium* orally can prevent *Salmonella* infection (Splichalova et al. 2021). Additionally, another study suggests that *Bifidobacterium* modifies gut microbiota, enhances intestinal barrier function, and regulates inflammatory cytokines (Arenas-Padilla et al. 2018). Finally, a microbiota analysis revealed that a diet supplemented with *Bifidobacterium animalis* significantly increased the relative abundances of beneficial bacteria. This study also demonstrated that using *Bifidobacterium* significantly improved growth performance and reduced diarrhea incidence (Pang et al. 2022).

Another probiotic that stimulates immunity is *Bacillus*. Their capacity to naturally generate spores that resume viability under favorable conditions gives them a significant edge over other probiotic species. *Bacillus* species frequently have the ability to activate the host immune system in chickens. According to research by (Mingmongkolchai and Panbangred 2018), *B. subtilis* may be involved in the enhancement of the IgA response, which is essential for protection against mucosal infections. In addition, Lee et al. (2015b) displayed immunological responses to *Eimera* spp. and *Clostridium perfringens*, the causal agents of necrotic enteritis. Khaksefidi and Ghoorchi (2006) showed that feeding *B. subtilis* to broiler chicks had a favorable effect on antibody generation against Newcastle disease. By boosting the number of intraepithelial lymphocytes and immunoglobulin-producing cells, *Bacillus* spores are said to aid in the development of the gut-associated lymphoid tissues (Molnár et al. 2011). According to a study by (Wang et al. 2018b), *B. subtilis* was able to reduce inflammation brought on by heat stress by producing more anti-inflammatory cytokines.

The use of *Bacillus* in aquaculture resulted in an improvement in immunological function overall, a rise in survival rates, and an increase in phagocytic, respiratory burst, catalase, nitric oxide synthase, acid phosphatase, amylase, and protease activities (Zhao et al. 2019). According to another report, adding *Bacillus subtilis* as a probiotic to a diet improved fertility, viability, and the quantity of normal fry generated (Ghosh et al. 2007). Finally, but certainly not least, the use of bacillus in fish has shown that it can reduce Vibrio populations and defend against *Aeromonas hydrophila* (Hostins et al. 2017; Devi et al. 2019). In horses, despite not being able to colonize the equine intestinal system, *Bacillus* organisms can survive in the large intestines of horses for up to two days and may even have positive effects by preventing the proliferation of harmful bacteria. Therefore, bacillus research on this species is not as frequent. According to research, bacillus has been effective against *Clostridium* spp., *E. coli*, *Salmonella* spp., *Rhodococcus equi*, and

Streptococcus equi. Moreover, horses are susceptible to developing intestinal bacterial overgrowth, enterocolitis, and diarrhea. Considering this, researchers looked into how *Bacillus subtilis* affects pathogenic elements in the equine digestive tract that frequently thrive despite antibiotic treatment (Burke and Morre 2017).

In ruminants, bacillus has been utilized in dairy cows where it enhanced milk IgA and lowered serum haptoglobin (Luan et al. 2015). According to a recent study, a combination of *Lactobacillus plantarum* and *Bacillus subtilis* used in Holstein calves as dietary supplements has been shown to have positive effects on the immune system because they boost the rate of T-cell transformation and decrease cortisol levels (Zhang et al. 2016). Last but not least, a study that was done to manage mastitis problem in dairy cows revealed that bacillus genera can limit the growth of mastitis pathogens more effectively against Gram-positive *S. aureus*, *S. epidermidis*, or *Trueperella pyogenes* than against Gram-negative *E. coli* or *Klebsiella* (Woodward et al. 1987). The use of *Bacillus* in swine revealed that piglets exposed to *Escherichia coli* had considerably higher serum IgG levels (Ahmed et al. 2014). Another study discovered that the use of a *Bacillus subtilis* mixture in sows before and throughout lactation could increase the number of piglets born overall and born alive, improve beneficial bacteria counts, and reduce the concentration of dangerous bacteria in the piglets' small intestine after birth (Baker et al. 2013). (Link et al. 2016) found that adding *Bacillus* to a breastfeeding piglet's diet decreased the diarrhea score.

Enterococci are common in the environment as commensal and pathogenic microorganisms, as well as gut symbiotic organisms. The *Enterococcus* spp. strains are well adapted to various food systems because of their tolerance to salts and acids. Additionally, a variety of *Enterococcus* strains have been found to produce bacteriocins and other antimicrobial substances. *Enterococcus* is a genus that has not yet been given the classification of "generally recognized as safe," although some of its species are employed as probiotics and feed additives to stop diarrhea or boost animal development (Franz et al. 2011). In chicken, Enterococcus was employed, and the outcome was an increase in IgA production (Beirão et al. 2018). In all other studies, *Enterococcus* was combined with various beneficial bacteria, and the results included a notable increase in the weight of immunological organs and antibody production, an improvement in antibody levels against the virus that causes Newcastle disease, and the induction of an anti-inflammatory response (Kabir et al. 2004; Hatab et al. 2016; Palamidi et al. 2016). A combination of *Enterococcus* and *Lactobacillus* strains had good effects in vitro, according to (Chaveerach et al. 2004), but only native bacteria obtained from adult chickens could protect young broilers from *Campylobacter* colonization. A more recent study found that the use of *Enterococcus faecium* can improve intestinal barrier damage caused by necrotic enteritis (Wu et al. 2019). Comparable results were observed in fish once *Enterococcus* was applied to their meal. All the studies found enhanced immunity and an increased survival rate. Particularly, Chang and Liu (2002) found that the administration of *Enterococcus* enhanced disease resistance against *Edwardsiella tarda* a well-known cause of hemorrhagic septicemia in fish. Later Wang et al. (2008) concluded that the use of *Enterococcus* influenced the immune system by increasing blood phagocyte respiratory burst activity. Lastly, Talpur et al. (2014) found that *Enterococcus* can enhance immunity and defense against infection by *Aeromonas hydrophila*.

In horses, *Enterococcus* is used in combination with other probiotics or prebiotics. Results from research by (Ward et al. 2004) on the decrease of *Salmonella* fecal shedding were encouraging. An oral gel product comprising *Lactobacillus* and *Enterococcus faecium* was given to the horses during the trial. The incidence of *Salmonella* shedding decreased as a consequence. On another study the use of *Enterococcus* in combination with Fructooligosaccharides triglyceride and cholesterol levels dropped, HDL and LDL concentrations in the treated horses were highest and lowest, respectively, and IgM levels were higher (Saeidi et al. 2021). In ruminants, the usage of *Enterococcus* is not so common. In a 2014 study conducted on dairy cows, it was shown that using *Enterococcus* and *Saccharomyces* together reduced the illness incidence of retained placenta, milk fever, metritis, endometritis, mastitis, ketosis, lameness, and other conditions (AlZahal et al. 2014). In several studies on pigs, it was discovered that adding *Enterococcus faecium* to the diet or feeding piglets a diet containing both *E. faecium* and other *Lactobacillus* strains greatly reduced postweaning diarrhea (Büsing and Zeyner 2015; Szabo et al. 2009; Vrotniakienė and Jatkauskas 2013). *Enterococcus faecium* treatment also shortens and decreases rotavirus shedding in pigs (Kreuzer et al. 2012). Other studies in pigs treated with *E. faecium* and challenged with *Salmonella typhimurium* DT104 showed that, despite the treated pigs' earlier and more intense development of humoral immunity, the excretion of *Salmonella* in feces and the colonization of organs increase without the presence of severe clinical signs of salmonellosis (Simon 2010; Szabo et al. 2009).

According to research on the yeast *Saccharomyces*, adding it to a diet can raise feed effectiveness, improve feed digestibility, boost animal performance, lower the quantity of harmful microorganisms, improve animal health, and lessen the negative environmental effects of farming animals. According to Bai et al. (2013), *Saccharomyces cerevisiae* and *Lactobacillus* probiotics increased the mRNA expression of TLR-2 and TLR-4 in the chicken's foregut. On another study made by Ajiguna et al. (2021) *Salmonella*-challenged broilers who received *Saccharomyces cerevisiae* as a dietary supplement gained weight and maintained their immune response. The same outcome found (Gingerich et al. 2021) where *Saccharomyces* reduced ***Salmonella enteritidis*** concentrations in the ceca. According to Paryad and Mahmoudi (2008), broiler chicks' blood cholesterol and heterophils/lymphocytes ratio both significantly decreased when *Saccharomyces cerevisiae* was introduced to their diet. Smialek et al. (2018) showed that a multispecies probiotic containing a mixture of Lactic Acid Bacteria and *Saccharomyces cerevisiae*, given to *Campylobacter*-challenged birds with as a supplement in their feed for the duration of the production cycle, was capable of lowering the *Campylobacter* population count in broiler ceca and feces. In this instance, the experimental birds' *Campylobacter* spp. count was around ten times lower than that of the control chickens.

Administration of *Saccharomyces cerevisiae* to fish enhanced development, improved immunological and antioxidant function, and improved disease resistance

(Abdel-Tawwab et al. 2008; El-Nobi et al. 2021). Another study found that administering *Saccharomyces cerevisiae* and *Bacillus* spp. reduced plasma cortisol levels, indicating improved stress tolerance and consequently better immunological function (Sutthi and van Doan 2020). Last but not least, a study on *Saccharomyces cerevisiae* found that some strains had potential as probiotics and aflatoxin B1 adsorbents when utilized in simulated fish digestive system settings (Pinheiro et al. 2020). In a study on the immunological response in horses, *Pediococcus* and *Saccharomyces boulardii* were mixed to examine their combined product. The study's findings demonstrated that the probiotic utilized raised neutrophil and immunoglobulin G concentrations. The outcomes showed that the probiotic administered had a positive impact on the system's immunity (Furr 2014). Another investigation on the administration of *Saccharomyces boulardii* in horses with acute enterocolitis revealed a substantial reduction in the severity of diarrheal sickness (Desrochers et al. 2005).

In ruminants, the administration of *Saccharomyces* to veal calves decreased occurrences of diarrhea and preserved a robust microbiota with Fecalibacterium as a dominant strain (Villot et al. 2019). In a different trial on calf rearing, providing calves with a synbiotic combination containing saccharomyces and prebiotics increased average daily growth and decreased the amount of *Escherichia coli* in the feces (Roodposhti and Dabiri 2012). Finally, Spaniol et al. (2014) discovered increased blood levels of TNF-α, IL-4, INF-γ, and serum globulins in dairy cows. A probiotic yeast applied to piglets has been shown in the majority of studies to have positive health effects. *Saccharomyces cerevisiae* may enhance mucosal immunity by increasing IgM and IgA activities against pathogens, promote intestinal development and function, adsorb mycotoxins, modulate gut microbiota, and lessen post-weaning diarrhea (Kogan and Kocher 2007; Shen et al. 2009; Jiang et al. 2015). Additionally, a live yeast supplement may boost the proliferation of gut cells in the mucus of pigs that have received it, enhancing their resistance to pathogenic invasion (Bontempo et al. 2006). Because the beta-D-glucans in yeast cell walls boost the activity of neutrophils and macrophages by attaching to their receptors and triggering cytokines, as well as raising the generation of antibodies, many positive benefits of yeast are proposed to stimulate immune-modulation (Kogan and Kocher 2007; Kim et al. 2017).

Probiotics on Intestinal Morphology and Digestibility Gut health and microbiota have a direct impact on an animal's ability to develop and function. In addition to becoming more effective at absorbing nutrients, a healthy gut is better resistant against harmful microbes. The animals benefit from a healthy microflora in a variety of ways, including by preventing the invasion of infections through bacterial competition or by occupying the attachment site in the gut and obstructing bacterial activity. Although the gut microflora is rather stable, several external conditions and an animal's state of health can affect it. Diet is the main element that influences the gut microbiota. Typically, probiotics are used to control the gut flora. Animal health and performance also depend on gut morphological factors since increased absorption area and nutrient absorption are strongly correlated with greater villus length to crypt depth ratios and higher villus height. Another indicator of better digestion is the number of goblet cells in the intestinal villi, since these cells decrease the likelihood of harmful bacteria adhering to the intestinal epithelium and increase mucin synthesis. Probiotics improve gut histomorphology.

Gao et al. (2017) fed *Lactobacillus plantarum* strain to broilers and found increased the proliferation of several intestinal *Lactobacillus* species and sped up the maturation of the intestinal microbiota. Earlier, similar findings were made by Lan et al. (2003), who discovered that *Lactobacillus* species maintained the natural balance of gut microorganisms and boosted the population of beneficial microflora. Palamidi et al. (2016) used *Lactobacillus* strains to examine the effect on digestibility and found significant improvement in the apparent total tract digestibility of lipids, crude protein, and ileal dry matter. In fish, *Lactobacillus* usage promoted intestinal microvilli length and growth (Pirarat et al. 2006). In another study, feeding fish exposed to cadmium with *Lactobacillus* reversed the alterations in their intestinal microbiota's composition and decreased the amount of flavobacterium and pseudomonas (Zhai et al. 2017). In two more investigations, *L. plantarum* significantly increased food intake, food absorption, and food conversion (Seenivasan et al. 2014; Sivagami and Ronald 2016). According to Suo et al. (2012), *Lactobacillus* caused an increase in the villus height of the ileum, jejunum, and duodenum in pigs. Another study on weaned pigs found that the villus height of the ileum was increased, the crypt depth in the duodenum was decreased, and the villus height to crypt depth ratio of the jejunum and ileum was also improved (Yi et al. 2018). Later, a different research came to the conclusion that the jejunum and ileum had a considerably higher ratio of the villus height to the crypt depth following *L. plantarum* treatment. The *L. plantarum* administration raised the mRNA abundance of pBD2 and pBD3 in the jejunum and ileum. The colonic microbiota's structure was also considerably altered by the *L. plantarum* therapy, with an increase in the phyla *Firmicutes*, *Actinobacteria*, and *Lactobacillus* (Wang et al. 2019b).

Bacillus species have same results as *Lactobacillus* on the intestinal morphology and digestibility. Hayashi et al. (2018) gave *B. subtilis* to broilers and discovered that better histologic modification was linked to the development of the defensive response in the ileum. Broilers' intestinal histology and microbiota are considerably improved by supplementing with *B. subtilis* alone or in conjunction with *S. boulardii*. Intestinal villus height, length, and goblet cell count all showed substantial improvements, with reduced crypt depth and much lower numbers of *Salmonella*, coliforms, and *E. coli* in the cecum (Rajput et al. 2013; Manafi et al. 2017). According to research made by Haque et al. (2021), the treatment of two strains of *Bacillus* resulted in a significantly greater amount of gut microbiota. According to histological data, probiotic administration throughout the nursing stages resulted in a favorable alteration in the gut morphological structures. Another study found that after 6 weeks of feeding, *B. cereus* and *B. subtilis* significantly improved the autochthonous gut microflora populations and stimulated a wide range of potential probiotics (Xia et al. 2020). *Bacillus* was administered to pigs, and Cai et al. (2015) found that this lengthened the villi of the duodenum and jejunum. In the ileum, Deng et al. (2020) discovered a rise in villus height and the villus height:crypt depth ratio. Piglets fed *B. subtilis* changed the composition of their gut microbiota by having more Firmicutes and less *Escherichia coli*. The effects of food supplementation with *Bacillus subtilis* and xylo-oligosaccharides in weaned pigs were also studied by Ding et al. (2021). By raising villus height and the ratio of villus height to crypt depth in the ileum, as well as colonic concentrations of butyrate, tryptamine, and cadaverine while lowering those of skatole, dietary supplementation with *B. subtilis* enhanced the intestinal morphology of weaned pigs.

Supplementing feed with *Saccharomyces cerevisiae* has been shown to increase feed digestibility and efficiency. Moreover, it has been demonstrated that adding yeast to meals can alter the intestinal microbiota of animals and prevent bacterial infections from colonizing the gastrointestinal system. After supplementation with *Saccharomyces*, Wang et al. (2016b) discovered a reduced ileal myeloperoxidase activity and a greater ratio of villus height to crypt depth in the ileum of broiler chickens. According to a different study by Muthusamy et al. (2011), adding saccharomyces to broiler chicken feed enhanced villus height in the jejunum, villus width in the ileum, and the number of goblet cells in the villi of the jejunum and ileum.

Probiotics on Sustainability Animal nutrition with probiotics has been recognized as a potential contributor to sustainability in the livestock industry. The use of probiotics in animal feed can help improve the overall health and productivity of livestock, reduce the use of antibiotics, and reduce greenhouse gas emissions from manure. Probiotics, as already noted, are live microorganisms that, when consumed, provide a health benefit to the host. In the context of animal nutrition, probiotics are added to feed in order to improve digestive health and overall gut function. This can lead to improved feed efficiency, reduced incidences of diseases, and better growth performance. For example, a study by Kim et al. (2015b) found that supplementing the diets of pigs with a probiotic improved their growth performance and reduced the incidence of diarrhea. One of the biggest challenges facing the livestock industry today is the increasing use of antibiotics in animal feed, due to the threat of diseases and other health issues. The overuse of antibiotics in animal feed has led to the development of antibiotic-resistant bacteria, which pose a serious threat to human health. By improving gut health and reducing the incidence of diseases, the use of probiotics in animal feed can help reduce the need for antibiotics. Another study by Pérez-Maldonado et al. (2017) found that supplementing broiler chicken diets with a probiotic reduced the need for antibiotics and improved the birds' overall health and growth performance. Another benefit of using probiotics in animal feed is the reduction of greenhouse gas emissions from manure. The use of probiotics can improve the overall digestive function of livestock, leading to better nutrient utilization and reduced waste. This, in turn, can lead to lower methane emissions from manure, as less feed is excreted in the feces. A study by Ying et al. (2017) found that supplementing the diets of dairy cows with a probiotic reduced methane emissions from manure by 11%. In conclusion, the use of probiotics in animal feed can play a significant role in improving the sustainability of the livestock industry. By improving animal health and reducing the use of antibiotics, as well as reducing greenhouse gas emissions from manure, the use of probiotics can help create a more sustainable and responsible industry.

3 Prebiotics

History Glenn Gibson and Marcel Roberfroid initially presented the idea of prebiotics in 1995. The first definition of prebiotic was given as "a non-digestible food ingredient that beneficially affects the host by selectively stimulating the growth and/or activity of one or a limited number of bacteria in the colon and thus improves host health." This criterion limits the substances that may be categorized as prebiotics in the carbohydrate class to a select few, including lactulose, GOS, and the short and long-chain fructans (FOS and inulin). So, 15 years later, at the sixth Meeting of the International Scientific Association of Probiotics and Prebiotics (ISAPP), prebiotics were defined as "a selectively fermented ingredient that results in specific changes in the composition and/or activity of the gastrointestinal microbiota, thus conferring benefit(s) upon host health" (Gibson et al. 2010).

Prebiotics can be divided into several groups based on their chemical structure and their target bacterial groups in the gut. Some common groups of prebiotics include:

- *Fructooligosaccharides (FOSs)*
- *Galactooligosaccharides (GOSs)*
- *Xylooligosaccharides (XOSs)*
- *Mannooligosaccharides (MOSs)*
- *Other Oligosaccharides*: Pectic oligosaccharide (POS) and trans-galactooligosaccharides (TOSs)

Mechanism of Action The majority of prebiotics are not digested or absorbed in the small intestine but instead are rapidly and efficiently fermented by bacteria in other sections of the alimentary canal, encouraging the growth of beneficial bacteria like *Lactobacillus* and *Bifidobacterium*. It is also noteworthy to note that a process known as "cross-feeding" occurs when the metabolism of dietary fibers, mostly prebiotics, from certain microorganisms indirectly promotes the growth of others, and the byproducts of fermentation serve as a fuel for the development of more bacteria. For instance, *Bifidobacteria* and *Lactobacilli*, which are the main fructan consumers, produce lactate and acetate during fermentation, which can be utilized by other bacteria as an energy source.

Fructooligosaccharide (FOS) A fructooligosaccharide is a special type of water-soluble carbohydrate. They include short fructose chains that the body is unable to break down. FOS is commonly used as a prebiotic in the animal diet. Some plants, such asparagus, onions, leeks, soybeans, wheat, tomatoes, and garlic, naturally contain FOS. It can provide food for probiotic microorganisms when included in a probiotic formula. According to studies, hazardous bacteria including *Salmonella*,

Clostridium perfringens, and other food-borne diseases can be suppressed with the use of FOS. Prebiotics like FOS can indirectly aid in preventing the growth of dangerous bacteria by promoting the growth of probiotic bacteria, maintaining a healthy balance of bacteria in the gut. FOS can also improve the body's absorption of calcium, which is good for bone health. Inulin is the most well-known fructooligosaccharide. Prebiotic additives like inulin are frequently utilized in the diets of farm animals and pets. This fructooligosaccharide promotes the growth and development of friendly bacterial species that live in the large intestine, demonstrating a positive impact on host health.

Research on the effects of inulin and fructooligosaccharide on the microbiota of the gastrointestinal tract, as well as the health and productivity of animals, has increased in recent years. FOS affect the intestinal bacterial microflora of chickens, which results in an increase in *Lactobacillus* populations and a decrease in the colonies of the following bacterial species: *Salmonella*, *Campylobacter*, and *Escherichia coli* (Xu et al. 2003). Recently, similar results were discovered when FOS was utilized in broiler chicken feeds. Pathogenic bacteria including *Helicobacter* and *Desulfovibrio* were shown to be greatly decreased (Shang et al. 2018). The shape of the intestines can be positively impacted by the inclusion of inulin and fructooligosaccharide in poultry diets. The small intestine has undergone positive alterations that have enhanced its growth and improved the intestinal villus system. Inulin supplementation of formulated broiler chicken feed resulted in an increased intestinal villus length and crypt depth (Rehman et al. 2007). According to Xu et al. (2003), fructooligosaccharide supplementation significantly improved intestinal morphology, particularly where there was a higher villus-to-crypts ratio. The same type of modification was seen by Shang et al. (2015), where fructooligosaccharide (FOS) supplementation resulted in a considerable increase in villus height, crypt depth, and total mucosa thickness in the ileum of broiler chickens.

Furthermore, studies have indicated that adding fructooligosaccharide to complex chicken diets can increase production and have an impact on the chickens' overall performance. Rebolé et al. (2010) investigated the impact of inulin on broiler chicken growth performance and reported that animals fed inulin-containing diets showed noticeably better final body weight gain. Supplementing laying hens with inulin and oligofructose has also been shown to have positive effects. Increased egg production and weekly increases in total egg weight were the results of this treatment (Chen et al. 2005).

Since the commencement of aquaculture industry, one of the key goals has been to enhance fish growth and feed efficiency, which might be accomplished by improving fish nutrition and metabolism. Improvements in fish nutrition will result in improved fish health and productivity, as well as a decrease in the environmental impact of aquaculture. The use of FOS as a feed supplement in aquaculture nutrition yields good results. Fish research has mostly focused on the effects of FOS on growth performance, immunological response, and illness resistance (Grisdale-Helland et al. 2008; Ye et al. 2011; Ortiz et al. 2013). There are several studies that indicate that administering FOS produces excellent results. Hu et al. (2019)

investigated the use of FOS in the shrimp nutrition and found that FOS increased microbial diversity and reduced various potential infections in shrimp intestines, including *Vibrio tubiashii*, *Vibrio parahaemolyticus*, and *Photobacterium damselae* like strains. Paz et al. (2019) reported similar findings in which FOS boosted circulating white blood cells and accelerated phagocytosis in fish infected with *Aeromonas hydrophila*. The addition of FOS resulted in not only a stronger immune response, but also improved performance parameters such as growth rate, ultimate body mass, weight gain, and feed efficiency. In two more trials, FOS supplementation was found to increase growth, nutritional utilization, immunological and antioxidant parameters, digestive enzyme activity, gut health, and disease resistance against *Aeromonas hydrophila* (Poolsawat et al. 2020; de Campos et al. 2022). All in all, the use of FOS has produced positive outcomes in aquaculture performance and immunological measures, particularly against *Aeromonas hydrophila*.

The use of FOS in swine diet has shown excellent results. Le Bourgot et al. (2016) discovered that FOS diet boosted anti-influenza IgA levels in postweaning pig serum, indicating that it can improve vaccination response and promote growth. Schokker et al. (2018) later discovered that daily oral treatment of FOS to suckling pigs had a distinct "bifidogenic" impact, with considerable changes in microbiota composition and diversity. With these data, Le Bourgot et al. (2019) showed that FOS generated a lasting alteration of the makeup of the fecal microbiota and elevated the *Prevotella* genus. More recent research has focused on the use of FOS in piglet diarrhea occurrences, and the results revealed that FOS reduced the incidence of pig diarrhea, which linked favorably with an increase in the abundance of *Lactobacillus* and *Bifidobacterium* in the gut (Zhang et al. 2022).

The use of FOS in companion animals is being studied for the benefits it can provide to the immune system and digestive health. Nowadays, it is critical to establish dietary methods targeted at positively affecting the intestinal health of companion animals. In one research, FOS was shown to reduce insulin resistance in obese dogs, which may aid in the prevention of type 2 diabetes (Respondek et al. 2008). In another study, FOS was found to lower blood cholesterol in dogs (Jeusette et al. 2004). Pinna et al. (2016) evaluated the effect of FOS on canine fecal microbiota metabolism and discovered that FOS reduced ammonia and resulted in greater concentrations of total volatile fatty acids. The study concluded that FOS may enhance canine intestinal microbiota metabolism.

Galactooligosaccharides (GOS) GOS are a type of prebiotic fiber, composed of short chains of galactose sugars. They are not digestible by animal digestive enzymes, but they serve as food for beneficial gut bacteria, specifically *Bifidobacteria*, promoting their growth and improving gut health. GOS are commonly used as a food ingredient and can be found in some dietary supplements. In animal nutrition, galactooligosaccharides (GOS) have been shown to have several potential benefits, including:

- Improved gut health: As prebiotics, GOS can help promote the growth of beneficial gut bacteria, which can improve gut health and reduce the risk of gastrointestinal disorders.
- Enhanced immune function: By promoting the growth of beneficial gut bacteria, GOS can also help boost the immune system, helping animals to better resist infections and other diseases.
- Improved nutrient absorption: By improving gut health and reducing inflammation, GOS may also enhance the absorption of nutrients from the diet, leading to improved growth and overall health in animals.
- Reduced risk of obesity: Some studies have suggested that GOS may help regulate metabolism and reduce the risk of obesity in animals. A study published in the British Journal of Nutrition in 2009 found that GOS supplementation reduced body weight gain and improved insulin sensitivity in rats, suggesting a potential role in preventing obesity and related metabolic disorders (Kim et al. 2009).

It is important to note that more research is needed to fully understand the impact of GOS on animal nutrition and health, and the effects may vary depending on the species and individual animal.

Galactooligosaccharides have been used in broiler nutrition to enhance gut health and improve performance. Supplementing the diets of broiler chickens with GOS has been shown to reduce gut inflammation, improve gut health, and enhance growth performance in several studies. For example, a study published by Ma et al. (2010) found that broiler chickens that were fed diets supplemented with GOS showed improved gut health and reduced gut inflammation and had better growth performance compared to a control group that did not receive GOS. Another study found that supplementing the diets of layer hens with GOS improved egg quality, reduced gut inflammation, and enhanced gut health compared to a control group that did not receive GOS. The GOS-supplemented hens had higher eggshell thickness and eggshell strength and lower levels of gut inflammation markers (Yang et al. 2014). Furthermore, Wang et al. (2012) discovered that supplementing broiler chicken diets with GOS improved gut health, decreased gut inflammation, and increased growth performance when compared to a control group that did not receive GOS. GOS supplementation was also observed to modify the gut microbiota, boosting the number of good bacteria such as *Lactobacillus* and *Bifidobacterium*. Zhang et al. (2013) discovered the same results, namely that supplementing broiler chicken diets with GOS improved gut health, decreased gut inflammation, and boosted growth performance when compared to a control group that did not receive GOS. These studies provide additional evidence of the potential benefits of GOS in chicken nutrition, but more research is needed to fully understand the effects and determine the most effective dosages for different species and conditions.

In aquaculture nutrition, they can be used as a prebiotic to promote the growth of beneficial gut bacteria in fish and shellfish. This can improve the digestive health and overall immunity of the cultured species, leading to better growth performance and survival rates. However, it is important to note that the nutritional requirements of different species of fish and shellfish may vary and the use of GOS in aquaculture diets should be carefully evaluated for each species based on its nutritional requirements and potential for interactions with other dietary components. In general, research has shown that the supplementation of GOS in the diets of fish and shellfish can have positive effects on gut health, digestion, and immunity. Studies on rainbow trout, for example, have demonstrated that adding GOS to the diet can improve gut microbiota, increase nutrient utilization and growth performance, and lower disease susceptibility. Similarly, GOS supplementation has been shown in experiments on Pacific white shrimp to improve the gut flora, increase growth and survival, and reduce illness incidence (Kim et al. 2016). Other studies have shown that GOS can be a beneficial supplement in the diets of other aquaculture species such as Atlantic salmon and tilapia. Particularly, Chen et al. (2017) found that the supplementation of GOS in the diet of Nile tilapia improved growth performance, increased the abundance of beneficial gut bacteria, and enhanced the non-specific immune responses of the fish. Specifically, the tilapia fed a diet containing GOS showed a significant increase in weight gain and feed efficiency compared to those fed a control diet without GOS. The study also found that GOS supplementation significantly altered the gut microbiota composition, increasing the relative abundance of beneficial bacteria such as *Lactobacillus* and *Bifidobacterium*, and reducing the relative abundance of potentially pathogenic bacteria. Furthermore, the study found that GOS supplementation improved the non-specific immune responses of Nile tilapia, as indicated by increased levels of lysozyme activity and phagocytic activity. Overall, the results of this study suggest that GOS can be a beneficial supplement for Nile tilapia, improving growth performance, gut health, and immunity. Same results were reported by Lee et al. (2015a) which showed that supplementation of GOS in the diet improved the growth performance of Atlantic salmon, with fish fed a diet containing GOS exhibiting a significant increase in weight gain and feed efficiency compared to those fed a control diet without GOS. The study also found that GOS supplementation led to an increase in the activity of digestive enzymes such as amylase, lipase, and protease, indicating improved digestive health in the fish. Furthermore, the study found that GOS supplementation altered the gut microbiota of Atlantic salmon, with the relative abundance of beneficial bacteria such as *Lactobacillus* and *Bifidobacterium* increased, while the relative abundance of potentially pathogenic bacteria was reduced.

Galactooligosaccharides have been studied for their potential benefits in swine nutrition. Some research has shown that GOS can improve gut health by promoting the growth of beneficial bacteria, leading to improved feed efficiency, weight gain, and overall health in swine. Other studies have also reported improved gut barrier function, reduced diarrhea and improved nutrient digestibility in swine fed diets containing GOS. However, the results of these studies are mixed, and more research is needed to fully understand the effects of GOS in swine nutrition and determine the optimal levels for use in pig diets. More specifically, a study conducted by Zhang et al. (2015) revealed that supplementing the diets of weanling pigs with galactooligosaccharides had effects on growth performance, with pigs fed diets containing GOS having improved average daily gain and feed conversion ratio compared to control pigs; nutrient digestibility, with pigs fed diets containing GOS having increased digestibility of crude protein, crude fat, and dry matter and in fecal bacterial populations, where the researchers found changes in the populations of fecal bacteria, with an increase in the numbers of *Bifidobacteria* and *Lactobacilli*. Sun et al. (2016) later explored the effects of supplementing weaned pig diets with galactooligosaccharides (GOS) on gut health and gut microbiota in pigs challenged with *E. coli*. Pigs fed GOS-containing diets demonstrated improved indicators of gut health, including lower diarrhea scores, improved intestinal morphology, and lower levels of pro-inflammatory cytokines in the gut. Furthermore, there was evidence that feeding GOS to pigs changed the makeup of their gut microbiota, with an increase in the numbers of helpful bacteria like *Lactobacilli* and *Bifidobacteria* and a decrease in the numbers of harmful bacteria like *E. coli*. Another study investigated the effects of GOS supplementation on growth performance, nutrient digestibility, and gut microbiota of growing pigs. The results showed that GOS supplementation improved the average daily gain and average daily feed intake of the pigs. The apparent total tract digestibility of dry matter and crude protein was also increased. In terms of gut microbiota, the study found that GOS supplementation led to an increase in the population of beneficial bacteria such as *Lactobacillus* and a decrease in the population of harmful bacteria such as *Escherichia coli* (Cui et al. 2019). Same results were found by Ma et al. (2020) on the gut environment and gut microbiota of growing-finishing pigs. The results showed that GOS supplementation increased the populations of beneficial bacteria such as *Lactobacillus* and *Bifidobacterium* in the gut of the pigs and decreased the population of harmful bacteria such as *Escherichia coli*. In terms of gut environment, the study found that GOS supplementation reduced the pH of the hindgut and increased the concentrations of short-chain fatty acids, particularly acetate and propionate. Overall, the results of this study suggest that GOS supplementation has the potential to improve the gut environment and modulate the gut microbiota in a beneficial direction in growing-finishing pigs. To sum up, galactooligosaccharides (GOS) can improve gut health and modulate gut microbiota in swine. GOS supplementation has resulted in improved feed efficiency, weight gain, and gut health in weanling and growing pigs, with increased growth performance and nutrient digestibility, and changes in the populations of fecal bacteria, such as an increase in the numbers of *Bifidobacteria* and *Lactobacilli* and a decrease in harmful bacteria like *E. coli*. The gut environment was also improved, with decreased pH of the hindgut and increased concentrations of short-chain fatty acids. More research is needed to fully understand the effects of GOS in swine nutrition.

Several studies have explored the use of galactooligosaccharides as a supplement in ruminant nutrition. The results of these studies have been mixed, with some finding positive effects on growth performance, feed efficiency, and gut health in ruminants, while others have found no significant effects. Rondina et al. (2017) investigated the use of GOS in dairy heifers' nutrition. The results of this study showed that dietary supplementation with GOS improved growth performance and feed efficiency in dairy heifers, as well as increased total tract digestibility of dry matter, organic matter, and neutral detergent fiber. Ma et al. (2017) used galactooligosaccharides in lambs' nutrition. In this study, lambs showed improved growth performance and feed efficiency, as well as increased nutrient digestibility. The researchers also found changes in ruminal fermentation, with an increase in the populations of beneficial bacteria such as *Lactobacillus* and a decrease in the populations of harmful bacteria such as *Escherichia coli*. Last, Lu et al. (2017) investigated the effects of GOS in sheep nutrition. The results of this study showed that dietary supplementation with GOS improved growth performance and feed efficiency in sheep, as well as increased nutrient digestibility and rumen pH. These studies suggest that GOS supplementation may have potential benefits for ruminants, such as improved growth performance, feed efficiency, and gut health. However, more research is needed to fully understand the effects of GOS in ruminant nutrition and determine the optimal levels for use in ruminant diets.

Xylooligosaccharides (XOS) Xylooligosaccharides are short chains of xylose sugars that can act as prebiotics in the gut. They are a type of oligosaccharide, which are carbohydrates made up of a small number of sugar units, typically two to ten. XOS are fermented by certain strains of bacteria in the gut, producing short-chain fatty acids (SCFAs) that can have beneficial effects on the host. XOS have been shown to increase the populations of beneficial bacteria such as *Bifidobacteria* and *Lactobacilli* in the gut, which can help to maintain gut health and improve overall gut microbiome diversity. They have also been reported to improve gut health and digestive function, boost the immune system, and promote mineral absorption. XOS supplementation has been shown to improve growth performance and nutrient digestibility in livestock animals and has potential applications in companion animal diets (Van Laere et al. 2010). Overall, XOS is a promising prebiotic with multiple potential health benefits. However, more research is needed to fully understand the effects of XOS on gut health and the optimal levels for use in animal diets.

Xylooligosaccharides are a type of prebiotic that have been studied for their potential benefits in chicken nutrition. XOS are derived from xylose, a sugar found in many plant materials and have been shown to promote the growth of beneficial bacteria in the gut. This can lead to improved gut health, which can have a positive impact on a chicken's overall health and performance. Studies have investigated the effects of XOS supplementation on various aspects of chicken nutrition, including growth performance, nutrient digestibility, and gut health. In one study, XOS supplementation was found to improve the growth performance and feed conversion efficiency of broiler chickens (Wang et al. 2019c). Another study found that XOS supplementation improved the digestibility of nutrients, such as crude protein and crude fat, in chickens (Zhao et al. 2018). In addition to improving growth performance and nutrient digestibility, XOS supplementation has also been found to improve gut health in chickens. For example, one study found that XOS supplementation reduced the levels of harmful bacteria in the gut of broiler chickens, while promoting the growth of beneficial bacteria (Ren et al. 2018). This can lead to improved gut barrier function, which can help to reduce the risk of enteric infections, such as *Escherichia coli* and *Salmonella* spp. Overall, these findings suggest that

XOS may be a useful addition to chicken diets for improving growth performance, nutrient digestibility, and gut health.

In aquaculture feeding XOS can improve gut health and growth performance in fish. XOS are oligosaccharides, which are short-chain carbohydrates digested by gut bacteria to create short-chain fatty acids (SCFAs). SCFAs have been proven to improve gut health by increasing the abundance of beneficial gut bacteria, decreasing the quantity of potentially pathogenic bacteria, and enhancing gut barrier function. One study investigated the effects of XOS supplementation on growth performance and gut microbiota of Pacific white shrimp (*Litopenaeus vannamei*). The results showed that XOS supplementation significantly improved the weight gain and feed efficiency of the shrimp compared to the control group. Additionally, XOS supplementation altered the gut microbiota composition, increasing the relative abundance of beneficial bacteria such as *Lactobacillus* and *Bifidobacterium*, and reducing the relative abundance of potentially pathogenic bacteria (Zhang et al. 2019). Another study investigated the effects of XOS supplementation on the growth performance, non-specific immunity, and gut microbiota of Pacific white shrimp. The results showed that XOS supplementation significantly improved weight gain and feed efficiency, and enhanced non-specific immune responses, as indicated by increased levels of hemolymph phenoloxidase activity and superoxide dismutase activity. Furthermore, XOS supplementation altered the gut microbiota composition, increasing the relative abundance of beneficial bacteria such as *Lactobacillus* and *Bifidobacterium*, and reducing the relative abundance of potentially pathogenic bacteria (Liu et al. 2019). One more study on the use of XOS in aquaculture nutrition is made by Chen et al. (2018) showed that xylooligosaccharides supplementation improved the growth performance, increased the activities of digestive enzymes and altered the gut microbiota composition of Nile tilapia. Specifically, the study found that supplementing the diet with XOS increased the weight gain, specific growth rate, and feed efficiency of Nile tilapia. The activities of digestive enzymes, such as trypsin, lipase, and amylase, were also enhanced in tilapia fed with xylooligosaccharides. The study also observed that the gut microbiota of tilapia fed with XOS was more diverse and had a higher proportion of beneficial bacteria, such as *Lactobacillus*. In conclusion, XOS have shown potential as a beneficial supplement for aquaculture, improving growth performance and gut health in fish and shrimp species. Further research is needed to fully understand the optimal concentration and effects of XOS in aquaculture, and to determine if the benefits observed in laboratory studies can be replicated in commercial-scale aquaculture systems. However, the results of the studies suggest that XOS have potential as a prebiotic for the promotion of gut health and growth performance in aquaculture.

Xylooligosaccharides have been proven in studies to increase swine growth performance, nutritional digestibility, and gut health. XOS supplementation can enhance the number of helpful gut bacteria like *Lactobacillus* and *Bifidobacterium* while decreasing the number of dangerous bacteria like *E. coli* and *Salmonella* spp. This can result in better gut health and a lower risk of enteric infections. One study conducted by Li et al. (2017b) found that XOS supplementation significantly improved the average daily gain and feed conversion efficiency of growing-finishing pigs. In addition, XOS supplementation was found to improve nutrient digestibility, including crude protein and gross energy. Another study by Chen et al. (2017) found that XOS supplementation improved gut health by reducing gut inflammation and increasing the number of beneficial bacteria. This was associated with improved growth performance, including higher average daily gain and improved feed conversion efficiency. These findings suggest that XOS can play an important role in improving the health and performance of swine. Last Zeng et al. (2018) found that supplementation of xylooligosaccharides improved the growth performance and nutrient digestibility of weaning pigs. The authors also found that xylooligosaccharides supplementation had a positive effect on the intestinal health of the pigs, as indicated by improved gut barrier function and reduced oxidative stress. From all these information, it can be deduced that Xylooligosaccharides (XOS) supplementation has a positive effect on the growth performance and gut health of swine. XOS supplementation can also be associated with improved gut barrier function and reduced oxidative stress in weaning pigs.

In ruminants, studies have shown that supplementing their diet with XOS can improve the growth performance and feed efficiency of the animals. For example, a study by Sun et al. (2017) showed that supplementation of XOS in the diet of beef cattle increased average daily gain and feed efficiency. Another study by Li et al. (2019a) found that adding XOS to the diet of dairy cows increased milk yield and improved feed conversion efficiency. XOS also have positive effects on gut health in ruminants. By selectively promoting the growth of beneficial bacteria, such as *Lactobacillus* and *Bifidobacterium*, XOS can help to maintain a healthy gut microbiota, reduce the risk of digestive disorders and improve animal health. A study by Liu et al. (2018) found that supplementing the diet of dairy cows with XOS reduced the incidence of subacute ruminal acidosis and improved feed digestibility. To sum up, xylooligosaccharides have been found to have positive effects on the growth performance and gut health of ruminants.

In companion animals, several studies have investigated the effects of XOS supplementation in their diets. In one study, the addition of XOS to the diets of dogs was found to increase the population of *Bifidobacteria* and *Lactobacilli* in the gut, which are considered to be beneficial bacteria (Couturier et al. 2011). In another study, the addition of XOS to the diets of cats was shown to improve gut health, as measured by an increased population of beneficial bacteria and a reduction in gut pH (Chou et al. 2018). XOS supplementation has also been shown to improve nutrient utilization in companion animals. In the same studies as before, the addition of XOS to the diets of dogs was found to increase the digestibility of protein and fat (Couturier et al. 2011). In the other study, the addition of XOS to the diets of cats was shown to improve the digestibility of starch and dietary fiber (Chou et al. 2018). In conclusion, Xylooligosaccharides are a promising nutritional supplement for companion animals, as they have been shown to improve gut health, support the immune system, and increase nutrient utilization.

Mannooligosaccharides (MOS) Are a type of prebiotic that are made up of short chains of mannose, a type of sugar. They are naturally found in some plants and have been found to have a number of health benefits when consumed. They are composed of short chains of mannose sugars and have a molecular weight of less than 1000 Daltons, which makes them highly soluble in water. The chain length of MOS varies and ranges from two to six mannose residues. MOS is attracting increasing attention as a dietary ingredient for animals because of its potential to improve gut health and overall performance. MOS works by selectively promoting the growth and activity of beneficial bacteria in the gut, thereby helping to maintain a healthy gut microbiome. This is important in animal nutrition, as the gut microbiome has been found to play a key role in many aspects of animal health and performance, including the digestion and absorption of nutrients, the immune response, and the prevention of disease. In addition to their prebiotic effects, MOS has also been found to have a number of other benefits in animal nutrition, including improved growth performance, increased nutrient digestibility, and reduced incidence of gut-related diseases. As a result, MOS is being increasingly used as a dietary ingredient in animal feed, particularly for poultry and swine. These findings suggest that MOS has the potential to play an important role in improving the health and performance of animals and may provide a promising alternative to other types of prebiotics and antibiotics.

Mannooligosaccharides (MOSs) are a type of prebiotic that are used in chicken nutrition to improve gut health and performance. MOS is a type of oligosaccharide derived from mannan, a naturally occurring polysaccharide found in the cell walls of yeast and other fungi. MOS is known to have several beneficial effects on gut health and function, including promoting the growth of beneficial bacteria and reducing the growth of pathogenic bacteria. Studies have shown that MOS supplementation in chickens can improve growth performance, feed conversion efficiency, and immune function. For example, a study conducted by Hong et al. (2017) found that supplementing chicken diets with MOS improved average daily gain and feed conversion efficiency compared to control groups. The authors also found that MOS supplementation improved immune function, as indicated by increased antibody production, and improved disease resistance. Another study by Tan et al. (2015) found that MOS supplementation in chickens was associated with an increased population of beneficial bacteria in the gut, including *Lactobacillus* and *Bifidobacterium* species. The authors also found that MOS supplementation reduced the presence of harmful bacteria, including *Escherichia coli* and *Salmonella* spp. These findings suggest that MOS can play an important role in maintaining gut health and reducing the risk of disease in chickens. In addition to improving gut health and performance, MOS supplementation has also been shown to have positive effects on nutrient digestibility and utilization in chickens. For example, a study conducted by Wang et al. (2017a) found that MOS supplementation improved nutrient digestibility, including crude protein and gross energy. The authors also found that MOS supplementation was associated with improved gut function, including increased intestinal villi length and reduced gut inflammation. These findings suggest that MOS can play an important role in optimizing nutrient utilization and improving overall health and performance in chickens.

In the nutrition of aquaculture MOS have been widely studied for their benefits to gut microbiota and digestive health, leading to better growth performance, feed utilization, and disease resistance in aquaculture species. Studies have shown that MOS supplementation in aquaculture diets can improve growth performance and feed utilization in fish and crustaceans. For example, a study by Wang et al. (2017b) found that MOS supplementation in the diet of grass carp improved growth performance and nutrient utilization. Another study by Zhang et al. (2019c) reported that MOS supplementation in the diet of white shrimp improved feed utilization and growth performance. Gut health is critical for the overall health and performance of aquaculture species, and MOS supplementation has been shown to improve gut health and reduce the risk of disease in fish and crustaceans. For instance, a study by Li et al. (2018a) found that MOS supplementation in the diet of black carp improved gut health and reduced the risk of disease. Another study by Chen et al. (2019b) reported that MOS supplementation in the diet of common carp improved gut microbiota, leading to improved gut health and disease resistance. In conclusion, MOS supplementation in aquaculture diets has been shown to improve growth performance, feed utilization, and gut health in fish and crustaceans. The benefits of MOS supplementation in aquaculture nutrition have been demonstrated in multiple studies, providing evidence for its use in the aquaculture industry.

In swine nutrition MOS can improve gut health and performance including promoting the growth of beneficial bacteria and reducing the growth of pathogenic bacteria. Studies have shown that MOS supplementation in swine can improve growth performance, feed conversion efficiency, and nutrient digestibility. For example, a study conducted by Li et al. (2018b) found that supplementing swine diets with MOS improved average daily gain and feed conversion efficiency compared to control groups. The authors also found that MOS supplementation improved nutrient digestibility, including crude protein and gross energy. Another study by Chen et al. (2019a) found that MOS supplementation in swine was associated with an increased population of beneficial bacteria in the gut, including *Lactobacillus* and *Bifidobacterium* species. The authors also found that MOS supplementation reduced the presence of harmful bacteria, including *Clostridium* and *Escherichia coli*. These findings suggest that MOS can play an important role in maintaining gut health and reducing the risk of disease in swine. In addition to improving gut health and performance, MOS supplementation has also been shown to have positive effects on gut morphology and function in swine. For example, a study conducted by Zeng et al. (2019) found that MOS supplementation improved gut barrier function and reduced oxidative stress. The authors also found that MOS supplementation was associated with improved gut morphology, including increased intestinal villi length and reduced gut inflammation. These findings suggest that MOS can play an important role in optimizing gut function and improving overall health and performance in swine.

MOS has been also shown to have a number of potential benefits in ruminant nutrition, including improved gut health, increased feed efficiency, and enhanced animal performance. One of the key mechanisms by which MOS works is by supporting the growth of beneficial gut bacteria and inhibiting the growth of harmful bacteria. This helps to maintain a healthy gut microflora and reduce the risk of digestive disorders, such as diarrhea, that can impact animal performance. MOS has also been shown to improve feed efficiency in ruminants by reducing the amount of feed that is passed undigested through the gut, thus increasing the overall nutrient utilization. Several studies have investigated the effects of MOS on ruminant performance, and the results have been generally positive. In one study, the effect of mannanoligosaccharides on performance, fecal characteristics, and cecal volatile fatty acid concentrations in young bulls was investigated. The results showed that mannanoligosaccharide supplementation improved feed efficiency, had a positive effect on fecal characteristics, and increased the concentration of propionic acid in the cecum (Rodehutscord et al. 2001). Furthermore, De Boever et al. (2005) discovered that feeding MOS to lambs enhanced feed conversion rates and boosted live weight growth. Another study, by Bélanger et al. (2010), found that MOS supplementation enhanced feed efficiency and milk output in dairy cows. In conclusion, the use of Mannooligosaccharides in ruminant nutrition has shown promising results in terms of improved gut health, increased feed efficiency, and enhanced animal performance.

Other Oligosaccharides Pectic-oligosaccharides and trans-galactooligosaccharides are two types of dietary oligosaccharides that have received attention for their potential use in animal nutrition. These oligosaccharides are non-digestible by the host animal and thus they can reach the gut microbiome where they may provide various benefits. Pectic-oligosaccharides (POSs) are oligomers derived from the breakdown of pectin; a type of polysaccharide found in plant cell walls. These oligosaccharides are known for their prebiotic properties, meaning that they selectively promote the growth and activity of beneficial gut microbes. They have been shown to improve gut health by increasing the populations of *Bifidobacteria* and *Lactobacilli*, and by reducing the populations of harmful bacteria such as *E. coli* and *Salmonella*. For example, a study by Pan et al. (2012) investigated the effects of pectic-oligosaccharides supplementation in weanling pigs and found that they improved gut health and reduced the populations of harmful bacteria in the gut.

Similarly, trans-galactooligosaccharides (TOSs) are prebiotic compounds derived from lactose. They have been shown to increase the populations of *Bifidobacteria* and *Lactobacilli* in the gut and to reduce the populations of harmful bacteria. In addition, they have been shown to improve gut health by improving the gut barrier function and reducing gut inflammation. Fan et al. (2017) investigated the effects of TOS supplementation in broiler chickens and found that it improved gut health and increased the populations of beneficial bacteria in the gut. All in all, POS and TOS have shown promise as potential dietary supplements in animal nutrition. However, more research is needed to fully understand their effects and to determine the optimal levels of supplementation for different species and production systems.

Prebiotics on Sustainability The use of prebiotics in animal nutrition has been shown to have a positive impact on animal health, productivity, and sustainability in animal production. One of the benefits of using prebiotics in animal nutrition is the improvement of gut health and digestive function. The growth of beneficial bacteria in the gut helps to maintain a healthy gut environment and to support the digestion and absorption of nutrients. This can lead to improved feed efficiency, which in turn can reduce the amount of feed required to produce a unit of animal product, and ultimately reduce the environmental impact of animal production. For example, a study by O'Doherty et al. (2017) found that supplementing the diets of dairy cows with a prebiotic improved feed efficiency and reduced the amount of nitrogen excreted in manure. Another benefit of using prebiotics in animal nutrition is the reduction of the use of antibiotics. Antibiotics are often used in animal production to prevent and treat diseases, but the overuse of antibiotics can lead to the development of antibiotic-resistant bacteria, which pose a serious threat to human health. The use of prebiotics in animal feed has been shown to improve gut health and reduce the incidence of diseases, which can help to reduce the need for antibiotics. For example, a study by Kumprecht et al. (2015) found that supplementing the diets of broiler chickens with a prebiotic reduced the need for antibiotics and improved the birds' overall health. The use of prebiotics can also have a positive impact on the environment by reducing greenhouse gas emissions from manure. The growth of beneficial bacteria in the gut can lead to improved nutrient utilization, reducing the amount of waste excreted in feces, and ultimately reducing the amount of methane produced from manure. A study by Wang et al. (2017) found that supplementing the diets of beef cattle with a prebiotic reduced methane emissions from manure by 9%. To sum up, the use of prebiotics in animal nutrition can contribute significantly to sustainability in animal production. By improving animal health and reducing the use of antibiotics, as well as reducing greenhouse gas emissions from manure, the use of prebiotics can help create a more sustainable and responsible industry.

4 Paraprobiotics and Postbiotics

History The concept of paraprobiotics and postbiotics as a means of promoting health in the gut is relatively new, but the use of probiotics and other gut-microbe-related interventions is a long-standing field of study. The terms "paraprobiotics" and "postbiotics" have been introduced to describe new approaches to gut health. Paraprobiotics are a class of functional food ingredients that are similar to probiotics but lack the capacity to colonize the host gut. They are typically derived from prebiotic compounds or inactivated probiotics and act by selectively promoting the growth and activity of beneficial gut bacteria, thereby modulating the gut microbiome and improving gut health. Unlike probiotics, paraprobiotics are not living organisms, making them more stable and easier to handle during storage and processing. These compounds have been shown to enhance gut health by increasing the abundance of beneficial bacteria and reducing the growth of harmful bacteria, thereby improving gut functionality and overall animal health (Wang et al. 2019a).

Postbiotics refer to the metabolic by-products of probiotics, some examples of postbiotics include short-chain fatty acids (SCFAs) such as acetate, propionate, and

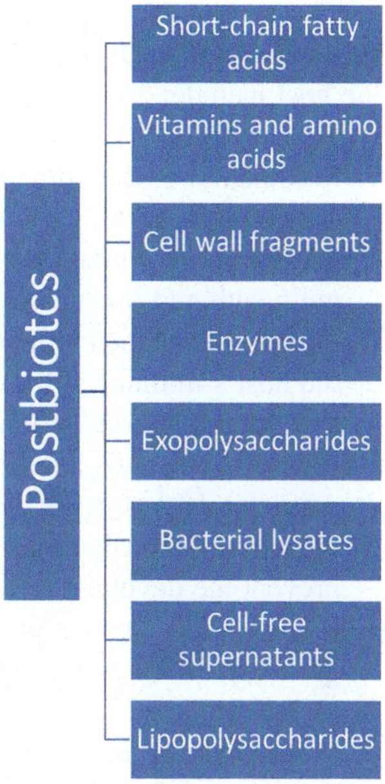

Fig. 3 Types of postbiotics

butyrate, bacteriocins, lipopolysaccharides, and host-derived peptides that are produced during the fermentation of probiotics. They play a crucial role in maintaining gut health by regulating the balance of gut microbiota and influencing the host's immune system. Unlike probiotics, postbiotics do not require live bacteria to be present for their benefits to be realized. Postbiotics have been shown to have a range of health benefits, including improved gut health, reduced inflammation, and improved immune function. They are also being investigated for their potential use in the treatment of a variety of diseases, including inflammatory bowel disease, irritable bowel syndrome, and colorectal cancer. Moreover, postbiotics are considered to be a promising new area of research in the field of gut health, as they offer many potential benefits without the concerns associated with live probiotics. However, more research is needed to fully understand the mechanisms by which postbiotics work and to determine the best methods for producing and delivering them in supplement form (Kim et al. 2015a) (Fig. 3).

Mechanism of Action Paraprobiotics work by altering the gut microbiota and leading to various health benefits. The mode of action of paraprobiotics can vary depending on the specific microorganism and the targeted animal species. However, some of the common mechanisms of action include stimulation of the immune system and inhibition of inflammation. Additionally, paraprobiotics can also modulate the gut environment, by producing organic acids and other metabolic by-products that can improve gut pH and affect gut permeability. These changes can enhance the stability and diversity of the gut microbiome, leading to improved gut health and animal performance (Siciliano et al. 2021).

The mechanism of action of postbiotics can vary depending on the specific substance produced, but some of the common ways that postbiotics can influence the host include:

1. Modulation of the gut environment: Certain metabolic by-products produced by gut bacteria, such as short-chain fatty acids (SCFAs) and lactic acid, can change the pH of the gut and modulate the gut environment. This can lead to improved gut health and prevent the growth of harmful bacteria.
2. Stimulation of the immune system: Some postbiotics have been shown to activate the host's immune system, leading to an increase in the production of cytokines and other immune system components. This can enhance the body's ability to fight off infections and prevent disease.
3. Antimicrobial activity: Some postbiotics have antimicrobial properties and can prevent the growth of harmful bacteria in the gut. This can improve gut health and prevent the development of digestive disorders.
4. Regulation of gene expression: Some postbiotics have been shown to influence gene expression in the host, leading to changes in metabolism and other physiological processes. This can lead to improved health and performance in animals.
5. Modulation of the gut microbiome: Postbiotics can influence the composition of the gut microbiome, leading to increased diversity and stability of gut bacteria. This can have a positive impact on gut health and animal performance.

The exact mechanism of action of postbiotics can depend on the specific substance produced, the dose and duration of treatment, and the target animal species. Additionally, further research is needed to fully understand the mechanisms by which postbiotics exert their effects (Żółkiewicz et al. 2020; Wegh et al. 2019) (Fig. 4).

Use of Para-postbiotics in the Poultry Nutrition The use of paraprobiotics and postbiotics in poultry nutrition has been a growing area of research in recent years due to their potential to improve the health and performance of poultry birds, as well as their promised ability to replace antibiotics (Abd El-Ghany 2020). In the study by Shu et al. (2018), the effects of non-viable yeast on growth performance, digestive enzyme activity, and gut microbiota of broilers were evaluated. The study found that the supplementation of non-viable yeast in the diet of broilers significantly improved their average daily gain and feed conversion ratio. The supplementation also increased the activity of digestive enzymes such as amylase, trypsin, and lipase, indicating improved digestion and nutrient utilization. Furthermore, the gut microbiota of broilers fed the non-viable yeast-supplemented diet showed increased abundance of beneficial bacteria, such as *Lactobacillus* and *Bifidobacterium*, and reduced abundance of pathogenic bacteria, such as *Escherichia coli*. These results suggest that the supplementation of non-viable yeast in the diet of broilers has beneficial effects on their growth performance, digestive function, and gut health. In another study, the authors investigated the effects of feeding broiler chickens with non-viable yeast cell wall on growth performance, gut health, and immune function.

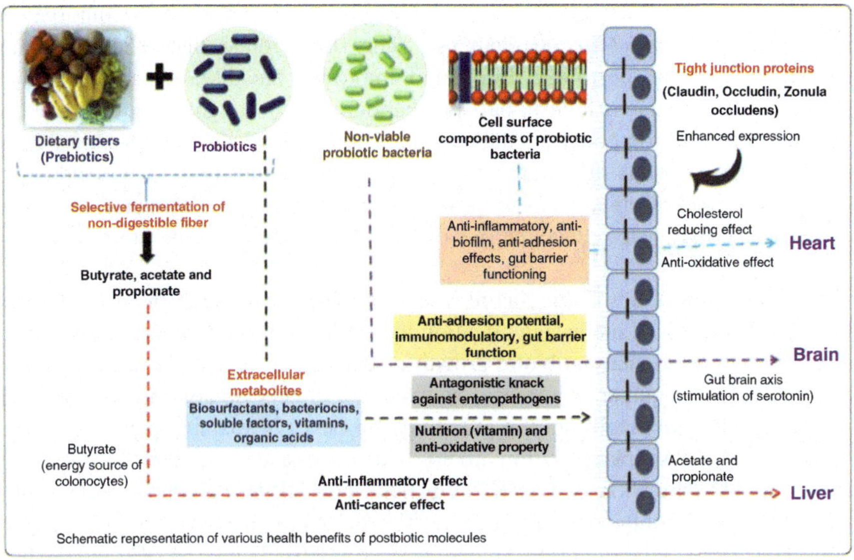

Fig. 4 Schematic representation of various benefits of postbiotic molecules

They found that the supplementation of non-viable yeast cell wall improved feed conversion ratio, enhanced gut health (by reducing the levels of harmful bacteria and increasing the populations of beneficial bacteria in the gut) and boosted the immune function of the broilers. The results of this study suggest that non-viable yeast cell wall may have potential as a nutritional supplement for improving the growth performance and gut health of broiler chickens (Liu et al. 2020). Furthermore, Lallemand Animal Nutrition (2022) made a hydrolyzed yeast that has been specifically designed to support animal performance, digestive care, and feed palatability while contributing to the feed protein balance. This hydrolized yeast can be used as a functional protein source due to its superior kinetics of protein digestibility, which starts at 77% at the beginning of the intestinal phase and increases to 94% after 48 hours. Last, a study that investigated the use of paraprobiotics in layer hen nutrition found that supplementing the diet of layer hens with non-viable yeast improved egg production and egg quality, while positively altering the gut microbiota. The authors concluded that non-viable yeast has potential as a dietary supplement in layer hen nutrition (Yong et al. 2019).

The same results were seen when postbiotics were delivered to layer chickens. As an example, consider the research by Zhang et al. (2019b). The scientists gave layer hens a meal enriched with postbiotics generated from lactic acid bacteria and discovered that the hens outperformed control birds in terms of growth performance, egg quality, and gut microbiome. The findings of this study imply that postbiotics may benefit the health and performance of layer hens. On another study made by Abd El-Ghany et al. (2022) found that the use of a postbiotic compound produced by stabilized non-viable *Lactobacilli* in the feed and drinking water of broiler chickens can have a positive impact on their health, growth performance, immunity, and resistance to *Escherichia coli* challenge. The study found that feed and water treatments with the postbiotic compound improved the disease picture, growth performance, immune response, bursa of Fabricius/body weight ratio, and reduced the intestinal coliform count in challenged chickens when compared to non-treated chickens. These results suggest that the use of the postbiotic compound in either dry or aqueous form is effective in improving the health and performance of broiler chickens and may help reduce the risk of colisepticemic disease in these birds.

Use of Para-postbiotics in the Swine Nutrition The use of paraprobiotics in swine nutrition has received increasing attention in recent years, as they offer a promising alternative to traditional antibiotic growth promoters. In one study made by Zhou et al. (2015) the authors aimed to evaluate the effects of non-viable yeast cell products on the gut microflora and growth performance of weaning pigs. The results showed that the supplementation of non-viable yeast cell products significantly increased the population of *Lactobacillus* and reduced the population of coliform bacteria in the gut of the pigs. The pigs fed the diet supplemented with high levels of non-viable yeast cell products had the highest average daily gain (ADG) and the best feed conversion ratio (FCR) compared to the control group. The authors concluded that the supplementation of non-viable yeast cell products improved the gut microflora and enhanced the growth performance of weaning pigs. Same findings were reported by Wang et al. 2018a who investigated the effects of a non-viable yeast-based product on the gut health and performance of weaning pigs. The results showed that supplementing the pigs' diet with the non-viable yeast-based product improved their weight gain and feed efficiency, as well as reduced the incidence of diarrhea. The study also demonstrated that the supplementation of the non-viable yeast-based product modulated the gut microbiota, leading to an increased abundance of beneficial bacteria and a reduced abundance of pathogenic bacteria. Finally, a study (Kim et al. 2019) evaluated the effects of a non-viable yeast product on the growth performance and gut microbiota of growing-finishing pigs. The results showed that supplementing the pigs' diet with the non-viable yeast product improved their average daily gain and feed efficiency, compared to the control group. Furthermore, the gut microbiota of the pigs fed with the non-viable yeast was found to be more diverse and had a higher proportion of beneficial bacteria compared to the control group.

As a relatively new area of research, there have not been many studies on the use of postbiotics in swine nutrition. One study that investigated the use of postbiotics, evaluated the effects of *Bacillus*-derived postbiotics on growth performance, gut morphology, and gut microbiota in growing pigs. The results showed that supplementation of postbiotics improved the average daily gain and feed efficiency of the pigs. In addition, the study found that the postbiotics treatment improved gut morphology, including villus height and crypt depth, and positively influenced the gut microbiota by increasing the abundance of beneficial bacteria and decreasing the abundance of pathogenic bacteria. The authors concluded that postbiotics could be a promising alternative to antibiotics for improving gut health and performance in growing pigs (Zhou et al. 2019b).

Use of Para-postbiotics in the Aquaculture Nutrition Paraprobiotics are increasingly being studied and used in aquaculture nutrition. The goal of using paraprobiotics in aquaculture is to promote the growth and health of fish and other aquatic species, while also reducing the use of antibiotics and other harmful chemicals. Paraprobiotics work by selectively promoting the growth of beneficial bacteria in the gut of aquatic animals, thereby modulating the gut microbiome and improving gut health. Studies have shown that the use of paraprobiotics in aquaculture can lead to improved growth performance, feed efficiency, and disease resistance in fish and other aquatic species. One study, for example, found that supplementing the diets of juvenile sea bass with a paraprobiotic mixture led to improved growth performance and feed efficiency, as well as a reduction in mortality rates (Ozkoc et al. 2016). Another study investigated the effect of a paraprobiotic containing *Lactobacillus reuteri* on the growth performance, gut microbiota, and immune response of tilapia. The results showed that the fish fed with the paraprobiotic had higher weight gain, improved feed conversion rate, and better gut microbiota compared to the control group. Additionally, the fish fed with the paraprobiotic showed an enhanced immune response, as indicated by higher levels of IgM and lysozyme activity. The authors concluded that the use of the *L. reuteri*-containing paraprobiotic has the potential to improve the growth performance, gut health, and immunity of tilapia in aquaculture (Li et al. 2019b). A third study looked at the effect of a paraprobiotic containing *Bacillus subtilis* on the growth performance, gut microbiota, and disease resistance of rainbow trout. The results showed that the dietary supplementation of the paraprobiotic significantly improved the weight gain, feed conversion ratio, and specific growth rate of the rainbow trout. Additionally, the paraprobiotic supplementation altered the gut microbiota composition, leading to an increase in the abundance of beneficial bacteria and a decrease in the abundance of harmful bacteria. Furthermore, the paraprobiotic supplementation increased the total antioxidant capacity, lysozyme activity, and interleukin-1β levels, indicating an improvement in the immune response of the rainbow trout. In conclusion, the results of the study suggest that the supplementation of a paraprobiotic containing *Bacillus subtilis* can improve the growth performance, gut microbiota, and disease resistance in rainbow trout. Zhou et al. (2019a) to sum up, the use of paraprobiotics in aquaculture is a promising approach for promoting the growth and health of fish and other aquatic species, while also reducing the use of antibiotics and other harmful chemicals. Further research is needed to fully understand the mechanisms of action and optimal use of paraprobiotics in aquaculture.

While there is a growing body of research on the use of postbiotics in other fields such as human and animal health, the use of postbiotics in aquaculture nutrition is still an emerging area of research with limited data available. Further studies are needed to establish the efficacy of postbiotics in enhancing the health and growth of aquaculture species. There is one study by Dong et al. (2020) who aimed to evaluate the effects of probiotic byproducts on growth performance, immune response, and gut microbiota of tilapia (*Oreochromis niloticus*). The study used four groups of tilapias, a control group and three treatment groups that were fed diets supplemented with different levels of postbiotics for 56 days. The results showed that the growth performance of tilapia fed diets containing postbiotics was significantly higher than the control group. The feed utilization efficiency of tilapia was also improved with increasing levels of postbiotics in the diet. The results also indicated that the postbiotics had a positive effect on the immune response of tilapia, as the serum lysozyme activity and phagocytic rate of the fish were significantly increased. In terms of gut microbiota, the results showed that the postbiotics significantly affected the abundance of certain bacteria in the gut of tilapia. The study found that the abundance of beneficial bacteria, such as *Lactobacillus* and *Bifidobacterium*, was increased, while the abundance of harmful bacteria, such as *Aeromonas* and *Escherichia coli*, was reduced. Overall, the results of this study suggest that the use of postbiotics in tilapia diets can improve growth performance, enhance immune response, and positively affect gut microbiota composition.

Use of Para-postbiotics in the Ruminant Nutrition The use of paraprobiotics in ruminant nutrition has gained considerable interest in recent years due to their potential to enhance animal performance, improve gut health, and modulate the rumen microbiome. A study by Cai et al. (2020) investigated the effects of a non-viable yeast cell product on growth performance, rumen fermentation and antioxidant capacity in beef cattle. The results showed that the beef cattle fed the yeast cell product had improved average daily gain, feed conversion ratio, and rumen pH compared to the control group. The study also showed that the antioxidant capacity of the beef cattle fed the yeast cell product was increased, indicating improved rumen health, and reduced oxidative stress. Another study by Moura et al. (2017) investigated the effects of supplementing non-viable yeast cell wall extracts on performance and ruminal acidosis in dairy cows. The results showed that supplementation with non-viable yeast cell wall extracts improved dry matter intake and milk yield, increased rumen pH, and decreased lactate concentration. The study also found that supplementation with non-viable yeast cell wall extracts reduced the occurrence of subacute ruminal acidosis (SARA), a common metabolic disorder in dairy cattle. These results suggest that the use of non-viable yeast cell wall extracts can improve the performance and health of dairy cattle by maintaining ruminal pH and reducing the risk of SARA. Prior to this study van der Meijden et al. (2015) also reported that non-viable yeast cell wall extracts supplementation reduced the risk of ruminal acidosis and improved health status in dairy cows. Furthermore, non-viable yeast cell wall extracts supplementation had no negative effect on production performance.

In ruminants, postbiotics have been shown to have positive effects on feed utilization, digestion, and gut health. One study conducted by Tian et al. (2019) investigated the effect of a fermented product containing *Bacillus subtilis* on dairy cows (a type of postbiotic). The study found that feeding the fermented product to the cows improved feed utilization, leading to increased milk production. Additionally, the fermented product was found to improve the health of the rumen, leading to a reduction in subacute ruminal acidosis (SARA), which is a common issue in dairy cattle. Another study conducted by Rauch et al. (2017) looked at the effect of feeding a yeast extract product to dairy cattle. The study found that feeding the yeast extract improved feed utilization and led to a reduction in the production of methane, which is a significant contributor to greenhouse gas emissions. In addition, feeding the yeast extract improved gut health, leading to a reduction in the occurrence of digestive issues. Finally, a study by Liu et al. (2017) investigated the effects of dietary supplementation with a *Bacillus subtilis* fermented product on the growth performance, blood characteristics, and gut microbiota in lambs. The results showed that lambs fed the *Bacillus subtilis* fermented product had improved average daily weight gain and feed conversion efficiency compared to the control group. The study also observed changes in the gut microbiota, with increased levels of beneficial bacteria such as *Lactobacillus* and *Bifidobacterium*. These findings suggest that postbiotics have potential to enhance growth performance and improve gut health in ruminants.

Use of Para-postbiotics in the Equine Nutrition In equine, the use of paraprobiotics is a new area of researching. One study that investigated the use of non-viable paraprobiotics in horse nutrition is "Effects of dietary non-viable yeast cell wall product on the digestive function, faecal parameters and gut microbiota of horses." This study was carried out to evaluate the effects of a non-viable yeast cell wall product on the digestive function, fecal parameters and gut microbiota of horses. The results showed that the dietary supplementation of the non-viable yeast cell wall product led to an improvement in fecal parameters and increased numbers of beneficial gut bacteria. The authors concluded that the use of non-viable yeast cell wall product as a dietary supplement may have positive effects on the gut health and digestive function of horses (Rousseau et al. 2018).

Unfortunately, there is a limited amount of research on the use of postbiotics in equine nutrition. The is one study by Flaminio et al. (2019), who investigated the effects of a postbiotic mixture (containing lactic acid bacteria byproducts) on the gut microbiota, digestive functions, and performance of horses. The study found that supplementation with the postbiotic mixture improved gut microbiota diversity and reduced the abundance of potential pathogenic bacteria. Additionally, horses supplemented with the postbiotic mixture had improved feed utilization and better performance compared to the control group. This study suggests that postbiotic supplementation can have positive effects on the gut microbiota, digestive function, and performance of horses. However, more research is needed to fully understand the potential benefits of postbiotics in equine nutrition.

Use of Para-postbiotics in the Companion Animal Nutrition The use of paraprobiotics in companion animal nutrition is an area of ongoing research. The aim is to utilize the by-products of probiotic bacteria to improve the digestive health and overall wellness of pet animals. The use of these paraprobiotics may help to enhance the gut microbiota, reduce digestive disorders, and improve the nutrient utilization in companion animals. One study that investigated the use of paraprobiotics in companion animal nutrition is made by Li et al. (2017a). The study was conducted on dogs with idiopathic chronic diarrhea and aimed to evaluate the effects of a non-viable yeast extract on the gut microbiota, antioxidant status, and clinical signs. The results showed that the supplementation of non-viable yeast extract significantly improved the gut microbiota diversity and reduced the severity of clinical signs in dogs with idiopathic chronic diarrhea. This study demonstrates the potential benefits of using non-viable paraprobiotics in companion animal nutrition and highlights the importance of further research in this field. By utilizing the paraprobiotics, pet owners may be able to improve the health and wellness of their animals and reduce the incidence of digestive disorders.

Postbiotics, the metabolites produced by probiotics, have gained increasing attention in companion animal nutrition for their potential benefits to animal health and well-being. The use of postbiotics in companion animal nutrition has been evaluated in several studies, showing promising results in terms of improving gut health, boosting the immune system, and enhancing nutrient utilization. One study evaluated the effects of postbiotics on dogs' gut health and found that supplementing with postbiotics can lead to an improvement in gut microbiota diversity and a reduction in harmful bacteria populations. This results in a more balanced gut microbiome, which can have a positive impact on digestive health, immune system function, and overall well-being (Kim et al. 2017a). Another study investigated the use of postbiotics in cats and found that supplementing with postbiotics can improve digestive health and reduce the risk of digestive disorders, such as diarrhea and vomiting. Additionally, the study reported a positive effect of postbiotics on cats' immune system, as well as improved nutrient utilization, which can result in better weight management and overall health (Kim et al. 2017b). In conclusion, the use of postbiotics in companion animal nutrition has shown promising results in terms of improving gut health, boosting the immune system, and enhancing nutrient utilization. Further research is needed to fully understand the potential benefits of postbiotics in companion animal nutrition and to develop more effective and safer postbiotic-based products for companion animals.

Para-postbiotics on Sustainability Paraprobiotics and postbiotics are two alternative approaches to traditional probiotics that have been shown to have a positive impact on sustainability in animal production. Paraprobiotics are defined as inactivated cells or cell fractions that have health-promoting effects. By using paraprobiotics in animal feed, it is possible to improve gut health and reduce the need for antibiotics in animal production, leading to a more sustainable industry. For example, a study by Duranti et al. (2015) found that supplementing the diets of broiler chickens with a paraprobiotic reduced the need for antibiotics and improved gut health. Postbiotics, on the other hand, are metabolic by-products of probiotic bacteria that have health-promoting effects. They are also non-viable and are produced by the fermentation of probiotic bacteria in food or feed. Like paraprobiotics, postbiotics can improve gut health and reduce the need for antibiotics in animal production, leading to a more sustainable industry. For example, a study by Zebeli et al. (2017) found that supplementing the diets of dairy cows with a postbiotic improved feed efficiency and reduced the amount of nitrogen excreted in manure. The use of non-viable paraprobiotics and postbiotics in animal production can also have a positive impact on the environment. By improving gut health and reducing the need for antibiotics, these compounds can reduce the amount of waste excreted in feces and the amount of greenhouse gas emissions produced from manure. This can help to reduce the environmental impact of animal production and create a more sustainable industry. In conclusion, the use of paraprobiotics and postbiotics in animal production has the potential to contribute significantly to sustainability. By improving gut health, reducing the need for antibiotics, and reducing the environmental impact of animal production, these compounds can help create a more sustainable livestock industry (Fig. 4).

5 Genetically Modified Microorganisms

Genetically modified microorganisms (GMOs) are microorganisms, such as bacteria, yeasts, or fungi, whose genetic material has been altered through genetic engineering techniques. This can be done to modify their properties, such as their growth rate, metabolism, or resistance to antibiotics or other environmental stressors. GMOs have a wide range of applications, including the production of industrial chemicals, the degradation of toxic waste, and the production of vaccines and other medical products. The use of GMOs in biotechnology has also led to the development of new crops with improved traits, such as increased yield or resistance to pests and diseases. One reference on the subject of GMOs is the book edited by Stemke (2004). This book provides a comprehensive overview of the science and technology of GMOs, as well as the regulatory and ethical issues surrounding their use. It covers topics such as the safety and risk assessment of GMOs, the social and economic implications of GMO technology, and the role of GMOs in sustainable agriculture and bioremediation.

The history of genetically modified microorganisms (GMOs) dates back to the 1970s, when scientists first developed the tools and techniques necessary to manipulate the genetic material of microorganisms. One of the key milestones in this history was the discovery of restriction enzymes, which allowed scientists to cut DNA at specific points and recombine it with DNA from other organisms. In 1973, Herbert Boyer and Stanley Cohen used restriction enzymes to create the first recombinant DNA molecule, which marked the birth of genetic engineering. In the following years, scientists made rapid progress in this field and developed techniques for introducing foreign genes into bacteria and other microorganism (Cohen et al. 1973). One of the earliest applications of GMOs was in the production of human insulin for medical use. In 1978, Genentech and Eli Lilly announced the successful production of human insulin in bacteria, which represented a major breakthrough in the treatment of diabetes. This was followed by the production of other medical products, such as human growth hormone and interferon, using GMOs (Baeshen et al. 2014). Another important application of GMOs was in the field of bioremediation, where microorganisms were used to clean up toxic waste and pollutants in the environment. This led to the development of new strains of bacteria with improved degradation capabilities, which could be used to clean up contaminated sites and reduce environmental damage. In conclusion, the history of GMOs spans several decades and encompasses a wide range of scientific and technological advances. GMOs have played a significant role in the development of new medical products and the cleanup of contaminated environments, and they continue to be an important tool for advancing biotechnology and improving health and well-being.

The administration of GMOs in animal nutrition has gained increasing attention in recent years, as scientists seek new ways to improve the efficiency and sustainability of animal production. GMOs have the potential to enhance animal health and productivity by providing new sources of vitamins, minerals, and other essential nutrients, as well as improving the digestibility of feed and reducing the impact of animal waste on the environment. One of the key applications of GMOs in animal nutrition is the production of transgenic crops and feeds, which can be enriched with essential nutrients and other beneficial compounds. For example, researchers have developed genetically modified soybeans that produce higher levels of essential amino acids, such as lysine and methionine, which are critical for the growth and development of livestock. This can help to reduce the need for supplementing feed with synthetic additives and improve the overall quality of the feed. Another important application of GMOs in animal nutrition is the production of probiotics. The administration of genetically modified probiotics in animal nutrition has become a promising approach in the effort to improve the health and productivity of livestock. One of the main benefits of genetically modified probiotics is that they can be designed to produce specific compounds or enzymes that can enhance animal health and performance. For example, a study by Kudriavtsev et al. (2015) found that supplementing pig diets with a genetically modified probiotic strain of *Lactobacillus plantarum* that produces high levels of lactic acid significantly improved growth performance and reduced the incidence of diarrhea in piglets. Another important benefit of genetically modified probiotics is that they can be designed to be more stable and effective in the harsh conditions of the gut. This can help to ensure that the probiotic strains are able to survive and thrive in the gut, leading to improved gut health and better performance outcomes. A study by Fan et al. (2020) found that supplementing broiler chicken diets with a genetically modified probiotic strain of *Bacillus subtilis* improved gut health and feed efficiency, resulting in improved growth performance and reduced manure production. Last, a very promising study made by Ritter et al. (2018) used a genetically modified antimicrobial peptide called "Microcin J25." The study results suggest that certain mutants of microcin J25 have increased specificity for pathogenic *Salmonella* species compared to human commensal *Escherichia coli*, meaning that these mutants may be more effective in treating infections caused by *Salmonella* while having less impact on the normal bacteria in the human gut. In conclusion, the administration of genetically modified probiotics in animal nutrition has the potential to play a significant role in improving the efficiency and sustainability of animal production. By improving gut health and boosting the immune system, genetically modified probiotics can help to promote sustainable and responsible animal agriculture that meets the needs of farmers, consumers, and the environment.

6 Conclusion

In conclusion, the use of probiotics, prebiotics, paraprobiotics, postbiotics, and genetically modified microorganisms (GMOs) in animal nutrition has the potential to greatly improve the health and productivity of farm animals. However, it is important to note that the effects of these additives on animal health and performance can vary depending on various factors, such as the type and dose of the additive, the age and health status of the animal, and the composition of the diet. Therefore, it is important to carefully evaluate the potential benefits and risks of each additive, and to use them in a responsible and sustainable manner.

The use of probiotics, prebiotics, paraprobiotics, postbiotics, and GMOs in animal nutrition can contribute to sustainability in several ways. Firstly, the administration of probiotics and prebiotics can improve gut health, reduce the incidence of diseases, and increase feed efficiency in farm animals. This can result in better utilization of feed resources, and reduced waste and emissions, thereby contributing to sustainability. Furthermore, the use of paraprobiotics and postbiotics can improve gut health, enhance nutrient utilization, and reduce inflammation in farm animals. This can lead to improved growth performance and feed efficiency, as well as reduced reliance on antibiotics and other drugs, which can contribute to sustainability by reducing the environmental impact of animal production and promoting animal welfare. In addition, the use of GMOs in animal nutrition can produce medical products and feed additives that can improve the growth performance and feed efficiency of farm animals. This can reduce the amount of feed resources required to produce a given amount of meat, eggs, or dairy products, and contribute to sustainability by reducing the environmental impact of animal production. Overall, the use of probiotics, prebiotics, paraprobiotics, postbiotics, and GMOs in animal nutrition has the potential to contribute to sustainability in animal production, but careful evaluation and responsible use are necessary to ensure that the benefits are realized, and the risks are minimized.

References

Abd El-Ghany WA (2020) Paraprobiotics and postbiotics: contemporary and promising natural antibiotics alternatives and their applications in the poultry field. Open Vet J 10(3):323–330. https://doi.org/10.4314/ovj.v10i3.11

Abd El-Ghany WA, Fouad H, Quesnell R, Sakai L (2022) The effect of a postbiotic produced by stabilized non-viable lactobacilli on the health, growth performance, immunity, and gut status of colisepticaemic broiler chickens. Trop Anim Health Prod 54(5):286. https://doi.org/10.1007/s11250-022-03300-w

Abdel-Moneim AE, Elbaz AM, Khidr RE, Badri FB (2019) Effect of in ovo inoculation of Bifidobacterium spp. on growth performance, thyroid activity, ileum histomorphometry and microbial enumeration of broilers. Probiotics Antimicrob Proteins 12(3):873–882. https://doi.org/10.1007/s12602-019-09613-x

Abdel-Tawwab M, Abdel-Rahman AM, Ismael NEM (2008) Evaluation of commercial live bakers' yeast, Saccharomyces cerevisiae as a growth and immunity promoter for fry Nile tilapia, *Oreochromis niloticus* (L.) challenged in situ with Aeromonas hydrophila. Aquaculture 280: 185–189

Agazzi A, Invernizzi G, Bontempo V, Savoini G (2015) The beneficial role of probiotics in monogastric animal nutrition and health. J Dairy Vet Anim Res 2(4):00041. https://doi.org/10.15406/jdvar.2015.02.00041

Ahmed ST, Hoon J, Mun H, Yang C (2014) Evaluation of *Lactobacillus* and *Bacillus*-based probiotics as alternatives to antibiotics in enteric microbial challenged weaned piglets. Afr J Microbiol Res 8:96–104

Ajiguna JC, Prakasita VC, Nahak TEM, Tabbu CR, Santosa CM, Wahyuni AETH (2021) The role of synbiotics (commercial product) as a substitute for antibiotic growth promotor (AGP) in the performance and blood values of cobb-strain broilers challenged with Salmonella enteritidis. Adv Biol Res 15:59–66

Al-Saiady MY (2010) Effect of probiotic bacteria on immunoglobulin G concentration and other blood components of newborn calves. J Anim Vet Adv 9:604–609

AlZahal O, McGill H, Kleinberg A, Holliday JI, Hindrichsen IK, Duffield TF, McBride BW (2014) Use of a direct-fed microbial product as a supplement during the transition period in dairy cattle. J Dairy Sci 97:7102–7114

Apas AL, Dupraz J, Ross R, Gonzalez SN, Arena ME (2010) Probiotic administration effect on fecal mutagenicity and microflora in the goat's gut. J Biosci Bioeng 110:537–540

Arenas-Padilla M, Duarte-Gutiérrez JL, Mata-Haro V (2018) Bifidobacterium animalis ssp. lactis BB12 induces IL-10 through cell membrane-associated components via TLR2 in swine. J Appl Microbiol 125:1881–1889. https://doi.org/10.1111/jam.14069

Arora M, Baldi A (2015) Regulatory categories of probiotics across the globe: A review representing existing and recommended caterorization. Indian J Med Microbiol 33:S2–S10

Baeshen NA, Baeshen MN, Sheikh A, Bora RS, Ahmed MM, Ramadan HA, Saini KS, Redwan EM (2014) Cell factories for insulin production. Microb Cell Factories 13:141. https://doi.org/10.1186/s12934-014-0141-0

Bai S, Wu A, Ding X, Lei Y, Bai J, Zhang K, Chio J (2013) Effects of probiotic-supplemented diets on growth performance and intestinal immune characteristics of broiler chickens. Poult Sci 92(3):663–670. https://doi.org/10.3382/ps.2012-02813

Baker AA, Davis E, Spencer JD, Moser R, Rehberger T (2013) The effect of a Bacillus- based direct-fed microbial supplemented to sows on the gastrointestinal microbiota of their neonatal piglets. J Anim Sci 91:3390–3399

Barba-Vidal E, Castillejos L, Roll VFB, Cifuentes-Orjuela G, Moreno Muñoz JA, Martín-Orúe SM (2017) The probiotic combination of Bifidobacterium longum subsp. infantis CECT 7210 and Bifidobacterium animalis subsp. Lactis BPL6 reduces pathogen loads and improves gut health of weaned piglets orally challenged with Salmonella typhimurium. Front Microbiol 8:1570. https://doi.org/10.3389/fmicb.2017.01570

Batista LHC, Cidrini IA, Prados LF, Cruz AAC, Torrecilhas JA, Siqueira GR, Resende FD (2022) A meta-analysis of yeast products for beef cattle under stress conditions: Performance, health and physiological parameters. Anim Feed Sci Technol 283:115182. https://doi.org/10.1016/J.ANIFEEDSCI.2021.115182

Beirão BC, Ingberman M, Fávaro C Jr (2018) Effect of an *Enterococcus faecium* probiotic on specific IgA following live Salmonella enteritidis vaccination of layer chickens. Avian Pathol 47:325–333

Bélanger G, Lacasse P, Brousseau P (2010) Mannan oligosaccharides, fructo-oligosaccharides and lactic acid bacteria supplementation of lactating dairy cow diets: effects on feed intake and lactation performance. Can J Anim Sci 90(3):507–515

Bermudez-Brito M, Plaza-Diaz J, Munoz-Quezada S, Gomez-Llorente C, Gil A (2012) Probiotic mechanisms of action. Ann Nutr Metab 61:160–174. https://doi.org/10.1159/000342079

Beukers AG, Zaheer R, Cook SR, Chaves AV, Ward MP, Tymensen L, Morley PS, Hannon S, Booker CW, Read RR, McAllister TA (2018) Comparison of antimicrobial resistance genes in feedlots and urban wastewater. Can J Vet Res 82(1):24–38

Bhogoju S, Nahashon S (2022) Recent advances in probiotic application in animal health and nutrition: A review. Agricultur 12(2):304. https://doi.org/10.3390/agriculture12020304

Bonos E, Giannenas I, Sidiropoulou E, Stylianaki I, Tzora A, Skoufos I, Barbe F, Demey V, Christaki E (2021) Effect of Bacillus pumilus supplementation on performance, intestinal morphology, gut microflora and meat quality of broilers fed different energy concentrations. Anim Feed Sci Technol 274. https://doi.org/10.1016/J.ANIFEEDSCI.2021.114859

Bontempo V, Giancamillo AD, Savoini G, Dell'Orto V, Domeneghini C (2006) Live yeast dietary supplementation acts upon intestinal morpho-functional aspects and growth in weanling piglets. Anim Feed Sci 129:224–236

Bouchard DS, Seridan B, Saraoui T, Rault L, Germon P, Gonzalez-Moreno C, Nader-Macias FM, Baud D, François P, Chuat V, Chain F, Langella P, Nicoli J, Le Loir Y, Even S (2015) Lactic acid bacteria isolated from bovine mammary microbiota: potential allies against bovine mastitis. PLoS One 10(12):e0144831. https://doi.org/10.1371/journal.pone.014483

Burke ML, Morre SA (2017) *Bacillus subtilis* strain PB6 demonstrates growth inhibition toward equine-specific bacterial pathogens. J Equine Vet 58:84–88

Büsing K, Zeyner A (2015) Effects of oral *Enterococcus faecium* strain DSM 10663 NCIMB 10415 on diarrhoea patterns and performance of sucking piglets. Benefic Microbes 6:41–44. https://doi.org/10.3920/BM2014.0008

Cai L, Indrakumar S, Kiarie E, Kim IH (2015) Effects of a multi-strain *Bacillus* species–based direct-fed microbial on growth performance, nutrient digestibility, blood profile, and gut health in nursery pigs fed corn–soybean meal–based diets. J Anim Sci 93(9):4336–4342. https://doi.org/10.2527/jas.2015-9056

Cai Y, Wang L, Chen J, Wang D (2020) The effects of yeast cell products on growth performance, rumen fermentation, and antioxidant capacity in beef cattle. Livest Sci 236:104332

Castanon JIR (2007) History of the use of antibiotic as growth promoters in European poultry feeds. Poult Sci 86:2466–2471. https://doi.org/10.3382/ps.2007-00249

Castex M, Chim L, Pham D, Lemaire P, Wabete N, Nicolas JL, Schmidely P, Mariojouls C (2008) Probiotic P. acidilactici application in shrimp Litopenaeus stylirostris culture subject to vibriosis in New Caledonia. Aquaculture 275:1–4. https://doi.org/10.1016/j.aquaculture.2008.01.011

Castex M, Lemaire P, Wabete N, Chim L (2010) Effect of probiotic Pediococcus acidilactici on antioxidant defences and oxidative stress of Litopenaeus stylirostris under Vibrio nigripulchritudo challenge. Fish Shellfish Immunol 28(4):622–631. https://doi.org/10.1016/J.FSI.2009.12.024

Chang C-I, Liu W-Y (2002) An evaluation of two probiotic bacterial strains, *Enterococcus faecium* SF68 and Bacillus toyoi, for reducing edwardsiellosis in cultured European eel, Anguilla Anguilla L. J Fish Dis 25:311–315

Chaveerach P, Lipman LJ, van Knapen F (2004) Antagonistic activities of several bacteria on *in vitro* growth of 10 strains of *Campylobacter jejuni/coli*. Int J Food Microbiol 90:43–50

Chen YC, Nakthong C, Chen TC (2005) Improvement of laying hen performance by dietary prebiotic chicory oligofructose and inulin. Int J Poultry Sci 4:103–108

Chen J, Wang X, Sun Y, Wang Y, Liu C, Li Y (2017) Xylooligosaccharides improve growth performance, nutrient utilization, and gut health of growing-finishing pigs. J Anim Sci 95 (7):3185–3193

Chen Y, Li J, Zhang S, Huang Y, Mao X (2018) Xylooligosaccharides supplementation improved growth performance and altered gut microbiota composition of Nile tilapia (*Oreochromis niloticus*). Fish Shellfish Immunol 78:74–80

Chen J, Zhang X, Li Y, Yang H (2019a) Effects of mannooligosaccharides on gut microbiota, gut health, and disease resistance in common carp. Aquac Res 51(2):391–400

Chen J, Li Y, Han M, Wei Y, Wang Y, Li Y (2019b) Mannooligosaccharides improve growth performance, nutrient utilization, and gut health of growing-finishing pigs. J Anim Sci 97(7): 3185–3193

Chiang ML, Chen HC, Chen KN, Lin YC, Lin YT, Chen MJ (2015) Optimizing production of two potential probiotic Lactobacilli strains isolated from piglet feces as feed additives for weaned piglets. Asian Australas J Anim Sci 28(8):1163e70

Chou S, Hsu T, Shih C, Cheng W (2018) Xylooligosaccharides supplementation in cat diets affects gut microbiota, digestibility, and fecal quality. J Anim Sci 96(6):2288–2297

Cohen SN, Chang AC, Boyer HW, Helling RB (1973) Construction of biologically functional bacterial plasmids in vitro. Proc Natl Acad Sci 70(11):3240–3244

Collado MC, Gueimonde M, Salminem S (2010) Probiotics in adhesion of pathogens: mechanisms of action. In: RRWatson VRP (ed) Bioactive foods in promoting health: probiotics and prebiotics, 1st edn. Academic Press, Elsevier, London, pp 353–370. https://doi.org/10.1016/B978-0-12-374938-3.00023-2

Couturier M, Thivichon-Prince B, Lener V, Poucheret P, Millet Y (2011) Prebiotic effect of xylooligosaccharides in dogs. Anim Feed Sci Technol 168:1–10

Cui Y, Du Y, Fan M, Wu J, Fan L (2019) Effects of galactooligosaccharides supplementation on growth performance, nutrient digestibility, and gut microbiota of growing pigs. J Anim Sci Technol 61(7):559

Cutting SM (2011) Bacillus probiotics. Food Microbiol 28(2):214–220. https://doi.org/10.1016/j.fm.2010.03.007

D'Amelio P, Sassi F (2017) Gut microbiota, immune system, and bone. Calcif Tissue Int 102(4): 415–425. https://doi.org/10.1007/s00223-017-0331-y

Dalloul RA, Lillehoj HS, Shellem TA, Doerr JA (2003) Enhanced mucosal immunity against Eimeria acervulina in broilers fed a lactobacillus-based probiotic. Poultry Sci. 82:62–66

De Boever JL, Van Nuffel A, Demeyer D, De Campeneere S (2005) The effects of mannan oligosaccharides on growth performance, nutrient utilization, and gut microflora of lambs. Anim Sci 81(2):297–303

de Campos CM, Zanuzzo FS, Gimbo RY, Favero GC, Soares MP, Pilarski F, Urbinati EC (2022) Dietary inulin modulated the cortisol response and increased the protection against pathogens in juvenile pacu (*Piaractus mesopotamicus*). Aquac Res 53:860–869. https://doi.org/10.1111/are.15626

Deng Q, Odhiambo JF, Farooq U, Lam T, Dunn SM, Ametaj BN (2015) Intravaginal lactic acid bacteria modulated local and systemic immune responses and lowered the incidence of uterine infections in periparturient dairy cows. PLoS One 10:e0124167

Deng B, Wu J, Li X, Zhang C, Xiaonimg M, Xu Z (2020) Effects of *Bacillus subtilis* on growth performance, serum parameters, digestive enzyme, intestinal morphology, and colonic microbiota in piglets. AMB Expr 10:212. https://doi.org/10.1186/s13568-020-01150-z

Desrochers AM, Dolente BA, Roy MF, Boston R, Carlisle S (2005) Efficacy of *Saccharomyces boulardii* for treatment of horses with acute enterocolitis. J Am Vet Med Assoc 227:954–959. https://doi.org/10.2460/javma.2005.227.954

Devi G, Harikrishnan R, Paray BA, Al-Sadoon MK, Hoseinifar SH, Balasundaram C (2019) Effect of symbiotic supplemented diet on innate-adaptive immune response, cytokine gene regulation and antioxidant property in Labeo rohita against Aeromonas hydrophila. Fish Shellfish Immunol 89:687–700

Dinan TG, Quigley EM, Ahmed SM, Scully P, O'Brien S, O'Mahony L, O'Mahony S, Shanahan F, Keeling PW (2006) Hypothalamic-pituitary gut axis dysregulation in irritable bowel syndrome: plasma cytokines as a potential biomarker? Gastroenterology 130:304–311. https://doi.org/10.1053/j.gastro.2005.11.033

Ding H, Zhao X, Azad MAK, Ma C, Gao Q, He J, Kong X (2021) Dietary supplementation with: Bacillus subtilis and xylo-oligosaccharides improves growth performance and intestinal morphology and alters intestinal microbiota and metabolites in weaned piglets. Food and Function 12(13):5837–5849. https://doi.org/10.1039/d1fo00208b

Dong L, Tang J, Liu L, Liu X, Li X (2020) The effects of probiotic byproducts on growth performance, immune response, and gut microbiota of tilapia (*Oreochromis niloticus*). Aquaculture 523:736334

Duranti M, Gioacchini G, Mancuso G, Gabrielli O, Scisciola M (2015) A randomized controlled trial of a bacterial-derived paraprobiotic in broiler chickens. Animal nutrition 1(3):150–154

El-Nobi G, Hassanin M, Khalil AA, Mohammed AY, Amer SA, Montaser MM, El-Sharnouby ME (2021) Synbiotic effects of saccharomyces cerevisiae, mannan oligosaccharides, and β-glucan on innate immunity, antioxidant status, and disease resistance of nile tilapia, *Oreochromis niloticus*. Antibiotics 10(5):567. https://doi.org/10.3390/antibiotics10050567

Fan Q, Wang J, Li Y, Zhang Q (2017) Effects of trans-galactooligosaccharides supplementation on growth performance, gut morphology, gut microbiota, and immunological responses in broiler chickens. Poult Sci 96(3):818–825

Fan Y, Zhang L, Wang L, Yang X, Zhao X (2020) The impact of a genetically modified Bacillus subtilis on gut health, feed efficiency, and growth performance in broiler chickens. Poult Sci 99(9):5054–5060

FDA (2012) The judicious use of medically important antimicrobial drugs in food-producing animals (Guidance for Industry, #209) online. Retrieved from https://www.fda.gov/downloads/animalveterinary/guidancecomplianceenforcement/guidanceforindustry/ucm216936.pdf

Fernández L, Martínez R, Pérez M, Arroyo R, Rodríguez JM (2019) Characterization of Lactobacillus rhamnosus MP01 and Lactobacillus plantarum MP02 and assessment of their potential for the prevention of gastrointestinal infections in an experimental canine model. Front Microbiol 10:1117. https://doi.org/10.3389/fmicb.2019.01117

Flaminio M, Prenesti E, Spanghero M, Novara A (2019) Effects of a mixture of fermented lactic acid bacteria by-products on the gut microbiota, digestive functions and performance of horses. Anim Feed Sci Technol 250:72–80

Franz CMAP, Huch M, Abriouel H, Holzapfel W, Gálvez A (2011) *Enterococci* as probiotics and their implications in food safety. J Food Microbiol 151:125–140. https://doi.org/10.1016/j.ijfoodmicro.2011.08.014

Furr M (2014) Orally administered Pediococcus acidilactici and saccharomyces boulardii–based probiotics alter select equine immune function parameters. J Equine Vet Sci 34:1156–1163

Gao P, Ma C, Sun Z (2017) Feed-additive probiotics accelerate yet antibiotics delay intestinal microbiota maturation in broiler chicken. Microbiome 5:91

Geigerova M, Vlkova E, Bunesova V, Rada V (2016) Persistence of bifidobacteria in the intestines of calves after administration in freeze-dried form or in fermented milk. Czech. J Anim Sci 61: 49e56

Geng X, Dong XH, Tan BP, Yang QH, Chi SY, Liu HY, Liu XQ (2012) Effects of dietary probiotic on the growth performance, non-specific immunity and disease resistance of cobia, *Rachycentron canadum*. Aquac Nutr 18:46–55

Ghosh S, Sinha A, Sahu C (2007) Effect of probiotic on reproductive performance in female livebearing ornamental fish. Aquac Res 38:518–526. https://doi.org/10.1111/j.1365-2109.2007.01696.x

Giang HH, Viet TQ, Ogle B, Lindberg JE (2011) Effects of supplementation of probiotics on the performance nutrient digestibility and fecal microflora in growing- finishing pigs. Asian Australas J Anim Sci 24(5):655e61

Giannenas I, Papadopoulos E, Tsalie E, Triantafillou E, Henikl S, Teichmann K, Tontis D (2012) Assessment of dietary supplementation with probiotics on performance, intestinal morphology and microflora of chickens infected with Eimeria tenella. Vet Parasitol 188(1–2):31–40. https://doi.org/10.1016/j.vetpar.2012.02.017

Giannenas I, Karamaligas I, Margaroni M, Pappas I, Mayer E, Encarnação P, Karagouni E (2014) Effect of dietary incorporation of a multi-strain probiotic on growth performance and health status in rainbow trout (*Oncorhynchus mykiss*). Fish Physiol Biochem. https://doi.org/10.1007/s10695-014-0010-0

Gibson GR, Scott KP, Rastall RA, Tuohy KM, Hotchkiss AT, Dubert-Ferrandon A, Gareau MG, Murphy EF, Saulnier D, Loh G, Macfarlane S, Delzenne NM, Ringel Y, Kozianowski G, Dickmann RS, Lenoir-Wijnkoop I, Walker C, Buddington RK (2010) Dietary prebiotics: current status and new definition. Food Sci Technol Bull: Functional Foods 7:1–19. https://doi.org/10.1616/1476-2137.15880

Gingerich E, Frana T, Logue CM, Smith DP, Pavlidis HO, Chaney WE (2021) Effect of feeding a postbiotic derived from saccharomyces cerevisiae fermentation as a preharvest food safety hurdle for reducing Salmonella enteritidis in the ceca of layer pullets. J Food Prot 84:275–280

Glenn G, Roberfroid M (1995) Dietary modulation of the human colonic microbiota: introducing the concept of prebiotics. J Nutr 125:1401–1412

Grisdale-Helland B, Helland SJ, Gatlin DM III (2008) The effects of dietary supplementation with mannanoligosaccharide, fructooligosaccharide or galactooligosaccharide on the growth and feed utilization of Atlantic salmon (Salmo salar). Aquaculture 283:163–167

Groves PJ, Williamson SL, Ahaduzzaman M, Diamond M, Ngo M, Han A, Sharpe SM (2021) Can a combination of vaccination, probiotic and organic acid treatment in layer hens protect against early life exposure to Salmonella typhimurium and challenge at sexual maturity? Vaccine 39: 815–824

Haghighi HR, Gong J, Gyles CL (2006) Probiotics stimulate production of natural antibodies in chickens. Clin Vaccine Immunol 13:975–980

Haghighi HR, Abdul-Careem MF, Dara RA, Chambers JR, Sharif S (2008) Cytokine gene expression in chicken cecal tonsils following treatment with probiotics and Salmonella infection. Vet Microbiol 126:225–233

Haque MM, Hasan NA, Eltholth MM, Saha P, Mely SS, Rahman T, Murray FJ (2021) Assessing the impacts of in-feed probiotic on the growth performance and health condition of pangasius (*Pangasianodon hypophthalmus*) in a farm trial. Aquac Rep 20. https://doi.org/10.1016/j.aqrep.2021.100699

Hasan MT, Jang WJ, Tak JY, Lee BJ, Kim KW, Hur SW, Han HS, Kim BS, Do HM, Kim SK, Kong IS (2018) Effects of Lactococcus lactis subsp. Lactis i2 with β-glucooligosaccharides on growth, innate immunity and streptococcosis resistance in olive flounder (*Paralichthys olivaceus*). J Microbiol Biotechnol 28:1433–1442

Hassanein MS, Soliman NK (2010) Effect of probiotic (Saccharomyces cerevisiae) adding to diets on intestinal microflora and performance of hy-line layers hens. J Am Sci 6:159–169

Hassani MHS, Jourdehi AY, Zelti AH, Masouleh AS, Lakani FB (2020) Effects of commercial superzist probiotic on growth performance and hematological and immune indices in fingerlings *Acipenser baerii*. Aquacult Int 28:377–387

Hatab MH, Elsayed MA, Ibrahim NS (2016) Effect of some biological supplementation on productive performance, physiological and immunological response of layer chicks. J Radiat Res Appl Sci 9:185–192

Hayashi RM, Lourenço MC, Kraieski AL, Araujo RB, GonzalezEsquerra R, Leonardecz E, da Cunha AF, Carazzolle MF, Monzani PS, Santin E (2018) Effect of feeding Bacillus subtilis spores to broilers challenged with Salmonella enterica serovar Heidelberg Brazilian strain UFPR1 on performance, immune response, and gut health. Front Vet Sci 5:13. https://doi.org/10.3389/fvets.2018.00013

Hedman HD, Vasco KA, Zhang L (2020) A review of antimicrobial resistance in poultry farming within low-resource settings. Animals 10(8):1264. https://doi.org/10.3390/ani10081264

Herfel TM, Jacobi SK, Lin X, Jouni ZE, Chichlowski M, Stahl CH, Odle J (2013) Dietary supplementation of Bifidobacterium longum strain AH1206 increases its cecal abundance and elevates intestinal interleukin-10 expression in the neonatal piglet. Food Chem Toxicol 60:116–122

Hollister A, Cheeke P, Robinson A (1989) Effects of water administrated probiotics and acidifiers on growth, feed conversion and enteritis mortality of weanling rabbits. J Appl Rabbit Res 12: 143–147

Hong Y, Zhang Q, Li X, Fan M, Li J, Ma L (2017) Effects of mannooligosaccharides on growth performance, nutrient utilization, and gut health of broiler chickens. Poult Sci 96(11): 3680–3687

Hostins B, Lara G, Decamp O, César DE, Wasielesky W (2017) Efficacy and variations in bacterial density in the gut of Litopenaeus vannamei reared in a BFT system and in clear water supplemented with a commercial probiotic mixture. Aquaculture 480:58–64

Hu X, Yang HL, Yan YY (2019) Effects of fructooligosaccharide on growth, immunity and intestinal microbiota of shrimp (*Litopenaeus vannamei*) fed diets with fish meal partially replaced by soybean meal. Aquac Nutr 25(1):194–204. https://doi.org/10.1111/anu.12843

Jahan N, Islam SMM, Rohani MF, Hossain MT, Shahjahan M (2021) Probiotic yeast enhances growth performance of rohu (Labeo rohita) through upgrading hematology, and intestinal microbiota and morphology. Aquaculture 545:737243

Jeusette I, Grauwels M, Cuvelier C, Tonglet C, Istasse L, Diez M (2004) Hypercholesterolaemia in a family of rough collie dogs. J Small Anim Pract 45:319–324

Jiang Z, Wei S, Wang Z, Zhu C, Hu S, Zheng C, Hu H, Wang L et al (2015) Effects of different forms of yeast Saccharomyces cerevisiae on growth performance, intestinal development, and systemic immunity in earlyweaned piglets. J Anim Sci Biotechnol 6:47–54

Julliand V, de Fombelle A, Varloud M (2006) Starch digestion in horses: the impact of feed processing. Livest Sci 100:44–52

Kabir SML, Rahman MM, Rahman MB, Rahman MM, Ahmed SU (2004) The dynamics of probiotics on growth performance and immune response in broilers. Int J Poult Sci 3:361–364. https://doi.org/10.3923/ijps.2004.361.364

Khaksefidi A, Ghoorchi T (2006) Effect of probiotic on performance and immunocompetence in broiler chicks. J Poult Sci 43:296–300. https://doi.org/10.2141/jpsa.43.296

Khalafalla MM, Ibrahim SA, Zayed MM, Awad MN, Mohamed RA (2020) Effect of a dietary mixture of beneficial bacteria on growth performance, health condition, chemical composition, and water quality of nile Tilapia, *Oreochromis niloticus* fingerlings. J Aquat Food Prod Technol 29:823–835

Kim DW, Cho SB, Lee HJ, Chung WT, Kim KH, Hwangbo J, Nam IS, Cho YI, Yang MP, Chung IB (2007) Comparison of cytokine and nitric oxide induction in murine macrophages between whole cell and enzymatically digested Bifidobacterium sp. obtained from monogastric animals. J Microbiol 45:305–310

Kim TI, Lee YH, Kim HJ, Kim YH, Lee KH, Lee YK (2009) Galactooligosaccharides reduce body weight gain and improve insulin sensitivity in rats. Br J Nutr 101(4):541–548

Kim D, Beck BR, Heo SB, Kim J, Kim HD, Lee SM, Kim Y, Oh SY, Lee K, Do HK, Lee KH, Holzapfel WH, Song SK (2013) Lactococcus lactis BFE920 activates the innate immune system of olive flounder (*Paralichthys olivaceus*), resulting in protection against Streptococcus iniae infection and enhancing feed efficiency and weight gain in large-scale field studies. Fish Shellfish Immunol 35(5):1585–1590. https://doi.org/10.1016/j.fsi.2013.09.008

Kim HY, Bae JW, Kim HJ (2015a) The roles of postbiotics in gut health and diseases. J Microbiol Biotechnol 25(2):149–162

Kim YJ, Kim JH, Kim JK, Kim SK (2015b) Effects of a probiotic on growth performance, intestinal microflora, and immune response of weaning pigs. J Anim Sci 93(2):587–592

Kim JS, Lee JH, Lee SH, Kim IH (2016) The effects of galactooligosaccharides on growth performance, digestive enzyme activity, and gut microbiota of rainbow trout (Oncorhynchus mykiss). Fish Shellfish Immunol 56:209–214

Kim JS, Hosseindoust A, Lee SH, Choi YH, Kim MJ, Lee JH, Kwon IK, Chae BJ (2017) Bacteriophage cocktail and multi-strain probiotics in the feed for weanling pigs: effects on intestine morphology and targeted intestinal coliforms and clostridium. Animal 11:45–53

Kim JH, Kim SH, Kim SH (2017a) The effect of probiotic-derived metabolites on the gut microbiota, immune system, and metabolic functions of dogs. Front Microbiol 8:1166

Kim JH, Kim SH, Kim SH, Lee SJ (2017b) The effect of probiotic-derived metabolites on digestive health and nutrient utilization in cats. J Anim Sci Technol 59(5):1–7

Kim HJ, Lee SJ, Kang SH, Kim SW, Kim HJ (2019) Effects of a non-viable yeast product on growth performance and gut microbiota of growing-finishing pigs. Animals 9(7):555

Kobayashi H, Kanmani P, Ishizuka T, Miyazaki A, Soma J, Albarracin L, Suda Y, Nochi T, Aso H, Iwabuchi N, Xiao JZ, Saito T, Villena J (2017) Development of an in vitro immunobiotic evaluation system against rotavirus infection in bovine intestinal epitheliocytes. Benef Microbes 8:309–321

Kogan G, Kocher A (2007) Role of yeast cell wall polysaccharides in pig nutrition and health protection. Livest Sci 109:161–165

Kreuzer S, Janczyk P, Assmus J, Schmidt MF, Brockmann GA, Nockler K (2012) No beneficial effects evident for *Enterococcus faecium* NCIMB 10415 in weaned pigs infected with *Salmonella enterica* serovar Typhimurium DT104. Appl Environ Microbiol 78:4816–4825. https://doi.org/10.1128/AEM.00395-12

Kudriavtsev B, Lo Leggio L, Succi M (2015) Probiotics and their impact on gut health, immunity and metabolism. Front Vet Sci 2:48

Kumprecht I, Vágner M, Konečná M (2015) The effects of a prebiotic mixture on broiler chicken performance, gut microflora and litter quality. Br Poult Sci 56(1):32–39

Lallemand Animal Nutrition (2022) Lallemand's hydrolyzed yeast offers unmatched protein digestibility kinetics. (online) https://lallemandanimalnutrition.com/en/europe/whats-new/lallemands-hydrolyzed-yeast-offers-unmatched-protein-digestibility-kinetics/

Lan PT, Binh le T, Benno Y (2003) Impact of two probiotic lactobacillus strains feeding on fecal lactobacilli and weight gains in chicken. J Gen Appl Microbiol 49:29–36

Le Bourgot C, Ferret-Bernard S, Blat S, Apper E, le Huërou-Luron I (2016) Short-chain fructooligosaccharide supplementation during gestation and lactation or after weaning differentially impacts pig growth and IgA response to influenza vaccination. J Funct Foods 24:307–315. https://doi.org/10.1016/J.JFF.2016.04.018

Le Bourgot C, Ferret-Bernard S, Apper E, Taminiau B, Cahu A, Le Normand L, Respondek F, Le Huërou-Luron I, Blat S (2019) Perinatal short-chain fructooligosaccharides program intestinal microbiota and improve enteroinsular axis function and inflammatory status in high-fat diet-fed adult pigs. FASEB J: official publication of the Federation of American Societies for Experimental Biology 33(1):301–313. https://doi.org/10.1096/fj.201800108R

Lee JH, Kim JS, Kim IH (2015a) The effects of galactooligosaccharides on growth performance, digestive enzyme activity, and gut microbiota of Atlantic salmon (Salmo salar L.). Fish Shellfish Immunol 46:32–36

Lee KW, Kim DK, Lillehoj HS, Jang SI, Lee SH (2015b) Immune modulation by Bacillus subtilis-based direct-fed microbials in commercial broiler chickens. Anim Feed Sci Technol 200(1): 76–85. https://doi.org/10.1016/j.anifeedsci.2014.12.006

Lee SH, Beck BR, Hwang S-H, Song SK (2021) Feeding olive flounder (*Paralichthys olivaceus*) with Lactococcus lactis BFE920 expressing the fusion antigen of Vibrio OmpK, and FlaB provides protection against multiple Vibrio pathogens: a universal vaccine effect. Fish Shellfish Immunol 114:253–262

Lerminiaux NA, Cameron ADS (2019) Horizontal transfer of antibiotic resistance genes in clinical environments. Can J Microbiol 65(1):34–44. https://doi.org/10.1139/cjm-2018-0275

Li L, Xu CL, Ji C, Ma Q, Hao K, Jin ZY, Li K (2006) Effects of a dried Bacillus subtilis culture on egg quality. Poult Sci 85:364–368

Li X, Li Y, Li Y, Li W, Li X, Liu H (2017a) Effect of a non-viable yeast extract on the gut microbiota, antioxidant status and clinical signs in dogs with idiopathic chronic diarrhea. Anim Sci J 88(4):647–655

Li Y, Zhang L, Ma X, Jiang Y, Cheng X, Han Y (2017b) Effects of xylooligosaccharides on growth performance, nutrient digestibility, and gut microbiota in growing-finishing pigs. J Anim Sci Biotechnol 8(1):12

Li Y, Zeng R, Liu Y, Wang J (2018a) Mannooligosaccharides improve gut health and disease resistance in black carp. Aquac Res 49(11):1538–1547

Li Y, Zhang X, Han M, Wei Y, Wang Y, Li Y (2018b) Effects of mannooligosaccharides on growth performance, nutrient digestibility, and gut microbiota in growing-finishing pigs. J Anim Sci Biotechnol 9(3):14

Li Y, Yang Q, Yang B, Li X, Gao F (2019a) Effects of Xylooligosaccharides on the growth performance, ruminal fermentation and microflora, and fatty acid composition of dairy cows. J Dairy Sci 102(4):3614–3624

Li Y, Li J, Wang S, Liu H (2019b) The effect of a paraprobiotic containing *Lactobacillus reuteri* on the growth performance, gut microbiota and immune response of tilapia (*Oreochromis niloticus*). Fish Shellfish Immunol 93:587–594

Lilly DM, Stillwell RH (1965) Probiotics. Growth promoting factors produced by microorganisms. Science 147:747–748

Link R, Reichel P, Kyzeková P (2016) The influence of probiotics on reproductive parameters of sows and health of their sucklings. Folia Vet 60:43–46

Liu H, Zhang J, Zhang SH, Yang FJ, Thacker PA, Zhang GL (2014) Oral administration of Lactobacillus fermentum I5007 favors intestinal development and alters the intestinal microbiota in formula-fed piglets. J Agric Food Chem 62:860e6

Liu Y, Geng Y, Li Y, Li J, Sun L (2017) Effects of dietary supplementation with Bacillus subtilis fermented product on growth performance, blood characteristics, and gut microbiota in lambs. J Anim Sci Biotechnol 8(1):61

Liu Q, Liu S, Zhang X, Wang Y, Hu Y, Chen X (2018) Effect of xylooligosaccharides supplementation on feed intake, digestibility, ruminal fermentation and microbial populations in dairy cows. Anim Feed Sci Technol 238:1–8

Liu Y, Ni P, Zhang J, Xia Y, Zhang Y (2019) The effect of Xylooligosaccharides on the growth performance, non-specific immunity and gut microbiota of juvenile Pacific white shrimp (Litopenaeus vannamei). Front Microbiol 10:2509

Liu Y, Cheng X, Yang Z, Liu Z (2020) Effects of non-viable yeast cell wall on growth performance, gut health and immune function of broilers. Anim Sci J 91(4):918–926

Lu C, Chen J, Li Y, Li Y, Li J, Zhang H (2017) Effects of dietary galactooligosaccharides supplementation on growth performance, nutrient digestibility, blood parameters and rumen fermentation in sheep. Anim Feed Sci Technol 235:128–135

Luan S, Duersteler M, Galbraith EA, Cardoso FC (2015) Effects of direct-fed Bacillus pumilus 8G-134 on feed intake, milk yield, milk composition, feed conversion, and health condition of pre- and postpartum Holstein cows. J Dairy Sci 98:6423–6432. https://doi.org/10.3168/jds.2015-9512

Ma L, Wang J, Lu L, Sun X, Zhang X (2010) The effect of galactooligosaccharides on growth performance, nutrient utilization and gut microflora of broilers. J Anim Physiol Anim Nutr 94(2):250–256

Ma L, Zhang Y, Ma L, He Y, Zhou Q (2017) Effects of galactooligosaccharides supplementation on ruminal fermentation, nutrient digestibility, and growth performance of lambs. Anim Feed Sci Technol 230:31–39

Ma W, Liu Y, Wu J, Wang X, Fan L (2020) Effects of galactooligosaccharides on gut environment and gut microbiota of growing-finishing pigs. J Anim Sci Technol 62(3):378

Madej JP, Stefaniak T, Bednarczyk M (2015) Effect of in ovo-delivered prebiotics and synbiotics on lymphoid-organs' morphology in chickens. Poult Sci 94:1209–1219

Maldonado NC, Chiaraviglio J, Bru E, De Chazal L, Santos V, Nader-Macias MEF (2017) Effect of milk fermented with lactic acid bacteria on diarrheal incidence, growth performance and microbiological and blood profiles of newborn dairy calves. Proteins, Probiotics Antimicrob. https://doi.org/10.1007/s12602-017-9308-4

Malvisi M, Stuknyte M, Magro G, Minozzi G, Giardini A, De Noni I, Piccinini R (2016) Antibacterial activity, and immunomodulatory effects on a bovine mammary epithelial cell line exerted by nisin A-producing Lactococcus lactis strains. J Dairy Sci 99:2288–2296. https://doi.org/10.3168/jds.2015-10161

Mamza SA, Egwu GO, Mshelia GD (2010) Beta-lactamase Escherichia coli and Staphylococcus aureus isolated from chickens in Nigeria. Vet Ital 46:155–165
Manafi M, Khalaji S, Hedayati M, Pirany N (2017) Efficacy of Bacillus subtilis and bacitracin methylene disalicylate on growth performance, digestibility, blood metabolites, immunity, and intestinal microbiota after intramuscular inoculation with Escherichia coli in broilers. Poult Sci 96:1174–1183. https://doi.org/10.3382/ps/pew347
Marshall-Jones ZV, Baillon MLA, Croft JM, Butterwick RF (2006) Effects of Lactobacillus acidophilus DSM13241 as a probiotic in healthy adult cats. Am J Vet Res 67:1005–1012. https://doi.org/10.2460/ajvr.67.6.1005
Matsuguchi T, Takagi A, Matsuzaki T, Nagaoka M, Ishikawa K, Yokokura T, Yoshikai Y (2003) Lipoteichoic acids from Lactobacillus strains elicit strong tumor necrosis factor alpha-inducing activities in macrophages through toll-like receptor 2. Clin Diagn Lab Immunol 10:259–266
Mattarelli P, Biavati B (2018) Chapter 2 - species in the genus Bifidobacterium. In: Mattarelli P, Biavati B, Holzapfel WH, Wood BJB (eds) The Bifidobacteria and related organisms. Academic Press, pp 9–48. https://doi.org/10.1016/B978-0-12-805060-6.00002-8
McFarland LV (2015) From yaks to yogurt. Clin Infect Dis 60(2):S85–S90. https://doi.org/10.1093/cid/civ054
Medeiros MAN, de Oliveira DCN, Rodrigues DP, de Freitas DRC (2011) Prevalence and antimicrobial resistance of salmonella in chicken carcasses at retail in 15 Brazilian cities. Rev Panam Salud Publica 30:555–560
Mingmongkolchai S, Panbangred W (2018) Bacillus probiotics: an alternative to antibiotics for livestock production. J Appl Microbiol 124(6):1334–1346. https://doi.org/10.1111/jam.13690
Molnár A, Podmaniczky B, Kürti P, Tenk I, Glávits R, Virág G, Szabo Z (2011) Effect of different concentrations of Bacillus subtilis on growth performance, carcase quality, gut microflora and immune response of broiler chickens. Br Poult Sci 52:658–665
Moura D, Abreu M, de Brito G, de Carvalho AP, Almeida AM, de Freitas JM (2017) Non-viable yeast cell wall extracts improve performance and reduce ruminal acidosis in dairy cows. J Dairy Sci 100(10):8233–8243
Mussatto SI, Mancilha IM (2007) Non-digestible oligosaccharides: a review. Carbohydr Polym 68: 587–597
Muthusamy N, Haldar S, Ghosh TK, Bedford MR (2011) Effects of hydrolysed *Saccharomyces cerevisiae* yeast and yeast cell wall components on live performance, intestinal histomorphology and humoral immune response of broilers. Br Poult Sci 52(6):694–703. https://doi.org/10.1080/00071668.2011.633072
Muzaffar K, Jan R, Ahmad Bhat N, Gani A, Ahmed Shagoo M (2021) Chapter 25 - commercially available probiotics and prebiotics used in human and animal nutrition. In: Dhanasekaran D, Sankaranarayanan A (eds) Advances in Probiotics. Academic Press, pp 417–435
Nataraj BH, Ali SA, Behare PV, Yadav H (2020) Postbiotics-parabiotics: the new horizons in microbial biotherapy and functional foods. Microb Cell Factories 19:168. https://doi.org/10.1186/s12934-020-01426-w
O'Doherty JV, Fitzgerald GJ, O'Mahony M (2017) Supplementation of prebiotics to high-forage diets: effects on digestion, rumen fermentation and microbial populations in lactating dairy cows. Animal nutrition 3(2):110–117
Odoi H (2016) Isolation and characterization of multi-drug resistant pseudomonas aeruginosa from clinical, environmental and poultry litter sources in Ashanti region of Ghana [thesis]. Kwame Nkrumah University of Science and Technology, Kumasi, Ghana
Oetzel GR, Emery KM, Kautz WP, Nocek JE (2007) Direct-fed microbial supplementation and health and performance of pre- and postpartum dairy cattle: A field trial. J Dairy Sci 90:2058–2068
Ong IM, Gonzalez JG, McIlwain SJ, Sawin EA, Schoen AJ, Adluru N, Alexander AL, Yu JJ (2018) Gut microbiome populations are associated with structure-specific changes in white matter architecture. Transl Psychiatry 8:6. https://doi.org/10.1038/s41398-017-0022-5

Ortiz LT, Rebolé A, Velasco S, Rodrı'guez ML, Treviño J, Tejedor JL, Alzueta C (2013) Effects of inulin and fructooligosaccharides on growth performance, body chemical composition and intestinal microbiota of farmed rainbow trout (Oncorhynchus mykiss). Aquac Nutr 19:475–482

Ozkoc MH, Demirci S, Kucuk O (2016) The effects of a live yeast probiotic on growth performance, feed utilization and some hematological parameters of sea bass (Dicentrarchus labrax L.). Aquac Res 47(4):1226–1233

Palamidi I, Fegeros K, Mohnl M, Abdelrahman WHA, Schatzmayr G, Theodoropoulos G, Mountzouris KC (2016) Probiotic form effects on growth performance, digestive function, and immune related biomarkers in broilers. Poult Sci 95:1598–1608. https://doi.org/10.3382/ps/pew052

Palma E, Tilocca B, Roncada P (2020) Antimicrobial resistance in veterinary medicine: an overview. Int J Mol Sci 21(6):1914. https://doi.org/10.3390/ijms21061914

Pambuka SR, Sjofjan O, Radiati LE (2014) Effect of liquid probiotics mixed culture supplements through drinking water on laying hens performance and yolk cholesterol. J World's Poult Res 4: 5–9

Pan Y, Yin Y, Li Y, Li D (2012) Effects of pectic oligosaccharides on growth performance, intestinal morphometry, and gut microflora in weanling pigs. Anim Feed Sci Technol 176(1): 78–85

Pang J, Liu Y, Kang L, Ye H, Zang J, Wang J, Han D (2022) Bifidobacterium animalis promotes the growth of weaning piglets by improving intestinal development, enhancing antioxidant capacity, and modulating gut microbiota. Appl Environ Microbiol. https://doi.org/10.1128/aem.01296-22

Paryad A, Mahmoudi M (2008) Effect of different levels of supplemental yeast (Saccharomyces cerevisiae) on performance, blood constituents and carcass characteristics of broiler chicks. Afr J Agric Res 3(12):835–842

Patel S, Shukla R, Goyal A (2015) Probiotics in valorization of innate immunity across various animal models. J Funct Foods 14:549–561. https://doi.org/10.1016/j.jff.2015.02.022

Pavlovic N, Stankov K, Mikov M (2012) Probiotics—interactions with bile acids and impact on cholesterol metabolism. Appl Biochem Biotechnol 168:1880–1895. https://doi.org/10.1007/s12010-012-9904-4

Paz A d L, da Silva JM, da Silva KMM, Val AL (2019) Protective effects of the fructooligosaccharide on the growth performance, hematology, immunology indicators and survival of tambaqui (Colossoma macropomum, Characiformes: Serrasalmidae) infected by Aeromonas hydrophila. Aquac Rep 15. https://doi.org/10.1016/J.AQREP.2019.100222

Pérez-Maldonado R, Gómez-Basauri V, del Amor FM, Madruga M (2017) Use of a probiotic in broiler chicken diets: effects on growth performance, gut morphology and immunity. Anim Nutr 3(3):185–189

Pinheiro REE, Rodrigues AMD, Lima CE, Santos JTO, Pereyra CM, Torres AM, Cavaglieri LR, Lopes JB, Muratori MCS (2020) Saccharomyces cerevisiae as a probiotic agent and a possible aflatoxin B1 adsorbent in simulated fish intestinal tract conditions. Arquivo Brasileiro de Medicina Veterinaria e Zootecnia 72(3):862–870. https://doi.org/10.1590/1678-4162-11280

Pinna C, Vecchiato CG, Zaghini G, Grandi M, Nannoni E, Stefanelli C, Biagi G (2016) In vitro influence of dietary protein and fructooligosaccharides on metabolism of canine fecal microbiota. BMC Vet Res 12:53. https://doi.org/10.1186/s12917-016-0672-1

Pirarat N, Kobayashi T, Katagiri T, Maita M, Endo M (2006) Protective effects and mechanisms of a probiotic bacterium Lactobacillus rhamnosus against experimental Edwardsiella tarda infection in tilapia (*Oreochromis niloticus*). Vet Immunol Immunopathol 113:339–347

Poolsawat L, Li X, Yang H, Yang P, Chowdhury MAK, Abdullateef Y, Leng X (2020) The potentials of fructooligosaccharide on growth, feed utilization, immune and antioxidant parameters, microbial community and disease resistance of tilapia (*Oreochromis niloticus* × O. aureus). Aquac Res 51:4430–4442. https://doi.org/10.1111/are.14786

Poppy GD, Rabiee AR, Lean IJ, Sanchez WK, Dorton KL, Morley PS (2012) A meta-analysis of the effects of feeding yeast culture produced by anaerobic fermentation of *Saccharomyces cerevisiae* on milk production of lactating dairy cows. J Dairy Sci 95:6027–6041

Praharaj I, John SM, Bandyopadhyay R, Kang G (2015) Probiotics, antibiotics and the immune responses to vaccines. Philos Trans R Soc B Biol Sci 370:20140144

Qin S, Huang Z, Wang Y, Pei L, Shen Y (2022) Probiotic potential of lactobacillus isolated from horses and its therapeutic effect on DSS-induced colitis in mice. Microb Pathog 165. https://doi.org/10.1016/j.micpath.2021.105216

Rahman Z, Mamun A, Ahmad I, Rashid I (2019) Influence of probiotics on the growth performance of sex reversed nile tilapia (*Oreochromis niloticus*, Linnaeus, 1758) fry. J Aquacult Res Dev 10: 8–14

Rajput IR, Li LY, Xin X, Wu BB, Juan ZL, Cui ZW, Li WF (2013) Effect of saccharomyces boulardii and Bacillus subtilis B10 on intestinal ultrastructure modulation and mucosal immunity development mechanism in broiler chickens. Poult Sci 92(4):956–965. https://doi.org/10.3382/ps.2012-02845

Rauch B, Mosenthin R, Zijlstra RT (2017) Impact of yeast cell wall derivatives on rumen metabolism and methane emissions in dairy cattle. Anim Feed Sci Technol 231:48–58

Rebolé A, Ortiz LT, Rodríguez ML, Alzueta C, Treviño J, Velasco S (2010) Effects of inulin and enzyme complex, individually or in combination, on growth performance, intestinal microflora, cecal fermentation characteristics, and jejunal histomorphology in broiler chickens fed a wheat- and barley-based diet. Poult Sci 89(2):276–286. https://doi.org/10.3382/ps.2009-00336

Rehman H, Rosenkranz C, Böhm J, Zentek J (2007) Dietary inulin affects the morphology but not the sodium-dependent glucose and glutamine transport in the jejunum of broilers. Poultry Sci 86:118–122

Ren J, Wang Y, Zhang Y, Gao J, Li D (2018) Effects of xylooligosaccharides on growth performance, nutrient digestibility, and gut microbiota in broiler chickens. Animal 12(2): 267–274

Respondek F, Swanson KS, Belsito KR, Vester BM, Wagner A, Istasse L, Diez M (2008) Short-chain fructooligosaccharides influence insulin sensitivity and gene expression of fat tissue in obese dogs. J Nutr 138(9):1712–1718. https://doi.org/10.1093/jn/138.9.1712

Ritter SC, Yang ML, Kaznessis YN, Hackel BJ (2018) Multispecies activity screening of microcin J25 mutants yields antimicrobials with increased specificity toward pathogenic Salmonella species relative to human commensal Escherichia coli. Biotechnol Bioeng 115(10): 2394–2404. https://doi.org/10.1002/bit.26772

Rodehutscord M, Böhm J, Eder K (2001) Mannanoligosaccharides in diets for young bulls: effects on performance, faecal characteristics and caecal volatile fatty acid concentrations. J Anim Physiol Anim Nutr 85(5):241–247

Rondina D, Bicalho RC, Vasconcelos JL, Bicalho ML, Waibl H (2017) Effects of dietary supplementation with galactooligosaccharides on growth performance, digestibility, and ruminal fermentation of dairy heifers. J Dairy Sci 100(4):2847–2854

Roodposhti PM, Dabiri N (2012) Effects of probiotic and prebiotic on average daily gain, fecal shedding of *Escherichia coli*, and immune system status in newborn female calves. Asian-Australas J Anim 25:1255–1261

Rousseau B, Dall'Amico D, Delezie E, Toussaint-Ravenel C (2018) Effects of dietary non-viable yeast cell wall product on the digestive function, faecal parameters and gut microbiota of horses. J Anim Physiol Anim Nutr 102(1):e171–e180

Sadat Hoseini Madani N, Adorian TJ, Ghafari FH, Hoseinifar SH (2018) The effects of dietary probiotic Bacilli (*Bacillus subtilis* and *Bacillus licheniformis*) on growth performance, feed efficiency, body composition and immune parameters of whiteleg shrimp (*Litopenaeus vannamei*) postlarvae. Aquac Res 49:1926–1933

Saeidi E, Mansoori Yarahmadi H, Fakhraei J, Mojahedi S (2021) Effect of feeding Fructooligosaccharides and *Enterococcus faecium* and their interaction on digestibility, blood, and immune parameters of adult horses. J Equine Vet 99. https://doi.org/10.1016/j.jevs.2021.103410

Santini C, Baffoni L, Gaggia F, Granata M, Gasbarri R, Di Gioia D, Biavati B (2010) Characterization of probiotic strains: an application as feed additives in poultry against Campylobacter jejuni. Int J Food Microbiol 141(1):S98–S108. https://doi.org/10.1016/j.ijfoodmicro.2010.03.039

Schokker D, Fledderus J, Jansen R, Vastenhouw SA, de Bree FM, Smits MA, Jansman AA (2018) Supplementation of fructooligosaccharides to suckling piglets affects intestinal microbiota colonization and immune development. J Anim Sci 96(6):2139–2153

Seenivasan C, Radhakrishnan S, Shanthi R, Muralisankar T, Saravana BP (2014) Effect of Lactobacillus sporogenes on survival, growth, biochemical constituents, and energy utilization of freshwater prawn *Macrobrachium rosenbergii* post larvae. J Basic Appl Zool 67:19–24

Selvaraj R, Bogar T (2019) Effect of probiotic supplemented feed on growth performance of molly fish (*Poecilia sphenops*) in recirculating aquaculture system. Pharma Innov J 8:533–537

Shang Y, Regassa A, Kim JH, Kim WK (2015) The effect of dietary fructooligosaccharide supplementation on growth performance, intestinal morphology, and immune responses in broiler chickens challenged with Salmonella enteritidis lipopolysaccharides. Poult Sci 94(12):2887–2897. https://doi.org/10.3382/ps/pev275

Shang Y, Kumar S, Thippareddi H, Kim WK (2018) Effect of dietary fructooligosaccharide (FOS) supplementation on ileal microbiota in broiler chickens. Poult Sci 97(10):3622–3634. https://doi.org/10.3382/ps/pey131

Shareef AM, Al-Dabbagh ASA (2009) Effect of probiotic (Saccharomyces cerevisiae) on performance of broiler chicks. Iraqi J Vet Sci 23:23–29

Shehta A, Omran H, Kiroloss F, Azmi M (2019) Effect of probiotic on growth performance and frequency of diarrhea in neonatal buffalo calves. Adv Anim Vet Sci 7:876e81

Shen YB, Piao XS, Kim SW, Wang L, Liu P, Yoon I, Zhen YG (2009) Effects of yeast culture supplementation on growth performance, intestinal health, and immune response of nursery pigs. J Anim Sci 87:2614–2624

Shu J, Xie M, Liu C, Ruan Y (2018) Effects of non-viable yeast on growth performance, digestive enzyme activity and gut microbiota of broilers. Poult Sci 97(6):1953–1962

Siciliano RA, Reale A, Mazzeo MF, Morandi S, Silvetti T, Brasca M (2021) Paraprobiotics: A new perspective for functional foods and nutraceuticals. Nutrients 13(4):1225. https://doi.org/10.3390/nu13041225

Simon O (2010) An interdisciplinary study on the mode of action of probiotics in pigs. J Animal Feed Sci 19:230–243. https://doi.org/10.22358/jafs/66284/2010

Sivagami M, Ronald K (2016) Effect of different probiotic enriched diets on growth performance of *Cirrhinus mrigala* fingerlings. Int J Fish Aquacult Sci 6:87–97

Smialek M, Burchardt S, Koncicki A (2018) The influence of probiotic supplementation in broiler chickens on population and carcass contamination with campylobacter spp.—field study. Res Vet Sci 118:312–316

Spaniol JS, Oltramari CE, Locatelli M, Volpato A, Campigotto G, Stefani LM, Da Silva AS (2014) Influence of probiotic on somatic cell count in milk and immune system of dairy cows. Comp Clin Pathol 24(3):677–681. https://doi.org/10.1007/s00580-014-1966-y

Splichalova A, Donovan SM, Tlaskalova-Hogenova H, Stranak Z, Splichalova Z, Splichal I (2021) Monoassociation of preterm germ-free piglets with Bifidobacterium animalis subsp. lactis BB-12 and its impact on infection with Salmonella typhimurium. Biomedicine 9:183

Stemke DJ (2004) Genetically modified microorganisms. In: Parekh SR (ed) The GMO handbook. Humana Press, Totowa, NJ. https://doi.org/10.1007/978-1-59259-801-4_4

Strompfová V, Kubašová I, Lauková A (2017) Health benefits observed after probiotic Lactobacillus fermentum CCM 7421 application in dogs. Appl Microbiol Biotechnol 101:6309–6319. https://doi.org/10.1007/s00253-017-8425-z

Sun X, Zhang C, Wang Y, Li J, Wang H, Chen Y, Ma Y (2016) Galactooligosaccharides improve gut health and modulate gut microbiota in weaned pigs challenged with Escherichia coli. Can J Microbiol 62(10):726–732

Sun X, Li Q, Li S, Li Y, Zhang X, Wu Z (2017) Effects of Xylooligosaccharides on growth performance, nutrient utilization, ruminal fermentation, and serum biochemistry in beef cattle. J Anim Sci 95(5):1985–1992

Sun D, Jeannot K, Xiao Y, Knapp CW (2019) Editorial: horizontal gene transfer mediated bacterial antibiotic resistance. Front Microbiol 10:1933. https://doi.org/10.3389/fmicb.2019.01933

Suo C, Yin Y, Wang X, Lou X, Song D, Wang X, Gu Q (2012) Effects of *Lactobacillus plantarum* ZJ316 on pig growth and pork quality. BMC Vet Res 8:89. https://doi.org/10.1186/1746-6148-8-89

Sutthi N, van Doan H (2020) Saccharomyces crevices and bacillus spp. effectively enhance health tolerance of Nile tilapia under transportation stress. Aquaculture 528:735766. https://doi.org/10.1016/j.aquaculture.2020.735527

Szabo I, Wieler LH, Tedin K, Scharek-Tedin L, Taras D, Hensel A et al (2009) Influence of a probiotic strain of *Enterococcus faecium* on *Salmonella enterica* serovar typhimurium DT104 infection in a porcine animal infection model. Appl Environ Microbiol 75:2621–2628. https://doi.org/10.1128/AEM.01515-08

Talpur AD, Munir MB, Mary A, Hashim R (2014) Dietary probiotics and prebiotics improved food acceptability, growth performance, haematology and immunological parameters and disease resistance against Aeromonas hydrophila in snakehead (Channa striata) fingerlings. Aquaculture 426–427:14–20

Tan X, Liu X, Fan M, Ma L, Li J (2015) Effects of mannooligosaccharides on growth performance, gut microbiota, and intestinal morphology in broiler chickens. J Anim Sci 93(9):4724–4733

Tian B, Liu X, Fan D, Liu H, Li X (2019) The effects of a fermented product of Bacillus subtilis on dairy cow performance, rumen fermentation, and microbial populations. Anim Feed Sci Technol 243:76–83

Tian Z, Cui Y, Lu H, Wang G, Ma X (2020) Effect of long-term dietary probiotic *Lactobacillus reuteri* 1 or antibiotics on meat quality, muscular amino acids and fatty acids in pigs. Meat Sci 171:108234. https://doi.org/10.1016/j.meatsci.2020.108234

Van den Bogaard AE, Stobberingh EE (2000) Epidemiology of resistance to antibiotics: links between animals and humans. Int J Antimicrob Agents 14:327–335. https://doi.org/10.1016/s0924-8579(00)00145-x

Van der Meijden E, Beumer RR, Tamminga S (2015) Non-viable yeast cell wall extract improves health status and reduces the risk of ruminal acidosis in dairy cows. Animal 9(9):1557–1563

Van Laere KMJ, Hartemink R, Bosveld M, Schoterman MHC, Kolsteren P (2010) Xylooligosaccharides (XOS) as prebiotics for humans. Crit Rev Food Sci Nutr 50(6):561–574

Villot C, Ma T, Renaud D, Ghaffari M, Gibson D, Skidmore A, Chevaux E, Guan L, Steele M (2019) Saccharomyces cerevisiae boulardii CNCM I-1079 affects health, growth, and fecal microbiota in milk-fed veal calves. J Dairy Sci 102:7011e25

Vrotniakienė V, Jatkauskas J (2013) Effects of probiotics dietary supplementation on diarrhea incidence, fecal shedding of *Escherichia coli* and growth performance in post-weaned piglets. Veterinarija ir Zootechnika 63:85

Wang YB, Tian ZQ, Yao JT, fen Li W (2008) Effect of probiotics, Enteroccus faecium, on tilapia (*Oreochromis niloticus*) growth performance and immune response. Aquaculture 277:203–207

Wang A, Yu H, Gao X, Li X, Qiao S (2009) Influence of Lactobacillus fermentum I5007 on the intestinal and systemic immune responses of healthy and E coli challenged piglets. Ant Van Leeuwenhoek 96:89e98

Wang J, Sun X, Ma L, Zhang Y, Zhang X (2012) Effect of galactooligosaccharides on growth performance, gut microbiota and digestive enzyme activities of broilers. J Appl Microbiol 113(1):32–40

Wang X, Farnell YZ, Peebles ED, Kiess AS, Wamsley KG, Zhai W (2016a) Effects of prebiotics, probiotics, and their combination on growth performance, small intestine morphology, and resident Lactobacillus of male broilers. Poult Sci 95:1332–1340. https://doi.org/10.3382/ps/pew030

Wang W, Li Z, Han Q, Guo Y, Zhang B, D'Inca R (2016b) Dietary live yeast and mannan-oligosaccharide supplementation attenuate intestinal inflammation and barrier dysfunction induced by *Escherichia coli* in broilers. Br J Nutr 116:1878–1888

Wang Z, Kebreab E, Cotta MA (2017) The effects of a prebiotic on methane production, ruminal fermentation, and nutrient digestion in beef cattle. Animal nutrition 3(3):190–197

Wang J, Tan X, Liu X, Fan M, Ma L, Li J (2017a) Effects of mannooligosaccharides on nutrient digestibility, gut health, and performance of broiler chickens. Poult Sci 96(6):1868–1875

Wang J, Liu Y, Li X, Liu J, Li Y, Guo Y (2017b) Effects of mannooligosaccharides on growth performance, nutrient utilization, and gut health of grass carp. Aquac Res 48(6):4323–4330

Wang Q, Wang ZH, Niu WX, Liu WQ, Jia CY, Ma XH (2018a) Effects of a non-viable yeast-based product on the gut health and performance of weanling pigs. Asian Australas J Anim Sci 31(11): 1647–1653

Wang W, Yan F, Hu J, Amen O, Cheng H (2018b) Supplementation of Bacillus subtilis-based probiotic reduces heat stress-related behaviors and inflammatory response in broiler chickens. J Anim Sci 96:1654–1666

Wang B, Li Y, Li Y, Zhang H (2019a) Paraprobiotics as modulators of gut microbiota: A review of their properties, benefits, and applications. Front Microbiol 10:1735

Wang Q, Sun Q, Qi R, Wang J, Qiu X, Liu Z, Huang J (2019b) Effects of *Lactobacillus plantarum* on the intestinal morphology, intestinal barrier function and microbiota composition of suckling piglets. J Anim Physiol Anim Nutr 103:1908–1918. https://doi.org/10.1111/jpn.13198

Wang Y, Song Y, Cai H, Guo X, Wang Y, Li X (2019c) Xylooligosaccharides improve the growth performance and modulate the gut microbiota of broiler chickens. Poult Sci 98(7):3317–3323

Wang J, Wang WW, Qi GH, Cui CF, Wu SG, Zhang HJ, Xu L, Wang J (2021) Effects of dietary Bacillus subtilis supplementation and calcium levels on performance and eggshell quality of laying hens in the late phase of production. Poult Sci 100:100970

Ward MP, Alinovi CA, Couetil LL, Glickman T, Wu CC (2004) A randomized clinical trial using probiotics to prevent Salmonella fecal shedding in hospitalized horses. J Equine Vet Sci 24:242–247. https://doi.org/10.1016/j.jevs.2004.05.004

Wegh CAM, Geerlings SY, Knol J, Roeselers G, Belzer C (2019) Postbiotics and their potential applications in early life nutrition and beyond. Int J Mol Sci 20(19):4673. https://doi.org/10.3390/ijms20194673

Woodward WD, Besser TE, Ward AC, Corbeil LB (1987) In vitro growth inhibition of mastitis pathogens by bovine teat skin normal flora. Can J Vet Res 51:27–31

Wu Y, Zhen W, Geng Y, Wang Z, Guo Y (2019) Pretreatment with probiotic *Enterococcus faecium* NCIMB 11181 ameliorates necrotic enteritis-induced intestinal barrier injury in broiler chickens. Sci Rep 9:10256

Xia Y, Wang M, Gao F, Lu M, Chen G (2020) Effects of dietary probiotic supplementation on the growth, gut health, and disease resistance of juvenile Nile tilapia (*Oreochromis niloticus*). Anim Nutr 6:69–79

Xu ZR, Hu CH, Xia MS, Zhan XA, Wang MQ (2003) Effects of dietary fructooligosaccharide on digestive enzyme activities, intestinal microflora and morphology of male broilers. Poultry Sci. 82:1030–1036

Yang M, Yang J, Yang Y, Wang Q, Fan H, Wang Y, Zhang X (2014) Effects of dietary galactooligosaccharides on egg quality, intestinal microbiota, and antibody production in laying hens. J Anim Sci 92(12):5275–5282

Ye J-D, Wang K, Li F-D, Sun Y-Z (2011) Single or combined effects of fructo- and mannan oligosaccharide supplements and Bacillus clausii on the growth, feed utilization, body composition, digestive enzyme activity, innate immune response and lipid metabolism of the Japanese flounder (*Paralichthys olivaceus*). Aquac Nutr 17:e902–e911

Yi H, Wang L, Xiong Y, Wen X, Wang Z, Yang X, Gao K, Jiang Z (2018) Effects of *Lactobacillus reuteri* LR1 on the growth performance, intestinal morphology, and intestinal barrier function in weaned pigs. J Anim Sci 96(6):2342–2351. https://doi.org/10.1093/jas/sky129

Ying Y, Jin L, Hou Y, Chen Y, Li L (2017) Effect of probiotics on growth performance, nutrient utilization, rumen fermentation, and methane emission in dairy cows. Anim Nutr 3(1):15–20

Yong J, Wang Q, Zhang X, Wang H (2019) Effects of dietary non-viable yeast supplementation on egg production, egg quality and intestinal microbiota in laying hens. Animal Nutrition 5(5): 362–369

Yu D, Xia Y, Ge L, Tan B, Chen S (2021) Effects of Lactococcus lactis on the intestinal functions in weaning piglets. Front Nutr 8. https://doi.org/10.3389/fnut.2021.713256

Zebeli Q, Ametaj BN, Drochner W (2017) The effects of a postbiotic on feed efficiency, nutrient utilization and manure excretion in lactating dairy cows. Animal nutrition 3(3):213–219

Zeng R, Yin Y, Liu J, Zhang W, Chen Y, Li J (2018) Effects of Xylooligosaccharides on growth performance, nutrient digestibility, and intestinal health in weaning pigs. Animals 8(7):94

Zeng R, Li Y, Han M, Wei Y, Wang Y, Li Y (2019) Effects of mannooligosaccharides on growth performance, nutrient digestibility, and intestinal health in weaning pigs. Animals 9(7):94

Zhai Q, Yu L, Li T, Zhu J, Zhang C, Zhao J, Zhang H, Chen W (2017) Effect of dietary probiotic supplementation on intestinal microbiota and physiological conditions of Nile tilapia (*Oreochromis niloticus*) under waterborne cadmium exposure. Antonie Leeuwenhoek 110: 501–513

Zhang Y, Wang J, Sun X, Ma L, Zhang X (2013) Effects of galactooligosaccharides on growth performance, nutrient utilization, and gut microflora of broilers. J Anim Sci 91(5):1636–1642

Zhang Y, Gao L, Li J, Li X, Li Y, Li Y, Ma Y (2015) Effects of dietary galactooligosaccharides on growth performance, nutrient digestibility, fecal bacterial populations and fecal organic acid concentrations of weanling pigs. Livest Sci 178:87–93

Zhang R, Zhou M, Tu Y, Zhang NF, Deng KD, Ma T, Diao QY (2016) Effect of oral administration of probiotics on growth performance, apparent nutrient digestibility and stress related indicators in Holstein calves. J Anim Physio An N 100:33–38

Zhang J, Ni P, Xia Y, Liu Y, Zhang Y (2019) The effect of Xylooligosaccharide supplementation on the growth performance and gut microbiota of juvenile Pacific white shrimp (Litopenaeus vannamei). Front Microbiol 10:2902

Zhang L, Chen J, Li Y, Wang J (2019b) Postbiotics improve growth performance, egg quality and gut microbiota in layer hens. Anim Nutr 5(6):503–509

Zhang X, Yang H, Li Y, Chen J (2019c) Effects of mannooligosaccharides on growth performance, feed utilization, and gut health of white shrimp. Aquac Res 50(5):564–572

Zhang Z, Zhang G, Zhang S, Zhao J (2022) Fructooligosaccharide reduces weanling pig diarrhea in conjunction with improving intestinal Antioxidase activity and tight junction protein expression. Nutrients 14(3):512. https://doi.org/10.3390/nu14030512

Zhao J, Wang X, Li X, Zhang Y, Li Y, Wang Y (2018) The effects of xylooligosaccharides on growth performance, nutrient digestibility, and gut microbiota in broiler chickens. Poult Sci 97(5):1367–1373

Zhao C, Zhu J, Hu J, Dong X, Sun L, Zhang X, Miao S (2019) Effects of dietary Bacillus pumilus on growth performance, innate immunity and digestive enzymes of giant freshwater prawns (Macrobrachium rosenbergii). Aquac Nutr 25:712–720

Zheng J, Wittouck S, Salvetti E, Franz CMAP, Harris HMB, Mattarelli P, O'Toole PW, Pot B, Vandamme P, Walter J, Watanabe K (2020) A taxonomic note on the genus Lactobacillus: Description of 23 novel genera, emended description of the genus Lactobacillus Beijerinck 1901, and union of Lactobacillaceae and Leuconostocaceae. Int J Syst Evol Microbiol 70(4): 2782–2858. https://doi.org/10.1099/ijsem.0.004107

Zhou H, Li Y, Liu H, Lu CD (2015) Non-viable yeast cell products improve the gut microflora and enhance growth performance in weaning pigs. Anim Feed Sci Technol 206:10–18

Zhou D, Fan D, Liu H, Li X (2019a) The effect of a paraprobiotic containing Bacillus subtilis on growth performance, gut microbiota, and disease resistance in rainbow trout (Oncorhynchus mykiss). Fish Shellfish Immunol 94:607–615

Zhou L, Lu X, Chen Y, Li Y (2019b) Effects of Bacillus-derived postbiotics on growth performance, intestinal morphology, and gut microbiota in growing pigs. J Anim Sci Biotechnol 10(1):27

Żółkiewicz J, Marzec A, Ruszczyński M, Feleszko W (2020) Postbiotics-a step beyond pre- and probiotics. Nutrients 12(8):2189. https://doi.org/10.3390/nu12082189

Novel Plant Extracts as Food Additives

Anastasia Kyriakoudi and Ioannis Mourtzinos

Abstract

Concerns and potential health risks related to the use of synthetic food additives have renewed the interest of the scientific community, food industry, and consumers toward replacing them with natural alternatives. Plant extracts represent an interesting approach based on their natural origin. Different plant materials as well as by-products and wastes from the agro-industrial sector constitute rich sources of valuable phytochemicals, including polyphenols, flavonoids, tannins, alkaloids, and terpenoids. A variety of extraction techniques can be employed for the recovery of these valuable compounds. During the last decades, emphasis is given on "green" techniques including ultrasound- and microwave- assisted extraction as well as supercritical fluid extraction and pressurized fluid extraction, along with the use of alternative solvents such as hydrophilic and hydrophobic deep eutectic solvents and aqueous solutions of cyclodextrins, which can reduce the negative health and environmental impact. The obtained extracts, either after purification or as such, could find applications as natural food antioxidants, preservatives, colorants, flavorings, etc., or for the fortification of active packaging materials, extending the shelf-life, maintaining the overall quality and add value to the food product. All the above are summarized in Fig. 1. The chapter covers the major recent analytical achievements in the light of which challenges and applications of plant extracts in the food industry are critically discussed.

A. Kyriakoudi · I. Mourtzinos (✉)
Laboratory of Food Chemistry and Biochemistry, Department of Food Science and Technology, School of Agriculture, Faculty of Agriculture, Forestry and Natural Environment, Aristotle University of Thessaloniki (AUTH), Thessaloniki, Greece
e-mail: ankyria@agro.auth.gr; mourtzinos@agro.auth.gr

G. Arsenos, I. Giannenas (eds.), *Sustainable Use of Feed Additives in Livestock*,
https://doi.org/10.1007/978-3-031-42855-5_9

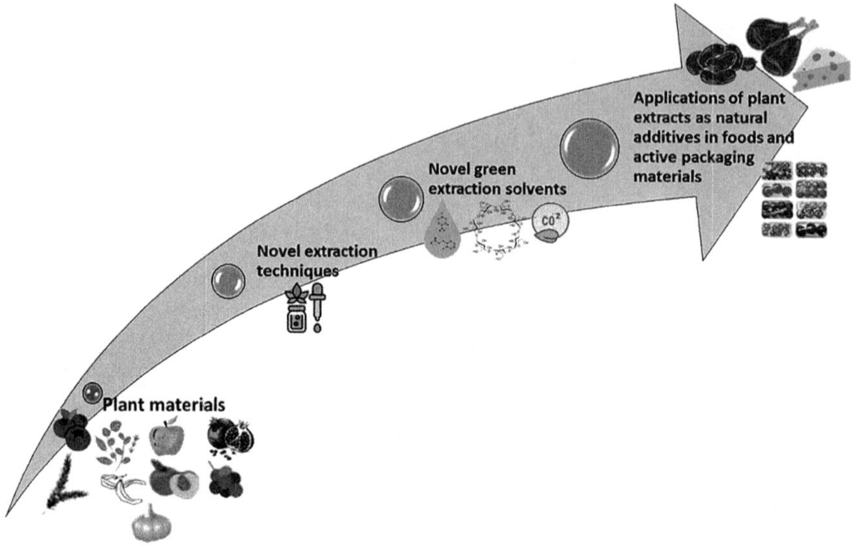

Fig. 1 Graphical representation of the main steps involved in the preparation of plant extracts as well as their applications in food products and/or active packaging materials as natural additives

Keywords

Plant materials · Food products · Food packaging · Active materials · Extraction techniques · Solvents

1 Plant Materials Used for the Preparation of Extracts for Food Applications

1.1 Incorporation of Plant Extracts in Food Products

A variety of plant materials such as fruits, vegetables, herbs, spices, and flowers, as well as by-products and wastes derived from the agrifood sector (e.g., peels, leaves, seeds, and pomace) that constitute rich sources of phenolic compounds, flavonoids, carotenoids, stilbenes, tannins, etc., are currently utilized for the preparation of novel extracts for applications by the food industry (Mir et al. 2022). It is worth mentioning that certain natural extracts, such as those obtained from rosemary, grape seed, and skin as well as olive pulp, are generally recognized as safe (GRAS) and can be applied in food products as antioxidants (EFSA 2008, FDA Notices No. 446, 93, 459). Below there is an overview of the plant materials that have been used for the preparation of extracts with potential applications in the food industry as natural additives. Emphasis is given on studies that have been published during the last decade.

Fruits constitute rich sources of various functional compounds such as phenolic acids, flavonoids, tannins, and proanthocyanidins. Extracts from different fruits have been used in various food products as natural antioxidants, antimicrobials, color enhancers, etc. More specifically, a lyophilized aqueous extract from *Nitraria retusa* fruits was used in beef patties (Mariem et al. 2014). Its addition was found to inhibit lipid oxidation and to improve microbiological stability during storage of the patties at 4 °C for 9 days. Bellucci et al. (2021) used a lyophilized aqueous red pitaya (or red dragon) extract in pork patties that were stored at 2 °C for 18 days. The authors suggested that the addition of this extract resulted in the delay of oxidative and color deterioration. A lyophilized aqueous black mulberry extract was used in beef patties stored under aerobic and vacuum packaging conditions for 15 days of chilled storage (Turan and Şimşek 2021). The addition of this extract resulted in the extension of the shelf-life of the patties by inhibiting lipid oxidation, discoloration, and microbial growth. Abdel-Naeem et al. (2022) used kiwi and pineapple extracts as well as their combination, in spent hen meat patties. The authors suggested that the incorporation of these extracts into the patties resulted in an enhancement in tenderness as well as in an improvement to all sensory attributes.

Vegetables, that are rich in phenolic compounds, carotenoids, tannins, and vitamins C and E, are also used for the preparation of extracts. In particular, the antioxidant effects of broccoli extracts were evaluated on goat meat nuggets (Banerjee et al. 2012) as well as on fresh ground beef (Kim et al. 2013). A dry potato extract was added in fresh and pre-cooked beef patties (Colle et al. 2019). Its addition resulted in improved water holding capacity, shelf life, and sensory characteristics of the patties during retail storage.

Extracts from various herbs and spices also find applications in the food industry. For example, rosemary and cloves extracts as well as their combination were found to inhibit microbial growth and lipid oxidation of raw chicken meat during storage at 4 °C for 15 days (Zhang et al. 2016). The addition of a powder sage extract on turkey meatballs that were packed in a cold modified atmosphere and stored at 4 °C for 9 days was found to delay the formation of lipid-derived products of oxidation (Gantner et al. 2018). Moreover, the addition of rosemary and oregano extracts on precooked meat hamburgers was found to inhibit lipid oxidation without altering their organoleptic traits (color, taste, aroma) (Manhani et al. 2018). Park et al. (2018) found that the fortification of milk and yogurt with a red ginseng powder increased its radical-scavenging activity whereas a reduction in the H_2O_2-induced DNA damage was observed. The effects of lyophilized rosemary and sage extracts on the oxidative stability of poultry pátê stored at 4 °C have been also examined (Bianchin et al. 2020).

Considering the European Union's (EU) action plan for the circular economy toward reducing food waste through the reuse, recovery, and recycling of materials and energy, the exploitation of agroindustrial by-products and wastes as sources of functional ingredients that could be used as natural food additives is a promising approach (Galanakis 2012). In this frame, peels, which are considered one of the main by-products of food processing, constitute a rich source of valuable bioactive compounds. They are reported to often possess increased antioxidant potential compared to the pulp. Various peel extracts have found applications as antioxidant and antimicrobial agents in food products. For example, an aqueous pomegranate peel extract has been used to grind goat meat and nuggets in order to examine its effect on its shelf-life (Devatkal et al. 2014). A concentrated lyophilized aqueous pomegranate peel extract has been applied to beef meatballs and was found to exhibit a lipid and protein retarding capacity (Turgut et al. 2016). The effect of a pomegranate peel extract on the phenolic content, fatty acid composition, and meat quality of broiler chickens has been investigated (Saleh et al. 2017). A tomato peel extract has been found to retard lipid peroxidation of lamp meat upon refrigerator storage (Alaa et al. 2019). Vasilopoulos et al. (2022) examined the effect of dietary supplementation with aqueous and cyclodextrin pomegranate peel extracts on broilers' performance parameters, welfare, and meat characteristics. An aqueous pomegranate peel extract has been also used in ground buffalo (*Bubalus bubalis*) meat to examine its storage stability at 4 °C for 20 days (Ghimire et al. 2022). Moreover, aqueous extracts of almond skins were used to examine their effect on the shelf-life of pork patties, however, only pH, weight loss and L^* values were found to be affected (Timón et al. 2022).

Leaves constitute important parts of the plants since they are responsible for the photosynthesis and oxygen production. Apart from chlorophyll, which is responsible for the green color of the leaves, they also contain carotenoids and anthocyanins that exert yellow to red colors during autumn. Various studies report the use of leaf extracts in different food products. In particular, an aqueous-ethanolic olive leaf extract was added in minced pork meat which was then stored at −18 °C under white fluorescent light to simulate retail display conditions, for 180 days (Botsoglou et al. 2014). The authors concluded that the addition of the extract was found to delay lipids and protein oxidation. The incorporation of methanolic extracts of oregano and sweet basil leaves was found to enhance the oxidative stability of ravioli with a dairy-based padding when stored at −18 °C for 4 months (Chinprahast et al. 2018).

Seeds constitute an inedible part of various fruits and vegetables and are considered as wastes by the food industry. Being rich in various bioactive compounds, seed extracts are widely used as a source of natural ingredients for the food industry. In particular, aqueous-ethanolic extracts from defatted grape seeds derived from two Greek grape varieties, namely "Moschofilero" and "Agiorgitiko," were used for the fortification of yogurts, which were found to exhibit higher antiradical and antioxidant activity compared to control ones, even after 3–4 weeks of cold storage (Chouchouli et al. 2013). Pateiro et al. (2018) added guarana seed extracts to pork patties that were stored at 2 °C for 18 days. These extracts were found to enhance the color of the patties, to protect meat against oxidation but were not found to have antimicrobial activity. The incorporation of a lyophilized Bordeaux grape seed extract in a Petit Suisse cheese did not affect its sensorial characteristics compared to the cheese manufactured without the grape seed extract and presented higher inhibition activity against angiotensin-converting enzyme, which controls blood pressure by regulating the volume of fluids in the body (Deolindo et al. 2019). The addition of a grape seed extract in ground meat patties (85% beef and 15% pork back fat) packaged under a high-oxygen modified atmosphere improved fresh meat color and minimized premature browning (Yang et al. 2022).

The spreading of a tomato pomace extract, a major by-product during tomato processing, on lamp meat, was found to inhibit redness decrease upon modified atmosphere packaging (Andres et al. 2017). Dried olive pomace, a by-product of olive oil separation, was incorporated into the conventional diet of dairy cows and was found to improve the nutritional and nutraceutical properties as well as to modify the aroma of milk and derived cheese (Castellani et al. 2017). No adverse effects were observed on the shelf-life of lamp meat during storage in retail sale conditions after the incorporation of dried red grape pomace, the main by-product generated by the wine industry, in the lamp diet (Guerra-Rivas et al. 2016). Demirkol and Tarakci (2018) suggested that an Isabella grape (*Vitis labrusa* L.) pomace extract could be used as a functional ingredient in yogurt. The addition of an aqueous-ethanolic blueberry pomace extract on pork patties was found to delay lipid oxidation during chilled storage (Peiretti et al. 2020).

1.2 Incorporation of Plant Extracts in Food Active Packaging Materials

Phytochemicals derived from plant materials can be used not only directly in food products but also in packaging materials in order to increase their shelf-life and quality. Taking into consideration that lately, there is an increased consumer's demand for slightly processed foods, the interest of the scientific community has turned to innovative food packaging technologies. The development of active packaging materials through their reinforcement with antioxidant and/or antimicrobial agents is a promising approach to improve food quality and increase the stability of sensitive food products. Considering that the use of synthetic additives has been linked to potential toxicological risks, the use of natural food additives derived from a variety of plant materials as well as agricultural by-products and wastes is a promising alternative and constitutes one of the current emerging trends in food packaging research field (Valdés et al. 2014).

In this view, extracts from various fruits, vegetables, herbs, and spices as well as flowers find also applications for the reinforcement of active packaging materials, e.g. edible films. In particular, an anthocyanin-rich red raspberry (*Rubus strigosus*) aqueous-ethanolic extract was incorporated in edible soy protein isolate films (Wang et al. 2012). The authors suggested that the obtained film had enhanced tensile strength and elongation at break, increased water swelling ratio as well as lower water solubility and water vapor permeability. Moreover, multifunctional bio-nanocomposite films based on konjac glucomannan/chitosan enriched with a mulberry anthocyanin extract for active food packaging showed UV–vis light barrier and pH-sensitive properties as well as antioxidant and antibacterial bioactivity (Sun et al. 2020). Alizadeh-Sani et al. (2021) prepared pH-responsive color indicator films by blending barberry anthocyanins with methylcellulose/chitosan nanofiber for real-time monitoring of meat freshness. The color indicator film showed color changes in response to pH alterations and ammonia gas, being suitable for indicating variations in food pH and the formation of volatile nitrogen compounds. An aqueous-ethanolic extract rich in anthocyanins derived from *Phyllanthus reticulatus* fruit was used for the preparation of smart biodegradable films based on chitosan/methylcellulose for monitoring the freshness of fish fillet (Gasti et al. 2021). The water solubility, moisture retention capacity, and water vapor transmission rate of the films were found to be enhanced along with an increase in the anthocyanin content. Moreover, the prepared films showed a pink color in acidic pH while yellowish in basic pH solution and exhibited strong antioxidant activity as well as antibacterial activity against *Staphylococcus aureus*, *Pseudomonas aeruginosa,* and *Escherichia coli*. Extracts from various vegetables have been also incorporated into active food packaging films. More specifically, active chitosan/PVA films loaded with anthocyanins from *Brassica oleracea* (red cabbage) as time–temperature indicators were prepared for intelligent food packaging (Pereira et al. 2015). The obtained films were applied on pasteurized milk exhibiting changes in the coloration of the film indicating alterations in its chemical composition. Koosha and Hamedi (2019) reported the preparation of intelligent chitosan/PVA nanocomposite films containing black carrot anthocyanins as pH indicators, with improved mechanical, thermal, and antibacterial properties. A purple-fleshed sweet potato extract was incorporated into a chitosan matrix for the development of antioxidant and intelligent pH-sensing packaging films (Yong et al. 2019). The authors suggested that the obtained films exhibited reduced elongation at break and were pH-sensitive due to the presence of anthocyanins. An anthocyanin-rich extract derived also from purple potato was used for the development of an intelligent/active food packaging film based on 2,2,6,6-tetramethylpiperidine-1-oxyl radical-oxidized bacterial cellulose for shelf-life extension of shrimp (Wen et al. 2021). Ma et al. (2017) reported the design of Tara gum/polyvinyl alcohol-based colorimetric NH_3 indicator films with the incorporation of curcumin for intelligent packaging in order to monitor shrimp spoilage. The obtained films exhibited a visible color change in an NH_3 environment whereas higher relative humidity in the environment accelerated the color change. Roselle anthocyanins have been used for the reinforcement of an intelligent pH film based on biodegradable polymers (i.e., starch, polyvinyl alcohol) in order to monitor pork freshness (Zhang et al. 2019a). The design of a colorimetric film by incorporating anthocyanins from fresh red rose (*Rosa rugosa*) petals in polyvinyl alcohol/okra mucilage polysaccharide (PVA/OMP) was designed for shrimp freshness monitoring (Kang et al. 2020). The prepared film exhibited improved mechanical and barrier properties, showed distinguishable color changes at pH 2–12, and was highly sensitive to volatile ammonia.

Additionally, a variety of agro-industrial by-products have been incorporated into edible films. More specifically, an anthocyanin-rich black plum peel extract was used for the development of multifunctional food packaging films (Zhang et al. 2019b). The prepared film was found to have antioxidant, ethylene scavenging, antimicrobial, and pH-sensitive properties. A banana peel extract prepared using 80% (v/v) aqueous ethanol was incorporated in chitosan films for the coating of apples in order to maintain their storage quality (Zhang et al. 2020). Extracts prepared from plant leaves have been also used for the fortification of edible films. Rambabu et al. (2019) incorporated a mango leaf extract in a chitosan-based film. The obtained film showed enhanced tensile strength and antioxidant activity. An aqueous amaranthus leaf extract, rich in phenolic compounds and betalains, was incorporated in an intelligent food packaging film developed with polyvinyl alcohol (PVA) and gelatin in order to monitor the freshness of fish and chicken meat (Kanatt 2020). This was accomplished by the color change of the extract under basic pH due to the presence of betalains. Martiny et al. (2020) reinforced a carrageenan-based active packaging film with an olive leaf extract for lamp meat preservation. Results showed that the addition of this extract increased the thickness and elongation at the break of the film whereas it was found to exhibit an antimicrobial capacity during the storage of the lamp meat at 7 °C for 2 days. Edible coating and films enriched with grape seed extract (0.1%) were applied on whole-grain cookies with grape and Aronia pomace (Molnar et al. 2023). The authors postulated that the addition of this extract to the film increased its thickness, its resistance to water vapor as well as its phenolic content resulting in improvement of the freshness degree, the postharvest quality, and sensory properties during six months of storage. A black rice bran extract rich in anthocyanins was used for the preparation of an intelligent film based on chitosan/oxidized chitin nanocrystals for seafood spoilage monitoring (Wu et al. 2019). The prepared films exhibited UV-barrier and antioxidant properties and were found to be pH dependent. Santos et al. (2022) reported the preparation of an edible active film based on gelatin and *Malpighia emarginata* (Acerola cherry) waste aqueous-ethanolic (34%, v/v) extract and its application in beef patties. It was shown that the active film reinforced with 4% of the extract, showed adequate physical and barrier properties, and it was protective in color changes and oxidation of lipids and proteins in frozen beef patties during storage (−18 °C, 60 days). A grape pomace extract along with guar gum was used for the dip coating treatment of minimally processed pomegranate arils to increase their shelf-life (Saurabh et al. 2022). A sea buckthorn pomace extract was added to a potato starch film in order to study its effect on the quality and spoilage bacteria of beef jerky sold in supermarkets (Guo et al. 2020). The authors suggested that the addition of this extract inhibited the growth of common spoilage microorganisms in beef jerky. A pectin-fish gelatin film containing olive antioxidants, namely hydroxytyrosol and 3,4-dihydroxyphenylglycol, which were extracted and purified from a by-product obtained from producing olive oil was used to study its effect on the preservation of raw beef meat during refrigerated storage (Bermúdez-Oria et al. 2019). Results showed that lipid oxidation, in terms of TBARS, was reduced in the meat samples that were wrapped with the prepared film.

2 Novel Extraction Techniques of High-Added Value Compounds from Plant Materials

Extraction yield of the targeted compound(s) is crucial for the selection of the extraction method. It usually depends on various factors such as the solvent, the apparatus used for the extraction, the solid:solvent ratio, the temperature, the duration, etc. Conventional extraction approaches, e.g., maceration, even though are still commonly used, they present certain disadvantages, including prolonged extraction duration, and low extraction yield. Novel extraction techniques, such as ultrasound- and microwave-assisted extraction as well as supercritical fluid extraction, that use lower amounts of solvents, are environmentally friendly, and result in high extraction yields (Farooq et al. 2022), have emerged.

2.1 Ultrasound-Assisted Extraction

Ultrasounds constitute a special type of sound wave beyond human hearing, usually between 20 kHz and 100 MHz. Ultrasound waves pass through a medium causing compression and expansion that create a phenomenon known as cavitation. The latter one describes the production, growth, and collapse of air bubbles. The advantages of ultrasound-assisted extraction (UAE) involve faster energy transfer, reduced extraction temperature, and equipment size (Chemat et al. 2017). UAE is an effective extraction technique for valuable compounds from plant materials. For example, Sharma et al. (2020) employed ultrasounds for the preparation of a defatted *Moringa oleifera* seed meal extract that was used as a natural antimicrobial coating, against *E. coli* and *Bacillus cereus*, for raw chicken sausages. UAE of polyphenols from the seeds of *Allium senescens* L. using water, as well as their application as natural antioxidants in Harbin dry sausages, was also reported (Qin et al. 2021). Moreover, phenolic compounds from pomegranate peels were extracted with UAE using water as a solvent and were added to chicken burgers as antioxidants (Ordaz-Rodríguez et al. 2022). The authors suggested that this extract was found to delay meat oxidation more efficiently than ascorbic acid. In the same frame, extracts rich in bioactive compounds obtained from pomegranate peels using UAE with 70% ethanol, were also employed as antioxidant agents to improve the shelf life of soybean and mustard edible oils (Rashid et al. 2022). Pollini et al. (2022) employed UAE for the extraction of polyphenols from apple pomace to be used as fortifying agents for beef burgers.

2.2 Microwave-Assisted Extraction

Microwave-assisted extraction (MAE) is also considered as an advanced technique for the extraction of phytochemicals from a variety of plant materials exploiting microwave energy. Microwaves are electromagnetic fields in the frequency range between 300 mHZ and 300 GHz. Electromagnetic energy is transformed to heat following ionic conduction and dipole rotation phenomena. MAE offers several advantages compared to conventional extraction techniques, including reduced extraction duration, and increased extraction efficiency and yield (Azmir et al. 2013). Microwaves were employed for the preparation of extracts from sea buckthorn pomace and seeds as antioxidants to stabilize canola oil (Patra et al. 2022). The use of microwaves was also reported for the extraction of polyphenols from blackcurrant by-products for possible uses of the extracts in active packaging (Alchera et al. 2022). The authors applied the prepared maltodextrin pad reinforced with the obtained extract to raspberries and reported an extension of their shelf life during storage at 4 °C for 13 days.

2.3 Supercritical Fluid Extraction

A supercritical state is a distinctive state, besides solid, liquid, and gas, that can be attained if a substance is subjected to temperature and pressure conditions beyond its critical point. The latter one is defined as the characteristic temperature and pressure above which distinctive gas and liquid phases do not exist. The obtained supercritical fluid possesses physicochemical properties (e.g., viscosity, surface tension) that make it suitable as an extraction solvent with reduced duration and higher yields. Carbon dioxide (CO_2), with a critical temperature close to room temperature (31 °C) and a low critical pressure (74 bars), is considered an ideal solvent for supercritical fluid extraction (SFE). Its only disadvantage is its low polarity, which makes it ideal for extracting lipids and non-polar compounds but unsuitable for extracting polar ones. This limitation can be tackled using a chemical modifier (cosolvent) to enhance its polarity (Azmir et al. 2013). Wild thyme (*Thymus serpyllum* L.) by-product extracts prepared using SFE were applied as natural antioxidants in ground pork patties (Šojić et al. 2020). The authors postulated that the prepared extracts were found to prevent color degradation and lipid and protein oxidation in the pork patties as well as to prolong their shelf-life. Moreover, extracts from tamarillo (*Solanum betaceum* Sendtn) epicarp obtained using SFE with CO_2 and ethanol as cosolvent were applied as antioxidants in cooked beef meat (Castro-Vargas et al. 2013).

3 Novel Solvents for the Extraction of High-Added Value Compounds from Plant Materials

3.1 Deep Eutectic Solvents

Apart from novel extraction techniques, the use of environmentally friendly, easily available, non-toxic, biodegradable, recyclable, economic, and nonflammable solvents is of utmost importance for the recovery of valuable compounds from plant materials for industrial applications (Socas-Rodríguez et al. 2021). In this frame, deep eutectic solvents (DESs), which can be defined as mixtures consisting of at least two components, one as the hydrogen bond donor (HBD) and another one as the hydrogen bond acceptor (HBA), are a promising alternative. The individual components of such mixtures interact with each other through the formation of a network of hydrogen bonds to create a system that exists in liquid form under ambient conditions. DESs present certain advantages such as biodegradability, biocompatibility, low cost, and easy preparation, simply by mixing the ingredients in different molar ratios upon heating and stirring. DESs that are synthesized by components of natural origin, such as organic acids (e.g., acetic acid, malic acid, and citric acid) and sugars (e.g., glucose, fructose, and sucrose) are called natural deep eutectic solvents (NADESs) (Sekharan et al. 2021). Moreover, except for hydrophilic NADESs that are used for the extraction of polar compounds (e.g., phenolic compounds), the use of hydrophobic natural deep eutectic solvents (HNADESs) has recently emerged for the extraction of non-polar compounds (e.g., carotenoids) (Cao and Su 2021). Since NADESs are prepared using starting materials of natural origin (e.g., sugars, organic acids, and amino acids), the obtained extracts could be used directly in food applications as ready-to-use extracts, eliminating the need for solvent removal (Rente et al. 2022). Selected examples of the use of DESs for the extraction of phytochemicals from different plant materials are given below. In particular, Kyriakidou et al. (2021) prepared pomegranate peel extracts using choline chloride- and glycerol-based DESs and incorporated them in chitosan films. Velásquez et al. (2021) used NADESs composed of different ratios of lactic acid, glycerol, tartaric acid, glucose, choline chloride, and citric acid, combined with ultrasounds, in order to extract anthocyanins from Chilean berries and incorporate them in edible films. Choline chloride- and acetic acid-based DESs coupled to ultrasounds were also employed for the extraction of phenolic compounds from *Pistacia lentiscus* L. fruits (Tebbi et al. 2023). A green extraction approach, using different hydrophobic natural deep eutectic solvents based on terpenes (i.e., menthol and thymol) and fatty acids (i.e., decanoic acid and dodecanoic acid) at different molar ratios, was also reported by Kyriakoudi et al. (2022) for the recovery of lycopene from tomato fruits for potential food industrial applications.

3.2 Aqueous Solutions of Cyclodextrins

Apart from DESs, another novel environmentally friendly system for the extraction of valuable compounds from plant materials is the aqueous solutions of cyclodextrins (CDs). The latter ones are cyclic oligosaccharides composed of six to eight glucopyranose molecules linked by α-1,4-glycosidic bonds. Cyclodextrins are relatively inexpensive substances that with their hydrophobic cavity can be used for the extraction of hydrophilic and hydrophobic compounds through the formation of inclusion complexes. In each case, the selection of the appropriate type of cyclodextrin is a critical parameter toward increasing the extraction efficiency. More specifically, α-CDs are more suitable for smaller molecules. The most commonly used β-CDs are suitable for medium-molecular weight compounds whereas γ-CDs are suitable for high-molecular weight compounds. The use of different types of CDs has been reported in literature for the extraction of various types of plant materials. For example, the formation of an inclusion complex between an olive leaf extract, an agricultural by-product rich in oleuropein, and with β-CD upon mixing of the components in aqueous media and subsequent freeze-drying, was reported in literature (Mourtzinos et al. 2007). The authors concluded that the obtained solid complex of olive leaf extract/β-CD could be used either to fortify foods or as a food supplement due to its increased stability. Moreover, Mourtzinos et al. (2016) employed ecofriendly extraction solvents based on water and glycerol along with 2-hydroxypropyl-β-cyclodextrin as an enhancer of the extraction for the extraction of polyphenols from olive leaves. The authors postulated that the obtained aqueous extract containing glycerol and cyclodextrins could be used as raw material/ingredient for several end users in the food, cosmetic, and pharmaceutical industries. In the same frame, Mourtzinos et al. (2018) prepared extracts rich in polyphenols from the onion (*Allium cepa*) solid wastes using ecofriendly solvents, such as water and glycerol as well as 2-hydroxypropyl-β-cyclodextrin as a cosolvent for the increase of the extraction yield. The authors added this extract as a food colorant in a yogurt matrix and concluded that it was found to be a stable natural colorant and that it could replace synthetic coloring agents. The simultaneous extraction and encapsulation of cornelian cherry polyphenols and iridoids using β-CD in order to enhance the ultrasound-assisted extraction efficiency of cornelian cherry bioactives were also reported in the literature (Popović et al. 2021). The authors suggested that due to the polyphenol encapsulation within β-CD, the obtained extract exhibited increased solubility in water, higher antioxidant activity, and prolonged release of anthocyanins from the dried powder making it suitable for various applications in food and pharmaceutical industries. Additionally, the use of aqueous solutions of β-cyclodextrin for the recovery of bioactive compounds (e.g., caffeine and chlorogenic acid) from coffee pulp, a by-product of coffee production was also reported by Loukri et al. (2020).

4 Overview

Nowadays, there is a continuously increasing demand both from consumers and industries for "clean label" foodstuffs. In this view, plant extracts prepared using novel extraction techniques and solvents, from different plant materials, constitute a promising and sustainable approach. Various parts of plants, as well as extracts obtained from agro-industrial by-products and wastes, are of particular interest as sources of natural antioxidants, antimicrobial, coloring, and flavoring agents. Such extracts can be either added directly in food products or they can be incorporated into active packaging materials in order to increase the shelf-life and improve the quality of food products, avoiding the use of synthetic additives.

Acknowledgments IM and AK acknowledge support of this work by the project "Chitosan-based nanoparticles and films loaded with agrifood by-product extracts for novel food applications-ChitoFoods" (Grant Agreement No. 3308), which is implemented under the frame of the "2nd

Call for H.F.R.I.'s Research Projects to Support Faculty Members & Researchers" funded by the Hellenic Foundation for Research and Innovation (HFRI).

References

Abdel-Naeem HHS, Abdelrahman AG, Imre K, Morar A, Herman V, Yassien NA (2022) Improving the structural changes, electrophoretic pattern, and quality attributes of spent hen meat patties by using kiwi and pineapple extracts. Foods 11:3430

Alaa A, Abdulhakim A, Rawaa T (2019) Potato and tomato peel extract – a natural antioxidant for retarding lipid peroxidation in lamb meat (Awassi) refrigerator storage. JFNA 2:1–5

Alchera F, Ginepro M, Giacalone G (2022) Microwave-assisted extraction of polyphenols from blackcurrant by-products and possible uses of the extracts in active packaging. Foods 11:2727

Alizadeh-Sani M, Tavassoli M, Mohammadian E, Ehsani A, Khaniki GJ, Priyadarshi R, Rhim JW (2021) pH-responsive color indicator films based on methylcellulose/chitosan nanofiber and barberry anthocyanins for real-time monitoring of meat freshness. Int J Biol Macromol 166:741–750

Andres AI, Petron MJ, Delgado-Adamez J, Lopez M, Timon M (2017) Effect of tomato pomace extracts on the shelf-life of modified atmosphere-packaged lamb meat. J Food Process Preserv 41:13018

Azmir J, Zaidul ISM, Rahman MM, Sharif KM, Mohamed A, Sahena F, Jahurul MHA, Ghafoor K, Norulaini NAN, Omar AKM (2013) Techniques for extraction of bioactive compounds from plant materials: a review. J Food Eng 117:426–436

Banerjee R, Verma AK, Das AK, Rajkumar V, Shewalkar AA, Narkhede HP (2012) Antioxidant effects of broccoli powder extract in goat meat nuggets. Meat Sci 91:179–184

Bellucci ERB, Munekata PES, Pateiro M, Lorenzo JM, da Silva Barretto AC (2021) Red pitaya extract as natural antioxidant in pork patties with total replacement of animal fat. Meat Sci 171:108284

Bermúdez-Oria A, Rodríguez-Gutiérrez G, Rubio-Senent F, Fernández-Prior A, Fernández-Bolaños J (2019) Effect of edible pectin-fish gelatin films containing the olive antioxidants hydroxytyrosol and 3,4-dihydroxyphenylglycol on beef meat during refrigerated storage. Meat Sci 148:213–218

Bianchin M, Pereira D, Almeida JF, Moura C, Pinheiro RS, Heldt LFS, Haminiuk CWI, Carpes ST (2020) Antioxidant properties of lyophilized rosemary and sage extracts and its effect to prevent lipid oxidation in poultry pátê. Molecules 6:5160

Botsoglou E, Govaris A, Ambrosiadis I, Fletouris D, Papageorgiou G (2014) Effect of olive leaf (*Olea europaea* L.) extracts on protein and lipid oxidation in cooked pork meat patties enriched with n-3 fatty acids. J Sci Food Agric 30:227–234

Cao J, Su E (2021) Hydrophobic deep eutectic solvents: the new generation of green solvents for diversified and colorful applications in green chemistry. J Clean Prod 314:127965

Castellani F, Vitali A, Bernardi N, Marone E, Palazzo F, Grotta L, Martino G (2017) Dietary supplementation with dried olive pomace in dairy cows modifies the composition of fatty acids and the aromatic profile in milk and related cheese. J Dairy Sci 100:8658–8669

Castro-Vargas HI, Benelli P, Ferreira SRS, Parada-Alfonso F (2013) Supercritical fluid extracts from tamarillo (*Solanum betaceum* Sendtn) epicarp and its application as protectors against lipid oxidation of cooked beef meat. J Supercrit Fluids 76:17–23

Chemat F, Rombaut N, Sicaire AG, Meullemiestre A, Fabiano-Tixier AS, Abert-Vian M (2017) Ultrasound assisted extraction of food and natural products. Mechanisms, techniques, combinations, protocols and applications. A review Ultrason Sonochem 34:540–560

Chinprahast N, Saengbamroong P, Wongtreenetkul S, Asawapuengrung W (2018) Antioxidant activities of local oregano and sweet basil leaves and application in a natural antioxidant enriched ravioli. Int Food Res J 25:1996–2004

Chouchouli V, Kalogeropoulos N, Konteles SJ, Karvela E, Makris DP, Karathanos VT (2013) Fortification of yoghurts with grape (*Vitis vinifera*) seed extracts. LWT - Food Sci Technol 53: 522–529

Colle MC, Richard RP, Smith DM, Colle MJ, Loucks WI, Gray SJ, Reynolds ZD, Sutton HA, Nasados JA, Doumit ME (2019) Dry potato extracts improve water holding capacity, shelf life, and sensory characteristics of fresh and precooked beef patties. Meat Sci 149:156–162

Demirkol M, Tarakci Z (2018) Effect of grape (*Vitis labrusca* L.) pomace dried by different methods on physicochemical, microbiological and bioactive properties of yoghurt. LWT - Food Sci Technol 97:770–777

Deolindo CTP, Monteiro PI, Santos JS, Cruz AG, da Silva MC, Granato D (2019) Phenolic-rich petit Suisse cheese manufactured with organic Bordeaux grape juice, skin, and seed extract: technological, sensory, and functional properties. LWT 115:108493

Devatkal SK, Thorat P, Manjunatha M (2014) Effect of vacuum packaging and pomegranate peel extract on quality aspects of ground goat meat and nuggets. J Food Sci Technol 51:2685–2691

European Food Safety Authority (2008) Use of rosemary extracts as a food additive-scientific opinion of the panel on food additives, flavourings, processing aids and materials in contact with food. EFSA J 2008(6):1–29

Farooq S, Mir SA, Shah MA, Manickavasagan A (2022) Extraction techniques. In: Mir SA, Manickavasagan A, Shah (eds) Plant extracts: applications in the food industry, 1st edn. Elsevier Inc., Amsterdam, pp 23–37

Galanakis CM (2012) Recovery of high added-value components from food wastes: conventional, emerging technologies and commercialized applications. Trends Food Sci Technol 26:68–87

Gantner M, Brodowska M, Górska-Horczyczaka E, Wojtasik-Kalinowskaa I, Najdab A, Pogorzelskaaand E, Godziszewska J (2018) Antioxidant effect of sage (*Salvia officinalis* L.) extract on Turkey meatballs packed in cold modified atmosphere. CYTA – J Food 16:628–636

Gasti T, Dixit S, D'souza OJ, Hiremani VD, Vootla SK, Masti SP, Chougale RB, Malabadi RB (2021) Smart biodegradable films based on chitosan/methylcellulose containing Phyllanthus reticulatus anthocyanin for monitoring the freshness of fish fillet. Int J Biol Macromol 187:451–461

Ghimire A, Paudel N, Poudel R (2022) Effect of pomegranate peel extract on the storage stability of ground buffalo (*Bubalus bubalis*) meat. LWT 154:112690

Guerra-Rivas C, Vieira C, Rubio B, Martínez B, Gallardo B, Mantecón AR, Lavín P, Manso T (2016) Effects of grape pomace in growing lamb diets compared with vitamin E and grape seed extract on meat shelf life. J Integr Agric 21:2445–2455

Guo Z, Han L, Yu QL, Lin L (2020) Effect of a sea buckthorn pomace extract-esterified potato starch film on the quality and spoilage bacteria of beef jerky sold in supermarket. Food Chem 326:127001

Kanatt SR (2020) Development of active/intelligent food packaging film containing Amaranthus leaf extract for shelf life extension of chicken/fish during chilled storage. Food Packag Shelf Life 24:100506

Kang S, Wang H, Xia L, Chen M, Li L, Cheng J, Li X, Jiang S (2020) Colorimetric film based on polyvinyl alcohol/okra mucilage polysaccharide incorporated with rose anthocyanins for shrimp freshness monitoring. Carbohydr Polym 229:115402

Kim SJ, Min SC, Shin HJ, Lee YJ, Cho AR, Kim SY, Han J (2013) Evaluation of the antioxidant activities and nutritional properties of ten edible plant extracts and their application to fresh ground beef. Meat Sci 93:715–722

Koosha M, Hamedi S (2019) Intelligent chitosan/PVA nanocomposite films containing black carrot anthocyanin and bentonite nanoclays with improved mechanical, thermal and antibacterial properties. Prog Org Coat 127:338–347

Kyriakidou A, Makris DP, Lazaridou A, Biliaderis CG, Mourtzinos I (2021) Physical properties of chitosan films containing pomegranate peel extracts obtained by deep eutectic solvents. Foods 10:1262

Kyriakoudi A, Tsiouras A, Mourtzinos I (2022) Extraction of lycopene from tomato using hydrophobic natural deep eutectic solvents based on terpenes and fatty acids. Foods 11:2645

Loukri A, Tsitlakidou P, Goula A, Assimopoulou AN, Kontogiannopoulos KN, Mourtzinos I (2020) Green extracts from coffee pulp and their application in the development of innovative brews. Appl Sci 10:6982

Ma Q, Du L, Wang L (2017) Tara gum/polyvinyl alcohol-based colorimetric NH3 indicator films incorporating curcumin for intelligent packaging. Sens Actuators B Chem 244:759–766

Manhani MR, Nicoletti MA, Barretto ACDS, Riveres De Jesus G, Munhoz CC, De Abreu GR, Zaccarelli-Magalhães J, Fukushima AR (2018) Antioxidant action of rosemary and oregano extract in pre-cooked meet hamburger. Food Sci Nutr 9:806–817

Mariem C, Sameh M, Nadhem S, Soumaya Z, Najiba Z, Raoudha EG (2014) Antioxidant and antimicrobial properties of the extracts from *Nitraria retusa* fruits and their applications to meat product preservation. Ind Crop Prod 55:295–303

Martiny TR, Raghavan V, Moraes CC, Rosa GS, Dotto GL (2020) Bio-based active packaging: carrageenan film with olive leaf extract for lamb meat preservation. Foods 9:1759

Mir SA, Manickavasagan A, Shah MA (2022) Sources of plant extracts. In: Mir SA, Manickavasagan A, Shah (eds) Plant extracts: applications in the food industry, 1st edn. Elsevier Inc., pp 1–22

Molnar J, Novotni D, Kurek M, Galić K, Iveković D, Bionda H, Ščetar M (2023) Characteristics of edible films enriched with fruit by-products and their application on cookies. Food Hydrocoll 135:108191

Mourtzinos I, Salta F, Yannakopoulou K, Chiou A, Karathanos VT (2007) Encapsulation of olive leaf extract in β-cyclodextrin. J Agric Food Chem 55:8088–8094

Mourtzinos I, Anastasopoulou E, Petrou A, Grigorakis S, Makris D, Biliaderiss CG (2016) Optimization of a green extraction method for the recovery of polyphenols from olive leaf using cyclodextrins and glycerin as co-solvents. J Food Sci Technol 53:3939–3947

Mourtzinos I, Prodromidis P, Grigorakis S, Makris DP, Biliaderis CG, Moschakis T (2018) Natural food colorants derived from onion wastes: application in a yoghurt product. Electrophoresis 39: 1975–1983

Ordaz-Rodríguez SB, Abadía-García L, Huerta-Manzanilla EL, Mendoza-Sánchez M (2022) Ultrasound-assisted extraction of pomegranate peel antioxidants: a green process to obtain a meat preservative. J Food Process Preserv 46:16712

Park H, Lee M, Kim KT, Park E, Paik HD (2018) Antioxidant and antigenotoxic effect of dairy products supplemented with red ginseng extract. J Dairy Sci 101:8702–8710

Pateiro M, Vargas FC, Chincha AAIA, Sant'Ana AS, Strozzi I, Rocchetti G, Barba FJ, Domínguez R, Lucini L, Amaral Sobral PJ, Lorenzo JM (2018) Guarana seed extracts as a useful strategy to extend the shelf life of pork patties: UHPLC-ESI/QTOF phenolic profile and impact on microbial inactivation, lipid and protein oxidation and antioxidant capacity. Food Res Int 114:55–63

Patra BR, Borugadda VB, Dalai AK (2022) Microwave-assisted extraction of sea buckthorn pomace and seed extracts as a proactive antioxidant to stabilize edible oils. Bioresour Technol 17:100970

Peiretti PG, Gai F, Zorzi M, Aigotti R, Medana C (2020) The effect of blueberry pomace on the oxidative stability and cooking properties of pork patties during chilled storage. J Food Pross Preserv 44:44:14520

Pereira VA, Pereira VA, de Arruda INQ, Stefani R (2015) Active chitosan/PVA films with anthocyanins from Brassica oleracea (red cabbage) as time–temperature indicators for application in intelligent food packaging. Food Hydrocoll 43:180–188

Pollini L, Blasi F, Ianni F, Grispoldi L, Moretti S, Di Veroli A, Cossignani L, Cenci-Goga BT (2022) Ultrasound-assisted extraction and characterization of polyphenols from apple pomace, functional ingredients for beef burger fortification. Molecules 27:1933

Popović BM, Blagojević B, Latković D, Četojević-Simin D, Kucharska AZ, Parisi F, Lazzara G (2021) A one step enhanced extraction and encapsulation system of cornelian cherry (*Cornus mas* L.) polyphenols and iridoids with β-cyclodextrin. LWT 141:110884

Qin L, Yu J, Zhu J, Kong B, Chen Q (2021) Ultrasonic-assisted extraction of polyphenol from the seeds of *Allium senescens* L. and its antioxidative role in Harbin dry sausage. Meat Sci 172: 108351

Rambabu K, Bharath G, Banat F, Show PL, Cocoletzi HH (2019) Mango leaf extract incorporated chitosan antioxidant film for active food packaging. Int J Biol Macromol 126:1234–1243

Rashid R, Masoodi FA, Wani SM, Manzoor S, Gull A (2022) Ultrasound assisted extraction of bioactive compounds from pomegranate peel, their nanoencapsulation and application for improvement in shelf life extension of edible oils. Food Chem 385:132608

Rente D, Bubalo MC, Panić M, Paiva A, Caprin B, Radojčić Redovniković I, ARC D (2022) Review of deep eutectic systems from laboratory to industry, taking the application in the cosmetics industry as an example. J Clean Prod 380:135147

Saleh H, Golian A, Kermanshahi H, Mirakzehi MT (2017) Effects of dietary α-tocopherol acetate, pomegranate peel, and pomegranate peel extract on phenolic content, fatty acid composition, and meat quality of broiler chickens. J Appl Anim Res 45:629–636

Santos EN, Sousa TCA, Neto DCS, Grisi CVB, Ferreira VCS, Silva FAP (2022) Edible active film based on gelatin and Malpighia emarginata waste extract to inhibit lipid and protein oxidation in beef patties. LWT 154:112837

Saurabh CK, Gupta S, Variyar PS (2022) Development of grape pomace extract based edible coating for shelf life extension of pomegranate arils. Food Meas 16:590–597

Sekharan TR, Chandira RM, Tamilvanan S, Rajesh SC, Venkateswarlu BS (2021) Deep eutectic solvents as an alternate to other harmful solvents. Biointerface Res Appl Chem 12:847–860

Sharma P, Wichaphon J, Klangpetch W (2020) Antimicrobial and antioxidant activities of defatted *Moringa oleifera* seed meal extract obtained by ultrasound-assisted extraction and application as a natural antimicrobial coating for raw chicken sausages. Int J Food Microbiol 332:108770

Socas-Rodríguez B, Torres-Cornejo MV, Álvarez-Rivera G, Mendiola JA (2021) Deep eutectic solvents for the extraction of bioactive compounds from natural sources and agricultural by-products. Appl Sci 11:4897

Šojić B, Tomović V, Kocić-Tanackov S, Kovačević DB, Putnik P, Mrkonjić Z, Đurović S, Jokanović M, Ivić M, Škaljac S, Pavlić B (2020) Supercritical extracts of wild thyme (*Thymus serpyllum* L.) by-product as natural antioxidants in ground pork patties. LWT 130:109661

Sun J, Jiang H, Wu H, Tong C, Pang J, Wu C (2020) Multifunctional bionanocomposite films based on konjac glucomannan/chitosan with nano-ZnO and mulberry anthocyanin extract for active food packaging. Food Hydrocoll 107:105942

Tebbi SO, Debbache-Benaida N, Kadri N, Kadi R, Zaidi S (2023) A novel strategy to improve the recovery of phenolic compounds from *Pistacia lentiscus* L. fruits using design-based statistical modeling for ultrasound-deep eutectic solvents extraction and the evaluation of their antioxidant potential. Sustain Chem Pharm 31:100933

Timón M, Andrés AI, Sorrentino L, Cardenia V, Petrón MJ (2022) Effect of phenolic compounds from almond skins obtained by water extraction on pork patty shelf life. Antioxidants 11:2175

Turan E, Şimşek A (2021) Effects of lyophilized black mulberry water extract on lipid oxidation, metmyoglobin formation, color stability, microbial quality and sensory properties of beef patties stored under aerobic and vacuum packaging conditions. Meat Sci 178:108522

Turgut SS, Soyer A, Işıkçı F (2016) Effect of pomegranate peel extract on lipid and protein oxidation in beef meatballs during refrigerated storage. Meat Sci 116:126–132

U.S. Department of Health and Human Services GRAS Notices No. 446 Red Grape Pomace Extract. Accessed January 21, 2023., from https://www.accessdata.fda.gov/scripts/fdcc/index.cfm?set=GRASNotices

U.S. Department of Health and Human Services GRAS Notices No. 459 Olive Pulp Extract. Accessed January 21, 2023., from https://www.accessdata.fda.gov/scripts/fdcc/index.cfm?set=GRASNotices

U.S. Department of Health and Human Services GRAS Notices No. 93 Grape Seed Extract and Grape Skin Extract. Accessed January 21, 2023., from https://www.accessdata.fda.gov/scripts/fdcc/index.cfm?set=GRASNotices

Valdés A, Mellinas AC, Ramos M, Garrigós MC, Jiménez A (2014) Natural additives and agricultural wastes in biopolymer formulations for food packaging. Front Chem 2:1–10

Vasilopoulos S, Dokou S, Papadopoulos GA, Savvidou S, Christaki S, Kyriakoudi A, Dotas V, Tsiouris V, Bonos E, Skoufos I, Mourtzinos I, Giannenas I (2022) Dietary supplementation with pomegranate and onion aqueous and cyclodextrin encapsulated extracts affects broiler performance parameters, welfare and meat characteristics. Poultry 1:74–93

Velásquez P, Bustos D, Montenegro G, Giordano A (2021) Ultrasound-assisted extraction of anthocyanins using natural deep eutectic solvents and their incorporation in edible films. Molecules 26:984

Wang S, Marcone M, Barbut S, Lim LT (2012) The impact of anthocyanin-rich red raspberry extract (ARRE) on the properties of edible soy protein isolate (SPI) films. J Food Sci 77:C497–C505

Wen Y, Liu J, Jiang L, Zhu Z, He S, He S, Shao W (2021) Development of intelligent/active food packaging film based on TEMPO-oxidized bacterial cellulose containing thymol and anthocyanin-rich purple potato extract for shelf life extension of shrimp. Food Packag Shelf Life 29:100709

Wu C, Sun J, Zheng P, Kang X, Chen M, Li Y, Ge Y, Hu Y, Pang J (2019) Preparation of an intelligent film based on chitosan/oxidized chitin nanocrystals incorporating black rice bran anthocyanins for seafood spoilage monitoring. Carbohydr Polym 222:115006

Yang XY, Xu BC, Lei HM, Luo X, Zhu LX, Zhang YM, Mao YW, Liang RR (2022) Effects of grape seed extract on meat color and premature browning of meat patties in high-oxygen packaging. J Integr Agric 21:2445–2455

Yong H, Wang X, Bai R, Miao Z, Zhang X, Liu J (2019) Development of antioxidant and intelligent pH-sensing packaging films by incorporating purple-fleshed sweet potato extract into chitosan matrix. Food Hydrocoll 90:216–224

Zhang H, Wu J, Guo X (2016) Effects of antimicrobial and antioxidant activities of spice extracts on raw chicken meat quality. Food Sci Hum Wellness 5:39–48

Zhang J, Zou X, Zhai X, HuangXW JC, Holmes M (2019a) Preparation of an intelligent pH film based on biodegradable polymers and roselle anthocyanins for monitoring pork freshness. Food Chem 272:306–312

Zhang X, Liu Y, Yong H, Qin Y, Liu J, Liu J (2019b) Development of multifunctional food packaging films based on chitosan, TiO_2 nanoparticles and anthocyanin-rich black plum peel extract. Food Hydrocoll 94:80–92

Zhang W, Li X, Jiang W (2020) Development of antioxidant chitosan film with banana peels extract and its application as coating in maintaining the storage quality of apple. Int J Biol Macromol 154:1205–1214

The Use of Feed and Food Additives in United States

Prafulla Kumar and Ajay Srivastava

Abstract

The feed additives market in the United States is a large and dynamic industry that plays a crucial role in supporting animal health, welfare, and productivity.

Additives are used for flavor and appeal, preparation, processing, freshness as well as safety and some additives have been in use for centuries for preservation, such as salt and sugar.

The United States Food Additives Market is expected to grow at a CAGR of 3.5% during 2023–2028 as per a study by Mordor Intelligence. The market size of feed additives varies by livestock species, with poultry and swine being the largest consumers of feed additives. The use of feed additives is subject to regulations by the FDA and AAFCO, which help to ensure the safety and efficacy of these products.

Feed additives may be classified as nutritional, zootechnical, sensory, coccidiostat or histomonostats, mycotoxin binders, and detoxifiers or other feed additives like antioxidants, enzymes, and binders which are used for a variety of purposes in animal nutrition. Some novel additives are being developed based on robust research and safety backgrounds for animal and human supplementation to serve the much-needed market demand such as Advanced Digestion Enhancing Protein Plus Technology (ADEPPT), which is ovoceutin and transferrin patented complex, CurCos-Zn for animals and patents have been applied for the same as well. These additives are used to supplement animal diets and address specific nutritional deficiencies or health challenges. Understanding the specific nutritional needs of different animal species, as well as the various feed additive options available, is critical for animal producers and nutritionists to maximize animal health and performance while also ensuring product quality and safety.

P. Kumar (✉) · A. Srivastava
RD LifeSciences, LLC, Clearwater, USA, Florida

G. Arsenos, I. Giannenas (eds.), *Sustainable Use of Feed Additives in Livestock*,
https://doi.org/10.1007/978-3-031-42855-5_10

Keywords

Food additive · Feed additive · Nutrition · Safety · Supplement

1 Food and Feed Additive

Food and feed additives in the United States are defined by the FDA as substances added to food or animal feed for a specific purpose, such as to improve their nutritional value, taste, texture, or shelf life. The FDA regulates the use of these additives and has established guidelines and regulations for their safe and effective use.

In the United States, food and feed additives are defined by the Food and Drug Administration (FDA) as substances that are added to food or animal feed for a specific purpose, such as to improve the nutritional value, taste, texture, or shelf life of the product. The FDA defines food additives as "*any substance, the intended use of which results or may reasonably be expected to result, directly or indirectly, in its becoming a component or otherwise affecting the characteristics of any food*" (FDA 2021a). Feed additives are defined as "*any substance, the intended use of which results or may reasonably be expected to result, directly or indirectly, in its becoming a component or otherwise affecting the characteristics of any animal food*" (FDA 2021b).

Feed additive may be one ingredient or combination of ingredients added to a basic feed mix to fulfill a specific need.

Usually, feed additive is used in microquantities and requires careful handling and mixing.

The regulation of feed additives in the United States is overseen by several federal and state agencies, including the Food and Drug Administration (FDA) and the Association of American Feed Control Officials (AAFCO).

The FDA is responsible for the approval and regulation of food and feed additives in the United States. The agency has established regulations and guidelines for the use of feed additives, including the generally recognized as safe (GRAS) list, which outlines substances that are considered safe for use in animal feed based on scientific evidence and expert opinion. The FDA also regulates the use of veterinary drugs in animal feed, including the approval of new drugs and the establishment of tolerance levels for residues in animal products.

The AAFCO is a nongovernmental organization that works with state and federal regulatory agencies to establish and maintain standards for animal feed and feed ingredients. The AAFCO provides guidance and recommendations on the use of feed additives, including the establishment of feed additive definitions, labeling requirements, and model regulations. The AAFCO also works with the FDA and other agencies to review and approve new feed additives and to establish guidelines for the safe and effective use of these substances in animal feed.

In summary, the FDA and AAFCO play important roles in the regulation of feed additives in the United States. The FDA is responsible for the approval and regulation of feed additives and veterinary drugs, while the AAFCO provides guidance and recommendations on the use of feed additives and works with regulatory agencies to establish and maintain standards for animal feed and feed ingredients.

2 Types of Feed Additives in United States

Feed additives are classified based on their intended use and mode of action. The following are the broad categories (Fig.1) of feed additives:

(a) **Nutritional feed additives**: These are additives that are used to supplement or enhance the nutrient content of animal diets, such as vitamins, minerals, and amino acids.
(b) **Zootechnical feed additives**: These are additives that are used to improve animal performance, such as growth promotion, feed efficiency, and disease prevention. Examples include probiotics, prebiotics, and acidifiers.
(c) **Sensory feed additives**: These are additives that are used to enhance the visual appeal or palatability of animal feed, such as flavors, sweeteners, and pigments.
(d) **Coccidiostats and histomonostats**: These are feed additives that are used to control specific diseases in poultry, such as coccidiosis and histomoniasis.
(e) **Mycotoxin binders and detoxifiers**: These are feed additives that are used to mitigate the harmful effects of mycotoxins in animal feed.
(f) **Other feed additives**: These include antioxidants, enzymes, and binders, which are used for a variety of purposes in animal nutrition.

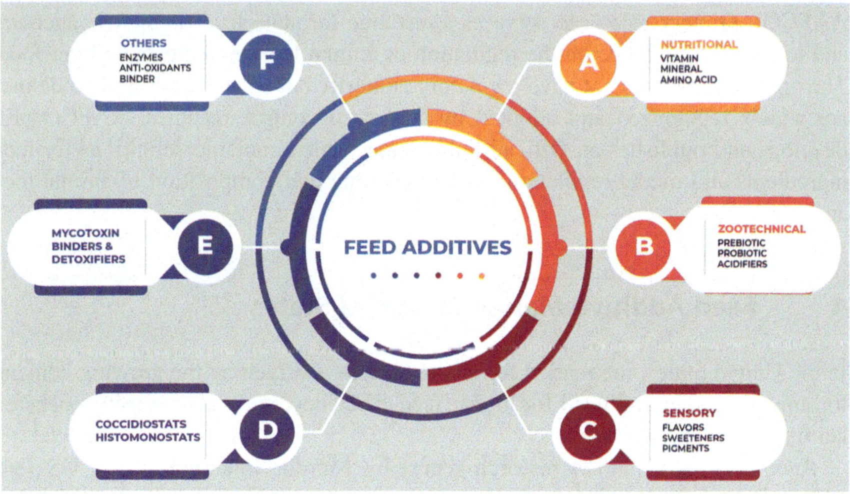

Fig. 1 Types of feed additives

The classification of feed additives is important for both regulatory purposes and for animal producers and nutritionists to select the appropriate additives to meet specific animal nutrition and health needs (Fig. 1).

3 Feed Additive Regulation in United States

Feed additives are regulated in the United States by the FDA. It has primary legal responsibility for determining their safe use in dogs, cats, horses, chickens, turkeys, pigs, sheep, and fish. The Food and Drug Administration's (FDA) Center of Veterinary Medicine monitors and establishes standards for feed contaminants, approves safe food additives, and manages the FDA's medicated feed and pet food programs. The approval process for feed additives involves a scientific evaluation of safety and efficacy data submitted by the manufacturer. The use of certain antibiotics as growth promoters in animal feed has been restricted in recent years due to concerns over the development of antibiotic-resistant bacteria.

Anything that can be added to food or not, its safety has to be assessed through a stringent approval process. The Food Safety and Inspection Service (FSIS) of the US Department of Agriculture shares responsibility with FDA for the safety and food additives used in meat, poultry, and egg products.

Labeling guidelines for all feed additives are given by AAFCO, which is the Association of American Feed Control Officials. AAFCO is a nonprofit organization that provides a forum for developing and implementing uniform and equitable laws, regulations, standards, definitions, and enforcement policies related to animal feeds in the United States. AAFCO does not have direct regulatory authority over feed additives in the United States. However, AAFCO provides guidance and recommendations on the use of feed additives and the labeling of animal feed, which are often used as the basis for regulation by state feed control agencies. AAFCO's primary role is to serve as a resource for state feed regulatory agencies and to provide guidance on the regulation of animal feed ingredients and pet food. The organization does not have regulatory authority but its standards and guidelines are widely recognized and adopted by state feed control agencies. AAFCO also develops and publishes an Official Publication, which contains definitions for feed ingredients and model regulations for the labeling and composition of animal feed and pet food.

4 Feed Additive Market in United States

In the United States, the market for feed additives is driven by the growing demand for animal protein, the need for higher feed efficiency, and the increasing focus on animal health and welfare.

According to a market research report by Mordor Intelligence, the US feed additives market is expected to grow at a compound annual growth rate (CAGR) of 3.5% from 2023 to 2028 (Mordor Intelligence 2021). The major factors driving the growth of the market include the increasing demand for meat and dairy products, the growing awareness about animal health and welfare, and the need to reduce the use of antibiotics in animal production.

The US feed additives market is highly competitive, with a large number of domestic and international players operating in the market. Some of the key players in the market include Archer Daniels Midland Company, Cargill, BASF, and Evonik Industries.

The feed additives market size in the United States varies depending on the livestock sector. Feed additives are used in different amounts and for different purposes depending on the specific needs of each livestock sector. The major livestock sectors in the United States include poultry, swine, cattle, and aquaculture.

(a) **Poultry:** The poultry sector is the largest consumer of feed additives in the United States. Poultry feed additives are used to improve feed efficiency, increase growth rate, and enhance the overall health of the birds. According to a market research report by Global Market Insights, the United States poultry feed additives market was valued at over $1.5 billion in 2018 and is projected to grow at a compound annual growth rate (CAGR) of 4.5% from 2019 to 2025 (Global Market Insights 2019).
(b) **Pigs:** The pig sector is another major consumer of feed additives in the United States. Pig feed additives are used to improve feed efficiency, promote growth and development, and enhance the immune system of the animals. According to the same report by Global Market Insights, the US pig feed additives market was valued at over $900 million in 2018 and is projected to grow at a CAGR of 3.5% from 2019 to 2025 (Global Market Insights 2019).
(c) **Cattle:** The cattle sector is also a significant consumer of feed additives in the United States. Cattle feed additives are used to improve feed efficiency, increase growth rate, and enhance the overall health of the animals. According to the same report by Global Market Insights, the US cattle feed additives market was valued at over $750 million in 2018 and is projected to grow at a CAGR of 3% from 2019 to 2025 (Global Market Insights 2019).
(d) **Aquaculture:** The aquaculture sector is a smaller but growing consumer of feed additives in the United States. Aquaculture feed additives are used to improve feed efficiency, increase growth rate, and enhance the overall health of fish and other aquatic animals. According to the same report by Global Market Insights, the US aquaculture feed additives market was valued at over $80 million in 2018 and is projected to grow at a CAGR of 5% from 2019 to 2025 (Global Market Insights 2019).

The US feed additives market is expected to grow in the coming years, driven by the increasing demand for animal protein, the need for higher feed efficiency, and the focus on animal health and welfare. Amino acids and enzymes are among the most widely used feed additives in the United States, and the market is highly competitive, with several major players operating in the market.

The feed additives market size in the United States varies depending on the livestock sector. The poultry sector is the largest consumer of feed additives, followed by swine, cattle, and aquaculture.

5 Nutritional Feed Additives

5.1 Vitamins

Vitamins are organic molecules that are needed in minute amounts for various physiological functions. The name vitamin originated from the term vital amine and refers to a group of compounds having a specific role in metabolism. Vitamins function as enzyme precursors, or co-enzymes, in different metabolic processes. Most vitamins need to be provided to the animal through diet.

Vitamins may be classified based on their solubility as fat or water-soluble vitamins

5.1.1 Fat-Soluble Vitamins

In animals, the fat-soluble vitamins are the same as in humans. These include:

(a) Vitamin A: This vitamin is important for vision, immune function, and cell growth. It is found in animal products such as liver, fish, and dairy products. Vitamin A supplementation is considered as a viable option to strengthen the immune system in neonates
(b) Vitamin D: This vitamin is important for bone health and helps the body absorb calcium and phosphorus. It is produced in the skin of animals when exposed to sunlight and is also found in some animal products such as fatty fish.
(c) Vitamin E: This vitamin is an antioxidant that helps protect cells from damage. It is found in vegetable oils and some animal products such as eggs and liver.
(d) Vitamin K: This vitamin is important for blood clotting and bone health. It is synthesized by gut bacteria in some animals and is also found in some animal products such as liver and egg yolks.

Since fat-soluble vitamins are stored in the body's fatty tissues, excess intake can lead to toxicity. Therefore, it is important for animals to receive adequate amounts of these vitamins through their diet, but not excessive amount as excessive storage can be toxic for some vitamins like vitamin A and vitamin D.

5.1.2 Water-Soluble Vitamins

Water-soluble vitamins are a group of vitamins that dissolve in water and are not stored in the body to the same extent as fat-soluble vitamins. They are absorbed in the small intestine and are transported to tissues throughout the body, where they are used for various functions. The water-soluble vitamins that are important for animals include:

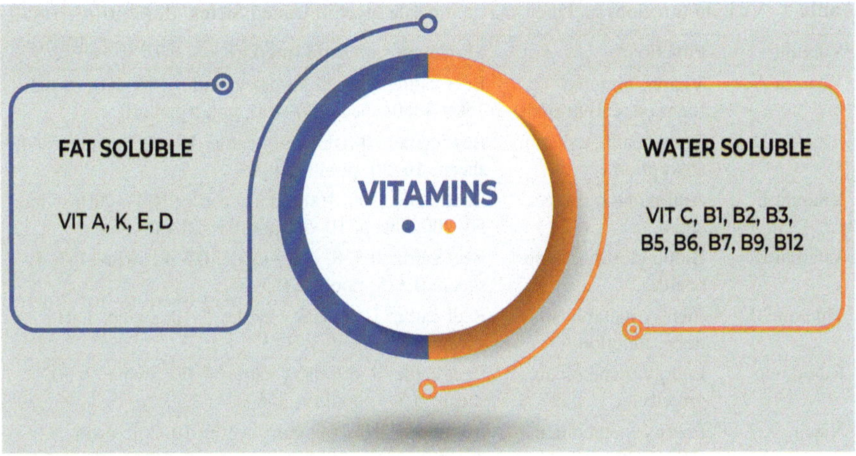

Fig. 2 Types of vitamins

(a) Vitamin B1 (thiamine): This vitamin is important for energy metabolism and nerve function.
(b) Vitamin B2 (riboflavin): This vitamin is important for energy metabolism and growth.
(c) Vitamin B3 (niacin): This vitamin is important for energy metabolism and skin health.
(d) Vitamin B5 (pantothenic acid): This vitamin is important for energy metabolism.
(e) Vitamin B6 (pyridoxine): This vitamin is important for protein metabolism and nerve function.
(f) Vitamin B7 (biotin): This vitamin is important for energy metabolism and skin health.
(g) Vitamin B9 (folic acid): This vitamin is important for DNA synthesis and cell growth.
(h) Vitamin B12 (cobalamin): This vitamin is important for red blood cell production.
(i) Vitamin C (ascorbic acid): This vitamin is an antioxidant that helps protect cells from damage and is important for collagen synthesis (Fig. 2 and Table 1).

5.2 Amino Acids

These are molecules that combine to form protein. Amino acids are the building block of protein. When proteins are broken or digested, amino acids are left. Amino acids play a great role in growth to production in animals. There are 20 unique amino acids, which are made from carbon, nitrogen, oxygen, hydrogen, and sometimes

Table 1 Vitamin functions and their maximum safe level in United States

Vitamin	Function	Maximum safe level (mg/kg)
Vitamin A	Vision, immune function, cell growth	Beef cattle: 500–2500; dairy cattle: 100–500; swine: 1500-3000; sheep: 50–400; poultry: 4–10
Vitamin D	Bone health, calcium absorption	Beef cattle: 10–100; dairy cattle: 10–100; swine: 50–80; sheep: 10–50; poultry: 1–2
Vitamin E	Antioxidant, protect cells	Beef cattle: 100–1000; dairy cattle: 100–1000; swine: 50–200; sheep: 10–50; poultry: 20–200
Vitamin K	Blood clotting, bone health	Beef cattle: 0.5–8; dairy cattle: 0.5–8; swine: 0.3–3; sheep: 0.5–5; poultry: 0.5–5
Thiamin	Energy metabolism, nerve function	Beef cattle: 1–10; dairy cattle: 1–10; swine: 1–10; sheep: 1–10; poultry: 1–3
Riboflavin	Energy metabolism, growth	Beef cattle: 4–60; dairy cattle: 4–60; swine: 2–30; sheep: 0.5–5; poultry: 3–6
Niacin	Energy metabolism, skin health	Beef cattle: 10–200; dairy cattle: 10–200; swine: 20–200; sheep: 10–50; poultry: 30–100
Pantothenic acid	Energy metabolism	Beef cattle: 10–100; dairy cattle: 10–100; swine: 10–100; sheep: 5–50; poultry: 3–10
Pyridoxine	Protein metabolism, nerve function	Beef cattle: 2–20; dairy cattle: 2–20; swine: 1–10; sheep: 1–10; poultry: 2–20
Biotin	Energy metabolism, skin health	Beef cattle: 0.1–1; dairy cattle: 0.1–1; swine: 0.05–0.2; sheep: 0.1–1; poultry: 0.02–0.1
Folic acid	DNA synthesis, cell growth	Beef cattle: 0.5–5; dairy cattle: 0.5–5; swine: 0.5–5; sheep: 0.5–5; poultry: 0.5–5
Vitamin B12	Red blood cell production	Beef cattle: 0.01–0.1; dairy cattle: 0.01–0.1; swine: 0.01–0.1; sheep: 0.01–0.1; poultry: 0.02–0.1

Guidance for Industry #235: Current Good Manufacturing Practice Requirements for Food for Animals. AAFCO (2021). Official Publication. Association of American Feed Control Officials.

sulfur in different configurations. These amino acids combine in different chains to create single protein in the animal body.

Amino acids are classified as essential or nonessential amino acids based on animals' ability to synthesize them in the body.

(a) **Nonessential amino acids**—Can be synthesized by the animal, so no need to be provided by the diet.
(b) **Essential amino acids**—Cannot be synthesized by the animal so it is mandatory to supply in the diet.

Whether essential or nonessential, animals need a sufficient amount of amino acids to meet their metabolic requirements. Industrial amino acids can be made through fermentation (L form of amino acid) or through chemical synthesis (DL form or racemic mixture of amino acids). However, pharmaceutical grade of amino acids can be manufactured through an enzymatic process. All amino acids used in protein synthesis in the body must be in L-configuration (Table 2).

Table 2 Nutritional classification of amino acids

Essential		Non-essential	
Common core	*Additional species related requirement*		*Conditionally nonessential*
Lysine	Arginine(cats, poultry, fish)	Cysteine	Glutamine
Histidine	Taurine (cats)	Tyrosine	Glycine
Leucine		Arginine	Serine
Isoleucine		Proline	Alanine
Valine			Asparagine
Methionine			Aspartate
Threonine			
Tryptophan			
Phenylalanine			

Amino acids for ruminants—Ruminants are able to synthesize many amino acids on their own through microbial fermentation in the rumen, but some essential amino acids must be provided in the diet in order to support optimal growth and production. The essential amino acids for ruminants include:

Essential amino acids for ruminants—methionine, lysine, histidine, phenylalanine, threonine, tryptophan, valine, isoleucine, and leucine.

Nonessential amino acids for ruminants—alanine, arginine, aspartic acid, cysteine, glutamic acid, glycine, proline, serine, and tyrosine (Licitra, G., Hernandez, T. M., & Van Soest, P. J. 1996)

Amino acids for pigs—Pigs require a balanced supply of essential and nonessential amino acids for optimal growth and production.

The essential amino acids for pigs include lysine, methionine, methionine + cystine, threonine, tryptophan, valine, isoleucine, leucine, arginine, and phenylalanine + tyrosine.

Lysine is considered the most limiting amino acid in swine diets and is often supplemented to ensure adequate amino acid balance.

In addition to these essential amino acids, pigs also have specific requirements for nonessential amino acids such as alanine, arginine, aspartic acid, cysteine, glutamic acid, glycine, proline, serine, and tyrosine, which are important for the synthesis of structural proteins and other physiological functions.

Amino acids for poultry—Poultry require a balanced supply of essential and nonessential amino acids for optimal growth and production.

The essential amino acids for poultry include arginine, histidine, isoleucine, leucine, lysine, methionine, methionine + cystine, phenylalanine + tyrosine, threonine, tryptophan, and valine.

Lysine is often considered the most limiting amino acid in poultry diets and is often supplemented to ensure adequate amino acid balance. Methionine is another important amino acid for poultry, as it is a precursor to cysteine and is important for feather growth.

Table 3 Amino acids and related products allowed as feed additives in United States

Additives	Remarks
L-methionine	Product should contain a minimum of 98.5% L-isomer of a-amino-4-(methylthio)butanoic acid. L-methionine is produced by *Escherichia coli* followed by enzymatic conversion to L-methionine.
DL-methionine hydroxy analog calcium	Product should contain a minimum 97% of a racemic mixture of 2-hydroxy-4-(methylthio)butanoic acid calcium salt.
DL-methionine	Product should contain a minimum of 99% racemic 2-amino-4-(methylthio)butanoic acid.
Glycine	Product should contain a minimum 97% amino acetic acid.
L-lysine	Product should contain minimum of 95% L-2, 6-diaminohexanoic acid.
L-threonine	Product contains a minimum of 95% of L-2-amino-3hydroxybutanoic acid.
DL-tryptophan	Product should contain a minimum of 97% racemic 2 amino-3-(3indolyl)-propionic acid.
L-tryptophan	Product should contain a minimum of 97% L-2-amino-3(3′ indolyl)-propionic acid.
Taurine	Product should contain minimum 97% 2-aminoethanesulfonic acid.
L-arginine	Product should contain a minimum of 98% L-2-amino-5-guanidyl-valeric acid
DL arginine	Product should contain a minimum of 98% racemic 2-amino-5-guanidyl-valeric acid
L-tyrosine	Product should contain a minimum of 98% L-2-amino-3-(4-hydroxyphenyl) propionic acid.
L-lysine liquid	Product should contain minimum of 50% L-2, 6-diaminohexanoic acid by weight in water solution.

AAFCO (2021)

In addition to these essential amino acids, poultry also have specific requirements for nonessential amino acids such as alanine, aspartic acid, glutamic acid, glycine, proline, serine, tyrosine, and cysteine (conditionally essential), which are important for the synthesis of structural proteins and other physiological functions.

It is important to note that the optimal amino acid requirements for poultry may vary depending on the specific animal breed, age, sex, and production stage (Table 3).

5.3 Minerals

These are inorganic elements that are important for the animal body, and their physiological and metabolic function. Minerals constitute about 4% of animal body. The functions of minerals are varied and range to regulate cellular function, immunity, acid-base balance, structural support, and expression and regulation of genes and enzymes. Minerals may be classified into two groups based on the amounts needed in the diet.

5.3.1 Macro Minerals

Macrominerals are essential minerals required in relatively large amounts by farm animals. Those minerals are required in large quantities in the diet (>0.01%).

They play a variety of important roles in the animal's body, including:

(a) Calcium: Calcium is an essential component of bones and teeth and is important for muscle function, nerve transmission, blood clotting, and enzyme activity.
(b) Phosphorus: Phosphorus is also an important component of bones and teeth, and is involved in energy metabolism, cell signaling, and DNA synthesis.
(c) Sodium: Sodium is involved in regulating fluid balance and acid-base balance in the body and is important for nerve and muscle function.
(d) Chloride: Chloride is involved in regulating fluid and electrolyte balance and is important for acid–base balance in the body.
(e) Magnesium: Magnesium is involved in enzyme activity, muscle function, and bone development.
(f) Potassium: Potassium is involved in regulating fluid balance, nerve function, and muscle function.
(g) Sulfur: Sulfur is important for the synthesis of amino acids and proteins (Fig. 3).

5.3.2 Micro Minerals

Microminerals, also known as trace minerals, are essential minerals required in relatively small amounts by farm animals. Required in trace amounts (<0.01%).

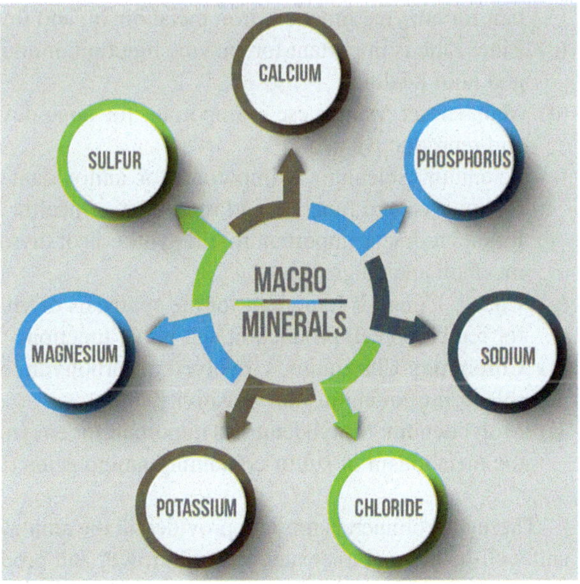

Fig. 3 Macrominerals

Fig. 4 Microminerals

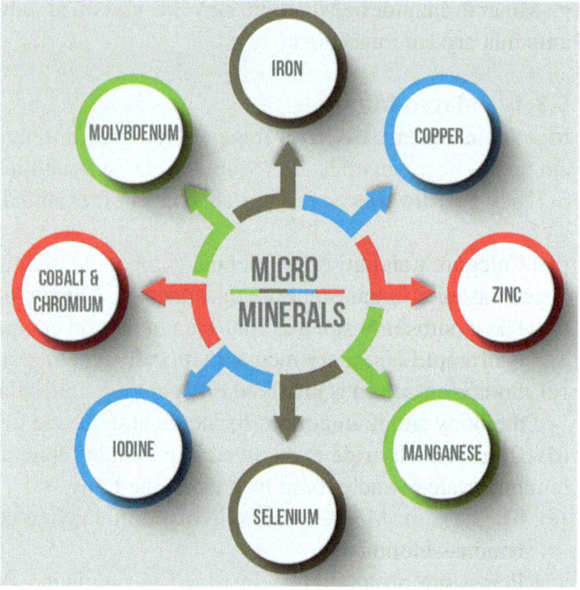

They play a variety of important roles in the animal's body, including:

(a) Iron: Iron is an essential component of hemoglobin, which is the protein in red blood cells that carries oxygen throughout the body. Iron is also important for enzyme function and immune system function.
(b) Copper: Copper is involved in the formation of connective tissue and is important for enzyme function, iron metabolism, and immune system function.
(c) Zinc: Zinc is important for enzyme function, immune system function, and skin and hoof health.
(d) Manganese: Manganese is important for bone development, enzyme function, and reproductive health.
(e) Selenium: Selenium is important for antioxidant function and is involved in immune system function and reproductive health.
(f) Iodine: Iodine is important for the synthesis of thyroid hormones, which regulate metabolism and growth.
(g) Cobalt: Cobalt is important for the synthesis of vitamin B12, which is essential for red blood cell production and nerve function.
(h) Chromium: Chromium is involved in carbohydrate metabolism and may play a role in improving insulin sensitivity.
(i) Molybdenum: Molybdenum is important for enzyme function and is involved in the metabolism of sulfur-containing amino acids (Fig. 4).

These microminerals must be provided in the animal's diet to support their health and well-being, and to ensure optimal growth and productivity.

Minerals cannot be added to the diet in their elemental forms and to be added as salts that are combined with other minerals (Table 4).

6 Zootechnical Feed Additives

6.1 Prebiotics

Prebiotics are nondigestible food ingredients that selectively promote the growth of beneficial microorganisms in the gut, such as bifidobacteria and lactobacilli. They serve as a substrate for the gut microbiota, promoting their growth and activity. Some examples of prebiotics include inulin, fructo-oligosaccharides (FOS), and galacto-oligosaccharides (GOS).

Prebiotics are often added to functional foods and dietary supplements and they have been shown to have health benefits such as improving gut health, boosting the immune system, and preventing certain diseases.

(a) **Inulin**—Inulin is one of the most commonly used and most effective prebiotics It is a soluble fiber. In farm animals, inulin is used as a feed additive for its prebiotic effects on the gut microbiota. Inulin supplementation in animal feed can improve gut health by promoting the growth of beneficial bacteria, such as bifidobacteria and lactobacilli, which can improve digestion and absorption of nutrients. This, in turn, can lead to improved feed efficiency, growth performance, and overall health of the animals.
Inulin can also be used to replace added sugars in animal feed, as it can have a sweetening effect and enhance palatability. Additionally, it can help to increase the viscosity of feed and reduce dustiness, improving handling and storage characteristics.
A common source of inulin is inulin is chicory root: one of the richest sources of inulin, with up to 60% inulin content, artichokes, garlic, onions, asparagus, etc.
It is intended as a source of soluble, fermentable fiber. It must not contain less than 90% of inulin on a dry matter basis (AAFCO 2021).

(b) **Fructo-oligosaccharides (FOS)**—FOS (fructo-oligosaccharides) are a type of prebiotic fiber derived from the breakdown of naturally occurring sugars such as fructose and glucose.
FOS (fructooligosaccharides) can be derived from a variety of sources, including:
   (i). Chicory root: One of the most commonly used sources of FOS, as it is a rich source of inulin, which can be hydrolyzed to produce FOS.
   (ii). Fruits: Some fruits, such as bananas, contain small amounts of FOS.
   (iii). Vegetables: Some vegetables, such as onions and garlic, also contain small amounts of FOS.
   (iv). Grains: Some grains, such as barley and wheat, can be used as a source of FOS.

Table 4 Approximate minerals requirements in different animal species

Species	Ca	P	K	Salt	S	Co	Cu	Fe	I	Mg	Mn	Se	Zn
	%					Ppm							
Swine	0.6	0.4	0.3	0.2	–	–	10	80	0.04	400	40	0.3	50
Dairy	0.5	0.4	0.7	0.25	0.2	0.1	10	100	0.10	600	20	0.3	40
Beef	0.4	0.3	–	0.5	–	0.1	6	–	–	–	10	0.3	–
Poultry	3.0	0.7	0.4	0.2	–	–	4	80	0.35	500	55	0.3	50
Sheep	0.4	0.3	–	0.5	0.1	0.07	5	–	0.39	600	–	0.3	100
All-species average	1.0	0.4	0.5	0.4	0.2	0.1	7	90	0.20	600	30	0.3	60

Total Diet Basis- Adapted from NAS-NRC recommendations
AAFCO (2021), *Ca* calcium, *P* Phosphorus, *K* Potassium, *S* Sulfur, *Co* Cobalt, *Fe* Iron, *I* Iodine, *Mg* Magnesium, *Mn* Manganese, *Se* Selenium, *Zn* Zinc

Synthesis: FOS can also be synthesized from fructose and glucose through chemical or enzymatic means.

The final product for FOS must contain a minimum of 80% fructo-ligosaccharide on a dry basis (AAFCO 2021)

(c) **Galacto-oligosaccharides (GOS)**—Galacto-oligosaccharides are a type of prebiotic fiber that is similar in structure to human milk oligosaccharides (HMOs). They are composed of short chains of galactose and glucose molecules and are naturally found in small amounts in some plant-based foods, such as legumes, lentils, and chickpeas.
GOS (galacto-oligosaccharides) are used in animal feed as a prebiotic ingredient to promote gut health and enhance overall health and performance. Some potential benefits of using GOS in animal feed include:

Improved gut health, enhanced digestion, reduced risk of disease: By promoting gut health, GOS can help to reduce the risk of digestive and other diseases in farm animals and improve immune functions.

6.2 Probiotics

Probiotics are live microorganisms that, when administered in adequate amounts, confer a health benefit to the host animal. Probiotics are often used in farm animal production to improve gut health and overall performance. Some of the commonly used probiotics in the United States for farm animals include:

(a) *Lactobacillus acidophilus*
(b) *Bifidobacterium* spp.
(c) *Streptococcus* spp.
(d) *Enterococcus* spp.
(e) *Bacillus subtilis*
(f) *Bacillus licheniformis*
(g) *Saccharomyces cerevisiae*
(h) *Aspergillus oryzae*

Each of these probiotics has been shown to have different beneficial effects on the animal's health and performance. For example, *Lactobacillus acidophilus* and *Bifidobacterium* spp. are commonly used to promote gut health and improve immune function, while *Saccharomyces cerevisiae* has been shown to improve feed efficiency and reduce the incidence of digestive disorders (Table 5).

6.3 Acidifiers

Acidifiers are a type of feed additive that are commonly used in animal nutrition to promote animal health and performance. They are organic acids or inorganic salts

Table 5 Probiotics, uses and its mechanism of action

Probiotic	Species	Use	Mechanism of action
Lactobacillus acidophilus	Poultry, swine, ruminants, equine	Promotes gut health and nutrient absorption, supports immune function, reduces pathogen colonization	Competitive exclusion, production of lactic acid and antimicrobial compounds, modulation of host immune response
Bifidobacterium spp.	Swine, ruminants	Improves gut health and nutrient absorption, reduces inflammation, and enhances immune function	Competitive exclusion, production of antimicrobial and anti-inflammatory compounds, modulation of host immune response
Streptococcus spp.	Poultry, swine, ruminants	Improves growth performance, supports immune function, reduces pathogen colonization	Competitive exclusion, production of lactic acid and antimicrobial compounds, modulation of host immune response
Enterococcus spp.	Poultry, swine	Improves growth performance, supports immune function, reduces pathogen colonization	Production of antimicrobial compounds, modulation of host immune response
Bacillus subtilis	Poultry, swine, ruminants	Improves growth performance, enhances immune function, reduces pathogen colonization	Production of antimicrobial compounds, modulation of host immune response, stimulation of digestive enzymes
Bacillus licheniformis	Poultry, swine	Improves growth performance, enhances immune function, reduces pathogen colonization	Production of antimicrobial compounds, stimulation of digestive enzymes
Saccharomyces cerevisiae	Poultry Swine Ruminants	Improves gut health and nutrient absorption, enhances immune function, improves rumen fermentation	Production of organic acids and B vitamins, stimulation of digestive enzymes, modulation of host immune response
Aspergillus oryzae	Cattle Poultry	Improves gut health and nutrient absorption, enhances immune function, improves rumen fermentation	Production of digestive enzymes and organic acids, modulation of host immune response

that are added to feed or drinking water to lower the pH of the animal's gastrointestinal tract, thereby creating an environment that is more conducive to beneficial microorganisms and less favorable to harmful bacteria. Some of the commonly used acidifiers in feed additives in the United States include:

(a) Formic acid
(b) Acetic acid
(c) Propionic acid

Table 6 Acidifiers, mechanism of action, and usage in species

Acid	Usage	Animal species	Mechanism of action
Formic acid	Preservation of feed, reduction of pathogen load, and improvement of growth performance	Poultry, swine, ruminants	Acidification of feed, reduction of bacterial load, improvement of gut health, and nutrient absorption
Acetic acid	Preservation of feed, reduction of pathogen load, and improvement of growth performance	Poultry, swine, ruminants	Acidification of feed, reduction of bacterial load, improvement of gut health, and nutrient absorption
Propionic acid	Preservation of feed, reduction of pathogen load, and improvement of growth performance	Ruminants, swine	Acidification of feed, reduction of bacterial load, improvement of rumen fermentation, and gut health
Butyric acid	Improvement of gut health and nutrient absorption, reduction of pathogen load, and enhancement of immune function	Poultry, swine	Stimulation of gut development, modulation of gut microflora, and improvement of gut barrier function
Lactic acid	Improvement of gut health and nutrient absorption, reduction of pathogen load, enhancement of immune function	Poultry, swine, ruminants	Acidification of feed, reduction of bacterial load, stimulation of digestive enzymes, and modulation of gut microflora
Citric acid	Preservation of feed, reduction of pathogen load, and improvement of growth performance	Poultry, swine	Acidification of feed, reduction of bacterial load, stimulation of digestive enzymes, improvement of gut health, and nutrient absorption

(d) Citric acid
(e) Lactic acid
(f) Fumaric acid
(g) Benzoic acid
(h) Phosphoric acid (Table 6)

Following are usage guidelines for some common acidifiers in the United States by AAFCO (2021).

Ammonium formate—Food additive ammonium formate may be safely used in the manufacture of complete pig feeds in the United States. It is intended for use as a feed acidifying agent to lower the pH in complete pig feed at levels not to exceed 1.2% of the complete feed.

To ensure safe use of the feed additive, formic acid, formate salts from all added sources cannot exceed 1.2% of complete feed when multiple sources of formic acid and its salts are used in combination.

Benzoic acid—This feed additive is used as a feed acidifying agent in complete pig feed at levels not to exceed 0.5% of the complete feed.

Formic acid—Formic acid is manufactured by heating carbon dioxide with sodium hydroxide under pressure and decomposing the resulting sodium formate with sulfuric acid. The additive is used as a feed acidifier to lower the pH in complete swine and poultry feed at levels not to exceed 1.2% of the complete feed. The additive should be free from methyl alcohol not to exceed 1000 ppm and moisture not to exceed 15%.

Feed grade sodium formate—The food additive feed grade sodium formate may be safely used in the manufacturing of complete swine and poultry feed in the United States. The additive is manufactured by the reaction of 99% formic acid with 50% sodium hydroxide in water to produce a solution made up of at least 20.5% sodium salt of formic acid and not more than 61% formic acid. The additive is used as an acidifying agent to lower the pH, in complete swine and poultry feed at levels not to exceed 1.2% of the complete feed.

6.4 Mycotoxin Detoxifier

Mycotoxins are toxic secondary metabolites produced by certain fungi, and they can contaminate food and feed, posing a risk to human and animal health. Mycotoxin detoxifiers are substances or compounds that can bind, degrade, or transform mycotoxins to reduce their toxic effects. In the United States, mycotoxin detoxifiers are classified as either feed additives or drugs, depending on their intended use and regulatory approval.

Mycotoxin binders are substances added to animal feed to reduce the toxic effects of mycotoxins. Mycotoxins are toxic secondary metabolites produced by certain fungi, and they can contaminate food and feed, posing a risk to human and animal health. Binder decontaminates mycotoxins in the feed by binding them strongly enough to prevent toxic interaction with consuming animals and to prevent mycotoxin absorption across the digestive tract. In the United States, mycotoxin binders are classified as feed additives and are regulated by the FDA.

There are several types of mycotoxin binders available in the United States, including clays, activated carbon, and yeast cell walls. Clays, such as bentonite and montmorillonite, are natural mineral products that can bind mycotoxins by electrostatic forces and hydrogen bonding. Activated carbon, also known as activated charcoal, is a form of carbon that has been treated with oxygen to make it more porous and increase its surface area. Activated carbon can bind mycotoxins by adsorption, which is the process of attracting and holding molecules to a surface. Yeast cell walls, such as those derived from *Saccharomyces cerevisiae*, can bind mycotoxins by ion exchange and physical adsorption.

They work by binding the mycotoxins in the animal's gastrointestinal tract, preventing their absorption and reducing their toxic effects. Mycotoxin binders are classified into different categories based on their chemical nature and mode of action.

Some of the common types of mycotoxin binders include clays, activated carbons (charcoal), and yeast cell walls.

Clays are natural mineral products that can bind mycotoxins by electrostatic forces and hydrogen bonding. They can be further classified as smectites or nonsmectites based on their crystal structure. Smectite clays, such as bentonite and montmorillonite, have a layered structure that allows them to expand and contract, which makes them more effective at binding mycotoxins. Nonsmectite clays, such as kaolin and zeolite, have a different crystal structure and are less effective at binding mycotoxins (Diaz et al. 2003).

Activated carbons are forms of carbon that have been treated with oxygen to make them more porous and increase their surface area. They can bind mycotoxins by adsorption, which is the process of attracting and holding molecules to a surface. Activated carbons can be further classified as nonpolar or polar based on their surface chemistry. Nonpolar-activated carbons are more effective at binding nonpolar mycotoxins, while polar-activated carbons are more effective at binding polar mycotoxins (Kumar and Singh 2015).

Yeast cell walls, such as those derived from *Saccharomyces cerevisiae*, can bind mycotoxins by ion exchange and physical adsorption. They contain various active components, such as mannans and beta-glucans, which can bind different types of mycotoxins (Magnoli et al. 2011).

In conclusion, mycotoxin binders are an important tool in reducing the toxic effects of mycotoxins in animal feed. Mycotoxin binders can be classified into different categories based on their chemical nature and mode of action. In the United States, mycotoxin binders are regulated by the FDA and are classified as feed additives. The types of mycotoxin binders available include clays, activated carbon, and yeast cell walls.

7 Sensory Feed Additives

7.1 Flavors and Sweeteners

Here are a wide variety of flavors and sweeteners available in the United States for use in food and beverage products. Flavors are substances that impart taste and aroma to food and beverages, while sweeteners are substances that provide a sweet taste. Some of the common flavors and sweeteners available in the United States include:

7.1.1 Flavors

(a) **Natural flavors**: derived from natural sources, such as fruits, vegetables, and herbs.
(b) **Artificial flavors**: created through chemical processes to mimic the taste and aroma of natural flavors.
(c) **Organic flavors**: made from organic ingredients and processed according to organic standards (USDA 2019).

(d) **Non-GMO flavors**: made from ingredients that have not been genetically modified.

7.1.2 Sweeteners

(a) **Sugar**: the most common sweetener, derived from sugar cane or sugar beets.
(b) **High fructose corn syrup (HFCS):** a corn-based sweetener that is widely used in processed foods and beverages.
(c) **Artificial sweeteners**: low-calorie or zero-calorie sweeteners that are used as sugar substitutes, such as aspartame, sucralose, and stevia (FDA 2018).
(d) **Natural sweeteners**: derived from natural sources, such as honey, agave, and maple syrup.
(e) **Sugar alcohols**: low-calorie sweeteners that occur naturally in some fruits and vegetables, such as xylitol and erythritol.
(f) The choice of flavor and sweetener depends on the specific application, consumer preferences, and regulatory requirements (Table 7).

7.2 Color Additives

In the United States, the use of coloring agents as feed additives in livestock and poultry diets is regulated by the Food and Drug Administration (FDA) under the Federal Food, Drug, and Cosmetic Act. The FDA allows the use of certain approved coloring agents in animal feed, but only for the purpose of improving the appearance of animal products, such as meat and eggs. The use of coloring agents for any other purpose, such as for nutritional benefits or for growth promotion, is not allowed. Additionally, the FDA requires that all coloring agents used in animal feed are safe for consumption and properly labeled (Table 8).

7.3 Pigments

Pigment feed additives are a type of feed additive that are used in animal nutrition to enhance the color of meat, egg yolks, and skin or feathers. They are commonly used in poultry and swine diets to improve the visual appeal and market value of animal products.

Some of the commonly used pigment feed additives in the United States include:

(a) Carotenoids (e.g., astaxanthin, lutein, and zeaxanthin)
(b) Canthaxanthin
(c) Riboflavin
(d) Turmeric
(e) Paprika

Carotenoids are the most commonly used pigments in animal nutrition and can be found naturally in many plant-based foods. Canthaxanthin, on the other hand, is a

Table 7 Common flavoring agents for feed in United States (AAFCO 2021)

Common name	Botanical name	Limitation
Aloe	*Aloe perryi, Aloe barbadensis, Aloe ferox, Aloe africana, Aloe spicata*	
Althea root and flowers	*Althea officinalis*	
Amyris (West Indian sandalwood)	*Amyris balsamifera* L.	
Artemisia (wormwood)	*Artemisia* spp.	Finished food thujone free[a]
Benzoin resin	Styrax bezoin Dryander, Styrax paralleloneurus, Styrax tonkinensis	
Blackberry bark	Rubus, section Eubatus	
Boronia flowers	Barosma betulina Bartl et Wendl. *Barosma crenulata* (L.) Hook. Or Barosma serratifolia wild	
Cajeput	*Melaleuca leucadendron* L	
Camphor tree	*Cinnamomum camphora*	Safrole free
Cascara sagrada	*Rhamnus purshiana* DC	
Cassie flowers	Acacia farnesiana(L.) wild	
Castor oil	Ricinus communis L.	
Catechu, black	Acacia catechu wild	
Cedar, white (Arborvitae), leaves and twigs	Thuja occidentalis L.	Finished food thujone free
Cherry pits	*Prunus avium* L or *Prunus cerasus* L.	Not to exceed 25 ppm prussic acid
Cherry-laurel leaves	*Prunus laurocerasus* L.	Not to exceed 25 ppm prussic acid
Chestnut leaves	Castanea dentata	
Dill, Indian	Anethum sowa Roxb.	
Dragon's blood	Daemonorops spp.	
Eucalyptus globulus leaves	Eucalyptus globulus Labill	
Fir "Pine" needles and twigs	Abies sibirica Ledeb. Abies alba, A. sachalinensis, A. mayrina	
Gambir	Uncaria gambir Roxb	
Genet flowers	*Spartium junceum*	
Gentian rhizome	*Gentiana lutea* L.	
Hyacinth flowers	*Hyacinthus orientalis* L.	
Lungmoss (lungwort)	Sticta pulmonacea ach	
Maple, mountain	Acer spicatum Lam	
Mimosa (black wattle)	Acacia decurrens Wild.var. dealbata	
Oak, white, chips	*Quercus alba* L.	
Passion flower	*Passiflora incarnata* L.	
Pennyroyal, American	*Hedeoma pulegioides* L	

(continued)

Table 7 (continued)

Common name	Botanical name	Limitation
Pennyroyal, European	*Mentha pulegium*	
Pine dwarf, needles, and twigs	*Pinus mugo*	
Quassia	*Picrasma excelsa*	
Rhubarb root	Rheum officinale	
Sandalwood, white, yellow, or East Indian	Santalum album L.	
Tagetes (Marigold)	*Tagetes patula* and other spp.	
Turpentine	Pinus palustris mill other pinus spp	
Walnut husks, hulls, leaves, and green nuts	*Juglans nigra* L.	
Yucca santa	*Eriodictyon californicum*	
Yucca Joshua tree	*Yucca brevifolia*	
Yucca, Mohave	Yucca schidigera Roezl	

Table 8 Some common color additives in United States

Additive	Classification	Limitation
FD&C blue no – 1, 2, green no – 3, red no – 3, 40, yellow no – 5, 6	Color additive	As per GMP
Astaxanthin	Color additive for orange to red color	As per GMP, for fish use. Not exceed 72 gm per ton of finished feed
Caramel	Color additive for brown color	As per GMP
Carrot oil	Color additive for orange color	As per GMP, carrot oil should not contain more than 25 ppm of hexane.
Cochineal extract	Color additive	As per GMP
Corn endosperm oil	Color additive	Chicken feed
Dehydrated beets	Color additive	As per GMP
Paprika	Color additive	As per GMP
Saffron	Color additive	As per GMP
Synthetic iron oxide	Color additive	As per GMP
Tagetes (Aztec Marigold)	Color additive	As per GMP
Titanium Dioxide	Color additive	As per GMP
Tomato lycopene	Color additive	As per GMP
Turmeric	Color additive	As per GMP
Ultramarine blue	Color additive	As per GMP
Beta carotene	Color additive	As per GMP

synthetic pigment that is commonly used in salmonid diets to improve the color of salmon flesh.

It is important to note that the use of pigment feed additives in animal feed is subject to regulations by the Food and Drug Administration (FDA) and the

Table 9 Pigment and their limit in various animal species

Pigment	Approved usage	Maximum limit	Animal species
Beta-carotene	Coloration	15 mg/kg	Poultry
Canthaxanthin	Coloration	8 mg/kg	Poultry
Capsanthin	Coloration	10 mg/kg	Poultry
Lutein	Coloration	40 mg/kg	Poultry
Xanthophylls	Coloration	40 mg/kg	Poultry
Astaxanthin	Coloration	25 mg/kg	Salmonid fish
Canthaxanthin	Coloration	50 mg/kg	Salmonid fish
Lutein coloration	Coloration	30 mg/kg	Salmonid fish
Zeaxanthin	Coloration	20 mg/kg	Salmonid fish
Apo-ester	Coloration	100 mg/kg	Egg-laying chicken

Association of American Feed Control Officials (AAFCO), and should be used in accordance with approved labeling and dosage guidelines.

Following are some pigment and their limit in few animal species (Table 9).

Overall, pigment feed additives are a valuable tool in animal nutrition that can help to improve the visual appeal and market value of animal products, while also supporting animal health and performance.

8 Coccidiostats and Histomonostats

Coccidiostats and histomonostats are feed additives that are used to control specific diseases in poultry, such as coccidiosis and histomoniasis.

Some common coccidiostats and histomonostats feed additives in the United States include:

(a) Amprolium: A coccidiostat that acts by interfering with the uptake of thiamine by the coccidian parasite.
(b) Decoquinate: A coccidiostat that works by blocking the electron transport system in the coccidian parasite.
(c) Lasalocid: A coccidiostat that affects the transport of ions across the cell membrane of the coccidian parasite.
(d) Monensin: A coccidiostat that interferes with the coccidian parasite's ability to transport ions across cell membranes.
(e) Narasin: A coccidiostat that affects the coccidian parasite's ability to produce energy by disrupting the electron transport system.
(f) Nitarsone: A histomonostat that is used to control histomoniasis in poultry by killing the causative agent, Histomonas meleagridis.

The use of coccidiostats and histomonostats is regulated by the FDA, and these additives are subject to strict withdrawal times to ensure that residues do not persist

Table 10 Common coccidiostats and histomonostats used in United States

Additives	Mode of action	Use	Withdrawal time
Amprolium	Thiamine uptake	Coccidiosis	24 hours
Decoquinate	Electron transport	Coccidiosis	5 days
Lasalocid	Ion transport	Coccidiosis	5 days
Monensin	Ion transport	Coccidiosis	5 days
Narasin	Electron transport	Coccidiosis	5 days
Nitarsone	Histomonas meleagridis	Histomoniasis	5 days for meat

in animal products. Proper management and biosecurity practices are also important for controlling coccidiosis and histomoniasis in poultry.

Here is a table of some common coccidiostats and histomonostats used in the United States, along with their withdrawal times (Table 10):

It is important to note that withdrawal times may vary depending on factors such as dosage and the specific animal species being treated. It is essential for animal producers and veterinarians to adhere to withdrawal times to prevent residues from these additives in animal products and ensure food safety.

9 Other Feed Additives

9.1 Anti-Oxidants

Antioxidants are natural or synthetic compounds that can prevent or repair damage to cells and tissues caused by free radicals, which are unstable molecules that can harm cells and contribute to the development of various diseases. Some common antioxidants found in animals include vitamin C, vitamin E, beta-carotene, selenium, and polyphenols. These antioxidants can be obtained through a variety of foods, including fruits, vegetables, and grains, as well as through dietary supplements formulated for animals. Antioxidants are believed to play an important role in promoting the health and longevity of animals.

Antioxidants can be classified into several groups based on their chemical structure and properties.

The following is a brief overview of some of the major classes of antioxidants:

Vitamins: Vitamins such as vitamin C (ascorbic acid), vitamin E (tocopherols and tocotrienols), and vitamin A (beta-carotene) are important antioxidants that can be obtained through diet. They help protect cells from oxidative damage by scavenging free radicals and reactive oxygen species (ROS) in the body (Foyer and Noctor 2005).

Carotenoids: Carotenoids are a class of pigments found in plants, algae, and some bacteria. Some examples of carotenoids include lycopene, zeaxanthin, and astaxanthin. Carotenoids act as antioxidants by neutralizing ROS and protecting cells from oxidative damage (Biesalski et al. 2009).

Flavonoids: Flavonoids are a group of polyphenolic compounds found in many fruits, vegetables, and herbs. Some common flavonoids include quercetin, kaempferol, and catechins. Flavonoids have been shown to exhibit antioxidant, anti-inflammatory, and anti-cancer properties (Zafra-Stone et al. 2007).

Enzymatic antioxidants: Enzymatic antioxidants are naturally occurring enzymes in the body that can neutralize ROS and prevent oxidative damage. Examples include superoxide dismutase, catalase, and glutathione peroxidase (Foyer and Noctor 2005).

Synthetic antioxidants: Synthetic antioxidants are compounds that are not naturally occurring in the body and are added to food or supplements to prevent oxidation. Some common synthetic antioxidants include butylated hydroxyanisole (BHA) and butylated hydroxytoluene (BHT) (Biesalski et al. 2009).

These are just a few examples of the many different types of antioxidants. The classification of antioxidants can be complex and there is ongoing research to better understand their mechanisms of action and potential health benefits.

Antioxidants offer several potential benefits to farm animals. Some of these benefits include:

Improved immune function: Antioxidants can help support the immune system by neutralizing harmful free radicals and reducing oxidative stress. This can improve the animal's ability to fight off infections and diseases (Mumpton and Hanninen 2011).

Reduced oxidative stress: Antioxidants can reduce the level of oxidative stress in farm animals, which can help prevent or mitigate the negative effects of various stressors such as heat stress, transportation, and weaning (Gupta and Sharma 2014).

Better growth performance: Antioxidants can improve growth performance in farm animals, which can lead to better meat or milk production. For example, studies have shown that supplementing with vitamin E and selenium can improve weight gain and feed conversion efficiency in pigs (Kim et al. 2011) and lambs, respectively.

Improved reproductive performance: Antioxidants can also improve reproductive performance in farm animals. For example, supplementing with vitamin E and selenium has been shown to improve semen quality and fertility in boars (Brouwer et al. 2002) and bulls (Ganaie et al. 2017), respectively.

One powerful synthetic antioxidant is made from chelated zinc and curcumin for use in animal feed. CurCos-Zn is a proprietary complex developed by RDL. Chelated zinc is a highly bioavailable form of zinc that is easily absorbed and utilized by animals. Zinc is an essential mineral that plays a critical role in numerous physiological processes, including immune function, skin and coat health, and digestive health.

The combination of chelated zinc and curcumin in CurCos-Zn provides a powerful tool for supporting animal health and well-being. The chelated zinc ensures that animals receive an optimal level of bioavailable zinc, while the curcumin provides additional health benefits through its antioxidant and anti-inflammatory properties. This makes CurCos-Zn an excellent choice for use in animal feed, particularly for supporting immune function and reducing inflammation.

Similarly, there is one novel antioxidant, which is made from chitosan and ovo-transferrin to help support a healthy gut, ADEPPT. It supports antioxidant mechanisms and offers additional prebiotic support to good bacteria in the gut.

Overall, antioxidants can play an important role in promoting the health and productivity of companion and farm animals.

9.2 Enzymes

Enzymes are proteins made up of amino acids or their derivatives, which catalyzes a defined chemical reaction. Enzymes are widely used in to animal food industry. It helps to improve animal performance and production economics through a variety of mechanisms. It breaks down substrates in feed and reduces the antinutritional substances, such as hydrolysis of phytic acids to improve phosphorus availability to reduce the antinutritional factors that cause performance losses in the structure of feed raw materials, and to improve performance. In the United States, feed enzymes are regulated at both the federal and state levels. At the federal level US Food and Drug Administration (FDA) Canter for Veterinary Medicine (CVM) Division on Animal Feed regulates animal food and each US state maintains and upholds its own state feed law and regulation unless the enzyme is generally recognized as safe (GRAS) for an intended use in accordance of federal law. The American Feed Control Officials (AAFCO) maintains a list of enzymes and source organisms that CVM accepts for use in animal feed.

Enzymes may be classified as per substrate they hydrolyze.

9.2.1 Carbohydrases

Carbohydrases are enzymes that catalyze the breakdown of carbohydrates into simpler sugars or other carbohydrates. They are commonly used in various industrial applications, including food and feed processing, biofuel production, and pharmaceuticals. The classification of carbohydrases is based on the type of reaction they catalyze and the substrate they act upon. Some of the major classes of carbohydrases include:

(a) **Amylases**: catalyze the hydrolysis of starch and glycogen into simpler sugars, such as glucose, maltose, and maltotriose (Garg et al. 2016).
(b) **Cellulases**: catalyze the hydrolysis of cellulose into glucose and other simple sugars (Bhat 2000).
(c) **Xylanases**: catalyze the hydrolysis of xylan, a major component of plant cell walls, into xylose and other simple sugars (Bhat 2000).
(d) **Pectinases**: catalyze the hydrolysis of pectin, a component of plant cell walls, into simple sugars and other by-products (Bhat 2000).
(e) **Lactases**: catalyze the hydrolysis of lactose, a disaccharide found in milk, into glucose and galactose (Bhatia et al. 2002).
(f) **Invertases**: catalyze the hydrolysis of sucrose into glucose and fructose (Bhatia et al. 2002).

Carbohydrases are widely used in the food and feed industry to improve the nutritional value and processing characteristics of various products. They are also used in biofuel production to break down complex carbohydrates into simple sugars that can be fermented into ethanol or other biofuels.

9.2.2 Lipases

Lipases are a group of enzymes that catalyze the hydrolysis of fats and oils into glycerol and fatty acids. They are widely used in various industrial applications, including food and feed processing, biofuel production, and pharmaceuticals. The classification of lipases is based on the type of reaction they catalyze and the substrate they act upon. Some of the major classes of lipases include:

(a) Carboxyl ester lipases (CEL) catalyze the hydrolysis of carboxylic ester bonds, such as those found in triglycerides and phospholipids (Hussain et al. 2016).
(b) Secretory lipases catalyze the hydrolysis of dietary fats and oils in the stomach and small intestine (Lowe 2002).
(c) Extracellular lipases catalyze the hydrolysis of fats and oils in the extracellular environment, such as in soil and aquatic ecosystems (Bornscheuer 2002).
(d) Intracellular lipases catalyze the hydrolysis of fats and oils within cells, such as in adipose tissue and liver cells (Bornscheuer 2002).
(e) Fungal lipases catalyze the hydrolysis of fats and oils, and are widely used in various industrial applications (Bhange et al. 2011).

Lipases are widely used in the food and feed industry to improve the nutritional value and processing characteristics of various products. They are also used in biofuel production to break down complex lipids into simpler fatty acids that can be converted into biodiesel or other biofuels.

9.2.3 Proteases

Proteases, also known as peptidases or proteolytic enzymes, are a group of enzymes that catalyze the hydrolysis of peptide bonds in proteins, leading to their breakdown into smaller peptides or amino acids. They are widely used in various industrial applications, including food and feed processing, biofuel production, and pharmaceuticals. The classification of proteases is based on the type of reaction they catalyze and the substrate they act upon. Some of the major classes of proteases include:

(f) **Serine proteases** catalyze the hydrolysis of peptide bonds using a serine residue in the active site, and include enzymes such as trypsin and chymotrypsin (Rawlings et al. 2018).
(g) **Cysteine proteases** catalyze the hydrolysis of peptide bonds using a cysteine residue in the active site, and include enzymes such as papain and bromelain (García-Carreño et al. 1993).

(h) **Aspartic proteases** catalyze the hydrolysis of peptide bonds using an aspartic acid residue in the active site, and include enzymes such as pepsin and renin (Rawlings et al. 2018).
(i) **Metalloproteases** catalyze the hydrolysis of peptide bonds using a metal ion, such as zinc or calcium, in the active site, and include enzymes such as matrix metalloproteases and carboxypeptidases (Rawlings et al. 2018).

Proteases are widely used in the food and feed industry to improve the nutritional value and processing characteristics of various products. They are also used in biofuel production to break down proteins into smaller peptides and amino acids that can be used as a nutrient source for microorganisms.

9.2.4 Oxidoreductase

Oxidoreductases are a class of enzymes that catalyze oxidation-reduction reactions, involving the transfer of electrons between molecules or ions. They are involved in many metabolic processes, including energy production, biosynthesis, and detoxification. The classification of oxidoreductases is based on the type of reaction they catalyze and the electron donor and acceptor molecules involved. Some of the major classes of oxidoreductases include:

(j) Dehydrogenases catalyze the transfer of hydride ions (H-) from a substrate to an electron acceptor, such as NAD+ or FAD. Examples of dehydrogenases include lactate dehydrogenase and alcohol dehydrogenase (Stryer 1995).
(k) Oxidases catalyze the transfer of electrons from a substrate to an oxygen molecule, producing hydrogen peroxide or water as a by-product. Examples of oxidases include cytochrome c oxidase and xanthine oxidase (Stryer 1995).
(l) Peroxidases catalyze the transfer of electrons from a substrate to a peroxide molecule, producing water or a different peroxide as a by-product. Examples of peroxidases include catalase and glutathione peroxidase (Stryer 1995).
(m) Reductases catalyze the transfer of electrons from an electron donor, such as NADH or FADH2, to a substrate. Examples of reductases include succinate dehydrogenase and NADPH-cytochrome P450 reductase (Stryer 1995).
(n) Oxidoreductases are widely used in various industrial applications, including food and feed processing, biofuel production, and pharmaceuticals. They are also involved in various physiological processes, including metabolism and energy production.

9.2.5 Phosphatases

Phosphatases are a class of enzymes that catalyze the hydrolysis of phosphoric acid ester bonds, leading to the removal of a phosphate group from a substrate molecule. They play a crucial role in various physiological processes, including signal transduction, energy metabolism, and nucleic acid synthesis. The classification of phosphatases is based on the type of substrate they act upon and the specific mechanism of action. Some of the major classes of phosphatases include:

(o) Acid phosphatases catalyze the hydrolysis of phosphate ester bonds in acidic environments, such as lysosomes and endosomes. Examples of acid phosphatases include tartrate-resistant acid phosphatase and prostatic acid phosphatase (Van Etten 1990).
(p) Alkaline phosphatases catalyze the hydrolysis of phosphate ester bonds in alkaline environments, such as the small intestine and bone tissue. Examples of alkaline phosphatases include intestinal alkaline phosphatase and tissue-nonspecific alkaline phosphatase (Millán 2013).
(q) Protein tyrosine phosphatases catalyze the dephosphorylation of tyrosine residues in proteins and play a crucial role in signal transduction pathways. Examples of protein tyrosine phosphatases include receptor protein tyrosine phosphatases and nonreceptor protein tyrosine phosphatases (Tonks 2006).
(r) Phosphatases are widely used in various industrial and medical applications, including enzyme assays, diagnostics, and therapeutics. They are also involved in various physiological processes, including metabolism and energy production.

Microbial enzymes are produced from nonpathogenic and nontoxicogenic strains (Table 11).

Novel Enzyme Blends: Novel enzyme blends are a type of enzyme formulation that is designed to address specific challenges in animal nutrition, such as improving feed efficiency, increasing nutrient availability, and enhancing overall animal health. These blends are typically made up of a combination of various enzymes, each with a specific function, and are used in the production of animal feed. One example of a novel enzyme blend is RDL's Zymace, which is designed to improve the digestibility of feed in dairy cows. R&D Lifesciences is a leading manufacturer of science-driven ingredients for the animal feed industry. Zymace contains a blend of enzymes, produced from direct fed *Aspergillus oryzae* fermentation extract including proteases, amylases, xylanases, and cellulases, as well as prebiotics that support gut health. Studies have shown that adding Zymace to dairy cow feed can improve milk production by up to 4.4 pounds per day per cow, while also reducing feed costs and improving feed efficiency.

The enzymes in Zymace are derived from natural sources and are carefully selected and blended to provide a targeted effect on specific feed components.

Zymace is just one example of the innovative and science-driven products offered by RDL, a company committed to developing high-quality, evidence-based ingredients to support the animal feed industry.

9.3 Binders

Binders are a type of feed additive that are commonly used in animal nutrition to improve the quality and consistency of feed pellets or granules. They are generally made from natural or synthetic materials that have a high binding capacity, such as clays, starches, and proteins.

Table 11 Allowed list of some enzymes and their typical substrate in the United States

Classification/ name	Source organism	Typical substrate	Function
Carbohydrases			
Alpha-amylase	*Animal pancreatic tissue, Aspergillus niger Aspergillus oryzae, Bacillus amyloliquefaciens, Bacillus licheniformis, Bacillus subtilis, Rhizopus neveus*	Corn silage, corn, corn feed meal, corn gluten feed, soybean meal, wheat, wheat middlings, barley, grain sorghum, pea, oat, tapioca, millet, rice	Hydrolyzes starch
Maltogenic alpha-amylase	*Bacillus subtilis* containing a *Bacillus stearothermophilus* gene for maltogenic alpha-amylase	See alpha-amylase	Hydrolyzes starch with production of maltose
Beta-amylase	Barley malt	See alpha-amylase	Hydrolyzes starch with production of maltose
Cellulase	*Aspergillus niger,* var. Humicola insolens *Trichoderma longibrachiatum* (also known as *T. reesei* or *T. viride*)	Corn, barley, wheat, wheat bran, rye, grain sorghum	Break down cellulose
Alpha galactosidase	*Aspergillus niger, Mortierella vinacea, Saccharomyces* sp.	Sweet lupin, soybean meal	Hydrolyzes oligosaccharides
Beta-glucanase	*Aspergillus niger, Aspergillus aculeatus, Bacillus lentus, Bacillus subtilis, Humicola insolens, Paenibacillus lentus, Talaromyces funiculosus*	Wheat, barley, canola meal, wheat by-product, oat groats, rye, triticale, grain sorghum.	Hydrolyzes beta-glucans, a type of nonstarch polysaccharide
Beta-glucosidase	*Aspergillus niger*, var.	Plant cell wall constituents	Hydrolyzes cellulose starch with production of glucose
Hemicellulase	Aspergillus aculeatus, *Aspergillus niger, Bacillus lentus, Bacillus subtilis, Paenibacillus lentus*	Corn, soybean meal, guar meal, barley, rye, grain, sorghum, wheat, oats, peas, lentils	Break down hemicellulose
Invertase	*Aspergillus niger* var. *Saccharomyces* sp.	Sucrose-containing products and by-products	Hydrolyzes sucrose to glucose and fructose
Lactase	*Aspergillus niger*, var. *Aspergillus oryzae*, var. *Candida pseudotropicalis* kluyveromyces	Lactose-containing products and by-products	Hydrolyzes lactose to glucose and galactose

(continued)

Table 11 (continued)

Classification/ name	Source organism	Typical substrate	Function
Beta-mannanase	*Aspergillus niger*, var. *Bacillus lentus*, *Paenibacillus lentus*	Corn, soybean meal, guar, meal, copra meal	Hydrolyzes beta-mannans, a component of hemicellulose
	Bacillus subtilis	Distillers dried grains with solubles	Hydrolyzes beta-mannans, a component of hemicellulose
Pectinase	*Aspergillus aculeatus*, *Aspergillus niger*, *Rhizopus oryzae*	Corn, wheat	Break down pectin
Xylanase	*Aspergillus niger*, var. *Aspergillus oryaze* expressing a Thermomyces lanuginosus xylanase gene, *Bacillus lentus*, *Bacillus subtilis* var. *Humicola insolens*, *Paenibacillus lentus*, *Talaromyces funiculosus*, *Trichoderma longibrachiatum*	Corn, barley, rye, wheat, grain sorghum, triticale, oats	Hydrolyzes xylans, a component of hemicellulose
Oxidoreductases			
Catalases	*Aspergillus niger*, var. *Micrococcus lysodeikticus*	Hydrogen peroxide	Produces water and oxygen from hydrogen peroxide
Glucose oxidase	*Aspergillus niger*, var.	Glucose	Degrades glucose to hydrogen peroxide and gluconic acid
Phosphatases	*Aspergillus niger*, var. *Aspergillus oryzae*, var. expressing the *Peniophora lycii* phytase gene phytase canola (*Brassica napus* expressing the *Aspergillus niger* phytase gene) *Pichia pastoris* expressing a phytase from a risk group 1 *Escherichia coli* Schizosaccharomyces pombe expressing an *Escherichia coli* strain B phytase gene		Hydrolyzes phytate and increase the digestibility of phytin bound phosphorus in poultry and swine diets.

(continued)

Table 11 (continued)

Classification/ name	Source organism	Typical substrate	Function
	Trichoderma reesei expressing an altered phytase gene from a risk group 1 *Escherichia coli* *Trichoderma reesei* expressing an altered phytase gene from a *Buttiauxella* sp.		
	Talaromyces funiculosus		Hydrolyzes phytate and increases the digestibility of phytin-bound phosphorus in poultry and swine diets.

Some of the commonly used binders in feed additives in the United States include:

(a) Bentonite clay
(b) Lignin sulfonate
(c) Guar gum
(d) Soy protein concentrate
(e) Corn gluten meal
(f) Wheat gluten
(g) Pectin

Binders can help to reduce feed waste and improve animal performance by promoting better feed intake and digestibility. They can also help to minimize the presence of fine particles or dust in the feed, which can lead to respiratory issues in animals.

Overall, binders are a useful tool in animal nutrition that can help to improve feed quality, reduce waste, and support animal health and performance.

Some of the common binders and anti-caking agents given by AAFCO are listed here.

(a) **Lignin sulfonate**—it is either one or a combination of ammonium calcium, magnesium, or sodium salt of the extract of spent sulfite liquor derived from the sulfite digestion of wood of abaca (*Musa textilis*) or Sisal (*Agave sisalana*) either liquid or dry form. It is being used as a pelleting aid in an amount not to exceed 4% of the finished pellet. It can also be used as a binding aid in liquid form in the flaking of feed used in feeds as liquid lignin sulfonate, in an amount not exceeding 11% of the molasses.

(b) **Sodium salt of fatty acids**—It is obtained by the neutralization of feed grade vegetable origin free fatty acids, or saponification of vegetable oil, or combination. Sodium hydroxide is used in the neutralization or saponification reactions. The resulting sodium salt is used as binder and/or lubricant in the pelleted and flaked feed as is basis not to exceed 5.5 lb./ton.

(c) **Potassium salt of fatty acids**—Here potassium hydroxide is used in the neutralization or saponification reactions. Potassium salt is used as a binder or lubricant in the pelleted feed and flaked feed. It has to be used in as-is basis not to exceed 15.5 lb./ton in animal feed.

(d) **Bentonite**—It is a naturally occurring mineral consisting of aluminum silicate and montmorillonite. It may contain calcium or sodium as the predominant available or exchange ion. It can be used as an anti-caking agent and pelleting aid in the amount not to exceed 2% of the total ration. It is not prohibited in medicated animal feed for the same purposes and at the same levels when it can be demonstrated that it does not interfere with the bioavailability of the medicament to animals and the analysis of feed.

(e) **Sodium bentonite**—It is a naturally occurring mineral consisting primarily of the trilayered hydrous aluminum silicate, montmorillonite characterized by a sodium exchange or available ion content of not less than 1% and not more than 2% of the air-dried material. It is intended to be used in nonmedicated animal feed as an anticaking agent and pelleting aid in an amount not exceeding 2% of the total ration. To reduce seepage in silage amount would not exceed 1% sodium bentonite.

(f) **Castor oil**—It is triglyceride obtained by extraction of oils from seeds of the castor bean plant, *Ricinus communis*. It consists predominantly of triglyceride ester of fatty acids. Ricinoleic acid should be guaranteed for not less than 87%. Castor oil may be safely used as an anti-caking agent, a releasing agent, and as a diluent in animal feeds at levels not exceeding 250 ppm in complete feed.

(g) **Pyrophyllite**—It is an aluminum silicate monohydrate and may be safely used as the sole anti-caking aid, blending agent, pelleting aid, and carrier in animal feed in an amount not to exceed 2% in complete animal feed.

(h) **Silicon dioxide**—The food additive silicon dioxide is intended to be used in feed components as an anti-caking agent and/or grinding aid as well. In no case, the amount should be exceeded by 2% by the weight of the finished feed.

(i) **Perlite**—It is a powdered form of a glassy volcanic rock, consisting essentially fused sodium potassium aluminum silicate. It is used as a filter aid or pressing aid in the processing of food and feed ingredients and may also use as an anti caking agent. It may not exceed 4% by weight of the product in which it is present as a processing aid (Table 12).

Table 12 Some common binders their approved usage, and limits in animal species in United States (AAFCO 2021)

Binder/thickening agent	Approved usage	Maximum limit	Animal species
Agar	Thickener	Not specified	Poultry, swine, ruminants
Alfalfa meal	Binder	Not specified	Poultry, swine
Bentonite	Anti-caking agent, mycotoxin binder	2%	Poultry, swine, ruminants
Calcium alginate	Pelleting aid, feed binder	Not specified	Poultry, swine, ruminants
Calcium carbonate	Anti-caking agent, pellet binder	Not specified	Poultry, swine, ruminants
Calcium lignosulfonate	Pelleting aid, feed binder	Not specified	Poultry, swine, ruminants
Carboxymethylcellulose (CMC)	Pelleting aid, feed binder	1.5%	Poultry, swine, ruminants
Carrageenan	Thickener	Not specified	Poultry, swine, ruminants
Gelatin	Binder, emulsifier	Not specified	Poultry, swine, ruminants
Guar gum	Thickener, binder	Not specified	Poultry, swine, ruminants
Lignin sulfonate	Pelleting aid, feed binder	Not specified	Poultry, swine, ruminants
Magnesium silicate	Anti-caking agent, mycotoxin binder	2%	Poultry, swine, ruminants
Maltodextrin	Binder, energy source	Not specified	Poultry, swine, ruminants
Molasses	Binder, energy source	Not specified	Poultry, swine, ruminants
Pectin	Thickener	Not specified	Poultry, swine, ruminants
Soy protein concentrate	Pellet binder, emulsifier	Not specified	Poultry, swine, ruminants
Wheat flour	Binder	Not specified	Poultry, swine

10 Conclusion

Feed additives are an important component of animal nutrition in the United States. They serve a variety of purposes, including improving feed efficiency, promoting growth and development, maintaining good health, and enhancing animal performance. From prebiotics and probiotics to enzymes and amino acids, there are a wide variety of feed additives available to meet the specific nutritional needs of different species and production systems. Amino acids, enzymes, mycotoxin binders, and color additives are also essential feed additives that play a significant role in enhancing animal nutrition and health in the United States. Amino acids are the building blocks of protein and must be supplemented in animal diets to meet their nutritional requirements. Enzymes improve nutrient absorption and utilization, particularly in nonruminant animals. Mycotoxin binders help reduce the negative effects of mycotoxins on animal health and productivity, while color additives improve the visual appeal of animal feed. Flavor additives play a crucial role in the feed industry by enhancing the palatability and acceptance of animal feed, thereby promoting better feed intake and improving animal performance. With the increasing demand for high-quality feed and the need for better animal nutrition, the use of flavor additives has become more important than ever before. The incorporation of flavors can mask unpleasant tastes and smells, encourage feed intake, and reduce stress in animals. As a result, they can contribute to better health and well-being of animals, as well as higher productivity and profitability for producers.

These additives can help maintain optimal animal health, growth, and productivity, leading to better quality animal products and overall profitability in the animal production industry.

Through ongoing research and development, the industry continues to innovate and improve the effectiveness and sustainability of feed additives, ultimately contributing to the production of safe, affordable, and high-quality animal protein for consumers.

References

Biesalski HK, Dragsted LO, Elmadfa I et al (2009) Bioactive compounds: definition and assessment of activity. Nutrition 25(11–12):1202–1205

Brouwer F, Beerda B, Hoving-Bolink A et al (2002) Effect of organic selenium supplementation on the distribution of selenium in boar semen and its effects on sperm motility. Anim Reprod Sci 74(1–2):43–50

FDA (2021a) Food Additives & Ingredients. https://www.fda.gov/food/food-ingredients-packaging/food-additives-ingredients

FDA (2021b) Animal & Veterinary. https://www.fda.gov/animal-veterinary

Foyer CH, Noctor G (2005) Redox homeostasis and antioxidant signaling: a metabolic interface between stress perception and physiological responses. Plant Cell 17(7):1866–1875

Ganaie AH, Sharma RK, Singh AK et al (2017) Effect of vitamin E and selenium supplementation on semen quality and reproductive performance of crossbred bulls. Theriogenology 94:72–77

Global Market Insights (2019)

Gupta A, Sharma RK (2014) Role of antioxidants in health and production of livestock. Vet World 7(11):946–951

Kim SW, Mateo RD, Yin YL et al (2011) Effects of dietary supplementation of antimicrobial peptide-A3 and Ovotransferrin on growth performance, nutrient digestibility, intestinal microbiota, and intestinal morphology of weanling pigs challenged with Escherichia coli K88. J Anim Sci 89(12):4141–4149

Mordor Intelligence (2021) USA feed additives market - growth, trends, COVID-19 impact, and forecasts (2021–2026)

Mumpton FA, Hanninen OO (2011) Antioxidants in animal nutrition. In: Antioxidants in food, vitamins and supplements: prevention and treatment of disease. Academic Press, Cambridge, pp 631–651
Zafra-Stone S, Yasmin T, Bagchi M, Chatterjee A, Vinson JA, Bagchi D (2007) Berry anthocyanins as novel antioxidants in human health and disease prevention. Mol Nutr Food Res 51(6): 675–683

Mycotoxin Detoxifier References

Diaz DE, Hagler WM Jr, Hopkins BA, Whitlow LW (2003) Effect of clay type on in vitro deactivation of aflatoxin. J Agric Food Chem 51(15):4439–4443
Kumar R, Singh R (2015) Mycotoxin binders: a review. Int J Curr Microbiol App Sci 4(1):482–491
Magnoli AP, Monge MP, Miazzo RD, Dalcero AM (2011) Mycotoxin binders in animal feed: a review. Mycotoxin Res 27(4):197–200

Flavors and Sweeteners References

Association of American Feed Control Officials (2021) AAFCO
FDA (2018) Additional information about high-intensity sweeteners permitted for use in food in the United States. https://www.fda.gov/food/food-additives-petitions/additional-information-about-high-intensity-sweeteners-permitted-use-food-united-states
USDA (2019) Organic standards. https://www.ams.usda.gov/grades-standards/organic-standards Association of American feed control officials, 2021

Enzymes References

Bhange K, Chaturvedi V, Bhatt R (2011) Fungal lipases: production, properties and applications. Res J Biotechnol 6(8):93–99
Bhat MK (2000) Cellulases and related enzymes in biotechnology. Biotechnol Adv 18(5):355–383
Bhatia Y, Mishra S, Bisaria VS (2002) Microbial β-glucosidases: cloning, properties, and applications. Crit Rev Biotechnol 22(4):375–407
Bornscheuer UT (2002) Microbial carboxyl esterases: classification, properties and application in biocatalysis. FEMS Microbiol Rev 26(1):73–81
García-Carreño FL, Dimes LE, Haard NF (1993) Substrate specificity of some marine proteases. Mar Biotechnol 1(1):60–66
Garg G, Singh N, Bhatia A (2016) A review on amylases, cellulases and lipases production by fungi on solid state fermentation. Int J Curr Microbiol App Sci 5(9):648–666
Hussain I, Choudhary MI, Atta-Ur-Rahman. (2016) Carboxyl ester lipases: classification, characterization and applications. Lipids Health Dis 15(1):1–11
Lowe ME (2002) The triglyceride lipases of the pancreas. J Lipid Res 43(12):2007–2016
Millán JL (2013) Alkaline phosphatases: structure, substrate specificity and functional relatedness to other members of a large superfamily of enzymes. Purinergic Signal 9(3):299–314
Rawlings ND, Barrett AJ, Finn R (2018) Twenty years of the MEROPS database of proteolytic enzymes, their substrates and inhibitors. Nucleic Acids Res 46(D1):D363–D371

Stryer L (1995) Biochemistry. W.H. Freeman and Company, New York
Tonks NK (2006) Protein tyrosine phosphatases: from genes, to function, to disease. Nat Rev Mol Cell Biol 7(11):833–846
Van Etten RL (1990) Acid phosphatases: structure, function, and regulation. Adv Enzymol Relat Areas Mol Biol 63:301–379

Part III

Biochemistry and General Effects

The Effects of Feed Additives on Farm Animals Under Heat Stress Conditions

Panagiotis Sakkas

Abstract

Heat stress **(HS)** challenges animal productivity, welfare, product quality, and reproductive success. Upon exposure to temperatures above their thermoneutral zone, animals' thermoregulatory mechanisms are activated to facilitate heat loss, which disrupt the acid balance causing alkalosis and acidosis in pigs and poultry. In ruminant species, the reduction of dry matter intake in combination with altered feeding patterns, increased respiration rates, and reduced volatile fatty acid absorption may lead to rumen acidosis. Feed intake declines to reduce the heat load, and feed efficiency is impaired, primarily due to the induction of oxidative stress and inflammation, especially in the gastrointestinal tract. Concurrently, HS increases the production of reactive oxygen species **(ROS)**, leading to the formation of oxidized products of lipid and protein metabolism, which damages enterocytes reducing nutrient digestion and absorption. Local inflammation, which may become systemic due to increased gut permeability, alters post-absorptive nutrient utilization and penalizes anabolic functions. Offering certain feed additives, isolated or in combination, incorporated in feed or water, may partly mitigate the adverse effects of HS in livestock and poultry species. These include (1) phytogenics, (2) vitamins, (3) microminerals, (4) electrolytes. Other additive solutions which do not fall within these categories are also assessed.

Keywords

Heat stress · Oxidative stress · Inflammation · Leaky gut · Additives · Phytogenics · Electrolytes · Vitamins · Microminerals

P. Sakkas (✉)
CCPA Group, Janze, France

G. Arsenos, I. Giannenas (eds.), *Sustainable Use of Feed Additives in Livestock*,
https://doi.org/10.1007/978-3-031-42855-5_11

Abbreviations

FI	Feed intake
HSP	Heat shock protein
HS	Heat stress
PSM	Plant secondary metabolite
ROS	Reactive oxygen species
TJ	Tight junction

1 Introduction

There is a consensus that climate change is leading to global warming, and rising temperatures are expected to increase heat waves' intensity, duration, and frequency. Exposure to high temperatures causes substantial economic losses in all livestock production systems (St-Pierre et al. 2003; Key et al. 2014), which are difficult to estimate and include losses attributed to (1) reduced product yield, (2) deteriorated product quality (meat, milk, and eggs), (3) reduced reproductive success, (4) increased morbidity and mortality and veterinary costs associated directly with HS and indirectly due to increased susceptibility to clinical and subclinical diseases, especially those affecting the gastrointestinal tract, (5) epigenetic effects as offspring of heat-stressed animals might express altered productive potential, and (6) lifelong penalties in performance persisting beyond the period of HS exposure in animals with a long production cycle. The expectation is that with increasing global temperatures, economic losses in the livestock sector will be more pronounced in the future, further necessitating the implementation of appropriate measures against its deleterious effects on livestock farming sustainability and animal welfare. Feed additives may improve the animal's thermal tolerance, i.e., capacity to produce under HS conditions, along with appropriate feed and husbandry strategies and adaptation of feed formulation standards. These additives may be incorporated in feed premixes, top dressed, or offered in water-soluble forms during heat exposure, offered during the whole production period in warm climates, or preventively when heat waves occur in temperate regions. Commercially available additives primarily aim at counteracting the two principal effects of HS, namely the disruption of acid–base balance and oxidative stress and inflammation, particularly in the intestinal tract. Over the course of this chapter, it will become apparent that there is no single additive solution against the harmful effects of HS. Often, cost-effective additive strategies incorporate multiple additive components, which display synergistic and/or additive effects owing to their complementary activities. The most prominent additive components include (a) phytogenic products consisting of plant extracts, oleoresins, and their containing plant secondary metabolites (PSMs), which may either directly affect thermal nociception and/or act as potent upregulators of endogenous antioxidant enzyme production and as inhibitors of pro-inflammatory

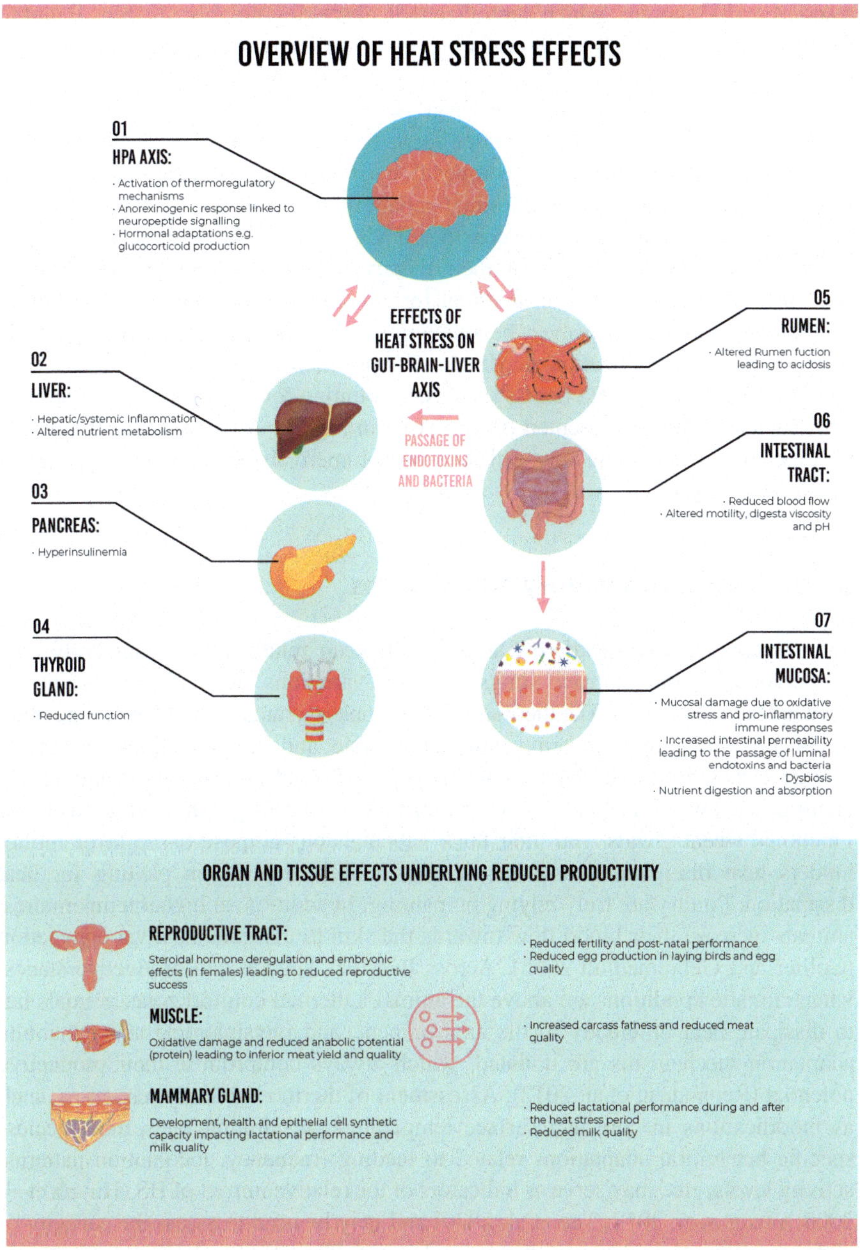

Fig. 1 Heat stress (HS) reduces feed intake in order to reduce internal heat load. The activation of thermoregulatory mechanisms may lead to respiratory alkalosis and acidosis and reduce the blood flow to internal organs to facilitate heat dissipation via convection. Oxidative stress, inflammation and dysbiosis occur in the intestine due to hypoxia, affecting nutrient digestion and absorption. Increased gut permeability leads to the increased passage of luminal endotoxins and bacteria. Both local and systemic pro-inflammatory immune responses alter nutrient metabolism via altered hormonal signalling (such as reduced thyroid function, increased glucocorticoid production, hyperinsulinemia). Metabolic adaptations penalise anabolic processes to meet the increased nutrient requirements of the activated immune system and to reduce heat load. As a result, product yield is reduced, product quality is compromised(meat, milk and eggs) and reproductive functions are penalised

responses, (b) vitamins, which neutralize radical species, notably vitamins E and C, (c) microminerals, which act as co-factors in the production of endogenous antioxidant enzymes such as Se and Zn, and (d) salts and their consisting electrolytes, which aim at correcting acid–base balance. Since HS insults intestinal health and function, the use of other additives may confer some benefits owing to their effect on the stabilization of the microbiota composition or the improvement in gut barrier function and amelioration of intestinal inflammation. Although amino acids and their intermediate metabolites are considered additives, it is beyond the scope of this chapter to review the effects of their dietary supplementation or other dietary nutrients modified during macronutrient formulation to support the productivity of heat-stressed animals. However, betaine will be assessed due to the extensive body of research on its utilization in HS conditions. In the following sections, some common key aspects of HS across species, accounting for the compromised productive efficiency will be presented (summarised in Fig. 1). Afterward, the effects and the efficacy of some of the most well-known, commercialized additives targeting HS across livestock species will be assessed.

2 Thermoregulatory Mechanisms

Farm animals are homeotherms as they can keep relatively constant body core temperature within narrow limits despite variations in ambient conditions (Renaudeau et al. 2012). The thermal comfort zone is a range in ambient temperature in which maintaining body temperature is possible, and an animal does not need to change its metabolic rate (Sejian et al. 2013). In terms of heat dissipation, animals are panting and sweating. Dairy cows are primarily sweating, while pigs have few functional sweat glands, and their thick subcutaneous adipose tissue layer further hinders heat dissipation. As a result, pigs mainly depend on panting for heat dissipation. Poultry are truly relying on panting. In addition, at higher temperatures, animals increase their blood flow towards the skin to dissipate heat via convection (Collier and Gebremedhin 2015). Across livestock species and productive stages, when climatic conditions are above the animal's thermal comfort zone, animals fail to dissipate heat efficiently to the environment, and physiological and metabolic adaptation mechanisms are initiated, which always compromise their productive potential (Renaudeau et al. 2012). Assessment of thermoregulatory responses, such as modifications in core and surface temperature, respiration rates, and species-specific behavioral adaptations related to feeding frequency, locomotion patterns, activity levels, etc., may serve as indicators of the relative impact of HS (Huynh et al. 2005; Wang et al. 2018; Tao et al. 2020) and may be used to assess the capacity of additive solutions to improve thermal tolerance. The effects of HS are more pronounced when high ambient temperatures are accompanied by elevated humidity levels as the physiological capacity to dissipate heat by evaporative heat losses is compromised (Beale et al. 2018). Indices used for characterizing the climatic environment may include some combination of temperature, humidity, radiation, air velocity, and convection, and responses to high ambient temperature can be modified by a variation in these other external parameters according to the different species, their productive stage, and the husbandry system. Thermal indexes were first used to assess the severity of HS in livestock animals and may serve as a forecast system to determine the possible threat or danger to the animals due to climatic conditions and implement appropriate interventions, such as strategic additive supplementation. Thermal indexes are a well-developed concept for dairy cattle, while for other animal species, relative humidity is often taken into account when defining the optimum temperature ranges, effectively being adjusted to lower desirable ranges at higher relative humidity values (Sejian et al. 2018b).

3 Upper Critical Temperatures and Losses in Production Performance

All livestock species at all productive stages are affected by HS. However, the effects are less important at earlier stages of ontogenic development in growing animals due to their lower upper critical temperature thresholds. The range and magnitude of production losses to HS depend on its (1) intensity, (2) duration ranging from hours to the overall production period in tropical and subtropical conditions, (3) whether temperatures approach the thermoneutral conditions over part of the day or not (cyclic vs. constant), (4) the age and/or weight of the animal upon HS exposure, (5) the genetic background, particularly in relation to the productive potential (within the same genetic line and between different genetic lines), and (6) the diet it is offered. Although HS effects may differ according to intrinsic and extrinsic factors, one of the first responses across species to HS exposure, which accounts for most losses in performance, is a reduction in voluntary FI. Modifications in feeding behavior aim to reduce the internal heat production deriving from exothermic processes such as nutrient ingestion, digestion, and metabolism (dietary-induced thermogenesis), compromising productive processes (growth, gestation, lactation, and egg production) and facilitating heat dissipation. As presented in later parts of this chapter, production losses are also partially attributed to poor feed efficiency. A meta-analysis in broilers suggests that wet bulb temperatures above 25.5 or 25.7° C, and a dry bulb temperature of 32.0° C were considered the threshold conditions for potential production losses of 36.5% in the weight gain of broiler chickens (Moura et al. 2015). In growing-finishing pigs, FI is reduced at an accelerating rate above the pigs' thermoneutral zone, which varies according to pig weight; larger pigs have a lower capacity to dissipate heat due to a lower surface area-to-mass ratio and higher backfat thickness (Renaudeau et al. 2011; Mayorga et al. 2019). In the case of finishing beef and if the stressful thermal conditions last for a relatively short period, the compensatory gain may be observed after the end of this period. However, it has been demonstrated that beef cattle under HS show slower growth rates (Summer et al. 2019). In dairy cows, FI begins to decline at temperatures of 25–26°C and reduces more rapidly above 30°C in temperate climatic conditions (Tao et al. 2020). Sows in the lactation stage are particularly sensitive to HS owing to their high FI and increased internal heat production to support lactation; respiration rates increase significantly even when comparing sows exposed at 18°C and 22°C, while rectal temperatures significantly rise above 27°C (Quiniou et al. 2000). Layers are vulnerable to HS as they must maintain a long production cycle (50 to 70 weeks). A meta-analysis indicated that FI decreases by approximately 10% from temperatures as low as 24°C, and production traits such as egg mass, egg production rate, egg weight, and shell strength are reduced by 5% in temperatures ranging between 24 and 30 °C along with increases in mortality and body weight loss (Mignon-Grasteau et al. 2014).

4 Acid–Base Balance

The activation of thermoregulatory mechanisms disturbs the animal's acid–base balance, leading to respiratory alkalosis and acidosis in poultry and pigs, and acidosis in ruminants. Disruptions of acid–base balance impact both FI and feed efficiency. In poultry and pigs, the events that undermine the acid–base balance disruption are initiated by increased panting to dissipate heat, which results in increased clearance of blood CO_2. As a result, partial CO_2 pressure is decreased, which in turn causes a decrease in the concentration of carbonic acids (H_2CO_3) and hydrogen ions (H^+). To counteract alkalosis, the kidneys increase bicarbonate ion (HCO_3^-) excretion and reduce H^+ excretion, thus increasing blood pH, leading to alkalosis (Teeter et al. 1985; Patience et al. 2005; Liu et al. 2018a).

On the other hand, continuous depletion of HCO_3^- ions along with renal Cl^- reabsorption as a metabolic compensation for blood alkalosis might lead to metabolic acidosis. Metabolic acidosis leads to the replacement of K^+ anions by hydrogen ions (H^+) intracellularly leading to increased plasma K^+ concentrations (hyperkalemia) (Borges et al. 2007; Liu et al. 2018a). Reduced blood levels of Na^+ and K^+ due to their increased excretion via the kidney have been consistently reported in heat-stressed broilers (Borges et al. 2007; Beckford et al. 2020; Livingston et al. 2022). Increased HCO_3^- losses, along with reduced carbonic anhydrase activity, which is an enzyme responsible for the formation of HCO_3^-, renders HCO_3^- unavailable for eggshell mineralization, negatively affecting eggshell quality (Mack et al. 2013; Barrett et al. 2019; Kim et al. 2020). In pigs, the concentration of Cl^+ has been reported to decrease (Ortega et al. 2022) or not.

A particular problem caused by decreased DM intake in dairy cows is the increased risk of acidosis. Essentially, rumination time is decreased under HS, leading to reduced DM intake. Due to the redistribution of blood to the periphery, digestion products such as volatile fatty acids are absorbed less efficiently, and their total rumen content may increase, reducing the pH. The reduced saliva production due to decreased rumination means that HCO3- production and the buffering capacity are also reduced, while the increased respiration due to panting increases CO_2 production. Collectively these events, along with altered feeding behavior, increase the risk of acidosis. The reduction in rumen pH impairs fiber digestion efficiency; rumen fibrolytic bacteria are the most affected when rumen pH drops below 6.0. The culmination of these events contributes to decreasing feed utilization during HS (Conte et al. 2018).

5 Oxidative Stress and Inflammation

Apart from reduced productivity due to decreased FI, feed efficiency is also typically impaired primarily due to HS-induced oxidative stress and inflammation. Upon HS exposure, there is an increase in mitochondrial energy generation (Akbarian et al. 2016), causing an imbalance between pro-oxidant and antioxidant systems, defined as the presence of reactive species, in excess of the available antioxidant capacity of animal cells. Increased production of reactive oxygen species (ROS) leads to an impairment of mitochondrial function, which is responsible for the production of most cellular energy via oxidative phosphorylation (Surai et al. 2019b). Mitochondrial dysfunction caused by HS is the basis of oxidative stress. ROS overproduction can damage proteins, lipids, and DNA, reducing energy generation efficiency and increasing ROS production in the mitochondria (Hu et al. 2019). The NF-E2-related factor 2 (Nrf-2 /Keap1) and nuclear factor kappa-light-chain-enhancer of activated B cells/inhibitory κB protein (Nf-κB/IkB) systems are considered to be the two major master regulators of the stress response (Surai et al. 2019b). To counteract increased ROS production, Nrf-2 is activated, which drives the production of endogenous antioxidant enzymes, such as superoxide dismutase (SOD), catalase (CAT), and glutathione peroxidase (GPx), improving cell survival under stress conditions. The upregulation of Nrf-2 pathway and its target gene HO-1 expression enhances barrier function, increasing the expression of tight junctions (TJ) proteins under hypoxic conditions (Liu et al. 2017c). In addition, heat shock factors (HSF) are activated, and in particular, HSF1, which induces the production of heat shock proteins (HSP). It is believed that the Nrf-2 and HSP represent two molecular responses with some overlapping and protective functions. Their activation downregulates pro-inflammatory type of responses by inhibiting Nf-κB activation, which is responsible for the transcriptional induction of pro-inflammatory cytokines, chemokines, and other inflammatory mediators (Ahmed et al. 2017; Liu et al. 2017a; Lian et al. 2020). Although in the short term, HS may increase Nrf-2 encoded endogenous antioxidants in an attempt to neutralize increasing ROS production, a subsequent decrease may occur in multiple tissues compromising the antioxidant system's capacity in poultry (Akbarian et al. 2016; Surai et al. 2019b), pigs (Cui et al. 2019), and ruminants (Abdelnour et al. 2019). Endogenous and exogenous nonenzymatic antioxidants such as glutathione and vitamin E provide an additional level of defense, which is nonetheless insufficient in HS conditions. The increased concentration of oxidation products of protein and lipid metabolism in various tissues, organs, and animal products and a proinflammatory immune profile reflects the prevailing pro-oxidant inflammatory milieu. Oxidative inflammation is detrimental to intestinal health and function but also affects multiple organs, such as the liver, spleen, reproductive tract, mammary gland, and tissues with high anabolic potential, such as muscles. Oxidative inflammation induces metabolic changes underscored by an altered immune and hormonal milieu. The upregulation of Nrf-2 antioxidant enzyme production, the limitation of protein and lipid oxidation, and the direct or indirect inhibition of the NF-kb pathways are targets of multiple additive solutions across species and livestock production systems. The efficacy of additive solutions in this respect is often evaluated by their effects on endogenous antioxidant enzyme activities, and oxidative and inflammatory status systemically and in tissues.

6 Intestinal Tract Effects

The effects of HS on the gastrointestinal tract of livestock and poultry species are of particular importance in terms of production losses and, as such is, the primary target of numerous additive solutions. As previously mentioned, during HS, there is a redistribution of blood to the periphery to facilitate heat dissipation (Lambert 2009). The shift of visceral blood flow toward the peripheral circulation during HS induces hypoxic conditions in the intestine and renders it particularly susceptible to oxidative stress and damages the mucosal tight junction **(TJ)** barrier (Rostagno 2020; Ortega and Szabó 2021). Altered protein expression of TJ proteins such as claudin and reducing occludin and zonula occludens 1 (ZO-1) leads to intestinal hyperpermeability in poultry (Ruff et al. 2020), pigs (Pearce et al. 2013; Gabler et al. 2018), and ruminants (Fontoura et al. 2022). As a result, intestinal antigens may infiltrate via the intestinal lumen to the systemic circulation and stimulate a local immune reaction causing a "leaky gut syndrome" leading to an increased endotoxin concentration which in turn elicits systemic inflammation and an acute phase protein response (Quinteiro-Filho et al. 2012; Pearce et al. 2012; Cui et al. 2019). In poultry, HS-induced intestinal inflammation is characterized by lymphoplasmacytic infiltration across segments of the small intestine (Quinteiro-Filho et al. 2010). Increased pathogen translocation due to increased intestinal permeability in the spleen and liver of broilers has also been observed (Alhenaky et al. 2017; Quinteiro-Filho et al. 2017). Intestinal inflammation impacts intestinal morphology and is accompanied by reduced villus length, increased crypt depth, and decreased mucosal absorptive surface area (Mitchell and Carlisle 1992; Garriga et al. 2006; Cui et al. 2018), although this effect appears less prominent in ruminants (Koch et al. 2019). Alterations in digestive enzyme activities (Pearce et al. 2012; Al-Zghoul et al. 2019) and nutrient transporter functions (Pearce et al. 2013; Sun et al. 2015) disturb nutrient digestion and absorption in monogastric (Morales et al. 2016; de Souza et al. 2016; Kim et al. 2020). Furthermore, changes in the intestinal environment, such as pH, transit time, digesta viscosity, and internal temperature, may further affect nutrient digestibility (Bonnet et al. 1997; Morales et al. 2016; de Souza et al. 2016; Kim et al. 2020). The insult to gut health and function also deregulates intestinal microbiota composition in pigs and poultry, increasing pathogenic while reducing beneficial bacterial populations (He et al. 2019; Cao et al. 2021; Patra and Kar 2021; Xia et al. 2022). In broilers, it may increase birds' predisposition to diseases detrimental for poultry farming, such as necrotic enteritis (Tsiouris et al. 2018).

7 Endocrine and Metabolic Effects

HS-induced oxidative inflammation is accompanied by hormonal alterations, which drive metabolic adaptations to ensure the animal's survival to the detriment of productive efficiency. In order to reduce internal heat production under high temperatures, lipogenesis is favored, and lipolysis is hindered, while protein breakdown is increased and protein synthesis is decreased (Belhadj Slimen et al. 2016). The energetic efficiency for lipid deposition is greater than that for protein deposition, and lipolysis generates more metabolic heat than carbohydrates and protein. Due to reduced fatty acid oxidation in HS conditions (Zhao et al. 2018b), glucose becomes the favored fuel for maintenance and growth functions. The increased protein breakdown facilitates glucose production via gluconeogenesis (Belhadj Slimen et al. 2016) to satisfy the increased energetic requirements of the immune system (Calder et al. 2007) as well as the production of immune system mediators (Iseri and Klasing 2013) and proteinaceous molecules such as acute phase proteins (Gruys et al. 2005; Kampman-van De Hoek et al. 2015) and HSP (Zulkifli et al. 2018). As a result, besides the obvious reduction of carcass yield due to reduced growth rates, growing livestock and poultry species have increased carcass fatness (Gonzalez-Rivas et al. 2020). In lactating sows and cows, the blunted lipolytic response hinders adipose tissue mobilization despite their weight loss, and they rely more on glucose to cover their energy requirements (Rhoads et al. 2013). The fact that heat-stressed lactating mammals fail to enlist this "shift" in post-absorptive energetic metabolism, even though they eat below requirements and are in a state of negative energy balance, directly impacts feed efficiency (Conte et al. 2018).

The previous metabolic adaptations to HS are orchestrated by a series of neuroendocrine responses (Smith and Vale 2006). In particular, HS activates the hypothalamic–pituitary–adrenal axis (HPA), elevating the levels of glucocorticoids hormones cortisol and corticosterone in livestock and avian species, respectively (Star et al. 2008; Hao et al. 2014; Quinteiro-Filho et al. 2017; Sejian et al. 2018a, 2018b). Glucocorticoids improve vasodilation and therefore heat loss, but also cause proteolysis and antagonize insulin and insulin growth factor while favoring lipogenesis by upregulating lipoprotein synthase activity (Scanes 2016; Gonzalez-Rivas et al. 2020). In reproducing livestock and poultry species increased glucocorticoid production increases energy deposition at the expense of reproductive functions as it may lead to decreased steroidogenesis and, as a result, impaired ovarian development and function (Shini et al. 2009; Oguntunji and Alabi 2010; Ross et al. 2017; Wang et al. 2017; Wolfenson and Roth 2019). Importantly, chronically elevated levels of glucocorticoids mediate pro-oxidative functions leading to increased mitochondrial metabolism (Spiers et al. 2015). Moreover, they suppress adaptive immune responses by causing the involution of immune organs (e.g., bursa of Fabricius, thymus, and spleen) and reducing B cell production and T-cell proliferation and function in poultry (Shini et al. 2010; Honda et al. 2015; He et al. 2018) and compromise immunocompetence in mammalian species (Bagath et al. 2019; Dahl et al. 2020). HS also affects thyroid function, causing principally a reduction in circulating thyroid hormone triiodothyronine (T3), which aims at reducing metabolic rate and, therefore, heat production to facilitate the animal's adaptation to the stressor (Star et al. 2008; Mullur et al. 2014; Beckford et al. 2020; Li et al. 2021). Reduced T3 further redirects the available energy toward fat deposition while sparing protein anabolism. Finally, increased circulating insulin levels commonly observed under chronic HS conditions may also account for the decreased circulating glucose concentration to satisfy the increased immune system requirements, the blunted lipolytic activity, and the increased lipogenesis observed in pigs and ruminants (Belhadj Slimen et al. 2016). Paradoxically, despite the increased insulin levels, which are typically observed during HS, there is reduced mammary protein synthesis (Rhoads et al. 2013). It has been suggested that hyperinsulinemia, along with increased endotoxin levels severely compromise sow ovarian function and fertility (Mayorga et al. 2020).

8 Product Quality

Besides lower carcass yields and increased carcass fatness, meat quality is also affected. In pigs and broilers, there is a greater incidence of pale color, lower water-holding capacity (WHC), and increased cook and drip losses and may lead to a higher frequency of pale soft exudative meat and increased lipid peroxidation may lead to shorter half-life (Zaboli et al. 2019; Gonzalez-Rivas et al. 2020). Broilers breast tissue, which has a high anabolic capacity, is affected the most by HS, and decreases in the proportion of breast muscle have been previously recorded (Zhang et al. 2012). Pectoral muscles have a high mitochondrial content and, consequently, more elevated ROS-producing potential and susceptibility to oxidative stress (Kikusato and Toyomizu 2019). Under HS conditions, a greater incidence of breast myopathies such as white stripping (Aslam et al. 2021) and decreased breast muscle protein synthesis and amino acid uptake (Ma et al. 2018) have been reported. Furthermore, meat safety issues arise as an increased frequency of carcass contamination by *Escherichia coli* and *Salmonella* has been reported during the hot season in pigs and poultry (Gregory 2010). In beef cattle, if they are slaughtered during or shortly after the end of the HS period, they have lower carcass weight, lower fat thickness, and worse meat quality in terms of pH, tenderness, and color (Summer et al. 2019) while increased development of a dark firm and dry meat (DFD) is typically observed across meat producing ruminants species (Gregory 2010). As previously stated, eggshell quality is also severely affected primarily due to blood alkalosis. In addition to defective egg mineralization, egg weight, and egg mass, Haugh units, yolk color are impacted by HS due to reduced FI, endocrine changes such as increased corticosterone secretion and reduced steroidogenesis (Franco-Jimenez et al. 2007; Oguntunji and Alabi 2010). Besides the reduced FI and reduced milk yield in dairy ruminants, there are effects on milk composition. In general, the lactose content is not considered to be affected (Cowley et al. 2015). However, a decrease in milk fat has been reported during summer (Bernabucci et al. 2015). The effects of HS are particularly pronounced on the protein fraction of the milk, which is independent of the parallel reduction in FI (Cowley et al. 2015), and primarily affect the casein fraction with adverse effects on cheese-producing properties (Bernabucci et al. 2015; Cowley et al. 2015). Increased mastitis incidence and somatic cell counts (Summer et al. 1999; Bernabucci et al. 2015), along with changes in mineral content, pH, and coagulation properties, further deteriorate cheese processing properties and yield (Summer et al. 2019).

9 Reproductive Success and Offspring Performance

HS affects sow and gilt reproductive performance and is typically manifested as seasonal infertility (Prunier et al. 1997). Increased temperatures have been shown to decrease farrowing rates, extend the weaning to estrus interval, and delay the onset of puberty (Bertoldo et al. 2009). Overall, HS exposure in gestating sows negatively affects early embryo development and reduces embryo viability and the number of piglets born alive (Ross et al. 2015). It has been suggested that intestinal hyperpermeability drives endotoxemia and alters insulin signaling, which affects ovarian function (Mayorga et al. 2020). In cows, the reproductive tract, and in particular, the ovarian components (i.e., follicles, oocytes, CL), and preimplantation embryos are susceptible to elevated temperatures. During HS, multiple reproductive processes are impaired, including oocyte competence, embryonic growth, gonadotropin secretion, ovarian follicular growth, steroidogenesis, development of the corpus luteum, and uterine endometrial responses. As a result, conception rates are severely compromised (Wolfenson and Roth 2019). Broiler male breeder and boar fertility are also affected by HS, primarily affecting sperm fertility and quality (Wettemann et al. 1976; Fouad et al. 2016). In laying hens, besides the direct effects on eggshell quality due to reduced ionized Ca and bicarbonate availability which affect the eggshell formation, there are direct effects of oxidative stress on inflammatory pathways, which in turn affect follicular cell apoptosis and reduce follicle numbers, decreasing the egg laying rate (Li et al. 2020). In both ruminant species and in sows, maternal HS during gestation affects the growth and physiology of the developing offspring. In dairy cattle, male calves born to heat-stressed mothers increased morbidity associated with impaired immunocompetence, and female calves produce less milk during their first lactation (2 years after the HS exposure) (Tao et al. 2012, 2014; Monteiro et al. 2016). It has been demonstrated that exposure to HS in gestating sows results in offspring that tend to accrete more body fat during later stages of growth (60 kg<), which in turn affects their feed efficiency and carcass quality (Johnson et al. 2015). Mammary development of multiparous dry cows exposed to HS is also affected, causing depressed mammary epithelial cell proliferation, potentially reducing the total number of cells in the mammary gland capable of producing milk (Tao et al. 2011).

10 Phytogenic Additives

Phytogenic feed additives, also known as phytobiotics or phytochemicals, may refer to plant extracts, essential oils, oleoresins, or dried plant material that contain plant secondary metabolites (PSMs), which generate biological activities. PSMs belong to a wide range of organic compounds, including polyphenols, carotenoids, lignans, coumarins, phytosterols, and S-containing compounds (Crozier et al. 2007). Their pharmacokinetics depend on the chemical structure and molecular weight but in general terms, upon ingestion, they exert local effects in the intestinal tract modulating epithelial, endocrine, and immune cell function. Certain PSM notably belong to polyphenols, interact with intestinal microbiota changing their profile, benefiting the growth of beneficial bacteria (Mithul Aravind et al. 2021). Furthermore, ingested PSM (parent metabolites) undergo metabolism by large intestinal and colonic microflora leading to the production of smaller molecules (Phase II metabolites) that can be absorbed and enter the circulation reaching distal organs where they can exert their biological activities (Oteiza et al. 2018; Kawabata et al. 2019). Phase II, PSM metabolites may possess additional biological activities in comparison to parent metabolites (Oteiza et al. 2018). As a result, it is often challenging to disassociate observed biological effects following PSM ingestion between parent and phase II metabolites (Ozdal et al. 2016; Williamson et al. 2018). Since phytogenic additives, even when based on a single plant, often contain a multitude of PSMs, observed effects may be derived primarily from the principal component but also from secondary PSMs existing in the product in lower concentrations. In general, commercially available phytogenic additives consist of combinations of extracts from different plants to exploit potential synergistic activities of PSMs when offered in combination, although antagonistic effects are known to exist (Chen et al. 2022b). Finally, to obtain desired effects the optimum doses of phytogenics need to be identified *in vitro* and *in vivo* trials, while the form (i.e., protected vs nonprotected) and phytochemical profile (flavonoids, saponins, alkaloids, etc.) may impact their bioavailability in target tissues.

Since there are multiple plant species containing PSMs (Abd El-Hack et al. 2020), the focus here will be on some of the most researched PSMs in the context of HS. Because of their efficacy in mediating relevant bioactivities, they represent the principal component of several HS-specific commercialized phytogenic additive solutions. There are numerous articles published over the last 15 years assessing the efficacy of phytogenic additives in poultry species, while evidence is accumulating in pigs and ruminants. One of the most essential aspects of phytogenic feed additive formulation is the capacity of the principal active components to modulate Nrf-2 and Nf-kb pathways, effectively re-enforcing endogenous antioxidant systems and downregulating inflammatory responses (Wu et al. 2014; Li et al. 2018; Alharbi et al. 2022). Among PSM, polyphenols, which include both flavonoid and non-flavonoid compounds, display potent antioxidant activities (Da Silveira et al. 2014; Gessner et al. 2017; Oteiza et al. 2018) and are present in the majority of HS targeted commercial phytogenic additives. Beyond their antioxidant and anti-inflammatory functions, PSMs also possess secondary activities that are also considered when formulating additive solutions. These include their ability to modulate gut transit time (Mendel et al. 2017), to increase digestive enzyme secretion (Kalpana and Srinivasan 2004; Yilmazer-Musa et al. 2012), to mitigate HS in target tissues other than the intestine according to their pharmacokinetics and relevant biological activities (Gupta and Pandey 2019), to modify microbiota composition (Feng et al. 2018; Dingeo et al. 2020), to alter thermal nociception and increase thermal tolerance (Zmrhal et al. 2018; Inagaki et al. 2019), and to alter metabolic responses (Cottrell et al. 2020).

10.1 Poultry

There are numerous studies that have investigated the effects of various phytogenic additives in broilers and laying hens, displaying their potency for mitigating the impact of HS. Certain polyphenols such as curcumin, resveratrol, and catechins have attracted considerable attention owing to their potent antioxidant and anti-inflammatory activities (Hu et al. 2019). Curcumin has a limited distribution in plant species and is the principal polyphenol found in the rhizome Turmeric (*Curcuma longa L.*). On the other hand, resveratrol is widely distributed among plants and is mainly extracted from red grapes (Tian and Liu 2020). Catechins are typically found in great tea extracts from *Camellia sinensis* and *C. assamica*, notably epigallocatechin gallate (EGCG) and epigallocatechin (EGC) (Cai et al. 2018). In heat-stressed broilers, the consumption of these polyphenols mainly improves feed efficiency and meat quality (drip loss, pH), while increases in FI have also been reported. Upon dietary intake, they have been shown to exert antioxidant activities in heat-sensitive tissues and organs as well as systemically, as indicated by increased endogenous antioxidant enzyme activities and reduced lipid peroxidation (Sahin et al. 2010b; Liu et al. 2014; Zhang et al. 2015a, 2015b, 2017a, 2018b; Luo et al. 2018). They have immunomodulatory effects, counteracting the immunosuppressive effects of HS on the bursa, thymus, and spleen weight (Liu et al. 2014; Zhang et al. 2018b). HS increases the expression of both HSP70 and HSP90, while supplementation with resveratrol, EGCG, and curcumin downregulates their expression in a variety of tissues including bursa, thymus, spleen, breast muscle, and the jejunum which is indicative of increased thermal tolerance (Orhan et al. 2013; Liu et al. 2014, 2016; Zhang et al. 2015a, 2015b; Yang et al. 2019). Curcumin supplementation has been shown to increase mitochondrial biogenesis–related genes in the liver and intestine (Zhang et al. 2015a, 2018a). Their dietary intake may alter neuroendocrine responses as reflected by reduced corticosterone levels (Park et al. 2016; Zhang et al. 2017a, 2018b; Yang et al. 2019). Their supplementation may improve intestinal histological features (Rajput et al. 2013; Liu et al. 2016; Zhang et al. 2017b) and intestinal epithelial barrier function via modifications in the mRNA expression of TJ-related genes (Zhang et al. 2017b). Finally, the dietary intake of all three metabolites has been shown to improve meat oxidative stability under conditions of HS (Zhang et al. 2015a; Zhao et al. 2021, 2022).

In laying hens, phytogenic consumption may result in reduced body weight loss and/or increased weight gain, higher egg laying rate, improved feed efficiency, and egg oxidative stability and quality (eggshell characteristics such as breaking force and thickness, Haugh units, yolk color). In addition to their intestinal tract effects, PSMs may improve liver oxidative status, which is of particular importance considering its importance in fat synthesis and metabolism (Zaefarian et al. 2019; Emami et al. 2021). In heat-stressed laying hens, curcumin supplementation, in particular, has been shown to reduce inflammatory responses in the liver by downregulating proinflammatory cytokine gene expression levels and protein expression of NF-κB (Nawab et al. 2019), possibly via reduced translocation of intestinal endotoxins in the circulation. Furthermore, dietary curcumin supplementation has recently been shown to reduce laying hen corticosterone levels in HS conditions, while increasing steroidal hormone production, positively influencing follicular development and ovulation (Liu et al. 2020). Effects extend to egg quality characteristics with supplemented birds producing eggs with increased eggshell thickness, eggshell strength, and albumen height (Liu et al. 2020). A well-known synergism exists between curcumin and baicalin, a potent antioxidant polyphenol found in *Scutellaria baicalensis* which attenuates inflammatory pathways under conditions that challenge liver health and function (Xu et al. 2021; Yang et al. 2021). Their combined supplementation may exert superior hepatoprotective effects, in comparison to monotherapy with the respective metabolites (Wang et al. 2021), owing to their synergistic actions on multiple signaling pathways regulating the pro-inflammatory response and antioxidant transcription factors (Wang et al. 2021). Plant extract associations of *C. longa* and *S. baicalensis* have also been shown to downregulate intestinal inflammatory response in heat-stressed broilers (Varmuzova et al. 2015) and improve multiple aspects of thermal tolerance in laying hens, including liver health (Giannenas et al. 2022). Other phytogenics, rich in catechins such as EGCG has been shown to improve oxidative stress biomarkers in the liver and intestine due to their effects on transcription factors (Sahin et al. 2010b; Orhan et al. 2013) in addition to improving egg oxidative status (Wang et al. 2020). Dietary supplementation with resveratrol did not improve performance in heat-stressed quails (Sahin et al. 2010a) despite reducing the degree of oxidative stress, assessed by measuring vitamin E blood concentration levels.

10.2 Livestock

Unlike poultry, where there is abundant evidence, fewer studies have investigated the effects of phytogenic additive supplementation in heat-stressed pigs. Dietary supplementation with essential oils to heat-stressed Chinese mini-pigs improved growth performance and increased glucose absorption (Song et al. 2009, 2010). In another study, dietary supplementation with oregano essential oil reduced cortisol and norepinephrine concentrations and improved the performance of growing-finishing pigs to a greater degree than betaine supplementation in HS conditions and resulted in improved DM, nitrogen, and gross energy digestibility (Lan and Kim

2018). Cinnamon supplementation improved insulin sensitivity in HS pigs, although there were no other effects on the physiological, metabolic, or biochemical responses to HS (Cottrell et al. 2020). Dietary supplementation with alkaloids may yield benefits in thermoregulatory responses of heat-stressed growing pigs independently of anti-inflammatory and antioxidant effects; their dietary supplementation has been shown to reduce respiration rates and core temperature and colonic permeability (Le et al. 2020). Of particular interest in terms of its thermoregulatory activities in pigs and ruminants (see below) is capsaicin, an alkaloid found in chili peppers and responsible for their burning and irritant effect (Fattori et al. 2016). Capsaicin is a prototypical transient receptor potential cation channel subfamily V member 1 (TRVP-1) agonist, which serves as one of the principal thermosensors for the thermoregulatory system being conserved across mammalian species (Garami et al. 2020). Activation of TRPV-1 by capsaicin lowers body temperature and induces hypothermia via peripheral vasodilation (Inagaki et al. 2019) and may increase the animal's upper critical temperature threshold levels and therefore it is thermal tolerance. Studies in growing-finishing pigs have shown that capsaicin-containing additives may improve feed efficiency in short- and long-term HS exposure although effects on respiration rates were limited under short-term HS exposure (Biggs et al. 2020; Kroscher et al. 2022).

Although phytogenic additive supplementation has multiple beneficial effects in gestating and lactating sows (Chen et al. 2022a) there have only been a few investigations in an HS context. Results confirm their potent effects at improving oxidative stress status, and reproductive and lactational performance. Supplementation of lactating sows has been shown to improve FI, DM digestibility, piglet weaning weight, and ADG, while decreasing backfat loss, serum cortisol level, as well piglet diarrhea (Liu et al. 2017b). Phytogenic supplementation of primiparous sows during gestation and lactation has been shown to alleviate oxidative stress by reducing the concentration of protein and lipid oxidation products decreasing stillbirth rates, weaning to estrous intervals, and back fat losses (Papatsiros et al. 2022). Importantly, phytogenic additive administration during late gestation and lactation may alleviate the oxidative stress in both sows and their offspring, increasing colostrum yield and milk fat content, while improving offspring intestinal barrier integrity and increasing weaning weight (Zhang et al. 2020).

Studies investigating the effects of capsaicin-containing additives in HS conditions in dairy cows have consistently shown increased milk production (Oh et al. 2015; Abulaiti et al. 2021; An et al. 2022). On the other hand, FI and rectal temperatures have not been consistently affected by their supplementation (Oh et al. 2015; An et al. 2022), which may be related to the dose of capsaicin, the timing of its administration, and the modalities of the different HS models employed across studies. In addition to lactational performance, supplementation with capsaicin has been shown to improve reproductive performance in estrous synchronized cows during summer; supplementation increased the estrus response, ovulatory follicle size, ovulation rate, and pregnancy rates (Abulaiti et al. 2021). In ruminants, flavonoids have attracted a lot of attention in mitigating the effects of stressors on ruminant health (Olagaray and Bradford 2019). Citrus extract supplementation in HS cows has generally been associated with improved mammary health, somatic cell counts, and altered locomotory activity patterns indicative of greater thermal tolerance (Havlin and Robinson 2015). *Radix bupleuri* extract supplementation has been shown to reduce increase FI, reduce body temperature, and improve production variables such as milk, protein, and fat yield (Pan et al. 2014). Offering a polyherbal phytogenic additive formulation improved productive performance while mediating immunomodulatory effects (Shan et al. 2018). Dietary supplementation with phytogenic additives has been also demonstrated as an efficient method to enhance the thermal tolerance of growing ruminants. Dietary supplementation of heat-stressed calves with resveratrol-rich grape seed extract has been shown to improve performance. In addition, supplemented calves had reduced respiratory rates and improved antioxidant activities, endocrine function, pro-inflammatory status, and hematological parameters compared to control calves (Urkmez and Biricik 2022a, 2022b). Results obtained in heat-stressed lambs investigating the use of phytogenic additives are promising; offering a polyherbal phytogenic additive improved performance, oxidative status, and carcass quality (Hashemzadeh et al. 2022) while dietary supplementation with the flavonoid naringenin simutaneously improved growth rates, feed efficiency, and antioxidant status and heightened humoral immune responses (Alhidary and Abdelrahman 2016).

11 Microminerals and Vitamins

11.1 Micromineral Supplementation

Sufficient dietary supplementation with microminerals is critical for farm animal thermal tolerance. The initial levels of the antioxidant defense system involve the production of water-soluble antioxidant enzymes, which requires the presence of selenium (Se), manganese (Mn), copper (Cu), and zinc (Zn); Se is a precursor of GP-x and other selenoproteins, while Mn, Cu, and Zn are essential parts of SOD (Surai 2014; Lauridsen 2019). These may be supplemented in organic forms or inorganic forms. A general increase in the supplementation levels of microminerals is not sustainable from an ecological standpoint, mainly due to their low retention and increased excretion leading to environmental pollution and is often confounded with regulations in regard to their maximum dietary incorporation levels (Gebreeyessus and Zewge 2019; Vats et al. 2022). On the other hand, offering them in organic forms may confer superior biological activities in comparison to inorganic forms, offered either isolated or in combination, at levels well below EU-regulated maximum incorporation limits.

Se acts as cofactor to antioxidant enzymes through its incorporation in selenoproteins, notably GPx. The latter is a component of glutathione pathways, which detoxify lipid peroxides protecting cellular and subcellular membranes against ROS damage (Pecoraro et al. 2022). In addition, Se intake affects thyroid function as it converts thyroxin (T4) into active T3 regulating metabolism (Shakeri et al. 2020), which is penalized under HS conditions as previously described. Both inorganic selenium, such as selenite or selenate, and more recently sources of organic selenium, such as yeast selenium, selenomethionine, selenocysteine, hydroxy-selenomethionine, are routinely incorporated into premixes to satisfy Se requirements as its concentration levels in natural ingredients are insufficient to cover nutrient requirements. Organic forms are more bioavailable than inorganic forms, conferring superior biological activities and allowing for better tissue distribution and accumulation as well as maternal transfer to the progeny via the eggs or milk (Surai and Fisinin 2014; Chen et al. 2016; Sun et al. 2021; Wang et al. 2022). The maximum supplementation doses of selenium currently permitted in the EU are 0.5 mg/kg for inorganic Se and 0.2 mg/kg for organic Se sources. Multiple studies in heat-stressed poultry have shown that Se supplementation at a range of 0.2–0.5 mg/kg with inorganic forms, well within the legal levels of dietary supplementation, improves the performance of layers, broilers, and egg quality characteristics (Habibian et al. 2015). In the case of ruminants, the absorption of inorganic forms is influenced by rumen conditions, while organic forms undergo considerably less alteration in the rumen, resulting in better availability (Surai et al. 2019a; Arshad et al. 2021). In dairy cows, supplementation with organic forms has been shown to be more effective than inorganic forms at improving their antioxidant status and performance in heat-stress conditions (Calamari et al. 2011; Sun et al. 2019). A study investigating maternal Se supply in heat-stressed sows showed that short-term supplementation during late gestation and lactation did not affect fetal growth regardless of environmental conditions. On the other hand, it improved piglet pre-weaning survival, colostrum, and milk composition, as well as maternal selenium antioxidant status and immunoglobulin transfer were observed at both thermoneutral and hot ambient conditions (Chen et al. 2019b). Supplementation of heat-stressed growing pigs with Se reduced proinflammatory gene expression, in the jejunum protecting against intestinal barrier disruption (He et al. 2022). Nonetheless, short-term supranutritional doses (1 mg/kg of feed) do not seem efficient in re-enforcing antioxidant defenses or attenuating other physiological HS responses besides a reduction in rectal temperatures (Liu et al. 2018b).

Zinc (Zn) is an essential mineral with a wide array of biological functions in addition to its role in free radicals suppression by activating endogenous antioxidative enzymes, including GSH-Px, SOD, glutathione S-transferase, and HO-1. Maximum dietary inclusion levels are regulated in the European Union (150 mg/kg and 120 mg/kg feed for pigs and poultry). There are several commercially available organic zinc sources (e.g., Zn protein, Zn amino acid, or Zn picolinate) or inorganic zinc sources ($ZnCl_2$, $ZnSO_4$, or ZnO). Dietary inclusion levels in the range of 40–60 mg/kg of feed have been shown to exert beneficial effects on poultry thermal tolerance. Studies in broilers, laying hens, and quails demonstrate improved antioxidant status by increasing SOD and GSP-Px activities and reduced levels of lipid peroxidation (Sahin and Kucuk 2003; Rao et al. 2016; Saleh et al. 2018). Zhu et al. found that high-level Zn supplementation in maternal breeder diets (110 mg/kg of feed), regardless of Zn sources, alleviated the negative effect of maternal heat stress on hatch chick quality and survivability, subsequent performance during the starter period and meat quality in comparison to nonsupplemented diets (Zhu et al. 2017). Greater biological antioxidant activities and immunomodulatory effects have been observed following organic in comparison to inorganic forms in heat-stressed broilers with a concentration (Akhavan-Salamat and Ghasemi 2019). Commercially available amino acid complexed Zn (120 mg/kg of feed) has been shown to improve intestinal barrier function and villi architecture in pigs following exposure to acute short-term (12h) HS (Pearce et al. 2015). In a separate study, dietary supplementation of the same product at the same level reduced inflammatory cytokine levels in heat-stressed pigs, which may be linked to its effect on intestinal barrier function (Mayorga et al. 2018). In steers, effects on intestinal morphology were more pronounced following partial substitution of $ZnSO_4$ by Zn AA complex at a total supplementation level of 75 mg/kg (Opgenorth et al. 2021). In cows, the same dose of this Zn complex improved mammary epithelium integrity (Weng et al. 2018).

In addition to Zn and Se, other microminerals may have higher physiological requirements in HS conditions. Studies have shown that dietary levels of Cu (El-Kassas et al. 2018) and Mg (El-Kassas et al. 2018; Estevez and Petracci 2019) may also influence thermal tolerance. Commercially available chelated-trace element additives may include multiple microminerals and their supplementation has been shown to mitigate some of the effects of HS (Saleh et al. 2020; Baxter et al. 2020). The effects of chromium (Cr) supplementation in HS conditions have been extensively reviewed elsewhere (Bin-Jumah et al. 2020). Although positive effects surrounding its use have been observed, Cr additive supplementation has not been yet authorized for use in the EU, and its utilization is tightly regulated in the USA by the Food and Drug Administration (FDA) (Spears 2019).

11.2 Vitamin Supplementation

Vitamin E (a-tocopherol) is a nonenzymatic lipid-soluble, chain-breaking exogenous antioxidant that protects cellular membranes against lipid peroxyl radicals. In the presence of vitamin E, peroxyl radicals react with α-tocopherol instead of lipid hydroperoxide, stopping peroxyl radical production, and preventing further oxidation of polyunsaturated fatty acids in cell membranes (Lee and Han 2018). It has been argued that antioxidant activities may in fact account for the majority of the observed effects following its dietary supplementation (Traber and Atkinson 2007). Its potent immunomodulatory functions are related to its antioxidant functions in immune cell membranes, which are rich in polyunsaturated fatty acids and display high metabolic activities, and are thus susceptible to oxidative damage under conditions of chronic inflammation (Lewis et al. 2019), as in the case of chronic HS exposure. Vitamin E is routinely included in all premixes, commercialized as α-tocopheryl-acetate or α-tocopheryl-succinate, as it is insufficiently and variably present in feed ingredients. Across livestock species, dietary vitamin E inclusion levels may vary significantly. There is a wide discrepancy between what is recommended by the NRC, breeders, and vitamin suppliers and what is applied in commercial practice. Herein, we will assess the efficacy of offering supranutritional vitamin E supplementation levels, i.e. levels way beyond what is considered as the requirement according to different sources. It is noteworthy that vitamin E supplementation represents a significant cost in animal premixes; therefore, administration of supranutritional doses is challenging from an economic standpoint. Recent literature reviews in heat-stressed poultry indicate that supranutritional levels between 150 and 500 mg/kg of feed have exerted positive effects on feed efficiency, an endogenous antioxidant enzyme, and lipid peroxidation status, with effects extending on egg mass, quality, and oxidative status in laying hens (Horváth and Babinszky 2018; Shakeri et al. 2020; Abdel-Moneim et al. 2021). Studies on the effects of supranutritional levels of vitamin E in isolation that is, without manipulating the inclusion level of other additive components, in heat-stressed mammalian species are scarce. The addition of 200 IU of vitamin E/kg of feed over 14 days did not alleviate oxidative stress in growing pigs under HS and did not impact rectal temperature and respiration rate (Liu et al. 2018a).

Another vitamin that has been shown to be efficient in mitigating HS in poultry is vitamin C (l-ascorbic acid). Among its multiple biological roles, including hormone biosynthesis (Shakeri et al. 2020), vitamin C functions act as a water-soluble antioxidant acting as a scavenger of ROS. Importantly, it works in conjunction with vitamin E as it regenerates membrane-bound α-tocopherol from its oxidized form of α-tocopheroxyl radicals, improving vitamin E recycling and utilization efficiency (Idamokoro et al. 2020). It has been argued that vitamin C synthesis, especially in poultry, may be inadequate to cover requirements in conditions that generate oxidative stress, and its circulating concentrations have been shown to be substantially decreased under HS conditions. Previous literature reviews have established that dietary supplementation with vitamin C at doses ranging from 150 to 250 mg/kg in broilers and 250–500 mg/kg in laying hens across studies improved HS responses (Horváth and Babinszky 2018; Shakeri et al. 2020). Although physiological requirements for vitamin C seem to be sufficiently covered by endogenous synthesis in mammalian species, decreases in plasma vitamin C concentration have been observed in lactating cows (Padilla et al. 2006) and its supplementation may be beneficial under stressful situations (Akinmoladun 2022). Nonetheless, there is scarce evidence of dietary vitamin C efficacy in heat-stressed ruminants and pigs, and further research is required (Akinmoladun 2022; Li et al. 2022).

Of particular interest for heat stressed ruminants is nicotinic acid (niacin; B_3). It has long been known to cause intense skin flushing, which increases peripheral heat loss (Chen et al. 2019a). Rungruang et al. (2014) demonstrated that the blood niacin concentration of dairy cows is decreased by HS, which suggests that a greater dietary supplementation level may be required. Overall, in the majority of performed studies, niacin seems to effectively reduce dairy cow body temperature. Feeding 12 g/d of rumen-protected niacin increased free plasma niacin levels and evaporative heat loss during peak thermal load and was associated with a small but detectable reduction in rectal and vaginal temperatures in dairy cows experiencing a mild thermal load (Zimbelman et al. 2010). The trial was repeated under commercial farm conditions and similarly, reduced vaginal temperatures in lactating dairy cows supplemented with 12 g/d of rumen-protected niacin (Zimbelman et al. 2013). These results were not validated in a subsequent dose-response study (Rungruang et al. 2014), which nonetheless demonstrated an increase in water intake, which is known to be limiting for high-producing dairy cows experiencing heat stress (Collier et al. 2019). The study of Pineda et al. (2016) observed a decrease in vaginal temperatures upon supplementing 15 g/d to heat-stressed cows, having no effect in cows raised under thermoneutral conditions. Although milk yield seems to be relatively unaffected by niacin supplementation, increased milk fat percentage (Zimbelman et al. 2010; Pineda et al. 2016) and total fat yield (Lohölter et al. 2013) have been observed.

Although supplementation of other vitamins belonging to the B complex may be of interest due to their relevant biological roles, more research is required to determine their optimum supplementation levels under stress conditions (Akinyemi and Adewole 2021), not to mention under HS conditions.

12 Electrolytes and Dietary Electrolyte Balance

The efficacy of dietary supplementation with electrolytes to alleviate HS in poultry has long been investigated. In broilers, to compensate for the increased excretion via the kidney of minerals during alkalosis, it is beneficial to include increased electrolyte levels containing K^+, Na^+, and Cl^-, while maintaining an increased dietary electrolyte balance (DEB), which is expressed as $DEB = (K^+ + Na^+ - Cl^-)$ (Mongin 1981; Borges et al. 2007; Sayed and Downing 2015). Providing increased levels of individual or multiple electrolytes such as sodium bicarbonate ($NaHCO_3$), sodium chloride (NaCl), ammonium chloride (NH_4Cl), potassium chloride (KCl), potassium sulfate (K_2SO_4), especially in the form of water-soluble additives is a commonly employed strategy during HS exposure in poultry (Mushtaq et al. 2013). As a result of their efficacy, several commercial additive solutions contain electrolytes. Their supplementation is crucial to increase water intake and consequently FI and growth rates and promotes body water retention, assisting heat dissipation and limiting dehydration (Ahmad et al. 2005). The comparative efficacy of dietary electrolytes is not equal; higher growth rates have been recorded in heat-stressed broilers provided with $NaHCO_3$ compared to those provided with Na_2CO_3 when the DEB was identical (Balnave and Muheereza 1997).

Similarly, dietary supplementation with $NaHCO_3$ was the most effective among different Na salts in improving performance and reducing mortality in diets with a constant DEB value (Ahmad et al. 2005, 2006). There are implications arising from maintaining a higher electrolyte balance, and the dietary supplementation levels with different electrolytes containing Na and K may differentially affect water intake responses in growing broilers and turkeys (Mushtaq et al. 2013) and, as a consequence, adversely impact litter quality parameters and footpad dermatitis development (Mushtaq et al. 2013; Swiatkiewicz et al. 2017). As a result, targeted electrolyte supplementation in water-soluble additive solutions during the period of heat stress exposure is advisable, especially in temperate climates faced with heat waves. In heat-stressed laying hens, dietary supplementation with $NaHCO_3$ is considered important as it partially restores defective eggshell quality due to improved eggshell mineralization (Balnave and Muheereza 1997). The majority of studies investigating electrolyte supplementation in the form of $NaHCO_3$ and KCl have been shown to improve dairy cow thermotolerance as indicated by restored acid-base balance, and lowered core temperature, while increasing FI and milk yield (Coppock et al. 1982; Tucker et al. 1988; West et al. 1991, 1992). Although there are numerous studies manipulating DEB in pigs, only a few have investigated the application of electrolytes in HS conditions. In growing-finishing pigs raised in heat stress conditions, feed intake (FI) and weight gain increased linearly with increasing dietary energy balance (DEB) up to 400, and this was accompanied by concomitant increases in blood pH, HCO_3, total CO_2, and Na concentration (Haydon et al. 1990). Although feeding sows diets with an increased DEB levels resulted in increased weaning piglet weight, effects were independent of season (warm vs cold) and DEB levels were inconsequential in relation to other sow performance parameters (Dove and Haydon 1994).

13 Betaine

It is beyond the scope of this paper to review macronutrient formulation approaches in HS conditions, such as modifications in protein and energy density, EAA profile, fat, and fiber levels, etc., as it is an active field of research in its own right. However, particular attention has to be drawn to the dietary levels of methionine, choline, and betaine, which are the primary sources of methyl donors in animals, and of B complex vitamins which act as cofactors in one-carbon metabolism (McFadden et al. 2020; Fu et al. 2022). Although there are studies that have investigated the effects of dietary supplementation of methionine in heat-stressed broilers (Brake et al. 1998; Del Vesco et al. 2015) and dairy cows (Pate et al. 2020) and choline in lactating and growing ruminants (Habeeb et al. 2017; Holdorf and White 2021) recording positive effects, the vast majority of research across farm animals has focused on betaine supplementation.

Betaine is a modified amino acid consisting of glycine with three methyl groups (trimethylglycine). It is present in various dietary ingredients and is also endogenously produced by choline (Eklund et al. 2005). Betaine, as a dietary additive, may be offered in feed but also in water due to its high solubility as betaine monophosphate, anhydrous betaine, and betaine hydrochloride (Alagawany et al. 2022). Upon dietary intake, it is readily absorbed by the small intestine of monogastric (Kettunen et al. 2001) and ruminants (Nakai et al. 2013) and enters hepatic cells. Its main functions are (1) to serve as a source of methyl groups in one-carbon metabolism and (2) to act as an organic osmolyte and chemical chaperone regulating cell volume and protecting cells from stress (Day and Kempson 2016).

Its methyl donor function relates to the re-methylation of homocysteine to methionine. Briefly, the transmethylation of betaine in the methionine cycle catalyzes homocysteine to form methionine principally in the mitochondria of liver and kidney cells (Obeid 2013). This reaction replenishes the pool of methionine, rendering it available to exert its functions related to protein synthesis and for methyl group transfer to synthesize numerous substances such as creatine, phosphatidylcholine, carnitine, adrenaline, methyl purines as well as methylated amino acids (Eklund et al. 2005). Via the transsulfuration pathway, homocysteine is irreversibly converted to cysteine, which in turn leads to the formation of the endogenous antioxidant glutathione (GSH) the non-protein amino acid taurine, and other metabolites which are increasingly required in oxidative stress conditions (McFadden et al. 2020; Rakhshandeh et al. 2020; Uyanga et al. 2022). Under increased oxidative stress, betaine's contribution to the detoxification of homocysteine is significant as hyperhomocysteinemia induces oxidative stress and apoptosis (Zhao et al. 2018a) with pronounced impacts on reproductive function (Rizzo and Sciorsci 2019). Effects related to its methyl donor functions have to be considered *in tandem* with the supplementation levels of other metabolites participating in one-carbon metabolism. At least in broilers offering adequate methionine and choline levels may limit some of the expected beneficial effects attributed to betaine supplementation (Sakomura et al. 2013; Mahmoudi et al. 2018; Park and Kim 2019; Sahebi Ala et al. 2019; Sahebi-Ala et al. 2021). Its second role relates to its osmolytic activity, as intact betaine inhibits cellular dehydration during heat stress conditions by minimizing water loss against the prevailing osmotic gradient maintaining cellular integrity, and protecting proteins against heat denaturation (Lever and Slow 2010; Day and Kempson 2016). Importantly, accumulation of intact betaine intracellularly occurs in the intestine and liver, which are majorly affected by HS (Kettunen et al. 2001; Clow et al. 2008; Le et al. 2020).

There is a large body of scientific studies addressing the effects of betaine administration in heat-stressed broilers and laying hens and is an active field of research. Beneficial effects of betaine administration in broilers and laying hens are not consistently observed across studies likely due to differences in their dietary supplementation levels, basal diet nutrient characteristics such as protein level and methyl donor supply, and the modality of the HS models used. In reviewing the literature, Ratriyanto and Mosethin (2018) demonstrated that dietary supplementation between 500 and 2000 mg/kg of feed may improve multiple parameters such as: (1) broiler and hen performance, (2) broiler carcasses characteristics such as increased breast meat yield and reduced abdominal fat composition, (3) eggshell and egg quality, (4) improved cockerel breeder semen quality and egg hatchability, (5) decreased respiration rates and rectal temperatures, (6) improved nutrient digestibility, (7) decreased rectal temperatures and respiration rates, and maintenance of electrolyte levels and blood gas concentrations, (8), improved gut barrier integrity and antioxidant status, and (9) hormonal function reflected by decreased corticosterone concentration and increased thyroid function (Ratriyanto and Mosenthin 2018).

Although results from studies in broilers generally support its utilization in HS conditions, results in growing-finishing pigs are conflicting. In a study investigating short-term effects of betaine supplementation (1250 mg/kg of feed) in heat-stressed growing pigs, reduced respiration rates, gut permeability, and decreased rates of intestinal associated injury were reported (Kellner et al. 2016). In late-finishing pigs housed in high environmental temperatures, dietary betaine supplementation (2000 mg/kg of feed) did not alter carcass composition but significantly reduced FI and ADG, whereas betaine supplemented at 625, 1250, and 1875 mg/kg of feed, elicited a quadratic decrease in carcass yield and did not improve pig performance (Mendoza et al. 2017b). In a companion study, although betaine (1000 mg/kg of feed) apparently alleviated some of the effects of HS, performance remained unaffected (Mendoza et al. 2017a). Beneficial effects have been observed when offering 1000 mg/kg of betaine although essential oil supplementation was more effective in alleviating HS effects (Lan and Kim 2018). In gestating sows, betaine supplementation has been shown to yield positive results. This is likely related to its ability to reduce homocysteine concentrations and its activities as a methyl donor, thereby sparing the conversion of methionine to homocysteine. The concentration of homocysteine is linked with reduced reproductive outcomes in multiple species (Rizzo and Sciorsci 2019). Adding betaine to summer gestation diets in a large field study increased litter size, especially in sows with more parities (van Wettere et al. 2012). In pigs, the homocysteine concentration is age-dependent. In fact, plasma homocysteine levels are lower in nulliparous than in multiparous sows, which may explain these results. It has been illustrated that betaine supplementation (3000 mg/kg of feed) in lactating sows during the summer months from 2d before farrowing until weaning d21 of lactation over the hot summer increased FI and reduced weaning to estrous intervals (Cabezón et al. 2016). A subsequent study offering betaine (2200 mg/kg of feed) reduced weaning to estrous intervals and was associated with a larger follicle size (Cabezón et al. 2017).

Dietary rumen unprotected betaine largely escapes microbial degradation and is available for uptake in the small intestine (McFadden et al. 2020). A lower efficacy of dietary betaine supplementation in HS conditions compared to thermoneutral conditions has been previously observed in dairy cows. Milk production and composition were similar between mid-lactation cows that had consumed betaine supplements to the control group (Hall et al. 2016), although feeding the same quantities improved performance in thermoneutral conditions. In stark contrast, Zhang et al. (2014) offered 15 g/day to heat-stressed dairy cows for 8 weeks which increased FI, milk production, lactose, and milk protein content (Zhang et al. 2014). Moreover, antioxidant capacity was improved based on the increase of plasma cortisol, GPx, SOD, and malondialdehyde levels. A similar improvement in milk yield by feeding rumen unprotected betaine (15 g/d) was also observed in heat-stressed lactating cows under a partial grazing system (Dunshea et al. 2019). Finally, a recent study (Shah et al. 2020) showed improved milk yield, protein, and fat content and digestibility at a dose of 15 g/d but not at 30 g/d. From the aforementioned results, it is evident that heat-stressed dairy cows may benefit from lower levels of betaine supplementation than in thermoneutral conditions, but the biological reason behind this phenomenon warrants further investigation.

14 Probiotics, Prebiotics, and Postbiotics

Both in monogastric and ruminant species, the utilization of probiotics, prebiotics, or postbiotics may yield positive effects. As previously described, HS impairs intestinal health and function and leads to the development of dysbiosis, which is defined as a deregulation of gut microbiota composition. It is typically characterized by the proliferation of pathogenic genera and decreased intestinal abundance of beneficial species. In both broilers and pigs beneficial species as *Lactobacillus* and *Bifidobacterium* are reduced (Song et al. 2014; Zhang et al. 2017b; Xia et al. 2022), while increased viable counts of opportunistic pathogens are observed (Ringseis and Eder 2022 for a recent review). The use of probiotics and prebiotics, which are covered in detail in Chapter 8, has been shown to partially restore the effects of HS on microbiota composition and intestinal barrier function and integrity in heat-stressed broilers. The readers are referred to the recent reviews for further information on the topic (Patra and Kar 2021; Ringseis and Eder 2022). Overall, they benefit intestinal health by promoting the growth of beneficial commensal bacteria, inhibiting the growth of pathogens, degrading bacterial antigens, strengthening the gut barrier, downregulating inflammatory processes, and stimulating gut immunity (Cristofori et al. 2021). Observed effects mediated by probiotic strains may extend to alterations of feeding behavior and expression of thermoregulatory mechanisms (e.g., rises in skin temperature) by affecting the microbiota-gut-brain axis (Serviento et al. 2022). Postbiotics, which consist of the metabolites and bioactive compounds produced from live probiotic organisms that have been inactivated by different methods (Zhong et al. 2022), may also mediate some of the effects of probiotics and promote heat stress tolerance (Humam et al. 2020). Their application consists of a relatively novel field of research in the context of animal feed additives in HS conditions and further research is required.

In ruminants, yeast products are commercially available and have been long researched in the context of HS. These may be yeast cultures that consist of fermentation by-products, which contain compounds that affect the growth of various types of rumen bacteria and protozoa. In contrast, active live dry yeast are products that, by definition, must contain >15 billion live yeast cells/g. The most common source is *Saccharomyces cerevisiae*. Their effect is assumed dependent on the yeast cell being alive in the rumen to mediate its effects (Poppy et al. 2012). The activities of live yeast or yeast cultures are strongly linked to subclinical rumen acidosis conditions (rumen pH <6.25), with feeding high-concentrate diets being a predisposing factor. Low rumen pH is known to inhibit FI and cell wall digestion, reducing the diet's energy value, especially that of the forage component. Typical symptoms include alterations of the volatile fatty acid profile in rumen fluid with low acetate-to-propionate ratios and accumulation of lactic acid. Since ruminal acidosis is one of the hallmarks of HS in dairy cows' studies with supplementation of yeast cultures and dry yeast in HS conditions have been carried. The favorable effects are associated with the growth stimulation of cellulolytic bacteria (capacity to remove oxygen from the ruminal environment) thereby increasing the potential to enhance fiber digestion, enhancing the growth of lactic acid utilizing bacteria decreasing its concentration in ruminal fluid and, therefore, improve the ability to mitigate a decline of rumen pH after feeding (Kumprechtová et al. 2019). Bach et al. (2007) suggested that one of the mechanisms by which live yeast increases the ruminal pH and reduces the risk of subacute ruminal acidosis is by altering feeding patterns; yeast-supplemented cows show increased meal frequency, which is thought to alleviate the acid load in the rumen. DeVries and Chevaux (2014) showed that offering live yeast indeed resulted in longer rumination time, smaller meal sizes, and more meals per day. In a more recent study (Dias et al. 2018a), it has been shown that supplemented cows had 1.2 more meals per day with the same meal duration. Rumen modifiers such as yeast products may result in a more consistent fermentation pattern that improves volatile fatty acid production and fiber digestibility, and allow a quicker return to eating. Although, an amelioration of rumen acidosis would allow for an increased FI and milk yield, these effects are not consistently observed. In the majority of trials, an increased efficiency of feed conversion to milk has been reported, likely related to improved rumen function and altered feeding patterns. An increase in rumen pH (Moallem et al. 2009; Perdomo et al. 2020) and an alteration of feeding behavior patterns have been confirmed in HS conditions (Perdomo et al. 2020). Increased rumen acetate concentration (Perdomo et al. 2020) and reduced concentrations of lactate and butyrate proportions have been observed (Salvati et al. 2015). One study has demonstrated improved digestibility (Perdomo et al. 2020) of both NDF and CP, likely due to improved rumen fermentation and microbial protein synthesis and increased nutrient flow to the small intestine and dairy cow performance. However, this was not the case in other studies assessing nutrient digestibility (Moallem et al. 2009; Dias et al. 2018b). The greater plasma niacin concentration in yeast-fed animals is a plausible explanation for the reduced rectal and skin temperatures and respiration rate. The vasodilatory effects of niacin support the increased sweating rate or reduced rectal temperatures observed (Salvati et al. 2015; Dias et al. 2018b). A lower acute phase protein response (Perdomo et al. 2020) and increased glucose concentrations have also been observed in response to feeding yeast cultures, which may be attributed to a reduction of elevated immune system requirements caused by acidosis (Dias et al. 2018b), or greater synthesis of propionate, a precursor for glucose synthesis (Nasiri et al. 2018).

15 Concluding Remarks and Perspectives

The effects of HS on animal biology is a field of rigorous investigation due to the magnitude of economic losses expected to increase further in the future, challenging the sustainability of animal production. Along with our increased understanding of its consequences, there has been extensive research on the application of additive solutions to mitigate its deleterious effects. In this chapter, I have provided a brief overview of the research conducted on some of the most widely used commercially available additives authorized for global use, which mitigate the effects of heat stress (HS) in various farm animal species at different stages of production. Since different additives act on different levels, specific dietary combinations may confer a multi-layered protection against the myriad insults of HS. Although there are studies investigating combined supplementation of additives, these usually consist of combinations between vitamins and microminerals. It is arguably difficult to capture the optimum combinations and doses of feed additives simultaneously, not to mention establishing appropriate levels of co-supplementation, due to experimental limitations. Regardless of the approach taken, successful additive solutions should limit oxidative stress and inflammation and safeguard intestinal health and function. Currently, there are multiple cost-effective additive solutions tailored to different species exposed to different modalities of HS. Technological developments and research advances in the field will inevitably lead to the formulation of even more efficient feed additive solutions.

References

Abd El-Hack ME, Abdelnour SA, Taha AE et al (2020) Herbs as thermoregulatory agents in poultry: An overview. Sci Total Environ 703:1–13. https://doi.org/10.1016/j.scitotenv.2019.134399

Abdel-Moneim AME, Shehata AM, Khidr RE et al (2021) Nutritional manipulation to combat heat stress in poultry – A comprehensive review. J Therm Biol 98:102915. https://doi.org/10.1016/j.jtherbio.2021.102915

Abdelnour SA, Abd El-Hack ME, Khafaga AF et al (2019) Stress biomarkers and proteomics alteration to thermal stress in ruminants: A review. J Therm Biol 79:120–134. https://doi.org/10.1016/j.jtherbio.2018.12.013

Abulaiti A, Ahmed Z, Naseer Z et al (2021) Effect of capsaicin supplementation on lactational and reproductive performance of Holstein cows during summer. Anim Prod Sci 61:1321–1328. https://doi.org/10.1071/AN20439

Ahmad T, Sarwar M, Mahr-Un-Nisa et al (2005) Influence of varying sources of dietary electrolytes on the performance of broilers reared in a high temperature environment. Anim Feed Sci Technol 120:277–298. https://doi.org/10.1016/j.anifeedsci.2005.02.028

Ahmad T, Mushtaq T, Mahr-Un-Nisa et al (2006) Effect of different non-chloride sodium sources on the performance of heat-stressed broiler chickens. Br Poult Sci 47:249–256. https://doi.org/10.1080/00071660600753342

Ahmed SMU, Luo L, Namani A et al (2017) Nrf2 signaling pathway: Pivotal roles in inflammation. Biochim Biophys Acta - Mol Basis Dis 1863:585–597. https://doi.org/10.1016/j.bbadis.2016.11.005

Akbarian A, Michiels J, Degroote J et al (2016) Association between heat stress and oxidative stress in poultry; mitochondrial dysfunction and dietary interventions with phytochemicals. J Anim Sci Biotechnol 7:37. https://doi.org/10.1186/s40104-016-0097-5

Akhavan-Salamat H, Ghasemi HA (2019) Effect of different sources and contents of zinc on growth performance, carcass characteristics, humoral immunity and antioxidant status of broiler chickens exposed to high environmental temperatures. Livest Sci 223:76–83. https://doi.org/10.1016/j.livsci.2019.03.008

Akinmoladun OF (2022) Stress amelioration potential of vitamin C in ruminants: a review. Trop Anim Health Prod 54:24. https://doi.org/10.1007/s11250-021-03026-1

Akinyemi F, Adewole D (2021) Environmental stress in chickens and the potential effectiveness of dietary vitamin supplementation. Front Anim Sci 2:775311. https://doi.org/10.3389/fanim.2021.775311

Alagawany M, Elnesr SS, Farag MR et al (2022) Betaine and related compounds: chemistry, metabolism and role in mitigating heat stress in poultry. J Therm Biol 104:103168

Alharbi KS, Afzal O, Almalki WH et al (2022) Nuclear factor-kappa B (NF-κB) inhibition as a therapeutic target for plant nutraceuticals in mitigating inflammatory lung diseases. Chem Biol Interact 354:109842. https://doi.org/10.1016/j.cbi.2022.109842

Alhenaky A, Abdelqader A, Abuajamieh M, Al-Fataftah AR (2017) The effect of heat stress on intestinal integrity and Salmonella invasion in broiler birds. J Therm Biol 70:9–14. https://doi.org/10.1016/j.jtherbio.2017.10.015

Alhidary IA, Abdelrahman MM (2016) Effects of naringin supplementation on productive performance, antioxidant status and immune response in heat-stressed lambs. Small Rumin Res 138: 31–36. https://doi.org/10.1016/j.smallrumres.2016.03.030

Al-Zghoul MB, Alliftawi AR e S, KMM S, Jaradat ZW (2019) Expression of digestive enzyme and intestinal transporter genes during chronic heat stress in the thermally manipulated broiler chicken. Poult Sci 98:4113–4122. https://doi.org/10.3382/ps/pez249

An Z, Zhang X, Gao S, et al (2022) Effects of Capsicum Oleoresin Supplementation on Lactation Performance, Plasma Metabolites, and Nutrient Digestibility of Heat Stressed Dairy Cow

Arshad MA, Ebeid HM, Hassan F-U (2021) Revisiting the effects of different dietary sources of selenium on the health and performance of dairy animals: a review. Biol Trace Elem Res 199: 3319–3337. https://doi.org/10.1007/s12011-020-02480-6

Aslam MA, İpek E, Riaz R et al (2021) Exposure of broiler chickens to chronic heat stress increases the severity of white striping on the pectoralis major muscle. Trop Anim Health Prod 53:502. https://doi.org/10.1007/s11250-021-02950-6

Bach A, Iglesias C, Devant M (2007) Daily rumen pH pattern of loose-housed dairy cattle as affected by feeding pattern and live yeast supplementation. Anim Feed Sci Technol 136:146–153. https://doi.org/10.1016/j.anifeedsci.2006.09.011

Bagath M, Krishnan G, Devaraj C et al (2019) The impact of heat stress on the immune system in dairy cattle: A review. Res Vet Sci 126:94–102. https://doi.org/10.1016/j.rvsc.2019.08.011

Balnave D, Muheereza SK (1997) Improving eggshell quality at high temperatures with dietary sodium bicarbonate. Poult Sci 76:588–593. https://doi.org/10.1093/ps/76.4.588

Barrett NW, Rowland K, Schmidt CJ et al (2019) Effects of acute and chronic heat stress on the performance, egg quality, body temperature, and blood gas parameters of laying hens. Poult Sci 98:6684–6692. https://doi.org/10.3382/ps/pez541

Baxter MFA, Greene ES, Kidd MT et al (2020) Water amino acid-chelated trace mineral supplementation decreases circulating and intestinal HSP70 and proinflammatory cytokine gene expression in heat-stressed broiler chickens. J Anim Sci 98:skaa049. https://doi.org/10.1093/JAS/SKAA049

Beale PK, Marsh KJ, Foley WJ, Moore BD (2018) A hot lunch for herbivores: physiological effects of elevated temperatures on mammalian feeding ecology. Biol Rev 93:674–692. https://doi.org/10.1111/brv.12364

Beckford RC, Ellestad LE, Proszkowiec-Weglarz M et al (2020) Effects of heat stress on performance, blood chemistry, and hypothalamic and pituitary mRNA expression in broiler chickens. Poult Sci 99:6317–6325. https://doi.org/10.1016/j.psj.2020.09.052

Belhadj Slimen I, Najar T, Ghram A, Abdrrabba M (2016) Heat stress effects on livestock: Molecular, cellular and metabolic aspects, a review. J Anim Physiol Anim Nutr (Berl) 100: 401–412. https://doi.org/10.1111/jpn.12379

Bernabucci U, Basiricò L, Morera P et al (2015) Effect of summer season on milk protein fractions in Holstein cows. J Dairy Sci 98:1815–1827. https://doi.org/10.3168/jds.2014-8788

Bertoldo M, Grupen CG, Thomson PC et al (2009) Identification of sow-specific risk factors for late pregnancy loss during the seasonal infertility period in pigs. Theriogenology 72:393–400. https://doi.org/10.1016/j.theriogenology.2009.03.008

Biggs ME, Kroscher KA, Zhao LD et al (2020) Dietary supplementation of artificial sweetener and capsicum oleoresin as a strategy to mitigate the negative consequences of heat stress on pig performance. J Anim Sci 98:1–12. https://doi.org/10.1093/jas/skaa131

Bin-Jumah M, Abd El-Hack ME, Abdelnour SA et al (2020) Potential use of chromium to combat thermal stress in animals: a review. Sci Total Environ 707:135996. https://doi.org/10.1016/j.scitotenv.2019.135996

Bonnet S, Geraert PA, Lessire M et al (1997) Effect of high ambient temperature on feed digestibility in broilers. Poult Sci 76:857–863. https://doi.org/10.1093/ps/76.6.857

Borges SA, Da Silva AVF, Maiorka A (2007) Acid-base balance in broilers. Worlds Poult Sci J 63: 73–81. https://doi.org/10.1079/WPS2005128

Brake J, Balnave D, Dibner JJ (1998) Optimum dietary arginine:lysine ratio for broiler chickens is altered during heat stress in association with changes in intestinal uptake and dietary sodium chloride. Br Poult Sci 39:639–647. https://doi.org/10.1080/00071669888511

Cabezón FA, Schinckel AP, Richert BT et al (2016) Effect of betaine supplementation during summer on sow lactation and subsequent farrowing performance. Prof Anim Sci 32:698–706. https://doi.org/10.15232/pas.2016-01532

Cabezón FA, Stewart KR, Schinckel AP, Richert BT (2017) Effects of betaine and heat stress on lactation and postweaning reproductive performance of sows. Prof Anim Sci 33:241–253. https://doi.org/10.15232/pas.2016-01571

Cai ZY, Li XM, Liang JP et al (2018) Bioavailability of tea catechins and its improvement. Molecules 23:10–13. https://doi.org/10.3390/molecules23092346

Calamari L, Petrera F, Abeni F, Bertin G (2011) Metabolic and hematological profiles in heat stressed lactating dairy cows fed diets supplemented with different selenium sources and doses. Livest Sci 142:128–137. https://doi.org/10.1016/j.livsci.2011.07.005

Calder PC, Dimitriadis G, Newsholme P (2007) Glucose metabolism in lymphoid and inflammatory cells and tissues. Curr Opin Clin Nutr Metab Care 10:531–540. https://doi.org/10.1097/MCO.0b013e3281e72ad4

Cao C, Chowdhury VS, Cline MA, Gilbert ER (2021) The microbiota-gut-brain axis during heat stress in chickens: a review. Front Physiol 12:752265. https://doi.org/10.3389/fphys.2021.752265

Chen J, Han JH, Guan WT et al (2016) Selenium and vitamin E in sow diets: I. Effect on antioxidant status and reproductive performance in multiparous sows. Anim Feed Sci Technol 221:111–123. https://doi.org/10.1016/j.anifeedsci.2016.08.022

Chen J, Yang Z, Dong G (2019a) Niacin nutrition and rumen-protected niacin supplementation in dairy cows: an updated review. Br J Nutr 122:1103–1112. https://doi.org/10.1017/S0007114519002216

Chen J, Zhang F, Guan W et al (2019b) Increasing selenium supply for heat-stressed or actively cooled sows improves piglet preweaning survival, colostrum and milk composition, as well as maternal selenium, antioxidant status and immunoglobulin transfer. J Trace Elem Med Biol 52: 89–99. https://doi.org/10.1016/j.jtemb.2018.11.010

Chen J, Huang Z, Cao X et al (2022a) Plant-derived polyphenols in sow nutrition: An update. Anim Nutr. https://doi.org/10.1016/j.aninu.2022.08.015

Chen X, Li H, Zhang B, Deng Z (2022b) The synergistic and antagonistic antioxidant interactions of dietary phytochemical combinations. Crit Rev Food Sci Nutr 62:5658–5677. https://doi.org/10.1080/10408398.2021.1888693

Clow KA, Treberg JR, Brosnan ME, Brosnan JT (2008) Elevated tissue betaine contents in developing rats are due to dietary betaine, not to synthesis. J Nutr 138:1641–1646. https://doi.org/10.1093/jn/138.9.1641

Collier RJ, Gebremedhin KG (2015) Thermal biology of domestic animals. Annu Rev Anim Biosci 3:513–532. https://doi.org/10.1146/annurev-animal-022114-110659

Collier RJ, Baumgard LH, Zimbelman RB, Xiao Y (2019) Heat stress: physiology of acclimation and adaptation. Anim Front 9:12–19. https://doi.org/10.1093/af/vfy031

Conte G, Ciampolini R, Cassandro M et al (2018) Feeding and nutrition management of heat-stressed dairy ruminants. Ital J Anim Sci 17:604–620. https://doi.org/10.1080/1828051X.2017.1404944

Coppock CE, Grant PA, Portzer SJ et al (1982) Lactating dairy cow responses to dietary sodium, chloride, and bicarbonate during hot weather. J Dairy Sci 65:566–576. https://doi.org/10.3168/jds.S0022-0302(82)82234-0

Cottrell JJ, Furness JB, Wijesiriwardana UA et al (2020) The effect of heat stress on respiratory alkalosis and insulin sensitivity in cinnamon supplemented pigs. Animals 10:1–13. https://doi.org/10.3390/ani10040690

Cowley FC, Barber DG, Houlihan AV, Poppi DP (2015) Immediate and residual effects of heat stress and restricted intake on milk protein and casein composition and energy metabolism. J Dairy Sci 98:2356–2368. https://doi.org/10.3168/jds.2014-8442

Cristofori F, Dargenio VN, Dargenio C et al (2021) Anti-Inflammatory and Immunomodulatory Effects of Probiotics in Gut Inflammation: A Door to the Body. Front Immunol 12:178. https://doi.org/10.3389/fimmu.2021.578386

Crozier A, Jaganath IB, Clifford MN (2007) Phenols, polyphenols and tannins: an overview. Plant Second Metab Occur Struct Role Hum Diet, 1–24. https://doi.org/10.1002/9780470988558.ch1

Cui Y, Hao Y, Li J et al (2018) Proteomic changes of the porcine skeletal muscle in response to chronic heat stress. J Sci Food Agric 98:3315–3323. https://doi.org/10.1002/jsfa.8835

Cui Y, Wang C, Hao Y et al (2019) Chronic heat stress induces acute phase responses and serum metabolome changes in finishing pigs. Animals 9(7):395. https://doi.org/10.3390/ani9070395

Da Silveira E, Sá RDC, Andrade LN, De Oliveira RDRB, De Sousa DP (2014) A review on anti-inflammatory activity of phenylpropanoids found in essential oils. Molecules 19:1459–1480. https://doi.org/10.3390/molecules19021459

Dahl GE, Tao S, Laporta J (2020) Heat Stress Impacts Immune Status in Cows Across the Life Cycle. Front Vet Sci 7:1–15. https://doi.org/10.3389/fvets.2020.00116

Day CR, Kempson SA (2016) Betaine chemistry, roles, and potential use in liver disease. Biochim Biophys Acta - Gen Subj 1860:1098–1106. https://doi.org/10.1016/j.bbagen.2016.02.001

de Souza LFA, Espinha LP, de Almeida EA et al (2016) How heat stress (continuous or cyclical) interferes with nutrient digestibility, energy and nitrogen balances and performance in broilers. Livest Sci 192:39–43. https://doi.org/10.1016/j.livsci.2016.08.014

Del Vesco AP, Gasparino E, De Oliveira GD et al (2015) Effects of methionine supplementation on the expression of oxidative stress-related genes in acute heat stress-exposed broilers. Br J Nutr 113:549–559. https://doi.org/10.1017/S0007114514003535

DeVries TJ, Chevaux E (2014) Modification of the feeding behavior of dairy cows through live yeast supplementation. J Dairy Sci 97:6499–6510. https://doi.org/10.3168/jds.2014-8226

Dias ALG, Freitas JA, Micai B et al (2018a) Effects of supplementing yeast culture to diets differing in starch content on performance and feeding behavior of dairy cows. J Dairy Sci 101:186–200. https://doi.org/10.3168/jds.2017-13240

Dias JDL, Silva RB, Fernandes T et al (2018b) Yeast culture increased plasma niacin concentration, evaporative heat loss, and feed efficiency of dairy cows in a hot environment. J Dairy Sci 101: 5924–5936. https://doi.org/10.3168/jds.2017-14315

Dingeo G, Brito A, Samouda H et al (2020) Phytochemicals as modifiers of gut microbial communities. Food Funct 11:8444–8471. https://doi.org/10.1039/d0fo01483d

Dove CR, Haydon KD (1994) The effect of various diet nutrient densities and electrolyte balances on sow and litter performance during two seasons of the year. J Anim Sci 72:1101–1106. https://doi.org/10.2527/1994.7251101x

Dunshea FR, Oluboyede K, DiGiacomo K et al (2019) Betaine improves milk yield in grazing dairy cows supplemented with concentrates at high temperatures. Animals 9:3–9. https://doi.org/10.3390/ani9020057

Eklund M, Bauer E, Wamatu J, Mosenthin R (2005) Potential nutritional and physiological functions of betaine in livestock. Nutr Res Rev 18:31–48. https://doi.org/10.1079/nrr200493

El-Kassas S, Abdo SE, El-Naggar K et al (2018) Ameliorative effect of dietary supplementation of copper oxide nanoparticles on inflammatory and immune reponses in commercial broiler under normal and heat-stress housing conditions. J Therm Biol 78:235–246. https://doi.org/10.1016/j.jtherbio.2018.10.009

Emami NK, Jung U, Voy B, Dridi S (2021) Radical response: effects of heat stress-induced oxidative stress on lipid metabolism in the avian liver. Antioxidants 10:1–15. https://doi.org/10.3390/antiox10010035

Estevez M, Petracci M (2019) Benefits of magnesium supplementation to broiler subjected to dietary and heat stress: improved redox status, breast quality and decreased myopathy incidence. Antioxidants 8(10):456. https://doi.org/10.3390/antiox8100456

Fattori V, Hohmann MSN, Rossaneis AC et al (2016) Capsaicin: current understanding of its mechanisms and therapy of pain and other pre-clinical and clinical uses. Molecules 21(7):844

Feng W, Ao H, Peng C (2018) Gut microbiota, short-chain fatty acids, and herbal medicines. Front Pharmacol 9:1–12. https://doi.org/10.3389/fphar.2018.01354

Fontoura ABP, Javaid A, Sáinz de la Maza-Escolà V et al (2022) Heat stress develops with increased total-tract gut permeability, and dietary organic acid and pure botanical supplementation partly restores lactation performance in Holstein dairy cows. J Dairy Sci 105:7842–7860. https://doi.org/10.3168/jds.2022-21820

Fouad AM, Chen W, Ruan D et al (2016) Impact of heat stress on meat, egg quality, immunity and fertility in poultry and nutritional factors that overcome these effects: A review. Int J Poult Sci 15:81–95. https://doi.org/10.3923/ijps.2016.81.95

Franco-Jimenez DJ, Scheideler SE, Kittok RJ et al (2007) Differential effects of heat stress in three strains of laying hens. J Appl Poult Res 16:628–634. https://doi.org/10.3382/japr.2005-00088

Fu R, Wang Q, Kong C et al (2022) Mechanism of action and the uses betaine in pig production. J Anim Physiol Anim Nutr (Berl) 106:528–536. https://doi.org/10.1111/jpn.13633

Gabler NK, Koltes D, Schaumberger S et al (2018) Diurnal heat stress reduces pig intestinal integrity and increases endotoxin translocation. Transl Anim Sci 2:1–10. https://doi.org/10.1093/tas/txx003

Garami A, Shimansky YP, Rumbus Z et al (2020) Hyperthermia induced by transient receptor potential vanilloid-1 (TRPV1) antagonists in human clinical trials: Insights from mathematical modeling and meta-analysis. Pharmacol Ther 208:107474. https://doi.org/10.1016/j.pharmthera.2020.107474

Garriga C, Hunter RR, Amat C et al (2006) Heat stress increases apical glucose transport in the chicken jejunum. Am J Physiol - Regul Integr Comp Physiol 290:195–201. https://doi.org/10.1152/ajpregu.00393.2005

Gebreeyessus GD, Zewge F (2019) A review on environmental selenium issues. SN Appl Sci 1:1–19. https://doi.org/10.1007/s42452-018-0032-9

Gessner DK, Ringseis R, Eder K (2017) Potential of plant polyphenols to combat oxidative stress and inflammatory processes in farm animals. J Anim Physiol Anim Nutr (Berl) 101:605–628. https://doi.org/10.1111/jpn.12579

Giannenas I, Sakkas P, Papadopoulos GA et al (2022) The association of Curcuma and Scutellaria plant extracts improves laying hen thermal tolerance and egg oxidative stability and quality under heat stress conditions. Front Vet Sci 9:957847. https://doi.org/10.3389/fvets.2022.957847

Gonzalez-Rivas PA, Chauhan SS, Ha M et al (2020) Effects of heat stress on animal physiology, metabolism, and meat quality: a review. Meat Sci 162:108025. https://doi.org/10.1016/j.meatsci.2019.108025

Gregory NG (2010) How climatic changes could affect meat quality. Food Res Int 43:1866–1873. https://doi.org/10.1016/j.foodres.2009.05.018

Gruys E, Toussaint MJM, Niewold TA, Koopmans SJ (2005) Acute phase reaction and acute phase proteins. J Zhejiang Univ Sci 6(B):1045–1056. https://doi.org/10.1631/jzus.2005.B1045

Gupta A, Pandey AK (2019) Plant Secondary Metabolites With Hepatoprotective Efficacy. Nutraceuticals and Natural Product Pharmaceuticals. Elsevier, In, pp 71–104

Habeeb AAM, Gad AE, Atta MAA, Abdel-Hafez MAM (2017) Evaluation of rumen-protected choline additive to diet on productive performance of male Zaraibi growing goats during hot summer season in Egypt. Trop Anim Health Prod 49:1107–1115. https://doi.org/10.1007/s11250-017-1292-x

Habibian M, Sadeghi G, Ghazi S, Moeini MM (2015) Selenium as a feed supplement for heat-stressed poultry: a review. Biol Trace Elem Res 165:183–193. https://doi.org/10.1007/s12011-015-0275-x

Hall LW, Dunshea FR, Allen JD et al (2016) Evaluation of dietary betaine in lactating Holstein cows subjected to heat stress. J Dairy Sci 99:9745–9753. https://doi.org/10.3168/jds.2015-10514

Hao Y, Feng Y, Yang P et al (2014) Nutritional and physiological responses of finishing pigs exposed to a permanent heat exposure during three weeks. Arch Anim Nutr 68:296–308. https://doi.org/10.1080/1745039X.2014.931522

Hashemzadeh F, Rafeie F, Hadipour A, Rezadoust MH (2022) Supplementing a phytogenic-rich herbal mixture to heat-stressed lambs: Growth performance, carcass yield, and muscle and liver antioxidant status. Small Rumin Res 206:106596. https://doi.org/10.1016/j.smallrumres.2021.106596

Havlin JM, Robinson PH (2015) Intake, milk production and heat stress of dairy cows fed a citrus extract during summer heat. Anim Feed Sci Technol 208:23–32. https://doi.org/10.1016/j.anifeedsci.2015.06.022

Haydon KD, West JW, McCarter MN (1990) Effect of dietary electrolyte balance on performance and blood parameters of growing-finishing swine fed in high ambient temperatures. J Anim Sci 68:2400–2406. https://doi.org/10.2527/1990.6882400x

He SP, Arowolo MA, Medrano RF et al (2018) Impact of heat stress and nutritional interventions on poultry production. Worlds Poult Sci J 74:647–664. https://doi.org/10.1017/S0043933918000727

He J, He Y, Pan D et al (2019) Associations of gut microbiota with heat stress-induced changes of growth, fat deposition, intestinal morphology, and antioxidant capacity in ducks. Front Microbiol 10:1–13. https://doi.org/10.3389/fmicb.2019.00903

He Y, Liu Y, Tang J et al (2022) Selenium exerts protective effects against heat stress-induced barrier disruption and inflammation response in jejunum of growing pigs. J Sci Food Agric 102:496–504. https://doi.org/10.1002/jsfa.11377

Holdorf HT, White HM (2021) Effects of rumen-protected choline supplementation in Holstein dairy cows during electric heat blanket-induced heat stress. J Dairy Sci 104:9715–9725. https://doi.org/10.3168/jds.2020-19794

Honda BTB, Calefi AS, Costola-De-Souza C et al (2015) Effects of heat stress on peripheral T and B lymphocyte profiles and IgG and IgM serum levels in broiler chickens vaccinated for Newcastle disease virus. Poult Sci 94:2375–2381. https://doi.org/10.3382/ps/pev192

Horváth M, Babinszky L (2018) Impact of selected antioxidant vitamins (Vitamin A, E and C) and micro minerals (Zn, Se) on the antioxidant status and performance under high environmental temperature in poultry. A review. Acta Agric Scand A Anim Sci 68:152–160. https://doi.org/10.1080/09064702.2019.1611913

Hu R, He Y, Arowolo MA et al (2019) Polyphenols as potential attenuators of heat stress in poultry production. Antioxidants 8:1–11. https://doi.org/10.3390/antiox8030067

Humam AM, Loh TC, Foo HL et al (2020) Dietary supplementation of postbiotics mitigates adverse impacts of heat stress on antioxidant enzyme activity, total antioxidant, lipid peroxidation, physiological stress indicators, lipid profile and meat quality in broilers. Animals 10:1–21

Huynh TTT, Aarnink AJA, Verstegen MWA et al (2005) Effects of increasing temperatures on physiological changes in pigs at different relative humidities. J Anim Sci 83:1385–1396. https://doi.org/10.2527/2005.8361385x

Idamokoro EM, Falowo AD, Oyeagu CE, Afolayan AJ (2020) Multifunctional activity of vitamin E in animal and animal products: a review. Anim Sci J 91:e13352. https://doi.org/10.1111/asj.13352

Inagaki H, Kurganov E, Park Y et al (2019) Oral gavage of capsaicin causes TRPV1-dependent acute hypothermia and TRPV1-independent long-lasting increase of locomotor activity in the mouse. Physiol Behav 206:213–224. https://doi.org/10.1016/j.physbeh.2019.04.015

Iseri VJ, Klasing KC (2013) Dynamics of the systemic components of the chicken (Gallus gallus domesticus) immune system following activation by Escherichia coli; implications for the costs of immunity. Dev Comp Immunol 40:248–257. https://doi.org/10.1016/j.dci.2013.02.005
Johnson JS, Sanz Fernandez MV, Patience JF et al (2015) Effects of in utero heat stress on postnatal body composition in pigs: Ii. Finishing phase. J Anim Sci 93:82–92. https://doi.org/10.2527/jas.2014-8355
Kalpana P, Srinivasan K (2004) Digestive Stimulant action of spices.pdf. Indian J Med Res 119: 167–169
Kampman-van De Hoek E, Sakkas P, Gerrits WJJ et al (2015) Induced lung inflammation and dietary protein supply affect nitrogen retention and amino acid metabolism in growing pigs. Br J Nutr 113:414–425. https://doi.org/10.1017/S0007114514003821
Kawabata K, Yoshioka Y, Terao J (2019) Role of intestinal microbiota in the bioavailability and physiological functions of dietary polyphenols. Molecules 24:370. https://doi.org/10.3390/molecules24020370
Kellner TA, Baumgard LH, Prusa KJ et al (2016) Does heat stress alter the pig's response to dietary fat? J Anim Sci 94:4688–4703. https://doi.org/10.2527/jas.2016-0756
Kettunen H, Tiihonen K, Peuranen S et al (2001) Dietary betaine accumulates in the liver and intestinal tissue and stabilizes the intestinal epithelial structure in healthy and coccidia-infected broiler chicks. Comp Biochem Physiol - A Mol Integr Physiol 130:759–769. https://doi.org/10.1016/S1095-6433(01)00410-X
Key N, Sneeringer S, Marquardt D (2014) Climate change, heat stress, and U.S. Dairy Production. SSRN Electron J. https://doi.org/10.2139/ssrn.2506668
Kikusato M, Toyomizu M (2019) Differential effects of heat stress on oxidative status of skeletal muscle with different muscle fibre compositions in broiler chicken. J Anim Feed Sci 28:78–82. https://doi.org/10.22358/jafs/102830/2019
Kim DH, Lee YK, Lee SD et al (2020) Changes in production parameters, egg qualities, fecal volatile fatty acids, nutrient digestibility, and plasma parameters in laying hens exposed to ambient temperature. Front Vet Sci 7:1–11. https://doi.org/10.3389/fvets.2020.00412
Koch F, Thom U, Albrecht E et al (2019) Heat stress directly impairs gut integrity and recruits distinct immune cell populations into the bovine intestine. Proc Natl Acad Sci U S A 116: 10333–10338. https://doi.org/10.1073/pnas.1820130116
Kroscher KA, Fausnacht DW, McMillan RP et al (2022) Supplementation with artificial sweetener and capsaicin alters metabolic flexibility and performance in heat-stressed and feed-restricted pigs. J Anim Sci 100:1–13. https://doi.org/10.1093/jas/skac195
Kumprechtová D, Illek J, Julien C et al (2019) Effect of live yeast (Saccharomyces cerevisiae) supplementation on rumen fermentation and metabolic profile of dairy cows in early lactation. J Anim Physiol Anim Nutr (Berl) 103:447–455. https://doi.org/10.1111/jpn.13048
Lambert GP (2009) Stress-induced gastrointestinal barrier dysfunction and its inflammatory effects. J Anim Sci 87:E101–E108. https://doi.org/10.2527/jas.2008-1339
Lan R, Kim I (2018) Effects of feeding diets containing essential oils and betaine to heat-stressed growing-finishing pigs. Arch Anim Nutr 72:368–378. https://doi.org/10.1080/1745039X.2018.1492806
Lauridsen C (2019) From oxidative stress to inflammation: Redox balance and immune system. Poult Sci 98:4240–4246. https://doi.org/10.3382/ps/pey407
Le HH, Shakeri M, Suleria HAR et al (2020) Betaine and isoquinoline alkaloids protect against heat stress and colonic permeability in growing pigs. Antioxidants 9:1–15. https://doi.org/10.3390/antiox9101024
Lee GY, Han SN (2018) The role of vitamin E in immunity. Nutrients 10:1–18. https://doi.org/10.3390/nu10111614
Lever M, Slow S (2010) The clinical significance of betaine, an osmolyte with a key role in methyl group metabolism. Clin Biochem 43:732–744. https://doi.org/10.1016/j.clinbiochem.2010.03.009

Lewis ED, Meydani SN, Wu D (2019) Regulatory role of vitamin E in the immune system and inflammation. IUBMB Life 71:487–494. https://doi.org/10.1002/iub.1976

Li YR, Li GH, Zhou MX et al (2018) Discovery of natural flavonoids as activators of Nrf2-mediated defense system: Structure-activity relationship and inhibition of intracellular oxidative insults. Bioorganic Med Chem 26:5140–5150. https://doi.org/10.1016/j.bmc.2018.09.010

Li GM, Liu LP, Yin B et al (2020) Heat stress decreases egg production of laying hens by inducing apoptosis of follicular cells via activating the FasL/Fas and TNF-α systems. Poult Sci 99:6084–6093. https://doi.org/10.1016/j.psj.2020.07.024

Li H, Zhang Y, Li R et al (2021) Effect of seasonal thermal stress on oxidative status, immune response and stress hormones of lactating dairy cows. Anim Nutr 7:216–223. https://doi.org/10.1016/j.aninu.2020.07.006

Li Q, Yang S, Chen F et al (2022) Nutritional strategies to alleviate oxidative stress in sows. Anim Nutr 9:60–73. https://doi.org/10.1016/j.aninu.2021.10.006

Lian P, Braber S, Garssen J et al (2020) Beyond heat stress: Intestinal integrity disruption and mechanism-based intervention strategies. Nutrients 12:1–31. https://doi.org/10.3390/nu12030734

Liu LL, He JH, Xie HB et al (2014) Resveratrol induces antioxidant and heat shock protein mRNA expression in response to heat stress in black-boned chickens. Poult Sci 93:54–62. https://doi.org/10.3382/ps.2013-03423

Liu L, Fu C, Yan M et al (2016) Resveratrol modulates intestinal morphology and HSP70/90, NF-κB and EGF expression in the jejunal mucosa of black-boned chickens on exposure to circular heat stress. Food Funct 7:1329–1338. https://doi.org/10.1039/c5fo01338k

Liu T, Zhang L, Joo D, Sun SC (2017a) NF-κB signaling in inflammation. Signal Transduct Target Ther 2:17023. https://doi.org/10.1038/sigtrans.2017.23

Liu WC, Yun HM, Pi SH, Kim IH (2017b) Supplementing lactation diets with herbal extract mixture during summer improves the performance of sows and nursing piglets. Ann Anim Sci 17:835–847. https://doi.org/10.1515/aoas-2016-0084

Liu Y, Bao Z, Xu X et al (2017c) Extracellular signal-regulated kinase/nuclear factor-erythroid2-like2/heme oxygenase-1 pathway-mediated mitophagy alleviates traumatic brain injury-induced intestinal mucosa damage and epithelial barrier dysfunction. J Neurotrauma 34:2119–2131. https://doi.org/10.1089/neu.2016.4764

Liu F, Celi P, Chauhan SS et al (2018a) A short-term supranutritional Vitamin E supplementation alleviated respiratory alkalosis but did not reduce oxidative stress in heat stressed pigs. Asian-Australasian J Anim Sci 31:263–269. https://doi.org/10.5713/ajas.17.0256

Liu F, Celi P, Cottrell JJ et al (2018b) Effects of a short-term supranutritional selenium supplementation on redox balance, physiology and insulin-related metabolism in heat-stressed pigs. J Anim Physiol Anim Nutr (Berl) 102:276–285. https://doi.org/10.1111/jpn.12689

Liu M, Lu Y, Gao P et al (2020) Effect of curcumin on laying performance, egg quality, endocrine hormones, and immune activity in heat-stressed hens. Poult Sci 99:2196–2202. https://doi.org/10.1016/j.psj.2019.12.001

Livingston ML, Pokoo-Aikins A, Frost T et al (2022) Effect of heat stress, dietary electrolytes, and Vitamins E and C on growth performance and blood biochemistry of the broiler chicken. Front Anim Sci 3:1–11. https://doi.org/10.3389/fanim.2022.807267

Lohölter M, Meyer U, Rauls C et al (2013) Effects of niacin supplementation and dietary concentrate proportion on body temperature, ruminal pH and milk performance of primiparous dairy cows. Arch Anim Nutr 67:202–218. https://doi.org/10.1080/1745039X.2013.793048

Luo J, Song J, Liu L et al (2018) Effect of epigallocatechin gallate on growth performance and serum biochemical metabolites in heat-stressed broilers. Poult Sci 97:599–606. https://doi.org/10.3382/ps/pex353

Ma B, He X, Lu Z et al (2018) Chronic heat stress affects muscle hypertrophy, muscle protein synthesis and uptake of amino acid in broilers via insulin like growth factor-mammalian target of rapamycin signal pathway. Poult Sci 97:4150–4158. https://doi.org/10.3382/ps/pey291

Mack LA, Felver-Gant JN, Dennis RL, Cheng HW (2013) Genetic variations alter production and behavioral responses following heat stress in 2 strains of laying hens. Poult Sci 92:285–294. https://doi.org/10.3382/ps.2012-02589

Mahmoudi M, Azarfar A, Khosravinia H (2018) Partial replacement of dietary methionine with betaine and choline in heat-stressed broiler chickens. J Poult Sci 55:28–37. https://doi.org/10.2141/jpsa.0170087

Mayorga EJ, Kvidera SK, Horst EA et al (2018) Effects of zinc amino acid complex on biomarkers of gut integrity and metabolism during and following heat stress or feed restriction in pigs. J Anim Sci 96:4173–4185. https://doi.org/10.1093/jas/sky293

Mayorga EJ, Renaudeau D, Ramirez BC et al (2019) Heat stress adaptations in pigs. Anim Front 9: 54–61. https://doi.org/10.1093/af/vfy035

Mayorga EJ, Ross JW, Keating AF et al (2020) Biology of heat stress; the nexus between intestinal hyperpermeability and swine reproduction. Theriogenology 154:73–83. https://doi.org/10.1016/j.theriogenology.2020.05.023

McFadden JW, Girard CL, Tao S et al (2020) Symposium review: one-carbon metabolism and methyl donor nutrition in the dairy cow. J Dairy Sci 103:5668–5683. https://doi.org/10.3168/jds.2019-17319

Mendel M, Chłopecka M, Dziekan N, Karlik W (2017) Phytogenic feed additives as potential gut contractility modifiers—A review. Anim Feed Sci Technol 230:30–46. https://doi.org/10.1016/j.anifeedsci.2017.05.008

Mendoza SM, Boyd RD, Ferket PR, van Heugten E (2017a) Effects of dietary supplementation of the osmolyte betaine on growing pig performance and serological and hematological indices during thermoneutral and heat-stressed conditions. J Anim Sci 95:5040–5053. https://doi.org/10.2527/jas2017.1905

Mendoza SM, Boyd RD, Zier-Rush CE et al (2017b) Effect of natural betaine and ractopamine HCL on whole-body and carcass growth in pigs housed under high ambient temperatures. J Anim Sci 95:3047–3056. https://doi.org/10.2527/jas2017.1622

Mignon-Grasteau S, Moreri U, Narcy A et al (2014) Robustness to chronic heat stress in laying hens: A meta-analysis. Poult Sci 94:586–600. https://doi.org/10.3382/ps/pev028

Mitchell MA, Carlisle AJ (1992) The effects of chronic exposure to elevated environmental temperature on intestinal morphology and nutrient absorption in the domestic fowl (Gallus domesticus). Comp Biochem Physiol -- Part A Physiol 101:137–142. https://doi.org/10.1016/0300-9629(92)90641-3

Mithul Aravind S, Wichienchot S, Tsao R et al (2021) Role of dietary polyphenols on gut microbiota, their metabolites and health benefits. Food Res Int 142:110189. https://doi.org/10.1016/j.foodres.2021.110189

Moallem U, Lehrer H, Livshitz L et al (2009) The effects of live yeast supplementation to dairy cows during the hot season on production, feed efficiency, and digestibility. J Dairy Sci 92:343–351. https://doi.org/10.3168/jds.2007-0839

Mongin P (1981) Recent advances in dietary anion-cation balance: applications in poultry. Proc Nutr Soc 40:285–294. https://doi.org/10.1079/pns19810045

Monteiro APA, Tao S, Thompson IMT, Dahl GE (2016) In utero heat stress decreases calf survival and performance through the first lactation. J Dairy Sci 99:8443–8450. https://doi.org/10.3168/jds.2016-11072

Morales A, Pérez M, Castro P et al (2016) Heat stress affects the apparent and standardized ileal digestibilities of amino acids in growing pigs. J Anim Sci 94:3362–3369. https://doi.org/10.2527/jas.2016-0571

Moura DJ, Vercellino RA, Santos JPA, Vale MM (2015) Heat stress impact on weight gain in broiler chickens: A meta-analytical study of environmental factor that impact production losses. ASABE 1st Climate Change Symposium: Adaptation and Mitigation, 124–126. https://doi.org/10.13031/cc.20152143764

Mullur R, Liu YY, Brent GA (2014) Thyroid hormone regulation of metabolism. Physiol Rev 94: 355–382. https://doi.org/10.1152/physrev.00030.2013

Mushtaq MMH, Pasha TN, Mushtaq T, Parvin R (2013) Electrolytes, dietary electrolyte balance and salts in broilers: An updated review on growth performance, water intake and litter quality. Worlds Poult Sci J 69:789–802. https://doi.org/10.1017/S0043933913000810

Nakai T, Sato T, Teramura M et al (2013) The effect of a continuous supply of betaine on the degradation of betaine in the rumen of dairy cows. Biosci Biotechnol Biochem 77:666–669. https://doi.org/10.1271/bbb.120839

Nasiri AH, Towhidi A, Shakeri M et al (2018) Effects of live yeast dietary supplementation on hormonal profile, ovarian follicular dynamics, and reproductive performance in dairy cows exposed to high ambient temperature. Theriogenology 122:41–46. https://doi.org/10.1016/j.theriogenology.2018.08.013

Nawab A, Li G, An L et al (2019) Effect of curcumin supplementation on TLR4 mediated non-specific immune responses in liver of laying hens under high-temperature conditions. J Therm Biol 84:384–397. https://doi.org/10.1016/j.jtherbio.2019.07.003

Obeid R (2013) The metabolic burden of methyl donor deficiency with focus on the betaine homocysteine methyltransferase pathway. Nutrients 5(9):3481–3495. https://doi.org/10.3390/nu5093481

Oguntunji AO, Alabi OM (2010) Influence of high environmental temperature on egg production and shell quality: A review. Worlds Poult Sci J 66:739–750. https://doi.org/10.1017/S004393391000070X

Oh J, Giallongo F, Frederick T et al (2015) Effects of dietary Capsicum oleoresin on productivity and immune responses in lactating dairy cows. J Dairy Sci 98:6327–6339. https://doi.org/10.3168/jds.2014-9294

Olagaray KE, Bradford BJ (2019) Plant flavonoids to improve productivity of ruminants – A review. Anim Feed Sci Technol 251:21–36. https://doi.org/10.1016/j.anifeedsci.2019.02.004

Opgenorth J, Abuajamieh M, Horst EA et al (2021) The effects of zinc amino acid complex on biomarkers of gut integrity, inflammation, and metabolism in heat-stressed ruminants. J Dairy Sci 104:2410–2421. https://doi.org/10.3168/jds.2020-18909

Orhan C, Tuzcu M, Gencoglu H et al (2013) Epigallocatechin-3-gallate exerts protective effects against heat stress through modulating stress-responsive transcription factors in poultry. Br Poult Sci 54:447–453. https://doi.org/10.1080/00071668.2013.806787

Ortega ADSV, Szabó C (2021) Adverse effects of heat stress on the intestinal integrity and function of pigs and the mitigation capacity of dietary antioxidants: A review. Animals 11:1–16. https://doi.org/10.3390/ani11041135

Ortega ADSV, Babinszky L, Ozsváth XE et al (2022) The effect of heat stress and vitamin and micro-mineral supplementation on some mineral digestibility and electrolyte balance of pigs. Animals 12:386. https://doi.org/10.3390/ani12030386

Oteiza PI, Fraga CG, Mills DA, Taft DH (2018) Flavonoids and the gastrointestinal tract: Local and systemic effects. Mol Aspects Med 61:41–49. https://doi.org/10.1016/j.mam.2018.01.001

Ozdal T, Sela DA, Xiao J et al (2016) The reciprocal interactions between polyphenols and gut microbiota and effects on bioaccessibility. Nutrients 8:1–36. https://doi.org/10.3390/nu8020078

Padilla L, Matsui T, Kamiya Y et al (2006) Heat stress decreases plasma vitamin C concentration in lactating cows. Livest Sci 101:300–304. https://doi.org/10.1016/j.livprodsci.2005.12.002

Pan L, Bu DP, Wang JQ et al (2014) Effects of Radix Bupleuri extract supplementation on lactation performance and rumen fermentation in heat-stressed lactating Holstein cows. Anim Feed Sci Technol 187:1–8. https://doi.org/10.1016/j.anifeedsci.2013.09.008

Papatsiros VG, Katsogiannou EG, Papakonstantinou GI et al (2022) Effects of phenolic phytogenic feed additives on certain oxidative damage biomarkers and the performance of primiparous sows exposed to heat stress under field conditions. Antioxidants 11:593. https://doi.org/10.3390/antiox11030593

Park JH, Kim IH (2019) The effects of betaine supplementation in diets containing different levels of crude protein and methionine on the growth performance, blood components, total tract nutrient digestibility, excreta noxious gas emission, and meat quality of the broiler chic. Poult Sci 98:6808–6815. https://doi.org/10.3382/ps/pez412

Park JH, Pi SH, Kim IH (2016) Wachstum, blutparameter, nährstoffverdaulichkeit und fleischqualität von broilern nach Fütterung eines futters mit zusatz eines scutellaria baicalenses-extraktes. Eur Poult Sci 80:1–10. https://doi.org/10.1399/eps.2016.155

Pate RT, Luchini D, Murphy MR, Cardoso FC (2020) Effects of rumen-protected methionine on lactation performance and physiological variables during a heat stress challenge in lactating Holstein cows. J Dairy Sci 103:2800–2813. https://doi.org/10.3168/jds.2019-17305

Patience JF, Umboh JF, Chaplin RK, Nyachoti CM (2005) Nutritional and physiological responses of growing pigs exposed to a diurnal pattern of heat stress. Livest Prod Sci 96:205–214. https://doi.org/10.1016/j.livprodsci.2005.01.012

Patra AK, Kar I (2021) Heat stress on microbiota composition, barrier integrity, and nutrient transport in gut, production performance, and its amelioration in farm animals. J Anim Sci Technol 63:211–247. https://doi.org/10.5187/JAST.2021.E48

Pearce SC, Mani V, Boddicker RL et al (2012) Heat stress reduces barrier function and alters intestinal metabolism in growing pigs. J Anim Sci 90:257–259. https://doi.org/10.2527/jas.52339

Pearce SC, Mani V, Boddicker RL et al (2013) Heat Stress Reduces Intestinal Barrier Integrity and Favors Intestinal Glucose Transport in Growing Pigs. PLoS One 8:1–9. https://doi.org/10.1371/journal.pone.0070215

Pearce SC, Sanz Fernandez MV, Torrison J et al (2015) Dietary organic zinc attenuates heat stress–induced changes in pig intestinal integrity and metabolism. J Anim Sci 93:4702–4713. https://doi.org/10.2527/jas.2015-9018

Pecoraro BM, Leal DF, Frias-De-Diego A et al (2022) The health benefits of selenium in food animals: a review. J Anim Sci Biotechnol 13:1–11. https://doi.org/10.1186/s40104-022-00706-2

Perdomo MC, Marsola RS, Favoreto MG et al (2020) Effects of feeding live yeast at 2 dosages on performance and feeding behavior of dairy cows under heat stress. J Dairy Sci 103:325–339. https://doi.org/10.3168/jds.2019-17303

Pineda A, Drackley JK, Garrett J, Cardoso FC (2016) Effects of rumen-protected niacin on milk production and body temperature of middle and late lactation Holstein cows. Livest Sci 187:16–23. https://doi.org/10.1016/j.livsci.2016.02.004

Poppy GD, Rabiee AR, Lean IJ et al (2012) A meta-analysis of the effects of feeding yeast culture produced by anaerobic fermentation of Saccharomyces cerevisiae on milk production of lactating dairy cows. J Dairy Sci 95:6027–6041. https://doi.org/10.3168/jds.2012-5577

Prunier A, Messias De Bragança M, Le Dividich J (1997) Influence of high ambient temperature on performance of reproductive sows. Livest Prod Sci 52:123–133. https://doi.org/10.1016/S0301-6226(97)00137-1

Quiniou N, Renaudeau D, Dubois S, Noblet J (2000) Influence of high ambient temperatures on food intake and feeding behaviour of multiparous lactating sows. Anim Sci 70:471–479. https://doi.org/10.1017/S1357729800051821

Quinteiro-Filho WM, Ribeiro A, Ferraz-de-Paula V et al (2010) Heat stress impairs performance parameters, induces intestinal injury, and decreases macrophage activity in broiler chickens. Poult Sci 89:1905–1914. https://doi.org/10.3382/ps.2010-00812

Quinteiro-Filho WM, Gomes AVS, Pinheiro ML et al (2012) Heat stress impairs performance and induces intestinal inflammation in broiler chickens infected with Salmonella Enteritidis. Avian Pathol 41:421–427. https://doi.org/10.1080/03079457.2012.709315

Quinteiro-Filho WM, Calefi AS, Cruz DSG et al (2017) Heat stress decreases expression of the cytokines, avian β-defensins 4 and 6 and Toll-like receptor 2 in broiler chickens infected with Salmonella Enteritidis. Vet Immunol Immunopathol 186:19–28. https://doi.org/10.1016/j.vetimm.2017.02.006

Rajput N, Muhammad N, Yan R et al (2013) Effect of dietary supplementation of curcumin on growth performance, intestinal morphology and nutrients utilization of broiler chicks. J. Poult. Sci. 50:44–52

Rakhshandeh A, De Lange CFM, Htoo JK, Rakhshandeh AR (2020) Immune system stimulation increases the irreversible loss of cysteine to taurine, but not sulfate, in starter pigs. J Anim Sci 98:1–8. https://doi.org/10.1093/jas/skaa001

Rao SVR, Prakash B, Raju MVLN et al (2016) Effect of supplementing organic forms of zinc, selenium and chromium on performance, anti-oxidant and immune responses in broiler chicken reared in tropical summer. Biol Trace Elem Res 172:511–520. https://doi.org/10.1007/s12011-015-0587-x

Ratriyanto A, Mosenthin R (2018) Osmoregulatory function of betaine in alleviating heat stress in poultry. J Anim Physiol Anim Nutr (Berl) 102:1634–1650. https://doi.org/10.1111/jpn.12990

Renaudeau D, Gourdine JL, St-Pierre NR (2011) Meta-analysis of the effects of high ambient temperature on growth performance of growing-finishing pigs. J Anim Sci 89:2220–2230. https://doi.org/10.2527/jas.2010-3329

Renaudeau D, Collin A, Yahav S et al (2012) Adaptation to hot climate and strategies to alleviate heat stress in livestock production. Animal 6:707–728. https://doi.org/10.1017/S1751731111002448

Rhoads RP, Baumgard LH, Suagee JK, Sanders SR (2013) Nutritional interventions to alleviate the negative consequences of heat stress. Adv Nutr 4:267–276. https://doi.org/10.3945/an.112.003376

Ringseis R, Eder K (2022) Heat stress in pigs and broilers: role of gut dysbiosis in the impairment of the gut-liver axis and restoration of these effects by probiotics, prebiotics and synbiotics. J Anim Sci Biotechnol 13:1–16. https://doi.org/10.1186/s40104-022-00783-3

Rizzo A, Sciorsci RL (2019) Role of homocysteine metabolism in animal reproduction: A review. Res Vet Sci 122:29–35. https://doi.org/10.1016/j.rvsc.2018.11.011

Ross JW, Hale BJ, Gabler NK et al (2015) Physiological consequences of heat stress in pigs. Anim Prod Sci 55:1381–1390. https://doi.org/10.1071/AN15267

Ross JW, Hale BJ, Seibert JT et al (2017) Physiological mechanisms through which heat stress compromises reproduction in pigs. Mol Reprod Dev 84:934–945. https://doi.org/10.1002/mrd.22859

Rostagno MH (2020) Effects of heat stress on the gut health of poultry. J Anim Sci 98:1–9. https://doi.org/10.1093/jas/skaa090

Ruff J, Barros TL, Tellez G et al (2020) Research note: evaluation of a heat stress model to induce gastrointestinal leakage in broiler chickens. Poult Sci 99:1687–1692. https://doi.org/10.1016/j.psj.2019.10.075

Rungruang S, Collier JL, Rhoads RP et al (2014) A dose-response evaluation of rumen-protected niacin in thermoneutral or heat-stressed lactating Holstein cows. J Dairy Sci 97:5023–5034. https://doi.org/10.3168/jds.2013-6970

Sahebi Ala F, Hassanabadi A, Golian A (2019) Effects of dietary supplemental methionine source and betaine replacement on the growth performance and activity of mitochondrial respiratory chain enzymes in normal and heat-stressed broiler chickens. J Anim Physiol Anim Nutr (Berl) 103:87–99. https://doi.org/10.1111/jpn.13005

Sahebi-Ala F, Hassanabadi A, Golian A (2021) Effect of replacement different methionine levels and sources with betaine on blood metabolites, breast muscle morphology and immune response in heat-stressed broiler chickens. Ital J Anim Sci 20:33–45. https://doi.org/10.1080/1828051X.2020.1868358

Sahin K, Kucuk O (2003) Zinc supplementation alleviates heat stress in laying Japanese quail. J Nutr 133:2808–2811. https://doi.org/10.1093/jn/133.9.2808

Sahin K, Akdemir F, Orhan C et al (2010a) Effects of dietary resveratrol supplementation on egg production and antioxidant status. Poult Sci 89:1190–1198. https://doi.org/10.3382/ps.2010-00635

Sahin K, Orhan C, Tuzcu M et al (2010b) Epigallocatechin-3-gallate prevents lipid peroxidation and enhances antioxidant defense system via modulating hepatic nuclear transcription factors in heat-stressed quails. Poult Sci 89:2251–2258. https://doi.org/10.3382/ps.2010-00749

Sakomura NK, Barbosa NAA, Longo FA et al (2013) Effect of dietary betaine supplementation on the performance, carcass yield, and intestinal morphometrics of broilers submitted to heat stress. Rev Bras Cienc Avic 15:105–111. https://doi.org/10.1590/S1516-635X2013000200005

Saleh AA, Ragab MM, Ahmed EAM et al (2018) Effect of dietary zinc-methionine supplementation on growth performance, nutrient utilization, antioxidative properties and immune response in broiler chickens under high ambient temperature. J Appl Anim Res 46:820–827. https://doi.org/10.1080/09712119.2017.1407768

Saleh AA, Eltantawy MS, Gawish EM et al (2020) Impact of dietary organic mineral supplementation on reproductive performance, egg quality characteristics, lipid oxidation, ovarian follicular development, and immune response in laying hens under high ambient temperature. Biol Trace Elem Res 195:506–514. https://doi.org/10.1007/s12011-019-01861-w

Salvati GGS, Morais Júnior NN, Melo ACS et al (2015) Response of lactating cows to live yeast supplementation during summer. J Dairy Sci 98:4062–4073. https://doi.org/10.3168/jds.2014-9215

Sayed MAM, Downing J (2015) Effects of dietary electrolyte balance and addition of electrolyte–betaine supplements in feed or water on performance, acid–base balance and water retention in heat-stressed broilers. Br Poult Sci 56:195–209. https://doi.org/10.1080/00071668.2014.995594

Scanes CG (2016) Biology of stress in poultry with emphasis on glucocorticoids and the heterophil to lymphocyte ratio. Poult Sci 95:2208–2215. https://doi.org/10.3382/ps/pew137

Sejian V, Naqvi SMK, Ezeji T et al (2013) Environmental stress and amelioration in livestock production. Springer, Berlin, pp 1–569. https://doi.org/10.1007/978-3-642-29205-7

Sejian V, Bhatta R, Gaughan JB et al (2018a) Review: adaptation of animals to heat stress. Animal 12:S431–S444. https://doi.org/10.1017/S1751731118001945

Sejian V, Krishnan G, Amitha JP (2018b) Thermal indices and heat stress assessment in livestock: way forward. Biomed J Sci Tech Res 3:2975–2976. https://doi.org/10.26717/bjstr.2018.03.000839

Serviento AM, Castex M, Renaudeau D, Labussière E (2022) Effect of live yeast supplementation and feeding frequency in male finishing pigs subjected to heat stress. Br J Nutr. https://doi.org/10.1017/S0007114522002513

Shah AM, Ma J, Wang Z et al (2020) Betaine supplementation improves the production performance, rumen fermentation, and antioxidant profile of dairy cows in heat stress. Animals 10:634. https://doi.org/10.3390/ani10040634

Shakeri M, Oskoueian E, Le HH, Shakeri M (2020) Strategies to combat heat stress in broiler chickens: unveiling the roles of selenium, vitamin E and vitamin C. Vet Sci 7:1–9. https://doi.org/10.3390/VETSCI7020071

Shan CH, Guo J, Sun X et al (2018) Effects of fermented Chinese herbal medicines on milk performance and immune function in late-lactation cows under heat stress conditions. J Anim Sci 96:4444–4457. https://doi.org/10.1093/jas/sky270

Shini S, Shini A, Huff GR (2009) Effects of chronic and repeated corticosterone administration in rearing chickens on physiology, the onset of lay and egg production of hens. Physiol Behav 98:73–77. https://doi.org/10.1016/j.physbeh.2009.04.012

Shini S, Huff GR, Shini A, Kaiser P (2010) Understanding stress-induced immunosuppression: Exploration of cytokine and chemokine gene profiles in chicken peripheral leukocytes. Poult Sci 89:841–851. https://doi.org/10.3382/ps.2009-00483

Smith SM, Vale WW (2006) The role of the hypothalamic-pituitary-adrenal axis in neuroendocrine responses to stress. Dialogues Clin Neurosci 8:383–395. https://doi.org/10.31887/dcns.2006.8.4/ssmith

Song X, Xu J, Wang T, Liu F (2009) Chinese medicine granule affects the absorption and transport of glucose in porcine small intestinal brush border membrane vesicles under heat stress. Asian-Australasian J Anim Sci 22:246–253. https://doi.org/10.5713/ajas.2009.80358

Song X, Xu J, Wang T, Liu F (2010) Traditional Chinese medicine decoction enhances growth performance and intestinal glucose absorption in heat stressed pigs by up-regulating the

expressions of SGLT1 and GLUT2 mRNA. Livest Sci 128:75–81. https://doi.org/10.1016/j.livsci.2009.11.002
Song J, Xiao K, Ke YL et al (2014) Effect of a probiotic mixture on intestinal microflora, morphology, and barrier integrity of broilers subjected to heat stress. Poult Sci 93:581–588. https://doi.org/10.3382/ps.2013-03455
Spears JW (2019) Boron, chromium, manganese, and nickel in agricultural animal production. Biol Trace Elem Res 188:35–44. https://doi.org/10.1007/s12011-018-1529-1
Spiers JG, Chen HJC, Sernia C, Lavidis NA (2015) Activation of the hypothalamic-pituitary-adrenal stress axis induces cellular oxidative stress. Front Neurosci 9:1–6. https://doi.org/10.3389/fnins.2014.00456
Star L, Decuypere E, Parmentier HK, Kemp B (2008) Effect of single or combined climatic and hygienic stress in four layer lines: 2. Endocrine and oxidative stress responses. Poult Sci 87: 1031–1038. https://doi.org/10.3382/ps.2007-00143
St-Pierre NR, Cobanov B, Schnitkey G (2003) Economic losses from heat stress by US livestock industries1. J Dairy Sci 86:E52–E77. https://doi.org/10.3168/jds.S0022-0302(03)74040-5
Summer A, Formaggioni P, Tosi F et al (1999) Effects of the hot-humid climate on rennet-coagulation properties of milk produced during summer of 1998 and relationship with the housing systems in the rearing of Italian Friesian cows. Ann Fac Med Vet Univ Parma 19: 167–179
Summer A, Lora I, Formaggioni P, Gottardo F (2019) Impact of heat stress on milk and meat production. Anim Front 9:39–46. https://doi.org/10.1093/af/vfy026
Sun X, Zhang H, Sheikhahmadi A et al (2015) Effects of heat stress on the gene expression of nutrient transporters in the jejunum of broiler chickens (Gallus gallus domesticus). Int J Biometeorol 59:127–135. https://doi.org/10.1007/s00484-014-0829-1
Sun LL, Gao ST, Wang K et al (2019) Effects of source on bioavailability of selenium, antioxidant status, and performance in lactating dairy cows during oxidative stress-inducing conditions. J Dairy Sci 102:311–319. https://doi.org/10.3168/jds.2018-14974
Sun H, Zhao L, Xu ZJ et al (2021) Hydroxy-selenomethionine improves the selenium status and helps to maintain broiler performances under a high stocking density and heat stress conditions through a better redox and immune response. Antioxidants 10(10):1542. https://doi.org/10.3390/antiox10101542
Surai PF (2014) Polyphenol compounds in the chicken/animal diet: from the past to the future. J Anim Physiol Anim Nutr (Berl) 98:19–31. https://doi.org/10.1111/jpn.12070
Surai PF, Fisinin VI (2014) Selenium in poultry breeder nutrition: an update. Anim Feed Sci Technol 191:1–15. https://doi.org/10.1016/j.anifeedsci.2014.02.005
Surai PF, Kochish II, Fisinin VI, Juniper DT (2019a) Revisiting oxidative stress and the use of organic selenium in dairy cow nutrition. Animals 9:462. https://doi.org/10.3390/ani9070462
Surai PF, Kochish II, Fisinin VI, Kidd MT (2019b) Antioxidant defence systems and oxidative stress in poultry biology: an update. Antioxidants 8:1–36. https://doi.org/10.3390/antiox8070235
Swiatkiewicz S, Arczewska-Wlosek A, Jozefiak D (2017) The nutrition of poultry as a factor affecting litter quality and foot pad dermatitis – an updated review. J Anim Physiol Anim Nutr (Berl) 101:e14–e20. https://doi.org/10.1111/jpn.12630
Tao S, Bubolz JW, do Amaral BC et al (2011) Effect of heat stress during the dry period on mammary gland development. J Dairy Sci 94:5976–5986. https://doi.org/10.3168/jds.2011-4329
Tao S, Monteiro APA, Thompson IM et al (2012) Effect of late-gestation maternal heat stress on growth and immune function of dairy calves. J Dairy Sci 95:7128–7136. https://doi.org/10.3168/jds.2012-5697
Tao S, Monteiro APA, Hayen MJ, Dahl GE (2014) Short communication: Maternal heat stress during the dry period alters postnatal whole-body insulin response of calves. J Dairy Sci 97: 897–901. https://doi.org/10.3168/jds.2013-7323

Tao S, Orellana Rivas RM, Marins TN et al (2020) Impact of heat stress on lactational performance of dairy cows. Theriogenology 150:437–444. https://doi.org/10.1016/j.theriogenology.2020.02.048

Teeter RG, Smith MO, Owens FN et al (1985) Chronic heat stress and respiratory alkalosis: occurrence and treatment in broiler chicks. Poult Sci 64:1060–1064. https://doi.org/10.3382/ps.0641060

Tian B, Liu J (2020) Resveratrol: a review of plant sources, synthesis, stability, modification and food application. J Sci Food Agric 100:1392–1404. https://doi.org/10.1002/jsfa.10152

Traber MG, Atkinson J (2007) Vitamin E, antioxidant and nothing more. Free Radic Biol Med 43: 4–15. https://doi.org/10.1016/j.freeradbiomed.2007.03.024

Tsiouris V, Georgopoulou I, Batzios C et al (2018) Heat stress as a predisposing factor for necrotic enteritis in broiler chicks. Avian Pathol 47:616–624. https://doi.org/10.1080/03079457.2018.1524574

Tucker WB, Harrison GA, Hemken RW (1988) Influence of dietary cation-anion balance on milk, blood, urine, and rumen fluid in lactating dairy cattle. J Dairy Sci 71:346–354. https://doi.org/10.3168/jds.S0022-0302(88)79563-6

Urkmez E, Biricik H (2022a) Grape seed extract supplementation in heat-stressed preweaning dairy calves: I. Effects on antioxidant status, inflammatory response, hematological and physiological parameters. Anim Feed Sci Technol 292:115421. https://doi.org/10.1016/j.anifeedsci.2022.115421

Urkmez E, Biricik H (2022b) Grape seed extract supplementation in heat-stressed preweaning dairy calves: II. Effects on growth performance, blood metabolites, hormonal responses, and fecal fermentation parameters. Anim Feed Sci Technol 292:115422. https://doi.org/10.1016/j.anifeedsci.2022.115422

Uyanga VA, Oke EO, Amevor FK et al (2022) Functional roles of taurine, L-theanine, L-citrulline, and betaine during heat stress in poultry. J Anim Sci Biotechnol 13:1–20. https://doi.org/10.1186/s40104-022-00675-6

van Wettere WHEJ, Herde P, Hughes PE (2012) Supplementing sow gestation diets with betaine during summer increases litter size of sows with greater numbers of parities. Anim Reprod Sci 132:44–49. https://doi.org/10.1016/j.anireprosci.2012.04.007

Varmuzova K, Matulova ME, Gerzova L et al (2015) Curcuma and Scutellaria plant extracts protect chickens against inflammation and Salmonella Enteritidis infection. Poult Sci 94:2049–2058. https://doi.org/10.3382/ps/pev190

Vats P, Kaur UJ, Rishi P (2022) Heavy metal-induced selection and proliferation of antibiotic resistance: A review. J Appl Microbiol 132:4058–4076. https://doi.org/10.1111/jam.15492

Wang XJ, Liu L, Zhao JP et al (2017) Stress impairs the reproduction of laying hens: An involvement of energy. Worlds Poult Sci J 73:845–855. https://doi.org/10.1017/S0043933917000794

Wang WC, Yan FF, Hu JY et al (2018) Supplementation of Bacillus subtilis-based probiotic reduces heat stress-related behaviors and inflammatory response in broiler chickens. J Anim Sci 96:1654–1666. https://doi.org/10.1093/jas/sky092

Wang J, Jia R, Celi P et al (2020) Green tea polyphenol epigallocatechin-3-gallate improves the antioxidant capacity of eggs. Food Funct 11:534–543. https://doi.org/10.1039/c9fo02157d

Wang X, Chang X, Zhan H et al (2021) Curcumin combined with Baicalin attenuated ethanol-induced hepatitis by suppressing p38 MAPK and TSC1/ eIF-2α/ATF4 pathways in rats. Food Biosci 40:100851. https://doi.org/10.1016/j.fbio.2020.100851

Wang W, Kang R, Liu M et al (2022) Effects of different selenium sources on the laying performance, egg quality, antioxidant, and immune responses of laying hens under normal and cyclic high temperatures. Animals 12(8):1006

Weng X, Monteiro APA, Guo J et al (2018) Effects of heat stress and dietary zinc source on performance and mammary epithelial integrity of lactating dairy cows. J Dairy Sci 101:2617–2630. https://doi.org/10.3168/jds.2017-13484

West JW, Mullinix BG, Sandifer TG (1991) Changing dietary electrolyte balance for dairy cows in cool and hot environments. J Dairy Sci 74:1662–1674. https://doi.org/10.3168/jds.S0022-0302(91)78329-X

West JW, Haydon KD, Mullinix BG, Sandifer TG (1992) Dietary cation-anion balance and cation source effects on production and acid-base status of heat-stressed cows. J Dairy Sci 75:2776–2786. https://doi.org/10.3168/jds.S0022-0302(92)78041-2

Wettemann RP, Wells ME, Omtvedt IT et al (1976) Influence of elevated ambient temperature on reproductive performance of boars. J Anim Sci 42:664–669. https://doi.org/10.2527/jas1976.423664x

Williamson G, Kay CD, Crozier A (2018) The bioavailability, transport, and bioactivity of dietary flavonoids: a review from a historical perspective. Compr Rev Food Sci Food Saf 17:1054–1112. https://doi.org/10.1111/1541-4337.12351

Wolfenson D, Roth Z (2019) Impact of heat stress on cow reproduction and fertility. Anim Front 9:32–38. https://doi.org/10.1093/af/vfy027

Wu KC, McDonald PR, Liu J, Klaassen CD (2014) Screening of natural compounds as activators of the keap1-nrf2 pathway. Planta Med 80:97–104. https://doi.org/10.1055/s-0033-1351097

Xia B, Wu W, Fang W et al (2022) Heat stress-induced mucosal barrier dysfunction is potentially associated with gut microbiota dysbiosis in pigs. Anim Nutr 8:289–299. https://doi.org/10.1016/j.aninu.2021.05.012

Xu J, Li S, Jiang L et al (2021) Baicalin protects against zearalenone-induced chicks liver and kidney injury by inhibiting expression of oxidative stress, inflammatory cytokines and caspase signaling pathway. Int Immunopharmacol 100:108097. https://doi.org/10.1016/j.intimp.2021.108097

Yang M, Lu Y, Zhu Q et al (2019) Effect of tea polyphenol on serum hormone, serum enzyme activity and antioxidant-related gene expression in chinese yellow chicken under heat. Stress. https://doi.org/10.20944/preprints201908.0009.v1

Yang JY, Li M, Zhang CL, Liu D (2021) Pharmacological properties of baicalin on liver diseases: a narrative review. Pharmacol Reports 73:1230–1239. https://doi.org/10.1007/s43440-021-00227-1

Yilmazer-Musa M, Griffith AM, Michels AJ et al (2012) Inhibition of α-Amylase and α-Glucosidase Activity by Tea and Grape Seed Extracts and their Constituent Catechins. J Agric Food Chem 60:8924–8929. https://doi.org/10.1021/jf301147n.Inhibition

Zaboli G, Huang X, Feng X, Ahn DU (2019) How can heat stress affect chicken meat quality? - A review. Poult Sci 98:1551–1556. https://doi.org/10.3382/ps/pey399

Zaefarian F, Abdollahi MR, Cowieson A, Ravindran V (2019) Avian liver: the forgotten organ. Animals 9:1–23. https://doi.org/10.3390/ani9020063

Zhang ZY, Jia GQ, Zuo JJ et al (2012) Effects of constant and cyclic heat stress on muscle metabolism and meat quality of broiler breast fillet and thigh meat. Poult Sci 91:2931–2937. https://doi.org/10.3382/ps.2012-02255

Zhang L, An WJ, Lian H et al (2014) Effects of dietary betaine supplementation subjected to heat stress on milk performances and physiology indices in dairy cow. Genet Mol Res 13:7577–7586. https://doi.org/10.4238/2014.September.12.25

Zhang J, Hu Z, Lu C et al (2015a) Effect of various levels of dietary curcumin on meat quality and antioxidant profile of breast muscle in broilers. J Agric Food Chem 63:3880–3886. https://doi.org/10.1021/jf505889b

Zhang JF, Hu ZP, Lu CH et al (2015b) Dietary curcumin supplementation protects against heat-stress-impaired growth performance of broilers possibly through a mitochondrial pathway. J Anim Sci 93:1656–1665. https://doi.org/10.2527/jas.2014-8244

Zhang C, Zhao X, Wang L et al (2017a) Resveratrol beneficially affects meat quality of heat-stressed broilers which is associated with changes in muscle antioxidant status. Anim Sci J 88:1569–1574. https://doi.org/10.1111/asj.12812

Zhang C, Zhao XH, Yang L et al (2017b) Resveratrol alleviates heat stress-induced impairment of intestinal morphology, microflora, and barrier integrity in broilers. Poult Sci 96:4325–4332. https://doi.org/10.3382/ps/pex266

Zhang J, Bai KW, He J et al (2018a) Curcumin attenuates hepatic mitochondrial dysfunction through the maintenance of thiol pool, inhibition of mtDNA damage, and stimulation of the mitochondrial thioredoxin system in heat-stressed broilers. J Anim Sci 96:867–879. https://doi.org/10.1093/jas/sky009

Zhang JF, Bai KW, Su WP et al (2018b) Curcumin attenuates heat-stress-induced oxidant damage by simultaneous activation of GSH-related antioxidant enzymes and Nrf2-mediated phase II detoxifying enzyme systems in broiler chickens. Poult Sci 97:1209–1219. https://doi.org/10.3382/ps/pex408

Zhang W, Heng J, Kim SW et al (2020) Dietary enzymatically-treated Artemisia annua L. supplementation could alleviate oxidative injury and improve reproductive performance of sows reared under high ambient temperature. J Therm Biol 94:102751. https://doi.org/10.1016/j.jtherbio.2020.102751

Zhao G, He F, Wu C et al (2018a) Betaine in inflammation: Mechanistic aspects and applications. Front Immunol 9:1–13. https://doi.org/10.3389/fimmu.2018.01070

Zhao L, McMillan RP, Xie G et al (2018b) Heat stress decreases metabolic flexibility in skeletal muscle of growing pigs. Am J Physiol - Regul Integr Comp Physiol 315:R1096–R1106. https://doi.org/10.1152/ajpregu.00404.2017

Zhao F, Wang X, Li Y et al (2021) Effects of dietary supplementation with epigallocatechin gallate on meat quality and muscle antioxidant capacity of broilers subjected to acute heat stress. Animals 11:1–10. https://doi.org/10.3390/ani11113296

Zhao Y, Li Z, Wang X et al (2022) Resveratrol attenuates heat stress-induced impairment of meat quality in broilers by regulating the Nrf2 signaling pathway. Animals 12:1–9. https://doi.org/10.3390/ani12151889

Zhong Y, Wang S, Di H et al (2022) Gut health benefit and application of postbiotics in animal production. J Anim Sci Biotechnol 13:1–12. https://doi.org/10.1186/s40104-022-00688-1

Zhu YW, Li WX, Lu L et al (2017) Impact of maternal heat stress in conjunction with dietary zinc supplementation on hatchability, embryonic development, and growth performance in offspring broilers. Poult Sci 96:2351–2359. https://doi.org/10.3382/ps/pew481

Zimbelman RB, Baumgard LH, Collier RJ (2010) Effects of encapsulated niacin on evaporative heat loss and body temperature in moderately heat-stressed lactating Holstein cows. J Dairy Sci 93:2387–2394. https://doi.org/10.3168/jds.2009-2557

Zimbelman RB, Collier RJ, Bilby TR (2013) Effects of utilizing rumen protected niacin on core body temperature as well as milk production and composition in lactating dairy cows during heat stress. Anim Feed Sci Technol 180:26–33. https://doi.org/10.1016/j.anifeedsci.2013.01.005

Zmrhal V, Lichovníková M, Hampel D (2018) The effect of phytogenic additive on behavior during mild-moderate heat stress in broilers. Acta Univ Agric Silvic Mendelianae Brun 66:939–944. https://doi.org/10.11118/actaun201866040939

Zulkifli I, Akmal AF, Soleimani AF et al (2018) Effects of low-protein diets on acute phase proteins and heat shock protein 70 responses, and growth performance in broiler chickens under heat stress condition. Poult Sci 97:1306–1314. https://doi.org/10.3382/ps/pex436

Feed Additives as Antiviral Agents

A. R. Yasmin, M. N. Sohaimi, S. N. A. Azaman, S. H. Nur-Fazila, and F. M. A. Amirul

Abstract

Viral infection is a major devastating challenge in the livestock industry. Hardy to completely eradicate the virus, the emergence of new variants accompanied by high morbidity and mortality further substantiates the problems in poultry, ruminant, and swine farms. Notifiable diseases caused by virus infection such as avian influenza (AI), infectious bursal disease (IBD), Newcastle disease (ND), foot and mouth disease (FMD), Aujeszky's disease, classical swine fever (CSF),

A. R. Yasmin (✉) · M. N. Sohaimi
Department of Veterinary Laboratory Diagnosis, Faculty of Veterinary Medicine, Universiti Putra Malaysia, Selangor, Malaysia

Laboratory of Vaccine and Biomolecules, Institute of Bioscience, Universiti Putra Malaysia, Selangor, Malaysia
e-mail: noryasmin@upm.edu.my; fitriahsohaimi@upm.edu.my

S. N. A. Azaman
Aquatic Animal Health and Therapeutics Laboratory (AquaHealth), Universiti Putra Malaysia, Selangor, Malaysia

Centre of Foundation Studies for Agricultural Sciences, Universiti Putra Malaysia, Serdang, Selangor, Malaysia
e-mail: s_norani@upm.edu.my

S. H. Nur-Fazila
Laboratory of Vaccine and Biomolecules, Institute of Bioscience, Universiti Putra Malaysia, Selangor, Malaysia

Department of Veterinary Pathology and Microbiology, Faculty of Veterinary Medicine, Universiti Putra Malaysia, UPM Serdang, Selangor, Malaysia
e-mail: nurfazila@upm.edu.my

F. M. A. Amirul
Department of Preclinical Science, Faculty of Veterinary Medicine, Universiti Malaysia Kelantan, Kota Bharu, Malaysia
e-mail: amirul.ma@umk.edu.my5

G. Arsenos, I. Giannenas (eds.), *Sustainable Use of Feed Additives in Livestock*,
https://doi.org/10.1007/978-3-031-42855-5_12

and other viral diseases are controlled by vaccination, proper husbandry, and strict biosecurity. The nutritional factors also play a pivotal role in ensuring optimum growth and immunity of the livestock as well as safeguarding sustainable food security as a protein source to human. Continuous global demand requires crucial maintenance of the good quality and value of livestock-based protein. It could be achieved by keeping the livestock in constant good health and by receiving proper and adequate nutrition. In order to fulfill the needs, feed additives are designed to be supplemented to animal feeds to improve the feed taste, stabilize gut microbes and to compensate nutritional deficiencies. In addition, feed additives function as antiviral, antioxidant, anticancer and exert immunostimulatory effects in combating pathogen. In this chapter, antiviral effects of feed additives are discussed by categories, namely, nutritional additives, zootechnical additives, natural additives, technological additives, and sensory additives. Most of the antiviral effects of the additives are studied for the usage in livestock, while some examples of additives are based on human virus studies, which could be potentially utilized for further usage in future in livestock.

Keywords

Antiviral · Feed additives · Immunostimulatory · Livestock

1 Introduction

Feed additives are an important product that contains chemical or natural substances, microorganisms, or mixtures that function to improve the feed taste, increase consumption, stabilize gut microbes, and to enhance the supplementation of nutrient deficiencies in livestock (Górniak et al. 2018). The usage of feed additives is pivotal in the supplementation of good nutrition to the livestock, thus enhancing the immunity of the livestock in combating pathogens particularly caused by viral infection (Thirumdas et al. 2021). Notifiable viral diseases such as avian influenza (AI), infectious bursal disease (IBD), Newcastle disease (NDV), foot and mouth disease (FMD), Aujeszky's disease and classical swine fever (CSF), and other viral diseases are major problems in farms (WOAH 2020). Combined effects of proper husbandry, herd health program includes vaccination, strict biosecurity together with supplementation of nutrition with good quality feed additives will ensure optimum health of the livestock. Antiviral effects of feed additives are discussed based on categorical value, namely, nutritional additives, zootechnical additives, natural additives, technological additives, and sensory additives as shown in Table 1.

1.1 Nutritional Additives

Nutritional additives are macronutrients and micronutrients that are required in livestock feed to enrich the nutrients and restore the loss of nutrients or prevent

Table 1 Categories of feed additives and their examples

Categories of feed additives	Examples
Nutritional additives	• Vitamins and minerals • Polysaccharide • Peptide/amino acid • Fatty acid
Zootechnical additives	• Probiotics • Prebiotics • Synbiotics
Natural additives	• Essential oils • Algae
Technological additives	• Preservatives • Antioxidant • Emulsifiers
Sensory additives	• Flavorings • Colorings

Note: These feed additives exert antiviral properties and immunomodulatory effects following several viral infections

nutrient deficiencies. Therapeutic antiviral and immune booster functions of micronutrients such as vitamins and minerals as well as macronutrients such as polysaccharides, peptides, and fatty acids will be discussed in this section.

1.1.1 Vitamins and Minerals

In poultry production, diseases like Newcastle diseases, avian influenza, and infectious bursal disease virus (IBDV) are controlled by vaccination. Studies have shown that the addition of nutritional additives such as vitamins and minerals inhibits viral activity and promotes the immune system in combating viral infection.

Vitamin E or tocopherol is a potent antioxidant that is also capable of enhancing immunity. In Newcastle disease virus (NDV), vitamin E has been shown to ameliorate the oxidative stress effects evidenced by histopathological changes in the gastrointestinal tract of the experimental chicken (Rehman et al. 2018).

Vitamin D is a fat-soluble secosteroid that is important for supporting the immune system and increasing certain minerals' intestinal absorption such as magnesium, calcium, and phosphate. In infectious bursal disease virus (IBDV) infection, a strong innate response was shown following vitamin D supplementation by the production of proinflammatory cytokines and several toll-like receptors that generate a good antiviral activity of the supplement (Jaime et al. 2020).

Selenium is a trace mineral and essential for metabolism, production, immune response, and improved cognition. Deficiency of selenium in livestock leads to immunosuppression, infertile, low milk yields, and reduced weight gains (Hefnawy and Tórtora-Pérez 2010). Supplementation of broiler chicken with selenium has increased serum antibody titer following highly pathogenic avian influenza infection, H5N1 suggestive of protective immunity against the infection (Liao et al. 2012).

In the swine industry, the usages of nutritional additives such as vitamins and minerals also have shown a positive effect in combating viral infection.

Supplementation of vitamin D to weaned pigs has reduced intestinal damage caused by porcine epidemic diarrhea (PEDV) and protected further inflammatory lesions on the gastrointestinal tract of the pigs (Yang et al. 2019).

Supplementation of selenium minerals also has shown antiviral effects against porcine circovirus type 2 (PCV2), a causative agent of postweaning multisystemic wasting syndrome (PMWS). PCV2 replication was inhibited and anti-oxidant effects of selenium were demonstrated following in vitro study (Pecoraro et al. 2022).

Moreover, the usage of vitamins and minerals as antiviral and immune boosters has also been demonstrated in many human in vitro and murine models. Vitamin A exerts antiviral properties and has been shown in influenza A virus, measles, SARS-CoV2, and HIV (Semba et al. 2005; Mullin 2011; Sarohan (2020)). While vitamin D has been shown to have positive effects in inhibiting viral infection of HIV, hepatitis C virus, SARS-CoV2, and influenza virus (Abu-Mouch et al. 2011; Grant et al. 2020; Akimbekov et al. 2020). Similarly, Mitchell et al. (2017), Reboul (2017) and Colunga Biancatelli et al. (2020) reported that vitamin E has antiviral effects on the influenza virus, hepatitis B virus, and RSV. Supplementation of vitamin C also had shown antiviral properties following infection with influenza virus, SARS-CoV 2, and HIV (Banerjee and Kaul 2010; Brinkevich et al. 2012; Carr 2020). Combinations of vitamin A and E have been shown to have substantial antioxidant effects and enhance the immune system in the influenza virus (Tantcheva et al. 2003). In addition, the administration of minerals like zinc and copper also has demonstrated antiviral effects in HIV, HSV, hepatitis C virus, human papillomavirus, influenza, and measles [(Lazarczyk et al. 2008; Awotiwon et al. 2017; Alpert 2017; Gupta et al. 2019; Vincent et al. 2018; Shah et al. 2019; Read et al. 2019).

1.1.2 Polysaccharide (Carbohydrate Origin)

Polysaccharide or glycan is a large and polymeric carbohydrate nutrient such as fructan, glycogen, starch, chitin, hyaluronic acid, and cellulose that contain a long chain of small sugar molecules or monosaccharide units. The chain is linked by a glycosidic bond with a formula known as (C6H10O5). Functionally, this macromolecule is essential for energy storage as well as structural organization for plant and animals. In addition to that, therapeutic effects such as antiviral, antineoplastic, and antioxidants of polysaccharides have been demonstrated together with immunomodulatory effects (Chang et al. 2015).

Fructans such as inulin originated from natural resources such as garlic, onion, and herbal plants (Dobrange et al. 2019) exert great antiviral effects in highly pathogenic avian influenza virus (HPAI from H5- and H-7 subtypes), swine-origin influenza (S-OIV), and human influenza type H1N1-type IV studies (Pleschka et al. 2009). Fructans from inulin were able to inactivate all of these viruses in cell culture assays. It also interferes with these viruses to bind in red blood cells (RBCs) in hemagglutination assay, thus inhibiting the viral entry into the cells by interfering with the viral HN proteins to bind with the cellular receptors on the RBCS (Hayashi et al. 2012). In another study, fruit-origin resveratrol oligosaccharides demonstrated antiviral properties by inhibiting SARS-COV2 infection in human lung cells in vitro study. In vitro and in vivo murine model studies inoculated with herpes simplex virus type 2 (HSV-2), fructan demonstrated a strong immunomodulatory effect by inducing productions of immunostimulatory cytokines and viral replication inhibitor (Lee et al. 2011). Moreover, fructan is able to produce anti-influenza A neutralizing antibodies following influenza A inoculation in mice in vivo. In another study by He et al. (2020), polysaccharide is able to promote viral clearance and interfere with the viral replication process.

Meanwhile, other polysaccharide such as sulfated polysaccharides has been shown to exert antiviral properties. Sulfated polysaccharides such as galactan, galactofucan, fucoid, and ulvan that could be acquired from herbal plants could interfere with the viral replication process. Interaction of sulfated polysaccharide with herpes simplex virus, hepatitis virus, influenza virus, coxsackievirus B3, rotavirus, HIV, and enterovirus causes failure of viral adsorption into the target cells (Chen and Huang 2018; He et al. 2020). Induction of T cells and cytokines following sulfated polysaccharide administration following infection with herpes simplex virus, SARS-COV2, and HIV clearly indicates that sulfated polysaccharide could enhance immune response and combat viral infections (Tuvaanjav et al. 2016; Lee and Han 2018).

1.1.3 Peptide

Peptides or proteins are macromolecules that contain a chain of amino acids that serve for defense, organization, development, and transportation. Proteins such as lactoferrin (originated from milk) and lectin exert antiviral effects and activate innate immunity as an initial defense mechanism. In HIV and herpesvirus infection, lactoferrin is able to inhibit viral attachment in the susceptible cells and cease the viral replication (Jenssen 2005; Macan et al. 2019). In review paper by Wakabayashi et al. (2014), lactoferrin is demonstrated as a competent antiviral against several human and animal viruses that lead to respiratory illness (respiratory syncytia virus, parainfluenza, influenza A, and avian influenza) such as common cold or gastrointestinal illness (rotavirus, feline calicivirus, and murine norovirus) such as diarrhea as well as in herpes virus infection. In this virus infection, lactoferrin serves as an antiviral by preventing attachment and entry to the target cells.

Lectin, on the other hand, could bind to the envelope of the virus and alter the conformational structure which further inhibits viral attachment into the cells has been shown in HIV, SARS-CoV, and hepatitis C virus (Tuvaanjav et al.,2016; Keyaerts et al. 2007; Mazalovska and Kouokam 2018). The immunomodulatory effect of lectin also has been shown in herpesvirus infection by increasing the division of natural killer lymphocytes and inhibiting the infection (Wetprasit et al. 2000).

1.1.4 Fatty Acid

Fatty acids or lipids such as phospholipids, monoglycerides, waxes, fats, and others are important for cell membranes' structural components, signaling, and storing energy. Lipid is also beneficial to a pathogen such as a virus as lipid is needed to support their infection. While some fatty acids also exert antiviral effects and enhance the host immune system. Unsaturated fatty acids, namely, linoleic, oleic, or arachidonic were shown to promote the loss of infectivity ability of several envelope viruses such as arbovirus, paramyxovirus, and herpesviruses in animal cells. The disruption of these viruses' envelope evidenced by electron microscope is believed to be the mechanism of fatty acids as antiviral (Kohn et al. 1980). Other than that, other types of fatty acids (such as PUFA, lipoxin, palmitate, lauric acid, arachidonic acid, and others) have been demonstrated to have antiviral effects and induce immune defense system in the interaction with several viruses such as SARS-CoV-2, MERS-CoV, SARS-CoV, arenavirus, herpes simplex virus, respiratory syncytia virus, influenza A and hepatitis C virus (Bartolotta et al. 2001; Hilmarsson et al. 2007; Martín-Acebes et al. 2012; Schönfeldt et al. 2016). PUFA or polyunsaturated fatty acids such as omega 3 or 6 are able to cause release of prostaglandins, an inhibitory of viral replication. PUFA also inhibits and prevents the expression of nuclear export factors, thus impairing the viral activity during the early stage of infection. Cytokines produced by macrophages and the synergistic interaction of PUFA with interferon suggest the role of PUFA in the enhancement of immunity (Sheridan et al. 2014).

1.2 Zootechnical Additives

Probiotics have been defined as living nonpathogenic microbes that beneficially affect the host either from bacteria or fungi (Fuller 1992). The probiotics can be classified into colonizing species such as *Lactobacillus* and *Enterococcus* spp. and free-flowing non-colonizing species, such as *Bacillus* spp. (spores) and *Saccharomyces cerevisiae* (Huyghebaert et al. 2011). The mode of action depends on the probiotic strain, which involves the production of specific metabolites either in the form of short organic fatty acids, lactic acid, hydrogen peroxide (H_2O_2), or intermediary metabolites with antimicrobial activity, interaction with receptor sites and stimulation of the immune system (Madsen et al. 2001; Sherman et al. 2009).

The majority of probiotics belong to the lactic acid bacteria (LAB) group and are highly abundant in several fermented foods of plant-based or animal-based origins (Zhao et al. 2019; Muhialdin et al. 2021). Probiotics include bacteria such as *Lactobacillus acidophilus, L. amylovorus, L. brevis, L. bulgaricus, L. casei, L. cellobiosus, L. crispatus, L. curvatus, L. delbrueckii* spp. *bulgaris, L. fermentum, L. gallinarum, L. helveticus, L. johnsonii, L. lactis, L. paracasei, L. plantarum, L. reuteri, L. rhamnosus; Streptococcus thermophilus, Lactococcus lactis, Leuconostoc mesenteroides, Pediococcus pentosaceus, P. acidilactici, Bifidobacterium adolescentis, B. animalis, B. bifidum, B. breve, B. essensis, B. infantis, B. laterosporum, B. thermophilum, B. longum, Propionibacterium acidipropionici, P. freudenreichii, P. jensenii, P. thoenii, Enterococcus fecalis, E. faecium, Bacillus alcalophilus, B. cereus, B. clausii, B. coagulans, B. subtilis, Escherichia coli, Sporolactobacillus inulinus*; as well as yeast such as *Saccharomyces boulardii and S. cerevisiae* (Bron et al. 2012; Saad et al. 2013).

Several studies reported that probiotics isolated from different fermented foods are effectively used as antiviral against virus infection by improving the immune cell function of the host (Muhialdin et al. 2021). Briefly, the mechanism of action was reported to be due to the stimulation of the immune system function via enhancing natural killers cell toxicity, enhancing the production of pro-inflammatory cytokines, and increasing the cytotoxic of T lymphocytes ($CD3^+$, $CD16^+$, $CD56^+$) (Muhialdin et al. 2021).

1.2.1 Probiotics

Probiotics utilize carbohydrates in the host to synthesize antiviral metabolites, which subsequently inhibit viral infections since these metabolites form a micro-environment that is not favorable to viral replication (Wang et al. 2022). In addition, this pathway exerts antiviral effects by preventing the virus attachment to host cells or directly inactivating the viruses. Probiotics form a defensive barrier against the invasion of viruses by enhancing the tight junctions (TJs) between intestinal epithelial cells (IECs).

Probiotics can also inhibit viruses directly by interacting with viruses or competing for the cellular receptors to inhibit virus entry into host cells (Lievin-Le and Servin 2014). In other studies, probiotics enhance the chemical barrier to maintain host health by producing antiviral metabolites or stimulating goblet cells to produce mucin (Wang et al. 2022). Recent studies have confirmed that probiotics secrete extracellular proteins, weaken the virus attachment, and protect the intestinal cells as reported by Liu et al. (2020).

1.2.1.1 Transmissible Gastroenteritis Virus (TGEV)

Probiotic *Bacillus subtilis* OKB105 inhibits the entry of transmissible gastroenteritis virus (TGEV) into the intestinal porcine epithelial cell line (IPEC-J2) by competing for the entry receptors (Wang et al. 2013; Wang et al. 2017).

1.2.1.2 Vesicular Stomatitis Virus (VSV)

Previous studies have revealed that probiotics can inhibit the infection of vesicular stomatitis virus (VSV) in vitro with a reduction of viral titer in cell monolayer (Botic et al. 2007). A possible mechanism of inhibiting VSV replication is that probiotics compete with the virus for cell binding and interference with virus attachment or entry. Another possible mechanism is that probiotics can trap VSV specifically or nonspecifically.

1.2.1.3 Human Immunodeficiency Virus (HIV)

The *Lactobacillus* genus is a probiotic that plays protective roles in the vaginal mucosa. Few studies have demonstrated that metabolites released by *Lactobacillus acidophilus* that is H_2O_2 are toxic to HIV agents (Knezevic et al. 2005).

1.2.1.4 Swine Influenza Virus

Another study revealed that *Enterococcus faecium* NCIMB 10415 was efficient to inhibit swine influenza virus (SIV) infection by the direct interaction between the SIV and probiotics (Wang et al. 2013).

1.2.1.5 Influenza A Virus

Probiotic bacteria obtained from fermented foods were tested in vitro and in vivo methods for analysis of antiviral activity. Lactic acid bacteria (LAB) as probiotics have been mostly studied using live bacteria that improve survival rates or partial protection against influenza virus (Jung et al. 2017). *Enterococcus faecium L3* from dairy product is effective against influenza A virus for subtype H3N2 and H1N1 by stimulation of interferon production and enhance humoral immune response based on testing in MDCK cell line and in female Balb/c mice (Ermolenko et al. 2019).

In previous work, kimchi known as Korean fermented cabbage was also produced in *Lactobacillus casei* DK128 probiotic, which confers broad protection against influenza A virus with 100% survival in treated mice. Alveolar macrophages were increased in the lung and bronchoalveolar lavage fluid (BALF) cells and maintained in the lungs of the treated mice whereas naïve mice showed a significantly lower level of macrophages in the lungs after viral infection (Jung et al. 2017). Alveolar macrophages appear to play a critical role in conferring protection mediated by prior treatment with heat-killed LAB. Upon influenza virus infection, alveolar macrophages were depleted to lower levels correlating with disease and mortality (Smith and Smith 2016). In addition, previous studies demonstrated that alveolar macrophages play a critical role in controlling lung viral loads and protecting against influenza virus infections (Tumpey et al. 2005; Fleming-Dutra et al. 2013). The immune pathway was performed by stimulation of rapid induction of IgG1 and IgG2a antibodies, induction of innate immune cells, and cytokines from infected mice (Jung et al. 2017).

Furthermore, heat-killed *E. faecalis* protected mice suppress influenza virus and enterovirus infections (Chen et al. 2017). It involves the activation of MCP-1/CCR2 pathway, which might act as a key mediator in the improved antiviral immune response. In another study, the mice treated with *Bifidobacterium bifidum* produced antibodies, IL-4, IL-12, and IFN-γ, and protected from the challenge of H1N1 influenza virus (Mahooti et al. 2019). IL-4 can induce Th2 immune responses, while IFN-γ modulates Th1 immune responses. The balance between Th1 and Th2 is important for the homeostasis of the host immune system.

1.2.1.6 Low Pathogenic Avian Influenza

The probiotics, *L. salivarius, L. johnsonii, and L. reuteri* were effectively used for antiviral and immunostimulatory effects against the LPAIV subtype H9N2 (Alqazlan et al. 2020). Studies in chickens revealed *Leuconostoc mesenteroides* YML003 showed antiviral activity against H9N2 LPAIV in vitro and in vivo (Seo et al. 2012). Another probiotic originated from camel milk *Lactiplantibacillus plantarum* KAU007 is highly effective against the LPAI (H9N2). As confirmed by the hemagglutination assay, KAU007 showed potent antiviral activity against H9N2 and vigorous antioxidant activity. The cell free culture supernatant (CFCS) showed a dose-dependent reduction in the levels of IL-6 and IFN-γ (Rather et al. 2022).

1.2.1.7 Rotavirus

Previous study demonstrated that *Lacticaseibacillus rhamnosus* GG (LGG) activates the TLR3 signaling pathway and effectively treat diarrhea induced by rotavirus infection (Aoki-Yoshida et al. 2016). This mechanism was induced by other probiotics such as *Ligilactobacillus (Lg.) salivarius* FFIG35 and FFIG58, which displayed antiviral effects (Indo et al. 2021).

In human cases, the underlying mechanism against rotavirus infections is an immune enhancement, as certain strains of lactobacilli promote immunological responses (Steyer et al. 2022). Increased anti-rotavirus-specific IgA caused a reduced intestinal microbiota imbalance, enhanced colonization of probiotics, and reduced the incidence of diarrhea (Binns et al. 2007).

1.2.1.8 Newcastle Disease Virus

The utilization of probiotics as feed additives caused the destruction of virus particles, blocking the virus infection into the cells and inhibiting viral proliferation in the cells (Wang et al. 2010). It was demonstrated that five probiotics and their metabolites could reduce the viral titer of NDV with inhibition of the replication rate of NDV significantly in a dose-dependent manner ($P < 0.01$). In another study, chicken fed with probiotics containing *Lactobacillus reuteri* PIA16 (*L. reuteri* PIA16) induced high antibody titer in broilers following immunization (Gonmei et al. 2019). The most likely reasons for *Lactobacillus* to increase the antibody titer in the present study might be due to the competitive exclusion of pathogens through competition of receptor sites, production of volatile fatty acids that are inhibitory of certain enteric pathogens, production of bacteriocins or competition with pathogens, and native flora for limiting nutrients or stimulation of a host innate immune response (Waititu et al. 2014).

1.2.2 Prebiotics

Prebiotics can be defined as non-digestible feed ingredients with selective effects on the intestinal microbiota. Prebiotics include fructans, oligosaccharides, arabino oligosaccharides, isomaltooligosaccharides, xylooligosaccharides, resistant starch, lactosucrose, lactobionic acid, galactomannan, psyllium, polyphenols, and polyunsaturated fatty acids (Olaimat et al. 2020; Gibson et al. 2017; Guarino et al. 2020; Davani-Davari et al. 2019). The health benefits of prebiotics to the GI tract including inhibition of pathogens and stimulation of the immune system are due to their ability to modulate the composition and activity of human microbiota (Davani-Davari et al. 2019).

1.2.2.1 SARS-COV-2

Prebiotics have a good potential effect against COVID-19 by enhancing probiotics growth and survivability. It has a direct effect on gastrointestinal symptoms via ACE enzyme blockage (Yeh et al. 2018).

1.2.2.2 Influenza A Virus

Mice fed with inulin or high fiber, HFD showed prolonged survival and exhibited reduced pulmonary resistance, mild airway constriction, and enhanced elasticity with better lung function (Trompette et al. 2018). In addition, dietary fiber influences antiviral immunity by alteration of $CD8^+$ T cell metabolism to accelerate and enhance effector function.

In another work, the *Houttuynia cordata* polysaccharide (HCP) was tested in mice and induced antiviral activity against the influenza A virus. HCP given orally alleviated lung injury and intestinal dysfunction caused by H1N1 infection. It is shown that HCP relieved the intestinal barrier damage and restored intestinal microbiota homeostasis by regulating the composition, diversity, and relative abundances of functional genes of intestinal microbiota. The decrease of TLRs and IL-1β levels and the increase of IL-10 concentration after HCP treatment contributed to inhibiting the inflammation (Chen et al. 2019).

Feed supplemented with mushroom-derived active hexose correlated compound (AHCC) modulates immunity and increases survival in response to a broad spectrum of acute infections, including influenza virus infection (Nogusa et al. 2009). The mice exhibited virus clearance and low weight loss compared to controls.

1.2.3 Synbiotic

Synbiotic is a mixture of probiotics and prebiotics that beneficially affects the host by activating the metabolism of one or a limited number of health-promoting bacteria and/or stimulating their growth selectively, improving the host's welfare (Ghahri et al. 2013). The synergistic effects of prebiotics and probiotics can be useful in stimulating beneficial bacteria and improving the health of the gut. In order to preserve gut microbiota and to promote host innate defenses, the administration of synbiotics (combinations of prebiotics, probiotics, and immunomodulators elements) as an alternative approach for promoting performance and immune responses in modern poultry husbandry is widely accepted (Talebi et al. 2015).

1.2.3.1 Newcastle Disease Virus

Application of synbiotin containing both probiotic (*Enterococcus faecium*) and prebiotic (fructo-oligosaccharides) in chickens enhances antibody responses following vaccination against ND, AI, IB, and IBD but it is more effective in live than killed vaccines and could be used as a feed additive adjuvant for improving innate and acquired immune responses in chickens (Khan et al. 2014).

1.3 Natural Feed Additives

Natural additives in animal feed are able to improve productivity and performance by enhancing digestibility, maintaining and stabilizing beneficial microflora in the gut, and finally can improve the quality of animal products and positively influence the environment (Arowolo and He 2018; Steiner and Syed 2015). Natural feed additives come from a variety of natural sources including plants, minerals, animals, and microorganisms. Another term used referring to the compounds obtained naturally from the environment is phytogenic. These feed additives originate from plants and algae, consisting of essential oils, extracted from different plant parts which include many different bioactive ingredients (Singh and Gaikwad 2020).

1.3.1 Essential Oil

Essential oils (EOs) are a complex mixture of different volatile and nonvolatile compounds produced by plants or algae dependent on environmental conditions. They are secondary metabolites produced at various parts of the plant including roots, stems, leaves, flowers, and seeds. Generally, EOs play an important role in protecting the plant in response to external agents, such as herbivores, pathogens, pollinators, climate changes, and environmental stresses (for example drought, high temperature, and UV radiation) (Bakkali et al. 2008; Regnault-Roger et al. 2012). The use of essential oil as feed additives was reviewed extensively by Stevanović et al. (2018). The main advantage of using essential oils as an antiviral agent gives the synergistic effect that would enhance and amplify their activity (Stevanović et al. 2018). The presence of various compounds in the essential oil can target different parts of the virus and its replication cycle which results in a more potent antiviral effect than if the individual compounds were used alone.

Essential oils have been studied for their potential antiviral activity against various animal viruses, including avian influenza, porcine reproductive and respiratory syndrome virus (PRRSV), and equine arteritis virus (EAV). Avian influenza virus (AVI) causes avian-flu or bird-flu and has been found in all of Eurasia among land birds, such as chicken, turkeys, ducks, geese, and wild waterfowl. Several EOs from plants such as *Melissa officinalis* (Pourghanbari et al. 2016), eucalyptus, peppermint (Kumosani et al. 2017), and *Zataria multiflora Boiss* (Shayeganmehr et al. 2018) showed strong antiviral activity against the avian influenza virus when tested on different land birds. The antiviral properties of these oils have been attributed to their high concentration of monoterpenoids. *Melissa officinalis* or lemon balm EO showed inhibition of AVI through different virus replication phases (pre- and post-infection), with the highest antiviral activity at the early phase of infection; during direct interaction or entry phase with the host cells (Pourghanbari et al. 2016). Meanwhile, a blend of EOs between eucalyptus and peppermint could prevent the direct interaction of the virus with host cells which later prevents the infection. In the study, 84.6% reduction in viral titer at 1.5 min contact time compared to the control virus deprived of contact with vaporized EOs (Kumosani et al. 2017). This implies the application of safe EOs in the reduction of air-suspended influenza viruses in closed systems harboring domestic animals as well as humans.

Another mechanism of EOs is shown by *Zataria multiflora Boiss*. EOs reduced viral replication in the trachea of broiler chickens when supplemented with EOs either before or after the H9N2 challenge (Shayeganmehr et al. 2018).

Several in vivo and in vitro studies reported the effectiveness of essential oils in reducing and preventing the infectivity of PPRSV on swine. EOs that are identified to have antiviral activity against PPRSV are eucalyptus, thyme, lemongrass, oregano, clove, cinnamon, and rosemary (Wani et al. 2021). Based on in vitro studies done by Fabros et al. (2018), cinnamon essential oil showed inhibition of PPRSV replication by more than 40% in post-infection assay. Furthermore, research by Kaewprom et al. (2017) demonstrated that hydrosols of *Thymus vulgaris* and *Nepeta cataria* have antiviral activities against PRRSV by blocking viral attachment, adsorption, replication, and release. They also reported that the oils were not only effective against PPRSV but also proven effective against feline coronavirus, which caused Feline infectious peritonitis (FIP) (Catella et al. 2021; Kaewprom et al. 2017).

Next, an in vivo study done on African swine fever virus (ASFV) of pig supplemented with a blend formulation of equal quantity of essential oil *Eucalyptus globulus*, *Pinus sylvestris* (pine), *and Lavandula latifolia* (lavender) showed deactivation of ASFV in a challenged group of pigs. No symptoms appeared when the pigs were given a blend of essential oil through their drinking water. The improvement of pig humoral immune system was also shown by enhancement of IgG level and reduction of IgM level. This study also confirmed that essential oils were well-tolerated and safe for use in animals (Babikian et al. 2021).

In the case of equine arteritis virus (EAV), an in vitro study was done on several essential oils such as *Origanum vulgare* (oregano), and *Thymus vulgaris* for EAV and the results showed a promising effect in reducing the virus titer with longer incubation period produced a significant effect (Blank et al. 2017; Pensel et al. 2014). They reported that the compounds mainly from the phenolic group of essential oils such as caffeic acid, p-coumaric acid, rosmarinic acid, quercetin, carnosic acid, kaempferol, and apigenin might interact with molecules on the viral surface, preventing the adsorption and penetration into cells, which caused the disintegration of the EAV (Blank et al. 2019).

Apart from the effect of essential oil in animals, essential oils also had been proven for their antiviral properties against human viruses. For example, tea tree oil has been found to have antiviral activity against herpes simplex virus type 1 (HSV-1) in vitro, and eucalyptus oil has been shown to have activity against influenza A virus in vitro (Schnitzler et al. 2001).

Lastly, it is important to note that while essential oils have shown promising results in laboratory studies and animal models, their effectiveness as antiviral additives for livestock has not been extensively studied and more research is needed to fully understand their potential. Additionally, the safety and efficacy of essential oils as antiviral additives for livestock can vary depending on the specific oil and the method of application.

1.3.2 Algae

Macro- and microalgae have been investigated for their potential as a source of antiviral agents for animal and human viruses. The antiviral component of algae is derived from a single extracted compound or whole cell extract. Microalgae compounds known to have antiviral activity are lectin, polysaccharides, pigments, flavonoids, polyphenols, and glycolipids (Carbone et al. 2021). Each of these compounds exhibits antiviral activity through various mechanisms. For example, lectin is known as a carbohydrate-binding protein that binds to the high glucan structure of the viral envelope and prevents the viral binding on human glycoprotein at the surface of the membrane (Botos and Wlodawer 2003). This microalgae compound found in *Oscillatoria agardhii*, *Nostoc ellipsosporum*, *Scytonema varium*, and *Microcystis* sp. was demonstrated could prevent HIV infection when tested in vitro (T. Sato and Hori 2009; Y. Sato et al. 2007).

Another compound such as polysaccharide has been tested for the prevention of human cytomegalovirus (HCMV) and mumps virus (MuV). HCMV causes an infected person to have a compromised immune system, pneumonitis, and possible liver failure. The use of polysaccharides from microalgae was to block the sites for virus attachment on the cells by interacting with the positively charged domains of the virus glycoprotein envelope and creating a non-reversible complex. *Porphyridium* sp., *Chlorella* sp., *Spirulina* sp., and *Nostoc* sp. are strains frequently studied to produce polysaccharides (Costa et al. 2021). The sulfate polysaccharide from *Spirulina platensis* was proven to reduce the replication of two animal viruses, Vaccinia virus (VV) and Ectromelia virus (ECTV) that caused smallpox in cattle (Huleihel et al. 2001).

Pigments from microalgae are one the most extensively studied for their potential of bioactivities to be used in both the industrial and pharmaceutical industries (Christaki et al. 2015; Pangestuti et al. 2020; Pignolet et al. 2013). The main pigment substances that showed the most interesting results are astaxanthin, carotenoids, phycobiliproteins, and Pheophorbide a. In a study on Herpes simplex virus type 1 (HSV-1), it was found that *D. salina* extract and *H. pluvialis* extract can reduce its activity ranging from 50% to 85%. The content *D. salina* extract might have higher carotenoid content (Santoyo et al. 2012). The mechanism of antiviral activity by carotenoids from microalgae is likely to be multifactorial, involving a combination of direct antiviral effects and modulation of host immune responses (Reynolds et al. 2021; Schnitzler et al. 2001).

Other microalgae compounds that showed antiviral activity include flavonoids, polyphenols, protein, and peptides. In a study, flavonoids from methanolic extract of microalgae cyanobacterium *Geitlerinema* sp. showed potent antivirals against hepatitis C virus by reducing the ATPase activity and consequently RNA helicase and virus replication (Agrawal 2011; Mustopa et al. 2016). A study done using HPLC on *Spirulina platensis* protein showed the antiviral effect against nuclear polyhedrosis virus (NPV), a double-stranded DNA-enveloped virus from the Baculoviridae family that causes hemodynamic septic disease in silkworms (Babu et al. 2005).

Besides producing compounds with anti-viral activity, microalgae also can be used as a bioreactor to develop a vaccine for animals. For example, a study attempted to create a vaccine by using microalgae against infectious bursal disease virus (IBDV), a non-enveloped RNA virus of the Birnaviridae family causing an immunosuppressive disease in poultry using a transformation technique. In the study, the protein VP2 of IBDV was expressed in *Chlorella pyrenoidosa* via Agrobacterium tumefaciens, which induced neutralizing antibodies, and conferred protection in young chickens against the action of the virus (Reddy et al. (2017) (Fig. 1).

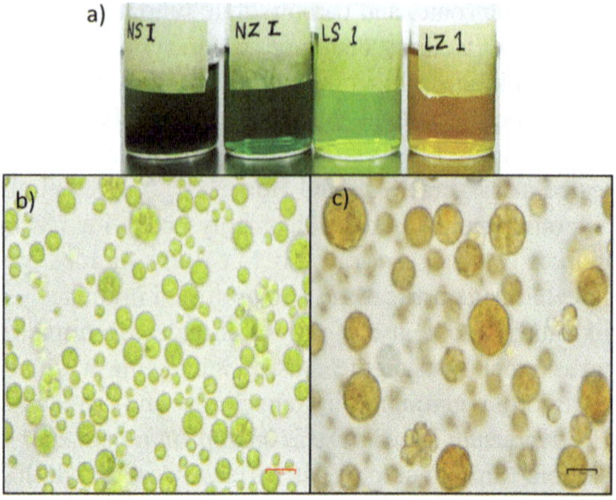

Fig. 1 *Chlorella* sp. produce various pigment content under different culture condition. (**a**) Pigment extracted from microalgae *Chlorella* cultured at different light intensity. (**b**) *Chlorella* sp. under normal condition, (**c**) *Chlorella* sp. under stress condition

It is worth noting that the specific mechanisms by which microalgal metabolites exhibit their antiviral activity can vary depending on the type of virus, the microalgal species, and the concentration and form of the metabolites used. Further research is needed to better understand the precise mechanisms involved and to optimize their antiviral efficacy.

1.4 Technological Additives

A technological additive is any substance added to feed for a technological purpose, it favorably affects the characteristics of feed. Technological additives such as preservatives are substances that allow feeds to be stored for a long time without spoiling. They are used to improve or stabilize the physical structure of the feed during feed production without a direct biological effect on animal production. Preservatives are listed as sodium benzoate, erythorbic acid, sodium benzoate, and nitrite oxide.

1.4.1 Preservatives

1.4.1.1 Sodium Benzoate

Sodium benzoate is a salt of benzoic acid that is well soluble in water, tasteless, and odorless. It is a widely used food preservative due to its bacteriostatic, fungistatic, and virustatic properties with immunomodulatory potential (Yadav et al. 2016). Benzoic acid derivatives have been reported to possess anti-influenza virus activities. Benzoic acid, termed NC-5, has potent anti-influenza activity in vivo and in vitro that is able to inhibit influenza A viruses by suppressing the nucleoprotein (NP) and matrix protein 1 (M1) expression levels during the late stages of viral biosynthesis and inhibit neuraminidase (NA) activity, which may influence virus release (Guo et al. 2019). Green tea extract with supplements of 2% citric acid, 0.1%

sodium benzoate, and 0.2% ascorbic acid as anti-oxidant demonstrated potent viral inactivating activity that could be formulated as safe and environmentally friendly personal hygiene against viral infections. Besides human use, it could also be used for protecting animals and livestock from viral infections and reducing zoonotic transmissions (Lee et al. 2018). Dietary supplements of bovine lactoferrin have antibacterial, antiviral, and antioxidant properties and enhance the immune system response (Duran et al. 2002).

1.4.1.2 Erythorbic Acid

Erythorbic acid is a stereoisomer of ascorbic acid (Vitamin C) that use as an antioxidant has potent enhancer of nonheme-iron absorption. It is also known as isoascorbic acid; erythorbic acid is a natural product, vegetable-derived food additive produced from sucrose. Dehydroascorbic acid, an oxidized form of ascorbic acid exhibits strong antiviral activity by inhibiting herpes simplex virus type 1 (HSV-1) multiplication probably at the assembly process of progeny virus particles after the completion of viral DNA replication (Furuya et al. 2008). Ascorbic acid decreases susceptibility genes expression, including mitochondrial antiviral signaling (MAVS) and interferon regulatory factor 3 (IRF3), and increases expression of NF-κB. These in conjunction induce type I interferons (IFNs) and elicit innate antiviral response in H1N1 virus-induced mice (Cai et al. 2015).

1.4.1.3 Nitrite Oxide

Nitric oxide composed of nitrogen (N) and oxygen (O) is a gas that increases blood flow by relaxing blood vessels. Antiviral effect of NO was demonstrated by inhibition of reovirus replication due to macrophage priming following virus infection for protection of cells against virus-induced replication and cytopathic effects. This protection is possibly mediated by the cytostatic effects of NO on the host cell (Pertile et al. 1996). Recent advances in life science involve usage of gaseous NO in the development of antiviral therapies was demonstrated by inhibiting viral enzymes through nitrosylation and disrupt the viral DNA during replication in the host. (Garren et al. 2021).

1.5 Sensory Additives

The term of sensory feed additives defines additives that improve or modify the visual properties of food obtained from animals. The additives should not adversely affect the product stability or organoleptic and nutritional properties of the food.

1.5.1 Colorings and Flavorings

Coloring additives include both synthetic substances and substances derived from natural sources while flavoring substance contains desired and pleasant aroma to mask unpleasant odors in such a way that the animals show a preference for their feed. Flavonoids are types of natural substances with phenolic structures isolated from a variety of plants. Flavonoids have antioxidant, anti-inflammatory, anticancer, and antiviral activities. Although most of the research or applications of flavonoids are focused on human diseases, flavonoids also show potential applicability against porcine virus infection (Zhang et al. 2022). Naturally occurring plant flavonoids are a promising class of antiviral agents to inhibit the African swine fever virus (ASFV), which causes highly fatal disease in pigs and is a major threat to the swine industry. Several flavonoids were able to significantly inhibit ASFV infection in vitro by targeting different stages of the viral life cycle (Hakobyan et al. 2016, 2019; Arabyan et al. 2018). Furthermore, kaempferol is a promising anti-ASFV agent and has a distinct antiviral mechanism compared to other anti-ASFV flavonoids (Arabyan et al. 2021).

Phytogenic feed additives are plant-based feed additives or botanicals that are derived from herbs, spices, and other plants and their extracts that are used in poultry nutrition. They have antimicrobial, antifungal, antiviral, antioxygenic, antiparasitic, and insecticidal properties. Their benefits include increased feed intake, stimulation of digestion, increased growth performance, reduced incidence of disease, improved reproductive parameters, feed efficiency, profitability, and reduced poultry house emissions (Murugesan et al. 2015). Polyphenols are active substances against various types of viral infections, against common viral infections of influenza, herpes, hepatitis, rotavirus, and coronavirus (Chojnacka et al. 2021). Essential oils (EOs) and aromatic plants are well known to exert antibacterial, antifungal, and antiviral activity in swine and poultry (Windisch et al. 2008).

Flavoring of essential oils (EOs) and extracts of rose geranium (*Pelargonium graveolens*) and petals of rose (*Rosa damascena*) have beneficial antimicrobial and antiviral properties and they can be used as natural preservatives (Androutsopoulou et al. 2021). Plant phytochemicals, in particular flavonoids and polyphenols, contain an abundant pool of potent antiviral molecules (e.g., vitexin), a flavonoid isolated from pink coral tree *Erythrina speciosa* leaves, demonstrated antiviral activity against herpes simplex virus type 1 (HSV-1) and hepatitis A virus-H10 (HAV H10) (Fahmy et al. 2020).

Tangeretin is a polymethoxylated flavone found in citrus fruit peels that inhibits viral entry into cells by blocking viral fusion (Tang et al. 2018), and citrus extracts are active against avian influenza virus (AIV), Newcastle disease virus (NDV), infectious bursal disease virus (IBDV) in different environments (Komura et al. 2019). The efficacy of tangeretin against seven other VHF-causing arenaviruses suggests that this compound could be used to develop an effective therapeutic to treat infection and disease caused by the Lassa virus and related viruses (Tang et al. 2018). Moreover, vaccines containing citrus-derived molecules were more efficient in stimulating the immune system with fewer side effects (Pennisi et al. 2017). Maltodextrins are plant polysaccharides that has been used as vaccine by using nanoparticles technology in influenza virus infection. Lactic acid also inhibited influenza A infection replication (Miyazaki 2017). Citric acid is another potent natural antimicrobial with antiviral activity that inhibits the foot and mouth disease virus (FMDV) (Hong et al. 2015).

References

Abu-Mouch S, Fireman Z, Jarchovsky J, Zeina A-R, Assy N (2011) Vitamin D supplementation improves sustained virologic response in chronic hepatitis C (genotype 1)-naïve patients. World J Gastroenterol: WJG 17(47):5184–5190

Agrawal AD (2011) Pharmacological activities of flavonoids: A review. Int J Pharm Sci Nanotechnol (IJPSN) 4(2):1394

Akimbekov NS, Ortoski RA, Razzaque MS (2020) Effects of sunlight exposure and vitamin D supplementation on HIV patients. J Steroid Biochem Mol Biol 105664:105664

Alpert PT (2017) The role of vitamins and minerals on the immune system. Home Health Care Manag Pract 29(3):199–202

Alqazlan N, Alizadeh M, Boodhoo N, Taha-Abdelaziz K, Nagy E, Bridle B, Sharif S (2020) Probiotic Lactobacilli limit avian influenza virus subtype H9N2 replication in chicken Cecal tonsil mononuclear cells. Vaccines (Basel) 8(4):605

Androutsopoulou C, Christopoulou SD, Hahalis P, Kotsalou C, Lamari FN, Vantarakis A (2021) Evaluation of essential oils and extracts of rose geranium and rose petals as natural preservatives in terms of toxicity, antimicrobial, and antiviral activity. Pathogens 10(4):494

Aoki-Yoshida A, Saito S, Fukiya S, Aoki R, Takayama Y, Suzuki C, Sonoyama K (2016) Lactobacillus rhamnosus GG increases toll-like receptor 3 gene expression in murine small intestine ex vivo and in vivo. Benefic Microbes 7(3):421–429

Arabyan E, Hakobyan A, Kotsinyan A, Karalyan Z, Arakelov V, Arakelov G et al (2018) Genistein inhibits African swine fever virus replication in vitro by disrupting viral DNA synthesis. Antivir Res 156:128–137

Arabyan E, Hakobyan A, Hakobyan T, Grigoryan R, Izmailyan R, Avetisyan A, Karalyan Z, Jackman JA, Ferreira F, Elrod CC, Zakaryan H (2021) Flavonoid library screening reveals Kaempferol as a potential antiviral agent against African swine fever virus. Front Microbiol 12: 736780

Arowolo MA, He J (2018) Use of probiotics and botanical extracts to improve ruminant production in the tropics: A review. Animal Nutrition 4(3):241–249

Awotiwon AA, Oduwole O, Sinha A, Okwundu CI (2017) Zinc supplementation for the treatment of measles in children. Cochrane Database Syst Rev 6(6):CD011177

Babikian HY, Jha RK, Truong QL, Nguyen LT, Babikyan Y, Nguyen HT et al (2021) Novel formulation with essential oils as a potential agent to minimize African swine fever virus transmission in an in vivo trial in swine. Veterinary World 14:1853–1866

Babu SM, Gopalaswamy G, Chandramohan N (2005) Identification of an antiviral principle in Spirulina platensis against Bombyx mori nuclear Polyhedrosis virus (BmNPV). Indian J Biotechnol 4:384–388

Bakkali F, Averbeck S, Averbeck D, Idaomar M (2008) Biological effects of essential oils-a review. Food Chem Toxicol 46(2):446–475

Banerjee D, Kaul D (2010) Combined inhalational and oral supplementation of ascorbic acid may prevent influenza pandemic emergency: A hypothesis. Nutrition 26(1):128–132

Bartolotta S, Garcí C, Candurra NA, Damonte EB (2001) Effect of fatty acids on arenavirus replication: inhibition of virus production by lauric acid. Arch Virol 146(4):777–790

Binns C, Lee A, Harding H, Gracey M, Barclay D (2007) The CUPDAY study: prebiotic-probiotic milk product in 1–3-year-old children attending childcare centres. Acta Paediatr 96:1646–1650

Blank DE, Corrêa RA, Freitag RA, Cleff MB, Hübner SDO (2017) Anti-equine arteritis virus activity of ethanolic extract and compounds from Origanum vulgare. Semina: Ciências Agrárias 38(2):759

Blank DE, Hübner SDO, Alves GH, Cardoso CAL, Freitag RA, Cleff MB (2019) Chemical composition and antiviral effect of extracts of Origanum vulgare. Adv Biosci Biotechnol 10(07):188–196

Botic T, Klingberg TD, Weingartl H, Cencic A (2007) A novel eukaryotic cell culture model to study antiviral activity of potential probiotic bacteria. Int J Food Microbiol 115:227–234

Botos I, Wlodawer A (2003) Cyanovirin-N: a sugar-binding antiviral protein with a new twist. Cell Mol Life Sci CMLS 60(2):277–287

Brinkevich SD, Boreko EI, Savinova OV, Pavlova NI, Shadyro OI (2012) Radical-regulating and antiviral properties of ascorbic acid and its derivatives. Bioorg Med Chem Lett 22(7): 2424–2427

Bron PA, Van Baarlen P, Kleerebezem M (2012) Emerging molecular insights into the interaction between probiotics and the host intestinal mucosa. Nat Rev Microbiol 10:66–78

Cai Y, Li YF, Tang LP, Tsoi B, Chen M, Chen H, Chen XM, Tan RR, Kurihara H, He RR (2015) A new mechanism of vitamin C effects on A/FM/1/47(H1N1) virus-induced pneumonia in restraint-stressed mice. Biomed Res Int 2015:675149

Carbone DA, Pellone P, Lubritto C, Ciniglia C (2021) Evaluation of microalgae antiviral activity and their bioactive compounds. Antibiotics 10(6):746

Carr AC (2020) A new clinical trial to test high-dose vitamin C in patients with COVID-19. Crit Care 24(1):1–2

Catella C, Camero M, Lucente MS, Fracchiolla G, Sblano S, Tempesta M et al (2021) Virucidal and antiviral effects of Thymus vulgaris essential oil on feline coronavirus. Res Vet Sci 137:44–47

Chang H-C, Bayeva M, Taiwo B, Palella FJ Jr, Hope TJ, Ardehali H (2015) High cellular iron levels are associated with increased HIV infection and replication. AIDS Res Hum Retrovir 31(3): 305–312

Chen L, Huang G (2018) The antiviral activity of polysaccharides and their derivatives. Int J Biol Macromol 115:77–82

Chen MF, Weng KF, Huang SY, Liu YC, Tseng SN, Ojcius DM, Shih SR (2017) Pretreatment with a heat-killed probiotic modulates monocyte chemoattractant protein-1 and reduces the pathogenicity of influenza and enterovirus 71 infections. Mucosal Immunol 10(1):215–227

Chen MY, Li H, Lu XX, Ling LJ, Weng HB, Sun W, Chen DF, Zhang YY (2019) Houttuynia cordata polysaccharide alleviated intestinal injury and modulated intestinal microbiota in H1N1 virus infected mice. Chin J Nat Med 17(3):187–197

Chojnacka K, Skrzypczak D, Izydorczyk G, Mikula K, Szopa D, Witek-Krowiak A (2021) Antiviral properties of polyphenols from plants. Foods 10(10):2277

Christaki E, Bonos E, Florou-Paneri P (2015) Chapter 14 - Innovative microalgae pigments as functional ingredients in nutrition. In: Kim S-K (ed) Handbook of marine microalgae. Academic Press, Boston, pp 233–243

Colunga Biancatelli RML, Berrill M, Marik PE (2020) The antiviral properties of vitamin C. Expert Rev Anti-Infect Ther 18(2):99–101

Costa JAV, Lucas BF, Alvarenga AGP, Moreira JB, De Morais MG (2021) Microalgae polysaccharides: an overview of production, characterization, and potential applications. Polysaccharides 2(4):759–772

Davani-Davari D, Negahdaripour M, Karimzadeh I, Seifan M, Mohkam M, Masoumi SJ, Berenjian A, Ghasemi Y (2019) Prebiotics: definition, types, sources, mechanisms, and clinical applications. Foods 8(3):92

Dobrange E, Peshev D, Loedolff B, Van den Ende W (2019) Fructans as immunomodulatory and antiviral agents: the case of Echinacea. Biomol Ther 9:615

Duran R, Deschler C, Precigou S, Goulas P (2002) Degradation of Chlorophenols by Phanerochaete Chrysosporium: effect of 3, 4-dichlorophenol on extracellular peroxidase activities. Appl Microbiol Biotechnol 59:284–288

Ermolenko EI, Desheva YA, Kolobov AA, Kotyleva MP, Sychev IA, Suvorov AN (2019) Anti–influenza activity of enterocin B in vitro and protective effect of bacteriocinogenic enterococcal probiotic strain on influenza infection in mouse model. Probiotics Antimicrob Proteins 11(2): 705–712

Fabros D Jr, Kankeaw U, Ruansit W, Tonlek B, Theenongsang S, Charerntantanakul W (2018) Evaluation of antiviral potential of cinnamon essential oil and its derived benzimidazole against porcine reproductive and respiratory syndrome virus. J Agricultural Res 35(2 Suppl):21–31

Fahmy NM, Al-Sayed E, Moghannem S, Azam F, El-Shazly M, Singab AN (2020) Breaking down the barriers to a natural antiviral agent: antiviral activity and molecular docking of erythrina speciosa extract, fractions, and the major compound. Chem Biodivers 17:e1900511

Fleming-Dutra KE, Taylor T, Link-Gelles R, Garg S, Jhung MA, Finelli L, Jain S, Shay D, Chaves SS, Baumbach J, Hancock EB, Beall B, Bennett N, Zansky S, Petit S, Yousey-Hindes K, Farley MM, Gershman K, Harrison LH, Ryan P et al (2013) Effect of the 2009 influenza A (H1N1) pandemic on invasive pneumococcal pneumonia. J Infect Dis 207(7):1135–1143

Fuller R (1992) History and development of probiotics. In: Probiotics. Springer, Netherlands, pp 1–8

Furuya A, Uozaki M, Yamasaki H, Arakawa T, Arita M, Koyama AH (2008) Antiviral effects of ascorbic and dehydroascorbic acids in vitro. Int J Mol Med 22(4):541–545

Garren MR, Ashcraft M, Qian Y, Douglass M, Brisbois EJ, Handa H (2021) Nitric oxide and viral infection: recent developments in antiviral therapies and platforms. Appl Mater Today 22: 100887

Ghahri H, Toloei T, Soleimani B (2013) Efficacy of antibiotic, probiotic, prebiotic and synbiotic on growth performance, organ weights, intestinal histomorphology and immune response in broiler chickens. Glob J Anim Sci Res 1(1):25–41

Gibson GR, Hutkins R, Sanders ME, Prescott SL, Reimer RA, Salminen SJ, Scott K, Stanton C, Swanson KS, Cani PD, Verbeke K, Reid G (2017) Expert consensus document: The International Scientific Association for Probiotics and Prebiotics (ISAPP) consensus statement on the definition and scope of prebiotics. Nat Rev Gastroenterol Hepatol 14(8):491–502

Gonmei G, Sapcota D, Saikia GK, Deka P, Mahanta JD, Kalita N, Saikia BN, Talukdar JK (2019) Studies on immune response to Newcastle disease virus in broiler chickens fed with Lactobacillus reuteri PIA16 isolated from the gut of indigenous chicken of Assam. India Vet World 12(8):1251–1255

Górniak W, Cholewińska P, Konkol D (2018) Feed additives produced on the basis of organic forms of micronutrients as a means of biofortification of food of animal origin. J Chem 2018: 8084127

Grant WB, Lahore H, McDonnell SL, Baggerly CA, French CB, Aliano JL, Bhattoa HP (2020) Evidence that vitamin D supplementation could reduce risk of influenza and COVID-19 infections and deaths. Nutrients 12:988

Guarino MPL, Altomare A, Emerenziani S, Di Rosa C, Ribolsi M, Balestrieri P, Iovino P, Rocchi G, Cicala M (2020) Mechanisms of action of prebiotics and their effects on gastrointestinal disorders in adults. Nutrients 12(4):1037

Guo M, Ni J, Yu J, Jin J, Ma L, Zhang H, Wang D, Zhang X, Dou J, Zhou C (2019) Antiviral activity of benzoic acid derivative NC-5 against influenza A virus and its neuraminidase inhibition. Int J Mol Sci 20:6261

Gupta S, Read SA, Shackel NA, Hebbard L, George J, Ahlenstiel G (2019) The role of micronutrients in the infection and subsequent response to hepatitis C virus. Cell 8(6):603

Hakobyan A, Arabyan E, Avetisyan A, Abroyan L, Hakobyan L, Zakaryan H (2016) Apigenin inhibits African swine fever virus infection in vitro. Arch Virol 161:3445–3453

Hakobyan A, Arabyan E, Kotsinyan A, Karalyan Z, Sahakyan H, Arakelov V, Nazaryan K, Ferreira F, Zakaryan H (2019) Inhibition of African swine fever virus infection by genkwanin. Antivir Res 167:78–82

Hayashi K, Umetsu R, Chijimatsu T, Hayashi T, Miyake S, Lee JB (2012) Anti-influenza A virus effects of fructan from Welsh onion (Allium fistulosum L.). Food Chem 134:2164–2168

He X, Fang J, Guo Q, Wang M, Li Y, Meng Y, Huang L (2020) Advances in antiviral polysaccharides derived from edible and medicinal plants and mushrooms. Carbohydr Polym 229:115548

Hefnawy AEG, Tórtora-Pérez JL (2010) The importance of selenium and the effects of its deficiency in animal health. Small Rumin Res 89(2–3):185–192

Hilmarsson H, Traustason B, Kristmundsd'ottir T, Thormar H (2007) Virucidal activities of medium-and long-chain fatty alcohols and lipids against respiratory syncytial virus and parainfluenza virus type 2: comparison at different pH levels. Arch Virol 152(12):2225–2236

Hong JK, Lee KN, You SH, Kim SM, Tark D, Lee HS, Ko YJ, Seo MG, Park JH, Kim B (2015) Inactivation of foot-and-mouth disease virus by citric acid and sodium carbonate with deicers. Appl Environ Microbiol 81(21):7610–7614

Huleihel M, Ishanu V, Tal J, Arad S (2001) Antiviral effect of red microalgal polysaccharides on Herpes simplex and Varicella zosterviruses. J Appl Phycol 13(2):127–134

Huyghebaert G, Ducatelle R, Van Immerseel F (2011) An update on alternatives to antimicrobial growth promoters for broilers. Vet J (London, England: 1997) 187(2):182–188

Indo Y, Kitahara S, Tomokiyo M, Araki S, Islam MA, Zhou B, Albarracin L, Miyazaki A, Ikeda-Ohtsubo W, Nochi T, Takenouchi T, Uenishi H, Aso H, Takahashi H, Kurata S, Villena J, Kitazawa H (2021) Ligilactobacillus salivarius strains isolated from the porcine gut modulate innate immune responses in epithelial cells and improve protection against intestinal viral-bacterial superinfection. Front Immunol 12:652923

Jaime J, Vargas-Bermúdez DS, Yitbarek A, Reyes J, Rodríguez-Lecompte JC (2020) Differential immunomodulatory effect of vitamin D (1, 25 (OH) 2 D3) on the innate immune response in different types of cells infected in vitro with infectious bursal disease virus. Poult Sci 99(9): 4265–4277

Jenssen H (2005) Anti-herpes simplex virus activity of lactoferrin/lactoferricin–an example of antiviral activity of antimicrobial protein/peptide. Cell Mol Life Sci CMLS 62(24):3002–3013

Jung YJ, Lee YT, Ngo VL, Cho YH, Ko EJ, Hong SM, Kim KH, Jang JH, Oh JS, Park MK, Kim CH, Sun J, Kang SM (2017) Heat-killed Lactobacillus casei confers broad protection against influenza A virus primary infection and develops heterosubtypic immunity against future secondary infection. Sci Rep 7(1):17360

Kaewprom K, Chen Y-H, Lin C-F, Chiou M-T, Lin C-N (2017) Antiviral activity of Thymus vulgaris and Nepeta cataria hydrosols against porcine reproductive and respiratory syndrome virus. Thai J Vet Med 47(1):25–33

Keyaerts E, Vijgen L, Pannecouque C, Van Damme E, Peumans W, Egberink H et al (2007) Plant lectins are potent inhibitors of coronaviruses by interfering with two targets in the viral replication cycle. Antivir Res 75(3):179–187

Khan RU, Rahman ZU, Javed I, Muhammad F (2014) Effect of vitamins, protein level and probiotics on immune response of moulted male broiler breeders. J Anim Physiol Anim Nutr 98(4):620–627

Knezevic A, Stepanovic S, Cupic M, Jevtovic D, Ranin J, Jovanovic T (2005) Reduced quantity and hydrogen-peroxide production of vaginal Lactobacilli in HIV positive women. Biomed Pharmacother 59:521–533

Kohn A, Gitelman J, Inbar M (1980) Interaction of polyunsaturated fatty acids with animal cells and enveloped viruses. Antimicrob Agents Chemother 18(6):962–968

Komura M, Suzuki M, Sangsriratanakul N, Ito M, Takahashi S, Alam MS, Ono M, Daio C, Shoham D, Táchira K (2019) Inhibitory effect of grapefruit seed extract (GSE) on avian pathogens. J Vet Med Sci 81(3):466–472

Kumosani T, Obeid Y, Shaib H, Abualnaja K, Moselhy S, Iyer A et al (2017) Standardization of a protocol for quantitative evaluation of anti-aerosolized influenza virus activity by vapors of a chemically-characterized essential oil blend. J Prev Infect Cntrol 03:1–8

Lazarczyk M, Pons C, Mendoza J-A, Cassonnet P, Jacob Y, Favre M (2008) Regulation of cellular zinc balance as a potential mechanism of EVER-mediated protection against pathogenesis by cutaneous oncogenic human papillomaviruses. J Exp Med 205(1):35–42

Lee GY, Han SN (2018) The role of vitamin E in immunity. Nutrients 10(11):1614

Lee JB, Fukai T, Hayashi K, Hayashi T (2011) Characterization of fructan from Chikuyo-Sekko-To, a Kampo prescription, and its antiherpetic activity in vitro and in vivo. Carbohydr Polym 85: 408–412

Lee YH, Jang YH, Kim YS, Kim J, Seong BL (2018) Evaluation of green tea extract as a safe personal hygiene against viral infections. J Biol Eng 12:1

Liao X, Lu L, Li S, Liu S, Zhang L, Wang G, Lia A, Luo X (2012) Effects of selenium source and level on growth performance, tissue selenium concentrations, antioxidation, and immune functions of heat-stressed broilers. Biol Trace Elem Res 150(1–3):158–165

Lievin-Le MV, Servin AL (2014) Anti-infective activities of lactobacillus strains in the human intestinal microbiota: from probiotics to gastrointestinal anti-infectious biotherapeutic agents. Clin Microbiol Rev 27:167–199

Liu Q, Yu Z, Tian F, Zhao J, Zhang H, Zhai Q, Chen W (2020) Surface components and metabolites of probiotics for regulation of intestinal epithelial barrier. Microb Cell Factories 19:23

Macan AM, Harej A, Cazin I, Klobuˇcar M, Stepani'c V, Paveli'c K et al (2019) Antitumor and antiviral activities of 4-substituted 1, 2, 3-triazolyl-2, 3- dibenzyl-L-ascorbic acid derivatives. Eur J Med Chem 184:111739

Madsen K, Cornish A, Soper P, McKaigney C, Jijon H, Yachimec C, Doyle J, Jewell L, De Simone C (2001) Probiotic bacteria enhance murine and human intestinal epithelial barrier function. Gastroenterology 121(3):580–591

Mahooti M, Abdolalipour E, Salehzadeh A, Mohebbi SR, Gorji A, Ghaemi A (2019) Immunomodulatory and prophylactic effects of Bifidobacterium bifidum probiotic strain on influenza infection in mice. World J Microbiol Biotechnol 35:91

Martín-Acebes MA, V'azquez-Calvo A, Caridi F, Saiz J-C, Sobrino F (2012) Lipid involvement in viral infections: present and future perspectives for the design of antiviral strategies. Lipid Metabolism 2012:291–322

Mazalovska M, Kouokam JC (2018) Lectins as promising therapeutics for the prevention and treatment of HIV and other potential coinfections. Biomed Res Int 2018:3750646

Mitchell CA, Ramessar K, O'Keefe BR (2017) Antiviral lectins: selective inhibitors of viral entry. Antivir Res 142:37–54

Miyazaki T (2017) Protective effects of lactic acid bacteria on influenza A virus infection. AIMS Allergy Immunol 1:138–142

Muhialdin BJ, Zawawi N, Abdull Razis AF, Bakar J, Zarei M (2021) Antiviral activity of fermented foods and their probiotics bacteria towards respiratory and alimentary tracts viruses. Food Control 127:108140

Mullin GE (2011) Vitamin A and immunity. Nutr Clin Pract 26(4):495–496

Murugesan GR, Syed B, Haldar S, Pender C (2015) Phytogenic feed additives as an alternative to antibiotic growth promoters in broiler chickens. Front Vet Sci 2:21

Mustopa AZ, Umami RN, Putri PH, Susilaningsih D, Farida H (2016) Identification of bioactive compound from microalga BTM 11 as Hepatitis C virus RNA helicase inhibitor. Jurnal Biologi Indonesia 11(2)

Nogusa S, Gerbino J, Ritz BW (2009) Low-dose supplementation with active hexose correlated compound improves the immune response to acute influenza infection in C57BL/6 mice. Nutr Res (New York, NY) 29(2):139–143

Olaimat AN, Aolymat I, Al-Holy M, Ayyash M, Abu Ghoush M, Al-Nabulsi AA, Osaili T, Apostolopoulos V, Liu SQ, Shah NP (2020) The potential application of probiotics and prebiotics for the prevention and treatment of COVID-19. NPJ Sci Food 4:17

Pangestuti R, Suryaningtyas IT, Siahaan EA, Kim S-K (2020) Cosmetics and cosmeceutical applications of microalgae pigments. In: In. Springer International Publishing, pp 611–633

Pecoraro BM, Leal DF, Frias-De-Diego A, Browning M, Odle J, Crisci E (2022) The health benefits of selenium in food animals: a review. J Animal Sci Biotechnol 13:58

Pennisi M, Russo G, Ravalli S, Pappalardo F (2017) Combining agent based-models and virtual screening techniques to predict the best citrus-derived vaccine adjuvants against human papilloma virus. BMC Bioinform 18:544

Pensel PE, Maggiore MA, Gende LB, Eguaras MJ, Denegri MG, Elissondo MC (2014) Efficacy of essential oils of Thymus vulgaris and Origanum vulgare on Echinococcus granulosus. Interdiscip Perspect Infect Dis 2014:1–12

Pertile TL, Karaca K, Sharma JM, Walser MM (1996) An antiviral effect of nitric oxide: inhibition of reovirus replication. Avian Dis 40(2):342–348

Pignolet O, Jubeau S, Vaca-Garcia C, Michaud P (2013) Highly valuable microalgae: biochemical and topological aspects. J Ind Microbiol Biotechnol 40(8):781–796

Pleschka S, Stein M, Schoop R, Hudson JB (2009) Anti-viral properties and mode of action of standardized Echinacea purpurea extract against highly pathogenic avian Influenza virus (H5N1, H7N7) and swine-origin H1N1 (S-OIV). Virol J 6:197

Pourghanbari G, Nili H, Moattari A, Mohammadi A, Iraji A (2016) Antiviral activity of the oseltamivir and Melissa officinalis L. essential oil against avian influenza A virus (H9N2). Virusdisease 27(2):170–178

Rather IA, Kamli MR, Sabir JSM, Ali S (2022) Evaluation of Lactiplantibacillus plantarum KAU007 against low-pathogenic avian influenza virus (H9N2). Pathogens 11:1246

Read SA, Obeid S, Ahlenstiel C, Ahlenstiel G (2019) The role of zinc in antiviral immunity. Adv Nutr 10(4):696–710

Reboul E (2017) Vitamin E bioavailability: mechanisms of intestinal absorption in the spotlight. Antioxidants 6(4):95

Reddy PH, Johnson AMA, Kumar JK, Naveen T, Devi MC (2017) Heterologous expression of infectious bursal disease virus VP2 gene in Chlorella pyrenoidosa as a model system for molecular farming. Plant Cell, Tissue Organ Cult (PCTOC) 131(1):119–126

Regnault-Roger C, Vincent C, Arnason JT (2012) Essential oils in insect control: low-risk products in a high-stakes world. Annu Rev Entomol 57(1):405–424

Rehman ZU, Che L, Ren S, Liao Y, Qiu X, Yu S, Sun Y, Tan L, Song C, Liu W, Ding Z, Munir M, Nair V, Meng C, Ding C (2018) Supplementation of vitamin E protects chickens from Newcastle disease virus-mediated exacerbation of intestinal oxidative stress and tissue damage. Cell Physiol Biochem 47:1655–1666

Reynolds D, Huesemann M, Edmundson S, Sims A, Hurst B, Cady S et al (2021) Viral inhibitors derived from macroalgae, microalgae, and cyanobacteria: A review of antiviral potential throughout pathogenesis. Algal Res 57:102331

Saad N, Delattre C, Urdaci M, Schmitter JM, Bressollier P (2013) An overview of the last advances in probiotic and prebiotic field. LWT Food Sci Technol 50:1–16

Santoyo S, Jaime L, Plaza M, Herrero M, Rodriguez-Meizoso I, Ibañez E, Reglero GJJ (2012) Antiviral compounds obtained from microalgae commonly used as carotenoid sources. J Appl Phycol 24:731–741

Sarohan AR (2020) COVID-19: endogenous retinoic acid theory and retinoic acid depletion syndrome. Med Hypotheses 144:110250

Sato T, Hori K (2009) Cloning, expression, and characterization of a novel anti-HIV lectin from the cultured cyanobacterium, *Oscillatoria agardhii*. Fish Sci 75(3):743–753

Sato Y, Okuyama S, Hori K (2007) Primary structure and carbohydrate binding specificity of a potent anti-HIV lectin isolated from the filamentous Cyanobacterium Oscillatoria agardhii. J Biol Chem 282(15):11021–11029

Schnitzler P, Schön K, Reichling J (2001) Antiviral activity of Australian tea tree oil and eucalyptus oil against herpes simplex virus in cell culture. Pharmazie 56(4):343–347

Schönfeldt HC, Pretorius B, Hall N (2016) Bioavailability of nutrients. In: Caballero B, Finglas P, Toldr'a FF (eds) The encyclopedia of food and health, vol 1, pp 401–406

Semba RD, Ndugwa C, Perry RT, Clark TD, Jackson JB, Melikian G et al (2005) Effect of periodic vitamin A supplementation on mortality and morbidity of human immunodeficiency virus–infected children in Uganda: A controlled clinical trial. Nutrition 21(1):25–31

Seo BJ, Rather IA, Kumar VJR, Choi UH, Moon MR, Lim JH, Park YH (2012) Evaluation of Leuconostoc mesenteroides YML003 as a probiotic against low-pathogenic avian influenza (H9N2) virus in chickens. J Appl Microbiol 113:163–171

Shah KK, Verma R, Oleske JM, Scolpino A, Bogden JD (2019) Essential trace elements and progression and management of HIV-1 infection. Nutr Res 71:21–29

Shayeganmehr A, Vasfi Marandi M, Karimi V, Barin A, Ghalyanchilangeroudi A (2018) Zataria multiflora essential oil reduces replication rate of avian influenza virus (H9N2 subtype) in challenged broiler chicks. Br Poult Sci 59(4):389–395

Sheridan DA, Bridge SH, Crossey MM, Felmlee DJ, Fenwick FI, Thomas HC et al (2014) Omega-3 fatty acids and/or fluvastatin in hepatitis C prior non-responders to combination antiviral therapy–a pilot randomised clinical trial. Liver Int 34(5):737–747

Sherman PM, Ossa JC, Johnson-Henry K (2009) Unravelling mechanism of action of probiotics. Nutr Clin Pract 21:10–14

Singh J, Gaikwad DS (2020) Phytogenic feed additives in animal nutrition. In: Natural bioactive products in sustainable agriculture. Springer, Singapore, pp 273–289

Smith AM, Smith APA (2016) Critical, Nonlinear Threshold Dictates Bacterial Invasion and Initial Kinetics During Influenza. Sci Rep 6(38703):2016. https://doi.org/10.1038/srep38703

Steiner T, Syed B (2015) Phytogenic feed additives in animal nutrition. In: Medicinal aromatic plants of the world: scientific, production, commercial utilization aspects. Springer, Dordrecht, pp 403–423

Stevanović Z, Bošnjak-Neumüller J, Pajić-Lijaković I, Raj J, Vasiljević M (2018) Essential oils as feed additives—future perspectives. Molecules 23(7):1717

Steyer A, Mičetić-Turk D, Fijan S (2022) The efficacy of probiotics as antiviral agents for the treatment of Rotavirus gastrointestinal infections in children: an updated overview of literature. Microorganisms 10(12):2392

Talebi A, Amani A, Pourmahmod M, Saghaei P, Rezaie R (2015) Synbiotic enhances immune responses against infectious bronchitis, infectious bursal disease, Newcastle disease and avian influenza in broiler chickens. Vet Res Forum 6(3):191–197

Tang K, He S, Zhang X, Guo J, Chen Q, Yan F, Banadyga L, Zhu W, Qiu X, Guo Y (2018) Tangeretin, an extract from citrus peels, blocks cellular entry of arenaviruses that cause viral hemorrhagic fever. Antivir Res 160:87–93

Tantcheva LP, Stoeva ES, Galabov AS, Braykova AA, Savov VM, Mileva MM (2003) Effect of vitamin E and vitamin C combination on experimental influenza virus infection. Methods Find Exp Clin Pharmacol 25(4):259–264

Thirumdas R, Kothakota A, Pandiselvam R, Bahrami A, Barba FJ (2021) Role of food nutrients and supplementation in fighting against viral infections and boosting immunity: A review. Trends Food Sci Technol 110:66–77

Trompette A, Gollwitzer ES, Pattaroni C, Lopez-Mejia IC, Riva E, Pernot J, Ubags N, Fajas L, Nicod LP, Marsland BJ (2018) Dietary fiber confers protection against flu by shaping $Ly6c^-$ patrolling monocyte hematopoiesis and $CD8^+$ T cell metabolism. Immunity 48(5):992–1005.e8

Tumpey TM, García-Sastre A, Taubenberger JK, Palese P, Swayne DE, Pantin-Jackwood MJ, Schultz-Cherry S, Solórzano A, Van Rooijen N, Katz JM, Basler CF (2005) Pathogenicity of influenza viruses with genes from the 1918 pandemic virus: functional roles of alveolar macrophages and neutrophils in limiting virus replication and mortality in mice. J Virol 79(23):14933–14944

Tuvaanjav S, Shuqin H, Komata M, Ma C, Kanamoto T, Nakashima H, Yoshida T (2016) Isolation and antiviral activity of water-soluble Cynomorium songaricum Rupr. Polysaccharides. J Asian Nat Prod Res 18(2):159–171

Vincent M, Duval RE, Hartemann P, Engels-Deutsch M (2018) Contact killing and antimicrobial properties of copper. J Appl Microbiol 124(5):1032–1046

Waititu SM, Yitbarek A, Matini E, Echeverry H, Kiarie E, Rodriguez-Lecompte JC, Nyachoti CM (2014) Effect of supplementing direct-fed microbials on broiler performance, nutrient digestibilities, and immune responses. Poult Sci 93(3):625–635

Wakabayashi H, Oda H, Yamauchi K, Abe F (2014) Lactoferrin for prevention of common viral infections. J Infect Chemother 20(11):666–671

Wang Z, Zhang P, Fu W, Zhang Y, Li T, Pan B, Wei P (2010) Wei sheng wu xue bao Acta microbiologica Sinica 50(12):1664–1669

Wang Z, Chai W, Burwinkel M, Twardziok S, Wrede P, Palissa C, Esch B, Schmidt MF (2013) Inhibitory influence of Enterococcus faecium on the propagation of swine influenza A virus in vitro. PLoS One 8(1):e53043

Wang X, Hu W, Zhu L, Yang Q (2017) Bacillus subtilis and surfactin inhibit the transmissible gastroenteritis virus from entering the intestinal epithelial cells. Biosci Rep 37(2):BSR20170082

Wang Y, Moon A, Huang J, Sun Y, Qiu HJ (2022) Antiviral effects and underlying mechanisms of probiotics as promising antivirals. Front Cell Infect Microbiol 12:928050

Wani AR, Yadav K, Khursheed A, Rather MA (2021) An updated and comprehensive review of the antiviral potential of essential oils and their chemical constituents with special focus on their mechanism of action against various influenza and coronaviruses. Microb Pathog 152:104620

Wetprasit N, Threesangsri W, Klamklai N, Chulavatnatol M (2000) Jackfruit lectin: properties of mitogenicity and the inhibition of herpesvirus infection. Jpn J Infect Dis 53(4):156–161

Windisch W, Schedle K, Plitzner C, Kroismayr A (2008) Use of phytogenic products as feed additives for swine and poultry. J Anim Sci 86:140–148

World organization of animal health (WOAH) (2020). http://www.animalhealthsurveillance.agriculture.gov.ie/notifiablediseases

Yadav A, Kumar A, Das M, Tripathi A (2016) Sodium benzoate, a food preservative, affects the functional and activation status of splenocytes at non cytotoxic dose. Food Chem Toxicol 88: 40–47

Yang J, Tian G, Chen D, Zheng P, Yu J, Mao X, He J, Luo Y, Luo J, Huang Z, Wu A, Yu B (2019) Dietary 25-Hydroxyvitamin D3 supplementation alleviates porcine epidemic diarrhea virus infection by improving intestinal structure and immune response in weaned pigs. Animals 9:627

Yeh TL, Shih PC, Liu SJ, Lin CH, Liu JM, Lei WT, Lin CY (2018) The influence of prebiotic or probiotic supplementation on antibody titers after influenza vaccination: a systematic review and meta-analysis of randomized controlled trials. Drug Des Devel Ther 12:217–230

Zhang X, Chen S, Li X, Zhang L, Ren L (2022) Flavonoids as potential antiviral agents for porcine viruses. Pharmaceutics 14:1793

Zhao W, Liu Y, Latta M, Ma W, Wu Z, Chen P (2019) Probiotics database: A potential source of fermented foods. Int J Food Prop 22(1):198–217

Food and Feed Additives to Counteract Mycotoxin Toxicity in Human and Animals

Daniela E. Marin and Ionelia Taranu

Abstract

Mycotoxins are secondary metabolites of fungi that frequently contaminate a broad category of substrates such as cereals, cereal products, medicinal plants, legumes, fruits, and spices. Mycotoxins are among the most common and increasingly well-known natural contaminants with proven toxic effects (hepatotoxicity, nephrotoxicity, immunotoxicity, carcinogenesis, teratogenesis, etc.) in both animals and humans. The elimination of mycotoxins from contaminated food and feed is an unresolved issue and once the contamination has occurred, there are few strategies that can be adopted to limit the adverse effects on human and farm animals.

To date, nutritional approaches used for adsorption, inactivation, or reduction of the mycotoxin concentration and toxicity are the most promising, given their ability to meet safety standards, and nutritional efficacy, on the one hand, and lower cost on the other.

A large number of dietary nutrients (proteins, fats, vitamins, and minerals) but also food and feed additives (enzymes, medicinal and aromatic plants, agri-food residues, probiotics, etc.) can interact with mycotoxins and reduce their effects. Certain characteristics are required for the selection of a food and feed additive with anti-mycotoxic activity, such as the safety of use, the stability of the complex between mycotoxin and additive in the gastrointestinal tract at different pH levels, the rapid degradation of mycotoxins into non-toxic metabolites, etc.

The present chapter reviews the botanical additives tested as mitigation agents for the reduction of the negative effect of mycotoxins using in vitro and in vivo approaches.

D. E. Marin (✉) · I. Taranu
National Institute for Research and Development for Biology and Animal Nutrition, IBNA, Balotesti, Romania
e-mail: daniela.marin@ibna.ro

G. Arsenos, I. Giannenas (eds.), *Sustainable Use of Feed Additives in Livestock*,
https://doi.org/10.1007/978-3-031-42855-5_13

Keywords

Mycotoxins · Animals · Human · Botanicals · Bioactive compounds · Waste

1 Introduction

Mycotoxins are secondary metabolites of fungi that frequently contaminate a broad category of substrates such as cereals, cereal products, medicinal plants, legumes, fruits, and spices (Taranu et al. 2005). They are among the most common and increasingly well-known natural contaminants with proven toxic effects (hepatotoxicity, nephrotoxicity, immunotoxicity, carcinogenesis, teratogenesis, etc.) in both animals and humans (Marin et al. 2016).

The elimination of mycotoxins from contaminated food and feed is an unresolved issue and once the contamination has occurred, there are few strategies that can be adopted to limit the adverse effects on human and farm animals. The use of physical (sorting, dehulling, extrusion, milling, washing, heating, irradiation, and use of adsorbents) and chemical (ammonia, sodium hydroxide, hydrochloric acid, butylated hydroxytoluene, butylated hydroxy-anisole, and oltipraz) methods to decontaminate mycotoxins should be limited due to the apparition of toxic residual products, health and environmental concerns (Meng et al. 2020) as well as interactions with food/feed nutrients or/and interference with the organoleptic properties of food and feed (Çelik 2020).

Biological decontamination methods, such as biodegradation and transformation of mycotoxins by either microorganisms, or agro-industrial by-products containing bioactive compounds efficient in adsorption and mitigating/alleviation of mycotoxin effects are the most promising, given their ability to meet safety standards, nutritional efficacy, on the one hand, and lower cost on the other (Adebo et al. 2017; Palade et al. 2020).

Recently, the use of botanicals as a natural alternative in reducing mycotoxin contamination has gained an increasing interest, as they are generally considered as natural, safe, and environmentally friendly alternative sources of bioactive compounds (Makhuvele et al. 2020).

Many studies have shown that diets rich in vegetables and fruits contain in addition vitamins and pro-vitamins, a large number of phytochemical compounds, such as polyphenols, flavonoids, terpenoids, and plant pigments with important antimycotic properties (Jideani et al. 2021).

The use of botanicals in the mitigation of mycotoxins effects targets: (i) to prevent the growth of mycotoxigenic fungi, mycotoxin synthesis, and mycotoxin contamination of food and feed; (ii) to be efficient in mycotoxin detoxification without any alteration of the nutritional value of food and feed; (iii) to counteract toxic effects of mycotoxins into the animal body, by stimulating the organism self-defence. According to EFSA, these substances called "interaction agents" which are different from "detoxifying agents (bio-transforming or adsorbing agents)" do not interact directly with mycotoxins, such as antioxidants which interact with the action of mycotoxins, but they are especially immunostimulants and their role in the mitigation of the toxic effects of mycotoxins cannot be neglected.

This review aimed to gather the literature studies that have investigated the capacity of botanicals to mitigate the mycotoxin effect in animals and humans both in vivo and in vitro. Although many plant extracts have been shown to be effective in detoxifying mycotoxins in vitro in abiotic systems, these results have rarely been validated by in vitro and in vivo exposure.

2 Use of Bioactive Compounds from Botanicals for Reduction of the Mycotoxins Effect in Animals

Phytobiotics are often included in animal diets as detoxifying/mitigating mycotoxin agents due to their antioxidant, anti-inflammatory, and immunostimulatory potential (Holanda and Kim 2021). Tables 1, 2, and 3 present several examples of different active phytocompounds tested for their capacity to alleviate the toxic effects of mycotoxins, especially on liver. For example, polyphenols from tea have important health benefits being immunostimulants, antioxidant, and anti-inflammatory. Thus, tea polyphenols administered to shrimp intoxicated for 20 days with 1.2 mg aflatoxin B1 (AFB1)/kg feed restored the muscle structure and chemical composition affected by the exposure to the toxin; also, they decreased the concentration of AFB1-albumin-adducts and reduced the liver AFB1 induced liver injury (Divakaran and Tacon 2000). A decrease in the concentration of the AFB1-albumin-adducts, accompanied by a decrease in AFM1 levels in urine or faeces was also observed in human volunteers accidentally exposed to AFB1 natural contaminated food (Tang et al. 2008) or in rats exposed to 100 μg AFB1/kg b.w after administration of tea polyphenols (Lu et al. 2017). Another polyphenol, curcumin (turmeric) derived from the rhizome of *Curcuma longa* with also hepatoprotective properties had the capacity to modulate the activity of the hepatic enzyme of phase-I and phase-II and to reduce AFB1-induced oxidative stress in chickens (Muhammad et al. 2018). In the same line, it worth mentioning phillygenin (PHI) a lignan derived primarily from *Forsythia suspensa* with medicinal properties which exert its hepatoprotective, antioxidant, and anti-inflammatory effect by inhibiting NF-κB and activating the Nrf2 signalling pathway in chickens intoxicated with AFB1 (Guo et al. 2022).

Cyanidin-3-*O*-glucoside (Cy3G) is the most known anthocyanin belonging to the flavonoids family and it is found mainly in red to blue fruits (Olivas-Aguirre et al. 2016). In vitro studies carried out either on hepatocytes cells (HepG2) or epithelial intestinal cells ($CaCO_2$) treated with AFB1 and ochratoxin A (OTA) have shown that Cy3G inhibits the toxic effects induced by the mycotoxins as cytotoxicity, DNA fragmentation, inhibition of DNA and protein synthesis (Guerra et al. 2005). In vivo studies performed in rats showed that Cy3G reduced the nitrosative stress induced by OTA by decreasing the expression of nitric oxide synthases (iNOS and eNOS), the concentration of nitric oxide as well as the activity of dimethyl arginine dimethylamino-hydrolase (DDAH) (Sorrenti et al. 2012; Pauletto et al. 2021). Resveratrol, a powerful antioxidant found in grapes is responsible for the reduction of expression

Table 1 Effect of different pure active compounds in alleviating the toxic effects induced by mycotoxins

No crt.	I.Specie/sex/age//II. Cell line	No. of individuals	Treatment duration	Mycotoxin concentration/way of administration	Type of active compounds	Effect of extracts	Authors
1	HepG2, CaCO2	–	24 h–72 h	AFB1 (0-100 μM), OTA (0-100 μM)	*Cyanidin-3-O-b -glucopyranoside (25; 50 mM)*	Inhibits the cytotoxicity of AFB1 and OTA in HepG2 and CaCO2 cells; counteracts the inhibition of DNA and protein synthesis induced by mycotoxins, reduced the DNA fragmentation and inhibition of caspase-3 activity	Guerra et al. (2005)
2	Shrimp (*Penaeus vannamei*)	150	20 day	AFB1 1.2 mg/kg	Tea polyphenols (0.4%–0.64%)	Inhibits the expansion of muscle fibre spaces and inflammation; protective effect against the decrease of muscle nutrients and changes in protein composition of shrimp caused by exposure to AFB1; decrease AFB1-albumin-adducts; promote the elimination of AFB1 in faeces; inhibit liver injury induced by AFB1	Divakaran and Tacon (2000)
3	Sprague–Dawley rats, male, 12 weeks old	42	14 days	0.5 mg OTA/kg b. w./day	Lycopene (5 mg/kg/day)	Decreases in tail length, tail moment, and tail intensity	Aydin et al. (2013)
4	Albino rats, male, 12 weeks old	30	1 month	0.1 mg/kg b.w	Rutin (50 mg/kg)	Decrease TBARs, SOD, GSH in liver; total thiols, total lipids and catalase	El-Sawi and Al-Seeni (2009)
5	Sprague–Dawley rats, male	12	20 day	100μgAFB1/kg b.w	Oxidized tea polyphenols 200 mg/kg b.w	Inhibits the absorption of complexed AFB1, decrease concentration of AFB1 adducts in serum, promote the elimination of AFB1 in faeces;	Lu et al. (2017)

6	Wistar rats, six weeks old, male	36	28 days	200 µg/kg b.w	Vanillic acid (0.48–48.0 mg/kg b.w)	Antimutagenicity against AFB1	Nilnumkhum et al. (2017)
7	Sprague-Dawley rats, male	24	14 days	0.5 mg OTA/kg b. w./day	Lycopene (5 mg/kg/day)	Increase GPx activity and GSH levels, and decreased apoptotic cell death in both cortex and medulla	Palabiyik et al. (2013)
8	Bovine hepatocyte cell line (BFH12)	–	48 h–64 h	3.6 µM AFB1	Resveratrol (10-30 mM)	Reverse the AFB1-dependent cytotoxicity, increased the NQO1 enzymatic activity, reduced CYP3A activity	Pauletto et al. (2021)
9	Fisher 344 rats, male, 4 week old	15	3 weeks	1.75 mg AFB1 /kg b. w, p.o. administration after 3 weeks	Curcumin (8 mg/kg; 80 mg/kg b.w)	Reduction of AFB1-DNA adducts; decrease of plasma AST and ALT activities, reduction of glutathione S-transferases	Poapolathep et al. (2015)
10	Sprague-Dawley rats, male	40	4 weeks	200 ppb OTA	*Cyanidin 3-O-ß-D-glucoside 1 g/kg feed*	Decreased expression of iNOS, eNOS, and DDAH-1; decrease nitrite levels and DDAH activity	Sorrenti et al. (2012)
11	Human volunteers	120	3 months	AFB1 natural contaminated food	Green tea polyphenols (500 mg/day, 1000 mg/day)	Reduction in AFB–AA adducts levels in serum, reduction in AFM1 levels in urine	Tang et al. (2008)
12	Ducks, male, 1 day old	45	10 weeks	0.2 mg/kg AFB1	Resveratrol 500 mg/kg diet	Decrease the mRNA expression of CYP1A1 and CYP1A2 ($p < 0.05$) and the production of AFB1-DNA adducts in ileum, reduced DNA damage and oxidative stress via the Nrf2/ Keap1 and NF- B/NLRP3 signalling pathways	Yang et al. (2022)

(continued)

Table 1 (continued)

No crt.	I.Specie/sex/age//II. Cell line	No. of individuals	Treatment duration	Mycotoxin concentration/way of administration	Type of active compounds	Effect of extracts	Authors
13	HepG2 cells	–	1 h	0–60 μM PAT	6-gingerol 10 μM	Reduce the DNA strand breaks and micronuclei formation caused by PAT; suppressed PAT-induced intracellular ROS formation and 8-OHdG level; reduce GSH depletion	Yang et al. (2011)
14	Ross 308 broilers	288	42 days	2 mg AFB1/kg feed; 2 mg OTA/kg feed	Berberine 200-600 mg/kg feed)	Improve growth performance, liver function, and antioxidant status of broilers fed diets contaminated with AFB and OTA	Malekinezhad et al. (2021)
15	Broiler chickens	60	26 days	0.5 mg·AFB1/kg bw	Piperine 60 ppm	Reduced DNA damage in peripheral blood cells, the number of erythrocytes with micronuclei, and polychromatic-to-normochromatic erythrocyte ratio	da Silva et al. (2016)

Table 2 Effect of different medicinal plant and spices in alleviating the toxic effects induced by mycotoxins

No crt.	I.Specie/sex/ age//II. Cell line	No. of animals/ group	Type of nutritional approach	Duration of treatment	Mycotoxin concentration/way of administration	Effect of extracts	Authors
1	Sprague-Dawley albino rats, male	80	*Camellia sinensis* leaves, *Carum carvi* seeds, *Alpinia galanga* rhizomes, *Boswellia serrata* resins, and *Cinchona officinalis* bark aqueous extracts	4 weeks	1 mg AFB1/kg b.w	Ameliorate the parameters of oxidative stress (MDA, GR, G-6PDH, vit C) in kidney and reverse the inhibitory action of AFB1 on markers of energy metabolism in the hearts	Soni et al. (1993)
2	Rats, male	36	*Alchornea cordifolia* ethanolic leaves extract	21 days	150 μg AFB1/kg diet	Alleviate biochemical parameters and body weight gain affected by AFB1 exposure; significant improvement of histological lesions of liver and kidney	*Baptiste et al. (2017)*
3	Sprague-Dawley albino rats of / male	72	*Moringa oleifera* leaf extract	1 month	0.7 g AFB1/kg b.w, i. p., four times weekly for 1 month	Decrease genetic alterations, sperm abnormalities, and biochemical destruction induced by AFB1	Farag et al. (2018)
4	Sprague-Dawley rats, male, three-month-old	60	*Calendula officinalis* ethanolic extract	6 weeks	2.5 mg AFB1/kg diet	Improve the biochemical parameters, inflammatory cytokines, decrease the oxidative stress, and improve the histological pictures in the liver of rats fed AFs-contaminated diet	Hamzawy et al. (2013)
5	Rats	40	*Coleus forskohlii* roots extract	2 weeks	1,5 mg AFB1, single dose, i.p	Increase TBARS, CAT; trigger cytoplasmic regeneration of hepatocytes	Atef et al. (2012)
6	Chicks, male, one day old	720	Chamomile flower extract and thyme oil extract	28 days	0.5 mg/kg diet AFB and 0.25 mg/kg diet OTA	Increase average daily gain, average daily feed intake, and decreased feed conversion ratio in the birds fed OTA-AFB toxin diets	Nazarizadeh et al. (2019)
7	Rat hepatoma cells (H4IIE-luc)		*Amaranthus hybridus* aqueous extract		0.025–50 mmol/L AFB1; 0.1-200 mmol/ L FB1	Improve of cell viability accompanied with a significant decrease in DNA damage and genotoxic effects	Ibrahim et al. (2015)

(continued)

Table 2 (continued)

No crt.	I.Specie/sex/age//II. Cell line	No. of animals/group	Type of nutritional approach	Duration of treatment	Mycotoxin concentration/way of administration	Effect of extracts	Authors
8	Sprague-Dawley rats, male	48	*Monanthotaxis caffra*, methanolic leaf extract	10 days	1 mgAFB1/kg b.w	Decrease serum levels of aspartate aminotransferase, alanine aminotransferase, lactate dehydrogenase, and creatinine; improve hepatocellular injuries induced by AFB1	Makhuvele et al. (2022)
9	Layers hens (white Leghorn), 25-week-old	28	*Aloe vera* extract (100–300 ppm)	28 days	1575 mg AFB1/kg b.w	Decrease the concentration of AFB1 residues and lipid peroxidation in egg yolk	Mohajer et al. (2021)
10	Sprague-Dawley albino rats, male	80	*Zingiber officinale* Roscoe rhizome, *Cinnamomum zeylanicum* bark, *Trigonella foenum graecum* seeds, *Camellia sinensis* leaves, *Salvia officinalis* leaves aqueous extract	3 weeks	1 mg AFB1/kg, single dose, i.p	Decrease the MDA level, increase the antioxidant level in kidney and glucose and glycogen contents in heart	Mohamed and Metwally (2009)
11	Ross 308 broilers, one-day-old chicks	192	Milk thistle whole plant powder	35 days	AF contaminated diet	Decrease alanine aminotransferase activity, increase the antibody titre	Afshin et al. (2017)
12	Ross-308 broiler chickens, 43 g	216	Licorice extract (3 g/kg feed; 6 g/kg feed)	42 days	0.48 mg AFB1/kg feed	Alleviate alkaline phosphatase, aspartate aminotransferase, and alanine transaminase activities, malondialdehyde concentration in breast meat and liver pathological damages	Rashidi et al. (2020)

13	Wistar albino male (200-250 g) rats	36	*Ixora coccinea* (Rubiaceae) and *Rhinacanthus nasutus*(Acanthaceae) roots extract *Spilanthes ciliata* (Asteraceae) whole plants (100, 200, 300 mg/kg)	72 h	1.5 mg/kg single dose, i.p	Decrease of the activity of the serum enzymes and enhance GSH in liver.	Shyamal et al. (2010)
14	Piglets (landrace large Yorkshire) female	9	Green tea extracts (Sunphenon) and coumarin	3 weeks	Ex vivo incubation with AFB1	Coumarin enhances AFB1 detoxification in liver and GST activity in the intestine. Sunphenon enhances AFB1 detoxification in intestine.	Tulayakul et al. (2007)
15	Broilers, 1 day old	480	Chinese gallnut	42 days	60 μg AFB1/kg diet (days 1 to 21);120 μg AFB1/kg diet (days 22 to 42);	Decrease the relative weight of liver and kidney, decreased the level of ALT, GGT, SOD, GSH-Px, GST	Zhang et al. (2022)
16	Albino rats, adult, male	36	Ginseng (100 mg/kg b. w)	5 days/ week, 4 weeks	250 mg AFB1/kg b.w	Improve the renal function affected by AFB1 as resulted from histological and immunohistochemical analyses.	Zidan et al. (2015)
17	Ross 308 broiler chicks, male, one day old	216	Milk thistle seed	21 days	250 ppb AFB1, 500 ppb AFB	Increased HDL and glucose; decrease AST	Amiridumari et al. (2013)
18	Broiler chicks, one day old	240	Milk thistle (10 g/kg diet)	5 weeks	80 mg/kg (first week); 520 mg/kg	Increased antibody titre against Newcastle disease, infectious bronchitis and infectious bursal diseases	Chand et al. (2011)
19	Ross 308 broiler chicks, male, one day old	216	Milk thistle seed	21 days	250 ppb AFB1, 500 ppb AFB1	Reduction of ALT and AST, increase the titre of specific antibody decreased by the toxin	Amiri Dumari et al. (2014)

(continued)

Table 2 (continued)

No crt.	I.Specie/sex/age//II. Cell line	No. of animals/group	Type of nutritional approach	Duration of treatment	Mycotoxin concentration/way of administration	Effect of extracts	Authors
20	Ross 308 broiler chicks, male, 7 days old	136	Milk thistle	21 days	0.5 ppm AFB1, 2 ppm AFB1	Improved performances in aflatoxin-challenged chicks, decrease of ileal populations of *Escherichia coli*, Salmonella, Klebsiella; increased villi height and villi height-to-crypt depth ratio in aflatoxicated birds	Jahanian et al. (2017)
21	Broilers, one day old	240	Polyherbal feed supplement (Growell) 0.35 g/kg of feed	6 weeks	0.2 ppm AFB1; 0.2 ppm OTA	Increase in haemoglobin values, reduce the alterations due to mycotoxins	Kalorey et al. (2005)
22	Broilers, one day old	240	Milk thistle (10 g/kg diet)	5 weeks	80 mg AFB1/kg (first week); 520 mg AFB1/kg	Improved body weight gain and feed intake	Muhammad et al. (2012)
23	Ross 308 broiler chicks	216	*Silybum marianum* seeds, 0.5%, 1%	5 weeks	250 ppb AFB1, 500 ppb AFB1	Decrease uric acid, glucose, AST and GGT enzymes	Fani et al. (2014)
24	Ross 308 broiler chicks	216	*Silybum marianum* seeds, 0.5%, 1%		250 ppb AFB1, 500 ppb AFB1	Improve weights of carcass and internal organs; reduce negative effects of AFB1 on liver morphology	Fani makki et al. (2013)
25	Broilers, 14 day old	21	*Silybum marianum*	35 days	0.8 mg AFB1/kg diet	Increase of BW gain and feed intake, restore the ALT activity affected by AFB1 exposure	Tedesco et al. (2004)
26	Broiler chicks, male, one day old	250	*Curcuma longa* (turmeric)	21 days	2 mg AFB1/kg diet	Decrease feed intake and body weight gain compared with control chicks; unable to reduce the severity of the lesions at the liver level.	Dos Anjos et al. (2015)

27	Broiler chickens, one day old	1225	Algae-based antioxidant	44 days	Naturally contaminated corn with FB1 620-11,700 ppb, FB2 133–5630 ppb, ZEA 1429 ppb	Enhanced body gain; no effect of AFB1 and algae supplement on villus height, crypt depth or on carcass, breast and fat yield	Bortoluzzi et al. (2016)
28	HepG2 cells		Polyphenol enriched cocoa extract		AFB1 (1-30 mM), OTA (50-200 mM)	Not effective against AFB1 but it increased the cell viability and reduced the amounts of ROS in cells treated with OTA or mixtures of AFB1 + OTA	Corcuera et al. (2012)
29	HepG2 cells; Wistar rats male	60	Ginger phenolic rich extract	28 days	AFB1 200 mg/kg b.w	Inhibits the production of ROS, DNA strand break, and cytotoxicity induced by AFB1; reduced AFB1 effects on serum markers of liver damage; reduced the lipid peroxidation and enhance the antioxidant enzymes activities	Vipin et al. (2017)
30	Ross 308 male broilers	160	*Zataria multiflora Boiss. 20 g/kg*	35 days	AFB1 1000 ppb	Improve birds performance and liver architecture	Fani makki et al. (2016)

Table 3 Effect of different vegetables, legumes, and cereals in alleviating the toxic effects induced by mycotoxins

No crt.	Specie/cell line	No. of animals	Type of nutritional approach	Treatment duration	Mycotoxin concentration / way of administration	Effect of extracts	Authors
1	Sprague-Dawley rats, male, 33 month old	120	Garlic, cabbage, and onion oil-soluble extracts	15 days	3 mg/kg diet, p.o	Improve the parameters of the oxidative stress (glutathione, malondialdehyde, and superoxide dismutase) in both kidney and liver	Abdel-Wahhab and Aly (2003)
2	Albino rats (*Rattus norvegicus*) 2-month-old male	35	*Dialium guineense* pulp phenolic extract	6 weeks	20 mg/day	AFB1-mediated elevation of oxidative stress biomarkers; malondialdehyde, conjugated dienes, lipid hydroperoxides, protein carbonyl, and percentage DNA fragmentation were decreased by D. guineense phenolic extract	Adeleye et al. (2014)
3	Rats, male	35	*Parkia biglobosa* phenolic extract	6 weeks	167 μg/kg b.w	Enhance antioxidant enzyme activities, decrease lipid peroxidation, protein oxidation, and DNA fragmentation	Ajiboye et al. (2014)
4	Wistar rats, six weeks old, male and females	120	Carrot, ginger, garlic ethanolic extracts	70 days	200 mg/kg	Garlic extract improved growth rate	Ewuola and Emerue (2021)
5	Chicks, 2 days old	180	Pomegranate peel and clove powders	30 days	2 mgAFB1/kg diet; 2mgOTA/kg diet	Increase of body weight, a decrease of mortality	Albahadly (2015)

6	Wistar rats, six weeks old, male	42	Purple rice (*Oryza sativa*) husk extract	28 days	200 mg/kg b.w	Antimutagenicity against AFB1	Nilnumkhum et al. (2017)
7	Chicken hepatocyte		Palm kernel cake	72 h	5 mM	Reduce the cellular lipid peroxidation and enhance antioxidant enzyme production in AFB1-treated chicken hepatocytes; improve the viability of AFB1-treated hepatocytes, up-regulate oxidative stress tolerance genes and down-regulate pro-inflammatory and apoptosis-associated genes.	Oskoueian et al. (2015)
8	Lung adenocarcinoma cell line (A549), rats, Wistar albino rats, male	48	*Sorghum bicolor* extracts (5-10 mg/mL - in vitro experiments; 2-10 mg/kg b. w - in vivo experiments)	28 days	50mgAF/kg b. w	Cytotoxic in adenocarcinoma cell line (A549) inhibits oxidative and nitrosative stress, inflammation, and apoptosis, maintain the normal histological structure of liver and kidney of rats treated with AFB1.	Owumi et al. (2022)

and activity of CYP1A1, CYP1A2, and CYP3A cytochromes in vitro and in vivo experiments (Pauletto et al. 2021; Yang et al. 2022). Resveratrol has an important antioxidant effect via Nrf2/Keap1 signalling pathway (Yang et al. 2022). Other bioactive compounds such as curcumin, gingerol, or lycopene have also a role in reducing the oxidative stress induced by mycotoxins by increasing the activity of antioxidant enzymes (e.g. GPx) and glutathione levels, suppressing mycotoxin-induced intracellular ROS formation and 8-OHdG level and reducing GSH depletion ((Yang et al. 2011; Poapolathep et al. 2015; Palabiyik et al. 2013). Besides cyanidin-3-o-glucoside, red and blue fruits contained another important anthocyanin cyanidin 3-O-galactoside (Cy3Gal) with the same capacity in counteracting the induced cytotoxicity of AFB1 (Liang et al. 2021). Cy3Gal reduces DNA fragmentation and restores the DNA and protein synthesis decreased by the exposure to mycotoxins (Guerra et al. 2005). In rats exposed to ochratoxin, administration of a diet containing Cy3Gal for 4 weeks had a protective effect on mycotoxin toxicity by decreasing the expression of iNOS, eNOS, and dimethyl-arginine dimethyl-amino-hydrolase (DDAH), the nitrite concentration and DDAH activity (Sorrenti et al. 2012).

Lycopene is a red carotenoid found in tomatoes and other red fruits and vegetables (Imran et al. 2020). Lycopene protects rats against DNA damage induced by Ochratoxin A by decreasing the tail length, tail moment, and tail intensity in the kidney and liver cells as assessed using the alkaline comet assay (Aydin et al. 2013) decreased apoptotic cell death in the kidney and restore the antioxidant response by increasing the GPx activity and GSH levels (Palabiyik et al. 2013).

Berberine derived from *Berberis* sp., *Mahonia aquifolium,* etc., and piperine derived from black pepper (*Piper nigrum Linn.*) are alkaloids with multiple pharmacological functions (Malekinezhad et al. 2021; Umaya et al. 2021). Different concentrations of barberine (200 mg/kg, 400 mg/kg, and 600 mg/kg) were analysed for their capacity to reduce the toxic effects of aflatoxin B1 (AFB) and ochratoxon (OTA) in broilers. The dose of 600 mg berberine/kg proved to be the most effective in counteracting the toxicity of AFB1 and OTA by improving growth performance, liver function, and antioxidant status of broilers fed diets contaminated with AFB and OTA, while piperine in concentration of 60 ppm inhibited subacute toxicity of AFB1 in broiler (da Silva et al. 2016). It was shown that *Zataria multiflora* Boiss. containing a complex of bioactive compounds (alpha-pinene, alpha-thujene, thymol, cis-sabinene hydrate, paracymene, cineole, myrcene, sabinene, and carvacrol) was able to improve performance and liver architecture in birds exposed to aflatoxins due to the antioxidant and free radical scavenging properties (Umaya et al. 2021).

Ethanolic extracts of cabbage, garlic, or onion administered to rats significantly improved growth rate as well as the parameters of the oxidative stress (glutathione, malondialdehyde, and superoxide dismutase) in both kidney and liver of intoxicated animals (Abdel-Wahhab and Aly 2003; Ewuola and Emerue 2021). From cereals, sorghum and rice extracts were tested as counteracting agents for mycotoxin toxicity showing important antioxidant, anti-inflammatory, and anti-apoptotic activities (Nilnumkhum et al. 2017; Owumi et al. 2022), maintaining at the same time the normal histological structure of liver and kidney in rats (Owumi et al. 2022).

Medicinal plants are used in traditional medicine as an alternative to conventional therapy, representing safer choices, and sometimes, the only effective treatment. The medicinal plants, their extracts or chemical compounds (polyphenols, carotenoids, sterols and terpenoids, quinoid compounds, tannins, vitamins, and minerals) have important therapeutic anti-inflammatory, antioxidant, anti-microbial, immunomodulatory effects that recommend their use in different pathologies. Medicinal or aromatic plants were used also as natural alternatives for the mitigation of the mycotoxin toxic effects on animal performances, biochemistry, oxidative stress, and immune or histopathological parameters affected by the exposure to mycotoxins. These bioactive compounds can provide protection against cell destruction due to their ability to bind free radicals by electron donation, metal chelation, and inhibition of lipoxygenases(Łuczaj and Skrzydlewska 2005). An important number of studies have analysed the anti-mycotoxin role of silymarin, a bioactive compound from *Silybum marianum* (milk thistle) known for its important antioxidant properties, which protects the cell membrane from radical-induced damage (Wellington and Jarvis 2001). Studies using milk thistle plant (powder or seeds) showed its ability to counteract the toxic effect of AFB, as a contaminant of broilers diet (Afshin et al. 2017; Amiridumari et al. 2013; Amiri Dumari et al. 2014; Chand et al. 2011; Jahanian et al. 2017; Muhammad et al. 2012; Fani et al. 2014; Fani makki et al. 2013; Tedesco et al. 2004). Milk thistle improved the performances (body weight gain and feed intake), (Fani makki et al. 2013; Muhammad et al. 2012; Tedesco et al. 2004) and the biochemistry parameters, by decreasing the activity of hepatic enzymes alkaline phosphatase, aspartate aminotransferase, and alanine transaminase in aflatoxin-challenged chicks (Jahanian et al. 2017; Amiri Dumari et al. 2014). Administration of milk thistle was able to counteract the effects of the aflatoxin on the immune response by decreasing inflammation and increasing the antibody titre against different diseases (e.g. Newcastle disease, infectious bronchitis, infectious bursal diseases) (Chand et al. 2011; Afshin et al. 2017; Amiri Dumari et al. 2014).

A very recent study by Masouri and col. demonstrated that *Mentha piperita* and its derivatives can be used to suppress aflatoxin effects on the liver, bone, and meat quality and to improve the performance of Japanese quails. Similar results were obtained from in vitro tests. AFB1 incubated for 72 h with medicinal plants was eliminated in the highest percentage by peppermint (81%) (Masouri et al. 2022).

Powder leaves of *Moringa oleifera* and *Eclipta prostata* are used traditionaly as hepatoprotective agents due to their high contents in bioactive compounds as wedelolactone, dimethyl wedelolactone, triterpenes and flovonoids. Suplimentation of poultry feed with these powders resulted in liver protection against AFB1 through the stimulation of hepatocytes regeneration, of the regulation of the hepatic microsomal drug-metabolizing enzyme levels, and of the antioxidant capacity (Umaya et al. 2021).

As shown in Table 1, a large variety of extracts, essential oils, or bioactive compounds from medicinal and aromatic plants (*Calendula officinalis*, *Chamomile*, *Berberine*, *Aloe vera,* Licorice, Ginseng, Boswellia, Salvia, Thyme, etc.), spices (Cinnamomum, cacao, etc.) were used in order to assess their anti-mycotoxin effects in vitro on bacteria (*Salmonella*), eukaryote cell cultures (hepatic cells-HepG2, lung adenocarcinoma cell line (A549), or in vivo on laboratory animals (rats) or farm animals (broilers, hens, and piglets). The main effect of medicinal plant extracts was to protect cells from the consequences of oxidative and nitrosative stress by inhibiting the formation of reactive intracellular oxygen species, decreasing lipid peroxidation, increased glutathione concentration, increased activity of antioxidant enzymes catalase, and superoxide dismutase (Soni et al. 1993; Atef et al. 2012; Mohamed and Metwally 2009). Some extracts were responsible for the significant improvement of histological lesions of the liver and kidney (Baptiste et al. 2017; Hamzawy et al. 2013; Makhuvele et al. 2022; Zidan et al. 2015). Furthermore, some essential oils from medicinal plants (fennel, cardamom, anise, chamomile, celery, cinnamon, thyme, taramira, oregano, and rosemary) were investigated in the attempt to reduce the occurrence of mycotoxins (e.g. OTA). The results showed their efficacy in reducing the fungus and OTA production as well as the expression levels of the genes responsible for the OTA biosynthesis (El Khoury et al. 2016). Astragalus polysaccharide, the major bioactive compound from Astragalus, a medicinal plant, used for over 2000 years in traditional Chinese medicine, reduced significantly cytotoxicity, apoptosis, and pro-inflammatory cytokine expressions induced by OTA (1.5 μg/ml) in porcine alveolar macrophages at a concentration of 20 μg/ml (Li et al. 2021).

3 Use of Bioactive Compounds from Agro-Industrial Wastes (by-Products) for Reduction of the Mycotoxins Effect in Animals

Agro-industry generates every year through plant processing a wide range of waste/by-products rich in biologically active compounds (antioxidants, polyunsaturated fatty acids, vitamins, minerals, polyphenols, etc.), whose concentration is sometimes higher than in the original raw material. These residues are worth to be used successfully in animal nutrition rather than being discarded on land and represent a harmful issue for the environment.

An interesting study carried out in the Laboratory of Animal Biology from INCDBNA-IBNA Balotesti scanned in vitro a series of agro-industrial waste/by-products in order to evaluate their potential to adsorb mycotoxins (Palade et al. 2020). Eight food by-products (grape seed meal, seabuckthorn meal, white and red skin potatoes, apple, carrot, beetroot, and celery) were assessed for their capacity to bind mycotoxins aflatoxin B1 and zearalenone (ZEA). Grape seed (GSM) and seabuckthorn (SBM) meals proved the highest mycotoxin binding capacity. The affinity of GSM and SBM for the two by-products was dependent on different factors such as time, concentration, temperature, or pH as well as their content in bioactive compounds. Both are rich in polyphenols, vitamins, minerals, and fatty acids with important anti-inflammatory and antioxidant properties at different levels (Shah et al. 2021). GSM had a higher affinity towards AFB1, while SBM for ZEA.

Overall, our study showed that these two by-products represent a natural, organic alternative to commercial mycotoxin adsorbents.

Winery industry produces an enormous amount of waste by-products that are currently under-used (Fabbri et al. 2015); grape seed meal (GSM), the residue left after oil extraction, is the second waste derived from grape after grape pomace (Dwyer et al. 2014). Due to their high content of bioactive compounds, grape pomace, grape seed meal, or other wine by-products have been used lately as food antioxidants (Ozvural and Vural 2014) or dietary supplements for cancer and cardiovascular disease prevention (Lamuela-Raventos and de la Torre-Boronat 1999; Veskoukis et al. 2012). Many studies have shown recently that grape waste can be used successfully in animal feeding. Besides its potential, as an alternative in case of feed shortages, the rich content in bioactive compounds can bring added value.

Recent studies performed in the Laboratory of Animal Biology from INCDBA-IBNA Baloteşti reported that grape waste (GSM) has a high potential to reduce the toxic effects of mycotoxins, acting both as a mycotoxin binder as well as a mitigating agent in AFB1 toxicity. Inclusion of 8% GSM in the diet of weaned piglets exposed to 320 ppm AFB1 increased the total antioxidant status as well as the activity of the enzymes involved in the antioxidant defence: catalase, superoxide dismutase, and glutathione peroxidase in mesenteric lymph nodes, responsible for gut immunity of intoxicated animals (Marin et al. 2020). Also, GSM was able to improve the inflammatory response, compromised by the exposure to the toxin, by decreasing the inflammatory markers increased by AFB1 (MAP kinases, metalloproteinases, and inflammatory cytokines). Furthermore, GSM has shown an important potential to modulate the piglet's microbiome, through a shift of colonic bacterial populations affected by the exposure to AFB1 (Grosu et al. 2019) as compared with the individual AFB1 or GSM treatments. The sequencing of colon-content microbiota using the Illumina MiSeq platform has shown that the relative abundance of Bacteroidetes and Proteobacteria was observed in animals receiving a diet containing GSM and contaminated with AFB1, accompanied by a decrease of abundance *Firmicutes phylum* in a synergic manner. AFB1 and GSM have a synergistic action on *Prevotella*, *Campylobacter,* and *Lactobacillus* bacteria, while an antagonistic effect was observed on *Lachnospira* (Grosu et al. 2019).

Liver is the main target for AFB1 in humans and animals. Feeding piglets with a diet contaminated with 320 ppm AFB1 led to a decrease of animal performance and liver histological damage. The addition of 8% GSM by-products into the AFB1 contaminated diet ameliorates histological liver injury and oxidative stress by decreasing MAPK (mitogen-activated protein kinase) and NF-kB (nuclear factor kB) signalling pathway and inhibiting of NF-kB signalling pathway overexpressed by AFB1 contaminated diet (Taranu et al. 2020). The capacity of GSM to reduce the excessive ROS production and oxidative stress comes from its high content in bioactive compounds, especially the polyphenols, rich in hydroxyl groups which together with the aromatic rings to which they are linked, confer their antioxidant potential (Chuang and McIntosh 2011).

Based on the in vitro results showing different affinity of GSM and SB for mycotoxins binding (Palade et al. 2020), a nutritional trial was carried out on piglets in order to investigate if a mix between the two by-products (GSM + SB) get supplementary efficacy in counteracting the toxic effect induced by dietary ochratoxin A (479 ppb) and aflatoxin B1 (62 ppb) than each individually both at the kidney and liver level (Popescu et al. 2021). The results showed that the mixture of mycotoxins induced an upregulation of cytochromes concentrations CYP2E1 and CYP3A29 in the liver, while a downregulation of cytochromes CYP1A2, CYP2E1, and CYP3A29 was observed in kidney. When the enzymatic activities of different cytochromes were analysed, it was observed that the exposure to mycotoxins induced a significant increase in the activity of CYP1A2, CYP2E1, and CYP3A29 in the kidney, while their activity was decreased in the liver. The diet containing the mix of waste (GSM + SB) tended to normalize these parameters to the control values, which suggests that this waste could represent valuable alternatives to inorganic or synthetic additives for counteracting mycotoxins effects (Popescu et al. 2021). In the same context, it is worth mentioning the study of Gutzwiller and collaborators who proved that dried apple pomace, a waste from apple juice production rich in pectin and other bioactive compounds added in a diet of pig (8%) contaminated with 3100 ppb DON and 65 ppb ZEA had the potential to restore the pig performance (Gutzwiller et al. 2007). Also, the dietary supplementation with glucomannan polymer (0.1, 0.2%) was effective in reversing the changes in serum biochemistry parameters, haematology and biliary IgA concentrations in broiler chicken (Swamy et al. 2002) fed with *Fusarium* contaminated diet.

Another important issue related to the contamination of food or feed with mycotoxins is represented by the possible transfer of mycotoxins in organs and meat. Indeed, many studies have shown that the chronic consumption of feed contaminated with mycotoxin can result into the accumulation of mycotoxins in organs, especially in organs involved in toxins metabolization and excretion (Pleadin et al. 2021). In order to analyse if a combination of agro-industrial wastes can decrease the concentration of mycotoxins in the liver of piglets exposed to zearalenone, 24 weanling piglets were randomly distributed into four groups (6 piglets/group): 1) control group, 2) ZEA group (piglets fed a diet contaminated with 290 μg/kg ZEA), 3) by-product group (piglets fed a diet supplemented with 5% flaxseed meal and 5% mixture of grapeseed meal and sea buckthorn meal in a ratio of 3:1), 4) ZEA + by-product group (piglets fed a diet supplemented with 5% flaxseed meal and 5% mixture of grapeseed meal and sea buckthorn meal in a ratio of 3:1 and contaminated with 290 μg/kg ZEA). Diets were administered ad libitum for 1 months. After the experimental period, piglets were euthanized, and samples of liver and intestinal content were collected for mycotoxin analysis. The results showed that exposure to a diet contaminated with zearalenone of piglets for 11 months resulted into an increase of concentration of ZEA in the liver (140.47 ± 15.4 ppb) vs control (0.92 ± 0.2 ppb) and in intestinal content (111.3 ± 20 ppb) vs control (0.91 ± 0.4 ppb). Feeding piglets with a diet that included the mix of waste resulted in a decrease of ZEA concentration in the liver of intoxicated piglets from 140.47 ± 15.4 ppb in the ZEA group to 70.7 ± 20 ppb in the

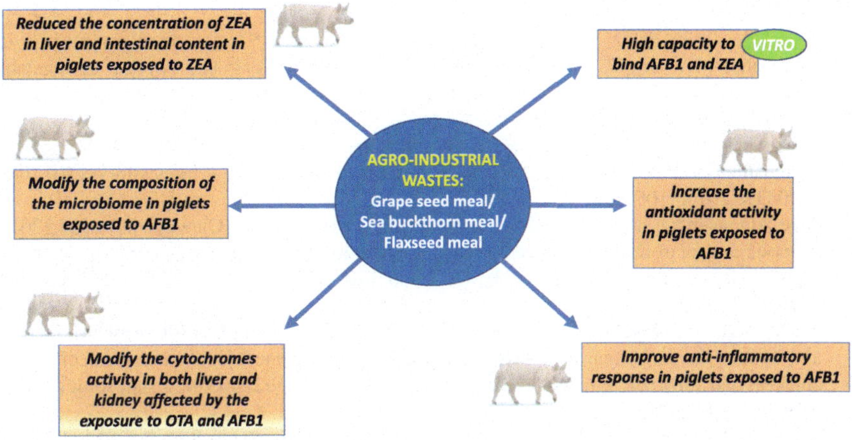

Fig. 1 Schematic presentation of some effects of agro-industrial wastes in alleviating the toxic effect of different mycotoxins

liver of piglets from ZEA+ mix of the waste group. Also, ZEA concentration in intestinal content decreased from 111.3 ± 20 ppb in the ZEA group to 44.4 ± 14 ppb in piglets from ZEA+ mix of waste group. These results represent additional evidence that these by-products can represent a valuable alternative to adsorbents used for counteracting the mycotoxin effects. A schematic presentation of some effects of agro-industrial wastes in alleviating the toxic effect of different mycotoxins is presented in Fig. 1.

4 Conclusion

Many literature studies investigated the capacity of different plants (medicinal plants, spices, fruits, legumes, cereals, etc.) to reduce the toxicity induced by mycotoxins. The results have shown that they can improve animal performance and health due to their rich content in bioactive compounds such as polyphenols, triterpene, vitamins, minerals, and unsaturated fatty acids with important antioxidant and anti-inflammatory properties.

However, other studies concerning their anti-mycotoxic properties are needed as not all plant bioactive compounds are suitable as food/feed additives due to concerns related to their flavours or colour (grapes, pepper, coffee, tea, or herbs) anti-nutritional factors. Another important aspect is related to the possible contamination of plants themselves with mycotoxins, as indicated by various studies. Also, due to the toxic properties of some bioactive compounds as coumarin, precautions should be taken before their use as an anti-mycotoxic agent.

References

Abdel-Wahhab MA, Aly SE (2003) Antioxidants and radical scavenging properties of vegetable extracts in rats fed aflatoxin-contaminated diet. J Agric Food Chem 51(8):2409–2414. https://doi.org/10.1021/jf0209185

Adebo OA, Njobeh PB, Gbashi S, Nwinyi OC, Mavumengwana V (2017) Review on microbial degradation of aflatoxins. Crit Rev Food Sci Nutr 57(15):3208–3217. https://doi.org/10.1080/10408398.2015.1106440

Adeleye AO, Ajiboye TO, Iliasu GA, Abdussalam FA, Balogun A, Ojewuyi OB et al (2014) Phenolic extract of Dialium guineense pulp enhances reactive oxygen species detoxification in aflatoxin B_1 hepatocarcinogenesis. J Med Food 17(8):875–885. https://doi.org/10.1089/jmf.2013.0157

Afshin M, Afzali N, Mojtahedi M, Mohammadi A (2017) Effects of milk thistle seeds, whole plant and extract on performance, carcass characteristics and some blood parameters of broiler chickens fed aflatoxin B1 contaminated diet. Anim Prod 19(2):403–414. https://doi.org/10.22059/jap.2017.62269

Ajiboye TO, Adeleye AO, Salau AK, Ojewuyi OB, Adigun NS, Sabiu S et al (2014) Phenolic extract of Parkia biglobosa fruit pulp stalls aflatoxin B1 – mediated oxidative rout in the liver of male rats. Rev Bras 24(6):668–676. https://doi.org/10.1016/j.bjp.2014.10.010

Albahadly H (2015) Activity of pomegranate peels and clove powders in detoxification of aflatoxin B1 and ochratoxin A from contaminated poultry diet. J Plant Pathol Microb 6:10

Amiri Dumari M, Sarir H, Fani Makki O, Afzali N (2014) Effect of Milk thistle (Silybum marianum L.) on biochemical parameters and immunity of broiler chicks fed aflatoxin B1 after three weeks. IJT 8(26):1098–1103

Amiridumari H, Sarir H, Afzali N, Fanimakki O (2013) Effects of milk thistle seed against aflatoxin B1 in broiler model. J Res Med Sci 18(9):786–790

Atef A, Hassan N, El-Mokhttar M, Mogda MM et al (2012) Efficacy of use of forskolin plant extract in control of toxic effects of aflatoxicosis in food. Am Sci 8(11):225–236

Aydin S, Palabiyik SS, Erkekoglu P, Sahin G, Başaran N, Giray BK (2013) The carotenoid lycopene protects rats against DNA damage induced by Ochratoxin A. Toxicon: Official J Int Soc Toxinol 73:96–103. https://doi.org/10.1016/j.toxicon.2013.07.004

Baptiste A, Manda P, Jean N, Aïssata D, Béatrice S, Dano S (2017) The assessment of Alchornea cordifolia activity in aflatoxicosis reduction in rat. J Antioxid Act 1:23–33. https://doi.org/10.14302/issn.2471-2140.jaa-17-1630

Bortoluzzi C, Schmidt JM, Bordignon HLF, Fülber LM, Layter JR, Fernandes J (2016) Efficacy of yeast derived glucomannan or algae-based antioxidant or both as feed additives to ameliorate mycotoxicosis in heat stressed and unstressed broiler chickens. Livest Sci 193:193. https://doi.org/10.1016/j.livsci.2016.09.005

Çelik K (2020) The efficacy of mycotoxin-detoxifying and biotransforming agents in animal nutrition, p 271–84

Chand N, Muhammad D, Durrani F, Qureshi M, Sahibzada S (2011) Protective effects of Milk thistle (Silybum marianum) against aflatoxin B1 in broiler chicks. Asian Australasian J Anim Sci 24:1011–1018. https://doi.org/10.5713/ajas.2011.10418

Chuang CC, McIntosh MK (2011) Potential mechanisms by which polyphenol-rich grapes prevent obesity-mediated inflammation and metabolic diseases. Annu Rev Nutr 31:155–176. https://doi.org/10.1146/annurev-nutr-072610-145149

Corcuera LA, Amézqueta S, Arbillaga L, Vettorazzi A, Touriño S, Torres JL et al (2012) A polyphenol-enriched cocoa extract reduces free radicals produced by mycotoxins. Food Chem Toxicol: An International Journal Published for the British Industrial Biological Research Association 50(3–4):989–995. https://doi.org/10.1016/j.fct.2011.11.052

da Silva Cardoso V, Vermelho AB, Ribeiro de Lima CA, Mendes de Oliveira J, Freire de Lima ME, Pinto da Silva LH et al (2016) Antigenotoxic effect of Piperine in broiler chickens intoxicated with aflatoxin B1. Toxins (Basel). 8(11):316. https://doi.org/10.3390/toxins8110316

Divakaran S, Tacon A (2000) Studies on the toxicity of shrimp (Penaeus vannamei) fed diets dosed with aflatoxin B1 to humans. J Aquat Food Prod Technol 9:115–120. https://doi.org/10.1300/J030v09n03_09

Dos Anjos FR, Ledoux DR, Rottinghaus GE, Chimonyo M (2015) Efficacy of adsorbents (bentonite and diatomaceous earth) and turmeric (Curcuma longa) in alleviating the toxic effects of aflatoxin in chicks. Br Poult Sci 56(4):459–469. https://doi.org/10.1080/00071668.2015.1053431

Dwyer K, Hosseinian F, Rod M (2014) The market potential of grape waste alternatives. Journal of. Food Res 3(2):91. https://doi.org/10.5539/jfr.v3n2p91

El Khoury R, Atoui A, Verheecke C, Maroun R, El Khoury A, Mathieu F (2016) Essential oils modulate gene expression and Ochratoxin A production in Aspergillus carbonarius. Toxins (Basel). 8(8):242. https://doi.org/10.3390/toxins8080242

El-Sawi NM, Al-Seeni MN (2009) Assessment of flavonoids as Rutin for detoxification of T-2 toxin. J Appl Anim Res 35(1):57–60. https://doi.org/10.1080/09712119.2009.9706985

Ewuola E, Emerue P (2021) Mitigating dietary aflatoxins in Wistar rat using selected phyto-antioxidant sources. Niger J Anim Prod 48:38–52. https://doi.org/10.51791/njap.v48i2.2931

Fabbri A, Bonifazi G, Serranti S (2015) Micro-scale energy valorization of grape marcs in winery production plants. Waste Manag (New York, NY) 36:156–165. https://doi.org/10.1016/j.wasman.2014.11.022

Fani Makki O, Afzali N, Omidi A (2013) Effect of different levels of Silybum marianum seeds on growth rate, carcass variables and liver morphology of broilers contaminated with aflatoxin B1. Poult Sci J 1(2):105–116. Gorgan University, GUASNR

Fani Makki O, Omidi A, Nik H, Hasheminejad S, Hosseini SM (2016) Anti-aflatoxin B1 effects of Shirazi thyme (Zataria multiflora) in broilers: evaluation of performance and liver histopathology. Vet Sci Dev 6:36–40. https://doi.org/10.4081/vsd.2016.6090

Fani O, Omidi A, Afzali N, Sarir H, M F, A S. (2014) Efficacy of Silybum Marianum seeds in ameliorating the toxic effects of aflatoxin B1 in broilers. Iranian J Toxicol 8:977–982

Farag I, Roshdy H, Radwan H, Ghaly I, Salah S, Abdel-Rahim E et al (2018) Ameliorative role of ethanolic extract of Moringa oleifera leaf on aflatoxin B 1 -induced genotoxicity and biochemical alterations in rats. J Arab Soc Med Res 13:60. https://doi.org/10.4103/jasmr.jasmr_33_17

Grosu IA, Pistol GC, Taranu I, Marin DE (2019) The impact of dietary grape seed meal on healthy and aflatoxin B1 afflicted microbiota of pigs after weaning. Toxins (Basel) 11(1):25. https://doi.org/10.3390/toxins11010025

Guerra MC, Galvano F, Bonsi L, Speroni E, Costa S, Renzulli C et al (2005) Cyanidin-3-O-beta-glucopyranoside, a natural free-radical scavenger against aflatoxin B1- and ochratoxin A-induced cell damage in a human hepatoma cell line (Hep G2) and a human colonic adenocarcinoma cell line (CaCo-2). Br J Nutr 94(2):211–220. https://doi.org/10.1079/bjn20051425

Guo J, Yan WR, Tang JK, Jin X, Xue HH, Wang T et al (2022) Dietary phillygenin supplementation ameliorates aflatoxin B(1)-induced oxidative stress, inflammation, and apoptosis in chicken liver. Ecotoxicol Environ Saf 236:113481. https://doi.org/10.1016/j.ecoenv.2022.113481

Gutzwiller A, Czeglédi L, Stoll P, Bruckner L (2007) Effects of fusarium toxins on growth, humoral immune response and internal organs in weaner pigs, and the efficacy of apple pomace as an antidote. J Anim Physiol Anim Nutr 91(9–10):432–438. https://doi.org/10.1111/j.1439-0396.2006.00672.x

Hamzawy MA, El-Denshary ESM, Hassan NS, Mannaa FA, Abdel-Wahhab MA (2013) Dietary supplementation of Calendula officinalis counteracts the oxidative stress and liver damage resulted from aflatoxin. ISRN Nutr 2013:538427. https://doi.org/10.5402/2013/538427

Holanda DM, Kim SW (2021) Mycotoxin occurrence, toxicity, and detoxifying agents in pig production with an emphasis on Deoxynivalenol. Toxins (Basel). 13(2):171. https://doi.org/10.3390/toxins13020171

Ibrahim M, Pieters R, Abdel-Aziem S, Walt A, Bezuidenhout C, Abdel-Wahhab PM (2015) Protective effects of Amaranthus hybridus against aflatoxin B1 and fumonisin B1-induced genotoxicity in H4IIE-luc cells. Hepatoma Res 1:136–146. https://doi.org/10.4103/2394-5079.167377

Imran M, Ghorat F, Habib Ur R, Aslam F, Heydari M, Shariati MA et al (2020) Lycopene as a natural antioxidant used to prevent human health disorders. Antioxidants 9. https://doi.org/10.3390/antiox9080706

Jahanian E, Mahdavi AH, Asgary S, Jahanian R (2017) Effects of dietary inclusion of silymarin on performance, intestinal morphology and ileal bacterial count in aflatoxin-challenged broiler chicks. J Anim Physiol Anim Nutr 101(5):e43–e54. https://doi.org/10.1111/jpn.12556

Jideani AIO, Silungwe H, Takalani T, Omolola AO, Udeh HO, Anyasi TA (2021) Antioxidant-rich natural fruit and vegetable products and human health. Int J Food Prop 24(1):41–67. https://doi.org/10.1080/10942912.2020.1866597

Kalorey DR, Kurkure NV, Ramgaonkar JS, Sakhare PS, Warke S, Nigot NK (2005) Effect of Polyherbal feed supplement "Growell" during induced Aflatoxicosis, Ochratoxicosis and combined Mycotoxicoses in broiler. Asian-Australas J Anim Sci 18(3):375–383. https://doi.org/10.5713/ajas.2005.375

Lamuela-Raventos RM, de la Torre-Boronat MC (1999) Beneficial effects of white wines. Drugs Exp Clin Res 25(2–3):121–124

Li L, Sun X, Zhao D, Dai H (2021) Pharmacological applications and action mechanisms of phytochemicals as alternatives to antibiotics in pig production. Front Immunol 12:798553. https://doi.org/10.3389/fimmu.2021.798553

Liang Z, Liang H, Guo Y, Yang D (2021) Cyanidin 3-O-galactoside: A natural compound with multiple health benefits. Int J Mol Sci 22(5):2261. https://doi.org/10.3390/ijms22052261

Lu H, Liu F, Zhu Q, Zhang M, Li T, Chen J et al (2017) Aflatoxin B1 can be complexed with oxidised tea polyphenols and the absorption of the complexed aflatoxin B1 is inhibited in rats. J Sci Food Agric 97(6):1910–1915. https://doi.org/10.1002/jsfa.7994

Łuczaj W, Skrzydlewska E (2005) Antioxidative properties of black tea. Prev Med 40(6):910–918. https://doi.org/10.1016/j.ypmed.2004.10.014

Makhuvele R, Naidu K, Gbashi S, Thipe VC, Adebo OA, Njobeh PB (2020) The use of plant extracts and their phytochemicals for control of toxigenic fungi and mycotoxins. Heliyon 6(10): e05291-e:e05291. https://doi.org/10.1016/j.heliyon.2020.e05291

Makhuvele R, Foubert K, Hermans N, Pieters L, Verschaeve L, Elgorashi E (2022) Protective effects of methanolic leaf extracts of Monanthotaxis caffra against aflatoxin B1-induced hepatotoxicity in rats. Onderstepoort J Vet Res 89(1):e1–e6. https://doi.org/10.4102/ojvr.v89i1.1968

Malekinezhad P, Ellestad LE, Afzali N, Farhangfar SH, Omidi A, Mohammadi A (2021) Evaluation of berberine efficacy in reducing the effects of aflatoxin B1 and ochratoxin A added to male broiler rations. Poult Sci 100(2):797–809. https://doi.org/10.1016/j.psj.2020.10.040

Marin D, Motiu M, Pistol G, Gras M, Israel-Roming F, Calin L et al (2016) Diet contaminated with ochratoxin A at the highest level allowed by EU recommendation disturbs liver metabolism in weaned piglets. World Mycotoxin J 9:1–10. https://doi.org/10.3920/WMJ2015.1993

Marin DE, Bulgaru CV, Anghel CA, Pistol GC, Dore MI, Palade ML et al (2020) Grape seed waste counteracts aflatoxin B1 toxicity in piglet mesenteric lymph nodes. Toxins. 12(12):800

Masouri L, Bagherzadeh-Kasmani F, Mehri M, Rokouei M, Masouri B (2022) Mentha piperita as a promising feed additive used to protect liver, bone, and meat of Japanese quail against aflatoxin B(1). Trop Anim Health Prod 54(5):254. https://doi.org/10.1007/s11250-022-03257-w

Meng D, Garba B, Ren Y, Yao M, Xia X, Li M et al (2020) Antifungal activity of chitosan against Aspergillus ochraceus and its possible mechanisms of action. Int J Biol Macromol 158:1063–1070. https://doi.org/10.1016/j.ijbiomac.2020.04.213

Mohajer A, Seifi S, Rezaie S, Khaniki G, Samarghandian S, Farkhondeh T et al (2021) Effect of aloe Vera extract on reducing aflatoxin B1 in eggs of laying hen and egg yolk oxidative stability. Biointerface Res Appl Chem 11:12680–12688. https://doi.org/10.33263/BRIAC115.1268012688

Mohamed A, Metwally N (2009) Antiaflatoxigenic activities of some plant aqueous extracts against aflatoxin-B1 induced renal and cardiac damage. J Pharmacol Toxicol 4:1–16. https://doi.org/10.3923/jpt.2009.1.16

Muhammad D, Chand N, Khan S, Sultan A, Mushtaq M, Rafiullah. (2012) Hepatoprotective role of milk thistle (Silybum marianum) in meat type chicken fed aflatoxin B1 contaminated feed. Pak Vet J 32:443–446

Muhammad I, Wang H, Sun X, Wang X, Han M, Lu Z et al (2018) Dual role of dietary curcumin through attenuating AFB(1)-induced oxidative stress and liver injury via modulating liver phase-I and phase-II enzymes involved in AFB(1) bioactivation and detoxification. Front Pharmacol 9:554. https://doi.org/10.3389/fphar.2018.00554

Nazarizadeh H, Mohammad Hosseini S, Pourreza J (2019) Effect of plant extracts derived from thyme and chamomile on the growth performance, gut morphology and immune system of broilers fed aflatoxin B1 and ochratoxin A contaminated diets. Ital J Anim Sci 18(1):1073–1081. https://doi.org/10.1080/1828051X.2019.1615851

Nilnumkhum A, Punvittayagul C, Chariyakornkul A, Wongpoomchai R (2017) Effects of hydrophilic compounds in purple rice husk on AFB1-induced mutagenesis. Mol Cell Toxicol 13(2): 171–178. https://doi.org/10.1007/s13273-017-0018-5

Olivas-Aguirre FJ, Rodrigo-García J, Martínez-Ruiz NDR, Cárdenas-Robles AI, Mendoza-Díaz SO, Álvarez-Parrilla E et al (2016) Cyanidin-3-O-glucoside: physical-chemistry, Foodomics and health effects. Molecules 21(9):1264. https://doi.org/10.3390/molecules21091264

Oskoueian E, Abdullah N, Zulkifli I, Ebrahimi M, Karimi E, Goh YM et al (2015) Cytoprotective effect of palm kernel cake phenolics against aflatoxin B1-induced cell damage and its underlying mechanism of action. BMC Complement Altern Med 15(1):392. https://doi.org/10.1186/s12906-015-0921-z

Owumi SE, Kazeem AI, Wu B, Ishokare LO, Arunsi UO, Oyelere AK (2022) Apigeninidin-rich Sorghum bicolor (L. Moench) extracts suppress A549 cells proliferation and ameliorate toxicity of aflatoxin B1-mediated liver and kidney derangement in rats. Sci Rep 12(1):7438. https://doi.org/10.1038/s41598-022-10926-1

Ozvural EB, Vural H (2014) Which is the best grape seed additive for frankfurters: extract, oil or flour? J Sci Food Agric 94(4):792–797. https://doi.org/10.1002/jsfa.6442

Palabiyik SS, Erkekoglu P, Zeybek ND, Kizilgun M, Baydar DE, Sahin G et al (2013) Protective effect of lycopene against ochratoxin A induced renal oxidative stress and apoptosis in rats. Exp Toxicol Pathol: Official Journal of the Gesellschaft fur Toxikologische Pathologie 65(6): 853–861. https://doi.org/10.1016/j.etp.2012.12.004

Palade LM, Dore MI, Marin DE, Rotar MC, Taranu I (2020) Assessment of food by-Products' potential for simultaneous binding of aflatoxin B1 and Zearalenone. Toxins (Basel). 13(1):2. https://doi.org/10.3390/toxins13010002

Pauletto M, Giantin M, Tolosi R, Bassan I, Barbarossa A, Zaghini A et al (2021) Discovering the protective effects of resveratrol on aflatoxin B1-induced toxicity: A whole transcriptomic study in a bovine hepatocyte cell line. Antioxidants (Basel) 10(8):1225. https://doi.org/10.3390/antiox10081225

Pleadin J, Lešić T, Milićević D, Markov K, Šarkanj B, Vahčić N et al (2021) Pathways of mycotoxin occurrence in meat products: A review. PRO 9(12):2122

Poapolathep S, Imsilp K, Machii K, Kumagai S, Poapolathep A (2015) The effects of curcumin on aflatoxin B1- induced toxicity in rats. Biocontrol Sci 20(3):171–177. https://doi.org/10.4265/bio.20.171

Popescu RG, Avramescu S, Marin DE, Ţăranu I, Georgescu SE, Dinischiotu A (2021) The reduction of the combined effects of aflatoxin and Ochratoxin A in piglet livers and kidneys by dietary antioxidants. Toxins. 13(9):648

Rashidi N, Khatibjoo A, Taherpour K, Akbari-Gharaei M, Shirzadi H (2020) Effects of licorice extract, probiotic, toxin binder and poultry litter biochar on performance, immune function, blood indices and liver histopathology of broilers exposed to aflatoxin-B1. Poult Sci 99(11): 5896–5906. https://doi.org/10.1016/j.psj.2020.08.034

Shah R, Idate A, Ugale V, Sharma P (2021) Comprehensive review on sea buckthorn: Biological activity and its potential uses. Pharma Innov 10:942–953. https://doi.org/10.22271/tpi.2021.v10.i5l.6325

Shyamal S, Latha P, Suja S, Shine V, Anuja G, Sini S et al (2010) Hepatoprotective effect of three herbal extracts on aflatoxin B1-intoxicated rat liver. Singap Med J 51(4):326

Soni KB, Rajan A, Kuttan R (1993) Inhibition of aflatoxin-induced liver damage in ducklings by food additives. Mycotoxin research 9(1):22–26. https://doi.org/10.1007/bf03192228

Sorrenti V, Di Giacomo C, Acquaviva R, Bognanno M, Grilli E, D'Orazio N et al (2012) Dimethylarginine Dimethylaminohydrolase/nitric oxide synthase pathway in liver and kidney: protective effect of Cyanidin 3-O-β-D-glucoside on Ochratoxin-A toxicity. Toxins 4(5): 353–363

Swamy HV, Smith TK, MacDonald EJ, Boermans HJ, Squires EJ (2002) Effects of feeding a blend of grains naturally contaminated with fusarium mycotoxins on swine performance, brain regional neurochemistry, and serum chemistry and the efficacy of a polymeric glucomannan mycotoxin adsorbent. J Anim Sci 80(12):3257–3267. https://doi.org/10.2527/2002.80123257x

Tang L, Tang M, Xu L, Luo H, Huang T, Yu J et al (2008) Modulation of aflatoxin biomarkers in human blood and urine by green tea polyphenols intervention. Carcinogenesis 29(2):411–417. https://doi.org/10.1093/carcin/bgn008

Taranu I, Marin DE, Bouhet S, Pascale F, Bailly J-D, Miller JD et al (2005) Mycotoxin Fumonisin B1 alters the cytokine profile and decreases the Vaccinal antibody titer in pigs. Toxicol Sci 84(2):301–307. https://doi.org/10.1093/toxsci/kfi086

Taranu I, Hermenean A, Bulgaru C, Pistol GC, Ciceu A, Grosu IA et al (2020) Diet containing grape seed meal by-product counteracts AFB1 toxicity in liver of pig after weaning. Ecotoxicol Environ Saf 203:110899. https://doi.org/10.1016/j.ecoenv.2020.110899

Tedesco D, Steidler S, Galletti S, Tameni M, Sonzogni O, Ravarotto L (2004) Efficacy of silymarin-phospholipid complex in reducing the toxicity of aflatoxin B1 in broiler chicks. Poult Sci 83(11):1839–1843. https://doi.org/10.1093/ps/83.11.1839

Tulayakul P, Dong KS, Li JY, Manabe N, Kumagai S (2007) The effect of feeding piglets with the diet containing green tea extracts or coumarin on in vitro metabolism of aflatoxin B1 by their tissues. Toxicon: Official Journal of the International Society on Toxinology 50(3):339–348. https://doi.org/10.1016/j.toxicon.2007.04.005

Umaya SR, Vijayalakshmi YC, Sejian V (2021) Exploration of plant products and phytochemicals against aflatoxin toxicity in broiler chicken production: present status. Toxicon: Official Journal of the International Society on Toxinology. 200:55–68. https://doi.org/10.1016/j.toxicon.2021.06.017

Veskoukis ASKA, Nikolaidis MG, Stagos D, Aligiannis N, Halabalaki M, Chronis K, Goutzourelas N, Skaltsounis L, Kouretas D (2012) The antioxidant effects of a polyphenol-rich grape pomace extract in vitro do not correspond in vivo using exercise as an oxidant stimulus. Oxidative Med Cell Longev 2012:14. https://doi.org/10.1155/2012/185867

Vipin KRR, Kurrey NK, Appaiah KA, Venkateswaran G (2017) Protective effects of phenolics rich extract of ginger against aflatoxin B(1)-induced oxidative stress and hepatotoxicity. Biomed Pharmacother 91:415–424. https://doi.org/10.1016/j.biopha.2017.04.107

Wellington K, Jarvis B (2001) Silymarin: a review of its clinical properties in the management of hepatic disorders. BioDrugs: Clinical Immunotherapeutics, Biopharmaceuticals and Gene Therapy 15(7):465–489. https://doi.org/10.2165/00063030-200115070-00005

Yang G, Zhong L, Jiang L, Geng C, Cao J, Sun X et al (2011) 6-Gingerol prevents Patulin-induced genotoxicity in HepG2 cells. Phytother Res 25(10):1480–1485. https://doi.org/10.1002/ptr.3446

Yang H, Wang Y, Yu C, Jiao Y, Zhang R, Jin S et al (2022) Dietary resveratrol alleviates AFB1-induced ileum damage in ducks via the Nrf2 and NF-κB/NLRP3 signaling pathways and CYP1A1/2 expressions. Agriculture 12(1):54

Zhang ZF, Xi Y, Wang ST, Zheng LY, Qi Y, Guo SS et al (2022) Effects of Chinese gallnut tannic acid on growth performance, blood parameters, antioxidative status, intestinal histomorphology, and cecal microbial shedding in broilers challenged with aflatoxin B1. J Anim Sci 100(4): skac099. https://doi.org/10.1093/jas/skac099

Zidan R, Elnegris H, Wahdan R (2015) Evaluating the protective role of ginseng against histological and biochemical changes induced by aflatoxin B1 in the renal convoluted tubules of adult male albino rats. J Clin Exp Pathol 5(6):253. https://doi.org/10.4172/2161-0681.1000253

Part IV

Specific Applications in Livestock by Animal Groups

Plant-Based Feed Additives for Livestock and Poultry in Southern Africa

Lyndy J. McGaw

Abstract

Southern Africa has a wealth of plant species, many with beneficial properties. In rural areas in particular, smallholder farmers have a strong tradition of using plants to support animal production as well as health. Such plants, as well as others still to be investigated, need to be evaluated for their ability to be used as feed additives in livestock and poultry production. Feed additives for production animals and poultry may be defined as compounds incorporated into the diet for specified purposes, including improvement of animal wellbeing, enhancement of the culture environment, improvement of quality of the animal as a final product, and enhancement of chemical and physical characteristics of the feed. A growing body of research has provided evidence on the beneficial effects of feed additives on growth performance, immune responses, feed utilization ability, and final product quality, including feed additives derived from plants. The global search for alternative feed additives for ruminants has become essential due to restrictions on the use of antibiotics and synthetic growth promoters in the livestock industry, as well as extensive pressure on farmers to reduce methane emissions while concomitantly maintaining production levels. Increasing interest in the use of plants as feed additives in southern Africa has led to a rising number of studies being conducted in recent years, providing interesting leads for further research and development.

Keywords

Phytogenic feed additives · Plant-based · Animal production · Poultry · Southern Africa

L. J. McGaw (✉)
Phytomedicine Programme, Department of Paraclinical Sciences, Faculty of Veterinary Science, University of Pretoria, Pretoria, South Africa
e-mail: lyndy.mcgaw@up.ac.za

G. Arsenos, I. Giannenas (eds.), *Sustainable Use of Feed Additives in Livestock*,
https://doi.org/10.1007/978-3-031-42855-5_14

1 Introduction

Southern Africa is host to an extraordinary level of plant diversity, with well over 30,000 species of higher plants (Van Wyk et al. 2009). Coupled with this is a rich cultural heritage demonstrated by an array of different cultures in the region. In common with global trends, there is a growing need in southern Africa for enhanced livestock productivity to meet the food requirements of the expanding population.

Global concerns are mounting concerning the levels of greenhouse gas (GHG) emissions from farm animals, particularly ruminants. Sub-Saharan Africa is not presently considered a major emitter of greenhouse gases (GHGs), but it is considered a possible future 'hotspot' for such emissions owing to the large livestock population spread over extensive dry areas characterized by feeds of low digestibility (Seketeme et al. 2022). One mitigation strategy is to research the use of enhancing feed digestibility by providing feed additives that improve quality of the diet and alter the rumen microbiome using specific forages containing plant secondary metabolites that decrease populations of methanogens, or methanogenesis (Seketeme et al. 2022).

Antimicrobial resistance (AMR) is another major concern relevant to animal production. The extensive use of antibiotics in animals as both therapeutic veterinary drugs and growth promoters is believed to be a significant factor contributing to emergence and spread of AMR (Gnanou and Sanders 2000). Although the use of in-feed antibiotics is not yet banned in southern Africa, as is the case in several other regions of the world, there remains a need to find alternative feed additives to replace antibiotics in the animal industry to enhance productivity and utilize natural resources for improved farming.

Ethnoveterinary medicine is widely used in rural areas of South Africa and neighbouring countries, and the use of plants as treatments for various ailments and to support growth features prominently in such practices (McGaw and Eloff 2008; McGaw et al. 2020). Additionally, there are many southern African plants known to have beneficial nutritional qualities as well as biological activity that may support livestock and poultry growth. A recent review highlighted the use of tannin-rich plants in particular as alternative supplementation for ruminants (Gxasheka et al. 2015). It was cautioned that the potential adverse effects of this class of compounds need to be balanced with the beneficial properties (Gxasheka et al. 2015).

Several research groups in South Africa target the investigation of plant extracts as additives to animal feed to enhance growth and nutritional status of production animals and poultry. This chapter provides an outline of work undertaken in this field to optimize use of the country's biodiversity in enhancing livestock and poultry production, whether it is on a commercial scale or for the benefit of small-scale rural farmers.

2 Plant-Based Feed Additives in Livestock Species

2.1 Cattle

In southern Africa, communal beef production is reliant on natural pasture, where environmental conditions are ideal for survival of pre-parasitic stages of gastrointestinal (GI) nematodes (Marufu et al. 2008). Various nematode species, particularly *Haemonchus contortus*, commonly affect cattle in communal farming areas of South Africa. The effects of high nematode infections, together with minimal management of cattle in communal areas, effectively result in considerable economic losses (Van Wyk et al. 1999). Nematode infections in the gastrointestinal tract reduce feed consumption and utilization, growth, meat yield and quality, resulting in blood loss and sometimes death.

Indigenous Nguni and non-descript crossbred cattle dominate communal beef production in South Africa. The Nguni breed is becoming more important internationally owing to its resistance to ticks, tick-borne diseases and nematodes, with meat quality comparable to those of Continental and British breeds (Strydom et al. 2008). As they are largely reliant on natural pasture, Nguni cattle often are not able to meet their nutritional requirements and they may lose condition in the dry season, becoming more susceptible to nematode infections and tick infestations, especially during the warmer, wetter seasons (Marufu et al. 2008).

Communal cattle farmers are often not able to afford or gain access to commercial anthelmintics. Supplementation with tannin-rich feeds can be a cost-effective complementary or alternative control strategy to reduce worm burdens in cattle in the communal areas. In 2009, Xhomfulana et al. compared nematode worm burdens in Nguni and crossbred cattle supplemented with leaf meal of *Acacia karroo (Vachellia karroo)*, which has a high protein and condensed tannin content. In this study, nematode worm burdens in Nguni and crossbred cattle supplemented with *A. karroo* leaf meal were compared. Results indicated that Nguni cattle fed *A. karroo* leaf meal had the lowest egg loads and worm burdens (Xhomfulana et al. 2009). It was concluded that supplementing cattle with *A. karroo* could reduce nematode burdens in addition to improving the nutritional status, growth performance and carcass traits of Nguni steers reared on natural pasture. Recommendations for further studies referred to the potential for development of a commercial anthelmintic product based on *A. karroo* foliage, which could be used in areas where cattle are farmed on high roughage diets, and where nematodes are endemic (Xhomfulana et al. 2009).

In another cattle study, it was noted that, although other studies have explored the possibility of using indigenous leguminous browse legume tree leaf meals and seed cakes as easily available and affordable supplements, the use of invasive alien legume plant leaf meals has not been investigated and this warrants further investigation (Dezah et al. 2021). Black wattle (*Acacia mearnsii*) is one of the most widespread and abundant invasive alien trees in South Africa. Although their leaves contain a rich diversity of phytonutrients, the utilization of *Acacia* leaf meals is limited by the presence of polyphenolic compounds. Polyphenolic compounds can be reduced to acceptable levels that do not affect nutrient utilization and growth performance of animals by mixing them with non-polyphenolic feeds. Dezah et al. (2021) explored the potential of *Acacia mearnsii* leaf meal (AMLM) as a ruminant feed to elucidate its efficacy in enhancing beef production and quality in smallholder areas. This complements the efforts by the National Agricultural Marketing Council of South Africa to intensify smallholder beef through the use of locally available and often underutilized browse legume species (Marandure et al. 2017). The objective of the study was to compare the effect of feeding finishing diets to crossbred steers containing either *Medicago sativa* hay or *A. mearnsii* leaf meals as substitutes for *Glycine max* on growth performance, carcass traits, and beef quality. Replacement of *G. max* with *M. sativa* did not affect growth performance or meat quality. (Dezah et al. 2021). In contrast, replacement with *A. mearnsii* resulted in low growth performance of the steers. *Medicago sativa* can be recommended as an alternative protein source for beef production in smallholder farming, but further research on optimum inclusion levels of *A. mearnsii* is needed (Dezah et al. 2021).

In 2009, Mapiye et al. reported on a study aiming to determine the effect of supplementing the diet of Nguni steers with *Acacia karroo* leaf meal on growth performance, blood chemistry, and carcass characteristics. It was reported that *A. karroo* supplementation improved the growth performance, nutritional status, and carcass traits of Nguni steers raised on natural pasture (Mapiye et al. 2009). The authors were of the opinion that *A. karroo* could be an economically viable alternative to commercial protein supplements for resource-poor beef cattle farmers in semi-arid areas. It was recommended that further studies should evaluate the meat quality of Nguni steers supplemented with *A. karroo* (Mapiye et al. 2009).

This suggested study was conducted by the same research group (Mapiye et al. 2010) when the meat quality of Nguni steers supplemented with *Acacia karroo* leaf meal was investigated. Based on the fat and cholesterol analysis of meat in this study, supplementing grazing steers with *A. karroo* could produce beef with similar health advantages to those finished on natural pasture. Supplementing natural pasture with *A. karroo* was reported to produce beef of similar quality to natural pasture but with a higher protein content and fresher appearance (Mapiye et al. 2010). Beef consumers require safe, healthier, and palatable meat, and thus research on fatty acid (FA) profiles and sensory quality attributes of beef produced from Nguni cattle supplemented with *A. karroo* leaf meal could be valuable (Mapiye et al. 2010).

A further study conducted by Mapiye et al. (2011a) investigated the fatty acid composition of beef derived from Nguni cattle supplemented with *A. karroo* leaf meal. The study was motivated by the recommendations of nutritionists who suggest that a reduction in total fat intake, especially saturated fatty acids (SFA) and trans fatty acids, may decrease risk of cardio-vascular diseases and some cancers (USDA and HHS 2010). In addition to reducing fat intake, nutritionists recommend the increased intake of polyunsaturated fatty acids (PUFA), particularly n-3 PUFA rather than n-6 PUFA (Griffin 2008). Beef from Nguni steers supplemented with *A. karroo* leaf meal had significantly higher alpha-linolenic acid content and an enhanced n-6/n-3 ratio than beef from steers receiving sunflower cake and control diets (Mapiye et al. 2011a). It was concluded that beef from cattle supplemented with *A. karroo* could provide health benefits to consumers (Mapiye et al. 2011a).

Mapiye et al. (2011b) went on to review the value of utilizing *Acacia karroo* in animal feed, as it is one of the most widespread and abundant indigenous tree legume species in southern Africa. *A. karroo* has positive attributes such as drought tolerance, high growth rates, ability to grow in infertile soils and resistance to large temperature variations. It has been considered to be a severe encroacher, adversely affecting land utilization and rangeland productivity (Nyamukanza and Scogings 2008). *A. karroo* leaves have desirable fatty acid profiles, and high nitrogen (N) and mineral concentrations. It was concluded by Mapiye et al. 2011b) that *A. karroo* leaf meal is a low-cost crude protein supplement that reduces worm burdens (most likely owing to the content of condensed tannins) and improves growth performance, protein status, carcass attributes, meat protein levels and appearance, as well as the proportion of desirable fatty acids in beef derived from cattle raised on rangeland.

Agricultural waste products have also been investigated by South African researchers in terms of their potential application as animal feed supplements. In a review article, Tayengwa and Mapiye (2018) explored the sustainable utilization of citrus and winery by-products as dietary supplements in ruminant production in low- to middle-income countries producing these by-products. These by-products have the potential to improve various indicators such as health and welfare, nutrient digestibility, rumen fermentation parameters, and growth performance, as well as carcass and meat quality attributes of ruminant animals (Tayengwa and Mapiye 2018). It was envisaged that a sustainable food production system could be attained through utilizing citrus and winery by-products to improve food, nutrition, and income security for resource-poor, vulnerable populations in low- to middle-income countries. Future studies were recommended to determine the effects of feeding citrus and winery by-products on various parameters in ruminants including meat fatty acid profiles, shelf life, and microbiological quality (Tayengwa and Mapiye 2018).

As a follow-up to the earlier review, Tayengwa et al. (2020) compared the growth performance, carcass, and meat quality effects of feeding 150 g/kg dry matter (DM) of dried citrus pulp (DCP) or dried grape pomace (DGP) as alternative dietary fibre sources to steers. Feeding the DGP and DCP diets resulted in better ($p \leq 0.05$) average daily gain and live weight compared to the control diet. The findings suggested that DGP is a better fibre source than DCP, causing enhanced growth performance, carcass attributes, and economic viability of feedlot steers (Tayengwa et al. 2020).

Another study by the same group compared the nutrient intake, digestibility, and utilization effects of feeding either 150 g/kg of dried citrus pulp (DCP) or grape pomace (DGP) as alternative dietary fibre sources to steers under feedlot conditions (Tayengwa et al. 2021a). Feeding DGP as an alternative fibre source to wheat bran improved nutrient intake, retention, and efficiency of N utilization but also reduced apparent nutrient digestibility compared to DCP. It was suggested that DGP may be a better fibre substitute for wheat bran than DCP in beef diets (Tayengwa et al. 2021a). It was also reported that finishing steers on diets containing DGP or DCP, compared to the control, increased proportions of total conjugated linoleic acid, n-3 and n-6 PUFA, and reduced concentrations of alcohols, aldehydes, and ketones, but did not affect the sensory attributes of the beef except for a slight reduction in tenderness (Tayengwa et al. 2021b).

In a valuable review of the potential of invasive Australian *Acacia* (IAA) species whose leaves are rich in nutrients and bioactive phytochemicals, Uushona et al. (2022) recommended a paradigm shift. It was suggested that *Acacia* could be valorised as feeds, rumen modifiers, nematode control agents, and biopreservatives in ruminant meat production. The authors noted that supplementing IAA leaf meals in grain ruminant diets either has neutral or positive effects on growth performance and carcass traits, as well as storing and eating quality of meat (Uushona et al. 2022). Generally, polyphenol-rich feeds are also capable of reducing bloat, and nitrogen and methane (CH_4) emissions to the environment. More work is required to determine inclusion rates of IAA leaves in forage diets for optimum ruminant health and production, meat shelf-life enhancement, and reductions in greenhouse gas emissions (Uushona et al. 2022).

Nyambali et al. (2022) conducted a study in Nguni cattle heifers to explore the economic viability of utilizing the invasive alien plant *Opuntia ficus-indica* (spineless cactus) cladodes as a supplementary feed. The impact of including cactus in the diet on animal growth performance and carcass characteristics was investigated. Carcass traits of heifers fed cactus diets and those fed non-cactus diets were comparable, and the higher economic returns from cactus inclusion supports the use of cactus-supplemented diets, especially during times of drought when commercial feed prices rise (Nyambali et al. 2022).

In a project undertaken by Chingala and co-workers, the effects of feeding Malawi Zebu steers with diets containing baobab (*Adansonia digitata*) seed meal and white thorn tree (*Vachellia polyacantha*) leaf meal as alternative protein sources to soybean meal on various parameters were investigated. It was reported that steers fed the baobab diet had higher microbial nitrogen supply, comparable feed conversion ratio (FCR) and nutrient digestibility (amongst other factors) to the soybean diet, indicating that baobab seed meal could be an alternative protein source to soybean meal for beef production (Chingala et al. 2019a). Additionally, steers fed the baobab diet resulted in the highest gross profits, followed by those fed the *Vachellia* and soybean diets, respectively (Chingala et al. 2019b). In total, feeding baobab and *Vachellia* diets improved gross profit and also produced beef comparable to the soybean diet (Chingala et al. 2019b).

In 2009, Hassen et al. investigated the nutritive value of *Ziziphus mucronata*, a valuable fodder tree in the drier parts of Africa, which is known to have nutritious leaves and edible fruit. New leaves of *Z. mucronata* were collected from tips of branches of the current season's growth in autumn, spring, and summer. In spring, the concentrations of Ca and Mn tended to be low but Ca, K, Zn, and Mn levels were generally sufficient to meet maintenance requirements of ruminants, regardless of the season. It appeared that P was generally deficient in *Z. mucronata* foliage in summer and autumn samples, and Cu was deficient throughout the year. It was concluded that *Z. mucronata* leaves have a relatively high nutritive value, with spring foliage having the best nutritive value (Hassen et al. 2009).

Several studies have also been undertaken to investigate the impact of southern African plant species on reducing gas and methane production. Hassen et al. (2016) researched the potential of six browse species (high, medium, and low condensed tannin concentrations) collected from the Kalahari Desert as anti-methanogenic additives to an *Eragrostis trichophora*-based substrate. The plant species studied were *Acacia luederitzii*, *Acacia erioloba*, *Acacia haematoxylon*, *Acacia mellifera*, *Monechma incanum,* and *Olea europaea*. The dry forage matter of the selected species was incubated with *Eragrostis trichophora* in a 30:70 (w/w) ratio together with buffered rumen fluid at 39°C for 48 h. Gas and methane production were determined at different time intervals and the volatile fatty acids concentration was tested after 48 h. *Acacia luederitzii* and *M. incanum* foliage decreased methane production by more than 50%, but also decreased rumen fermentation parameters such as volatile fatty acids concentration, and digestibility. It was concluded that tannin extracts from *A. luederitzii* could perhaps be used as a dietary alternative to reduce methane production, but optimum levels of inclusion that do not compromise rumen fermentation efficiency and overall digestibility of the diet need to be determined (Hassen et al. 2016).

In a similar study, Akanmu and Hassen (2017) determined the effect of selected medicinal plant extracts to reduce in vitro methane production while enhancing in vitro organic matter digestibility. The anti-methanogenic activities of leaf extracts of *Azadirachta indica*, *Aloe vera*, *Carica papaya*, *Moringa oleifera*, *Piper betle*, *Tithonia diversifolia*, *Jatropha curcas,* and *Moringa oleifera* pods were studied at different doses. It was concluded that methanol extracts of *A. indica*, *C. papaya*, *J. curcas*, *M. oleifera* (leaves and pods), and *T. diversifolia* reduced methane production, and may potentially manipulate rumen conditions, improve feed digestibility, and reduce enteric methane emission from ruminants, but results need to be confirmed in vivo (Akanmu and Hassen 2017).

In an MSc dissertation, Davis (2019) tested seven South African plant species for their ability to modify rumen fermentation characteristics in vitro. The plant species were *Combretum erythrophyllum*, *Dodonaea viscosa*, *Erythrina lysistemon*, *Heteropyxis natalensis*, *Sclerocarya birrea*, *Searsia chirindensis,* and *Halleria lucida*. Results indicated that plant extracts from *C. erythrophyllum* and *H. lucida* when used at 50 mg/kg, and plant extract from *S. chirindensis* at 50 or 300 mg/kg had promising potential as ruminant diet additives to improve the rate of gas production of poor-quality roughage-based diets (Davis 2019). This may in turn improve energy availability to the ruminant, but further studies were recommended to refine the optimal dose rate of these plant extracts for reducing methane production without negatively altering digestibility and volume of gas produced (Davis 2019).

2.2 Small Ruminants

Goats are a major source of meat, milk, manure and income for resource-poor smallholder farmers in southern Africa. The animals are predominantly reared in smallholder areas and commonly rely on natural pastures and crop residues, but availability of this feed is seasonally limited, compromising feed quantity and quality (Smith et al. 2005). Supplementation of the goat diet, particularly doelings and pregnant does, is highly useful to improve their condition during dry seasons. Smallholder farmers may include soya bean meal, sunflower cake, and cotton seed meal (CSM) as conventional supplements but these may be expensive or unavailable (Madzimure et al. 2011). The velvet bean (*Mucuna pruriens*) is a drought-tolerant, high-yielding legume and may be useful as a potential cheap protein supplement for small ruminants (Mapiye et al. 2007). Madzimure et al. (2014) evaluated supplementation of different levels of velvet bean seed meal (VBM; 0, 12, 24, and 36%) to Mashona goat doelings in terms of their feed intake (FI) and growth. The results of the study indicated that high inclusion levels of VBM negatively affected growth of young goats, which is most likely due to the presence of anti-nutritional factors.

Dietary protein sources in animal nutrition are becoming increasingly expensive and challenging to access (Gebregiorgis et al. 2011). *Moringa oleifera* grows throughout the tropics and has nutritious properties with a crude protein range of 23–40% (Mendieta-Araica et al. 2011), making it an ideal protein supplement. Moyo et al. (2012) investigated the effect of supplementing crossbred Xhosa lop-eared goat castrates with *M. oleifera* leaves in terms of the growth performance, as well as carcass and non-carcass characteristics. Feeding *M. oleifera* leaves or sunflower seed cake improved the growth performance and carcass characteristics of goats in an almost similar way, indicating that *M. oleifera* could be used as an alternative protein supplement to sunflower seed cake in goats. However, in South Africa the cost of *M. oleifera* leaves is higher than that of sunflower cake seed, which has been attributed to its scarcity and claims of its promising medicinal and nutritive properties (Moyo et al. 2012). Research on improving cultivation of *M. oleifera* is needed to increase production of the leaf meal, which will lower the cost for more economic and readily available animal feed supplementation.

In another study in goats, Gumede et al. (2022) ascertained the immunomodulatory effect of *Moringa oleifera* leaves supplementing the diets of BaPedi goats following vaccination with blanthrax vaccine. In this study, goats were subcutaneously vaccinated with blanthrax vaccine, a combined vaccine protecting against anthrax and blackleg. Three experimental diets were formulated by replacing a conventional supplement of lucerne with 0–50% of *M. oleifera*. Results indicated that only monocyte counts, platelet counts, and mean corpuscular volume (MCV) were significantly different ($p < 0.05$) amongst the 11 blood profiles observed (Moyo et al. 2012). Significant differences ($p < 0.05$) were noted in terms of body weight gain (BWG), growth rate (GR), and metabolic weight gain (MWG), but differences were not significant ($p > 0.05$) for feed intake (FI) and feed conversion ratio (FCR) (Moyo et al. 2012). It was suggested that *M. oleifera* leaves can be used as a feed supplement at 20 and 50% inclusion levels without adversely affecting growth performance and blood parameters.

Marume et al. (2012a) determined the effect of dietary supplementation with fresh *Acacia karroo* on experimental haemonchosis in four-month-old Xhosa lop-eared goats. The results demonstrated that consumption of fresh *A. karroo* leaves modulated infection level by reducing the establishment of *H. contortus* in the goats. In an associated study, supplementation of goat feed with *A. karroo* positively influenced fatty acid composition and sensory attributes of chevon derived from goats infected with *H. contortus* (Marume et al. 2012b).

A recent study published by Uushona et al. (2023) reported on the effects of substituting increasing levels of *Acacia mearnsii* leaf meal (AMLM) for *Triticum aestivum* bran in lamb finishing diets on growth, carcass, and meat quality attributes. It was found that AMLM could replace up to 100 g/kg dry matter of *T. aestivum* bran in lamb finisher diets without negatively affecting meat production and quality (Uushona et al. 2023).

Mahachi et al. (2020) contributed a useful review on the possibility of using the well-known plant sericea lespedeza (SL; *Lespedeza juncea* var. *sericea*) as a natural helminth suppressant, feed component, and meat preservative for the improvement of small ruminant health, meat production, and shelf life. Sericea lespedeza contains crude protein, minerals, vitamins, fibre amino acids, and diverse physiologically active substances including sterols, flavanols, and condensed tannins, which contribute nutritional, antimicrobial, antioxidant, anti-bloat, and helminth-suppressing properties. The review highlighted the potential of feeding SL to small ruminants as a sustainable mechanism to control nematode infections, and to enhance meat production and shelf life, while concomitantly reducing greenhouse gas emissions (Mahachi et al. 2020). However, it was cautioned that further research is necessary to determine optimal feeding strategies and doses (Mahachi et al. 2020).

The same research group went on to evaluate the effects of feeding incremental levels of sericea lespedeza (SL; 0, 62.5, 125, 187.5, and 250 g/kg diet) instead of lucerne on meat production and quality, and oxidative shelf life of meat from feedlot lambs infected or non-infected with *Haemonchus contortus* (Mahachi et al. 2023). It was reported that feeding SL did not affect meat physico-chemical, fatty acid, colour, and protein shelf-life profiles ($p > 0.05$). It appeared that substituting lucerne with SL up to 187.5 g/kg in lamb feedlot diets therefore had no effect on meat production and quality but it was able to enhance lipid stability regardless of nematode infection (Mahachi et al. 2023).

Du Toit et al. (2020) investigated the efficacy of *Lespedeza cuneata* in terms of helping to reduce ruminant methane (CH_4) emissions. The effect of feeding different levels of *L. cuneata* hay on the feed intake and enteric methane emissions of sheep fed a basal diet of subtropical *Eragrostis curvula* hay was investigated. It was determined that inclusion of *L. cuneata* improved diet digestibility, and led to increased concentrations of crude protein, non-fibre carbohydrates, and neutral detergent fibre (Du Toit et al. 2020). Substituting *E. curvula* hay with 60% *L. cuneata* (on a dry matter basis) resulted in the greatest reduction in CH_4 production of 21.4%, compared with a diet of 100% *E. curvula*, indicating that *L. cuneata* has promising potential to reduce CH_4 yield, and also to possibly increase production from sheep by improving diet digestibility and through improved dry matter intake (Du Toit et al. 2020).

2.3 Pigs

Smallholder pig production in southern Africa is constrained by feed shortages (D'Mello 1995). The demand for cereals to feed the ever-growing human population in southern Africa makes it imperative to identify alternative feedstuffs for feeding pigs. In a study conducted in pigs, Khanyile et al. (2014) proposed that leguminous leaf meals are potential substitutes for protein ingredients, and that inclusion of these forages also increases the dietary fibre content in pig feed. Inclusion levels of tannin-rich leaf meals should, however, not occur to such an extent that the pigs are prevented from consuming enough nutrients needed for growth (Khanyile et al. 2014). Tannins have several negative effects in animals, including depressing feed intake and reducing nutrient digestibility, and these factors need to be taken into account. Khanyile et al. (2014) assessed the nutritive value of leaf meals of selected leguminous *Acacia* species leaf meals. Leaves of five dominant species, namely *Acacia nilotica*, *Acacia nigrescens*, *Acacia tortilis*, *Acacia robusta,* and *Acacia xanthophloea*, were harvested and analysed for chemical and physical properties. Although the crude protein content of *A. xanthophloea* and *A. tortilis* was similar, the latter species was the most abundant in the area and had the lowest water holding and swelling capacity, as well as moderate levels of condensed tannins (Khanyile et al. 2014). In determining the optimum inclusion level of *A. tortilis* leaf meal in finishing pigs, animals were randomly allocated to six diets comprising 0, 50, 100, 150, 200, and 250 g/kg inclusion of *A. tortilis* leaf meal. Using piece-wise regression (broken-stick analyses), it was observed that *A. tortilis* leaf meal can be included up to 129 g/kg DM in finishing pig feeds, without negatively affecting the gain:feed (G:F) ratio, which was calculated by dividing average daily gain by average daily feed intake. Interestingly, it was found that the ability of pigs to utilize leaf-meal-based diets improved with duration of exposure to such diets (Khanyile et al. 2014).

A follow-up study to that conducted by Khanyile et al. (2014) was initiated by Ndou et al. (2015) from the same research group. In this investigation, the response of metabolites to increasing levels of *A. tortilis* leaf meal in growing pigs was assessed, as there was postulated to be a difference in supplementing leaf meals in growing and finishing pigs. The potential of using nutritionally related blood metabolites to estimate optimum levels of *A. tortilis* leaf meal inclusion was assessed (Ndou et al. 2015). It was found that inclusion of leaf meal of *A. tortilis* did not influence blood uric acid concentrations. Feed intake and average daily gain were constrained once levels of *A. tortilis* in diets of grower pigs exceeded 66.9 and 64.8 g/kg dry matter (DM), respectively. Optimum levels of leaf meals at which serum iron and total protein reached threshold values were at 60.0 and 63.2 g/kg DM, respectively.

The use of *Vachellia* (previously known as *Acacia*) leaves in the diets of pigs in the southern parts of Africa has been explored, but most such studies have focused on growth performance (Khanyile et al. 2014; Ndou et al. 2015), not paying attention to the quality attributes of the pork. Khanyile et al. (2020) determined growth performance, and carcass characteristics as well as fatty acid composition of pigs fed diets containing graded levels of *Vachellia (Acacia) tortilis* leaf meal (VTLM) in finishing pigs. It was reported that the inclusion of the leaf meal reduced the total SFA and total n-6 PUFA but increased the proportion of total monounsaturated fatty acid (MUFA) and n-3 PUFA of pork meat. The optimum MUFA and total *n-3* fatty acid were obtained at approximately 70 g/kg DM of *V. tortilis* leaf meal inclusion level.

The effect of dietary inclusion of *Moringa oleifera* leaf meal (MOLM) on feed conversion ratio (FCR) of finisher pigs, as well as physico-chemical meat quality, fatty acid (FA) composition, and pork shelf life was investigated by Mukumbo et al. (2014). Pigs were allocated to one of four dietary treatments, containing 0, 2.5, 5, or 7.5% MOLM. A significant reduction in intramuscular fat and stearic acid content was recorded as levels of MOLM increased. However, the other fatty acid profiles, ratios, and health lipid indices were not significantly different across treatments. Including 2.5 and 5% of MOLM in finisher pig feed had no detrimental effects on the feed conversion efficiency, carcass characteristics or physico-chemical meat quality, and interestingly significantly prolonged acceptability of pork odour and colour over ten days of refrigerated storage. Inclusion of 7.5% MOLM, however, resulted in poorer feed conversion efficiency. While including MOLM in finisher pig feed significantly improved n-3 content and n-6:n-3 feed ratio, these improvements were not shown in fatty acid composition of meat and subcutaneous fat. This may be a result of reduced levels of dietary fatty acid incorporation in favour of saturated and monounsaturated fatty acids incorporation from de novo lipogenesis, which becomes more dominant as growth progresses. Therefore, Mukumbo et al. (2014) advised that inclusion of MOLM in feed from an earlier age may have a more pronounced effect and thus be worth investigating.

Marula (*Sclerocarya birrea* subsp. *caffra*) nut cake is a by-product of oil extraction from marula fruits (Mdziniso et al. 2016) in southern Africa. Hlongwana et al. (2021) surmised that pig production systems could benefit from utilizing the discarded marula nut cake. The study aimed to determine the response in nitrogen (N) balance in slow-growing pigs supplemented with incremental levels of marula nut cake. It was concluded that the nut cake reduced nitrogen excretion, potentially minimizing ammonia volatilization, and thus it could be used as an alternative protein source for slow-growing pigs (Hlongwana et al. 2021).

Mabena et al. (2022) evaluated the effects of varying levels of marula (*Sclerocarya birrea* A. Rich) nut cake (ANC) on nutrient digestibility, growth performance, and carcass characteristics in pigs. It was concluded that ANC could replace soybean meal in the diet of growing pigs at less than 150 g/kg inclusion level (Mabena et al. 2022).

3 Poultry

The poultry industry is globally the most rapidly growing agricultural subsector (Mottet and Tempio 2017). In sub-Saharan Africa, as is the case in most developing countries, poultry demand is high as a major source of protein (Steinfeld et al. 2006). Antibiotic growth promoters are employed for disease preventative purposes and to improve growth rate and feed conversion rates in poultry production. However, although they are useful in preventing subclinical infections and promoting growth as in-feed antibiotics, they are also responsible in part for the worldwide scourge of antimicrobial resistance. The banning of antibiotic growth promoters has stimulated research on alternatives to antibiotic feed additives, including the use of phytogenic feed additives. These are considered to be safer compared to synthetic antibiotic growth promoters as they are natural products and commonly residue free although verification in each case is required (Dhama et al. 2014). There is a need to evaluate the use of plants that may be used in intensive poultry production to enable farmers to meet the demand for poultry products, given current restrictions on antibiotic growth promoters. Additionally, low-cost plant-based additives for chicken raising in rural areas may be developed and recommended for use to enhance food security for local people.

In a project investigating the potential of plants used in avian ethnomedicine for development of phytogenic feed additives, Jambwa et al. (2022a) conducted an ethnoveterinary survey in three districts of Zimbabwe. After surveying 146 smallholder farmers, 36 plant species belonging to 22 families were documented to be used in treating and managing various poultry ailments (Jambwa et al. 2022a). This project went on to investigate the antibacterial, antioxidant, and anti-lipoxygenase activities, and the cytotoxicity of selected plant species identified in the preliminary survey (Jambwa et al. 2023). Three reference American Type Culture Collection (ATCC) strains (*Staphylococcus aureus*, *Escherichia coli*, *Salmonella* Enteritidis) as well as two clinical strains isolated from chickens (*Escherichia coli* and *Salmonella* Gallinarum) were included in the antibacterial assays. The study concluded that *Senna singueana* had a variety of useful bioactivities, supporting further investigation of this plant species for development of phytogenic poultry feed additives. Other plant species with interesting biological activities include *Erythrina abyssinica*, *Bobgunnia madagascariensis,* and *Aloe greatheadii* (Jambwa et al. 2023).

Jambwa et al. (2022b) focussed on *Senna singueana* to assess the antibacterial, anti-lipoxygenase (anti-inflammatory) and antioxidant activities, and in vitro safety of fractions and isolated compounds from the leaf material. Bioassay-guided fractionation led to the isolation and identification of a known bioactive compound, luteolin (Jambwa et al. 2022b). Identification of chemical compounds in bioactive plant species, particularly those constituents with useful biological activity, is valuable for potential future use of these compounds as chemical markers in commercial preparations developed as antibiotic feed additive alternatives.

In a project investigating the value of selected African plants as potential poultry feed additives, Olawuwo et al. (2022a) assessed the mineral content and other useful properties of selected plants. Macro- and micro-mineral concentrations in *Morinda lucida*, *Acalypha wilkesiana*, and *Ficus exasperata* compared to standard broiler feed were determined using inductively coupled plasma mass spectrometry (ICP-MS), and inductively coupled plasma optical emission spectroscopy (ICP-OES). Degradation of nitrogen (N) and carbon (C) over specific intervals were evaluated by Dumas dry oxidation. Antioxidant efficacy in terms of radical scavenging activities was investigated using chemical assays and levels of phytochemical constituents were evaluated using standard methods. The research revealed that leaves of *F. exasperata*, *M. lucida*, and *A. wilkesiana* have significant amounts of macro- and micro-minerals, sugars, organic acids and anions, all nutritional requirements of poultry (Olawuwo et al. 2022a). Organic acids decrease intestinal loads of *Escherichia coli* and *Salmonella* species in broilers. Minerals and organic acids are essential antimicrobial and antioxidant constituents of poultry diets for the maintenance of gut flora health, and their body metabolic, enzymatic and antioxidant defence. High antioxidant activity was correlated with high total C and C/N ratio of the plant powders. Appreciable levels of total phenolics and flavonoids were present in the extracts, and these compounds are commonly associated with useful biological activities.

Further research targeted antibacterial and anti-biofilm activities as well as in vitro toxicity of the chosen medicinal plant leaf extracts against pathogens implicated in poultry diseases (Olawuwo et al. 2022b). Contamination of poultry products with a range of zoonotic pathogens is concerning to food safety and public health. Additionally, there is increased consumer demand for organic poultry products, further motivating studies in the area of phytogenic poultry feed additives. Many pathogens such as *Salmonella* spp. and *Campylobacter* spp. form biofilms, which further exacerbate diseases in poultry as well as contributing to resistance to antimicrobial treatment. Biofilms are generally complex biological structures that comprise many bacterial cells surrounded by layers of substances produced by the bacteria to form a protective barrier. The selected plant extracts had varying degrees of antimicrobial activity against bacteria and fungi relevant to poultry disease, further supporting the potential value of the plant species as candidates for future research as natural feed additives with protective attributes against relevant poultry pathogens.

Most poultry feed available to scavenging birds is of poor quality and mostly in short supply during dry seasons (Mwale et al. 2008). This, together with price increases of conventional poultry feed resources, especially protein sources, has spurred research on low-cost and locally available indigenous feed resources. A potential locally available feed resource may be derived from the multipurpose baobab (*Adansonia digitata*) tree, which is native to semi-arid areas in Zimbabwe and seeds are mostly available throughout the dry season. The seeds are known to be rich in protein, energy, amino acids (lysine and thiamine), vitamin C, calcium, and iron (Mwale et al. 2008). Baobab seed cake is also known to contain anti-nutritional factors including saponins, tannins, oxalate, and phytate, but levels of these compounds are generally assumed to be below established toxic levels in poultry (Nkafamiya et al. 2007). The effect on growth performance of guinea fowl keets after feeding graded levels of baobab seed cake (0, 5, 10, and 15%) was subsequently investigated by Mwale et al. (2008). Feed intake and body weight gain of keets were high in birds fed control and 5% baobab seed cake diets, and it was concluded that further research should evaluate the effects of high baobab seed cake inclusion levels in adult guinea fowl diets and other poultry species, including assessing yield and quality of meat and eggs of adult birds (Mwale et al. 2008).

The effects of increasing dietary inclusion levels of baobab (*Adansonia digitata*) seed oilcake (BSOC) on the growth performance, yield, and carcass characteristics in broiler chicks were assessed by Chisoro et al. (2018). It was concluded that including 5% BSOC could improve growth performance, as well as reducing feed costs, along with a reduction in the gross margin in the grower stage of broilers (Chisoro et al. 2018).

A series of studies on the effects of using *Moringa oleifera* leaf meal (MOLM) as an additive to broiler feed have been undertaken in southern Africa. Wapi et al. (2013) investigated physico-chemical shelf-life indicators of meat from broilers and found that including MOLM as an additive affected colour and pH of broiler meat, and also stability of these parameters during storage. The pH and colour (lightness) of the meat were stable during storage until Day 6 when a decrease was noted, while drip loss increased with storage time (Wapi et al. 2013). Nkukwana et al. (2014a) examined the effects of dietary supplementation of MOLM as a growth promoter in broiler chickens on the growth performance, digestibility, digestive organ size, and carcass yield. The results supported supplementation of MOLM up to 25 g per kg of feed as this did not alter efficiency of nutrient utilization, but enhanced growth performance (Nkukwana et al. 2014a). It was also found that supplementation of MOLM up to 5% of the bird's dry matter intake improved the fatty acid profile in addition to reducing lipid oxidation in broiler breast meat (Nkukwana et al. 2014b). In a related study, it was reported that ground *M. oleifera* leaves may be incorporated in the broiler diet as a phytogenic feed additive up to concentrations of 25 g/kg with significant positive effects on intestinal morphology, digesta pH, and digestive organ size (Nkukwana et al. 2015). Broiler dietary supplementation of MOLM also had positive effects on proximate composition and shelf-life quality indicators of broiler breast meat (Nkukwana et al. 2016). In comparing the effects of MOLM, a probiotic and an organic acid on growth, weights of digestive organs, and meat quality, Nduku et al. (2020) concluded that MOLM induced effects comparable with the probiotic and organic acid. All of the treatments improved growth performance, digestive organ size and meat quality characteristics when included in broiler diets as alternatives to antibiotic growth promoters (Nduku et al. 2020).

Mabusela et al. (2018) set out to determine how the partial supplementation of *Moringa oleifera* whole seed meal (MOWSM) would affect layer performance, egg quality, and egg fatty acid profile. Although MOWSM inclusion improved yolk colour, improved the fatty acid profile, and maintained external egg quality, the deleterious effect that it had on layer performance indicated that it may not be fed to early-lay hens at 1, 3, and 5% MOWSM (Mabusela et al. 2018).

4 Conclusion

With its rich plant and cultural diversity, southern Africa has much potential to utilize natural resources to enhance animal health and production. In recent years, a proliferation of studies has emerged from the region, reporting on results of evaluating various plant species for beneficial effects as feed additives in livestock and poultry. Many of these have been addressed in this chapter but more still remain to be evaluated and compared with one another in terms of methods used and outcomes reported. A variety of factors were assessed in the reported studies, including effects on growth performance, feed conversion ratios, nematode burden, and emission of greenhouse gases to name a few. The most common plant species investigated as feed supplements were those belonging to the *Acacia* (*Vachellia*) genus, including *A. karroo* and *A. tortilis*. A common thread noted in many of the studies was that further work was recommended to be done on a larger scale, with assessment of more parameters as a means of confirming in vivo practicality, efficacy, and safety of the plant-based feed additives. The growing interest in this field ensures that much future research will be undertaken in this area in the southern African region.

References

Akanmu AM, Hassen A (2017) The use of certain medicinal plant extracts reduced *in vitro* methane production while improving *in vitro* organic matter digestibility. Anim Prod Sci 58:900–908

Chingala G, Raffrenato E, Dzama K, Hoffman LC, Mapiye C (2019a) Intake, digestibility, rumen protein synthesis, and growth performance of Malawi Zebu steers fed diets containing rangeland-based protein source. Trop Anim Health Prod 51:199–204

Chingala G, Raffrenato E, Dzama K, Hoffman LC, Mapiye C (2019b) Carcass and meat quality attributes of Malawi Zebu steers fed *Vachellia polyacantha* leaves or *Adansonia digitata* seed as alternative protein sources to *Glycine max*. S Afr J Anim Sci 49:395–402

Chisoro P, Nkukwana TT, Mupangwam JF, Mabusela TP (2018) Feed intake, growth performance and carcass traits of broilers fed diets with various inclusion levels of baobab seed oilcake. S Afr J Anim Sci 48:284–294

D'Mello JFP (1995) Leguminous leaf meals in non-ruminant nutrition. In: D'Mello JPF, Devendra P (eds) Tropical legumes in animal nutrition. CAB International, Wallingford, pp 247–281

Davis JM (2019) *In vitro* methane and gas production characteristics of *Eragrostis curvula* feed supplemented with seven south African plant extract additives. MSc dissertation, University of Pretoria, South Africa

Dezah TJ, Chikwanha OC, Dzama K, Strydom PE, Mapiye C (2021) Comparison of *Acacia mearnsii* and *Medicago sativa* as substitutes for *Glycine max* in cattle finisher diets: effects on beef production and quality. Scientific African 11:e00673

Dhama K, Tiwari R, Khan RU, Chakraborty S, Gopi M, Karthic K, Saminathan M, Desingu PA, Sunkara LT (2014) Growth promoters and novel feed additives improving poultry production and health, bioactive principles and beneficial applications: the trends and advances–a review. Int J Pharmacol 10:129–159

Du Toit CJL, van Niekerk WA, Meissner HH, Erasmus LJ, Coertze RJ (2020) Methane emissions from sheep fed *Eragrostis curvula* hay substituted with *Lespedeza cuneate*. Anim Prod Sci 60: 1777–1784

Gebregiorgis F, Negesse T, Nurfeta A (2011) Feed intake and utilization in sheep fed graded levels of dried moringa (*Moringa stenopetala*) leaf as a supplement to Rhodes grass hay. Trop Anim Health Prod 44:517–517. https://doi.org/10.1007/s11250-011-9927-9

Gnanou JC, Sanders P (2000) Antibiotic resistance in bacteria of animal origin: methods in use to monitor resistance in EU countries. Int J Antimicrob Agents 15:311–322

Griffin BA (2008) How relevant is the ratio of dietary n-6 to n-3 polyunsaturated fatty acids to cardiovascular disease risk? Evidence from the OPTILIP study. Curr Opin Lipidol 19:57–62

Gumede L, Tyasi TL, Chitura T, Mbatha KR (2022) Immune and growth response of indigenous Pedi goats vaccinated with blanthrax to an inclusion of *Moringa oleifera* (drumstick tree) in *Cenchrus ciliaris* (Buffel grass) hay-based diet. Adv Anim Vet Sci 10:573–581

Gxasheka M, Tyasi TL, Qin N, Lyu Z-C (2015) An overview of tannins rich plants as alternative supplementation on ruminant animals: a review. Int J Agric Res Rev 3:343–349

Hassen A, Rethman NFG, van Niekerk WA (2009) A note on the potential nutritive value of *Ziziphus mucronata* (buffalo thorn) foliage during different seasons. Afr J Range Forage Sci 26: 103–105

Hassen A, Theart JJF, van Niekerk WA, Adejoro FA, Gemeda BS (2016) *In vitro* methane and gas production characteristics of *Eragrostis trichophora* substrate supplemented with different browse foliage. Anim Prod Sci 56:634–640

Hlongwana FY, Thabethe F, Thomas RS, Chimonyo M (2021) Nitrogen balance in slow-growing Windsnyer pigs fed on incremental levels of amarula (*Sclerocarya birrea* subsp. *caffra*) nut cake. Trop Anim Health Prod 53:364

Jambwa P, Katsande S, Matope G, McGaw LJ (2022a) Ethnoveterinary remedies used in avian complementary medicine in selected communal areas in Zimbabwe. Planta Med 88:313–323. https://doi.org/10.1055/a-1529-8618

Jambwa P, Makhubu FN, Matope G, Fouche G, McGaw LJ (2022b) Bioassay guided fractionation of *Senna singueana* and its potential for development of poultry phytogenic feed additives. Frontiers in veterinary science, section veterinary. Pharmacol Toxicol 8:800272. https://doi.org/10.3389/fvets.2021.800272

Jambwa P, Nkadimeng SM, Mudimba TN, Matope G, McGaw LJ (2023) Antibacterial and anti-inflammatory activity of plant species used in traditional poultry ethnomedicine in Zimbabwe: a first step to developing alternatives to antibiotic poultry feed additives. J Ethnopharmacol 300: 115687. https://doi.org/10.1016/j.jep.2022.115687

Khanyile SP, Ndou SP, Chimonyo M (2014) Influence of *Acacia tortilis* leaf meal-based diets on growth performance of pigs. Livest Sci 167:211–218

Khanyile M, Mapiye C, Thabethe F, Ncobela CN, Chimonyo M (2020) Growth performance, carcass characteristics and fatty acid composition of finishing pigs fed on graded levels of *Vachellia tortilis* leaf meal. Livest Sci 241:104259

Mabena PM, Ratsaka MM, Nkukwana TT, Malebana IMM, Nkosi BD (2022) Growth performance, nutrient digestibility and carcass characteristics of pigs fed diets containing amarula (*Sclerocarya birrea* A. Rich) nut cake as replacement to soybean meal. Trop Anim Health Prod 54:8–18

Mabusela SP, Nkukwana TT, Mokoma M, Muchenje V (2018) Layer performance, fatty acid profile and the quality of eggs from hens supplemented with *Moringa oleifera* whole seed meal. S Afr J Anim Sci 48:234–243

Madzimure J, Musimurimwa C, Chivandi E, Gwiriri L, Mamhare E (2011) Milk yield and quality in Guernsey cows fed cottonseed cake-based diets partially substituted with baobab (*Adansonia digitata* L.) seed cake. Trop Anim Health Prod 43:77–82

Madzimure J, Mutema N, Chimonyo M, Bakare AG, Mapiye C (2014) Performance of Mashona doelings supplemented with different levels of velvet bean (*Mucuna pruriens* L. DC. Var. *utilis*) seedmeal. Trop Anim Health Prod 46:901–904

Mahachi LN, Chikwanha OC, Katiyatiya CLF, Marufu MC, Aremu AO, Mapiye C (2020) Sericea lespedeza (*Lespedeza juncea* var. *sericea*) for sustainable small ruminant production: feed, helminth suppressant and meat preservation capabilities. Anim Feed Sci Technol 270:114688

Mahachi LN, Chikwanha OC, Katiyatiya CLF, Marufu MC, Aremu AO, Mapiye C (2023) Meat production, quality, and oxidative shelf-life of *Haemonchus*-parasitised and non-parasitised lambs fed incremental levels of sericea lespedeza substituted for lucerne. Meat Sci 195:109009

Mapiye C, Mwale M, Mupangwa JF, Mugabe PH, Poshiwa X, Chikumba N (2007) Utilisation of legumes as livestock feed in Zimbabwe. Tropical Grasslands 41:84–91

Mapiye C, Chimonyo M, Dzama K, Strydom PE, Muchenje V, Marufu MC (2009) Nutritional status, growth performance and carcass characteristics of Nguni steers supplemented with *Acacia karroo* leaf-meal. Livest Sci 126:206–214

Mapiye C, Chimonyo M, Dzama K, Muchenje V, Strydom PE (2010) Meat quality of Nguni steers supplemented with *Acacia karroo* leaf-meal. Meat Sci 84:621–627

Mapiye C, Chimonyo M, Dzama K, Hugo A, Strydom PE, Muchenje V (2011a) Fatty acid composition of beef from Nguni steers supplemented with *Acacia karroo* leaf-meal. J Food Compos Anal 24:523–528

Mapiye C, Chimonyo M, Marufu MC, Dzamac K (2011b) Utility of *Acacia karroo* for beef production in southern African smallholder farming systems: a review. Anim Feed Sci Technol 164:135–146

Marandure T, Mapiye C, Makombe G, Dzama K (2017) Indicator-based sustainability assessment of the smallholder beef cattle production system in South Africa. Agroecol Sustain Food Syst 41:3–29

Marufu MC, Chimonyo M, Dzama K, Mapiye C (2008) Egg loads of gastrointestinal strongyles in Nguni cattle on communal rangelands in the semi-arid areas. A poster presented at the world Association for Animal Production. In: The south African Society for Animal Science and the all-Africa Association of Animal Production 10th world conference on animal production, 23–-28 November 2008, Cape Town, South Africa. Book of abstracts. Wageningen Academic Publishers, The Netherlands, p 188

Marume U, Chimonyo M, Dzama K (2012a) Influence of dietary supplementation with *Acacia karroo* on experimental haemonchosis in indigenous Xhosa lop-eared goats of South Africa. Livest Sci 144:132–139

Marume U, Hugo A, Chimonyo M, Muchenje V, Dzama K (2012b) Effect of dietary supplementation with *Acacia karroo* leaves on fatty acid profiles and consumer sensory attributes of Xhosa lop-eared goats under artificial haemonchosis. Anim Prod Sci 52:1099–1106

McGaw LJ, Eloff JN (2008) Ethnoveterinary use of southern African plants and scientific evaluation of their medicinal properties. J Ethnopharmacol 119:559–574

McGaw LJ, Famuyide IM, Khunoana ET, Aremu AO (2020) Ethnoveterinary botanical medicine in South Africa: a review of research from the last decade (2009 to 2019). J Ethnopharmacol 257: 112864

Mdziniso MP, Dlamini AM, Khumalo GZ, Mupangwa JF (2016) Nutritional evaluation of Marula (*Sclerocarya birrea*) seed cake as a protein supplement in dairy meal. J Appl Life Sci Int 4:1–11

Mendieta-Araica B, Sporndly R, Reyes-Sánchez N, Sporndly E (2011) Moringa (*Moringa oleifera*) leaf meal as a source of protein in locally produced concentrates for dairy cows fed low protein diets in tropical areas. Livest Sci 137:10–17

Mottet A, Tempio G (2017) Global poultry production: current state and future outlook and challenges. Worlds Poult Sci J 73:245–256

Moyo B, Masika BP, Muchenje V (2012) Effect of supplementing crossbred Xhosa lop-eared goat castrates with *Moringa oleifera* leaves on growth performance, carcass and noncarcass characteristics. Trop Anim Health Prod 44:801–809. https://doi.org/10.1007/s11250-011-9970-6

Mukumbo FE, Maphosa V, Hugo A, Nkukwana TT, Mabusela TP, Muchenje V (2014) Effect of Moringa oleifera leaf meal on finisher pig growth performance, meat quality, shelf life and fatty acid composition of pork. S Afr J Anim Sci 44. https://doi.org/10.4314/sajas.v44i4.9

Mwale M, Mupangwa JF, Mapiye C, Saina H, Chimvuramahwe J (2008) Growth performance of Guinea fowl keets fed graded levels of baobab seed cake diets. Int J Poult Sci 7:429–432

Ndou SP, Khanyile M, Chimonyo M (2015) Growth performance and nutrition-related serum metabolites in growing pigs fed on *Acacia tortilis* leaf meal. Livest Sci 182:22–27

Nduku XP, Mabusela SP, Nkukwana TT (2020) Growth and meat quality of broiler chickens fed Moringa oleifera leaf meal, a probiotic and an organic acid. S Afr J Anim Sci 50:710–718

Nkafamiya II, Osemeahon SA, Dahiru D, Umaru HA (2007) Studies on the chemical composition and physico-chemical properties of the seeds of baobab *(Adansonia digitata)*. Afr J Biotechnol 6:756–759

Nkukwana TT, Muchenje V, Pieterse E, Masika PJ, Mabusela TP, Hoffman LC, Dzama K (2014a) Effect of *Moringa oleifera* leaf meal on growth performance, apparent digestibility, digestive organ size and carcass yield in broiler chickens. Livest Sci 161:139–146

Nkukwana TT, Muchenje V, Masika PJ, Hoffman LC, Dzama K, Descalzo AM (2014b) Fatty acid composition and oxidative stability of breast meat from broiler chickens supplemented with *Moringa oleifera* leaf meal over a period of refrigeration. Food Chem 142:255–261

Nkukwana TT, Muchenje V, Masika PJ, Mushonga B (2015) Intestinal morphology, digestive organ size and digesta pH of broiler chickens fed diets supplemented with or without *Moringa oleifera* leaf meal. S Afr J Anim Sci 45:362–370

Nkukwana TT, Muchenje V, Masika PJ, Pieterse E, Hoffman LC, Dzama K (2016) Proximate composition and variation in colour, drip loss and pH of breast meat from broilers supplemented with *Moringa oleifera* leaf meal over time. Anim Prod Sci 56:1208–1216

Nyambali A, Mndela M, Tjelele TJ, Mapiye C, Strydom PE, Raffrenato E, Dzama K, Muchenje V, Mkhize NR (2022) Growth performance, carcass characteristics and economic viability of Nguni cattle fed diets containing graded levels of *Opuntia ficus-indica*. Agriculture 12:1023

Nyamukanza CC, Scogings PF (2008) Sprout selection and performance of goats fed *Acacia karroo* coppices in the false Thornveld of the Eastern Cape, South Africa. S Afr J Anim Sci 38:83–90

Olawuwo OS, Abdalla MA, Mühling KH, McGaw LJ (2022a) Proximate analysis of nutrients and *in vitro* radical scavenging efficacy in selected medicinal plant powders with potential for use as poultry feed additives. S Afr J Bot 146:103–110

Olawuwo OS, Famuyide IM, McGaw LJ (2022b) Antibacterial and antibiofilm activity of selected medicinal plant leaf extracts against pathogens implicated in poultry diseases. Front Vet Sci Sect Vet Pharmacol Toxicol 9:820304. https://doi.org/10.3389/fvets.2022.820304

Seketeme M, Madibela OR, Khumoetsile T, Rugoho I (2022) Ruminant contribution to enteric methane emissions and possible mitigation strategies in the southern African development community region. Mitig Adapt Strateg Glob Chang 27:47

Smith T, Mlambo V, Sikosana JLN, Maphosa V, Mueller-Harvey I, Owen E (2005) *Dichrostachys cinerea* and *Acacia nilotica* fruits as dry season feed supplements for goats in a semi-arid environment: summary of a DFID funded project in Zimbabwe. Anim Feed Sci Technol 122: 149–157

Steinfeld H, Wassenaar T, Jutzi S (2006) Livestock production systems in developing countries: status, drivers, trends. Revue Scientifique et Technique-Office International des Epizooties 25: 505–516

Strydom PE, Frylinck L, Van der Westhuizen J, Burrow HM (2008) Growth performance, feed efficiency and carcass and meat quality of tropically adapted breed types from different farming systems in South Africa. Aust J Exp Agric 48:599–607

Tayengwa T, Mapiye C (2018) Citrus and winery wastes: promising dietary supplements for sustainable ruminant animal nutrition, health, production, and meat quality. Sustainability 10: 3718. https://doi.org/10.3390/su10103718

Tayengwa T, Chikwanha OC, Dugan MER, Mutsvangwa T, Mapiye C (2020) Influence of feeding fruit by-products as alternative dietary fibre sources to wheat bran on beef production and quality of Angus steers. Meat Sci 161:107969

Tayengwa T, Chikwanha OC, Raffrenato E, Dugan MER, Mutsvangwa T, Mapiye C (2021a) Comparative effects of feeding citrus pulp and grape pomace on nutrient digestibility and utilization in steers. Animal 15:100020

Tayengwa T, Chikwanha OC, Neethling J, Dugan MER, Mutsvangwa T, Mapiye C (2021b) Polyunsaturated fatty acid, volatile and sensory profiles of beef from steers fed citrus pulp or grape pomace. Food Res Int 139:109923

USDA and HHS (2010) Dietary guidelines for Americans 2010. Department of Health and Human Services (HHS)/Department of Agriculture (USDA) Available from http://www.cnpp.usda.gov/DGAs2010-DGACReport.htm

Uushona T, Chikwanha OC, Katiyatiya CLF, Tayengwa T, Strydom PE, Mapiye C (2022) Ruminant meat production and quality enhancement, nematode suppression and greenhouse gas emission mitigation: a sustainable paradigm for valorisation of *Acacia* leaves. Anim Feed Sci Technol 284:115187

Uushona T, Chikwanha OC, Katiyatiya CLF, Strydom PE, Mapiye C (2023) Production and meat quality attributes of lambs fed varying levels of *Acacia mearnsii* leaf-meal as replacement for *Triticum aestivum* bran. Meat Sci 196:109042

Van Wyk JA, Stenson MO, Van der Merwe JS, Vorster RJ, Viljoen G (1999) Antihelmintic resistance in South Africa: surveys indicate an extremely serious situation in sheep and goat farming. Onderstepoort J Vet Res 66:273–284

Van Wyk B-E, Van Oudtshoorn B, Gericke N (2009) Medicinal Plants of South Africa. Briza Publications, Pretoria

Wapi C, Nkukwana TT, Hoffman LC, Dzama K, Pieterse E, Mabusela T, Muchenje V (2013) Physico-chemical shelf-life indicators of meat from broilers given *Moringa oleifera* leaf meal. South African. J Anim Sci 43(Issue 5, Supplement 1):43

Xhomfulana V, Mapiye C, Chimonyo M, Marufu MC (2009) Supplements containing *Acacia karroo* foliage reduce nematode burdens in Nguni and crossbred cattle. Anim Prod Sci 49:646–653

Natural Alternatives to Anticoccidial Drugs to Sustain Poultry Production

Kelsilandia Aguiar-Martins, Caela Burrell, Damer P. Blake, and Virginia Marugan-Hernandez

Abstract

For many farm animal diseases natural compounds are emerging as alternatives to classical drugs and antibiotics. In this chapter, we provide a brief overview of coccidiosis in chickens caused by protozoan parasites from the genus *Eimeria*, outline the measures used to assess anticoccidial drug efficacy and summarise progress towards validation of natural alternatives. It is important to understand current measures of anticoccidial control and their limitations in order to design better, more environment-friendly and antimicrobial-free approaches. Recognising the optimal parameters required to evaluate anticoccidial efficacy in animal trials will encourage better and more consistent experimental design, improving comparison between studies and optimising animal use. Once these concepts have been developed, the chapter provides an extensive review of natural compounds with potential to reduce the occurrence of coccidiosis in chickens, including plant extracts, herbal formulations, essential oils and probiotics. Finally, a compilation of studies using in vitro models as alternatives to animals is summarised. In vitro studies cannot directly evaluate anticoccidial efficacy, but they can become important tools for pre-screening of novel alternative compounds, replacing and reducing animal use.

Keywords

Chicken coccidiosis · *Eimeria* · Anticoccidial efficacy · Natural compounds · In vitro models

K. Aguiar-Martins · C. Burrell · D. P. Blake (✉) · V. Marugan-Hernandez
Pathobiology and Populations Science, Royal Veterinary College, Hertfordshire, UK
e-mail: dblake@rvc.ac.uk

G. Arsenos, I. Giannenas (eds.), *Sustainable Use of Feed Additives in Livestock*,
https://doi.org/10.1007/978-3-031-42855-5_15

1 *Eimeria* Species Parasites: The Cause of Coccidiosis in Poultry

The enteric disease coccidiosis, caused by protozoan parasites of the genus *Eimeria*, is recognised as one of the principal intestinal disorders within poultry posing a significant risk to animal welfare and economic productivity. Until recently, it was believed only seven species of *Eimeria* could infect chickens, namely *Eimeria acervulina*, *Eimeria brunetti*, *Eimeria maxima*, *Eimeria mitis*, *Eimeria necatrix*, *Eimeria praecox* and *Eimeria tenella*, with names such as *Eimeria hagani* and *Eimeria mivati* considered *nomina dubia* (Vrba et al. 2011). All seven species have a global distribution and can be found wherever chickens are farmed (Clark et al. 2016). However, the recent identification of three new species named *Eimeria lata*, *Eimeria nagambi* and *Eimeria zaria* (previously Operational Taxonomic Units [OTUs] X, Y and Z) has required the taxonomy of the genus to be revisited (Blake et al. 2021b). Each *Eimeria* species replicates within restricted region(s) of the intestinal tract during the endogenous phase of the lifecycle, featuring multiple rounds of asexual replication (schizogony) followed by a single round of sexual replication (Fig. 1). The enteric site(s) of replication defines each species, although overlap between species limits value as a pathognomonic diagnostic (Johnson and Reid 1970). Following replication in the gut, unsporulated oocysts are shed into the

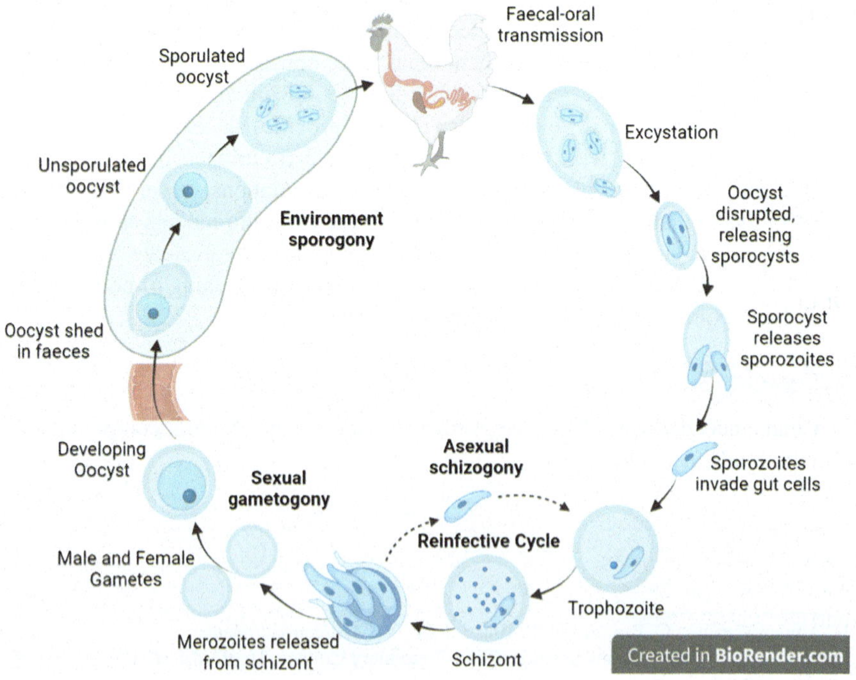

Fig. 1 The *Eimeria* lifecycle in chickens

environment initiating the exogenous phase of the lifecycle. Sporulation occurs in the environment where oocysts persist as a source of infection via faecal–oral transmission (Williams 1995; Velkers et al. 2012). Clinical coccidiosis can be characterised by malabsorption, enteritis and, in severe cases, mortality, translating to limited bird growth, increased intra-flock variation and reduced feed conversion efficiency.

Ingestion of a sporulated *Eimeria* oocyst initiates the endogenous phase of the lifecycle. The tough oocyst wall is disrupted mechanically during transit through the crop and gizzard, releasing four sporocysts. Two sporozoites are released from each sporocyst in the intestine following exposure to digestive enzymes and bile, after which each sporozoite continues to pass through the intestinal lumen until it attaches to and invades the epithelial cell layer. The location of invasion and subsequent replication varies for each *Eimeria* species (Blake et al. 2021b). After invasion the sporozoite develops into a trophozoite then undergoes schizogony (asexual multiple fission), producing multiple first-generation merozoites. Once mature, the merozoites rupture the schizont and leave the host cell before invading another epithelial cell and entering a second round of schizogony. *Eimeria* that infect chickens commonly undergo three or four rounds of schizogony in total prior to differentiating into gametes as sexual replication begins. Macrogametocytes, which develop into uninucleate macrogametes, are fertilised by motile, biflagellated microgametes produced in microgametocytes by multiple fission, resulting in the formation of diploid zygotes. Once fertilised the developing zygote forms a resistant wall and transforms into an unsporulated oocyst that escapes from the host cell to be excreted into the environment, starting the exogenous phase. The unsporulated oocyst sporulates in the environment, requiring warmth, oxygen and moisture to become a sporulated oocyst. The sporulated oocyst contains four sporocysts, each of which contains two sporozoites.

2 Why Is Coccidiosis a Problem?

Distinct *Eimeria* species can cause coccidiosis in all livestock and many non-production species but chickens are most economically important, representing a third of all meat produced globally with an estimated 68 billion chickens farmed in 2018 (Blake et al. 2020). The combined cost of control (prophylaxis and treatment), morbidity and mortality attributed to coccidiosis in chickens has recently been estimated to exceed £99 million in the UK alone, rising from £38 million in 1995 (Williams 1999). The economic burden of coccidiosis in chickens based on data from countries representing six continents indicated a global cost in excess of £10.4 billion in 2016 (Blake et al. 2020). Morbidity factors including poor bodyweight gain (BWG) and feed conversion ratio (FCR) represent the biggest cost components, but the wider effects of subclinical infections are hard to quantify (Blake et al. 2021a). Such consequences include increased intra-flock variation during poultry production and reduced welfare, commonly leading to increased culling, condemnation, and quality downgrades, in addition to increased risk factors to public health and food security (Blake and Tomley 2014). Rapid flock cycle schedules, introducing multiple batches of immunologically naïve and immature chickens to environments contaminated with *Eimeria* oocysts, ensure an ongoing threat of coccidiosis.

Indirect consequences of coccidiosis can include enteric dysbiosis, characterised by altered microbiome diversity and structure resulting in an imbalance in intestinal homeostasis. Dysbiosis associated with *Eimeria* infection can predispose towards successful incursion or exacerbation of other pathogens such as *Clostridium perfringens* and *Campylobacter jejuni*, posing additional risks to overall health and productivity (Madlala et al. 2021). Specifically, necrotic enteritis (NE) resulting from *C. perfringens* infection develops during concurrent *Eimeria* infection, influenced by the intestinal damage and change in enteric environment caused by coccidiosis (Macdonald et al. 2019). Similarly, the increased proclivity for opportunistic infections by naturally occurring gut bacteria can pose an increased risk to public health, influencing the carriage of foodborne zoonotic pathogens such as *Salmonella* Typhimurium and *C. jejuni* (Baba et al. 1982; Macdonald et al. 2019). Furthermore, wet litter resulting from dysbiosis poses a significant risk for pododermatitis, a leading cause of quality downgrades, culling and condemnation resulting from reduced BWG and locomotion (De Jong et al. 2014). Thus, the true cost of *Eimeria* infection and coccidiosis is likely to be far higher than current estimates suggest.

3 Current Measures to Control *Eimeria* and Prevent Coccidiosis

3.1 Husbandry

Good husbandry within poultry houses is an essential and economical means of reducing the threat of coccidiosis. Hygiene measures including restricting bird access to faeces, maintaining good (dry) litter quality and cleaning between flocks are key, while environmental measures controlling house temperature, ventilation and moisture levels are also beneficial (Blake and Tomley 2014; Stayer et al. 1995). Additional husbandry considerations aimed at reducing exposure to stressors, including overcrowding, overheating and debeaking, coupled with reducing dietary deficiencies and feed restriction, contribute to good overall immune health and support effective responses to *Eimeria* infection (Williams 1998).

3.2 Chemoprophylaxis

Good husbandry alone is not sufficient to effectively prevent the occurrence of coccidiosis. Routine chemoprophylaxis using anticoccidial compounds such as chemical anticoccidials and ionophores is necessary for sufficient control of *Eimeria* infection. Two broad categories of anticoccidial drugs are available, namely organic ionophores (derived from fermentation products) and synthetic chemicals (produced by chemical synthesis) (Chapman 1997). Ionophores such as monensin dominate the market (Blake et al. 2020), working by increasing the ion permeability of cell membranes to inhibit the parasite (Bakker 1979). From a regulatory standpoint there are notable distinctions in the classification of anticoccidial drugs under different governing bodies. Ionophores are classified as zootechnical feed additives in Europe, while they are regarded as polyether ionophorous antibiotics in the USA, meaning they are tightly regulated under legislation restricting the use of antibiotics in livestock feed (Chapman 2001; Noack et al. 2019; Shirley and Chapman 2005). The difference in regulation has increased pressure to reduce drug use in poultry production.

Benefits of using anticoccidial drugs include ease of administration, since most can be incorporated into feed or dispensed in drinking water (Attree et al. 2021). Ionophores have also been shown to possess antibacterial properties against Gram-positive bacteria including *C. perfringens*, offering protection against not only coccidiosis, but also the risk of necrotic enteritis (Lanckriet et al. 2010). However, routine chemoprophylaxis is not without drawbacks. Most importantly, anticoccidial drug resistance develops rapidly and is now ubiquitous within all *Eimeria* species. Reduced efficacy has been reported for all current anticoccidial drugs (Braunius 1982; Chapman 1997; Joyner 1970). It has also been suggested that routine use of ionophores such as salinomycin might co-select bacteria including *Enterococcus faecalis* and *Enterococcus faecium* for resistance to other medically relevant drugs such as erythromycin, tetracycline and ampicillin (Pikkemaat et al. 2022). Similarly, concerns regarding the risk of environmental contamination and selection for resistance, as well as risks of accidental ingestion and toxicity to non-target organisms, have been highlighted (Dorne et al. 2013; Mooney et al. 2020).

The longstanding use and ease of administration of chemoprophylactic drugs are a testament to their success within the poultry industry, but the rate of development of drug resistance coupled with legislative restrictions and consumer concern for chemical residues has discouraged the development of new anticoccidials (Chapman 2001, Chapman et al. 2005). The requirement for withdrawal periods prior to slaughter also leaves animals vulnerable to uncontrolled coccidiosis infections, posing significant health and welfare risks (Kadykalo et al. 2018; Peek and Landman 2011; Williams 1998). Consequently, these pressures have resulted in calls for alternative measures for the control of coccidiosis.

3.3 Vaccination

Anticoccidial vaccination is currently achieved by controlled administration of varied formulations of live wild-type (i.e. virulent) or attenuated *Eimeria* species parasites, inducing an immune response equivalent to that stimulated by a natural eimerian infection. A transmission blocking subunit vaccine containing affinity-purified gametocyte antigens extracted from *E. maxima* named CoxAbic is available in some parts of the world, although uptake has been limited (Sharman et al. 2010).

Considerable efforts have been applied to development of effective and scalable subunit or recombinant anticoccidial vaccines, but at the time of writing none have reached the market (Blake et al. 2021a; Shirley et al. 2005; Soutter et al. 2020, 2022).

Developed in the 1950s, the first generation of anticoccidial vaccines comprised live, wild-type, sporulated *Eimeria* oocysts dosed at low levels over several days in order to induce protective immunity comparable to natural exposure to the parasite (Williams 2002). Due to their ease of administration via spraying onto newly hatched chicks or onto their feed, supplementation into drinking water or via gels, coupled with their relatively low cost of production, live virulent vaccines are widely used within the global poultry industry (Blake et al. 2021a; Soutter et al. 2020). Nonetheless, correct administration requires uniform dosage across a flock and effective cycling to limit the occurrence of clinical disease and compromised flock performance (Blake et al. 2020, 2021a; Shirley et al. 2005).

Recognising the limitations surrounding live virulent vaccines, a second generation of anticoccidial vaccines was developed in the 1980s based upon live-attenuated *Eimeria* lines. These attenuated lines were mostly developed by selection for 'precocious' development, reducing the duration of the lifecycle, fecundity and pathogenicity while retaining immunogenicity (Jeffers 1975; Shirley et al. 2005). Although safer than their virulent predecessors, live-attenuated vaccines incur relatively high production costs and can be difficult to scale up due to their reduced fecundity, supporting uptake by much of the egg-laying and breeding sectors but more rarely in broilers (Blake et al. 2021a).

The use of live virulent and attenuated vaccines presents additional benefits by introducing drug-susceptible parasites into flocks and permitting decreased use of anticoccidial drugs, reducing selective pressures towards resistance and supporting the restoration of drug susceptibility (Chapman and Jeffers 2015). Nonetheless, several drawbacks are associated with the production and use of live anticoccidial vaccines. Production of parasites for use in vaccine formulations requires use of live chickens since *Eimeria* cannot productively replicate in vitro, raising production costs and ethical considerations. Further to this, as natural immunity to *Eimeria* is parasite species- or (in the case of *E. maxima* in some examples) strain-specific, multiple lines are required in each vaccine to provide comprehensive immune protection (Blake et al. 2021a). Consequently, vaccine development has taken a new direction in recent decades attempting to develop recombinant anticoccidial vaccines. Although numerous antigens have been defined and tested as potential vaccine candidates, the search for immunoprotective targets has been met with several frustrations. In particular, the complex nature of apicomplexan lifecycles coupled with the multitude of protein-coding genes within each *Eimeria* species genome has made it difficult for scientists to pinpoint specific antigens with the ability to stimulate a sufficient immunoprotective response (Lee et al. 2022).

4 Measures of Anticoccidial Efficacy

Efficacy of anticoccidial drugs, vaccines and alternative products for control of *Eimeria* in chickens can be assessed using a variety of parasitological, immunological, pathological and/or production measures. The choice of measure can be influenced by the scale of the study and proximity to market. Combining measures from two or more categories is preferred, offering a more robust assessment. Assessment of anticoccidial activity has traditionally been done using controlled or environmental infection of live chickens. Here, the chicken is the target system and not simply a model, offering directly relevant data. However, assessment of the outcomes of infection requires controlled conditions, excluding or controlling for inadvertent exposure to *Eimeria* or other pathogens.

4.1 Parasitological Measures

Eimeria follow host species-specific monoxenous lifecycles with faecal–oral transmission via the environment. Infection by ingestion of one or more sporulated *Eimeria* oocysts initiates multiple sequential rounds of asexual replication (schizogony), followed by a single round of sexual reproduction (gametogony), after which the in vivo phase is completed when unsporulated oocysts are shed with faeces (Blake and Tomley 2014). Eimerian lifecycles do not include dormant or long-lived encysted forms, unlike close relatives such as *Neospora caninum*, *Toxoplasma gondii* or *Sarcocystis* species (Dubey 2019). Understanding the lifecycle offers opportunities to quantify *Eimeria* replication. Oocysts shed in faeces are most easily accessible and can be quantified by microscopy. Oocysts can be enumerated directly from faeces, litter or other environmental substrates by counting in a smear or similar wet preparation (Lawal et al. 2016). Greater sensitivity can be achieved by flotation, either directly in a counting chamber such as a modified McMaster or mini-FLOTAC chamber, or by employing pre-count enrichment by flotation of a larger volume of faeces/substrate and sub-sampling into a counting chamber (Long et al. 1976; Bortoluzzi et al. 2018). Saturated sodium chloride solution (specific gravity [s. g.] 1.18–1.20) is most commonly used for oocyst flotation in samples collected from chickens, achieving an optimum balance between oocyst flotation and separation from faecal debris, although alternatives such as sugar (s.g. 1.28) can also be used and are often preferred in studies with ruminants (e.g. Ruiz et al. 2013). Measures of oocyst output include oocysts per gram (OPG) in faeces or litter, or total output over a defined period of time such as a day or the full patent period (Blake et al. 2005). OPG is most easily calculated, requiring collection of a small sample that can illustrate oocyst numbers at a single fixed time. However, OPG figures can vary due to faecal volume and time of sampling within the patent period. Collecting faecal material over a longer period of time is more robust, but requires suitable experimental accommodation such as individual cages on wire gridded floors. Microscopic enumeration of oocysts offers an easily accessible and cost-effective method to define *Eimeria* replication, but figures can be difficult to interpret if the time of infection is unknown (i.e. oocyst output varies between early, peak and late patent periods). Variation in oocyst output has also been noted between strains of some *Eimeria* species (Blake et al. 2005). Mixed infections with more than one *Eimeria* species can be challenging, with overlapping oocyst morphology and considerable variation in fecundity between some species (Long et al. 1976). Detection and diagnosis of *Eimeria* infection can also be achieved by microscopy using wet smears or tissue samples collected directly from the intestine during post-mortem (Long et al. 1976), but quantitation can be challenging if samples are not representative.

More recently, development of quantitative polymerase chain reaction (qPCR) assays targeting the *Eimeria* genus or each individual *Eimeria* species that infects chickens created tools to improve quantification (Blake et al. 2006; Vrba et al. 2010). Using qPCR to enumerate *Eimeria* genomes in intestinal contents, faeces, tissue or environmental samples reduces the requirement for specialist parasitological knowledge and can be less subjective than microscopy, but does require relatively expensive equipment that currently precludes routine use in the field. Samples preserved by freezing (as collected, or in 70% v/v + ethanol) or in storage buffers such as RNA later permit delayed analysis and support medium throughput screening. Key steps include normalisation to compare between samples, with examples including comparison to host reference genes (e.g. in tissue) or standardisation of input genomic DNA template (e.g. from faeces or litter). Quantitative PCR can be highly sensitive, capable of detecting as few as 10^{-1} (*Eimeria* genus specific, targeting the multicopy 5S rRNA gene) or 10^{1} (individual species; Blake et al. 2006; Vrba et al. 2010) genome copies. Quantitative PCR can also be highly reproducible, requiring smaller group sizes for statistical power than methods such as oocyst microscopy (Nolan et al. 2015).

4.2 Immunological Measures

Eimeria infections induce a robust inflammatory immune response that features T-helper (Th) 1, Th17 and T-regulatory cells (Kim et al. 2019; Lee et al. 2022). For species such as *Eimeria maxima* a complex balance has been described between interferon gamma (IFNγ) and interleukin (IL)-10 (Bremner et al. 2021), but the timing and magnitude of response can be highly variable. At the time of writing no immune parameter has been described as a reliable biomarker for *Eimeria* infection or replication, despite considerable effort (e.g. Bremner et al. 2021; Lee et al. 2022).

4.3 Pathological Measures

In the absence of an easily quantifiable measure to assess parasite replication or protective immune response on farm, alternative measures have included assessment of gross pathology at post-mortem. Consideration of intestinal pathology associated with eimerian infection in chickens found dead or sacrificed as sentinels of flock health has been standardised following description of a lesion scoring system for six of the seven established *Eimeria* species that infect chickens (Johnson and Reid 1970). All three of the most economically relevant *Eimeria* species were considered (*E. acervulina*, *E. maxima* and *E. tenella*), together with *E. brunetti* and *E. necatrix*. *Eimeria mivati* was included, but is now recognised to be *E. mitis* (Vrba et al. 2011), leaving *E. praecox* unrepresented. The situation has become more complex with the recent description of three new species (*E. lata*, *E. nagambi* and *E. zaria*), although lesions and gross pathology have now been reported (Blake et al. 2021a). The Johnson & Reid scoring system is used widely, providing a standardised and semi-quantitative protocol for assessment of pathognomonic pathology and lesions where a score of '0' indicates a healthy, apparently uninfected chicken, and '+4' represents heavy infection or dead due to coccidiosis (Johnson and Reid 1970). The protocol is straightforward to apply, permitting description of the infecting species and severity of infection, but can be subjective and requires consistent support from an experienced operator. Drawbacks to the use of lesion scoring to assess anticoccidial efficacy include the absence of opportunity for repeat measures and inconsistencies between operators undermining comparison between studies. It is also difficult to assess oocyst output or production parameters in chickens sampled for lesion scoring given that the scoring should be completed five to seven days post infection, early in or even before the patent period of oocyst excretion (Long et al. 1976; Blake et al. 2021b). Within small groups of chickens it has been suggested that lesion score is a poor proxy for pathogenicity (Johnson and Reid 1970), although larger cohort studies have described clear and significant associations between lesion scores and production parameters such as BWG (Boulton et al. 2018). Blood plasma colouration has been assessed in chickens fed maize-based diets as a proxy for gut leakage, linking intestinal damage and lesions to elevated plasma colouration due to leakage of carotenoids as measured by optical density between 450 and 490 nm (Hamzic et al. 2015), but the assay is not appropriate with non-maize (sometimes termed 'corn') diets.

4.4 Production Measures

Coccidiosis is primarily a disease of production animals, with the costs attributed to lost performance estimated to represent between 51 and 90% of the total cost of coccidiosis (e.g. £82.4 million in the UK in 2016; Blake et al. 2020). Thus, assessment of anticoccidial efficacy usually requires one or more measure of production performance to convince academics, poultry producers and industry professionals since this is directly relevant to profit.

4.4.1 Bodyweight Gain (BWG)

Assessment of bodyweight gain (BWG) requires two or more paired measurements of chicken weight, a key performance indicator for broiler chickens. The measure is easy to assess, but does require the same individuals to be weighed on each occasion. Unless the disease caused by infection is severe it can be difficult to detect a statistically significant difference within five to seven days of infection (i.e. the period targeted for lesion scoring), incurring a requirement for additional chickens to be tested over ten or more days if BWG and lesion scoring are to be considered (e.g. Soutter et al. 2022). A considerable difference in weight gain is expected between male and female chickens. Analysing weight gain separately in male and female chickens can reduce experimental variation, improving statistical power, but it is helpful to test both sexes to ensure the absence of a sex-linked effect.

4.4.2 Feed Conversion Ratio (FCR)

Feed conversion ratio (FCR) can be a sensitive measure of subclinical coccidiosis, sometimes referred to as coccidiasis (Williams 1986). Chickens that show no clinical signs of coccidiosis but fail to thrive, or require additional feed to match performance shown by uninfected contemporaries, can be identified by variation in FCR. FCR is a standard performance measure tracked by most commercial broiler producers, but requires accurate recording of feed consumed and minimal loss of feed to other sources (spillage, rodents, etc.).

4.4.3 Egg Production

Coccidiosis is primarily a disease of young, growing chickens, with greatest impact on fast-growing broiler breeds reared for meat, although species such as *E. brunetti* and *E. necatrix* can affect older chickens (McDougald et al. 1990). Nonetheless, egg-laying is rarely significantly affected by coccidiosis (Mayhew 1934), although modest losses can be attributed to layer and breeding stock (Blake et al. 2020). Greater losses in flock-level egg production can be attributed to mortality of replacement layer and breeding stock.

4.4.4 Mortality

Severe outbreaks of coccidiosis can feature extensive mortality, a measure that has historically been assessed in studies of anticoccidial efficacy (e.g. the Morehouse-Baron Method of evaluating anticoccidial activity; Morehouse and Baron 1970). However, mortality in experimental animals is now rarely assessed, becoming unnecessary given the range of alternatives available and ethical considerations of welfare and suffering.

4.4.5 Combined Measures

Attempts to assess multiple measures of anticoccidial efficacy and combine into a single composite score have included the anticoccidial index (ACI). The components included have varied between studies, but percentage survival, percentage relative weight gain, lesion score and oocyst count are common (Jeffers 1974; Chapman 1989). Use of an ACI can be informative, providing a global overview of parasitological, pathological and performance measures, but it is worth noting that oocyst counts performed during the five- to seven-day window for lesion scoring post infection are unlikely to be representative for *Eimeria* species with longer pre-patent periods such as *E. maxima* or *E. tenella*.

4.4.6 Novel Measures of Anticoccidial Effects

Eimeria infections have been associated with significant variation in colonisation by bacterial pathogens such as *Salmonella* Typhimurium, *C. jejuni* and *C. perfringens* (Arakawa et al. 1981; Macdonald et al. 2019; Ringenier et al. 2021), as well as disruption of enteric microbiomes (Macdonald et al. 2017). In the future testing of novel anticoccidial controls is likely to include microbiome analysis, assessing impact on the occurrence of specific bacterial pathogens and contribution to enteric dysbiosis.

4.5 Recommendations on Best Practices to Assess Anticoccidial Efficacy

The complex nature of coccidiosis highlights that a single measure of efficacy is unlikely to be reliable in evaluation of anticoccidial products. Instead, it is preferred to include measures of parasite replication, pathology induced by clinical challenge and performance under industry conditions. The use of such a combination can illustrate efficacy of parasite control, reduction in disease and protection of economic margin and welfare. Studies of performance under industry conditions commonly require large cohort sizes, incurring financial and ethical costs. For this reason, small-scale studies of parasite replication can provide a valuable primary measure that can be supplemented by assessment of pathology before large group size studies are required. The recent development of improved quantitative in vitro assays for parasite replication under anticoccidial selection can now provide a valuable pre-screening step for many antimicrobial and phytobiotic products before in vivo studies are initiated (Marugan-Hernandez et al. 2020; Sidiropoulou et al. 2020), as outlined below.

5 Phytochemicals and Essential Oils: New Opportunities for Anticoccidial Control in Poultry

Challenges faced in the routine application of anticoccidial drugs and vaccines in chickens have encouraged investigation of a range of natural alternatives, highlighting a selection of phytobiotics and other natural products. There is now increasing public and legislative pressure in many countries to replace established anticoccidial drugs with new, more socially acceptable alternatives. Opportunities include increased use of a range of 'natural' treatments including phytochemicals (plant extracts, herbal formulations and essential oils [EOs]) or probiotics as prophylactic or therapeutic options to control coccidiosis in poultry. In some examples these alternatives are already used widely to treat animals in developing countries, often driven by economic cost or cultural habits (Abbas et al. 2017). In most examples the mechanism of anticoccidial action is not known, but extensive studies have served to document the efficacy of several species of plants, probiotics and

Table 1 Examples of natural products tested to control *Eimeria* in chickens

Author	*Eimeria* species	Chickens used	Compound
Baron et al. (2022)	*Eimeria mivati*, *Eimeria maxima*, *Eimeria acervulina* and *Eimeria tenella*	70 broilers	Toltrazuril
Messaï and Redouane-Salah (2022)	*Eimeria tenella*	120 Hubbard-ISA broilers	*Artemisia herba alba Asso*
EzeC et al. (2022)	*Eimeria tenella* and *Eimeria maxima*	119 broilers	*Sacoglottis gabonensis* (*Magnoliophyta*, *Humiriaceae*)
Qaid et al. (2022a)	*Eimeria tenella*	150 Ross 308 broilers	Cinnamon bark (*Cinnamomum verum*)
Qaid et al. (2022b)	*Eimeria tenella*	225 broilers	*Cinnamomum verum* bark and *Rumex nervosus* leaves
Khorrami et al. (2022)	*Eimeria necatrix*, *Eimeria maxima*, *Eimeria acervulina* and *Eimeria tenella*	300 broilers	Pomegranate peel extract
Yadav et al. (2022)	*Eimeria maxima*	576 broilers	*Brassica* spp. *rapeseed*, canola meal and isothiocyanate
Pham et al. (2022)	*Eimeria maxima* and *Eimeria necatrix*	432 Arbor Acres broilers	Essential oils (carvacrol, thyme, hexanoic, benzoic and butyric acid.
Yang et al. (2022)	*Eimeria mivati*, *Eimeria maxima*, *Eimeria acervulina* and *Eimeria tenella*	105 broilers	*Akkermansia muciniphila*
Lima et al. (2021)	*E. acervulina*, *E. maxima* and *E. tenella*	900 broilers	*Mushrooms Agaricus subrufescens* and *Pleurotus ostreatus*
Fu et al. (2021)	*E. tenella*	630 broilers	*Ethanamizuril*
Nguyen et al. (2021)	*E. acervulina*, *E. maxima*, *E. tenella*, *Eimeria mitis* and *Eimeria praecox*	120 broilers	*Berberine*
Herrero-Encinas et al. (2021)	*Eimeria maxima*, *E. maxima MF*, *Eimeria acervulina*, *Eimeria tenella* and *Eimeria mivati*	400 broilers	*Olea europaea*
Abdelhady et al. (2021)	*Eimeria tenella*, *E. brunetti*, *E. hagani*, *E. mivati*, *E. acervulina*, *E. maxima*, *E. necatrix* and *E. praecox*	750 broilers	Atomonsin, Madramycin, Atozuril and Maxiban
Galli et al. (2021)	*Eimeria: E. acervulina*, *E. brunetti*, *E. maxima*, *E. necatrix*, *E. praecox*, *E. tenella* and *E. mitis*	480 broilers	Phytogenic blend (curcuminoids, cinnamaldehyde and glycerol monolaurate)
	E. tenella	150 broilers	*Cinnamomum verum*

(continued)

Table 1 (continued)

Author	*Eimeria* species	Chickens used	Compound
Qaid et al. (2021)			
Cheng et al. (2021)	*E. tenella*, *E. maxima*, *E. necatrix* and *E. acervulina*	108 broilers	*Bacillus licheniformis*
Akhter et al. (2021)	*Eimeria* spp.	80 Sonali chickens	Papaya (*Carica papaya*) leaves extract
Sharma et al. (2021)	*E. tenella*	25 Cobb chicks	Extract of Indian Gooseberry (*Phyllanthus emblica*)
Qaid et al. (2021)	*E. tenella*	150 broilers	*Rumex nervosus* leaf powder
Sidiropoulou et al. (2020)	*E. tenella* and *Eimeria acervulina*	180 broilers	Oregano and garlic essential oil
Song et al. (2020)	*E. tenella*	200 broilers	Herbal powder 'Shi Ying Zi'
Bo et al. (2020)	*E. tenella*	84 broilers	Decoquinate nanoliposome
Yong et al. (2020)	*E. tenella*	180 broilers	*Fructus Meliae toosendan* extract
Wang et al. (2020)	*E. tenella*	90 broilers	*Radix dichroae*
Hamilton et al. (2019)	*E. acervulina*, *E. maxima* and *E. tenella*	240 broilers	Green tea, apple cider vinegar and amprolium
El-Shazly et al. (2020)	*E. tenella*	90 broilers	*Fluoroquinolone lomefloxacin*
El-Sawah et al. (2020)	*E. tenella*, *E. necatrix*, *E. maxima* and *E. acervulina*	240 broilers	*Enterococcus faecium* and *diclazuril*
Vereecken et al. (2020)	*E. acervulina*, *E. maxima* and *E. tenella*,	150 broilers	*Nicarbazin* and *Monensin*
Granstad et al. (2020)	*E. acervulina*, *E. mitis*, *E. tenella* and *E. maxima*	308 broilers	*Bacillus subtilis* and *Saccharomyces cerevisiae*
Muraina et al. (2020)	*E. tenella*, *E. necatrix* and *E. brunetti*	25 broilers	*Khaya senegalensis* aqueous stem bark extract
Srinivasu et al. (2019)	*Eimeria* spp.	420 broilers	*Cocciban* herbal
Rasheed et al. (2020)	*E. acervulina* and *E. tenella*	1028 broilers	IL-10 antibody
Talghari et al. (2020)	*E. tenella*, *E. maxima*, *E. necatrix* and *E. acervulina*	300 broilers	Sodium bisulphate, *Monensin*
Pop et al. (2019)	*E. acervulina*, *E. tenella* and *E. maxima*	100 broilers	Herbal formula
Zhang et al. (2019)	*Eimeria* spp.	440 broilers	*Triazine*
Yang et al. (2019)	*E. tenella*	200 layers	*Bidens pilosa*

(continued)

Table 1 (continued)

Author	*Eimeria* species	Chickens used	Compound
Upadhaya et al. (2019)	*E. tenella* and *E. maxima.*	800 broilers	Vitamin D
Nabian et al. (2018)	*E. acervulina, E. maxima, E. tenella* and *E. necatrix*	54 broiler	Salinomycin and amprolium + ethopabate
Jarujareet et al. (2018)	*E. tenella*	85 broilers	Riboflavin supplement
Preetam et al. (2019)	*E. tenella*	420 broilers	*Cocciban* herbal
Ademola et al. (2019)	*Eimeria* spp.	96 broilers	*Pleurotus ostreatus* extract
Hernandez-Patlan et al. (2019)	*Eimeria maxima*	120 broilers	*Bacillus*
Wang et al. (2018a)	*E. acervulina, E. maxima* and *E. tenella*	1344 broilers	*Bacillus subtilis* and zinc
Hayajneh et al. (2018)	*Eimeria* spp.	450 broilers	Apple cider vinegar
Lai et al. (2018)	*E. tenella*	2000 broilers	Methionine
Hong et al. (2018)	*E. tenella*	30 broilers	Ethanol extract of *Tribulus terrestris*
Galli et al. (2018)	*E. necatrix, E. acervulina* and *E. maxima*	60 layers	Curcumin
Ahad et al. (2018)	*E. tenella*	70 broilers	Fruit peel *Punica granatum*
Awais et al. (2018)	*E. tenella, E. acervulina, E. necatrix* and *E. maxima*	200 broilers	*L. bagasse*-derived polysaccharides
Wang et al. (2018b)	*E. tenella*	270 broilers	*Areca* nut

trace elements reporting effects on BWG, lesions scores and/or oocyst outputs (Table 1).

Plant-based compounds are the natural products most commonly tested for control of coccidiosis. It has been suggested that plant components can constrain the disease through three mechanisms: directly targeting, inhibiting or killing one or more stages of the eimerian lifecycle, regulating the host immune response, or indirectly by regulation of the enteric microbiome (Muthamilselvan et al. 2016). Plants or their secondary metabolites with anti-parasitic properties are already well-known sources of drugs for treatment of humans; in some cases, these botanicals have been used as an alternative to treat animal diseases, including coccidiosis. Examples include artemisinin, derived from extracts of sweet wormwood (*Artemisia annua*) and used to treat malaria, but also shown to be effective against *Eimeria*. When tested in birds infected with *E. tenella*, artemisinin improved BWG and reduced lesion scores, although the same effect was not observed in chickens infected with *E. acervulina* (Allen et al. 1997).

Approximately 400,000 species of plants exist in the world; however, only 10% of these have been explored as medicines, prompting considerable interest in the potential of the remaining 90% for drug discovery (Volenzo and Odiyo 2020). Bioactive components can be extracted or purified using a range of techniques, highlighting that standardisation of the active ingredients in phytotherapy remains a challenge. Variation can arise due to differences in plant genetics, climatic and growing factors, the time of harvesting, and the extraction methods (Miliauskas et al. 2004). Extraction of active plant materials or secondary metabolites commonly requires an appropriate solvent and standard extraction procedure, but the choice of method is influenced by the type of plant, the solvents used, pH of the solvent, temperature, and the intended use of the final product (Abubakar and Haque 2020). While derivatives of several plants have been shown to influence the outcome of coccidiosis, the extraction methods required were often expensive or the plant itself was not widely available (Awais et al. 2018). In other examples, the correct dosage may not be known or incidental negative clinical parameters might constrain commercial use of certain plants. Other challenges include international divergence in regulations, terminologies and classification of herbal medicines, limiting dissemination and application around the world (Alostad et al. 2018). Nonetheless, considerable efforts have been expended searching for botanical products that can be used to control coccidiosis, as discussed here.

5.1 Plant Extracts

'Plant extract' is a general term used to describe the extracted product of a solid/liquid separation operation: the solid material (plant) is submitted to a fluid (solvent), where the active component of interest is solubilised in the solvent. Optionally, this solution can pass through a second step of separation to obtain a more solid extract (Abubakar and Haque 2020). Plant extracts are relatively easy to manipulate and are commonly included in poultry diets to improve health (Akyildiz and Muzaffer 2016). Several plant extracts have been explored as sources of control for *Eimeria*, with some examples discussed below.

5.1.1 *Rumex nervosus*

Rumex nervosus, commonly known as Othrob, is a shrub belonging to the Polygonaceae family; it is native to some Eastern African countries and the Arabian Peninsula. Different parts of this plant have been explored as natural medicines with diverse pharmacological properties. However, leaf extracts have attracted most attention in the poultry industry due to anticoccidial and anti-inflammatory activities. Challenge studies have shown a limited anticoccidial effect, reducing lesion scores and oocyst output while improving weight gain (Qaid et al. 2021). *Rumex nervosus* is common in many developing countries where coccidiosis incurs a notable economic impact, encouraging its application as a cheap alternative to supress the effects of *Eimeria* infection.

5.1.2 *Punica granatum*

Pomegranate peel is a rarely explored by-product from the pomegranate (*Punica granatum*) industry. It is rich in phenolic compounds and has been associated with strong antioxidant, anti-inflammatory and anti-parasitic activities, including against *Eimeria* parasites. Extracts of pomegranate peel have been tested in a range of animals including mice, rabbits and poultry, demonstrating varied levels of efficacy for treatment of coccidiosis (Dkhil 2013). Multiple divergent methods of extraction and manipulation have been described, possibly explaining the variation between reports. In chickens, pomegranate extract has shown anticoccidial activity, lowering oocyst counts and reducing intestinal lesions; however, the dose of extract required has been associated with negative side effects on growth indices (Khorrami et al. 2022).

5.1.3 *Carica papaya*

Carica papaya belongs to the family Caricaceae. It grows widely in tropical and subtropical lowland regions around the world, being the third most cultivated tropical crop worldwide (Chávez-Pesqueira and Núñez-Farfán 2017). The leaf of the plant is rich in saponins and papain, both of which can be used to treat intestinal parasites (Rahmasari and Wibowo 2019). Saponin is known to bind cholesterol, altering the integrity of parasite surface membranes and eventually resulting in death (Wang et al. 1998). It has also been shown that an extract of this plant can reduce sporozoite invasion of intestinal epithelial cells, possibly due to proteolytic destruction by papain (El-Shall et al. 2022). Challenge studies have revealed potential using the leaf of the plant to boost growth, reduce oocyst output and improve haematological parameters in Sonali chickens, although the outcome was not equivalent to the synthetic drug toltrazuril (Akhter et al. 2021).

5.1.4 *Bidens pilosa*

Bidens pilosa is an Asteraceae plant that originated from South America but is now widely distributed around the world. This herb has been attributed with a range of therapeutic properties, being rich in several polyacetylene, polyacetylene glycoside, flavonoid, flavone glycoside, aurone, chalcone, okanin glycoside, phenolic acid, terpene, pheophytin and fatty acid compounds (Xuan and Khanh 2016). Phytochemicals derived from *B. pilosa* have been reported to modulate gastrointestinal diseases, immunity and bacteria in humans (Bartolome et al. 2013). More recently, the plant has gained attention for control of coccidiosis. The extract of the plant can inhibit the occurrence of disease and decrease the severity of *Eimeria* infection in chickens. It has been suggested that regulation of the gut microbiota during *Eimeria* infection might be the main contribution of this herbal product. Treatment using *B. pilosa* has been reported to reduce villus destruction during *Eimeria* infection and support a healthy enteric microbiota (Memon et al. 2021; Chang et al. 2016).

5.1.5 *Phyllanthus emblica*

Phyllanthus emblica, commonly called the Indian gooseberry, is native to tropical and southern Asia. Extracts of *P. emblica* leaves and fruit have been described as treatment for a diverse range of diseases such as cancer, osteoporosis, neurological disorders, hypertension, parasitic and other infectious disorders (Variya et al. 2016). Treatment of chickens during eimerian infection has been shown to inhibit subsequent oocyst sporulation, reducing infectivity and improving weight gain in other members of the flock. The identity of the active compound(s) remains unknown (Sharma et al. 2021). Another plant from the same family, *Emblica officinalis*, had previously been shown to enhance BWG and offer immunostimulatory properties in broiler chickens (Kaleem et al. 2014).

5.1.6 *Sacoglottis gabonensis*

Sacoglottis gabonensis is a large tree from the Humiriaceae family, common in tropical countries around the world. Extracts of the stem bark from this tree are rich in flavonoids, alkaloids, saponins, tannins, glycosides, terpenoids and phenolic compounds (Ejikeme et al. 2014). A recent study in chickens infected with *E. tenella* and *E. maxima* indicated that continuous use of *S. gabonensis* stem bark extract can improve haemato-biochemical indices, clinical signs and significantly improve survivability (EzeC et al. 2022). Toxicity tests in this same study suggest that the extract is a relatively safe product that could be included in livestock and poultry diets to control coccidiosis.

5.2 Herbal Formulations and Derivatives

Herbal formulations can be prepared by several extraction methods and may include one or more plants containing one or more active substances. The beneficial, and often synergistic, properties of different herbs are widely recognised in traditional Asian medicine.

5.2.1 Shi Ying Zi

Shi Ying Zi powder is a traditional Chinese herbal medicine containing *Cnidium monnieri*, *Taraxacum mongolicum*, and sodium chloride. These herbs are commonly used in combination to treat infectious diseases. A recent study comparing the effect of this medicine with the ionophore monensin and sulfamlopyrazine (also known as sulfachloropyrazine sodium) found comparable efficacy in the prevention and treatment of *E. tenella* infection in broiler chickens, reducing the occurrence of haemorrhage and total oocyst output in faeces (Song et al. 2020).

5.2.2 *Fructus Meliae toosendan*

Also known as *Chuan Lian Zi* in Chinese medicine, *Fructus Meliae toosendan* is the mature fruit of *Melia toosendan*, a plant found primarily in the southwest region of China. This traditional Chinese herbal medicine has been tested for diverse uses in human treatment including anti-cancer therapy, but has only recently been tested to control coccidiosis in chickens. *Fructus Meliae toosendan* extract was tested during *E. tenella* challenge, inhibiting oocyst sporulation and improving the enteric microbiota of chickens as well as improving some clinical signs. Although *M. toosendan* cytotoxic and deeper clinical parameters were not fully assessed yet in chickens, it seemed to prevent coccidial infection in broilers (Yong et al. 2020).

5.2.3 Berberine

Berberine is a compound that can be extracted from several plants, especially Berberis (*Berberis vulgaris*). It is a traditional Chinese medicine with antimicrobial activities against several infectious agents including viruses, bacteria, fungi, parasitic protozoans and helminths (Dkhil 2013). Recently, berberine has received more attention due to its anti-parasitic potential against coccidiosis. A diet supplemented with this medicine reduced faecal oocyst shedding in chickens infected with various *Eimeria* species (*E. acervulina*, *E. maxima*, *E. mitis*, *E. praecox* and *E. tenella*) (Nguyen et al. 2021). It is suggested that berberine can regulate access to important trace elements during *Eimeria* replication, reducing oocyst output and clinical signs (Huang et al. 2022).

5.3 Current Herbal Commercial Products

Several formulations of herbal products are currently marketed to control or treat coccidiosis. Examples include Cocciban® (Indian Herbs Specialities Pvt. Ltd), a herbal formula containing *Azadirachta indica* leaves, *Nicotiana tabacum* leaves, *Calotropis procera* flowers and *Trachyspermum ammi* seeds. One study has suggested Cocciban® exhibits efficacy comparable to synthetic anticoccidial drugs as salinomycin and dinitolmide, reducing oocyst excretion and improving health in terms of haemato-biochemical, faecal and histopathological parameters (Srinivasu et al. 2019). Another recent commercial formula, Norponin XO2®, has also been shown to control coccidiosis at a level equivalent to an ionophore. Produced from a blend of *Yucca schidigera* and *Trigonella foenum-graecum*, plants rich in several steroidal saponins, a recent evaluation of this product revealed that supplementation during infection with *E. tenella* performed at a level equivalent to monensin (Benarbia et al. 2022).

5.4 Essential Oils

Essential oils (EOs) are obtained from different species of aromatic plants following a wide range of methods of extraction (Aziz et al. 2018). EOs can contain over 200 substances of volatile and non-volatile nature. They have applications in multiple industries and have been investigated for their efficacy as antimicrobial, anti-inflammatory and anti-viral agents. The first study evaluating effects of EOs after infection with *Eimeria tenella* was reported by (Giannenas et al. 2003). Here, oregano EO (300 mg/kg) supported BWG and feed intake at levels comparable to the unchallenged group, higher than the infected group but lower than the group treated with lasalocid (75 mg/kg). More recently, oregano and citrus EOs (500 mg/kg) did not improve performance in chickens experimentally infected with *Eimeria* spp. (challenge using a Coccivac® at 25× the recommended dose) (Gordillo Jaramillo et al. 2021). Similarly, oregano EO (50, 24 or 12 mg/kg) also failed to improve BWG and did not reduce oocyst output compared to uninfected controls (Bozkurt et al. 2016). However, intestinal morphology was improved and total antioxidant status was increased. In the same experiment monensin sodium (50 mg/kg) did improve these parameters. In a similar experiment published by Mohiti-Asli and Ghanaatparast-Rashti (2015), supplementation with oregano EO (500 ppm) increased bodyweight and improved FCR during *Eimeria* challenge (50× Livacox® vaccine recommended dose), and reduced lesion scores and oocyst outputs, similar to those observed with diclazuril supplementation (200 ppm). Further support has been published describing supplementation with Colombian oregano EO (100 ppm) resulting in improved bodyweight and reduced FCR when compared with challenged as well as unchallenged controls (Betancourt et al. 2019).

In other studies *Psidium guajava* EO (1 and 5 mg/kg) was shown to exert a beneficial effect against chicken coccidiosis supporting increased antioxidant enzyme activities (i.e. glutathione peroxidase), although the OPG was lower in chickens receiving diclazuril (1 mg/kg) and challenged with mixed *Eimeria* spp. compared to chickens that received EO at any of the doses. No differences were observed in FCR between treated groups (Langerudi et al. 2022). Chang et al. (2021) showed that continuous supplementation with garlic EO could significantly reduce clinical signs of coccidiosis, including caecal lesions and shed oocysts, while also increasing BWG and exhibiting an anticoccidial index similar to chickens supplemented with diclazuril (0.06 mL/L). In this study, garlic EO (0.06 mL/L) also improved immune function in chickens, showing an increase on the biochemical index for IgM, IgG and IgA. Feed supplementation with encapsulated thymol and carvacrol EO (60–120 mg/kg) improved production parameters such as BWG and FCR after challenge with *Eimeria* spp. (25× Coccivac® vaccine recommended dose) at a level comparable to chickens treated with salinomycin (60 mg/kg) (Lee et al. 2020), also reducing volatile fatty acids and increasing catalase activity.

Blends of EOs have also been tested for efficacy in control of coccidiosis. A feed additive composed of eugenol, thymol and piperine blends of EOs (CRINA Poultry DSM [Dutch State Mines]: 15, 30 or 50 mg/kg) and vitamin D3 (50, 69 or 138 ug/kg; respectively) improved BWG during the broiler growing phase in comparison to controls. Groups with higher doses of CRINA and vitamin D3 showed increased BWG and reduced FCR. All treatments reduced intestinal pathology caused by *Eimeria* challenge (300–500 purified sporulated oocysts of *E. tenella* and *E. maxima*) (Upadhaya et al. 2019). A blend of eucalyptus and peppermint EOs (Chemist Werner Krull) improved weight gain after challenge with eight *Eimeria* spp. (a total of 1.76×10^5 oocysts per bird of equal number of sporulated oocysts of *E. acervulina*, *E. brunetti*, *E. hagani*, *E. maxima*, *E. mivati*, *E. necatrix*, *E. praecox* and *E. tenella*) and provided a growth promotion effect in unchallenged groups; FCR was also improved when challenged chickens were supplemented with EO, preventing mortality and reducing intestinal lesions and output oocysts (Barbour et al. 2015). A mixture of essential oils (Essential – Oligo Basics Agroind – 1.5–2 kg/ton) reduced intestinal lesions after challenge with *Eimeria* spp. Feed efficiency and growth rate were also improved before challenge and during the initial seven days post challenge. Supplementation increased the apparent metabolisable energy (AME) as well as villi height (Murakami et al. 2014). A blend of EOs containing carvacrol, cineole, camphor and thymol (Herba) did not improve performance in a study where other treatments including salinomycin, a multienzyme, probiotics and prebiotics all showed good results (Bozkurt et al. 2014). A blend of three EOs (oregano, laurel leaf and lavender; 50 mg/kg) significantly improved performance of chickens challenged with a mixture of *Eimeria* species (35 × 10^4 sporulated oocysts from field isolates of *E. acervulina*, *E. maxima*, *E. tenella*, *E. mitis*, *E. brunetti* and *E. praecox*), as well as reducing oocyst outputs; however, this was lower than in groups supplemented with monensin (100 mg/kg) (Bozkurt et al. 2012).

6 Probiotics

Probiotics can enhance intestinal health by favouring beneficial bacteria, with the host benefitting from their immunostimulatory and antioxidant properties. However, specific mechanisms of action of probiotics remain unknown. There are several publications describing the use of probiotics (live microorganisms) to reduce the impact of chicken coccidiosis.

Giannenas et al. (2012) evaluated different probiotics individually (*Enterococcus faecium*, *Bifidobacterium animalis*, *Lactobacillus reuteri*, *Bacillus subtilis*) and a multi-species probiotic mix PoultryStar® (BIOMIN GmbH, Austria) for their ability to reduce the effects of coccidiosis. All the groups exhibited better (lower) FCR in comparison to the un-supplemented infected control. At day 21, seven days after challenge with *E. tenella* (4000 oocysts by oral gavage), all supplemented groups showed a significant increase in body weight compared to the un-supplemented infected group. However, at day 42, only the uninfected control, the anticoccidial-treated control (lasalocid) and PoultryStar® remained significantly heavier. In a similar experiment (Giannenas et al. 2014), *E. faecium*, *B. animalis* and *Lactobacillus salivarius* (at a 6:3:1 ratio) were added via drinking water or via feed. Seven days after challenge with *E. acervulina* (5000 oocysts), *E. maxima* (2000 oocysts) and *E. tenella* (2000 oocysts), the un-supplemented infected group showed the lowest BWG values, with all the probiotic-treated chickens showing similar levels to the anticoccidial-treated control (lasalocid), although the latter exhibited lower lesion scores and oocyst numbers than the probiotic-treated groups.

Some of these probiotics have been evaluated in further studies. Addition of *B. subtilis* to the chicken diet improved feed efficiency, reduced lesions and enhanced intestinal immunity and epithelial barrier integrity seven days after infection with *E. maxima* (1000 oocysts) at a level comparable or better than antimicrobials (virginiamycin or bacitracin methylene disalicylate) (Park et al.

2020). The probiotic *E. faecium* also supported higher bodyweights, and lower lesion scores and oocyst counts than un-supplemented infected controls (El-Sawah et al. 2020) when chickens were challenged with 25,000 mixed species oocysts including *E. tenella* (80%), *E. necatrix* (10%), *E. maxima* (5%) and *E. acervulina* (5%). A synergetic effect was found when combined with the anticoccidial diclazuril; nevertheless, the probiotic did not ease clinical signs when used as a therapeutic after coccidiosis had been detected.

Probiotic mixtures are becoming popular in the poultry industry. Protexin, a probiotic preparation containing *Bifidobacterium bifidum*, *E. faecium*, *Lactobacillus acidophilus*, *Lactobacillus delbrueckii*, *Lactobacillus plantarum*, *Lactobacillus rhamnosus* and *Streptococcus thermophiles* (Hilton Pharmaceutical (Pvt) Ltd., Pakistan) supported increased bodyweight gain and lower FCR, and reduced oocyst shedding, caecal lesions and mortality after challenge with 25,000 *E. tenella* oocysts, alone or in combination with the anticoccidial diclazuril, compared with the un-supplemented control (Memon et al. 2022). Other probiotic combinations such as Antox® (*Saccharomyces cerevisiae*) and En-florax® (*E. faecium*, *Lactobacillus casei*, *L. plantarum* and *Pediococcus acidilactici*) were also evaluated. After a challenge with 2000 *E. tenella* oocysts, both probiotic combinations supported similar feed intake, FCR and BWG than chickens supplemented with amprolium. Lesion scores and oocyst outputs were lower than for un-supplemented controls, but not as good as the amprolium-supplemented group (Ogwiji et al. 2022).

7 In Vitro Studies

As discussed, there has been a significant recent increase in research intended to discover new extracts, compounds or products effective against coccidiosis in chickens, but commercialisation requires extensive assessments of testing and validation. Safety and efficacy studies for new anticoccidial drugs are commonly designed using large numbers of chickens to determine parasitological and pathological parameters (e.g. parasite replication, enteric lesion score, serum carotenoids as a measure of gut leakage), with complementary studies to evaluate commercial parameters such as BWG and FCR. The World Association for Advancement of Veterinary Parasitology published guidance for evaluating efficacy of anticoccidial drugs in chickens and other poultry nearly 20 years ago (Holdsworth, Conway et al. 2004), and the U.S. Food and Drug Administration periodically releases revised guidelines for industry anticoccidial investigations with standards and specific protocols (https://www.fda.gov/media/81837/download), although meeting these are not a legal requirement. In these guidelines there are three different levels of studies in experimental animals:

- **Battery studies (cages):** to evaluate the efficacy of new compounds under investigation. Effects are assessed under controlled laboratory conditions, usually in White Leghorn (layer) chickens.

- **Floor-pen studies:** simulated conditions representative of genetic stock used in contemporary poultry production (e.g. Cobb500™, broilers).
- **Multi-location commercial-scale field studies:** to confirm the feeding level selected for the compound is effective under commercial production conditions. This type of study is conducted under industrial conditions using commercial-scale facilities with a variety of production systems, conditions and chicken breeds.

Such studies are expensive, time consuming and ethically challenging. The availability of in vitro models to pre-screen novel anticoccidial compounds would allow the evaluation of larger numbers of compounds in a more time- and cost-effective way. Several studies using in vitro evaluation of potential compounds for effects in the parasite itself have been published. However, many are semi-quantitative, time consuming or subjective to the person performing the experiment.

7.1 Evaluation of Oocyst Destruction

Rahmani et al. (2021)) evaluated the destructive effects of *Pistacia lentiscus* vegetable oil or diclazuril on a mixture of oocysts of *Eimeria* species isolated from field samples. Oocysts (15,000/well in triplicate) were incubated in 96 well plates at room temperature (RT) for 24 hours with different concentrations of the oil. After incubation, the percentage of oocysts destroyed was estimated visually and by measuring absorbance at 273 nm. A dose-dependent effect was observed (Lethal Dose $[LD]_{50} = 5.86$ and 3.79 mg/mL for each substance). Similarly, secreted compounds of the yeast *Meyerozyma guilliermondii* were assessed for damage caused in oocyst walls. Here, batches of 100,000 *E. tenella* oocysts were exposed to yeast extracts and incubated at 30°C with mixing at 200 rpm for 24 and 48 hours, after which damage was observed microscopically. Integrity was retained in 61% of oocysts at 24 hours and decreased 1.3-fold by 48 hours (Dantan-Gonzalez et al. 2015). In another study, sporulated oocysts (4×10^5 with approximately 32.05% *E. acervulina*, 26.92% *E. tenella*, 15.35% *E. mitis* and 14.10% and 11.53% *E. maxima*) from a field isolate were incubated with olive pulp at ambient temperature for 24 hours. Destruction (LC_{50}) was estimated and absorbance measured at 273 nm (Debbou-Iouknane et al. 2019). Remmal et al. (2011)) evaluated oocyst destruction caused by EOs (purchased from Seema International) by microscopic counting at 273 nm absorbance. Oocysts (3×10^5) were incubated with different EO concentrations for 24 hours. Lethal concentration $(LC)_{50}$ was less than 1 mg/mL for EOs from artemisia, tea tree, thyme and clove.

7.2 Evaluation of Effects in Oocyst Sporulation

Desalegn and Ahmed (2020) evaluated the effects of *Aloe debrana* and *Aloe pulcherrima* leaf gel infusions on oocyst sporulation in a mixed species *Eimeria* isolate. Batches of 1500 oocysts/mL were incubated with 10, 15, 25 or 30% w/v gel infusions in 1% potassium dichromate. The highest doses inhibited sporulation by >79% and >69% for each of the substances, respectively. Gadelhaq et al. (2018) evaluated the effects of two herbal extracts of garlic and *Moringa oleifera* on sporulation in comparison with diclazuril. Non-characterised oocysts from a field isolate propagated in chickens were incubated with either extract for 48 hours at 25–19°C, but neither showed any effect on oocyst sporulation. Habibi et al. (2016) evaluated the influence of extracts from *Biarum bovei, Nectaroscordum tripedale, Dorema aucheri, Cichorium intybu* or *Prangos ferulaceae* leaves, with sporulation inhibited from 8.51 to 29.69%.

7.3 Evaluation of Effects on Free Sporozoites

Daneshmand et al. (2021) evaluated effects of sugarcane extract (Polygain) on free sporozoites after incubation at 37°C for 14 hours, finding a significant decrease in sporozoite numbers of different species (*E. acervulina*, *E. maxima*, *E. brunetti*, *E. tenella*, *E. necatrix* and *E. mitis*) after the treatment.

7.4 Evaluation of Effects on Sporozoite Invasion

In 2011, Khalafalla et al. (2011) evaluated the effects of curcumin (diferuloymethane) on sporozoite morphology, viability (using trypan blue) and ability to invade in Madin-Darby bovine kidney (MDBK) cells. Exposure resulted in deformation, swelling and loss of viability at concentrations in excess of >25 μM. Infectivity of fluorescent-labelled sporozoites with 10 μM 5(6)-carboxyfluorescein diacetate *N*-succinimidyl ester measured by flow cytometry showed reductions of 41.6% and 72.8% for 100 and 200 μM concentrations, respectively. Burt et al. (2013) evaluated the effects of 4 phytochemicals (betaine, carvacrol, curcumin and *Echinacea purpurea* extract) on sporozoites added to treated MDBK cell monolayers and incubated for 2, 4 or 20 hours at 37°C, then stained with haematoxylin-eosin to calculate the number of invaded cells. Carvacrol, curcumin and *E. purpurea* extract only showed a reduced invasion percentage after 2 hours of incubation, but not after 4 and 20 hours. Ten EOs extracted from Thai indigenous plants were evaluated for oocysticidal properties, assessing levels of sporulation and oocyst degeneration. Inhibition of invasion in MDBK cells was also assessed after 1 hour pre-incubation in each EO followed by cells fixing 24 hours after infection with Bouin solution and staining with Periodic Acid Schiff. *Boesenbergia pandurata* and *Ocimum basilicum* EOs reduced sporulation as well as cell invasion by 70% (Jitviriyanon et al. 2016). Effects of probiotics have been also evaluated in vitro by comparing efficacy of sporozoite invasion. Sporozoites of *E. tenella* were labelled with 5(6)-carboxyfluorescein diacetate *N*-succinimidyl ester to evaluate invasion levels by microscopy. Invasion levels in MDBK cells were reduced up to 80% after treatment with different bacterial strains (*Bifidobacterium animalis*,

Enterococcus faecium, *Lactobacillus reuteri*, *L. salivarius* and *Bacillus subtilis*) (Hessenberger et al. 2016). Teng et al. (2020) evaluated the effects of nitro compounds on sporozoite invasion. Infected cells were incubated at 40°C for 72 hours, then monolayers were fixed with iced methanol and labelled with a mouse anti-sporozoite antibody. Quantification was done by microscopy. Balta et al. (2021) evaluated the effects of an organic acid mixture on inhibition of invasion of MDBK and chicken lung epithelial cells (CLEC213). Invasion levels were estimated by DAPI (4′,6-diamidino-2-phenylindole) staining and microscopy, finding inhibition of 50–80%.

7.5 Quantitative Methods

In 2014, Jenkins et al. (2014) used quantitative polymerase chain reaction (qPCR) and semi-qPCR to quantify inhibition of *E. tenella* sporozoite invasion in MDBK cell monolayers using classical anticoccidial compounds (salinomycin and monensin). Thabet et al. (2015) used similar protocols to evaluate the efficacy of different classical anticoccidials (monensin, maduramicin, slainomycin and lasalocid). This same technique was used to evaluate the effects of allicin (Alnassan et al. 2015) with further refinement in 2017 by Thabet et al. (2017). Tests of natural compounds using this approach illustrated a range of levels of inhibition for nutmeg oil (35.5–49.5%), cinnamon oil (none) and glabridin (14.1–81.7%) (Thabet et al. 2022). Felici et al. (2020) used qPCR to define significant levels of inhibition of sporozoite invasion (>20%) when sporozoites were exposed to thymol and carvacrol blends and more in particular when these were combined with saponins. Marugan-Hernandez et al. (2020) have compared qPCR and florescent methods to evaluate inhibition of sporozoite invasion and also endogenous development in MDBK. This study provided detailed microscopical studies to evaluate the quality of the schizonts and merozoites generated by in vitro methods as well as protein expression profiles along the development.

8 Conclusions

A range of prophylactic and therapeutic measures are currently available to control coccidiosis in chickens, but gaps in our knowledge of each product are emphasised by the host of challenges associated with each approach. While it is unlikely that a single all-purpose solution will suffice to control coccidiosis across all poultry sectors, alternatives that can be upscaled for mass application and aimed at working within current and possible future legislative restrictions and consumer concerns are required to support future food security and agro-economy. The drive towards natural prophylactics and remedies for use in livestock and poultry production is not limited to coccidiosis, gaining considerable traction across much of the farming sector to address ‘drug-free’ requirements from consumers. Many naturally existing therapies can also boast the benefits of affordable manufacture, ease of administration and scalability while providing efficacious alternatives to drugs, facilitating the reduction of drug-resistant parasite populations.

References

Abbas Z, Khan SM, Alam J, Khan SW, Abbasi AM (2017) Medicinal plants used by inhabitants of the Shigar Valley Baltistan region of Karakorum range-Pakistan. J Ethnobiol Ethnomed 13(1). https://doi.org/10.1186/s13002-017-0172-9

Abdelhady AY, El-Safty SA, Hashim M, Ibrahim MA, Mohammed FF, Elbaz AM, Abdel-Moneim A-ME (2021) Comparative evaluation of single or combined anticoccidials on performance, antioxidant status, immune response, and intestinal architecture of broiler chickens challenged with mixed Eimeria species. Poult Sci 100(6):101162. https://doi.org/10.1016/j.psj.2021.101162

Abubakar A, Haque M (2020) Preparation of medicinal plants: basic extraction and fractionation procedures for experimental purposes. J Pharm Bioallied Sci 12(1):1

Ademola IO, Ojo PO, Odeniran PO (2019) Pleurotus ostreatus extract inhibits Eimeria species development in naturally infected broiler chickens. Trop Anim Health Prod 51(1):109–117. https://doi.org/10.1007/s11250-018-1665-9

Ahad S, Tanveer S, Malik TA, Nawchoo IA (2018) Anticoccidial activity of fruit peel of Punica granatum L. Microb Pathog 116:78–83. https://doi.org/10.1016/j.micpath.2018.01.015

Akhter M, Aziz F, Hasan M, Islam R, Parvez M, Sarkar S, Meher M (2021) Comparative effect of papaya (Carica papaya) leaves' extract and Toltrazuril on growth performance, hematological parameter, and protozoal load in Sonali chickens infected by mixed Eimeria spp. J Adv Vet Anim Res 8(1):1. https://doi.org/10.5455/javar.2021.h490

Akyildiz S, Muzaffer D (2016) Application of plant extracts as feed additives in poultry nutrition. Scientific Papers (LIX) 2393–2260

Allen P, Lydon J, Danforth H (1997) Effects of components of Artemisia annua on coccidia infections in chickens. Poult Sci 76(8):1156–1163

Alnassan AA, Thabet A, Daugschies A, Bangoura B (2015) In vitro efficacy of allicin on chicken *Eimeria tenella* sporozoites. Parasitol Res 114(10):3913–3915

Alostad AH, Steinke DT, Schafheutle EI (2018) International comparison of five herbal medicine registration systems to inform regulation development: United Kingdom, Germany, United States of America, United Arab Emirates and Kingdom of Bahrain. Pharm Med 32(1):39–49

Arakawa A, Baba E, Fukata T (1981) *Eimeria tenella* infection enhances Salmonella typhimurium infection in chickens. Poult Sci 60(10):2203–2209

Attree E, Sanchez-Arsuaga G, Jones M, Xia D, Marugan-Hernandez V, Blake D, Tomley F (2021) Controlling the causative agents of coccidiosis in domestic chickens; an eye on the past and considerations for the future. CABI Agric Biosci 2:37. https://doi.org/10.1186/s43170-021-00056-5

Awais MM, Akhtar M, Anwar MI, Khaliq K (2018) Evaluation of Saccharum officinarum L. bagasse-derived polysaccharides as native immunomodulatory and anticoccidial agents in broilers. Vet Parasitol 249:74–81. https://doi.org/10.1016/j.vetpar.2017.11.012

Aziz ZAA, Ahmad A, Setapar SHM, Karakucuk A, Azim MM, Lokhat D, Rafatullah M, Ganash M, Kamal MA, Ashraf GM (2018) Essential oils: extraction techniques, pharmaceutical and therapeutic potential - A review. Curr Drug Metab 19(13):1100–1110

Baba E, Fukata T, Arakawa A (1982) Establishment and persistence of Salmonella typhimurium infection stimulated by *Eimeria tenella* in chickens. Res Vet Sci 33(1):95–98. https://doi.org/10.1016/S0034-5288(18)32366-X

Bakker EP (1979) Ionophore antibiotics. In: Mechanism of action of antibacterial agents. Springer, Berlin, pp 67–97. https://doi.org/10.1007/978-3-642-46403-4_5

Balta I, Marcu A, Linton M, Kelly C, Stef L, Pet I, Ward P, Pircalabioru GG, Chifiriuc C, Gundogdu O, Callaway T, Corcionivoschi N (2021) The in vitro and in vivo anti-virulent effect of organic acid mixtures against *Eimeria tenella* and Eimeria bovis. Sci Rep 11(1):16202

Barbour EK, Bragg RR, Karrouf G, Iyer A, Azhar E, Harakeh S, Kumosani T (2015) Control of eight predominant Eimeria spp. involved in economic coccidiosis of broiler chicken by a chemically characterized essential oil. J Appl Microbiol 118(3):583–591

Baron LF, da Fonseca FN, Maciag SS, Bellaver FAV, Ibeli AMG, Mores MAZ, de Almeida GF, Guterres SS, Bastos APA, Paese K (2022) Toltrazuril-loaded polymeric Nanocapsules as a promising approach for the preventive control of coccidiosis in poultry. Pharmaceutics 14(2): 392. https://doi.org/10.3390/pharmaceutics14020392

Bartolome AP, Villaseñor IM, Yang W-C (2013) Bidens pilosa L. (Asteraceae): botanical properties, traditional uses, Phytochemistry, and pharmacology. Evid Based Complement Alternat Med 2013:1–51

Benarbia MEA, Gaignon P, Manoli C, Chicoteau P (2022) Saponin-Rich Plant Premixture supplementation is as efficient as Ionophore Monensin supplementation under experimental Eimeria spp challenge in broiler chicken. Front Vet Sci 9:946576

Betancourt L, Hume M, Rodriguez F, Nisbet D, Sohail MU, Afanador-Tellez G (2019) Effects of Colombian oregano essential oil (Lippia origanoides Kunth) and Eimeria species on broiler production and cecal microbiota. Poult Sci 98(10):4777–4786

Blake DP, Tomley FM (2014) Securing poultry production from the ever-present Eimeria challenge. Trends Parasitol 30(1):12–19

Blake DP, Hesketh P, Archer A, Carroll F, Shirley MW, Smith AL (2005) The influence of immunizing dose size and schedule on immunity to subsequent challenge with antigenically distinct strains of Eimeria maxima. Avian Pathol 34(6):489–494

Blake DP, Hesketh P, Archer A, Shirley MW, Smith AL (2006) Eimeria maxima: the influence of host genotype on parasite reproduction as revealed by quantitative real-time PCR. Int J Parasitol 36(1):97–105

Blake DP, Knox J, Dehaeck B, Huntington B, Rathinam T, Ravipati V, Ayoade S, Gilbert W, Adebambo A, Jatau I, Raman M, Parker D, Rushton J, Tomley F (2020) Re-calculating the cost of coccidiosis in chickens. Vet Res 51:115

Blake DP, Marugan-Hernandez V, Tomley FM (2021a) Spotlight on avian pathology: Eimeria and the disease coccidiosis. Avian Pathol 50(3):209–213. https://doi.org/10.1080/03079457.2021.1912288

Blake DP, Vrba V, Xia I, Jatau S, Spiro M, Nolan GU, Tomley F (2021b) Genetic and biological characterisation of three cryptic Eimeria operational taxonomic units that infect chickens (Gallus gallus domesticus). Int J Parasitol 51:621

Bo R, Dai X, Huang J, Wei S, Liu M, Li J (2020) Evaluation of optimum conditions for decoquinate nanoliposomes and their anticoccidial efficacy against diclazuril-resistant *Eimeria tenella* infections in broilers. Vet Parasitol 283:109186. https://doi.org/10.1016/j.vetpar.2020.109186

Bortoluzzi C, Paras KL, Applegate TJ, Verocai GG (2018) Comparison between McMaster and mini-FLOTAC methods for the enumeration of Eimeria maxima oocysts in poultry excreta. Vet Parasitol 254:21–25

Boulton K, Nolan MJ, Wu Z, Psifidi A, Riggio V, Harman K, Bishop SC, Kaiser P, Abrahamsen MS, Hawken R, Watson KA, Tomley FM, Blake DP, Hume DA (2018) Phenotypic and genetic variation in the response of chickens to *Eimeria tenella* induced coccidiosis. Genet Sel Evol 50(1):63

Bozkurt M, Selek N, Kucukyilmaz K, Eren H, Guven E, Catli AU, Cinar M (2012) Effects of dietary supplementation with a herbal extract on the performance of broilers infected with a mixture of Eimeria species. Br Poult Sci 53(3):325–332

Bozkurt M, Aysul N, Kucukyilmaz K, Aypak S, Ege G, Catli AU, Aksit H, Coven F, Seyrek K, Cinar M (2014) Efficacy of in-feed preparations of an anticoccidial, multienzyme, prebiotic, probiotic, and herbal essential oil mixture in healthy and Eimeria spp.-infected broilers. Poult Sci 93(2):389–399

Bozkurt M, Ege G, Aysul N, Aksit H, Tuzun AE, Kucukyilmaz K, Borum AE, Uygun M, Aksit D, Aypak S, Simsek E, Seyrek K, Kocer B, Bintas E, Orojpour A (2016) Effect of anticoccidial monensin with oregano essential oil on broilers experimentally challenged with mixed Eimeria spp. Poult Sci 95(8):1858–1868

Braunius WW (1982) Coccidiosis in broilers: the effective use of anticoccidial drugs. Worlds Poult Sci J 38(3):176–185. https://doi.org/10.1079/WPS19820013

Bremner A, Kim S, Morris KM, Nolan MJ, Borowska D, Wu Z, Tomley F, Blake DP, Hawken R, Kaiser P, Vervelde L (2021) Kinetics of the cellular and transcriptomic response to Eimeria maxima in relatively resistant and susceptible chicken lines. Front Immunol 12:653085

Burt SA, Tersteeg-Zijderveld MH, Jongerius-Gortemaker BG, Vervelde L, Vernooij JC (2013) In vitro inhibition of *Eimeria tenella* invasion of epithelial cells by phytochemicals. Vet Parasitol 191(3–4):374–378

Chang CL-T, Yang C-Y, Muthamilselvan T, Yang W-C (2016) Field trial of medicinal plant, Bidens pilosa, against eimeriosis in broilers. Sci Rep 6(1):24692. https://doi.org/10.1038/srep24692

Chang LY, Di KQ, Xu J, Chen YF, Xi JZ, Wang DH, Hao EY, Xu LJ, Chen H, Zhou RY (2021) Effect of natural garlic essential oil on chickens with artificially infected *Eimeria tenella*. Vet Parasitol 300:109614

Chapman HD., et al. (2005). Guidelines for evaluating the efficacy and safety of live anticoccidial vaccines, and obtaining approval for their use in chickens and turkeys. Avian Pathol 34(4):279–290. https://doi.org/10.1080/03079450500178378

Chapman HD (1989) Sensitivity of field isolates of *Eimeria tenella* to anticoccidial drugs in the chicken. Res Vet Sci 47(1):125–128

Chapman HD (1997) Biochemical, genetic and applied aspects of drug resistance in Eimeria parasites of the fowl. Avian Pathol 26:221–244

Chapman HD (2001) Use of anticoccidial drugs in broiler chickens in the USA: analysis for the years 1995 to 1999. Poult Sci 80(5):572–580. https://doi.org/10.1093/PS/80.5.572

Chapman HD, Jeffers TK (2015) Restoration of sensitivity to salinomycin in Eimeria following 5 flocks of broiler chickens reared in floor-pens using drug programs and vaccination to control coccidiosis. Poult Sci 94(5):943–946

Chávez-Pesqueira M, Núñez-Farfán J (2017) Domestication and Genetics of Papaya: a Review. Front Ecol Evol 5. https://doi.org/10.3389/fevo.2017.00155

Cheng Y-H, Horng Y-B, Chen W-J, Hua K-F, Dybus A, Yu Y-H (2021) Effect of fermented products produced by bacillus licheniformis on the growth performance and Cecal microbial Community of Broilers under Coccidial challenge. Animals 11(5):1245. https://doi.org/10.3390/ani11051245

Clark EL, Macdonald SE, Thenmozhi V, Kundu K, Garg R, Kumar S, Ayoade S, Fornace KM, Jatau ID, Moftah A, Nolan MJ, Sudhakar NR, Adebambo AO, Lawal IA, Alvarez Zapata R, Awuni JA, Chapman HD, Karimuribo E, Mugasa CM, Namangala B, Rushton J, Suo X, Thangaraj K, Srinivasa Rao AS, Tewari AK, Banerjee PS, Dhinakar Raj G, Raman M, Tomley FM, Blake DP (2016) Cryptic Eimeria genotypes are common across the southern but not northern hemisphere. Int J Parasitol 46:537–544

Daneshmand A, Young P, Campbell B, Kheravii SK, Sharma NK, Afshari R, Dias DA, Flavel M, Kitchen B, Wu SB (2021) In vitro inhibitory activities of sugarcane extract on avian Eimeria sporozoites. Int J Parasitol Drugs Drug Resist 17:1–4

Dantan-Gonzalez E, Quiroz-Castaneda RE, Cobaxin-Cardenas M, Valle-Hernandez J, Gama-Martinez Y, Tinoco-Valencia JR, Serrano-Carreon L, Ortiz-Hernandez L (2015) Impact of Meyerozyma guilliermondii isolated from chickens against Eimeria sp. protozoan, an in vitro analysis. BMC Vet Res 11:278

De Jong IC, Gunnink H, Van Harn J (2014) Wet litter not only induces footpad dermatitis but also reduces overall welfare, technical performance, and carcass yield in broiler chickens. J Appl Poult Res 23(1):51–58. https://doi.org/10.3382/JAPR.2013-00803

Debbou-Iouknane N, Nerin C, Amrane M, Ghemghar M, Madani K, Ayad A (2019) In vitro Anticoccidial activity of olive pulp (Olea europaea L. var. Chemlal) extract against Eimeria oocysts in broiler chickens. Acta Parasitol 64(4):887–897

Desalegn AY, Ahmed MR (2020) Anticoccidial activity of aloe debrana and Aloe pulcherrima leaf gel against Eimeria oocysts. J Parasitol Res 2020:8524973

Dkhil MA (2013) Anti-coccidial, anthelmintic and antioxidant activities of pomegranate (Punica granatum) peel extract. Parasitol Res 112(7):2639–2646

Dorne JLCM, Fernández-Cruz ML, Bertelsen U, Renshaw DW, Peltonen K, Anadon A, Feil A, Sanders P, Wester P, Fink-Gremmels J (2013) Risk assessment of coccidostatics during feed cross-contamination: animal and human health aspects. Toxicol Appl Pharmacol 270(3): 196–208. https://doi.org/10.1016/J.TAAP.2010.12.014

Dubey JP (2019) Coccidiosis in livestock, poultry, companion animals, and humans. CRC Press, Taylor & Francis Group, Boca Raton, USA

Ejikeme CM, Ezeonu CS, Eboatu AN (2014) Determination of physical and phytochemical constituents of some tropical timbers indigenous to Niger Delta area of Nigeria. European Scientific Journal, ESJ, p 10

El-Sawah AA, Aboelhadid SM, El-Nahass EN, Helal HE, Korany AM, El-Ashram S (2020) Efficacy of probiotic enterococcus faecium in combination with diclazuril against coccidiosis in experimentally infected broilers. J Appl Microbiol 129(4):1020–1028

El-Shall NA, Abd El-Hack ME, Albaqami NM, Khafaga AF, Taha AE, Swelum AA, El-Saadony MT, Salem HM, El-Tahan AM, AbuQamar SF (2022) Phytochemical control of poultry coccidiosis: a review. Poult Sci 101(1):101542

El-Shazly KA, El-Latif AA, Abdo W, El-Morsey A, El-Aziz MIA, El-Mogazy H (2020) The anticoccidial activity of the fluoroquinolone lomefloxacin against experimental *Eimeria tenella* infection in broiler chickens. Parasitol Res 119(6):1955–1968. https://doi.org/10.1007/s00436-020-06692-6

EzeC P, ObiC F, IdikaI K, NwosuC O (2022) Evaluation of the anti-coccidial effect of aqueous Sacoglottis gabonensis (Magnoliophyta, Humiriaceae) stem bark extract in broilers experimentally infected with mixed Eimeria species. Span J Agric Res 20(1):e0503–e0503. https://doi.org/10.5424/sjar/2022201-18620

Felici M, Tugnoli B, Ghiselli F, Massi P, Tosi G, Fiorentini L, Piva A, Grilli E (2020) In vitro anticoccidial activity of thymol, carvacrol, and saponins. Poult Sci 99(11):5350–5355

Fu Y, Zhou J, Zhang L, Fei C, Wang X, Wang M, Wang C, Zhang K, Xue F, Liu Y (2021) Pharmacokinetics and anticoccidial activity of ethanamizuril in broiler chickens. Vet Parasitol 289:109318. https://doi.org/10.1016/j.vetpar.2020.109318

Gadelhaq SM, Arafa WM, Abolhadid SM (2018) In vitro activity of natural and chemical products on sporulation of Eimeria species oocysts of chickens. Vet Parasitol 251:12–16

Galli GM, Da Silva AS, Biazus AH, Reis JH, Boiago MM, Topazio JP, Migliorini MJ, Guarda NS, Moresco RN, Ourique AF, Santos CG, Lopes LS, Baldissera MD, Stefani LM (2018) Feed addition of curcumin to laying hens showed anticoccidial effect and improved egg quality and animal health. Res Vet Sci 118(10)101–106. https://doi.org/10.1016/j.rvsc.2018.01.022

Galli GM, Petrolli TG, Aniecevski E, Santo AD, Leite F, Griss LG, Dazuk V, Boiago MM, dos Santos HV, Simões CADP (2021) Phytogenic blend protective effects against microbes but affects health and production in broilers. Microb Pathog 152:104590. https://doi.org/10.1016/j.micpath.2020.104590

Giannenas I, Florou-Paneri P, Papazahariadou M, Christaki E, Botsoglou NA, Spais AB (2003) Effect of dietary supplementation with oregano essential oil on performance of broilers after experimental infection with *Eimeria tenella*. Arch Tierernahr 57(2):99–106

Giannenas I, Papadopoulos E, Tsalie E, Triantafillou E, Henikl S, Teichmann K, Tontis D (2012) Assessment of dietary supplementation with probiotics on performance, intestinal morphology and microflora of chickens infected with *Eimeria tenella*. Vet Parasitol 188(1–2):31–40

Giannenas I, Tsalie E, Triantafillou E, Hessenberger S, Teichmann K, Mohnl M, Tontis D (2014) Assessment of probiotics supplementation via feed or water on the growth performance, intestinal morphology and microflora of chickens after experimental infection with Eimeria acervulina, Eimeria maxima and *Eimeria tenella*. Avian Pathol 43(3):209–216

Gordillo Jaramillo FX, Kim DH, Lee SH, Kwon SK, Jha R, Lee KW (2021) Role of oregano and citrus species-based essential oil preparation for the control of coccidiosis in broiler chickens. J Anim Sci Biotechnol 12(1):47

Granstad S, Kristoffersen AB, Benestad SL, Sjurseth SK, David B, Sørensen L, Fjermedal A, Edvardsen DH, Sanson G, Løvland A (2020) Effect of feed additives as alternatives to in-feed antimicrobials on production performance and intestinal Clostridium perfringens counts in broiler chickens. Animals 10(2):240. https://doi.org/10.3390/ani10020240

Habibi H, Firouzi S, Nili H, Razavi M, Asadi SL, Daneshi S (2016) Anticoccidial effects of herbal extracts on *Eimeria tenella* infection in broiler chickens: in vitro and in vivo study. J Parasit Dis 40(2):401–407

Hamilton M, Ma X, McCrea BA, Carrisosa M, Macklin KS, Zhang C, Wang X, Hauck R (2019) Influence of Eimeria spp. infection on chicken Jejunal microbiota and the efficacy of two alternative products against the infection. Avian Dis 64(2):123. https://doi.org/10.1637/0005-2086-64.2.123

Hamzic E, Bed'Hom B, Juin H, Hawken R, Abrahamsen MS, Elsen JM, Servin B, Pinard-van der Laan MH, Demeure O (2015) Large-scale investigation of the parameters in response to Eimeria maxima challenge in broilers. J Anim Sci 93(4):1830–1840

Hayajneh FMF, Jalal M, Zakaria H, Abdelqader A, Abuajamieh M (2018) Anticoccidial effect of apple cider vinegar on broiler chicken: an organic treatment to measure anti-oxidant effect. Pol J Vet Sci 21(2):361–369

Hernandez-Patlan D, Solis-Cruz B, Pontin KP, Hernandez-Velasco X, Merino-Guzman R, Adhikari B, López-Arellano R, Kwon YM, Hargis BM, Arreguin-Nava MA (2019) Impact of a bacillus direct-fed microbial on growth performance, intestinal barrier integrity, necrotic enteritis lesions, and Ileal microbiota in broiler chickens using a laboratory challenge model. Front Vet Sci 6. https://doi.org/10.3389/fvets.2019.00108

Herrero-Encinas J, Menoyo D, Blanch M, Pastor JJ, Rochell SJ (2021) Response of broiler chickens fed diets supplemented with a bioactive olive pomace extract from Olea europaea to an experimental coccidial vaccine challenge. Poult Sci 100(2):575–584. https://doi.org/10.1016/j.psj.2020.11.027

Hessenberger S, Schatzmayr G, Teichmann K (2016) In vitro inhibition of *Eimeria tenella* sporozoite invasion into host cells by probiotics. Vet Parasitol 229:93–98

Holdsworth PA, Conway DP, McKenzie ME, Dayton AD, Chapman HD, Mathis GF, Skinner JT, Mundt HC, Williams RB, P. World Association for the Advancement of Veterinary (2004) World Association for the Advancement of Veterinary Parasitology (WAAVP) guidelines for evaluating the efficacy of anticoccidial drugs in chickens and turkeys. Vet Parasitol 121(3–4):189–212

Hong S, Moon M-N, Im E-K, Won J-S, Yoo J-H, Kim O (2018) Anti-coccidial activity of the ethanol extract of Tribulus terrestris fruits on *Eimeria tenella*. Lab Anim Res 34(1):44. https://doi.org/10.5625/lar.2018.34.1.44

Huang Z, Zhang X, Zhu Q, Cao F, Liu W, Shi P, Yang X (2022) Effect of Berberine on copper and zinc levels in chickens infected with *Eimeria Tenella*. Mol Biochem Parasitol 249:111478

Jarujareet W, Shigenoki Y, Taira K, Ooi HK (2018) *Eimeria tenella* oocyst excretion and riboflavin supplement in infected chicken. J Vet Med Sci 80(9):1392–1394. https://doi.org/10.1292/jvms.18-0219

Jeffers TK (1974) Eimeria acervulina and E. maxima: incidence and anticoccidial drug resistance of isolants in major broiler-producing areas. Avian Dis 18(3):331–342

Jeffers TK (1975) Attenuation of *Eimeria tenella* through selection for precociousness. J Parasitol 61(6):1083–1090

Jenkins MC, O'Brien CN, Fuller L, Mathis GF, Fetterer R (2014) A rapid method for determining salinomycin and monensin sensitivity in *Eimeria tenella*. Vet Parasitol 206(3–4):153–158

Jitviriyanon S, Phanthong P, Lomarat P, Bunyapraphatsara N, Porntrakulpipat S, Paraksa N (2016) In vitro study of anti-coccidial activity of essential oils from indigenous plants against *Eimeria tenella*. Vet Parasitol 228:96–102

Johnson J, Reid WM (1970) Anticoccidial drugs: lesion scoring techniques in battery and floor-pen experiments with chickens. Exp Parasitol 28(1):30–36

Joyner LP (1970) Coccidiosis: problems arising from the development of anticoccidial drug resistance. Exp Parasitol 28(1):122–128. https://doi.org/10.1016/0014-4894(70)90077-9

Kadykalo S, Roberts T, Thompson M, Wilson J, Lang M, Espeisse O (2018) The value of anticoccidials for sustainable global poultry production. Int J Antimicrob Agents 51(3): 304–310. https://doi.org/10.1016/J.IJANTIMICAG.2017.09.004

Kaleem QM, Akhtar M, Awais MM, Saleem M, Zafar M, Iqbal Z, Muhammad F, Anwar MI (2014) Studies on Emblica officinalis Derived tannins for their Immunostimulatory and protective activities against coccidiosis in industrial broiler chickens. Sci World J 2014:1–10

Khalafalla RE, Muller U, Shahiduzzaman M, Dyachenko V, Desouky AY, Alber G, Daugschies A (2011) Effects of curcumin (diferuloylmethane) on *Eimeria tenella* sporozoites in vitro. Parasitol Res 108(4):879–886

Khorrami P, Gholami-Ahangaran M, Moghtadaei-Khorasgani E (2022) The efficacy of pomegranate peel extract on Eimeria shedding and growth indices in experimental coccidiosis in broiler chickens. Veterinary Medicine & Sci 8(2):635–641. https://doi.org/10.1002/vms3.714

Kim WH, Chaudhari AA, Lillehoj HS (2019) Involvement of T cell immunity in avian coccidiosis. Front Immunol 10:2732

Lai A, Dong G, Song D, Yang T, Zhang X (2018) Responses to dietary levels of methionine in broilers medicated or vaccinated against coccidia under *Eimeria tenella*-challenged condition. BMC Vet Res 14(1):140. https://doi.org/10.1186/s12917-018-1470-8

Lanckriet A, Timbermont L, de Gussem M, Marien M, Vancraeynest D, Haesebrouck F, Ducatelle R, van Immerseel F (2010) The effect of commonly used anticoccidials and antibiotics in a subclinical necrotic enteritis model. Avian Pathol J W.V.P.A 39(1):63–68. https://doi.org/10.1080/03079450903505771

Langerudi MT, Youssefi MR, Tabari MA (2022) Ameliorative effect of Psidium guajava essential oil supplemented feed on chicken experimental coccidiosis. Trop Anim Health Prod 54(2):120

Lawal JR, Jajere SM, Ibrahim UI, Geidam YA, Gulani IA, Musa G, Ibekwe BU (2016) Prevalence of coccidiosis among village and exotic breed of chickens in Maiduguri. Nigeria Vet World 9(6):653–659

Lee JW, Kim DH, Kim YB, Jeong SB, Oh ST, Cho SY, Lee KW (2020) Dietary encapsulated essential oils improve production performance of coccidiosis-vaccine-challenged broiler chickens. Animals (Basel) 10(3):481

Lee Y, Lu M, Lillehoj HS (2022) Coccidiosis: recent Progress in host immunity and alternatives to antibiotic strategies. Vaccines (Basel) 10(2):215

Lima GA, Barbosa BFS, Araujo RGAC, Polidoro BR, Polycarpo GV, Zied DC, Biller JD, Ventura G, Modesto IM, Madeira AMBN (2021) Agaricus subrufescens and Pleurotus ostreatus mushrooms as alternative additives to antibiotics in diets for broilers challenged with Eimeria spp. Br Poult Sci 62(2):251–260. https://doi.org/10.1080/00071668.2020.1837344

Long PL, Joyner L, Millard B, Norton C (1976) A guide to laboratory techniques used in the study and diagnosis of avian coccidiosis. Folia Vet Lat 6(3):201–217

Macdonald SE, Nolan MJ, Harman K, Boulton K, Hume DA, Tomley FM, Stabler RA, Blake DP (2017) Effects of *Eimeria tenella* infection on chicken caecal microbiome diversity, exploring variation associated with severity of pathology. PLoS One 12(9):e0184890

Macdonald SE, Van Diemen PM, Martineau H, Stevens MP, Tomley FM, Stabler RA, Blake DP (2019) Impact of *eimeria tenella* coinfection on campylobacter jejuni colonization of the chicken. Infect Immun 87(2):e00772-18. https://doi.org/10.1128/IAI.00772-18/SUPPL_FILE/IAI.00772-18-S0001.PDF

Madlala T, Okpeku M, Adeleke MA (2021) Understanding the interactions between Eimeria infection and gut microbiota, towards the control of chicken coccidiosis: A review. Parasite 28:48. https://doi.org/10.1051/parasite/2021047

Marugan-Hernandez V, Jeremiah G, Aguiar-Martins K, Burrell A, Vaughan S, Xia D, Randle N, Tomley F (2020) The growth of *Eimeria tenella*: characterization and application of quantitative methods to assess Sporozoite invasion and endogenous development in cell culture. Front Cell Infect Microbiol 10:579833

Mayhew RL (1934) Studies on coccidiosis: VI. Effect of early attack on egg production. Poult Sci 13:148–154

McDougald L, Fuller A, McMurray B (1990) An outbreak of Eimeria necatrix coccidiosis in breeder pullets: analysis of immediate and possible Long-term effects on performance. Avian Dis 34:485–487

Memon FU, Yang Y, Lv F, Soliman AM, Chen Y, Sun J, Wang Y, Zhang G, Li Z, Xu B (2021) Effects of probiotic and Bidens pilosa on the performance and gut health of chicken during induced *Eimeria tenella* infection. J Appl Microbiol 131(1):425–434

Memon FU, Leghari IH, Rajput N, Gadahi JA, Sahito JZA, Yang Y, Baig MB, Laghari F, Memon HA, Si H (2022) Immunomodulatory and ameliorative effects of probiotic in combination with diclazuril on broilers under coccidia infection. J Appl Microbiol 132(4):3181–3188

Messaï A, Redouane-Salah S (2022) Dietary use of Artemisia Herba Alba Asso as A potential Coccidiostat against Cæcal coccidiosis: Haematological parameters variations. Trop Anim Health Prod 54:28. https://doi.org/10.1007/s11250-021-03038-x

Miliauskas G, Venskutonis PR, van Beek TA (2004) Screening of radical scavenging activity of some medicinal and aromatic plant extracts. Food Chem 85(2):231–237

Mohiti-Asli M, Ghanaatparast-Rashti M (2015) Dietary oregano essential oil alleviates experimentally induced coccidiosis in broilers. Prev Vet Med 120(2):195–202

Mooney D, Richards K, Danaher M, Grant J, Gill L, Mellander P-E, Coxon C (2020) An investigation of anticoccidial veterinary drugs as emerging organic contaminants in groundwater. Sci Total Environ Elsevier BV 746:141116. https://doi.org/10.1016/j.scitotenv.2020.141116

Morehouse NF, Baron RR (1970) Coccidiosis: evaluation of coccidiostats by mortality, weight gains, and fecal scores. Exp Parasitol 28:25–29

Muraina IA, Gotep JG, Tanko JT, Onyiche TE, Aladelokun OB, Dapuliga CC, Forcados GE, Ozele N, Samuel AL, Oladipo OO (2020) Anticoccidial effects of Khaya senegalensis aqueous stem bark extract on broiler chickens experimentally infected with Eimeria species. Trop Anim Health Prod 52(3):1249–1255. https://doi.org/10.1007/s11250-019-02125-4

Murakami AE, Eyng C, Torrent J (2014) Effects of functional oils on coccidiosis and apparent metabolizable energy in broiler chickens. Asian-Australas J Anim Sci 27(7):981–989

Muthamilselvan T, Kuo T-F, Wu Y-C, Yang W-C (2016) Herbal remedies for coccidiosis control: A review of plants, compounds, and Anticoccidial actions. Evid Based Complement Alternat Med 2016:1–19

Nabian S, Arabkhazaeli F, Seifouri P, Farahani A (2018) Morphometric analysis of the intestine in experimental coccidiosis in broilers treated with Anticoccidial drugs. Iran J Parasitol 13(3): 493–499

Nguyen BT, Flores RA, Cammayo PLT, Kim S, Kim WH, Min W (2021) Anticoccidial activity of Berberine against Eimeria-infected chickens. Korean J Parasitol 59(4):403–408

Noack S, Chapman HD, Selzer PM (2019) Anticoccidial drugs of the livestock industry. Parasitol Res 118(7):2009. https://doi.org/10.1007/S00436-019-06343-5

Nolan MJ, Tomley FM, Kaiser P, Blake DP (2015) Quantitative real-time PCR (qPCR) for *Eimeria tenella* replication - implications for experimental refinement and animal welfare. Parasitol Int 64(5):464–470

Ogwiji M, Jatau ID, Natala JA, Sani D, Habibu B, Andamin AD, Kyari S, Gasaliyu KA (2022) Comparative effect of dietary supplements on the performance and severity of experimental *Eimeria tenella* infection in broiler chickens. Trop Anim Health Prod 54(3):191

Park I, Lee Y, Goo D, Zimmerman NP, Smith AH, Rehberger T, Lillehoj HS (2020) The effects of dietary Bacillus subtilis supplementation, as an alternative to antibiotics, on growth performance, intestinal immunity, and epithelial barrier integrity in broiler chickens infected with Eimeria maxima. Poult Sci 99(2):725–733

Peek HW, Landman WJM (2011) Coccidiosis in poultry: Anticoccidial products, vaccines and other prevention strategies. Veterinary Quarterly Taylor & Francis Group 31(3):143–161. https://doi.org/10.1080/01652176.2011.605247

Pham VH, Abbas W, Huang J, He Q, Zhen W, Guo Y, Wang Z (2022) Effect of blending encapsulated essential oils and organic acids as an antibiotic growth promoter alternative on growth performance and intestinal health in broilers with necrotic enteritis. Poult Sci 101(1): 101563. https://doi.org/10.1016/j.psj.2021.101563

Pikkemaat MG, Rapallini MLBA, Stassen JHM, Alewijn M, Wullings BA (2022) Ionophore resistance and potential risk of ionophore driven co-selection of clinically relevant antimicrobial resistance in poultry. (Report/Wageningen Food Safety Research; No. WFSR 2022.005)

Pop LM, Varga E, Coroian M, Nedişan ME, Mircean V, Dumitrache MO, Farczádi L, Fülöp I, Croitoru MD, Fazakas M (2019) Efficacy of a commercial herbal formula in chicken experimental coccidiosis. Parasites Vectors 12(1):343. https://doi.org/10.1186/s13071-019-3595-4

Preetam VC, Srinivas G, Rao DS (2019) Effect of Cocciban herbal coccidiostats on hematobiochemical, fecal parameters and cecal histopathology of broiler chicken. Trop Anim Health Prod 51(6):1375–1381. https://doi.org/10.1007/s11250-019-01831-3

Qaid MM, Al-Mufarrej SI, Azzam MM, Al-Garadi MA, Alqhtani AH, Fazea H, E., Suliman, G. M., & Alhidary, I. A. (2021) Effect of Rumex nervosus leaf powder on the breast meat quality, carcass traits, and performance indices of *Eimeria tenella* oocyst-infected broiler chickens. Animals 11(6):1551

Qaid MM, Al-Garadi MA, Al-abdullatif AA, Alharthi AS, Al-Mufarrej SI (2022a) Efficacy of Rumex nervosus leaves or Cinnamomum verum bark as natural growth promoters on the growth performance, immune responsiveness, and serum biochemical profile of broiler chickens. Ital J Anim Sci 21(1):792–801. https://doi.org/10.1080/1828051x.2022.2065941

Qaid MM, Al-Mufarrej SI, Azzam MM, Al-Garadi MA, Alqhtani AH, Al-abdullatif AA, Hussein EO, Suliman GM (2022b) Dietary cinnamon bark affects growth performance, carcass characteristics, and breast meat quality in broiler infected with *Eimeria tenella* oocysts. Animals 12(2):166. https://doi.org/10.3390/ani12020166

Rahmani A, Ahmed Laloui H, Zaak H, Selmania A, Oufroukh K, Chareb N, Klikha A, Ghalmi F (2021) Effect of Pistacia lentiscus L. vegetable oil on growth performance and coccidiosis in broiler chickens: in vitro and in vivo assessment. Acta Parasitol 66(4):1151–1157

Rahmasari FV, Wibowo FA (2019) Effectiveness test of papaya leaves extract (Carica papaya L.) as Antihelmintics of Ascaridia galli worm. J Berk Ked 15(2):97

Rasheed MS, Tiwari UP, Oelschlager ML, Smith BN, Jespersen JC, Escobar J, Olmeda-Geniec N, Dilger RN (2020) Dietary supplementation with anti–IL-10 antibody during a severe Eimeria challenge in broiler chickens. Poult Sci 99(12):6493–6502. https://doi.org/10.1016/j.psj.2020.09.012

Remmal A, Achahbar S, Bouddine L, Chami N, Chami F (2011) In vitro destruction of Eimeria oocysts by essential oils. Vet Parasitol 182(2–4):121–126

Ringenier M, Caekebeke N, De Meyer F, Van Limbergen T, Eeckhaut V, Ducatelle R, Van Immerseel F, Dewulf J (2021) A field study on correlations between macroscopic gut health scoring, histological measurements and performance parameters in broilers. Avian Pathol 50(6): 500–506

Ruiz A, Matos L, Munoz MC, Hermosilla C, Molina JM, Andrada M, Rodriguez F, Perez D, Lopez A, Guedes A, Taubert A (2013) Isolation of an Eimeria ninakohlyakimovae field strain (Canary Islands) and analysis of its infection characteristics in goat kids. Res Vet Sci 94(2): 277–284

Sharma UNS, Fernando DD, Wijesundara KK, Manawadu A, Pathirana I, Rajapakse RPVJ (2021) Anticoccidial effects of Phyllanthus emblica (Indian gooseberry) extracts: potential for controlling avian coccidiosis. Vet Parasitol: Regional Studies and Reports 25:100592

Sharman PA, Smith NC, Wallach MG, Katrib M (2010) Chasing the golden egg: vaccination against poultry coccidiosis. Parasite Immunol 32(8):590–598

Shirley MW, Chapman HD (2005) Eight decades of research on Eimeria in poultry. In Proceedings of the IXth International Coccidiosis Conference, 23–29 September 2005, pp. 29–35

Shirley MW, Smith AL, Tomley FM (2005) The biology of avian Eimeria with an emphasis on their control by vaccination. Adv Parasitol 60:285–330

Sidiropoulou E, Skoufos I, Marugan-Hernandez V, Giannenas I, Bonos E, Aguiar-Martins K, Lazari D, Blake DP, Tzora A (2020) In vitro Anticoccidial study of oregano and garlic essential oils and effects on growth performance, fecal oocyst output, and intestinal microbiota in vivo. Front Vet Sci 7:420

Song X, Li Y, Chen S, Jia R, Huang Y, Zou Y, Li L, Zhao X, Yin Z (2020) Anticoccidial effect of herbal powder Shi Ying Zi in chickens infected with *Eimeria tenella*. Animals 10(9):1484. https://doi.org/10.3390/ani10091484

Soutter F, Werling D, Tomley FM, Blake DP (2020) Poultry coccidiosis: design and interpretation of vaccine studies. Front Vet Sci 7:101. Frontiers Media S.A. https://doi.org/10.3389/fvets.2020.00101

Soutter F, Werling D, Nolan M, Kuster T, Attree E, Marugan-Hernandez V, Kim S, Tomley FM, Blake DP (2022) A novel whole yeast-based subunit Oral vaccine against *Eimeria tenella* in chickens. Front Immunol 13:809711

Srinivasu B, Preetam VC, Srinivas G, Rao DS (2019) Effect of Cocciban herbal coccidiostats on hematobiochemical, fecal parameters and cecal histopathology of broiler chicken. Trop Anim Health Prod 51(6):1375–1381. https://doi.org/10.1007/s11250-019-01831-3

Stayer PA, Pote LM, Keirs RW (1995) A comparison of Eimeria oocysts isolated from litter and fecal samples from broiler houses at two farms with different management schemes during one growout. Poult Sci 74(1):26–32. https://doi.org/10.3382/ps.0740026

Talghari M, Behnamifar A, Rahimi S, Karimi Torshizi MA, Beckstead R, Grimes JL (2020) The effect of sodium bisulfate and coccidiostat on intestinal lesions and growth performance of Eimeria spp.–challenged broilers. Poult Sci 99(10):4769–4775. https://doi.org/10.1016/j.psj.2020.06.060

Teng PY, Fuller AL, Kim WK (2020) Evaluation of nitro compounds as feed additives in diets of Eimeria-challenged broilers in vitro and in vivo. Poult Sci 99(3):1320–1325

Thabet A, Alnassan AA, Daugschies A, Bangoura B (2015) Combination of cell culture and qPCR to assess the efficacy of different anticoccidials on *Eimeria tenella* sporozoites. Parasitol Res 114(6):2155–2163

Thabet A, Zhang R, Alnassan AA, Daugschies A, Bangoura B (2017) Anticoccidial efficacy testing: in vitro *Eimeria tenella* assays as replacement for animal experiments. Vet Parasitol 233:86–96

Thabet A, Alzuheir I, Alnassan AA, Daugschies A, Bangoura B (2022) In vitro activity of selected natural products against *Eimeria tenella* sporozoites using reproduction inhibition assay. Parasitol Res 121(1):335–344

Upadhaya SD, Cho SH, Chung TK, Kim IH (2019) Anti-coccidial effect of essential oil blends and vitamin D on broiler chickens vaccinated with purified mixture of coccidian oocyst from *Eimeria tenella* and Eimeria maxima. Poult Sci 98(7):2919–2926. https://doi.org/10.3382/ps/pez040

Variya BC, Bakrania AK, Patel SS (2016) Emblica officinalis (Amla): a review for its phytochemistry ethnomedicinal uses and medicinal potentials with respect to molecular mechanisms. Pharmacol Res 111:180–200. https://doi.org/10.1016/j.phrs.2016.06.013

Velkers FC, Bouma A, Stegeman JA, de Jong MC (2012) Transmission of a live Eimeria acervulina vaccine strain and response to infection in vaccinated and contact-vaccinated broilers. Vaccine 30(2):322–328

Vereecken M, Dehaeck B, Berge AC, Marien M, Geerinckx M, De Gussem K (2020) Synergistic effect of a combination of nicarbazin and monensin against coccidiosis in the chicken caused by Eimeria spp. Avian Pathol 49(4):389–393. https://doi.org/10.1080/03079457.2020.1756226

Volenzo T, Odiyo J (2020) Integrating endemic medicinal plants into the global value chains: the ecological degradation challenges and opportunities. Heliyon 6(9):e04970. https://doi.org/10.1016/j.heliyon.2020.e04970

Vrba V, Blake DP, Poplstein M (2010) Quantitative real-time PCR assays for detection and quantification of all seven Eimeria species that infect the chicken. Vet Parasitol 174(3–4): 183–190

Vrba V, Poplstein M, Pakandl M (2011) The discovery of the two types of small subunit ribosomal RNA gene in Eimeria mitis contests the existence of E. mivati as an independent species. Vet Parasitol 183(1–2):47–53

Wang Y, McAllister TA, Newbold CJ, Rode LM, Cheeke PR, Cheng K-J (1998) Effects of Yucca schidigera extract on fermentation and degradation of steroidal saponins in the rumen simulation technique (RUSITEC). Anim Feed Sci Technol 74(2):143–153. https://doi.org/10.1016/S0377-8401(98)00137-0

Wang D, Zhou L, Li W, Zhou H, Hou G (2018a) Anticoccidial effects of areca nut (Areca catechu L.) extract on broiler chicks experimentally infected with *Eimeria tenella*. Exp Parasitol 184:16–21. https://doi.org/10.1016/j.exppara.2017.11.002

Wang X, Kiess AS, Peebles ED, Wamsley KGS, Zhai W (2018b) Effects of Bacillus subtilis and zinc on the growth performance, internal organ development, and intestinal morphology of male broilers with or without subclinical coccidia challenge. Poult Sci 97(11):3947–3956. https://doi.org/10.3382/ps/pey262

Wang L, Guo Z, Gong Z, Cai J, Yang F, Wei X, Niu B (2020) Efficacy of an Oral solution prepared from the ultrasonic extract of Radix dichroae roots against *Eimeria tenella* in broiler chickens. Evid Based Complement Alternat Med 2020:1–8. https://doi.org/10.1155/2020/3870902

Williams RB (1986) On the aetiology of intestinal intussusception in the fowl. Avian Pathol 15(2): 301–304

Williams RB (1995) Epidemiological studies of coccidiosis in the domesticated fowl (Gallus gallus): II. Physical condition and survival of Eimeria acervulina oocysts in poultry-house litter. Appl Parasitol 36(2):90–96

Williams RB (1998) Epidemiological aspects of the use of live anticoccidial vaccines for chickens. Int J Parasitol 28(7):1089–1098. https://doi.org/10.1016/S0020-7519(98)00066-6

Williams RB (1999) A compartmentalised model for the estimation of the cost of coccidiosis to the world's chicken production industry. Int J Parasitol 29(8):1209–1229. https://doi.org/10.1016/S0020-7519(99)00086-7

Williams RB (2002) Fifty years of anticoccidial vaccines for poultry (1952-2002). Avian Dis 46(4): 775–802

Xuan TD, Khanh TD (2016) Chemistry and pharmacology of Bidens pilosa: an overview. J Pharm Investig 46(2):91–132

Yadav S, Teng P-Y, Singh AK, Choi J, Kim WK (2022) Influence of Brassica spp. rapeseed and canola meal, and supplementation of bioactive compound (AITC) on growth performance, intestinal-permeability, oocyst shedding, lesion score, histomorphology, and gene expression of broilers challenged with E. maxima. Poult Sci 101(2):101583. https://doi.org/10.1016/j.psj.2021.101583

Yang W-C, Yang C-Y, Liang Y-C, Yang C-W, Li W-Q, Chung C-Y, Yang M-T, Kuo T-F, Lin C-F, Liang C-L (2019) Anti-coccidial properties and mechanisms of an edible herb, Bidens pilosa, and its active compounds for coccidiosis. Sci Rep 9(1):2896. https://doi.org/10.1038/s41598-019-39194-2

Yang W-Y, Chou C-H, Wang C (2022) The effects of feed supplementing *Akkemansia muciniphila* on incidence severity and gut microbiota of necrotic enteritis in chickens. Poult Sci 101(4):101751. https://doi.org/10.1016/j.psj.2022.101751

Yong T, Chen M, Li Y, Song X, Huang Y, Chen Y, Jia R, Zou Y, Li L, Yin L (2020) Anticoccidial effect of Fructus Meliae toosendan extract against *Eimeria tenella*. Pharm Biol 58(1):636–645. https://doi.org/10.1080/13880209.2020.1784234

Zhang M, Li X, Zhao Q, She R, Xia S, Zhang K, Zhang L, Wang X, Wang M, Liu Y (2019) Anticoccidial activity of novel triazine compounds in broiler chickens. Vet Parasitol 267:4–8. https://doi.org/10.1016/j.vetpar.2019.01.006

Feed Additives to Combat Intestinal Diseases in Antibiotic-Free Poultry Farming

Vasileios Tsiouris, Tilemachos Mantzios, Konstantinos Kiskinis, and Paschalis Fortomaris

Abstract

A rapid rise in the world population has led to an increased demand for animal protein, a large amount of which is covered by the poultry industry. However, poultry protein must be produced in a way that promotes birds' health and welfare and simultaneously is ethical and environmentally friendly. Traditionally, antimicrobial growth promoters (AGPs) have been widely used to improve performance, health, and welfare, modulate the intestinal microbiota, and control enteric pathogens in birds. Nevertheless, widespread scientific concern for the risk of developing cross-resistance pathogens to antibiotics has resulted in the restriction and even complete ban of AGPs for therapeutic and prophylactic purposes in poultry. In addition, the augmented consumer demand for "organic" poultry products further compels the poultry industry to avoid the use of AGPs and to apply more environmental and poultry-friendly management practices, such as free-range, all vegetable feeding, outdoor grassing, and slow-growth hybrids. Considering all the above, the prohibition or voluntary exclusion of AGPs in poultry feed and the change of management practices disturbed the ecosystem of the gastrointestinal tract and dramatically increased the incidence of economically important intestinal diseases. The intestinal tract is not only responsible for digestion and absorption of nutrients but also acts as a metabolic, endocrinal, and immunological organ. Digestive tract infections are a major concern in the poultry industry and result in growth retardation, increased

V. Tsiouris (✉) · T. Mantzios · K. Kiskinis · P. Fortomaris
School of Veterinary Medicine, Aristotle University of Thessaloniki, Thessaloniki, Greece
e-mail: biltsiou@vet.auth.gr

G. Arsenos, I. Giannenas (eds.), *Sustainable Use of Feed Additives in Livestock*,
https://doi.org/10.1007/978-3-031-42855-5_16

mortality, deterioration of birds' health and welfare, and increased risk of contamination of poultry products intended for human consumption. All these adaptations in poultry farming bring new challenges and demand a holistic approach through novel feed strategies in order to minimize the impact of digestive tract infections on poultry health and welfare as well as to guarantee food safety and security for a growing human population. A wide range of feed additives have been suggested, including eco-friendly compounds, such as polyphenols in herbal extracts, essential oils, phytogenic feed additives, organic and/or inorganic acids, prebiotics, probiotics, competitive exclusion products, hen egg antibodies, and exogenous enzymes. The aim of this chapter is to provide the current knowledge of novel feed additives with the potential to combat intestinal diseases in antibiotic-free poultry farming.

Keywords

Probiotics · Prebiotics · Organic acids · Inorganic acids · Phytogenics · Zoonoses · Necrotic enteritis · Colibacillosis · Salmonellosis · Coccidiosis · Feed additives · Alternatives · Poultry · Gut health

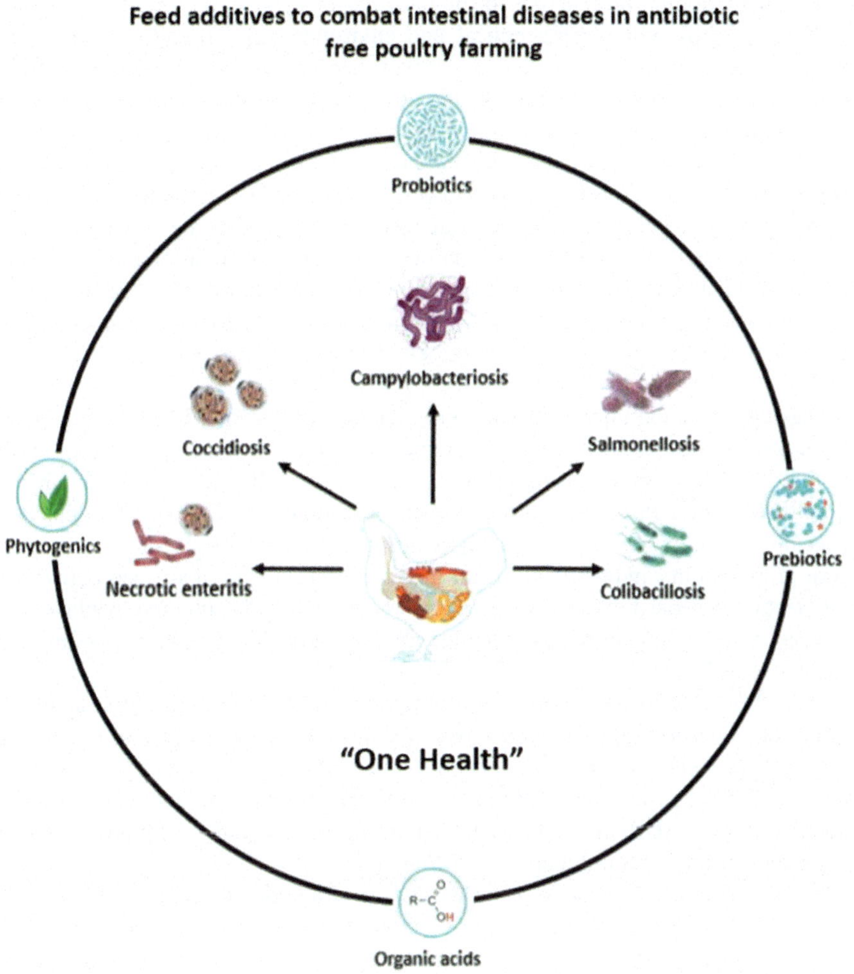

1 Introduction

The global human population is estimated to reach 9.6 billion by 2050 (Kiarie et al. 2019). Rapid global population growth and rising incomes in developing countries have led to increased demand for protein, which is one of the most expensive and limiting feed ingredients (Kim et al. 2019a). The poultry industry can play a key role in alleviating poverty as well as food scarcity as it can provide high-quality protein and nutrients for human consumption through meat and eggs (Raza et al. 2019). Global poultry production has tripled to an annual production of 90 million tons of chicken meat and 1.1 trillion eggs (http://www.fao.org/faostat/) over the past two decades (Kiarie et al. 2019).

However, poultry protein must be produced in a way that promotes the health and welfare of birds while being ethical and environmentally friendly. The poultry industry faces an ongoing challenge to develop management strategies to optimize chicken performance while minimizing food safety concerns. For example, the poultry industry has widely used significant amounts of antibiotic growth promoters (AGPs) to improve performance, health, and welfare, regulate gut microbiota, and control enteric pathogens in poultry. However, the misuse of antibiotics has led to an increase in the development of antibiotic resistance and antibiotic residues in meat, with serious concerns for human and animal health and the environment. Public and consumer concerns have necessitated consumer discontinuation and legislative regulation and/or restrictions on the use of antibiotics for therapeutic and prophylactic uses in poultry production and livestock in general (Kiarie et al. 2019; Zhu et al. 2021).

Taking all of the above into account, banning or voluntarily excluding AGPs from poultry feed and changing management practices have disrupted the gastrointestinal tract ecosystem and dramatically increased the incidence of economically important enteric diseases. Gastrointestinal infections, such as necrotic enteritis, coccidiosis, campylobacteriosis, salmonellosis, and colibacillosis, are a major concern in the poultry industry and result in growth retardation, poor feed conversion efficiency, increased mortality, deterioration of birds' health and welfare, and increased risk of contamination in poultry products intended for human consumption (Kiarie et al. 2019).

The shift away from antibiotic use in poultry has led industry and academia to find alternatives to AGPs to improve bird health, welfare, and performance as well as prevent and control enteric pathogens in poultry. In addition, increased consumer demand for "organic" poultry products is further forcing the poultry industry to avoid the use of AGPs and to implement more environmentally and poultry-friendly management practices (Abdelli et al. 2021).

Considering the shift away from the use of antibiotics in poultry feed and the impact of the aforementioned diseases on poultry, public, and environmental health, a "One Health" approach should be adopted by the scientific international community and poultry industry in order to promote performance, health, and welfare of poultry and to minimize public health issues. For this reason, a wide range of alternative environmental-friendly feed additives have been proposed, such as probiotics, prebiotics, organic and/or inorganic acids, and phytogenic feed additives (Abdelli et al. 2021; Zhu et al. 2021). The aim of this chapter is to provide current knowledge on new feed additives with the potential to combat intestinal diseases and improve intestinal health in poultry.

2 Necrotic Enteritis

Necrotic enteritis (NE) is a disease of high economic impact (Van der Sluis 2000), which affects the health and welfare of broilers and also poses a threat to public health (Van Immerseel et al. 2004a, b). Worldwide losses from this disease were estimated at $6 billion in 2015, and thus it is commonly described as a $6 billion disease (Wade and Keyburn 2015). The ban of AGPs in diets of poultry significantly affected the intestinal ecosystem and disrupted its homeostasis, allowing the overgrowth of *Clostridium perfringens*, leading to the outbreak of the disease (Williams 2005; McDevitt et al. 2006). NE represents a typical example of disease syndrome, which is considered a gut health issue and depends on interactions between host, infectious, nutritional, and managerial factors (Williams 2005; McDevitt et al. 2006). Particularly, factors such as higher intestinal viscosity increase the average retention time of the intestinal content and the amount of undigested material in the intestinal tract, which gives the *C. perfringens* more time and substrate to colonize the small intestine, to proliferate and to produce toxins (Waldenstedt et al. 2000). Conversely, high concentrations of lactic acid produced by bacteria can reduce the pH to levels that are low enough to inhibit the growth of *C. perfringens* (McReynolds et al. 2007).

Despite our present understanding of the disease, and identification of *C. perfringens* as the etiological agent, the predisposing factors, which are essential for the outbreak of the disease, are numerous, but many are ill-defined and experimental results have been contradictory (Williams 2005; McDevitt et al. 2006; Dahiya 2007). The best-known predisposing factor for NE is the mucosal damage, caused by coccidian infection (Williams 2005; Timbermont et al. 2011). On the contrary, attenuated anticoccidial vaccination, which also causes mild intestinal lesions, showed a significant protective effect against subclinical experimental NE in broiler chicks (Tsiouris et al. 2013). The protective effect of anticoccidial vaccine against NE was attributed to the stimulation of nonspecific and specific immunity mechanisms, as a result of local inflammation. In addition, the reduction of severe coccidiosis lesions in the mucosa limits the mucus secretion and the available nutrients for *C. perfringens* proliferation in the gastrointestinal tract. Finally, the control of coccidiosis discourages the attachment of *C. perfringens* to the intestinal mucosa (Timbermont et al. 2011).

A thorough understanding of the pathogenesis and interaction between nutrients, intestinal mucosa, intestinal microbiota, and *C. perfringens* is essential to prevent and reduce the incidence of NE in poultry. Several alternatives to in-feed antibiotics such as probiotics, prebiotics, organic and/or inorganic acids, and phytogenic feed additives have been employed to control the incidence of NE in poultry. These approaches rely on the positive modulation of host immune response, nutritional manipulation, intestinal microbiota interaction, and pathogen reduction (Fathima et al. 2022a; Revington 2002). The following section provides an update on feed additives that could contribute to combat NE in poultry.

2.1 Probiotics

Probiotics are feed additives that have attained popularity in livestock nutrition after the ban of antibiotic growth promoters. In particular, probiotics are portrayed as live microorganisms that benefit the host when fed in adequate amounts (Shehata et al.

2022). Among others, probiotics are considered one of the most promising alternative approaches, to combat the withdrawal of AGPs in poultry, since they can enhance nutrient digestibility and counteract the growth of pathogenic bacteria such as *Escherichia coli*, *Salmonella* spp., *Clostridium* spp., and *Campylobacter* spp. (Shehata et al. 2022; Fathima et al. 2022b; Kulkarni et al. 2022). To that end, lactic acid-producing bacteria such as *Lactobacillus* spp., *Bifidobacterium* spp., *Enterococcus faecalis*, and *Streptococcus thermophilus* are gaining popularity since they have been found to be highly efficient in positively modulating host immune fractions, exclude the colonization of pathogenetic bacteria in the intestine, and are also resistant to gastric and bile salts (Shehata et al. 2022). In particular, Wang et al. (2017a, b) reported that dietary *Lactobacillus johnsonii* BS15 improved the performance (in terms of body weight [BW] gain and feed conversion ratio [FCR]) and intestinal immunity in Cobb 500 broiler chicks, which were experimentally induced in an NE model. In a second study, in an effort to better understand the immune regulatory mechanism of *L. johnsonii* BS15, Wang et al. (2018) demonstrated that the addition of this probiotic strain may prevent subclinical NE in broilers, by stimulating intestinal immunity as well as by boosting blood parameters related to the immunity. In another study, Shojadoost et al. (2022) also demonstrated immunomodulatory effects of four different species of *Lactobacilli* (*L. johnsonii*, *Ligilactobacillus salivarius*, *Limosilactobacillus reuteri*, and *L. crispatus*) in experimentally challenged broilers. In particular, researchers revealed that the application of the above probiotic mixture accompanied by significant alterations in interferon (IFN)-γ, interleukin (IL)-1β, IL-2, IL-12, IL-17, and transforming growth factor-beta (TGF-β) gene transcription in the duodenum and jejunum as well as to subtle changes in the frequency of CD8+ T cells and B cells in the cecal tonsil of the treated chickens (Shojadoost et al. 2022). On the other hand, Zhao et al. (2022) investigated the effect of dietary *Bacillus licheniformis* H2 in NE-challenged Ross 308 male broiler chicks, through RNA and 16S rDNA amplicon sequencing. Researchers demonstrated that the application of *B. licheniformis* H2 significantly reduced the relative abundance of *C. perfringens* in the intestine and ameliorated the adverse effects induced by NE in the liver and the ileum of the challenged birds. Furthermore, probiotic application enhanced the function of the intestinal barrier and the epithelial renewal, whereas it also ameliorated the inflammatory response and increased the immunity in broilers with subclinical NE (Zhao et al. 2022). Furthermore, dietary *Butyricicoccus pullicaecorum* during NE challenge significantly reduced the number of birds with necrotic lesions in their intestines (Eeckhaut et al. 2016). Currently, several approaches have been utilized for the delivery of probiotics in poultry, including in ovo application (Beck et al. 2019; Pender et al. 2017), inclusion in feed and/or water, as well as chick spraying with a probiotic solution (Luan et al. 2019). However, to prevent failures and obtain practical advantages, it is essential to choose probiotics carefully in terms of dose and type (Shehata et al. 2022).

2.2 Prebiotics

Prebiotics are defined as "selectively fermented feed ingredients that cause specific changes in the composition and activity of the gastrointestinal microbiota, thus conferring benefits upon host health" (Gibson et al. 2010). Indigestible compounds, such as mannan-oligosaccharides (MOS), β-glucans, galactooligosaccharides, and fructooligosaccharides (FOS), are widely used in poultry diets, since they resist exposure to gastric acid, support the growth of beneficial gut microbiota, enhance the intestinal structure, and improve the feed conversion and efficiency, through the synthesis of metabolites from their fermentation in the intestine. These metabolites include short-chain fatty acids, primarily acetate, propionate, and butyrate, which are directly absorbed from the hindgut and used as an energy source, thereby promoting weight gain and performance (Fathima et al. 2022a, b).

A good source of MOS, D-mannose, glucans, and methyl-D-mannose is the yeast cell wall. In chickens, supplementation with yeast cell wall extract promotes the development of the beneficial gut microbiome and controls the activation of proinflammatory pathways, and to that end seems a promising approach in ameliorating the adverse effects of NE in poultry. In particular, M'Sadeq et al. (2015) reported that adding yeast cell wall extract (derived from *Saccharomyces cerevisiae*) in the diet of experimentally challenged broilers ameliorated the adverse effects of NE, in terms of the improved feed intake, weight gain, livability, and lesion score in the intestine. Similarly, Tian et al. (2016) reported the high capacity of yeast-derived β-glucans in improving the body weight, feed efficiency, and antibody levels against *C. perfringens*, as well as villus height and villus height/crypt depth ratio, in experimentally challenged Arbor Acres male broilers. Finally, researchers reported that dietary β-glucans reduced *C. perfringens* counts and NE-lesions in the intestine of birds (Tian et al. 2016).

2.3 Organic Acids

Acidifiers are feed additives used in the poultry industry to lower the pH of feed or water and to stimulate the function and physiology of the gastrointestinal tract of birds. Organic acids, such as acetic, propionic, and lactic acid, are commonly used in poultry feed as a means of controlling the growth of harmful bacteria, thus promoting gut health and improving feed efficiency (Yang et al. 2018; Pearlin et al. 2020; Ebeid et al. 2022). For microorganisms, organic acids can act either as a source of carbon and energy or as antimicrobial agents, depending on the concentration of the acid, its ability to enter the cell, and the capacity of the organism to metabolize the acid (Pearlin et al. 2020; Yang et al. 2018). Organic acids have antimicrobial activity against various microorganisms, including bacteria, fungi, viruses, and parasites, making them useful for food preservation and for controlling the growth of pathogenic microorganisms (Mantzios et al. 2023; Ebeid et al. 2022; Nguyen et al. 2020; Pearlin et al. 2020; Yang et al. 2018; Emami et al. 2017). The mechanism of the antibacterial action of organic acids involves several key steps. Organic acids lower the pH of the surrounding environment, thus creating an acidic environment that is hostile to bacteria (Dibner and Buttin 2002). In addition, organic acids can penetrate the bacterial cell membrane and disrupt its integrity, leading to loss of essential nutrients, while it can also interfere with cellular metabolic processes, such as respiration and energy production, leading to inhibition of the activity of specific enzymes and eventual to cell death (Scicutella et al. 2021; Khan and Iqbal 2016; Dibner and Buttin 2002).

In poultry species, dietary organic acids and their salts demonstrate high efficiency in counteracting pathogens such as *C. perfringens* and *Eimeria* spp., and to that end could be promising alternatives for the control of NE. In particular, dietary addition of various combinations of short-chain fatty acids (SCFAs) with medium-chain fatty acids (MCFAs) and/or phenolic compounds effectively ameliorated the adverse effects of NE in male Ross 308 broilers that were exposed under a subclinical NE challenge (Kumar et al. 2022). More specifically, challenged birds that fed organic acid blends had significantly reduced intestinal permeability (decreased fluorescein isothiocyanate-dextran (FITC-d) from gut to serum), pathogenetic *Bacteroides* spp. counts, and numerically reduced *C. perfringens* loads in the ceca (Kumar et al. 2022). Additionally, Pham et al. (2022) showed that a blend of encapsulated organic acids and essential oils improved the FCR, reduced FITC-d levels in the serum, minimized gross lesions in the gut, and *C. perfringens* loads in the ceca and livers in experimentally challenged broiler chicks (Pham et al. 2022). Furthermore, dietary supplementation with benzoic acid, *Enterococcus faecium*, and an essential oil complex modulated the gut health and supported the recovery (by reducing gross lesion score and modulating microbial composition) of laying hens that were experimentally challenged by *Eimeria* spp. and *C. perfringens* (Zhang et al. 2023). Furthermore, a blend of microencapsulated organic acids and botanicals (citric, sorbic acids, thymol, and vanillin) reduced NE via specific signaling pathways (by modulating the T-cell receptor, tumor necrosis factor [TNF], and nuclear factor-κB [NF-κB] signaling) in experimentally challenged broilers (Swaggerty et al. 2022).

2.4 Phytogenic Feed Additives

Phytogenic feed additives (PFAs) or phytobiotics are natural bioactive compounds derived from plants. Based on their origin, they can generally be classified as herbs (flowering, non-woody, and nonpersistent plants) or spices (nonleaf parts of plants with intensive taste or smell including seeds, fruits, bark, or root). Based on their form and the process used to produce active compounds, PFAs can further be classified into essential oils (EOs) (extracted by steam distillation, hydrodistillation, hydrodiffusion, or solvent extraction) and oleoresins (extracted by nonaqueous solvents). Polyphenols, the main active compound of PFAs, vary in composition and concentration, depending on the type and the part of the plant, the geographical origin, the harvesting season, environmental factors, storage conditions, and processing techniques (extraction, distillation, and stabilization) (Windisch et al.

2008). PFAs have been used since ancient times and even today in aromatherapy, cosmetics, and analgesics as well as therapeutic approaches to various diseases in humans and animals (Bilia et al. 2014). Currently, PFAs, as well as their derivatives EOs, have gained rising interest from the scientific community as natural growth promoters in the poultry industry, imitating the action of AGPs, and thus proving various beneficial effects such as antimicrobial, antioxidant, anti-inflammatory, immunomodulatory, and digestion-stimulating activities (Omonijo et al. 2018; Suresh et al. 2018; Zhai et al. 2018).

Numerous studies have reported the antimicrobial activity of PFAs, which is not the result of a specific mode of action but a combinational effect of their bioactive compounds on different cell targets. The antimicrobial actions of EOs are attributed to their hydrophobic nature, which is responsible for their ability to penetrate through the cell wall, disrupting the bacterial structures and increasing the bacterial cell membrane permeability. Specifically, the main key steps of the PFA activity include: (a) destruction of the cell wall, (b) damage to membrane proteins or/and the cytoplasmic membrane, (c) coagulation of the cytoplasm, (d) increased permeability, which leads to leakage of cell contents, (e) reduction of the proton motive force, (f) decrease in ATP synthesis and hydrolysis of the ATP leading to reduction in the intracellular ATP pool (Bhavaniramya et al. 2019).

The antibacterial efficacy of thymol and carvacrol was supported by the results of a study when these natural compounds were used as feed additives in broiler chickens challenged by *C. perfringens.* Additionally, the above-mentioned *phytogenics* promoted birds' gut health, by supporting the abundance of *Lactobacillus* spp. in the gut (Du et al. 2015). The dietary commercial blend comprised 5% carvacrol, 3% cinnamaldehyde, and 2% capsicum oleoresin downregulated IFN- and IL-6 cytokines in broilers that were reared under nonchallenging conditions (Pirgozliev et al. 2019). In broilers under challenging conditions such as NE, Lee et al. 2013 showed that a mixture of capsicum and turmeric oleoresins reduced intestinal IL-8, lipopolysaccharide-induced TNF-α factor, IL-17A, and IL-17F mRNA levels (Lee et al. 2013). Similarly, the expression of pro-inflammatory cytokines was reduced by the supplementation of *Allium hookeri* in lipopolysaccharide (LPS)-induced young broiler chickens (Lee et al. 2017) and thyme powder in broilers without any challenge (Hassan and Awad 2017). The dietary inclusion of *Capsicum* and *Curcuma longa* oleoresins in the basal diet of three commercial broiler breeds reduced the negative impact of the NE challenge model, in terms of improved performance and intestinal lesions (Kim et al. 2015). In another study, the supplementation of a PFA blend (thyme and anise) improved the feed efficiency and reduced pathogenetic bacteria in the intestines of birds, such as *C. perfringens* and *E. coli.* Additionally, the PFA blend significantly reduced the gross lesions induced by NE in the intestines of birds (Cho et al. 2014). Microcapsules of a blend of EOs (thyme, peppermint, savory, and black pepper) were also demonstrated to improve growth parameters and the body's antioxidant status. Histologically, birds in the EO group revealed enhanced villus height, crypt depth, and villous height and crypt depth (VH:CD) ratio compared to those in the untreated challenged group (Moharreri et al. 2021). Similar histological findings were recorded with the supplementation of another EO blend (thyme, anise, oregano, carvacrol, yucca extracts, and cinnamaldehyde) in the basal diet of broiler chicks. This blend also elevated the blood protein and globulin concentration in the serum of broilers (Abudabos et al. 2018). Finally, in a current study, Ibrahim et al. (2021b) noted an increase in the expression levels of occludin and Jam-2 tight-junction proteins, in the intestine of broilers, that received garlic nanohydrogel in their basal diet during an NE challenge model (Table 1).

3 Coccidiosis

Coccidiosis is considered the most significant parasitic disease of poultry, caused by coccidian of the genus *Eimeria* spp. Several stages of the coccidian life cycle are evolving in the intestinal tract and cause severe damage to the intestinal mucosa, and subsequently diarrhea, growth retardation, mortality, wet litter, reduced egg production, and predisposition to other diseases such as salmonellosis, campylobacteriosis, and necrotic enteritis (Noack et al. 2019; Williams 2005). It is globally recognized as a major pathogen for the poultry industry, infecting annually over 60 billion chickens (Shirley et al. 2005), with significant health, welfare, and financial consequences. The annual global cost of coccidiosis in the poultry industry has been estimated at over US $3 billion (Noack et al. 2019). However, a recent recalculation of annual global cost of coccidiosis catapulted to the amount of 10.36 billion Great Britain pounds (GBP) (Chapman et al. 2013; Blake et al. 2020). This economic impact of coccidiosis in the poultry industry is attributed to the impaired digestion of nutrients, growth retardation, and increased mortality, along with the cost of control of coccidiosis by the use of anticoccidials or coccidiostatic drugs, vaccines, and herbs (Blake et al. 2020). Moreover, it is recognized as a major welfare concern for birds since it can significantly affect health status and cause diarrhea and mortality. Therefore, both poultry industry and consumer organizations are trying to alleviate its impact (Giannenas et al. 2023; Taylor et al. 2022; Tsiouris et al. 2021).

The control of poultry coccidiosis has been successfully achieved by the use of anticoccidial drugs for almost a century. However, the poultry industry is facing several challenges regarding legislative restrictions to limit chemoprophylaxis of coccidiosis (Tsiouris et al. 2021). Almost simultaneously with the discovery of anticoccidial drugs, the occurrence of resistance or decreased sensitivity to anticoccidial drugs by *Eimeria* spp. was observed (Noack et al. 2019; Chapman et al. 2013; Abbas et al. 2011a). In addition, anticoccidial residues in poultry meat, eggs, and their by-products have been reported and raise a public health concern and an increasing consumers preference for "zero" use of anticoccidial drugs in poultry. Thus, there is an urgent need for the poultry industry to look for alternative anticoccidial compounds, which should be more environmentally friendly and acceptable by consumers compared to conventional anticoccidial drugs (Giannenas et al. 2023; Taylor et al. 2022; Tsiouris et al. 2021). Feed additives such as probiotics, prebiotics, (in)organic acids, and phytogenic feed additives (PFAs) can

Table 1 In vivo experimental studies on the efficacy of probiotics, prebiotics, (in)organic acids, and phytogenics to combat necrotic enteritis in poultry

Type of additive	Type of bird	Dosage	Challenge	Effect	References
Probiotics					
Lactobacillus johnsonii BS15	Cobb 500 broiler chicks	1×10^6 CFU/g	*C. perfringens*	↑ body weight gain, ↓ FCR, ↑ intestinal immunity	Wang et al. (2017a, b)
Lactobacillus johnsonii, *Ligilactobacillus salivarius*, *Limosilactobacillus reuteri*, and *L. crispatus*	Cobb 500 broiler chicks (male)	1×10^7–1×10^8 CFU	*C. perfringens*	↓ gut NE lesions, ↑ villus/crypt ratio, ↑ *Actinobacteria*, *Lactobacillaceae*, and *Firmicutes*	Shojadoost et al. (2022)
Limosilactobacillus reuteri and *Ligilactobacillus salivarius*	Cobb 500 broiler chicks (female) and Hubbard M99 (male)	250–1000 μL (10^7 CFU/mL)	*C. perfringens*	↑ weight gain, ↓ gut NE lesions	Vieco-Saiz et al. (2022)
Bacillus licheniformis H2	Ross 308 broilers (male)	1.0×10^6 CFU/g	*C. perfringens*	↓ relative abundance of *C. perfringens* in the ileum, ↓ pathological damage in the ileum and liver, ↑ intestinal barrier function, ↑ epithelial renewal, ↓ energy consumption, ↑ enteral nutrition absorption, ↓ inflammatory response, ↑ immunity	Zhao et al. (2022)
Butyricicoccus pullicaecorum	Ross 308 broiler chicks	10^9 CFU/g	*C. perfringens*	↓ number of birds with NE lesions	Eeckhaut et al. (2016)
Multistrain probiotic (*Bacillus subtilis* and *B. velezensis*)	Ross 308 broilers (male)	$1–2 \times 10^9$ CFU	*C. perfringens*	↑ performance and improved gut health, ↑ duodenal villi height, and villi height/crypt depth ratio in the duodenum and jejunum, ↓ relative liver weights	Ramlucken et al. (2020)
Bacillus amyloliquefaciens BLCC1-0238	Arbor Acres broiler chicks	2×10^5 CFU/g	*C. perfringens* CVCC2030	↑ growth performance, ↓ overall mortality, ↓ intestinal mucosal	Zhang et al. (2022)

(continued)

Table 1 (continued)

Type of additive	Type of bird	Dosage	Challenge	Effect	References
				lesions, ↑ expression levels of claudin-3 and peroxisome proliferator-activated receptor-gamma coactivator-1α, ↓ expression levels of interferon-γ, IL-10, and secretory immunoglobulin A	
Prebiotics					
Yeast β-glucans	Arbor Acres broilers (male)	200 mg/kg	*C. perfringens* type A CVCC 49	↑ BW, ↑ feed efficiency, ↑ antibody levels against *C. perfringens*, ↑ villi height and villi height/crypt depth ratio, ↓ intestinal *C. perfringens* numbers, ↓ gut NE lesions, ↑ gene expression of Cath-2, AvBD-4, and AvBD-10 (early infection stage), ↑ gene expression of Cath-1, Cath-2, and AvBD-1 (later infection stage), ↓ AvBD-10 and LEAP-2 mRNA levels (later infection stage)	Tian et al. (2016)
Yeast cell wall extract (*Saccharomyces cerevisiae*)	Ross 308 broilers (male)	200–800 mg/kg	*C. perfringens* A EHE-NE18	↑ feed intake, ↑ weight gain, ↑ livability, ↓ NE-lesions	M'Sadeq et al. (2015)
Organic acids					
(A) a blend of SCFA, MCFA, and a phenolic compound, (B) a blend of free and buffered SCFA with MCFA, and (C) a blend of free and buffered	Ross 308 broilers	A: 0.5–1.5 g/kg B: 1.0–2.5 g/kg C: 1.0–2.0 g/kg	*C. perfringens*	↓ FITC-d from gut to serum, ↓ *Bacteroides*, ↓ *C. perfringens* loads, ↑ acetate and butyrate concentrations in the ceca	Kumar et al. (2022)

SCFA with a high concentration of MCFA					
A blend of encapsulated organic acids with essential oils (butyric acid, benzoic, hexanoic carvacrol, and thyme)	Arbor Acres broiler chicks (male)	200–800 mg/kg	*C. perfringens*	↑ growth performance, ↓ FCR, ↓ NE lesions, ↓ serum concentration of FITC-d, ↓ *C. perfringens* counts in the ceca and liver	Pham et al. (2022)
Butyric and valeric acid esters (tributyrin, monovalerin)	Broiler chicks	Tributyrin: Feed = 0.5 kg/metric ton, water = 1 kg/1000 L Bacitracin: Feed = 55 g/metric ton.	*C. perfringens*	↓ NE lesions, ↑ livability	Hofacre et al. (2020)
Commercial products 70% benzoic acid (99.5% of purity), 5% of EO (thymol: carvacrol = 1: 1), 5% of *Enterococcus faecium* (EF, 2 × 108 CFU/kg diet) and 20% of its own carrier (50% silica and 50% dextrin)	Lohmann gray hens	2.0 g/kg	*C. perfringens*	↓ NE lesions, ↑ recovery of laying hens after coccidia and *C. perfringens* challenge	Zhang et al. (2023)
Phytogenics					
Blend of star anise and thyme	Ross 308 broilers	250 mg/kg	*C. perfringens*	↑ growth performance, ↓ blood TC, ↓ *C. perfringens* and *E. coli* proliferation in small and large intestine	Cho et al. (2014)
Garlic extract (nanohydrogel)	Ross 308 broilers (male)	400 mg/kg	*C. perfringens*	↑ body weight gain, ↓ FCR, ↓ intestinal permeability, ↓ *C. perfringens* counts, ↓ mortality rates ↓ gross lesions in the intestine	Ibrahim et al. (2021b)

(continued)

Table 1 (continued)

Type of additive	Type of bird	Dosage	Challenge	Effect	References
Blend of capsicum and *Curcuma longa* oleoresins	Cobb, Ross 308, Hubbard broilers (male)	4 mg of each oleoresin/kg	*C. perfringens*	↑ BW, ↓ gross lesions in the intestine, beneficial alterations in the gut microbiome	Kim et al. (2015)
Blend of carvacrol and thymol	Cobb 500 broilers (male)	60–240 mg/kg	*C. perfringens*	↓ intestinal lesions, ↓ *E. coli* counts in the ileum, ↑ *Lactobacillus* spp. in the ceca	Du et al. (2015)
Blend of thyme, peppermint savory, black pepper (encapsulated)	Ross 308 broilers (male)	0.5–2 kg/ton	*C. perfringens*	↑ total feed intake, ↓ FCR, ↑ body antioxidant status, ↑ ileum morphostructure, improved intestinal microbial population,	Moharreri et al. (2021)
Blend of capsicum, turmeric (oleoresins)	Ross 308	4 mg/kg	*C. perfringens*	↑ BW, ↓ intestinal lesions, ↓ serum α-toxins, ↓ IL-8, TNF-α factor, IL-17A and IL17F mRNA levels	Lee et al. (2013)

Abbreviations: *IL*, interleukin; *NE*, necrotic enteritis; *FCR*, feed conversion ratio; *BW*, body weight; *FITC-d*, fluorescein isothiocyanate-dextran; blood; *TC*, total number of white blood cells; *TNF-α*, tumor necrosis factor-alpha; *SCFA*, short-chain fatty acid; *MCFA*, medium chain fatty; *EO*, essential oil; *CFU*, colony-forming unit play a critical role to control coccidiosis and promote performance, health, and welfare of poultry.

3.1 Probiotics

Probiotics, such as *Lactobacillus* spp., *Pediococcus* spp., *Enterococcus* spp., *Bacillus* spp., and yeast *Saccharomyces cerevisiae,* are currently used in poultry ratios. The mode of action of probiotics may include the production of antibacterial compounds (short-chain fatty acids, hydrogen peroxide, bacteriocins, etc.), competition for nutrients and adhesion sites on host cells, interaction with the intestinal microbiome, enhancement of the digestive enzyme's activity, and stimulation of the host cellular and humoral immunity (El-Sawah et al. 2020; Pajarillo et al. 2015). In recent decades, many reports have proved that probiotic bacteria were effective anticoccidiosis candidates either in vitro or in vivo (Giannenas et al. 2023; Tierney et al. 2004; Dalloul et al. 2003; Kasornpikul et al. 2009; Giannenas et al. 2014a; Hessenberger et al. 2016). In particular, the supplementation of poultry diets with *Lactobacillus plantarum* P8 ameliorated the adverse effect of coccidiosis, by reducing mortality, oocyst shedding, oxidative stress, inflammation, and intestinal permeability in broilers experimentally challenged with multiple *Eimeria* species (*E. tenella*, *E. necatrix*, *E. maxima*, and *E. acervulina*). Metagenomic analysis revealed that *L. plantarum* P8 reduced the abundance of *E. tenella*, highlighting its role in the regulation of gut microbiota in challenged birds (Wang et al. 2021). Similarly, Mohsin et al. (2022) demonstrated the potential of *L. plantarum* to combat *E. tenella* infection, by enhancing the cell-mediated and humoral immune response, performance, antioxidant enzymes, tight junction proteins, and serum chemistry in experimentally challenged broiler chicks (Mohsin et al. 2022). In another study, the combination of probiotic blend (90% *Bifidobacterium* species and remaining *Lactobacillus* spp., *Saccharomyces* spp., *Streptococcus* spp., and *Enterococcus* spp.) and organic zinc ameliorated the adverse effects of *E. tenella* in experimentally challenged Japanese quails, in terms of mortality, growth performance, lesion score in the intestine, oocyst shedding, gross, and histological lesions (Saeeda et al. 2023). Additionally, the supplementation of a host-specific probiotic, containing *Bifidobacterium bifidum, E. faecium, Lactobacillus acidophilus, Lactobacillus delbrueckii, L. plantarum, Lactobacillus rhamnosus, and Streptococcus. thermophiles,* enhanced the immune status of chickens and inhibited the negative effects of *Eimeria* spp. infection by improving the BW and FCR, reducing the oocyst shedding, cecal lesions, and mortality while in combination with diclazuril exhibited a synergistic effect by improving growth performance (Memon et al. 2022). Hence, it can be concluded that the competitive exclusion between *Eimeria* spp. and bacterial probiotics for adhesion sites/receptors negatively affects the penetration, replication, and shedding of oocysts, and therefore probiotics could act as efficient and safe alternatives for the control of coccidiosis in poultry (El-Sawah et al. 2020; Dalloul et al. 2003).

Another very interesting research topic is the investigation of the combination of a probiotic with a coccidiosis vaccine. In particular, the combination of a single probiotic *Bacillus subtilis* or a multibacteria commercial probiotic product (PoultryStar, BIOMIN GmbH, Austria) containing *Enterococcus* spp., *Bifidobacterium* spp., *Pediococcus* spp., and *Lactobacillus* spp., with coccidiosis vaccines, decreased the intestinal lesion score and oocyst per gram (OPG) after *Eimeria* spp. infection and resulted in a significant improvement in the production performance in experimentally challenged broiler chicks. These protective effects may be closely related to the effect on the intestinal microbiota, such as *Romboutsia, Blautia*, and *Butyricicoccus* (Cai et al. 2022; Ritzi et al. 2016).

3.2 Prebiotics

The use of prebiotics for the control of coccidiosis in broiler chicks showed promising results since they improved their performance and reduced the coccidiosis lesions (Giannenas et al. 2023). In a recent study by Giannenas et al. (2023), a multistrain yeast fraction (*Cyberlindnera jadinii* and *Saccharomyces cerevisiae*) prebiotic product plus ionophore anticoccidial drug (salinomycin) resulted in a significant reduction in oocyst excretion and coccidiosis lesions in the intestinal mucosa alongside with an improvement in body weight, feed conversion ratio, and intestinal microbiota. The positive effect against coccidiosis could be attributed to the interaction of the multistrain yeast fraction prebiotic product with the intestinal microbiota, depending on the severity of coccidiosis lesions. In particular, the more severe the coccidiosis lesions, the more significant the effect on the intestinal microbiota (Giannenas et al. 2023). In addition, mannan-oligosaccharide (MOS), a starch material extracted from the cell component of the yeast, *Saccharomyces cerevisiae*, improved the growth performance, reduced the oocyst count per gram of feces, and ameliorated the coccidian lesions in broiler chicks challenged with *E. tenella* (Chand et al. 2016).

The application of prebiotics xylo-oligosaccharides (XOS) in poultry diets improves growth performance and birds' gut health, by stimulating the intestinal microbiome and altering its composition (Craig et al. 2020), and specifically by increasing the beneficial bacteria and their metabolites such as SCFA (Shokryazdan et al. 2017). Lin et al. (2022) reported that the addition of XOS in diets of broilers alleviated the depression in growth performance and nutrient utilization, caused by *Eimeria* spp. challenge. In addition, by changes in the gene expression of claudin-1, supplemental XOS showed their potential to alleviate the negative impact of *Eimeria* spp. in the cecal fermentation pattern (Lin et al. 2022). Similarly, Craig et al. (2020) showed the prebiotic potential of XOS, in broilers experimentally challenged by 12× the recommended dose of the Paracox 8 vaccine. In particular, the supplementation of XOS increased the feed intake, nitrogen digestibility, digestibility of several amino acids, and the concentration of arabinose and xylose in experimentally challenged birds (Craig et al. 2020). In conclusion, all the above responses are speculated as modes of action by which prebiotics alleviated *Eimeria* spp. infections in poultry.

3.3 Organic and Inorganic Acids

In previous experimental studies, adding organic acids to poultry feed effectively controlled coccidiosis, through mechanisms that reduced the virulence of coccidian and ameliorated the host inflammatory response (Balta et al. 2021). Balta et al. (2021) reported that feeding a blend of citric acid, sodium citrate, maltodextrin, silica, malic acid, sodium chloride, citrus extract, and olive extract in broilers challenged by *Eimeria* spp. significantly reduced the number of *E. tenella* oocysts in ceca and excreted feces. The acidic blend also increased the manganese superoxide dismutase (Mn-SOD) and SCFA levels, and affected the immune response by reducing the reactive oxygen species and ameliorating the inflammatory oxidative stress in experimentally challenged birds. Researchers demonstrated that improved gross lesion scores were associated with significantly increased levels of immunoglobulin M (IgM) and immunoglobulin A (IgA) in the intestinal mucosa of birds that received the organic acid blend, revealing thus the potential of organic acids to stimulate the immune system of birds (Balta et al. 2021). Additionally, individual acids, mainly SCFA (butyric, acetic, and benzoic), have shown preventive effects against coccidiosis and demonstrated a reduction in the severity of lesion scores caused by the parasite in intestines, a fact that may be attributed to the lowering of pH in intestines, that negatively affect the virulence of oocysts (Mustafa et al. 2021; Adhikari et al. 2020; Aristimunha et al. 2016). Abbas et al. (2011b) reported that acetic acid exhibited almost equivalent results with amprolium, in ameliorating the adverse effects of *E. tenella* in growth performance and gross lesions of experimentally challenged broiler chicks. In another study, Nouri (2022) reported the potential of a blend of encapsulated organic acids (mainly lactic, acetic, benzoic, formic, and citric acids) as a safe alternative for the control of *Eimeria* spp. infection in broilers, especially in combination with diclazuril treatment. In particular, the blend of encapsulated organic acids improved the European production efficiency factor, survival rate, oocyst output per gram feces, sporulation percentage, cecal lesion score, immunity, and bloody diarrhea score, highlighting its anticoccidial and immunogenic properties (Nouri 2022). Considering the high capacity of organic acids in modulating the gastrointestinal environment and improving gut health in birds, further in vivo and in-field studies are proposed, in order to determine the possible maximum safe levels of individual organic acids or blends, to be used as an alternative to chemotherapeutic drugs for coccidian infections in poultry.

3.4 Phytogenic Feed Additives

A promising group of natural dietary alternatives to anticoccidial drugs is aromatic plant-derived substances, the so-called PFAs. Giannenas et al. (2003) in an effort to investigate the effect of dietary supplementation of oregano essential oil in broiler chicks challenged by *E. tenella* observed an improvement in the performance and a partial protection against coccidiosis, since the phytogenic product reduced both the oocyst excretion and coccidiosis lesions in the intestinal mucosa (Giannenas et al. 2003). In addition, Ali et al. (2019) observed that Hubbard broilers infected by *Eimeria* spp. exhibited higher feed intake (FI) and BW when ginger and/or garlic were supplemented to their diets. Both phytogenics lowered the overall oocyst number, mortality rates, and moderated the lesion score in the intestines of challenged birds (Ali et al. 2019). Similarly, Fatemi et al. (2017) noted that dietary *Artemisia annua* ethanolic extract improved the growth performance and decreased the number of oocysts per gram in fecal samples in broilers challenged by *Eimeria* spp. (Fatemi et al. 2017). Furthermore, the inclusion of artemisinin or *Artemisia annua* leaves in the diet of Chinese yellow broilers challenged by *E. tenella* reduced diarrhea and lesions in the ceca of birds (Jiao et al. 2018). Finally, both treatments promoted apoptosis and decreased the inflammatory response in experimentally challenged birds. In another study, the inclusion of oregano and garlic essential oils in the diet of birds improved the BW, FCR, reduced fecal oocyst excretion, and lowered the count of *C. perfringens* (Sidiropoulou et al. 2020). The possible synergistic anticoccidial effect of a polyherbal extract product was investigated by Tsiouris et al. (2021). In particular, a polyherbal extract consisted *Berberis aristata, Polygonum aviculare, Syzygiumaromaticum, Holarrhena antidysenterica,* and *Allium sativum* improved the performance, reduced the number of oocyst and coliform counts, and ameliorated coccidiosis lesions in Ross 308 female broiler chicks that were experimentally challenged by *Eimeria* spp. Additionally, the polyherbal extract improved the intestinal mucosal architecture in terms of jejunum villus height, crypt depth, and the number of villus goblet cells, while also improved the antioxidant capacity and increased the storage duration of poultry breast and thigh meat (Tsiouris et al. 2021). In conclusion, the dietary inclusion of curcumin extract in the basal diet of Cobb 500 broilers challenged by *Eimeria* spp. improved the FCR but decreased the lesion score and oocyst shedding in the intestine (Yadav et al. 2020) (Table 2).

4 Campylobacteriosis

Bacteria of the genus *Campylobacter* spp., mainly *C. jejuni*, rank first among the causative agents of human bacterial gastroenteritis worldwide and pose a severe health and economic burden (Taha-Abdelaziz et al. 2023; Al Hakeem et al. 2022). *Campylobacter* spp. are commensal (or non)microorganisms found in poultry, cows, pigs, sheep, pets, and wildlife. However, poultry species represent the main reservoir and therefore handling and consumption of raw and/or undercooked poultry meat is one of the most important routes for human infection (EFSA 2021; Kozu Clarke and Ajlouni 2021; Thames and Sukumaran 2020). It is reported that the higher body temperature of chicks is favorable for thermotolerant *Campylobacter* spp., which rapidly transmitted in the flock and colonizes the intestinal tract of birds resulting in

Table 2 In vivo experimental studies on the efficacy of probiotics, prebiotics, (in)organic acids, and phytogenics to combat coccidiosis in poultry

Type of additive	Type of bird	Dosage	Challenge	Effect	References
Probiotics					
A commercial probiotic product (blend of *B. bifidum*, *E. faecium*, *L. acidophilus*, *L. delbrueckii*, *L. plantarum*, *L. rhamnosus*, and *St. thermophiles*)	Cobb broiler chicks	1 g/kg of feed	*E. tenella*	↑ BW, ↓ FCR, ↓ mortality, ↓ oocyst shedding (day 4, 5, 6, and 7 post-infection), ↓ gross lesions in ceca, ↑ IL-10 and IgM in serum	Memon et al. (2022)
B. subtilis (and *Eimeria* vaccine)	Mahuang chickens	1.25×10^9 CFU/kg feed	*Eimeria* spp.	↑ performance, ↓ *Eimeria* spp. lesion score, ↓ oocyst shedding	Cai et al. (2022)
A commercial multistrain probiotic (PoultryStar: *Enterococcus*, *Bifidobacterium*, *Pediococcus*, and *Lactobacillus* species) and coccidiosis vaccine (Immucox I)	Cobb 500 broiler chicks (male)	20 mg/bird	*E. acervulina* and *E. maxima*	↑ BW, ↑ protective effect against the challenge, ↓ gross lesions in the duodenum	Ritzi et al. (2016)
Lactobacillus plantarum P8	Arbor Acres broilers (male)	1×10^7 CFU/g	*E. tenella, E. necatrix, E. maxima* and *E. acervulina*	↑ performance, ↓ mortality, ↓ oocyst shedding, ↑ intestinal barrier, ↑ intestinal morphology, ↓ intestinal permeability, ↓ oxidative stress, ↓ inflammation, ↓ oxidative stress, ↓ pro-inflammation, ↓ abundance of *Eimeriidae*, and *E. tenella*	Wang et al. (2021)

(continued)

Table 2 (continued)

Type of additive	Type of bird	Dosage	Challenge	Effect	References
Lactobacillus plantarum	Broiler chicks	1×10^8 CFU	*E. tenella* (Beijing strain)	↑ cell-mediated and humoral immune response, ↑ performance, ↑ antioxidant enzymes (PepT1), ↑ tight junction proteins, ↑ serum chemistry (AST, ALT, and LDH)	Mohsin et al. (2022)
A probiotic blend (90% *Bifidobacterium* species and remaining *Lactobacillus*, *Saccharomyces*, *Streptococcus*, and *Enterococcus*) and organic zinc (OZn)	Japanese quails	OZn: 50 mg/kg, probiotic: 6×10^9 CFU/g feed	*E. tenella*	↑ growth performance, ↓ lesion score in the intestine, ↓ mortality, ↓ oocysts per gram, ↓ histological dimensions of cecum	Saeeda et al. (2023)
Prebiotics					
Multistrain yeast fraction product (MSYF) and salinomycin	Ross 308 broiler chicks (male)	MSYF: 0.4 g/kg feed, salinomycin: 60 mg/kg feed	*E. acervulina*, *E. maxima*, and *E. tenella*	↑ growth performance, ↑ Lactobacilli counts in jejunum and ceca, ↑ villus height in duodenum and jejunum, ↓ lesion score in the intestine, ↓ oocyst shedding, ↓ eimerian DNA concentrations, ↑ bone ash content in both femur and tibia, ↑ bone comprehensive strength,	Giannenas et al. (2023)
Mannan-oligosaccharide (MOS)	Ross 308 broiler	0.8 g/kg feed	*E. tenella*	↑ weight gain, ↑ feed intake, ↑ FCR, ↓ oocysts per gram, ↓	Chand et al. (2016)

	chicks (male)			pinpoint hemorrhages, thickness of cecal wall, bloody fecal contents, and mucoid contents in the cecum	
Xylo-oligosaccharides (XOS)	Cobb 500 broiler chicks	0.5 g/kg XOS	*E. acervulina*, *E. maxima*, and *E. tenella*	Alleviated the depression in growth performance and nutrient utilization from the *Eimeria* challenge. In addition, supplemental XOS reversed the gene expression changes of claudin-1, and also showed the potential to alleviate the negative cecal fermentation pattern induced by *Eimeria* infection.	Lin et al. (2022)
Xylo-oligosaccharides (XOS)	Ross 308 broiler chicks (male)	0.025% XOS	*E. acervulina, E. brunetti HP, E. maxima CP, E. maxima MFP, E. mitis HP, E. necatrix HP, E. praecox HP* and *E. tenella*	↑ feed intake, ↑ nitrogen digestibility, ↑ digestibility of several amino acids, ↑ of arabinose and xylose concentration, ↑ short-chain fatty acid production	Craig et al. (2020)
Organic and inorganic acids					
Water: Blend of formic acid, acetic acid, and ammonium formate) feed: Blend of encapsulated butyrate encapsulated MCFAs, organic acids mainly sorbic acid, and phenolic compound	Ross 308 broiler chicks	Water: 1.5/1 L, feed: 0.15% - 0.1%	*E acervulina*, *E. maxima*, and *E. tenella*	↑ BW, ↑ average body weight gain, ↓ FCR, ↓ lesion score in the intestine	Mustafa et al. (2021)

(continued)

Table 2 (continued)

Type of additive	Type of bird	Dosage	Challenge	Effect	References
Encapsulated organic acids (comprised 15% of lactic acid, 10% of acetic acid, 5% of benzoic acid, 5% of formic acid, and 5% of citric acid)	Ross 308 broiler chicks	0.1%	*Eimeria* spp.	↑ European production efficiency factor, ↑ survival rate, ↓ oocyst output per gram feces, ↓ sporulation percentage, ↓ cecal lesion score, ↑ immunity, ↓ bloody diarrhea, ↑ anticoccidiosis index,	Nouri (2022)
Organic acid blend (maltodextrin, sodium chloride, citric acid, sodium citrate, silica, malic acid, citrus extract and olive extract)	Ross 308 broiler chicks	0.5%	*E. tenella*	↓ *E. tenella* levels in the ceca and excreted feces, ↓ inflammatory oxidative stress, ↑ immune response (↓ ROS, ↑ Mn-SOD, and SCFA levels), ↑ IgA and IgM, ↓ lesions in ceca in cecal tissue	Balta et al. (2021)
Acetic acid	Hubbard Al-Noor broiler chicks	3%	*E. tenella*	↑ performance, ↓ pathogenic effects associated with coccidiosis, ↓ cecal lesion score, ↓ pH of the cecal content	Abbas et al. (2011a, b)
Phytogenics					
Artemisinin (ART)/ *Artemisia annua* leaves (LAA)	Chinese yellow broiler chicks	ART: 100 mg/kg or LAA: 5%	*E. tenella*	↓ diarrhea, ↓ lesions in ceca ART: ↓ Bcl-2 expression, ↑ expression levels of Bax and cleaved caspase-3, ↓ mRNA expressions of NF-κB and interleukin-17A in ceca LAA: ↓ clinical symptoms, ↑	Jiao et al. (2018)

				apoptosis, and ↓ inflammatory response	
Blend of *Holarrhena antidysenterica*, *Berberis aristata*, *Syzygium aromaticum*, *Polygonum aviculare*, and *Allium sativum*	Ross 308 broiler chicks (female)	1–2 g/kg	*E acervulina*, *E. maxima*, and *E. tenella*	↑ performance, ↑ antioxidant status, ↓ oocyst and coliform counts, ↑ *Lactobacillus* spp. Counts	Tsiouris et al. (2021)
Blend of oregano and garlic EOs	Ross 308 broiler chicks (male)	Premix (1 g/kg feed) containing the oregano (50 g/kg premix) and garlic (5 g/kg premix)	*E. tenella (Wisconsin)*	↑ BW, ↓ FCR, ↓ fecal oocyst excretion, ↓ total anaerobe counts, ↓ *C. perfringens* counts, ↑ *E. coli* and *Enterobacteriaceae* counts	Sidiropoulou et al. (2020)
Artemisia annua extract	Ross 308 broiler chicks	0.4% wheat bran with 2000 ppm *Artemisia annua* extract (AE) (contained 5 ppm artemisinin)	*E. acervulina*, *E. necatrix*, *E. tenella*	↑ BW, ↓ oocysts per gram	Fatemi et al. (2017)
Curcumin	Cobb 500 broiler chicks (male)	100–200 mg/kg	*Eimeria* spp.	↑ antioxidant status, ↓ lesion score, ↓ oocyst shedding	Yadav et al. (2020)
Oregano oil	Cobb 500 broiler chicks	300 mg/kg	*E. tenella*	↑ BW, ↓ FCR, ↓ fecal oocyst excretion, ↓ diarrhea, ↓ lesions in ceca, ↓ mortality rates	Giannenas et al. (2003)

(continued)

Table 2 (continued)

Type of additive	Type of bird	Dosage	Challenge	Effect	References
Garlic/ginger	Hubbard broiler chicks	15 g/kg or 5 g/kg	*Eimeria* spp.	↑ BW, ↑ FI, ↓ fecal oocyst excretion, ↓ diarrhea, ↓ lesions in ceca, ↓ mortality rates	Ali et al. (2019)

Abbreviations: *IL*, interleukin; *FCR*, feed conversion ratio; *BW*, body weight; *IgM*, immunoglobulin M; *IgA*, immunoglobulin A; *ROS*, reactive oxygen species; *SCFA*, short-chain fatty acids; *AST*, aspartate aminotransferase; *ALT*, alanine transaminase; *LDH*, lactate dehydrogenase; *MCFA*, medium chain fatty; *EO*, essential oil; *CFU*, colony-forming unit; *Mn-SOD*, manganese superoxide dismutase high populations, approximately 1×10^9 CFU/g in the ceca (Awad et al. 2018; Hansson et al. 2018).

Despite the early connection of *Campylobacter* spp. with clinical manifestations in ostriches (pathological lesions in the liver and intestines) and laying hens (*Vibrionic hepatitis*), in general, microorganisms are not associated with clinical signs in poultry under natural conditions (Awad et al. 2018). However, in the last decades, an increased number of reports linked *Campylobacter hepaticus* infection with diarrhea, increased mortality, reduced egg production, and spotty lesions in the liver in free-ranged layers (spotty liver disease/syndrome) (Phung et al. 2019). Additionally, current studies elucidate the ability of *Campylobacter* spp., especially *C. jejuni*, to induce a robust inflammatory response and clinical disease, including diarrhea, mortality, reduced performance, and uniformity in experimentally challenged broiler chicks (Awad et al. 2018). *Campylobacter* spp. infection is also linked with increased footpad lesions and hock burns due to degraded gut health, which leads to diarrhea and increased moisture in the litter. It is evident that all the above compromises poultry's health and welfare status, whereas the even frequent isolation of antibiotic-resistant *C. jejuni* strains from both humans and poultry highlights the interventions that should be applied in a "One Health" approach (Schiaffino et al. 2019).

For *Campylobacter* spp., there are many recommendations for the control measures that should be applied to reduce the incidence of the microorganism in the poultry production chain (Koutsoumanis et al. 2020; Hansson et al. 2018; Awad et al. 2018). However, implementing control strategies at the farm level is recognized as a vital intervention for controlling poultry and human infection. It is reported that lowering by 3 log10 the *Campylobacter* spp. counts in the ceca of broilers could reduce the incidence of human infection that is attributable to broiler meat by 58% (Koutsoumanis et al. 2020). In addition, the high consumer demand for organic products and additional rules of production that the European Union (EU) bounds for this product type category highlight the urgent demand for alternative biocides that could control bacterial infections in poultry and simultaneously be safe for the consumer and the environment (Lassen et al. 2022). To that end, feed additives could support the effort of other interventions, like biosecurity measures, vaccination, and improved hygiene in slaughter, in reducing *Campylobacter* spp. counts in the final product for human consumption. The inclusion of dietary additives, such as probiotics, prebiotics, (in)organic acids, and phytogenic feed additives (PFAs), is getting attention, as they can effectively stimulate the immune system, modify the physicochemical properties of intestinal content, and improve the structure of and the abundance of the beneficial intestinal microbiome (Ismail et al. 2023, Jahan et al. 2022, Al Hakeem et al. 2022; Aljazzar et al. 2022; Helmy et al. 2022; Lassen et al. 2022; Ibrahim et al. 2020; Kim et al. 2019a, b; Chacher et al. 2017; Eeckhaut et al. 2016). However, until now, none of these strategies, by themselves, can provide an effective, reliable, and practical intervention measure to prevent *Campylobacter* spp. colonization in broiler chicks (Tsiouris et al. 2019). The following section provides an update on feed additives that could contribute to the control of *Campylobacter* spp. in poultry.

4.1 Probiotics

The mechanisms underlying the beneficial effects of the probiotic application in poultry are not fully elucidated. It is generally reported that, after colonization, a probiotic strain interacts with the host's intestinal microbiome by producing biocide compounds (such as organic acids and bacteriocins) and by stimulating the gut immune system (Abd El-Hack et al. 2020; Park et al. 2016). Additionally, probiotic accumulation alters physicochemical characteristics of the gut contents (pH and viscosity), thus promoting the absorption of nutrients and creating an unfavorable microenvironment for the growth of pathogenetic bacteria such as *Salmonella* spp. and *Campylobacter* spp. (Abd El-Hack et al. 2020; Saint-Cyr et al. 2016). In particular, for *Campylobacter* spp., substantial advancements have been recently achieved (Ismail et al. 2023; Šimunović et al. 2022; Helmy et al. 2022; Abd El-Hack et al. 2020). However, the efficacy of probiotics on *C. jejuni* loads in broilers varied in degree, and more research is needed to improve this approach.

Lactobacilli probiotics have shown a high capacity to stimulate immunity and induce competitive exclusion in *C. jejuni* colonization in broilers. Saint-Cyr et al. (2017) reported that the oral ingestion of *Lactobacillus salivarius* every 2 days significantly reduced (2.81 log CFU/g) *C. jejuni* loads in experimentally challenged broiler chicks. In addition, taxonomic analysis suggests that the application of *L. salivarius* limited the impact of *C. jejuni* infection on *Anaerotruncus* sp. decrease and *Subdoligranulum* sp. increase (Saint-Cyr et al. 2017). Similarly, Nishiyama et al. (2014) reported that oral administration of *Lactobacillus gasseri* SBT2055 for over 14 days significantly reduced *C. jejuni* counts (8 log CFU/g) in the ceca of experimentally challenged to white leghorn chicks.

Other probiotic microorganisms such as species of *Butyricicoccus* spp., *Bacillus* spp., and *E. coli* Nissle 1917 have also shown a potential to combat *Campylobacter* spp. infections in poultry. Ismail et al. (2023) reported the preventive potential of dietary *Bacillus amyloliquefaciens* in encapsulated nanoparticles in terms of improved performance, reduced cecal colonization, and fecal shedding of *C. jejuni* and beneficial alterations in the gut microbiome of experimentally challenged broiler chicks. In addition, researchers reported that applying the probiotic *B. amyloliquefaciens* increased mRNA expression levels of digestive enzymes encoding genes (AMY2a, PNLIP, CELA1, and CCK) and the expression of barrier functions-linked genes, including DEFB1, FABP-2, and MUC-2. In another study, dietary *Butyricicoccus pullicaecorum*, a butyrate-producing probiotic strain, also effectively reduced the abundance of *Campylobacter* spp. by 1.5 log10 gene copies/g cecal content in Ross 308 broiler chicks aged 40 days (Eeckhaut et al. 2016). In addition, Šimunović et al. (2022) reported that oral administration of probiotic *Bacillus subtilis* PS-216 significantly reduced the colonization of *C. jejuni* and improved the weight gain in experimentally challenged broiler chicks. Helmy et al. (2022) reported the high capacity of probiotic *E. coli* Nissle 1917, when supplemented in an encapsulated form in drinking water, in reducing *C. jejuni* counts (approximately 2.0 log) in the ceca of experimentally challenged specific pathogen-free (SPF) chickens. In addition, researchers reported the high capacity of the probiotic strain in experimentally challenged birds in terms of improved gut histomorphology (increased villous height, crypt depth, and villous height/crypt depth ratio in the jejunum and ileum), enhanced immunity (increased *C. jejuni*-specific and total IgA and IgY antibodies in serum), as well as the expression of several cytokines and chemokines, which activate the Th1, Th2, and Th17 pathways (Helmy et al. 2022).

Multispecies probiotics have shown increased potential than individuals in controlling and decreasing *C. jejuni* loads in poultry (Al Hakeem et al. 2022). The daily oral administration of an avian-specific commercial probiotic mixture (*Enterococcus faecium, Pediococcus acidilactici, Bifidobacterium animalis, Lactobacillus salivarius*, and *Lactobacillus reuteri*) resulted in an extreme reduction in *C. jejuni* counts (6 log CFU/g) in the cecal content in experimentally challenged broiler chicks, at 35 days of age (Ghareeb et al. 2012). Furthermore, Ştef et al. (2015) reported that the administration of four probiotic strains (*Lactobacillus paracasei* JR, *L. rhamnosus* 15b, Y *L. lactis*, and *L. lactis* FOA) in different combinations and different periods induced significant changes in the intestinal mucosa histology (smaller goblet cells and reduced presence of leukocyte infiltration in the chorion) and effectively inhibited the development of *C. jejuni* (Ştef et al. 2015). In an in-field study, Smialek et al. (2018) tested the efficiency of a commercial probiotic mixture (*L. lactis, Carnobacterium divergens, Lactobacillus casei, Lactobacillus plantarum*, and *Saccharomyces cerevisiae*) on intestinal counts and carcass contamination of broilers with *Campylobacter* spp. Researchers showed that using the probiotic mixture might decrease the pollution level in the birds' environment and the extent of *Campylobacter* spp. invasion in the gastrointestinal tract of birds. Thus, it can be supported that probiotic beneficial effects are not only limited to lowering *Campylobacter* spp. counts in the gastrointestinal tract of birds but also to reducing the spread of the pathogen in the environment (farm, slaughterhouse, vehicles, and personnel) (Smialek et al. 2018).

In conclusion, it is essential to note that various parameters should be considered when evaluating the efficiency of a single probiotic bacterial strain or mixture against *C. jejuni*. In particular, the probiotic dose, age, feed composition, sex, hybrid of birds, and experimental protocol, as well as the host-microbiome composition, could significantly affect the activity of a probiotic and are considered important factors for the variability of the results among previous investigations.

4.2 Prebiotics

Prebiotics, such as mannan-oligosaccharides, β-glucans, and fructans, are widely used in the poultry industry as they have been found to beneficially influence the gut microbiome and the host. At the same time, prebiotics expresses high synergy when combined with probiotics, essential oils, and other alternatives (Taha-Abdelaziz et al. 2023; Jahan et al. 2022; Chacher et al. 2017).

Mannan-oligosaccharides (MOSs: primarily derived from outer cell membranes of *Saccharomyces cerevisiae* yeast) are widely applied in poultry feed. They are resistant to hydrolysis by digestive enzymes, and can reduce the growth of pathogenetic bacteria, mainly by enhancing the growth of beneficial bacteria such as *Lactobacilli* and *Bifidobacterium* spp. (Jahan et al. 2022; Kim et al. 2019a, b; Chacher et al. 2017). In addition, MOSs are rich in mannoproteins, mannan, and glucan and can inhibit gastrointestinal colonization of pathogens by obstructing their attachment to epithelial cells of the intestine. In particular, it is reported that by binding the type-1 fimbriae and inhibiting lectin of Gram-negatives such as *Salmonella* spp., *E. coli*, and *C. jejuni*, MOSs allows the movement of the pathogen through the intestinal tract without permitting its colonization (Jahan et al. 2022; Kim et al. 2019a, b; Chacher et al. 2017). Additionally, by increasing the mucin secretion in the intestine, MOSs provide an additional barrier for pathogenic bacteria to interact and attach to intestinal epithelial cells (Jahan et al. 2022; Chacher et al. 2017). Rostami et al. (2021) reported that adding MOS to the poultry diet restored the performance, mortality rate, cecal coliform counts, and blood IL-6 and IFN-γ concentrations that is induced by *C. jejuni* infection in experimentally challenged male broiler chicks. Additionally, the MOS supplementation reduced the ileal and cecal counts of *C. jejuni* but resulted in significantly higher concentrations of IgG and IgA in the serum of birds (Rostami et al. 2021). On the other hand, Munoz et al. (2023) reported that the dietary yeast cell wall did not have any impact on growth performance, innate immune response, cecal colonization, carcass yield, or *C. jejuni* prevalence after processing, in broilers experimentally challenged at day 21 (Munoz et al. 2023). Similarly, Froebel et al. (2019) and Baurhoo et al. (2007) demonstrated that including *Saccharomyces*-derived prebiotics in the diet of broiler chicks reduced the colonization of *C. jejuni* by <1 log. In addition, Tsiouris et al. (2019) reported that adding 1% whey to broiler diets did not affect the performance, welfare, and cecal *C. jejuni* counts after the experimental challenge. Thus, for *Campylobacter* spp., it may be concluded that the supplementation of prebiotics alone does not appear to offer the best protection. In this context, previous investigations have shown promising results in combining prebiotics with other active ingredients such as probiotics, essential oils, and organic acids. However, further in vivo studies are proposed to elucidate their synergies against *C. jejuni* colonization in experimentally challenged poultry species (Shehata et al. 2022).

4.3 Organic and Inorganic Acids

According to the scientific opinion published by EFSA 2020, "Update and review of control options for Campylobacter in broilers at primary production," adding organic acids, chlorine-based biocides, or hydrogen peroxide to the drinking water of broilers could reduce the risk of *Campylobacter*-positive flocks up to 55% (Koutsoumanis et al. 2020). This recommendation is fully supported by previous epidemiological investigations in Great Britain, France, and Spain, which reveal that the sanitation of drinking water by organic or inorganic acids is strongly correlated with a reduced percentage of *Campylobacter*-positive flocks (Allain et al. 2014; Torralbo et al. 2014; Ellis-Iversen et al. 2009). In this context, Mantzios et al. (2023)

demonstrated the in vitro anti-*Campylobacter* capacity of nine commercial water acidifiers, organic acid, and glycerides blended with EU approval for application in the drinking water of poultry. Among the tested products, researchers highlighted the high capacity of hydrogen peroxide-based products in inhibiting the growth of the tested *C. jejuni* and *C. coli* strains under extremely low concentrations (Mantzios et al. 2023).

Despite the promising results of the in vitro and field data, the application of organic and/or inorganic acids to experimentally challenged birds seems to have limited effect on the colonization of *C. jejuni* in the ceca of birds at slaughter age (Szott et al. 2022; Mortada et al. 2020). In particular, Szott et al. (2022) tested a blend of sorbic acid, benzoic acid, propionic acid, and acetic acid in broilers, which were challenged by *C. jejuni*. Researchers reported that adding organic acids consistently reduced significant *C. jejuni* loads in cloacal swabs during the early stage of infection, whereas at the end of the study, there were no significant differences in *C. jejuni* loads of cecal and colon contents compared to the positive control group (Szott et al. 2022). Similarly, Mortada et al. (2020) reported that an organic acid-based commercial product (formic acid, cinnamaldehyde) significantly reduced the proliferation of *C. jejuni* under the in vitro tests but did not alter *C. jejuni* loads in ceca or on the carcass of birds at 42 days of age (Mortada et al. 2020). On the other hand, Lassen et al. (2022) in a systematic literature review of products with potential application for use in the control of *Campylobacter* spp. in organic and free-range broilers concluded that a blend of organic acids could be a potential candidate for reducing *Campylobacter* spp. in broilers. In the same context, Ebrahimi et al. (2016) demonstrated that combining butyric acid (salts and esters of butyric acid) glycerides with protected organic acids in the feed could reduce *C. jejuni* counts in the cecum at 42 days of age (Ebrahimi et al. 2016). In addition, Skoufos et al. (2019) reported that dietary ferric tyrosine significantly reduced *C. jejuni* counts (2 log CFU/g) in the ceca of broilers that were experimentally challenged after exposure to contaminated litter. Wagle et al. (2017) observed reductions of 2.5 log CFU/g of a four-strain mixture of *C. jejuni* when β-resorcylic acid were orally given to chickens for the first 14 days of life. Finally, Skånseng et al. (2010) reported that the supplementation of broiler diets with formic acid and sorbate could prevent the colonization of *C. jejuni*, whereas adding formic acid alone was insufficient to prevent *C. jejuni* colonization (Skånseng et al. 2010).

In summary, it can be reported that the efficacy of organic and inorganic acids in broilers can be affected by several factors, such as the type of organic acid, concentration, growth stage, feed ingredients, microbial population, and environmental factors. These factors may explain the recorded variations in the previous investigations.

4.4 Phytogenic Feed Additives

The anti-*Campylobacter* activity of PFAs is early evident by previous in vivo investigations, although the effectiveness of treatments varies considerably. Ibrahim et al. (2020) demonstrated that the supplementation of *Glycyrrhiza glabra* extract in the basal diet of broiler chicks under *C. jejuni* challenge improved the body weight gain, the FCR, and reduced the mortality rates. The tight junction expression of occludin and JAM-2, as well as the expression of mucin-2, was increased compared to the control group. Finally, *C. jejuni* counts in the fecal samples on the seventh day post-infection were significantly reduced in the phytogenic product group. Reduced *C. jejuni* counts were also reported when thymol and carvacrol were added in the diet of broiler chickens (Arsi et al. 2014; Szott et al. 2020). Nooreh et al. (2021) stated that dietary *Ferulago angulata* extract in *C. jejuni*-challenged broilers improved the FCR, the mortality rates, and decreased the *C. jejuni* counts as well as coliforms in the ileum and cecum. Finally, this extract boosted the immune response, by enhancing the IgG, IgA, IL-6, and gamma-interferon concentrations under *Campylobacter* spp. challenge. Aljazzar et al. (2022) reported that dietary eugenol and trans-cinnamaldehyde downregulated mRNA expression levels of *flaA, virB11*, and *WlaN* virulence genes of *C. jejuni*, at the 7th and 14th days post-infection, in experimentally challenged Ross 308 broiler chicks. Finally, in Hy-Line laying hens experimentally challenged by *C. hepaticus* strain HV10, dietary sanguinarine and chelerythrine extracts reduced miliary lesions on the liver, the lesion score, and improved the growth performance (Quinteros et al. 2021) (Table 3).

5 Salmonellosis

Salmonella spp. belong to the Enterobacteriaceae family, which are known intracellular, gram-negative motile bacilli with peritrichous flagella. The name "Salmonella" is derives from an American scientist, Daniel E. Salmon, who isolated the enteric pathogen from the pig's intestine. *Salmonella* spp. cause localized or systemic infections and can cause devastating damage to the poultry industry, and simultaneously pose a significant public health issue. Each year, approximately 48 million people contract a foodborne disease, costing the US economy around $90 billion in healthcare expenses. According to the Centers for Disease Control and Prevention, *Salmonella* causes 1.35 million infections per year in the USA (CDC 2021), whereas *Salmonella* spp. in poultry products causes $2.8 billion in financial losses to the US poultry industry (Egg-STAT-Ic; Scharff 2020). In particular, salmonellosis is a major zoonotic pathogen, with poultry meat and eggs being the primary source of infection (Ruvalcaba-Gómez et al. 2022; El-Saadony et al. 2022). It is estimated that 41% of chicken meat and one in 20,000 eggs are contaminated by *Salmonella* spp. (Park et al. 2015; Thames and Sukumaran 2020).

Salmonellosis causes significant losses to poultry, and thus several strategies to control and reduce its spread at the farm level have been attempted. For several years the use of antibiotics has been the main strategy for the control of *Salmonella* spp. However, the use of antibiotics has been questioned since antibiotic resistance or decreased sensitivity of *Salmonella* spp. has been observed and poses a threat to public health. Over the last years, the use of feeding-based strategies as an alternative to antibiotics for *Salmonella* prophylaxis in poultry has been recommended. Among

Table 3 In vivo experimental studies on the effect of probiotics, prebiotics, (in)organic acids, and phytogenics to combat *Campylobacter* spp. infections in poultry

Type of additive	Type of bird	Dosage	Challenge	Effect	References
Probiotics					
Lactobacillus salivarius SMXD51	Ross PM3 chicks	10^7 CFU/mL	*C. jejuni*	↓ *C. jejuni* loads (2.81 log CFU/g)	Saint-Cyr et al. (2017)
Lactobacillus gasseri SBT2055	White leghorn chicks	1×10^8 CFU in 100 μL	*C. jejuni* 81–176	↓ *C. jejuni* counts (8 log CFU/g) in the ceca	Nishiyama et al. (2014)
Bacillus amyloliquefaciens	Ross 308 broilers (male)	2.5–7.5 × 10^5 CFU/g	*C. jejuni*	↑ growth performance, ↓ cecal colonization and fecal shedding of *C. jejuni*, ↑ *Bifidobacterium* and *Lactobacillus* species, ↓ *Clostridium* species and *Enterobacteriaceae*, ↑ mRNA expression levels of digestive enzymes encoding genes (AMY2a, PNLIP, CELA1, and CCK), ↑ expression of barrier functions-linked genes including DEFB1, FABP-2, and MUC-2	Ismail et al. (2023)
E. coli Nissle 1917	SPF chickens	9.6×10^8 –1× 10^9 CFU/bird	*C. jejuni* (multiple strains)	↓ 2.0–2.5 log *C. jejuni* colonization in the cecum, ↑ villous height, crypt depth and villous height/crypt depth ratio in the jejunum and ileum, ↑ *C. jejuni*-specific antibodies in serum, ↑ IgA and IgY antibodies, ↑ cytokines and chemokines (Th1, Th2, and Th17 pathways)	Helmy et al. (2022)
Bacillus subtilis PS-216	Broiler chicks	2.5×10^6 CFU/ mL water	*C. jejuni*	↓ *Campylobacter* counts in ceca, ↑ weight gain	Šimunović et al. (2022)

(continued)

Table 3 (continued)

Type of additive	Type of bird	Dosage	Challenge	Effect	References
Commercial probiotic mixture (*Enterococcus faecium*, *Pediococcus acidilactici*, *Bifidobacterium animalis*, *Lactobacillus salivarius*, and *Lactobacillus reuteri*)	Ross 308 broiler chicks	2–20 mg/bird per day	*C. jejuni*	↓ *C. jejuni* counts (6 log CFU/g) cecal content	Ghareeb et al. (2012)
Prebiotics					
Yeast cell wall (YCW)	Ross 708 broilers (male)	200–800 g/ton	*C. jejuni*	The addition of YCW during a *C. jejuni* challenge did not have an impact on growth performance, innate immune response, cecal colonization, carcass yield, or *C. jejuni* prevalence after processing	Munoz et al. (2023)
Whey	Ross 308 broiler chicks	1%	*C. jejuni* MB 4185 (KC 40)	The use of whey did not significantly affect the performance and cecal *C. jejuni* counts	Tsiouris et al. (2019)
Mannan-oligosaccharide	Ross 308 broilers (male)	2 g/kg	*C. jejuni*	↓ *C. jejuni* counts in the ceca and ileum, ↑ serum IgG and IgA concentrations	Rostami et al. (2021)
Saccharomyces-derived prebiotic refined functional carbohydrates (RFC) with yeast culture	Cobb broiler chicks (male)	Feed: 50–100 g/ton Water: 500 ppm	*Campylobacter* spp.	↑ ADWG, ↓ cecal *Campylobacter* counts	Froebel et al. (2019)
Organic acids and inorganic acids					
A blend of sorbic acid, benzoic acid, propionic acid, and acetic acids	Ross 308 broiler chicks	480 mmol/L	*C. jejuni* strain BfR-CA-14430	↓ *Campylobacter* loads in cloacal swabs (2 log). However, at the end of the trial, no significant differences were detected in *Campylobacter* loads of cecal and colon contents	Szott et al. (2022)

the-
se

Commercial product (blend of formic acid, cinnamaldehyde, and a permeabilizing complex)	Cobb 500 broilers (male)	0.5 kg/ton	*C. coli*	↓ proliferation of *Campylobacter* under the in vitro tests but did not alter *Campylobacter* loads in ceca or on the carcass of birds at 42 days of age	Mortada et al. (2020)
Several combinations of medium-chain fatty acids (MCFAs), monoglycerides (MGs) of MCFA and MG-MCFA+OA.	Ross 308 broiler chicks	Several concentrations depending on group	*C. jejuni*	↓ cecal *Campylobacter* colonization compared to control at 35 days, but only the group treated with monoglycerides of MCFA, maintained the reduction at 42 days.	Gracia et al. (2016)
Combinations of formic acid and/or potassium sorbate	Ross 308 broiler chicks	0.1–2%	*C. jejuni* C484	↓ colonization of *C. jejuni*	Skånseng et al. (2010)
Caprylic acid	Broiler chickens	0.7–1.4%, 0.35%, or 0.7%	*C. jejuni* (5-wild-type strain mixture)	↓ cecal *Campylobacter* colonization	Solis de los Santos et al. (2009)
Ferric tyrosine	Ross 308 broilers (male)	0.02–0.20 g/kg	*C. jejuni*	↓ *C. jejuni* counts (0. 20 g/kg: 2 log10) in the ceca, ↑ weight gain and feed efficiency	Skoufos et al. (2019)
β-resorcylic acid	Cobb 500 broiler chicks	0.25–1%	*C. jejuni* (four-strain mixture)	↓ *C. jejuni* counts in cecal content (0.5%: ~2.5 log CFU/g and 1%: 1.7 log CFU/g)	Wagle et al. (2017)
Phytogenics					
Ferulago angulate extract	Ross 308 broiler chicks	200–400 mg/kg	*C. jejuni*	↑ growth performance, ↑ immune status (↑ IgG, IgA, IL-6, and gamma interferon), ↓ C. *jejuni* counts only in the ileum	Nooreh et al. (2021)
Thymol/carvacrol/blend of thymol, carvacrol	Ross 308 broilers	0.25–2%	*C. jejuni*	↓ *C. jejuni* counts	Arsi et al. (2014)

(continued)

Table 3 (continued)

Type of additive	Type of bird	Dosage	Challenge	Effect	References
Glycyrrhiza glabra (licorice) extract	Ross 308 broilers	0.25–2 g/kg	*C. jejuni* (ATCC 33291)	↑ body weight gain, ↓ FCR, ↓ expression of genes coding for tight junction proteins occludin and junctional adhesion molecules, ↑ expression of GLP-2 gene, ↓ TLR-4, ↓ IL-1β, ↓ mean log10 count of *C. jejuni* in cecal samples (2 g/kg: 2.28 log10 CFU/g),	Ibrahim et al. (2020)
Blend of eugenol and trans-cinnamaldehyde	Ross 308 broilers	400 mg/kg	*C. jejuni*	↑ growth performance variables, ↓ *C. jejuni* fecal loads, ↓ *C. jejuni* virulence genes (flaA, virB11, and wlaN) expressions, ↑ immunostimulant and anti-inflammatory activities	Aljazzar et al. (2022)
Carvacrol	Ross 308 broilers	120 mg/kg	*C. jejuni*	↓ *C. jejuni* counts (1.17 log10) in cloacal swabs. However, carvacrol did not successfully reduce *Campylobacter* cecal colonization at the end of the trial.	Szott et al. (2020)
Commercial product (blend of sanguinarine and chelerythrine)	Hy-Line laying hens	100–200 mg/kg	*C. hepaticus* (HV10)	↑ BW, ↑ FI, ↓ lesion score, ↑ egg mass, ↓ IL-8 in blood	Quinteros et al. (2021)

Abbreviations: *IL*, interleukin; *FCR*, feed conversion ratio; *BW*, body weight; *MCFA*, medium-chain fatty acids; *IgM*, immunoglobulin M; *OPG*, oocysts per gram; *FI*, feed intake; *ADWG*, average daily weight gain; *IgG*, immunoglobulin G; *TLR-4*, Toll-like receptor 4; *SPF*, specific pathogen-free; *CFU*, colony-forming unit feeding-based strategies, the use of probiotics, prebiotics, (in)organic acids, and phytogenic feed additives (PFAs) is an essential approach to combat *Salmonella* spp. in poultry (Ruvalcaba-Gómez et al. 2022; El-Saadony et al. 2022).

5.1 Probiotics

Probiotics seem an attractive alternative for the control of salmonellosis in poultry since they have been successfully used to inhibit the growth of pathogenic bacteria, are safe for consumption by birds, are bile-resistant, and can perhaps be modified to produce a variety of antimicrobial peptides that are now recognized. The mechanisms underlying the probiotic's efficacy are related to the strengthening of the intestinal microbiota and the production of antimicrobial factors such as bacteriocins, hydrogen peroxide, lactic acid, and short-chain fatty acids (Ruvalcaba-Gómez et al. 2022). Wang et al. (2023) reported the probiotic potential of *Lactobacillus fermentum* 1.2133, both in vitro and in vivo, in broiler chicks that were experimentally challenged by *Salmonella pullorum* CVCC533 at the 14th day of age. The dietary application of *L. fermentum* significantly regulated the intestinal flora of challenged birds by reducing the number of *Salmonella* and aerobic bacteria in the chicks' intestinal content, and by improving the abundance of *Lactobacilli* species. Additionally, *L. fermentum* resulted in increased expression of avian β-defensin-2 (AvBD2) mRNA in the small intestine of chickens. It is reported that defensins, a class of the host endogenous antimicrobial peptides, play a key role in innate immunity and the cellular barrier (Yi et al. 2014), and their expression in birds that received probiotics, could be an additional mechanism of probiotic action (Wang et al. 2023). Forkus et al. (2017) engineered probiotic *E. coli* Nissle 1917 to express and secrete the antimicrobial peptide, microcin J25. When researchers tested *E. coli* Nissle 1917, under in vitro experiments and in an animal model of 300 turkeys, they reported that *Salmonella enterica* carriage was 97% lower in the treated group than in the positive control. Additionally, Shao et al. (2022) reported the high capacity of *Enterococcus faecium* NCIMB 11181, in ameliorating the adverse effects of *Salmonella typhimurium* when added to diets of Arbor Acres male broilers that were experimentally challenged. In particular, *E. faecium* NCIMB 11181 reduced the intestinal colonization of *S. typhimurium* and its translocation to visceral organs, whereas diminished the damage in the gut morphological structure as well as the dysbiosis, which is caused by *S. typhimurium* infection. In another study, the administration of the probiotic *C. butyricum* HJCB998 reduced counts of *Salmonella* spp., *C. perfringens*, and *E. coli* in the ceca of broiler chicks (Yang et al. 2012). Additionally, the supplementation of broilers fed with *C. butyricum* HJCB998 resulted in increased IgA, IgG, and complement component-3 concentrations in the serum, indicating its promoting effect on birds' immune function (Yang et al. 2012).

5.2 Prebiotics

Another promising feeding-based strategy for the control of salmonellosis is the use of *prebiotics* (Micciche et al. 2018). There are several categories of prebiotics, such as fructooligosaccharides (FOS), mannan-oligosaccharides (MOS), and galactooligosaccharides (Ruvalcaba-Gómez et al. 2022), and some of them have effectively ameliorated the adverse effects of *Salmonella* infection in previous in vivo investigations. In particular, Adhikari et al. (2018) reported the supplementation of poultry diets with FOS could impair the pathogenesis of *Salmonella enteritidis*, in experimentally challenged white leghorn laying hens, by modulating humoral immunity. Specifically, birds that were challenged and received FOS by their diet had reduced *S. enteritidis* cecal counts and upregulation of TLR-4, IFN-γ, and IgA compared to birds that were only challenged (Adhikari et al. 2018). Similarly, Wu et al. (2020) reported that supplementing the diet with prebiotic trehalose alleviated the adverse effects of *S. typhimurium* in experimentally challenged Arbor Acres male broiler chicks. More specifically, the prebiotic trehalose increased the abundance of *Lactobacilli* and decreased *S. typhimurium* counts in ceca, whereas exhibited anti-inflammatory effects by downregulation of IL-1β and lipopolysaccharide-induced tumor necrosis factor-α (LITAF), and upregulation of IL-10 and IFN-α (Wu et al. 2020). In another study, Santana et al. (2020) reported the beneficial effect of lactulose in the intestinal development of turkeys in an *S. enteritidis* challenge model. In conclusion, possible explanations for the protective effect of prebiotics against salmonellosis, apart from enhancing the growth of beneficial bacteria of intestinal microbiota, could be the immune-stimulative effect on the gut-associated lymphoid tissue and increased levels of secretory IgA in the intestinal mucosa (Ruvalcaba-Gómez et al. 2022; El-Saadony et al. 2022).

The addition of whey in poultry diets has been used for the control of intestinal pathogens, such as *Salmonella* spp. and *C. perfringens* in poultry, although chicks' digestive system is not able to degrade its main component, lactose (Corrier et al. 1990; De Loach et al. 1990; McReynolds et al. 2007). In particular, the use of whey in poultry diets decreased the number of *Salmonella*-positive birds by 90%, and the intestinal carriage of *S. typhimurium* by more than 2 $\log^{10}$ in chicks fed with whey compared to controls (De Loach et al. 1990). The proposed mechanism is that lactose acts as a growth substrate for beneficial bacteria of intestinal microbiota, such as *Lactobacillus* spp. and *Bifidobacterium* spp., which ferment it to lactic acid and volatile fatty acids. Possible mechanisms for the effect of lactic acid bacteria on intestinal pathogens include the production of bacteriocins, the stimulation of adaptive immunity, the alteration of the cecal microbiome, and the antimicrobial activity of the produced lactic acid (Gülşen et al. 2002; Neal-McKinney et al. 2012). Finally, in another study, El-Shall et al. (2020) reported the high efficiency of a commercial prebiotic product containing 2.6 Beta LevaFructan alongside with *Salmonella* vaccine, in ameliorating the adverse effects of *Salmonella* in terms of mortality rate, fecal shedding, and *S. enteritidis* translocation in liver, spleen, heart, and cecum in experimentally challenged birds.

5.3 Organic Acids

Commercial products, blends of organic and/or inorganic acids, exhibit promising efficiency in counteracting the growth of critical zoonotic bacteria, such as *Salmonella* spp. under in vitro tests (Mantzios et al. 2023; Kovanda et al. 2019). In particular, the anti-*Salmonella* activity of organic acids is not fully elucidated; however, it is mainly attributed to the nonionized form of organic acids, that can pass through the *Salmonella* cell wall and express their hazardous effects. Additionally, it is reported that the application of organic acids may decrease pH of the upper gastrointestinal tract, thus forming a barrier for the invasion and the colonization of *Salmonella* spp. in the intestine (El-Saadony et al. 2022; Lawhon et al. 2002; Van Immerseel et al. 2004a, b). The potential of organic acids as alternatives to combat *Salmonella* spp. infections in poultry is early evident in previous in vivo investigations (Wang et al. 2019; Bourassa et al. 2018; Milbradt et al. 2014). In particular, Bourassa et al. (2018) investigated the effect of various concentrations of formic and propionic acids, in both feed and water, in different periods, in broilers experimentally challenged by *S. typhimurium* at the early stage of life. Researchers concluded that the addition of formic acid from the sixth week resulted in no recovery of *S. typhimurium* from the ceca of birds. In contrast, Adhikari et al. (2020) found that the addition of a buffered formic acid and sodium formate mixture in the feed of broilers had no significant effect on the colonization of *S. typhimurium* in broiler chickens (Adhikari et al. 2020). Furthermore, Aljumaah et al. (2020) reported that a commercial organic acid-based product (a blend of short- and medium-chain fatty acids and a low dose of β1-4 mannobiose) marginally influenced the overall performance of broiler chicks challenged by *S. typhimurium* ATCC 14028. However, the acidic blend significantly increased cecal acetic and butyric acid concentrations and resulted in alterations in bacterial communities in the intestine (Aljumaah et al. 2020). In addition, Wang et al. (2019) conducted a study to determine the ameliorative potential of organic acids when added to the drinking water of broiler chicks, which were experimentally challenged by *S. pullorum*. The supplementation of the commercial organic acid blend promoted the performance of birds. Following metabolomic analysis, researchers showed a possible metabolic mechanism of organic acids, to alleviate stress responses mediated by steroid hormones. Finally, Milbradt et al. (2014) documented that a dietary mixture of short- and medium-chain fatty acids (acetic acid, formic acid, propionic acid, sorbic acid, polysorbate, silicon dioxide, coconut oil, and ammonium hydroxide) reduced *Salmonella* spp. and *Enterobacteriaceae* and increased *Lactobacillus* spp. in both crop and cecum and increased butyric acid concentration in cecal contents in commercial turkeys challenged with *S. enteritidis*.

5.4 Phytogenic Feed Additives

PFAs and their derivatives, especially essential oils, have gained rising interest from the scientific community as natural growth promoters in the poultry industry. They are natural plant-derived compounds or plant extracts with antimicrobial, antioxidant, anti-inflammatory, immunomodulatory, and digestion-stimulating activities (Omonijo et al. 2018; Zhai et al. 2018). Several essential oils, such as cinnamaldehyde, eugenol, thymol, and carvacrol, have been successfully used for the control of *Salmonella* spp. since they can significantly reduce the colonization and counts of *Salmonella* spp. in the intestinal tract of broiler chicks and layer hens (El-Saadony et al. 2022; Kollanoor-Johny et al. 2012). Moharreri et al. (2022) reported that the supplementation of diets with a microencapsulated blend of essential oil (thyme, summery savory, peppermint, and black pepper seed) improved the BW, FI, and FCR in broilers, which were experimentally challenged by *S. enteritidis*. The phytogenic blend also reduced the lipid peroxidation and enhanced the antioxidant capacity in the serum, breast meat, and total phenolic contents of birds' breast. Additionally, the gene expression analysis in samples from infected birds that received the phytogenic blend showed increased expression of occludin and SOD and decreased expression of inducible nitric oxide synthase (iNOS). The phytogenic blend also improved the histological structure of the ileum, by increasing the villus height, villus width, villus-to-crypt (V:C) ratio, and the number of goblet cells. Finally, the blend reduced *S. enteritidis* counts in the ileum and cecum of birds. A similar trial was performed by using Cobb broiler chicks, which were infected by *S. enteritidis* and received dietary supplements with a commercial phytogenic product (blend of fennel, melissa balm, peppermint, anise, oak, clove, and thyme). The microbiota analysis of the cecal content revealed that the inclusion of the phytogenic blend in broiler diets decreased the number of *Salmonella* spp., *E. coli*, and *C. perfringens* and increased the number of total *Lactobacillus* spp. in the ceca of birds (Wati et al. 2015). In another study, laying hens infected by *Salmonella gallinarum* in a fowl typhoid model showed lower mortality rates in the group that received a chestnut extract in the basal diet compared to the positive control group (Casanova et al. 2021). Ibrahim et al. (2021a, b) noted that broilers infected by *S. typhimurium* and fed a diet that was supplemented with encapsulated thymol upregulated the expression of digestive enzymes *AMY2A, PNLIP*, and *CCK*, the expression of biomarkers *Muc-2, FABP2, IL-10*, and downregulated the expression of *IL-2* and *IL-6*. Finally, thymol increased total *Lactobacillus* spp. counts and decreased *S. typhimurium* counts in broilers experimentally challenged by *S. typhimurium*. Recently, garlic and onion extract chitosan nanoparticles inhibited the colonization of *S. typhimurium* and simultaneously improved the synthesis of intestinal microbiota in birds (Enoka et al. 2021) (Table 4).

6 Colibacillosis

Avian pathogenic *Escherichia coli* (APEC) infections in poultry cause several pathological conditions collectively known as "colibacillosis" and are characterized by a triad of lesions of "perihepatitis, pericarditis and airsacculitis," accompanied by septicemia and increased mortality (Fancher et al. 2020). APEC infections are associated with major economic losses due to mortality, meat and egg production

Table 4 In vivo experimental studies on the efficacy of probiotics, prebiotics, (in)organic acids, and phytogenics to combat *Salmonella* infections in poultry

Type of additive	Type of bird	Dosage	Challenge	Effect	References
Probiotics					
Lactobacillus fermentum 1.2133	Roman broiler chicks	2.5–5 × 10^8 CFU/chick	*S. pullorum* CVCC533	↓ *S. pullorum* and aerobic bacteria in the chicken duodenum, ileum, and cecum, ↑ species abundance of lactobacilli, ↓ intestinal damage, ↑ avian β-defensin	Wang et al. (2023)
Enterococcus faecium	Ross 308 broiler chicks (eggs)	16 × 10^9 CFU/egg	*S. enteritidis*	↓ *S. enteritidis*-positive chicks	De Oliveira et al. (2014)
E. coli Nissle 1917 (Microcin J25)	Hybrid Converter Breed Tom turkey poults	10E7 CFU EcN (J25)/bird	*S. enterica*	↓ (97%) *S. enterica* carriage	Forkus et al. (2017)
Clostridium butyricum	Lingnan yellow broiler chicks (male)	2–3 × 10^7 CFU *C. butyricum*/kg of diet	–	↑ BW, ↑ ADG, ↑ IgA and IgG in the serum, ↑ complement component-3 in the serum, ↓ *Salmonella*, ↓ *E. coli*, ↓ *C. perfringens* in ceca, ↑ cecal *Lactobacillus* and *Bifidobacterium* counts, ↑ *C. butyricum* counts	Yang et al. (2012)
Enterococcus faecium NCIMB 11181	Arbor Acres broilers chicks (male)	4 × 108 CFU/kg of diet	*S. typhimurium*	↓ *S. typhimurium* colonization and translocation, ↓ gut morphological structure damage and intestinal cell apoptosis induced by *S. typhimurium* infection, ↑ villous height, ↓ intestinal TUNEL-positive cell numbers, ↑ anti-*Salmonella* antibodies in intestinal	Shao et al. (2022)

(continued)

Table 4 (continued)

Type of additive	Type of bird	Dosage	Challenge	Effect	References
				mucosa and serum of the infected birds, ↓ dysbiosis, ↑ *Lachnospiraceae* and *Alistipes*, ↓ *Barnesiella* abundance	
Prebiotics					
Fructooligosaccharides (FOS)	White Leghorn W-36 laying hens	0.1%	*S. enteritidis*	↓ fecal *S. enteritidis* counts, ↓ *S. enteritidis*-positive ovaries, ↑ TLR-4, IFN-γ, and IgA,	Adhikari et al. (2018)
Trehalose dihydrate	Arbor Acres broiler chicks (male)	1–5% w/w	*S. typhimurium*	↓ FCR, ↓ *S. typhimurium* counts in ceca, ↑ abundance of lactobacilli, ↓ inflammation, ↓ adverse effects on serum levels of AST, TG, and albumin and globulin ratio, ↓ intestinal damages, maintained the integrity of cecal epithelial cells, ↓ IL-1β and LITAF, ↑ IL-10 and IFN-α.	Wu et al. (2020)
2.6 Beta LevaFructan	Cobb broiler chicks	100 g on 1000 mL/ 0.5 mL per liter of drinking water)	*S. enteritidis*	↓ negative effect of live vaccine growth performance, ↓ mortality rate, ↓ fecal shedding, ↓ *S. enteritidis* in liver, spleen, heart and cecum.	El-Shall et al. (2020)
Lactulose	Buta turkeys	0.003 mL kg^{-1} (body weight)	*S. enteritidis*	↑ intestinal development (in the prebiotic control group)	Santana et al. (2020)
Organic acids					
Formic	Cobb 500 broiler chicks (male)	4 kg/ton	*S. typhimurium*	No recovery of *S. typhimurium* from ceca, ↓ transmission of	Bourassa et al. (2018)

				S. typhimurium from shedders to sentinels	
Commercial organic acid blend (short- and medium chain-fatty acids and a low dose of β1-4 mannobiose)	Ross 308 broiler chicks	3.0 g/kg	*S. typhimurium* ATCC# 14028	↑ increased cecal acetic and butyric acids concentrations, minor alterations of intestinal bacterial communities.	Aljumaah et al. (2020)
Commercial organic acid blend (formic acid, acetic acid, and ammonium formate)	Arbor Acre broiler chicks (male)	1–1.5 L/ 1000 L	*S. pullorum* (CVCC521, C79–5)	↑ performance, modulate the systemic metabolic perturbation caused by challenge, ↓ stress responses mediated by steroid hormone, ↑ antioxidant or immune defense, and modified intestinal microbiota metabolism.	Wang et al. (2019)
Commercial organic acid blend (acetic acid, formic acid, propionic acid, sorbic acid, polysorbate, silicon dioxide, coconut oil, and ammonium hydroxide)	British United Turkey poults	2 g/kg of feed	*S. enteritidis*	↓ *Enterobacteriaceae* abundance in crop and cecum, ↑ butyric acid, ↓ *S. enteritidis* colonization ceca, ↓ contamination risk of meat during slaughter.	Milbradt et al. (2014)
Commercial organic acid preparations (a blend based on fumaric acid, calcium format, calcium propionate, and potassium sorbate; and a bend based on citric acid, calcium formate, calcium butyrate, and calcium lactate)	Ross 308 broiler chicks	0.06% and 0.10%, respectively	–	↓ *E. coli* and *salmonella* spp. counts	Hassan et al. (2010)
Phytogenics					
Thymol (encapsulated)	Ross 308 broilers (male)		*S. typhimurium*	Reduction of *S. typhimurium* loads and virulence	Ibrahim et al. (2021a, b)
		150 mg/kg		↑ BW, ↓ FCR, retention of N and crude fiber, ↑ fecal moisture	Wati et al. (2015)

(continued)

Table 4 (continued)

Type of additive	Type of bird	Dosage	Challenge	Effect	References
Blend of fennel, melissa balm, peppermint, anise, oak, clove, and thyme	Cobb 400 broiler chicks (male)		*S. enteritidis* 0363P (ATCC 14028) and *E. coli* (ATCC 25922)	content, ↓ digesta transit time, ↓ *Salmonella* spp. *E. coli* and *Clostridium* spp. counts, ↑ *Lactobacillus* spp. counts, ↑ humoral immune response, ↓ heterophil-to-lymphocyte ratio, ↑ ALP	
Blend of thyme, summer savory, peppermint, black pepper seed (encapsulated)	Ross 308 broilers (male)	0.5–2 kg/ton	*S. enteritidis*	↑ BW, ↑ FI, ↓ FCR, ↑ antioxidant status, ↑ ileal morpho structure, ↑ intestinal microbial population, ↑ occludin, ↑ SOD	Moharreri et al. (2022)
Quebracho wood (*S. lorentzii*, >78% condensed tannins) or chestnut (*C. sativa*, >84% hydrolysable tannins)	Cobb 500 broilers	0.1% w/w	*S. enteritidis* INTA86	↓ *S. enteritidis* excretion	Casanova et al. (2021)
Chestnut (*C. sativa*, >84% hydrolysable tannins)	Lohmann Brown Classic laying hens	0.1%	*S. gallinarum* INTA91 S	↓ mortality	Casanova et al. (2021)

Abbreviations: *IL*, interleukin; *FCR*, feed conversion ratio; *BW*, body weight; *IgM*, immunoglobulin M; *IgA*, immunoglobulin A; *FI*, feed intake; *TLR-4*, Toll-like receptor 4, *IFN-γ*, interferon gamma, *LITAF*, lipopolysaccharide-induced tumor necrosis factor-alpha factor; *IFN-α*, interferon alpha; *ALP*, alkaline phosphatase; *SOD*, superoxide dismutase; *CFU*, colony-forming unit; *ADG*, average daily gain losses, carcass condemnation at slaughter, costs associated with treatment, and prophylaxis as well as compromised animal welfare in industrial poultry production (Tarabees et al. 2019). Young birds are more susceptible to severe infections compared to adult birds. Bacteria can be transferred among humans, animals, and the environment (Kathayat et al. 2021; Christensen et al. 2021; Hu et al. 2022).

APEC is a member of chickens' intestinal microbiota and resident of the upper respiratory tract, mainly in the pharynx and trachea. For several years, APEC has been considered a secondary pathogen requiring predisposing factors such as poor ventilation and hygiene, high ammonia levels, extreme house temperatures, immunosuppression, and other infections, to result in pathological conditions (Nolan et al. 2013). However, according to Nolan et al. (2019), APEC could have a primary role since *E. coli* was isolated from colibacillosis without obvious predisposing factors. In addition, several APEC outbreaks have been related to single clones such as *E. coli* ST95 and *E. coli* ST117 (Christensen et al. 2021; Ronco et al. 2017).

For several years the control of APEC infections has relied on antibiotics, which led to the emergence of antibiotic-resistant *E. coli* and resurgence of the public and food safety authorities' concerns. In particular, concerns have been raised that resistant strains or plasmids could be transferred to humans via the food chain and could increase the risk for extraintestinal infections and impaired their treatment. Therefore, there is a huge need to consider the "One Health" perspective (Kathayat et al. 2021; Christensen et al. 2021; Hu et al. 2022). The adaption of vaccination strategies, strict biosecurity measures, and the control of immunosuppressive agents and/or stress factors could be preventative to the odds of APEC infections (Kathayat et al. 2021). However, feed additives such as probiotics, prebiotics, organic acids, and phytogenics have been widely studied during the last decade (Wang et al. 2017a, b; Emami et al. 2017; Hashem et al. 2022; Ibrahim et al. 2022) aiming to develop an alternative to antibiotics for the control of infectious agents, such as APEC, in poultry.

6.1 Probiotic

The supplementation of probiotic mixtures in poultry diets influences the intestinal microbiota structure, including counts of APEC, leading to increased production, health, and welfare (Kathayat et al. 2021). Probiotics such as *Lactobacillus* spp. produce lactic acid, an organic acid that is known for its inhibitory action against pathogenic *E. coli* (Watkins et al. 1982). The enrichment of poultry microbiome with *Lactobacilli* species could be a promising approach to reducing the incidence of APEC colonization. Wang et al. (2017a, b) reported that dietary *Lactobacillus plantarum* B1 significantly increased the BW and significantly decreased *E. coli* counts in the cecal content of broiler chicks that were experimentally challenged by *E. coli* K88. In addition, researchers stated that *L. plantarum* B1 increases the population of cecal lactic acid bacteria, as well as improves intestinal mucosal immunity by increasing the ileal mucosal secretory IgA concentration and reducing the IFN-γ, IL-2, IL-4, and TNF-α levels in the ileum (Wang et al. 2017a, b).

Similarly, Ding et al. (2019) demonstrated that the dietary addition of *L. plantarum* 15-1 improved intestinal health by increasing the levels of SCFAs and mitigating the damage caused by *E. coli* O78 in experimentally challenged broilers. In a current study, Ye et al. (2021) investigated the effects of a probiotic mixture, including both oxybiotic and anaerobic microbes, on growth performance and intestinal microbiota of partridge shank broiler chicks. Researchers reported that the supplementation of the probiotic mixture in the feed improved the growth performance by increasing the BW, average daily weight gain (ADWG), and FCR and positively influencing the intestine's morphological characteristics. Researchers noted that the abundance of harmful bacteria in the phyla of *Firmicutes*, *Euryarchaeota*, and *Ruminococcus* was dramatically decreased in the chicks that received the probiotic mixture, while the abundance of beneficial bacteria in the phyla of *Actinobacteria* and *Eremiobacterota* was significantly increased (Ye et al. 2021). In another study, the administration of a multistrain probiotic containing *Bacillus subtilis*, *Clostridium butyricum*, and *Lactobacillus plantarum* improved the performance, affected the synthesis of cecal microbiota, and reduced the APEC O78 translocation to visceral organs in experimentally challenged broiler chicks (Tarabees et al. 2019). Furthermore, Tarabees et al. (2020) demonstrated that the addition of *Enterococcus faecalis*-1 significantly improved immunological response and growth performance, and increased the total *Enterococcus* counts in the ceca, while significantly decreased the mortality, and *E. coli* O78 visceral invasion in experimentally challenged broilers. In another study, Kathayat et al. (2022) investigated in vitro and in vivo effects of *Lacticaseibacillus rhamnosus* GG against APEC infection in broiler chicks. The oral administration of *L. rhamnosus* GG reduced the colonization (~1.6 logs) of APEC and moderated the APEC-induced alterations in the microbial community in the ceca of experimentally challenged broilers. In particular, *L. rhamnosus* GG decreased the abundance of *Proteobacteria*, particularly those belonging to *Enterobacteriaceae* (*Escherichia-Shigella*) family. Applying a different approach, Li et al. (2021) investigated the effect of the in ovo inoculation of a single probiotic strain (*Lactobacillus animalis*, *Lactobacillus reuteri*, or *Lactobacillus rhamnosus*) on the incidence of APEC infection in broilers, and evaluated the virulence and antimicrobial resistance properties of APEC isolates. There were no significant differences between the control and treatments when screening isolates for APEC-related genes (*iroN, ompT, hlyF, iss*, and *iutA*) by polymerase chain reaction (PCR). Consequently, researchers concluded that the in ovo inoculation of a single probiotic strain did not confer protection against APEC strains in broilers.

However, the in ovo probiotic ingestion is an up-and-coming technique that suits modern poultry production requirements and was previously effective in combatting *S. enteritidis* in experimentally challenged broilers that received *E. faecium,* both in ovo and through the diet (De Oliveira et al. 2014). However, to better understand the potential of in ovo probiotic ingestion, to combat APEC infection in poultry, additional trials using individual probiotics or combinations of probiotics should be conducted.

6.2 Competitive Exclusion

The use of competitive exclusion cultures may significantly reduce intestinal colonization with challenge strains of APEC, as well as the excretion and transmission of APEC. In particular, a commercial competitive exclusion product (Aviguard®, Microbial Developments Ltd.) reduced the excretion and transmission of an extended-spectrum cephalosporin-resistant strain, in the absence of antimicrobial treatment (Ceccarelli et al. 2017). Similarly, Chantziaras et al. (2018) found that the *E. coli* population originating from Aviguard® was predominant and largely prevented the colonization and transmission of both the bacteriologically fit and the bacteriologically nonfit enrofloxacin-resistant strains. Furthermore, the administration of Aviguard® resulted in a substantial decrease in cecal extended-spectrum β-lactamases and AmpC-producing *E. coli* counts in young chickens (Methner and Rösler 2020).

6.3 Prebiotics

Prebiotics, such as fructooligosaccharides (FOS), inulin, galactooligosaccharides (GOS), and other undigested starches, are frequently added in poultry diets, for their beneficial effect on gut microbiome populations, as they serve as nutrients' source for beneficial bacteria in the intestine. Indeed, bacteria such as *Bifidobacterium* utilize prebiotics by fermentation that produces short-chain fatty acids (SCFAs), which are known for their inhibitory effect against *E. coli* (Shastak et al. 2015). β-Mannan is a nonstarch polysaccharide (NSP), usually found in soybean meal, sesame meal, palm kernel meal, copra meal, and guar meal (Dhawan and Kaur, 2007). It is reported that β-1,4-mannan-oligosaccharides potentially limit pathogenic bacteria to adherence to the gut as a result of saturation of binding sites (a mannose-specific binding site) on the bacterial surface. Thus, bacteria such as APEC that have a mannose-specific binding site on their surface and create bonds by β-1-4-MOS or D-mannose in the intestinal lumen are expelled by the bird's physiological mechanical defense mechanisms such as mucus secretion and peristalsis (Oyofo et al. 1989). The reason for the prebiotic effect of β-1-4-MOS is that it can be used as nutrients by *Bifidobacteria* and *Lactobacilli* but not, for example, by *C. perfringens* and *E. coli*, thereby promoting healthful intestinal microflora (Asano et al. 2001). Tarabees et al. (2019) reported that the dietary addition of the prebiotic isomaltooligosaccharide significantly improved the growth performance and modulated the intestinal microbiota of broiler chickens challenged with APEC O78. In particular, the addition of isomaltooligosaccharide in diets of birds reduced mortality percentage and *E. coli* recovery rates from the liver and spleen, while increased the carcass weight as well as the total lactobacilli and total lactobacilli-enterococci populations in the ceca of challenged birds (Tarabees et al. 2019). In addition, Ding et al. (2019) reported that dietary FOS improved growth performance and increased valeric acid and total SCFAs in ceca, of male Arbor Acres broilers, which were experimentally challenged by *E. coli* O78.

6.4 Organic Acids

The capacity of individual organic acids or commercial blends to inhibit *E. coli* under in vitro tests was previously reported (Kovanda et al. 2019) and could be a strong indication for their potential application as alternatives to control APEC infections in poultry (Mantzios et al. 2023). The ability of the undissociated form of organic acids to diffuse across *E. coli* cellular membranes is theorized to be the main factor contributing to their toxicity, whereas their power to modify the gastrointestinal environment and to stimulate the immune system and gut microbiome in birds has indirect beneficial effects, detailed in previous in vivo investigations (Pham et al. 2022; Warsito et al. 2021; Emami et al. 2017; Giannenas et al. 2014b). In particular, Giannenas et al. (2014b) showed that 0.03 or 0.1% dietary benzoic acid decreased the total coliforms and increased lactic acid bacteria in the ceca of turkey poults. Additionally, organic acid blends effectively decreased mortality and bacterial counts in the gut of turkeys suffering from poult enteritis and mortality syndrome (Roy et al. 2002). Warsito et al. (2021) reported that a combination of acidifier-dextrose can increase hen day production and decrease FCR in laying hens infected with APEC. In another study, Emami et al. (2017) investigated the effect of commercial organic acid blends on male broilers challenged with *E. coli* K88. Similarly, researchers noted that the addition of organic acids to diets of ETEC-challenged birds improved growth performance, ileal morphology, immune responses, and increased cecal *Lactobacilli* counts (Emami et al. 2017). Surprisingly, Hassan et al. (2010) reported that the dietary addition of two commercial mixtures of organic acids (Galliacid® and Biacid®) in Ross 308 broiler chicks was more efficient than the antibiotic growth promoter (Enramycin) in decreasing intestinal *E. coli* and *Salmonella* spp. counts. Pham et al. (2022) reported that dietary coated essential oil and organic acid mixture (benzoic, butyric, hexanoic acid, thyme, and carvacrol) supplementation could mildly alleviate induced gut injury and inflammation by *E. coli* O78 in experimentally challenged broiler chicks. Researchers stated that dietary coated essential oil and organic acid mixture improved feed conversion efficiency, reduced gross lesion scores and cecal *E. coli* counts, and increased intestinal goblet cells and serum IgG concentration in experimentally challenged birds (Pham et al. 2022).

6.5 Phytogenic Feed Additives

Previous in vitro investigations have observed the high capacity of individual or blends of PFAs against several APEC strains, and to that end could be promising alternatives for the control of APEC infections in the field (Wan et al. 2016; Shen et al. 2020). However, further in vivo challenge studies are proposed in order to elucidate the pharmacokinetics of these bioactive compounds and their role in combating APEC infections in various poultry species. Khishtan and Beski (2020) stated that adding chamomile flower in the diet of broilers under the *E. coli* challenge improved the FCR, the protein digestibility, the histological profile (longer villi and lower crypt depth), and the serum protein profile including total serum proteins, albumin, globulin, and aspartate aminotransferase (AST) in contrast with the control group of the trial. Similarly, the inclusion of encapsulated eugenol displayed improved performance parameters in broilers infected by *E. coli* O78. Additionally, the eugenol-supplemented group performed enhanced gut barrier integrity by the upexpression of cathelicidins-2, β-defensin-1, MUC-2, JAM-2, occludin, CLDN-1, and FABP-2 genes and modulation of cytokine genes (IL-1β, TNF-α, IL-6, IL-8, and IL-10) expression (Ibrahim et al. 2022). Elmowalid et al. (2019) concluded that extracts of garlic and ginger in the basal diet of Sasso broiler chickens infected by *E. coli* O78 resulted in the reduction of NO production and in increased expression of cytogenes IL-1β, IL-6, and IFN-γ in both groups compared with the positive control group. Additionally, protection percentages against *E. coli* 078 infection were very high in both supplemented groups, especially in the garlic-supplemented group in which *E. coli* counts were not detected in the vital organs of birds compared to the nonsupplemented group where high CFU counts of *E. coli* were detected in the vital organs of birds with high mortality rates (Elmowalid et al. 2019). In another study, the dietary lemongrass improved performance parameters, increased IgM, IgG, and IgA levels, lysozyme values, and superoxide dismutase activities in Japanese quails. Additionally, lemongrass supplementation led to lower *E. coli* and *Salmonella* spp. counts but higher *Lactobacillus* spp. counts in the ceca of birds (Alagawany et al. 2021). Finally, the inclusion of propolis in the basal diet of layers infected by *E. coli* reduced the inflammation markers TNF-α, IL-1β, and the plasma corticosterone concentration in the blood as well as the apoptotic factor (Foxo3) and the MDA concentration. The treated group showed improved performance parameters, immune response, and total antioxidant capacity concentrations (Abbas et al. 2020) (Table 5).

7 Conclusions and Perspectives

Optimal gut health is an increasingly important topic in poultry and involves a number of physiological, microbiological, and physical functions that work together to maintain intestinal homeostasis and resistance to environmental and infectious stresses (Kogut 2022). The intestine is a vital organ and is not only responsible for the digestion and absorption of nutrients. In addition, the gut is responsible for the interface with the outside world and acts as a metabolic, endocrine, and immune organ.

The implementation of the new legislation on the use of antibiotics is a factor that has had a huge impact on the performance, gut health, and welfare of birds as well as the profit of poultry farmers and the hygiene status of poultry products. In addition, increased consumer demand for organic poultry products has forced the poultry industry to seek more environmentally and poultry-friendly management practices, e.g., free-range farming, cage-free farming, and access to pasture. However, these management practices increase the risk of poultry and human pathogens and could disturb the balance of the intestinal ecosystem.

Table 5 In vivo experimental studies on the efficacy of probiotics, prebiotics, (in)organic acids, and phytogenics to combat APEC infections in poultry

Type of additive	Type of bird	Dosage	Challenge	Effect	References
Probiotics					
Lactobacillus plantarum B1	Arbor Acres broiler chicks	2×10^9 CFU/kg	*E. coli* K88	↑ BW, ↓ *E. coli* bacteria counts in ceca, ↑ cecal lactic acid bacteria counts, and improved intestinal mucosal immunity (↑ IgA, ↓ IFN-γ, IL-2, IL-4, and TNF-α)	Wang et al. (2017a, b)
Lactobacillus plantarum 15-1	Arbor Acres broilers (male)	1×10^8 CFU/kg	*E. coli* O78	Improved intestinal health, ↑ acetic acid and total SCFAs in ceca, ↓ histological damage caused by *E. coli.*	Ding et al. (2019)
Lactobacillus plantarum	Hubbard broiler chicks	1.6×10^{12} CFU/kg (1 g/kg feed)	*E. coli O78*	↑ growth performance, ↓ hepatic enzymes and renal function tests, ↑ serum total proteins, ↑ immunological parameters, ↑ antioxidant enzymes, ↓ DNA damage, ↓ histopathological changes within intestinal, hepatic, and renal, tissues.	Hashem et al. (2022)
Enterococcus faecalis-1	Cobb 500 broiler chicks	0.5 mL (1 × 108 CFU)/ chicken	*E. coli* O78	↓ mortality, and *E. coli* O78 visceral invasion, ↑ immunological response, ↑ growth performance, ↑ total *Enterococcus* counts in the ceca.	Tarabees et al. (2020)
Bacillus subtilis MORI 91, *Clostridium butyricum* M7, *Lactobacillus plantarum* K34 (multistrain probiotic)	Cobb 500 broiler chicks	2×10^8 CFU/g, 2.06×10^8 CFU/g, 2×10^8 CFU/g	*E. coli* O78	↑ BW, ↓ mortality, ↓ *E. coli* O78 recovery rates in the liver and spleen	Tarabees et al. (2019)
Lacticaseibacillus rhamnosus GG	Leghorn (SPF) chickens	10^8 CFU/chicken	*E. coli* O78	↓ colonization (~1.6 logs) of *E. coli* O78, ↓ *E. coli*-induced alterations in the microbial community in the ceca, ↓ the abundance of *Proteobacteria*, particularly those belonging to	Kathayat et al. (2022)

				Enterobacteriaceae (*Escherichia-Shigella*) family.	
Prebiotics					
Isomaltooligosaccharide (IMO)	Cobb 500 broiler chicks	0.5 g/kg	*E. coli* O78	↑ BW, ↓ mortality, ↓ *E. coli* O78 recovery rates in the liver and spleen, ↑ total lactobacilli and total lactobacilli-enterococci populations in the ceca	Tarabees et al. (2019)
Fructooligosaccharides (FOS)	Arbor Acres broilers (male)	5 g/kg	*E. coli* O78	↑ growth performance, ↑ valeric acid and total SCFAs in ceca	Ding et al. (2019)
Organic acids					
Commercial organic acid preparations (various mixtures of formic and propionic acid and their salts on different carriers)	Ross 308 broilers (male)	0.2–0.4%, 0.005–0.1%, 0.05–0.1%	*E. coli* K88	↑ growth performance, ↑ ileal morphology, ↑ immune responses, ↑ cecal *Lactobacilli*, ↓ cecal *E. coli*.	Emami et al. (2017)
Coated essential oil and organic acid mixture (benzoic, butyric, hexanoic acid, thyme, and carvacrol)	Arbor Acres broilers (male)	500 mg/kg feed	*E. coli* O78	↑ feed conversion efficiency, ↓ gross lesion scores, ↓ cecal *E. coli* counts, ↑ intestinal goblet cells, ↑ IgG concentration, ↓ IFN-γ mRNA, ↓ *Bacteroidetes* and genus *Lactobacillus* abundance.	Pham et al. (2022)
Benzoic acid	Nicholas 300 turkey poults (male)	300–1000 mg/kg	–	↓ total coliforms and ↑ lactic acid bacteria in the ceca	Giannenas et al. (2014a, b)
Commercial organic acid preparations (a blend based on fumaric acid, calcium format, calcium propionate, and potassium sorbate; and a bend based on citric acid, calcium formate, calcium butyrate, and calcium lactate)	Ross 308 broiler chicks	0.06% and 0.10%, respectively	–	↓ *E. coli* and *Salmonella* spp. counts	Hassan et al. (2010)

(continued)

Table 5 (continued)

Type of additive	Type of bird	Dosage	Challenge	Effect	References
Commercial organic acid blend (primarily propionic acid)	British United Turkey poults	0.625–2.5%	Poult enteritis and mortality syndrome	↓ mortality, ↓ bacterial content, maintenance of packed cell volume and hemoglobin content	Roy et al. (2002)
Phytogenics					
Eugenol	Ross 308 broiler chicks	100–400 mg/kg	*E. coli* O78	↑ BW, ↑ FI, ↑ *Lactobacillus* spp. counts, ↓ FCR, ↓ *E. coli* counts, ↑ gut barrier integrity	Ibrahim et al. (2022)
Chamomile extract	Ross 308 broiler chicks	5–10 g/kg (or Lt)	*E. coli*	↑ BW, ↓ FCR, ↑ nutrient digestibility, ↓ negative impact of *E. coli*, improved protein profile of serum	Khishtan and Beski (2020)
Garlic (*Allium sativum*) and ginger (*Zingiber officinale*)	Sasso broiler chicks	10–15% v/v, and 15 g/kg	*E. coli* O78	↑ immune response, mortality	Elmowalid et al. (2019)

Abbreviations: *SCFAs*, short-chain fatty acids; *IgA*, immunoglobulin A; *IFN-γ*, interferon gamma; *IL*, interleukin; *FCR*, feed conversion ratio; *BW*, body weight; *FI*, feed intake; *TNF-α*, tumor necrosis factor-alpha; *IgA*, immunoglobulin A; *CFU*, colony-forming unit

Therefore, the poultry industry should adopt a holistic approach through innovative strategies to control gut health disorders and infectious diseases in order to guarantee food safety for a growing human population, to minimize their impact on health and welfare of poultry, and to enhance the sustainability of the poultry industry. The manipulation of gut microflora, mucosa, and immunity using alternatives to antibiotic feed additives, such as phytogenic feed additives, competitive exclusion products, probiotics, prebiotics, organic/inorganic acids, and enzymes, is a very promising strategy. These additives have growth-promoting, antimicrobial, antioxidant and immunostimulating effects and can partially compensate the use of antibiotics. Despite their beneficial effect on the poultry, human, and environmental health, none of them can completely replace the use of antibiotics. For this reason, it is necessary to find the appropriate combination so that a synergistic action is observed, which will contribute to gut homeostasis and improved poultry meat quality.

Therefore, there is an urgent need for further research to scientifically elucidate their mechanisms of action and to establish safe, economic, and effective supplementation programs, such as dosage regimen, optimal combination, duration, and timing of administration as well as possible toxicity effect and antagonism between product categories. Finally, it would be of great value to evaluate them both in intensive and organic farming systems as well as in fast- and slow-growing breeds of birds.

References

Abbas RZ, Iqbal Z, Blake D et al (2011a) Anticoccidial drug resistance in fowl coccidia: the state of play revisited. World Poult Sci J 67:337–350

Abbas RZ, Munawar SH, Manzoor Z et al (2011b) Anticoccidial effects of acetic acid on performance and pathogenic parameters in broiler chickens challenged with *Eimeria tenella*. Pesqui Vet Bras 31:99–103

Abbas AO, Alaqil AA, El-Beltagi HS et al (2020) Modulating laying hens productivity and immune performance in response to oxidative stress induced by E. coli challenge using dietary propolis supplementation. Antioxidants 9:0893

Abd El-Hack ME, El-Saadony MT, Shafi ME et al (2020) Probiotics in poultry feed: a comprehensive review. J Anim Physiol Anim Nutr 104:1835–1850

Abdelli N, Solà-Oriol D, Pérez JF (2021) Phytogenic feed additives in poultry: achievements, prospective and challenges. Animals 11:3471

Abudabos AM, Alyemni AH, Dafalla YM et al (2018) The effect of phytogenics on growth traits, blood biochemical and intestinal histology in broiler chickens exposed to Clostridium perfringens challenge. J Appl Anim Res 46:691–695

Adhikari P, Cosby DE, Cox NA et al (2018) Effect of dietary fructooligosaccharide supplementation on internal organs Salmonella colonization, immune response, ileal morphology, and ileal immunohistochemistry in laying hens challenged with Salmonella Enteritidis. Poult Sci 97: 2525–2533

Adhikari P, Yadav S, Cosby DE et al (2020) Research note: effect of organic acid mixture on growth performance and Salmonella Typhimurium colonization in broiler chickens. Poult Sci 99:2645–2649

Al Hakeem WG, Fathima S, Shanmugasundaram R et al (2022) Campylobacter jejuni in poultry: pathogenesis and control strategies. Microorganisms 10(11):2134

Alagawany M, El-Saadony MT, Elnesr SS et al (2021) Use of lemongrass essential oil as a feed additive in quail's nutrition: its effect on growth, carcass, blood biochemistry, antioxidant and immunological indices, digestive enzymes and intestinal microbiota. Poult Sci 101:172

Ali M, Chand N, Khan RU et al (2019) Anticoccidial effect of garlic (Allium sativum) and ginger (Zingiber officinale) against experimentally induced coccidiosis in broiler chickens. J Appl Anim Res 47:79–84

Aljazzar A, Abd El-Hamid MI, El-Malt RMS et al (2022) Prevalence and antimicrobial susceptibility of Campylobacter species with particular focus on the growth promoting, immunostimulant and anti-Campylobacter jejuni activities of eugenol and trans-cinnamaldehyde mixture in broiler chickens. Animals 12:905

Aljumaah MR, Alkhulaifi MM, Abudabos AM et al (2020) Organic acid blend supplementation increases butyrate and acetate production in Salmonella enterica serovar Typhimurium challenged broilers. PLoS One 15:e0232831

Allain V, Chemaly M, Laisney MJ et al (2014) Prevalence of and risk factors for Campylobacter colonisation in broiler flocks at the end of the rearing period in France. Br Poult Sci 55:452–459

Aristimunha PC, Rosa AP, Boemo LS et al (2016) A blend of benzoic acid and essential oil compounds as an alternative to antibiotic growth promoters in broiler diets. J Appl Poult Res 25: 455–463

Arsi K, Donoghue AM, Venkitanarayanan K et al (2014) The efficacy of the natural plant extracts, thymol and Carvacrol against Campylobacter colonization in broiler chickens. J Food Saf 34: 321–325

Asano I, Nakamura Y, Hoshino et al (2001) Use of mannooligosaccharides from coffee mannan by intestinal bacteria. Nippon Nogeikagaku Kaishi 75:1077–1083

Awad WA, Hess C, Hess M (2018) Re-thinking the chicken-Campylobacter jejuni interaction: a review. Avian Pathol 47(4):352–363

Balta I, Marcu A, Linton M et al (2021) The in vitro and in vivo anti-virulent effect of organic acid mixtures against Eimeria tenella and Eimeria bovis. Sci Rep 11:16202

Baurhoo B, Phillip L, Ruiz-Feria C (2007) Effects of purified lignin and mannan oligosaccharides on intestinal integrity and microbial populations in the ceca and litter of broiler chickens. Poult Sci 86(6):1070–1078

Beck CN, McDaniel CD, Wamsley KGS et al (2019) The potential for inoculating Lactobacillus animalis and Enterococcus faecium alone or in combination using commercial in ovo technology without negatively impacting hatch and post-hatch performance. Poult Sci 98(23): 7050–7062

Bhavaniramya S, Vishnupriya S, Al-Aboody MS et al (2019) Role of essential oils in food safety: antimicrobial and antioxidant applications. Grain Oil Sci Technol 2:49–55

Bilia AR, Guccione C, Isacchi B et al (2014) Essential oils loaded in nanosystems: a developing strategy for a successful therapeutic approach. Evid Based Complement Alternat Med. 2014: 651593

Blake DP, Knox J, Dehaeck B et al (2020) Re-calculating the cost of coccidiosis in chickens. Vet Res 51:115

Bourassa DV, Wilson KM, Ritz CR et al (2018) Evaluation of the addition of organic acids in the feed and/or water for broilers and the subsequent recovery of Salmonella Typhimurium from litter and ceca. Poult Sci 97(1):64–73

Cai H, Luo S, Zhou Q et al (2022) Effects of Bacillus subtilis and coccidiosis vaccine on growth indices and intestinal microbiota of broilers. Poult Sci 101:102091

Casanova NA, Redondo LM, Redondo EA et al (2021) Efficacy of chestnut and quebracho wood extracts to control Salmonella in poultry. J Appl Microbiol 131:135–145

Ceccarelli D, van Essen-Zandbergen A, Smid B et al (2017) Competitive exclusion reduces transmission and excretion of extended-spectrum-β-lactamase-producing Escherichia coli in broilers. Appl Environ Microbiol 83:e03439–e03416

Center for Disease Control and Prevention. Salmonella. https://www.cdc.gov/Salmonella/index.html. Accessed 30 November 2021

Chacher MFA, Kamran Z, Ahsan U et al (2017) Use of mannan oligosaccharide in broiler diets: an overview of underlying mechanisms. World Poult Sci J 73(4):831–844
Chand N, Faheem H, Khan RU et al (2016) Anticoccidial effect of mananoligosacharide against experimentally induced coccidiosis in broiler. Environ Sci Pollut Res 23:14414–14421
Chantziaras I, Smet A, Filippitzi ME et al (2018) The effect of a commercial competitive exclusion product on the selection of enrofloxacin resistance in commensal E. coli in broilers. Avian Pathol 47:443–454
Chapman HD, Barta JR, Blake D et al (2013) A selective review of advances in coccidiosis research. Adv Parasitol 83:93–171
Cho JH, Kim HJ, Kim IH (2014) Effects of phytogenic feed additive on growth performance, digestibility, blood metabolites, intestinal microbiota, meat color and relative organ weight after oral challenge with Clostridium perfringens in broilers. Livest Sci 160:82–88
Christensen H, Bachmeier J, Bisgaard M (2021) New strategies to prevent and control avian pathogenic Escherichia coli (APEC). Avian Pathol 50(5):370–338
Corrier D, Hinton A, Ziprin R et al (1990) Effect of dietary lactose on Salmonella colonization of market-age broiler chickens. Avian Dis 34:668–676
Craig AD, Khattak F, Hastie P et al (2020) The similarity of the effect of carbohydrase or prebiotic supplementation in broilers aged 21 days, fed mixed cereal diets and challenged with coccidiosis infection. PLoS One 15:e0229281
Dahiya J (2007) Nutritional strategies to control Clostridium perfringens in gastrointestinal tract of broiler chickens (PhD Thesis). Department of Animal and Poultry Science, University of Saskatchewan, Saskatoon, Canada
Dalloul RA, Lillehoj HS, Shellem TA et al (2003) Intestinal immunomodulation by vitamin A deficiency and lactobacillus-based probiotic in Eimeria acervulina-infected broiler chickens. Avian Dis 47:1313–1320
De Loach JR, Oyofo BA, Corrier DE et al (1990) Reduction of Salmonella typhimurium concentration in broiler chickens by milk or whey. Avian Dis 34:389–392
De Oliveira J, Van der Hoeven-Hangoor E, Van de Linde I et al (2014) In ovo inoculation of chicken embryos with probiotic bacteria and its effect on posthatch Salmonella susceptibility. Poult Sci 93:818–829
Dhawan S, Kaur J (2007) Microbial mannanases: an overview of production and applications. Crit Rev Biotechnol 27:197–216
Dibner JJ, Buttin P (2002) Use of organic acids as a model to study the digestibility, immune response and intestinal morphology of male broilers fed phosphorus deficient diets supplemented with microbial phytase and organic acids. Livest Sci 157:506–513
Ding S, Wang Y, Yan W et al (2019) Correction: effects of Lactobacillus plantarum 15–1 and fructooligosaccharides on the response of broilers to pathogenic Escherichia coli O78 challenge. PLoS One 14(9):e0222877
Du E, Gan L, Li Z et al (2015) In vitro antibacterial activity of thymol and carvacrol and their effects on broiler chickens challenged with Clostridium perfringens. J Anim Sci Biotechnol 6:58
Ebeid TA, Tůmová E, Ketta M et al (2022) Recent advances in the role of feed restriction in poultry productivity: part II- carcass characteristics, meat quality, muscle fibre properties, and breast meat myopathies. World Poult Sci J 2022:989–1005
Ebrahimi H, Rahimi S, Khaki P et al (2016) The effects of probiotics, organic acid, and a medicinal plant on the immune system and gastrointestinal microflora in broilers challenged with Campylobacter jejuni. Turk J Vet Anim Sci 40:329–336
Eeckhaut V, Wang J, Van Parys A et al (2016) The probiotic Butyricicoccus pullicaecorum reduces feed conversion and protects from potentially harmful intestinal microorganisms and necrotic enteritis in broilers. Front Microbiol 7:1416
Egg-STAT-Ic About Eggs. https://www.usda.gov/media/blog/2020/04/14/egg-stat-ic-about-eggs. Accessed 29 January 2023
Ellis-Iversen J, Jorgensen F, Bull S et al (2009) Risk factors for Campylobacter colonization during rearing of broiler flocks in Great Britain. Prev Vet Med 89:178–184

Elmowalid GA, Abd E-HMI, Abd El-Wahab AM et al (2019) Garlic and ginger extracts modulated broiler chicks innate immune responses and enhanced multidrug resistant Escherichia Coli O78 clearance. Comp Immunol Microbiol Infect Dis 66:101334

El-Saadony MT, Salem HM, El-Tahan AM et al (2022) The control of poultry salmonellosis using organic agents: an updated overview. Poult Sci 101:101716

El-Sawah AA, Aboelhadid SM, El-Nahass EN et al (2020) Efficacy of probiotic Enterococcus faecium in combination with diclazuril against coccidiosis in experimentally infected broilers. J Appl Microbiol 129:1020–1028

El-Shall NA, Awad AM, El-Hack MEA et al (2020) The simultaneous administration of a probiotic or prebiotic with live Salmonella vaccine improves growth performance and reduces fecal shedding of the bacterium in Salmonella-challenged broilers. Animals 10:70

Emami NK, Daneshmand A, Naeini SZ et al (2017) Effects of commercial organic acid blends on male broilers challenged with E. coli K88: performance, microbiology, intestinal morphology, and immune response. Poult Sci 96(9):3254–3263

Enoka VIL, Kikuvi GM, Ndungu PW (2021) Effect of garlic and onion extract chitosan nanoparticles on selected intestinal bacterial flora in indigenous rainbow rooster chicken in Kenya. AIMS Mol Sci 8:98–116

European Food Safety Authority EFSA (2021) The European Union One Health 2019 Zoonoses Report. EFSA J 19:e06406

Fancher CA, Zhang L, Kiess AS et al (2020) Avian pathogenic Escherichia coli and Clostridium perfringens: challenges in no antibiotics ever broiler production and potential solutions. Microorganisms 8:1533

Fatemi A, Keramat A, Razavi SM (2017) Anticoccidial effects of artemisia annua ethanolic extract: prevention, simultaneous challenge-medication, and treatment. Parasitol Res 116:2581–2589

Fathima S, Hakeem WGA, Shanmugasundaram R et al (2022a) Necrotic enteritis in broiler chickens: a review on the pathogen, pathogenesis, and prevention. Microorganisms 10:1958

Fathima S, Shanmugasundaram R, Adams D et al (2022b) Gastrointestinal microbiota and their manipulation for improved growth and performance in chickens. Foods 11:1401

Forkus B, Ritter S, Vlysidis M et al (2017) Antimicrobial probiotics reduce Salmonella enterica in Turkey gastrointestinal tracts. Sci Rep 7:40695

Froebel L, Jalukar S, Lavergne T et al (2019) Administration of dietary prebiotics improves growth performance and reduces pathogen colonization in broiler chickens. Poult Sci 98(12): 6668–6676

Ghareeb K, Awad WA, Mohnl M et al (2012) Evaluating the efficacy of an avian-specific probiotic to reduce the colonization of *Campylobacter jejuni* in broiler chickens. Poult Sci 91(8): 1825–1832

Giannenas I, Florou-Paneri P, Papazahariadou M et al (2003) Effect of dietary supplementation with oregano essential oil on performance of broilers after experimental infection with Eimeria tenella. Arch Tierernahr 57:99–106

Giannenas I, Tsalie E, Triantafillou E et al (2014a) Assessment of probiotics supplementation via feed or water on the growth performance, intestinal morphology and microflora of chickens after experimental infection with Eimeria acervulina, Eimeria maxima and Eimeria tenella. Avian Pathol 43:209–216

Giannenas I, Papaneophytou C, Tsalie E et al (2014b) Dietary supplementation of benzoic acid and essential oil compounds affects buffering capacity of the feeds, performance of Turkey poults and their antioxidant status, pH in the digestive tract, intestinal microbiota and morphology. Anim Biosci 27:225–236

Giannenas I, Tsiouris V, Bonos E et al (2023) Effect of a multi-strains yeast fractions product plus anticoccidial drug on performance, anticoccidial index, intestinal health, and bone mineralization of broiler chickens infected with Eimeria spp. Comp Clin Pathol 32:15–28

Gibson GR, Scott KP, Rastall RA et al (2010) Dietary prebiotics: current status and new definition. Food Sci Technol Bull Funct Foods 7(1):1–19

Gracia MI, Millán C, Sánchez J et al (2016) Efficacy of feed additives against Campylobacter in live broilers during the entire rearing period: Part B. Poult Sci 95:886–892

Gülşen N, Coskun B, Umucalilar HD et al (2002) Effect of lactose and dried whey supplementation on growth performance and histology of the immune system in broilers. Arch Anim Nutr 56: 131–139

Hansson I, Sandberg M, Habib I et al (2018) Knowledge gaps in control of Campylobacter for prevention of campylobacteriosis. Transbound Emerg Dis 65:30–48

Hashem MA, Hassan AEA, Abou-Elnaga HMM et al (2022) Modulatory effect of dietary probiotic and prebiotic supplementation on growth, immuno-biochemical alterations, DNA damage, and pathological changes in E. coli-infected broiler chicks. Front Vet Sci 9:964738

Hassan FAM, Awad A (2017) Impact of thyme powder (Thymus vulgaris L.) supplementation on gene expression profiles of cytokines and economic efficiency of broiler diets. Environ Sci Pollut Res 15:816–826

Hassan HMA, Mohamed MA, Youssef AW et al (2010) Effect of using organic acids to substitute antibiotic growth promoters on performance and intestinal microflora of broilers. Asian Australas J Anim Sci 11:1348–1353

Helmy YA, Closs G Jr, Jung K et al (2022) Effect of probiotic E. coli Nissle 1917 supplementation on the growth performance, immune responses, intestinal morphology, and gut microbes of Campylobacter jejuni infected chickens. Infect Immun 90(10):e0033722

Hessenberger S, Schatzmayr G, Teichmann K et al (2016) In vitro inhibition of Eimeria tenella sporozoite invasion into host cells by probiotics. Vet Parasitol 229:93–98

Hofacre CL, Mathis GF, Lumpkins BS et al (2020) Efficacy of butyric and valeric acid esters in a necrotic enteritis challenge model. Avian Dis 64:407–414

Hu X, Fu Y, Shi H et al (2022) Neglected resistance risks: cooperative resistance of antibiotic resistant bacteria influenced by primary soil components. J Hazard Mater 429:128229

Ibrahim D, Sewid AH, Arisha AH et al (2020) Influence of *Glycyrrhiza glabra* extract on growth, gene expression of gut integrity, and *Campylobacter jejuni* colonization in broiler chickens. Front Vet Sci 7:612063

Ibrahim D, Hassan AA, Badawi M et al (2021a) Thymol nanoemulsion promoted broiler chicken's growth, gastrointestinal barrier and bacterial community and conferred protection against Salmonella typhimurium. Sci Rep 11:7742

Ibrahim D, Ismail TA, Khalifa E et al (2021b) Supplementing garlic nanohydrogel optimized growth, gastrointestinal integrity and economics and ameliorated necrotic enteritis in broiler chickens using a Clostridium perfringens challenge model. Animals 11:2027

Ibrahim D, Fatma E, Metwally AS et al (2022) Dietary eugenol nanoemulsion potentiated performance of broiler chickens: orchestration of digestive enzymes, intestinal barrier functions and cytokines related gene expression with a consequence of attenuating the severity of E. coli O78 infection. Front Vet Sci 9:847580

Ismail H, Ibrahim D, El Sayed S et al (2023) Prospective application of nanoencapsulated Bacillus amyloliquefaciens on broiler chickens' performance and gut health with efficacy against Campylobacter jejuni colonization. Animals 13:775

Jahan AA, Gonzalez Ortiz G, Moss AF et al (2022) Role of supplemental oligosaccharides in poultry diets. World Poult Sci J 2022:615–639

Jiao JY, Yang YQ, Liu M et al (2018) Artemisinin and artemisia annua leaves alleviate eimeria tenella infection by facilitating apoptosis of host cells and suppressing inflammatory response. Vet Parasitol 254:172–177

Kasornpikul C, Chaiyasut C, Sirithanyalug B et al (2009) Effect of the probiotic Lactobacillus plantalum CMU-FP002 on oocyst shedding by broilers inoculated with Eimeria tenella. Avian Biol Res 2:157–159

Kathayat D, Lokesh D, Ranjit S et al (2021) Avian pathogenic Escherichia coli (APEC): an overview of virulence and pathogenesis factors, zoonotic potential, and control strategies. Pathogens 10:467

Kathayat D, Closs GJ, Helmy YA et al (2022) In vitro and in vivo evaluation of Lacticaseibacillus rhamnosus GG and bifidobacterium lactis Bb12 against avian pathogenic Escherichia coli and identification of novel probiotic-derived bioactive peptides. Probiotics Antimicrob Proteins 14: 1012–1028

Khan SH, Iqbal J (2016) Recent advances in the role of organic acids in poultry nutrition. J Appl Anim Res 44:359–369

Khishtan ATM, Beski SSM (2020) Delivery route of chamomile on the growth and subsequent physiology of broiler chickens under E. coli challenge. Iraqi J Agric Sci 51:1058–1073

Kiarie EG, Leung H, Akbari Moghaddam Kakhki R et al (2019) Utility of feed enzymes and yeast derivatives in ameliorating deleterious effects of coccidiosis on intestinal health and function in broiler chickens. Front Vet Sci 6:6473

Kim JE, Lillehoj HS, Hong YH et al (2015) Dietary Capsicum and Curcuma longa oleoresins increase intestinal microbiome and necrotic enteritis in three commercial broiler breeds. Res Vet Sci 102:150–158

Kim SA, Jang MJ, Kim SY et al (2019a) Potential for prebiotics as feed additives to limit foodborne Campylobacter establishment in the poultry gastrointestinal tract. Front Microbiol 10:91

Kim SW, Less JF, Wang L et al (2019b) Meeting global feed protein demand: challenge, opportunity, and strategy. Annu Rev Anim Biosci 7:221–243

Kogut MH (2022) Approaches to stimulate mucosal immunity in the intestinal tract of chickens. In: Seventh IHSIG International Conference on Poultry Intestinal Health, Cartagena, Colombia

Kollanoor-Johny A, Mattson T, Baskaran SA et al (2012) Reduction of Salmonella enterica serovar enteritidis colonization in 20-day-old broiler chickens by the plant-derived compounds trans-cinnamaldehyde and eugenol. Appl Environ Microbiol 78:2981–2987

Koutsoumanis K, Allende A, Alvarez-Ordóñez A et al (2020) Update and review of control options for Campylobacter in broilers at primary production. EFSA J 18:1–89

Kovanda L, Zhang W, Wei X et al (2019) In vitro antimicrobial activities of organic acids and their derivatives on several species of gram-negative and gram-positive bacteria. Molecules 24:3770

Kozu Clarke A, Ajlouni S (2021) Recommended practices to eliminate Campylobacter from live birds and chicken meat in Japan. Food Saf 9(3):57–74

Kulkarni RR, Gaghan C, Gorrell K et al (2022) Probiotics as alternatives to antibiotics for the prevention and control of necrotic enteritis in chickens. Pathogens 11:692

Kumar A, Toghyani M, Kheravii SK et al (2022) Organic acid blends improve intestinal integrity, modulate short-chain fatty acids profiles and alter microbiota of broilers under necrotic enteritis challenge. Anim Nutr 8(1):82–90

Lassen B, Helwigh B, Petersen KC et al (2022) Systematic review of products with potential application for use in the control of Campylobacter spp. in organic and free-range broilers. Acta Vet Scand. 64:24

Lawhon SD, Maurer R, Suyemoto M et al (2002) Intestinal short-chain fatty acids alter Salmonella typhimurium invasion gene expression and virulence through BarA/SirA. Mol Microbiol 46(5): 1451–1464

Lee SH, Lillehoj HS, Jang SI et al (2013) Dietary supplementation of young broiler chickens with Capsicum and turmeric oleoresins increases resistance to necrotic enteritis. Br J Nutr 110:840–847

Lee Y, Lee S, Gadde UD et al (2017) Dietary Allium hookeri reduces inflammatory response and increases expression of intestinal tight junction proteins in LPS-induced young broiler chicken. Res Vet Sci 112:149–155

Li T, Castañeda CD, Miotto J et al (2021) Effects of in ovo probiotic administration on the incidence of avian pathogenic Escherichia coli in broilers and an evaluation on its virulence and antimicrobial resistance properties. Poult Sci 100:100903

Lin Y, Teng PY, Olukosi OA (2022) The effects of xylo-oligosaccharides on regulating growth performance, nutrient utilization, gene expression of tight junctions, nutrient transporters, and cecal short chain fatty acids profile in Eimeria-challenged broiler chickens. Poult Sci 101: 102125

Luan SJ, Sun YB, Wang Y et al (2019) Bacillus amyloliquefaciens spray improves the growth performance, immune status, and respiratory mucosal barrier in broiler chickens. Poult Sci 98: 1403–1409
M'Sadeq SA, Wu SB, Choct M et al (2015) Use of yeast cell wall extract as a tool to reduce the impact of necrotic enteritis in broilers. Poult Sci 94:898–905
Mantzios T, Tsiouris V, Kiskinis K et al (2023) In vitro investigation of the antibacterial activity of nine commercial water disinfectants, acidifiers, and glyceride blends against the most important poultry zoonotic bacteria. Pathogens 12(3):381
McDevitt R, Brooker J, Acamovic T et al (2006) Necrotic enteritis: a continuing challenge for the poultry industry. World Poult Sci J 62:221–247
McReynolds J, Byrd J, Genovese K et al (2007) Dietary lactose and its effect on the disease condition of necrotic enteritis. Poult Sci 86:1656–1661
Memon FU, Leghari IH, Rajput N et al (2022) Immunomodulatory and ameliorative effects of probiotic in combination with diclazuril on broilers under coccidia infection. J Appl Microbiol. 132:3181–3188
Methner U, Rösler U (2020) Efficacy of a competitive exclusion culture against extended-spectrum β-lactamase-producing Escherichia coli strains in broilers using a seeder bird model. BMC Vet Res 16:143
Micciche AC, Foley SL, Pavlidis HO et al (2018) A review of prebiotics against salmonella in poultry: current and future potential for microbiome research applications. Front Vet Sci 5:191
Milbradt EL, Zamae JR, Araújo Júnior JP et al (2014) Control of Salmonella enteritidis in turkeys using organic acids and competitive exclusion product. J Appl Microbiol 117:554–563
Moharreri M, Vakili R, Oskoueian E et al (2021) Phytobiotic role of essential oil-loaded microcapsules in improving the health parameters in Clostridium perfringens-infected broiler chickens. Ital J Anim Sci 20:2075–2085
Moharreri M, Vakili R, Oskoueian E et al (2022) Effects of microencapsulated essential oils on growth performance and biomarkers of inflammation in broiler chickens challenged with Salmonella enteritidis. J Saudi Soc Agric Sci 21:349–357
Mohsin M, Zhang Z, Yin G (2022) Effect of probiotics on the performance and intestinal health of broiler chickens infected with Eimeria tenella. Vaccines 10:97
Mortada M, Cosby DE, Shanmugasundaram R et al (2020) In vivo and in vitro assessment of commercial probiotic and organic acid feed additives in broilers challenged with Campylobacter coli. J Appl Poult Res 29:435–446
Munoz LR, Bailey MA, Krehling JT et al (2023) Effects of dietary yeast cell wall supplementation on growth performance, intestinal Campylobacter jejuni colonization, innate immune response, villus height, crypt depth, and slaughter characteristics of broiler chickens inoculated with Campylobacter jejuni at day 21. Poult Sci 102:609
Mustafa A, Bai S, Zeng Q et al (2021) Effect of organic acids on growth performance, intestinal morphology, and immunity of broiler chickens with and without coccidial challenge. AMB Express 11:140
Neal-McKinney JM, Lu X, Duong T et al (2012) Production of organic acids by probiotic Lactobacilli can be used to reduce pathogen load in poultry. PLoS One 7:e43928
Nguyen DH, Seok WJ, Kim IH (2020) Organic acids mixture as a dietary additive for pigs—a review. Animals 10:952
Nishiyama K, Seto Y, Yoshioka K et al (2014) Lactobacillus gasseri SBT2055 reduces infection by and colonization of Campylobacter jejuni. PLoS One 9:e108827
Noack S, Chapman HD, Selzer PM (2019) Anticoccidial drugs of the livestock industry. Protozool Res 118:2009–2026
Nolan LK, Barnes HJ, Vaillancourt JP et al (2013) Colibacilliosis, Chap. 18. In: Swayne DE, Glisson JR, McDougald LR, Nolan LK, Suarez DL, Nair V (eds) Diseases of poultry, 13th edn. Wiley-Blackwell, Ames, IW, pp 751–805
Nolan LK, Logue CM, Barbeiri NL (2019) APEC: a global one health pathogen and the myths that complicate its control. In: Presentation at WVPAC 2019, Bangkok

Nooreh Z, Kamran T, Mohammad et al (2021) Effects of a dietary direct-fed microbial and ferulago angulata extract on growth performance, intestinal microflora, and immune function of broiler chickens infected with Campylobacter jejuni. Poult Sci 100:100942
Nouri A (2022) Anticoccidial and immunogenic effectivity of encapsulated organic acids and anticoccidial drugs in broilers infected with Eimeria spp. Sci Rep 12(17):060
Omonijo FA, Ni L, Gong J et al (2018) Essential oils as alternatives to antibiotics in swine production. Anim Nutr 4:126–136
Oyofo BA, Droleskey RE, Norman JO et al (1989) Inhibition by mannose of in vitro colonisation of chicken small intestine by Salmonella typhimurium. Poult Sci 68:1351–1356
Pajarillo EABJ, Chae P, Balolong MP et al (2015) Effects of probiotic Enterococcus faecium NCIMB 11181 administration on swine fecal microbiota diversity and composition using barcoded pyrosequencing. Anim Feed Sci Technol 201:80–88
Park S, Choi S, Kim H et al (2015) Fate of mesophilic aerobic bacteria and Salmonella enterica on the surface of eggs as affected by chicken feces, storage temperature, and relative humidity. Food Microbiol 48:200–205
Park YH, Hamidon F, Rajangan C et al (2016) Application of probiotics for the production of safe and high-quality poultry meat. Korean J Food Sci Anim Resour 36:567–576
Pearlin BV, Muthuvel S, Govidasamy P et al (2020) Role of acidifiers in livestock nutrition and health: a review. J Anim Physiol Anim Nutr 104:558–569
Pender CM, Kim S, Potter TD et al (2017) In ovo supplementation of probiotics and its effects on performance and immune-related gene expression in broiler chicks. Poult Sci 96:1052–1062
Pham VH, Abbas W, Huang J et al (2022) Dietary coated essential oil and organic acid mixture supplementation improves health of broilers infected with avian pathogenic Escherichiacoli. Anim Nutr 12:245–262
Phung C, Vezina B, Anwar A et al (2019) Campylobacter hepaticus, the cause of spotty liver disease in chickens: transmission and routes of infection. Front Vet Sci 6:505
Pirgozliev V, Mansbridge SC, Rose SP et al (2019) Immune modulation, growth performance, and nutrient retention in broiler chickens fed a blend of phytogenic feed additives. Poult Sci 98: 3443–3449
Quinteros JA, Scott PC, Wilson TB et al (2021) Isoquinoline alkaloids induce partial protection of laying hens from the impact of Campylobacter hepaticus (spotty liver disease) challenge. Poult Sci 100:101423
Ramlucken U, Ramchuran SO, Moonsamy G et al (2020) A novel Bacillus based multi-strain probiotic improves growth performance and intestinal properties of Clostridium perfringens challenged broilers. Poult Sci 99:331–341
Raza A, Bashir S, Tabassum R (2019) An update on carbohydrases: growth performance and intestinal health of poultry. Heliyon 5:e01437
Revington B (2002) Feeding poultry in the post-antibiotic era. In: Proceedings of the multi-state poultry meeting, Atlanta, GA
Ritzi MM, Abdelrahman W, Van-Heerden K et al (2016) Combination of probiotics and coccidiosis vaccine enhances protection against an Eimeria challenge. Vet Res 47:1–8
Ronco T, Stegger M, Olsen RH et al (2017) Spread of avian pathogenic Escherichia coli ST117 O78:H4 in Nordic broiler production. BMC Genomics 18:13
Rostami F, Taherpour K, Ghasemi HA et al (2021) Effects of Scrophularia striata hydroalcoholic extract in comparison to salinomycin on growth performance, intestinal health and immunity in broiler chickens following a mixed-species Eimeria challenge. Vet Parasitol 293:109417
Roy RD, Edens FW, Parkhurst CR et al (2002) Influence of a propionic acid feed additive on performance of turkey poults with experimentally induced poult enteritis and mortality syndrome. Poult Sci 81:951–957
Ruvalcaba-Gómez JM, Villagrán Z, Valdez-Alarcón JJ et al (2022) Non-antibiotics strategies to control salmonella infection in poultry. Animals 12:102
Saeeda K, Chand N, Khan NU et al (2023) Dietary organic zinc and probiotic alleviate induced Eimeria tenella infection in Japanese quails model of coccidiosis. Trop Anim Health Prod 55:37

Saint-Cyr MJ, Guyard-Nicodème M, Messaoudi S et al (2016) Recent advances in screening of anti-Campylobacter activity in probiotics for use in poultry. Front Microbiol 7:553
Saint-Cyr MJ, Haddad N, Taminiau B et al (2017) Use of the potential probiotic strain Lactobacillus salivarius SMXD51 to control Campylobacter jejuni in broilers. Int J Food Microbiol 247:9–17
Santana ES, Andrade MA, da Silveira Neto OJ et al (2020) Intestinal integrity and performance of turkeys subjected to inoculation of Salmonella enteritidis and a diet supplemented with lactulose. Pesqui Agropecu Bras 55
Scharff RL (2020) Food attribution and economic cost estimates for meat- and poultry-related illnesses. J Food Prot 83:959–967
Schiaffino F, Platts-Mills J, Kosek MN (2019) A One Health approach to prevention, treatment, and control of campylobacteriosis. Curr Opin Infect Dis 32:453–460
Scicutella F, Mannelli F, Daghio M et al (2021) Polyphenols and organic acids as alternatives to antimicrobials in poultry rearing: a review. Antibiotics 10:1010
Shao Y, Zhen W, Guo F et al (2022) Pretreatment with probiotics Enterococcus faecium NCIMB 11181 attenuated Salmonella typhimurium-induced gut injury through modulating intestinal microbiome and immune responses with barrier function in broiler chickens. J Anim Sci Biotechnol 13:130
Shastak Y, Ader P, Feuerstein D et al (2015) ß-Mannan and mannanase in poultry nutrition. World Poult Sci J 71:161–174
Shehata AA, Yalçın S, Latorre JD et al (2022) Probiotics, prebiotics, and phytogenic substances for optimizing gut health in poultry. Microorganisms 10:395
Shen C, Christensen L, Rasmussen P et al (2020) In vitro immunomodulatory effects of thymol and cinnamaldehyde in a pig intestinal epithelial cell line (IPEC-J2). J Appl Anim Nutr 8:127–134
Shirley MW, Smith AL, Tomley FM (2005) The biology of avian Eimeria with an emphasis on their control by vaccination. Adv Parasitol 60:285–330
Shojadoost B, Alizadeh M, Boodhoo N et al (2022) Effects of treatment with Lactobacilli on necrotic enteritis in broiler chickens. Probiotics Antimicrob Proteins 14:1110–1129
Shokryazdan P, Faseleh Jahromi M, Navidshad B et al (2017) Effects of prebiotics on immune system and cytokine expression. Med Microbiol Immunol 206:1–9
Sidiropoulou E, Skoufos I, Marugan-Hernandez V et al (2020) In vitro anticoccidial study of oregano and garlic essential oils and effects on growth performance, fecal oocyst output, and intestinal microbiota in vivo. Front Vet Sci 7:420
Šimunović K, Sahin O, Erega A et al (2022) Bacillus subtilis PS-216 Spores supplemented in broiler chicken drinking water reduce Campylobacter jejuni colonization and increases weight gain. Front Microbiol 13:910616
Skånseng B, Kaldhusdal M, Moen B, Gjevre AG et al (2010) Prevention of intestinal *Campylobacter jejuni* colonization in broilers by combinations of in-feed organic acids. J Appl Microbiol 109:1265–1273
Skoufos I, Tzora A, Giannenas I et al (2019) Evaluation of in-field efficacy of dietary ferric tyrosine on performance, intestinal health and meat quality of broiler chickens exposed to natural *Campylobacter jejuni* challenge. Livest Sci 221:44–51
Smialek M, Burchardt S, Koncicki A (2018) The influence of probiotic supplementation in broiler chickens on population and carcass contamination with Campylobacter spp. – field study. Res Vet Sci 118:312–316
Solis de los Santos F, Donoghue AM, Venkitanarayanan K et al (2009) The natural feed additive caprylic acid decreases Campylobacter jejuni colonization in market-aged broiler chickens. Poult Sci 88:61–64
Ştef L, Dumitrescu G, Simiz E et al (2015) The effect of probiotics on broiler growth and intestinal morphology when used to prevent Campylobacter jejuni colonization. Sci Pap Anim Sci Biotechnol 2015:43–50
Suresh G, Das RK, Brar SK et al (2018) Alternatives to antibiotics in poultry feed: molecular perspectives. Crit Rev Microbiol 44:318–335

Swaggerty CL, Byrd JA, Arsenault RJ et al (2022) A blend of microencapsulated organic acids and botanicals reduces necrotic enteritis via specific signaling pathways in broilers. Poult Sci 101(4): 101753

Szott V, Reichelt B, Alter T et al (2020) In vivo efficacy of carvacrol on Campylobacter jejuni prevalence in broiler chickens during an entire fattening period. Eur J Microbiol Immunol 10: 131–138

Szott V, Peh E, Friese A et al (2022) Antimicrobial effect of a drinking water additive comprising four organic acids on Campylobacter load in broilers and monitoring of bacterial susceptibility. Poult Sci 101:102209

Taha-Abdelaziz K, Singh M, Sharif S et al (2023) Intervention strategies to control Campylobacter at different stages of the food chain. Microorganisms 11:113

Tarabees R, Gafar KM, El-Sayed MS et al (2019) Effects of dietary supplementation of probiotic mix and prebiotic on growth performance, cecal microbiota composition, and protection against Escherichia coli O78 in broiler chickens. Probiotics Antimicrob Proteins 11:981–989

Tarabees R, El-Sayed MS, Shehata AA et al (2020) Effects of the probiotic candidate E. faecalis-1, the poulvac E. coli vaccine, and their combination on growth performance, caecal microbial composition, immune response, and protection against E. coli O78 challenge in broiler chickens. Probiotics Antimicrob Proteins 12:860–872

Taylor J, Walk C, Misiura M et al (2022) Quantifying the effect of coccidiosis on broiler performance and infection outcomes in the presence and absence of control methods. Poult Sci 101:101746

Thames HT, Sukumaran AT (2020) A review of salmonella and Campylobacter in broiler meat: emerging challenges and food safety measures. Foods 9:776

Tian X, Shao Y, Wang Z et al (2016) Effects of dietary yeast β-glucans supplementation on growth performance, gut morphology, intestinal Clostridium perfringens population and immune response of broiler chickens challenged with necrotic enteritis. Anim Feed Sci Technol 215: 144–155

Tierney J, Gowing H, Van Sinderen D et al (2004) In vitro inhibition of Eimeria tenella invasion by indigenous chicken Lactobacillus species. Vet Parasitol 122:171–182

Timbermont L, Haesebrouck F, Ducatelle R et al (2011) Necrotic enteritis in broilers: an updated review on the pathogenesis. Avian Pathol 40:341–347

Torralbo A, Borge C, Allepuz A et al (2014) Prevalence and risk factors of Campylobacter infection in broiler flocks from southern Spain. Prev Vet Med 114:106–113

Tsiouris V, Georgopoulou I, Batzios C et al (2013) The role of an attenuated anticoccidial vaccine on the intestinal ecosystem and on the pathogenesis of experimental necrotic enteritis in broiler chickens. Avian Pathol 42:163–170

Tsiouris V, Economou E, Lazou T et al (2019) The role of whey on the performance and campylobacteriosis in broiler chicks. Poult Sci 98:236–243

Tsiouris V, Giannenas I, Bonos E et al (2021) Efficacy of a dietary polyherbal formula on the performance and gut health in broiler chicks after experimental infection with Eimeria spp. Pathogens 10:524

Van der Sluis W (2000) Clostridial enteritis a syndrome emerging worldwide. World Poult 16:56–57

Van Immerseel F, De Buck J, De Smet I et al (2004a) Interactions of butyric acid and acetic acid-treated Salmonella with chicken primary cecal epithelial cells in vitro. Avian Dis 48:384–391

Van Immerseel F, De Buck J, Pasmans F et al (2004b) Clostridium perfringens in poultry: an emerging threat for animal and public health. Avian Pathol 33:537–549

Vieco-Saiz N, Belguesmia Y, Raspoet R et al (2022) Protective effects of novel lactobacillaceae strains isolated from chicken caeca against necrotic enteritis infection: in vitro and in vivo evidences. Microorganisms 10:152

Wade B, Keyburn AL (2015) The true cost of necrotic enteritis. World Poultry 2015:16–17

Wagle BR, Upadhyay A, Arsi K et al (2017) Application of β-Resorcylic acid as potential antimicrobial feed additive to reduce Campylobacter colonization in broiler chickens. Front Microbiol 8:599

Waldenstedt L, Elwinger K, Lunden A et al (2000) Intestinal digesta viscosity decreases during coccidial infection in broilers. Br Poult Sci 41:459–464

Wan MLY, Ling KH, Wang MF et al (2016) Green tea polyphenol epigallocatechin-3-gallate improves epithelial barrier function by inducing the production of antimicrobial peptide pBD-1 and pBD-2 in monolayers of porcine intestinal epithelial IPEC-J2 cells. Mol Nutr Food Res 60:1048–1058

Wang H, Ni X, Qing X et al (2017a) Probiotic enhanced intestinal immunity in broilers against subclinical necrotic enteritis. Front Immunol 8:1592

Wang S, Peng Q, Jia HM et al (2017b) Prevention of Escherichia coli infection in broiler chickens with Lactobacillus plantarum B1. Poult Sci 96(8):2576–2586

Wang H, Ni X, Qing X et al (2018) Probiotic Lactobacillus johnsonii BS15 improves blood parameters related to immunity in broilers experimentally infected with subclinical necrotic enteritis. Front Microbiol 9:49

Wang J, Dai D, Zhang HJ et al (2019) Organic acids modulate systemic metabolic perturbation caused by salmonella pullorum challenge in early-stage broilers. Front Physiol 10:1418

Wang Y, Lv X, Li X et al (2021) Protective effect of Lactobacillus plantarum P8 on growth performance, intestinal health, and microbiota in eimeria-infected broilers. Front Microbiol 12:705758

Wang M, Hu J, Yu H et al (2023) Lactobacillus fermentum 1.2133 display probiotic potential in vitro and protect against Salmonella pullorum in chicken of infection. Lett Appl Microbiol 76:ovac041

Warsito SH, Sabdoningrum EK, Tripalupi N et al (2021) The effect of acidifier-dextrose against hen day production and feed conversion ratio in laying hens infected with avian pathogenic Escherichia coli. Vet Med Int 26:6610778

Wati T, Tapan K, Syed GB et al (2015) Comparative efficacy of a phytogenic feed additive and an antibiotic growth promoter on production performance, caecal microbial population and humoral immune response of broiler chickens inoculated with enteric pathogens. Anim Nutr 1:213–219

Watkins BA, Miller BF, Neil DH (1982) In vivo inhibitory effects of Lactobacillus acidophilus against pathogenic Escherichia coli in gnotobiotic chicks. Poult Sci 61:1298–1308

Williams R (2005) Intercurrent coccidiosis and necrotic enteritis of chickens: rational, integrated disease management by maintenance of gut integrity. Avian Pathol 34:159–180

Windisch W, Schedle K, Plitzner C et al (2008) Use of phytogenic products as feed additives for swine and poultry. J Anim Sci 86:E140–E148

Wu YT, Yang WY, Wu SYH et al (2020) Modulations of growth performance, gut microbiota, and inflammatory cytokines by trehalose on Salmonella Typhimurium-challenged broilers. Poult Sci 99:4034–4043

Yadav S, Teng PY, dos Santos TS et al (2020) The effects of different doses of curcumin compound on growth performance, antioxidant status, and gut health of broiler chickens challenged with Eimeria species. Poult Sci 99:5936–5945

Yang CM, Cao GT, Ferket PR et al (2012) Effects of probiotic, Clostridium butyricum on growth performance, immune function, and cecal microflora in broiler chickens. Poult Sci 91:2121–2129

Yang X, Xin H, Yang C et al (2018) Impact of essential oils and organic acids on the growth performance, digestive functions and immunity of broiler chickens. Anim Nutr 4:388–393

Ye Y, Li Z, Wang P et al (2021) Effects of probiotic supplements on growth performance and intestinal microbiota of partridge shank broiler chicks. PeerJ 9:e12538

Yi G, Qu L, Liu J et al (2014) Genome-wide patterns of copy number variation in the diversified chicken genomes using next-generation sequencing. BMC Genomics 15:962

Zhai H, Liu H, Wang S et al (2018) Potential of essential oils for poultry and pigs. Anim Nutr 4: 179–186

Zhang H, Zhou Y, Xu H et al (2022) Bacillus amyloliquefaciens BLCC1–0238 alone or in combination with Mannan-Oligosaccharides alleviates subclinical necrotic enteritis in broilers. Probiotics Antimicrob Proteins 14:158–168

Zhang H, Li M, Zhang K et al (2023) Effect of benzoic acid, Enterococcus faecium, and essential oil complex on intestinal microbiota of laying hens under coccidia and Clostridium perfringens challenge. Poult Sci 102:102490

Zhao Y, Zeng Y, Zeng D et al (2022) Dietary probiotic supplementation suppresses subclinical necrotic enteritis in broiler chickens in a microbiota-dependent manner. Front Immunol 13: 855426

Zhu Q, Sun P, Zhang B et al (2021) Progress on gut health maintenance and antibiotic alternatives in broiler chicken production. Front Nutr 8:692839

The Effects of Feed Additives on the Immune System of Poultry

Alireza Seidavi, Babak Hosseintabar-Ghasemabad, and Ambra Rita Di Rosa

Abstract

The immune system is considered one of the most important influencing factors in improving the performance and health of poultry, and nutrition, in turn, plays an important role in balancing and strengthening the immune system. On the other hand, poultry diets are mainly formulated based on the minimum cost in order to achieve targeted performance goals, and attention to the criteria necessary to achieve ideal responses in the immune system is ignored. The increase in the world's population and the growth in the demand for quality and healthy food have led to the fact that a 2% increase in the annual production of food is necessary, and in parallel with this increase in production, the least pressure should be placed on the environment. To achieve such goals, it is absolutely necessary to optimize feedstuff with the help of various feed additives in order to improve health, increase the bioavailability of nutrients, and improve animal performance. The main key to achieving these goals is only by having productive birds in terms of innate and acquired immunity during the breeding period. The use of feed additives in poultry nutrition can be considered as an effective solution to strengthen body immunity and a reassuring and economical proposal. Knowledge and proper understanding of the characteristics and use of additives and their effects in poultry nutrition are the goals of many researchers' studies in recent years. Many researches have shown that feed additives can increase the bioavailability and digestibility of nutrients without negatively affecting the nutrient balance of poultry diets, and at the same time leading to the improvement

A. Seidavi (✉)
Department of Animal Science, Rasht Branch, Islamic Azad University, Rasht, Iran

B. Hosseintabar-Ghasemabad
Department of Animal Science, Faculty of Agriculture, University of Tabriz, Tabriz, Iran

A. R. Di Rosa
Department of Veterinary Sciences, University of Messina, Messina, Italy

G. Arsenos, I. Giannenas (eds.), *Sustainable Use of Feed Additives in Livestock*,
https://doi.org/10.1007/978-3-031-42855-5_17

of the health of the digestive system, and the quality and quantity of manufactured products have positive and promising effects. One of the beneficial and special effects that recently made the importance of a series of additives important in poultry research is their effect on the poultry immune system. The bird's immune system is a complex network of biological structures and processes, and a decrease in performance or a slight change in any of its aspects can affect the performance and health of the bird and increase the risk of contracting various diseases. Therefore, any nutritional solution with the approach of improving physiological and immunological function can help strengthen the immune system and at the same time lead to an increase in the production capacity of poultry flocks, which can be of interest to researchers and poultry industry activists. Functional additives in poultry nutrition have many varieties. This chapter provides readers with the latest and most recent information with the aim and focus on the types of additives that were able to be effective on the poultry immune system and subsequently improve the performance and other health parameters of the poultry flock. In this chapter, an introduction is mentioned about the poultry immune system and its importance on the production performance and health of the bird. Then, the review and introduction of additives in the nutrition of poultry models and their effects on parameters related to bird health have been discussed. The feed additives investigated and effective on the body's immune system in this season include all kinds of probiotics, prebiotics, synbiotics, medicinal plants, and phytogenic compounds, all of which are among the most valid additives. They are considered useful in poultry nutrition, which can directly or indirectly affect the poultry's immune system.

Keywords

Feed additive · Health · Immune system · Production

1 Introduction

The global pattern in most industries has shifted from production efficiency to public health (Ferket 2004). From a technical point of view, the use of feed additives has never been mandatory and does not always guarantee maximum productivity and economic profit, but the increase in consumer awareness and production standards has led to the growth of demand for healthy food products of animal origin, and subsequently, more pressure on activists. The poultry production industry was created to use feed additives (Pandey et al. 2019). On the other hand, intensive poultry farming always leads to the growth of a wide range of microorganisms that are constantly competing with and affecting the immune system (Seidavi et al. 2020). The use of growth-promoting antibiotics, which were very useful in the past in order to improve performance and protect animals against pathogenic microorganisms, today, due to bacterial resistance and disturbance in the health of poultry, the environment, and its consequences on human health faced a serious challenge. Therefore, in the first step in order to strengthen and balance the immune system and respond to the aforementioned concerns and challenges, the European Union banned the use of antibiotics to prevent diseases and also as growth promoters, and subsequently the ground was prepared for the use of alternative additives (Ferket 2004).

In general, using nutritional knowledge to improve health is a preventive and logical scientific solution that can help to balance and strengthen the immune system. Therefore, in the first step, providing sufficient amounts of nutrients and creating a balance in nutrients should be prioritized to achieve improved health and safety (Klasing 2007). In line with this goal, the use of feed additives to improve the bioavailability of nutrients can be very effective and helpful, because many researches have shown that nutrients have a great impact on the development of the immune system and the number of antibodies produced. On the other hand, poultry feed formulations are mainly adjusted based on production indicators such as the improvement of growth status, eggs and feed efficiency, and not much attention paid to the criteria necessary for responses of the immune system. In addition, it has been found that in conditions of illness, stress, and the acute stage of the immune system, the requirement for nutrients increase, and the proper functioning of the immune system is dependent on the improvement of metabolizable energy and amino acids, and any disturbance in the supply of nutrients and failure to comply with the ratio of nutrients can lead to a disturbance in the normal state of the bird's natural physiology and subsequently have a negative effect on the functioning of the immune system. A review of the scientific literature indicates that feed additives can play a constructive role in improving the availability of nutrients and animal health in acute conditions. However, the strategy and provision of auxiliary solutions through dietary supplements play a very important role in protecting the bird's health against the invasion of pathogenic agents, and any deficiency and disturbance in the absorption of nutrients can have irreparable effects on normal performance—leave the immune system (Klasing 2007; Olgun et al. 2021). In recent years, nutritionists have reported poultry nutritional requirement tables focusing on the effect of nutrients on the performance of the poultry's immune system, such as observing the ratio of amino acids, the ratio of omega-6 to omega-3, and the ratio of calcium to phosphorus and other nutrients, which all play a significant role, and they are safe in balancing. Using feed additives to help absorb and balance nutrients can play a positive role in improving the immune system. Because the nutrient is considered the food of the cell, better access to nutrients with the help of various feed additives plays an important role in the balance and improving immunity (Prabakar et al. 2016; Olgun et al. 2021).

All over the world, feed additives are widely used in the poultry industry for various purposes. A great variety of feed additives and, on the other hand, the different goals of manufacturers, have caused several classifications of these compounds to be presented by experts in recent years (Pirgozliev et al. 2019). It is certain that a number of additives undertake several tasks and approaches that can directly and indirectly affect the animal's immune system and health (Pandey et al. 2019). In general, the role of nutrients in the body's immune system is undeniable, and any use of feed additives to improve the absorption of nutrients and subsequent improvement and balance of immunity with the approach of nutrition knowledge is one of the goals of this chapter. In fact, this is an approach that can have a positive effect on improving the condition of the microbiota and microflora of the digestive system and also strengthen the vital signaling pathways in cells and immune-stimulating organs.

2 Probiotics and Immunity

The US National Food Association identifies probiotics as a source of living microorganisms, including bacteria, fungi, and yeasts. Probiotics are defined by the Food and Agriculture Organization (FAO) as living microorganisms whose consumption in sufficient quantities promotes the health of the host body.

Probiotics work in poultry by preserving natural intestinal microorganisms by competitive replacement mechanisms and antagonism, altering metabolism by increasing the activity of digestive enzymes and reducing the activity of bacterial enzymes and ammonia production, and improving feed intake and digestion, and immune system stimulation occurs (Abd El-Hack et al. 2020; Krysiak et al. 2021).

The effect of probiotics on increasing the production of antibodies in broilers has been reported in several studies (Lutful Kabir 2009). Khaksefidi and Ghoorchi (2006) reported an increase in antibody titers in response to injection of sheep red blood cells (SRBCs) at 7 and 14 days of age when consuming 50 mg of probiotics per kg of feed in broilers. In addition, Haghighi et al. (2005) showed an increase in antibodies in response to seven antigens in broilers. Dalloul et al. (2005) observed the effects of lactobacilli-based probiotics on the intestinal immune response during infection with *Imeriaacerolna*. The researchers observed positive effects of probiotics on the stimulation of some of the primary immune responses, such as the secretion of interferon-γ and interleukin-2 against *Eimeria acervulina*, which improved local defense against coccidiosis. Probiotics have been suggested as immunomodulators to reduce the expression of proinflammatory cytokines in the intestine and to increase IgA secretion and the integrity of the intestinal epithelial barrier and serum antibody levels to specific antigens (Haghighi et al. 2005; Hasted et al. 2021). Koenen et al. (2004) also reported an increase in antibody production using probiotics, suggesting that several probiotics may stimulate an adequate protective immune response to enhance resistance to pathogenic bacteria. Some strains of probiotics such as *Lactobacillus, Lactobacillus acidophilus, Bifidobacterium lactis* have been shown to be able to stimulate macrophages and neutrophil populations. *Lactobacillus* increases the efficiency of the population of helper T lymphocytes in mice (O'mahony et al. 2001). Kabir et al. (2004) observed high levels of antibody production in broilers receiving 2 mg of the probiotic protoxin in 10 liters of water. Some probiotics can stimulate a protective immune response by increasing resistance to microbial pathogens.

Khaksefidi and Ghoorchi (2006) using 50 mg of probiotic *Bacillus subtilis* per kg of diet observed improvements in antibody production against sheep red blood cells and Newcastle disease vaccine 10 days after injection. Increased antibody production due to probiotic consumption may be due to the accumulation of essential nutrients in stimulating immune cells to produce antibodies by probiotic microorganisms. Nahashon (1994) reported that feeding *Lactobacillus* culture media to laying hens increases the number of lymph follicles scattered in the ileum, which stimulates the mucosal immune system to produce antibody-stimulated immunoglobulin A. Lactic acid-producing bacteria activate a nonspecific immune response by increasing the activity of macrophages. Multiple strains produce live bacteria, and lactic acid induces the release of pro-inflammatory cytokines, tumor necrosis factor-α, and interleukin-6 in vitro, indicating nonspecific immune stimulation. Shoeib (1997) observed an increase in erythrocytes and leukocytes as well as a significant increase in the percentage of lymphocytes and monocytes with the addition of probiotics to the diet in broilers. Probiotics improve the immune system by stimulating lymphatic tissues (Kabir et al. 2004) and by altering the microbial population of the gastrointestinal tract (Jin et al. 1998). The role of probiotic bacterial strains, primarily lactobacilli, is to maintain a balance in the population of the intestinal microbial flora that prevents the spread of pathogenic microorganisms (Kiczorowska et al. 2017). This effect is due to the increased concentration of lactic acid and volatile fatty acids (VFAs) (Kiczorowska et al. 2017). *Lactobacillus* bacteria can increase the proportion of mucous membrane proteins involved in energy metabolism and their availability in the animal jejunum (Kiczorowska et al. 2017). Increasing the concentration of intestinal VFA and lactic acid bacteria (LAB) reduces the incidence of diarrhea (Kiczorowska et al. 2017). In addition, LAB fermentation leads to the production of higher amounts of lactic and acetic acids in the ileum and colon, which has an indirect effect on the concentration of propionic and butyric acids (Kiczorowska et al. 2017). Their presence lowers the intestinal pH and thus prevents the proliferation of pathogenic bacteria (Wang et al. 2016). Colonization of intestinal probiotic bacteria can contribute to various mucosal immune responses, including increased expression of TLR2, TLR9, and amplitude of nucleotide-binding oligomerization domain (NOD), as well as increased cytokine secretion and increased levels of IgG-producing slow cells (IgG). *Lactobacillus* can increase the number of intraepithelial lymphocytes (IELs) and IgA-producing cells in the gastrointestinal tract (Salah et al. 2012; Kiczorowska et al. 2017). These events lead to the development of intestinal mucosal resistance, which is favorably affected by increased interleukin (IL-6) secretion. It has now been well established that the observed beneficial effects of probiotics are exerted by lowering the pH through the production of VFAs and inhibiting the growth of harmful pathogenic bacteria (Khan and Naz 2013). Another mechanism is through competition for sites of adhesion to the intestinal epithelium that prevents the formation of colonies of pathogenic bacteria (Khan and Naz 2013). The protective effect of probiotics is associated with increased humoral and cellular immune responses through increased production of T lymphocytes, CD^{+} cells, and antibody-secreting cells, expression of pro- and anti-inflammatory cytokines, interleukins, and production of natural interferon-γ (IFN-γ) cells. Antibody and macrophage activation increases (Chichlowski et al. 2007; Khan and Naz 2013). Another method of action of probiotics is to reduce the activity of intestinal and fecal β-glucosidase and β-glucuronidase bacterial enzymes (Khan and Naz 2013). These enzymes are involved in the formation of toxins in the body (Khan and Naz 2013).

Culture of *Lactobacillus* may reduce β-glucosidase and β -glucuronidase activities by attaching itself along the intestine of poultry, thus preventing the colonization of bacteria with toxin-stimulating enzymes (Chichlowski et al. 2007; Khan and Naz 2013). Probiotics produce antimicrobials that prevent the localization of pathogenic bacteria in the animal's gut. These small antimicrobial molecules, known as bacteriocins, defensins, and cathelicidins, work to fight or prevent colonization by pathogenic bacteria. These protein complexes have an antagonistic effect against pathogenic bacteria (Chichlowski et al. 2007; Khan and Naz 2013).

Polyamine-derived piperidine, which is produced by the intestinal microflora as a result of amino acid degradation, has been shown to prevent the attachment of *Salmonella* and *Shigella* to intestinal epithelial cells (Chichlowski et al. 2007; Khan and Naz 2013). In general, in order to stimulate immunity, the intestine and its resident microflora play an important role in shaping the immune system (Khan and Naz 2013). Supplementation of probiotics additives in laying hens increases the cellularity of Payer's patches in the ileum, a sign of stimulation of the mucosal immune system that responds to antigenic stimuli and secretes immunoglobulin (IgA). Oral administration of probiotics mainly stimulates splenic natural killer (NK) cell activity and phagocytic activity (Khan and Naz 2013). Antibody titers in Newcastle, bronchitis, and influenza infections increase with probiotics alone or in combination (Khan and Naz 2013). The higher antibody titer seen in probiotic-treated birds has been attributed to the beneficial effect of microflora on the gut to maintain a healthy balance of immune cells (Khan and Naz 2013). Higgins et al. (2007) reported that *Lactobacillus*-based probiotic cultures improved the number of macrophages in the cecum as well as increased phagocytic activity against *Salmonella* enteritidis, indicating that probiotics have the ability to modulate the innate immunity of broilers.

3 Prebiotics and Immunity

Prebiotics increase digestion and absorption of nutrients by limiting the growth of pathogenic bacteria and stimulating the growth of beneficial bacteria, improving intestinal health and integrity, improving intestinal morphological parameters, and strengthening the intestinal immune system and lead to improved poultry performance and health (Yang et al. 2009). Prebiotics used against pathogens include fructooligosaccharides (FOS), xylooligosaccharides (XOS), mannan-oligosaccharides (MOS), galactooligosaccharides (GOS). Prebiotic compounds by agaro-oligosaccharides (AOS), arabinoxylans, cyclodextrins, FOS, inulin, isomaltose, lactose (for poultry), lactulose, MOS oligofructose, raffinose, stachyose, xylooligosaccharides (XOS), and β-galactooligosaccharides (GOS) the main prebiotics used in feeding poultry and pigs (Pirgozliev et al. 2019; Hasted et al. 2021; Ricke et al. 2020). The advantage of prebiotics over probiotics is that they stimulate the growth of beneficial bacteria that are already present in the host organism and adapt to all environmental conditions. Prebiotics are an indigestible compound that can selectively grow one or a number of bacteria at the end of the gastrointestinal tract (cecum and colon), stimulates, and thus improves host performance (McFarlane et al. 2019, 2022). Certain carbohydrates can have very specific activities. For example, sugars (monosaccharides) have been shown to regulate hormones, regulate fetal development, direct the movement of proteins and cells from tissues, and regulate the immune system (Ricke et al. 2020). It has been suggested that the regulation of immune mechanisms may be affected by the type of carbohydrate in the diet because dietary carbohydrates interact in the gastrointestinal tract with intestinal cells, bacteria, and the immune system (Ferket 2004; Al-Khalaifah 2018; Ricke et al. 2020). Mannan-oligosaccharides have beneficial effects on humoral immunity and immunoglobulin status. The humoral immune response is a nutritionally useful tool for disease resistance and is more active than the inflammatory response (Chacher et al. 2017). Savage (1996) showed an increase in blood immunoglobulin G and bile IgA in prebiotic-fed chickens. The increase in antibody response following the consumption of mannan-oligosaccharides is due to the improvement of the immune system's ability to resist antimicrobials of foreign microbes. Parts of the *Saccharomyces cerevisiae* cell wall are oligosaccharide-like and have strong antigenic properties. However, mannan-oligosaccharides increase humoral immunity against specific pathogenic bacteria by preventing the colonization of pathogenic bacteria, and of course, the possibility of these bacteria being detected by the cell provides immunity as an antigen.

In fact, fructooligosaccharides facilitate the secretion of immunoglobulin A into the intestinal mucosa, and as a result, pathogens become very unstable against the phagocytic action of intestinal lymphocytes. Mannan-oligosaccharides have been reported to stimulate the production of mannose-bonded lectins (an acute phase protein that promotes phagocytosis). It has been suggested that mannose in the superficial part of mannan-oligosaccharides stimulates the production of mannose-binding lectins by stimulating Toll-like receptor systems (Ferket 2004; Muhammad et al. 2020; Ricke et al. 2020; Rostami et al. 2022). It has also been suggested that mannan-oligosaccharides have a direct effect on intestinal immune cells by entering mast cells in intestinal surface glands. As a result, the synthesis of specific immunoglobulins in response to antigen in plasma and bile increases (Ferket 2004). MOS and FOS have been used to increase the body weight of farmed animals, feed efficiency, energy digestible, and intestinal microbiota population as well as reduce serum cholesterol levels (Nabizadeh 2012; Yaqoob et al. 2021). MOS, derived from the yeast cell wall of *Saccharomyces cerevisiae*, has been shown to be indigestible to nonruminant animals and to improve animal performance and health through a diversity of mechanisms such as preventing pathogen binding to the gastrointestinal tract (GIT), changing microbial populations and digestion, and helping boost immune function (Chee et al. 2010). Several mechanisms are responsible for the beneficial effects of MOS. They are also associated with improved intestinal health and enhanced mucosal immunity by increasing the number of goblet cells and villi length, increasing the population of beneficial bacteria such as lactobacilli and bifidobacteria in the intestines of monogastric animals, and decreasing the population of *Escherichia coli* and *Salmonella* (Spring et al. 2015). Various studies show that lactobacilli and bifidobacteria can increase mucin synthesis and secretion in the intestine (Smirnov et al. 2005). The immunomodulatory effects of yeast cell wall polysaccharides are related to the ability of macrophages to stimulate cytokine production as well as to improve the humoral immune parameter (immunoglobulin) and intrinsic (nonspecific) immune factors (lysozyme and complement) (Spring et al. 2015).

The immunosuppressive effect of prebiotics may be due to the direct interaction between prebiotics and intestinal immune cells, as well as the indirect effect of prebiotics through preferential microbial colonization—useful and microbial products that interact with immune cells. Administration of MOS to the feed mixture increases mucosal IgA secretion and humoral and cellular immune responses in chickens.

Inulin is another health-enhancing feed additive in the diet of monogastric animals, which is a hydrolysis and fermentation medium for beneficial intestinal microbiomes and increases the abundance of beneficial bacteria (Song et al. 2020). Long-chain fructan has been shown to improve serum lipid profile, stimulating the immune system and enhancing the productivity and bioavailability of minerals such as zinc, iron, and copper that can have a positive effect on immune system function. These fructans have a positive effect on the immune system by stimulating the production of cytokines, mononuclear cells, and phagocytic macrophages and by inducing the synthesis of immunoglobulin, especially IgA (Siwek et al. 2018; Abd El-Hack et al. 2021; Khomayezi and Adewole 2022). Research by Huang et al. (2015) has confirmed that dietary inulin at levels of 5–10 g/kg can enhance the intestinal immune function of young broilers when intestinal function is not fully developed. Increased intestinal acidity due to the use of prebiotics may help reduce infection in the intestines of chickens (Shehata et al. 2022). Chitosan, as a natural source of prebiotics, is a natural biopolymer that is formed by deacetylated chitin, a major component of fungal cell walls and the exoskeleton of arthropods. Chitosan has several benefits including antimicrobial and antioxidant properties (Shehata et al. 2022). In agriculture, horticulture, environmental sciences, industry, microbiology, and medicine, chitosan has also shown promising applications (Shehata et al. 2022). In addition, many studies have used chitosan as a mucosal adjuvant that increases IgA levels (Kumar 2000; Shehata et al. 2022). Yalçın et al. (2014) reported that baker's yeast-derived yeast cell wall was an effective prebiotic feed additive in broiler feed due to increased growth function, increased humoral immune response, and decreased abdominal fat. In another study of laying hens, Yalçın et al. (2014) concluded that the yeast cell wall has beneficial effects on the production of low-cholesterol eggs and improves the humoral immune response. Prebiotics improve the immune system by blocking the binding site of bacteria. That is, pathogenic bacteria can be supplied to the immune system as attenuated antigens.

4 Synbiotics and Immunity

When probiotics are used in combination with prebiotics, they are called synbiotics, which lead to the ability to further improve the survival of probiotics (Shehata et al. 2022). In fact, synbiotics are a mixture of both probiotics and prebiotics, which have more benefits than taking either alone. Probiotics, prebiotics, and synbiotics are now widely used worldwide (Scherezenmeir and De Verse 2001; Shehata et al. 2022). Recent research has shown that synbiotics improve the microbial environment of the gut, activate host immunity, and ultimately prevent the replacement of harmful bacteria (Bandyopadhyay and Mandal 2014). Basically, the intestinal microbiota as a hidden metabolic organ for the body's immunology can play an important role in the immune system, and the use of feed additives such as synbiotics as effective compounds in improving the intestinal microbiota can indirectly play an important role in improving the immune system (Shehata et al. 2022). Further research into the effectiveness of prebiotics and synbiotics, as well as how these products work, reveals that manufacturers' concern is to find suitable, reliable, and cost-effective prebiotics and synbiotics as feed additives that can be used successfully and continuously in animal husbandry in the future. These additives can improve the epithelial barrier by affecting the intestinal microbiota and also improve the integrity of strong bonds by controlling tight junction proteins and inflammatory signaling pathways in host animals (Shehata et al. 2022). In this regard, a product containing oligofructose and *Bifidobacterium* probiotic is a good example of an active synbiotic, whereas a mixture of oligofructose and *Lactobacillus* case probiotic does not have this ability (Scherezenmeir and De Verse 2001). Dibaji et al. (2014) stated that the addition of synbiotics reduced *Escherichia coli* and the total coliform population in intestines of broilers. In contrast, different levels of probiotics increased the number of lactobacilli in intestines of broilers. Concentrations of synbiotics higher than the recommended levels in the diet also increased the population of lactic acid bacteria in intestines of broilers. Because the body's beneficial microbes help develop a strong immune system, reduce morbidity and mortality, and improve digestion and feed conversion ratio (FCR), the use of this synbiotic improves bird performance and thus reduces production costs. However, in the final summary of the above report, the researchers believe that the use of prebiotics and synbiotics can be used as a useful tool in altering the intestinal microflora, interacting with the immune system, and ultimately improving the growth and efficiency of substances to be used orally in farm animals (Dibaji et al. 2014). The role of synbiotics in gastrointestinal physiology is to enhance avian health and safety, e.g., competitive exclusion, antimicrobial activity, ammonia reduction, production of vitamins and maintain intestinal pH, immune modulation (stimulation/tolerance), epithelial barrier integrity, enzyme production, and microbial diversity and balance. However, many of the mechanisms of effects of probiotics and prebiotics on the immune system can apply to synbiotics (Tayeri et al. 2018; Shehata et al. 2022). Tayeri et al. (2018) in their research studies compared the effects of antibiotics, probiotics, synbiotics, and prebiotics on the performance and carcass characteristics of broiler chickens. The results showed that prebiotics had the most beneficial effect on cecum microbiota, stimulating lactic and aerobic acid-producing bacteria and reducing *Escherichia coli*. Synbiotic supplementation feeding in broiler diets had the most beneficial effects compared to other treatments containing prebiotics, probiotics, antibiotics, and controls. The final conclusion indicated that the use looks for prebiotics, probiotics, and synbiotics instead of antibiotics, significant performance, and health benefits for broiler.

5 Phytogenics and Immunity

Phytogenics are a group of natural growth stimulators (NGPs) or nonantibiotic growth stimulators of plant origin, which can play an effective role in improving health and immunity due to their bioactive and antioxidant compounds (Windisch et al. 2008; Alloui et al. 2014; Kiczorowska et al. 2017). One of the factors that reduces the immunity of poultry is stress. This is due to technological, environmental, heat, cold, high ammonia, nutritional, mycotoxins, imbalance of vitamins, mineral, and bacterial and viral challenges, which can cause the production of free radicals and defects in antioxidant protection at the cellular level and ultimately lead to oxidative stress and severe immunological decline (Surai et al. 2017). Therefore, it is important to strengthen the body's antioxidant system, which is responsible for preventing damage caused by free radicals to biological molecules such as proteins, fats, and DNA and can maintain the immunity of poultry (Surai et al. 2019). In different plants, a wide range of active phytochemical compounds including flavonoids, terpenoids, lignans, sulfides, polyphenols, carotenoids, coumarins, saponins, plant sterols, curcumins, and phthalides have been identified. Some of these phytochemicals either inhibit nitrosation or the formation of DNA adducts or stimulate the activity of protective enzymes such as the phase II enzyme glutathione transferase (EC 2.5.1.18) (Craig 1999). For example, carotenoid pigments are effective antioxidants that quench free radicals and protect against cells from oxidative damage, as well as stimulate the immune system. It has also been proven that the mentioned compounds can strengthen the activity of lymphocytes, increase phagocytosis, and induce interferon production, and the combination of these events plays an important role in improving the body's immune function (Van den Biggelaar et al. 2020; Abdelli et al. 2021). Antioxidant activity is one of the important properties of phytogenic products, and it is very important in strengthening the immune system. Antioxidant ability destroys free radicals and plays an important role in preventing some diseases caused by free radicals (Miguel 2010; Pandey et al. 2019). Previous research has shown that the antioxidant activity is due to their ability to donate hydrogen or electrons to free radicals and to transfer unpaired electrons, which are the main mechanisms of protecting other biological molecules against oxidation (Pandey et al. 2019). Phytogenics can directly eliminate stress-related reactive oxygen species (ROS) production through inhibition of enzymes or trace metals chelating (Abd El-Hack et al. 2020). Likewise, they have the ability to activate antioxidant enzymes and inhibit pro-oxidant enzymes such as lipoxygenase and NADPH oxidase (Pandey et al. 2019; Makala 2022).

As a result, phytogenics help host animals from immune defense stress in critical situations and increase intestinal access to essential nutrients for absorption, thus helping animals to grow within the framework of genetic potential. By strengthening their immunity and improving their health, they can have better growth (Yitbarek 2015). Some medicinal plants that contain molecules with immune-stimulating properties are: echinacea (*Echinacea* spp.), licorice (*Glycyrrhiza* L.), and garlic (*Allium sativum* L.) (Kiczorowska et al. 2017). It has been proven that these plants can improve the activity of lymphocytes, macrophages, and natural killer (NK) cells. They increase phagocytosis or stimulate interferon synthesis and are effective in strengthening the immune system (Kiczorowska et al. 2017). Garlic is one of the phytogenics that is often used in feeding animals. This plant shows very strong antiviral, bactericidal, antifungal, and antiparasitic properties (Kiczorowska et al. 2017). Kumar et al. (2014) showed that the use of garlic oil (*Allium sativum*) leads to strengthening the immunity and improving the antioxidant properties of the host animal. It is effective in combating fungal infections of the skin and mucosa of the digestive and respiratory systems, and strengthens and coordinates the immune system in birds (Kiczorowska et al. 2017). Many researches in this field show that supplementing the feed mixture with 0.5–3% garlic can be used to improve intestinal health, increase energy and nutrient consumption, and stimulate the growth of broiler chickens (Kiczorowska et al. 2017).

Glycyrrhizin, a triterpenoid saponin with a sweet taste, is the main bioactive component of licorice radix (*Glycyrrhiza glabra* L) (Craig 1999; Alagawany et al. 2019; Reda et al. 2021). It has been reported that glycyrrhizin and its aglycone, glycyrrhetinic acid, induce interferon activity and increase the activity of natural killer cells (Gowthaman et al. 2021). Hatano et al. (1988) reported that chalcones in licorice have antiviral activity against HIV. The popular spice turmeric (*Curcuma longa* L.) is also used as a feed additive in livestock production to strengthen the immune system (Khan et al. 2012a; Kiczorowska et al. 2017). Its activity in reducing intestinal inflammation and its strong antiviral, antibacterial, and antifungal activity are successfully used in large-scale production because it can improve the health status of animals and stimulate more production efficiency (Kanani et al. 2016; Kiczorowska et al. 2017; Shohe et al. 2019; Mitra et al. 2022). Turmeric oleoresin increased the expression of genes related to the recruitment of neutrophils, the complement system and its regulatory proteins, chemokines, cytokines, and the processing and presentation of antigens (Kiczorowska et al. 2017; Olarotimi 2018). Similarly, ginger rhizome (*Zingiber officinalis*) used in broiler diets has been reported to stimulate growth performance and health (Kiczorowska et al. 2017; Asghar et al. 2021). The use of ginger and cinnamon at the level of 0.8% of the feed mixture significantly improved health parameters in birds and increased the level of hemoglobin and the number of red blood cells (RBCs) (Ademola et al. 2009; Khan et al. 2012b; Kiczorowska et al. 2017; Adekiya et al. 2020). Cinnamon (*Cinnamon zeylanicum*), as a phytogenic option, has a stimulating effect on the immune system, which can be attributed to the antioxidant property of the plant's active compounds (El-Hack et al. 2020; Ali et al. 2021). Cinnamon strengthens the phagocytic activity of macrophages and plays an important role in the body's immune defense system. Cinnamon oils have antioxidant activity in broilers and its extract can have protective effects on diseases caused by oxidative stress. Cinnamon has cinnamaldehyde and diterpenes, which have the anti-inflammatory and antioxidant potential along with modulating activity on the body's immune system. In addition, thyme (*Thymus vulgaris*), black seed (*Nigella sativa*), and anise (*Pimpinella anisum*), which are used as spices, increase the production and resistance of poultry to diseases and strengthen the immune system (Al-Beitawi et al. 2010). This issue was confirmed in the studies conducted by Yazdi et al. (2014) on broiler chickens, in which the addition of anise seeds at a level of 10 g/kg of diet increased the antibody titer against avian influenza virus. Soltan et al. (2008) found that supplementing poultry diet with anise seeds improved blood parameters and increased phagocytic activity and the number of lymphocytes. Purple coneflower (*Echinacea purpurea*) is one of the most important and popular medicinal plants, which can be a phytogenic immune enhancer due to its stimulating effect on the immune system (Kiczorowska et al. 2017; Saeed et al. 2018; Daryin et al. 2019). The Dehkordi and Fallah (2011) research shows that feeding *Echinacea purpura*, especially for a long time, may improve performance and increase the immune response in broilers. Daryin et al. (2019) also pointed out that repeated short-term use of water extract of *Echinacea* plant has stimulating effects on the immune system in laying hens.

Pepper (*Capsicum* spp.) is an important plant species and a good source with various phytochemical compounds, including vitamin C, phenolic compounds, flavonoids, and carotenoids, which have significant antioxidant activity (Shehata et al. 2022). Capsaicin is the main bioactive component of red pepper (*Capsicum*), which gives a spicy taste to food. Capsaicin has been associated with many biological effects, including reducing body fat, anti-inflammatory, anticancer, antioxidant, and intestinal motility modulation (Shehata et al. 2022; Abd El-Hack et al. 2020a, b). These actions are mostly due to its role as an agonist of the transient receptor potential vanilloid 1 (TRPV1), which is expressed in the mesenteric nervous system and colon epithelial cells. The anti-inflammatory effect of capsaicin is also related to its role in activating the peroxisome proliferator-activated receptor gamma (PPARγ). Experimental studies showed that capsaicin can reduce intestinal inflammation, which can involve not only the TRPV1 receptor but also PPARγ (Santos and Alvarez-Leite 2019; Shehata et al. 2022). Next, another plant that is considered as an immunogenic phytogenic is the flower extract of *Calendula officinalis* L. (marigold), which is antibacterial, anti-antitumor-promoting, and cicatrizing effects, in vitro anti-HIV activity, hypoglycemic effects, gastric emptying inhibitory activity, and gastroprotective effects (Muley et al. 2009; Shehata et al. 2022). *Calendula officinalis* triterpenoids are considered as the most important anti-inflammatory principles of its extract (Muley et al. 2009; Shehata et al. 2022). Considering that marigold contains triterpenoids, taraxasterol-3-O-myristate, and arnidiol-3-O-myristate, it was shown that it can moderate the stress damage caused by H_2O_2 and INF-γ + tumor necrosis factor-alpha (TNF-α) and be effective against intestinal inflammations and strengthening immunity (Dall'Acqua et al. 2016; Shehata et al. 2022). Rajput et al. (2012) reported that dietary supplementation with marigold flower extract increased antibody titer against Newcastle disease virus (NDV) and avian influenza virus (AIV) and improved growth performance of broiler chickens. Eugenol is a volatile phenolic compound of clove essential oil that is obtained from the buds and leaves of *Eugenia caryophyllata* (Shehata et al. 2022; Gürbüz and Korkmaz 2022). 1,2-Eugenol is the main component (70–90%) of clove oil and is responsible for clove aroma (Shehata et al. 2022). Clove (*Syzygium aromaticum*) oil has antimicrobial, antifungal, antiviral, antioxidant, anti-inflammatory, and anticancer properties (Han and Parker 2017; Shehata et al. 2022). The authors found that clove essential oil significantly modulated global gene expression and altered signaling pathways critical for inflammation, tissue regeneration, and cancer cell signaling processes (Suliman et al. 2021; Shehata et al. 2022). Research studies for poultry also showed that clove essential oil is an anti-inflammatory agent, modulating the immune system and tissue regeneration of birds. Birds fed with eugenol and garlic reduce the CLDN5 expression in birds and lead to improved intestinal health, reduced effects of necrotizing enteritis, and enhanced immunity (Shehata et al. 2022).

Quercetin is the most common flavonoid in nature (Saeed et al. 2017; Shehata et al. 2022). It is also found in fruits and vegetables such as onions, cabbage, and apples. Quercetin in onion peel has a higher bioavailability than in apple peel (Shehata et al. 2022). This compound is one of the most studied polyphenols that has various health-promoting properties; for example, antimicrobial, antioxidant, anti-inflammatory, and metabolic effects have been confirmed in many studies and research (Erlund 2004; Saeed et al. 2017). Quercetin induces its antibacterial activity as a DNA gyrase on various cellular targets, bacterial membrane and motility, type II fatty acid biosynthesis pathway, and D-alanine:D-alanine ligase enzyme inhibitor (Wu et al. 2008; Shehata et al. 2022). Abdel-Latif et al. (2021) reported that total coliforms and *Clostridium perfringens* were reduced in groups fed with quercetin supplementation (200–800 ppm). On the contrary, the number of lactobacilli increased due to the improvement of the intestinal microbiota environment in the groups supplemented with quercetin, and following this event, the health and immunity of birds were maintained. Suzuki and Hara (2009) reported that quercetin enhances intestinal barrier function through the accumulation of zonula occludens (ZO)-2, occludin, and claudin-1 by inhibiting protein kinase Cδ (PKCδ). The analysis of intestinal permeability in the colon of laboratory mice showed that quercetin partially inhibits the effects of TNF-α and IFN-γ and reduces the total resistance of the intestinal barrier (Amasheh et al. 2009; Shehata et al. 2022). Carrasco-Pozo et al. (2013) evaluated the protective effect of quercetin on ZO-1 and occludin in Caco-2 cells treated with indomethacin and rotenone (an environmental toxin). Quercetin treatment protected the translocation of ZO-1 and prevented the downregulation of ZO-1 and occludin expression. The authors hypothesized that the effects of quercetin may be due to its mitochondrial protective properties. Quercetin inhibited isoform-mixed protein kinase C (PKC) and phosphoinositide 3-kinase (P13K). Summarizing the results showed that quercetin had a modulating effect on the activity of different intracellular signaling molecules, which can regulate the integrity of tight junction (TJ) and play a positive role in improving and strengthening the body's immune system (Shehata et al. 2022).

Curcumin is the main active component of turmeric with a concentration of about 1–5% and as a food flavoring, which is extracted and isolated from the dry powdered rhizome of *Curcuma longa* Linn (Shehata et al. 2022). According to the Food and Drug Administration (FDA) and the World Health Organization (WHO), oral administration is safe for adults and 0–0.3 mg/kg is also based on acceptable daily intake (ADI) recommendation. Curcumin is very resistant to low pH and is not metabolized in the stomach. It is absorbed from the large intestine and is identified in the form of glucuronide conjugate and sulfate conjugate in the blood. In enterocytes and hepatocytes, it is metabolized to di-, tetra-, and hexahydrocurcumin by reductase enzyme (Shehata et al. 2022). Curcumin shows poor gastrointestinal absorption and low bioavailability, which is mainly attributed to insolubility in water and rapid metabolism and excretion (Stohs et al. 2020). In rats, about 75% of curcumin was excreted in feces, and very little was found in urine (Shehzad et al. 2010). Natural products such as piperine, in addition to nanoformulations, increased the bioavailability of curcumin (Han 2011). It is said that curcumin has various types of medicinal activities, including antioxidant, anti-inflammatory, anticancer, antidiabetic, and anti-HIV effects (Shehata et al. 2022). It was shown that curcumin induces several endogenous antioxidants in intestinal disorders and reduces mucosal damage in trinitrobenzene sulfonic acid-induced colitis in vivo. Some studies reported that curcumin has cell protective properties by inducing heme oxygenase protective protein (HO-1). Curcumin prevents TNF-α-induced decrease in zonula and occluden-1 (ZO-1) protein levels in Caco-2 cell layers (Shehata et al. 2022). The interaction between curcumin and gut microbiota is bidirectional (Shehata et al. 2022). On the one hand, intestinal microbiota enzymes play a role in curcumin metabolism through reduction, acetylation, hydroxylation, demethylation, and demethoxylation. On the other hand, curcumin modulates the intestinal microbiota, improves intestinal barrier models, and counteracts pro-inflammatory mediators, and the set of these events aims to improve the immune system (Jiang et al. 2006; Scazzocchio et al. 2020). Curcumin has its anticoccidial effect through its antioxidant effect on the immune system. Therefore, curcumin can be useful for the treatment of intestinal disorders through the following effects: It protects intestinal epithelial cells against H_2O_2-induced tight junction (TJ) disruption and barrier dysfunction through the HO-1 pathway (Shehata et al. 2022). It restores occludin enzyme and ZO-1 protein levels after H_2O_2 treatment (Shehata et al. 2022). Curcumin can also reduce the release of IL-1β secreted by lipopolysaccharides (LPS), induce intestinal epithelial cells (IECs) and macrophages and prevent the disintegration of tight junction proteins such as ZO-1, claudin-1, claudin-7, and actin filaments. In specific pathogen-free (SPF) chickens experimentally infected with *Eimeria maxima*, curcumin reduced intestinal isoprostane 8-iso-PGF2 and prostaglandin GF2 (Shehata et al. 2022).

Zaker-Esteghamati et al. (2020) in a review report on the effect of *Silybum marianum* and its derivatives stated that silymarin is part of polyphenols and has the ability to absorb and neutralize oxygen-free radicals. One of the main functions

Fig. 1 *Silybum marianum* farms in the country of Iran and Gilan province in 2022

of hemoglobin is to carry oxygen throughout the body and remove carbon dioxide. Therefore, by increasing hemoglobin, oxygen delivery to different tissues was done by feeding thistle, and as a result, the metabolic process and immunity level improved, and probably the bird's health will be more stable with the consumption of thistle and its derivatives (Fig. 1).

Khazaei et al. (2022) in their review report in connection with *Silybum marianum* stated that the seeds of this plant contain about 70–80% of silymarin flavonolignans and about 20–30% of polymeric and polyphenol compounds are oxidized (such as tannins). The use of different levels of milk thistle (0.5% and 1.5%) significantly improved the total antioxidant values of bird blood. Consumption of silymarin in quail diet increased the number of white blood cells, calcium, vitamin D_3, and albumin. Silymarin also reduced the relative weight of the bursa of fabricius and the spleen. Also, the increase in blood hemoglobin is one of the important characteristics of iron storage in red blood cells (Khazaei et al. 2022). The antioxidant function of silymarin is to prevent the formation of free radicals by specifically inhibiting enzymes that produce reactive oxygen species and improve the integrity of mitochondria in stress conditions and strengthen the immune system (Surai 2015). Silymarin reduces liver glutathione and plays an important role in liver protein production and cell repair. The antioxidant properties of silymarin are due to the polyphenolic structure with the methoxy group in one of the phenolic rings, which can reduce oxidative stress and protect cells from apoptosis. Also, silymarin increases the activity of pancreatic antioxidant enzymes, glutathione peroxidase, superoxide dismutase, and catalase. Aldose reductase inhibitors such as silymarin can reduce oxidative stress by inhibiting the protein kinase C enzyme (Khazaei et al. 2022). Gazak et al. (2007) reported that silymarin in milk thistle reduces free radicals caused by aflatoxin and increases cellular activity for protein synthesis and antibody production. By acting on transcription enzymes, silymarin accelerates the process of protein synthesis, which by entering the nucleus and acting on RNA polymerase 1 and rRNA transcription enzymes leads to an increase in ribosome formation. This event, in turn, accelerates the synthesis of protein and DNA and increases the construction process in the cytoplasm and as a result, the synthesis of structural and functional proteins increases. At least conceptually, this stimulation may enable the cell to cope with the reduction of transporters and enzymes that occur in many pathological conditions. Silymarin, together with silibinin, can increase the regeneration of liver cells and increase the production of enzymes needed for DNA synthesis. Silymarin can also increase the release of acetylcholine (Zaker-Esteghamati et al. 2020). Silybin has antibacterial activity against gram-positive bacteria, which is much stronger than silymarin, while it has no effect on gram-negative bacteria (Zaker-Esteghamati et al. 2020) (Fig. 2).

Feshanghchi et al. (2022) investigated the effects of milk thistle (*Silybum marianum*; MT), toxin binder (TB), and marine algae (*Spirulina platensis*; SP) on the performance, blood parameters, humoral immunity, and cecum microbiota of broilers exposed to aflatoxin-B1 (AFB1). Feeding diets contaminated with AFB1 led to a significant increase in AST and ALT activity. MT, TB, and SP powders significantly reduced AST and ALT blood activity in broilers. When diets infected with AFB1 were fed by adding MT, TB, and SP compounds to the diet, growth performance, immunological performance, and serum biochemical parameters of broiler chickens improved significantly and promisingly. Previous research also showed that marine algae have a wide range of therapeutic properties and biological activities, including immunomodulatory, antioxidant, and anti-inflammatory capabilities. Broilers fed with SP supplemented diet reduced the levels of oxidative stress leading to better antioxidant capacity. These researchers believe that the active combination of milk thistle and spirulina algae has the ability to protect the supply of glutathione and reduce oxidation, as well as have a direct effect on immune cells, and can increase the level of immunoglobulin, leading to the strengthening of the immune system (Feshanghchi et al. 2022).

Cynara scolymus is a medicinal plant containing quality antioxidants and the research showed that feeding the powder and extract of *C. scolymus* in broilers leads to an increase in Newcastle antibody titer and a decrease in blood pressure, and subsequently, the bird's immunity is strengthened (Zaker-Esteghamati et al. 2021a). In addition, diets containing *C. scolymus* increased the titer of antisheep red blood cells (SRBCs) in chickens under conditions of heat stress (Effati et al. 2014). Feeding *C. scolymus* in Japanese quail led to reduction of oxidative stress and improvement of safety and health-related parameters (Kh 2015). Tajodini et al. (2015) showed that feeding *C. scolymus* (1.5% powder) increased the antibody titer against Newcastle virus at the age of 42 days. Rouzmehr et al. (2014) also

Fig. 2 Capitol and seeds of *Silybum marianum* in the country of Iran and Gilan province in 2022

reported that, with the consumption of 150 g per ton of *C. scolymus* powder on days 28 and 42, an increase in the antibody titer against Gumboro and Newcastle disease was observed (Fig. 3).

Zaker-Esteghamati et al. (2021b) investigated the effects of adding a dry extract of *Cynara scolymus* (CS) and *Silybum marianum* (SM) in the diet of broilers. They

Fig. 3 *Cynara scolymus* plant in the country of Iran and Gilan province in 2022

showed that feeding 250 and 500 mg/kg of dry extract of CS and SM in the diet may have positive effects on the studied traits. The results of the present study showed that CS and SM supplements in aflatoxin-free diets had no effect on immune organs, but improved the antibody titer against influenza virus and SRBC. Also, dietary supplementation with CS and SM improved blood lipid profile and liver enzymes. There are reports that plant flavonoid compounds, including silymarin in SM, can strengthen the immune system and increase antibody production in animals due to their antibacterial properties and vitamin C-enhancing activity (Zaker-Esteghamati et al. 2021b). *Cynara scolymus* was also effective in strengthening the immune system of broilers due to its phenolic compounds and high antioxidant properties. Fallah Huseini et al. (2012) investigated the effect of 1.5% *Cynara scolymus* leaf meal and oregano extract (200 mg/kg in drinking water) on Ross strain broilers (Zaker-Esteghamati et al. 2021a). The results showed that the effect of these treatments on the antibody titer against NDV and heterophile on the lymphocyte was not significant. However, the percentage of heterophiles increased significantly with the application of the treatments, so that the control treatment and *Cynara scolymus* leaf meal 1.5% had the lowest and highest percentages of heterophile, respectively. The percentage of lymphocytes decreased with the use of 1.5% *Cynara scolymus* leaf meal in the diet (83.54%) compared to the control (84.65%). Therefore, it can be concluded that the use of *Cynara scolymus* is quite effective in correcting stress conditions and improving the immune system of broilers. But the

Fig. 4 Capitol of *Cynara scolymus* in the country of Iran and Gilan province in 2022

use of these compounds in normal conditions has no adverse effect on the immune system and can be used as a harmless food supplement in the diet of broilers (Zaker-Esteghamati et al. 2021a, b) (Fig. 4).

Green tea (*Camellia sinensis* L.) as a valuable medical plant contains polyphenolic compounds such as epi-phenolate gallate, tannin, caffeine, theophylline, and theobromine. Green tea as an antioxidant feed additive can be used in poultry nutrition in order to improve blood antioxidant status, poultry health, and strengthen immunity, but it is necessary to pay attention to the use of appropriate levels in order to prevent weight loss and performance (Pliego et al. 2022) (Fig. 5).

Several studies have shown that dietary green tea powder affects immunological parameters of broilers. Green tea catechins have many biological and biochemical effects, which are mainly attributed to their high antioxidant capacity. Epigallocatechin-3-gallate is the most biologically active substance in green tea (Seidavi et al. 2017). The effects of diet containing fish oil (1.5% and 2%) and green tea powder 1% and 1.5% as supplements on the immunity of broilers showed that the hemorrhagic response to influenza and Newcastle disease improved. If the powder (1 and 1.5% by weight) is used as a supplement for 42 days, it effectively prevents the growth of pathogens, improves performance, and reduces mortality. Investigating immunological parameters of broiler chickens fed with green tea powder showed that it led to increased resistance against influenza. Another study investigated the effect of green tea extract (GTE; 125, 250, 500, 1000, and 2000 mg/kg) on growth performance, serum lipid profile, and liver glutathione. The results of

Fig. 5 Tea (*Camellia sinensis* L.) gardens in Iran and Gilan province

this research confirmed the reduction of malondialdehyde in thigh muscle and the humoral immune response against Newcastle disease virus vaccines of broiler chickens. In addition, green tea supplementation increased the specific antibody titer against Newcastle virus vaccines at the age of 28 and 35 days in broilers. In another study, the effect of green tea (0.5 and 1 g per kg of diet) on growth performance, carcass characteristics, blood biochemical parameters, oxidative stability of meat, and the expression of interleukin-6 and interferon-γ genes in broilers was determined, and green tea extract showed a significant effect on liver enzymes and a decrease in malondialdehyde content. In broiler chickens, these plant compounds increased the expression of interleukin-6 and γ-interferon liver genes and the findings showed that green tea can be used not only to reduce abdominal fat and lipid metabolites, but also act as antioxidants. Green tea can greatly affect the intestinal microbiota of chickens by changing their composition. Green tea dietary supplement caused the proliferation of *Lactobacillus* and prevented the proliferation of *Escherichia coli* in the ileum and cecum (Seidavi et al. 2017). Seidavi et al. (2014) investigated the effect of fish oil supplements and green tea powder on the immunity of broiler chickens. They showed that the group fed with 2% fish oil plus 1% green tea showed a significant increase in total immunoglobulin production, particularly IgM production. Abbas et al. (2017) investigated the effect of green tea against coccidiosis in chickens and showed that the cellular and humoral immunity against coccidiosis increased and the reason was the increase in immune response and the level of total immunoglobulin, IgG, and IgM. Zarezadeh et al. (2013) reported that the addition of green tea and fish oil in the diet of broilers with the addition of 1.5% fish oil and 1.5% green tea stimulated the humoral immune response. Song et al. (2016) investigated the effects of nutritional supplements including *Enterococcus faecalis* and green tea extract on growth performance, immune system, and serum biochemical parameters of broiler chickens. The results of that experiment showed that the increase in IgG and IgA levels leads to the improvement in the immune system, the reduction in total cholesterol, and the increase in serum biochemical parameters. Polyphenols, especially catechins, which have antibacterial and antioxidant properties, are a valuable option for a variety of chronic immune disorders and for improving hemorrhoidal responses to influenza and Newcastle disease (Song et al. 2016; Seidavi et al. 2020).

Baghban-Kanani et al. (2019) showed that supplementing the diet of laying hens with *Artemisia annua* leaf powder increased glutathione peroxidase (GSH-Px) concentration and decreased malondialdehyde (MDA). It was previously shown that *Artemisia annua* extract increases GSH-Px activity and decreases MDA production in the liver of rats (Ryu et al. 1998). By supplementing the diet with *Artemisia annua*, Wan et al. (2016) showed an increase in GSH-Px activity and a decrease in MDA levels in the serum and liver of birds and proved that the plant has a good potential to improve the antioxidant status of poultry, which can corroborate the strengthening of the immune system. *Artemisia annua* contains phenolic compounds and flavonoids with superior antioxidant activity (Brisibe et al. 2009; Iqbal et al. 2012). However, *Artemisia annua* is one of the most important species among plants of the Artemisia family, which is well known for its antimalarial properties. This plant was used as a medicine with the support of the United Nations in all countries to save human from malaria (Han et al. 2005) (Fig. 6).

It was proved that artemisinin present in this plant, in addition to its antimalarial effect, can also be effective against various cancer cells such as leukemia and colon cancer (Baldi and Dixit 2008). The literature review indicates that *Artemisia annua* as a potential anticoccidial in the poultry industry can have an important role (Nahed et al. 2022). Reports show that the hydroalcoholic extract of *Artemisia annua* causes the destruction of the cystic stage of the *Giardia lamblia* parasite (Bahman et al. 2012). *Artemisia annua* can strengthen the body's immune system due to its artemisinin composition. Also, due to the antiparasitic effects of artemisinin base compounds and the high antioxidant capacity of crude extracts of this plant, its use in nutritional programs is expanding. Published research shows that artemisinin drugs have been effective against *Leishmania, Trypanosoma,* and some viruses that affect humans and animals. In animals, the antiparasitic activity of *Artemisia* spp. and artemisinin has been proven effective against several known parasites including *Fasciola*, *Eimeria*, *Trichostrongylus*, and *Haemonchus*. Apart from the nutritional value of *Artemisia annua* leaves, the benefits associated with other antioxidant compounds in the diet, such as vitamin E and other phenolic compounds, make this plant a natural additive with high potential, which can be effective in stimulation and balance of immunity (Cherian et al. 2013), because improving the antioxidant

Fig. 6 *Artemisia annua* plants in the country of Iran and Gilan province in 2022

status of broiler blood occurs through the removal of free radicals and hydrogen supply by flavonoids and phenolic compounds (Wan et al. 2016).

Reviewing the literature of the last few years shows that the use of *Astragalus* in nutrition of poultry can play an important role in improving the safety and health of birds (Liang et al. 2021; Liao et al. 2021; Qiao and Kyselov 2021). Xi et al. (2014) investigated the effects of stems and leaves in ultra-fined powder of *Astragalus* (SLASUP) and reported that SLASUP at the level of 1.5% can improve the immune function of chickens by increasing the antibody titer, content IL-2 and IFN-γ significantly. Stems and leaves of the mentioned source are clinically useful to strengthen the immune response and to prevent various infections such as Newcastle disease in the poultry farm.

Farag and Alagawany (2019) announced that *Astragalus membranaceus* is a member of the *Leguminosae* family and is widely used as an immune-modulating agent. It has been proven that *Astragalus* polysaccharides (APS) have impressive and promising biological activities that can improve immune system, antioxidant, antiviral, antimicrobial, and antiparasitic activities in poultry. These researchers studied the immune system-enhancing activities of *Astragalus membranaceus* in birds and reported that its feeding leads to the development of immune-modulating effects in broiler chickens and increased the weight of immune organs and increased IgG levels and improved liver function and antioxidant status (Awadallah 2019; Farag and Alagawany 2019).

Also, Shan et al. (2019) showed that the oral administration of the plant can increase the immune function of the intestinal mucosa of chickens and as a vaccine booster it can play a significant role in improving the health of the flock. Jiang et al. (2010) also believed that the immune function of chicken erythrocytes after being infected with infectious bursal disease virus (IBDV) was significantly reduced by the consumption of *Astragalus* plant powder and clearly increased the immune function of chicken erythrocytes. These researchers showed that probiotics together with the oral administration of the plant powder showed synergistic modulation effects on the body's immunity and intestinal microbiota.

6 Conclusion

In order to control and prevent all kinds of diseases, strengthening the poultry's immune system through the feeding of feed additives can be considered as a low-risk solution and an alternative to antibiotics. The review of the scientific literature in this chapter showed that additives such as probiotics, prebiotics, synbiotics, and phytogenics can increase the health and performance of poultry while stimulating the immune system. Probiotics, prebiotics, and synbiotics investigated through their effect on gut microbiota and microflora were able to balance and strengthen the immune system and subsequently improve the health of the herd. Due to the presence of bioactive compounds, the investigated phytogenics were able to affect the vital cell signaling pathways and lead to balance, stimulation, and strengthening of the poultry's immunity. In general, the formulation and use of natural additives using useful microbial and plant compounds as immune system stimulants can be a sustainable and environmentally friendly solution in the poultry industry, which improves performance and health and strengthens the system so that poultry flock safety is effective and accepted by activists of this sector.

Funding Financial support by Rasht Branch, Islamic Azad University, Rasht, Iran, grant number 17-16-1-462 is gratefully acknowledged.

References

Abbas A, Iqbal Z, Abbas RZ, Khan MK, Khan JA, Hussain K, Rizwan HM (2017) Immunomodulatory effects of *Camellia sinensis* against coccidiosis in chickens. J Anim Plant Sci 27(2)

Abd El-Hack ME, Abdelnour SA, Taha AE, Khafaga AF, Arif M, Ayasan T, Abdel-Daim MM (2020a) Herbs as thermoregulatory agents in poultry: an overview. Sci Total Environ 703: 134399

Abd El-Hack ME, El-Saadony MT, Shafi ME, Qattan SY, Batiha GE, Khafaga AF, Alagawany M (2020b) Probiotics in poultry feed: a comprehensive review. J Anim Physiol Anim Nutr 104(6): 1835–1850

Abd El-Hack ME, El-Saadony MT, Shafi ME, Alshahrani OA, Saghir SA, Al-Wajeeh AS, Abdel-Moneim AME (2021) Prebiotics can restrict *Salmonella* populations in poultry: a review. Anim Biotechnol 33(7):1668–1677

Abdel-Latif MA, Elbestawy AR, El-Far AH, Noreldin AE, Emam M, Baty RS, Abd El-Hamid HS (2021) Quercetin dietary supplementation advances growth performance, gut microbiota, and intestinal mRNA expression genes in broiler chickens. Animals 11(8):2302

Abdelli N, Solà-Oriol D, Pérez JF (2021) Phytogenic feed additives in poultry: achievements, prospective and challenges. Animals 11(12):3471

Adekiya AO, Agbede TM, Ejue WS, Aboyeji CM, Dunsin O, Aremu CO, Adesola OO (2020) Biochar, poultry manure and NPK fertilizer: sole and combine application effects on soil properties and ginger (*Zingiber officinale Roscoe*) performance in a tropical Alfisol. Open Agric 5(1):30–39

Ademola SG, Farinu GO, Babatunde GM (2009) Serum lipid, growth and hematological parameters of broilers fed garlic, ginger and their mixtures. World J Agric Sci 5(1):99–104

Alagawany M, Elnesr SS, Farag MR, Abd El-Hack ME, Khafaga AF, Taha AE, Dhama K (2019) Use of licorice (*Glycyrrhiza glabra*) herb as a feed additive in poultry: current knowledge and prospects. Animals 9(8):536

Al-Beitawi NA, El-Ghousein SS, Athamneh MZ (2010) Effect of adding crushed *Pimpinella anisum, Nigella sativa* seeds and *Thymus vulgaris* mixture to antibiotics-free rations of vaccinated and non-vaccinated male broilers on growth performance, antibody titer and haematological profile. Ital J Anim Sci 9(2):e43

Ali A, Ponnampalam EN, Pushpakumara G, Cottrell JJ, Suleria HA, Dunshea FR (2021) Cinnamon: a natural feed additive for poultry health and production - a review. Animals 11(7):2026

Al-Khalaifah HS (2018) Benefits of probiotics and/or prebiotics for antibiotic-reduced poultry. Poult Sci 97(11):3807–3815

Alloui MN, Agabou A, Alloui N (2014) Application of herbs and phytogenic feed additives in poultry production: a review. Glob J Anim Sci Res 2(3):234–243

Amasheh M, Grotjohann I, Amasheh S, Fromm A, Söderholm JD, Zeitz M, Schulzke JD (2009) Regulation of mucosal structure and barrier function in rat colon exposed to tumor necrosis factor alpha and interferon gamma in vitro: a novel model for studying the pathomechanisms of inflammatory bowel disease cytokines. Scand J Gastroenterol 44(10):1226–1235

Asghar MU, Rahman A, Hayat Z, Rafique MK, Badar IH, Yar MK, Ijaz M (2021) Exploration of *Zingiber officinale* effects on growth performance, immunity and gut morphology in broilers. Braz J Biol 83

Awadallah MRF (2019) The role of *Astragalus membranaceus* as immunomodulator in poultry. World's Poultry Science Association

Baghban-Kanani P, Hosseintabar-Ghasemabad B, Azimi-Youvalari S, Seidavi A, Ragni M, Laudadio V, Tufarelli V (2019) Effects of using *Artemisia annua* leaves, probiotic blend, and organic acids on performance, egg quality, blood biochemistry, and antioxidant status of laying hens. J Poult Sci 56(2):120–127

Bahman RE, Gholami S, Azadbakht M, Ziaei H (2012) Effect of Hydroalcoholic extract of *Artemisia annua* on cysts of *Giardia lamblia* in Invitro. J Maz Univ Med Sci 22(90):71–80

Baldi A, Dixit VK (2008) Yield enhancement strategies for artemisinin production by suspension cultures of *Artemisia annua*. Bioresour Technol 99(11):4609–4614

Bandyopadhyay B, Mandal NC (2014) Probiotics, prebiotics and synbiotics in health improvement by modulating gut microbiota: the concept revisited. Int J Curr Microbiol App Sci 3(3):410–420

Brisibe EA, Umoren UE, Brisibe F, Magalhäes PM, Ferreira JF, Luthria D, Prior RL (2009) Nutritional characterisation and antioxidant capacity of different tissues of *Artemisia annua* L. Food Chem 115(4):1240–1246

Carrasco-Pozo C, Morales P, Gotteland M (2013) Polyphenols protect the epithelial barrier function of Caco-2 cells exposed to indomethacin through the modulation of occludin and zonula occludens-1 expression. J Agric Food Chem 61(22):5291–5297

Chacher MFA, Kamran Z, Ahsan U, Ahmad S, Koutoulis KC, Din HQU, Cengiz Ö (2017) Use of mannan oligosaccharide in broiler diets: an overview of underlying mechanisms. World Poult Sci J 73(4):831–844

Chee SH, Iji PA, Choct M, Mikkelsen LL, Kocher A (2010) Functional interactions of manno-oligosaccharides with dietary threonine in chicken gastrointestinal tract. I. Growth performance and mucin dynamics. Br Poult Sci 51(5):658–666

Cherian G, Orr A, Burke IC, Pan W (2013) Feeding *Artemisia annua* alters digesta pH and muscle lipid oxidation products in broiler chickens. Poult Sci 92(4):1085–1090

Chichlowski M, Croom J, McBride BW, Havenstein GB, Koci MD (2007) Metabolic and physiological impact of probiotics or direct-fed-microbials on poultry: a brief review of current knowledge. Int J Poult Sci 6(10):694–704

Craig WJ (1999) Health-promoting properties of common herbs. Am J Clin Nutr 70(3):491S–499S

Dall'Acqua S, Catanzaro D, Cocetta V, Igl N, Ragazzi E, Giron MC, Montopoli M (2016) Protective effects of ψ taraxasterol 3-O-myristate and arnidiol 3-O-myristate isolated from Calendula officinalis on epithelial intestinal barrier. Fitoterapia 109:230–235

Dalloul RA, Lillehoj HS, Tamim NM, Shellem TA, Doerr JA (2005) Induction of local protective immunity to *Eimeria acervulina* by a *Lactobacillus*-based probiotic. Comp Immunol Microbiol Infect Dis 28(5–6):351–361

Daryin AI, Kerdyashov NN, Naumov AA, Shishkina TV (2019) Effect of *Echinacea purpurea* on biochemical composition of meat poultry hatching eggs. Volga Region Farmland 1:42–47

Dehkordi SH, Fallah V (2011) Enhancement of broiler performance and immune response by *Echinacea purpurea* supplemented in diet. Afr J Biotechnol 10(54):11280–11286

Dibaji SM, Seidavi A, Asadpour L, Da Silva FM (2014) Effect of a synbiotic on the intestinal microflora of chickens. J Appl Poult Res 23(1):1–6

Effati M, Samadi F, Dastar B, Azari MA, Hashemi SR (2014) Effects of different levels of artichoke (*Cynara scolymus*) on growth performance and immune responses of broilers underheat stress. Iran J Appl Anim Sci 4(2):399–403

El-Hack A, Mohamed E, Alagawany M, Abdel-Moneim AME, Mohammed NG, Khafaga AF, Elnesr SS (2020) Cinnamon (*Cinnamomum zeylanicum*) oil as a potential alternative to antibiotics in poultry. Antibiotics 9(5):210

Erlund I (2004) Review of the flavonoid's quercetin, hesperetin, and naringenin. Dietary sources, bioactivities, bioavailability, and epidemiology. Nutr Res 24(10):851–874

Fallah Huseini H, Kianbakht S, Heshmat R (2012) *Cynara scolymus* L. in treatment of hypercholesterolemic type 2 diabetic patients: a randomized double-blind placebo-controlled clinical trial. J Med Plants 11(41):58–65

Farag MR, Alagawany M (2019) The role of *Astragalus membranaceus* as immunomodulator in poultry. World Poult Sci J 75(1):43–54

Ferket P R (2004) Alternatives to antibiotics in poultry production: responses, practical experience and recommendations. In: Nutritional biotechnology in the feed and food industries. Proceedings of Alltech's 20th Annual Symposium: re-imagining the feed industry, Lexington, Kentucky, USA, 23–26 May 2004, pp 57–67

Feshanghchi M, Baghban-Kanani P, Kashefi-Motlagh B, Adib F, Azimi-Youvalari S, Hosseintabar-Ghasemabad B, Tufarelli V (2022) Milk thistle (*Silybum marianum*), marine algae (*Spirulina platensis*) and toxin binder powders in the diets of broiler chickens exposed to aflatoxin-b1: growth performance, humoral immune response and cecal microbiota. Agriculture 12(6):805

Gazak R, Walterova D, Kren V (2007) Silybin and silymarin-new and emerging applications in medicine. Curr Med Chem 14(3):315–338

Gowthaman V, Sharma D, Biswas A, Deo C (2021) Liquorice (*Glycyrrhiza glabra*) herb as a poultry feed additive: a review: Liquorice as poultry feed additive. Lett Anim Biol 1(2):14–20

Gürbüz M, Korkmaz BİO (2022) The anti-campylobacter activity of eugenol and its potential for poultry meat safety: a review. Food Chem 394:133519

Haghighi HR, Gong J, Gyles CL, Hayes MA, Sanei B, Parvizi P, Sharif S (2005) Modulation of antibody-mediated immune response by probiotics in chickens. Clin Vaccine Immunol 12(12): 1387–1392

Han HK (2011) The effects of black pepper on the intestinal absorption and hepatic metabolism of drugs. Expert Opin Drug Metab Toxicol 7(6):721–729

Han X, Parker TL (2017) Anti-inflammatory activity of clove (*Eugenia caryophyllata*) essential oil in human dermal fibroblasts. Pharm Biol 55(1):1619–1622

Han JL, Wang H, Ye HC, Liu Y, Li ZQ, Zhang Y, Li GF (2005) High efficiency of genetic transformation and regeneration of *Artemisia annua* L via agrobacterium tumefaciens-mediated procedure. Plant Sci 168(1):73–80

Hasted TL, Sharif S, Boerlin P, Diarra MS (2021) Immunostimulatory potential of fruits and their extracts in poultry. Front Immunol 12:1634

Hatano T, Yasuhara T, Miyamoto K, Okuda T (1988) Anti-human immunodeficiency virus phenolics from licorice. Chem Pharm Bull 36(6):2286–2288

Higgins JP, Higgins SE, Vicente JL, Wolfenden AD, Tellez G, Hargis B (2007) Temporal effects of lactic acid bacteria probiotic culture on *Salmonella* in neonatal broilers. Poult Sci 86(8): 1662–1666

Huang Q, Wei Y, Lv Y, Wang Y, Hu T (2015) Effect of dietary inulin supplements on growth performance and intestinal immunological parameters of broiler chickens. Livest Sci 180:172–176

Iqbal S, Younas U, Chan KW, Zia-Ul-Haq M, Ismail M (2012) Chemical composition of *Artemisia annua* L leaves and antioxidant potential of extracts as a function of extraction solvents. Molecules 17(5):6020–6032

Jiang H, Deng CS, Zhang M, Xia J (2006) Curcumin-attenuated trinitrobenzene sulphonic acid induces chronic colitis by inhibiting expression of cyclooxygenase-2. World J Gastroenterol 12(24):3848

Jiang J, Wu C, Gao H, Song J, Li H (2010) Effects of *Astragalus* polysaccharides on immunologic function of erythrocyte in chickens infected with infectious bursa disease virus. Vaccine 28(34): 5614–5616

Jin LZ, Ho YW, Abdullah N, Jalaludin S (1998) Growth performance, intestinal microbial populations, and serum cholesterol of broilers fed diets containing *Lactobacillus cultures*. Poult Sci 77(9):1259–1265

Kabir SL, Rahman MM, Rahman MB, Rahman MM, Ahmed SU (2004) The dynamics of probiotics on growth performance and immune response in broilers. Int J Poult Sci 3(5): 361–364

Kanani PB, Daneshyar M, Najafi R (2016) Effects of cinnamon (*Cinnamomum zeylanicum*) and turmeric (*Curcuma longa*) powders on performance, enzyme activity, and blood parameters of broiler chickens under heat stress. Poult Sci J 4(1):47–53

Kh M (2015) The effect of artichoke leaf powder on performance, internal organs weight and some blood parameters of Japanese quail poisoned with carbon tetrachloride. J Anim Sci Res 25(2): 57–67

Khaksefidi A, Ghoorchi T (2006) Effect of probiotic on performance and immunocompetence in broiler chicks. J Poult Sci 43(3):296–300

Khan RU, Naz S (2013) The applications of probiotics in poultry production. World Poult Sci J 69(3):621–632

Khan RU, Naz S, Javdani M, Nikousefat Z, Selvaggi M, Tufarelli V, Laudadio V (2012a) The use of turmeric (*Curcuma longa*) in poultry feed. World Poult Sci J 68(1):97–103

Khan RU, Naz S, Nikousefat Z, Tufarelli V, Javdani M, Qureshi MS, Laudadio V (2012b) Potential applications of ginger (*Zingiber officinale*) in poultry diets. World Poult Sci J 68(2):245–252

Khazaei R, Seidavi A, Bouyeh M (2022) A review on the mechanisms of the effect of silymarin in milk thistle (*Silybum marianum*) on some laboratory animals. Vet Med Sci 8(1):289–301

Khomayezi R, Adewole D (2022) Probiotics, prebiotics, and synbiotics: an overview of their delivery routes and effects on growth and health of broiler chickens. World Poult Sci J 78(1): 57–81

Kiczorowska B, Samolińska W, Al-Yasiry ARM, Kiczorowski P, Winiarska-Mieczan A (2017) The natural feed additives as immunostimulants in monogastric animal nutrition – a review. Ann Anim Sci 17(3):605–625

Klasing KC (2007) Nutrition and the immune system. Br Poult Sci 48(5):525–537

Koenen ME, Kramer J, Van Der Hulst R, Heres L, Jeurissen SHM, Boersma WJA (2004) Immunomodulation by probiotic *lactobacilli* in layer and meat type chickens. Br Poult Sci 45(3):355–366

Krysiak K, Konkol D, Korczyński M (2021) Overview of the use of probiotics in poultry production. Animals 11(6):1620

Kumar MNR (2000) A review of chitin and chitosan applications. React Funct Polymers 46(1): 1–27

Kumar M, Kumar V, Roy D, Kushwaha R, Vaiswani S (2014) Application of herbal feed additives in animal nutrition-a review. Int J Livest Res 4(9):1–8

Liang W, Li H, Zhou H, Wang M, Zhao X, Sun X, Zhang X (2021) Effects of *Taraxacum* and *Astragalus* extracts combined with probiotic *Bacillus subtilis* and *Lactobacillus* on *Escherichia coli* infected broiler chickens. Poult Sci 100(4):101007

Liao L, Li J, Li J, Huang Y, Wu Y (2021) Effects of *Astragalus* polysaccharides on intestinal morphology and intestinal immune cells of Muscovy ducklings infected with Muscovy duck reovirus. Poult Sci 100(1):64–72

Lutful Kabir SM (2009) The role of probiotics in the poultry industry. Int J Mol Sci 10(8): 3531–3546

Makala H (2022) Herbs and phytogenic feed additives in poultry nutrition. Vet Med Sci Pract 78(1): 11–18

McFarlane C, Ramos CI, Johnson DW, Campbell KL (2019) Prebiotic, probiotic, and synbiotic supplementation in chronic kidney disease: a systematic review and meta-analysis. J Ren Nutr 29(3):209–220

McFarlane C, Kelly JT, Conley M, Johnson DW, Campbell KL (2022) Consumers' perspectives and experiences of prebiotics and probiotics for gut health in chronic kidney disease. J Ren Nutr 33:116

Miguel MG (2010) Antioxidant and anti-inflammatory activities of essential oils: a short review. Molecules 15(12):9252–9287

Mitra S, Da Silva LE, Tuli HS, Dhama K (2022) Potential health benefits of dietary *Curcuma longa* (turmeric) supplementation on health and production of poultry: a mini-review. Indian Vet J 99(02):07–13

Muhammad SL, Sheikh IS, Bajwa MA, Mehmood K, Rashid N, Akhter MA, Rehman UA (2020) Effect of Mannan oligosaccharide (MOS) on growth, physiological and immune performance parameters of broiler chickens. Pak-Euro J Med Life Sci 3(2):76–85

Muley BP, Khadabadi SS, Banarase NB (2009) Phytochemical constituents and pharmacological activities of *Calendula officinalis* Linn (*Asteraceae*): a review. Trop J Pharm Res 8(5)

Nabizadeh A (2012) The effect of inulin on broiler chicken intestinal microflora, gut morphology, and performance. J Anim Feed Sci 21(4):725–734

Nahashon SN (1994) Effect of a direct-fed microbial on performance of single comb white leghorn chickens

Nahed A, Abd El-Hack ME, Albaqami NM, Khafaga AF, Taha AE, Swelum AA, Elbestawy AR (2022) Phytochemical control of poultry coccidiosis: a review. Poult Sci 101(1):101542

O'mahony L, Feeney M, O'halloran S, Murphy L, Kiely B, Fitzgibbon J, Collins JK (2001) Probiotic impact on microbial flora, inflammation and tumor development in IL-10 knockout mice. Aliment Pharmacol Therap 15(8):1219–1225

Olarotimi OJ (2018) Turmeric (*Curcuma longa*): an underutilized phytogenic additive in poultry nutrition. Turk J Agric Food Sci Technol 6(1):102–106

Olgun O, Abdulqader AF, Karabacak A (2021) The importance of nutrition in preventing heat stress at poultry. World Poult Sci J 77(3):661–678

Pandey AK, Kumar P, Saxena MJ (2019) Feed additives in animal health. In: Nutraceuticals in veterinary medicine. Springer, Cham, pp 345–362

Pirgozliev V, Rose SP, Ivanova S (2019) Feed additives in poultry nutrition. Bulg J Agr Sci 25(1): 8–11

Pliego AB, Tavakoli M, Khusro A, Seidavi A, Elghandour MM, Salem AZ, Rene Rivas-Caceres R (2022) Beneficial and adverse effects of medicinal plants as feed supplements in poultry nutrition: a review. Anim Biotechnol 33(2):369–391

Prabakar G, Pavulraj S, Shanmuganathan S, Kirubakaran A, Mohana N (2016) Early nutrition and its importance in poultry: a review. Indian J Anim Nutr 33(3):245–252

Qiao Y, Kyselov O (2021) Using *Astragalus* polysaccharide herbal supplement to improve disease resistance in poultry

Rajput N, Naeem M, Ali S, Rui Y, Tian W (2012) Effect of dietary supplementation of marigold pigment on immunity, skin and meat color, and growth performance of broiler chickens. Braz J Poult Sci 14:291–295

Reda FM, El-Saadony MT, El-Rayes TK, Farahat M, Attia G, Alagawany M (2021) Dietary effect of licorice (*Glycyrrhiza glabra*) on quail performance, carcass, blood metabolites and intestinal microbiota. Poult Sci 100(8):101266

Ricke SC, Lee SI, Kim SA, Park SH, Shi Z (2020) Prebiotics and the poultry gastrointestinal tract microbiome. Poult Sci 99(2):670–677

Rostami F, Taherpour K, Ghasemi HA, Gharaei MA, Shirzadi H (2022) Efficacy of *Scrophularia striata* hydroalcoholic extract and mannan-oligosaccharide on productive performance, intestinal bacterial community, and immunity in broiler chickens after infection with Campylobacter jejunum. Anim Feed Sci Technol 285:115217

Rouzmehr F, Mohit A, Koshsekeh M, Hassanzade M (2014) The effect of the additive containing artichoke extract (APC) on growth performance, blood cholesterol level, carcass characteristics and immune system of broiler chickens. Vet Clin Pathol 8(1):357–366

Ryu BK, Ahn BO, Oh TY, Kim SH, Kim WB, Lee EB (1998) Studies on protective effect of DA-9601, *Artemisia asiatica* extract, on acetaminophen-and CCI4-induced liver damage in rats. Arch Pharm Res 21(5):508–513

Saeed M, Naveed M, Arain MA, Arif M, Abd El-Hack ME, Alagawany M, Sun C (2017) Quercetin: nutritional and beneficial effects in poultry. World Poult Sci J 73(2):355–364

Saeed M, Babazadeh D, Arain MA, Naveed M, Shah QA, Kamboh AA, Chao S (2018) The use of chicoric acid from *Echinacea purpurea* as a feed additive in poultry nutrition. World Poult Sci J 74(1):69–78

Salah RB, Trabelsi I, Mansour RB, Lassoued S, Chouayekh H, Bejar S (2012) A new *Lactobacillus plantarum* strain, TN8, from the gastro intestinal tract of poultry induces high cytokine production. Anaerobe 18(4):436–444

Santos EAD, Alvarez-Leite JI (2019) Capsaicin: a potential therapy adjuvant for intestinal bowel disease. JDDD 2:8–16

Savage TF (1996) The effects of feeding a mannan oligosaccharide on immunoglobulins, plasma IgG and bile IgA, of Wrolstad MW male turkeys. Poult Sci 75:143–148

Scazzocchio B, Minghetti L, D'Archivio M (2020) Interaction between gut microbiota and curcumin: a new key of understanding for the health effects of curcumin. Nutrients 12(9):2499

Scherezenmeir J, De Verse M (2001) Probiotics, and synbiotics approaching a definition. Am J Clin Nutr 73(2 Suppl):361S–364S

Seidavi A, Asadpour L, Dadashbeiki M, Payan-Carreira R (2014) Effects of dietary fish oil and green tea powder supplementation on broiler chickens' immunity. Acta Sci Vet 42(1):1–13

Seidavi A, Dadashbeiki M, Asadpour L, Van den Hoven R, Alimohammadi-Saraei MH, Alise M, Santini A (2017) Dietary green tea powder affects the immunologic parameters of broiler chicks. Ital J Anim Sci 16(1):108–114

Seidavi A, Belali M, Elghandour MM, Adegbeye MJ, Salem AZM (2020) Potential impacts of dietary inclusion of green tea (*Camellia sinensis* L) in poultry feeding: a review. Agrofor Syst 94(4):1161–1170
Shan C, Sun B, Dalloul RA, Zhai Z, Sun P, Li M, Luan W (2019) Effect of the oral administration of *Astragalus* polysaccharides on jejunum mucosal immunity in chickens vaccinated against Newcastle disease. Microb Pathog 135:103621
Shehata AA, Yalçın S, Latorre JD, Basiouni S, Attia YA, Abd El-Wahab A, Tellez-Isaias G (2022) Probiotics, prebiotics, and phytogenic substances for optimizing gut health in poultry. Microorganisms 10(2):395
Shehzad A, Wahid F, Lee YS (2010) Curcumin in cancer chemoprevention: molecular targets, pharmacokinetics, bioavailability, and clinical trials. Arch Pharm 343(9):489–499
Shoeib HK (1997) Response of broiler chicks to probiotic (pronifer) supplementation. Assiut Vet Med J 36:103–116
Shohe A, Vidyarthi VK, Zuyie R (2019) Performance of broiler chicken on diet supplemented with turmeric powder (*Curcuma longa*). Livest Res Int 7(2):77–82
Siwek M, Slawinska A, Stadnicka K, Bogucka J, Dunislawska A, Bednarczyk M (2018) Prebiotics and synbiotics in ovo delivery for improved lifespan condition in chicken. BMC Vet Res 14(1): 1–17
Smirnov A, Perez R, Amit-Romach E, Sklan D, Uni Z (2005) Mucin dynamics and microbial populations in chicken small intestine are changed by dietary probiotic and antibiotic growth promoter supplementation. J Nutr 135(2):187–192
Soltan MA, Shewita RS, El-Katcha MI (2008) Effect of dietary anise seeds supplementation on growth performance, immune response, carcass traits and some blood parameters of broiler chickens. Int J Poult Sci 7(11):1078–1088
Song D, Wang YW, Hou YJ, Dong ZL, Wang WW, Li AK (2016) The effects of dietary supplementation of microencapsulated *Enterococcus faecalis* and the extract of *Camellia oleifera* seed on growth performance, immune functions, and serum biochemical parameters in broiler chickens. J Anim Sci 94(8):3271–3277
Song J, Li Q, Everaert N, Liu R, Zheng M, Zhao G, Wen J (2020) Effects of inulin supplementation on intestinal barrier function and immunity in specific pathogen-free chickens with *Salmonella* infection. J Anim Sci 98(1):skz396
Spring P, Wenk C, Connolly A, Kiers A (2015) A review of 733 published trials on Bio-Mos®, a mannan oligosaccharide, and Actigen®, a second-generation mannose rich fraction, on farm and companion animals. J Appl Anim Nutr 3
Stohs SJ, Chen O, Ray SD, Ji J, Bucci LR, Preuss HG (2020) Highly bioavailable forms of curcumin and promising avenues for curcumin-based research and application: a review. Molecules 25(6):1397
Suliman GM, Alowaimer AN, Al-Mufarrej SI, Hussein EO, Fazea EH, Naiel MA, Swelum AA (2021) The effects of clove seed (*Syzygiumaromaticum*) dietary administration on carcass characteristics, meat quality, and sensory attributes of broiler chickens. Poult Sci 100(3):100904
Surai PF (2015) Silymarin as a natural antioxidant: an overview of the current evidence and perspectives. Antioxidants 4(1):204–247
Surai PF, Kochish II, Fisinin VI (2017) Antioxidant systems in poultry biology: nutritional modulation of vitagenes. Eur J Poult Sci 81:1612–9199
Surai PF, Kochish II, Fisinin VI, Kidd MT (2019) Antioxidant defense systems and oxidative stress in poultry biology: an update. Antioxidants 8(7):235
Suzuki T, Hara H (2009) Quercetin enhances intestinal barrier function through the assembly of zonula occludens-2, occludin, and claudin-1 and the expression of claudin-4 in Caco-2 cells. J Nutr 139(5):965–974
Tajodini M, Samadi F, Hashemi SR, Hassani S, Ghasemnejad A (2015) Effects of different levels of artichoke (*Cynara scolymus* L) powder and vitamin E on performance and immune system response of broiler chickens. Iran J Med Aromat Plants Res 31(1):92–101

Tayeri V, Seidavi A, Asadpour L, Phillips CJ (2018) A comparison of the effects of antibiotics, probiotics, synbiotics and prebiotics on the performance and carcass characteristics of broilers. Vet Res Commun 42(3):195–207

Van den Biggelaar RH, Van Eden W, Rutten VP, Jansen CA (2020) Nitric oxide production and fc receptor-mediated phagocytosis as functional readouts of macrophage activity upon stimulation with inactivated poultry vaccines in vitro. Vaccine 8(2):332

Wan XL, Niu Y, Zheng XC, Huang Q, Su WP, Zhang JF, Wang T (2016) Antioxidant capacities of *Artemisia annua* L leaves and enzymatically treated *Artemisiaannua* L in vitro and in broilers. Anim Feed Sci Technol 221:27–34

Wang X, Farnell YZ, Peebles ED, Kiess AS, Wamsley KGS, Zhai W (2016) Effects of prebiotics, probiotics, and their combination on growth performance, small intestine morphology, and resident lactobacillus of male broilers. Poult Sci 95(6):1332–1340

Windisch W, Schedle K, Plitzner C, Kroismayr A (2008) Use of phytogenic products as feed additives for swine and poultry. J Anim Sci 86(suppl 14):E140–E148

Wu D, Kong Y, Han C, Chen J, Hu L, Jiang H, Shen X (2008) D-Alanine: D-alanine ligase as a new target for the flavonoid's quercetin and apigenin. Int J Antimicrob Agents 32(5):421–426

Xi N, Kang J, Hao L, Li R, Bao Y, Shi W (2014) Effects of ultrafine powder of the stem and leaf of Astragalus on immunity in chickens. Ital J Anim Sci 13(1):3022

Yalçın S, Yalçın S, Eser H, Şahin A, Yalçın SS (2014) Effects of dietary yeast cell wall supplementation on performance, carcass characteristics, antibody production and histopathological changes in broilers. Kafkas Univ Vet Fak Derg 20(5):757–764

Yang Y, Iji PA, Choct M (2009) Dietary modulation of gut microflora in broiler chickens: a review of the role of six kinds of alternatives to in-feed antibiotics. World Poult Sci J 65(1):97–114

Yaqoob MU, Abd El-Hack ME, Hassan F, El-Saadony MT, Khafaga AF, Batiha GE, Wang M (2021) The potential mechanistic insights and future implications for the effect of prebiotics on poultry performance, gut microbiome, and intestinal morphology. Poult Sci 100(7):101143

Yazdi FF, Ghalamkari G, Toghiani M, Modaresi M, Landy N (2014) Anise seed (*Pimpinella anisum* L) as an alternative to antibiotic growth promoters on performance, carcass traits and immune responses in broiler chicks. Asian Pac J Trop Dis 4(6):447–451

Yitbarek MB (2015) Phytogenics as feed additives in poultry production: a review. Int J Extensive Res 3:49–60

Zaker-Esteghamati H, Seidavi A, Bouyeh M (2020) A review on the effect of *Silybum marianum* and its derivatives on broilers under healthy and aflatoxicosis conditions: part 1: performance, carcass and meat characteristics, and intestinal microflora. World Poult Sci J 76(2):318–327

Zaker-Esteghamati H, Seidavi A, Bouyeh M (2021a) Effect of *Cynara scolymus* and its derivatives on broilers: an updated review. Anim Biotechnol 32(5):656–662

Zaker-Esteghamati H, Seidavi A, Bouyeh M (2021b) The effects of *Cynara scolymus* and *Silybum marianum* on growth, carcass and organ characteristics, immunity, blood constitutes, liver enzymes, jejunum morphology, and fatty acid profile of breast meat in broilers. Food Sci Nutr 9(12):6692–6706

Zarezadeh S, Rustaei AM, Alizadeh A, Mohammadi QM (2013) The effect of adding green tea and fish oil to the diet of broiler chickens on performance and humoral immune response against Newcastle. Anim Prod Res 1:1–13

Application of Aromatic Plants and Their Extracts in Livestock

Lizhi Jin, Ioannis Panitsidis, Stelios Vasilopoulos, Stella Dokou, Paraskevi Chantzi, Konstantina Vasilopoulou, Michalis K. Stefanakis, Praffula Kumar, and Ilias Giannenas

Abstract

Phytobiotic feed additives can be defined as plants/herbs derived products added to the feed in order to improve performance of animals. The most intensive use of such additives is being used in broiler chickens. The improvement of broiler performance and product quality may be due to phytobiotics' multi-directional effects, such as decreasing the gastrointestinal tract (GIT) microbial population through direct antimicrobial effects or indirectly through increasing mucous secretion in the gut and inhibiting bacterial adhesion to cell membranes, modifying the GIT bacterial population through the competitive exclusion of certain types of population, and making the environment favorable. Secondly, phytobiotic feed additives have great antioxidant, anti-inflammatory, and immunomodulatory effects. Those medicinal actions and their aromatic components justify their names as medicinal and aromatic plants. Different parts of aromatic and medicinal plants (flowers, fruits, roots, leaves, seeds, bark, and berries) or whole plants are utilized as sources of therapeutic compounds. For more than 30 centuries, Greeks, Egyptians, Indians, Sumerians, and Chinese were aware of the benefits of herbs, and today, a large portion of the world's population relies on indigenous traditional medicine. This chapter highlights the description of some

L. Jin
Meritech (Asia Pacific) Biotech Pte Ltd., Singapore, Singapore

I. Panitsidis · S. Vasilopoulos · S. Dokou · P. Chantzi · K. Vasilopoulou · M. K. Stefanakis · I. Giannenas (✉)
Laboratory of Nutrition, Faculty of Veterinary Medicine, Aristotle University of Thessaloniki, Thessaloniki, Greece
e-mail: igiannenas@vet.auth.gr

P. Kumar
Technical Sales and Research, Vets Plus, Inc., Menomonie, WI, USA

G. Arsenos, I. Giannenas (eds.), *Sustainable Use of Feed Additives in Livestock*,
https://doi.org/10.1007/978-3-031-42855-5_18

selected herbs and the potential application and health benefits of phytobiotics and their derived products in broiler chicken nutrition.

Keywords

Essential oils · Antioxidants · Antimicrobials · Livestock production

1 Introduction

The use of aromatic plants, their extracts, and essential oils as functional components in the pharmaceutical, food, and feed sectors is a result of rising consumer interest in products with natural origins as well as rising public concern over antibiotic growth promoters (Khodaei et al. 2021). In recent years, the use of essential oils in livestock and food industries has gained attention as a natural alternative to synthetic chemicals. Consumers are concerned about the use of synthetic additives in foods, and this has forced food processors to find ways to produce food products without the use of these additives. Therefore, there is a need in the food industry to find "clean label" products. In this context, the inclusion of medicinal plants into food products is an excellent strategy to produce functional foods, because plant-based extracts are rich in phytochemicals, with particular importance due to the health-beneficial effects. Therefore, it is reasonable to imagine that, over the next few decades, the uses of natural extracts from medicinal plants will rapidly increase (Soyadı and El-Ghany 2020).

Medicinal plants or medicinal herbs have been identified and used since ancient times to improve the sensory characteristics of food. The main compounds found in plants correspond to four major biochemical classes: Polyphenols, terpenes, glycosides and alkaloids. Plants synthesize these compounds for a variety of purposes, including protection of the plant against fungi and bacteria, defense against insects and attraction of pollinators and dispersal agents to favor the dispersion of seeds and pollens. Nowadays, there is also a growing interest in medicinal plants as natural alternatives to synthetic additives in foods because herbs and spices are generally recognized as safe (GRAS) and are excellent substitutes for chemical additives. The major activities of extracts and herbs from medicinal plants are antimicrobial, anti-inflammatory, bactericidal, antiviral, antifungal, and preservative for foods. The use of natural preservatives to increase the shelf life of food systems is a promising technology since many vegetal substances show antioxidant and antimicrobial properties. Taking into account all these considerations, recent changes in legislation controlling the use of animal feed additives and increased demand by consumers for healthier food products, if possible free of chemical additives, have stimulated interest in bioactive secondary metabolites from medicinal plants as alternative performance enhancers.

The plant extracts market consists of sales of plant extracts by entities (organizations, sole traders, and partnerships) that are used in natural pigments, traditional Chinese medicine, food additives, daily chemicals, medicines, and cosmetics. Plant extracts are screened through an extraction and separation process in which plants are used as raw materials. They can be solid, liquid, or soft extracts. The main types of plant extracts are spices, essential oils and natural extracts, flavors and fragrance, phytomedicines and herbal extracts, phytochemicals, and others. Phytochemicals refer to bioactive nutrient plant chemicals in fruits, vegetables, grains, and other plant foods that will provide desirable health benefits beyond basic nutrition to reduce the risk of major chronic diseases.

The global natural extracts market size to be valued at USD 18.95 billion by 2028 and is expected to grow at a compound annual growth rate (CAGR) of 8.9% during the forecast period. Factors such as consumer awareness, changing lifestyles, increasing adoption of natural ingredient-based products, and biotechnological advancements are driving the market. Moreover, macroeconomic factors such as growing consumer disposable incomes, R&D, imports and exports, growth of the agriculture industry, aging population, and population growth have positively influenced the demand for natural extracts over the past years. Natural extracts are gaining popularity in various industries, such as food and beverage, cosmetics, perfumery, pharmaceuticals, and nutraceuticals.

Increasing technological advancements have emerged as the key trend gaining popularity in the plant extract market. Companies operating in the plant extract sector are focused on developing new technological solutions to reinforce their position. Natural extracts have no detrimental side effects unlike most of the chemical-based ingredients used in personal care and cosmetics, pharmaceuticals, and food and beverage products. In addition, the matter of ecological and natural sourcing of ingredients has gained top priority among manufacturers, especially in North America and Europe. Numerous overall health benefits associated with natural extracts are anticipated to drive the product demand in pharmaceutical and nutraceutical applications. These factors are likely to significantly impact the demands for the product in the near future.

Essential oils have been shown to have a variety of effects on livestock, including improving their health, reducing stress, and increasing feed efficiency (Du et al. 2017). The food sectors attempt to avoid using potentially dangerous synthetic chemicals, while the livestock industry is urged to utilize natural growth promoters. These sectors are now searching for effective, safe, and affordable compounds with well-defined mechanisms of action and documented advantages.

The use of essential oils in livestock and food industries is a relatively new field of research, but it is rapidly growing due to increasing interest in natural alternatives to synthetic chemicals. Essential oils have been used in the food industry as a natural preservative to improve the shelf life of food products and as a natural flavor enhancer to replace synthetic flavorings. In livestock, essential oils have been used as a natural alternative to antibiotics and other synthetic chemicals to promote health and prevent disease (Florou-Paneri et al. 2019). Numerous active substances, including alkaloids, tannins, steroids, glycosides, resins, phenols, volatile oils, and flavonoids, are found in essential oils (EOs) (Amirifar et al. 2022). Therefore, components generated from plants have a great potential to meet such needs and contribute significantly to a wide range of industries by enhancing the productivity and health of livestock or by creating better products for human consumption.

In animal nutrition, the use of essential oils has a long history, dating back to ancient times. People have used various plants and herbs for their medicinal and aromatic properties to improve the health and well-being of livestock. For example, in ancient Greece and Rome, essential oils were used to treat various ailments in animals and to promote their health. Similarly, in traditional Chinese medicine, essential oils have been used for centuries to treat various diseases in animals and to enhance their performance. The exact essential oils used by ancient Greece and Rome in animal nutrition are not well documented. However, some of the most commonly used essential oils in ancient times included lavender, rosemary, thyme, fennel, and chamomile. These essential oils were used for their medicinal and aromatic properties to promote the health and well-being of livestock (Florou-Paneri et al. 2019).

Two of the most researched compounds of essential oils are carvacrol and thymol. Giannenas et al. (2018) have been found that these compounds have several functional properties. These properties include antibacterial, antioxidant, anti-inflammatory, and anticancer effects. The authors also mention that carvacrol and thymol have been shown to improve gut health and performance in livestock, making oregano oil which consists of these compounds, a potentially valuable feed additive for animal production.

2 Essential Oils

Essential oils are highly concentrated, volatile, odorous chemicals found in aromatic plants. These come from a variety of plant components, including the flowers, buds, seeds, leaves, twigs, bark, wood, fruits, and roots (Negi 2012). More than 17,000 plant species with EOs may be found in roughly 60 different plant families, with the Alliaceae, Apiaceae, Asteraceae, Lamiaceae, Myrtaceae, Poaceae, and Rutaceae being the most noteworthy (Hammer and Carson 2011). The secondary metabolites that make up the essential oils are intricate combinations of phenylpropenes and terpenes with low boiling points. Terpenes can be linear or cyclic, and various compounds, such as terpenes (with 10 carbon atoms), terpenes (with 15 carbon atoms), and diterpenes (with 20 carbon atoms), are produced based on the quantity and arrangement of the carbon atoms (Franz and Novak 2020; Asgari Lajayer et al. 2017). Eugenol, carvacrol, cinnamic acid, thymol, cinnamaldehyde, citral, and geraniol are some of the most vital EO components (Burt 2004) (Fig. 1).

All these compounds can be found in many common herbs, including those listed below.

Oregano (*Origanum vulgare* L.) is a rich source of various bioactive compounds including phenolic compounds, terpenes, and terpenoids. These compounds contribute to the antioxidant, antimicrobial, anti-inflammatory, and anticancer properties of oregano essential oil. Additionally, oregano essential oil has been used as a natural preservative in food industries due to its antimicrobial activity against a wide range

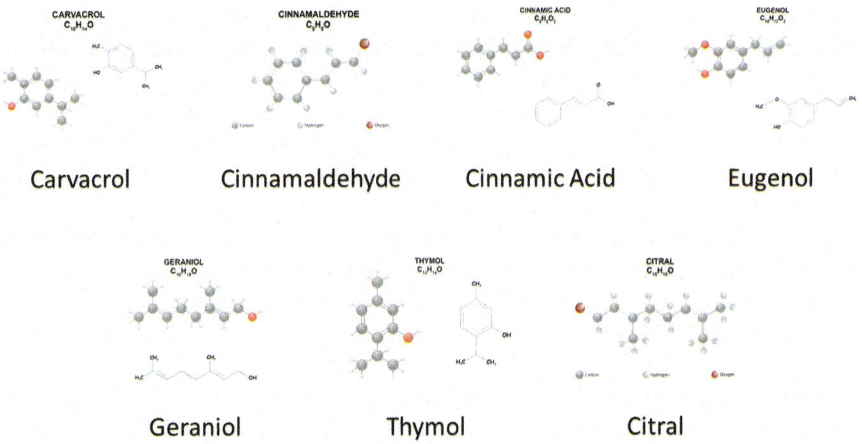

Fig. 1 Chemical composition of EOs compounds

of microorganisms. Overall, the oregano essential oil has promising potential for use in the food, pharmaceutical, and cosmetic industries as a natural source of bioactive compounds (Liu et al. 2017).

Turmeric oil is a well-known ingredient with antibacterial, anti-allergic, antimicrobial, antiparasitic, antifungal, anti-worm, and antiviral properties. It is gaining attention for its anticancer effects and thus is primarily used in food and beverage and healthcare applications (Goudarzi et al. 2020).

Rota et al. (2008) investigated the chemical composition and antimicrobial activity of essential oils derived from three Thymus species: *Thymus vulgaris*, *Thymus zygis*, and *Thymus hyemalis*. The results showed that all three essential oils had strong antimicrobial activity against several strains of bacteria and fungi, with Thymus vulgaris oil being the most effective. The authors attributed the antimicrobial activity to the presence of thymol and carvacrol, two major components found in all three oils. The study suggests that thyme essential oils could be potentially useful as natural antimicrobial agents in animal nutrition.

Ginger oil is an essential oil extracted from ginger rhizome at high concentrations after distillation. It has a distinct aroma described as warm and spicy and is often used in aromatherapy. Ginger oil finds wide applications in hair care and skin care products. For instance, in skincare, ginger oil inhibits cellular oxidation and helps soothe redness, reduce skin damage and aging, and restore skin color and radiance. Furthermore, it is widely used as a flavoring agent in beverages, confectionery, and bakery products, such as cookies, cakes, ginger snaps, and biscuits (Wang et al. 2019).

Nieto et al. (2018a, b) reviewed the antioxidant and antimicrobial properties of rosemary. This review gives an overview on the use of natural extract from rosemary as a preservative in foods. Their use is limited due to their negative organoleptic properties, such as odor and taste. However, different new extraction methods have been developed in order to get colorless and odorless rosemary extracts. Several studies have reported that bioactive compounds, present in rosemary extracts and essential oils, delay lipid oxidation and microbiological spoilage, extending the shelf life of food. Taking into account all these aspects, rosemary extracts could be used in functional foods, pharmaceutical products, plant products, and food preservation. The application of this natural extract can be complimented in different food systems such as meat, oils, and dressing.

As an alternative to synthetic antioxidants, natural extracts can be used, from plants such as grape, olive, sesame seed, tea, and soybean, with antioxidant properties. Antioxidant compounds are usually added at a moderate dosage level since high levels of inclusion may mechanistically cause adverse effects through pro-oxidative action.

EOs are produced by plants via secondary metabolic processes as signaling and defense chemicals. In general, EOs contribute significantly to a plant's defense against outside forces, such as direct and indirect plant responses to pathogens, herbivores, and pollinators, as well as responses to climatic effects and stress factors like drought, high temperatures, and ultraviolet (UV) light (Bakkali et al. 2008; Seow et al. 2014; El Asbahani et al. 2015). Terpenoids are generated by several biosynthetic routes, with two primary precursors, isopentenyl diphosphate (IPP) and its isomer, dimethylallyl diphosphate (DMAPP), synthesized by two distinct reaction chains in a plant cell. Sesquiterpenoids are produced by the cytoplasmic acetate-mevalonate route, which begins with the condensation of acetyl-CoA, whereas isoprene, monoterpenes, and diterpenes are produced by the plastidial methylerythritol phosphate (MEP) system, which requires pyruvate and glyceraldehydes 3-phosphate (Gershenzon and Dudareva 2007; Champagne and Boutry 2016). The monoterpenes and sesquiterpenes that are produced mainly have cyclic structures. Although a terpene synthase-independent mechanism also exists, it has been reported that a complicated pathway that arose for terpene biosynthesis in plants, where monoterpenes are generated in plastids and the cytosol by canonical monoterpene synthases, has emerged (Sun et al. 2016). The wide variety of terpene synthases, which may transform a phenyl diphosphate into various products through a variety of reaction cycles, is a result of the high heterogeneity in the chemical structure of terpenoid molecules. Essential oil aromatic molecules, which are less well-reported than terpenoids, are created via a different synthesis route (Degenhardt et al. 2009; Pavela and Benelli 2016).

3 Extraction and Determination of Essential Oils

The oils are usually extracted by two main processes: steam distillation and supercritical carbon dioxide extraction. In the first process, the plant material is combined with steam in a sizable container, which is often constructed of stainless steel. The plant material containing the required oils is infused with steam through an input, which causes the plant's aromatic molecules to release and transform into vapor. The condenser or condensation flask receives the vaporized plant components. Here, two

Fig. 2 Schematic representation of the extraction methods of essential oils

distinct pipelines allow hot water to leave and cold water to enter the condenser, respectively. This causes the vapor to return to liquid form and cool. The separator, a container beneath the condenser where the aromatic liquid by-product gathers, is where it falls from the condenser. The essential oil floats on top of the water since water and oil do not mix. From here, it is siphoned off. Similar to oils obtained by distillation, essential oils from the supercritical CO_2 extraction of plants are created. Thus, unlike the steam distillation method, the CO_2 extraction procedure may result in oils of greater quality since they were not damaged by the application of high heat. Heat doesn't affect any of the oil's components during CO_2 extraction. Pressurized carbon dioxide undergoes the following process, which results in it being liquid while still being in a gaseous form and becoming "supercritical." It is then pumped into a chamber that contains plant materials in this condition. The CO_2 acts as a solvent on the natural plant matter, drawing the oils and other compounds like pigment and resin from it because of the gas's liquid qualities. The liquid CO_2 then dissolves the essential oil component. The resultant oil is what remains when the CO_2 is returned to its normal pressure and returns to a gaseous condition. Following extraction, spectroscopic techniques such as UV and visible spectroscopy, infrared spectroscopy, mass spectroscopy, isotope-ratio mass spectrometry, nuclear magnetic resonance (NMR), Fourier transform infrared (FTIR), and Raman spectroscopy, as well as separation techniques such as gas chromatography (GC), chiral GC, liquid chromatography, thin layer chromatography, and thin layer chromatography, are used to determine the chemical composition of essential oils (Do et al. 2015) (Fig. 2).

4 Modes of Action

4.1 Immunomodulatory Action

Over the years, studies have shown that numerous compounds contained in essential oils, such as aldehydes, phenylpropanoid, and terpenes, have antimicrobial effects against a variety of pathogens. The major factor influencing different essential oils' potential usage in diverse applications is their antimicrobial effectiveness (Fokou et al. 2020). Because of this, the paragraphs below have been compiled with some of the main antibacterial, antifungal, and antiviral potentials of essential oils.

Numerous essential oils contain antibacterial activity among their diverse antimicrobial characteristics. The limitation of microbial growth is caused by a variety of modes of action. The capacity of EO to prevent or stop bacterial development is referred to as its antibacterial activity (Raut and Karuppayil 2014). Numerous studies have revealed that some herbs and essential oils, including cinnamon, clove, thyme, oregano, and rosemary, have potent antibacterial qualities (Dutra et al. 2019; Lorenzo-Leal et al. 2019). According to studies, the phenolic compounds in essential oils, such eugenol, carvacrol, and thymol, are primarily responsible for the antibacterial effects of EO on a variety of bacteria, including *Escherichia coli*, *Staphylococcus aureus*, *Bacillus cereus*, and *Staphylococcus pneumoniae* (Lopez-Romero et al. 2015; Marchese et al. 2016).

Essential oils have been shown to be beneficial in maintaining gut health in animal nutrition. They can stimulate the growth of beneficial gut microorganisms, inhibit the growth of pathogenic bacteria, and improve the absorption of nutrients in the gut. According to a study by Li et al. (2020), the use of essential oils as feed additives in piglets resulted in reduced inflammation and improved gut health. The study found that the addition of thymol, cinnamaldehyde, and their combination to the diet of weaned piglets improved intestinal morphology, increased the population of beneficial gut bacteria, and decreased the population of harmful bacteria. Furthermore, the essential oil treatments were found to enhance the immune response of the piglets. These results suggest that essential oils can be a useful tool in maintaining gut health and improving the overall health and performance of animals in animal production.

Furthermore, essential oils have been found to possess anti-inflammatory properties that can benefit animal nutrition. A study conducted by Hassan et al. (2020) found that the essential oil of oregano had anti-inflammatory effects on the small intestine of broiler chickens. The study showed that oregano essential oil reduced the levels of pro-inflammatory cytokines and increased the levels of anti-inflammatory cytokines in the intestine. This indicates that the essential oil of oregano can reduce inflammation in the small intestine of broiler chickens. The anti-inflammatory effect of essential oils can benefit animals by improving their gut health and reducing the risk of inflammation-related diseases. The use of essential oils as feed additives has become increasingly popular due to their many beneficial effects on animal health and performance.

Two of the most used techniques for determining the minimum inhibitory activity of essential oils are the Kirby-Bauer and broth microdilution procedures. By measuring the inhibitory zone created, the Kirby-Bauer test employs disks infused with a predetermined quantity of a possible antibacterial to identify bacterial resistance. In contrast, broth microdilution injects varied quantities of possible antimicrobials into the microculture while resazurin is used to assess bacterial viability, allowing a minimum inhibitory concentration (MIC) to be established (Yang et al. 2017). Oregano, basil, and rosemary from the Lamiaceae family as well as parsley, coriander, and anise from the Apiaceae family all showed notable antimicrobial efficacy against harmful microorganisms. According to the findings of one study, oregano essential oil is efficient against *Salmonella typhimurium*, *Yersinia enterocolitica*, and *Escherichia coli* (Liu et al. 2017). The same outcomes were observed with lavender essential oil, which has been extensively studied for its antimicrobial properties. According to the findings, lavender-based essential oil is effective against a variety of bacteria, including *Pseudomonas aeruginosa*, *Klebsiella pneumoniae*, *Aeromonas*, *Citrobacter*, and *Proteus* (Hossain et al. 2017; Yang et al. 2017). Additionally, research indicates that cinnamon bark essential oil has high antibacterial action against *Pseudomonas aeruginosa*, *Staphylococcus aureus*, *Escherichia coli*, *Klebsiella pneumoniae*, and *Staphylococcus aureus* (Firmino et al. 2018; Yang et al. 2019). Similar findings were made for peppermint essential oil, which was shown to be efficient and effective against a range of *Enterobacteriaceae*, *S. aureus*, and *Listeria monocytogenes* (Saeed et al. 2006; Liang et al. 2012).

Similar to antibacterial activity, determining the species of fungi is typically done using the Kirby-Bauer technique. Oral pills or topically applied lotions are the major treatments for fungal infections. Because animal and fungal cells are both eukaryotic, treating fungal infections is considerably more challenging than treating bacterial infections. The fungal cell must thus be the target of any antifungal medication, not the animal cell. As a result, several essential oils and each of their unique constituents have been thoroughly tested against a variety of fungi (Schroder et al. 2017; Nazzaro et al. 2017). It has been discovered that EO possesses antifungal action that plays a part in preventing cell-to-cell contact, reducing fungal growth, and obstructing the synthesis of mycotoxin (Prakash et al. 2012; Nazzaro et al. 2017). Numerous EOs have been investigated for their anti-*C. albicans* fungal action. Plants with powerful inhibitory effects included thyme, rosemary, and oregano (Liu et al. 2017; Ksouri et al. 2017). Moreover, various Ocimum EO species have been tested against various *Candida* species (Vieira et al. 2014). Another research tested 60 commercially available essential oils against *Aspergillus niger*, *Candida albicans*, and C*andida neoformans* with excellent results. These oils included cinnamon bark essential oil, lemon essential oil, peppermint essential oil, and others (Powers et al. 2018). By rupturing the fungal cell membrane, the essential oils of oregano and thyme, which were the subject of a different investigation, significantly inhibited the growth of fungal infections (Swamy et al. 2016). In general, it has been discovered that oregano affects spore development and breaks down the membranes of fungal cells. Additionally, it has been found that the action of cinnamon and clove essential oils dramatically reduced the mycelial development of three different fungal *Aspergillus* species (Hu et al. 2019) (Table 1).

Antivirals are drugs that treat viral infections by focusing on viral replication events, which inhibits viral replication, however they frequently aren't very successful (Gelzleichter 2013). Alternative therapeutic substances, such as essential oils, have been researched in addition to the small number of effective antiviral medications because several studies have demonstrated that they can have substantial antiviral activity against several RNA and DNA viruses by preventing viral replication (Tariq et al. 2019). Some chemical compounds detected in essential oils have demonstrated potent antiviral action, inhibiting viral reproduction and virus adsorption to host cells. It has been discovered that the essential oils from ginger, thyme, and hyssop are bioactive against the HSV (Schnitzler 2019). Additionally, a distinct investigation revealed that the H1N1 influenza A virus viral particles are highly resistant to essential oils extracted from the cinnamon bark, eucalyptus, carrot, and rosemary (Brochot et al. 2017). Nonetheless, various EOs will operate against viruses in different ways; typically, they target the nucleic acid polymerases. Thymol and phenylpropanoids have generally been found to be the chemicals primarily responsible for the antiviral action in HSV. As further studies must be carried out to give more confirming data in terms of efficacy and safety, EOs' antiviral characteristics may be a potential option in the future (Table 2).

4.2 Antioxidant Activity

Aromatic herbs and their essential oils are rich sources of phenolic compounds, such as eugenol, thymol, and carvacrol, which are natural antioxidants (Franz et al. 2010). Although the bioactivity is linked to aglycon structures, particularly catechol in aglycons, polyphenols often exist as glycosides (Sakakibara et al. 2003). It is believed that the high redox properties and chemical structure of these compounds, which can be accountable for the neutralization of free radicals, chelating transitional metals, and quenching singlet and triplet oxygen by delocalization or decomposing peroxides, are the reasons for their antioxidant activity (Chun et al. 2005). The phenolic antioxidants' favorable health functioning is linked to the aforementioned characteristics since they help to postpone the onset of a number of oxidative stress-related disorders, including cancer, diabetes, Alzheimer's, and cardiovascular conditions (Miron et al. 2010). In addition, plant phenols can operate as chain-breaking peroxyl radical scavengers, shielding highly unsaturated lipids in feed from oxidation process. One of the primary issues limiting the quality and acceptance of food products is lipid oxidation (Miguel 2010).

Recent studies suggest that EOs can also modulate the immune system through their antioxidant effects, which reduce oxidative stress and inflammation, and enhance the animal's ability to fight infections. This effect is attributed to the EOs' ability to scavenge free radicals and activate antioxidant enzymes, such as superoxide dismutase and catalase. Moreover, EOs can stimulate the production of cytokines, such as interleukin-1β and tumor necrosis factor-α, which play an

Table 1 Chemical composition of various essential oils and their antibacterial-antifungal activity against pathogens

Plant	Chemical compound	Inhibited microorganisms	References
Origanum vulgare	Carvacrol, thymol, γ-terpinene	*L. monocytogens, E. coli*, adeno virus, polio virus	Ultee and Smid (2001)
Origanum vulgare	Carvacrol	*Candida* spp.	Tampieri et al. (2005)
Origanum vulgare	Carvacrol, *p*-cymene, γ-terpinene	*Clostridium botulinum, C. perfringens, Listeria monocytogenes, E. coli, S. choleraesuis, S. typhimurium, S. aureus, B. subtilis, Pseudomonas aeruginosa, Shigella sonnei, Sarcina lutea, M. flavus, K. pneumoniae*	Amatiste et al. (2014) Béjaoui et al. (2013)
Cinnamomum zelancium	Cinnamaldehyde	*S. aureus, Streptococcus pyogenes, S. pneumoniae, B. cereus, E. coli, Klebsiella pneumoniae, Proteus mirabilis, P. aeruginosa, S. typhimurium, C. perfringens, Mycobacterium smegmatis*	Unlu et al. (2010)
Cinnamomum zelancium	Cinnamaldehyde	*Enterobacteriaceae*	Hood et al. (2003)
Coriandrum sativum	2E-Decenal, decanal, 2E-decen-1-ol, n-decanol	*S. aureus, Bacillus* spp., *E. coli, Salmonella typhi, K. pneumonia, P. mirabilis, P. aeruginosa*	Matasyoh et al. (2009) Begnami et al. (2010)
Thymus vulgaris	Thymol, linalool, carvacrol, 1,8-cineole, eugenol, camphor, camphene, α-pinene, borneol, β-pinene	*L. monocytogenes, E. coli, S. typhimurium, S. aureus, C. botulinum, C. perfringens, S. sonnei, S. lutea, M. flavus, C. albicans, L. monocytogenes, P. putida, S. putrefaciens*	Giordani et al. (2004) Amatiste et al. (2014) Béjaoui et al. (2013) Santurio et al. (2014)
Thymus mastichina	m-thymol, carvacrol, trans-caryophyllene	*B. thermosphacta, E. coli, L. innocua, L. monocytogenes, P. putida, S. typhimurium, S. putrefaciens*	Teixeira et al. (2013)
Allium sativum	Isothiocynate, Ethanolic compounds	*S. aureus, Enteriobacteriaceae, Candida albicans*	Grohs and Kunz (2000)
Ocimum basilicum	Estragole	*Aspergillus* spp., *F. moniliforme, Candida* spp.	Sakkas et al. (2016) Soliman and Badeaa (2002)

(continued)

Table 1 (continued)

Plant	Chemical compound	Inhibited microorganisms	References
Salvia officinalis	α-Thujone, camphor, 1,8-cineole, α-pinene	*S. aureus, P. stuartii, P. stuartii, E. coli, Shigella sonnei, Sarcina lutea, M. flavus, B. thermosphacta, E. coli, L. innocua, L. monocytogenes, Candida* spp.	Mulyaningsih et al. (2010) Jirovetz et al. (2007)
Salvia sclarea	Linalool, linalyl acetate, geranyl acetate, – ocimene acetate, caryophyllene oxide	*S. aureus, S. agalactiae, S. epidermis, E. coli, Proteus vulgaris, P. aeruginosa, K. pneumonia, E. faecalis, B. pumilus, B. subtilis, S. typhimurium, Candida* spp.	Jirovetz et al. (2007) Fraternale et al. (2005) Cui et al. (2015)
Rosmarinus officinalis	Camphor, camphene, limonene, geraniol, myrcene, linalool benzoylacetate, linalool, α-pinene, α-terpinolene, bornyl acetate, borneol	*E. coli, S. typhimurium, B. cereus, Bacillus subtilis, S. aureus, S. agalactiae, S. epidermidis, S. aureus, P. vulgaris, P. aeruginosa, K. pneumonia, E. faecalis, B. thermosphacta, L. innocua, L. monocytogenes, P. putida, S. typhimurium, S. putrefaciens, M. smegmatis*	Fu et al. (2007) Teixeira et al. (2013) Rodriguez Flores et al. (2014)

important role in immune responses. A study conducted by Cheng et al. (2020) showed that feeding broilers with a mixture of EOs improved their antioxidant capacity and immunity, as evidenced by increased levels of cytokines and antibody titers. Thus, EOs can be a valuable tool in animal nutrition to support immune function and overall health.

5 Use of Essential Oils in Food Preservation

By generating natural preservatives, essential oils can prolong the shelf life of food (Holley and Patel 2005). Microbes often impact all food items. Only a small percentage of these bacteria, nevertheless, have a positive impact on enhancing food flavor. Most of them produce toxins that cause food to deteriorate and lower the product's organoleptic quality. Therefore, a variety of food preservation techniques are utilized to prevent or manage food degradation. These techniques include pasteurization, cooling, nutritional limitation, acidification, fermentation, and chemical preservatives. These techniques are typically linked to lowering the organoleptic characteristics of meals and lowering consumer acceptance (Negi

Table 2 Chemical composition of various essential oils and their activity against viruses

Virus	Plant	Chemical compound	References
H1N1	*Cinnamomum zeylanicum, Citrus bergamia, Cymbopogon flexuosus, Thymus vulgaris, Lavandula ocinalis, Eucalyptus globulus, Pelargonium graveolens*	Carvacrol, eugenol, Santalol	Paulpandi et al. (2012) Vimalanathan and Hudson (2014) Brochot et al. (2017)
HSV	Star Anise, *Mentha suaveolens, Sinapis arvensis, Pulicaria vulgaris, Zataria multiflora, Eucalyptus caesia, Artemisia kermanensis, Rosmarinus ocinalis, Thymus capitatus*	β-Caryophyllene, Farnesol, trans-anethole, thymol, p-cymene, β-Pinene, limonene	Astani et al. (2011) Civitelli et al. (2014) Gavanji et al. (2015) Sharifi-Rad et al. (2017)
BVDV	Mexican oregano (*Lippia graveolens*), *Ocimum basilicum*	1, 8-cineole, camphor, carvacrol, thymol	Pilau et al. (2011) Kubiça et al. (2014)
Coxsackie virus	*Dysphania ambrosioides, Eucalyptus globulus bicostata*	–	Elaissi et al. (2012) El Mokni et al. (2019)
Respiratory syncytial virus	Patchouli, Mexican oregano (*Lippia graveolens*)	Carvacrol	Wei et al. (2012) Pilau et al. (2011)
Zika	*Ayapana triplinervis*	Thymohydroquinone dimethyl ether	Haddad et al. (2019)

2012). Numerous bioactive chemicals found in EOs, including terpinol, eugenol, myrtenol, thymol, camphor, and carvacrol, suggest substantial activity. Gram-positive bacteria seem to be more susceptible to essential oils than Gram-negative bacteria. In some circumstances, the combination of two or more chemicals causes higher activity than when they are used separately to have synergistic effects, resulting in a greater effect against pathogens (Burt 2004). Lipid oxidation, a complicated process that can impact the composition and behavior of meat during food processing, is one of the key elements in meat products that lowers meat quality (Domínguez et al. 2019). Several metabolites that are produced as a result of lipid oxidation have a significant impact on the quality of meat. Essential oils offer superior antioxidant properties to synthetic antioxidants, which can also benefit customers' overall health (Jayawardana et al. 2019). The creation of healthier meat products has been the industry's major emphasis in recent years. Two methods were used to achieve this goal: lowering the levels of unwanted compounds and raising the concentrations of the desired bioactive molecules (Domínguez et al. 2020) (Table 3).

Table 3 List of uses of essential oils as natural antioxidants in meat

Essential oil	Food	Result	References
Green tea	Chopped cooked ham	Reduction of color oxidation	Šojić et al. (2018)
Black pepper	Meat	Reduction of color oxidation Lipoprotein oxidation reduction	Heck et al. (2019)
Oregano vulgare	Lamb meat Chopped cooked ham	Reduction of color oxidation Lipid oxidation reduction Decrease Sensory degradation	Šojić et al. (2018) Fernandes et al. (2017)
Oregano vulgare	Egg	Improve egg shell and freshness Reduction of yolk lipid oxidation	Feng et al. (2021, 12:72) Radwan et al. (2008)
Oregano vulgare	Pork	Reduction of drip loss Lipid oxidation reduction	Cheng et al. (2017, 103–109)

6 Essential Oil Properties in Livestock Nutrition

6.1 Poultry

As a consequence of years of study, today's poultry performance factors are adjusted, resulting in quickly developing chickens with reduced feed intake and improved health. The livestock sector therefore works to improve the quality of the finished product after obtaining quantity since there is a significant demand for the final product to be healthier for the consumers' overall health, have better bioactive compounds, and be produced in a more sustainable manner. For this reason, nowadays, adding essential oils to feed or drinking water is a common practice in the production of chickens. Most EOs used in poultry are extracted from onion (*Allium cepa*), cinnamon (*Cinnamomum zeylanicum*), oregano (*Origanum vulgare*), garlic (*Allium sativum*), sage (*Salvia officinalis*), rosemary (*Rosmarinus officinalis*), turmeric (*Curcuma longa*), thyme (*Thymus vulgaris*), and clove (*Syzygium aromaticum*), among others (Gopi et al. 2014; Feizi et al. 2014).

The biggest concern to the chicken business is infection, and a number of potential risks might make it much more dangerous. One of the risks is immunosuppression, which can be brought on by a variety of factors, including mycotoxin, viral infections with immunosuppressive effects, failed vaccinations, and the overuse of antibiotics (Galal et al. 2016). Studies have demonstrated that herbal essential oils can boost IFN-γ, promote the generation of immunoglobulin, and enhance lymphocyte activity (Faramarzi et al. 2013; Gopi et al. 2014; Adaszyńska-Skwirzyńska and Szczerbińska 2017). With the addition of thyme oil to diet, significant levels of IgA were seen in another trial (Placha et al. 2014). Higher globulin levels also suggest

Table 4 Summary of studies with *Origanum vulgare* essential oil on FCR in broilers

Animal breed	Experimental period (day)	FCR improvement comparing with control (%)	References
Ross 307	42	4.6%	Calislar et al. (2009)
AA	44	4.3%	Hu et al. (2010)
AA	21	1.5%	SFR (2009)
Three-yellow	49	14.8%	Liu et al. (2008)
Cobb-500	35	3.5%	Giannenas (2004)
Cobb-500	42	2.6%	Giannenas et al. (2003)
Ross	41	4.5%	Waldenstedt et al. (2003)
Cobb	35	2%	Saini et al. (2003a)
Cobb	45	4.9%	Saini et al. (2003b)

superior resistance to the disease and the immunological response of birds, and significantly higher antibody titers against infectious bursa disease are further evidence of improved health brought on by thyme oil (Bovera et al. 2015). According to the results of another study, a combination of essential oils from *Oreganum aetheroleum* is superior to ciprofloxacin for the treatment of *Escherichia coli* in broiler chickens. Signs, fatalities, postmortem lesions, humoral and cell-mediated immune responses, as well as bacterial re-isolation, all decreased (Abd El-Ghany and Ismail 2014).

Due to their ability to stimulate growth, essential oils are regarded as growth enhancers in chicken feed. *Origanum* spp. contains a variety of potent chemical substances, including phenols. According to research that looked at the effect of oregano essential oil on chickens, groups fed the oregano oil grew more quickly and better FCR (Table 4) than birds fed a standard diet (Nieto et al. 2018a, b). Oregano essential oil effect on broilers' FCR, evaluated in various experimental studies, is presented in Table 4. Another potent essential oil is obtained from *Cinnamomum zeylanicum*. In order to determine the effects of clove bud oil, cinnamon bark oil, and ajwain seed oil in broilers, a study was carried out in India. The findings of this study demonstrated that adding cinnamon bark oil to the feed had more advantages for broiler chicks' intestinal health, antioxidant status, blood cholesterol, and immunology than utilizing antibiotic growth boosters (Chowdhury et al. 2018).

The gastrointestinal system is another area that essential oils have a favorable impact because they help the system restoration of the balance of bacteria and boost nutrient absorption, which is mostly attributed to terpenoids (Barbour et al. 2010). The effects of thyme oil, which has thymol and carvacrol among its principal ingredients, were investigated in an experiment on broiler chicks. In the broiler production system, the results showed that feeding thyme oil caused considerable weight growth, an improved feed conversion ratio, enhanced livability, and higher turnover (Wade et al. 2018). Additionally, it has also been speculated that thymol oil can safeguard the intestinal microvilli responsible for nutritional absorption while simultaneously encouraging the release of consumable endogenous enzymes and blood components (Hashemipour et al. 2013a).

Essential oils can benefit poultry nutrition in numerous areas besides their primary uses. They could be successful in combating parasites. In instance, lice, mites, and ticks are among the many ectoparasites that different essential oils are highly effective against. Different exposure methods, including immersion and direct physical contact with surfaces treated with essential oils, have demonstrated their effectiveness (Bakkali et al. 2008). The anti-inflammatory qualities of terpenoids and flavonoids, which are part of essential oils, decrease the absorption of prostaglandins (Krishan and Narang 2014).

6.2 Swine

The use of antibiotics has been banned or restricted in many countries of the western world due to spread of antimicrobial-resistant bacteria in both pigs and humans presenting a serious public health risk (Yang et al. 2015). Essential oils are a promising option as an alternative solution. When compared to antibiotics, essential oils are said to be more natural, less harmful, and devoid of residues (Gong et al. 2014).

Most of the studies focused on the growth-promoting features, such as feed intake, weight gain, and feed conversion ratio. The effects of essential oils on feed intake of pigs showed promising results. In the review by Franz et al. (2010), the change in feed intake compared to control ranged from 9% to 12%, while a recent analysis by Zeng et al. (2015) revealed a range of 3–19%. The growth-promoting property of Essential oils is primarily connected with actions on the gastrointestinal tract to raise feed palatability, promote digestive fluid production, improve intestinal shape, stabilize intestinal flora, and reduce inflammation (Steiner and Syed 2015). Improved palatability of animal feed stimulates appetite and enhances feed intake in many different ways. Cinnamaldehyde and thymol increased piglet apparent total tract digestibility of dry matter, crude protein, and energy (Li et al. 2012). Platel and Srinivasan (2004) proposed that the improved apparent digestibility of fat and nutrients could be attributed to increased bile and enzyme secretion, which was proposed as the primary mode for the digestive stimulant action of spices. Carvacrol, thymol, eugenol, and cinnamaldehyde were shown to be virtually entirely absorbed in the gut and small intestine of pigs within 2 hours (Michiels et al. 2008). Some essential oils may have a selective effect on the gut microbiota. A combination of carvacrol, cinnamaldehyde, and capsicum oleoresin enhanced the population of *Lactobacilli* and the ratio of *Lactobacilli* to *Enterobacteria* of early-weaned piglets (Manzanilla et al. 2006; Castillo et al. 2006). Lastly, essential oils may enhance immunological responses. Piglets' blood lymphocyte proliferation rate, phagocytosis rate, immunoglobulin IgG, IgA, IgM, C3, and C4 levels were enhanced by supplementing essential oils (Li et al. 2012). The addition of essential oils to weaned pigs improved both non-specific cellular and humoral immune responses, according to Halas et al. (2011).

6.3 Aquaculture

Aquaculture is one of the most promising industries because it offers nutritious and healthy protein sources to human consumers. It is also a profitable business and has become a popular activity for many farmers and aquaculturists around the world. However, in recent years aquaculture is facing a lot of challenges, including water elimination and pollution, high feed costs, deterioration of animal health and others that create a negative impact on production and economy (Abdel-Latif et al. 2022; Dawood et al. 2021). As a result of these stressful situations that aquaculture farming is facing, there has been a disease outbreak on fish population. Since recently, antibiotics and other chemical drugs were used widely to prevent these fish diseases resulting to pathogen resistance, microbial population imbalance, and environmental pollution (Zhang et al. 2022). The urge for safer and environmentally friendly alternatives in combination with the growing consumer demand for high-quality aquatic products has led to the investigation of herbal medicines and their derivatives, such as essential oils.

Essential oils (EOs) have become very popular and are known as natural antioxidants and immunostimulants. Their chemical structure is resistant to gastric acid and are also known to improve palatability by controlling appetite genes in fish. EOs have a powerful antibacterial effect against pathogenic bacteria which permit the beneficial bacteria to act and improve food digestion, gut health, and resistance to pathogens (Dawood et al. 2021). They are also known for their antiparasitic and antifungus activities. As a result, due to their therapeutic properties, low toxicity, minor side effects, and low risk of developing drug resistance, EOs have become a great alternative to antibiotics and chemical drugs in fish aquaculture (Zhang et al. 2022).

Because of all these beneficial properties of EOs in aquaculture, more and more researchers have focused on the application of EOs and their impact on fish performance and health. For example, Abdel-Latif et al. (2020) found that dietary supplementation with oregano essential oil common carp (*Cyprinus carpio L.*), improved growth performance indices and feed utilization parameters. In another study, Aanyu et al. (2018) showed that dietary limonene, which is the major component in the oil of citrus fruit peels, enhanced the growth of Nile tilapia. As mentioned before EOs oils are also known for their antibacterial, antiparasitic and antifungal effects. Giannenas et al. (2012) studied the effects of dietary supplementation with carvacrol or thymol derived from *T. vulgaris* EO on the intestinal microbiota of rainbow trout and discovered a significant decrease in gut's total anaerobic bacteria. Moreover, Ferreira et al. (2019) showed that the EO of *Mentha piperita* (peppermint) was effective enough to reduce the number of dinoflagellate *Piscinoodinium pillulare* in tambaqui juveniles. Finally, Madrid et al. (2015) discovered that the anti-oomycete activity of *Laureliopsis philippianna* EO against *Saprolegnia* spp. was satisfying, proving that it can be used as a remedy for this aquatic fungus.

All in all, EOs are a good alternative remedy against fish diseases. Its action mechanisms are primarily linked to their immunoenhancement and antioxidation effects on fish, as well as direct pathogen killing efficacies. However, additional research is required to fill the gaps and to enhance the knowledge that we have so far in practical applications (Zhang et al. 2022).

6.4 Ruminants

In accordance with all major livestock sectors, in ruminant farming the use of antibiotics for purposes of enhancing animal growth performance was banned, yet significant amounts are used therapeutically and for prophylaxis and metaphylaxis as well (Rochfort et al. 2008). Concurrently, consumers are correlating product higher quality with milk and meat originating from animals deprived of antibiotics, thus furtherly decreasing the use of antibiotics is under public demand (Centner 2016). Essential oils as plant-derive products can offer support to this collective action with possible economic and environmental benefits. Various essential oils were investigated with promising results in animal performance and health, rumen fermentation as well as product quality characteristics (Giannenas et al. 2011; Elcoso et al. 2019; Westphalen et al. 2021).

Essential oils have been reported to support milk and meat production resulting in higher production yields both in small and large ruminants (Dorantes-Iturbide et al. 2022; Orzuna-Orzuna et al. 2022). These effects could be attributed to several mechanisms. Various essential oils have been recorded to increase dry matter intake (DMI), although they transfuse different aroma and can affect feed palatability (Mucha and Witkowska, 2021). This increase could be attributed to the higher rate of fiber degradation resulting from the increased population of fiber-degrading microorganisms, such as *Fibrobacter succinogenes* and *Ruminococcus albus* (Zhou et al. 2019). Rumen pH was found unaltered by essential oils' bioactive compounds while ammonia-N and methane concentrations were found decreased (Khiaosa-Ard and Zebeli 2013). Bioactive compounds have also the ability to alter rumen epithelium by increasing villus length in the small intestine that can lead to nutrient absorption enhancement (Patra et al. 2019). At the same time studies have linked essential oils with increased serum immunoglobulins, suggesting immune system invigoration (Wu et al. 2021). All these chemical and structural alterations attributed to EOs result in increased animal productivity. Essential oils can also act as natural antioxidants supporting milk and meat oxidative stability (Kalaitsidis et al. 2021; Rivaroli et al. 2016).

Although the potential beneficial effects of ruminant supplementation with EOs are plenty, the available studies investigating their incorporation are significantly lower compared to monogastric animal sector and high variability of the results is present in the existing literature (Dorantes-Iturbide et al. 2022). A possible explanation is due to the difference in animals' production phase, the duration of supplementation and the application of in vitro experiments that can display different results in vivo (Dorantes-Iturbide et al. 2022) (Table 5).

Table 5 Herbal essential oils and their impacts on the physical performances and physiological responses of livestock animals

Animal	Plant	Compound	Results	References
Broiler	*Origanum vulgare*	–	Enhanced body weight gain, feed conversion ratio and lower lipid oxidation	Florou-Paneri et al. (2006)
Broiler	*Origanum vulgare*	–	Lower lipid oxidation	Giannenas et al. (2005)
Broiler	*Thymus vulgaris*, Cinnamon	–	Improved total proteins, feed intake, body weight gain and feed conversion ratio, reduced serum cholesterol	Al-Kassie (2009)
Broiler	*Origanum vulgare*, *Pimpinella anisum* and citrus	Carvacrol, anethol, and limonen	Enhanced body weight gain with decreased feed intake, significant effect on the total tract apparent digestibility of organic matter and the nitrogen corrected apparent metabolisable energy	Mountzouris et al. (2011)
Broiler	*Artemisia annua*	40 different flavonoids, phenolics, purines, and other compounds	Lower lipid oxidation in breast and thigh	Cherian et al. (2013)
Broiler	–	Carvacrol Thymol	Improved immunological response and performance, higher antioxidant and digestive enzyme activities, delayed lipid oxidation	Hashemipour et al. (2013b)
Broiler	*Thymus vulgaris*, *Glycyrrihiza galabra*	Thymol, glycyrritinic acid, flavonoids	Improved weight gain, heavier immune organs. Lower abdominal fat	Kalantar et al. (2017)
Broiler	*Lavandula angustifolia*	α-Pinene, caryophyllene, α-terpineol, borneol, eucarvone, linalool and other compounds	Positive effect on body weight, feed conversion ratio, positive impact on the gut microflora of the ileum	Kirkpinar et al. (2014)
Broiler	*Allium sativum*, *Zingiber officinale*	–	Improved oocysts shedding, lesion score and histopathology of the small intestines after induced infection with *Eimeria* spp.	Ali et al. (2019)
Broiler	*Lavandula angustifolia*	Linalool, α-pinene, caryophyllene, α-terpineol, terpinen-4-ol	Improved antioxidant status in serum and liver, improved growth performance, intestinal morphology and microflora	Barbarestani et al. (2020)

(continued)

Table 5 (continued)

Animal	Plant	Compound	Results	References
Nursery pigs	*Origanum vulgare*	Carvacrol Thymol	No results have demonstrated	Neill et al. (2006)
Pigs	*Allium sativum, Glycyrrhiza glabra, Thymus vulgaris, Carum cavi*	–	Improved average daily gain, feed conversion ratio and blood parameters	Czech et al. (2009)
Grower–finisher pigs	*Thymus vulgaris, Salvia rosmarinus, Origanum vulgare*	–	Increased average daily gain, digestibility of nitrogen and energy, enhanced the longissimus muscle area, reduced the *b** value, and the NH3 excretion	Yan et al. (2010)
Finisher pigs	*Origanum vulgare*	Carvacrol Thymol	Increased ADG, reduced body weight loss due to transportation	Zhang et al. (2015)
Weaned pigs	–	Thymol, cinnamaldehyde	Increased antioxidant capacity, greater villus height to crypt depth ratio, reduced *E. coli* in colon	Li et al. (2012)
Growing-finishing pigs	*Origanum vulgare*	–	Decreased lipid oxidation, enhanced oxidative stability, and improved sensory properties	Cheng et al. (2017)
Growing-finishing pigs	*Origanum vulgare*	–	Significantly less oxidation of lipids and greater concentrations of monounsaturated fatty acids	Chuan-Shang et al. (2017)
Weaning to finishing pigs	Microencapsulated complex essential oils	Thymol	Improved feed efficiency, feed intake, red blood cells and immunoglobulin G, fecal microflora, and meat quality	Oh et al. (2018)
Finisher pigs	Mexican oregano (*Lippia graveolens*)	Carvacrol, p-cymene, thymol, trans-caryophyllene, γ-terpinene, α-caryophyllene	Significant impact on carcass length and width, improved carcase performance, greatest levels of customer acceptance for color, flavor, texture, and overall, like following taste evaluation.	Janacua-Vidales et al. (2019)

Pigs	*Origanum vulgare*	–	Greater oxidative stability and after consumer test oregano groups achieved higher consistency scores and overall liking	Forte et al. (2017)
Holstein cows	*Medicago sativa*, capsicum, *Pimpinella anisum*	Cinnamaldehyde, eugenol	Decreased ammonia N concentrations and protozoal counts	Cardozo et al. (2006)
Ruminants	*Rheum officinale*, *Allium sativum*, *Frangula alnus*	–	Improved digestibility, decreased methane	García-González et al. (2008)
Crossbred bulls (Angus × Nellore)	*Ricinus communis*, *Origanum vulgare*, *Anacardium occidentale*	–	Increased bull performance, however, meat qualities did not alter much	Fugita et al. (2018)
Beef cattle	*Origanum vulgare*	–	Increased digestive ability by modulating epithelial development and microbiota	Zhang et al. (2021)
Sheep	*Origanum vulgare*	Carvacrol and thymol	Improved growth performance, modification of intestinal morphology and microflora	Jiale Sun et al. (2022)
Rainbow trout (*Oncorhynchus mykiss*)	*Mentha spicata*, *Thymus vulgaris*, *Salvia* sp.	1,8-cineole, carvacrol or Pulegone	Improved growth rates, feed utilization	Sönmez et al. (2015)
Nile tilapia (*Oreochromis niloticus*)	–	Cinnamaldehyde, thymol	Improved growth parameters and feed utilization, lysozyme activity, IgM, IgG levels, reduced oxidation	Amer et al. (2018)
Nile tilapia (*Oreochromis niloticus*)	–	Limonene and thymol	Improved growth performance, IGF-I, MUC, PEPT1, LPL, ALP, and CAT	Aanyu et al. (2018)
Common carp (*Cyprinus carpio* L.)	*Origanum vulgare*	Carvacrol, thymol	Improved growth rates, feed utilization, intestinal histomorphology, potential hepato-protective effects	Abdel-Latif et al. (2020)

(continued)

Table 5 (continued)

Animal	Plant	Compound	Results	References
Nile tilapia (*Oreochromis niloticus*)	–	Menthol	Improved growth rates, feed utilization, immune response, anti-inflammatory, and proinflammatory, enhanced resistance against chlorpyrifos toxicity	Dawood et al. (2020)

7 Conclusion

The essential oils have been used for their medicinal and aromatic properties in the food industry and animal nutrition for centuries. These natural compounds are derived from various plants and herbs and have been shown to have a number of functional properties, including antimicrobial, antioxidant, and anti-inflammatory effects. The use of essential oils as feed additives in animal nutrition has the potential to improve growth performance, digestive health, and disease resistance in livestock. However, it is important to note that the efficacy of essential oils as feed additives can vary depending on the type of oil, the dose, and the animal species. The use of essential oils in the food industry is also gaining popularity due to their potential to improve the sensory quality, safety, and shelf life of food products. Essential oils can be used as natural preservatives, flavorings, and fragrances in food products, and have been shown to have antimicrobial properties that can help prevent spoilage. Additionally, essential oils can also provide a source of antioxidants and other health-promoting compounds in food products (Florou-Paneri et al. 2019). While the use of essential oils in the food industry and animal nutrition is promising, it is important to conduct further research to fully understand their effects and to determine the optimal doses and combinations for various applications. This will help ensure the safe and effective use of essential oils in food and animal nutrition and help maximize their potential benefits.

To sum up, the use of essential oils as feed additives in animal production has been studied for several decades and has shown promising results. In poultry production, the use of essential oils has been shown to improve growth performance, feed efficiency, and gut health. For example, a study found that the addition of oregano essential oil to broiler chicken diets increased weight gain, improved feed efficiency, and reduced harmful bacteria in the gut (Bozkurt et al. 2007). In addition, the use of essential oils such as cinnamon, clove, and ginger has been shown to improve egg production and egg quality in laying hens (Ezenduka et al. 2015). In ruminant production, essential oils have also been shown to have positive effects on growth performance, feed efficiency, and gut health. A study found that the addition of cinnamon essential oil to the diets of dairy cows increased milk production and improved feed efficiency (Khanal et al. 2015). In addition, the use of essential oils such as oregano and thyme has been shown to reduce the incidence of ruminal acidosis and improve rumen function in cattle (Christie et al. 2013). The use of essential oils in animal production is also associated with various health benefits. For example, the use of essential oils such as oregano and cinnamon has been shown to have antibacterial, antifungal, and antiviral properties that can help improve the immunity and disease resistance of livestock (Abd El-Hack et al. 2015). In addition, the use of essential oils has been shown to have anti-inflammatory effects that can help reduce the incidence of various inflammatory diseases in livestock, such as mastitis in dairy cows (Caneschi et al. 2023). All in all, the use of essential oils as feed additives in animal production has shown promising results, with the potential to improve growth performance, feed efficiency, gut health, and immunity, and reduce the incidence of various diseases in livestock.

References

Aanyu M, Betancor MB, Monroig O (2018) Effects of dietary limonene and thymol on the growth and nutritional physiology of Nile tilapia (Oreochromis niloticus). Aquaculture 488:217–226. https://doi.org/10.1016/j.aquaculture.2018.01.036

Abd El-Ghany WA, Ismail M (2014) Tackling experimental colisepticaemia in broiler chickens using phytobiotic essential oils and antibiotic alone or in combination. Iran J Vet Res 15:110–115

Abd El-Hack ME, El-Saadony MT, Saad AM, Salem HM, Ashry NM, Ghanima MMA, El-Tarabily KA (2022) Essential oils and their nanoemulsions as green alternatives to antibiotics in poultry nutrition: a comprehensive review. Poult Sci 101(2):101584

Abdel-Latif HMR, Abdel-Tawwab M, Khafaga AF, Dawood MAO (2020) Dietary oregano essential oil improved the growth performance via enhancing the intestinal morphometry and hepato-renal functions of common carp (Cyprinus carpio L.) fingerlings. Aquaculture 526: 735432. https://doi.org/10.1016/j.aquaculture.2020.735432

Abdel-Latif HMR, Dawood MAO, Alagawany M, Faggio C, Nowosad J, Kucharczyk D (2022) Health benefits and potential applications of fucoidan (FCD) extracted from brown seaweeds in aquaculture: an updated review. Fish Shellfish Immunol 122:115–130. https://doi.org/10.1016/j.fsi.2022.01.039

Adaszyńska-Skwirzyńska M, Szczerbińska D (2017) Use of essential oils in broiler chicken production – a review. Ann Anim Sci 17:317–335. https://doi.org/10.1515/aoas-2016-0046

Ali M, Chand N, Khan RU (2019) Anticoccidial effect of garlic (Allium sativum) and ginger (Zingiber officinale) against experimentally induced coccidiosis in broiler chickens. J Appl Anim Res 47:79–84

Al-Kassie GAM (2009) Influence of two plant extracts derived from thyme and cinnamon on broiler performance. Pak Vet J 29:169–173

Amatiste S, Sagrafoli D, Giacinti G, Rosa G, Carfora V, Marri N, Rosati R (2014) Antimicrobial activity of essential oils against Staphylococcus aureus in fresh sheep cheese. Ital J Food Saf 3(3)

Amer SA, Metwally AE, Ahmed SAA (2018) The influence of dietary supplementation of cinnamaldehyde and thymol on the growth performance, immunity and antioxidant status of monosex Nile tilapia fingerlings (Oreochromis niloticus). Egypt J Aquat Res 44:251–256

Amirifar A, Hemati A, Asgari Lajayer B, Pandey J, Astatkie T (2022) Impact of various environmental factors on the biosynthesis of alkaloids in medicinal plants. In: Aftab T (ed) Environmental challenges and medicinal plants. Environmental challenges and solutions. Springer, Cham. https://doi.org/10.1007/978-3-030-92050-0_9

Asgari Lajayer B, Ghorbanpour M, Nikabadi S (2017) Heavy metals in contaminated environment: Destiny of secondary metabolite biosynthesis, oxidative status and phytoextraction in medicinal plants. Ecotoxicol Environ Saf 145:377–390. https://doi.org/10.1016/j.ecoenv.2017.07.035

Astani A, Reichling J, Schnitzler P (2011) Screening for antiviral activities of isolated compounds from essential oils. Evid Based Complement Altern Med 2011:1–8

Bakkali F, Averbeck S, Averbeck D, Idaomar M (2008) Biological effects of essential oils – a review. Food Chem Toxicol 46:446–475. https://doi.org/10.1016/j.fct.2007.09.106

Barbarestani SY, Jazi V, Mohebodini H, Ashayerizadeh A, Shabani A, Toghyani M (2020) Effects of dietary lavender essential oil on growth performance, intestinal function, and antioxidant status of broiler chickens. Livest Sci 233:103958

Barbour EK, Yaghi RH, Jaber LS, Shaib HA, Harakeh S (2010) Safety and antiviral activity of essential oil against avian influenza and Newcastle disease viruses. Int J Appl Res Vet Med 8: 60–64

Begnami AF, Duarte MCT, Furletti V, Rehder VLG (2010) Antimicrobial potential of Coriandrum sativum L. against different Candida species in vitro. Food Chem 118(1):74–77

Béjaoui A, Chaabane H, Jemli M, Boulila A, Boussaid M (2013) Essential oil composition and antibacterial activity of Origanum vulgare subsp. glandulosum Desf. at different phenological stages. J Med Food 16(12):1115–1120

Bovera F, Piccolo G, Gasco L, Marono S, Loponte R, Vassalotti G, Mastellone V, Lombardi P, Attia YA, Nizza A (2015) Yellow mealworm larvae (Tenebrio molitor, L.) as a possible alternative to soybean meal in broiler diets. Br Poult Sci 56:569–575. https://doi.org/10.1080/00071668.2015.1080815

Bozkurt M, Cekic O, Kucuk O, Caliskan M (2007) The effect of oregano essential oil on performance, digestive enzymes and gut microflora in broiler chickens. South Afr J Anim Sci 37(2):139–145. https://doi.org/10.4314/sajas.v37i2.6

Brochot A, Guilbot A, Haddioui L, Roques C (2017) Antibacterial, antifungal, and antiviral effects of three essential oil blends. Microbiology 6:e00459

Burt S (2004) Essential oils: their antibacterial properties and potential applications in foods—a review. Int J Food Microbiol 94:223–253. https://doi.org/10.1016/j.ijfoodmicro.2004.03.022

Calislar S, Gemci I, Kamalak A (2009) Effects of Orego-Stim® on broiler chick performance and some blood parameters. J Anim Vet Adv 8:2617–2620

Caneschi A, Bardhi A, Barbarossa A, Zaghini A (2023) Plant essential oils as a tool in the control of bovine mastitis: an update. Molecules 28(8):3425

Cardozo PW, Calsamiglia S, Ferret A, Kamel C (2006) Effects of alfalfa extract, anise, capsicum, and a mixture of cinnamaldehyde and eugenol on ruminal fermentation and protein degradation in beef heifers fed a high-concentrate diet. J Anim Sci 84(10):2801–2808. https://doi.org/10.2527/jas.2005-593

Castillo M, Martín-Orúe S, Roca M, Manzanilla E, Badiola I, Perez J (2006) The response of gastrointestinal microbiota to avilamycin, butyrate, and plant extracts in early-weaned pigs. J Anim Sci 84:2725–2734

Centner TJ (2016) Efforts to slacken antibiotic resistance: labeling meat products from animals raised without antibiotics in the United States. Sci Total Environ 563:1088–1094

Champagne A, Boutry M (2016) Proteomics of terpenoid biosynthesis and secretion in trichomes of higher plant species. Biochim Biophys Acta 1864:1039–1049

Cheng C, Zhang X, Xia M, Liu Z, Wei H, Deng Z, Wang C, Jiang S, Peng J (2017) Effect of oregano essential oil and benzoic acid supplementation to a low-protein diet on meat quality, fatty acid composition, and lipid stability of longissimus thoracis muscle in pigs. Lipids Health Dis 16(1):164. https://doi.org/10.1186/s12944-017-0535-1

Cheng Y-H, Hsiao F-S, Huang Y-C, Chen T-C (2020) Effects of essential oil supplementation on growth performance, antioxidation, and immune responses in broilers. Animals 10(1):107

Cherian G, Orr A, Burke IC, Pan W (2013) Feeding *Artemisia annua* alters digesta pH and muscle lipid oxidation products in broiler chickens. Poult Sci 92:1085–1090

Chowdhury S, Mandal GP, Patra AK, Kumar P, Samanta I, Pradhan S, Samanta AK (2018) Different essential oils in diets of broiler chickens: 2. Gut microbes and morphology, immune response, and some blood profile and antioxidant enzymes. Anim Feed Sci Technol 236:39–47. https://doi.org/10.1016/j.anifeedsci.2017.12.003

Christie WW, Fondevila M, McAllister TA, Cheng K-J, Yang WZ (2013) Essential oil supplementation to reduce the risk of ruminal acidosis and improve rumen fermentation and milk production in dairy cattle

Chuan-Shang C, Wei H, Peng J (2017) Effect of oregano essential oil supplementation to a reduced-protein diet on meat quality, fatty acid composition, and oxidative stability of longissimus thoracis muscle in growing-finishing pigs. J Anim Sci 95(4):346. https://doi.org/10.2527/asasann.2017.710

Chun S-S, Vattem DA, Lin Y-T, Shetty K (2005) Phenolic antioxidants from clonal oregano (*Origanum vulgare*) with antimicrobial activity against *Helicobacter pylori*. Process Biochem 40:809–816

Civitelli L, Panella S, Marcocci ME, De Petris A, Garzoli S, Pepi F, Vavala E, Ragno R, Nencioni L, Palamara A (2014) In vitro inhibition of herpes simplex virus type 1 replication by Mentha suaveolens essential oil and its main component piperitenone oxide. Phytomedicine 21:857–865

Cui H, Zhang X, Zhou H, Zhao C, Lin L (2015) Antimicrobial activity and mechanisms of Salvia sclarea essential oil. Bot Stud 56(1):1–8

Czech A, Kowalczuk E, Grela ER (2009) The effect of an herbal extract used in pig fattening on the animal's performance and blood components. Ann Univ Mariae Curie Sklodowska 27:25–33

Dawood MAO, El-Salam Metwally A, Elkomy AH, Gewaily MS, Abdo SE, Abdel-Razek MAS, Soliman AA, Amer AA, Abdel-Razik NI, Abdel-Latif HMR, Ahamad Pray B (2020) The impact of menthol essential oil against inflammation, immunosuppression, and histopathological alterations induced by chlorpyrifos in Nile tilapia. Fish Shellfish Immunol 102:316–325

Dawood MAO, El Basuini MF, Yilmaz S, Abdel-latif HMR, Alagawany M, Abdul Kari Z, Khairul Azhar Abdul Razab M, Khalidah Abdul Hamid N, Moonmanee T, Van Doan H (2021) Exploring the role of dietary herbal essential oils as natural antioxidative agents in aquaculture: a review. Animals 12(823):1–19

Degenhardt J, Köllner TG, Gershenzon J (2009) Monoterpene and sesquiterpene synthases and the origin of terpene skeletal diversity in plants. Phytochemistry 70:1621–1637

Do TKT, Hadji-Minaglou F, Antoniotti S, Fernandez X (2015) Authenticity of essential oils. Trends Anal Chem 66:146–157

Domínguez R, Pateiro M, Gagaoua M, Barba FJ, Zhang W, Lorenzo JM (2019) A comprehensive review on lipid oxidation in meat and meat products. Antioxidants 8:429

Domínguez R, Gullón P, Pateiro M, Munekata P, Zhang W, Lorenzo J (2020) Tomato as potential source of natural additives for meat industry. A review. Antioxidants 9:73

Dorantes-Iturbide G, Orzuna-Orzuna JF, Lara-Bueno A, Miranda-Romero LA, Mendoza-Martínez GD, Hernández-García PA (2022) Effects of a polyherbal dietary additive on performance, dietary energetics, carcass traits, and blood metabolites of finishing lambs. Metabolites 12:413. https://doi.org/10.3390/metabo12050413

Du J, Zhang Y, Zhang J, Wang Q, Li H (2017) Essential oils as feed additives for livestock: a review. Livest Sci 205:70–79

Dutra TV, Castro JC, Menezes JL, Ramos TR, do Prado IN, Machinski M, Mikcha JMG, de Abreu Filho BA (2019) Bioactivity of oregano (Origanum Vulgare) essential oil against Alicyclobacillus Spp. Ind Crop Prod 129:345–349

El Asbahani A, Miladi K, Badri W, Sala M, Addi EH, Casabianca H, El Mousadik A, Hartmann D, Jilale A, Renaud FNR (2015) Essential oils: from extraction to encapsulation. Int J Pharm 483: 220–243

El Mokni R, Youssef FS, Jmii H, Khmiri A, Bouazzi S, Jlassi I, Jaidane H, Dhaouadi H, Ashour ML, Hammami S (2019) The essential oil of Tunisian Dysphania ambrosioides and its antimicrobial and antiviral properties. J Essent Oil Bear Plants 22:282–294

Elaissi A, Rouis Z, Ben Salem NA, Mabrouk S, Ben Salem Y, Salah KBH, Aouni M, Farhat F, Chemli R, Harzallah-Skhiri F (2012) Chemical composition of 8 eucalyptus species' essential oils and the evaluation of their antibacterial, antifungal and antiviral activities. BMC Complement Altern Med. 12:81

Elcoso G, Zweifel B, Bach A (2019) Effects of a blend of essential oils on milk yield and feed efficiency of lactating dairy cows. Appl Anim Sci 35(3):304–311

Ezenduka CC, Obamuyi OO, Nwosu VO, Okeke IN (2015) Antibacterial and antioxidant activities of cinnamon (Cinnamomum verum), clove (Syzygium aromaticum) and ginger (Zingiber officinale) essential oils. Int J Food Sci Technol 50(9):1977–1982. https://doi.org/10.1111/ijfs.12833

Faramarzi S, Bozorgmehrifard MH, Khaki A, Moomivand H, Ezati MS, Rasoulinezhad S, Bahnamiri AJ, Dizaji B (2013) Study on the effect of Thymus vulgaris essential oil on humoral immunity and performance of broiler chickens after La Sota vaccination. Ann Biol Res 4:290–294

Feng J, Lu M, Wang J, Zhang H, Qiu K, Qi G, Wu S (2021) Dietary oregano essential oil supplementation improves intestinal functions and alters gut microbiota in late-phase laying hens. J Anim Sci Biotechnol 12(1):1–15

Feizi A, Bijanzad P, Asfaram H, Khiavi TM, Alimardan M, Hamzei S, Ghabel H, Faramarzi S (2014) Effect of thyme extract on hematological factors and performance of broiler chickens. Eur J Exp Biol 4:125–128

Fernandes R, Trindade M, Tonin F, Pugine S, Lima C, Lorenzo J, De Melo M (2017) Evaluation of oxidative stability of lamb burger with Origanum vulgare extract. Food Chem 233:101–109

Ferreira LC, da Cruz MG, Lima TBC, Serra BNV, Chaves FCM, Chagas EC, Ventura AS, Jerônimo GT (2019) Antiparasitic activity of Mentha piperita (Lamiaceae) essential oil against Piscinoodinium pillulare and its physiological effects on Colossoma macropomum (Cuvier, 1818). Aquaculture 512:734343. https://doi.org/10.1016/j.aquaculture.2019.734343

Firmino DF, Cavalcante TTA, Gomes GA (2018) Antibacterial and antibiofilm activities of *Cinnamomum* sp. essential oil and cinnamaldehyde: antimicrobial activities. Sci World J 2018:7405736

Florou-Paneri P, Giannenas I, Christaki E, Govaris A, Botsoglou NA (2006) Performance of chickens and oxidative stability of the produced meat as affected by feed supplementation with oregano, vitamin C, vitamin E and their combinations. Arch Geflugelkunde 70:232–240

Florou-Paneri P, Christaki E, Giannenas I (eds) (2019) Feed additives aromatic plants and herbs in animal nutrition and health. Academic Press. ISBN 9780128147009

Fokou JBH, Dongmo PMJ, Boyom FF (2020) Essential oil's chemical composition and pharmacological properties. In: Essential oils - oils of nature. IntechOpen, London

Forte C, Ranucci D, Beghelli D, Branciari R, Acuti G, Todini L (2017) Dietary integration with oregano (*Origanum vulgare* L.) essential oil improves growth rate and oxidative status in outdoor-reared, but not indoor-reared, pigs. J Anim Physiol Anim Nutr 101(5):352–361

Franz C, Novak J (2020) Sources of essential oils. In: Handbook of essential oils. CRC Press, Boca Raton, FL, pp 41–83

Franz C, Baser KHC, Windisch W (2010) Essential oils and aromatic plants in animal feeding—a European perspective: a review. Flavour Fragr J 25:327–340

Fraternale D, Giamperi L, Bucchini A, Ricci D, Epifano F, Genovese S, Curini M (2005) Composition and antifungal activity of essential oil of Salvia sclarea from Italy. Chem Nat Compd 41:604–606

Fu Y, Zu Y, Chen L, Shi X, Wang Z, Sun S, Efferth T (2007) Antimicrobial activity of clove and rosemary essential oils alone and in combination. Phytother Res 21(10):989–994

Fugita CA, do Prado Rodolpho M, Velandia VM, Bonafé Elton G, Barbosa CC, Ana G, Carlos S, do Prado Ivanor N (2018) Effect of the inclusion of natural additives on animal performance and meat quality of crossbred bulls (Angus × Nellore) finished in feedlot. Anim Prod Sci 58:2076–2083. https://doi.org/10.1071/AN16242

Galal AAAE, El-Araby IE, Hassanin O, Omar AE (2016) Positive impact of oregano essential oil on growth performance, humoral immune responses and chicken interferon alpha signaling pathway in broilers. Adv Anim Vet Sci 4:57–65. https://doi.org/10.14737/journal.aavs/2016/4.1.57.65

García-González R, López S, Fernández M, Bodas R, González JS (2008) Screening the activity of plants and spices for decreasing ruminal methane production in vitro. Anim Feed Sci Technol 147:36–52

Gavanji S, Sayedipour SS, Larki B, Bakhtari A (2015) Antiviral activity of some plant oils against herpes simplex virus type 1 in Vero cell culture. J Acute Med 5:62–68

Gelzleichter TR (2013) Chapter 7—Early characterization of biosimilar therapeutics. In: Plitnick LM, Herzyk DJ (eds) Nonclinical development of novel biologics, biosimilars, vaccines and specialty biologics. Academic Press, San Diego, CA, pp 185–210. ISBN 978-0-12-394810-6

Gershenzon J, Dudareva N (2007) The function of terpene natural products in the natural world. Nat Chem Biol 3:408–414

Giannenas I (2004) Effect of diet supplementation with ground oregano on performance of broiler chickens challenged with Eimeria tenella. Arch Geflugelkunde 68:247–252

Giannenas I, Florou-Paneri P, Papazahariadou M, Christaki E, Botsoglou NA, Spais AB (2003) Effect of dietary supplementation with oregano essential oil on performance of broilers after experimental infection with eimeria tenella. Arch Anim Nutr 57(2):99–106. https://doi.org/10.1080/0003942031000107299

Giannenas I, Florou-Paneri P, Botsoglou NA, Christaki E, Spais AB (2005) Effect of supplementing feed with oregano and (or) alpha-tocopheryl acetate on growth of broiler chickens and oxidative stability of meat. J Anim Feed Sci 14:521–535

Giannenas I, Skoufos J, Giannakopoulos C, Wiemann M, Gortzi O, Lalas S, Kyriazakis I (2011) Effects of essential oils on milk production, milk composition, and rumen microbiota in Chios dairy ewes. J Dairy Sci 94(11):5569–5577

Giannenas I, Triantafillou E, Stavrakakis S, Margaroni M, Mavridis S, Steiner T, Karagouni E (2012) Assessment of dietary supplementation with carvacrol or thymol containing feed additives on performance, intestinal microbiota and antioxidant status of rainbow trout (Oncorhynchus mykiss). Aquaculture 350–353:26–32. https://doi.org/10.1016/j.aquaculture.2012.04.027

Giannenas I, Bonos E, Christaki E, Florou-Paneri P (2018) Oregano: a feed additive with functional properties. Therap Foods 179–208. https://doi.org/10.1016/B978-0-12-811517-6.00006-4

Giordani R, Regli P, Kaloustian J, Mikail C, Abou L, Portugal H (2004) Antifungal effect of various essential oils against Candida albicans. Potentiation of antifungal action of amphotericin B by essential oil from Thymus vulgaris. Phytother Res 18:990–995

Gong J, Yin F, Hou R, Yin YL (2014) Review: Chinese herbs as alternatives to antibiotics in feed for swine and poultry production: potential and challenges in application. Can J Anim Sci 94: 223–241

Gopi M, Karthik K, Manjunathachar HV, Tamilmahan P, Kesavan M, Dashprakash M, Balaraju BL, Purushothaman MR (2014) Essential oils as a feed additive in poultry nutrition. Adv Anim Vet Sci 2:1–7. https://doi.org/10.14737/journal.aavs/2014.2.1.1.7

Goudarzi M, Namjooyan F, Saeedi M, Shafiee-Nick R (2020) Chemical composition, antioxidant activity, and antibacterial effect of turmeric (Curcuma longa L.) essential oil. Vet Res Forum 11(2):159–165

Grohs BM, Kunz B (2000) Use of spices for the stabilization of fresh portioned pork. Food Control 11(6):433–436

Haddad JG, Picard M, Bénard S, Desvignes C, Desprès P, Diotel N, El Kalamouni C (2019) Ayapana triplinervis essential oil and its main component Thymohydroquinone dimethyl ether inhibit Zika virus at doses devoid of toxicity in zebrafish. Molecules 24:3447

Halas V, Nochta I, Pásti Z, Szabó C, Tóthi R, Tossenberger J (2011) Cellular immune response of weaned pigs fed diet supplemented with an essential oil. Agric Conspec Sci 76:279–282

Hammer KA, Carson CF (2011) Antibacterial and antifungal activities of essential oils. In: Thormar H (ed) Lipids and essential oils as antimicrobial agents. Springer. https://doi.org/10.1002/9780470976623.ch11

Hashemipour H, Kermanshahi H, Golian A, Veldkamp T (2013a) Effect of thymol and carvacrol feed supplementation on performance, antioxidant enzyme activities, fatty acid composition, digestive enzyme activities, and immune response in broiler chickens. Poult Sci 92(8): 2059–2069

Hashemipour H, Kermanshahi H, Golian A, Raji A, Van Krimpen MM (2013b) Effect of Thymol+ carvacrol by next enhance 150® on intestinal development of broiler chickens fed CMC containing diet. Iran J Appl Anim Sci 3:567–576

Hassan S, Khalil MM, El-Badry AA, Aal SE (2020) Effect of oregano essential oil on growth performance, immune response and gut health of broiler chickens. Egypt Poult Sci J 40(1): 89–104

Heck RT, Saldaña E, Lorenzo JM, Correa LP, Fagundes MB, Cichoski AJ, de Menezes CR, Wagner R, Campagnol PCB (2019) Hydrogelled emulsion from chia and linseed oils: a promising strategy to produce low-fat burgers with a healthier lipid profile. Meat Sci 156:174–182

Holley RA, Patel D (2005) Improvement in shelf-life and safety of perishable foods by plant essential oils and smoke antimicrobials. Food Microbiol 22:273–292

Hood JR, Wilkinson JM, Cavanagh HM (2003) Evaluation of common antibacterial screening methods utilized in essential oil research. J Essent Oil Res 15(6):428–433

Hossain S, Heo H, De Silva BCJ, Wimalasena SHMP, Pathirana HNKS, Heo GJ (2017) Antibacterial activity of essential oil from lavender (*Lavandula angustifolia*) against pet turtle-borne pathogenic bacteria. Lab Anim Res 33(3):195–201. https://doi.org/10.5625/lar.2017.33.3.195

Hu WQ, Li QY, Xiong F, Deng XJ (2010) Effects of plant extract feed additive (Haolicao) on performance and economic benefits of broilers. Chin J Anim Sci 8:3

Hu F, Tu X-F, Thakur K, Hu F, Li X-L, Zhang Y-S, Zhang J-G, Wei Z-J (2019) Comparison of antifungal activity of essential oils from different plants against three fungi. Food Chem Toxicol 134:110821

Janacua-Vidales H, Peña-González E, Alarcon-Rojo AD, Ortega-Gutiérrez J, Aguilar-Palma N (2019) Determination of carcase yield, sensory and acceptance of meat from male and female pigs with dietary supplementation of oregano essential oils. Ital J Anim Sci 18(1):668–678. https://doi.org/10.1080/1828051X.2018.1553507

Jayawardana BC, Warnasooriya VB, Thotawattage G, Dharmasena V, Liyanage R (2019) Black and green tea (Camellia sinensis L.) extracts as natural antioxidants in uncured pork sausages. J Food Process Preserv 43:e13870

Jirovetz L, Wlcek K, Buchbauer G, Gochev V, Girova T, Stoyanova A et al (2007) Antifungal activities of essential oils of Salvia lavandulifolia, Salvia officinalis and Salvia sclarea against various pathogenic Candida species. J Essent Oil Bear Plants 10(5):430–439

Kalaitsidis K, Sidiropoulou E, Tsiftsoglou O, Mourtzinos I, Moschakis T, Basdagianni Z et al (2021) Effects of cornus and its mixture with oregano and thyme essential oils on dairy sheep performance and milk, yoghurt and cheese quality under heat stress. Animals 11(4):1063

Kalantar M, Hosseini SM, Yang LG, Raza SHA, Gui LS, Rezaie M, Khojastekey M, Wei DW, Khan R, Yasar S, Syed SF, Kachiwal AB, Elkhairey M, Lei QJ, Kaleri RR, El-Aziz AHA (2017) Performance, immune, and carcass characteristics of broiler chickens as affected by thyme and licorice or enzyme supplemented diets. Open J Anim Sci 7:105–109. https://doi.org/10.4236/ojas.2017.72009

Khanal RK, Lee JH, Park SH, Kim HK (2015) Effects of dietary cinnamon essential oil on growth performance, blood characteristics, and meat quality in growing-finishing pigs. Asian Australas J Anim Sci 28(10):1429–1436. https://doi.org/10.5713/ajas.15.0087

Khiaosa-Ard R, Zebeli Q (2013) Meta-analysis of the effects of essential oils and their bioactive compounds on rumen fermentation characteristics and feed efficiency in ruminants. J Anim Sci 91(4):1819–1830

Khodaei N, Nguyen MM, Mdimagh A, Bayen S, Karboune S (2021) Compositional diversity and antioxidant properties of essential oils: predictive models. LWT Food Sci Technol 138:110684. https://doi.org/10.1016/j.lwt.2020.110684

Kirkpinar F, Ünlü HB, Serdaroğlu M, Turp GY (2014) Effects of dietary oregano and garlic essential oils on carcass characteristics, meat composition, colour, PH and sensory quality of broiler meat. Br Poult Sci 55:157–166

Krishan G, Narang A (2014) Use of essential oils in poultry nutrition: a new approach. J Adv Vet Anim Res 1:156–162. https://doi.org/10.5455/javar.2014.a36

Ksouri S, Djebir S, Bentorki AA, Gouri A, Hadef Y, Benakhla A (2017) Antifungal activity of essential oils extract from Origanum floribundum Munby, Rosmarinus officinalis L. and Thymus ciliatus Desf. against Candida albicans isolated from bovine clinical mastitis. J Mycol Med 27:245–249

Kubiça TF, Alves SH, Weiblen R, Lovato LT (2014) In vitro inhibition of the bovine viral diarrhoea virus by the essential oil of Ocimum basilicum (basil) and monoterpenes. Braz J Microbiol 45: 209–214

Li P, Piao X, Ru Y, Han X, Xue L, Zhang H (2012) Effects of adding essential oil to the diet of weaned pigs on performance, nutrient utilization, immune response and intestinal health. Asian-Australas J Anim Sci 25:1617–1626

Li Y, Wang T, Wu X, Zhang H, Liu Q, Chen Y (2020) Effect of thymol, cinnamaldehyde and their combination on intestinal morphology, microbiota, and immune response of weaned pigs. Anim Nutr 6(2):208–214

Liang R, Xu S, Shoemaker CF, Li Y, Zhong F, Huang Q (2012) Physical and antimicrobial properties of peppermint oil nanoemulsions. J Agric Food Chem 60:7548–7555

Liu XC, Sun YN, Li HP (2008) Effects of plant extract feed additives on performance of yellow-feathered broilers. Chin J Anim Sci 44(6):3

Liu Q, Meng X, Li Y, Zhao C-N, Tang G-Y, Li H-B (2017) Antibacterial and antifungal activities of spices. Int J Mol Sci 18:1283. https://doi.org/10.3390/ijms18061283

Lopez-Romero JC, González-Ríos H, Borges A, Simões M (2015) Antibacterial effects and mode of action of selected essential oils components against Escherichia coli and Staphylococcus aureus. Evid Based Complement Alternat Med 2015:9

Lorenzo-Leal AC, Palou E, López-Malo A (2019) Evaluation of the efficiency of allspice, thyme and rosemary essential oils on two foodborne pathogens in in-vitro and on alfalfa seeds, and their effect on sensory characteristics of the sprouts. Int J Food Microbiol 295:19–24

Madrid A, Godoy P, González S, Zaror L, Moller A, Werner E, Cuellar M, Villena J, Montenegro I (2015) Chemical characterization and anti-oomycete activity of Laureliopsis philippianna essential oils against Saprolegnia parasitica and S. australis. Molecules 20(5):8033–8047. https://doi.org/10.3390/molecules20058033

Manzanilla E, Nofrarias M, Anguita M, Castillo M, Perez J, Martin-Orue S (2006) Effects of butyrate, avilamycin, and a plant extract combination on the intestinal equilibrium of early-weaned pigs. J Anim Sci 84:2743–2751

Marchese A, Orhan IE, Daglia M, Barbieri R, Di Lorenzo A, Nabavi SF, Gortzi O, Izadi M, Nabavi SM (2016) Antibacterial and antifungal activities of thymol: a brief review of the literature. Food Chem 210:402–414

Matasyoh JC, Maiyo ZC, Ngure RM, Chepkorir R (2009) Chemical composition and antimicrobial activity of the essential oil of Coriandrum sativum. Food Chem 113(2):526–529

Michiels J, Missotten J, Dierick N, Fremaut D, Maene P, De Smet S (2008) In vitro degradation and in vivo passage kinetics of carvacrol, thymol, eugenol and trans-cinnamaldehyde along the gastrointestinal tract of piglets. J Sci Food Agric 88:2371–2381

Miguel MG (2010) Antioxidant activity of medicinal and aromatic plants. A review. Flavour Fragr J 25:291–312

Miron TL, Gazi I, Plaza del Moral M (2010) Romanian aromatic plants as source of antioxidants. Innov Rom Food Biotechnol 6:18–24

Mountzouris KC, Paraskevas V, Tsirtsikos P, Palamidia I, Steiner T, Schatzmayr G (2011) Assessment of a phytogenic feed additive effect on broiler growth performance, nutrient digestibility and caecal microflora composition. Anim Feed Sci Technol 168:223–231

Mucha W, Witkowska D (2021) The applicability of essential oils in different stages of production of animal-based foods. Molecules 26:3798

Mulyaningsih S, Sporer F, Zimmermann S, Reichling J, Wink M (2010) Synergistic properties of the terpenoids aromadendrene and 1,8-cineole from the essential oil of Eucalyptus globulus against antibiotic-susceptible and antibiotic-resistant pathogens. Phytomedicine 17(13): 1061–1066

Nazzaro F, Fratianni F, Coppola R, Feo VD (2017) Essential oils and antifungal activity. Pharmaceuticals 10:86. https://doi.org/10.3390/ph10040086

Negi PS (2012) Plant extracts for the control of bacterial growth: efficacy, stability, and safety issues for food application. Int J Food Microbiol 156:7–17

Neill CR, Nelssen JL, Tokach MD, Goodband RD, Derouchey JM, Dritz SS, Groesbeck CN, Brown KR (2006) Effects of oregano oil on growth performance of nursery pigs. J Swine Health Prod 14(6):312–316

Nieto CA, Ortiz RE, Tellez GA (2018a) Effect of two chemotypes of oregano essential oil on broiler performance, nutrient balance, and lipid peroxidation of breast meat during storage. Cienc Anim Bras Goiânia 19:1–15. https://doi.org/10.1590/1809-6891v19e-47819

Nieto G, Ros G, Sánchez JC (2018b) Antioxidant and antimicrobial properties of rosemary (Rosmarinus officinalis L.): a review. Medicines 5(4):98. https://doi.org/10.3390/medicines5040098

Oh HJ, Kim IH, Song MH, Kwak WG, Yun W, Lee JH, Cho JH (2018) Effects of microencapsulated complex of organic acids and essential oils on growth performance, nutrient retention, blood profiles, fecal microflora, and lean meat percentage in weaning to finishing pigs. Can J Anim Sci 99:41–49. https://doi.org/10.1139/cjas-2018-0006

Orzuna-Orzuna JF, Dorantes-Iturbide G, Lara-Bueno A, Miranda-Romero LA, Mendoza-Martínez GD, Santiago-Figueroa I (2022) A meta-analysis of essential oils use for beef cattle feed: rumen fermentation, blood metabolites, meat quality, performance and, environmental and economic impact. Fermentation 8(6):254

Patra AK, Amasheh S, Aschenbach JR (2019) Modulation of gastrointestinal barrier and nutrient transport function in farm animals by natural plant bioactive compounds – a comprehensive review. Crit Rev Food Sci Nutr 59(20):3237–3266

Paulpandi M, Kannan S, Thangam R, Kaveri K, Gunasekaran P, Rejeeth C (2012) In vitro anti-viral effect of β-santalol against influenza viral replication. Phytomedicine 19:231–235

Pavela R, Benelli G (2016) Essential oils as ecofriendly biopesticides? Challenges and constraints. Trends Plant Sci 21:1000–1007

Pilau MR, Alves SH, Weiblen R, Arenhart S, Cueto AP, Lovato LT (2011) Antiviral activity of the Lippia graveolens (Mexican oregano) essential oil and its main compound carvacrol against human and animal viruses. Braz J Microbiol 42:1616–1624

Placha I, Takacova J, Ryzner M, Cobanova K, Laukova A, Strompfova V, Venglovska K, Faix S (2014) Effect of thyme essential oil and selenium on intestine integrity and antioxidant status of broilers. Br Poult Sci 55:105–114. https://doi.org/10.1080/00071668.2013.873772

Platel K, Srinivasan K (2004) Digestive stimulant action of spices: a myth or reality? Indian J Med Res 119(5):167

Powers CN, Osier JL, McFeeters RL, Brazell CB, Olsen EL, Moriarity DM, Satyal P, Setzer WN (2018) Antifungal and cytotoxic activities of sixty commercially-available essential oils. Molecules 23

Prakash B, Singh P, Kedia A, Dubey NK (2012) Assessment of some essential oils as food preservatives based on antifungal, antiaflatoxin, antioxidant activities and in vivo efficacy in food system. Food Res Int 49:201–208. https://doi.org/10.1016/j.foodres.2012.08.020

Radwan Nadia L, Hassan RA, Qota EM, Fayek HM (2008) Effect of natural antioxidant on oxidative stability of eggs and productive and reproductive performance of laying hens. Int J Poult Sci 7(2):134–150

Raut JS, Karuppayil SM (2014) A status review on the medicinal properties of essential oils. Ind Crops Prod 62:250–264

Rivaroli DC, Guerrero A, Velandia Valero M, Zawadzki F, Eiras CE, Campo MD, Sañudo C, Mendes Jorge A, Nunes do Prado I (2016) Effect of essential oils on meat and fat qualities of crossbred young bulls finished in feedlots. Meat Sci 121:278–284

Rochfort S, Parker AJ, Dunshea FR (2008) Plant bioactives for ruminant health and productivity. Phytochemistry 69(2):299–322

Rodriguez Flores C, Pennec A, Nugier-Chauvin C, Daniellou R, Herrera-Estrella L, Chauvin AL (2014) Chemical composition and antibacterial activity of essential oils extracted from plants cultivated in Mexico. J Mex Chem Soc 58(4):452–455

Rota MC, Herrera A, Martínez RM, Sotomayor JA, Jordán MJ (2008) Antimicrobial activity and chemical composition of Thymus vulgaris, Thymus zygis and Thymus hyemalis essential oils. Food Control 19(7):681–687

Saeed S, Naim A, Tariq P (2006) In vitro antibacterial activity of peppermint. Pak J Bot 38(3): 869–872

Saini R, Davis S, Dudley-Cash W (2003a) Oregano essential oil reduces necrotic enteritis in broilers. In: proceedings of the 52 th western poultry disease conference, Sacramento, California, pp 95–97

Saini R, Davis S, Dudley-Cash W (2003b) Oregano essential oil reduces the expression of coccidiosis in broilers. In: proceedings of the 52 th western poultry disease conference, Sacramento, California, pp 97–98

Sakakibara H, Honda Y, Nakagawa S, Ashida H, Kanazawa K (2003) Simultaneous determination of all polyphenols in vegetables, fruits, and teas. J Agric Food Chem 51(3):571–581. https://doi.org/10.1021/jf020926l

Sakkas H, Gousia P, Economou V, Sakkas V, Petsios S, Papadopoulou C (2016) In vitro antimicrobial activity of five essential oils on multidrug resistant gram-negative clinical isolates. J Intercult Ethnopharmacol 5:212–218

Santurio DF, de Jesus FPK, Zanette RA, Schlemmer KB, Fraton A, Fries LLM (2014) Antimicrobial activity of the essential oil of thyme and of thymol against Escherichia coli strains. Acta Sci Vet 42(1):1–4

Schnitzler P (2019) Essential oils for the treatment of herpes simplex virus infections. Chemotherapy 64(1):1–7. https://doi.org/10.1159/000501062

Schroder T, Gaskin S, Ross K, Whiley H (2017) Antifungal activity of essential oils against fungi isolated from air. Int J Occup Environ Health 23(3):181–186. https://doi.org/10.1080/10773525.2018.1447320

Seow YX, Yeo CR, Chung HL, Yuk HG (2014) Plant essential oils as active antimicrobial agents. Crit Rev Food Sci Nutr 54(5):625–644. https://doi.org/10.1080/10408398.2011.599504

Sharifi-Rad J, Salehi B, Schnitzler P, Ayatollahi SA, Kobarfard F, Fathi M, Eisazadeh M, Sharifi-Rad M (2017) Susceptibility of herpes simplex virus type 1 to monoterpenes thymol, carvacrol, p-cymene and essential oils of Sinapis arvensis L., Lallemantia royleana Benth. and Pulicaria vulgaris Gaertn. Cell Mol Biol (Noisy-le-Grand) 63(8):42–47. https://doi.org/10.14715/cmb/2017.63.8.10

Šojić B, Pavlić B, Zeković Z, Tomović V, Ikonić P, Kocić-Tanackov SD, Džinić N (2018) The effect of essential oil and extract from sage (Salvia officinalis L.) herbal dust (food industry by-product) on the oxidative and microbiological stability of fresh pork sausages. LWT Food Sci Technol 89:749–755

Soliman KM, Badeaa RI (2002) Effect of oil extracted from some medicinal plants of different mycotoxigenic fungi. Food Chem Toxicol 40:1669–1675

Sönmez AY, Bilen S, Albayrak M, Yılmaz S, Biswas G, Hisar O, Yanık T (2015) Effects of dietary supplementation of herbal oils containing 1,8-cineole, carvacrol or pulegone on growth performance, survival, fatty acid composition, and liver and kidney histology of rainbow trout (Oncorhynchus mykiss) fingerlings. Turk J Fish Aquat Sci 15:813–819

Soyadı Y, El-Ghany WA (2020) Phytobiotics in poultry industry as growth promoters, antimicrobials and Immunomodulators - a review. J World Poult Res. 10:571–579. https://doi.org/10.36380/jwpr.2020.65

Steiner T, Syed B (2015) Phytogenic feed additives in animal nutrition. In: Máthé Á (ed) Medicinal and aromatic plants of the world, vol 1. Springer, Dordrecht, The Netherlands, pp 403–423. ISBN 978-94-017-9810-5

Sun P, Schuurink RC, Caissard JC, Hugueney P, Baudino S (2016) My way: noncanonical biosynthesis pathways for plant volatiles. Trends Plant Sci 21:884–894

Sun J, Cheng Z, Zhao Y, Wang Y, Wang H, Ren Z (2022) Influence of increasing levels of oregano essential oil on intestinal morphology, intestinal flora and performance of Sewa sheep. Ital J Anim Sci 21(1):463–472. https://doi.org/10.1080/1828051X.2022.2048208

Swamy MK, Akhtar MS, Sinniah UR (2016) Antimicrobial properties of plant essential oils against human pathogens and their mode of action: an updated review. Evid Based Complement Altern Med 2016:3012462

Tampieri MP, Galuppi R, Macchioni F, Carelle MS, Falcioni L, Cioni PL, Morelli I (2005) The inhibition of Candida albicans by selected essential oils and their major components. Mycopathologia 159:339–345

Tariq S, Wani S, Rasool W, Bhat MA, Prabhakar A, Shalla AH, Rather MA (2019) A comprehensive review of the antibacterial, antifungal and antiviral potential of essential oils and their chemical constituents against drug-resistant microbial pathogens. Microb Pathog 2019:103580

Teixeira B, Marques A, Ramos C, Neng NR, Nogueira JM, Saraiva JA, Nunes ML (2013) Chemical composition and antibacterial and antioxidant properties of commercial essential oils. Ind Crop Prod 43:587–595

Unlu M, Ergene E, Unlu GV, Zeytinoglu HS, Vural N (2010) Composition, antimicrobial activity and in vitro cytotoxicity of essential oil from Cinnamomum zeylanicum Blume (Lauraceae). Food Chem Toxicol 48(11):3274–3280

Ultee A, Smid EJ (2001) Influence of carvacrol on growth and toxin production by Bacillus cereus. Int J Food Microbiol 64(3):373–378

Vieira PR, Morais SM, Bezerra FH, Ferreira PA, Oliveira ÍR, Silva MG (2014) Chemical composition and antifungal activity of essential oils from Ocimum species. Ind Crop Prod 55:267–271

Vimalanathan S, Hudson JB (2014) Anti-influenza virus activity of essential oils and vapors. Am J Essent Oils Nat Prod 2:47–53

Wade MR, Manwar SJ, Kuralkar SV, Waghmare SP, Ingle VC, Hajare SW (2018) Effect of thyme essential oil on performance of broiler chicken. J Entomol Zool Stud 6:25–28

Waldenstedt L, Inborr J, Hansson I, Elwinger K (2003) Effects of astaxanthin-rich algal meal (Haematococcus pluvalis) on growth performance, caecal campylobacter and clostridial counts and tissue astaxanthin concentration of broiler chickens. Anim Feed Sci Technol 108:119–132. https://doi.org/10.1016/S0377-8401(03)00164-0

Wang X, Zhang J, Xu Y, Chen J, Luo X (2019) Chemical composition, antioxidant, and antimicrobial activities of essential oil from ginger and garlic. J Food Qual 2019:1–9

Wei X, Cheng P, Feng W (2012) Study on the effect of anti-respiratory viruses of patchouli oil in vitro. J Pharm Clin Chin Mater Med 6:65–68

Westphalen MF, Carvalho PH, Oh J, Hristov AN, Staniar WB, Felix TL (2021) Effects of feeding rumen-protected capsicum oleoresin on growth performance, health status, and total tract digestibility of growing beef cattle. Anim Feed Sci Technol 271:114778

Wu JP, Zhou R, Liu LS, Casper DP, Lang X, Wang CL, Zhang LP, Wei S, Liu HB (2021) Growth performance, nutrient digestibility, blood parameters, and carcass characteristics by lambs fed an oregano and cobalt blend. Animal 15:100365

Yan L, Wang JP, Kim HJ, Meng QW, Ao X, Hong SM, Kim IH (2010) Influence of essential oil supplementation and diets with different nutrient densities on growth performance, nutrient digestibility, blood characteristics, meat quality and fecal noxious gas content in grower–finisher pigs. Livest Sci 128:115–122

Yang C, Chowdhury M, Huo Y, Gong J (2015) Phytogenic compounds as alternatives to in-feed antibiotics: potentials and challenges in application. Pathogens 4(1):137–156

Yang SK, Yusoff K, Mai CW, Lim WM, Yap WS, Lim SE, Lai KS (2017) Additivity vs synergism: investigation of the additive interaction of cinnamon bark oil and Meropenem in combinatory therapy. Molecules 22(11):1733. https://doi.org/10.3390/molecules22111733

Yang SK, Yusoff K, Ajat M, Thomas W, Abushelaibi A, Akseer R, Lim SHE, Lai KS (2019) Disruption of KPC-producing Klebsiella pneumoniae membrane via induction of oxidative stress by cinnamon bark (Cinnamomum verum J. Presl) essential oil. PLoS One 14(4): e0214326. https://doi.org/10.1371/journal.pone.0214326

Zeng Z, Zhang S, Wang H, Piao X (2015) Essential oil and aromatic plants as feed additives in non-ruminant nutrition: a review. J Anim Sci Biotechnol 6:1–10

Zhang T, Zhou YF, Zou Y, Hu XM, Zheng LF, Wei HK, Giannenas I, Jin LZ, Peng J, Jiang SW (2015) Effects of dietary oregano essential oil supplementation on the stress response, antioxidative capacity, and HSPs mRNA expression of transported pigs. Livest Sci 180:143–149

Zhang R, Wu J, Lei Y, Bai Y, Jia L, Li Z, Liu T, Xu Y, Sun J, Wang Y, Zhang K, Lei Z (2021) Oregano essential oils promote rumen digestive ability by modulating epithelial development and microbiota composition in beef cattle. Front Nutr 8:722557. https://doi.org/10.3389/fnut.2021.722557

Zhang W, Zhao J, Ma Y, Li J, Chen X (2022) The effective components of herbal medicines used for prevention and control of fish diseases. Fish Shellfish Immunol 126:73–83. https://doi.org/10.1016/j.fsi.2022.05.036

Zhou R, Wu J, Zhang L, Liu L, Casper DP, Jiao T et al (2019) Effects of oregano essential oil on the ruminal pH and microbial population of sheep. PLoS One 14(5):e0217054

Improving the Sustainability of Laying Hen Production Using Plant-Based Feed Additives

David Harrington, Matthew Leach, and Jan Dirk Van Der Klis

Abstract

Eggs are an affordable source of proteins, amino acids, minerals and vitamins. They contain an optimal balance of saturated and unsaturated fatty acids. Eggs provide a relatively inexpensive nutritious food source for a growing world population. Global egg production has increased from 50 to 85 million tonnes in the last 20 years and continues to increase. Increased number of rearing and laying facilities, bird selection to improve hen productivity under variable environmental conditions and the necessary increase in inputs like feed, energy, and water create challenges to meeting the burgeoning demand for sustainable production practices to help minimise environmental impact. While eggs have a relatively low environmental impact per kg of protein, compared to other livestock production, there is a continuous need to further reduce the environmental footprint. To this end, several factors need to be considered including economics, energy usage, global feedstuff transports, manure management, housing, length of laying cycle, production efficiency, sexing and good intestinal health, and disease management. Especially good intestinal health is essential to support sustainability measures, because it is central to nutrient utilisation, product quality and disease resistance of the hen, with feed additives having a key role to play. Amongst the variety of feed additives available for hens are plant-based additives that offer a potential advantage given their production is often rooted in sustainable practices. Documented effects of plant additives include enhancement of

D. Harrington (✉)
teamTWO Solutions, Silkeborg, Denmark
e-mail: david.harrington@teamtwo-solutions.com

M. Leach
Comparative Biology Centre, Newcastle University, Newcastle upon Tyne, UK

J. D. Van Der Klis
Arctium lappa Consultancy, Amersfoort, The Netherlands

G. Arsenos, I. Giannenas (eds.), *Sustainable Use of Feed Additives in Livestock*,
https://doi.org/10.1007/978-3-031-42855-5_19

feed digestion and nutrient utilisation, enhanced antioxidant and immune status, improved liveability, mineral utilisation, egg quality and production and reduced greenhouse gas emissions. Feed additives as a tool must have a clear benefit to improve efficiency/reduce environmental contaminants, while delivering a positive economic return. Of course, a return on investment is essential to ensure environmental sustainability of egg production.

Keywords

Laying hens · Plants · Phytogenics · Extracts · Environment · Sustainability, production

1 Introduction

Eggs are a highly nutritious and affordable food source that deliver both essential macro- and micronutrients and are amongst the lowest-cost animal source for proteins, vitamins and minerals (Nys and Sauveur 2004; Drewnowski 2010). This makes eggs an important food source for a growing global population which is predicted to reach 9.8 billion people by 2050, with the significant population growth occurring in the developing countries of South Asia and sub-Saharan Africa (Searchinger et al. 2018). Eggs are particularly important for developing nations where the selection of affordable protein can be limited.

1.1 Global Egg Production and Future Trends

Global egg production has increased during the last two decades almost twofold (Fig. 1) and continues to increase. Regionally, Asia has seen a 5% increase in production during this period, while EU-28 production contracted by a similar

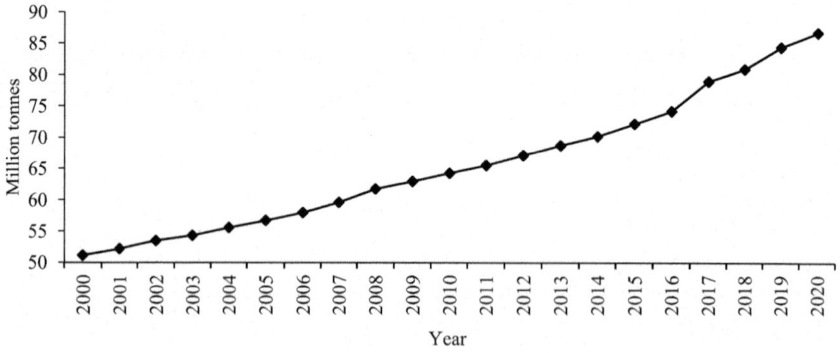

Fig. 1 Global egg production increase for the last 20 years (FAO 2022)

amount; Asia now accounts for approximately 62% of global shell-egg production, and Americas and EU-28 both account for 21% of global shell-egg (FAO 2022).

Estimates have suggested approximately 50 million additional laying chickens would have to go into production each year to meet the growing demand for eggs (Preisinger 2013), delivering approximately 1 million tonnes of additional eggs/year. Of course, this requires the construction of more rearing and laying facilities, genetic selection to improve hen productivity under variable environmental conditions and increased inputs of feed, energy, and water. In many countries, hen housing is moving towards non-cage systems, driven by a combination of societal, animal welfare and environmental concerns, as well as legislation to meet sustainability targets (Mench et al. 2011; EC 2021). For example, by 2025 it is estimated that 51% of egg production in the USA will be cage-free, an increase of 46% from 2011 (Mench et al. 2011; O'Keefe 2022), while the EU is intending to phase out cages from 2027 (EC 2021). However, the sustainability benefits of non-cage systems are still open to question, with concerns over air quality, economics, emissions and environmental footprint, while access to outdoor areas in some systems increases the risk to animal health (Xin et al. 2011; Molnár and Szőllősi 2020; de Olde et al. 2020). Alternative systems with access to orchards or increased backyard production have also been suggested as highly sustainable egg production systems, particularly for developing countries, although the economics of doing so are still considered challenging (Morris et al. 2018; Rocchi et al. 2019; Rajkumar et al. 2021; Soares et al. 2022). While there are differences between Northern and Southern hemispheres, urbanisation impacts on the development of new laying facilities with complaints of fine dust, emissions, and smell. This creates challenges to meeting the burgeoning demand for sustainable production practices.

It is also important to recognise that within the four dimensions of sustainability (societal, environmental, economic, enforcement (FAO 2012)), the relative importance of parameters can differ between developing and developed countries. For example, in Europe experts prioritise greenhouse gas (GHG) emissions and water usage as the most important environmental impact; in South America land-related impacts are the most important; and in Africa, it is also soil and land degradation, followed by land use and water usage (Paul et al. 2020). Consequently, when considering the environmental footprint, and potential solutions to reduce that footprint, regionality and stage of development are significant factors.

1.2 Egg Production and Its Environmental Footprint

Hen genotypes are capable of producing 500 eggs in a 100-week production cycle without comprising skeletal calcium, bone strength or welfare (Bain et al. 2016; Preisinger 2018; Anderson 2019; Dunn et al. 2021), with moulting practiced in some countries. End of lay or spent hens have traditionally been considered to have minimal value, either being rendered into protein meal for feed or turned into pet food There is a growing trend to use spent hens for human consumption or other valorised products (Fan and Wu 2022) and to add value to male chicks that are currently of limited use (Giersberg and Kemper 2018; Gautron et al. 2021) to enhance the sustainability profile of egg production. While a shift to hybrid strains that exhibit both strong laying characteristics and good carcass characteristics could bring added value, they produce less eggs and are less feed efficient which would likely adversely impact the environmental footprint of egg production and be economically unviable (Gautron et al. 2021).

1.2.1 GHG Emissions from Egg Production

Improved efficiencies in supply chain activities, changes in feed composition for both pullet laying hens and improvements in resource efficiencies, animal health, and productivity at the level of pullet and egg production have led to a substantial reduction in the environmental impact of cage egg production (Pelletier 2018). For example, in Canada, despite egg production increasing approximately 51% from 1962 to 2012, the total life GHG emissions were 57% lower.

While comparing life cycle assessments (LCA) of production systems is challenging (Costantini et al. 2021), egg production has lower GHG per kg protein versus beef, pork or chicken (Table 1), although GHG estimates have the potential to be skewed by geographical location or datapoints from producers with large impacts (Poore and Nemecek 2018). Furthermore, terrestrial acidification and eutrophication potential are also lower for egg production than other livestock production systems (Table 1). Within the egg production process, egg laying has by far the biggest environmental impact (79.0%) compared to rearing (19.1%) or packaging of eggs (1.9%) (Estrada-González et al. 2020). Countries are increasingly legislating for management of emissions. For example, in the European Union, Directive 2016/2284/EU requires the reduction in ammonia emissions by 19% per year from 2030, using 2005 as a reference (EC 2016).

1.2.2 Energy and Water Usage

Maintaining the optimum environment for the birds is crucial for efficient performance and egg laying while pullet rearing facilities require significant inputs of water, gas and energy to run (Estrada-González et al. 2020). While egg production has become more energy efficient, e.g. the Canadian egg industry saw reductions of 10%, 71% and 53% in energy land and water use, respectively, over 50 years (Pelletier 2018), the environmental impact from primary energy is still a

Table 1 Estimated global variation in GHG emissions, land use, terrestrial acidification and eutrophication potential for selected terrestrial livestock production systems per 1 kg/protein (adapted from Poore and Nemecek (2018))

Production	GHG (CO_2 eq kg^{-1})	Acidifying (SO_2 eq g^{-1})	Eutrophying (PO_4^{3-} eq g^{-1})	Freshwater (1000 L)	Land use (m^2)
Beef	498.9	1598.9	1511.6	174.2	1636.0
Pork	76.1	881.7	472.1	413.2	107.3
Poultry meat	56.9	591.3	281.1	81.9	70.6
Eggs	42.1	48.0	196.1	162.1	56.5

consideration. LCA of on-farm energy inputs suggest anywhere from 1.05 to 15% are from electricity (Leinonen and Kyriazakis 2016; Sasanya and Olaifa 2022). Despite improvements in efficiencies, concerns over the impact of energy generation on the climate and supply chain vulnerability to shock events (e.g. disease outbreaks or geopolitical events) (Hamilton et al. 2020) push for both further efficiency and a reduced reliance on single energy sources. Direct energy inputs in net zero energy barns (NZEB) (buildings incorporate renewable energy generation to offset non-renewable energy used) were found to be 25% lower than conventional houses, while environmental payback varied from approximately 1.4–20.7 years, depending on renewable energy sources (Li et al. 2021). Compared to chicken meat production, egg production has nearly twice the water requirement (and therefore greater environmental impact for water usage) due to the water requirement of the egg (Poore and Nemecek 2018).

1.2.3 Manure Management

The principal source of GHGs and environmental emissions in egg production are predominantly nitrogen and phosphorus-based, rather than carbon; hens contribute little to methane outputs (MacLeod 2011). Poultry house design and management are two key factors that affect ammonia emissions and efficiency of reduction (Hayes et al. 2013; Wang-Li et al. 2013). While on-farm manure mitigation strategies such as manure storage, air scrubbers or increase mechanical ventilation are considered amongst the most effective to reduce emissions (Ershadi et al. 2021), their economic feasibility and potential energy use have been questioned (de Olde et al. 2020).

1.2.4 Feed

Feeding hens has the greatest cost and environmental impact and is dependent on the origin of energy and protein sources (land use, deforestation and transport) and on the efficiency of nutrient utilisation (Pelletier 2018). There are several strategies to improve the sustainability of feed: (1) optimisation of nutrient balance; (2) maximise nutrient availability; (3) reduce nutrient excess or safety margin; and (4) alternative ingredients (MacLeod 2011; Leinonen and Kyriazakis 2016). Optimising the nutrient balance of the feed is actively undertaken in laying hens and diets are formulated to meet the needs of the hen at least cost and to maximise egg output (Bryden et al. 2021); excess nutrients and safety margins are substantially reduced (MacLeod 2011). Feed enzymes are routinely used to improve nutrient availability, while balancing the ration with synthetic amino acids is common practise. Alternative crops to soya meal such as beans, peas or rapeseed can be used where the environmental footprint is lower (Leinonen and Kyriazakis 2016) or the cereal is more cost effective. Dekker et al. (2013) suggested that using locally procured ingredients, even if a slight decrease in performance was observed, led to substantial reductions in the ecological impact of egg production through reductions in global warming potential, nitrogen deficit and occupation and energy use.

1.2.5 Disease Management

Good disease management and intestinal health are essential to support sustainability measures because it is central to the overall efficiency of egg production by the hen, with feed additives having a key role, particularly in alternative systems (Jeni et al. 2021). A wide selection of additives has been used in laying hens including enzymes, prebiotics, e.g. yeast, probiotics, organic acids and plant-based feed additives (Świątkiewicz et al. 2013). The application of feed additives based on either terrestrial or aquatic plants is a growing field in animal production, including laying egg production (Karásková et al. 2016).

2 Plant-Based Feed Additives

Approximately 60% of the plants used for medicinal or aromatic purposes are cultivated via agroforestry or controlled cultivation (science-based, specifically controlled and designed to optimise active phytocompounds), while approximately 20% are wild-harvested (Brinckmann et al. 2022). Seaweed is the only biomass that can be cultivated and harvested daily all year round (van den Burg et al. 2013); harvesting and cultivation are often seasonal. Controlled cultivation can challenge the ethos of sustainability by potentially disrupting ecosystems, despite the requirement of several regulatory authorities that wild species should be cultivated (Giannenas et al. 2020). Sustainable harvesting should be the most important conservation strategy for wild-harvested plant species as it contributes to the economy and long-term stability of local communities (Giannenas et al. 2020). Plant-based additives can also be manufactured by extraction of agricultural by-products, e.g., fruit pomace, and artichoke leaves and bracts (Sosnówka-Czajka and Skomorucha 2021; Ligas et al. 2021; Brunetti et al. 2022). Using by-products rich in phytocompounds of interest would both reduce the total amount of waste biomass produced by the agricultural industry (Brunetti et al. 2022) and contribute to sustainability and development of the circular economy.

There has been a significant focus on aromatic herbs for feed additives (Franz et al. 2020), but aquatic plants such as seaweeds and other non-aromatic plants have also been considered (Karásková et al. 2016; Mohammadi Gheisar and Kim 2018; Kulshreshtha et al. 2020; Michalak and Mahrose 2020) (Table 2). Plant-based additives can comprise the whole plant or sections of plant, e.g. seeds or leaves or plant extracts. Extracts, including essential oils, oleoresins and liquid extracts, can be prepared via several methods including distillation, supercritical CO_2 extraction and maceration. Essential oils and distillation extracts from aromatic plants, are one of the more predominant, complex mixtures of phytocompounds used in agriculture (Reddy 2019).

Table 2 Example of terrestrial and aquatic plant species, some of their principle phytocompound and reported effects

Common name	Scientific name	Family	Principle phytocompounds	Reported effect	References
Black cumin	*Nigella sativa L.*	Ranunculaceae	Thymoquinone, thymol, γ-Terpinene, β-caryophyllene	Anti-inflammatory, antioxidant, antitumour, antibacterial	Aksu et al. (2021), Ermumcu (2022)
Damask rose	*Rosa damascena*	Rosaceae	Stearoptene, nerol, geraniol, citronellol	Antimicrobial, antioxidant, antitussive, hypnotic	Boskabady et al. (2011)
Garlic	*Allium sativum*	Amaryllidaceae	Allicin, quercitin, gallic acid	Antifungal, antioxidant, antimicrobial, anti-schistosomal, anti-inflammatory	Metwally et al. (2018), Kothari et al. (2019)
Cananga	*Cananga odorata*	Annonaceae	Linalool, beta-caryophyllene, farnesene	Antimicrobial, insect repellent, antimelanogenic	Tan et al. (2015
Angelica	*Angelica archangelica*	Apiaceae	Alpha-pinene, alpha-phellandrene camphene	Anti-anxiety, anti-convulsant, antiviral activity, cholinesterase inhibitory, anti-inflammatory	Kaur and Bhatti (2021)
Anise	*Pimpinella anisum*	Apiaceae	Methylcavicol, anisaldehyde, aniscetone	Antimicrobial, antifungal, antiviral, antioxidant, muscle relaxant, analgesic, anticonvulsant	Shojaii and Abdollahi Fard (2012)
Caraway	*Carum carvi*	Apiaceae	D-carvone, limonene, dihydrocarvone	Anti-aflatoxigenic, antioxidant, antimicrobial	Mahboubi (2019)
Asian ginseng	*Panax ginseng*	Araliaceae	Ginsenosides, sterols, flavonoids	Immunomodulatory, anticarcinogenic, neurotrophic, antioxidant	Wee et al. (2011), Mehta et al. (2021)
Yucca	*Yucca schidigera*	Asparagaceae	5β-spirostanic saponin, resveratrol, yuccaol	Cytotoxic, anti-inflammatory, hypocholesterolaemic	Cheeke (2000), Patel (2012), Jiménez et al. (2021)
Camomile	*Matricaria recutita*	Asteraceae	Chamazulene, alpha-bisabolol, pharnesene, bisabolol oxide, apigenin	Antioxidant, antimicrobial, antiplatelet, antispasmodic, anxiolytic	McKay and Blumberg (2006)

(continued)

Table 2 (continued)

Common name	Scientific name	Family	Principle phytocompounds	Reported effect	References
Echinacea	*Echinacea purpurea*	Asteraceae	Caffeic acid, echinacoside, alkylamide, cichoric acid, α-pinene, limonene	Anti-inflammatory, antioxidant, antifungal, anti-osteoporotic	Burlou-Nagy et al. (2022)
Sweet wormwood	*Artemisia annua*	Asteraceae	Artemisinin	Anti-protozoal	Waizel-Bucay and Waizel-Haiat (2019)
Tarragon	*Artemisia dracunculus*	Compositae	Methyl chavicol, anethole, camphene, sabinene, anisol	Analgesic, hypnotic, antiepileptic, anti-inflammatory, antipyretic, anthelminthic	Ekiert et al. (2021)
Peacocks ear	*Padina pavonica*	Dictyotaceae	O-cresol, p-cresol, phlorotannin	Antioxidant, antimicrobial osteogenic	Ansari et al. (2019), Minetti et al. (2019), Hakim and Patel (2020)
Irish moss	*Chondrus crispus*	Gigartinaceae	Galactolipids, floridoside, isothionic acid, taurine, phenylalanine, l-citruline	Antimicrobial, antioxidant	Sangha et al. (2013), Rodríguez et al. (2021)
St John's wort	*Hypericum perforatum L.*	Hypericaceae	Hypercin, hyperforin, naphthodianthrones, quercitrin	Antidepressant, antioxidant, antipyretic, anxiolytic, anti-inflammatory	Raziq et al. (2016), Jarzębski et al. (2020)
Peppermint	*Mentha piperita*	Lamiaceae	Menthol, menthone, pinene, menthyl acetate	Carminative, anti-inflammatory, analgesic, antispasmodic	Sharifi-Rad et al. (2017)
Rosemary	*Rosmarinus officinalis L.*	Lamiaceae	1,8-cineole, camphor, apigenin, rosmadial	Antiseptic, vaso-dilator, hepatoprotector	Waizel-Bucay and Waizel-Haiat (2019)
Sage	*Salvia officinalis*	Lamiaceae	Thujone, cineol, camphor, bomeol	Antipyretic, antimicrobial, antidiarrheal	Waizel-Bucay and Waizel-Haiat (2019)
Oregano	*Origanum vulgare*	Lamiaceae	Carvacrol, thymol	Antimicrobial, antioxidant, anti-inflammatory, appetite stimulant	Waizel-Bucay and Waizel-Haiat (2019)
Fenugreek		Leguminosae	Elemene, campesterol caempferol	Mucolytic, antimicrobial, anti-inflammatory, analgesic	Waizel-Bucay and Waizel-Haiat (2019)

	Trigonella foenum-graecum				
Moringa	*Moringa oleifera*	Moringaceae	Quercetin, kaempferol, γ-tocopherol	Anti-inflammatory, antimicrobial larvicidal, hypocholesterolaemic	Ferreira et al. (2008)
Clove	*Syzygium aromaticum*	Myrtaceae	Eugenol, beta-caryophyllene, vanillin, kaempferol	Antipyretic, anxiolytic, antiemetic, antimicrobial, anti-inflammatory	Batiha et al. (2020)
Eucalyptus	*Eucalyptus* sp.	Myrtaceae	Cineole, phellandrene, piperitone	Antipyretic, antiseptic, antitussive	Waizel-Bucay and Waizel-Haiat (2019)
Plume-poppy	*Macleaya cordata*	Papaveraceae	Sanguinarine, chelerythrine, quercitin	Anti-inflammatory, anticarcinogenic, antimicrobial	Lin et al. (2017)
Pepper	*Piper nigrum*	Piperaceae	Limonene, sabinene, pinene, caryophyllene	Analgesic, antimicrobial, anti-inflammatory, antispasmodic	Waizel-Bucay and Waizel-Haiat (2019)
Lemongrass	*Cymbopogon citratus*	Poaceae	Neral, geranial, mycrene, genariol, citronellol, α-oxobisabolene	Antioxidant, dermatotoxic, antimalarial, anti-inflammatory, insecticidal	Oladeji et al. (2019)
Gu sui Bu	*Aglaomorpha fortunei/ Drynaria roosii*	Polypodiaceae	Naringin, esculetin, catechin, catechin, protocatechuic acid, campesterol	Anti-osteoporotic, anti-adipogenic	Jang et al. (2019)
Soap bark treee	*Quillaja saponaria Molina*	Quillajaceae	Quillaic acid, quercitrin, tanins	Cytotoxic, anti-inflammatory, antimicrobial, expectorant, antiparasitic	Waizel-Bucay and Waizel-Haiat (2019)
Bitter orange	*Citrus aurantium*	Rutaceae	Linalool, naringin, linoleic acid, limonene, p-coumarin	Antioxidant, antimicrobial, cytotoxic	Maksoud et al. (2021)
Star Anise	*Illicium verum Hook*	Schisandraceae	Trans-anethole, estragole, limonene, shikimic acid	Insecticidal, antibacterial, antifungal, anti-inflammatory	Wei et al. (2014), Sharafan et al. (2022)
Red string seaweed	*Sarcodiotheca gaudichaudii*	Solieriaceae	Carrageenan, alginate, fucosterol, fucoidans	Antiviral, antimicrobial, antiaging	Wells et al. (2017)

(continued)

Table 2 (continued)

Common name	Scientific name	Family	Principle phytocompounds	Reported effect	References
Oil-seed camellia	*Camellia oleifera*	Theaceae	Oleiferasaponin, oleiferaol, camelliaside, protocatechuate acid sesamin	Anti-inflammatory, hypolipidemic, antimelanogenic,	Lee and Yen (2006), Luan et al. (2020)
Ginger	*Zingiber officinale*	Zingiberaceae	6-gingerol, 8-gingerol, 10, gingerol	Antioxidant, anti-inflammatory, antimicrobial, anticancer, neuroprotective, antiemetic	Mao et al. (2019)
Turmeric	*Curcuma longa*	Zingiberaceae	Curcumin	Antimicrobial, antioxidant, anti-inflammatory, appetite stimulant	Memarzia et al. (2021)
Casssia	*Cinnamomum cassia*	Lauraceae	Cinnamaldehyde, trans-cinnamaledehyde, linalool, eugenol, cinncassin E	Antimicrobial, anticarcinogenic, antiviral, cytoprotective	Zhang et al. (2019)

2.1 Plants Are Complex

Plants and plant extracts (phytocompounds) are compositionally complex and typically comprise bioactive compounds including secondary plant metabolites, as well as vitamins, and polyunsaturated fatty acids (Franz et al. 2020). Secondary metabolites are produced for several reasons: response to environmental stressors, protection against pathogens or predators or attraction of pollinators. The composition of secondary metabolites can vary according to several parameters including time of year, location and soil condition (Reddy 2019). While secondary plant metabolites are often the focus of biological activity of plants, other compounds such as carbohydrates, e.g. carageens from red algae are also important (Menaa et al. 2021) and require further investigation in hens.

Phytocompounds generally exhibit broad range of activities in the hen including antifungal, antibacterial, antioxidant, antiviral and anti-inflammatory (Bakkali et al. 2008; de Armas-Ricard et al. 2019; Cueva et al. 2020). The diversity of all compounds in plant-based feed additives is high, especially if taking into consideration the inter- and infraspecific variation (Franz et al. 2020; Ligas et al. 2021).

2.1.1 Categorisation of Secondary Metabolites

The biologically active phytocompounds can be broadly categorised as follows:

Phenols are the largest group of secondary plant metabolites and range from simple compounds with a single aromatic ring to complex compounds, polyphenols. (Teoh 2015). Herbs and spices typically have the highest polyphenol content by weight (Pérez-Jiménez et al. 2010). Phenolics can be divided into soluble compounds, e.g. flavonoids, phenolic acids and insoluble compounds, e.g. condensed tannins and lignins. Oxygenated phenolics include alcohols (e.g. borneal), aldehydes (e.g. cinnamaldehyde) and ketones (e.g. carvone) (Christaki et al. 2020). In algae, the polyphenols are classified as phloroglucinols and phlorotannins, based on an aromatic phenyl ring with three hydroxyl groups (Menaa et al. 2021).

Flavonoids comprise an aromatic ring that absorbs ultraviolet radiation giving rise to plant pigments. Flavonoids are known for the antioxidant activity, e.g. quercetin, but they have several other activities including antimicrobial, anti-inflammatory, antitumour, and both oestrogenic and anti-oestrogenic activities (Teoh 2015).

Stilbenoids are a group of phenolic compounds that have a stilbene as a backbone but differ in the constituents on the carbon ring. Examples include resveratrol and gnetol, used in folk medicine for arthritis (Akinwumi et al. 2018).

Terpenes comprise a five carbon isoprene ring and the number of rings determines their classification: 1–6 and 8 rings, hemiterpenes, monoterpenes, sesquiterpenes, diterpenes, sesterpenes, triterpenes and tetraterpenes (Christaki et al. 2020). Linalool is a monoterpene and is the major component in the scent of orchids, artemisinin is a sesquiterpene (3 rings) and has antiprotozoal activity; carotenoids are tetraterpenes (Teoh 2015).

Triterpenoid and steroidal glycosides are similar in structure but synthesised through different pathways and are collectively known as saponins. Saponins contain both lipophilic (either a triterpenoid or steroidal moiety known as the aglycone) and hydrophilic groups (sugars). Examples include ginsenosides from ginseng or digitalis from foxglove (*Digitalis purpurea*) (Teoh 2015).

Alkaloids contain one or more nitrogen atoms, usually in a heterocyclic ring and are derived from amino acids. They are bitter in taste and many alkaloids act on the nervous system and are stimulants, e.g. morphine or cocaine (Teoh 2015).

3 The Role of Plants in Laying Hen Production

Authors have been circumspect in drawing over-arching conclusions on the benefits of plants in laying hen production (Bozkurt et al. 2014; Harrington et al. 2020). Inter-study comparisons can be confounded by numerous parameters including duration of study, age of birds, breed of bird, feed, source of plant, whole plant or extract thereof, extract type, composition, geographical location and housing. The challenge is to define a robust methodology that can facilitate reliable inter- and intra- study comparison to fully evaluate plant-based additives in hen production (Harrington et al. 2020). Nonetheless, a meta-analysis by Darmawan et al. (2022) concluded that phytocompounds (water, ethanol and petroleum ether extracts of leaf, bulb, seed, or peel) fed to 19–74-week-old laying hens favourably influenced productivity, egg quality, blood constituents, antioxidant and immunological parameters in laying hens.

Considering the potential benefits of phytocompounds from the perspective of sustainable egg production can shed new light on how these compounds might be used strategically. In the following sections, selected examples of plants or phytocompounds and their potential role in sustainable egg production will be given within the broad context of microbiota management; feed digestion and nutrient utilisation; mineral utilisation; egg quality and production; antioxidant and immune status; and reduced GHG emissions.

3.1 Gut Microbiota Management and Gut Health

The birds' microbiome includes the gut, skin, oral cavity, respiratory tract and reproductive tract (which differs between the upper and lower tract) (Kropáčková et al. 2017; Lee et al. 2019; Engel et al. 2020; Su et al. 2021; Wen et al. 2021; Kursa et al. 2022), but the gut microbiome is the most studied. The gut microbiota and its interaction with the host is essential for bird health, digestion and absorption of nutrients, pathogen exclusion and development of the immune system (Shang et al. 2018; Broom and Kogut 2018). The normal gut microbiota of the laying hen stabilizes in the mid- and late-phase of lay and is generally dominated by the phyla Proteobacteria, Firmicutes, Bacteroidetes, Fusobacteria and Actinobacteria. The microbial composition differs with intestinal location, bird genotype, age, stage of production and production system (Seidlerova et al. 2020; Khan et al. 2020; Wiersema et al. 2021; Joat et al. 2021) and can be influenced by several factors including diet, pathogens, antimicrobials, feed restriction, mycotoxins and vaccination (Kers et al. 2018; Artdita et al. 2021; Borey et al. 2022; Yuan et al. 2022). The composition of the gut microbiota can also affect the composition of the odour of poultry manure, for example hydrogen sulphide concentration (Huang et al. 2019). An unbalanced gut microbiota can induce inflammation, poor nutrient digestion, leaky gut or other gut-related pathologies, while early insults on the developing gut microbiota, e.g. from antibiotic use during rearing can affect later antibody responses during lay (Simon et al. 2016). In older hens, follicular degradation is partly associated with an unbalanced gut microbiota and increased oxidative stress (Wang et al. 2021). Forced moulting can lead to an increase in *Clostridium perfringens* and coliform counts with a concurrent drop in *Lactobacillus* spp. (Bozkurt et al. 2016).

3.1.1 Modulating the Gut Microbiota

Powdered extracts of leaves from Bilberry (*Vaccinium myrtillus*) and Walnut (*Juglans regia* L.) (rich in antioxidant and phenolic compounds, respectively) fed to 32-week-old hens for 4 weeks, reduced the level of Bacteroidetes and *Enterobacteriaceae* and increased the abundance of firmicutes and lactobacilli (Popescu et al. 2020). In breeder hens fed grape seed extract rich in procyanidin flavonoids, the relative abundance of *Bifidobacteriaceae*, *Lactobacilliaceae* and *Lachnospiraceae* were shown to increase after only 2 weeks (Grandhaye et al. 2020). Oregano oil fed to hens over a 7-week period was shown to increase numbers of lactobacilli and bifidobacteria while lowering levels of *Escherichia coli* and Salmonella (He et al. 2017). Hens fed thyme or curcumin (*Curcuma longa*) have also been reported to have lower intestinal coliform, enterococci, yeasts and fungi counts (Arpášová et al. 2013; da Rosa et al. 2020). Bolukbasi and Erhan (2007) saw no difference in coliform counts after feeding thyme to hens for 12 weeks, and nor did Boka et al. (2014) when hens were fed black cumin seeds (*Nigella sativa* L.). While Kim and Kang (2015) saw no change in *E. coli* counts in 28-week-old hens fed chlorella by-products, the authors did observe an increase in *Lactobacillus* in ileal contents. Abad et al. (2021) reported an increase in lactobacilli numbers in laying hens fed garlic extract. While feeding oregano essential oil to post-moult hens did not significantly alter the caecal microbiota, a trend in decreasing numbers of coliforms and increasing *Lactobacillus* spp. was observed (Bozkurt et al. 2016).

3.1.2 Reducing the Impact of Pathogens

Food borne pathogens: *Salmonella* Enteritidis infected hens fed feed supplemented with the red seaweeds *Chondrus crispus* or *Sarcodiotheca gaudichaudii*, had a significant reduction in caecal colonisation by *Salmonella* Enteritidis with concurrent increased levels of *Lactobacillus acidophilus* (Kulshreshtha et al. 2017). The reduction in Salmonellae could be related to interference with motility and virulence of *S.* Enteritidis, thereby blocking epithelial attachment or a probiotic effect (Kulshreshtha et al. 2017). A reduction in the number of *Salmonella*-positive tissue samples from *Salmonella*-infected hens fed trans-cinnamaldehyde, oregano or eugenol has also been reported, with birds remaining clear of salmonella for 30 days post inoculation (Ordonez et al. 2008; Upadhyaya et al. 2015; He et al. 2017). Zhao et al. (2022) suggested that eugenol could lessen *Salmonella* infections through the relief of inflammatory stress, stabilisation of the mucosal barrier and inhibition of bacterial adhesion and invasion.

Intestinal infections: Muthamilselvan et al. (2016) reviewed over 68 plants and phytocompounds reported to control protozoal infections in birds. Examples included artemisinin from *Artemisia annua*; condensed tannins from pine bark; papain from papaya (*Carica papaya*); essential oils extracted from oregano (*Origanum compactum*), rosemary (*Rosmarinus officinalis*), curcumin, Madeira vine (*Anredera cordifolia*); and blue evergreen hydrangea (*Dichroa febrifuga*); and saponins from *Yucca schidigera* and *Quillaja saponaria*. Several modes of action have been proposed to explain the mitigation of *Eimeria* infection in poultry including modulating host immunity, prebiotic activity, improving intestinal integrity and direct antiprotozoal activity via inhibition or impairment of parasite invasion, replication or development (Muthamilselvan et al. 2016; El-Shall et al. 2022). Cinnamaldehyde has been shown to significantly lower *Brachyspira intermedia* counts in hens (Verlinden et al. 2013), while tea tree (*Melaleuca alternifolia*) essential oil (TTO) has been suggested as an alternative to antibiotics. When used at 100 mg/kg feed, hens fed TTO had the same level of recovery from *Mycoplasma synoviae* infection as hens given tetracycline, oxytetracycline or chlortetracycline (Puvača et al. 2020).

3.1.3 Improving Intestinal Integrity

Numerous plant extracts and phytocompounds including crushed cumin seeds, sea buckthorn (*Hippophae rhamnoides*), oregano (*Origanum* spp.), red seaweed (*Chondrus crispus*), curcumin, polysaccharides from the seaweed *Enteromorpha* have been documented to improve intestinal morphology in birds such as increased villus height, increased goblet cell numbers or proliferation of lamina propria lymphatic follicles (Boka et al. 2014; He et al. 2017; Kulshreshtha et al. 2017; Dalal et al. 2018; Guo et al. 2020; Feng et al. 2021; Panaite et al. 2022). Polyphenols (magnolol and its isomer honokiol) from the root and bark of *Magnolia officinalis* were shown by Chen et al. (2022) to increase jejenul and ileal villus height and increase expression of genes associated with tight junction proteins such as occludin and zonula occludens (ZO)-1 and maintenance of the intestinal barrier. However, Lokaewmanee et al. (2014) found no significant change in intestinal morphology in hens fed 0.1% extracts from red clover (*Trifolium pratense*) and garlic (*Allium sativa*), while Mousa et al. (2019) also found no evidence of change in intestinal morphology after feeding up to 1% garlic or thyme.

3.1.4 Without Good Gut Health and Balanced Microbiota, Egg Production Is Unsustainable

Good gut health and the microbiota is the key to success of many initiatives for sustainability, and the evidence suggests plant-based additives are a useful tool to support the hen's microbiota. Reducing mortality has a significant impact on the environmental footprint of egg production; a 5% reduction in overall mortality can reduce energy demands and GHG emissions by 6–7% (Weeks et al. 2016). Improved gut health and a balanced gut microbiota can reduce disease susceptibility, lower the risk of infection from zoonotic pathogens, improve vaccination response, maintain or improve feed efficiency and lower the need for antibiotics. Increase in zoonotic infections such as Salmonella have been associated with environmental extremes such as heat or extreme rainfall (Morgado et al. 2021) and the changing global climate would suggest extreme weather events and therefore increased zoonotic outbreaks are more likely. Balancing the hens' microbiota to reduce pathogen load using phytocompounds, particularly when phytocompounds have been shown to boost antioxidant status during environmental stress, would seem a sensible approach.

There are some indications that hens in cage free systems with outdoor access have less a diverse gut microbiota and poorer intestinal morphology versus caged birds (Wiersema et al. 2021). Birds in conventional cages receive optimum nutrient balance and higher substrate availability versus cage-free, outdoor birds, contributing to a more diverse gut microbiota (Seidlerova et al. 2020). Furthermore, birds are transferred from rearing to laying accommodation around 16–18 weeks during a period when the microbiota also changes (Joat et al. 2021). Strategies for sustainable egg production should include phytocompounds to support a balanced, diverse microbiota in cage-free systems or at critical times, e.g. transfer or moulting.

Dai et al. (2022) proposed that focusing on intestinal microbiota could be a useful tool for managing egg quality and safety in the future. Certainly, as our understanding of the interaction between phytocompounds and the hen microbiota increases, this would be a promising strategy. It will also be essential to help reduce use of antibiotics and to use phytocompounds preventatively rather than therapeutically as antibiotics are used now.

3.2 Feed Digestion and Nutrient Utilisation

Herbs and spices can affect feed digestion and nutrient utilisation in multiple ways, e.g. via stimulation of: (1) the pancreas to produce lipase, amylase or proteases; (2) the liver to produce bile acids, which are important for fat digestion; (3) the differentiation of the enterocytes, resulting in a higher brush border enzyme activity; or (4) changing the rate of passage of feed through the intestinal tract, increasing the time available for digestion and absorption. The utilisation of absorbed nutrients is dependent on the health status of the hen. Stress and intestinal or systemic infections will reduce the efficiency of nutrient utilisation. Moreover, the egg production cycle causes a diurnal variation in nutrient requirements. Effects of herbs and spices as antioxidant, anti-inflammatory and immunomodulatory actives are dealt with in the following paragraphs.

Platel and Srinivasan (2004) reviewed the effects of spices on nutrient digestion in animals. They showed that capsaicin, curcumin, ginger and piperine stimulated the production of pancreatic enzymes, whereas bile flow was increased by fenugreek and mustard and brush border enzymes by coriander and onion. Unfortunately, to our knowledge, no systematic studies of phytocompounds on digestibility have been carried out with laying hens. Scientific publications on the effect of phytocompounds on digestibility in laying hens are mostly based on commercial products, not allowing to conclude on the effect of single actives.

Yu et al. (2019) showed in a dose response study of star anise (*Illicium verum*) oil (SAO) using 29-week-old Hyline brown laying hens that 400 mg SAO/kg was the optimal inclusion level to maximise the apparent digestibility of crude protein and first limiting amino acids lysine, methionine and threonine. They showed a significant quadratic effect of the SAO inclusion level on nutrient digestibility. Although the digestibility of fat reached a maximum at the same SAO inclusion level, this effect was only numerical. Abdel-Wareth and Lohakare (2020) tested the effect of peppermint (*Mentha piperit* L.) oil (PO) on nutrient digestibility in 32-week-old Bovans hens. They used increasing dose levels from 0 to 296 mg PO/kg, and showed a linear increase in the digestibility of crude protein and fat with PO inclusion level. Dietary supplementation of 100 mg/kg oregano oil (OA) stimulated the activity of amylase and trypsin in the small intestinal contents of 30-week-old Hyline laying hens, being significantly higher than 50 mg OA/kg and numerically higher than 150 mg OA/kg (He et al. 2017). They also observed a significantly higher GLUT2, SGLT1 and PepT1 gene expression in duodenal and jejunal tissue, indicating increased levels of intestinal transporters stimulating nutrient absorption. Dietary *Moringa oleifera* leaves supplementation in 64-week-old Hyline Brown laying hens decreased excreta ammonia concentration and improved feed efficiency (Abdel-Wareth and Lohakare 2021) with increasing dose levels up to an inclusion level of 6 g/kg. The maximum *Moringa oleifera* leaves inclusion was 9 g/kg.

The polyphenols magnolol and honokiol from *Magnolia officinalis* work synergistically in the hen to regulate hepatic lipid metabolism leading to lower liver triglyceride levels (Chen et al. 2022), important to help manage fatty liver syndrome in hens When nettle (*Urtica cannabina*) was fed to laying hens, the total n-3 PUFA concentration in the yolk was increased, while the ratio of n-6/n-3 was reduced in the nettle group (Zhang et al. 2020). Supplementing 29-week-old hens diet with the algae *Spirulina platensis* or turmeric at 0.3 and 0.6%, respectively, improved feed efficiency (Mobarez Samia et al. 2018), although in older hens (60 weeks) *Spirulina platensis* at up to 2% had no effect on FCR (Curabay et al. 2021).

3.2.1 Improving Feed Utilisation Can Reduce the Environmental Footprint

Feed has the largest impact on the environmental footprint of egg production, A laying hen at 90% production produces 0.05 Phosphorus, 0.3 Nitrogen and 3 kg CO_2 eq kg^{-1} protein (Flachowsky et al. 2018); as egg production decreases, so emissions increase. Improving nitrogen utilisation, for example, can help to lower the inevitable emissions increase as production drops. Increased feed use efficiency could have significant impact on lowering emissions (Turner et al. 2022). Alternative protein and energy sources are increasingly considered in egg production to both lower

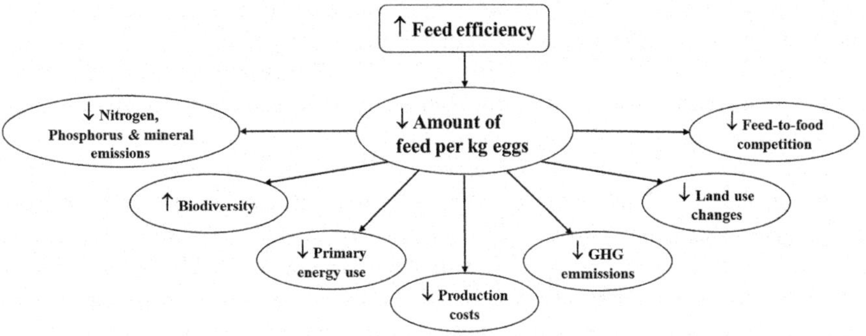

Fig. 2 Potential sustainability benefits deriving from feed efficiency improvements in laying hens (adapted from Zampiga et al. 2021)

environmental footprint and economic reasons. Abín et al. (2018) suggested the replacement of palm oil with cottonseed could reduce the carbon footprint by 2.3 kg CO_2 eq kg^{-1} eggs. Using phytocompounds to stimulate enzyme production and intestinal nutrient transporters could facilitate the use of alternative ingredients in layer feed. Stimulating improved nutrient digestion and improved feed conversion can lower reduce many aspects of the environmental footprint (Fig. 2). Where hybrid strains are used, some genotypes preferentially select lower protein diets, suggesting there is scope for phytocompounds to possibly improve feed efficiency in these genotypes.

3.3 Mineral Utilisation

Calcium for eggshell formation is obtained from the diet via absorption and mobilisation from the medullary bone. Compromises in calcium balance leads to bone weakness, low egg production, poor eggshell quality and reduced bird welfare. Abe et al. (1982) showed that the efficacy of 1-alpha hydroxylase in kidney homogenates from aged laying hens is reduced, resulting in reduced calcium absorption and eggshell quality. These observations were recently confirmed by Pampouille et al. (2022), who demonstrated that the serum content of 1,25 $(OH)_2D_3$ was reduced in laying hens with poor shell quality compared to hens with good shall quality, whereas the intermediate metabolite $25(OH)D_3$ was not different between both groups. This indicated impaired activation of $25(OH)D_3$ in the metabolically active form of vitamin D_3. Dietary supplementation with 1,25 $(OH)_2D_3$ as glycosides from *Solanum glaucophyllum* leaves, a plant that naturally contains this active substance, effectively improved calcium metabolism and shell quality in aged laying hens (Van der Klis and Pedrosa 2022). Abdel-Wareth and Lohakare (2020) showed that peppermint oil increased eggshell strength and thickness, while dietary *Moringa oleifera* leaves supplementation improved eggshell thickness between 64 and 72 weeks of age in Hyline Brown laying hens up to an inclusion level of 6 g/kg (Abdel-Wareth and Lohakare 2021).

Essential oils *Ferulago campestris* and marine extracts from the algae *Padina pavonica* increased osteoblast proliferation by increasing alkaline phosphatase, mitogen-activated protein kinase and stimulating mineralisation activity. This resulted in increased bone mineral density and bone strength (Sabbieti et al. 2011; Minetti et al. 2019). Maltanedienol in *P. pavonica* exerts a phytooestrogenic effect; it is similar in structure to oestradiol-17β and is thought to stimulate oestrogen receptors, thereby promoting calcium fixation (Cassar et al. 2013). Other plants and extracts, e.g. essential oils from fennel, garlic, oregano rosemary, sage, echinacea and *Drynaria roosii* have also been shown to increase bone strength, although there may be a threshold above which these essential oils become detrimental to bone health (Olgun 2016; Świątkiewicz et al. 2018; Huang et al. 2020), as indicated also for nutrient utilisation. Betaine, found in many plants including spinach and sugar beets, has been shown to improve bone strength and density by reducing plasma homocysteine concentrations; high homocysteine levels have been shown to increase bone resorption leading to lower bone strength (Maidin et al. 2021).

Horsetail (*Equisetum arvense*) is a plant rich in silica and flavonoids (Law and Exley 2011; Gründemann et al. 2014). When combined with *Spirulina platensis* (0.5 and 2% respectively) eggshell thickness, strength and yolk colour were significantly improved in 63-week-old hens (Tufarelli et al. 2021). Silica accelerates the process of calcification and deposition in bone and shell, improving bone matrix quality and facilitating bone mineralisation (Jugdaohsingh et al. 2004), and therefore shell quality. Spirulina has been shown to improve eggshell quality, also likely due to its mineral content (Selim et al. 2018).

3.3.1 Enhancing Mineral Utilisation Is a Key Tool for Sustainable Production

Improving mineral utilisation, and thereby bone strength and eggshell strength, is crucial for animal welfare and egg production, especially towards the end of the laying cycle where eggshell quality tends to decline. A reduction in wasted egg eggs due to cracked shells would contribute to overall increased production, reduced farm environmental footprint and potentially to reduced losses due to breakage during transport. On-farm egg wastage (due to grading and predominantly eggshell defects) has been estimated up to 0.5% (Singh et al. 2009; Redlingshöfer et al. 2017), while wastage due to distribution/packaging accounts for 1% of the total waste from primary production to consumption (Birney et al. 2017). The environmental impacts estimated for food wastage vary, but for example, in Turkey, national on-farm egg wastage was estimated at 3.57 Mt. CO_2 eq/year (Cakar et al. 2020), approximately 1.5 CO_2 eq kg^{-1} protein. Even a 10% reduction in egg wastage through eggshell improvements could yield a reduction of 0.15 CO_2 eq kg^{-1} per kg/protein based on these calculations.

Declining egg production and egg quality are the main drivers to replace laying flocks. Ensuring efficient mineral utilisation, and therefore skeletal integrity and eggshell quality, helps to keep birds in production longer, reducing the overall number of hens in the system and the need to rear more pullets. Pullet rearing contributes approximately 19% to the environmental impact of egg production (Estrada-González et al. 2020), and thus reduction in overall pullet numbers would contribute to a reduction in environmental footprint of this stage of production.

3.4 Egg Quality and Production

Egg quality and egg production tend to decline with age of bird, associated with reduced endocrine and liver function, yolk precursor, reduced synthesis of 1,25 $(OH)_2D_3$ by the kidney, aged shell glands and poor digestive function (Joyner et al. 1987; Wistedt et al. 2019; Gu et al. 2021a, b). Moulting also leads to atrophy of the reproductive organs and reduction in endocrine secretions (Gongruttananun et al. 2017; Socha et al. 2017). At a minimum, mitigating declining egg production rates and egg quality through production should be a target for the use of plant-based feed additives.

3.4.1 Egg Production

Essential oil from the chaste tree (*Vitex agnus-castus*), a tree common to the Mediterranean, is rich in 1,8-cineole, sabinene and beta-caryophyllene (a sesquiterpene) and has been used for human female reproductive health for centuries as a source of phytoestrogens (Daniele et al. 2005). The supplementation of *Vitex agnus-castus* to laying hen feed led to a 19% increase in egg production, 5% increase in eggshell weight and an increase in oestradiol-17β hormone from 222 to 435 pg/mL, in control and treated group respectively (El-Saadany et al. 2022). Rather than stimulating oestrogen receptors (flax seed, for example, contains phenols which act as oestrogenic alternatives (Rubilar et al. 2010)), it is suggested *V. agnus-castus* decreases levels of prolactin and rebalances endocrine hormone (van Die et al. 2012), leading to increased oestradiol-17β. Red pepper (*Capsicum frutescens*) and anise (*Pimpinella Anisum* L.) have been shown to increase oestrogen levels in 60-week-old hens (El-Maaty and Sherif 2016). Forty-seven-week-old hens fed a commercial preparation using sanguinarine and chelerythrine (alkaloid extracts from the poppy *Macleaya cordata*) for 12 weeks were determined to have increased serum concentrations of luteinising hormone, progesterone, follicle stimulating hormone and oestradiol, although egg production did not differ from controls (Guo et al. 2021). Flavonoid extracts from *Epimedium* (a group of plants used in Chinese medicine) have been demonstrated to restore egg production and improve development of fallopian tubes and follicles post-moulting, via the stimulation of oestrogen synthesis (Huo et al. 2020). Turmeric has also been shown to improve egg production (Laganá et al. 2011; Rahardja et al. 2015; Widjastuti et al. 2017; Mobarez Samia et al. 2018), although there might be threshold above which performance no longer benefits (Fawaz et al. 2022). Interestingly, Hassan (2016) reported no benefits of turmeric on egg production.

Tea and tea extracts have shown positive effects on egg production (Rizk 2017; Xia et al. 2018, 2022). While Wang et al. (2018) found no effect of 1% green tea powder on egg production, 2% green tea powder inclusion led to a negative impact on egg production, Zhu et al. (2020) reported that the type of tea extract (polyphenol or catechin) was significant for its effect on egg production; polyphenols increased egg production, while catechins had no effect. A combination of peppermint and thyme essential oil improved performance (increased egg production and egg mass and reduced FCR) in hens reared under low temperature conditions (6.8 ± 3 °C) (Akbari et al. 2016). Crushed black cumin seeds have been shown to increase egg production and eggshell thickness (Akhtar et al. 2003; Aydin et al. 2008; Mohammed and Al-Hameed 2021), although the effect is inconsistent (Hassan and Ragab 2007).

3.4.2 Egg Quality

Egg quality characteristics, other than flavour and odour include yolk quality (height and diameter), albumen height, Haugh units, egg mass and eggshell thickness are influence by phytocompounds. In their review of polyphenols used in the poultry industry, Abd El-Hack et al. (2022) concluded that when administered in accurate doses, polyphenols:

- Improved egg size and numbers
- Improved egg weight
- Improved egg colour intensity
- Improved eggshell quality
- Improved egg yolk colour intensity and fatty acid contents
- Increased yolk and albumen weights

A variety of plants and their extracts such as peppermint oil, *Moringa oleifera* leaves, black cumin, rosemary, clove bud powder, river saltbush (*Atriplex amnicola*) and *Yucca schidigera* extract increased eggshell thickness, and in some cases egg yolk colour (Aydin et al. 2008; Taheri Gandomani et al. 2014; Alagawany et al. 2018; Çufadar 2018; de Koning et al. 2019; Abdel-Wareth and Lohakare 2020, 2021). Leaves of *Artemisia annua* (a medicinal plant known to have antiprotozoal activity (Remmal et al. 2011; Fatemi et al. 2015) fed to hens was also shown to increase shell thickness (Baghban-Kanani et al. 2019). However, other authors report no benefits of some of these plants, e.g. peppermint oil or *Yucca schidigera* on improving eggshell thickness, although egg yolk cholesterol was reduced (Wang and Kim 2011; Dilawar et al. 2021). Similarly, *Spirulina platensis* did not influence eggshell thickness but did enhance yolk colour (Zahroojian et al. 2013). Peppermint oil as well as combinations of extracts from pomegranate (*Punica granatum*), *Thymus vulgaris*, and *Allium sativum*, and a combination of magnolia vine (*Schisandra. chinensis*) pomace, fermented pine needle (*Pinus densiflora*) extract and Chinese chive (*Allium tuberosum*) powder also increased egg quality in terms of Haugh units (Abou-Elkhair et al. 2018; Sharma et al. 2020; Abdel-Wareth and Lohakare 2020; Moon et al. 2021).

Rosemary or thyme/thymol have been shown to increase Haugh units (Abdel-Wareth 2016), while echinacea and oregano influence yolk colour and yolk index respectively (Radwan et al. 2008; Jahanian et al. 2017; Alagawany et al. 2017). However, other studies report no effect on egg composition with the same herbs or other such eucalyptus or Pudina (*Mentha spicata*) (Botsoglou et al. 2005; Florou-Paneri et al. 2005; Abd et al. 2008; Bozkurt et al. 2016; Devi et al. 2017).

3.4.3 Enhancing Egg Quality and Production Meets a Key Sustainability Objective: Efficiency

The potential for plants and plant extracts to improve outputs from egg production systems is especially interesting in those markets, e.g. developing countries, which might use native breeds with low production; phytocompounds with phytoestrogenic properties could help boost production, and thereby efficiency. A 10% increase in egg production could lead to a reduction of 0.24 kg CO_2 eq kg^{-1} (Weeks et al. 2016). Similarly, an increase of 10 weeks of production could preserve approximately 14 of potentially polluting nitrogen per kg egg protein from entering the environment based on Bain et al. (2016). Where moulting is practiced, phytoestrogens could help restore productivity. El-Laithy (2009) suggested parenteral administration of oestradiol-17β to restore egg production after moulting, a function that could potentially be fulfilled by flax seed or *Vitex agnus-castus*, for example. Replacement of hens at the end of production can contribute up to 24% of the environmental impact of egg production, hence extension of the laying cycle has been suggested as a key factor to lessen the environmental footprint (Abín et al. 2018).

Enhancing egg quality has a clear role in sustainability. The storage temperature of eggs and storage time as well as age of bird influence egg quality (de Menezes et al. 2012). Refrigerated (8 °C) eggs have better long-term quality than eggs stored at room temperature (25 °C) but egg quality still declines with storage duration, irrespective of temperature. Similarly, eggs from older hens tend to have poorer quality than those from younger hens. In addition to improving the egg antioxidant status and shelf-life, phytocompounds could also improve other measures of egg quality such as yolk quality and colour, characteristics that can enhance customer buying power. Improved eggshell strength can reduce egg losses both in the house and at later stages in the supply chain. As mentioned earlier, increased saleable eggs can lower the environmental footprint of egg production on farm and in the later stages of the supply chain, e.g. retail.

3.5 Antioxidant Status

The antioxidant status of the hen is a key factor in bird health and production. Stresses in commercial poultry result from environmental, nutritional, microbiological and management factors which negatively impact poultry health and production (Mishra and Jha 2019). For example, heat stress increases the rate of oxidative stress (Nidamanuri et al. 2017). In hens, the production of corticosterone in response to stress suppresses follicular development. This in turn increases energy intake and available energy is increased by decreasing the availability of the circulating yolk precursor preventing yolk deposition in follicles leading to lower egg production (Odihambo Mumma et al. 2006; Wang et al. 2017). Acute oxidative stress has been shown to directly reduce egg production (Abbas et al. 2022). The gastrointestinal tract (GIT) is continuously exposed to a plethora of challenges from feed ingredients, pathogens, mycotoxins etc. and is under constant stress (Mishra and Jha 2019) Given the critical role of the GIT in nutrient absorption and immune response and as a selective barrier, supporting a balanced antioxidant status in the GIT is essential.

Oxidative stress is a consequence of these stresses; an imbalance between reactive oxygen species (ROS) and antioxidants leading to elevated ROS (Schieber and Chandel 2014). ROS are by-products of aerobic metabolism, including immune system processes, and are chemically reactive for different biological targets. An excess of ROS can lead to damage of cell membranes, proteins and DNA. However, ROS are also important for signalling in physiological processes as well as cell growth regulation, synthesis of biological products, and energy production. ROS are also important in the immune system acting as cell signals and in phagocytosis; the immune system produces ROS to kill pathogens acting as cell signals and compounds. Oxidative stress can limit the immune system, leading to comprised immune defences (Biller-Takahashi et al. 2015). The nuclear factor (erythroid-derived 2)-like 2/Kelch-like ECH-associated protein 1 (Nrf2/Keap1) system plays a major role in maintaining the balance of ROS via antioxidant response elements which regulate a series of detoxifying and antioxidant enzymes (Qin and Hou 2017).

3.5.1 Affecting Antioxidant Status in the Bird

Many phytocompounds evaluated for inclusion as feed additives demonstrate antioxidant activity predominantly associated with the presence of polyphenols Essential oils from plant families such *Lamiaceae* have well documented antioxidant activity in vitro and in vivo in birds (Matkowski and Piotrowska 2006). Clove bud essential oil has been shown to have more than ten-fold higher antioxidant activity to inhibit 2–2-diphenyl-1-picric hydrazyl (DPPH) free radical than vitamin E (Taheri Gandomani et al. 2014). Supplementation of hen diets with combinations of thyme, mint, rosemary or dill, resulted in lower serum and liver glutathione peroxidase (GSH) activity malondialdehyde (MDA) concentrations and elevated serum super oxide dismutase (SOD) activity (El-Hack and Alagawany 2015; Alagawany and El-Hack 2015; Alagawany et al. 2017; Mousavi et al. 2017), while the algae *Spirulina platensis* and star anise (*Illicium verum* Hook. f.) also improved overall antioxidant capacity (Yu et al. 2018; Tufarelli et al. 2021). Wang et al. (2018) observed that hens fed tea polyphenols had improved magnum morphology, attributed to improved magnum health due to increased antioxidant activity; increases in albumen height and haugh units were also observed. Astaxanthin (ASTA), from the algae *Haematococcus pluvialis,* has been shown to increase total antioxidant capacity, SOD and GSH levels in the plasma, livers and egg yolks of hens under both normal and heat-stressed conditions, likely via upregulation of enzymes in the Nrf2 pathway (Magnuson et al. 2018; Gao et al. 2020; Heng et al. 2021). Generally, phytocompounds have low absorption in the small intestine, suggesting their effect on the animal is via indirect pro-oxidant properties, biotransformation, signalling transduction and gene expression regulation rather than direct antioxidant properties (Qin and Hou 2017).

Impact of external stressors: External stressors, e.g. temperature can induce oxidative stress. Chen et al. (2018) demonstrated that supplementation of eucalyptus leaves to feed did not significantly affect hen's total antioxidant capacity under normal conditions, yet when hens were exposed to ethanol-induced oxidative damage, serum antioxidant status was significantly improved, resulting in lower live damage. Akbarian et al. (2016) also observed in their review of dietary phytochemicals used in heat-stress birds, that many of the whole plants or extracts that contained polyphenols (flavonoids and lignans) only demonstrated a benefit when birds were heat-stressed, i.e. the plants were less, or ineffective, in non-heat stressed birds. Hens fed bitter apple fruit (*Citrullus colocynthis*) seeds had improved overall antioxidant capacity under both non-stressed and oxidative stressed conditions using the herbicide paraquat (paraquat toxicity leads to excessive superoxide radical production, triggering inflammation responses and oxidative stress) (Abbas et al. 2022).

Meat quality: Enhancement of meat quality in hens using plant extracts is unclear. In 50-week hens fed microalgae (*Aurantiochytrium* sp.), there was no significant change in meat quality (Liu et al. 2020a), while no improvement in carcass yield was recorded in 101-week old hens fed Sea Buckthorn (Kang et al. 2015). However, in 36-week-old hens fed polyphenolic extracts from hybrid Eucalyptus, meat pH was increased and drip loss decreased (Chen et al. 2018).

Mycotoxins: Mycotoxins such as zearalenone or aflatoxin B1 have been shown to lower gut microbiota diversity, impact ovarian function and compromise immunity and antioxidant status (Azzam and Gabal 1998; Rizzi et al. 2003; Nakade et al. 2016; Yuan et al. 2022). Oregano essential oil (400 mg/kg) has been shown to counter the reduction in antioxidant status (increased MDA and reduced SOD) observed in Japanese quail fed aflatoxin B1 2.5 mg/kg diet (Abdelrazek et al. 2015). Silymarin (a flavonoid isolated from milk thistle (*Silybum marianum*)) fed to hens at 100 mg/kg feed simultaneously with 0.05 ppm aflatoxin and 20 ppm fumonisin significantly reduced glutathione *S*-transferase activity and numerically increased thiobarbituric acid reactive substances (TBARS) compared to birds only receiving mycotoxins (Armanini et al. 2021).

Protozoal infections: The antioxidant properties of plants have been suggested as one reason for their apparent efficacy to reduce protozoal infections in birds (Naidoo et al. 2008). Over 68 plants and phytocompounds have shown efficacy against protozoa including plants from the *Lamiaceae* (e.g. *Origanum* spp. and *Mentha* spp.), *Artemisia* spp. (Asteraceae), *Moringa* spp. (*Moringaceae*), *Melia azadirachta* (*Meliaceae*) (Muthamilselvan et al. 2016). *Allium tripedale*, a species of wild onion, has been shown to produce similar results to the anticoccidial drug diclazuril in birds for both coccidial control and performance (Habibi et al. 2016).

3.5.2 Affecting Antioxidant Status in the Egg

Signs of oxidation in eggs include changes in colour and flavour and off-odour, i.e. indicators of reduced palatability and quality. In 78-week-old hens fed dried sage total tocopherol in egg yolk was observed to increase 2.3-fold versus control eggs from birds fed a basal diet containing 40 IU vitamin E. Loetscher et al. (2014) hypothesised that the improved antioxidant status in the study using natural antioxidants was related to a vitamin E sparing effect in the gut. Layers fed eucalyptus leaves, green tea water extracts, echinacea, oregano, rosemary or skullcap mushroom have shown evidence of reduced lipid oxidation in fresh egg yolks or prolonged storage (Botsoglou et al. 2005; Florou-Paneri et al. 2005; Radwan et al. 2008; Jahanian et al. 2017; Alagawany et al. 2017; Batista et al. 2017; Chen et al. 2018; Xia et al. 2022). Egg shelf-life was improved in eggs kept at both 5 °C and 21 °C for 28 days from 50-week hens fed pumpkin seed meal (Vlaicu and Panaite, 2022). The authors attributed this improvement in shelf life to delayed lipid oxidation and protein degradation.

Torki et al. (2018) reported a synergistic effect between chicory, rosemary and dill on egg yolk oxidative capacity when fed to hens, both in non-stressed and heat-stressed birds. Conversely, (Galobart et al. 2001) saw no improvement in markers of lipid oxidation in eggs from hens fed rosemary powder. Aqueous-alcoholic extract of roselle (*Hibiscus sabdariffa*) leaves also had no effect on antioxidant status of eggs from 35-week-old hens (Sarvestani et al. 2020).

3.5.3 Maintaining Antioxidant Status Is a Significant Contributor to Sustainable Egg Production

The antioxidant status of the hen is the basis for a healthy bird. It is crucial for the efficiency of the GIT, for immunity and therefore performance of the hen. The Nrf2/Keap1 system is linked to many other functions including antioxidant and anti-inflammatory, hence the modulation of this system by phytocompounds likely plays a role in the broad biological activity of phytocompounds (Qin and Hou 2017). Maintaining feed intake, health and production in the face of disease and enteric challenge will help maintain the low GHG emissions (less than 0.5 CO_2 eq kg^{-1}) per average daily feed intake reported for egg production (de Vries and de Boer 2010).

Improving egg quality and potentially egg shelf life via slower lipid oxidation using phytocompounds is important for the supply chain; eggs can be stored longer reducing wastage or there is potentially less adverse impact of variation in storage temperature. However, it cannot be assumed that elevated antioxidant capacity in the hen translates to improved antioxidant status in the egg (Harrington et al. 2020); older hens preferentially deposit antioxidants in the egg not body (Loetscher et al. 2014). Consequently, antioxidant status of the hen and egg should be considered separately when evaluating potential plant-based feed additives for both egg quality and using spent hens in the food-chain. Improving carcass quality of spent hens can further reduce GHG emissions footprint (Taylor et al. 2014).

Thermal stress can adversely impact hen oxidative status and performance, leading to reduced feed efficiency and egg production (Nidamanuri et al. 2017; Barrett et al. 2019; Li et al. 2020). Improving the thermotolerance of birds using phytocompounds (Samarasinghe et al. 2003; Radwan et al. 2008; Rajput et al. 2013; Fawaz et al. 2022) could reduce primary energy inputs for heating/cooling or counter the effects of climate are more pronounced, improved bird thermotolerance by maintaining hen feed and production efficiency. Some estimates suggest maintaining the house environment using primary energy contributes 5.14×10^{-3} CO_2 eq kg^{-1} due to primary energy usage to support birds (Estrada-González et al. 2020); using phytocompounds to improve hen thermotolerance could reduce this impact.

Feed insecurity associated with high feed cost and compromised feed safety, e.g. from mycotoxins, is a challenge for the poultry industry, particularly in developing countries (Ochieng et al. 2021) and is compounded by climatic extremes. The supplementation of hen diets with phytocompounds could help to counter the detrimental impact of mycotoxins on production, particularly where poor-quality cereals are used.

3.6 Immune Status

The energy demands of mounting an immune response requires a potential trade-off between body condition, egg-laying/egg composition and immune function if resources are limited (Hanssen et al. 2004; Cucco et al. 2010). Some nutrients can be considered essential for the immune system and their deficiency can influence immunocompetence and be deleterious for those tissues that are sensitive to nutrient levels. Other nutrients could be considered immunomodulatory whereby the intensity of immune response can be influenced by their supplementation rather than the absence of such nutrients comprising immune status (Koutsos and Klasing 2014). Laying activity in hens appears to bias the immune system toward long-term humoral response with strong innate responses (Koenen et al. 2002; Schmucker et al. 2021). Forced moulting, typically stimulated by a reduction in nutrient density, is practised to restore higher egg production and can negatively affect both innate and acquired immunity, with a decrease in the functional immune response of leucocytes (Nathan et al. 1977; Holt 1992a, b; Kogut 2009). Moulting can also increase entry of enteric foodborne pathogens into the reproductive tract, leading to contaminated eggs (Cogan and Humphrey 2003).

3.6.1 Phytocompounds Can Have Broad Ranging Effects on the Immune System

Antibody responses: Circulating immunoglobulin levels increase in hens given Fenugreek (*Trigonella foenum-graecum L.*), and red, blue-green algae powder (SPP) in water (Zadeh et al. 2015; Kulshreshtha et al. 2017; Nia et al. 2021). Humoral immunoglobulins also increase in hens fed dried rosemary, thyme, or clove bud or their corresponding essential oils (Radwan et al. 2008; Taheri Gandomani et al. 2014; Alagawany et al. 2017). (Gultepe et al. 2020) report that myrtle (*Myrtus communis* L.) increased IgY levels in older hens. *Yucca schidigera* extract and curcumin have also been shown to increase circulating IgY levels in hens (Liu et al. 2020b). Powders of the Chinese herbal medicine plants *Ligustrum lucidum* and *Schisandra chinensis* and blue-green algae *Spirulina platensis* have been shown to increase Newcastle Disease virus (NDV) vaccinal titres, including under heat stress (Ma et al. 2005; Nia et al. 2021). Conversely, when specific antibody titres were measured, e.g. to infectious bronchitis virus, NDV or infectious bursal disease, essential oils from oregano, laurel, sage, myrtle, fennel and citrus peel had no effect in hens 36–68 weeks of age (Özek 2012; Bozkurt et al. 2012). Dried Egyptian clover (*Trifolium alexandrinum*), a forage crop grown in sub-tropical regions, or seeds of *Nigella sativa* have shown no demonstrable effect on the immune system of the laying hen (Hassan and Ragab 2007; Ragab and Hassan 2007), despite reports of increased egg production (Akhtar et al. 2003).

Cellular immunity: Hens fed eucalyptus powder had significantly higher wattle swelling 24–72 h after injection with phytohemagglutinin compared to control hens, indicative of an elevated cell-mediated response (Abd et al. 2008). Eucalyptus, fennel, garlic and thyme have been shown to increase monocyte levels in hens (Abd et al. 2008; Ghasemi et al. 2010; Nasiroleslami and Torki 2010; Bozkurt et al. 2012). Eugenol (from clove oil), gingerol (from ginger), quercetin (found in many fruits, vegetables and tea) and capsaicin have also been indicated to have anti-inflammatory activity by, for example, lowering expression of pro-inflammatory cytokines or modulation of the Nf-κB pathway (Yang et al. 2020; Feng et al. 2021). Curcumin powder increased complement activity and reduce proinflammatory cytokine expression (IL-1β, IL-6 and TNF-α) and subsequently lower TLR4 protein expression in thermally stressed birds, perhaps as a dose response (Nawab et al. 2019, 2020; Mosayyeb Zadeh et al. 2020; Liu et al. 2020b; da Rosa et al. 2020). Interestingly, while *Spirulina platensis* stimulated an increase in humoral antibody response, there was no increase in cell mediated responses to phytohemagglutinin injection (Nia et al. 2021). The apparent heterogeneity in response to phytocompounds could relate to the type, route and level of immune stressor.

3.6.2 The Importance of Hen Immunity for Sustainability

Enhancing protective immune responses to pathogens in the absence of antibiotics for example, is important to maintain efficient, and therefore sustainable production. Vaccination is essential for maintaining animal health, particularly where reliance on chemical and antimicrobial tools is reduced. Boosting vaccinal antibody titres can help protect hens during production, reducing the impact of pathogen challenges including zoonotic organisms will ensure efficient and uninterrupted egg production. Similarly, bolstering immune responses in older hens using phytocompounds is important when longer laying cycles are considered or birds are moulted leading to comprise immunity.

As described earlier, modulating the gut and other microbiota either directly or indirectly, can reduce pathogen load. Modulating inflammatory responses to infection, e.g. coccidiosis or colibacillosis, is a key component of mitigating the impact of the pathogen on the host, either directly or via reduction in the impact of secondary infections. There is also mitigation of the energy cost of mounting immune responses, particularly if an infection is chronic. Plant-based additives can facilitate the hen into diverting resources back into egg laying.

Adoption of extensive poultry housing increases the risk of exposure of birds to a broader range of potential pathogen challenges requiring a more robust protective immune response either via natural immunity or vaccination. The challenge is to identify the specific immunomodulating properties of the plant of interest and use it strategically, given the potential other additives in use in the feed and immune stressors to be experienced.

3.7 Ammonia Reduction

While Swelum et al. (2021) proposed that plant feed additives be used for GHG mitigation due to their ability to increase 'birds' intestinal dignity', reduce unabsorbed and undigested nutrients and modulate microflora, some phytocompounds can have direct activity on ammonia. *Yucca schidigera*, a plant native to Southwestern USA and Mexico, has been shown to lower ammonia emissions from poultry manure by up to 73%, while supplementation in poultry diets can lower ammonia by to 40%, without adversely affecting performance (Gurbuz et al. 2011; Chepete et al. 2012; Matusiak et al. 2016; Alagawany et al. 2018). *Y. schidigera* contains steroidal saponins and glycoproteins which physically bind ammonia, thereby reducing free ammonia (Wu et al. 1994; Cheeke 2000). The saponins may also reduce urease activity during microbial decomposition in manure which would result in slow ammonia release (Saeed et al. 2018), although it is questionable how much yucca inhibits microbial degradation of urea is unclear (Makkar et al. 1999). There are other plants rich in saponins including fenugreek (*Trigonella foenum-graecum* L), a source of steroidal saponins, that warrant further investigation for ammonia reduction (Petit et al. 1995; Wina 2018). Recent work by Park et al. (2018) showed that laying hens fed fenugreek seed extract emitted 38% less ammonia than control birds. However, the authors attributed this reduction in ammonia to modification of the gut microbiota rather than binding of ammonia per se. A source of triterpenoid saponins, *Quillaja saponaria* Molina (a tree endemic to Chile) has also been evaluated for the reduction in ammonia in animal production, either alone or in combination with *Y. schidigera* (Güroy et al. 2016; Bartoš et al. 2016; Dang and Kim 2020; Abozeid et al. 2021; Bafundo et al. 2021), lowering ammonia production in commercial egg laying facilities by up to 63% (Harrington and Decap 2021).

3.7.1 Plant Extracts Can Be a Tool for Direct Reduction in GHGs to Lower Environmental Footprint

While GHG emissions can be reduced via modulation of the gut microbiota and nutrition, studies would support the case for saponin sources as potential in-feed mitigants to lower ammonia emissions (Adegbeye et al. 2020). Plants such as yucca and quillaja do not negatively impact performance, and indeed, in some instances have been shown to enhance performance. Saponins have other well-documented properties including anti-inflammatory and anti-protozoal activity (Cheeke 2000; Hristov et al. 2003). Consequently, they have benefits for sustainable production in addition to reducing GHG emissions, with the potential to modulate gut health and protozoal infections and reduce reliance on drugs.

4 Sustainability Requires a Multifaceted Approach

The discussion of what is sustainable in egg production has changed dramatically over the last 20 years; in 2002 cages were considered the most sustainable option; current thinking suggests hens should be reared cage-free (de Boer and Cornelissen 2002; EC 2021). Undoubtedly, the discussion of what comprises sustainable egg production will continue and evolve. The stakeholder groups are diverse and different perspectives, e.g. extremist views from activist groups, can often inform the debate without prudence to underlying science or fact (de Olde et al. 2020; Leroy et al. 2022); although under the societal pillar of sustainability (FAO 2012), it could be argued the extremist view has as much validity as mainstream. LCA of egg production is complex and the outcome can be influenced by numerous factors including location and the underlying data assumptions (Dekker et al. 2013; Abín et al. 2018; Estrada-González et al. 2020; Costantini et al. 2021).

4.1 The Phytocompound Cascade in Sustainability

The on-farm sustainability considerations for egg production are diverse and include energy usage, manure management, housing, feed, length of laying cycle, genetics, production efficiency, sexing and good intestinal health and disease management and require a multifaceted approach. Guillaume et al. (2022) identified feed conversion ratio (FCR), feed composition and manure management to be the most important factors influencing the total environmental impacts of egg production. LCA show a clear relationship between environmental footprint and feed and production efficiency of the hen. If the hen can produce more saleable eggs for lower inputs, the overall environment footprint (from primary production to household consumption) will lower. Feed additives based on phytocompounds offer the potential to influence the physiological efficiency of the hen and therefore positively influence the considerations of sustainability either directly or indirectly, via:

- enhancing gut health
- improving feed and mineral utilisation
- bolstering immune status
- improving production and egg quality
- reducing faecal ammonia
- enhancing food safety

Plant-based additives could be useful tools to help meet the demands of sustainable egg production if used strategically. For example:

- Intestinal health – grape seed or Bilberry leaves could support gut microbiota
- Disease management – red seaweed to reduce *Salmonella* and ensure food safety; blue-green algae *Spirulina platensis* to enhance immune responses and vaccinal titres; *Artemisia* spp. for protozoal challenges

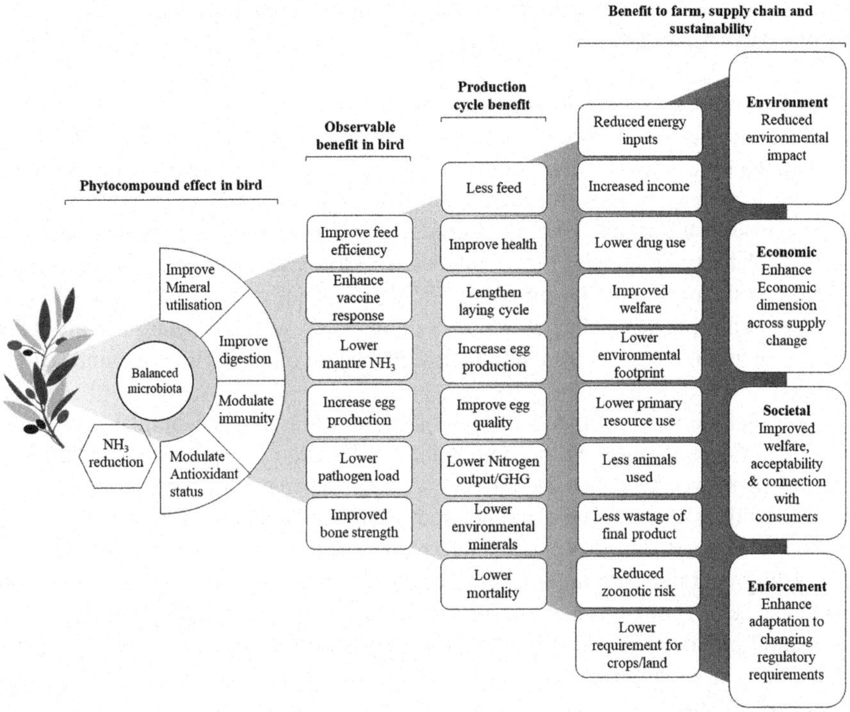

Fig. 3 The multiplying effect of impact of phytocompounds in the laying hen on dimensions of sustainability for egg production

- Production efficiency – *Solanum glaucophyllum* as a natural vitamin-D source throughout the cycle; pumpkin seed meal to improve egg quality and shelf-life
- Length of laying cycle – *Padina pavonica* to increase bone strength and mineralisation in older layers; Anise (*Pimpinella Anisum* L.) to maintain hormone levels
- Manure management – *Yucca schidigera* or *Quillaja saponaria* for ammonia reduction from manure
- Feed efficiency – *Origanum vulgare* to increase enzyme activity and intestinal nutrient transporters; *Illicium verum* to increase protein digestibility; *Magnolia officinalis* to support hepatic lipid metabolism and lower risk of fatty liver.

The ability to combine plants and their extracts to achieve specific outcomes is an exciting prospect. An example of the potential cascade of benefits from physiological effect in the hen to the influence on the four dimensions of sustainability (FAO 2012) is shown in Fig. 3.

4.2 Understanding Current Limitations and Nature Identical Compounds

The lack of consistency of efficacy can still present a barrier for commercial adoption of many interesting ingredients (Bozkurt et al. 2014; Harrington et al. 2020). Karásková et al. (2016) summarised the challenge well, '...long term studies will be crucial, proving mainly the efficacy of these additives, their safety with regard to animal health, the quality of animal products and environment, and, subsequently, their availability in terms of their anticipated regular use'. The potential to combine plants and their extracts to achieve specific performance or environmental outcomes is an exciting prospect but not all phytocompounds will act synergistically when used contemporaneously or at high inclusions, while other phytocompounds can be toxic.

Consideration of the plants as a biological resource to manufacture the feed additive is as important as the output for LCA; resource replenishment needs to exceed depletion. Questions often arise concerning manufacturing costs of plant-based additives; can they be manufactured in a cost-efficient, ethical and sustainable way that it is still economically viable to be used by the farmer? The impact of harvesting wild plants or the impact of cultivating plant species on the environment and local ecosystem also needs to be considered. Combining multiple plant extracts or natural phytocompounds to deliver a highly efficacious product can also potentially be cost prohibitive if production costs are high. Cost is one of the major concerns for sustainable egg production (de Olde et al. 2020).

Synthetic production of phytocompounds or nature identical (NI) is a growing area (Braga et al. 2019) enabling the production of fully synthetic plant-based feed additives. While NI phytocompounds appear to be as efficacious as their naturally derived counterparts, the larger diversity of phytocompounds in natural products is more effective for the target animal and environment (Reddy 2019; Rossi et al. 2020). However, the concept of 'natural' products and associated 'natural production' is strong and important to consumers (Grigoriadou et al. 2020) and how synthetic products would fit within this scenario needs further exploration.

5 Conclusions

A multifactorial, multistakeholder approach is needed to ensure eggs are produced sustainably, which includes the rearing phase. Big steps have been made by introducing hens that can produce over 500 eggs per hen housed. Although it is recognised that plant-based additives have a wide variety of functions that support sustainable egg production, the modes of action on single components and of synergies between them are currently only partly understood. To fully use their potential, additional research is needed in pullets and laying hens. Plant-based feed additives must have a clear benefit to improve efficiency/reduce environmental contaminants, while delivering a positive economic return.

References

Abad P, Arroyo-Manzanares N, Ariza JJ et al (2021) Effect of Allium extract supplementation on egg quality, productivity, and intestinal microbiota of laying hens. Animals 11:41. https://doi.org/10.3390/ani11010041

Abbas AO, Alaqil AA, Kamel NN, Moustafa ES (2022) Citrullus colocynthis seed ameliorates layer performance and immune response under acute oxidative stress induced by paraquat injection. Animals 12:945. https://doi.org/10.3390/ani12080945

Abd El-Hack ME, Salem HM, Khafaga AF et al (2022) Impacts of polyphenols on laying hens' productivity and egg quality: a review. J Anim Physiol Anim Nutr 1–20. https://doi.org/10.1111/jpn.13758

Abd AM, Motaal E, Ahmed AMH et al (2008) Productive performance and immunocompetence of commercial laying hens given diets supplemented with Eucalyptus. Int J Poult Sci 7:445–449

Abdelrazek HMA, Yusuf MS, Hassan MA et al (2015) Ameliorative effect of oregano essential oil on mycotoxins-induced immune impairments in growing japanese quail. Egypt Acad J Biol Sci C Physiol Mol Biol. 7:101–114. https://doi.org/10.21608/eajbsc.2015.13695

Abdel-Wareth AAA (2016) Effect of dietary supplementation of thymol, synbiotic and their combination on performance, egg quality and serum metabolic profile of Hy-Line Brown hens. Br Poult Sci 57:114–122. https://doi.org/10.1080/00071668.2015.1123219

Abdel-Wareth AAA, Lohakare JD (2020) Productive performance, egg quality, nutrients digestibility, and physiological response of Bovans brown hens fed various dietary inclusion levels of peppermint oil. Anim Feed Sci Technol 267:114554. https://doi.org/10.1016/j.anifeedsci.2020.114554

Abdel-Wareth AAA, Lohakare J (2021) Moringa oleifera leaves as eco-friendly feed additive in diets of Hy-line brown hens during the late laying period. Animals 11:1116. https://doi.org/10.3390/ani11041116

Abe E, Horikawa H, Masumura T et al (1982) Disorders of cholecaiciferol metabolism in old egg-laying hens. J Nutr 112:436–446

Abín R, Laca A, Laca A, Díaz M (2018) Environmental assesment of intensive egg production: a Spanish case study. J Clean Prod 179:160–168. https://doi.org/10.1016/j.jclepro.2018.01.067

Abou-Elkhair R, Selim S, Hussein E (2018) Effect of supplementing layer hen diet with phytogenic feed additives on laying performance, egg quality, egg lipid peroxidation and blood biochemical constituents. Anim Nutr 4:394–400. https://doi.org/10.1016/j.aninu.2018.05.009

Abozeid AM, Abdel-Rahim MM, Abouelenien F et al (2021) Quillaja saponaria and/or Yucca schidigera ameliorate water quality, growth performance, blood health, intestine and gills histomorphology of Nile tilapia, Oreochromis niloticus. Aquac Res 52:6117–6131. https://doi.org/10.1111/are.15474

Adegbeye MJ, Ravi Kanth Reddy P, Obaisi AI et al (2020) Sustainable agriculture options for production, greenhouse gasses and pollution alleviation, and nutrient recycling in emerging and transitional nations - an overview. J Clean Prod 242:118319. https://doi.org/10.1016/j.jclepro.2019.118319

Akbari M, Torki M, Kaviani K (2016) Single and combined effects of peppermint and thyme essential oils on productive performance, egg quality traits, and blood parameters of laying hens reared under cold stress condition (6.8 ± 3 °C). Int J Biometeorol 60:447–454. https://doi.org/10.1007/s00484-015-1042-6

Akbarian A, Michiels J, Degroote J et al (2016) Association between heat stress and oxidative stress in poultry; mitochondrial dysfunction and dietary interventions with phytochemicals. J Anim Sci Biotechnol 7:37. https://doi.org/10.1186/s40104-016-0097-5

Akhtar MS, Nasir Z, Abid AR (2003) Effect of feeding powdered Nigella sativa L. seeds on poultry egg production and their suitability for human consumption. Vet Arh 73:181–190

Akinwumi BC, Bordun K-AM, Anderson HD (2018) Biological activities of stilbenoids. Int J Mol Sci 19:792. https://doi.org/10.3390/ijms19030792

Aksu M, Ozkan G, Kiralan SS et al (2021) Composition and functionality of Nigella sativa essential oil. In: Fawzy Ramadan M (ed) Black cumin (Nigella sativa) seeds: chemistry, technology, functionality, and applications. Springer, Cham, pp 409–420

Alagawany M, El-Hack MEA (2015) The effect of rosemary herb as a dietary supplement on performance, egg quality, serum biochemical parameters, and oxidative status in laying hens. J Anim Feed Sci 24:341–347. https://doi.org/10.22358/jafs/65617/2015

Alagawany M, El-Hack M, Saeed M et al (2017) Effect of some phytogenic additives as dietary supplements on performance, egg quality, serum biochemical parameters and oxidative status in laying hens. Indian J Anim Sci 87:103–108

Alagawany M, Abd El-Hack ME, Farag MR et al (2018) Dietary supplementation of Yucca schidigera extract enhances productive and reproductive performances, blood profile, immune function, and antioxidant status in laying Japanese quails exposed to lead in the diet. Poult Sci 97:3126–3137. https://doi.org/10.3382/ps/pey186

Anderson KE (2019) Final report of the fortieth North Carolina layer performance and management test. North Carolina Cooperative Extension, Raleigh, NC

Ansari A, Alghanem S, Naeem M (2019) Brown alga Padina: a review. Int J Bot Stud 4:2455–2541

Armanini EH, Boiago MM, de Cécere BGO et al (2021) Protective effects of silymarin in broiler feed contaminated by mycotoxins: growth performance, meat antioxidant status, and fatty acid profiles. Trop Anim Health Prod 53:442. https://doi.org/10.1007/s11250-021-02873-2

Arpášová H, Kačániová M, Haščík P, Mellen M (2013) Effect of thyme essential oil addition on physical and microbiological quality of table eggs. J Microbiol Biotechnol Food Sci 2:1092–1106

Artdita CA, Zhuang Y-R, Liu T-Y et al (2021) The effect of feeding restriction on the microbiota and metabolome response in late-phase laying hens. Animals 11:3043. https://doi.org/10.3390/ani11113043

Aydin R, Karaman M, Cicek T, Yardibi H (2008) Black cumin (Nigella sativa L.) supplementation into the diet of the laying hen positively influences egg yield parameters, shell quality, and decreases egg cholesteroll. Poult Sci 87:2590–2595. https://doi.org/10.3382/ps.2008-00097

Azzam AH, Gabal MA (1998) Aflatoxin and immunity in layer hens. Avian Pathol 27:570–577. https://doi.org/10.1080/03079459808419386

Bafundo KW, Männer K, Duerr I (2021) The combination of quillaja and yucca saponins in broilers: effects on performance, nutrient digestibility and ileal morphometrics. Br Poult Sci 62:589–595. https://doi.org/10.1080/00071668.2021.1891523

Baghban-Kanani P, Hosseintabar-Ghasemabad B, Azimi-Youvalari S et al (2019) Effects of using Artemisia annua leaves, probiotic blend, and organic acids on performance, egg quality, blood biochemistry, and antioxidant status of laying hens. J Poult Sci 56:120–127. https://doi.org/10.2141/jpsa.0180050

Bain MM, Nys Y, Dunn IC (2016) Increasing persistency in lay and stabilising egg quality in longer laying cycles. What are the challenges? Br Poult Sci 57:330–338. https://doi.org/10.1080/00071668.2016.1161727

Bakkali F, Averbeck S, Averbeck D, Idaomar M (2008) Biological effects of essential oils - a review. Food Chem Toxicol 46:446–475

Barrett NW, Rowland K, Schmidt CJ et al (2019) Effects of acute and chronic heat stress on the performance, egg quality, body temperature, and blood gas parameters of laying hens. Poult Sci 98:6684–6692. https://doi.org/10.3382/ps/pez541

Bartoš P, Dolan A, Smutný L et al (2016) Effects of phytogenic feed additives on growth performance and on ammonia and greenhouse gases emissions in growing-finishing pigs. Anim Feed Sci Technol 212:143–148. https://doi.org/10.1016/j.anifeedsci.2015.11.003

Batiha GE-S, Alkazmi LM, Wasef LG et al (2020) Syzygium aromaticum L. (Myrtaceae): traditional uses, bioactive chemical constituents, pharmacological and toxicological activities. Biomolecules 10:E202. https://doi.org/10.3390/biom10020202

Batista NR, Garcia ERM, Oliveira CL et al (2017) Trace mineral sources and rosemary oil in the diet of brown laying hens: egg quality and lipid stability. Braz J Poult Sci 19:663–672. https://doi.org/10.1590/1806-9061-2016-0369

Biller-Takahashi JD, Takahashi LS, Mingatto FE, Urbinati EC (2015) The immune system is limited by oxidative stress: dietary selenium promotes optimal antioxidative status and greatest immune defense in pacu Piaractus mesopotamicus. Fish Shellfish Immunol 47:360–367. https://doi.org/10.1016/j.fsi.2015.09.022

Birney CI, Franklin KF, Davidson FT, Webber ME (2017) An assessment of individual foodprints attributed to diets and food waste in the United States. Environ Res Lett 12:105008. https://doi.org/10.1088/1748-9326/aa8494

Boka J, Mahdavi AH, Samie AH, Jahanian R (2014) Effect of different levels of black cumin (Nigella sativa L.) on performance, intestinal Escherichia coli colonization and jejunal morphology in laying hens. J Anim Physiol Anim Nutr 98:373–383. https://doi.org/10.1111/jpn.12109

Bolukbasi SC, Erhan MK (2007) Effect of dietary thyme (Thymus vulgaris) on laying hens performance and Escherichia coli (E. coli) concentration in feces. Int J Nat Eng Sci 1:55–58

Borey M, Bed'Hom B, Bruneau N et al (2022) Caecal microbiota composition of experimental inbred MHC-B lines infected with IBV differs according to genetics and vaccination. Sci Rep 12:9995. https://doi.org/10.1038/s41598-022-13,512-7

Boskabady MH, Shafei MN, Saberi Z, Amini S (2011) Pharmacological effects of Rosa damascena. Iran J Basic Med Sci 144:295–307

Botsoglou N, Florou-Paneri P, Botsoglou E et al (2005) The effect of feeding rosemary, oregano, saffron and alpha-tocopheryl acetate on hen performance and oxidative stability of eggs. South Afr J Anim Sci 35:143–151. https://doi.org/10.4314/sajas.v35i3.4053

Bozkurt M, Küçükyilmaz K, Çatli AU et al (2012) Performance, egg quality, and immune response of laying hens fed diets supplemented with mannan-oligosaccharide or an essential oil mixture under moderate and hot environmental conditions. Poult Sci 91:1379–1386. https://doi.org/10.3382/ps.2011-02023

Bozkurt M, Hippenstiel F, Abdel-Wareth AAA et al (2014) Effects of selected herbs and essential oils on performance, egg quality and some metabolic activities in laying hens: a review. Eur Poult Sci 78:1–15

Bozkurt M, Bintaş E, Kırkan Ş et al (2016) Comparative evaluation of dietary supplementation with mannan oligosaccharide and oregano essential oil in forced molted and fully fed laying hens between 82 and 106 weeks of age. Poult Sci 95:2576–2591. https://doi.org/10.3382/ps/pew140

Braga A, Rocha I, Faria N (2019) Microbial hosts as a promising platform for polyphenol production. In: Ahktar M, Swamy M, Sinniah U (eds) Natural bio-active compounds. Springer, Singapore, pp 71–103

Brinckmann JA, Kathe W, Berkhoudt K et al (2022) A new global estimation of medicinal and aromatic plant species in commercial cultivation and their conservation status. Econ Bot 76: 319–333. https://doi.org/10.1007/s12231-022-09554-7

Broom LJ, Kogut MH (2018) The role of the gut microbiome in shaping the immune system of chickens. Vet Immunol Immunopathol 204:44–51. https://doi.org/10.1016/j.vetimm.2018.10.002

Brunetti L, Leuci R, Colonna MA et al (2022) Food industry byproducts as starting material for innovative, green feed formulation: a sustainable alternative for poultry feeding. Molecules 27: 4735. https://doi.org/10.3390/molecules27154735

Bryden WL, Li X, Ruhnke I et al (2021) Nutrition, feeding and laying hen welfare. Anim Prod Sci 61:893–914. https://doi.org/10.1071/AN20396

Burlou-Nagy C, Bănică F, Jurca T et al (2022) Echinacea purpurea (L.) Moench: biological and pharmacological properties. a review. Plants 11. https://doi.org/10.3390/plants11091244

Cakar B, Aydin S, Varank G, Ozcan HK (2020) Assessment of environmental impact of food waste in Turkey. J Clean Prod 244:118846. https://doi.org/10.1016/j.jclepro.2019.118846

Cassar M, Shoemake C, Azzopardi LM et al (2013) De novo design of non-steroidal oestrogen receptor modulating molecules using maltanedienol as a lead molecule. Bioorg Med Chem. 11(7):1389–1396

Cheeke PR (2000) Actual and potential applications of Yucca schidigera and Quillaja saponaria saponins in human and animal nutrition. In: Oleszek W, Marston A (eds) Saponins in food, feedstuffs and medicinal plants. Springer, Dordrecht, The Netherlands, pp 241–254

Chen Y, Chen H, Li W et al (2018) Polyphenols in Eucalyptus leaves improved the egg and meat qualities and protected against ethanol-induced oxidative damage in laying hens. J Anim Physiol Anim Nutr 102:214–223. https://doi.org/10.1111/jpn.12680

Chen F, Zhang H, Zhao N et al (2022) Effects of magnolol and honokiol blend on performance, egg quality, hepatic lipid metabolism, and intestinal morphology of hens at late laying cycle. Animal 16:100532. https://doi.org/10.1016/j.animal.2022.100532

Chepete HJ, Xin H, Mendes LB et al (2012) Ammonia emission and performance of laying hens as affected by different dosages of Yucca schidigera in the diet. J Appl Poult Res 21:522–530. https://doi.org/10.3382/japr.2011-00420

Christaki E, Giannenas I, Bonos E, Florou-Paneri P (2020) Chapter 2 - Innovative uses of aromatic plants as natural supplements in nutrition. In: Florou-Paneri P, Christaki E, Giannenas I (eds) Feed additives. Academic Press, pp 19–34

Cogan TA, Humphrey TJ (2003) The rise and fall of Salmonella Enteritidis in the UK. J Appl Microbiol 94(Suppl):114S–119S. https://doi.org/10.1046/j.1365-2672.94.s1.13.x

Costantini M, Ferrante V, Guarino M, Bacenetti J (2021) Environmental sustainability assessment of poultry productions through life cycle approaches: a critical review. Trends Food Sci Technol 110:201–212. https://doi.org/10.1016/j.tifs.2021.01.086

Cucco M, Pellegrino I, Malacarne G (2010) Immune challenge affects female condition and egg size in the grey partridge. J Exp Zool Part Ecol Genet Physiol 313A:597–604. https://doi.org/10.1002/jez.635

Cueva C, Silva M, Pinillos I et al (2020) Interplay between dietary polyphenols and oral and gut microbiota in the development of colorectal cancer. Nutrients 12:625. https://doi.org/10.3390/nu12030625

Çufadar Y (2018) Effects of dietary different levels of rosemary essential oil on performance and eggshell quality parameters in laying hens. Selcuk J Agric Food Sci. 32:454–457. https://doi.org/10.15316/SJAFS.2018.122

Curabay B, Sevim B, Cufadar Y, Ayasan T (2021) Effects of adding spirulina platensis to laying hen rations on performance, egg quality, and some blood parameters. J Hell Vet Med Soc 72: 2945–2952. https://doi.org/10.12681/jhvms.27538

da Rosa G, Dazuk V, Alba DF et al (2020) Curcumin addition in diet of laying hens under cold stress has antioxidant and antimicrobial effects and improves bird health and egg quality. J Therm Biol 91:102618. https://doi.org/10.1016/j.jtherbio.2020.102618

Dai D, Qi G, Wang J et al (2022) Intestinal microbiota of layer hens and its association with egg quality and safety. Poult Sci 101:102008. https://doi.org/10.1016/j.psj.2022.102008

Dalal R, Kosti D, Tewatia B (2018) Effect of turmeric powder on egg quality, gut morphology, ecology and on immune system of laying hen: a review. J Entomol Zool Stud 6:978–982

Dang DX, Kim IH (2020) Effects of dietary supplementation of Quillaja saponin on growth performance, nutrient digestibility, fecal gas emissions, and meat quality in finishing pigs. J Appl Anim Res 48:397–401. https://doi.org/10.1080/09712119.2020.1813739

Daniele C, Coon JT, Pittler MH, Ernst E (2005) Vitex agnus castus. Drug Saf 28:319–332. https://doi.org/10.2165/00002018-200,528,040-00004

Darmawan A, Hermana W, Suci DM et al (2022) Dietary phytogenic extracts favorably influence productivity, egg quality, blood constituents, antioxidant and immunological parameters of laying hens: a meta-analysis. Animals 12:2278. https://doi.org/10.3390/ani12172278

de Armas-Ricard M, Ruiz-Reyes E, Ramírez-Rodríguez O (2019) Caffeates and caffeamides: synthetic methodologies and their antioxidant properties. Int J Med Chem 2019:1–15. https://doi.org/10.1155/2019/2592609

de Boer IJM, Cornelissen AMG (2002) A method using sustainability indicators to compare conventional and animal-friendly egg production systems. Poult Sci 81:173–181. https://doi.org/10.1093/ps/81.2.173

de Koning C, Barekatain R, Singh M, Drake K (2019) Saltbush (Atriplex nummularia and A. amnicola) as potential plants for free-range layer farms: consequences for layer performance, egg sensory qualities, and excreta moisture. Poult Sci 98:4555–4564. https://doi.org/10.3382/ps/pez294

de Menezes PC, de Lima ER, de Medeiros JP et al (2012) Egg quality of laying hens in different conditions of storage, ages and housing densities. Rev Bras Zootec 41:2064–2069. https://doi.org/10.1590/S1516-35982012000900014

de Olde EM, van der Linden A, olde Bolhaar LD, de Boer IJM (2020) Sustainability challenges and innovations in the Dutch egg sector. J Clean Prod 258:120974. https://doi.org/10.1016/j.jclepro.2020.120974

de Vries M, de Boer IJM (2010) Comparing environmental impacts for livestock products: a review of life cycle assessments. Livest Sci 128:1–11. https://doi.org/10.1016/j.livsci.2009.11.007

Dekker SEM, de Boer IJM, van Krimpen M et al (2013) Effect of origin and composition of diet on ecological impact of the organic egg production chain. Livest Sci 151:271–283. https://doi.org/10.1016/j.livsci.2012.11.013

Devi K, Palod J, Singh D et al (2017) Effect of graded levels of Pudina (Mentha arvensis) leaf powder supplementation on certain serum biochemical parameters in laying hens. Int J Livest Res 7:206–216. https://doi.org/10.5455/ijlr.20170520042932

Dilawar MA, Mun HS, Rathnayake D et al (2021) Egg quality parameters, production performance and immunity of laying hens supplemented with plant extracts. Animals 11:975. https://doi.org/10.3390/ani11040975

Drewnowski A (2010) The nutrient rich foods index helps to identify healthy, affordable foods. Am J Clin Nutr 91:1095S–1101S. https://doi.org/10.3945/ajcn.2010.28450D

Dunn IC, De Koning D-J, McCormack HA et al (2021) No evidence that selection for egg production persistency causes loss of bone quality in laying hens. Genet Sel Evol 53:11. https://doi.org/10.1186/s12711-021-00603-8

EC (2016) Directive (EU) 2016/2284 of the European Parliament and of the Council of 14 December 2016 on the reduction of national emissions of certain atmospheric pollutants, amending Directive 2003/35/EC and repealing Directive 2001/81/EC

EC (2021) European citizens' initiative: commission to propose phasing out of cages for farm animals. In: European Commission. https://ec.europa.eu/commission/presscorner/detail/en/ip_21_3297. Accessed 15 Oct 2022

Ekiert H, Świątkowska J, Knut E et al (2021) Artemisia dracunculus (tarragon): a review of its traditional uses, phytochemistry and pharmacology. Front Pharmacol 12

El-Hack MEA, Alagawany M (2015) Performance, egg quality, blood profile, immune function, and antioxidant enzyme activities in laying hens fed diets with thyme powder. J Anim Feed Sci. 24:127–133. https://doi.org/10.22358/jafs/65638/2015

El-Laithy S (2009) Effect of induced molting on egg quality and some blood constituents in Hy-Line hens. Ann Agric Sci Moshtohor 47:1–13

El-Maaty HA, Sherif SK (2016) The efficacy of stocking density and aromatic plant powder supplemented-diets for H&N laying hens strain during high environmental temperature. Egypt J Nutr Feeds 19:573–584

El-Saadany AS, Hanafy MM, Elkomy AE (2022) Flaxseed and Agnus-castuson vitex as a source of phytoestrogens and their impact on productive performance, some blood constituents, and blood oestradiol profile of aged laying hens. Ital J Anim Sci 21:821–830. https://doi.org/10.1080/1828051X.2022.2066578

El-Shall NA, Abd El-Hack ME, Albaqami NM et al (2022) Phytochemical control of poultry coccidiosis: a review. Poult Sci 101:101542. https://doi.org/10.1016/j.psj.2021.101542

Engel K, Pankoke H, Jünemann S et al (2020) Family matters: skin microbiome reflects the social group and spatial proximity in wild zebra finches. BMC Ecol 20:58. https://doi.org/10.1186/s12898-020-00326-2

Ermumcu MŞK (2022) Chapter 4 - Biological activities of black cumin (Nigella sativa) seed oil. In: Mariod AA (ed) Multiple biological activities of unconventional seed oils. Academic Press, pp 43–53

Ershadi SZ, Heidari MD, Dutta B et al (2021) Comparative life cycle assessment of technologies and strategies to improve nitrogen use efficiency in egg supply chains. Resour Conserv Recycl 166:105275. https://doi.org/10.1016/j.resconrec.2020.105275

Estrada-González IE, Taboada-González PA, Guerrero-García-Rojas H, Márquez-Benavides L (2020) Decreasing the environmental impact in an egg-producing farm through the application of LCA and lean tools. Appl Sci 10:1352. https://doi.org/10.3390/app10041352

Fan H, Wu J (2022) Conventional use and sustainable valorization of spent egg-laying hens as functional foods and biomaterials: a review. Bioresour Bioprocess 9:43. https://doi.org/10.1186/s40643-022-00529-z

FAO (2012) Sustainability assessment of food and agriculture systems guidelines (SAFA). FAO, Rome, Italy

FAO (2022) FAOSTAT. Food Agric. Organ. U. N. FAO, In. Accessed 13 Oct 2022

Fatemi A, Razavi SM, Asasi K, Torabi Goudarzi M (2015) Effects of Artemisia annua extracts on sporulation of Eimeria oocysts. Parasitol Res 114:1207–1211. https://doi.org/10.1007/s00436-014-4304-z

Fawaz M, Südekum K-H, Ahmed Hassan H, Abdel-Wareth A (2022) Productive, physiological and nutritional responses of laying hens fed different dietary levels of turmeric powder. J Anim Physiol Anim Nutr. https://doi.org/10.1111/jpn.13686

Feng J, Lu M, Wang J et al (2021) Dietary oregano essential oil supplementation improves intestinal functions and alters gut microbiota in late-phase laying hens. J Anim Sci Biotechnol 12:72. https://doi.org/10.1186/s40104-021-00600-3

Ferreira PMP, Farias DF, de Oliveira JTA, de Carvalho AFU (2008) Moringa oleifera: bioactive compounds and nutritional potential. Rev Nutr 21:431–437. https://doi.org/10.1590/S1415-52732008000400007

Flachowsky G, Meyer U, Südekum K-H (2018) Invited review: Resource inputs and land, water and carbon footprints from the production of edible protein of animal origin. Arch Anim Breed 61:17–36. https://doi.org/10.5194/aab-61-17-2018

Florou-Paneri P, Nikolakakis I, Giannenas I et al (2005) Hen performance and egg quality as affected by dietary oregano essential oil and alpha-tocopheryl acetate supplementation. Int J Poult Sci 4:449–454. https://doi.org/10.3923/ijps.2005.449.454

Franz CM, Baser KHC, Hahn-Ramssl I (2020) Chapter 3 - Herbs and aromatic plants as feed additives: aspects of composition, safety, and registration rules. In: Florou-Paneri P, Christaki E, Giannenas I (eds) Feed additives. Academic Press, pp 35–56

Galobart J, Barroeta AC, Baucells MD et al (2001) Effect of dietary supplementation with rosemary extract and alpha-tocopheryl acetate on lipid oxidation in eggs enriched with omega3-fatty acids. Poult Sci 80:460–467. https://doi.org/10.1093/ps/80.4.460

Gao S, Li R, Heng N et al (2020) Effects of dietary supplementation of natural astaxanthin from Haematococcus pluvialis on antioxidant capacity, lipid metabolism, and accumulation in the egg yolk of laying hens. Poult Sci 99:5874–5882. https://doi.org/10.1016/j.psj.2020.08.029

Gautron J, Réhault-Godbert S, Van de Braak TGH, Dunn IC (2021) Review: What are the challenges facing the table egg industry in the next decades and what can be done to address them? Animal 15:100282. https://doi.org/10.1016/j.animal.2021.100282

Ghasemi R, Zarei M, Torki M (2010) Adding medicinal herbs including garlic (Allium sativum) and thyme (Thymus vulgaris) to diet of laying hens and evaluating productive performance and egg quality characteristics. Am J Anim Vet Sci 5:6

Giannenas I, Sidiropoulou E, Bonos E et al (2020) Chapter 1 - The history of herbs, medicinal and aromatic plants, and their extracts: Past, current situation and future perspectives. In: Florou-Paneri P, Christaki E, Giannenas I (eds) Feed additives. Academic Press, pp 1–18

Giersberg MF, Kemper N (2018) Rearing male layer chickens: a German perspective. Agriculture 8:176. https://doi.org/10.3390/agriculture8110176

Gongruttananun N, Kochagate P, Poonpan K et al (2017) Effects of an induced molt using cassava meal on body weight loss, blood physiology, ovarian regression, and postmolt egg production in late-phase laying hens. Poult Sci 96:1925–1933. https://doi.org/10.3382/ps/pew457

Grandhaye J, Douard V, Rodriguez-Mateos A et al (2020) Microbiota changes due to grape seed extract diet improved intestinal homeostasis and decreased fatness in parental broiler hens. Microorganisms 8:1141. https://doi.org/10.3390/microorganisms8081141

Grigoriadou K, Krigas N, Lazari D, Maloupa E (2020) Chapter 4 - Sustainable use of Mediterranean medicinal-aromatic plants. In: Florou-Paneri P, Christaki E, Giannenas I (eds) Feed additives. Academic Press, pp 57–74

Gründemann C, Lengen K, Sauer B et al (2014) Equisetum arvense (common horsetail) modulates the function of inflammatory immunocompetent cells. BMC Complement Altern Med 14:283. https://doi.org/10.1186/1472-6882-14-283

Gu YF, Chen YP, Jin R et al (2021a) Age-related changes in liver metabolism and antioxidant capacity of laying hens. Poult Sci 100:101478. https://doi.org/10.1016/j.psj.2021.101478

Gu YF, Chen YP, Jin R et al (2021b) A comparison of intestinal integrity, digestive function, and egg quality in laying hens with different ages. Poult Sci 100:100949. https://doi.org/10.1016/j.psj.2020.12.046

Guillaume A, Hubatová-Vacková A, Kočí V (2022) Environmental impacts of egg production from a life cycle perspective. Agriculture 12(3):355

Gultepe EE, Iqbal A, Cetingul IS et al (2020) Effect of Myrtus communis L. plant extract as a drinking water supplement on performance, some blood parameters, egg quality and immune response of older laying hens. Kafkas Univ Vet Fak Derg 26:9–16. https://doi.org/10.9775/kvfd.2019.22058

Guo Y, Zhao Z-H, Pan Z-Y et al (2020) New insights into the role of dietary marine-derived polysaccharides on productive performance, egg quality, antioxidant capacity, and jejunal morphology in late-phase laying hens. Poult Sci 99:2100–2107. https://doi.org/10.1016/j.psj.2019.12.032

Guo S, Lei J, Liu L et al (2021) Effects of Macleaya cordata extract on laying performance, egg quality, and serum indices in Xuefeng black-bone chicken. Poult Sci 100:101031. https://doi.org/10.1016/j.psj.2021.101031

Gurbuz E, Balevi T, Kurtoglu V, Oznurlu Y (2011) Use of yeast cell walls and Yucca schidigera extract in layer hens' diets. Ital J Anim Sci 10:e26. https://doi.org/10.4081/ijas.2011.e26

Güroy B, Mantoğlu S, Merrifield DL, Guroy D (2016) Effects of dietary Nutrafito Plus on growth, haemotological parameters and total ammonia-nitrogen excretion of juvenile striped catfish Pangasianodon hypophthalmus. Aquac Res 47:1770–1777. https://doi.org/10.1111/are.12634

Habibi H, Firouzi S, Nili H et al (2016) Anticoccidial effects of herbal extracts on Eimeria tenella infection in broiler chickens: in vitro and in vivo study. J Parasit Dis Off Organ Indian Soc Parasitol 40:401–407. https://doi.org/10.1007/s12639-014-0517-4

Hakim MM, Patel IC (2020) A review on phytoconstituents of marine brown algae. Fut J Pharm Sci 6:129. https://doi.org/10.1186/s43094-020-00147-6

Hamilton H, Henry R, Rounsevell M et al (2020) Exploring global food system shocks, scenarios and outcomes. Futures 123:102601. https://doi.org/10.1016/j.futures.2020.102601

Hanssen SA, Hasselquist D, Folstad I, Erikstad KE (2004) Costs of immunity: immune responsiveness reduces survival in a vertebrate. Proc R Soc B Biol Sci 271:925–930

Harrington D, Decap S (2021) Mitigating the impact of ammonia emissions from poultry. Feed Addit Mag. 2021:3

Harrington D, Hall H, Wilde D, Wakeman W (2020) Chapter 11 - Application of aromatic plants and their extracts in the diets of laying hens. In: Florou-Paneri P, Christaki E, Giannenas I (eds) Feed additives. Academic Press, pp 187–203

Hassan SM (2016) Effects of adding different dietary levels of turmeric (Curcuma longa Linn) powder on productive performance and egg quality of laying hens. Int J Poult Sci 15:156–160

Hassan HA, Ragab MS (2007) Effects of using black seeds on egg production, egg quality and immune response in laying diets varying in their protein content. Fayoum J Agric Res Dev. 21: 206–224. https://doi.org/10.21608/fjard.2007.197559

Hayes M, Xin H, Li H et al (2013) Ammonia, greenhouse gas, and particulate matter concentrations and emissions of aviary layer houses in the midwestern USA. Trans ASABE 56:1921–1932. https://doi.org/10.13031/2013.41575

He X, Hao D, Liu C et al (2017) Effect of supplemental oregano essential oils in diets on production performance and relatively intestinal parameters of laying hens. Am J Mol Biol 7:73–85. https://doi.org/10.4236/ajmb.2017.71006

Heng N, Gao S, Chen Y et al (2021) Dietary supplementation with natural astaxanthin from Haematococcus pluvialis improves antioxidant enzyme activity, free radical scavenging ability, and gene expression of antioxidant enzymes in laying hens. Poult Sci 100:101045. https://doi.org/10.1016/j.psj.2021.101045

Holt PS (1992a) Effects of induced moulting on immune responses of hens. Br Poult Sci 33:165–175. https://doi.org/10.1080/00071669208417454

Holt PS (1992b) Effect of induced molting on B cell and CT4 and CT8 Tcell numbers in spleens and peripheral blood of white leghorn hens. Poult Sci 71:2027–2034. https://doi.org/10.3382/ps.0712027

Hristov AN, Ivan M, Neill L, McAllister TA (2003) Evaluation of several potential bioactive agents for reducing protozoal activity in vitro. Anim Feed Sci Technol 105:163–184. https://doi.org/10.1016/S0377-8401(03)00060-9

Huang C-B, Xiao L, Xing S-C et al (2019) The microbiota structure in the cecum of laying hens contributes to dissimilar H_2S production. BMC Genomics 20:770. https://doi.org/10.1186/s12864-019-6115-1

Huang J, Tong XF, Yu ZW et al (2020) Dietary supplementation of total flavonoids from Rhizoma Drynariae improves bone health in older caged laying hens. Poult Sci 99:5047–5054. https://doi.org/10.1016/j.psj.2020.06.057

Huo S, Li Y, Guo Y et al (2020) Improving effects of Epimedium flavonoids on the selected reproductive features in layer hens after forced molting. Poult Sci 99:2757–2765. https://doi.org/10.1016/j.psj.2019.12.053

Jahanian E, Jahanian R, Rahmani H-R, Alikhani M (2017) Dietary supplementation of Echinacea purpurea powder improved performance, serum lipid profile, and yolk oxidative stability in laying hens. J Appl Anim Res 45:45–51. https://doi.org/10.1080/09712119.2015.1091344

Jang S-A, Hwang Y-H, Kim T et al (2019) Anti-osteoporotic and anti-adipogenic effects of the water extract of Drynaria roosii Nakaike in ovariectomized mice fed a high-fat diet. Molecules 24:3051. https://doi.org/10.3390/molecules24173051

Jarzębski M, Jarzębski W, Baranowska HM et al (2020) Characterization of St. John's wort (Hypericum perforatum L.) and the impact of filtration process on bioactive extracts incorporated into carbohydrate-based hydrogels. Food Hydrocoll 104:105748. https://doi.org/10.1016/j.foodhyd.2020.105748

Jeni RE, Dittoe DK, Olson EG et al (2021) An overview of health challenges in alternative poultry production systems. Poult Sci 100:101173. https://doi.org/10.1016/j.psj.2021.101173

Jiménez GG, Durán AG, Macías FA, Simonet AM (2021) Structure, bioactivity and analytical methods for the determination of yucca saponins. Molecules 26:5251. https://doi.org/10.3390/molecules26175251

Joat N, Van TTH, Stanley D et al (2021) Temporal dynamics of gut microbiota in caged laying hens: a field observation from hatching to end of lay. Appl Microbiol Biotechnol 105: 4719–4730. https://doi.org/10.1007/s00253-021-11,333-8

Joyner CJ, Peddie MJ, Taylor TG (1987) The effect of age on egg production in the domestic hen. Gen Comp Endocrinol 65:331–336. https://doi.org/10.1016/0016-6480(87)90117-1

Jugdaohsingh R, Tucker KL, Qiao N et al (2004) Dietary silicon intake is positively associated with bone mineral density in men and premenopausal women of the Framingham Offspring cohort. J Bone Miner Res Off J Am Soc Bone Miner Res 19:297–307. https://doi.org/10.1359/JBMR.0301225

Kang HK, Kim J-H, Hwangbo J, Kim CH (2015) Effects of dietary supplementation of vitamin C and sea buckthorn on the performance and meat quality in old laying hens. Korean J Poult Sci 42:181–189. https://doi.org/10.5536/KJPS.2015.42.2.181

Karásková K, Suchý P, Straková E (2016) Current use of phytogenic feed additives in animal nutrition: a review. Czech J Anim Sci. 60:521–530. https://doi.org/10.17221/8594-CJAS

Kaur A, Bhatti R (2021) Understanding the phytochemistry and molecular insights to the pharmacology of Angelica archangelica L. (garden angelica) and its bioactive components. Phytother Res 35:5961–5979. https://doi.org/10.1002/ptr.7206

Kers JG, Velkers FC, Fischer EAJ et al (2018) Host and environmental factors affecting the intestinal microbiota in chickens. Front Microbiol 9:235. https://doi.org/10.3389/fmicb.2018.00235

Khan S, Moore RJ, Stanley D, Chousalkar KK (2020) The gut microbiota of laying hens and its manipulation with prebiotics and probiotics to enhance gut health and food safety. Appl Environ Microbiol 86:e00600–e00620. https://doi.org/10.1128/AEM.00600-20

Kim CH, Kang HK (2015) Effect of dietary supplementation with a chlorella by-product on the performance, immune response and metabolic function in laying hens. Eur Poult Sci 79. https://doi.org/10.1399/eps.2015.108

Koenen ME, Boonstra-Blom AG, Jeurissen SHM (2002) Immunological differences between layer- and broiler-type chickens. Vet Immunol Immunopathol 89:47–56. https://doi.org/10.1016/S0165-2427(02)00169-1

Kogut MH (2009) Impact of nutrition on the innate immune response to infection in poultry1. J Appl Poult Res 18:111–124. https://doi.org/10.3382/japr.2008-00081

Kothari D, Lee W-D, Niu K-M, Kim S-K (2019) The genus Allium as poultry feed additive: a review. Animals 9:1032. https://doi.org/10.3390/ani9121032

Koutsos EA, Klasing KC (2014) Chapter 17 - Factors modulating the avian immune system. In: Schat KA, Kaspers B, Kaiser P (eds) Avian immunology, 2nd edn. Academic Press, Boston, pp 299–313

Kropáčková L, Pechmanová H, Vinkler M et al (2017) Variation between the oral and faecal microbiota in a free-living passerine bird, the great tit (Parus major). PLoS One 12:e0179945. https://doi.org/10.1371/journal.pone.0179945

Kulshreshtha G, Rathgeber B, MacIsaac J et al (2017) Feed supplementation with red seaweeds, Chondrus crispus and Sarcodiotheca gaudichaudii, reduce Salmonella enteritidis in laying hens. Front Microbiol 8:567

Kulshreshtha G, Hincke MT, Prithiviraj B, Critchley A (2020) A review of the varied uses of macroalgae as dietary supplements in selected poultry with special reference to laying hen and broiler chickens. J Mar Sci Eng 8:536. https://doi.org/10.3390/jmse8070536

Kursa O, Tomczyk G, Adamska K et al (2022) The microbial community of the respiratory tract of commercial chickens and turkeys. Microorganisms 10:987. https://doi.org/10.3390/microorganisms10050987

Laganá C, Pizzolante CC, Saldanha E, de Moraes J (2011) Turmeric root and Annato seed in second-cycle layer diets: performance and egg quality. Braz J Poult Sci 13:171–176. https://doi.org/10.1590/S1516-635X2011000300002

Law C, Exley C (2011) New insight into silica deposition in horsetail (Equisetum arvense). BMC Plant Biol 11:112. https://doi.org/10.1186/1471-2229-11-112

Lee C-P, Yen G-C (2006) Antioxidant activity and bioactive compounds of tea seed (Camellia oleifera abel.) oil. J Agric Food Chem 54:779–784. https://doi.org/10.1021/jf052325a

Lee S, La T-M, Lee H-J et al (2019) Characterization of microbial communities in the chicken oviduct and the origin of chicken embryo gut microbiota. Sci Rep 9:6838. https://doi.org/10.1038/s41598-019-43,280-w

Leinonen I, Kyriazakis I (2016) How can we improve the environmental sustainability of poultry production? Proc Nutr Soc 75:265–273. https://doi.org/10.1017/S0029665116000094

Leroy F, Abraini F, Beal T et al (2022) Animal board invited review: Animal source foods in healthy, sustainable, and ethical diets – an argument against drastic limitation of livestock in the food system. Animal 16:100457. https://doi.org/10.1016/j.animal.2022.100457

Li G-M, Liu L-P, Yin B et al (2020) Heat stress decreases egg production of laying hens by inducing apoptosis of follicular cells via activating the FasL/Fas and TNF-α systems. Poult Sci 99:6084–6093. https://doi.org/10.1016/j.psj.2020.07.024

Li Y, Allacker K, Feng H et al (2021) Net zero energy barns for industrial egg production: an effective sustainable intensification strategy? J Clean Prod 316:128014. https://doi.org/10.1016/j.jclepro.2021.128014

Ligas B, Izydorczyk G, Mikula K et al (2021) Valorization of postextraction residues—analysis of the influence of new feed additives with micronutrients on eggs quality parameters. Poult Sci 100:101416. https://doi.org/10.1016/j.psj.2021.101416

Lin L, Liu Y, Huang J et al (2017) Medicinal plants of the genus Macleaya (Macleaya cordata, Macleaya microcarpa): a review of their phytochemistry, pharmacology, and toxicology. Phytother Res 32. https://doi.org/10.1002/ptr.5952

Liu B, Jiang J, Yu D et al (2020a) Effects of supplementation of microalgae (Aurantiochytrium sp.) to laying hen diets on fatty acid content, health lipid indices, oxidative stability, and quality attributes of meat. Foods 9:1271. https://doi.org/10.3390/foods9091271

Liu M, Lu Y, Gao P et al (2020b) Effect of curcumin on laying performance, egg quality, endocrine hormones, and immune activity in heat-stressed hens. Poult Sci 99:2196–2202. https://doi.org/10.1016/j.psj.2019.12.001

Loetscher Y, Kreuzer M, Messikommer RE (2014) Late laying hens deposit dietary antioxidants preferentially in the egg and not in the body. J Appl Poult Res 23:647–660. https://doi.org/10.3382/japr.2014-00973

Lokaewmanee K, Yamauchi K, Komori T, Saito K (2014) Eggshell quality, eggshell structure and small intestinal histology in laying hens fed dietary Pantoea-6® and plant extracts. Ital J Anim Sci 13:3163. https://doi.org/10.4081/ijas.2014.3163

Luan F, Zeng J, Yang Y et al (2020) Recent advances in Camellia oleifera Abel: a review of nutritional constituents, biofunctional properties, and potential industrial applications. J Funct Foods 75:104242. https://doi.org/10.1016/j.jff.2020.104242

Ma D, Shan A, Chen Z et al (2005) Effect of Ligustrum lucidum and Schisandra chinensis on the egg production, antioxidant status and immunity of laying hens during heat stress. Arch Anim Nutr 59:439–447. https://doi.org/10.1080/17450390500353499

MacLeod MG (2011) 19 - Environmental sustainability of egg production and processing. In: van Immersel F, Nys Y, Bain M (eds) Improving the safety and quality of eggs and egg products. Woodhead Publishing, pp 445–462

Magnuson AD, Sun T, Yin R et al (2018) Supplemental microalgal astaxanthin produced coordinated changes in intrinsic antioxidant systems of layer hens exposed to heat stress. Algal Res 33:84–90. https://doi.org/10.1016/j.algal.2018.04.031

Mahboubi M (2019) Caraway as important medicinal plants in management of diseases. Nat Prod Bioprospect 9:1–11. https://doi.org/10.1007/s13659-018-0190-x

Maidin MBM, McCormack HA, Wilson PW et al (2021) Dietary betaine reduces plasma homocysteine concentrations and improves bone strength in laying hens. Br Poult Sci 62:573–578. https://doi.org/10.1080/00071668.2021.1883550

Makkar HPS, Aregheore EM, Becker K (1999) Effect of saponins and plant extracts containing saponins on the recovery of ammonia during urea-ammoniation of wheat straw and fermentation kinetics of the treated straw. J Agric Sci 132:313–321. https://doi.org/10.1017/S0021859699006334

Maksoud S, Abdel-Massih RM, Rajha HN et al (2021) Citrus aurantium L. active constituents, biological effects and extraction methods. an updated review. Molecules 26:5832. https://doi.org/10.3390/molecules26195832

Mao Q-Q, Xu X-Y, Cao S-Y et al (2019) Bioactive compounds and bioactivities of ginger (Zingiber officinale Roscoe). Foods 8:185. https://doi.org/10.3390/foods8060185

Matkowski A, Piotrowska M (2006) Antioxidant and free radical scavenging activities of some medicinal plants from the Lamiaceae. Fitoterapia 77:346–353. https://doi.org/10.1016/j.fitote.2006.04.004

Matusiak K, Oleksy M, Borowski S et al (2016) The use of Yucca schidigera and microbial preparation for poultry manure deodorization and hygienization. J Environ Manag 170:50–59. https://doi.org/10.1016/j.jenvman.2016.01.007

McKay DL, Blumberg JB (2006) A review of the bioactivity and potential health benefits of chamomile tea (Matricaria recutita L.). Phytother Res 20:519–530. https://doi.org/10.1002/ptr.1900

Mehta SD, Rathore P, Rai G (2021) Ginseng: pharmacological action and phytochemistry prospective. In: Hano C, Chen J (eds) Ginseng - modern aspects of the famed traditional medicine. IntechOpen, pp 1–15

Memarzia A, Khazdair MR, Behrouz S et al (2021) Experimental and clinical reports on anti-inflammatory, antioxidant, and immunomodulatory effects of Curcuma longa and curcumin, an updated and comprehensive review. BioFactors Oxf Engl 47:311–350. https://doi.org/10.1002/biof.1716

Menaa F, Wijesinghe U, Thiripuranathar G et al (2021) Marine algae-derived bioactive compounds: a new wave of nanodrugs? Mar Drugs 19:484. https://doi.org/10.3390/md19090484

Mench JA, Sumner DA, Rosen-Molina JT (2011) Sustainability of egg production in the United States - The policy and market context. Poult Sci 90:229–240. https://doi.org/10.3382/ps.2010-00844

Metwally DM, Al-Olayan EM, Alanazi M et al (2018) Antischistosomal and anti-inflammatory activity of garlic and allicin compared with that of praziquantel in vivo. BMC Complement Altern Med 18:135. https://doi.org/10.1186/s12906-018-2191-z

Michalak I, Mahrose K (2020) Seaweeds, intact and processed, as a valuable component of poultry feeds. J Mar Sci Eng 8:620. https://doi.org/10.3390/jmse8080620

Minetti M, Bernardini G, Biazzo M et al (2019) Padina pavonica extract promotes in vitro differentiation and functionality of human primary osteoblasts. Mar Drugs 17:E473. https://doi.org/10.3390/md17080473

Mishra B, Jha R (2019) Oxidative stress in the poultry gut: potential challenges and interventions. Front Vet Sci 6:60. https://doi.org/10.3389/fvets.2019.00060

Mobarez Samia M, Rizk AM, Abdel Latif AM, El-Sayed OA (2018) Effect of supplementing diet with Spirulina platensis algae or turmeric on productive and reproductive performance of Golden Montazah layers. Egypt Poult Sci J. 38:109–125. https://doi.org/10.21608/epsj.2018.5596

Mohammadi Gheisar M, Kim IH (2018) Phytobiotics in poultry and swine nutrition – a review. Ital J Anim Sci 17:92–99. https://doi.org/10.1080/1828051X.2017.1350120

Mohammed HA, Al-Hameed SAM (2021) Effect of dietary black cumin seeds (Nigella stavia), garlic (Allium sativum) and lettuce leaves (Lactuca sativa) on performance and egg quality traits of native layer hens. IOP Conf Ser Earth Environ Sci 910:012001. https://doi.org/10.1088/1755-1315/910/1/012001

Molnár S, Szőllősi L (2020) Sustainability and quality aspects of different table egg production systems: a literature review. Sustainability 12:7884. https://doi.org/10.3390/su12197884

Moon S-G, Lee S-K, Lee W-D et al (2021) Effect of dietary supplementation of a phytogenic blend containing Schisandra chinensis, Pinus densiflora, and Allium tuberosum on productivity, egg quality, and health parameters in laying hens. Anim Biosci 34:285–294. https://doi.org/10.5713/ajas.20.0552

Morgado ME, Jiang C, Zambrana J et al (2021) Climate change, extreme events, and increased risk of salmonellosis: foodborne diseases active surveillance network (FoodNet), 2004–2014. Environ Health 20:105. https://doi.org/10.1186/s12940-021-00787-y

Morris SS, Beesabathuni K, Headey D (2018) An egg for everyone: pathways to universal access to one of nature's most nutritious foods. Matern Child Nutr 14:e12679. https://doi.org/10.1111/mcn.12679

Mosayyeb Zadeh A, Mirghelenj SA, Hasanlou P, Shakouri Alishah H (2020) Effects of turmeric (Curcuma longa) powder supplementation in laying hens' diet on production performance, blood biochemical parameters and egg quality traits. J Anim Physiol Anim Nutr. https://doi.org/10.1111/jpn.13746

Mousa B, Awad A, Alhamdani H et al (2019) Inclusion of garlic (Allium sativum.) and turmeric (Curcuma longa L.) powder to laying hens' diets on egg quality traits, bacterial population and intestinal histomorphology. Ann Trop Med Public Health. 22:231–240. https://doi.org/10.36295/ASRO.2019.221224

Mousavi A, Mahdavi AH, Riasi A, Soltani-Ghombavani M (2017) Synergetic effects of essential oils mixture improved egg quality traits, oxidative stability and liver health indices in laying hens fed fish oil. Anim Feed Sci Technol 234:162–172. https://doi.org/10.1016/j.anifeedsci.2017.10.001

Muthamilselvan T, Kuo T-F, Wu Y-C, Yang W-C (2016) Herbal remedies for coccidiosis control: a review of plants, compounds, and anticoccidial actions. Evid Based Complement Alternat Med 2016:e2657981. https://doi.org/10.1155/2016/2657981

Naidoo V, Mcgaw L, Bisschop S et al (2008) The value of plant extracts with antioxidant activity in attenuating coccidiosis in broiler chickens. Vet Parasitol 153:214–219. https://doi.org/10.1016/j.vetpar.2008.02.013

Nakade M, Balogh K, Zándoki E et al (2016) Effect of short-term aflatoxin exposure in combination with medicinal herb mixture on lipid peroxidation and glutathione redox system in laying hens. Columella J Agric Environ Sci 3:65–73. https://doi.org/10.18380/SZIE.COLUM.2016.3.1.65

Nasiroleslami M, Torki M (2010) Including essential oils of fennel (Foeniculum vulgare) and ginger (Zingiber officinale) to diet and evaluating performance of laying hens, white blood cell count and egg quality characteristics. Adv Environ Biol 4:341–345

Nathan DB, Heller ED, Perek M (1977) The effect of starvation on antibody production of chicks. Poult Sci 56:1468–1471. https://doi.org/10.3382/ps.0561468

Nawab A, Li G, An L et al (2019) Effect of curcumin supplementation on TLR4 mediated non-specific immune responses in liver of laying hens under high-temperature conditions. J Therm Biol 84:384–397. https://doi.org/10.1016/j.jtherbio.2019.07.003

Nawab A, Tang S, Li G et al (2020) Dietary curcumin supplementation effects on blood immunological profile and liver enzymatic activity of laying hens after exposure to high temperature conditions. J Therm Biol 90:102573. https://doi.org/10.1016/j.jtherbio.2020.102573

Nia EA, Mehri M, Shirmohammad F (2021) Effects of dietary Spirulina platensis algae powder on performance, egg quality, ovarian follicles and immune system in Lohmann LSL laying hens. Iran J Anim Sci. 51:349–359. https://doi.org/10.22059/ijas.2021.314088.653810

Nidamanuri AL, Murugesan S, Mahapatra RK (2017) Effect of heat stress on physiological parameters of layers - a review. Int J Livest Res 7:1–17

Nys YY, Sauveur B (2004) Valeur nutritionnelle des oeufs. Prod Anim Inst Natl Rech Agron 17:385–393

Ochieng PE, Scippo M-L, Kemboi DC et al (2021) Mycotoxins in poultry feed and feed ingredients from Sub-Saharan Africa and their impact on the production of broiler and layer chickens: a review. Toxins 13:633. https://doi.org/10.3390/toxins13090633

Odihambo Mumma J, Thaxton JP, Vizzier-Thaxton Y, Dodson WL (2006) Physiological stress in laying hens1. Poult Sci 85:761–769. https://doi.org/10.1093/ps/85.4.761

O'Keefe T (2022) US egg producers predict eggs not 100% cage free in 2040. In: WATTPoultry.com. https://www.wattagnet.com/articles/44100-us-egg-producers-predict-eggs-not-100-cage-free-in-2040. Accessed 25 Sep 2022

Oladeji OS, Adelowo FE, Ayodele DT, Odelade KA (2019) Phytochemistry and pharmacological activities of Cymbopogon citratus: a review. Sci Afr 6:e00137. https://doi.org/10.1016/j.sciaf.2019.e00137

Olgun O (2016) The effect of dietary essential oil mixture supplementation on performance, egg quality and bone characteristics in laying hens. Ann Anim Sci 16:1115–1125. https://doi.org/10.1515/aoas-2016-0038

Ordonez G, Llopis N, Penalver P (2008) Efficacy of eugenol against a Salmonella enterica serovar Enteritidis experimental infection in commercial layers in production. J Appl Poult Res 17:376–382

Özek K (2012) Effects of dietary herbal essential oil mixture and/or mannan-oligosaccharide supplementation on laying performance, some serum biochemical markers and humoral immunity in laying hens exposed to heat. Rev Méd Vét 163:153–159

Pampouille E, Gautron J, Même N, et al (2022) Existe-t-il des marqueurs metaboliques de la qualité de coquille chez la poule pondeuse. In: 14th Journées de la Recherche Avicole et Palmipèdes à Foie Gras, 9–10 Mars 2022

Panaite TD, Vlaicu PA, Saracila M et al (2022) Impact of watermelon rind and sea buckthorn meal on performance, blood parameters, and gut microbiota and morphology in laying hens. Agriculture 12:177. https://doi.org/10.3390/agriculture12020177

Park J-H, Kim Y-M, Kim I-H (2018) Egg production, egg quality, blood profiles, cecal microflora, and excreta noxious gas emission in laying hens fed with fenugreek (Trigonella foenum-graecum L.) seed extract. J Poult Sci 55:47–53. https://doi.org/10.2141/jpsa.0170011

Patel S (2012) Yucca: a medicinally significant genus with manifold therapeutic attributes. Nat Prod Bioprospect 2:231–234. https://doi.org/10.1007/s13659-012-0090-4

Paul BK, Butterbach-Bahl K, Notenbaert A et al (2020) Sustainable livestock development in low- and middle-income countries: shedding light on evidence-based solutions. Environ Res Lett 16: 011001. https://doi.org/10.1088/1748-9326/abc278

Pelletier N (2018) Changes in the life cycle environmental footprint of egg production in Canada from 1962 to 2012. J Clean Prod 176:1144–1153. https://doi.org/10.1016/j.jclepro.2017.11.212

Pérez-Jiménez J, Neveu V, Vos F, Scalbert A (2010) Identification of the 100 richest dietary sources of polyphenols: an application of the phenol-explorer database. Eur J Clin Nutr 64:112–120

Petit PR, Sauvaire YD, Hillaire-Buys DM et al (1995) Steroid saponins from fenugreek seeds: extraction, purification, and pharmacological investigation on feeding behavior and plasma cholesterol. Steroids 60:674–680. https://doi.org/10.1016/0039-128x(95)00090-d

Platel K, Srinivasan K (2004) Digestive stimulant action of spices: a myth or reality? Indian J Med Res 119:167

Poore J, Nemecek T (2018) Reducing food's environmental impacts through producers and consumers. Science 360:987–992. https://doi.org/10.1126/science.aaq0216

Popescu RG, Voicu SN, Gradisteanu Pircalabioru G et al (2020) Effects of dietary inclusion of bilberry and walnut leaves powder on the digestive performances and health of Tetra SL laying hens. Animals 10:823. https://doi.org/10.3390/ani10050823

Preisinger R (2013) Layer breeding for the future. In: Poultry world. https://www.poultryworld.net/poultry/layer-breeding-for-the-future/. Accessed 12 Oct 2022

Preisinger R (2018) Innovative layer genetics to handle global challenges in egg production. Br Poult Sci 59:1–6. https://doi.org/10.1080/00071668.2018.1401828

Puvača N, Lika E, Tufarelli V et al (2020) Influence of different tetracycline antimicrobial therapy of Mycoplasma (Mycoplasma synoviae) in laying hens compared to tea tree essential oil on table egg quality and antibiotic residues. Foods 9:612. https://doi.org/10.3390/foods9050612

Qin S, Hou D-X (2017) The biofunctions of phytochemicals and their applications in farm animals: the Nrf2/Keap1 system as a target. Engineering 3:738–752. https://doi.org/10.1016/J.ENG.2017.03.011

Radwan NL, Hassan RA, Qota EM, Fayek HM (2008) Effect of natural antioxidant on oxidative stability of eggs and productive and reproductive performance of laying hens. Int J Poult Sci 7: 134–150

Ragab MS, Hassan HA (2007) Effects of using dried Egyptian clover and orange peels as natural feed additives on egg production egg quality and immune response of laying hens. Fayoum J Agric Res Dev. 21:188–205. https://doi.org/10.21608/fjard.2007.197558
Rahardja DP, Hakim MR, Lestari VS (2015) Egg production performance of old laying hen fed dietary turmeric powder. Int J Anim Vet Sci 9:748–752
Rajkumar U, Rama Rao SV, Raju MVLN, Chatterjee RN (2021) Backyard poultry farming for sustained production and enhanced nutritional and livelihood security with special reference to India: a review. Trop Anim Health Prod 53:176. https://doi.org/10.1007/s11250-021-02621-6
Rajput N, Muhammad N, Yan R et al (2013) Effect of dietary supplementation of curcumin on growth performance, intestinal morphology and nutrients utilization of broiler chicks. J Poult Sci 50:44–52. https://doi.org/10.2141/jpsa.0120065
Raziq N, Saeed M, Shahid M et al (2016) Pharmacological basis for the use of Hypericum oblongifolium as a medicinal plant in the management of pain, inflammation and pyrexia. BMC Complement Altern Med 16:41. https://doi.org/10.1186/s12906-016-1018-z
Reddy DN (2019) Essential oils extracted from medicinal plants and their applications. In: Akhtar MS, Swamy MK, Sinniah UR (eds) Natural bio-active compounds: Volume 1: production and applications. Springer, Singapore, pp 237–283
Redlingshöfer B, Coudurier B, Georget M (2017) Quantifying food loss during primary production and processing in France. J Clean Prod 164:703–714. https://doi.org/10.1016/j.jclepro.2017.06.173
Remmal A, Achahbar S, Bouddine L et al (2011) In vitro destruction of Eimeria oocysts by essential oils. Vet Parasitol 182:121–126. https://doi.org/10.1016/j.vetpar.2011.06.002
Rizk YS (2017) Effect of dietary green tea and dried seaweed on productive and physiological performance of laying hens during late phase of production. Egypt Poult Sci J. 37:685–706. https://doi.org/10.21608/epsj.2017.7534
Rizzi L, Simioli M, Roncada P, Zaghini A (2003) Aflatoxin B1 and clinoptilolite in feed for laying hens: effects on egg quality, mycotoxin residues in livers, and hepatic mixed-function oxygenase activities. J Food Prot 66:860–865. https://doi.org/10.4315/0362-028x-66.5.860
Rocchi L, Paolotti L, Rosati A et al (2019) Assessing the sustainability of different poultry production systems: a multicriteria approach. J Clean Prod 211:103–114. https://doi.org/10.1016/j.jclepro.2018.11.013
Rodríguez M, Caleja C, Pereira E et al (2021) Red seaweeds as a source of nutrients and bioactive compounds: optimization of the extraction. Chemosensors 9:132. https://doi.org/10.3390/chemosensors9060132
Rossi B, Toschi A, Piva A, Grilli E (2020) Single components of botanicals and nature-identical compounds as a non-antibiotic strategy to ameliorate health status and improve performance in poultry and pigs. Nutr Res Rev. 33:218–234. https://doi.org/10.1017/S0954422420000013
Rubilar M, Gutiérrez C, Verdugo M et al (2010) Flaxseed as a source of functional ingredients. J Soil Sci Plant Nutr 10:373–377. https://doi.org/10.4067/S0718-95162010000100010
Sabbieti MG, Agas D, Maggi F et al (2011) Molecular mediators involved in Ferulago campestris essential oil effects on osteoblast metabolism. J Cell Biochem 112:3742–3754. https://doi.org/10.1002/jcb.23306
Saeed M, Arain MA, Naveed M et al (2018) Yucca schidigera can mitigate ammonia emissions from manure and promote poultry health and production. Environ Sci Pollut Res 25(35027–35):033. https://doi.org/10.1007/s11356-018-3546-1
Samarasinghe K, Wenk C, Silva KFST, Gunasekera JMDM (2003) Turmeric (Curcuma longa) root powder and mannanoligosaccharides as alternatives to antibiotics in broiler chicken diets. Asian Australas J Anim Sci 16:1495–1500
Sangha JS, Fan D, Banskota AH et al (2013) Bioactive components of the edible strain of red alga, Chondrus crispus, enhance oxidative stress tolerance in Caenorhabditis elegans. J Funct Foods 5:1180–1190. https://doi.org/10.1016/j.jff.2013.04.001

Sarvestani SS, Hosseini SM, Farhangfar SH (2020) Effect of aqueous-alcoholic extract of Hibiscus sabdariffa calyx and leaf calyx and leaf on performance, egg quality, immune system and antioxidant balance of laying hens. Iran J Appl Anim Sci 10:317–325

Sasanya BF, Olaifa O (2022) Analysis of energy consumption in poultry management for table egg production in Nigeria. Heliyon 8:e10053. https://doi.org/10.1016/j.heliyon.2022.e10053

Schieber M, Chandel NS (2014) ROS function in redox signaling and oxidative stress. Curr Biol 24: R453–R462. https://doi.org/10.1016/j.cub.2014.03.034

Schmucker S, Hofmann T, Sommerfeld V et al (2021) Immune parameters in two different laying hen strains during five production periods. Poult Sci 100:101408. https://doi.org/10.1016/j.psj.2021.101408

Searchinger T, Waite R, Hanson C et al (2018) Creating a sustainable food future: A menu of solutions to feed nearly 10 billion people by 2050 - final report. World Resources Institute

Seidlerova Z, Kubasova T, Faldynova M et al (2020) Environmental impact on differential composition of gut microbiota in indoor chickens in commercial production and outdoor, backyard chickens. Microorganisms 8:767. https://doi.org/10.3390/microorganisms8050767

Selim S, Hussein E, Abou-Elkhair R (2018) Effect of Spirulina platensis as a feed additive on laying performance, egg quality and hepatoprotective activity of laying hens. Eur Poult Sci 82. https://doi.org/10.1399/eps.2018.227

Shang Y, Kumar S, Oakley B, Kim WK (2018) Chicken gut microbiota: importance and detection technology. Front Vet Sci 5:254. https://doi.org/10.3389/fvets.2018.00254

Sharafan M, Jafernik K, Ekiert H et al (2022) Illicium verum (star anise) and trans-anethole as valuable raw materials for medicinal and cosmetic applications. Molecules 27:650. https://doi.org/10.3390/molecules27030650

Sharifi-Rad J, Sureda A, Tenore G et al (2017) Biological activities of essential oils: from plant chemoecology to traditional healing systems. Molecules 22:70. https://doi.org/10.3390/molecules22010070

Sharma MK, Dinh T, Adhikari PA (2020) Production performance, egg quality, and small intestine histomorphology of the laying hens supplemented with phytogenic feed additive. J Appl Poult Res 29:362–371. https://doi.org/10.1016/j.japr.2019.12.001

Shojaii A, Abdollahi Fard M (2012) Review of pharmacological properties and chemical constituents of Pimpinella anisum. ISRN Pharm 510:795. https://doi.org/10.5402/2012/510795

Simon K, Verwoolde MB, Zhang J et al (2016) Long-term effects of early life microbiota disturbance on adaptive immunity in laying hens. Poult Sci 95:1543–1554. https://doi.org/10.3382/ps/pew088

Singh RP, Beura CK, Mahapatra AS et al (2009) Assessment of quantitative losses of eggs between farm and household consumer. Indian J Poult Sci 44:239–242

Soares PR, Lopes MAR, Conceição MA et al (2022) Sustainable integration of laying hens with crops in organic farming. A review. Agroecol Sustain Food Syst 46:969–1001

Socha JK, Sechman A, Mika M, Hrabia A (2017) Effect of growth hormone on steroid concentrations and mRNA expression of their receptor, and selected egg-specific protein genes in the chicken oviduct during pause in laying induced by fasting. Domest Anim Endocrinol 61:1–10. https://doi.org/10.1016/j.domaniend.2017.05.001

Sosnówka-Czajka E, Skomorucha I (2021) Effect of supplementation with dried fruit pomace on the performance, egg quality, white blood cells, and lymphatic organs in laying hens. Poult Sci 100: 101278. https://doi.org/10.1016/j.psj.2021.101278

Su Y, Ge Y, Xu Z et al (2021) The digestive and reproductive tract microbiotas and their association with body weight in laying hens. Poult Sci 100:101422. https://doi.org/10.1016/j.psj.2021.101422

Swelum AA, El-Saadony MT, Abd El-Hack ME et al (2021) Ammonia emissions in poultry houses and microbial nitrification as a promising reduction strategy. Sci Total Environ 781:146978. https://doi.org/10.1016/j.scitotenv.2021.146978

Świątkiewicz S, Arczewska-Włosek A, Krawczyk J et al (2013) Effects of selected feed additives on the performance of laying hens given a diet rich in maize dried distiller's grains with solubles (DDGS). Br Poult Sci 54:478–485. https://doi.org/10.1080/00071668.2013.797563

Świątkiewicz S, Arczewska-Włosek A, Krawczyk J, Szczurek W, Puchała M, Józefiak D (2018) Effect of selected feed additives on egg performance and eggshell quality in laying hens fed a diet with standard or decreased calcium content. Ann Anim Sci 18:167–183

Taheri Gandomani V, Mahdavi AH, Rahmani HR et al (2014) Effects of different levels of clove bud (Syzygium aromaticum) on performance, intestinal microbial colonization, jejunal morphology, and immunocompetence of laying hens fed different n-6 to n-3 ratios. Livest Sci 167: 236–248. https://doi.org/10.1016/j.livsci.2014.05.006

Tan LTH, Lee LH, Yin WF et al (2015) Traditional uses, phytochemistry, and bioactivities of Cananga odorata (Ylang-ylang). Evid Based Complement Alternat Med 2015:e896314. https://doi.org/10.1155/2015/896314

Taylor RC, Omed H, Edwards-Jones G (2014) The greenhouse emissions footprint of free-range eggs. Poult Sci 93:231–237. https://doi.org/10.3382/ps.2013-03489

Teoh ES (2015) Secondary metabolites of plants. In: Medicinal orchids of Asia. Springer, pp 59–73

Torki M, Sedgh-Gooya S, Mohammadi H (2018) Effects of adding essential oils of rosemary, dill and chicory extract to diets on performance, egg quality and some blood parameters of laying hens subjected to heat stress. J Appl Anim Res 46:1118–1126. https://doi.org/10.1080/09712119.2018.1473254

Tufarelli V, Baghban-Kanani P, Azimi-Youvalari S et al (2021) Effects of horsetail (Equisetum arvense) and spirulina (Spirulina platensis) dietary supplementation on laying hens productivity and oxidative status. Animals 11:335. https://doi.org/10.3390/ani11020335

Turner I, Heidari D, Pelletier N (2022) Environmental impact mitigation potential of increased resource use efficiency in industrial egg production systems. J Clean Prod 354:131743. https://doi.org/10.1016/j.jclepro.2022.131743

Upadhyaya I, Upadhyay A, Kollanoor-Johny A et al (2015) In-feed supplementation of trans-cinnamaldehyde reduces layer-chicken egg-borne transmission of Salmonella enterica serovar Enteritidis. Appl Environ Microbiol 81:2985–2994

van den Burg SWK, Stuiver M, Veenstra FA et al (2013) A Triple P review of the feasibility of sustainable offshore seaweed production in the North Sea. Wageningen University & Research, Wageningen

Van der Klis JD, Pedrosa K (2022) Understanding the lack of 1,25(OH)2 vitamin D3 in aged laying hens. Int Poult Prod 30:27–28

van Die M, Burger H, Teede H, Bone K (2012) Vitex agnus-castus extracts for female reproductive disorders: a systematic review of clinical trials. Planta Med 79:562–575. https://doi.org/10.1055/s-0032-1,327,831

Verlinden M, Pasmans F, Mahu M et al (2013) In vitro sensitivity of poultry Brachyspira intermedia isolates to essential oil components and in vivo reduction of Brachyspira intermedia in rearing pullets with cinnamaldehyde feed supplementation. Poult Sci 92:1202–1207. https://doi.org/10.3382/ps.2012-02690

Vlaicu PA, Panaite TD (2022) Effect of dietary pumpkin (Cucurbita moschata) seed meal on layer performance and egg quality characteristics. Anim Biosci 35:236–246. https://doi.org/10.5713/ab.21.0044

Waizel-Bucay J, Waizel-Haiat S (2019) Las plantas con principios amargos y su uso medicinal. ¿Un futuro dulce? Otorrinolaringología 64:202–228

Wang JP, Kim IH (2011) Effect of caprylic acid and Yucca schidigera extract on production performance, egg quality, blood characteristics, and excreta microflora in laying hens. Br Poult Sci 52:711–717. https://doi.org/10.1080/00071668.2011.635638

Wang XJ, Liu L, Zhao JP et al (2017) Stress impairs the reproduction of laying hens: an involvement of energy. Worlds Poult Sci J 73:845–856. https://doi.org/10.1017/S0043933917000794

Wang X, Wang X, Wang J et al (2018) Dietary tea polyphenol supplementation improved egg production performance, albumen quality, and magnum morphology of Hy-Line Brown hens during the late laying period. J Anim Sci 96:225–235. https://doi.org/10.1093/jas/skx007

Wang J, Jia R, Gong H et al (2021) The effect of oxidative stress on the chicken ovary: involvement of microbiota and melatonin interventions. Antioxidants 10:1422. https://doi.org/10.3390/antiox10091422

Wang-Li L, Li Q, Chai L et al (2013) The national air emissions monitoring study's Southeast layer site: part III. Ammonia concentrations and emissions. Trans ASABE 56:1185–1197. https://doi.org/10.13031/trans.56.9673

Wee JJ, Mee Park K, Chung A-S (2011) Biological activities of ginseng and its application to human health. In: Benzie IFF, Wachtel-Galor S (eds) Herbal medicine: biomolecular and clinical aspects, 2nd edn. CRC Press/Taylor & Francis, Boca Raton, FL

Weeks CA, Lambton SL, Williams AG (2016) Implications for welfare, productivity and sustainability of the variation in reported levels of mortality for laying hen flocks kept in different housing systems: a meta-analysis of ten studies. PLoS One 11:e0146394. https://doi.org/10.1371/journal.pone.0146394

Wei L, Hua R, Li M et al (2014) Chemical composition and biological activity of star anise Illicium verum extracts against maize weevil, Sitophilus zeamais adults. J Insect Sci 14:80. https://doi.org/10.1093/jis/14.1.80

Wells ML, Potin P, Craigie JS et al (2017) Algae as nutritional and functional food sources: revisiting our understanding. J Appl Phycol 29:949–982. https://doi.org/10.1007/s10811-016-0974-5

Wen C, Li Q, Lan F et al (2021) Microbiota continuum along the chicken oviduct and its association with host genetics and egg formation. Poult Sci 100:101104. https://doi.org/10.1016/j.psj.2021.101104

Widjastuti T, Setiawan I, Abun (2017) The use of turmeric (Curcuma domestica Val) meal in the ration as feed additive on hen-day production and egg quality of Sentul chicken. Sci Pap Ser Anim Sci 60:131–135

Wiersema ML, Koester LR, Schmitz-Esser S, Koltes DA (2021) Comparison of intestinal permeability, morphology, and ileal microbial communities of commercial hens housed in conventional cages and cage-free housing systems. Poult Sci 100:1178–1191. https://doi.org/10.1016/j.psj.2020.10.052

Wina E (2018) The role of saponin as feed additive for sustainable poultry production. Indones Bull Anim Vet Sci 27:117. https://doi.org/10.14334/wartazoa.v27i3.1588

Wistedt A, Ridderstråle Y, Wall H, Holm L (2019) Age-related changes in the shell gland and duodenum in relation to shell quality and bone strength in commercial laying hen hybrids. Acta Vet Scand 61:14. https://doi.org/10.1186/s13028-019-0449-1

Wu Z, Sadik M, Sleiman FT et al (1994) Influence of yucca extract on ruminal metabolism in cows3. J Anim Sci 72:1038–1042. https://doi.org/10.2527/1994.7241038x

Xia B, Liu Y, Sun D et al (2018) Effects of green tea powder supplementation on egg production and egg quality in laying hens. J Appl Anim Res 46:927–931. https://doi.org/10.1080/09712119.2018.1431240

Xia B, Liu J, Sun D et al (2022) Changes in egg production, egg quality and egg nutritional values of laying hens in response to dietary supplementation with green tea water extracts. J Appl Anim Res 50:254–258. https://doi.org/10.1080/09712119.2022.2039155

Xin H, Gates RS, Green AR et al (2011) Environmental impacts and sustainability of egg production systems1. Poult Sci 90:263–277. https://doi.org/10.3382/ps.2010-00877

Yang JX, Maria TC, Zhou B et al (2020) Quercetin improves immune function in Arbor Acre broilers through activation of NF-κB signaling pathway. Poult Sci 99:906–913. https://doi.org/10.1016/j.psj.2019.12.021

Yu C, Wei J, Yang C et al (2018) Effects of star anise (Illicium verum Hook.f.) essential oil on laying performance and antioxidant status of laying hens. Poult Sci 97:3957–3966. https://doi.org/10.3382/ps/pey263

Yu C, Guo Y, Yang Z et al (2019) Effects of star anise (Illicium verum Hook. f.) essential oil on nutrient and energy utilization of laying hens. Anim Sci J 90:880–886. https://doi.org/10.1111/asj.13221

Yuan T, Li J, Wang Y et al (2022) Effects of zearalenone on production performance, egg quality, ovarian function and gut microbiota of laying hens. Toxins 14:653. https://doi.org/10.3390/toxins14100653

Zadeh JB, Kor NM, Olfati A (2015) The effects of different levels aqueous extract of Fenugreek (Trigonella foenum-graecum L.) on performance and immune response of laying hens. J Essent Oil Bear Plants 18:1476–1481. https://doi.org/10.1080/0972060X.2014.981594

Zahroojian N, Moravej H, Shivazad M (2013) Effects of dietary marine algae (Spirulina platensis) on egg quality and production performance of laying hens. J Agric Sci Technol 15:1353–1360

Zampiga M, Calini F, Sirri F (2021) Importance of feed efficiency for sustainable intensification of chicken meat production: implications and role for amino acids, feed enzymes and organic trace minerals. Worlds Poult Sci J 77:639–659. https://doi.org/10.1080/00439339.2021.1959277

Zhang C, Fan L, Fan S et al (2019) Cinnamomum cassia Presl: a review of its traditional uses, phytochemistry, pharmacology and toxicology. Molecules 24:3473. https://doi.org/10.3390/molecules24193473

Zhang J, Na T, Jin Y et al (2020) Thicker shell eggs with enriched n-3 polyunsaturated fatty acids and lower yolk cholesterol contents, as affected by dietary nettle (Urtica cannabina) supplementation in laying hens. Animals 10:1994. https://doi.org/10.3390/ani10111994

Zhao X, Zheng S, Wei S et al (2022) The protective effect and potential mechanisms of eugenol against Salmonella in vivo and in vitro. Poult Sci 101:101801. https://doi.org/10.1016/j.psj.2022.101801

Zhu YF, Wang JP, Ding XM et al (2020) Effect of different tea polyphenol products on egg production performance, egg quality and antioxidative status of laying hens. Anim Feed Sci Technol 267:114544. https://doi.org/10.1016/j.anifeedsci.2020.114544

Application of Feed Additives in the Diets of Turkeys

Mehmet Bozkurt, Ahmet Önder Üstündağ, Ahmet Engin Tüzün, and Metin Çabuk

Abstract

Over the last half-century, the poultry production industry witnessed a great dependency on the in-feed use of antimicrobials to promote health and productivity. However, the continuous and uncontrolled use led to the ban of growth-promoting antibiotics (AGPs) due to resistance and residue concerns. In addition, intensive farm management procedures of modern turkey rearing systems also result in physiological, biochemical and metabolic changes that are difficult to address. Consequently, turkey growers are faced with the challenge of maintaining productivity and preventing diseases on their farms. These challenges facilitated the development and evaluation of novel antibiotic alternatives, such as probiotics, prebiotics, synbiotics, organic acids, enzymes, yeast-based products, phytogenics, bacteriophages and algal derivates. The main purported mode of action of these feed additives was to manipulate gut microbiota to achieve optimal gut function and integrity along with balanced microbiota for better growth and improved health status of turkeys. However, reports of the effect of dietary supplementation with these additives on turkey performance traits are contradictory and inconclusive. In addition, a vast number of studies with broiler chickens have revealed distinctive characteristics of these additives under intestinal health challenges or any other management stressor, but

M. Bozkurt (✉) · A. Ö. Üstündağ
Faculty of Agriculture, Department of Animal Science, Aydın Adnan Menderes University, Aydın, Turkey
e-mail: mbozkurt@adu.edu.tr

A. E. Tüzün
Koçarlı Vocational School, Aydın Adnan Menderes University, Aydın, Turkey

M. Çabuk
Veterinary Faculty, Department of Animal Nutrition and Nutritional Diseases, Science and Technology Application and Research Center, Dokuz Eylül University, İzmir, Turkey

G. Arsenos, I. Giannenas (eds.), *Sustainable Use of Feed Additives in Livestock*,
https://doi.org/10.1007/978-3-031-42855-5_20

experiments with turkeys under such circumstances are considerably limited and still require elucidation. It is apparent that for a deeper understanding of the effectiveness of these additives purported to maintain gut health and production efficiency in turkey production, a more comprehensive approach is needed beyond the mere use of the additives. Therefore, the current review aims to compile results published with turkeys and suggest how the use of these novel feed additives may improve the overall health status of turkeys and production efficiency in a sustainable, economic, environmental- and animal welfare-friendly way while enabling safe food for consumers.

Keywords

Turkey feeding · Probiotic · Prebiotic · Enzyme · Phytogenic

1 Introduction

Commercial turkey production practices have evolved considerably over the last half-century. This has resulted from improvements in poultry genetics, nutrition, management and health maintenance, among other factors (Scott 1987; Grashorn and Bessei 2004; Huff et al. 2007; Chiang et al. 2008; Hafez and Awad 2021; Gous et al. 2022; Krautwald-Junghannsa and Janja 2022). Modern hybrid male and female turkeys weighing almost 18 kg and 14 kg, respectively, after a 112 days fattening period are the result of an intensive and continuous artificial selective process for commercial objectives, which has led to increased productivity (Huff et al. 2008; Hocking 2014). However, this continuous selection process also brought physiologic, biochemical and metabolic changes and immunodepression associated with the intensive farm management procedures of modern turkey production rearing systems (Hocking 2014; Huff et al. 2014; Erasmus 2018). For almost a half-century, turkey producers have achieved great success in addressing certain challenges, such as those hampering optimal bird health and productive output, by supplementation with AGPs (Leeson 1984; Firman and Kirn 1989; Waibel et al. 1991; Ferket 2013).

However, the emergence of antibiotic-resistant bacteria and the rise in antibiotic residue in edible meat have led to consumer concerns over their use in feed and have resulted in a move away from the use of antibiotics in livestock production worldwide (Neveling and Dicks 2021). The consideration of strict antimicrobial restrictions on their use in animal agriculture also led the turkey industry to seek efficient, cost-effective, sustainable and environmentally friendly alternatives to AGPs with no residue or resistant properties (Ferket 2013).

Consequently, a vast number of new feed additives have become continuously available in the market, with the purpose of manipulating the functionality, structure, integrity and microbial composition of the intestinal tract to manage intestinal health challenges and maintain optimal growth performance. Strategies to accomplish these crucial mechanisms of the digestive system through organic acids, probiotics, prebiotics, synbiotics, phytochemicals, yeast-based products and enzymes are showing promise in enhancing general health status and production efficiency in turkeys. Although this list includes the most researched alternatives, there are exceptional alternatives to antibiotics that have been discussed by other authors. Bacteriocins (Cole et al. 2006), antimicrobial peptides (Forkus et al. 2017), clays (Denli et al. 2009), and several trace minerals (Flores et al. 2021) have also been reviewed as antibiotic alternatives, but only organic acids, probiotics, prebiotics, yeast-based products, phytogenics and enzyme alternatives are discussed in the current literature review. These principal control strategies are likely to include combinations of products as replacements for AGPs such as synbiotics (Omidiwura et al. 2018; Czech et al. 2020). Thus, there is a great opportunity to demonstrate the efficacy of six kinds of alternative strategies to feed antibiotics in the turkey production chain, but adequate information on the use of exceptional strategies is significantly lacking.

Altogether, the available information indicates that these compounds could be used as dietary supplements with various applications, including antimicrobial, immune system stimulation, antiparasitic, anti-inflammatory and antioxidant actions (Hollister et al. 1994; Hafez and Hauck 2006; Dutta et al. 2008; Loeffler 2014), as well as several beneficial reflections on meat quality traits, including microbial and sensory actions (Botsoglou et al. 2003; Dewi et al. 2021). Furthermore, positive implications regarding environmental and turkey welfare outcomes through supplementation of these feed additives were demonstrated (Odetallah et al. 2002; Huff et al. 2011). Meeting these demands while still being profitable and successful is often a great challenge for turkey producers. This review provides a brief overview of performance-enhancing feed additive use in turkey production and introduces some examples of how dietary and other management factors interact with their mechanism of action.

2 Organic Acids

In the feeding of poultry, organic acids, known to have strong antibacterial effects, have been used as control agents to reduce the intraluminal concentration of coliform bacteria and other acid-intolerant organisms, such as Campylobacter and Salmonella, known to be involved in digestive disorders (Izat et al. 1990; Ricke 2003; Immerseel et al. 2006). Organic acids, so-called acidifiers, also have several additional effects on gastrointestinal tract (GIT) health and productive performance. These include a reduction in digesta pH, increased pancreatic secretion, and enhanced feed utilization, thereby increasing the digestion and absorption of nutrients (Patten and Waldroup 1988; Yang et al. 2018).

The majority of research studies related to the use of organic acids have been conducted on broiler chickens, often with inconsistent results (Günal et al. 2006; Hernández et al. 2006; Adewole et al. 2021). There are a limited number of studies with hybrid turkey toms; scant information is available on organic acids as feed additives in turkey diets, reporting antibacterial activity in the gut with some success but failure to promote body weight (BW), body weight gain (BWG), feed conversion rate (FCR) or meat yield.

In the study by Wajda et al. (2010), formic acid, mostly used as an acidifying agent in poultry and pig production, was administered to Big 6 turkey toms via the drinking water route over the entire experimental growth period (from 1 to 4 weeks of age—5 l/tonne, from 5 to 8 weeks of age—4 l/tonne, from 9 to 18 weeks of age—3 l/tonne). Body weight gain, FCR and mortality rate determined at 4, 8, 12 and 18 weeks of age were not affected by formic acid (Acidum formicum) intake. Slaughter characteristics (the percentage rate of carcass yield, breast muscles, thigh muscles, drumstick muscles, total leg muscles and abdominal fat), chemical composition (percentage of dry matter, total crude protein, fat and ash) and sensory properties (pH 24, brightness, water-holding capacity, palatability, juiciness and tenderness) of meat from turkeys determined at the end of the trial were comparable between birds treated with and without formic acid.

One strategy is to increase the effectiveness of organic acids by means of an antimicrobial in-feed preparation by combining them with phytogenic compounds with certain antibacterial activities (Basmacioğlu-Malayoğlu et al. 2016; Stefanello et al. 2020). Such a combination strategy showed benefits in broiler chickens in terms of growth performance indices, villus structure and absorptive surface area and reduced the pathogenic bacteria count in the intestines (Bozkurt et al. 2012; Weber et al. 2012; Gheisar et al. 2015). This synergistic mechanism of action between organic acids and plant essential oils, formerly demonstrated in broiler chickens, was tested on 400 male turkeys (Large White Big 6) from one day of age to 140 days of age (Mikulski et al. 2008). For this purpose, a four-phase basal feed mixture without any performance-enhancing feed additives was initially prepared and then supplemented with either a blend of organic acids (formic acid and propionic acid) or a mixture of organic acids (citric acid, fumaric acid, orthophosphoric acid and malic acid) and an essential oil blend from citrus fruits, cinnamon, oregano and thyme. Such a combination of specific plant bioactives with diverse antimicrobial and enzymatic activities and organic acids with strong antimicrobial properties, either alone or in combination, showed no notable effect on BWG or FCR in turkeys after a 140-day feeding period. When compared to the unsupplemented treatment, significant reductions were determined in crop pH with dietary supplementation of organic acids and blends of essential oils, but no such effect was observed for gizzard and caecal pH levels. The enzymatic activities of α-glucosidase, β-glucosidase, α-galactosidase, β-galactosidase and β-glucuronidase in the caecal content were significantly higher in birds fed the tested supplements than in those fed the unsupplemented control, while only α-glucosidase responded positively to the organic acid blend in ileal digesta. These improvements in enzymatic activities derived from the organic acid and essential oil supplementation do not seem to be reflected in nutrient utilization and, ultimately, the final growth performance of the turkeys.

In contrast to observations on the performance of hybrid turkeys reared under standard management conditions, organic acids demonstrated some beneficial effects in improving the performance of turkey poults at risk for potentially pathogenic bacteria. In this regard, successful results were illustrated by Roy et al. (2002), who used a commercial organic acid blend (primarily propionic acid; Myco Curb) to control poult enteritis and mortality syndrome (PEMS), which describes potentially lethal enteritis of young turkeys (Yu et al. 2000). Myco Curb, a mould inhibitor for processed feed ingredients and animal feeds, was maintained in the feed for the duration of the 3-week periods. At 6 days post-hatch, BUT Big 9 poults were given a 1-mL oral gavage of a 10% suspension of faeces from coronavirus-negative PEMS-positive poults. Myco Curb was supplemented into the diet at levels of 1.25 and 2.5%. Myco Curb delayed the onset of the initial mortality spike associated with the disease and reduced the cumulative mortality by more than 50%. The marked reduction in mortality in the poults treated with Myco Curb was attributed to decreased bacterial content of the gut and the maintenance of packed cell volume and haemoglobin content. However, the beneficial attributes to gut health and liveability generated by fortification of the diet with Myco Curb were not reflected in the growth performance indices of turkeys. Male and female poults had a significantly lower BW and higher FCR when they were given 2.5% Myco Curb in their feed, while no significant changes were observed at dietary inclusion levels of 0.625 and 1.25%.

Effects similar to those in the abovementioned work were reported in a study (Milbradt et al. 2014a) in which 1-day-old female commercial cross turkey poults (BUT Big 9) were fed diets supplemented with a mixture of short- and medium-chain fatty acids (acetic acid, formic acid, propionic acid, sorbic acid and vegetal fatty acids). Although the organic acid blends had no significant effects on the intestinal villus height, crypt depth, or villus: crypt ratio, negative effects on the early gain or feed intake were observed throughout the study. The same failure was witnessed in another experiment conducted for a 60-day period using white California turkey toms (Çelik et al. 2003). No significant effect of 2% dietary supplementation with a commercial organic acid blend (Biotronic® SE) comprised of formic acid (17%), ammonium format (14.1%), propionic acid (12.4%), ammonium propionate (%8.4) and carrier material (47.7%) on the BWG and FCR of turkeys was found.

Incompetence in improving BWG and FCR in turkey toms between 0 and 28 days of age with dietary organic acid administration was also reported in a more recent study (Milbradt et al. 2017), while a significant reduction in caecal numbers of *Salmonella enteritis* was observed. In a former study (Milbradt et al. 2014b), in-water treatment with a blend of organic acids (formic acid, acetic acid, ammonium format, mono- and diglyceride of unsaturated fatty acids and copper acetate) elicited significant reductions in the number of *Salmonella enteritidis* colony forming units in the caecum and crop of turkeys (BUT Big 9) that were challenged 72 and 24 hours prior to slaughter. However, it was noted that only constant treatment of organic acids in drinking water resulted in antibacterial activity, while the transient administration regimen failed to do so.

As demonstrated in a very recent study (Makowski et al. 2022), significant levels of decreases in faecal populations of *Escherichia coli* and *Clostridium perfringens* accompanying an increase in the gizzard pH were a manifestation of antimicrobial activity generated by butyric acid glycerides or coated sodium butyrate when added to turkey diets. The addition of butyric acid in different forms to turkey diets increased the butyric acid concentration in the caecal digesta and ileal protein digestibility and in turn improved FCR.

In brief, as presented in Table 1, almost all the published studies have unanimously concluded that organic acids, irrespective of their chemical characteristics and application procedures, may have the potential to be more effective in the field, where poults are subjected to repeated challenges regarding disease pressure, stocking density, litter quality, diet characteristics and management practices. The lack of reflection of the certain antimicrobial activity posed by organic acids in the growth performance of turkeys is noticeable.

3 Enzymes

The use of exogenous microbial enzymes as feed additives is a well-established practice in food-producing animal production, and their application is ubiquitous in commercial broiler and laying hen diets as a tool to enhance depolymerization of nonstarch polysaccharides (NSP), reduce digesta viscosity and modulate bird intestinal health (Bedford and Schulze 1998; Adeola and Cowieson 2011; Mahmood and Guo 2020). In particular, the application of phytase and, to a lesser extent, xylanases and ß-glucanases has been widely adopted, but many other enzyme classes can significantly improve the utilization of feed, such as proteases, pectinases, amylases, pentosanases, beta-mannanase, galactosidases, or cellulase. Evidently, supplementation with these different enzymes in poultry gives a wide range of effects on growth performance, nutrient digestibility and health (Slomonski 2011). However, in turkey feeding, there has been a lack of studies examining exogenous enzyme efficacy for over three decades (Ferket 1993, 2013; Nguyen et al. 2022).

In a study by Odetallah et al. (2002), two endoxylanase enzyme preparations (Natugrain and Lyxasan forte), with apparently different modes of action, were fed to turkey toms aged between 8 and 140 days, either alone or in combination (Natugrain Blend). Cumulative feed consumption was higher by 2 kg or more in birds treated with Natugrain Blend than in other treatments. Accordingly, Natugrain Blend markedly improved body weight at 84 and 112 days of age in comparison to the unsupplemented control diet. Lyxasan forte had the best feed/gain ratio throughout the experiment. The liveability and feather scores of birds were significantly improved in all three enzyme treatments compared to the control.

Ritz et al. (1995) found that pancreatic organ weight and pancreatic amylase activity were not consistently affected by diet in an experimental schedule in which male turkeys were administered commercial preparations of amylase (Avizyme TK0492-1®) and xylanase (TK0492-1®) from hatch to 8 weeks of age. Amylase activity within the intestinal chyme increased sporadically with dietary amylase supplementation compared to the control and xylanase-supplemented diets. The authors concluded that increased supplemental amylase activity levels may provide

Table 1 Broad influences of dietary supplementation with organic acids and enzymes on the body weight, feed intake, feed conversion ratio, mortality rate, meat yield, and digestive system development of growing turkeys

Bird genotype	Period (days)	Product specification	Supplemental level (g/kg)	Body weight	Feed intake	FCR[a]	Mort. rate	Meat yield	Dig.sys devlp[b].	References
Organic acids										
BUT big 9	21	Blend of acids	1.25 2.5	–[c] –	– ↓[f]	– –	– –	nd nd	↑ ↑	Roy et al. (2002)
California	60	Blend of acids	2.0	–	↓	nd	nd	nd	nd	Çelik et al. (2003)
BUT Big-6	140	Blend of acids	5.0	–	–	–	–	nd	nd	Mikulski et al. (2008)
BUT Big-6	126	Formic acid	3–5 ml/kg	–	–	–	–	–	nd	Wajda et al. (2010)
BUT Big 9	70	Blend of acids	2.0	–	↓	–	–	nd	↑	Milbradt et al. (2014a)
BUT Big 9	28	Blend of acids	2.0	–	–	–	nd	nd	↑	Milbradt et al. (2017)
BUT Big-6	105	Sodium butyrate But. a. Glyceride	3.3 3.4	↑[d] ↑	nd nd	↑ ↑	– –	nd nd	↑ ↑	Makowski et al. (2022)
Enzymes										
Nicholas	56	Amylase	1.1	nd[e]	nd	nd	nd	nd	↑	Ritz et al. (1995)
LW (BUT)	140	Lyxasan forte Natugrain blend	0.1 0.15	– –	– ↑	– –	– ↓	nd nd	nd nd	Odetallah et al. (2002)
LW (BUT)	17	Enzyme cocktail	0.2	↑	↑	↑	–	nd	↑	Santos, (2002)
Nicholas	21	Phytase	1000 U/kg	–	–	–	nd	nd	nd	Applegate et al. (2003)
BUT Big-6	112	Phytase	750 U/kg	–	–	–	nd	nd	nd	Brodacki et al. (2009)
BUT Big-6	56	Enzyme cocktail	1.0	–	–	–	nd	nd	↑	Juśkiewicz et al. (2010)

(continued)

Table 1 (continued)

Bird genotype	Period (days)	Product specification	Supplemental level (g/kg)	Body weight	Feed intake	FCR[a]	Mort. rate	Meat yield	Dig.sys devlp[b].	References
Diamond	119	Enzyme cocktail	1.0	↑	—	↑	nd	↑	nd	García-López et al. (2011)
BUT Big-6	56	Enzyme cocktail	1.0	↑	nd	↑	nd	nd	↑	Zduńczyk et al. (2013)
Nicholas	28	β-Mannanase	0.5	nd	nd	nd	nd	nd	↑	Ayoola et al. (2015)
Nicholas	21	Enzyme cocktail	0.2 ml/kg	—	—	—	nd	nd	—	Flores Maldonado (2018)

[a]FCR: feed conversion ratio (g feed/g gain); [b]Digestive system development expresses some metabolic, biochemical and morphologic aspects in the GIT including intestinal microflora enzymes, short chain fatty acids content in the ceca, viscosity of intestinal lumen, villus morphological characteristics and pH changes; [c]Unchanged (—); [d]Increased (↑); [e]Not determined (nd); [f]Decreased (↓)

more conclusive evidence of an additive effect of dietary amylase and endogenous amylase activity.

Findings determined from several studies indicated that the age and metabolic needs of the turkey influence enzyme activity levels within the pancreas and small intestine. In a study by Ritz et al. (1995), pancreatic amylase activity seemed to increase in a linear fashion with age and continued to follow this progressive rise throughout the 8-week period. This continued increase in activity contradicts the results reported by Krogdahl and Sell (1989), who observed a stabilization in pancreatic amylase after 21 days. The observations of McKnight (1997) are parallel to earlier establishments that exogenous xylanase supplementation in a wheat-based diet was only effective at a young age (up to 70 days of age).

The rearing system of turkeys was found to intervene in the efficacy of in-feed enzymes in a study by García-López et al. (2011). Addition of a commercial enzyme preparation (Allzyme-Vegpro; mixture of protease, amylase, cellulase, beta-galactosidase and pentosanases) to the diet resulted in a lack of bird response in a semi-confinement rearing system with daily access for 4 hours to grassland meadows, whereas enzyme supplementation improved the BWG and FCR of turkeys kept in conventional confinements.

Zduńczyk et al. (2013) examined the relationship between dietary fibre status and supplementation with an enzyme premix containing pectinase, cellulase, xylanase, glucanase, mannanase, galactanase, amylase and protease on endogenous intestinal enzyme secretion of Big 6 male turkeys. From 1 to 56 days of age, the enzyme premix was added to a soybean-based control diet and an experimental diet in which rapeseed meal was gradually replaced with soybean up to 180 g/kg level. The results indicated that, regardless of the diet type, enzyme supplementation tended to reduce ileal viscosity, decrease ammonia concentration, increase the glycolytic activities of the intestinal microflora enzymes α-glucosidase, α-galactosidase and β-galactosidase, decrease the activity of β-glucuronidase and increase BWG in turkeys. The authors suggested that supplementation with the NSP-degrading enzyme could maintain adequate endogenous glucosidase and galactosidase activity levels for digestion in the high fibre-containing diets with rapeseed meal inclusion while maintaining gut physiology parameters of turkeys similar to those fed the SBM-based diet. However, in the previous work by the same research team (Juśkiewicz et al. 2010), significant decreases were observed in the final body weight, small intestine and caecal tissue mass, caecal digesta mass and production of volatile fatty acids in the caeca when the same enzyme preparation was used. Of note, the experimental protocol was similar between the two studies except that sunflower meal was replaced with soybean meal instead of rapeseed meal.

Within the framework of this thesis, the magnitude of the response to frost damage during seed development and dietary energy density was examined in turkey toms by dietary supplementation with a natural blend of enzymes (Natugrain Blend®) containing endoxylanases and other enzymes (β-glucanase, hemicellulase, cellulase and protease) (Santos 2002). The results from two consecutive trials demonstrated that enzyme supplementation had positive effects on the nutrient utilization of different wheat sources and cultivation conditions. In addition, different sources of supplemental enzymes had variable effects according to the age of the birds. Phospholipase alleviated the adverse effect of dietary NSP by improving fat digestion and absorption in young turkeys, whereas endoxylanase was more effective in older birds, which have greater digestive capacity and a more mature gut microbial ecosystem.

Ayoola et al. (2015) carried out two trials to examine the effects of dietary supplementation with either β-mannanase, a commercial blend of xylanase, amylase and protease, or direct-fed microbes on the gut health of turkeys. The research results demonstrated that supplements can improve gut health in 42-day-old Nicholas turkey poults, as indicated by improved morphological development of the enteric mucosa and reduction in adherent ileal mucin secretion. Similarly, in a more recent comprehensive thesis study, interactive effects between enzyme supplementation and the NSP profile of the diet and particle size distribution were assessed by feeding toms from 1 to 21 days of age (Flores Maldonado 2018). It was determined that supplementation with a commercial enzyme cocktail (Rovabio Advance, Adisseo) containing arabinofuranosidases and xylanases may improve the digestibility of feed formulated with a low crude protein soybean meal and screened feeds by reducing the negative effects of NSP in the intestinal tract.

Accumulated research over the past three decades has shown that phytase enzymes have profound effects on health, performance and skeletal development, as they influence phosphorus utilization and retention and bone mineralization in poultry. Today, nearly all broiler and layer hen feed now contain phytase enzyme products formulated in place of inorganic phosphorus preparations such as mono- and dicalcium phosphate (Selle and Ravindran 2007; Cowieson et al. 2011). The application of phytase has been widely adopted in other poultry species but has been much less investigated in turkeys. In one of the two available studies with turkeys (Brodacki et al. 2009), phytase supplementation in diets (Ronozyme P; provides 750 FTU phytase enzyme per g feed) from 6 to 16 weeks of age did not influence the BWG and FCR of female turkeys. The lack of significant effects due to phytase (regardless of source) on the performance criteria of turkey toms aged between 10 and 21 days was reported by Applegate et al. (2003). The tibia and toe ash, however, were significantly affected by the phytase source (*Escherichia coli*-derived vs. Aspergillus-or Peniophora-derived phytases) and concentration (0, 250, 500,750 and 1000 U/kg), indicating significant increases in ash content with heightened supplemental phytase levels over 250 U/kg feed.

Barbour et al. (2002) demonstrated that digestion of pressed turkey feathers with an enzyme mixture (protease, lipase and amylase) prior to autoclaving can improve the protein and amino acid nutritional values. Responsiveness of growing turkeys to exogenous in-feed enzyme preparations in terms of productive performance is illustrated in Table 1.

In general, the available scientific literature has demonstrated that supplementation with appropriate enzymes in poultry diets is an effective way to overcome antinutritional effects that mostly stem from the high NSP content in grains. Although relatively limited data are available, turkey poults appear to respond in the same manner as broilers and layer hens. The addition of enzyme products containing xylanase, mostly in combination with other carbohydrases, allows for the inclusion of greater amounts of wheat in diets without loss of performance (Leeson et al. 1996). In addition, glucanases and galactosidase in combination with proteinase and pectinases may work synergistically to improve the nutritive value of diets containing significant levels of soybean meal. Of note, the effect appears to be age-related, with the greatest response generally shown by young birds at 0 to 7–10 weeks.

4 Probiotics

The concept of probiotic microorganisms as in-feed preparations has become an area of much interest in poultry production, particularly for broiler and layer chickens (Rojman and Deborah 2022). However, relatively few reports have been published concerning the use of probiotics in turkey production. The favourable effects of probiotics are thought to be attributed to their ability to regulate gut microbiota balance, which in turn plays a critical role in maintaining host health (Sekirov et al. 2010). Probiotics, which are viable microorganisms with no pathogenic properties, consisting of yeast or bacteria, have also been reported to be useful under several circumstances for proper maintenance of the structure and function of the turkey gut (Marteau et al. 2001; Soccol et al. 2013).

In one of the pioneer studies, Jiraphocakul et al. (1990) compared the effectiveness of dried *Bacillus subtilis* culture with feed-grade antibiotics in two consecutive trials using female and male turkeys grown for 16 and 20 weeks, respectively. In the former experiment, the dietary *B. subtilis* culture significantly increased *B. subtilis* counts in the crop and caecum but failed to influence intestinal Lactobacillus or *Escherichia coli* counts. BW and FCR were not significantly affected by feeding the *B. subtilis* culture in this experiment. In the latter experiment, significant improvements in BWG at 12 weeks and FCR at 20 weeks were observed in male birds receiving the *B. subtilis* culture. The authors concluded that the results of the two experiments differ somewhat with respect to BWG or FCR, probably due to the bird gender difference.

Unbeneficial results from the addition of probiotic preparation [a mixture of *Lactobacillus bulgaricus, Lactobacillus acidophilus, Streptococcus thermophilus, Lactococcus lactis* and *Propionibacterium* (0.1×10^9 CFU/per gram)] at 0.05% in the diets of hybrid commercial female turkeys were reported in a recent study by Lalev et al. (2020). Supplementation of the basal diet containing 10% silkworm meal as soybean meal replacement with these probiotics did not affect the performance and meat yield characteristics of the turkeys after a 74-day feeding period (from 56 to 130 days of age). However, fortification of basal turkey grower and finisher diets with a blend of probiotic microorganisms significantly decreased serum glucose and uric acid concentrations at the end of the grow-out period.

Supplemental levels of commercial *Bacillus cereus* var. *Bacillus toyoi* strain (Toyocerin®) were the subject of another study using male middle-heavy Big 9 turkey poults (Grela et al. 2009). Addition of Toyocerin probiotic preparation to the diet increased overall BWG (0 to 16 weeks), but significant enhancements in FCR in the initial 12-week growing period disappeared thereafter. The results regarding BWG and FCR indicated that the higher the supplemental probiotic dose (0.2×10^9 versus 1×10^9 *B. toyoi* CFU per kg of diet), the greater the benefit, suggesting a dose-dependent effectiveness. A similar pattern in the performance traits to that observed in the study of Grela et al. (2009) was described in a later study by Batkowska et al. (2015) in which Big 9 turkey males were fed diets supplemented with 1×10^9 *B. toyoi* CFU/kg instead of the 0.2×10^9 level. In partial agreement with these earlier determinations, in a more recent study (Dobrowolski et al. 2019), mixed probiotic preparations at three different supplemental levels exerted a beneficial effect on the histological structure of the small intestine; however, the observed effects were dose- and region dependent. The improvements determined in absorption surface area with probiotic treatment were not reflected in turkey growth performance at market age. Similarly, significant enhancements of intestinal microbial populations and gut morphology derived from probiotic supplementation were not paired with food utilization in turkeys, although the final BW increased with probiotics (Agboola et al. 2014).

The effects of adding a microbial preparation or zinc bacitracin, either alone or in combination, to the diet were studied in an initial study by Francis et al. (1978). The addition of either Lactobacillus or zinc bacitracin to the diet resulted in poults with numerically increased BW over poults fed the control diet. The authors noticed that the improvement in growth from the combination of the two supplements was not as great as when either was administered alone. Nevertheless, probiotics have been shown to be more effective as antimicrobial agents than zinc bacitracin, as evidenced by significantly decreased coliform and total aerobe counts in the feed and the digestive system of turkey poults.

Recently, probable synergism between probiotics and prebiotic preparations, so-called synbiotics, has been the subject of several studies. In one study (Omidiwura et al. 2018), turkeys receiving a bacterial culture (*B. subtilis*), either alone or in combination with a prebiotic preparation, for an 8-week period showed comparable BWG and FCR to untreated control birds. In later work, Czech et al. (2020) found that Big 6 turkeys experienced limited changes in plasma biochemicals, including pro- and antioxidant compounds, enzymes, minerals and hormones, when fed a diet supplemented with fodder yeast (*Yarrowia lipolytica* or *Saccharomyces cerevisiae*; 30 g/kg diet) and probiotics after a 112-day feeding period. In this study, the addition of probiotic (a mixture of *Bacillus licheniformis*-1.6×10^9 CFU/g and *Bacillus subtilis*-1.6×10^9 CFU/g; 0.5 g/kg diet) alone to a diet generally resulted in a lack of bird response.

In a subsequent study, the effectiveness of two commercial probiotic preparations [(3.2×10^{10} CFU/g *Bacillus* spp. (BioPlus, Biochem) and 1.0×10^{10} CFU/g *Enterococcus faecium* (Cylactin, DSM)] and a synbiotic preparation [(containing 2.0×10^9 CFU/g of *Lactobacillus* spp., 2.0×10^7 CFU/g of *Saccharomyces cerevisiae* yeast and 2% inulin)] was evaluated using Big 6 turkey toms challenged with dietary mycotoxin (Śliżewska et al. 2020). The basal feed contained wheat that was naturally contaminated with mycotoxin (ochratoxin, OTA), resulting in OTA concentrations in the phase diets ranging from 198 to 462 mg/kg, and was fed for a 15-week grow-out period. The synbiotic preparation modulated the intestinal microbiota, increasing beneficial bacteria while reducing the number of potential pathogens. However, the benefits obtained from both probiotic preparations were markedly inferior to those of synbiotic application. The benefits in intestinal health resulting from these preparations did not translate into productivity in toms as measured by BWG and FCR at 6 and 14 weeks of age. This means that mycotoxin-induced performance failures could not be ameliorated by probiotic and synbiotic preparations.

In a study comprising two consecutive experiments carried out by Torres-Rodriguez et al. (2007), the effects of a commercially available lactic acid bacterium as a probiotic alone and supplemented with lactose as a prebiotic were evaluated to determine its effects on turkey BW during the brooding and grow-out phases under commercial conditions. Regardless of whether the probiotic was administered in drinking water or feed, the combination of pro- and prebiotic caused increased BW at days 26 and 28 in the first experiment and at market age in the later trial compared with the control group. However, treatment with probiotic microorganisms alone was ineffective in significantly changing the growth rate of hybrid turkeys at any age.

In terms of intestinal development, it was determined that poults differed in the response to different commercial probiotic preparations incorporated at the recommended commercial dosage (Loeffler 2014). Supplementation with one of two commercial probiotic preparations between 0 and 11 days of age increased both the villus length and area as well as the crypt depth, but the remaining treatment failed to do so. The beneficial effects of dietary supplementation with commercial direct-fed microbes (Primalac, Star Labs Inc., containing primarily *Lactobacillus acidophilus* and *Lactobacillus casei*) were reported in an earlier study by Rahimi et al. (2009). In a factorial experimental design, female one-day-old turkey poults were fed mash or crumbled feed and were subjected to a *Salmonella* spp. challenge by oral gavage or not. According to measurements performed at day 21, regardless of the feed form and experimental Salmonella infection, the probiotic-fed birds showed increased goblet cell numbers, total goblet cell area, goblet cell mean size, and mucosal thickness and a greater number of segmented filamentous bacteria compared with the controls. The authors postulated that changes in intestinal morphology, as observed in their study, support the optimum gut health and function concept.

Tomaszewska et al. (2016) revealed that bone characteristics and mineralization in meat-type female turkeys interact with the supplementation dose of probiotics. They found that the influence of probiotic administration on bone mineral density, bone mineral concentration, bone tissue density and bone ash was dose-dependent. The investigated properties of long bones in female turkeys were positively affected by probiotic-supplemented diets in a dose-dependent manner. In general, a review of the available data on turkeys supports the concept that poultry gut health, structure and function can be improved by probiotics, but these amendments were barely coupled with growth performance. Different components of probiotics under various methods of administration are discussed in this section and summarized in Table 2.

5 Prebiotics

Many of the yeast products prepared from the yeast *Saccharomyces cerevisiae* are classified as prebiotics, and were proven to be a good option as alternatives to AGPs in broiler chickens (Rosen 2007a; Hooge and Connolly 2011), turkeys (Hooge 2004; Rosen 2007b) and laying hens (Salami et al. 2022). Prebiotics are expected to decrease the pH of intestinal contents, thus inhibiting the growth of harmful bacteria such as Clostridium and Salmonella and promoting the growth of *Lactobacillus* spp. and *Bifidobacterium* spp. found in the intestines (Ofek et al. 1977; Spring et al. 2000; Rehman et al. 2007). The above specific mechanism of action has been well documented with respect to short-chain fructo oligosaccharides (FOS) and the fructose polymer inulin (Rehman et al. 2009). In contrast to the mode of action of antibiotics, which limit or suppress the growth of common gram-positive microflora, mannan oligosaccharides (MOS) and other oligosaccharides can serve as decoy attachment sites for gram-negative pathogens, thereby preventing attachment onto enterocytes and subsequent enteric refection (Oyofo et al. 1989; Newman 1994; Parks et al. 2001). Moreover, over the past three decades, a great amount of information has been gathered on the antimicrobial potential of yeast-based products, particularly for MOS, FOS and the fructose polymer inulin to a lesser extent, with reflections on overall health status and productivity in poultry (Al-Khalaifah 2018; Al-Khalaifah et al. 2019; Reuben et al. 2021; Khomayezi and Adewole 2021; Salami et al. 2022). However, the data on the use of yeast cell derivates as natural feed additives as alternatives to AGPs in turkey diets published to date are inconsistent.

Firman et al. (2013) reported improved FCR and increased pectoralis major yield in turkey toms fed a diet containing yeast culture from 15 to 18 weeks. Bradley and Savage (1995) and Hayat et al. (1993) demonstrated that dietary supplementation with a yeast fermentation product improved the reproductive performance of some turkey breeder hens. They concluded that yeast culture inclusion in turkey breeder hen diets could improve productive efficiency, such as the hatchability of fertile eggs.

Based on the results of a study with a commercial strain of large white turkeys (Nicholas Turkeys), Fritts and Waldroup (2003) concluded that a yeast fermentation product (Bio-Mos®) might be considered as part of an overall feeding program to aid in overcoming the potential loss of AGPs. Data from this study indicated that FCR from 0 to 20 weeks of age was significantly improved by 0.10% Bio-Mos® compared to the negative control diet without performance enhancer feed additive, whereas BW, mortality, and breast meat yield were not significantly ($P < 0.05$) influenced by either Bio-Mos® or the AGPs. This was in agreement with previously published studies (Olsen 1996; Savage and Zakrzewska 1997), which reported that turkeys fed diets provisioned with Bio-Mos® had significantly improved FCR in the

Table 2 Responses to dietary administration with probiotics and prebiotics by growing turkeys as an expression of body weight, feed intake, feed conversion ratio, mortality rate, meat yield, and digestive system development

Bird genotype	Period (days)	Product specification	Supplemental level (g/kg)	Body weight	Feed intake	FCR[a]	Mort. rate	Meat yield	Dig. sys devel.[b]	References
Probiotics										
LW Nicholas	140	*Bacillus subtilis*	0.25–1.0	↑[c]	nd[f]	↓	–	nd	–	Jiraphocakul et al. (1990)
Nicholas-200	28	Blend of bacteria	2.5 mL/kg	↑	nd	nd	nd	nd	nd	Torres-Rodriguez et al. (2007)
BUT-9	126	*Bacillus toyoi*	0.2×10^9 cfu	↑	nd	–	nd	nd	nd	Grela et al. (2009)
BUT-9	126	*Bacillus toyoi*	0.2×10^9 cfu	↑	nd	↓	–	nd	nd	Batkowska et al. (2015)
BUT-6	56	*Bacillus subtilis*	0.5	–[d]	↑	–	nd	nd	↑	Omidiwura et al. (2018)
BUT-6	80	Blend of bacteria	0.5	–	nd	–	nd	nd	↑	Dobrowolski et al. (2019)
Hybrid turkeys	64	Blend of bacteria	0.5	↑	↑	–	nd	–	nd	Lalev et al. (2020)
BUT-6	105	Blend of bacteria	0.4	–	nd	–	–	nd	–	Śliżewska et al. (2020)
Prebiotics										
BUT-6	56	MOS	1.0	↑	nd	↓	nd	nd	nd	Savage and Zakrzewska (1996)
BUT-6	21	MOS	1.0	↑	nd	–	–	nd	↑	Fairchild et al. (2001)
Hybrid turkeys	126	MOS	1.0	↑	nd	–	–	nd	nd	Parks et al. (2001)
LW Nicholas	140	MOS	1.0	–	nd	↓	–	–	nd	Fritts and Waldroup, (2003)
	30	MOS								

(continued)

Table 2 (continued)

Bird genotype	Period (days)	Product specification	Supplemental level (g/kg)	Body weight	Feed intake	FCR[a]	Mort. rate	Meat yield	Dig. sys devel.[b]	References
BUT Big-9			1.0 4.0	— —	— —	— —	nd nd	nd nd	cs[g] cs	Juśkiewicz et al. (2003)
BUT-9	56	MOS Inulin	4.0 4.0	— —	— —	— —	nd nd	nd nd	↑ ↑	Zduńczyk et al. (2004a)
BUT-9	56	MOS	1.0 5.0	— —	— —	— —	nd nd	nd nd	cs cs	Zduńczyk et al. (2004b)
BUT-9	112	Inulin	0.1 1.0	— ↓[e]	nd nd	— —	nd nd	nd nd	— ↑	Juśkiewicz et al. (2005)
Hybrid turkeys	93	Lactose	1.0	↑	nd	nd	nd	nd	nd	Torres-Rodriguez et al. (2007)
BUT Big-6	70	MOS	1.0	—	↑	↑	nd	—	nd	Konca et al. (2009)
Hybrid Turkeys	63	Yeast fer. Product	0.625 2.50	↑ —	nd nd	— —	nd nd	nd nd	↑ —	Paiva et al. (2010)
LW Nicholas	126	Yeast fer. Product	1.25 2.50	— —	nd nd	↓ ↓	nd nd	↑ ↑	— —	Firman et al. (2013)
Hybrid turkeys	140	Yeast fer. product	1.25–0.625	—	—	—	nd	—	nd	Barasch (2012)
Nicholas-300	98	MOS + betaglucans	1.0	—	—	—	nd	nd	nd	Vahabi-Asil et al. (2017)

[a]FCR: feed conversion ratio (g feed/g gain); [b]Digestive system development expresses some metabolic, biochemical and morphologic aspects in the GIT including intestinal microflora enzymes, short chain fatty acids content in the ceca, viscosity of intestinal lumen, villus morphological characteristics and pH changes; [c]Increased (↑);[d] Unchanged (—); [e]Decreased (↓); [f]Not determined (nd); [g]Not decisive (cs)

absence of increased BWG. Significantly improved BWG and FCR in turkey poults fed to 8 weeks of age were reported in another study (Savage and Zakrzewska 1996).

In an early feeding study conducted at 3 weeks of age, Fairchild et al. (2001) reported that both Bio-Mos® and bambermycin significantly improved the BWG of turkey poults. Similarly, Parks et al. (2001) observed an improvement in early FCR (0–3 weeks) with the addition of Bio-Mos®, bambermycin, virginamycin and Bio-Mos® and bambermycin used in combination. The addition of Bio-Mos®, as well as the virginiamycin treatments, increased the overall BWG compared to the unsupplemented dietary treatments.

It is noteworthy that antibiotics and MOS improved the FCR of poults from 0 to 3 weeks of age, a period when gut microflora are not fully developed and stabilized. Therefore, these feed additives may provide advantages by stabilizing the gut microflora and limiting colonization by pathogens through the early stages of life.

Based on a meta-analysis of several trials, Hooge (2004) reported that dietary MOS can improve BWG and decrease mortality in turkeys compared to negative control diets, but dietary MOS supplementation does not always improve FCR. Based on the statistical similarity of the bird performance response to either MOS supplementation or antibiotic supplementation, Hooge (2004) suggested that MOS, a polysaccharide-protein complex derived from yeast, could replace AGPs.

In contrast to earlier studies reporting benefits on BWG and FCR of turkey toms, Valancony et al. (2001) observed no difference in slaughter weight or carcass yield of turkeys fed diets with added avilamycin antibiotic and Bio-Mos®; no mention was made regarding FCR. Final BW, feed intake and FCR in turkey toms at 56 days of age were comparable when treated with control, inulin, and MOS at 4 g/kg supplementation level (Zduńczyk et al. 2004a). Similarly, no significant effect of 0.1, 0.25 and 0.5% dietary inclusion of Bio-Mos® was observed on poult performance after an 8-week feeding experiment (Zduńczyk et al. 2004b). In the previous work of this research team (Juśkiewicz et al. 2003), feeding turkeys with diets supplemented with the same levels of Bio-Mos® for a 4-week period did not affect the productivity of the birds as measured by BWG, feed intake and FCR. In their following related study (Juśkiewicz et al. 2005), insignificant effects of fortifying prebiotic (inulin) in turkey diets with different supplementation levels on growth performance were determined though main indicators of gut health have remarkably enhanced.

Unbeneficial effects of feeding diets supplemented with Bio-Mos® and live yeast from 10 to 20 weeks of age on the growth performance of Big 6 male turkeys were reported by Konca et al. (2009). Likewise, Barasch (2012) reported that the inclusion of a yeast fermentation product (Original XPC™, Diamond V Mills) in the breeder hen diet during the first 2 weeks of laying did not have a carry-over effect on the performance of the male progeny compared to the unsupplemented control breeder hens. Another study provided evidence that supplementation with prebiotics containing the same amount of MOS and beta-glucans from *Saccharomyces cerevisiae* yeast in turkey diets with different levels of protein did not lead to any significant changes in performance, blood biochemistry, serum immunoglobulin concentrations, or haemagglutination-inhibition titres (Vahabi-Asil et al. 2017).

A similar response was seen in a study by Paiva et al. (2010), where turkey hens raised to 9 weeks of age were fed a diet with different levels of the yeast fermentation product. The birds receiving the prebiotic-supplemented diets had higher BWG at four and 6 weeks of age than the control birds. By 9 weeks of age, a significant difference was no longer observed. Likewise, supplementation with lactose as a prebiotic, either alone or in combination with lactic acid bacteria-based probiotics, was demonstrated to increase the BWG of poults by 15% and 17% after 26 and 28 days of treatment, respectively, regardless of the administration route via feed or drinking water (Torres-Rodriguez et al. 2007).

Scrutinization of the available literature on turkey toms reveals that the benefits reaped from the dietary treatment of yeast-cell wall products were notably ahead of improvements in growth performance in terms of mitigating the stress generated by management and health challenges (Juśkiewicz et al. (2006). This phenomenon was supported by two battery experiments conducted by Huff et al. (2007), where 1-week-old male poults and progeny of 33- and 40-week-old hens were subjected to cold stress and respiratory *E. coli* challenge. Poults were administered a commercial yeast extract feed supplement (Alphamune) for a 3-week period at 504 or 1008 g/t inclusion levels. Immunostimulation using yeast extract supplements protected poults from some of the production loss due to cold stress and *E. coli.*

Turkeys fed MOS during a specific challenge from *Salmonella typhimurium* had a decreased incidence of faecal contamination, whereas broilers fed MOS had reduced faecal counts of *Salmonella dublin* and *E. coli* (Spring 1996). Schoeni and Wong (1994) also reported a reduction in *Campylobacter jejuni* colonization when birds were fed MOS. As previously mentioned, Fairchild et al. (1999) observed improved performance of poults challenged with field isolates of *E. coli* and fed MOS. In agreement, dietary lactose, as a putative prebiotic, was found to be beneficial to intestinal health, as manifested in decreased caecal Salmonella colonization in turkey poults (Corrier et al. 1991; Hollister et al. 1994).

The success of feed supplementation with commercial yeast extract (Alphamune™) in improving the protection of turkeys against the development of Clostridial dermatitis, a production disease of commercial turkeys that is characterized by sudden mortality in market-aged male birds, was also demonstrated in another study (Huff et al. 2014). The results indicated that the addition of Alphamune™ to the feed at 1008 g/tonne may decrease the mortality rate in male poults at 2, 7 and 12 weeks of age, whereas there were no significant differences in mortality at week 16. The antimicrobial activity of another yeast cell derivate, inulin, was shown in a study by Juśkiewicz et al. (2005). They demonstrated that the supplementation of turkey diets with 1.0% inulin led to a significant reduction in *E. coli* populations, an insignificant but noticeable increase in *Bifidobacterium* and *Lactobacillus* counts, and the greatest SCFA production. However, the final BW of turkeys fed inulin was inferior to that of their untreated counterparts.

The association between the stress incurred due to management practice (i.e., moving birds to larger facilities three times per grow-out period and an increase in the turkey disease osteomyelitis complex) and increased colonization with foodborne pathogens was noted in several studies (Dutta et al. 2008; Huff et al.

2009, 2010). By means of healing this stress-induced health problem, it was demonstrated that Alphamune™ effectively protected turkeys from the immunosuppressive effects of incurred transport stress, as evidenced by decreased serum heterophil oxidative burst activity and increased serum corticosterone levels (Huff et al. 2011). The authors suggested that yeast derivates could be used as nutritional immunomodulators that may be applied prior to the transport of turkeys. However, the greater protective response to Alphamune™ in female turkeys than in males is noticeable.

Yeast derivates have also been shown to modulate caecal fermentation metabolites, which are accepted as reliable indicators of gut health, and the proper function and integrity of intestines (Juśkiewicz et al. 2003). As determined by Zduńczyk et al. (2004b), dietary supplementation with commercial yeast derivate (Bio-Mos®, 5 g/kg) reduced ammonia concentrations and enhanced volatile fatty acid concentrations, especially acetate and butyrate, in the caecal digesta. However, a significant enhancement of propionate, isobutyrate and isovalerate concentrations in the digesta was observed when the supplemental dosage of Bio-Mos® was halved (i.e., 2.5 g/kg). A similar MOS product (Bio-Mos®, 1.8 g/kg) showed a positive effect on early intestinal development in poults aged 9 and 11 days, increasing the villus height, villus surface area and crypt depth at both ages (Loeffler 2014).

Overall, a review of the available information stresses that management factors, including the brooding temperature during the starter period, challenge with enteric pathogens and litter quality and production stressors during the later stages of turkey production (e.g., social and psychological stress due to crowding and transportation stress), as well as hen and bird age and gender, should be considered when designing studies to evaluate the antimicrobial potential of yeast-based products as alternatives to AGPs. Table 2 also shows the growth performance responses of turkeys fed to various prebiotic preparations at different concentrations.

6 Phytogenics

Over the past two decades, plant-derived extracts have garnered interest for feed industry applications, with particular interest for pigs and poultry, along with the increase in popularity of the search for natural alternatives to in-feed antibiotics that are free from residue and resistant effects (Mountzouris et al. 2009; Wallace et al. 2010; Diaz-Sanchez et al. 2015; Giannenas et al. 2020). This is similar to the food industry's prioritization of high-quality, clean-label meat products for consumers (Botsoglou et al. 2003; Keokamnerd et al. 2008; Dewi et al. 2021). Scientific studies with poultry are mostly devoted to broiler chickens and, to a lesser extent, layer hens and findings obtained from studies over the last two decades have been summarized in comprehensive reviews (Bozkurt et al. 2014; Harrington et al. 2020; Jin et al. 2020). However, in turkey diets, the magnitude of the response to dietary supplementation with aromatic plants and their extracts has rarely been compiled. Fortunately, the recent review of Bozkurt and Tüzün (2020) provided excellent coverage of early research appreciating metabolic, physiologic and morphologic changes along with the growth performance traits, gut function, general health status, meat yield and meat quality aspects in response to dietary consumption of plant-derived chemicals. This pioneering review also provides background information for the present review paper. The results obtained from several studies conducted after the publication of the above review are presented below.

The objectives of a recent study carried out by Zumbaugh et al. (2020) were to determine the effect of commercially available plant bioactives (Digestarom®, Biomin) on the performance of turkey poults fed a reduced protein and amino acid diet over a 6-week brooding period. Digestarom® is a proprietary phytogenic feed additive consisting of a unique blend of herbs, essential oils and functional flavours. The results showed that formulating starter and grower diets to be nutrient-deficient negatively affected the growth of poults. However, as the birds fed Digestarom® showed significantly higher (3.5%) BWG than those fed the nonsupplemented and protein-deficient diets, its addition is thought to have helped compensate for the reduced protein and limiting amino acids. However, it had minimal effect on FCR and nutrient digestion parameters, including digestive enzyme activity, pancreatic enzyme gene expression and nutrient transporter gene expression.

Salmonella Heidelberg, an invasive pathogen in humans, is the leading cause of foodborne illness in ground and processed turkey meat (Bearson et al. 2017; CDC 2019). Therefore, continued efforts have been made to mitigate the threat induced by this enteropathogen in poultry products (Antunes et al. 2016; Dewi et al. 2021). For this concept, the antimicrobial efficacy of three plant-derived antimicrobials, lemongrass essential oil, citral and trans-cinnamaldehyde, against *S.* Heidelberg in ground turkeys was evaluated in a very recent study (Dewi et al. 2021). The results of this study indicated that the three tested plant-derived chemicals are effective against *S.* Heidelberg in ground turkey during refrigerated storage, indicating their potential use as interventions to mitigate Salmonella contamination in comminuted turkeys. However, the beneficial effects of these phytogenic compounds on the flavour and appearance of the raw turkey patties were not as great as the antimicrobial activity against *S.* Heidelberg.

The application of oregano products as a feed additive on the performance, health and quality of turkey meat was outlined in a very recent review by Bozakova and Ivanov (2022). The efficiency of ground oregano and its main active compounds in the prevention and treatment of protozoal diseases in turkeys was suggested by the authors.

According to the available results from a limited number of studies, it can be stated that phytogenic compounds have the potential to be considered as feed additives to promote growth, health and meat quality in turkeys, and render nutrients more available for digestion as well. However, the underlying mechanism and reasons for these improvements still require clarification.

7 Mycotoxin-Counteracting Strategies

The contamination of feed with mycotoxins is a problem of critical importance in continuing feed safety issues leading to economic losses in animal production (Wu 2007). Mycotoxins are secondary metabolites produced by toxigenic fungal species. The mycotoxins produced (mostly trichothecenes, zearalenone, deoxynivalenol, aflatoxins and fumonisins) can cause deleterious effects on animal and consequently human health after oral intake (Murugesan et al. 2015; Haquea et al. 2020). Mycotoxin-detoxifying agents supplemented as feed additives are still the most promising and are therefore most commonly used in turkey feed, as is the case in other poultry species (Kolosova and Stroka 2011; Murugesan et al. 2015). These detoxifiers can be classified as mycotoxin binders and mycotoxin modifiers (EFSA 2009). Mycotoxin binders can suppress or reduce the absorption of mycotoxins in the gut, resulting in the excretion of toxin-binder complexes in faeces, whereas mycotoxin modifiers modify their mode of action and transform the toxin into nontoxic metabolites (EFSA 2010; Devreese et al. 2013, 2014). Of note, evidence from previous studies has demonstrated that poults are rather tolerant to selected mycotoxins because infected birds show comparable performance traits and biochemical, histological and immunological measurements, which are regarded as indicators of mycotoxicosis, compared to uninfected healthy birds (Giambrone et al. 1985; Olsen et al. 1986; Morris et al. 1999; Quist et al. 2000; Rauber et al. 2007; Girish et al. 2008; Grimes et al. 2010; Xu et al. 2011).

Although a range of research studies show promising results for toxin binders and toxin modifier agents in broilers, layer hens and ducks (Murugesan et al. 2015), there is a dearth of research on turkey toms. The efficacy of bentonite clay, proven to be a mycotoxin-detoxifying agent, was tested in an in vitro model using a drug interaction model (Devreese et al. 2013). In the study, based on field observations and suggestions made by other authors that mycotoxin-detoxifying agents decrease or enhance the oral absorption of drugs, the interaction between the macrolide antibiotic tylosin and a bentonite toxin binder was investigated. The results indicated that bentonite significantly reduced tylosin passage and thus bound tylosin. Therefore, the authors suggested that the combined use of bentonite in the feed with tylosin in the feed or drinking water should be avoided, as this could lead to therapy failure and eventually enhance antibiotic resistance towards tylosin due to subtherapeutic plasma concentrations. On the other hand, the modified glucomannan binder did not alter the passage of tylosin significantly, indicating safe combined use.

In their follow-up study, Devreese et al. (2014) evaluated the effect of a commercial yeast-derived glucomannan mycotoxin binder agent (Mycosorb®, Alltech Inc.) on selected nonspecific parameters and specific toxicokinetic markers after feeding hybrid turkeys a diet naturally contaminated with Fusarium mycotoxins, mainly deoxynivalenol. The mycotoxin-detoxifying agent, Mycosorb, was supplemented at 2 kg/tonne in each rearing phase diet, including starter (0–3 weeks), grower (4–6 weeks), developer (7–9 weeks) and finisher (10–12 weeks). Except for the starter phase, no significant differences in BW, BWG, feed intake, or FCR were observed. The feeding of contaminated diets reduced the duodenal villus height and apparent villus surface area. This mycotoxin-induced negative effect was prevented by glucomannan supplementation. However, the feeding of contaminated diets elevated the total duodenal CD8+ T-lymphocyte counts, but this effect was not ameliorated by Mycosorb. Thus, the provision of a diet with glucomannan was able to counteract the negative effects of Fusarium mycotoxins on duodenal morphometry but was ineffective to counter the influx of specific toxicokinetic markers and decrease deoxynivalenol absorption after naturally occurring oral toxin intake for a 3-week period. Nonetheless, in terms of the growth performance parameters, the lack of response to mycotoxicosis or mycotoxin-detoxifier agents used is noticeable.

In a more recent study, the effectiveness of in-feed mycotoxin-binder agents in turkey poult diets containing low to moderate levels of mycotoxins (aflatoxin, deoxynivalenol and zearalenone) was determined by Tilley et al. (2017). Corn, wheat and barley with the naturally occurring mycotoxins mentioned above were used to make rations for feeding turkey hen poults to 6 weeks of age. The commercial mycotoxin-detoxifier agents used in this study were Biomin BioFix (2 lb/tonne), Kemin Kallsil (4 lb/tonne) and Nutriad UNIKE (3 lb/tonne). The feed additives lessened the detrimental effect of mycotoxicosis on nutrient utilization, which manifested as improvements in FCR. The physiological effect of feeding the additives was observed as reduced relative gizzard weight for both groups and a smaller increase in the relative kidney weight for the birds fed the mycotoxin feed. The authors considered that the feed additives used in the study alleviated the harmful effect of dietary mycotoxins to some degree.

Review of the above studies shows that feed additives with mycotoxin-binding and mycotoxin-modifying activities could detoxify some mycotoxins or provide a direct benefit to poults. Considering that turkey poults are rather tolerant to different mycotoxins, it was tentatively concluded by almost all the researchers that markedly higher naturally occurring or artificially infected levels of mycotoxins, compared to those applied in the turkey trials, are warranted to establish the actual potential of mycotoxin-detoxifier agents used.

8 Conclusion

The ban on nontherapeutic antibiotics adds to the challenge of maintaining flock health, productive output and welfare in modern turkey production. Scrutiny of results from available studies indicated that, in the absence of AGPs, using various alternative feed additives with antimicrobial activity could open up options to improve nutrient utilization, absorption, metabolism and modulation of host immunity, which in turn affect the gut health of turkeys. In this sense, prebiotics, probiotics, phytogenics, organic acids, yeast-based products and enzymes appear to be regarded as practical and sustainable tools to be utilized. However, such improvements in the function and health status of the digestive system barely coincided with improved production efficiency. In general, either no differences or somewhat beneficial trends can be found in the literature on the influence of feed additives on growth performance traits, including BW, BWG, FCR and mortality

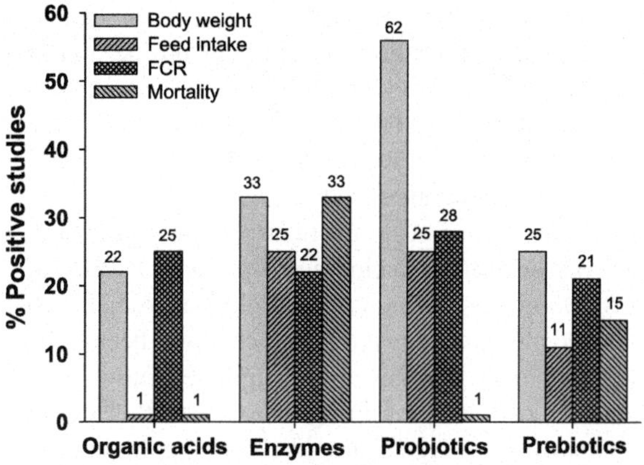

Fig. 1 The experiments cited in this chapter in which organic acids, enzymes, probiotics, and prebiotics significantly improved growth performance characteristics in growing turkeys

rate. The same is the case for digestive organ weight and length, meat yield and processed parts with economic importance. The mechanisms of the improvement of gut health and bird resilience due to these additives may also include improvement of intestinal health and utilization of nutrients by maintaining gut integrity and enhancing antioxidant capacity. However, as far as growth performance is concerned, the feed additives evaluated herein mostly resulted in a lack of bird response, probably due to their longer grow-out period (12–20 weeks) than broilers (5–7 weeks), allowing the turkeys more time to encounter pathogens and thus perform under less stressful conditions. In addition, the relative age and bird gender may play a role, although sufficient data are not available to validate this. Taken together, a review of the current studies provides supporting evidence that performance-enhancing feed additives can help modulate the immune system and improve intestinal development in young turkeys without compromising growth performance during times of disease challenge. The figure also provides an overview of the results from a large body of research and allows the reader to glean more effective conclusions on the usefulness of these performance enhancer feed additives (Fig. 1).

9 Further Directives

Microbiota modulation by using feed additives, particularly exogenous carbohydrase enzymes and protease, is of particular importance when feeding poults on starter diets when the young birds have an immature microbiota. Thus, controlling the antinutritional effects of NSP and detrimental effects of fermentative organisms proliferating in the small intestine due to dietary supplementation of soybean meal (almost 40%) could have significantly positive impacts in antibiotic-free and no-antibiotic-ever production systems.

On the other hand, late feeding programs with low protein levels using alternative protein sources to soybean, such as sunflower meal, rapeseed cake, cotton seed cake and DDGS, may gain relevance in physiology, microbiology, metabolic function and nutrient utilization, mainly because of their high fibre content, poor nutrient digestibility, unbalanced amino acid profile and anti-nutritional factors that limit their use in rations. Certain benefits of exogenous enzymes in overcoming these nutritional issues and their contributions to economic, environmental and sustainable agriculture practices remain to be elucidated for this late growth phase, where almost half of the cumulative feed is consumed.

While these issues are relatively straightforward, the control of bacterial and protozoan diseases, including *Histomonas meleagridis, Salmonella, Campylobacter, Listeria monocytogenes* and coccidiosis, remains challenging in turkey production. Due to the ban on AGPs, which have been successively used to control these diseases over decades, turkey growers now face the challenge of maintaining productivity and preventing diseases on their farms. Thus, whether these additive alternatives to AGPs are likely to be highly influential in coping with these enteropathogens merits further investigation through further studies.

References

Adeola O, Cowieson AJ (2011) Board-invited review: opportunities and challenges in using exogenous enzymes to improve nonruminant animal production. J Anim Sci 89:3189–3218

Adewole DI, Oladokun S, Santin E (2021) Effect of organic acids–essential oils blend and oat fiber combination on broiler chicken growth performance, blood parameters, and intestinal health. Anim Nutr 7:1039–1051

Agboola AF, Aroniyo I, Suberu S et al (2014) Dietary supplementation of probiotics and synbiotics on intestinal microbial populations and gut morphology of Turkey poults. Afr J Livest Ext 14: 13–20

Al-Khalaifah HS (2018) Benefits of probiotics and/or prebiotics for antibiotic-reduced poultry. Poult Sci 97:3807–3815. https://doi.org/10.3382/ps/pey160

Al-Khalaifah H, Al-Nasser A, Al-Surayee T et al (2019) Effect of dietary probiotics and prebiotics on the performance of broiler chickens. Poult Sci 98:4465–4479. https://doi.org/10.3382/ps/pez282

Antunes P, Mourao J, Campos J et al (2016) Salmonellosis: the role of poultry meat. Clin Microb Infect 22:110–121

Applegate TJ, Webel DM, Lei XG (2003) Efficacy of a phytase derived from Escherichia coli and expressed in yeast on phosphorus utilization and bone mineralization in Turkey poults. Poult Sci 82:1726–1732

Ayoola AA, Malheiros RD, Grimes JL et al (2015) Effect of dietary exogenous enzyme supplementation on enteric mucosal morphological development and adherent mucin thickness in turkeys. Front Vet Sci 2:1–8

Barasch IB (2012) The effects of yeast-derived feed additives and their potential as antibiotic alternative feed additives on the growth performance of Turkey toms raised to market age. Master Thesis. Graduate Faculty of North Carolina State University (USA)

Barbour GW, Werling M, Yersin AG et al (2002) The effect of enzyme predigestion on the nutritional quality of prepressed Turkey feather meal. Poult Sci 81:1032–1037

Basmacioğlu-Malayoğlu H, Özdemir P, Bağriyanik HA (2016) Influence of an organic acid blend and essential oil blend, individually or in combination, on growth performance, carcass parameters, apparent digestibility, intestinal microflora and intestinal morphology of broilers. Br Poult Sci 57:227–234. https://doi.org/10.1080/00071668.2016.1141171

Batkowska J, Brodacki A, Tomczyk G (2015) The influence of probiotic bacteria (Bacillus toyoi) on livability and performance of young meat-type turkeys. Braz J Poult Sci 17(4):433–438

Bearson BL, Bearson SMD, Looft T et al (2017) Characterization of a multidrug-resistant salmonella enterica Serovar Heidelberg outbreak strain in commercial turkeys: colonization, transmission, and host transcriptional response. Front Vet Sci 4:1–7

Bedford MR, Schulze H (1998) Exogenous enzymes for pigs and poultry. Nutr Res Rev 11:91–114

Botsoglou NA, Grigoropoulou SH, Botsoglou E et al (2003) The effects of dietary oregano essential oil and α-tocopheryl acetate on lipid oxidation in raw and cooked Turkey during refrigerated storage. Meat Sci 65:1193–1200

Bozakova N, Ivanov V (2022) Effects of oregano (*Origanum vulgare*) as a dietary supplement in Turkey farming – review. J Hyg Engin Des 38:219–224

Bozkurt M, Tüzün AE (2020) Application of aromatic plants and their extracts in diets of turkeys. In: Florou-Paneri P, Christaki E, Giannenas I (eds) Feed additives: aromatic plants and herbs in animal nutrition and health. Academic Press (an imprint of Elsevier), London, pp 205–220

Bozkurt M, Küçükyilmaz K, Çatli AU et al (2012) Effects of administering an essential oil mixture and an organic acid blend separately and combined to diets on broiler performance. Arch Geflügelk 76:81–87

Bozkurt M, Hippenstiel F, Abdel-Wareth AAA et al (2014) Effects of selected herbs and essential oils on performance, egg quality and some metabolic activities in laying hens-a review. Eur Poult Sci 78:1–15

Bradley GL, Savage TF (1995) The effect of autoclaving a yeast culture of Saccharomyces cerevisiae on Turkey poult performance and the retention of gross energy, and selected minerals. Anim Feed Sci Technol 55:1–7

Brodacki A, Batkowska J, Makarski B (2009) The effect of feeding with feedstuff containing microbial and plant phytase additive on performance of young slaughter turkeys. Acta Agricul Scand Sec A 59:137–141

CDC (2019) Outbreak of multidrug-resistant Salmonella infections linked to raw turkey products. 2018 Outbreaks. https://www.cdc.gov/salmonella/reading-07-18/index.html. Accessed July 2019

Çelik K, Erbil EI, Uzatici A et al (2003) The using of organic acids in California Turkey chicks and its effects on performance before pasturing. Int J Poult Sci 2:446–448

Chiang W, Booren A, Strasburg G (2008) The effect of heat stress on thyroid hormone response and meat quality in turkeys of two genetic lines. Meat Sci 80:615–622

Cole K, Farnell MB, Donoghue AM et al (2006) Bacteriocins reduce campylobacter colonization and alter gut morphology in Turkey poults. Poult Sci 85:1570–1575

Corrier DE, Hinton A Jr, Kubena LF et al (1991) Decreased salmonella colonization in Turkey poults inoculated with anaerobic cecal microflora and provided dietary lactose. Poult Sci 70: 1345–1350

Cowieson AJ, Wilcock P, Bedford MR (2011) Superdosing effects of phytase in poultry and other monogastrics. World Poult Sci J 67:225–235

Czech A, Merska-Kazanowska M, Całyniuk Z (2020) Redox status, biochemical parameters and mineral elements content in blood of Turkey hens fed a diet supplemented with Yarrowia lipolytica yeast and two Bacillus species. Animals 10:459. https://doi.org/10.3390/ani10030459

Denli M, Blandon JC, Guynot ME et al (2009) Effects of dietary AflaDetox on performance, serum biochemistry, histopathological changes and AF residues in broilers exposed to AFB1. Poult Sci 88:1444–1451

Devreese M, Pasmans F, De Backer P et al (2013) An in vitro model using the IPEC-J2 cell line for efficacy and drug interaction testing of mycotoxin detoxifying agents. Toxicol Vitro 27:157–163

Devreese M, Girgis GN, Si-T T et al (2014) The effects of feed-borne fusarium mycotoxins and glucomannan in Turkey poults based on specific and non-specific parameters. Food Chem Toxicol 63:69–75

Dewi G, Nair DVT, Peichel C et al (2021) Effect of lemongrass essential oil against multidrug-resistant salmonella Heidelberg and its attachment to chicken skin and meat. Poult Sci 100: 101116

Diaz-Sanchez S, D'Souza D, Biswas D et al (2015) Botanical alternatives to antibiotics for use in organic poultry production. Poult Sci 94:1419–1430

Dobrowolski P, Tomaszewska E, Klebaniuk R et al (2019) Structural changes in the small intestine of female turkeys receiving a probiotic preparation are dose and region dependent. Animal 13(12):2773–2781

Dutta V, Huff GR, Huff WE et al (2008) The effects of stress on respiratory disease and transient colonization of turkeys with Listeria monocytogenes Scott A. Avian Dis 52:581–589

EFSA (2009) Review of mycotoxin-detoxifying agents used as feed additives: mode of action, efficacy and feed/food safety. EFSA Support Publ 6(9):22E. https://doi.org/10.2903/sp.efsa.2009.EN-22

EFSA (2010) Statement on the establishment of guidelines for the assessment of additives from the functional group 'substances for reduction of the contamination of feed by mycotoxins'. EFSA J 8(7):1693. https://doi.org/10.2903/j.efsa.2010.1693

Erasmus MA (2018) Welfare issues in Turkey production. Adv Poult Welf Pages 3:263–291. https://doi.org/10.1016/B978-0-08-100915-4.00013-0

Fairchild AS, Grimes JL, Edens FW et al (1999) Effect of hen age, Bio-Mosand Flavomycin on susceptibility of Turkey poults to oral Escherichia coli challenge. In: Lyons TP, Jacques KA (eds) Proceedings of Alltech's 15th annual symposium, under the microscope: focal points for the new millenium. Biotechnology in the feed industry. Nottingham University Press, Nottingham, UK, pp 185–201

Fairchild AS, Grimes JL, Edens FW et al (2001) Effects of hen age, Bio-Mos, and Flavomycin on poult susceptibility to oral Escherichia coli challenge. Poult Sci 80:562–571

Ferket PR (1993) Practical use of feed enzymes for turkeys and broilers. J Appl Poult Res 2:75–81

Ferket PR (2013) Novel nutritional applications to optimize feed efficiency in turkeys. In: Paper presented at the 7th Turkey science and production conference. Proceedings, pp 1–11

Firman JD, Kirn BN (1989) Research note: effects of monensin and bambermycins on the performance of market turkeys. Poult Sci 68:1724–1726

Firman JD, Moore D, Broomhead J et al (2013) Effects of dietary inclusion of a Saccharomyces cerevisiae fermentation product on performance and gut characteristics of male turkeys to market weight. Int J Poult Sci 12:141–143. https://doi.org/10.3923/ijps.2013.141.143

Flores Maldonado KR (2018) Effect of feed quality, feed additives and litter amendments in a Turkey antibiotic free program on performance and health. Dissertation, University of North Carolina State

Flores KR, Fahrenholz A, Ferket PR et al (2021) Effect of methionine chelated Zn and Mn and corn particle size on large white male Turkey live performance and carcass yield. Poult Sci 100: 101444

Forkus B, Ritter S, Vlysidis M et al (2017) Antimicrobial probiotics reduce Salmonella enterica in Turkey gastrointestinal tracts. Sci Rep 7:40695

Francis C, Janky DM, Arafa AS et al (1978) Interrelationship of lactobacillus and zinc bacitracin in the diets of Turkey poults. Poult Sci 57:1687–1698

Fritts CA, Waldroup PW (2003) Evaluation of bio-Mos mannan oligosaccharide as a replacement for growth promoting antibiotics in diets for turkeys. Int J Poult Sci 2:19–22

García-López JC, Pinos-Rodríguez JM, García-Galicia IA et al (2011) Effects of enzyme and feeding system on Turkey performance. Arch Zootec 60:297–300

Gheisar MM, Hosseindoust A, Kim IH (2015) Evaluating the effect of microencapsulated blends of organic acids and essential oils in broiler chickens diet. J Appl Poult Res 24:511–519

Giambrone JJ, Diener UL, Davis ND et al (1985) Effects of aflatoxin on young turkeys and broiler chickens. Poult Sci 64:1678–1684

Giannenas I, Sidiropoulou E, Bonos E et al (2020) The history of herbs, medicinal and aromatic plants, and their extracts: past, current situation and future perspectives. In: Florou-Paneri P,

Christaki E, Giannenas I (eds) Feed additives: aromatic plants and herbs in animal nutrition and health. Academic Press (an imprint of Elsevier), London, pp 1–15
Girish CK, Smith TK, Boermans HJ et al (2008) Effects of feeding blends of grains naturally contaminated with fusarium mycotoxins on performance, hematology, metabolism, and immunocompetence of turkeys. Poult Sci 87:421–432
Gous RM, Fisher C, Tůmová E et al (2022) Dietary energy: protein ratio influences the efficiency of utilization of dietary protein by growing turkeys. Br Poult Sci. https://doi.org/10.1080/00071668.2022.211669
Grashorn MA, Bessei W (2004) Comparison of heavy Turkey breeds BUT Big 6 and hybrid Euro FP for fattening performance, slaughter yield and meat quality. Arch Geflügelk 68:2–7
Grela ER, Brodacki A, Batkowska J et al (2009) Influence of a probiotic of bacillus toyoi strain on performance of growing Turkey poults. Arch Geflügelk 73(3):160–166
Grimes JL, Koci MD, Stark CR et al (2010) Biological effect of naturally occurring mycotoxins fed to poults reared to 21 days of age. Int J Poult Sci 9:871–874
Günal M, Yayli G, Kaya O et al (2006) The effects of antibiotic growth promoter, probiotic or organic acid supplementation on performance, intestinal microflora and tissue of broilers. Int J Poult Sci 5:149–155. https://doi.org/10.3923/ijps.2006.149.155
Hafez MH, Awad AS (2021) Turkey production and health: current challenges. German J Vet Res 1:3–14
Hafez HM, Hauck R (2006) Efficacy of a herbal product against Histomonas meleagridis after experimental infection of Turkey poults. Arch Anim Nutr 60:436–442
Haquea A, Wanga Y, Shenc Z et al (2020) Mycotoxin contamination and control strategy in human, domestic animal and poultry: a review. Microb Pathog 142:104095. https://doi.org/10.1016/j.micpath.2020.104095
Harrington D, Hall H, Wilde D (2020) Application of aromatic plants and their extracts in the diets of laying hens. In: Florou-Paneri P, Christaki E, Giannenas I (eds) Feed additives: aromatic plants and herbs in animal nutrition and health. Academic Press (an imprint of Elsevier), London, pp 187–199
Hayat J, Savage TF, Mirosh LW (1993) The reproductive performance of two genetically distinct lines of medium white Turkey hens when fed breeder diets with and without a yeast culture containing Saccharomyces cerevisiae. Anim Feed Sci Technol 43:291–301
Hernández F, García V, Madrid J et al (2006) Effect of formic acid on performance, digestibility, intestinal histomorphology and plasma metabolite levels of broiler chickens. Br Poult Sci 47:50–56
Hocking PM (2014) Unexpected consequences of genetic selection in broilers and turkeys: problems and solution. Br Poult Sci 55:1–10
Hollister AG, Corrier DE, Nisbet DJ et al (1994) Comparison of effects of chicken cecal microorganisms maintained in continuous culture and provision of dietary lactose on cecal colonization by Salmonella typhimurium in Turkey poults and broiler chicks. Poult Sci 73:640–647
Hooge DM (2004) Turkey pen trials with dietary mannan oligosaccharide: meta-analysis, 1993-2003. Int J Poult Sci 3(3):179–188
Hooge DM, Connolly A (2011) Meta-analysis summary of broiler chicken trials with dietary Actigen® (2009-2011). Int J Poult Sci 10:819–824
Huff GR, Huff WE, Rath NC et al (2007) Differential effects of sex and genetics on behaviour and stress response of turkeys. Poult Sci 86:1294–1303
Huff GR, Huff WE, Rath NC et al (2008) Effects of Escherichia coli challenge and transport stress on hematology and serum chemistry values of three genetic lines of turkeys. Poult Sci 87:2234–2241
Huff GR, Dutta V, Huff WE et al (2009) Co-infection of market-age turkeys with Escherichia coli and listeria monocytogenes in two stress models. Avian Dis 53:495–501

Huff GR, Huff WR, Farnell MB et al (2010) Bacterial clearance, heterophil function, and hematological parameters of transport-stressed Turkey poults supplemented with dietary yeast extract. Poult Sci 89:447–456

Huff GR, Dutta V, Huff WE et al (2011) Effects of dietary yeast extract on Turkey stress response and heterophil oxidative burst activity. Br Poult Sci 52:446–455

Huff GR, Huff WE, Rath NC (2014) Effects of vitamin d and yeast extract supplementation on Turkey mortality and clostridial dermatitis incidence in a dexamethasone immunosuppression model. Avian Dis 58:572–578

Immerseel FV, Russell JB, Flythe MD et al (2006) The use of organic acids to combat Salmonella in poultry: a mechanistic explanation of the efficacy. Avian Pathol 35:182–188

Izat AL, Tidwell NM, Thomas RA et al (1990) Effects of a buffered propionic acid in diets on the performance of broiler chickens and on microflora of the intestine and carcass. Poult Sci 69: 818–826

Jin L-Z, Dersjant-Li Y, Giannenas I (2020) Application of aromatic plants and their extracts in diets of broiler chickens. In: Florou-Paneri P, Christaki E, Giannenas I (eds) Feed Addıtıves: aromatic plants and herbs in animal nutrition and health. Academic Press (an imprint of Elsevier), London, pp 159–179

Jiraphocakul S, Sullivan TW, Shahani KM (1990) Influence of a dried Bacillus subtilis culture and antibiotics on performance and intestinal microflora in turkeys. Poult Sci 69:1966–1973

Juśkiewicz J, Zduńczyk Z, Jankowski J (2003) Effect of adding mannan-oligosaccharide to the diet on the performance, weight of digestive tract segments, and caecal digesta parameters in young turkeys. J Anim Feed Sci 12:133–142

Juśkiewicz J, Jankowski J, Zduńczyk Z et al (2005) Performance and microbial status of turkeys fed diets containing different levels of inulin. Arch Geflügelk 69(4):175–180

Juśkiewicz J, Zduńczyk Z, Jankowski J (2006) Growth performance and metabolic response of the gastrointestinal tract of turkeys to diets with different levels of mannanoligosaccharide. Worlds Poult Sci J 62:612–625

Juśkiewicz J, Jankowski J, Lecewicz A et al (2010) Effect of diets with different contents of sunflower meal without or with exogenous enzymes supplementation on gastrointestinal tract response of growing turkeys. J Anim Feed Sci 19:468483

Keokamnerd T, Acton JC, Han IY et al (2008) Effect of commercial rosemary oleoresin preparations on ground chicken thigh meat quality packaged in a high-oxygen atmosphere. Poult Sci 87:170–179

Khomayezi R, Adewole D (2021) Probiotics, prebiotics, and synbiotics: an overview of their delivery routes and effects on growth and health of broiler chickens. World Poult Sci J 78(1): 57–81. https://doi.org/10.1080/00439339.2022.1988804

Kolosova A, Stroka J (2011) Substances for reduction of the contamination of feed by mycotoxins: a review. World Mycotox J 4:225–256

Konca Y, Kirkpinar F, Mert S (2009) Effects of mannan-oligo saccharides and live yeast in diets on the carcass, cut yields, meat composition and colour of finishing turkeys. Asian Aust J Anim Sci 22:550–556. https://doi.org/10.5713/ajas.2009.80350

Krautwald-Junghannsa ME, Janja S (2022) The influence of stocking density on behaviour, health, and production in commercial fattening turkeys - a review. Br Poult Sci 63:434–444. https://doi.org/10.1080/00071668.2022.2050673

Krogdahl A, Sell JL (1989) Influence of age on lipase, amylase, and protease activities in pancreatic tissue and intestinal contents of young turkeys. Poult Sci 68:1561–1568

Lalev M, Mincheva N, Oblakova M et al (2020) Effects of insect- and probiotic-based diets on turkeys' production, health, and immune parameters. Bulg J Agric Sci 26(6):1254–1265

Leeson S (1984) Growth and carcass characteristics of chicken and Turkey broilers fed diets containing flavomycin. Can J Anim Sci 64:971–976

Leeson S, Caston L, Yungblut D (1996) Adding roxazyme to wheat diets of chicken and Turkey broilers. J Appl Poult Res 5:167–172

Loeffler SL (2014) Effects of probiotic and prebiotic supplementation in Turkey poults on intestinal morphology and MUC2 gene expression. Dissertation, The Ohio State University

Mahmood T, Guo Y (2020) Dietary fiber and chicken microbiome interaction: where will it lead to? Anim Nutr 1:1–8

Makowski Z, Lipiński K, Mazur-Kuśnirek M (2022) The effects of sodium butyrate, coated sodium butyrate, and butyric acid glycerides on nutrient digestibility, gastrointestinal function, and fecal microbiota in turkeys. Animals 12:1836. https://doi.org/10.3390/ani12141836

Marteau PR, De Vrese M, Cellier CJ (2001) Protection from gastrointestinal diseases with the use of probiotics 1–3. Am J Clin Nutr 73:430S–436S

McKnight F (1997) Use of NSP enzymes in wheat based Turkey diets. Use of Natugrain NSP enzyme in poultry nutrition. In: Paper presented at the BASF technical symposium. BASF, Ludwigshafen, Germany. Proceedings, pp 50–72

Mikulski D, Zduńczyk Z, Jankowski J et al (2008) Effects of organic acids or natural plant extracts added to diets for turkeys on growth performance, gastrointestinal tract metabolism and carcass characteristics. J Anim Feed Sci 17:233–246

Milbradt EL, Okamoto AS, Rodrigues JCZ et al (2014a) Use of organic acids and competitive exclusion product as an alternative to antibiotic as a growth promoter in the raising of commercial turkeys. Poult Sci 93:1855–1861

Milbradt EL, Zamae JR, Araujo Junior JP et al (2014b) Control of salmonella enteritidis in turkeys using organic acids and competitive exclusion product. J Appl Microbiol 117:554–563

Milbradt EL, Okamoto AS, Padovani CR et al (2017) Use of organic acids and a competitive exclusion product as growth promoter and salmonella enteritidis control in commercial turkeys. Braz J Poult Sci 19:551–558

Morris CM, Li YC, Ledoux DR et al (1999) The individual and combined effects of feeding moniliformin, supplied by fusarium fujikuroi culture material, and deoxynivalenol in young Turkey poults. Poult Sci 78:1110–1115

Mountzouris KC, Paraskevas V, Fegeros K (2009) Phytogenic compounds in broiler nutrition. In: Steiner T (ed) Phytogenics in animal nutrition: natural concepts to optimize gut health and performance. Nottingham University Press, Nottingham, pp 97–101

Murugesan GR, Ledoux DR, Naehrer K et al (2015) Prevalence and effects of mycotoxins on poultry health and performance, and recent development in mycotoxin counteracting strategies. Poult Sci 94:1298–1315

Neveling DP, Dicks LMT (2021) Probiotics: an antibiotic replacement strategy for healthy broilers and productive rearing. Prob Antimicrob Proteins 13:1–11

Newman K (1994) Mannan-oligosaccharides: natural polymers with significant impact on the gastrointestinal microflora and the immune system. In: Lyons TP, Jacques KA (eds) Biotechnology in the feed industry. Proceeding of Alltech's Tenth Annual Symposium. Nottingham University Press, Nottingham, pp 167–175

Nguyen HT, Wu SB, Bedford MR et al (2022) Dietary soluble non-starch polysaccharide level and xylanase influence the gastrointestinal environment and nutrient utilisation in laying hens. Br Poult Sci 3:340–350

Odetallah NH, Parks CW, Ferket PR (2002) Effect of wheat enzyme preparation on the performance characteristics of tom turkeys fed wheat-based rations. Poult Sci 81:987–994

Ofek I, Mirelman D, Sharon N (1977) Adherence of Escherichia coli to human mucosal cells mediated by mannose receptors. Nature (London) 265:623–625

Olsen R (1996) Experience with mannan oligosaccharides in commercial Turkey production. Zootech Int 19:38–39

Olsen M, Mirocha CJ, Abbas HK et al (1986) Metabolism of high concentrations of dietary zearalenone by young male Turkey poults. Poult Sci 65:1905–1910

Omidiwura BRO, Agboola AF, Olaleye OO et al (2018) Influence of prebiotics, probiotics or synbiotics on performance, intestinal mucosal integrity and gut microbiota of Turkey poults. Niger J Anim Sci 20(2):64–75

Oyofo BA, DeLoach JR, Corrier DE et al (1989) Prevention of Salmonella typhimurium colonization of broiler with D-mannose. Poult Sci 68:1357–1360

Paiva DM, Walk CL, Lehman R et al (2010) Turkey response to the inclusion of a Saccharomyces cerevisiae fermentation product, original XPCTM, in antibiotic free diets following a coccidia vaccination. Poult Sci 89(Suppl 1):283–284

Parks CW, Grimes JL, Ferket PR et al (2001) The effect of mannan oligosaccharides, bambermycins, and virginianmycin on performance of large white male market turkeys. Poult Sci 80:718–723

Patten JD, Waldroup PW (1988) The use of organic acid in broiler diets. Poult Sci 67:1182–1187

Quist CF, Bounous DI, Kilburn JV et al (2000) The effect of dietary aflatoxin on wild Turkey poults. J Wild Dis 36:436–444

Rahimi S, Grimes JL, Fletcher O et al (2009) Effect of a direct-fed microbial (Primalac) on structure and ultrastructure of small intestine in Turkey poults. Poult Sci 88:491–503

Rauber RH, Dilkin P, Giacomini LZ et al (2007) Performance of Turkey poults fed different doses of aflatoxin in the diet. Poult Sci 86:1620–1624

Rehman H, Rosenkranz C, Böhm J et al (2007) Dietary inulin affects the morphology but not the sodium-dependent glucose and glutamine transport in the jejunum of broilers. Poult Sci 86:118–122

Rehman H, Vahjen W, Kohl-Parisini A et al (2009) Influence of fermentable carbohydrates on the intestinal bacteria and enteropathogens in broilers. Worlds Poult Sci J 65(1):75–90

Reuben RC, Sarkar SL, Ibnat H et al (2021) Novel multi-strain probiotics reduces Pasteurella multocida induced fowl cholera mortality in broilers. Sci Rep 11:8885. https://doi.org/10.1038/s41598-021-88299-0

Ricke SC (2003) Perspectives on the use of organic acids and short chain fatty acids as antimicrobials. Poult Sci 82:632–639

Ritz CW, Hulet RM, Self BB et al (1995) Endogenous amylase levels and response to supplemental feed enzymes in male turkeys from hatch to eight weeks of age. Poult Sci 74:1317–1322

Rojman K, Deborah A (2022) Probiotics, prebiotics, and synbiotics: an overview of their delivery routes and effects on growth and health of broiler chickens. World Poult Sci J 78(1):57–81

Rosen GD (2007a) Holo-analysis of the efficacy of Bio-Mos® in broiler nutrition. Br Poult Sci 48(1):21–26. https://doi.org/10.1080/00071660601050755

Rosen GD (2007b) Holo-analysis of the efficacy of Bio-Mos® in Turkey nutrition. Br Poult Sci 48(1):27–32. https://doi.org/10.1080/00071660601050730

Roy RD, Edens FW, Parkhurst CR et al (2002) Influence of a propionic acid feed additive on performance of Turkey poults with experimentally induced poult enteritis and mortality syndrome. Poult Sci 81:951–951

Salami SA, Ross SA, Patsiogiannis A et al (2022) Performance and environmental impact of egg production in response to dietary supplementation of mannan oligosaccharide in laying hens: a meta-analysis. Poult Sci 101:101745. https://doi.org/10.1016/j.psj.2022.101745

Santos AA (2002) Improvement of nutritional value of wheat by dietary enzyme supplementation for turkeys. Dissertation, University of North Carolina State

Savage TF, Zakrzewska EI (1996) The performance of male turkeys fed a starter diet containing a manna-oligosaccharide (Bio-Mos) from day old to eight weeks of age. In: Lyons TP, Jacques KA (eds) Biotechnology in The Feed Industry. Proceedings of Alltech's 12 Annual Symposium. Nottingham University Press, Nottingham, UK, pp 47–54

Savage TF, Zakrzewska EI (1997) The performance of male turkeys fed a starter diet containing a mannan oligosaccharide. Zootech Int 20:30–32

Schoeni JCL, Wong ACL (1994) Inhibition of Campylobacter jejuni colonization in chicks by defined competitive exclusion cultures. Appl Environ Microbiol 60:1191–1197

Scott M (1987) Energy, protein and amino acid requirements of turkeys. In: Nutrition of the Turkey. M.L. Scott of Ithaca, Ithaca, NY, pp 28–82

Sekirov I, Russell S, Antunes L (2010) Gut microbiota in health and disease. Physiol Rev 90:859–904

Selle PH, Ravindran V (2007) Review: microbial phytase in poultry nutrition. Anim Feed Sci Technol 135:1–41

Śliżewska K, Markowiak-Kopeć P, Sip A et al (2020) The effect of using new synbiotics on the Turkey performance, the intestinal microbiota and the fecal enzymes activity in turkeys fed Ochratoxin A contaminated feed. Toxins 12:578. https://doi.org/10.3390/toxins12090578

Slomonski BA (2011) Recent advances in research on enzymes for poultry diets. Poult Sci 90: 2013–2023

Soccol CR, de Souza P, Vandenberghe L, Rigon Spier M et al (2013) The potential of probiotics: a review. Food Technol Biotechnol 48:413–434

Spring P (1996) Effects of mannanoligosaccharide on different cecal parameters and on cecalconcentrations of enteric pathogens in poultry. Dissertation, Swiss Federal Institute of Technology, Zurich, Switzerland

Spring P, Wenk C, Dawson KA et al (2000) The effects of dietary mannaoligosaccharides on cecal parameters and the concentrations of enteric bacteria in the ceca of salmonella-challenged broiler chicks. Poult Sci 79:205–211

Stefanello C, Rosa DP, Dalmoro YK et al (2020) Protected blend of organic acids and essential oils improves growth performance, nutrient digestibility, and intestinal health of broiler chickens undergoing an intestinal challenge. Front Vet Sci 6:491

Tilley JEN, Grimes JL, Koci MD et al (2017) Efficacy of feed additives to reduce the effect of naturally occurring mycotoxins fed to Turkey hen poults reared to 6 weeks of age. Poult Sci 96: 4236–4244

Tomaszewska E, Kwiecień M, Dobrowolski P et al (2016) Dose-dependent effects of probiotic supplementation on bone characteristics and mineralisation in meat-type female turkeys. Anim Prod Sci 58:507–516

Torres-Rodriguez A, Higgins SE, Vicente JLS et al (2007) Effect of lactose as a prebiotic on Turkey body weight under commercial conditions. J Appl Poult Res 16:635–641

Vahabi-Asil O, Bouyeh M, Qotbi A et al (2017) Effects of a prebiotic on growth performance, blood parameters and immunity response of turkeys fed low protein diets. Eur Poult Sci 81:196. https://doi.org/10.1399/eps.2017.196

Valancony H, Humbert F, Rukelibuga J et al (2001) Comparison of some substitutes for antibiotic additives in diets for Turkey poults. Effects on production and on resistance to salmonella colonization. Sci Tech Avicoles 35:25–34

Waibel PE, Halvorson JC, Noll SL et al (1991) Influence of virginiamycin on growth and efficiency of large white turkeys. Poult Sci 70:837–847

Wajda S, Śmiecińska K, Jankowski J et al (2010) The efficacy of lactic acid bacteria Pediococcus acidilactici, lactose and formic acid as dietary supplements for turkeys. Pol J Vet Sci 13:45–51

Wallace RJ, Oleszek W, Franz C et al (2010) Dietary plant bioactives for poultry health and productivity. Br Poult Sci 51:461–487

Weber GM, Michalczuk M, Huyghebaert G et al (2012) Effects of a blend of essential oil compounds and benzoic acid on performance of broiler chickens as revealed by a meta-analysis of 4 growth trials in various locations. Poult Sci 91:2820–2828

Wu F (2007) Measuring the economic impacts of fusarium toxins in animal feeds. Anim Feed Sci Technol 137:363–374

Xu L, Eicher SD, Applegate TJ (2011) Effects of increasing dietary concentrations of corn naturally contaminated with deoxynivalenol on broiler and Turkey poult performance and response to lipopolysaccharide. Poult Sci 90:2666–2774

Yang X, Xin H, Yang C et al (2018) Impact of essential oils and organic acids on the growth performance, digestive functions and immunity of broiler chickens. Anim Nutr 4:388–393. https://doi.org/10.1016/j.aninu.2018

Yu M, Ismail MM, Qureshi MA et al (2000) Viral agents associated with poult enteritis and mortality syndrome: role of a small round virus and Turkey coronavirus. Avian Dis 44:297–304

Zduńczyk Z, Jankowski J, Juśkiewicz J et al (2004a) Response of young turkeys to dietscontaining flavomycyn, mannan-oligosacharide or inulin. Vet ir Zootech 25:94–96

Zduńczyk Z, Juśkiewicz J, Jankowski J et al (2004b) Performance and caecal adaptation of turkeys to diets without or with antibiotic and with different levels of mannan oligosaccharide. Arch Anim Nutr 58:367–378

Zduńczyk Z, Jankowski J, Juśkiewicz J et al (2013) Effect of different dietary levels of low-glucosinolate rapeseed (canola) meal and non-starch polysaccharide-degrading enzymes on growth performance and gut physiology of growing turkeys. Can J Anim Sci 93:353–362

Zumbaugh CA, Murugesan GR, Wong EA et al (2020) Evaluation of a phytogenic feed additive on performance, nutrient digestion, and absorption in Turkey poults. Anim Feed Sci Technol 267: 114575. https://doi.org/10.1016/j.anifeedsci.2020.114575

Effect of Phytochemical Feed Additives on Health Status, Milk Yield, and Quality Characteristics in Ruminants

E. Tsiplakou, A. Mavrommatis, A. I. Gelasakis, A. I. Kalogianni, and P. E. Simitzis

Abstract

Milk and dairy products possess remarkable nutritive and sensory properties retaining a prominent position in consumer preferences. However, the raising consumer demands for safe dairy products of high nutritional value has reshaped the direction of livestock production and food industry. Currently, research efforts are focused on the development of niche functional dairy products that fortify human health and are in harmony with the concept of sustainable production, green economy, environmental protection, and proper health and welfare status of farm animals. Innovative feeding strategies are therefore evaluated on an evidential basis and accordingly adapted to dairy ruminant production systems to improve milk yield and quality traits, without impairing their health and well-being. Incorporation of natural feed ingredients, such as essential oils, plant extracts, and agro-industrial by-products, into ruminants' diets can greatly

E. Tsiplakou · A. Mavrommatis
Laboratory of Nutritional Physiology and Feeding, Agricultural University of Athens, Athens, Greece

Department of Animal Science, School of Animal Biosciences, Agricultural University of Athens, Athens, Greece

A. I. Gelasakis · A. I. Kalogianni
Department of Animal Science, School of Animal Biosciences, Agricultural University of Athens, Athens, Greece

Laboratory of Anatomy and Physiology of Farm Animals, Agricultural University of Athens, Athens, Greece

P. E. Simitzis (✉)
Department of Animal Science, School of Animal Biosciences, Agricultural University of Athens, Athens, Greece

Laboratory of Animal Breeding and Husbandry, Agricultural University of Athens, Athens, Greece
e-mail: pansimitzis@aua.gr

G. Arsenos, I. Giannenas (eds.), *Sustainable Use of Feed Additives in Livestock*,
https://doi.org/10.1007/978-3-031-42855-5_21

improve the intrinsic milk quality, and enhance ruminants' gastrointestinal function, resistance against nematode infections, and their overall antioxidant and immunological status. A desirable manipulation of bacteria involved in the ruminal bio-hydrogenation and a decrease on the methane emissions are also evident, associated with the optimization of ruminal function and feeding efficiency, the minimization of environmental burden, and the reinforcement of productivity and profitability. However, the commercialization and widespread exploitation of natural feed additives in ruminants is a rather complicated task demanding further research to elucidate possible adverse effects on productivity and health status and standardize their use.

Keywords

Phytobiotics · Essential oils · Tannins · Saponins · Milk production · Milk composition · Methane mitigation

1 Introduction

Phytobiotics constitute a large group of secondary compounds with a remarkable diversity in their chemical structure and properties. Parameters, such as the utilized part of the plant (seeds, leaf, root, or bark), the harvesting period, geographical location, and method of isolation (cold expression, steam distillation, extraction with non-aqueous solvents, etc.) modify their activity (Dudareva et al. 2004). These natural compounds belong to diverse categories; however, in this chapter three main groups that are regularly added in the diets ruminants' diets are reviewed: (1) essential oils (EOs) typically composed of (mono- and sesqui-) terpenoids and phenylpropanoids, (2) tannins, and (3) saponins (Jouany and Morgavi 2007).

EOs include a mixture of phenolic compounds such as terpenes, terpenoids, and phenylpropanoids, which are responsible for their intense aroma. The most commonly used EOs in animals' diets are derived from oregano (carvacrol, thymol, p-cymene, γ-terpinene), clove (eugenol), coriander (linalool), ginger (α-pinene, cineole, borneol, geraniol, α-curcumene, camphene, and eucalyptol), rosemary (carnosic acid, carnosol, rosmadial, genkwanin, rosmarinic acid, 1,8-cineole, α-pinene, limonene, and camphor), mint (menthol), and thyme (thymol, carvacrol, p-cymene, γ-terpinene, linalool). Monoterpenes, such as carvacrol, thymol, linalool, and limonene, represent the most common metabolites of EOs (ca. 90%) and originate from an isoprene unit through the mevalonate pathway (Zwenger and Basu 2008). On the other hand, phenylpropanoids (cinnamaldehyde, eugenol, anethole, etc.) are compounds with a chain of three carbons bound to an aromatic ring of six carbons and are mainly derived from the phenylalanine that is synthetized by the shikimate metabolic pathway (De Cássia da Silveira e Sá et al. 2014). The hydrophobicity of EOs enables them to disrupt the structure of bacterial cell membrane, resulting in the leakage of ions and other cell components (Burt 2004).

Tannins are water-soluble polymeric polyphenols formed by quinones or flavonoids and bearing at least one hydroxyl substituent. They are divided into condensed (CT) and hydrolysable (HT) tannins and can be both found in shrubs, legumes, cereals, grains, and fruits (Patra and Saxena 2011; Tong et al. 2022). Tannins demonstrate antioxidant, antimicrobial, antiparasitic, and anti-inflammatory activity; however, their mode of action has not been sufficiently clarified yet (Tong et al. 2022). The main characteristic of these bioactive compounds is their capacity to bind with proteins, due to the presence of phenolic hydroxyl groups, and the formation of complexes. These complexes remain undigested in rumen, since necessary nutrients become inaccessible to fibrolytic bacteria which disrupt their metabolism (Frutos et al. 2004). As a result, dietary protein degradation in the rumen is delayed, amino acid bioavailability in the small intestine is increased and nitrogen excretion in urine is reduced (Jouany and Morgavi 2007). For years, tannins were considered as antinutritional compounds; however, the perspective of their application as rumen regulating factors and beneficial biological substances has currently gained ground.

Saponins are steroid or triterpenoid glycosides of high molecular weight that possess the ability to form stable foam in aqueous solutions (Hart et al. 2008). Steroid saponins are mainly found in wild plants (e.g., oats, capsicum peppers, tomato seeds), whereas triterpenoid saponins dominate in cultivated crops (e.g., alfalfa, soybean, chickpeas, beans, peanuts, lentils, ginseng roots, sunflower seeds, spinach, and tea leaves). Saponins exhibit various biological effects; among their beneficial effects is the regulation of permeability of cell membranes, the stimulation of the immune system, their hypocholesterolemic and anticarcinogenic properties, the antiprotozoal, antifungal, antiparasitic, and antiviral effects, and the improvement of feed efficiency. Nevertheless, they may have adverse effect on protein digestibility and on the reproduction in ruminants (Francis et al. 2002, Oleszek and Oleszek 2020).

2 Plant Feed Additives: Are the Allies Against the Livestock Environmental Impact?

The Paris Agreement's aim of limiting the increase in global temperature to 1.5 °C above pre-industrial levels demands the implementation of efficient and sustainable mitigation strategies with the intention to reduce global greenhouse gas (GHG) emissions mainly by suppressing methane (CH_4) produced by the agriculture sector (IPCC 2021). Mitigation of GHG emissions is a challenging task due to the predicted raised demand for meat and milk during the next decades (Beauchemin et al. 2020). To meet the challenge of reducing the GHG emissions in the livestock sector without jeopardizing digestive and metabolic processes and the ruminants' high production, research efforts have been focused on the in-depth understanding of the rumen function and compounds that could reinforce it.

The rumen microbiome constitutes a community of microorganisms that inhabit the rumen and is characterized by its high population density, extensive diversity (encompassing bacteria, archaea, protozoa, and fungi), and complexity of interactions (Huws et al. 2018). The unique habitat in the rumen promotes assembling a community of archaea; nearly all of these being methanogens (Patra et al. 2017). Rumen methanogens form CH_4 by utilizing H_2 and CO_2 produced by other fermentative rumen microbes as a result of complex carbohydrate degradation. Rumen CH_4 emission accounts not only about 17% of the global CH_4 emission (Knapp et al. 2014) but also for a loss of about 2–12% of the ingested feed energy (Johnson and Johnson 1995) reminding that CH_4 formation produced by ruminant sector compromise both the environment and the feed efficiency in ruminants.

Diets of ruminants are mainly based on cereal grains and other easily degradable by-products of high nutritional value to efficiently meet the nutritional demands of high-yielding animals in energy and protein. However, these practices lead to an impaired rumen ecosystem associated with a high prevalence of rumen acidosis, intestinal inflammation, and diarrhea (Zebeli et al. 2010). Till 2006, antibiotics were used to overcome these problems in the European Union. The ban of their dietary application has led to the continuous seeking on efficient, safe, and cost-effective substances with similar properties. At the same time, synthetic additives are nowadays questioned due to their suspected carcinogenic potential and the consumers' preference for natural, healthy, safe, and nutritious meat and dairy products. As a result, natural phytobiotics have been emerged as candidate substances in ruminant diets for improving their health status, growth performance, and the quality traits of the derived products (Simitzis 2017).

A wide variety of plant feed additives have been investigated as rumen habitat modifiers primarily focusing on methanogens suppression. EOs are secondary metabolites which are extracted from the volatile fraction of plants' distillation process. EOs have been reported to possess bioactive compounds which can inhibit rumen deamination and methanogenesis through the inhibition of hyper-ammonia-producing (HAP) bacteria (*Prevotella* spp., *Ruminobacter amylophilus*, etc.) and protozoa-methanogens at the stages of attachment and colonization, respectively, resulting in lower levels of ammonia, methane and acetate and higher concentrations of butyrate and propionate. Thus, the effect of EOs on rumen methanogenesis has been extensively assessed (Benchaar and Greathead 2011). In a study by Benetel et al. (2022), a variety of EOs (star anise, citronella, clove bud, staigeriana eucalyptus, globulus eucalyptus, ginger, ho wood, melaleuca, oregano, and white thyme) were in vitro assessed for their methano-inhibitory potential in the rumen. Among the aforementioned EOs, only oregano and white thyme EOs reduced methane production when included in the semi-automated gas production system at a level of 250 mg/L, without negatively affecting feed degradation. Similarly, Rofiq et al. (2021) investigated the potential of garlic, thyme, clove, orange peel, mint, and cinnamon EOs in rumen fermentation and methane production in vitro. The experiment resulted in a significant reduction of methane formation by ingesting orange peel and cinnamon oils (300 ppm) after 6 h incubation, while administration of clove oils (300 ppm) had the highest in vitro true digestibility of dry matter; in summary, all tested essential oils reduced in vitro rumen methane production and there was no negative effect on in vitro digestibility.

Jiménez-Ocampo et al. (2022) assessed the potential of orange EOs (OEO) on in vitro and in vivo beef heifers' rumen fermentation. In the in vitro trial, the addition of 0.25% OEO resulted in a significant reduction in CH_4 production, while no changes were observed in pH, and the proportions of volatile fatty acids (VFAs). In the in vivo trial, the gross energy consumption, apparent total tract digestibility, CH_4 emissions, and rumen valerate concentration were decreased by supplementation with 0.5% OEO in the heifers' diet. These outcomes suggest that EOs may negatively affect animal performance through the reduction of nutrients digestibility in vivo. These controversial attributes of EOs between in vitro and in vivo trials were confirmed in a recent study by Sari et al. (2022), who summarized the effect of garlic on rumen microbiome and methane formation. In that study, in 9 out of 13 in vitro experiments, garlic acid reduced the yield of methane emissions by 47% (range 9.8–96%), while in 11 out of 14 in vivo experiments the methane yield was suppressed by 12.8% (range 5–33%). The latter finding further supports the persistence mechanisms of rumen biochemistry in radical changes driven by a complex host-microbiome interplay which is neglected in in vitro studies. Moreover, it should be underlined that although the bioactive compounds of garlic appear to suppress methane formation, the in vivo trials provided conflicting results regarding animal performance, since in many cases the DM and/or neutral detergent fiber (NDF) digestibility were compromised (Sari et al. 2022). It can be concluded that in vitro approaches cannot always generate reliable data, while in vivo studies provide inconclusive and often controversial results due to discrepancies in dosage, chemical structure, diet or the form in which an agent is introduced into the animal nutrition (Cardozo et al. 2004, 2005).

Oregano essential oil is rich in carvacrol and thymol that exert a general antimicrobial activity in the rumen (Kolling et al. 2018). However, only two in vivo studies (Tekippe et al. 2011; Hristov et al. 2013) have shown a reduction of CH_4 formation of up to 40% in dairy cattle. In addition, Hristov et al. (2013) did not find any adverse effect of oregano leaves dietary inclusion (8.7, 18.9, and 28.2 g/kg DM) on feed efficiency and rumen pH or VFAs being accompanied by the methane reduction in dairy cows. Similarly, Kolling et al. (2018) reported a decrease in CH_4 production (g/kg digestible DMI) by using 560 g oregano extract/kg DM in lactating dairy cattle. In contrast, several other studies have not shown any significant effect of dietary supplementation with oregano on CH_4 production; for instance, lactating dairy cattle supplemented with oregano oil and carvacrol at 50 mg/kg DM did not express any anti-methanogenic effect (Benchaar 2020).

Although the main anti-methanogenic action mode of EOs lies on the disruption of bacterial membranes (i.e., alteration of permeability), it has been reported that their broad antimicrobial spectrum is likely attributed to synergistic mechanisms (Helander et al. 1998), which though remain unclear. This hypothesis justifies the concerns regarding potential adverse effects on the fibrolytic rumen microbes and the rumen degradative capacity. Considering methane mitigation through EOs inclusion, Cobellis et al. (2016) pointed out that the in vivo reduction of methane yield as an effect of EOs seems to be fictitious since it is more likely to be associated with the overall degradative suppression in the rumen due to their antimicrobial action. Inconsistencies in the available literature due to inter-studies variation regarding the experimental designs render impossible the suggestion of universally applicable protocols for the most effective integration of EOs in ruminants' diets. Thus, a network meta-analysis could holistically benchmark and synthetize the obtained literature outcomes regarding EOs and their prevailing compounds' effects on both in vivo methane formation and feed efficiency, to assess their realistic value for the reduction of ruminants' methane emissions.

The anti-methanogenic activity of plant feed additives is often accompanied by a reduction in organic matter (OM) digestibility and/or DMI, compromising animal performance as previously reported by Correddu et al. (2020), who reported lower VFAs production and digestibility when polyphenols-rich by-products were used in small ruminants' diets. In general, the reduction of ruminal methane yield by plant bioactive compounds, namely EOs, tannins, saponins, and other secondary metabolites is undoubtable; however, there is ample evidence suggesting that this CH_4 reduction is attributed to an overall suppression of rumen activity which can negatively affect animal performance. These natural rumen modifiers demonstrate strong potential for future eco-friendly dietary approaches; however, the current understanding of rumen biochemistry is inadequate to successfully implement such strategies at the farm level. At last but not least, holistic models should be developed for the assessment of the environmental footprint of the aforementioned strategies considering simultaneously the actual methane emissions and nitrogen excretion and the environmental decongestion impact due to the agro-industrial by-products valorization.

When the environmental impact of livestock production is concerned, nitrogen excretion is usually overlooked, despite its great contribution to the GHG. Over 70% of the ingested feed protein is degraded in the rumen to ammonia, while only a minor fraction of this ammonia is utilized by rumen microorganisms to produce microbial protein (Bach et al. 2005). Ammonia production constitutes not only a waste of dietary nitrogen but also an important environmental issue (Bach et al. 2005). EOs are able to suppress protein degradation in the rumen and increase by-pass protein flow to the gut of the host animal, therefore improving nitrogen efficiency. In this context, Zhou et al. (2020) reported that oregano EOs decreased the nitrogen ammonia concentration during in vitro trials, while the microbial protein synthesis was also improved. Cobellis et al. (2016) reported that the effects of EOs on rumen ammonia suppression were more consistent compared to those on methane emissions.

Although EOs and other bioactive compounds derivatives of plant materials have been extensively investigated as rumen modifiers, mostly focused on their anti-methanogenic properties, their on-farm scale implementation is accompanied by crucial limitations. The financial burden of their inclusion in animal diets, their questionable long-term effects on animal performance, and the competition between food-feed-nutraceutical industries are the main factors that obstruct their extensive application in farm animals. Thus, research is now focusing on the valorization of raw plant materials, agro-industrial by-products and leftovers, rich in bioactive compounds with a potential to favorably modify rumen functionality. Currently, the increasing availability of agro-industrial by-products (e.g., grape, olive, tomato, citrus, and myrtle residues) appears to serve the efforts for i) reduction of competition with human edible foods, ii) lowering of the feedstuffs cost, and iii) suppressing the environmental footprint and the water waste associated with crop cultivation (Correddu et al. 2020). Other bioactive compounds, such as polyphenols, flavonoids, and various secondary plant metabolites have also been reported as potential rumen function manipulators similar to the abovementioned EOs. Some of them are summarized in the study of Correddu et al. (2020), who estimated the effects of agro-industrial by-products on dairy small ruminants feeding efficiency. In that study it was found that pistachio hull, grape pomace, and exhausted myrtle berries reduced the ammonia concentration in the rumen indicating lower nitrogen excretion to the environment, while tomato silage supplementation in the goats' diet reduced the methane formation and emission.

3 Can Plant Feed Additives Promote Ruminants' Feed Efficiency Through Rumen Orchestration?

Considering that methane formation constitutes a significant energy loss in the rumen, it could be assumed that methane mitigation strategies could be simultaneously considered as the roadmap for higher feed efficiency. However, since the in vivo assessment of plant feed additives aiming to reduce methane emission often led to lower digestibility because of their non-specific antimicrobial effect in the rumen microbiomes, this section will strictly focus on the impact of plant bioactive compounds as feed efficiency promoters through advantageous changes in rumen fermentation.

In the study of Jiao et al. (2021), the inclusion of 4 g/day oregano EOs resulted in higher hemicellulose digestibility of corn silage in rams without significant improvement of DM, CP, and NDF digestibilities. Similarly, oregano EOs increased nutrient digestibility in dairy cows (Yuan and Tong 2005). Rosemary and lemongrass dietary supplementation in goats increased the OM digestibility and the concentration of propionate and glucose (Kholif et al. 2017). The study by Zhang et al. (2021) estimated the effect of dietary supplementation with EOs on cattle's rumen degradation potential. In that study, the morphology of rumen altered since the high EOs (260 mg/head/day) inclusion level increased the rumen papillae length. Under a functional perspective, amylase activity was decreased, whereas β-glucosidase and cellulase activities, and propionate and butyrate concentrations were significantly increased in the high inclusion level (Zhang et al. 2021). In summary, EOs supplementation increased the rumen digestive ability by modulating epithelial development and enzymatic degradative potential in beef cattle (Zhang et al. 2021). In a study by Yang et al. (2007), the inclusion of 5 g/kg DM of garlic EO in cows' total mixed ration (TMR) increased the total feed digestibility. Similarly, the dietary supplementation with 20–35 g garlic oil per goat per day improved acid detergent fiber (ADF) and lignin digestibility, total VFA, and feed conversion rate (FCR) (Okoruwa and Edoror 2019). Moreover, 5 g garlic oil in sheep diet increased feed digestibility and consequently feed efficiency by promoting the growth of anaerobic rumen fungi which promoted fiber digestion (Klevenhusen et al. 2011).

However, the aforementioned constitutes almost the majority of the limited positive outcomes of EOs on feed digestibility and rumen fermentation parameters in the existing literature. The antimicrobial properties of plant bioactive compounds and especially the secondary metabolites make difficult the improvement of feed efficiency since this is firmly linked to rumen microbiome structure and bacterial viability. Nevertheless, it has been evidenced that under rumen dysbiosis, EOs show promising results by rebalancing the rumen microbiome and sustaining its metabolism.

More specifically, in a study by Rivera-Chacon et al. (2022), the dietary supplementation of a phytogenic feed additive based on L-menthol, thymol, eugenol, mint oil, and cloves powder amended the ruminal fermentation profile and modulated the risk of subacute ruminal acidosis (SARA) in cattle. The phytogenic feed additive group had an increased ruminal pH compared to the control one, reduced time to pH < 5.8, decreased ruminal concentration of D-lactate, and tended to lower total lactate. Additionally, phytogenic feed additive group presented increased acetate, butyrate, isobutyrate, isovalerate, and the acetate to propionate ratio compared to the control. In the same context, Ahmed et al. (2022) assessed the potential of selected plant extracts to control severe subacute ruminal acidosis (SARA) in vitro. Cinnamon, grape seeds, orange, pomegranate peels, propolis, and guava extracts significantly increased ruminal pH. Furthermore, the addition of cinnamon extract and monensin decreased lactic acid concentration and VFA at 24 h incubation. The cinnamon extract significantly increased the acetate–propionate ratio, while monensin reduced it. Both cinnamon extract and monensin significantly increased ruminal pH at 6 and 24 h of incubation and reduced lactic acid concentration at 24 h compared to control, with corn used as substrate. The authors concluded that cinnamon extract can be used effectively as an alternative to antibiotics for the control of ruminal acidosis when corn is used as a basal diet.

Polyphenols and particularly condensed tannins activate the route of "bypass" protein; this is achieved through (i) the binding with proteins in the rumen, (ii) the reduction of protein fermentation rate to ammonia, (iii) the balance between nitrogen rumen absorption and urine excretion, and (iv) the increase on both the synthesis of microbial protein and the absorption of amino acids in abomasum and small intestine, where the pH values enable the separation of tannin–protein complexes (Ahmed et al. 2020; Correddu et al. 2020; Jouany and Morgavi 2007; Lillehoj et al. 2018; Mahachi et al. 2020; Piluzza et al. 2014; Tayengwa and Mapiye 2018; Theodorou et al. 2006). Likewise, saponins deteriorate protozoa (defaunation) and/or methanogenic archaea by forming complexes with sterols in the protozoal membrane surface leading to an increased microbial protein synthesis in the rumen that is associated with nitrogen and methane mitigation (Newbold et al. 2015). This effect is more profound when saponins are added into high-concentrate rather than high-forage diets, since their activity is strongly associated with rumen pH (5.5 instead of 7.0, respectively) (Poungchompu et al. 2009).

These bioactive compounds can therefore offer a natural means to support the more effective exploitation of dietary proteins, thus leading to improved feed efficiency and increased weight gain in ruminants (Lillehoj et al. 2018; Tayengwa and Mapiye 2018). However, tannins included in specific forages (i.e., *Lotus corniculatus*) exhibit antimicrobial action against rumen proteolytic bacteria through the binding with bacterial extracellular enzymes which impairs the synthesis of bacterial proteins (Jouany and Morgavi 2007; Min et al. 2003). Except for proteins, polyphenols also present secondarily affinity for carbohydrates, polysaccharides, nucleic acids, and minerals, forming complex substances which escape from the rapid fermentation in the rumen (Correddu et al. 2020; Min and Solaiman 2018; Piluzza et al. 2014). Condensed tannins reduce carbohydrates digestion by inhibiting the responsible enzymes and altering the nitrogen availability for the carbohydrates' fermentation processes (Kelln et al. 2021).

4 What About Phytobiotics and Animal Performance?

According to the existing literature, no effect of eugenol (Benchaar et al. 2012), cinnamaldehyde and eugenol (Tager and Krause 2011), cinnamaldehyde and garlic oil (Blanch et al. 2016), garlic or juniper berry EO (Yang et al. 2007), oregano leaves (Hristov et al. 2013; Tekippe et al. 2011), EO components mixture (thymol, eugenol, vanillin, guaiacol, and limonene) (Benchaar et al. 2006, 2007; Kung et al. 2008; Joch et al. 2019), blend of oregano, cinnamon, thyme, and orange peel EOs (Spanghero et al. 2009) and mixture of eugenol, geranyl acetate, and coriander oil (Santos et al. 2010) on milk yield and composition of dairy cattle is observed. However, according to the meta-analysis by Belanche et al. (2020), long-term exposure to a commercial blend of EOs (Agolin) resulted in a slight increase in milk yield at the level of 4%, while no effects on feed intake and milk composition were evident. Braun et al. (2019) suggested that the aforementioned effects of EOs on milk production in dairy cattle is related with an improved feed efficiency, nutrients absorption, and increased uptake of cations like calcium and ammonium as a result of EOs dietary supplementation.

In dairy ewes, orange peel EO (Kotsampasi et al. 2018), cornus extract, enriched with EOs of oregano and thyme (Kalaitsidis et al. 2021), thyme and celery seed mixture (Khattab et al. 2020), and EO components mixture (thymol, eugenol, vanillin, guaiacol, and limonene) (Giannenas et al. 2011) resulted in an increase in milk yield. However, no effect of anise, clove, and thyme EO (El-Essawy et al. 2021) or citral oil (Canaes et al. 2017) on milk yield was observed, while an increase in milk yield, protein, and fat levels was observed in dairy goats as a result of Boswellia sacra resin (Soltan et al. 2021) and rosemary or lemon grass (Kholif et al. 2017) dietary supplementation.

Polyphenols have been found to decrease feed intake in sheep in a type- or dose-dependent manner, whereas, in goats this adverse effect may be hindered by the proline-rich saliva enzymes, which inactivate the action of bioactive compounds such as the tannins, and the physiological ruminal differences (increased rumen volume and digestive rate) (Correddu et al. 2020; Kelln et al. 2021; Mahachi et al. 2020). High content of polyphenols in the feedstuffs, especially of proanthocyanidins and tannins, has undesirable effect in feed intake of ruminants due to their astringent taste and reduced palatability (Mahachi et al. 2020; Tayengwa and Mapiye 2018). This unfavorable effect of polyphenols is limited when polyphenol-rich feedstuffs are used in moderate levels or in the form of pellet and hay (Mahachi et al. 2020). In that case, the addition of polyethylene glycol (PEG) has been suggested as a countermeasure to neutralize the antinutritional effect of tannins.

In general, high tannin intake (>50 g/kg DM) has a clear negative effect on productivity as a result of the reduction in diet palatability, voluntary feed intake, and nutrients digestibility and alteration of the animal physiology (higher salivation rate, mucosal perturbations, etc.) (Serrano et al. 2009). At the same time, no significant effects on milk yield or milk composition is generally observed in dairy cattle as a result of tannin ingestion (Broderick et al. 2004; Benchaar et al. 2008; Dschaak et al. 2011; Aguerre et al. 2016; Focant et al. 2019), although a decrease in milk urea N and plasma urea N is reported (Tiemann et al. 2008; Aboagye et al. 2019). Tannin-containing forages may contribute to improved sheep wool quality associated with the increased protein absorption (Min and Solaiman 2018; Ramírez-Restrepo et al. 2005). Moreover, mimosine found in *Leucaena leucocephala* may act as a defleecing agent in wool sheep breeds, though it should be carefully used due to its high toxicity (Rochfort et al. 2008).

Effect of saponins on animal performance is generally inconsistent. Recent studies did not find any significant effects of dietary saponins on milk yield and composition in dairy cattle (Holtshausen et al. 2009), while the application of composite plant extract (CPE) rich in polyphenolics and saponins improved milk production in goats (Shilwant et al. 2022). In general, defaunation caused by saponins decreases fiber degradation leading to an improved nitrogen metabolism, especially in animals fed low protein diets (Firkins et al. 2007), although this improvement is not associated with enhanced animal productivity.

As already mentioned, there is an increasing interest over the secondary bioactive compounds of agro-industrial by-products as agents for improving animal products quality with minimal cost. In cows, substitution up to 30% of the diet by ensiled mixed tomato and apple pomace leads to a slight increase in milk yield without any other adverse effects (Abdollahzadeh et al. 2010). On the other hand, olive by-products dietary supplementation generally does not have a significant effect on milk yield and composition in dairy cattle, sheep, and goats, although fatty acid profile is positively modified, since an increase of monounsaturated fatty acids and a decrease of saturated fatty acids are observed (Tzamaloukas et al. 2021). Moreover, data concerning pomegranate by-products indicate that incorporation at the level of 8% is probably too high for lactating cows and should be lowered to the half with the intention to improve milk yield and composition (Shaani et al. 2015; Jami et al. 2012; Abarghuei et al. 2013).

5 Improvement of Ruminant's Product Characteristics Through Changes in Rumen Functional Routes: The Case of Biohydrogenation

Ruminant milk contains prominent proportions of saturated fatty acids (SFA), which have been related with increased risk of coronary heart disease incidence (CHD) in human. Conversely, polyunsaturated fatty acids (PUFA) which are also found in ruminants' milk in low concentrations, have been extensively investigated for their beneficial properties for human health. Consequently, it is of utmost importance for nutritionists to find novel and sustainable ways to produce animal products with high PUFA content (Kliem and Shingfield 2016). Although ruminants' diets are rich in unsaturated fatty acids (USFAs), the lipids leaving the rumen and insert the blood stream are in majority SFA. The former biotransformation is performed by rumen microbes via two main routes, lipolysis, and biohydrogenation (BH). Lipids passing into the rumen are firstly transformed by microbial lipases in a process called lipolysis. The microbial lipases hydrolyze the ester linkages in complex lipids, causing the release of FA. After lipolysis, USFAs go through BH by ruminal microbes. This process converts the USFA to SFA with the parallel formation of variable *trans* intermediates (Lashkari et al. 2019). Among these *trans* FAs produced, conjugated linoleic acids also known as rumenic acids (c9t11 CLA) have been reported to have anti-carcinogenic and other health-promoting properties in animal models and cell cultures (Crumb 2011). The c9t11 CLA is produced either in the rumen during the BH of C18:2n6 (LA; linoleic acid) or in the animal tissues by Δ9-desaturase using vaccenic acid (VA; C18:1 t11), an intermediate of ruminal BH of C18 USFA (Gunal et al. 2014). However, the principal route of c9t11 CLA formation is through the Δ9-desaturase action mainly in mammary gland using as a precursor molecule the VA escaping the rumen BH (Bichi et al. 2012). Hence, modulating rumen fermentation and the BH process to increase the ruminal flow of VA through dietary manipulations has received considerable attention in recent years.

In their study, Gunal et al. (2014) assessed the effect of anise oil, cedar wood oil, cinnamon oil, eucalyptus oil, and tea tree oil on rumen fermentation and BH in vitro. The inclusion of EOs reduced the concentrations of rumen C18:0 regardless of the dose level, while LA was substantially accumulated when EOs added at 500 mg/L. EOs tested in this study had no effects on VFA profile and consequently on the fermentation potential, but significantly inhibited the BH process which is related to higher levels of USFA reaching the ruminants tissues. Another study assessed the effect of dietary supplementation with *Coleus amboinicus* Lour. (CAL), a semi-succulent perennial plant in the family *Lamiaceae* with a pungent oregano-like flavor and odor rich in polyphenols on modulate ruminal BH of USFA (Yanza et al. 2022). CAL increased the PUFA proportion in the rumen of lambs, which was mainly attributable to the increase in C18:3n3 (LNA; linolenic acid) proportion since it would be able to escape from the rumen BH process. In the study of Zhu et al. (2012) the effect of ruminal infusion with 0.8 g garlic oil per day per goat on rumen fermentation and microbial populations was investigated. The infusion of garlic oil in goats' rumen increased the proportions of VA, CLA, and PUFA, while SFAs were significantly decreased (Zhu et al. 2012). Although the results of FAs indicated a severe inhibition of rumen BH, the relative abundance of *Butyrivibrio* group was not significantly suppressed. Thus, a low correlation was found between *B. proteoclasticus* abundance and C18:0 accumulation in the rumen (Zhu et al. 2012). The involvement of *Butyrivibrio* species on rumen biohydrogenation constitutes one of the oldest discoveries in the field of rumen microbiome-biochemistry interplay (Polan et al. 1964). More specifically, the rumen BH of dietary PUFAs to SFAs is discriminated in two steps. During the first step, *Butyrivibrio fibrisolvens* converts LA to trans C18:1 isomers (including the VA), while, within the second step, *Butyrivibrio proteoclasticus* further hydrogenate 18:1 isomers to stearic acid (C18:0) (McKain et al. 2010). However, the paradox in many relative studies is that *Butyrivibrio* species rarely correlated with the extent of accumulated biohydrogenation intermediates in the rumen. Indeed, in the study of Mavrommatis et al. (2021), *Flintibacter, Intestinimonas, Sporobacter,* and *Blautia* species in the rumen were more tightly correlated with milk ω3, PUFA, and CLA content than *Butyrivibrio* species, indicating a direct or indirect involvement of other species on rumen BH.

Based on the existing literature about the potential of EOs to manipulate the rumen BH, there were no consistent results or substantial changes in fatty acid profile especially under in vivo conditions indicating that although EOs have an impact on biohydrogenating microbes, they cannot be considered as a promising and reliable strategy to enrich ruminants' products with beneficial FAs. On the other hand, plant feed additives rich in tannins have a strong effect on rumen BH as indicated by both in vitro and in vivo trials (Vasta et al. 2019). More specifically in the review of Vasta et al. (2019), polyphenol-rich feedstuffs decrease the accumulation of C18:0 and increase that of VA and CLA in rumen digesta in vitro. These outcomes reflect the promising aspects of polyphenols/tannins on modulating rumen microbiome aiming to produce functional animal-food rich in bioactive fatty acids beneficial for human health. However, in these studies, analysis of the microbiota composition did not provide clear indications about which changes in the bacterial population were responsible for the effects on rumen BH. This could be attributed mainly to two reasons; firstly, the genomic footprint of microorganisms is not always strictly related to the metabolic activity of the microbes, and secondly, the research may be focused on well-known species involved in the BH process overlooking other species that may be also implicated.

Data about the effect of pure non-tannic polyphenols on BH and rumen microbiota composition are still scarce. However, some indications may be obtained by considering results obtained in studies adopting agro-industrial by-products such as those from the olive oil which are rich in both non-tannic and tannic compounds. A modulating effect on rumen BH has been reported for stoned olive pomace rich in olive polyphenols in vitro (Pallara et al. 2014). The general effect of HT and CT on the BH process is quite consistent with results observed in in vitro trials: an increase of BH intermediates (especially VA) at the expense of SA. Buccioni et al. (2015) reported that feeding ewes a diet supplemented with quebracho tannins (as a source of CT) or chestnut tannins (as a source of HT) increased the ruminal relative abundance of *B. fibrisolvens* and reduced *B. proteoclasticus* compared to control animals, and this effect was more evident with quebracho tannins. Considering the outcomes of Vasta et al. (2019) study, the inclusion of polyphenols through the valorization of agro-industrial by-products appears to be a promising strategy for rumen biohydrogenation manipulation toward the enrichment of dairy products with unsaturated FAs. As it can be concluded, tannins induce a decrease in rumen methanogenesis due to their effect on methanogens but also the indirect decrease in H2 production as a result of decreased fiber digestion and protozoal population in the rumen (Patra et al. 2017).

6 Can Plant Feed Additives Fortify Ruminants Health Status and Benefit Immunocompetence?

Anthelmintic resistance is an emerging threat for ruminants' health, adversely affecting their productivity, health and welfare status and increasing the production cost (Papadopoulos et al. 2012). Anthelmintic resistance and the consumer-driven demands for safe, sustainable, and environmentally friendly livestock production with a minimal use of pharmaceuticals, have highlighted the necessity for alternative treatment/prevention strategies to mitigate gastrointestinal nematodes in ruminants. Among these strategies, the addition of natural feed additives with bioactive compounds in ruminants' diets constitutes an emerging and promising alternative to reduce the parasitic burden.

Several bioactive compounds have been studied for their nematocidal activity in ruminants; among them, lipids, phenolics, alkaloids, terpenes, and essential oils have been extensively reviewed (Githiori et al. 2006; Hoste and Torres-Acosta 2011; Hoste et al. 2022; Mahachi et al. 2020; Piluzza et al. 2014; Rochfort et al. 2008; Tayengwa and Mapiye 2018). Alkaloids, glycosides, sequiterpenes, kosotoxins, anthraquinones, terpenoids, resins, tannins, and lactones have demonstrated a nematocidal effect against mixed gastrointestinal nematode infections in ruminants, whereas various plants and agro-industrial by-products containing thiosulfonates, anthraquinones, terpenes, terpenoids, anthocyanins, alkaloids, and tannins have shown an in vivo nematocidal activity against *Haemonchus contortus* (Githiori et al. 2006; Hoste et al. 2022; Tayengwa and Mapiye 2018). Similarly, tannin-rich forages such as *Hedysarum coronarium*, *Lotus pedunculatus*, *L. corniculatus*, *Lespedeza cuneata*, *Acalia nilotica, Onobrychis viciiofolia*, and tannins derived from quebracho, mimosa, and chestnut, reduced fecal egg counts (FECs) and improved small ruminants' resistance against nematode infections (Huang et al. 2018; Niezen et al. 2002; Piluzza et al. 2014). Silage and hay of sainfoin (*Onobrychis viciifolia*) reduced adult *H. contortus* and the FEC of *Cooperia curticei* in lambs (Heckendorn et al. 2006) and FEC of *H. contortus, Teladorsagia circumcincta*, and *Trichostrongylus colubriformis* in goats (Paolini et al. 2005). Also, pellet of sainfoin (*On. alba* subsp. *laconica* and *On. peloponnesiaca*) reduced the female worm fecundity of *H. contortus* and FEC in lambs, whereas carob pod (*Ceratonia siliqua*) exerted activity against eggs and adult worms of *H. contortus* attributed to its condensed tannins content (Saratsi 2021). Furthermore, the tannin-rich *Sericea lespedeza* hay and pellets exhibited time- and dose-dependent nematocidal activity against *H. contortus, T. colubriformis, T. circumcincta*, and *Eimeria* species in small ruminants (Mahachi et al. 2020). Chicory (*Cichorium intybus L.*) silage, containing sequiterpene lactones significantly reduced *Ostertagia ostertagi* in experimentally infected calves (Peña-Espinoza et al. 2016), whereas, limonene and eugenol efficiently reduced the FEC in sheep and goats (Tayengwa and Mapiye 2018).

Nematocidal activity of specific natural bioactive compounds is universally acceptable; however, the mode of action has not been fully understood. Moreover, their nematocidal efficiency varies both between the ruminant species (e.g., the proline-rich saliva enzymes of goats inactivate the tannins) and the nematode species (Huang et al. 2018; Quijada et al. 2015). It is suggested that bioactive compounds in phytochemicals either act directly via the inhibition of larvae development and egg hatching, or indirectly, via the upregulation of immunological response of animals associated with the improvement of protein efficiency (Ahmed et al. 2020; Jouany and Morgavi 2007; Mahachi et al. 2020; Waghorn and McNabb 2003). Condensed tannins and phenols impair the mechanism of oxidative phosphorylation, therefore resulting in the energy drain and the collapse of parasites. Also, tannins bind with glycoproteins around the epicuticle of worms leading to their malnourishment and death, whereas saponins and condensed tannins can cause structural changes in the cell membranes of the worm cuticles, adversely affecting their mobility, feed intake, and fecundity (Ahmed et al. 2020; Hoste et al. 2012; Jouany and Morgavi 2007; Mahachi et al. 2020; Martin et al. 2001; Waghorn and McNabb 2003). On the other hand, alkaloids hinder the glycose absorption from the gastrointestinal tract of parasites and impair their central nervous system function, thus leading to paralysis, while they reduce the nitrate generation and challenge the parasites' homeostasis due to their antioxidant effect (Martin et al. 2001). At last but not least, essential oils interfere with the mechanism of nematodes' embryogenesis inhibiting their egg hatching (Tayengwa and Mapiye 2018). Nevertheless, the applicability and systematic use of bioactive compounds as nematicides is hindered by the remarkably divergent results among the different ruminant species, the unstandardized concentration of bioactive compounds in different forages and feedstuffs, the incomplete neutralization of worm burdens, and the non-reproducible or controversial results between in vivo and in vitro studies (Rochfort et al. 2008; Waghorn and McNabb 2003). In the case of ectoparasites, the ectoparasiticide mode of action of natural bioactive compounds is not sufficiently documented; however, it is likely that they may exhibit a repellency effect via the odor or an acaricide effect via the inhibition of the mites egg hatching (Martin et al. 2001).

Part of the molecular multifunctionality of the natural bioactive compounds is their antioxidant capacity which improves the immune status and reduces oxidative stress of animals (Salem et al. 2012; Tayengwa and Mapiye 2018). Phenolic compounds have been found to in vitro modify immune status via the downregulation of the inflammatory response, since they reduce the production of cytokines and reactive oxygen species, as well as the functionality of cytotoxic

T-lymphocytes and natural killer cells (Provenza and Villalba 2010). The anti-inflammatory and antioxidant activity of polyphenols has been confirmed by in vivo studies in dairy cows, goats, and ewes (Correddu et al. 2020; Niderkorn and Jayanegara 2021; Salem et al. 2012). The aforementioned properties are attributed to their ability to chelate with free radicals, inhibit the enzyme actions associated with the mechanisms of oxidative stress, reinforce the functionality of antioxidant mechanisms, and prevent the lipid oxidation (Correddu et al. 2020; Niderkorn and Jayanegara 2021; Soldado et al. 2021; Tayengwa and Mapiye 2018). Tannins and proanthocyanidins increase the bioavailability of by-pass proteins favoring the adequate supply of arginine, glutamine, and cysteine which are critical amino acids for the biosynthesis of cell-mediated and humoral immune response elements in ruminants. In growing calves, a combination of essential oils, tannins, and saponins had beneficial effects on their health, reducing the levels of cortisol and glucose and increasing the concentration of total proteins, albumin, globulins, aspartate transaminase, and alanine transaminase, as well as the immune response (Lakhani et al. 2019). In dairy cows, the consumption of condensed tannins resulted in lower serum urea, increased values of antioxidant enzymes, and decreased lipid peroxidation, while it exhibited a probiotic effect in the rumen and the gastrointestinal tract, increasing the commensal bacteria and inhibiting the harmful ones (Provenza and Villalba 2010, Dey and De 2014). In lambs, condensed tannins increased the production of specific antibodies against *Ostertagia circumcincta* and *Trichostrongylus colubriformis* (Provenza and Villalba 2010). Finally, a significant reduction of *Escherichia coli* and *E. coli O157:H7* bacteria (Huang et al. 2018) and in the number of flies (Waghorn and McNabb 2003) in the ruminants fed tannin-rich feedstuffs has been observed.

Finally, condensed tannins exhibit a favorable action against bloat in ruminants at a rate of 5–50 g/kg dry matter (Ahmed et al. 2020; Rochfort et al. 2008; Tayengwa and Mapiye 2018; Waghorn and McNabb 2003). They restrict the formation of rumen foam through the formation of complexes with proteins which elude the ruminal degradation. Subsequently, the production of fermentation gases, which are entrapped in polysaccharide slime forming the rumen foam, decreases (Mahachi et al. 2020; O'Connell and Fox 2001; Tayengwa and Mapiye 2018; Waghorn and McNabb 2003).

7 Conclusion

Phytobiotics have the potential to positively manipulate gut microbiota and rumen fermentation. However, their effectiveness in ruminants has not yet been proved to be consistent and conclusive and some issues need to be addressed before their regular application. Although phytogenic feed supplements are generally recognized as safe (GRAS) for use and their accumulation in the body is not probable due to their exhalation (carbon dioxide) and rapid elimination (glucuronides by the kidneys), their use at high concentrations could induce cytotoxic effects on the living cells. At the same time, their beneficial effects should justify the additional cost of their use. Finally, our knowledge regarding their activities in animal organism is still rather limited and there is a strong need for research to better understand their absorption, distribution, bioavailability, metabolism, and excretion and establish appropriate protocols for their integration in ruminants' diets.

References

Abarghuei MJ, Rouzbehan Y, Salem AZM, Zamiri MJ (2013) Nutrient digestion, ruminal fermentation and performance of dairy cows fed pomegranate peel extract. Livest Sci 157:452–461. https://doi.org/10.1016/j.livsci.2013.09.007

Abdollahzadeh F, Pirmohammadi R, Fatehi F, Bernousi I (2010) Effect of feeding ensiled mixed tomato and apple pomace on performance of Holstein dairy cows. Slovak J Anim Sci 43:31–35

Aboagye IA, Oba M, Koenig KM, Zhao GY, Beauchemin KA (2019) Use of gallic acid and hydrolyzable tannins to reduce methane emission and nitrogen excretion in beef cattle fed a diet containing alfalfa silage. J Anim Sci 97(5):2230–2244. https://doi.org/10.1093/jas/skz101

Aguerre MJ, Capozzolo MC, Lencioni P, Cabral C, Wattiaux MA (2016) Effect of quebracho-chestnut tannin extracts at 2 dietary crude protein levels on performance, rumen fermentation, and nitrogen partitioning in dairy cows. J Dairy Sci 99(6):4476–4486. https://doi.org/10.3168/jds.2015-10745

Ahmed A, Tariq A, Habib S (2020) Plant phenolics in sustainable agriculture. In: Lone R, Shuab R, Kamili AN (eds) Plant phenolics in sustainable agriculture, vol 1. Springer Singapore, Singapore. https://doi.org/10.1007/978-981-15-4890-1

Ahmed MG, Al-Sagheer AA, El-Zarkouny SZ et al (2022) Potential of selected plant extracts to control severe subacute ruminal acidosis in vitro as compared with monensin. BMC Vet Res 18:356. https://doi.org/10.1186/s12917-022-03457-4

Bach A, Calsamiglia S, Stern MD (2005) Nitrogen metabolism in the rumen. J Dairy Sci 88(Suppl 1):E9–E21. https://doi.org/10.3168/jds.S0022-0302(05)73133-7

Beauchemin KA, Ungerfeld EM, Eckard RJ, Wang M (2020) Review: fifty years of research on rumen methanogenesis: lessons learned and future challenges for mitigation. Animal 14:s2–s16. https://doi.org/10.1017/S1751731119003100

Belanche A, Newbold CJ, Morgavi DP, Bach A, Zweifel B, Yáñez-Ruiz DR (2020) A meta-analysis describing the effects of the essential oils blend Agolin ruminant on performance, rumen fermentation and methane emissions in dairy cows. Animals 10(4):620

Benchaar C (2020) Feeding oregano oil and its main component carvacrol does not affect ruminal fermentation, nutrient utilization, methane emissions, milk production, or milk fatty acid composition of dairy cows. J Dairy Sci 103(2):1516–1527. https://doi.org/10.3168/jds.2019-17230

Benchaar C, Greathead H (2011) Essential oils and opportunities to mitigate enteric methane emissions from ruminants. Anim Feed Sci Technol 166-167:338–355. https://doi.org/10.1016/j.anifeedsci.2011.04.024

Benchaar C, Petit HV, Berthiaume R, Whyte TD, Chouinard PY (2006) Effects of addition of essential oils and monensin premix on digestion, ruminal fermentation, milk production, and milk composition in dairy cows. J Dairy Sci 89:4352–4364. https://doi.org/10.3168/jds.S0022-0302(06)72482-1

Benchaar C, Petit HV, Berthiaume R, Ouellet DR, Chiquette J, Chouinard PY (2007) Effects of essential oils on digestion, ruminal fermentation, rumen microbial populations, milk production, and milk composition in dairy cows fed alfalfa silage or corn silage. J Dairy Sci 90:886–897. https://doi.org/10.3168/jds.S0022-0302(07)71572-2

Benchaar C, Calsamiglia S, Chaves AV, Fraser GR, Colombatto D, McAllister TA, Beauchemin KA (2008) A review of plant-derived essential oils in ruminant nutrition and production. Anim Feed Sci Technol 145(1–4):209–228. https://doi.org/10.1016/j.anifeedsci.2007.04.014

Benchaar C, Lettat A, Hassanat F, Yang WZ, Forster RJ, Petit HV, Chouinard PY (2012) Eugenol for dairy cows fed low or high concentrate diets: effects on digestion, ruminal fermentation characteristics, rumen microbial populations and milk fatty acid profile. Anim Feed Sci Technol 178:139–150. https://doi.org/10.1016/j.anifeedsci.2012.10.005

Benetel G, Silva TDS, Fagundes GM, Welter KC, Melo FA, Lobo AAG, Muir JP, Bueno ICS (2022) Essential oils as in vitro ruminal fermentation manipulators to mitigate methane emission by beef cattle grazing tropical grasses. Molecules 27:2227. https://doi.org/10.3390/molecules27072227

Bichi E, Toral PG, Hervas G, Frutos P, Gomez-Cortes P, Juarez M, de la Fuente MA (2012) Inhibition of Δ9-desaturase activity with sterculic acid: effect on the endogenous synthesis of cis-9 18:1 and cis-9, trans-11 18:2 in dairy sheep. J Dairy Sci 95:5242–5252

Blanch M, Carro MD, Ranilla MJ, Viso A, Vazquez-Anon M, Bach A (2016) Influence of a mixture of cinnamaldehyde and garlic oil on rumen fermentation, feeding behaviour and performance of lactating dairy cows. Anim Feed Sci Technol 219:313–323. https://doi.org/10.1016/j.anifeedsci.2016.07.002

Braun HS, Schrapers KT, Mahlkow-Nerge K, Stumpff F, Rosendahl J (2019) Dietary supplementation of essential oils in dairy cows: evidence for stimulatory effects on nutrient absorption. Animal 13(3):518–523. https://doi.org/10.1017/S1751731118001696

Broderick GA, Udén P, Murphy ML, Lapins A (2004) Sources of variation in rates of in vitro ruminal protein degradation. J Dairy Sci 87(5):1345–1359. https://doi.org/10.3168/jds.S0022-0302(04)73284-1

Buccioni A, Pauselli M, Viti C, Minieri S, Pallara G, Roscini V, Rapaccini S, Trabalza Marinucci M, Lupi P, Conte G, Mele M (2015) Milk fatty acid composition, rumen microbial population, and animal performances in response to diets rich in linoleic acid supplemented with chestnut or quebracho tannins in dairy ewes. J Dairy Sci 98:1145–1156. https://doi.org/10.3168/jds.2014-8651

Burt S (2004) Essential oils: their antibacterial properties and potential applications in foods—A review. Int J Food Microbiol 94:223–253

Canaes TS, Zanferari F, Maganhe BL, Takiya CS, Silva TH, Del Valle TA, Rennó FP (2017) Increasing dietary levels of citral oil on nutrient total tract digestibility, ruminal fermentation, and milk composition in Saanen goats. Anim Feed Sci Technol 229:47–56. https://doi.org/10.1016/j.anifeedsci.2017.05.002

Cardozo PW, Calsamiglia S, Ferret A, Kamel C (2004) Effects of natural plant extracts on ruminal protein degradation and fermentation profiles in continuous culture. J Anim Sci 82(11):3230–3236. https://doi.org/10.2527/2004.82113230x

Cardozo PW, Calsamiglia S, Ferret A, Kamel C (2005) Screening for the effects of natural plant extracts at different pH on in vitro rumen microbial fermentation of a high-concentrate diet for beef cattle. J Anim Sci 83(11):2572–2579. https://doi.org/10.2527/2005.83112572x

Cobellis G, Trabalza-Marinucci M, Yu Z (2016) Critical evaluation of essential oils as rumen modifiers in ruminant nutrition: a review. Sci Total Environ 545–546:556–568. https://doi.org/10.1016/j.scitotenv.2015.12.103

Correddu F, Lunesu MF, Buffa G, Atzori AS, Nudda A, Battacone G, Pulina G (2020) Can agro-industrial by-products rich in polyphenols be advantageously used in the feeding and nutrition of dairy small ruminants? Animals 10(1):1–25. https://doi.org/10.3390/ani10010131

Crumb DJ (2011) Conjugated linoleic acid (CLA) – an overview. Int J Appl Res Nat Prod 4:12–18

De Cássia da Silveira e Sá R, Andrade LN, Dos Reis Barreto de Oliveira R, De Sousa DP (2014) A review on anti-inflammatory activity of phenylpropanoids found in essential oils. Molecules 19:1459–1480. https://doi.org/10.3390/molecules19021459

Dey A, De PS (2014) Influence of condensed tannins from Ficus bengalensis leaves on feed utilization, milk production and antioxidant status of crossbred cows. Asian Australas J Anim Sci 27(3):342–348. https://doi.org/10.5713/ajas.2013.13295

Dschaak CM, Williams CM, Holt MS, Eun JS, Young AJ, Min BR (2011) Effects of supplementing condensed tannin extract on intake, digestion, ruminal fermentation, and milk production of lactating dairy cows. J Dairy Sci 94(5):2508–2519. https://doi.org/10.3168/jds.2010-3818

Dudareva N, Pichersky E, Gershenzon J (2004) Biochemistry of plant volatiles. Plant Physiol 135: 1893–1902. https://doi.org/10.1104/pp.104.049981

El-Essawy AM, Anele UY, Abdel-Wahed AM, Abdou AR, Khattab IM (2021) Effects of anise, clove and thyme essential oils supplementation on rumen fermentation, blood metabolites, milk yield and milk composition in lactating goats. Anim Feed Sci Technol 271:114760. https://doi.org/10.1016/j.anifeedsci.2020.114760

Firkins JL, Yu Z, Morrison M (2007) Ruminal nitrogen metabolism: perspectives for integration of microbiology and nutrition for dairy. J Dairy Sci 90(Suppl 1):E1–E16. https://doi.org/10.3168/jds.2006-518

Focant M, Froidmont E, Archambeau Q, Dang Van QC, Larondelle Y (2019) The effect of oak tannin (Quercus robur) and hops (Humulus lupulus) on dietary nitrogen efficiency, methane emission, and milk fatty acid composition of dairy cows fed a low-protein diet including linseed. J Dairy Sci 102(2):1144–1159. https://doi.org/10.3168/jds.2018-15479

Francis G, Kerem Z, Makkar HPS, Becker K (2002) The biological action of saponins in animal systems: a review. Br J Nutr 88(6):587–605. https://doi.org/10.1079/bjn2002725

Frutos P, Hervas G, Giraldez FJ, Mantecon AR (2004) Review. Tannins and ruminant nutrition. Span J Agric Res 2(2):191–202. https://doi.org/10.5424/sjar/2004022-73

Giannenas I, Skoufos J, Giannakopoulos C, Wiemann M, Gortzi O, Lalas S, Kyriazakis I (2011) Effects of essential oils on milk production, milk composition, and rumen microbiota in Chios dairy ewes. J Dairy Sci 94(11):5569–5577. https://doi.org/10.3168/jds.2010-4096

Githiori JB, Athanasiadou S, Thamsborg SM (2006) Use of plants in novel approaches for control of gastrointestinal helminths in livestock with emphasis on small ruminants. Vet Parasitol 139(4):308–320. https://doi.org/10.1016/j.vetpar.2006.04.021

Gunal M, Ishlak A, AbuGhazaleh AA, Khattab W (2014) Essential oils effect on rumen fermentation and biohydrogenation under *in vitro* conditions. Czech J Anim Sci 59:450–459

Hart KJ, Yanez-Ruiz DR, Duval SM, McEwan NR, Newbold CJ (2008) Plant extracts to manipulate rumen fermentation. Anim Feed Sci Technol 147(1–3):8–35. https://doi.org/10.1016/j.anifeedsci.2007.09.007

Heckendorn F, Häring DA, Maurer V, Zinsstag J, Langhans W, Hertzberg H (2006) Effect of sainfoin (Onobrychis viciifolia) silage and hay on established populations of Haemonchus contortus and Cooperia curticei in lambs. Vet Parasitol 142(3–4):293–300. https://doi.org/10.1016/j.vetpar.2006.07.014

Helander IM, Alakomi H-L, Latva-Kala K, Mattila-Sandholm T, Pol I, Smid EJ, Gorris LGM, von Wright A (1998) Characterization of the action of selected essential oil components on gram-negative bacteria. J Agric Food Chem 46:3590–3595

Holtshausen L, Chaves AV, Beauchemin KA, McGinn SM, McAllister TA, Odongo NE, Cheeke PR, Benchaar C (2009) Feeding saponin-containing Yucca schidigera and Quillaja saponaria to decrease enteric methane production in dairy cows. J Dairy Sci 92(6):2809–2821. https://doi.org/10.3168/jds.2008-1843

Hoste H, Torres-Acosta JFJ (2011) Non chemical control of helminths in ruminants: adapting solutions for changing worms in a changing world. Vet Parasitol 180(1–2):144–154. https://doi.org/10.1016/j.vetpar.2011.05.035

Hoste H, Martinez-Ortiz-De-Montellano C, Manolaraki F, Brunet S, Ojeda-Robertos N, Fourquaux I, Sandoval-Castro CA (2012) Direct and indirect effects of bioactive tannin-rich tropical and temperate legumes against nematode infections. Vet Parasitol 186(1–2):18–27. https://doi.org/10.1016/j.vetpar.2011.11.042

Hoste H, Meza-Ocampos G, Marchand S, Sotiraki S, Sarasti K, Blomstrand BM, Morgan ER (2022) Use of agro-industrial by-products containing tannins for the integrated control of gastrointestinal nematodes in ruminants. Parasite 29. https://doi.org/10.1051/parasite/2022010

Hristov AN, Lee C, Cassidy T, Heyler K, Tekippe JA, Varga GA, Corl B, Brandt RC (2013) Effect of Origanum vulgare L. leaves on rumen fermentation, production, and milk fatty acid composition in lactating dairy cows. J Dairy Sci 96:1189–1202. https://doi.org/10.3168/jds.2012-5975

Huang Q, Liu X, Zhao G, Hu T, Wang Y (2018) Potential and challenges of tannins as an alternative to in-feed antibiotics for farm animal production. Anim Nutr 4(2):137–150. https://doi.org/10.1016/j.aninu.2017.09.004

Huws SA, Creevey CJ, Oyama LB, Mizrahi I, Denman SE, Popova M, Muñoz-Tamayo R, Forano E, Waters SM, Hess M, Tapio I, Smidt H, Krizsan SJ, Yáñez-Ruiz DR, Belanche A, Guan L, Gruninger RJ, McAllister TA, Newbold CJ, Roehe R, Dewhurst RJ, Snelling TJ, Watson M, Suen G, Hart EH, Kingston-Smith AH, Scollan ND, do Prado RM, Pilau EJ, Mantovani HC, Attwood GT, Edwards JE, NR ME, Morrisson S, Mayorga OL, Elliott C, Morgavi DP (2018) Addressing global ruminant agricultural challenges through understanding the rumen microbiome: past, present, and future. Front Microbiol 9:2161. https://doi.org/10.3389/fmicb.2018.02161

IPCC (2021) "Summary for policymakers," in climate change 2021: the physical science basis. In: Masson-Delmotte V, Zhai P, Pirani A, Connors SL, Péan C, Berger S et al (eds) Contribution of working group I to the sixth assessment report of the intergovernmental panel on climate change. Cambridge University Press, Cambridge, MA

Jami E, Shabtay A, Nikbachat M, Yosef E, Miron J, Mizrahi I (2012) Effects of adding a concentrated pomegranate-residue extract to the ration of lactating cows on in vivo digestibility and profile of rumen bacterial population. J Dairy Sci 95:5996–6005. https://doi.org/10.3168/jds.2012-5537

Jiao T, Wu J, Casper DP, Davis DI, Brown MA, Zhao S, Liang J, Lei Z, Holloway B (2021) Feeding sheep cobalt and oregano essential oil alone or in combination on ruminal nutrient digestibility, fermentation, and fiber digestion combined with scanning electron microscopy. Front Vet Sci 8: 639432. https://doi.org/10.3389/fvets.2021.639432

Jiménez-Ocampo R, Montoya-Flores MD, Pamanes-Carrasco G, Herrera-Torres E, Arango J, Estarrón-Espinosa M, Aguilar-Pérez CF, Araiza-Rosales EE, Guerrero-Cervantes M, Ku-Vera JC (2022) Impact of orange essential oil on enteric methane emissions of heifers fed bermudagrass hay. Front Vet Sci. 9:863910. https://doi.org/10.3389/fvets.2022.86391

Joch M, Kudrna V, Hakl J, Božik M, Homolka P, Illek J, Tyrolova Y, Výborná A (2019) In vitro and in vivo potential of a blend of essential oil compounds to improve rumen fermentation and performance of dairy cows. Anim Feed Sci Technol 251:176–186. https://doi.org/10.1016/j.anifeedsci.2019.03.009

Johnson KA, Johnson DE (1995) Methane emissions from cattle. J Anim Sci 73:2483–2492

Jouany JP, Morgavi DP (2007) Use of 'natural' products as alternatives to antibiotic feed additives in ruminant production. Animal 1(10):1443–1466. https://doi.org/10.1017/S1751731107000742

Kalaitsidis K, Sidiropoulou E, Tsiftsoglou O, Mourtzinos I, Moschakis T, Basdagianni Z, Vasilopoulos S, Chatzigavriel S, Lazari D, Giannenas I (2021) Effects of cornus and its mixture with oregano and thyme essential oils on dairy sheep performance and milk, yoghurt and cheese quality under heat stress. Animals 11(4):1063. https://doi.org/10.3390/ani11041063

Kelln BM, Penner GB, Acharya SN, McAllister TA, Lardner HA (2021) Impact of condensed tannin-containing legumes on ruminal fermentation, nutrition, and performance in ruminants: A review. Can J Anim Sci 101(2):210–223. https://doi.org/10.1139/cjas-2020-0096

Khattab MSA, Kholif AE, Abd El Tawab AM, Shaaban MM, Hadhoud FI, El-Fouly HA, Olafadehan OA (2020) Effect of replacement of antibiotics with thyme and celery seed mixture on the feed intake and digestion, ruminal fermentation, blood chemistry, and milk lactation of lactating Barki ewes. Food Funct 11(8):6889–6898. https://doi.org/10.1039/D0FO00807A

Kholif AE, Matloup OH, Morsy TA, Abdo MM, Abu Elella AA, Anele UY, Swanson KC (2017) Rosemary and lemongrass herbs as phytogenic feed additives to improve efficient feed utilization, manipulate rumen fermentation and elevate milk production of Damascus goats. Livest Sci 204:39–46. https://doi.org/10.1016/j.livsci.2017.08.001

Klevenhusen F, Zeitz J, Duval S, Kreuzer M, Soliva C (2011) Garlic oil and its principal component diallyl disulfide fail to mitigate methane, but improve digestibility in sheep. Anim Feed Sci Technol 166:356–363

Kliem KE, Shingfield KJ (2016) Manipulation of milk fatty acid composition in lactating cows: opportunities and challenges. Eur J Lipid Sci Technol 118:1661–1683

Knapp JR, Laur GL, Vadas PA, Weiss WP, Tricarico JM (2014) Invited review: enteric methane in dairy cattle production: quantifying the opportunities and impact of reducing emissions. J Dairy Sci 97:3231–3261

Kolling GJ, Stivanin SCB, Gabbi AM, Machado FS, Ferreira AL, Campos MM, Tomich TR, Cunha CS, Dill SW, Pereira LGR, Fischer V (2018) Performance and methane emissions in dairy cows fed oregano and green tea extracts as feed additives. J Dairy Sci 101:4221–4234. https://doi.org/10.3168/jds.2017-13841

Kotsampasi B, Tsiplakou E, Christodoulou C, Mavrommatis A, Mitsiopoulou C, Karaiskou C, Sossidou E, Fragioudakis N, Kapsomenos I, Bampidis VA, Christodoulou V, Zervas G (2018) Effects of dietary orange peel essential oil supplementation on milk yield and composition, and blood and milk antioxidant status of dairy ewes. Anim Feed Sci Technol 245:20–31. https://doi.org/10.1016/j.anifeedsci.2018.08.007

Kung L, Williams P, Schmidt RJ, Hu W (2008) A blend of essential plant oils used as an additive to alter silage fermentation or used as a feed additive for lactating dairy cows. J Dairy Sci 91:4793–4800. https://doi.org/10.3168/jds.2008-1402

Lakhani N, Kamra DN, Lakhani P, Alhussien MN (2019) Immune status and haemato-biochemical profile of buffalo calves supplemented with phytogenic feed additives rich in tannins, saponins and essential oils. Trop Anim Health Prod 51(3):565–573. https://doi.org/10.1007/s11250-018-1727-z

Lashkari S, Bonefeld Petersen M, Krogh JS (2019) Rumen biohydrogenation of linoleic and linolenic acids is reduced when esterified to phospholipids or steroids. Food Sci Nutr 8(1): 79–87. https://doi.org/10.1002/fsn3.1252

Lillehoj H, Liu Y, Calsamiglia S, Fernandez-Miyakawa ME, Chi F, Cravens RL et al (2018) Phytochemicals as antibiotic alternatives to promote growth and enhance host health. Vet Res 49(1):1–18. https://doi.org/10.1186/s13567-018-0562-6

Mahachi LN, Chikwanha OC, Katiyatiya CLF, Marufu MC, Aremu AO, Mapiye C (2020) Sericea lespedeza (Lespedeza juncea var. sericea) for sustainable small ruminant production: feed, helminth suppressant and meat preservation capabilities. Anim Feed Sci Technol 270 (September):114688. https://doi.org/10.1016/j.anifeedsci.2020.114688

Martin M, Mathias E, McCorkle CM (2001) Ethnoveterinary medicine: an annotated bibliography of community animal healthcare. Medicine. https://doi.org/10.3362/9781780441146

Mavrommatis A, Skliros D, Flemetakis E, Tsiplakou E (2021) Changes in the rumen Bacteriome structure and enzymatic activities of goats in response to dietary supplementation with Schizochytrium spp. Microorganisms 9:1528. https://doi.org/10.3390/microorganisms9071528

McKain N, Shingfield KJ, Wallace RJ (2010) Metabolism of conjugated linoleic acids and 18:1 fatty acids by ruminal bacteria: products and mechanisms. Microbiology 156:579–588

Min BR, Solaiman S (2018) Comparative aspects of plant tannins on digestive physiology, nutrition and microbial community changes in sheep and goats: a review. J Anim Physiol Anim Nutr 102(5):1181–1193. https://doi.org/10.1111/jpn.12938

Min BR, Barry TN, Attwood GT, McNabb WC (2003) The effect of condensed tannins on the nutrition and health of ruminants fed fresh temperate forages: a review. Anim Feed Sci Technol 106(1–4):3–19. https://doi.org/10.1016/S0377-8401(03)00041-5

Newbold CJ, de la Fuente G, Belanche A, Ramos-Morales E, McEwan NR (2015) The role of ciliate protozoa in the rumen. Front Microbiol 6:1313. https://doi.org/10.3389/fmicb.2015.01313

Niderkorn V, Jayanegara A (2021) Opportunities offered by plant bioactive compounds to improve silage quality, animal health and product quality for sustainable ruminant production: a review. Agronomy 11(1). https://doi.org/10.3390/agronomy11010086

Niezen JH, Charleston WAG, Robertson HA, Shelton D, Waghorn GC, Green R (2002) The effect of feeding sulla (Hedysarum coronarium) or lucerne (Medicago sativa) on lamb parasite burdens and development of immunity to gastrointestinal nematodes. Vet Parasitol 105(3):229–245. https://doi.org/10.1016/S0304-4017(02)00014-6

O'Connell JE, Fox PF (2001) Significance and applications of phenolic compounds in the production and quality of milk and dairy products: a review. Int Dairy J 11(3):103–120. https://doi.org/10.1016/S0958-6946(01)00033-4

Okoruwa M, Edoror O (2019) Effect of garlic oil supplementation on intake, digestibility, performance and rumen function of goats fed silage based-diet. Eur J Nutr Food Saf 10:104–115

Oleszek M, Oleszek W (2020) Saponins in food. In: Handbook of dietary phytochemicals. Springer. https://doi.org/10.1007/978-981-13-1745-3_34-1

Pallara G, Buccioni A, Pastorelli R, Minieri S, Mele M, Rapaccini S, Messini A, Pauselli M, Servili M, Giovannetti L, Viti C (2014) Effect of stoned olive pomace on rumen microbial communities and polyunsaturated fatty acid biohydrogenation: an in vitro study. BMC Vet Res 10:271. https://doi.org/10.1186/s12917-014-0271-y

Paolini V, De La Farge F, Prevot F, Dorchies P, Hoste H (2005) Effects of the repeated distribution of sainfoin hay on the resistance and the resilience of goats naturally infected with gastrointestinal nematodes. Vet Parasitol 127(3–4):277–283. https://doi.org/10.1016/j.vetpar.2004.10.015

Papadopoulos E, Gallidis E, Ptochos S (2012) Anthelmintic resistance in sheep in Europe: a selected review. Vet Parasitol 189(1):85–88. https://doi.org/10.1016/j.vetpar.2012.03.036

Patra AK, Saxena J (2011) Exploitation of dietary tannins to improve rumen metabolism and ruminant nutrition. J Sci Food Agric 91(1):24–37. https://doi.org/10.1002/jsfa.4152

Patra A, Park T, Kim M, Yu Z (2017) Rumen methanogens and mitigation of methane emission by anti-methanogenic compounds and substances. J Anim Sci Biotechnol 8:13. https://doi.org/10.1186/s40104-017-0145-9

Peña-Espinoza M, Thamsborg SM, Desrues O, Hansen TVA, Enemark HL (2016) Anthelmintic effects of forage chicory (Cichorium intybus) against gastrointestinal nematode parasites in experimentally infected cattle. Parasitology 143(10):1279–1293. https://doi.org/10.1017/S0031182016000706

Piluzza G, Sulas L, Bullitta S (2014) Tannins in forage plants and their role in animal husbandry and environmental sustainability: a review. Grass Forage Sci 69(1):32–48. https://doi.org/10.1111/gfs.12053

Polan CE, McNeill JJ, Tove SB (1964) Biohydrogenation of unsaturated fatty acids by rumen bacteria. J Bacteriol 88:1056–1064

Poungchompu O, Wanapat M, Wachirapakorn C, Wanapat S, Cherdthong A (2009) Manipulation of ruminal fermentation and methane production by dietary saponins and tannins from mangosteen peel and soapberry fruit. Arch Anim Nutr 63(5):389–400. https://doi.org/10.1080/17450390903020406

Provenza FD, Villalba JJ (2010) The role of natural plant products in modulating the immune system: an adaptable approach for combating disease in grazing animals. Small Rumin Res 89(2–3):131–139. https://doi.org/10.1016/j.smallrumres.2009.12.035

Quijada J, Fryganas C, Ropiak HM, Ramsay A, Mueller-Harvey I, Hoste H (2015) Anthelmintic activities against Haemonchus contortus or Trichostrongylus colubriformis from small ruminants are influenced by structural features of condensed tannins. J Agric Food Chem 63(28):6346–6354. https://doi.org/10.1021/acs.jafc.5b00831

Ramírez-Restrepo CA, Barry TN, López-Villalobos N, Kemp PD, Harvey TG (2005) Use of Lotus corniculatus containing condensed tannins to increase reproductive efficiency in ewes under commercial dryland farming conditions. Anim Feed Sci Technol 121(1–2):23–43. https://doi.org/10.1016/j.anifeedsci.2005.02.006

Rivera-Chacon R, Castillo-Lopez E, Ricci S, Petri RM, Reisinger N, Zebeli Q (2022) Supplementing a phytogenic feed additive modulates the risk of subacute rumen acidosis, rumen fermentation and systemic inflammation in cattle fed acidogenic diets. Animals 12:1201. https://doi.org/10.3390/ani12091201

Rochfort S, Parker AJ, Dunshea FR (2008) Plant bioactives for ruminant health and productivity. Phytochemistry 69(2):299–322. https://doi.org/10.1016/j.phytochem.2007.08.017

Rofiq MN, Negara W, Martono S, Gopar RA, Boga M (2021) Potential effect of some essential oils on rumen methane reduction and digestibility by in vitro incubation technique. IOP Conf Ser Earth Environ Sci 905(1):012138

Salem AZM, López S, Robinson PH (2012) Plant bioactive compounds in ruminant agriculture - impacts and opportunities. Anim Feed Sci Technol 176(1–4):1–4. https://doi.org/10.1016/j.anifeedsci.2012.07.001

Santos MB, Robinson PH, Williams P, Losa R (2010) Effects of addition of an essential oil complex to the diet of lactating dairy cows on whole tract digestion of nutrients and productive performance. Anim Feed Sci Technol 157:64–71. https://doi.org/10.1016/j.anifeedsci.2010.02.001

Saratsi A (2021) The contribution of Mediterranean flora against gastrointestinal nematodes in small ruminants. Agric Univ Athens. https://doi.org/10.12681/eadd/50128

Sari NF, Ray P, Rymer C, Kliem KE, Stergiadis S (2022) Garlic and its bioactive compounds: implications for methane emissions and ruminant nutrition. Animals 12:2998. https://doi.org/10.3390/ani12212998

Serrano J, Puupponen-Pimia R, Dauer A, Aura AM, Saura-Calixto F (2009) Tannins: current knowledge of food sources, intake, bioavailability and biological effects. Mol Nutr Food Res 53(Suppl. 2):S310–S329. https://doi.org/10.1002/mnfr.200900039

Shaani Y, Eliyahu D, Mizrahi I, Yosef E, Ben-Meir Y, Nikbachat M, Solomon R, Mabjeesh SJ, Miron J (2015) Effect of feeding ensiled mixture of pomegranate pulp and drier feeds on digestibility and milk performance in dairy cows. J Dairy Res 83:35–41. https://doi.org/10.1017/S0022029915000618

Shilwant S, Hundal JS, Singla M, Patra AK (2022) Ruminal fermentation and methane production in vitro, milk production, nutrient utilization, blood profile, and immune responses of lactating goats fed polyphenolic and saponin-rich plant extracts. Environ Sci Pollut Res 1–13:10901. https://doi.org/10.1007/s11356-022-22931-y

Simitzis PE (2017) Enrichment of animal diets with essential oils—a great perspective on improving animal performance and quality characteristics of the derived products. Medicines 4:35

Soldado D, Bessa RJB, Jerónimo E (2021) Condensed tannins as antioxidants in ruminants—effective-ness and action mechanisms to improve animal antioxidant status and oxidative stability of products. Animals 11(11). https://doi.org/10.3390/ani11113243

Soltan YA, Morsy AS, Hashem NM, Sallam SM (2021) Boswellia sacra resin as a phytogenic feed supplement to enhance ruminal fermentation, milk yield, and metabolic energy status of early lactating goats. Anim Feed Sci Technol 277:114963. https://doi.org/10.1016/j.anifeedsci.2021.114963

Spanghero M, Robinson PH, Zanfi C, Fabbro E (2009) Effect of increasing doses of a microencapsulated blend of essential oils on performance of lactating primiparous dairy cows. Anim Feed Sci Technol 153:153–157. https://doi.org/10.1016/j.anifeedsci.2009.06.004

Tager LR, Krause KM (2011) Effects of essential oils on rumen fermentation, milk production, and feeding behavior in lactating dairy cows. J Dairy Sci 94:2455–2464. https://doi.org/10.3168/jds.2010-3505

Tayengwa T, Mapiye C (2018) Citrus and winery wastes: promising dietary supplements for sustainable ruminant animal nutrition, health, production, and meat quality. Sustainability (Switzerland) 10(10):1–22. https://doi.org/10.3390/su10103718

Tekippe JA, Hristov AN, Heyler KS, Cassidy TW, Zheljazkov VD, Ferreira JFS, Karnati SK, Varga GA (2011) Rumen fermentation and production effects of Origanum vulgare L. leaves in lactating dairy cows. J Dairy Sci 94:5065–5079. https://doi.org/10.3168/jds.2010-4095

Theodorou MK, Kingston-Smith AH, Winters AL, Lee MRF, Minchin FR, Morris P, MacRae J (2006) Polyphenols and their influence on gut function and health in ruminants: a review. Environ Chem Lett 4(3):121–126. https://doi.org/10.1007/s10311-006-0061-2

Tiemann TT, Lascano CE, Wettstein HR, Mayer AC, Kreuzer M, Hess HD (2008) Effect of the tropical tannin-rich shrub legumes Calliandra calothyrsus and Flemingia macrophylla on methane emission and nitrogen and energy balance in growing lambs. Animal 2(5):790–799. https://doi.org/10.1017/S1751731108001791

Tong Z, He W, Fan X, Guo A (2022) Biological function of plant tannin and its application in animal health. Front Vet Sci 8(January):1–7. https://doi.org/10.3389/fvets.2021.803657

Tzamaloukas O, Neofytou MC, Simitzis PE (2021) Application of olive by-products in livestock with emphasis on small ruminants: implications on rumen function, growth performance, milk and meat quality. Animals 11(2):531. https://doi.org/10.3390/ani11020531

Vasta V, Daghio M, Cappucci A, Buccioni A, Serra A, Viti C, Mele M (2019) Invited review: plant polyphenols and rumen microbiota responsible for fatty acid biohydrogenation, fiber digestion, and methane emission: experimental evidence and methodological approaches. J Dairy Sci 102(5):3781–3804. https://doi.org/10.3168/jds.2018-14985

Waghorn GC, McNabb WC (2003) Consequences of plant phenolic compounds for productivity and health of ruminants. Proc Nutr Soc 62(2):383–392. https://doi.org/10.1079/PNS2003245

Yang WZ, Benchaar C, Ametaj BN, Chaves AV, He ML, McAllister TA (2007) Effects of garlic and juniper berry essential oils on ruminal fermentation and on the site and extent of digestion in lactating cows. J Dairy Sci 90:5671–5681. https://doi.org/10.3168/jds.2007-0369

Yanza YR, Szumacher-Strabel M, Lechniak D et al (2022) Dietary Coleus amboinicus Lour. Decreases ruminal methanogenesis and biohydrogenation, and improves meat quality and fatty acid composition in longissimus thoracis muscle of lambs. J Animal Sci Biotechnol 13(1):5. https://doi.org/10.1186/s40104-021-00654-3

Yuan LG, Tong HM (2005) Oregano oil and Flavomycin in dairy cattle production application test. Chin J Vet Med 41:20–22

Zebeli Q, Mansmann D, Steingass H, Ametaj BN (2010) Balancing diets for physically effective fibre and ruminally degradable starch: a key to lower the risk of sub-acute rumen acidosis and improve productivity of dairy cattle. Livest Sci 127:1–10. https://doi.org/10.1016/j.livsci.2009.09.003

Zhang R, Wu J, Lei Y, Bai Y, Jia L, Li Z, Liu T, Xu Y, Sun J, Wang Y, Zhang K, Lei Z (2021) Oregano essential oils promote rumen digestive ability by modulating epithelial development and microbiota composition in beef cattle. Front Nutr 8:722557. https://doi.org/10.3389/fnut.2021.722557

Zhou R, Wu J, Lang JX, Liu L, Casper DP, Wang C, Zhang L, Wei S (2020) Effect of oregano essential oil on in vitro ruminal fermentation, methane production, and ruminal microbial community. J Dairy Sci 103:2303–2314

Zhu Z, Mao S, Zhu W (2012) Effects of ruminal infusion of garlic oil on fermentation dynamics, fatty acid profile and abundance of bacteria involved in biohydrogenation in rumen of goats. Anim Biosci 25(7):962–970. https://doi.org/10.5713/ajas.2011.11442

Zwenger S, Basu C (2008) Plant terpenoids: applications and future potentials. Biotechnol Mol Biol Rev 3:1–7

Feed Additives as Dietary Tools to Improve Welfare Status in Ruminants

Soumela Savvidou, Maria-Anastasia Karatzia, and Basiliki Kotsampasi

Abstract

Animal welfare is intrinsically associated with animal health and productivity and has been established as an important public awareness subject, leading to a demand for high welfare standards by all relevant professionals and consumers alike.

This chapter will discuss the interconnection of farm animal welfare with nutrition, and particularly the potential of feed additives to improve ruminant welfare status, regardless of health status, productive state, breed, and on-farm conditions, and especially the potential of feed additives to improve welfare status of ruminants. The inclusion of feed additives, such as probiotics, prebiotics, enzymes, essential oils, toxin binders, phytogenics, and others in ruminant nutrition, has been proven to be a functional dietary tool towards a considerable ruminant welfare level improvement. Within this concept, the inclusion of the abovementioned feed additives in ruminant nutrition leads to:

- Support of natural feeding behavior
- Environmental stress alleviation
- Rumen fermentation enhancement
- Methane and ammonia mitigation
- Improvement of production efficiency
- Antiparasitic, antiviral, and anthelminthic action
- Metabolic disease prevention
- Subclinical mastitis abatement
- Antioxidant status improvement
- Immunomodulation
- Anti-inflammatory action

S. Savvidou (✉) · M.-A. Karatzia · B. Kotsampasi
Research Institute of Animal Science, ELGO-DEMETER, Paralimni Giannitsa, Greece

G. Arsenos, I. Giannenas (eds.), *Sustainable Use of Feed Additives in Livestock*,
https://doi.org/10.1007/978-3-031-42855-5_22

Thus, it leads, indirectly, to an improvement in the overall well-being of animals.

Feeding behavior affects nutritional costs, which in turn generate significant repercussions on total production expenses. Consequently, improving production efficiency can have a considerable impact on the profitability of ruminant industry, where alternative nutritional strategies are constantly pursued and implemented, in compliance to consumer desire for animal products that are produced in an ethical aspect and within the One Welfare concept.

Keywords

Feed additives · Probiotics · Prebiotics · Enzymes · Essential oils · Toxin binders · Phytogenics · Welfare · Ruminants

1 Introduction

Animal welfare is intrinsically associated with animal health and productivity and has been established as an important public awareness subject, leading to a rigorous demand for high welfare standards by all relevant professionals and consumers alike. This chapter will discuss the interconnection of animal welfare with another fundamental aspect of animal production, nutrition, and especially the potential capacity of feed additives to improve the welfare status of ruminants.

Nutrition is recognized as the most fundamental welfare need for all animals, regardless of health status, productive state, breed, and on-farm conditions. Even though significant advances in nutritional practices have been achieved in recent years, nutrition is still perceived as the most critical welfare issue in most cattle, sheep, and goat farms.

For ruminant livestock, food preference and behavior, that indicates the level of their welfare, is developed through an intricate process combining taste, smell, sight, and satisfaction of physiological requirements. Additionally, the interplay between flavor and postingestive feedback affects nutritional behavior dynamically. Diversity of feed sources and feed choice, in association with grazing behavior, should also be considered as a critical aspect of animal welfare.

Feeding behavior affects nutritional costs, which in turn exhibit significant repercussions on total production expenses. Consequently, improving production efficiency can have a considerable impact on the profitability of ruminant industry, where alternative nutritional strategies are constantly pursued and implemented, in compliance to consumer desire for animal products that are produced in an ethical aspect and within the One Welfare concept.

Feed additives are ingredients that are applied in small amounts in animal nutrition. They have been reported to enhance the efficacy of feed acceptance, nutrient utilization, growth, and health status, and thus indirectly the overall well-being of the animals. Different types of feed additives will be discussed in this chapter, such as probiotics, prebiotics, essential oils, toxin binders, enzymes, and phytogenics, all of them reported to modify the digestion kinetics and immune function in ruminant animals leading to accelerated production efficiency, health, and welfare.

To conclude, this chapter will provide an overview of the current knowledge, limitation, and prospects for feed additives applied to ruminant enterprise as dietary tools to improve welfare status.

2 Feed Additives

The ruminal microbiome, i.e., the community of prokaryotic and eukaryotic microorganisms that inhabits the rumen, is characterized as a complex and dynamic ecosystem, comprised of a collection of bacteria, archaea, protozoa, and fungi, in high population density and diversity. The rumen microbiome plays a crucial role in digestive processes of complex fibrous ingredients by producing enzymes (fibrolytic, glycolytic, lipolytic, and proteolytic) that facilitate volatile fatty acids as a main product of feed metabolism and fermentation. Lately, it is gradually acknowledged that the microorganisms colonizing the host's digestive system can directly or indirectly impact brain operation and cognitive behavior (Kraimi et al. 2019). Kelsey and Colpoys (2018) suggest that calves provided with probiotic bacteria showed a reduced stress-related behavior. The main driver of modulating rumen microbiome, and thus impacting animal production and wellbeing is diet. Even though feed additives (probiotics, yeast cell walls, prebiotics, essential oils, toxin binders, enzymes, and phytogenics) are not considered as dietary essential to ruminants, they have been reported to modify digestion kinetics in the rumen and enhance the efficacy of feed acceptance, nutrient utilization, growth, and health status, and thus indirectly the overall well-being of animals.

3 Probiotics: Mode of Actions—Effects on Ruminants

In 2013, the International Scientific Association for Probiotics and Prebiotics (ISAPP) reviewed the term "probiotic," defined in 2001 by an Expert Consultation of the Food and Agricultural Organization of the United Nations and the World Health Organization (FAO/WHO), and the scientific literature related to the topic, and it reiterated, actually, the proposed definition of the term "probiotic" as "Live microorganisms that, when administered in adequate amounts, confer a health benefit on the host."

Probiotics have been extensively applied in the livestock industry as one of the antimicrobial substitutes owing to the banned or rigorously restricted preventative and growth promoting usage of antibiotics. As for the ruminant nutrition, incorporation of probiotics is not as extensive as in monogastric animals' nutrition due mainly to aspects such as viability in the rumen and interactions with ruminal microbiota and the host. Considering that, mode of action of probiotics can be faced as two distinct general mechanisms, one occurring in the rumen and the other in the

Table 1 Some of the most commonly used probiotics for ruminants

Probiotic microorganisms	References
Saccharomyces cerevisiae	Chaucheyras-Durand et al. (2008), Elghandour et al. (2020), Marden et al. (2008), Maamouri and Ben Salem (2021)
Lactic acid bacteria	Chiquette et al. (2008), Weimer (2015)
Aspergillus oryzae	Jouany et al. (1998), Mathieu et al. (1996)
Bacillus	Uyeno et al. (2015)
Enterococcus	Ma et al. (2020)

lower gut. Anyhow, there have been controversial scientific outcomes, with some of the studies reporting positive results (Nocek et al. 2003; Nocek and Kautz 2006) while others, little or no effect (Oetzel et al. 2007, AlZahal et al. 2014a, b).

The majority of probiotic bacteria in animal nutrition are classified as lactic acid-producing or lactic acid-utilizing bacteria, including species of *Lactobacillus*, *Bifidobacterium*, *Streptococcus*, *Enterococcus*, *Megasphaera,* and *Propionibacterium.* Fungal species of *Saccharomyces* and *Aspergillus* are also widely applied as probiotics in ruminant diets. Commercial probiotic products may be composed of either single- or multispecies microorganisms. Table 1 presents some of the most commonly used probiotics for ruminants, such as *Saccharomyces cerevisiae* (Chaucheyras-Durand et al. 2008; Elghandour et al. 2020), lactic acid bacteria (LAB) (Chiquette et al. 2008; Weimer 2015); *Aspergillus oryzae* (Jouany et al. 1998; Mathieu et al. 1996); and *Bacillus* and *Enterococcus* (Uyeno et al. 2015), all of which found to confer a substantial impact on ruminant health.

4 Proposed Mode of Action in the Lower Gut

The probiotic mode of action proposed in scientific literature (Anee et al. 2021; Ban and Guan 2021) includes:

- Inhibition of pathogenic adhesion and competition for nutrients in the gastrointestinal tract, positively impacting beneficial bacteria in the gut
- Secretion of defensins/bacteriocins that negatively affect pathogenic bacteria
- Competitive exclusion of pathogenic microorganisms, by binding with receptors in the intestinal epithelium cell (IEC) or mucus layer
- Enhancement of gut barrier function by structuring a dense mucus layer through mucin glycoprotein secretion by mucus-producing cells that facilitate to reduce intracellular permeability to pathogens
- Decrease of luminal pH, by activating lactic and acetic acid production, toxic to pathogenic bacteria
- Stimulation of host immunity by activating T- and antibody-producing B-cells through the dendritic cell, found in the mucosal lamina propria, surface epithelium, and Peyer's patches

- Stimulation of expression and turnover of neurotransmitters (e.g., serotonin), modulating enteric nervous and central nervous system bidirectional communication and interaction and thus, decreasing stress and improving animal behavior (Kraimi et al. 2019)

5 Proposed Mode of Action in the Rumen

Probiotic bacteria have been found to beneficially influence rumen fermentation and prevent subacute ruminal acidosis (Kumprechtová et al. 2019; Philippeau et al. 2017), a metabolic disorder of ruminants fed high-concentrate diets, where a depression of ruminal pH below 5.6 at least 3 h/day is observed, with most common symptoms being lethargy, elevated respiratory and pulse rate, increased temperature, diarrhea, weight loss, and decreased feed intake. Lactic acid bacteria (*Lactobacillus* and *Enterococcus*) have been found to boost beneficial ruminal microorganisms adjusting to the presence of lactic acid in the rumen (Yoon and Stern 1995), and stimulating lactic acid-utilizing bacteria, exerting a probiotic effect.

6 Proposed Mode of Action of Yeast Probiotics

The inclusion of fungal probiotic microorganisms, especially yeast products (e.g., *Saccharomyces cerevisiae* and *Saccharomyces boulardii*) in the diet, is related to the stimulation of rumen microbial flora, in particular of the beneficial cellulolytic microorganisms, which compete with lactic acid bacteria for substrate and enhance anaerobiosis by sifting oxygen present in the rumen (Khan et al. 2016; Vibhute et al. 2011; Chiquette 2009; Marden et al. 2008), as well as to the increase in fiber degradation and microbial protein flow to the small intestine and to modification of rumen fermentation and volatile fatty acid (VFA) production (Pinloche et al. 2013; Jiang et al. 2017).

7 Effects of Probiotics on Ruminants

Improved zootechnical performance parameters such as daily weight gain, feed efficiency, and milk production are some of the well-documented effects of probiotics on ruminants (Kirankumar et al. 2022). Enhanced daily weight gain has been reported in several studies, for example, when a probiotic mixture of *Lactobacillus reuteri* DDL 19, *Lactobacillus alimentarius* DDL 48, *Enterococcus faectum* DDE 39, and *Bifidobacterium bifidum* DDBA provided to goats increased their standard body weight by 9% (Apás et al. 2010). Du et al. 2018 suggest that stimulation of growth hormone/insulin-like growth factor-1 (GH/IGF-1) by probiotic *Bacillus amyloliquefaciens* C-1 is responsible for improved growth performance and regulation of gut microbiota in beef calves. Higher milk production in lactating ruminants provided with probiotics was also reported. Specifically, enhanced milk production was reported when a probiotic mixture of *Bacillus subtilis*, *Saccharomyces cerevisiae*, and *Enterococcus faecalis* (5 × 1011 CFU/day) was delivered to dairy goats (Ma et al. 2020), when a mixture of *Bacillus licheniformis* and *Bacillus subtilis* (2.56 × 10^9 CFU/day) was administered (Kritas et al. 2006), and when a mixture of *Enterococcus faecium* (5 × 10^9 CFU) and *Saccharomyces cerevisiae* (2 × 10^9 CFU) was provided to dairy cattle (by 2.3 L per cow each day; Nocek and Kautz 2006). Increased milk yield was also reported when lactating dairy cows (AlZahal et al. 2014a, b; Kumprechtová et al. 2019), lactating dairy goats (Stella et al. 2007), and lactating dairy sheep (Mavrommatis et al. 2020) were provided with *Saccharomyces cerevisiae*. However, some studies reported no impact on production parameters after yeast addition in diets of dairy cows (Poppy et al. 2012; Elghandour et al. 2022).

Crucial for the health performance, rumen fermentation can be reinforced by probiotics, mainly through regulation of the rumen microbial ecosystem and the production of antimicrobial substances that can eliminate zoonotic pathogenic bacteria and control ammonia production. *Rhodopseudomonas palustris* has been applied as a potential probiotic on lactating cows' nutrition, and proved to positively impact the viability and growth of rumen microorganisms and boost microbial fermentation towards nonglucogenic energy supply (Chen et al. 2020). Administration of *Megasphaera elsdenii*, a lactate-utilizing bacterium, was reported to enhance ruminal butyrate production and feed intake in neonatal calves (Muya et al. 2015). Addition of yeasts in steers' diet enhanced the population of both carbohydrate-digesting bacteria (*Ruminococcus albus, Ruminococcus champanellensis, Ruminococcus bromii,* and *Ruminococcus obeum*) and lactate-utilizing bacteria (*Megasphaera elsdenii, Desulfovibrio desulfuricans,* and *Desulfovibrio vulgaris*), and the ability to scavenge oxygen within the rumen (Ogunade et al. 2019). However, there is an inconsistency in the probiotic bacteria effects on ruminal fermentation ability. For instance, when dairy calves were administered with *Lactobacillus plantarum* and *Bacillus subtilis,* they did not change any ruminal fermentation characteristics (Zhang et al. 2017). Strain, dose, frequency, and time/duration of supplementation are some of the factors that may affect the response of ruminants to probiotics (Buntyn et al. 2016; Amin and Mao 2021).

Health performance is directly related to the immune system in ruminants and can also be affected by probiotic use. Health issues could be overcome through administration of probiotics. Incidence of diarrhea, a very common problem faced by calves, has been reduced by administration of a mixture of lactic acid bacteria (*Lactobacillus acidophilus*, *Lactobacillus salivarius*, and *Lactobacillus plantarum*) (Signorini et al. 2012). A compound probiotic (consisting of *Lactobacillus plantarum*, *Pediococcus acidilactici*, *Pediococcus pentosaceus*, and *Bacillus subtilis*) decreased fecal scores and consequently medicine treatments and improved the health status of newborn calves (Wang et al. 2022). According to Cao et al. (2007), mastitis, a very challenging problem in dairy cows, can be treated by *Staphylococcus aureus*, and more recently Alawneh et al. (2020) proposed an alternative treatment or prevention strategy, a *Lactobacillus*-based, post-milking teat spray that has been found to potentially improve teat end sphincter functions and udder health. A mixture of *Saccharomyces cerevisiae* and *Lactococcus* species has been also found to lower the mastitis-related inflammation by a mechanism of decreasing serum concentration of tumor necrosis factor-α (TNF-α), interleukin-6 (IL-6), and IL-1β in dairy cows (Gao et al. 2020). Ruminal acidosis, as mentioned above, could be also prevented via administration of probiotic bacteria that are capable of utilizing lactic acid in the rumen, either in calves (Krehbiel et al. 2003) or in dairy cows (Aikman et al. 2011; Mazon et al. 2020). *Saccharomyces cerevisiae* is popular for its probiotic effect yeast, can stabilize rumen pH, and reduce susceptibility to acidosis in dairy animals (Chaucheyras-Durand et al. 2008; Marden et al. 2008; Maamouri and Ben Salem 2021).

Lifecycle of ruminants can be very demanding due to stressful events encountered, i.e., pathogenic bacteria and disease, environmental factors, handling anxiety, nutritional stress, relocation actions, and novel environments. Mitigation of some of these negative effects associated with stress is faced by supplementation of probiotics (Broadway et al. 2015). Supplemented *Saccharomyces cerevisiae* to ruminants showed to alleviate some negative effects associated heat stress (Burdick Sanchez et al. 2015), acute inflammatory stress induced by lipopolysaccharide (Finck et al. 2014), diarrheal symptoms of colostrum-deprived dairy calves challenged with *Escherichia coli* (Galvão et al. 2005).

8 Prebiotics' Mode of Actions: Effects on Ruminants

Prebiotics are defined as nondigestible, nonviable substrates that selectively stimulate the growth and/or activity of bacteria in the gastrointestinal tract, and consequently enable the host organism to defend against pathogens and modulate the immune system, conferring a health benefit to the host (Gibson and Roberfroid 1995, Markowiak and Śliżewska 2017).

The most used prebiotics fed to ruminants belong to the group of the carbohydrate-based oligosaccharides:

Fructooligosaccharides (FOS): They consist of one glucose molecule connected to numerous fructose molecules via β-(2-1) or β-(2-6) bonds.
Galactooligosaccharides (GOS): They are created from lactose and connect to several galactose molecules via β-(1-3) and β-(1-4) linkages.
Xylooligosaccharides (XOS).
Mannooligosaccharides (MOS): They are products of cell walls of *S. cerevisiae* and contain repeating mannose units via α-(1-2) and α-(1-3) linkages.
Other oligosaccharides: Pectic oligosaccharides (POS) and transgalactooligosaccharides (TOS)

Lately, noncarbohydrate compounds (e.g., polyphenols and polyunsaturated fatty acids) and polysaccharides (e.g., xylans, pectins, inulins, and β-glucans that are polysaccharides of glucose that differ in their arrangement of β-glycosidic bonds) have also been characterized and used as prebiotic feed additives in ruminants' nutrition (Moorthy et al. 2020; Costantini et al. 2017; Scott et al. 2020).

Prebiotics have been applied to calves' nutrition, especially during the preweaning period in order to improve their health and performance (Ghosh and Mehla 2012). Both MOS and FOS are found to enhance the nutrient absorption and consequently the feed efficiency and yields (Cangiano et al. 2020). Supplementing MOS and polysaccharides to preweaning calves increased their average daily gain, feed efficiency (Heinrichs et al. 2003; Ghosh and Mehla 2012; Roodposhti and Dabiri 2012; Heinrichs et al. 2013), and growth (He et al. 2017), which according to Michalak et al. (2021) is attributed to the fact that MOS have the ability to improve the epithelium of ruminants (Alves Costa et al. 2019) by increasing the length of the rumen papillae and the height of the jejunum villi, which is possibly associated to the increase in substrate availability by MOS-utilizing bacteria. Similarly, inclusion of short-chain FOS in the calf milk replacer enhanced the growth performance of preruminant calves (Grand et al. 2013), and as for a commercial product Bio-Mos®, when added in milk or milk replacer fed to dairy calves also improved daily weight gain and body weight in the preweaning period.

As stated by Chang et al. (2022), application of oligosaccharides in ruminant nutrition might also be accountable for the positive manipulation of rumen fermentation by increasing concentrations of volatile fatty acids (VFAs) and bacterial protein and reducing that of ammonia nitrogen (NH_3–N) (Li et al. 2011, 2018), as well as by specifically promoting the proliferation of particular bacterial species. Additionally, Cangiano et al. (2020) suggested that prebiotic supplementation stimulates the growth of ciliated protozoa and cellulolytic bacteria that stabilize the rumen pH. Li et al. (2011) showed that nondigestible oligosaccharides (isomalto-oligosaccharides, MOS and FOS) enhanced rumen fermentation in sheep possibly through boosting the proliferation of some beneficial bacteria.

FOS have been reported to positively impact the health status of ruminants, by supporting their immune system and by reducing intensity of diarrhea and mortality rates, especially in calves (Cangiano et al. 2020; Michalak et al. 2021). The high-risk preweaning period that is characterized by gastrointestinal infections, microbial dysbiosis, diarrhea, dehydration, and eventually early calf mortality might be prevented or even treated by prebiotic supplementation. Lower fecal scores and low diarrhea incidence were reported after weaning when calves were provided with GOS, and it was attributed by Senevirathne et al. (2019) to the improved intestinal microbial community, which was able to prevent pathogens. Likewise, and more consistently, lower mean fecal scores were found in MOS-supplemented Holstein calves (Hill et al. 2009; Morrison et al. 2010; Ghosh and Mehla 2012; Heinrichs et al. 2013). Immune response to a potential infection was promoted in calves fed β-glucans; Kim et al. (2021) noted an increased production of total serum immunoglobulin A (IgA) (both bacterial-specific and viral-specific IgA) as well as increased neutrophils, neutrophil:lymphocyte ratio, and serum lactoferrin when challenged with a live bacterial and viral vaccine. In addition, Eicher et al. (2011) increased gene expression of toll-like receptor-4 (TLR4) and interleukin-12 (IL12) in the lung. MOS are also reported to enhance health traits, and when supplemented in ruminants' nutrition are found to have an immune-boosting effect (Wismar et al. 2010; Raza et al. 2022).

However, the results for the efficacy of prebiotic supplementation in ruminants' nutrition in the literature seem to be inconsistent and even contradictory as there are studies (da Silva et al. 2012; Froehlich et al. 2017; Tóth et al. 2020) suggesting that prebiotic supplementation has no effect on calves' growth performance and health. Definitely, further studies are needed to investigate the potential role of prebiotics in ruminant health.

9 Essential Oils

Essential oils (EOs) are naturally produced by several plants that are generally localized in temperate to warm countries, like the Mediterranean and tropical countries, representing an important part of the traditional pharmacopeia (Bakkali et al. 2008). They are aromatic oily liquids obtained from various plant materials (flowers, buds, seeds, leaves, twigs, bark, herbs, wood, fruits, and roots) (Burt 2004) by hydrodistillation or steam distillation (Bakkali et al. 2008; Shaaban et al. 2012). An estimated 3,000 EOs are known, of which about 300 are commercially important as flavor and fragrance agents (Van de Braak and Leijten 1999). EOs are a mixture of low-molecular-weight molecules that include terpenes (monoterpenes and sesquiterpenes), alcohols, aldehydes, and ketones, which, in addition to their bioactive molecules, are also responsible for the aromatic fragrance that these materials frequently exhibit (Jean 2008). Essential oils are considered safe for human and animal consumption and are categorized as generally recognized as safe (GRAS; FDA 2004).

Some examples of EOs from aerial parts (leaves and flowers) include oregano oil (*Origanum vulgare*) and thyme oil (*Thymus vulgaris*), which both contain thymol at proportions of 63.3% and 19.5–54.1%, respectively. Other EOs derived from leaves and flowers are basil oil (*Ocimum basilicum* L.), with an estragole content, dill oil (*Anethum graveolens* L.), which contains phellandrene. EOs can also be obtained from fruit parts, such as anise oil (*Pimpinella anisum* L.), coriander oil (*Coriandrum sativum* L.), lemon oil (*Citrus limon*), mountain pride oil (*Heracleum persicum*), and nutmeg oil (*Myristica fragrans*), with the main content of each EO in the form of trans-anethole, at 76.9–93.7%, linalool, at 72.2–87.5%, limonene, at 65.6–69.9%, hexyl butyrate, at 56.6%, and α-pinene, at 22.2%. EOs can also be obtained from other parts of plants such as seeds, rhizomes, tree bark, tubers, and buds. *Nigella sativa* L. and *Carum carvi* L. are examples of medicinal plants where EOs are extracted from seeds as black cumin seed oil and caraway oil, respectively. The primary active compound in black cumin seed oil is para-cymene, while caraway oil contains carvone. Other EOs such as turmeric oil (1,8-cineole) are taken from the rhizome *Curcuma longa* L. Cinnamon oil (E-cinnamaldehyde, 97.7%) is derived from the bark of the *Cinnamomum zeylanicum* tree, while clove oil (eugenol 88.6%) was extracted from buds of *Eugenia caryophyllata*, and garlic oil (diallyl disulphide, 53%) was extracted from *Allium sativum* tubers (Ramdani et al. 2023).

10 EO Effects on Ruminants' Fermentation and Production

EOs and their bioactive metabolites have been reported to exert various biological effects, when given to animals with the diet, such as antimicrobial, anti-inflammatory, antioxidant, and antiparasitic effects, whereas they are also known to modulate ruminal fermentation in ruminants (Pessoa et al. 2002; Evangelista et al. 2022; Akram et al. 2021; Zhang et al. 2022; Štrbac et al. 2022). The effect of dietary EO supplementation has been mostly studied on monogastric animals, while recent research has been developed in ruminants, with the majority of them conducted in vitro.

Many studies both in vitro and in vivo have revealed that EOs and their metabolites, due to their antimicrobial activity, manly inhibiting methanogenic archaea responsible for methane production, can affect ruminal microbiota, thus altering ruminal fermentation process towards better utilization of energy and protein of the feed. Since inefficiency in rumen fermentation leads to energy and protein losses in the form of methane and ammonia gas production, dietary supplementation of EOs could decrease in methane and ammonia emissions that would favor both for the animals and the environment.

EOs possess a significant influence on protein metabolism and reduce ammonia production by inhibiting the deamination of amino acids (AAs), possibly through the suppression of hyperammonia-producing (HAP) bacteria at the level of adhesion and colonization (Benchaar and Greathead 2011; McIntosh et al. 2003). Several EOs (oregano, cinnamon, eucalyptus, rosemary, clove oil, garlic oil, and peppermint oil) have already been tried in vitro and in vivo in animals to reduce methane and ammonia production (Baraz et al. 2018; Cobellis et al. 2015, 2016a, b; Hamdani et al. 2019; Tomkins et al. 2015; Zhou et al. 2020; Patra and Yu 2012). These effects are highly dependent on EO type, source, applied dose, and the mixture of EOs. Effects can also be transient due to an adaptation of the microbiota to survive EOs' antimicrobial effects. Considering the multiple effects of EOs on rumen microbiota, it has been proposed that EOs could improve the feed efficiency and nutrient utilization in ruminants.

In small ruminants, a meta-analysis was conducted by Dorantes-Iturbide et al. (2022) using 74 peer-reviewed publications in the data set so as to evaluate the effects of dietary supplementation with essential oils on productive performance, carcass characteristics, nutrient digestibility, ruminal parameters, serum metabolites, and meat and milk quality. The results of the study revealed that EOs can be used as natural growth promoters in small ruminants, while at the same time, they improve feed intake and feed efficiency, and nutrient digestibility, as well as milk production and quality, meat quality, and shelf-life. Furthermore, dietary supplementation with EOs improves ruminal fermentation and reduces environmental impact by increasing ruminal propionate concentration and by reducing methane emissions, ruminal ammonia nitrogen concentration, and the number of total protozoa and methanogens. However, these results depend on the EO dosage rate and type and duration of animal feeding.

The effects of inclusion of various EOs in ruminants' diets on animal performance and health have also been well reviewed by Akram et al. (2021). In sheep, dietary inclusion of essential oil from thyme, clove, orange peel, rosemary, garlic, or EO mixtures increased rumen fermentation and N metabolism, decreased ruminal ammonia nitrogen, improved protein digestibility and total digestibility of nutrients, increased milk yield and milk fat, and improved animal antioxidant status. However, no effects on dry matter intake, average growth rate, and performance were observed. In goats, *Callistemon viminalis* oil supplementation improved dry matter intake, nutrient digestibility, and nitrogen utilization, whereas no effects were observed when given rosemary EOs, other than increased immunity status. Juniper oil addition in goats' diet, although did not affect feed intake and performance, increased feed efficiency and improved their antioxidant status.

In bovines, inclusion of EOs of oregano, thyme, and coriander or blends and mixtures of EOs increased nutrients digestibility, nitrogen utilization, and microbial protein synthesis and decreased methane production, as well as ammonium nitrogen. Moreover, it improved milk yield in cows and growth rate in calves, as well as feed efficiency. Positive effects were also observed in animal health status regarding immunity improvement, amelioration of calves' diarrhea incidents, and improvement of udder health and immunity (Akram et al. 2021).

However, other in vivo studies on ruminants have shown controversial effects. Chaves et al. (2008) observed no change in ruminal ammonia concentration in lambs supplemented with cinnamaldehyde or carvacrol. Likewise, Yang et al. (2010a, b) observed no change in ruminal ammonia concentration in feedlot cattle supplemented with various levels of either eugenol or cinnamaldehyde. Feeding cinnamon oil, or its metabolite cinnamaldehyde, or eugenol to dairy cows had no effect on ruminal protein degradation and ammonia concentration (Benchaar et al. 2008, 2012, 2015; Benchaar 2016), whereas Tekippe et al. (2011) reported that pulse dosing 500 g/day of oregano leaves (90.8% carvacrol of total essential oil) in the rumen of dairy cows increased ruminal ammonia concentration. Thus, although results may suggest some tendencies and beneficial effects, inconsistencies regarding the efficacy of EOs, which are probably linked to the nature of compounds and other factors, such as infection, nutritional status, environment, and diet composition, must be evaluated with caution.

11 Health-Promoting Effects

Meta-analysis of Dorantes-Iturbide et al. (2022) revealed that dietary supplementation with EOs reduces oxidative stress in small ruminants, as indicated by the increase in total antioxidant capacity (TAC), catalase (CAT), and sodium dismutase (SOD) activity. According to authors, EOs reduce reactive oxygen species (ROS) in blood serum due to the negative correlation between TAC and blood, as well as due to the reported effect that CAT, SOD, and glutathione peroxidase (GPx) antioxidant enzymes can reduce oxidative stress because they convert ROS into other compounds less harmful to biological macromolecules in the organism.

Regarding EO control of pathogens, according to Benchaar et al. (2008), EOs and its main constituents can inhibit several foodborne pathogens including *E. coli* O157:H7, *S. aureus*, *Listeria monocytogenes,* and *Salmonella* spp. EOs that exhibit the highest antimicrobial activity against ruminal bacteria are also generally most potent against pathogens. However, the broad-spectrum antimicrobial activity of these compounds may make it difficult to use them to specifically target pathogens within the ruminant digestive tract and even more may also increase the sensitivity of bacteria to other antimicrobials. EOs from peppermint were shown to exhibit activity against *Giardia* (Vidal et al. 2007), a protozoan parasite that is highly prevalent in cattle (Olson et al. 2004). It is possible that EOs, or their active components, may also have activity against other parasites that reside in the intestine such as *Cryptosporidium*, coccidia or nematodes. Effects of EOs on parasites in the lower digestive tract would be dependent on the ability of the antimicrobial components that they contain to remain active after passage through the rumen. The extent to which EOs escape from the rumen and flow to the lower digestive tract is under investigation. The effects of EOs as anthelmintics are related to the interaction of these compounds with the structure of the parasite. This occurs when lipophilic compounds, such as essential oil constituents, can break or damage the cell membrane of the parasite, thus affecting membrane permeability and leading to some enzyme and nutrient losses. It is also possible that these EOs inhibit cell growth and differentiation, a very rapid process of worm egg embryogenesis (Ramdani et al. 2023).

Although the conventional therapy against mastitis still includes treatment with antibiotics, their use is under criticism, due to the development of antimicrobial resistance, as shown for *Streptococcus agalactiae* and *S. aureus*, causing antibiotic resistance in humans, as well. An alternative complementary approach could be EO use. In vitro studies have demonstrated the antibacterial activity of EOs against several pathogens that can cause mastitis, such as, among the others, *L. monocytogenes*, *Salmonella typhimurium*, *E.coli*, and *S. aureus* (Burt 2004). Their natural origin, coupled with lower side effects and the limited development of resistance after prolonged use, has identified EOs as promising therapeutic agents against mastitis, both in vitro and in vivo (Kher et al. 2019). Limited research has been conducted in vivo to study the effectiveness of EOs in treating mastitis. The intramammary infusion and external application on the quarter of 10% of a mixture of *Thymus vulgaris* and *Lavandula angustifolia* caused a substantial decrease in the bacterial colony count, after four days of treatment. The most potent antibacterial activity was achieved by massaging the udder with the mixture of EOs (Abboud et al. 2015). Another study was conducted to evaluate the effect of *Origanum vulgare* by intra-mammary infusion of 0.9 mL of EO applied twice a day for three days. *S. aureus* and *E. coli* were not detected in milk after the treatment (Chao et al. 2015). The intramammary infusion of sage EO to ewes affected with subclinical mastitis resulted in a significant decrease in somatic cell count 24 and 48 h post-treatment. On the other hand, Lefevre et al. treated 55 cases of mastitis with a mixture of EOs containing *Thymus vulgaris, Rosmarinus verbenone,* and *Laurus nobilis* (1.5% each in 10 mL of sunflower oil), and 45 mastitis with a combination of

Thymus vulgaris and *Rosmarinus verbenone* (6% of each in sunflower oil or water). The results showed that the recovery rate was only 40% (Kammerer et al. 2009).

12 Toxin Binders and Modifiers

Toxin binders or mycotoxin binders and modifiers refer to feed additives added to monogastric feeds or ruminant rations that are able to bind or modify harmful substances, mostly mycotoxins or endotoxins (lipopolysaccharides), within animals' gastrointestinal tracts.

13 Mycotoxins and Their Effect on Ruminants

Mycotoxins are secondary metabolic products from molds belonging to the *Aspergillus*, *Penicillium*, and *Fusarium* genera. Mycotoxin contamination represents a worldwide problem for various agricultural commodities both pre- and post-harvest. More than 300 secondary metabolites have been identified although only around 30 have true toxic properties, which are of some concern. Two groups of toxinogenic (mycotoxin-producing) fungi can be distinguished. The first one consists of fungi (such as *Fusarium*), which invade their substrate and produce mycotoxins on growing plants before harvesting: this is the category of field (pre-harvest) toxins. *Aflatoxins* and *Fusarium* toxins are included in this group. The other group contains fungi, which produce toxins after harvesting and during crop storage and transportation. These toxins are named storage (or post-harvest) toxins and ochratoxin A belongs to this group. The mycotoxins that are of some concern for the safety of animal feed in the European Union include aflatoxin B1 (AFB1), ergot sclerotia, zearalenone (ZEN), deoxynivalenol (DON), ochratoxin A (OTA), fumonisins (especially fumonisin B1, FB1), T-2 and HT-2 toxins, the maximum levels of which have been recommended by European Union (Commission recommendation 2006/576/EC3) (Boudergue et al. 2009).

Since ruminant diets depend mostly on cereals, protein feeds, their by-products as well as hay and grass, grass/legume, and silages, the risk of mycotoxin exposure is higher compared to swine and poultry that have less varied diets. Evidence suggests the greatest exposure for ruminants to some regulated mycotoxins (aflatoxins, trichothecenes, ochratoxin A, fumonisins, and zearalenone) and to many other secondary metabolites (Gallo et al. 2015).

When livestock ingest one or more mycotoxins, the effect on health could be acute, meaning evident signs of disease are present or even causing death. However, acute manifestation of mycotoxicosis is rare under farm conditions. The effects of mycotoxin ingestion are mainly chronic, implying hidden disorders with reduced ingestion, productivity, and fertility. Ruminants are considered less susceptible to mycotoxins than monogastrics. This is based on the assumption that the rumen microbiota degrades and inactivates mycotoxins and is regarded to be the first line of defense against mycotoxins such as ZEN, OTA, and T-2 toxin, thus protecting the animal. However, a number of mycotoxins resist rumen degradation, such as AFB1, fumonisin, and patulin, causing distinct clinical signs of intoxication. Moreover, due to their complex diet, dairy cows may be exposed to a varying number of mycotoxins, originating from different feed materials such as roughage and concentrates. Exposure to these complex mixtures of mycotoxins may result in unexpected health risks (Gallo et al. 2015; Fink-Gremmels 2008; Phuong 2015). Effects of mycotoxins on animals are diverse and range from carcinogenicity, hepatoxicity, and neurotoxicity to impaired reproduction, digestive disorders, changes in animal growth, feed intake reduction or feed refusal, alteration in nutrient absorption and metabolism, decrease in productive performance, increased risk of mastitis, and effects on the endocrine system, as well as inflammatory and immunological suppression, all of which impair animal health status and wellbeing. Clinical signs can be seen at high levels of mycotoxin contamination but more frequently we observe subclinical effects. Already moderate levels of mycotoxins, especially during chronic exposure, can negatively affect animals. Mycotoxins influence the immune system, the integrity of the gut barrier, and act as predisposing factors for disease. Hepatic enzymes are typically increased and prothrombin time can be prolonged. Particular attention should be given to young animals, as their rumen is not yet fully developed, so they are more susceptible to liver damage caused by mycotoxins. Next to preruminating calves, cows in the transition period are particularly susceptible, as the negative energy balance is exacerbated by the presence of molds and mycotoxins in their diet. In both cases, typically, these animals are also given the lowest quality feedstuffs in the farm, as they are not in production (Gallo et al. 2015; Fink-Gremmels 2008; Mostrom and Jacobsen 2011; Rodrigues 2014).

The presence of mycotoxins in feeds could pose a risk to consumers if these toxins and/or their metabolites are excreted and accumulated in animal products such as milk and meat, and for this reason their presence especially in milk must be regularly monitored. The majority of mycotoxins found in feeds do not cause problems either because they are not excreted into milk or because they are excreted as a less toxic metabolite. However, aflatoxin M1 (AFM1) is most prevalent in feedstuffs and recognized as the most hazardous due to its genotoxic and carcinogenic potential. Ruminants, after consuming feed contaminated with AFB1, transformed AFB1 mainly into AFM1, which is then excreted into milk. The amount of AFM1 found in milk represents normally 1–2% of the ingested AFB1. However, it can reach 6% in high-producing cows. AFM1 is toxic and carcinogenic and it is classified by the International Agency of Research on Cancer as class 2B, possible human carcinogens. Because of this concern, the AFM1 level in milk is regulated in several countries; the European Union limit of 0.05 g/kg is one of the lowest in the world (EFSA 2004).

14 Definition of Toxin Binders and Modifiers

The Commission regulation (EC) No 386/2009 defines these groups of feed additives as "substances for reduction of the contamination of feed by mycotoxins: substances that can suppress or reduce the absorption or promote the excretion of mycotoxins or modify their mode of action.". These feed additives may act by reducing the bioavailability of the mycotoxins or by degrading them or transforming them into less toxic metabolites (Boudergue et al. 2009). Therefore, according to their mode of action, these feed additives are classified to adsorbing agents (binders) and biotransforming agents (modifiers).

Toxin-absorbing agents are compounds that have the capacity to bind the mycotoxins in contaminated feed without dissociating in the gastrointestinal tract of the animal. In this way the toxin-adsorbing agent complex passes through the animal and is eliminated via feces. This prevents or minimizes exposure of animals to mycotoxins. Mycotoxin-adsorbing agents can be silica-based inorganic compounds or carbon-based organic polymers (Boudergue et al. 2009).

The inorganic adsorbing agents include natural clay products as well as synthetic polymers. These agents include aluminosilicates, the largest class of mycotoxin-sequestering agents. Within this group, there are two important subclasses: the phyllosilicate subclass and the tectosilicate subclass. The phyllosilicate subclass includes bentonites, montmorillonites, smectites, kaolinites, and illites and the tectosilicate subclass includes zeolites. Bentonites are generally impure clay consisting mostly of montmorillonite. Montmorillonite is a layered silicate, which adsorbs organic substances either on its external surfaces or within its interlaminar spaces, whereas the modified montmorillonite nanocomposite (MMN) is a new absorptive additive, developed with nanomodification techniques. Zeolites have large pores that provide space for large cations such as sodium, potassium, and calcium of which clinoptilolite is a natural zeolite whose main application is the adsorption of heavy metals from aqueous solutions. Moreover, the hydrated sodium calcium aluminosilicate (HSCAS) is perhaps the most studied mycotoxin-sequestering agent among the mineral clays. It is a naturally occurring and heat-processed calcium montmorillonite that is commonly used as an anticaking additive in animal feed. Other categories of toxins binders are activated carbon, yeast cell walls (derived from *Saccharomyces cerevisiae*), which are also used as a dietary mycotoxin-adsorbing agent, bacteria (strains of lactic acid bacteria such as *Lactobacillus rhamnosus* strain GG and *Lactobacillus rhamnosus* strain LC-705 that are used to remove mycotoxins), micronized fibers, obtained from different plant materials such as cereals (wheat, barley, and oat), pea hulls, apple, and bamboo. They are constituted mainly of cellulose, hemicelluloses and lignin, polymers (cholestyramine and polyvinylpyrrolidone), and additives resulting from the modification and activation of diatomaceous earth, which is a natural material extracted from a quarry with a maximum of 70% silicon dioxide (Boudergue et al. 2009).

Biotransforming agents (mycotoxin modifiers), such as bacteria/fungi or enzymes, are another strategy to control mycotoxicosis in animals (Boudergue et al. 2009). These agents biodegrade or biotransform mycotoxins into less toxic metabolites. They can be divided into four classes: bacteria, yeasts, fungi, and enzymes (Boudergue et al. 2009). They act in the intestinal tract of animals prior to the absorption of mycotoxins (Devreese et al. 2013).

15 Effects of Dietary Toxin Binders and Modifiers on Ruminants

Various studies on ruminants have been contacted to evaluate the effects of toxin binders on animals. Most of them consider bentonite and zeolite clay effects on performance and health of both sheep and cows. According to studies these clays increase feed intake, nutrient digestibility, protein and carbohydrate metabolism and digestion, rumen pH, and volatile fatty acids in the rumen, whereas decrease ammonia concentration (Mehany and Hegazy 2020; Saleh et al. 1999; Salem et al. 2001; Colling et al. 1979; May and Barker 1988; Osman and Soliman 2021; Marrero et al. 1987; Ivan et al. 2001; Nikkhah et al. 2000; Khachlouf et al. 2018; Mohsen et al. 2017). Feed conversion ratio improved in most cases of clay inclusion in the feed of lactating cows and sheep, as well as of lambs and calves (Mehany and Hegazy 2020; Mohsen et al. 2017; Nowar et al. 1993). On the other hand, Buldo and Fehring (2004) found no effect of dietary inclusion of zeolites on feed efficiency. Moreover, the addition of clays in ruminants' diet could increase milk yield and in some cases improve milk chemical composition (Mehany and Hegazy 2020; Salem et al. 2001; Khachlouf et al. 2018; Abd El-Baki et al. 2001; Ella 2007; Bampidis et al. 2014; Osman and Soliman 2021; Pooyan et al. 2020).

Moreover, Pooyan et al. (2020) have shown that dietary bi-partial toxin binder improved the health of lactating cows, when the feed was exposed to multiple mycotoxin contamination, as suggested by the lower rectal temperature, respiratory rate, and heartbeat. Subacute ruminal acidosis (SARA) is a serious metabolic disease of cows as a result of consuming diets high in grain and other highly fermentable carbohydrates that although increases milk production, it increases the risk of subacute ruminal acidosis. Even without clinical symptoms, this metabolic disorder is dangerous and can have many detrimental effects. SARA is defined as periods of moderately depressed ruminal pH, from about 5.5 to 5.0. SARA may be associated with laminitis and other health problems resulting in decreased production (Krause and Oetzel 2006). Supplementation with toxin bidders has been proved to protect animals from this metabolic syndrome and alleviate symptoms. Clay mineral-based supplementation of cattle has been proved to have beneficial effects on liver function, even during stress conditions of subacute ruminal acidosis (Humer and Kröger 2019). According to Lean (2007) and Sulzburger (2016), clay supplementation as an additive to cereal grains for lactating cows buffers rumen pH and reduces the time of rumen pH below 5.6 after grain feeding. Cows fed clay tended to have higher milk yield and higher 3.5% fat-corrected milk (FCM), which confirmed that clays could alleviate symptoms of gastrointestinal stress caused by changes in pH levels. Also, Valpotic and Gracner (2017) concluded that zeolite used as a feed additive improved health and economical parameters in cattle production.

Several binders/adsorbents have been proved to be effective on mycotoxins in terms of reducing the carryover of AFM1 in milk and the effect on animal health. Summing up the results of various studies in dairy ruminants, it could be concluded that the inclusion in the feed of several types of clay binders (bentonite, aluminosilicate, montmorillonite, vermiculite, and nontronite) with or without live yeast, yeast culture, has the potential to reduced transfer of dietary AFB1 as AFM1 in milk, when animal consumed AFB1-contaminated feeds (Pietri et al. 2009; Kissell et al. 2013; Kutz et al. 2009; Xiong et al. 2015; Jiang et al. 2018; Sulzberger et al. 2017; Mugerwa et al. 2015).

However, it seems that the efficacy of these adsorbents to bind aflatoxin (AF) can be affected by the ratio of adsorbent to mycotoxin, pH, and temperature. Sumantri et al. (2012) reported no effect of inclusion of bentonite in the diet of cows fed with AFB1-contaminated feed. Ogunade et al. (2016), despite no reduction in AFM1 concentration after feeding a diet containing sodium bentonite and *S. cerevisiae* fermentation product, reported a reduction of the time required to reduce AFM1 in milk to safe levels after AFB1-contaminated feed withdrawal.

In contrast to inorganic and organic binders, studies on in vivo efficacy of modifiers or detoxifiers (i.e., probiotics and enzymes) are scarce and are limited in broilers. The application of probiotics and enzymes in mycotoxin mitigation in farm animals is limited because microbes/enzymes are mostly active in their own environment. Moreover, it is very difficult to evaluate the mechanism of biotransformation and the identification of metabolites and their toxicities (Boudergue et al. 2009). Nevertheless, they offer an alternative environmentally friendly strategy for minimizing livestock exposure to aflatoxins, to be in practice.

16 Enzymes

Exogenous fibrolytic enzymes (EFEs) have been widely and successfully applied in monogastric animal nutrition, yet for ruminants' production, exogenous enzyme supplementation in diets has recently gained acknowledgment and challenged the scientific community to further study enzymes as feed additives in ruminants' nutrition. Fiber digestibility is limited in the rumen ecosystem, and applying EFEs in ruminant diets can potentially improve cell wall digestion and the efficiency of feed utilization (Reddy et al. 2016).

EFEs used in ruminants' nutrition can be categorized into three primary categories based on the specific substrates on which their enzyme activity occurs: fibrolytic, amylolytic, and proteolytic enzymes (Durge et al. 2022). Exogenous fibrolytic enzymes can be further classified according to their ability to act as cellulase, which hydrolyze the fiber of plant cell wall to glucose, cellobiose, or cellooligosaccharides with simultaneous action of three enzymes: endoglucanases, exoglucanases, and β-glucosidases (Zhang and Lynd 2004). EFEs may hydrolyze feed directly, and act synergistically with ruminal microbes to enhance feed digestion (McAllister et al. 2001), indirectly boost the attachment and numbers of cellobiose- and glucose-utilizing bacteria in the rumen (Nsereko et al. 2002), and work synergistically with the large intestine microflora (Beauchemin et al. 2004).

The most common EFEs, used in ruminants' nutrition are xylanase (X), cellulase-xylanase (C-X), ferulic acid esterase (FAE), cellulase-FAE (C-FAE), endoglucanase-xylanase (En-X), and amylase-exogenous proteolytic enzyme (A-EPE), and some nonfibrolytic activity EFE such as amylase and protease or bacteria. Ruminant enzyme additives are derived primarily from four bacterial (*Bacillus subtilis, Lactobacillus acidophilus, Lactobacillus plantarum*, and *Streptococcus faecium*) and three fungal (*Aspergillus oryzae, Trichoderma reesei*, and *Saccharomyces cerevisiae*) species (McAllister et al. 2001)

A meta-analysis on the impact of dietary application of exogenous fibrolytic enzymes on the performance of dairy cows (Arriola et al. 2017) showed that application of EFEs did not impact dry matter intake and feed efficiency but had a tendency to enhance the total-tract dry matter digestibility and neutral detergent fiber digestibility (NDFD) by quite small amounts (1.36 and 2.30%, respectively, or <0.31 standard deviation units). Also, application of EFEs increased yields of milk (0.83 kg/day), 3.5% fat-corrected milk (0.55 kg/day), milk protein (0.03 kg/day), and milk lactose (0.05 kg/day) by moderate to small amounts (<0.30 standard deviation units). The authors also suggest that there is a superiority of the C-X enzymes as for their beneficial effects, compared to the rest enzymes examined. Another related meta-analysis (Tirado-González Tirado-González et al. 2018) suggests that although EFEs may enhance the performance parameters (average daily gain, feed conversion, feed intake, in vivo dry matter, and digestibility in low-forage (F:C < 50%) grass-based diets) of dairy cows and beef cattle, the reaction varies according to the appropriate dosage mixture of cellulases and xylanases to the diet. There are unknown interactions between supplements of exogenous fibrolytic enzymes (EFEs) and cell walls of feedstuff in ruminal conditions. A commercial product, Viscozyme®, containing a GH74 endoglucanase, GH71 α-1,3-glucanase, GH5 mannanase, GH7 cellobiohydrolase, GH28 pectinase, and esterases, enhanced ruminal saccharification of barley straw (Badhan et al. 2018).

In a cycle of in vitro fermentation tests, Eun and Beauchemin (2007) found that four feed enzyme products (FE), endoglucanase, exoglucanase, xylanase, and protease, that varied in enzymatic activities on the degradation of alfalfa hay and corn silage may increase in vitro neutral detergent fiber (NDF) digestibility and might be capable of improving lactational performance of dairy cows. When lactating cows were provided with an exogenous phytase plus cellulase formulation, a reduction of the fecal nutrient excretion was observed, as well as increase of the fiber and dry matter (DM) digestibility (Knowlton et al. 2007). A recombinant fibrolytic enzyme (RFE; xylanase XYL10C) selected specifically for forage-fed ruminants on ruminal pH did not improve fiber digestion, but increased ruminal NH_3–N concentration, which might suggest that it increased ruminal proteolytic activity (Ran et al. 2019).

Understanding the complexity of the rumen microbial ecosystem and the nature of its interactions with plant cell walls is crucial to exogenous enzyme usage for improving plant cell wall digestibility and thus the efficiency of feed utilization in ruminants. The inconsistency of the results might be attributed to the variability of the enzyme formulations available, their different mode of action, the synergism between exogenous enzyme and rumen microflora, the supplementation dosage, the suitable mode and time of delivery, and the substrate feed specificity, as well as the host itself (for example, stage of lactation of dairy cows).

17 Phytogenics

Phytogenics, also referred to as phytobiotics, are defined as natural bioactive compounds that are extracted from various plants and incorporated into animal feed as plant feed additives (PFAs), alternative to the controversial chemical additives. They have been considered to have positive effects on animal growth and health due to their antimicrobial, antiviral, antifungal, and antioxidative properties. They have been used traditionally as complementary or alternative medicines to improve human health or cure human diseases (Gadde et al. 2017; Yang et al. 2015).

This category of natural additives includes essential oils (EOs), saponins, flavonoids, and tannins as the main biologically active compounds obtained from various parts of plants (Bodas et al. 2012; Salem et al. 2014; Ebeid et al. 2020). These bioactive compounds are referred to as phytochemicals or plant secondary metabolites (PSMs). Except essential oils that have already been discussed previously, there are numerous plant extracts that can be fed to livestock reared under both conventional and organic conditions (Hrelia and Angeloni 2020; Bampidis and Robinson 2006; Brenes et al. 2016). Furthermore, these PFAs are considered natural, harmless, and residue-free; hence, they could meet both consumer expectations and the need of the feed industry to find valid alternatives to synthetic agents.

Several PFAs such as flavonoids (e.g., polyphenolics), phenolic acids, terpenoids (e.g., terpenes), alkaloids, stilbenoids, epicatechin, hesperidin, and quercetin have been recognized in various plants, which have been demonstrated to exert antioxidant, anti-inflammatory, antiparasitic activities and improve rumen fermentation (Tohge et al. 2013).

Flavonoids are polyphenolic compounds and, although nearly 5000 flavonoids have been identified, the most significant ones are divided into 12 subclasses and include the following: anthocyanidins, aurone, chalcone, coumarin, dihydrochalcone, dihydroflavonol, flavan-3,4-diol, flavan-3-ol, flavanones, flavones, flavonols, and isoflavones (Hahlbrock 1981; Harborne 1984). Tannins are an important group of flavonoids (polyphenolics compounds) in terrestrial plants, broadly categorized into two major compounds, condensed tannins (CTs) and hydrolyzable tannins (HTs). CTs have been used extensively in ruminants due to their abilities to decrease methane (CH_4) emissions, shift protein digestion from the rumen to the small intestine, improve the maternal environment and reproductive efficiency (i.e., ovulation, scanning, pregnancy, and fecundity rates), support early embryonic survival, enhance embryo and fetal development, lambing rates, and lamb survival from birth to weaning, among other numerous applications. Similar to this, the ability of CT to lower gastrointestinal parasite burdens is particularly intriguing given growing worries about pharmaceutical antiparasitic resistance in grazing ruminants as a result of their ongoing, sometimes unnecessary, treatment with the macrocyclic lactone ivermectin (Tedeschi et al. 2021).

Due to their propensity to be broken down in the rumen and ingested by ruminants, hydrolyzable tannins have frequently been thought of as potentially harmful, antinutritional plant secondary metabolites (Tedeschi et al. 2021). However, recently studies have reported that HTs exert beneficial effects on ruminants. The inclusion in cattle diet of HTs from various sources (chestnut, tannic acid, and gallic acid) decreases environmental impact of ruminants (lower CH_4 and ammonia emissions), without decreasing animal performance (Aboagye et al. 2019). Moreover, HT extracted from chestnut could be considered as an alternative for parasitic control in cattle, since it was found that it causes a decrement in motility with the subsequent death of larvae 3 (L_3) of *Haemonchus contortus* (Acevedo-Ramírez et al. 2020).

The antiparasitic effect of tannins relies on reducing gastrointestinal nematodes by increasing the host response to parasites. The capability of tannins to bind to proteins is able to protect them from rumen degradation and improve protein flow and amino acid absorption in the small intestine. Increased protein supply in the small intestine has been seen to enhance host homeostasis and immune response to helminths (Ramdani et al. 2023).

Many in vitro studies have indicated the potential of tannins to exert antioxidant effects. The in vivo antioxidant activities of tannins were also demonstrated in different animal tissues. Inclusion of forage containing CT improved the antioxidant status of both cattle and sheep by increasing serum antioxidant activity. Quebracho tannins in lamb diets improved the antioxidant status of muscle, liver, and plasma (Huang et al. 2018). López-Andrés et al. (2013) found that quebracho tannins were not degraded or absorbed in the gastrointestinal tract but increased the antioxidant capacity of liver and plasma in sheep, which demonstrated that CT may indirectly affect antioxidant status in animal tissues. It has been speculated that dietary tannins may spare other nutritive antioxidants during the digestive process, or they may protect proteins, carbohydrates, and lipids in the digestive tract from oxidative damage during digestion. However, the antioxidant mechanism of tannins in animal tissues is unknown (Huang et al. 2018).

Approximately 30,000 terpenoids compounds have been identified. Among these, saponins are classified into triterpene or steroidal glycosides, which are found in many plant species, having many different biophysicochemical properties. Most studies in ruminants have focused only on saponins' nutritional aspects to prevent digestive disorders when used as feed additives. Plant extract is the typical form adopted to achieve the desired phytochemical compound, and five saponin-rich plants have been consistently examined: *Camellia sinensis* L., *Quillaja saponaria*, and *Sapindus rarak* DC.—lerak and *Sapindus saponaria* L.—soapberry with a triterpenoid structure; and *Yucca schidigera* with a steroidal nucleus (Tedeschi et al. 2021). According to Patra and Saxena (2009), saponins appear to have a major antiprotozoal (defaunation) action in the rumen, which may improve the efficiency of microbial protein production and protein flow to the duodenum. Moreover, saponins may reduce methane generation directly by lowering methanogen populations and activities (such as the rate of methanogenesis or expression of methane-producing genes). The rumen metabolism may be altered favorably or unfavorably by saponins' selective effects on particular rumen bacteria and fungus. While saponins' ammonia adsorption and modification of digesta transit in the rumen have also been linked to altered rumen metabolism, their physiological effects are probably insignificant in comparison to microbiological ones. It has been discovered that saponins' effects on rumen fermentation are inconsistent. These differences seem to be connected to the chemical makeup and dosage of saponins, diets, and microbial populations.

Depending on the diets and the saponins involved, the antiprotozoal properties of saponins would be anticipated to have a mixed influence on ruminant productivity. It is commonly accepted that eliminating or suppressing protozoa will improve ruminant performances, especially when fed a low-protein diet, despite the possibility that defaunation would reduce fiber digestibility in the rumen. The effect of defaunation depends on the balance between the energy and protein requirements of animals and nutrients supplied by diets. Therefore, the impact of saponins is influenced by feed composition (Patra and Saxena 2009). Furthermore, Rochfort et al. (2008) in their review have indicated the immunostimulant and antivirus potential of saponins.

18 Feed Additives as a Tool Towards Ruminant Welfare

Animal welfare has been defined in different ways, but a concise approach that encompasses all aspects is that it characterizes animals that are healthy, while having their needs met. This definition emphasizes the significance of maintaining good health and fulfilling the physical and emotional needs of animals in order to attain high standards of welfare. In a society that is increasingly demanding ethical and sustainable livestock production practices, with consumers who are rejecting products made under unacceptable conditions, it is essential to integrate animal welfare into sustainable livestock production practices.

Any effort to increase the efficiency of production must prioritize animal welfare, along with other measures of sustainability. While there are various strategies to improve animal health, implement novel management practices and feeding protocols, decrease greenhouse emissions and highlight product quality in ruminant production, some of these strategies may come at the expense of animal welfare. Therefore, it is crucial to adopt approaches that not only promote sustainable productivity but also enhance animal welfare, such as feed supplementation and improving animal health (Llonch et al. 2017). Within that concept, the inclusion of the abovementioned feed additives in ruminant nutrition has been proven to be a functional dietary tool towards ruminant welfare boost (Fig. 1).

The gut microbiota is an essential factor for animal welfare, although there is still limited evidence of its direct influence on ruminants. However, probiotics are

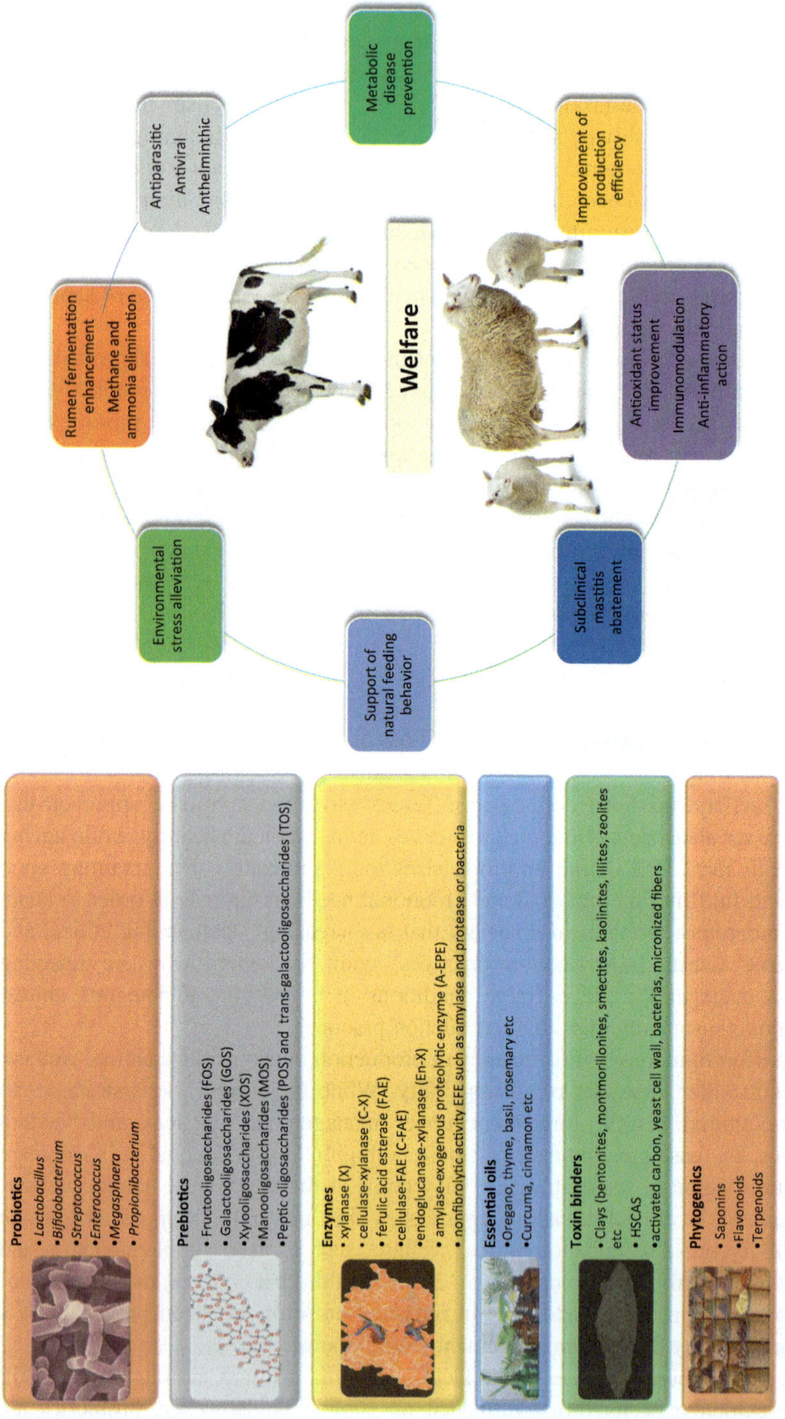

Fig. 1 Schematic illustration of feed additive effects on ruminants

frequently used to manipulate the microbiota in order to improve productivity by impacting physiological processes. These processes can affect behavioral functions related to animal welfare, such as stress, anxiety, and changes in memory, cognition, and social behavior.

In ruminants that suffer from acidosis, the microbiota-gut-brain axis is suspected to modify feeding behavior, through rumen microbiota affecting impact on appetite during pathological states. In cattle with subacute acidosis, ruminal microbiota is altered, leading to a drop in feed intake and rumination time. The probiotic *Saccharomyces cerevisiae*, commonly used in ruminants, has been shown to protect against physiological changes induced by acidosis and also instigate behavioral changes, such as reducing the minimum interval between meals and increasing time spent ruminating. While there is limited information about the impact of the microbiota-gut-brain axis on behavior in ruminants, it indicates that it could have significant effects on their welfare status that are not yet fully recognized. The influence of the gut microbiota on behavior can be added to its influence on health through its role in the immune response, making it a crucial factor in achieving animal welfare.

A possible alternative approach for manipulating the microbiota is to incorporate prebiotics into the diet. Research involving germ-free animals, as well as interventions such as the provision of pro- or prebiotics, and alterations of their diet, has shown that conversions to the gut microbiota (GUT-M) are associated with changes in various behavioral patterns that affect animal welfare directly in a rather distinct manner. Providing prebiotic supplementation can be beneficial before or during stressful events, such as handling or transport of ruminants, to prevent activation of the hypothalamic-pituitary-adrenal axis and associated anxiety-like behaviors that depress the welfare level of animals (Kraimi et al. 2019).

The host-protecting function of most prebiotics in ruminants can be inhibited due to their degradation in the rumen, yet their activity towards promoting gut health especially, during the stressful period of weaning in young animals, must be noted (Gaggìa et al. 2010). Keeping welfare at satisfying levels during this time is a significant assistance towards calves, lambs, and kids for reaching their future full productive potential (Cangiano et al. 2020). Apart from their beneficial effects on health and growth, pro- and prebiotic prophylactic action is correlated with a reduction in mortality and morbidity in ruminants, which are two of the main markers of welfare (Ortiz-Pelaez et al. 2008; Radzikowski 2017).

By promoting the growth of beneficial bacteria in the gut, eubiotics such as probiotics, prebiotics, and essential oils can help improve nutrient absorption, reduce inflammation, and support immune function. This, in turn, can help reduce the incidence of gastrointestinal disorders and infections, thereby reducing the need for antibiotics in cattle, sheep, and goats (Gupta et al. 2019; Pulina et al. 2017).

In addition to their potential to improve rumen fermentation, increase feed digestibility, and enhance animal performance, essential oils have also been shown to reduce methane and ammonia production (Arteaga-Wences et al. 2021). These effects may be attributed to the antimicrobial and antioxidant properties of essential oils, which can improve gut health and nutrient absorption (Cobellis et al. 2016a, b). Results on methane production reduction can be characterized as varied, depending on the type and dosage of essential oil used, as well as the composition of the animal's diet (Belanche et al. 2020; Torres et al. 2020; Giannenas et al. 2013).

Apart from nutrient absorption enhancement, essential oils appear to also act as appetizers and flavoring agents, improving diet acceptance and preference for ruminants, thus promoting the manifestation of natural feeding behavior and respecting animal welfare (Durmic and Blache 2012). Some essential oils exhibit a significant antibacterial effect against bacteria isolated from small ruminant mastitis, showing promising therapeutic properties towards subclinical and clinical mastitis and a path towards related stress alleviation (Abdalhamed et al. 2018).

Within the scope of alleviating climate change effects on ruminants, the supplementation of dairy cattle rations with toxin binders such as bentonite and zeolite has shown promising results against heat stress. Through the incorporation of toxin binders in their nutrition, heat-stressed cows maintain their feed intake and reproductive parameters at optimal level, thus safeguarding their welfare (Devreese et al. 2013; Kihal et al. 2022; Mehany and Hegazy 2020).

Moreover, the use of toxin binders towards the reduction of aflatoxins in milk and mycotoxins in animal feed, both as regulatory and preventive measures, can ensure animal health and nutrition, further translating into impacts relevant to ruminant welfare (Katsoulos et al. 2016; Mantovani et al. 2022).

Improving the efficiency of ruminant production is essential for its future success, particularly in light of the price volatility that often affects both dairy and meat markets. Rather than focusing on increasing milk production, it may be beneficial to explore alternative approaches that enable the production of the same amount of milk using fewer cows. This could significantly increase herd efficiency, farm profitability, and animal health and welfare.

A promising approach for improving ruminant production efficiency is the use of feed enzymes, which can enhance feed degradation and maximize starch digestibility. By increasing the availability of energy and nutrients from feed, feed enzymes can help to match the requirements of high-yielding dairy cows and fast-growing beef cattle, leading to increased milk production and growth rates without increasing the number of cows or total feed consumption. Improved feed efficiency using feed enzymes can help reduce feed costs and improve profitability for farmers, while also reducing the environmental impact of ruminant production by reducing greenhouse gas emissions and nutrient excretion. The utilization of essential oil compounds can provide further advantages such as enhancing the taste and increasing feed intake, promoting the secretion of digestive enzymes, and providing antioxidant, antimicrobial, antifungal, and antiparasitic effects. Overall, feed enzymes are a promising ecofriendly tool for improving animal welfare, efficiency, and sustainability of ruminant production (Bonneau and Laarveld 1999; McGrath et al. 2018).

Regarding the wider group of phytogenics or phytochemicals, it is widely accepted and supported by research that herbal bioactive compounds have been found effective in improving ruminant health and productivity, especially in stressful situations that compromise animal welfare. Supplementing lambs under heat stress with tannin or naringin can enhance performance, antioxidant status, and immune responses. The same herbal extracts have been reported to improve feed intake and mitigate the adverse effects of heat stress on performance in transition dairy cows and cows suffering from subclinical mastitis. In heat-stressed cows, phytogenics and phytochemicals reduce body core temperature and modulate heat stress conditions of lactating dairy cows, as evidenced by lower rectal temperature and respiration rate due to the induction of oxidative stress (Flachowsky and Lebzien 2012; Hashemzadeh et al. 2022; Karásková et al. 2015).

Sheep feeding behavior can also be modulated through administration of phytochemical mixtures. In particular, animals appear to be devoting more time to eating and therefore having less available time to exhibit abnormal behaviors and oral nonnutritive behaviors, which are indicators of poor welfare (Ahmed et al. 2021).

19 Limitations, Prospects, and Conclusions

Any diet supplementation of feed additives in ruminants' nutrition towards promoting health, welfare, and production could yield profits. Studies have clearly demonstrated that dietary interventions applied in neonatal ruminants, rather than in later development stages, can effectively manipulate gut microbiota and subsequently boost the metabolism and the immune response. Weaning period is similarly very critical for the lasting effects on rumen microbiota establishment. A better understanding of host genetics and physiology, the molecular mechanisms associated with microbial populations, and the interactions between them would better elucidate the mechanism of action of feed additives on gastrointestinal development and microbial colonization. Current research on the host specificity of feed additives is still considered limited. Apparently, these dietary supplements have potential to develop and improve in terms of their host specificity and effectiveness. Yet again, metabolic and toxicological studies will be required for evaluating their favorable consequences for the animals in ruminant farming and consumers.

As an increasing public awareness trend, high welfare standards in animal husbandry practices are persistently demanded by animal scientists, veterinarians, farmers, and consumers. Animal welfare is conventionally associated with animal health and production, though there is also an association of animal welfare with other fundamental aspects of animal production, such as nutrition, and especially feed additives, which are reported to have the capacity to improve welfare status of ruminants. So, several critical aspects of animal welfare, directly or indirectly, affect production efficiency and may have a considerable impact on the profitability of ruminant industry, which constantly pursue and implement alternative nutritional strategies, which comply with consumer desire for animal products that are produced in an ethical approach.

References

Abboud M, El Rammouz R, Jammal B, Sleiman M (2015) In vitro and in vivo antimicrobial activity of two essential oils Thymus vulgaris and Lavandula angustifolia against bovine Staphylococcus and Streptococcus mastitis pathogen. Middle East J Agric Res 4:975–983

Abd El-Baki SM, Soliman ES, Youssed NA (2001) Effect of Tafla clay on performance of lactating cows fed different levels of concentrate feed mixture with sulphuric acid-urea treated rice straw. Egypt J Nutr Feeds 4(special issue):337–347

Abdalhamed AM, Zeedan GSG, Abou Zeina HAA (2018) Isolation and identification of bacteria causing mastitis in small ruminants and their susceptibility to antibiotics, honey, essential oils, and plant extracts. Vet World 11(3):355

Aboagye IA, Oba M, Koenig KM, Zhao GY, Beauchemin KA (2019) Use of gallic acid and hydrolyzable tannins to reduce methane emission and nitrogen excretion in beef cattle fed a diet containing alfalfa silage1,2. J Anim Sci 97(5):2230–2244. https://doi.org/10.1093/jas/skz101

Acevedo-Ramírez PMC, Hallal-Calleros C, Flores-Pérez I, Alba-Hurtado F, Mendoza-Garfias MB, Barajas R (2020) Nematicidal effect and histological modifications induced by hydrolysable tannin extract on the third-stage infective larvae of *Haemonchus contortus*. Biology 9(12):442. https://doi.org/10.3390/biology9120442

Ahmed E, Batbekh B, Fukuma N, Kand D, Hanada M, Nishida T (2021) A garlic and citrus extract: impacts on behavior, feed intake, rumen fermentation, and digestibility in sheep. Anim Feed Sci Technol 278:115007

Aikman PC, Henning PH, Humphries DJ, Horn CH (2011) Rumen pH and fermentation characteristics in dairy cows supplemented with *Megasphaera elsdenii* NCIMB 41125 in early lactation. J Dairy Sci 94(6):2840–2849. https://doi.org/10.3168/jds.2010-3783

Akram M, Asghar M, Jalal H (2021) Essential oils as alternatives to chemical feed additives for maximizing livestock production. J Hellenic Vet Med Soc 72(1):2595–2610. https://doi.org/10.12681/jhvms.26741

Alves Costa, N., Pansani, A. P., de Castro, C. H., Basile Colugnati, D., Xaxier, C. H., Guimarães, K. C., et al (2019). Milk restriction or oligosaccharide supplementation in calves improves compensatory gain and digestive tract development without changing hormone levels. PLoS One, 14(3), e0214626.

Alawneh JI, James AS, Phillips N, Fraser B, Jury K, Soust M, Olchowy TW (2020) Efficacy of a Lactobacillus-based teat spray on udder health in lactating dairy cows. Frontiers in veterinary science 7:584436

AlZahal O, Dionissopoulos L, Laarman AH, Walker N, McBride BW (2014a) Active dry *Saccharomyces cerevisiae* can alleviate the effect of subacute ruminal acidosis in lactating dairy cows. J Dairy Sci. 97(12):7751–7763. https://doi.org/10.3168/jds.2014-8212

AlZahal O, McGill H, Kleinberg A, Holliday JI, Hindrichsen IK, Duffield TF et al (2014b) Use of a direct-fed microbial product as a supplement during the transition period in dairy cattle. J Dairy Sci. 97(11):7102–7114. https://doi.org/10.3168/jds.2014-8248

Amin AB, Mao S (2021) Influence of yeast on rumen fermentation, growth performance and quality of products in ruminants: a review. Anim Nutr 7(1):31–41

Anee IJ, Alam S, Begum RA et al (2021) The role of probiotics on animal health and nutrition. J Basic Appl Zool 82:52. https://doi.org/10.1186/s41936-021-00250-x

Apás AL, Dupraz J, Ross R, González SN, Arena ME (2010) Probiotic administration effect on fecal mutagenicity and microflora in the goat's gut. J Biosci Bioeng. https://doi.org/10.1016/j.jbiosc.2010.06.005

Arriola KG, Oliveira AS, Ma ZX, Lean IJ, Giurcanu MC, Adesogan AT (2017) A meta-analysis on the effect of dietary application of exogenous fibrolytic enzymes on the performance of dairy cows. J Dairy Sci 100(6):4513–4527

Arteaga-Wences YJ, Estrada-Angulo A, Ríos-Rincón FG, Castro-Pérez BI, Mendoza-Cortéz DA, Manriquez-Núñez OM et al (2021) The effects of feeding a standardized mixture of essential

oils vs monensin on growth performance, dietary energy and carcass characteristics of lambs fed a high-energy finishing diet. Small Ruminant Res 205:106557
Badhan A, Ribeiro GO Jr, Jones DR, Wang Y, Abbott DW, Di Falco M et al (2018) Identification of novel enzymes to enhance the ruminal digestion of barley straw. Bioresour Technol 260:76–84
Bakkali F, Averbeck S, Averbeck D, Idaomar M (2008) Biological effects of essential oils – a review. Food Chem Toxicol 46(2):446–475. https://doi.org/10.1016/j.fct.2007.09.106
Bampidis VA, Robinson PH (2006) Citrus by-products as ruminant feeds: a review. Anim Feed Sci Technol 128:175–217
Bampidis VA, Christodoulou V, Theophilou N, Kotsampasi V (2014) Effect of dietary palygorskite on performance and blood parameters of lactating Holstein cows. Appl Clay Sci 91–92:25–29. https://doi.org/10.1016/j.clay.2014.02.006
Ban Y, Guan LL (2021) Implication and challenges of direct-fed microbial supplementation to improve ruminant production and health. J Anim Sci Biotechnol 12:109. https://doi.org/10.1186/s40104-021-00630-x
Baraz H, Jahani-Azizabadi H, Azizi O (2018) Simultaneous use of thyme essential oil and disodium fumarate can improve in vitro ruminal microbial fermentation characteristics. Vet Res Forum 9:193–198
Beauchemin KA, Colombatto D, Morgavi DP, Yang WZ, Rode LM (2004) Mode of action of exogenous cell wall degrading enzymes for ruminants. Can J Anim Sci 84:13–22
Belanche A, Newbold CJ, Morgavi DP, Bach A, Zweifel B, Yáñez-Ruiz DR (2020) A meta-analysis describing the effects of the essential oils blend agolin ruminant on performance, rumen fermentation and methane emissions in dairy cows. Animals 10(4):620
Benchaar C (2016) Diet supplementation with cinnamon oil, cinnamaldehyde, or monensin does not reduce enteric methane production of dairy cows. Animal 10:418–425. https://doi.org/10.1017/S175173111500230X
Benchaar C, Greathead H (2011) Essential oils and opportunities to mitigate enteric methane emissions from ruminants. Anim Feed Sci Technol 166:338–355
Benchaar C, Calsamiglia S, Chaves AV, Fraser GR, Colombatto D, McAllister TA, Beauchemin KA (2008) A review of plant-derived essential oils in ruminant nutrition and production. Anim Feed Sci Technol. 145:209–228
Benchaar C, Lettat A, Hassanat F, Yang WZ, Forster RJ, Petit HV, Chouinard PY (2012) Eugenol for dairy cows fed low or high concentrate diets: effects on digestion, ruminal fermentation characteristics, rumen microbial populations and milk fatty acid profile. Anim Feed Sci Technol 178(3–4):139–150. https://doi.org/10.1016/j.anifeedsci.2012.10.005
Benchaar C, Hassanat F, Petit HV (2015) Dose–response to eugenol supplementation to dairy cow diets: methane production, N excretion, ruminal fermentation, nutrient digestibility, milk production, and milk fatty acid profile. Anim Feed Sci Technol 209:51–59. https://doi.org/10.1016/j.anifeedsci.2015.07.027
Bodas R, Prieto N, Garcia-Gonzalez R et al (2012) Manipulation of rumen fermentation and methane production with plant secondary metabolites. Anim Feed Sci Technol 176:78–93. https://doi.org/10.1016/j.anifeedsci.2012.07.010
Bonneau M, Laarveld B (1999) Biotechnology in animal nutrition, physiology and health. Livest Prod Sci 59(2–3):223–241
Boudergue C, Burel C, Dragacci S, Favrot, MC, Fremy, JM, Massimi, C., ... & Avantaggiato, G. (2009) Review of mycotoxin-detoxifying agents used as feed additives: mode of action, efficacy and feed/food safety. EFSA Supporting Publications 6(9):22E
Brenes A, Viveros A, Chamorro S, Arija I (2016) Use of polyphenol-rich grape by-products in monogastric nutrition. A review. Anim Feed Sci Technol 211:1–17
Broadway PR, Carroll JA, Sanchez NCB (2015) Live yeast and yeast cell wall supplements enhance immune function and performance in food-producing livestock: a review. Microorganisms 3(3):417–427. https://doi.org/10.3390/microorganisms3030417
Buldo RG, Fehring A (2004) Effect of addition a natural zeolite on productivity of dairy calves after weaning. Archivos-de-Medicinaeterinaria 36(2):197–201

Buntyn JO, Schmidt TB, Nisbet DJ, Callaway TR (2016) The role of direct-fed microbials in conventional livestock production. Annu Rev Anim Biosci 4:335–355
Burdick Sanchez NC, Carroll JA, Broadway PR, McBride ML, Ortiz XA, Collier JL et al (2015) OmniGen-AF alters rectal temperature (RT) and leukocyte profiles in dairy cows exposed to heat stress (HS) following acute activation of the stress axis. In: Proceeding of the Joint Annual Meeting of the American Society of Animal Science, Orlando, FL, USA
Burt S (2004) Essential oils: Their antibacterial properties and potential applications in foods—a review. Int J Food Microbiol 94:223–253. https://doi.org/10.1016/j.ijfoodmicro.2004.03.022
Cangiano LR, Yohe TT, Steele MA, Renaud DL (2020) Invited review: strategic use of microbial-based probiotics and prebiotics in dairy calf rearing. Appl Anim Sci 36(5):630–651
Cao LT, Wu JQ, Xie F, Hu SH, Mo Y (2007) Efficacy of nisin in treatment of clinical mastitis in lactating dairy cows. J Dairy Sci. https://doi.org/10.3168/jds.2007-0153
Chang M, Wang F, Ma F, Jin Y, Sun P (2022) Supplementation with galacto-oligosaccharides in early life persistently facilitates the microbial colonization of the rumen and promotes growth of preweaning Holstein dairy calves. Anim Nutr 10:223–233
Chaucheyras-Durand F, Walker ND, Bach A (2008) Effects of active dry yeasts on the rumen microbial ecosystem: past, present and future. Anim Feed Sci Technol. https://doi.org/10.1016/j.anifeedsci.2007.04.019
Chao G, Bao G, Cao Y, Yan W, Wang Y, Zhang X et al (2015) Prevalence and diversity of enterotoxin genes with genetic background of Staphylococcus aureus isolates from different origins in China. Int J Food Microbiol 211:142–147
Chaves AV, Stanford K, Gibson LL, McAllister TA, Benchaar C (2008) Effects of carvacrol and cinnamaldehyde on intake, rumen fermentation, growth performance, and carcass characteristics of growing lambs. Anim Feed Sci Technol 145:396–408
Chen YY, Wang YL, Wang WK, Zhang ZW, Si XM, Cao ZJ, Li SL, Yang HJ (2020) Beneficial effect of *Rhodopseudomonas palustris* on in vitro rumen digestion and fermentation. Benef Microbes. https://doi.org/10.3920/BM2019.0044
Chiquette J (2009) The role of probiotics in promoting dairy production. WCDS Adv Dairy Technol. 21:143–157
Chiquette J, Allison MJ, Rasmussen MA (2008) *Prevotella bryantii* 25A used as a probiotic in early-lactation dairy cows: effect on ruminal fermentation characteristics, milk production, and milk composition. J Dairy Sci. https://doi.org/10.3168/jds.2007-0849
Cobellis G, Petrozzi A, Forte C, Acuti G, Orrù M, Marcotullio MC, Aquino A, Nicolini A, Mazza V, Trabalza-Marinucci M (2015) Evaluation of the effects of mitigation on methane and ammonia production by using Origanum vulgare L. and Rosmarinus officinalis L. essential oils on in vitro rumen fermentation systems. Sustainability 7:12856–12869
Cobellis G, Trabalza-Marinucci M, Marcotullio MC, Yu Z (2016a) Evaluation of different essential oils in modulating methane and ammonia production, rumen fermentation, and rumen bacteria in vitro. Anim Feed Sci Technol 215:25–36
Cobellis G, Trabalza-Marinucci M, Yu Z (2016b) Critical evaluation of essential oils as rumen modifiers in ruminant nutrition: a review. Sci Total Environ 545:556–568
Colling DP, Britton RA, Farlin SD, Nielson MK (1979) Effect of adding sodium bentonite to high grain diets for ruminants. J Anim Sci 48(3):641
Costantini L, Molinari R, Farinon B, Merendino N (2017) Impact of omega-3 fatty acids on the gut microbiota. Int J Mol Sci 18:2645–2662
da Silva JT, Bittar CMM, Ferreira LS (2012) Evaluation of mannan-oligosaccharides offered in milk replacers or calf starters and their effect on performance and rumen development of dairy calves. Rev Bras Zootec. 41:746–752
Devreese M, De Backer P, Croubels S (2013) Different methods to counteract mycotoxin production and its impact on animal health. Vlaams Diergeneeskundig Tijdschrift 82(4)
Dorantes-Iturbide G, Orzuna-Orzuna JF, Lara-Bueno A, Mendoza-Martínez GD, Miranda-Romero LA, Lee-Rangel HA (2022) Essential oils as a dietary additive for small ruminants: a meta-

analysis on performance, rumen parameters, serum metabolites, and product quality. Vet Sci 2022(9):475. https://doi.org/10.3390/vetsci9090475

Du R, Jiao S, Dai Y, An J, Lv J, Yan X, Wang J, Han B (2018) Probiotic *Bacillus amyloliquefaciens* C-1 improves growth performance, stimulates GH/IGF-1, and regulates the gut microbiota of growth-retarded beef calves. Front Microbiol. https://doi.org/10.3389/fmicb.2018.02006

Durge A, Tiwari J, Nair PM, Singh AK, Yadav SK, Chouraddi R et al (2022) A review on the role of exogenous fibrolytic enzymes in ruminant nutrition. Curr J Appl Sci Technol 41(36):45–58

Durmic Z, Blache D (2012) Bioactive plants and plant products: Effects on animal function, health and welfare. Anim Feed Sci Technol 176(1–4):150–162

Ebeid HM, Kholif AE, Chrenkova M, Anele UY (2020) Ruminal fermentation kinetics of Moringa oleifera leaf and seed as protein feeds in dairy cow diets: in sacco degradability and protein and fiber fractions assessed by the CNCPS method. Agrofor Syst 94:905–915. https://doi.org/10.1007/s10457-019-00456-7

EFSA Journal (2004) Opinion of the Scientific Panel on Contaminants in the Food Chain on a request from the Commission related to Aflatoxin B1 as undesirable substance in animal feed (Request N° EFSA-Q-2003-035). EFSA J 39:1–27

Eicher SD, Patterson JA, Rostagno MH (2011) β-Glucan plus ascorbic acid in neonatal calves modulates immune functions with and without Salmonella enterica serovar Dublin. Vet Immunol Immunopathol 142(3–4):258–264

Elghandour MMY, Tan ZL, Abu Hafsa SH, Adegbeye MJ, Greiner R, Ugbogu EA, Cedillo Monroy J, Salem AZM (2020) *Saccharomyces cerevisiae* as a probiotic feed additive to non and pseudo-ruminant feeding: a review. J Appl Microbiol. https://doi.org/10.1111/jam.14416

Elghandour MM, Abu Hafsa SH, Cone JW, Salem AZ, Anele UY, Alcala-Canto Y (2022) Prospect of yeast probiotic inclusion enhances livestock feeds utilization and performance: an overview. Biomass Convers Biorefinery 2022:1–13. https://doi.org/10.1007/s13399-022-02562-6

Ella AAA (2007) Effect of dry yeast and/or bentonite as feed additives on the productive performance of lactating ewes and its offsprings. Egypt J Nutr Feeds. 10(1):81–98

Eun JS, Beauchemin KA (2007) Enhancing in vitro degradation of alfalfa hay and corn silage using feed enzymes. J Dairy Sci 90(6):2839–2851

Evangelista AG, Corrêa JAF, Pinto ACSM, Luciano FB (2022) The impact of essential oils on antibiotic use in animal production regarding antimicrobial resistance—a review. Crit Rev Food Sci Nutr 62:5267–5283. https://doi.org/10.1080/10408398.2021.1883548

FDA (2004) Food and Drug Administration of the US, 21 CFR 184. Online, Available. at: http://www.cfsan.fda.gov/eafus.html

Finck DN, Ribeiro FRB, Burdick NC, Parr SL, Carroll JA, Young TR et al (2014) Yeast supplementation alters the performance and health status of receiving cattle. Prof Anim Sci 30(3):333–341

Fink-Gremmels J (2008) The role of mycotoxins in the health and performance of dairy cows. Vet J 176(1):84–92

Flachowsky G, Lebzien P (2012) Effects of phytogenic substances on rumen fermentation and methane emissions: a proposal for a research process. Anim Feed Sci Technol 176(1-4):70–77

Froehlich KA, Abdelsalam KW, Chase C, Koppien-Fox J, Casper DP (2017) Evaluation of essential oils and prebiotics for newborn dairy calves. J Anim Sci. 95:3772–3782

Gadde U, Kim WH, Oh ST, Lillehoj HS (2017) Alternatives to antibiotics for maximizing growth performance and feed efficiency in poultry: a review. Anim Health Res Rev 18:26–45

Gaggìa F, Mattarelli P, Biavati B (2010) Probiotics and prebiotics in animal feeding for safe food production. Int J Food Microbiol 141:S15–S28

Galvão KN, Santos JE, Coscioni A, Villaseñor M, Sischo WM, Berge ACB (2005) Effect of feeding live yeast products to calves with failure of passive transfer on performance and patterns of antibiotic resistance in fecal Escherichia coli. Reprod Nutr Dev 45(4):427–440

Gao J, Liu YC, Wang Y, Li H, Wang XM, Wu Y et al (2020) Impact of yeast and lactic acid bacteria on mastitis and milk microbiota composition of dairy cows. AMB Exp 10:22–12. https://doi.org/10.1186/s13568-020-0953-8

Gallo A, Giuberti G, Frisvad JC, Bertuzzi T, Nielsen KF (2015) Review on mycotoxin issues in ruminants: Occurrence in forages, effects of mycotoxin ingestion on health status and animal performance and practical strategies to counteract their negative effects. Toxins 7(8):3057–3111

Ghosh S, Mehla RK (2012) Influence of dietary supplementation of prebiotics (mannanoligosaccharide) on the performance of crossbred calves. Trop Anim Health Prod 44: 617–622

Giannenas I, Bonos E, Christaki E, Florou-Paneri P (2013) Essential oils and their applications in animal nutrition. Med Aromat Plants 2(140):2167–0412

Gibson GR, Roberfroid MB (1995) Dietary modulation of the human colonic microbiota: introducing the concept of prebiotics. J Nutr 125(6):1401–1412

Grand E, Respondek F, Martineau C, Detilleux J, Bertrand G (2013) Effects of short-chain fructooligosaccharides on growth performance of preruminant veal calves. J Dairy Sci 96(2): 1094–1101

Gupta RC, Srivastava A, Lall R (eds) (2019) Nutraceuticals in veterinary medicine, vol 120. Springer, Cham, Switzerland

Hahlbrock K (1981) Flavonoids. In: Stumpf PK, Conn EE (eds) The Biochemistry of plants: a comprehensive treatise; secondary plant products. Academic Press, San Diego, CA, pp 425–456

Hamdani H, Chami N, Oukhouia M, Jabeur I, Sennouni C, Remmal A (2019) Effect of a thymol-based additive on rumen fermentation, on methane emissions in eructed gas and on milk production in Holstein cows. Livest Res Rural Dev 31

Harborne JB (1984) Phytochemical methods: a guide to modern techniques of plant analysis. Springer, New York, NY

Hashemzadeh F, Rafeie F, Hadipour A, Rezadoust MH (2022) Supplementing a phytogenic-rich herbal mixture to heat-stressed lambs: growth performance, carcass yield, and muscle and liver antioxidant status. Small Rumin Res 206:106596

He ZX, Ferlisi B, Eckert E, Brown HE, Aguilar A, Steele MA (2017) Supplementing a yeast probiotic to pre-weaning Holstein calves: feed intake, growth and fecal biomarkers of gut health. Anim Feed Sci Technol. 226:81–87

Heinrichs AJ, Jones CM, Heinrichs BS (2003) Effects of mannan oligosaccharide or antibiotics in neonatal diets on health and growth of dairy calves. J Dairy Sci 86:4064–4069

Heinrichs AJ, Heinrichs BS, Jones CM (2013) Fecal and saliva IgA secretion when feeding a concentrated mannan oligosaccharide to neonatal dairy calves. Prof Anim Sci 29:457–462

Hill SR, Hopkins BA, Davidson S, Bolt SM, Diaz DE, Brownie C, Brown T, Huntington GB, Whitlow LW (2009) The addition of cottonseed hulls to the starter and supplementation of live yeast or mannanoligosaccharide in the milk for young calves. J Dairy Sci 92(2):790–798

Hrelia S, Angeloni C (2020) New mechanisms of action of natural antioxidants in health and disease. Antioxidants 9:344

Huang Q, Liu X, Zhao G, Hu T, Wang Y (2018) Potential and challenges of tannins as an alternative to in-feed antibiotics for farm animal production. Anim Nutr 4(2):137–150

Humer E, Kröger I (2019) Supplementation of a clay mineral-based product modulates plasma metabolomic profile and liver enzymes in cattle fed grain-rich diets. Animal 13:1214–1223

Ivan ML, Neill RA, Jalaludin S (2001) Efficacy of bentonite on rumen fermentation and duodenal low of dietary components in sheep fed palm kernel cake by-product. Anim Feed Sci Technol 92(20):127–135

Jean B (2008) Pharmacognosy phytochemistry, medicinal plants, 2nd edn. Lavoisier, Paris, France, p 934

Jiang Y, Ogunade IM, Qi S, Hackmann TJ, Staples CR, Adesogan AT (2017) Effects of the dose and viability of *Saccharomyces cerevisiae*. 1. Diversity of ruminal microbes as analyzed by Illumina MiSeq sequencing and quantitative PCR. J Dairy Sci. 100(1):325–342. https://doi.org/10.3168/jds.2016-11263

Jiang Y, Ogunade IM, Kim DH, Li X, Pech-Cervantes AA, Arriola KG, Oliveira AS, Driver JP, Ferraretto LF, Staples CR et al (2018) Effect of adding clay with or without a Saccharomyces

cerevisiae fermentation product on the health and performance of lactating dairy cows challenged with dietary aflatoxin B1. J Dairy Sci. 101:3008–3020
Jouany JP, Mathieu F, Senaud J, Bohatier J, Bertin G, Mercier M (1998) Effect of Saccharomyces cerevisiae and Aspergillus oryzae on the digestion of nitrogen in the rumen of defaunated and refaunated sheep. Anim Feed Sci Technol. https://doi.org/10.1016/S0377-8401(98)00194-1
Kammerer M, Le Guenic M, Roussel P, Linclau O, Cartaud G, Tainturier D, Larrat M, Bareille N (2009) Le traitement des mammites cliniques de la vache laitière par des huiles essentielles. Innov Agron. 4:79–83
Karásková K, Suchý P, Straková E (2015) Current use of phytogenic feed additives in animal nutrition: a review. Czech J Anim Sci 60(12):521–530
Katsoulos PD, Karatzia MA, Boscos C, Wolf P, Karatzias H (2016) In-field evaluation of clinoptilolite feeding efficacy on the reduction of milk aflatoxin M1 concentration in dairy cattle. J Anim Sci Technol 58(1):1–7
Kelsey AJ, Colpoys JD (2018) Effects of dietary probiotics on beef cattle performance and stress. J Vet Behav. 27:8–14. https://doi.org/10.1016/j.jveb.2018.05.010
Khachlouf K, Hamed H, Gdoura R, Gargouri A (2018) Effects of zeolite supplementation on dairy cow production and ruminal parameters: a review. Ann Anim Sci 18:857–877
Khan RU, Shabana N, Kuldeep D, Karthik K, Ruchi T, Mutassim MA (2016) Direct-fed microbial: beneficial applications, modes of action and prospects as a safe tool for enhancing ruminant production and safeguarding health. Int J Pharm. 12:220–231
Kher MN, Sheth NR, Bhatt VD (2019) In vitro antibacterial evaluation of terminalia chebula as an alternative of antibiotics against bovine subclinical mastitis. Anim Biotechnol. 30:151–158
Kihal A, Marquès C, Rodríguez-Prado M, Jose-Cunilleras E, Calsamiglia S (2022) Effect of diet supplementation with the mycotoxin binder montmorillonite on the bioavailability of vitamins in dairy cows. Toxins 14(1):26
Kim HS, Whon TW, Sung H, Jeong YS, Jung ES, Shin NR et al (2021) Longitudinal evaluation of fecal microbiota transplantation for ameliorating calf diarrhea and improving growth performance. Nat Commun 12(1):161
Kirankumar N, Naresh Kumar M, Harmeet Singh D, Santosh R, Namita R, Matthias H, Anil KP (2022) Impact of probiotics on dairy production efficiency. Front Microbiol 13. https://doi.org/10.3389/fmicb.2022.805963
Kissell L, Davidson S, Hopkins BA, Smith GW, Whitlow LW (2013) Effect of experimental feed additives on aflatoxin in milk of dairy cows fed aflatoxin-contaminated diets. J Anim Physiol Anim Nutr (Berl) 97:694–700
Knowlton KF, Taylor MS, Hill SR, Cobb C, Wilson KF (2007) Manure nutrient excretion by lactating cows fed exogenous phytase and cellulase. J Dairy Sci 90(9):4356–4360
Kraimi N, Dawkins M, Gebhardt-Henrich SG, Velge P, Rychlik I, Volf J (2019) Influence of the microbiota-gut-brain axis on behavior and welfare in farm animals: a review. Physiol Behav. 210:112658. https://doi.org/10.1016/j.physbeh.2019.112658
Krause KM, Oetzel GR (2006) Understanding and preventing subacute ruminal acidosis in dairy herds: a review. Anim Feed Sci Technol 126(3–4):215–236
Krehbiel C, Rust S, Zhang G, Gilliland S (2003) Bacterial direct-fed microbials in ruminant diets: performance response and mode of action. J Anim Sci. https://doi.org/10.2527/2003.8114_suppl_2E120x
Kritas SK, Govaris A, Christodoulopoulos G, Burriel AR (2006) Effect of Bacillus licheniformis and Bacillus subtilis supplementation of ewe's feed on sheep milk production and young lamb mortality. J Veterinary Med Ser A 53(4):170–173
Kumprechtová D, Illek J, Julien C, Homolka P, Jančík F, Auclair E (2019) Effect of live yeast (*Saccharomyces cerevisiae*) supplementation on rumen fermentation and metabolic profile of dairy cows in early lactation. J Anim Physiol Anim Nutr 103(2):447–455. https://doi.org/10.1111/jpn.13048

Kutz RE, Sampson JD, Pompeu LB, Ledoux DR, Spain JN, Vazquez-Anon M, Rottinghaus GE (2009) Efficacy of soils, NovasilPlus, and MTB-100 to reduce Aflatoxin M1 levels in milk of early to mid-lactation dairy cows fed Aflatoxin B1. J Dairy Sci. 92:3959–3963

Lean S (2007) Ruminal acidosis—aetiopathogenesis, prevention and treatment. Aust Vet Assoc 61: 7–43

Li GH, Ling BM, Qu MR, You JM, Song XZ (2011) Effects of several oligosaccharides on ruminal fermentation in sheep: an in vitro experiment. Rev Med Vet 162(4):192–197

Li Z, Bai H, Zheng L, Jiang H, Cui H, Cao Y, Yao J (2018) Bioactive polysaccharides and oligosaccharides as possible feed additives to manipulate rumen fermentation in Rusitec fermenters. Int J Biol Macromol 109:1088–1094

Llonch P, Haskell MJ, Dewhurst RJ, Turner SP (2017) Current available strategies to mitigate greenhouse gas emissions in livestock systems: an animal welfare perspective. Animal *11*(2): 274–284

López-Andrés P, Luciano G, Vasta V, Gibson TM, Biondi L, Priolo A, Mueller-Harvey I (2013) Dietary quebracho tannins are not absorbed, but increase the antioxidant capacity of liver and plasma in sheep. Br J Nutr 110(4):632–639

Ma ZZ, Cheng YY, Wang SQ, Ge JZ, Shi HP, Kou JC (2020) Positive effects of dietary supplementation of three probiotics on milk yield, milk composition and intestinal flora in Sannan dairy goats varied in kind of probiotics. J Anim Physiol Anim Nutr 104(1):44–55

Maamouri O, Ben Salem M (2021) The effect of live yeast Saccharomyces cerevisiae as probiotic supply on growth performance, feed intake, ruminal pH and fermentation in fattening calves. Vet Med. Sci. 8:398–404. https://doi.org/10.1002/vms3.631

Mantovani A, Aquilina G, Cubadda F, Marcon F (2022) Risk-benefit assessment of feed additives in the one health perspective. Front Nutr 128

Marden JP, Julien C, Monteils V, Auclair E, Moncoulon R, Bayourthe C (2008) How does live yeast differ from sodium bicarbonate to stabilize ruminal pH in high yielding dairy cows? J Dairy Sci. 91:3528–3535

Markowiak P, Śliżewska K (2017) Effects of probiotics, prebiotics, and symbiotics on human health. Nutrients 9(9):1021

Marrero D, Martinez RO, Rivas JL (1987) Effect of the inclusion of bentonite on the performance and ruminal fermentation of calves fed wheat. Cuban J Agric Sci 21(3):253–258

Mathieu F, Jouany JP, Sénaud J, Bohatier J, Bertin G, Mercier M (1996) The effect of *Saccharomyces cerevisiae* and *Aspergillus oryzae* on fermentations in the rumen of faunated and defaunated sheep; protozoal and probiotic interactions. Reprod Nutr Dev. https://doi.org/10.1051/rnd:19960305

Mavrommatis A, Mitsiopoulou C, Christodoulou C, Karabinas D, Nenov V, Zervas G et al (2020) Dietary supplementation of a live yeast product on dairy sheep milk performance, oxidative and immune status in peripartum period. J Fungi 6(4):334. https://doi.org/10.3390/jof6040334

May PJ, Barker DJ (1988) Sodium bentonite in high grain diets for young cattle. Proc Austr Soc Anim Prod 17:439

Mazon G, Campler MR, Holcomb C, Bewley JM, Costa JHC (2020) Effects of a *Megasphaera elsdenii* oral drench on reticulorumen pH dynamics in lactating dairy cows under subacute ruminal acidosis challenge. Anim Feed Sci Tech. 261:114404. https://doi.org/10.1016/j.anifeedsci.2020.114404

McAllister TA, Hristov AN, Beauchemin KA, Rode LM, Cheng K-J (2001) Enzymes in ruminants diets. In: Bedford MR, Partridge GG (eds) Enzymes in farm animal nutrition. CABI, Wiltshire, UK, pp 273–229

McGrath J, Duval SM, Tamassia LF, Kindermann M, Stemmler RT, de Gouvea VN et al (2018) Nutritional strategies in ruminants: a lifetime approach. Res Vet Sci 116:28–39

McIntosh FM, Williams P, Losa R, Wallace RJ, Beever DA, Newbold CJ (2003) Effects of essential oils on ruminal microorganisms and their protein metabolism. Appl Environ Microbiol. 69(8): 5011–5014

Mehany AA, Hegazy MM (2020) Effect of toxin binder supplemented to Friesian cows rations on stress states and their reproductive performance under summer climatic conditions in delta region. تأثير مضادات السموم الفطرية المضافة لعلائق الأبقار الفريزيان على حالة الاجهاد الحرارى و الأداء التناسلى لها تحت الظروف المناخية الصيفية لأقليم الدلتا. J Anim Poult Prod 11(10):389–397

Michalak M, Wojnarowski K, Cholewińska P, Szeligowska N, Bawej M, Pacoń J (2021) Selected alternative feed additives used to manipulate the rumen microbiome. Animals 11(6):1542

Mohsen MK, Abdel-raouf EM, Gaafar HM, Mesbah RA (2017) productive performance of early weaning friesian calves supplemented with bee pollen and black seeds. Egyptian Journal of Agricultural Research 95(2):913–930

Moorthy M, Chaiyakunapruk N, Jacob SA, Palanisamy UD (2020) Prebiotic potential of polyphenols, its effect on gut microbiota and anthropometric/clinical markers: a systematic review of randomised controlled trials. Trends Food Sci Technol. 99:634–649

Morrison SJ, Dawson S, Carson AF (2010) The effects of mannan oligosaccharide and Streptococcus faecium addition to milk replacer on calf health and performance. Livest Sci 131(2–3): 292–296

Mostrom MS, Jacobsen BJ (2011) Ruminant mycotoxicosis. Veterinary Clinics: Food Animal Practice 27(2):315–344

Mugerwa S, Kabirizi J, Zziwa E (2015) Effect of supplementing lactating goats fed on aflatoxin contaminated feed with calcium bentonite and activated charcoal on Aflatoxin M1 concentration, excretion and carryover in milk. Uganda J Agric Sci 16:83–92

Muya MC, Nherera FV, Miller KA, Aperce CC, Moshidi PM, Erasmus LJ (2015) Effect of Megasphaera elsdenii NCIMB 41125 dosing on rumen development, volatile fatty acid production and blood β-hydroxybutyrate in neonatal dairy calves. J Anim Physiol Anim Nutr 99(5): 913–918. https://doi.org/10.1111/jpn.12306

Nikkhah AM, Goodarzi, Ashtiania MSRM (2000) Effect of various levels of zeolites on milk yield, milk composition and pH of rumen and frces. Iran J Agric Sci 31(2):221–225

Nocek JE, Kautz WP (2006) Direct-fed microbial supplementation on ruminal digestion, health, and performance of pre- and postpartum dairy cattle. J Dairy Sci. 89(1):260–266. https://doi.org/10.3168/jds.S0022-0302(06)72090-2

Nocek JE, Kautz WP, Leedle JAZ, Block E (2003) Direct-fed microbial supplementation on the performance of dairy cattle during the transition period1. J Dairy Sci. 86(1):331–335. https://doi.org/10.3168/jds.S0022-0302(03)73610-8

Nowar MSK, Al-Shawabkeh, Khour HN (1993) Effect of feeding farm animals with Jordanian clay deposits containing montmorillonite. 1-Effect of fatting lambs' performance, with special reference to blood hematology, liver and kidney functions, and parasitological and serological examinations. Zagz J Agric Res 20(2A):651–667

Nsereko VL, Beauchemin KA, Morgavi DP, Rode LM, Furtado AF, McAllister TA, Iwassa AD, Yang WZ, Wang Y (2002) Effect of fibrolytic enzyme preparation from Trichoderma Longibrachiatum on the rumen microbial population of dairy cows. Can J Microbiol 48:14–20

Oetzel GR, Emery KM, Kautz WP, Nocek JE (2007) Direct-fed microbial supplementation and health and performance of pre- and postpartum dairy cattle: a field trial. J Dairy Sci. 90(4): 2058–2068. https://doi.org/10.3168/jds.2006-484

Ogunade IM, Arriola KG, Jiang Y, Driver JP, Staples CR, Adesogan AT (2016) Effects of 3 sequestering agents on milk Aflatoxin M1 concentration and the performance and immune status of dairy cows fed diets artificially contaminated with Aflatoxin B1. J Dairy Sci. 99:6263–6273

Ogunade IM, Lay J, Andries K, McManus CJ, Bebe F (2019) Effects of live yeast on differential genetic and functional attributes of rumen microbiota in beef cattle. J Anim Sci Biotechnol *10*(1):1–7

Olson ME, O'Handley RM, Ralston BJ, McAllister TA, Thompson RCA (2004) Update on Cryptosporidium and Giardia infections in cattle. Trends Parasitol. 20:185–191

Ortiz-Pelaez A, Pritchard D, Pfeiffer D, Jones E, Honeyman P, Mawdsley J (2008) Calf mortality as a welfare indicator on British cattle farms. Vet J. 176:177–181. https://doi.org/10.1016/j.tvjl.2007.02.006
Osman AA, Soliman SA (2021) Effects of bentonite supplementation on milk yield, milk composition, digestibility and nutritive values in Holstein cows. J Anim Poult Fish Prod 10(1):21–25. https://doi.org/10.21608/japfp.2021.178519
Patra A, Saxena J (2009) The effect and mode of action of saponins on the microbial populations and fermentation in the rumen and ruminant production. Nutr Res Rev *22*(2):204–219. https://doi.org/10.1017/S0954422409990163
Patra AK, Yu Z (2012) Effects of essential oils on methane production and fermentation by, and abundance and diversity of, rumen microbial populations. Appl Environ Microbiol 78:4271–4280
Pessoa LM, Morais SM, Bevilaqua CML, Luciano JHS (2002) Anthelmintic activity of essential oil of Ocimum gratissimum Linn. and eugenol against Haemonchus contortus. Vet Parasitol 109(1–2):59–63
Philippeau C, Lettat A, Martin C, Silberberg M, Morgavi DP, Ferlay A et al (2017) Effects of bacterial direct-fed microbials on ruminal characteristics, methane emission, and milk fatty acid composition in cows fed high- or low-starch diets. J Dairy Sci. 100(4):2637–2650. https://doi.org/10.3168/jds.2016-11663
Phuong NH, Thieu NQ, Ogle B, Pettersson H (2015). Aflatoxins, fumonisins and zearalenone contamination of maize in the southeastern and central highlands provinces of Vietnam. Agriculture, 5(4), 1195–1203
Pietri A, Bertuzzi T, Piva G, Binder EM, Schatzmayr DA, Rodrigues I (2009) Aflatoxin transfer from naturally contaminated feed to milk of dairy cows and the efficacy of a mycotoxin deactivating product. Int J Dairy Sci. 4:34–42
Pinloche E, McEwan N, Marden J-P, Bayourthe C, Auclair E, Newbold CJ (2013) The effects of a probiotic yeast on the bacterial diversity and population structure in the rumen of cattle. PLoS One. 8(7):e67824. https://doi.org/10.1371/journal.pone.0067824
Pooyan H, Towhidi A, Dehghan banadaky, M., Farzaneh, M., Hasan, J. (2020) Effects of toxin binder BIOTOX on productive performance and health status in Holstein cows. Anim Prod 22(2):313–324. https://doi.org/10.22059/jap.2020.286480.623428
Poppy GD, Rabiee AR, Lean IJ, Sanchez WK, Dorton KL, Morley PS (2012) A meta-analysis of the effects of feeding yeast culture produced by anaerobic fermentation of Saccharomyces cerevisiae on milk production of lactating dairy cows. J Dairy Sci *95*(10):6027–6041. https://doi.org/10.3168/jds.2012-5577
Pulina G, Francesconi AHD, Stefanon B, Sevi A, Calamari L, Lacetera N et al (2017) Sustainable ruminant production to help feed the planet. Ital J Anim Sci 16(1):140–171
Radzikowski D (2017) Effect of probiotics, prebiotics and synbiotics on the productivity and health of dairy cows and calves. World Sci News 78:193–198
Ramdani D, Yuniarti E, Jayanegara A, Chaudhry AS (2023) Roles of essential oils, polyphenols, and saponins of medicinal plants as natural additives and anthelmintics in ruminant diets: a systematic review. Animals 13(4):767. https://doi.org/10.3390/ani13040767
Ran T, Saleem AM, Shen Y, Ribeiro GO Jr, Beauchemin KA, Tsang A et al (2019) Effects of a recombinant fibrolytic enzyme on fiber digestion, ruminal fermentation, nitrogen balance, and total tract digestibility of heifers fed a high forage diet. J Anim Sci 97(8):3578–3587
Raza M, Yousaf MS, Ahmad J, Rashid MA, Majeed KA, Tahir SK, Ashraf S, Numan M, Khalid A, ur Rehman H (2022) Prebiotics supplementation modulates pre-weaning stress in male cattle calves by improving growth performance, health scores and serum biomarkers. Czech J Anim Sci 67(3):102–113
Reddy PRK, Raju J, Reddy AN, Ramadevi A, Reddy PPR, Raju J (2016) Recent trends in supplementation of exogenous fibrolytic enzymes in ruminant nutrition: a review. Indian J Nat Sci 7(38):11700–11708

Rochfort S, Parker AJ, Dunshea FR (2008) Plant bioactives for ruminant health and productivity. Phytochemistry 69(2):299–322

Rodrigues I (2014) A review on the effects of mycotoxins in dairy ruminants. Anim Prod Sci 54 (9):1155–1165

Roodposhti PM, Dabiri N (2012) Effects of probiotic and prebiotic on average daily gain, fecal shedding of *Escherichia coli*, and immune system status in newborn female calves. Asia-Austral J Anim Sci. 25:1255–1261

Saleh MSA, El-Raouf EM, Mohsen MK, Salem AY (1999) Bentonites supplementation to concentrate ration for lactating buffaloes. Egypt J Nutr Feed 2:67–78

Salem FAF, El-Amary H, Hassanin SH (2001) Effect of Bentonite supplementation on nutrients digestibility, rumen fermentation. Some blood physiological parameters and performance growing Lambs. Egypt J Nutr Feed 4:179–191

Salem AZM, Kholif AE, Elghangour MMY et al (2014) Effect of increasing levels of seven tree species extracts added to a high concentrate diet on in vitro rumen gas output. Anim Sci J 85: 853–860. https://doi.org/10.1111/asj.12218

Scott KP, Grimaldi R, Cunningham M, Sarbini SR, Wijeyesekera A, Tang MLK et al (2020) Developments in understanding and applying prebiotics in research and practice—an ISAPP conference paper. J Appl Microbiol. 128:934–949

Senevirathne ND, Anderson JL, Metzger L (2019) Growth performance, nutrient utilization, and health of dairy calves supplemented with condensed whey solubles. J Dairy Sci 102(9): 8108–8119

Shaaban HAE, El-Ghorab AH, Shibamoto T (2012) Bioactivity of essential oils and their volatile aroma components: review. J Essent Oil Res. 24:203–212

Signorini ML, Soto LP, Zbrun MV, Sequeira GJ, Rosmini MR, Frizzo LS (2012) Impact of probiotic administration on the health and fecal microbiota of young calves: a meta-analysis of randomized controlled trials of lactic acid bacteria. Res Vet Sci. https://doi.org/10.1016/j.rvsc.2011.05.001

Stella AV, Paratte R, Valnegri L, Cigalino G, Soncini G, Chevaux E et al (2007) Effect of administration of live *Saccharomyces cerevisiae* on milk production, milk composition, blood metabolites, and faecal flora in early lactating dairy goats. Small Ruminant Res. 67(1):7–13. https://doi.org/10.1016/j.smallrumres.2005.08.024

Štrbac F, Bosco A, Maurelli MP, Ratajac R, Stojanović D, Simin N, Orčić D, Pušić I, Krnjajić S, Sotiraki S, Saralli G, Cringoli G, Rinaldi L (2022) Anthelmintic properties of essential oils to control gastrointestinal nematodes in sheep—in vitro and in vivo studies. Vet Sci. 9:93. https://doi.org/10.3390/vetsci9020093

Sulzberger SA, Melnichenko S, Cardoso FC (2017) Effects of clay after an aflatoxin challenge on aflatoxin clearance, milk production, and metabolism of Holstein cows. J Dairy Sci. 100:1856–1869

Sulzburger SA (2016) Effect of clay supplementation on rumen environment, metabolism, inflammation, and performance in dairy cows. Master's Thesis, University of Illinois at Urbana, Champaign, IL, USA, 13 October, 120p

Sumantri I, Murti TW, van der Poel AFB, Boehm J, Agus A (2012) Carry-over of aflatoxin B1-feed into aflatoxin M1-milk in dairy cows treated with natural sources of aflatoxin and bentonite. J Indones Trop Anim Agric 37:271–277

Tedeschi LO, Muir JP, Naumann HD, Norris AB, Ramírez-Restrepo CA, Mertens-Talcott SU (2021) Nutritional aspects of ecologically relevant phytochemicals in ruminant production. Front Vet Sci. 8:628445. https://doi.org/10.3389/fvets.2021.628445

Tekippe JA, Hristov AN, Heyler KS, Cassidy TW, Zheljazkov VD, Ferreira JFS, Karnati SK, Varga GA (2011) Rumen fermentation and production effects of Origanum vulgare L. leaves in lactating dairy cows. J Dairy Sci. 94:5065–5079. https://doi.org/10.3168/jds.2010-4095

Tirado-González DN, Miranda-Romero LA, Ruíz-Flores A, Medina-Cuéllar SE, Ramírez-Valverde R, Tirado-Estrada G (2018) Meta-analysis: effects of exogenous fibrolytic enzymes in ruminant diets. J Appl Anim Res 46(1):771–783

Tohge T, Watanabe M, Hoefgen R, Fernie AR (2013) The evolution of phenylpropanoid metabolism in the green lineage. Crit Rev Biochem Mol Biol. 48:123–152. https://doi.org/10.3109/10409238.2012.758083

Tomkins NW, Denman SE, Pilajun R, Wanapat M, McSweeney CS, Elliott R (2015) Manipulating rumen fermentation and methanogenesis using an essential oil and monensin in beef cattle fed a tropical grass hay. Anim Feed Sci Technol 200:25–34

Torres MG, Soriano R, Peralta JJ, Alejos JI, Sánchez P, Arias L et al (2020) Challenges of livestock: climate change, animal welfare and agroforestry. Large Anim Rev 26(1):39–45

Tóth S, Kovács M, Bóta B, Szabó-Fodor J, Bakos G, Fébel H (2020) Effect of mannanoligosaccharide (MOS) and inulin supplementation on the performance and certain physiological parameters of calves reared on milk replacer. J Appl Anim Res. 48:228–234

Uyeno Y, Shigemori S, Shimosato T (2015) Effect of probiotics/prebiotics on cattle health and productivity. Microbes Environ. https://doi.org/10.1264/jsme2.ME14176

Valpotic H, Gracner D (2017) Zeolite clinoptilolite nanoporous feed additive for animals of veterinary importance: potentials and limitations. Period Biol. 119:159–172

Van de Braak SAAJ, Leijten GCJJ (1999) Essential oils and oleoresins: a survey in the Netherlands and other major markets in the European Union. Rotterdam, CBI, Centre for the Promotion of Imports from Developing Countries, p 116

Vibhute VM, Shelke RR, Chavan SD, Nage SP (2011) Effect of probiotics supplementation on the performance of lactating crossbred cows. Vet World. 4:557–561

Vidal F, Vidal JC, Gadelha APR, Lopes CS, Coelho MGP, Monteiro-Leal LH (2007) Giardia lamblia: the effects of extracts and fractions from Mentha × piperita Lin. (Lamiaceae) on trophozoites. Exp Parastiol. 115:25–31

Wang H, Yu Z, Gao Z, Li Q, Qiu X, Wu F et al (2022) Effects of compound probiotics on growth performance, rumen fermentation, blood parameters, and health status of neonatal Holstein calves. J Dairy Sci 105(3):2190–2200. https://doi.org/10.3168/jds.2021-20721

Weimer PJ (2015) Redundancy, resilience, and host specificity of the ruminal microbiota: implications for engineering improved ruminal fermentations. Front Microbiol. https://doi.org/10.3389/fmicb.2015.00296

Wismar R, Brix S, Frøkiær H, Lærke HN (2010) Dietary fibers as immunoregulatory compounds in health and disease. Ann N Y Acad Sci 1190(1):70–85

Xiong JL, Wang YM, Nennich TD, Li Y, Liu JX (2015) Transfer of dietary aflatoxin B1 to milk aflatoxin M1 and effect of inclusion of adsorbent in the diet of dairy cows. J Dairy Sci. 98:2545–2554

Yang WZ, Ametaj BN, Benchaar C, Beauchemin KA (2010a) Dose response to cinnamaldehyde supplementation in growing beef heifers: ruminal and intestinal digestion. J Anim Sci 88:680–688

Yang WZ, Ametaj BN, He ML, Benchaar C, Beauchemin KA (2010b) Cinnamaldehyde in feedlot cattle diets: intake, growth performance, carcass characteristics, and blood metabolites. J Anim Sci. 88:1082–1092

Yang C, Chowdhury MAK, Huo Y, Gong J (2015) Phytogenic compounds as alternatives to in-feed antibiotics: potentials and challenges in application. Pathogens 4(1):137–156. https://doi.org/10.3390/pathogens4010137

Yoon IK, Stern MD (1995) Influence of direct-fed microbials on ruminal microbial fermentation and performance of ruminants - a review. Asian Austral J Anim. 8(6):533–555. https://doi.org/10.5713/ajas.1995.553

Zhang YHP, Lynd LR (2004) Toward an aggregated understanding of enzymatic hydrolysis of cellulose: noncomplexed cellulase systems. Biotechnol Bioeng 88(7):797–824

Zhang R, Dong X, Zhou M, Tu Y, Zhang N, Deng K, Diao Q (2017) Oral administration of Lactobacillus plantarum and Bacillus subtilis on rumen fermentation and the bacterial community in calves. Anim Sci J 88(5):755–762. https://doi.org/10.1111/asj.12691

Zhang L, Gao F, Ge J, Li H, Xia F, Bai H, Piao X, Shi L (2022) Potential of aromatic plant-derived essential oils for the control of foodborne bacteria and antibiotic resistance in animal production: a review. Antibiotics 11:1673

Zhou R, Wu J, Lang X, Liu L, Casper DP, Wang C, Zhang L, Wei S (2020) Effects of oregano essential oil on in vitro ruminal fermentation, methane production, and ruminal microbial community. J Dairy Sci 103:2303–2314

Application of Feed Additives to Improve Pork Meat Quality

Zhongxin Zhou, Hong-Kui Wei, Chuanshang Cheng, Gene Jin, and Jian Peng

Abstract

Every aspect of pork production has an impact on meat quality, including genetic selection, nutritional control and pre-slaughter stress. Pre-slaughter transport is one of the major stress sources in pigs. Transport stress affects the animal's neuroendocrine system and blood biochemical parameters, and also has a serious impact on meat quality. The decrease in meat quality induced by transport stress may be due to the fact that transport stress increases hypothalamo-pituitary-adrenal (HPA) axis activity; promotes the release of glucocorticoid; disrupts intracellular calcium ion homeostasis affecting the energy metabolism level of muscle, rapid accumulation of lactic acid in muscle and changes in pH; as well as causes the destruction of the body's antioxidant system leading to oxidative stress in muscle tissue, ultimately resulting in oxidative damage to muscle tissue. Measures for alleviating transport stress to improve meat quality include breeding stress-resistant varieties, improving transport conditions and nutrient regulation strategies. Nutrition regulation strategy is the most critical. Oregano essential oils (OEOs) improve meat quality and prolong the shelf life of pork of finishing pigs under transport stress. Supplementation of OEOs in the low-protein and amino-acid-supplemental diets increased tenderness, overall acceptance, intramuscular fat content, percentage of n-3 polyunsaturated fatty acids (n-3 PUFAs) and percentage of monounsaturated fatty acids of *longissimus thoracis* muscles, and reduced drip loss of *longissimus thoracis* muscles.

Z. Zhou (✉) · H.-K. Wei · C. Cheng · J. Peng
Department of Animal Nutrition and Feed Science, College of Animal Sciences & Technology, Huazhong Agricultural University, Wuhan, China
e-mail: zhongxinzhou@mail.hzau.edu.cn

G. Jin
Meritech Bioengineering Co., Ltd, Guangzhou, China

G. Arsenos, I. Giannenas (eds.), *Sustainable Use of Feed Additives in Livestock*,
https://doi.org/10.1007/978-3-031-42855-5_23

Keywords

Pork quality · Transport stress · Oregano essential oil · *Longissimus thoracis* muscles · Pigs

1 Factors Affecting Pork Quality

Every aspect of pork production has an impact on meat quality, with genetic selection, nutritional control and pre-slaughter stress reduction playing an important role in regulating meat quality. Pre-slaughter transport is one of the major sources of stress in pigs. Reducing the stimulation of pigs during pre-slaughter transport and regulating pre-slaughter transport procedures can not only improve meat quality but also meet animal welfare requirements.

1.1 Genetic Factors

With the application of molecular biology in genetic breeding and the construction of high-density genetic maps of pigs, it is possible to search for master and candidate genes that regulate meat quality traits at the molecular level. It has been shown that the halothane gene and the sour meat gene are the two main effect genes for meat quality traits and have been used as marker genes for genetic selection (Larzul et al. 1997).

The recessive pure gene for halothane causes porcine stress syndrome (PSS), hence the name, and is an important marker gene for meat quality. Lawrie showed that pigs recessive and purely homozygous for the halothane gene had calmodulin in the mitochondria that facilitates the release of Ca^{2+} and stimulates glycogenolysis, which in turn affects tethered hydration (Lawrie 2006). De Smet slaughtered transport-stressed pigs and found significant differences in pork pH and drip loss between pure and heterozygous genotypes 40 min after slaughter (De Smet et al. 1996). GisPert used different transport densities and rest periods before slaughter and did a fasting treatment before slaughter and found that the dominant pure heterozygous genotype for the halothane gene had the highest final pH in the muscle and the recessive pure heterozygous genotype had the highest lactate content (GisPert et al. 2000).

The Rendement Napole gene (RN-gene), also known as the acid meat gene, is found in Hampshire and Hampshire pedigree pig populations and is also known as the Hampshire gene, another master effect gene affecting meat quality, located on chromosome 15 of the pig (Monin and Sellier 1985). Studies have confirmed that the muscle glycogen content of RN-dominant pigs is more than 70% higher than normal pigs, probably due to an increase in glycogen synthesis as a result of the suppressed activity of adenosine monophosphate (AMP)-activated protein kinase (AMPK) (Le Roy et al. 1990). This results in a higher amount of post-slaughter lactic acid production, a decrease in muscle pH and an increase in muscle acidity, causing a reduction in tethering power. The acid meat gene often determines the final pH of pork to be lower than normal muscle, resulting in altered meat flavour and reduced palatability (Rosenvold and Andersen 2003). Therefore, the acid meat gene is also one of the main factors affecting the marketing and processing of meat products.

1.2 Nutritional Factors

The nutritional level of the diet affects meat quality to a certain extent, and meat quality can be improved by regulating the nutritional level of the diet (D'Souza et al. 2003). Protein levels have a strong influence on meat tenderness. It was found that increasing the protein level of the diet increased leanness, reduced carcass backfat and reduced intramuscular fat levels, but meat tenderness was reduced (Davey et al. 1976). Low-protein diets, on the other hand, increase intramuscular fat content and the rate of protein metabolism in the body, the amount of collagen synthesis is reduced and collagen is unable to form cross-linked structures, thus improving meat tenderness (Goerl et al. 1995). Energy levels affect the degree of collagen cross-linking, with low energy levels increasing the degree of collagen cross-linking and thus leading to reduced meat tenderness (Aberle et al. 1981). The source of fatty acids in the diet affects the fatty acid composition of the carcass. High levels of unsaturated fatty acids result in soft carcasses and oxidative rancidity of fat, resulting in off-flavours and reduced meat quality, and Dverlan showed that the long-term addition of fish oil to the diet significantly increased the polyunsaturated fatty acid (PUFA) content of fat and increased fat off-flavours and odours, more significantly with increasing amounts of fish oil (Irie and Sakimoto 1992). Mg plays an important role in energy and protein metabolism and is a Ca^{2+} antagonist that reduces neuromuscular excitability, decreases glycolysis, raises muscle endpoint pH, improves tethering and reduces pale soft exudative (PSE) meat production (Cernak et al. 2000). The addition of Mg to the diet reduced muscle overstrain in pigs, reduced extravasation of sarcoplasm from myogenic fibres and reduced dark, firm and dry (DFD) meat production (Kietzman 1985). Vitamin E (VE) significantly reduces lipid peroxidation, prevents the oxidation of Fe^{2+} to Fe^{3+} in haemoglobin, enhances the stability of oxyhaemoglobin and prolongs the retention of natural meat colour in fresh meat (Swaran 2011). Therefore, in pig production, the provision of appropriate dietary protein and energy levels, fat sources, and vitamin and mineral content plays an important role in the regulation of meat quality.

1.3 Pre-Slaughter Transport Stress

Due to the increased scale and intensification of farm production, animals need to be transported from the farm to the slaughterhouse. The transport process can affect the animal's performance, hormone levels, blood biochemical indicators, immune levels and the quality of livestock products due to crushing, bumps and changes in ambient temperature. Dairy cows lose up to 2% of their milk production and weight after transport. Goats lose up to 10% of their body weight (Kannan et al. 2000), laying hens lose egg production (Arthington et al. 2003) and pigs experience a reduction in feed intake (Del Barrio et al. 1993). Transport can increase serum levels of creatine kinase (CK) and lactate dehydrogenase (LDH) in pigs (Warriss et al. 1998) and can affect both the humoral and cellular immune systems (Buckham Sporer et al. 2007). In pigs, serum cortisol concentrations increased significantly at transport densities less than 0.5 m^2/100 kg (Blackshaw et al. 1997). Rempel further demonstrated that pre-slaughter transport increased muscle contraction and diastole frequency, disrupted Ca^{2+} homeostasis, and increased glycolysis, resulting in increased lactate production, gradual accumulation of H^+, decreased muscle pH and increased energy metabolism (Rempel et al. 1995). Hambrech found that prolonged transport had a serious impact on muscle, increasing the glycolytic potential of the animal and the concentration of lactate in the muscle, resulting in a decrease in pH and tethering power of the longest dorsal muscle, which is predominantly glycolytic, for 24 h after slaughter, reducing muscle brightness and muscle tenderness, increasing electrical conductivity (EC) and predisposing the animal to PSE meat (Hambrecht et al. 2005). Warriss found that prolonged transport resulted in an increased probability of DFD meat in pigs (Warriss et al. 1998). Therefore, improving pre-slaughter transport is one of the important factors in improving pork quality.

2 The Concept of Transport Stress and Its Hazards

Animals are subjected to a variety of external factors during transfer or transport to the slaughterhouse, including loading and unloading, shaking of transport vehicles, wet and crowded spaces, mixing, high temperatures and unfamiliar environments. The effects of these stressors, singly or in combination, on the animals are often referred to as transport stress (Fazio and Ferlazzo 2003).

Transport stress affects the animal's neuroendocrine system and blood biochemical parameters (Fazio and Ferlazzo 2003). It also has a serious impact on meat quality (Warriss 2003). Short periods of stress tend to produce PSE meat and long periods of stress tend to produce DFD meat (Guardia et al. 2005). Transport stress leads to an increase in mortality and weight loss and suggests that transport causes weight loss of 40–60 g/kg and increases mortality to 0.1–0.4% (Lambooij 2000; Odore et al. 2004). Therefore, weight loss, mortality and the production of poor-quality meat due to transport stress can cause significant economic losses to animal production.

2.1 Effects of Transport Stress on the HPA Axis and Glucocorticoids

Transport stress affects the neuroendocrine system. The hypothalamo-pituitary-adrenal (HPA) axis and the sympathetic adreno-medullary axis (SAM) are the most important endocrine axes in the body's response to stress. The HPA axis has a response centre in the paraventricular nucleus of the hypothalamus, while the SAM axis belongs to the noradrenergic system and has a response centre in the hypothalamic basket (Freeman 2008). Stress induces excitation of the endocrine axis and promotes the secretion of adrenaline and corticotropin-releasing factor (CRF) (Elenkov et al. 2000). CRF activates the pituitary gland to release adrenocorticotropic hormone (ACTH) into the peripheral circulation, which is transported to the adrenal glands via the bloodstream to promote the synthesis and release of glucocorticoids, which ultimately activate the systemic adaptive stress response (Vanitallie 2002). In turn, the activity of the HPA axis is regulated by negative feedback from glucocorticoids, which bind to either the mineralocorticoid receptor (MR) or the glucocorticoid receptor (GR) and inhibit the over-activation of the HPA axis caused by excessive release of CRF from the hypothalamus (Kapoor and Winn 2006).

The various stressors are converted into nerve impulses by peripheral receptors, which project upwards to the nociceptors via the basket spot-norepinephrine neurons (locus coeruleus-norepinephrine, LC-NE), causing an emotional response, and downwards to the sympathetic-adrenal medulla, causing the release of catecholamine hormones into the blood. At the same time, LC-NE connects to CRF neurons, initiating the HPA axis stress response (Stratakis and Chrousos 1995). Pérez found that serum glucocorticoid concentrations in pigs arriving at the slaughterhouse after transport were significantly altered by the duration of rest before slaughter, with pigs resting for 3 h before slaughter having significantly lower glucocorticoid levels than those slaughtered directly without rest (Pérez-Campo et al. 1998). However, as the transport time was extended, the body slowly adapted to the transport environment through its own adjustment, and the levels of glucocorticoids and ACTH tended to decrease, but this trend was related to the intensity of stress suffered by the pigs during the transport process. Chai showed that high-density transport could activate the HPA axis, and plasma glucocorticoid concentrations increased dramatically.

2.2 Effect of Transport Stress on the Redox Status of the Animal Organism

Oxidative stress (OS) refers to the excessive production of free radicals in the body when the living organism is exposed to external stimuli, which exceeds the body's ability to scavenge free radicals and causes oxidative damage to the body with an excess of free radicals and their oxidation end products. The main manifestation of this is the massive production of free radicals, which react with the unsaturated structures of cellular phospholipid membranes to produce large amounts of lipid peroxides such as malondialdehyde (MDA), damaging cell structure and function (Onmaz et al. 2011). It has been reported that after prolonged (12 h) transport of horses, oxidative stress occurs in vivo, with a decrease in the activity of antioxidant enzymes such as GSH-Px in serum and a significant increase in MDA levels (Onmaz et al. 2011). The results showed that transport stress had an effect on the antioxidant capacity of animal serum. The above results suggest that the transport process promotes oxidative stress in the animal organism and reduces the antioxidant capacity of the organism, resulting in oxidative damage to tissues and organs.

2.3 Effects of Transport Stress on the Animal's Intestinal Tract

2.3.1 Effect of Transport Stress on Morphological Characteristics of the Intestinal Tract

The intestine is an organ for the digestion and absorption of nutrients, as well as an important immune organ. When animals are exposed to stress, the intestine is damaged due to ischaemia (Radwan et al. 2010), oedema of the intestinal mucosa and loss of mucosal epithelium (Boros et al. 1995). Studies have shown that the gut is the earliest to experience ischaemia and the latest to recover than other tissues and organs (Secchi et al. 2000), making it the initiating organ in the animal's response to stress.

2.3.2 Alteration of Intestinal Flora by Transport Stress

A large number of hygienic communities exist in the gut, which provides colonisation sites and nutrition for microorganisms. It has been found that transport stress alters the balance of intestinal flora and increases the number of pathogenic bacteria in animals (Eisenhofer et al. 1997). It may be that stressful conditions release catecholamines from the enteric nervous system that affect the status and behaviour of beneficial intestinal bacteria fixation and pathogenic microorganisms (Monnikes et al. 2001). Stress reduces the ability of gastric acid production (Tache 1999) and increases the pH of gastric juice, which does not effectively kill pathogenic microorganisms such as *Escherichia coli*, *Salmonella* and *Campylobacter* from food sources, leading to pathogen colonisation in the gut and disruption of the flora balance. The pathogens multiply and release enterotoxins and reactive oxygen species (ROS), causing a series of adverse reactions in the gastrointestinal tract (Ferguson and Warner 2008). Thus, transport stress is an important factor in the disruption of the intestinal flora.

2.3.3 Transport Stress Induced Oxidative Stress in the Intestine

In addition, the cellular ATP supply is reduced, the cellular membrane Ca^{2+} pump is dysfunctional and the intracellular Ca^{2+} content is elevated, activating Ca^{2+}-dependent proteins that catalyse the conversion of xanthine dehydrogenase to xanthine oxidase. During blood reperfusion, oxygen molecules enter in large quantities with the blood, and xanthine is oxidised to uric acid under the action of xanthine oxidase, and a large number of free radicals are generated (Nilsson et al. 1994). Free radicals react with unsaturated fatty acids on phospholipid membranes to produce peroxidation products such as MDA, causing peroxidative chain reactions that cause oxidative damage to the intestine (Zimmerman and Granger 1994). Thus, when transport stress occurs, it is also accompanied by oxidative damage to intestinal tissues.

2.3.4 Transport Stress Disrupts Intestinal Barrier Function and Induces an Inflammatory Response in the Intestine

The intestinal barrier function is a combination of mechanical, immune, chemical and biological barriers. Free radicals generated during transport stress act on the intestinal epithelial mucosa and cause damage to the intestinal barrier function. Free radicals attack sulfhydryl groups in proteins and DNA, resulting in inactivation of functional proteins and reduced protein expression (Tanida et al. 2011). Tight junctions are the most important type of connection between epithelial cells (Turner 2009; Marchiando et al. 2010). Tests have shown that lipid oxidants can induce depolymerisation of tight junction proteins in vitro (Musch et al. 2006), disrupting intercellular tight junctions (Chen-Quay et al. 2009). In animals, oxidative stress has been found to phosphorylate the tyrosines of tight junction proteins and impair tight junction function (Yao et al. 2010). When the tight junctions are impaired, harmful substances in the gut (e.g. endotoxins secreted by Gram-negative bacteria) enter the bloodstream via the intestinal paracellular pathway and bind to specific receptors to activate the intestinal immune system, prompting the immune system to release inflammatory factors such as tumour necrosis factor-α (TNF-α), interleukin-6 (IL-6) and other inflammatory factors, causing an inflammatory response in the intestine (Suzuki 2013). Thus, transport stress induces an inflammatory response in the intestine, mainly through free radicals that damage intestinal tight junctions, which in turn affects the barrier function of the intestine.

3 Mechanisms of Transport Stress on Meat Quality

Stressors during animal transport are an important factor in causing physiological abnormalities and a decline in meat quality. The decrease in meat quality is mainly manifested in the rapid decrease in pH and the high incidence of PSE in meat after slaughter. Analysis of the causes reveals the fact that transport stress increases HPA axis activity; promotes the release of glucocorticoid; disrupts intracellular calcium ion homeostasis, affecting the energy metabolism level of muscle, rapid accumulation of lactic acid in muscle and changes in pH; as well as causes the destruction of the body's antioxidant system leading to oxidative stress in muscle tissue. In addition, stress may increase the inflammatory response and promote the release of free radicals, ultimately resulting in oxidative damage to muscle tissue. The possible causes are detailed below.

3.1 Effects of Transport Stress and Intestinal Inflammation on the Activity of the Endocrine System

TNF-α, produced by macrophages, T-cells and monocytes, is the first inflammatory factor to appear in the inflammatory response and induces the production and release of other inflammatory factors (e.g. IL-1, IL-6, IL-8) with a half-life of only 2 min in serum (Rubenfeld et al. 2005). It has a variety of biological effects, such as mediating immune responses, inhibiting polymorphonuclear (PMN) apoptosis, inducing PMN to produce endogenous platelet-activating factor, stimulating acute-phase protein secretion by hepatocytes, activating B-lymphocytes etc. (Forman and Torres 2003). It was found that in the intestine, the production of inflammatory factors TNF-α and IL-6 can damage the endoplasmic reticulum and mitochondria of enterocytes, causing apoptosis of intestinal epithelial cells and further disrupting the barrier function of the intestine (Suzuki et al. 2011; Forman and Torres 2003). At the same time inflammatory factors act on phagocytic neutrophils, inducing further oxidative stress (Sugino et al. 1987). Thus, oxidative stress causes inflammation in the intestine, further exacerbating intestinal oxidative stress.

CRF is expressed in both neurocentral and peripheral tissues, mainly by neurons in the paraventricular nucleus of the hypothalamus and synthesised and released by intestinal mast cells, intestinal chromophores and macrophages in peripheral tissues. corticotropin-releasing factor receptors (CRFRs) are found in the central nervous system as well as in many peripheral tissues (Grammatopoulos and Chrousos 2002) and in response to stressors, HPA axis function is activated and CRF neurons in the paraventricular nucleus of the hypothalamus are activated. The CRF neurons in the paraventricular nucleus increase the secretion of CRF, which reaches the pituitary via the pituitary portal circulation. CRF binding to CRFRs in the anterior pituitary activates the adenylate cyclase system, stimulating the pituitary to release ATCH, which enters the bloodstream and activates ATCH receptors in the adrenal cortex, ultimately leading to the synthesis and release of glucocorticoids and regulating the effects of stress on the body (Lightman 2008).

Both exogenous and endogenous cytokines have a regulatory effect on the HPA axis. The administration of IL-6 to rats stimulates the secretion of CRF and glucocorticoids (John and Buckingham 2003). And a variety of cytokines can stimulate HPA axis activity (Rivier 1993). Glucocorticoids were elevated in rats injected with lipopolysaccharides (LPS), but glucocorticoid levels did not change significantly in cytokine-deficient rats, suggesting that cytokines have an important regulatory role in stress hormones (Shixuya et al. 1997). Rats were restrained for 24 h and elevated plasma CRF was found to be accompanied by elevated levels of hypothalamic IL-6 mRNA expression (Mastorakos et al. 1994). When oxidative stress occurs in the gut, tight junction protein function is impaired and large toxic substances enter the bloodstream, stimulating the immune system and causing an inflammatory response in the gut and elevated levels of inflammatory cytokines such as TNF-α and IL-6 in the blood. TNF-α and IL-6 can activate the HPA axis by acting at CRF sites in the hypothalamus (Haddad et al. 2003), promoting the release of ACTH from the pituitary. This in turn increases the release of adrenal glucocorticoids, which reduce the release of inflammatory factors, control intestinal inflammation and restore intestinal homeostasis through immunosuppression (Borovikova et al. 2000; Gay et al. 2008). Thus, in stressful situations, the gut undergoes oxidative stress and releases inflammatory factors, which activate the HPA axis and increase circulating hormone levels.

It has been shown that transport stress activates the HPA axis and increases the release of glucocorticoids, which promote muscle degradation. Yoshioka and Guzik found that transport stress increased blood and saliva cortisol levels and increased pH in meat (45 min) (Yoshioka et al. 2004; Guzik et al. 2006). Subsequent studies also found that transport stress increased blood cortisol levels and PSE meat by 40%. More recent studies have found that cortisol affects carcass quality, with high levels of cortisol increasing hot carcass weight and carcass fat thickness but reducing meat flavour (Dokmanovic et al. 2015). However, it remains to be further confirmed whether, in the case of transport stress, it introduces increased inflammation by affecting intestinal stress, which in turn affects the HPA axis and the release of glucocorticoids, ultimately leading to a reduction in meat quality.

3.2 Effect of Glucocorticoids on Meat Quality

3.2.1 Effect of Glucocorticoids on Intracellular Ca^{2+} Concentrations

Intracellular Ca^{2+} is involved in the regulation of most intracellular physiological functions, including muscle contraction, energy metabolism, gene expression and apoptosis. The balance of intracellular Ca^{2+} concentration relies on a complex and elaborate Ca^{2+} transport and clearance mechanism. The intracellular Ca^{2+} concentration is maintained at the nM level in the resting state and at the mM level in the extracellular environment.

Calcium ions play an important role in mediating apoptotic signalling (Lee et al. 2002). When cells are stimulated by the external environment, the influx of calcium ions from the extracellular environment or the efflux of calcium from the endoplasmic reticulum via the IP3 receptor on the endoplasmic reticulum membrane will increase the intracellular calcium content, and the mitochondria will take up the excess calcium ions and redistribute them, resulting in mitochondrial calcium overload, leading to mitochondrial damage, release of cytochrome C, activation of the calcium-dependent protein caspase and induction of apoptosis (Breckenridge et al. 2003). The addition of the calcium-dependent nucleic acid endonuclease blocker zinc or the Ca^{2+}/K^{+} ion channel blocker verapamil was found to inhibit apoptosis in a cadmium-induced apoptosis model. The above studies suggest that an increase in intracellular Ca^{2+} may be a cause of apoptosis.

Glucocorticoids are steroidal substances that have good cell membrane permeability and can freely enter and exit the cell membrane and cytoplasm, where they bind to the glucocorticoid receptor (GR) and participate in a variety of life activities and maintain life stability (Chrousos and Kino 2007). In hippocampal cells, activated GR dimers bind to DNA and regulate the release of Ca^{2+} from L-type calcium channels (Chameau et al. 2007). In a rat restraint stress model, blood levels of glucocorticoids are elevated and promote binding to GR and upregulate the endo plasmic reticulum calcium release channel proteins IP3R1 and IP3R2, thereby regulating changes in cytoplasmic calcium ion concentrations (Lencesova et al. 2004). The ability of glucocorticoids to promote increased calcium ion concentrations has also been observed in C2C12 myogenic cells and mouse dorsal root ganglion cells (Passaquin et al. 1998). In pork production it has also been found that apoptosis and poor pork quality due to transport stress are associated with intracellular Ca^{2+} overload, with excess Ca^{2+} not being cleared in time, resulting in disturbed endoplasmic reticulum Ca^{2+} concentrations, not only in live pigs but also in post-slaughter pork, and the more lean the pork, the more severe the situation (Mclennan and Phillips 1992). It can be hypothesised that transport stress activates the HPA axis and promotes the secretion of glucocorticoids, which penetrate the cell membrane via the blood circulation and, by binding to GR, form activated receptor complexes that upregulate IP3R or downregulate SarcoEndoplasmic Reticulum Calcium ATPase (SERCA) through transcriptional activation, resulting in the influx of calcium ions from the endoplasmic reticulum calcium pool into the cytoplasm. And it cannot be absorbed de novo into the endoplasmic reticulum, resulting in elevated intracellular calcium ion levels, mitochondrial uptake of intracytoplasmic calcium ions, mitochondrial calcium ion overload, release of cytochrome C, activation of calcium-ion-dependent nucleic acid endonucleases, and induction of apoptosis, tissue damage and spillage of cell contents, leading to the production of poor-quality meat.

3.2.2 Effect of Glucocorticoids on Muscle Energy Metabolism

Glucocorticoids and insulin are two important hormones in the regulation of energy in the body and play opposite regulatory roles. Glucocorticoids act mainly in the liver, muscle and adipose tissue. Excessive secretion of glucocorticoids can interfere with the action of insulin (Rosmond 2003). Studies have shown that glucocorticoids cause insulin resistance in the liver and skeletal muscle by directly inhibiting insulin secretion from pancreatic beta-cells or by interfering with the binding of insulin to its receptors and by attenuating the insulin activity of the bound receptors (Lambillotte et al. 1997). Stress-induced elevation of glucocorticoids diminished insulin action, compromised myoglycogen synthesis, enhanced gluconeogenesis in the liver and increased glycolytic rate in muscle tissue supplied mainly through the glycolytic pathway (Hambrecht et al. 2005).

In order to maintain the stability of the body's internal environment and increase the chances of individual survival, HPA axis activity is enhanced and glucocorticoid secretion is increased under stress (Sapolsky et al. 2000). Rempel further demonstrated that transport stress increases the frequency of muscle contraction and diastole in animals, disrupting Ca^{2+} homeostasis (Rempel et al. 1995). Hambrech found that prolonged transport increased muscle glycolytic potential and lactate concentration, reduced muscle tenderness and muscle brightness, and increased electrical conductivity, resulting in a decrease in pH and tethering power in the longest dorsal muscles, which are predominantly glycolytic fibres, for 24 h after slaughter, making them susceptible to PSE meat being produced (Hambrecht et al. 2005). The regulation of gluconeogenesis by glucocorticoids in response to stress causes pH abnormalities that are a major cause of reduced meat quality (Geers et al. 1994), with lower pH producing PSE meat and higher pH producing DFD meat (Warriss et al. 1998; Pérez et al. 2002). Thus, glucocorticoids can influence meat quality by modulating the way muscle energy is supplied, controlling the accumulation of muscle lactate and the rate of pH reduction.

3.3 Effect of Oxidative Stress on Meat Quality in Animal Muscle

Transport stress increases the number of free radicals in the blood, liver, muscle and other tissues and organs of the body, which attack cell membranes and produce large amounts of MDA, a lipid peroxidation product, and affect the normal levels of SOD, CAT and GSH-Px in the tissues and organs. This leads to oxidative damage in all tissues and organs.

3.3.1 Effect of Muscle Oxidative Stress on Muscle Tethered Hydrodynamics

Myocyte membranes and subcellular membranes are rich in polyunsaturated fatty acids, and their phospholipid structure is susceptible to attack by free radicals; once the reaction is initiated, a chain reaction occurs, resulting in the production of a large number of lipid peroxidation end products such as MDA and acrolein, the normal structure of the cell membrane is disrupted and the cell contents are exuded (Nakaya et al. 1987). The water content of muscle is about 75%, of which 4–5% is in the bound state, with free-state water present in the sarcoplasm and intercellular spaces. Free radical attack on the myocyte membrane causes rupture and damage to the cell membrane, resulting in loss of free-state water from the sarcoplasm and intercellular spaces, which leads to a reduction in muscle tethering and increased drip loss (Simitzis et al. 2010). Therefore, the attack on myocyte membranes by free radicals generated by muscle oxidation is an important cause of reduced muscle tethering power, which affects meat quality.

3.3.2 Effect of Muscle Oxidative Stress on Meat Colour

The colour of pork is mainly determined by the status of myoglobin. Ferrous haemoglobin in myoglobin plays a major role in colour rendering. Oxygen has a strong affinity for ferrous haemoglobin and when oxygen is not bound to it, myoglobin is dark red, and when bound to oxygen, myoglobin is bright red (Pisula 1996). It has been found that the oxidation of pigmentsPigments in muscle is positively correlated with the degree of lipid oxidation , due to free radicals generated during fat oxidation either directly promoting pigmentPigments oxidation or indirectly acting by damaging pigmentPigments degradation systems (Faustman et al. 1989), thus affecting the colour of the meat. When muscle is under oxidative stress, Fe^{2+} in ferrous haemoglobin is oxidised to Fe^{3+}, myoglobin becomes denatured myoglobin (methemoglobin) and the meat colour appears dark brown (Daiwen et al. 2002).

In addition, some of the light absorbed by the muscle is absorbed by myosin and becomes coloured, while much of the rest is reflected back. The more the light is reflected back, the paler the muscle appears (Daiwen et al. 2002). In contrast, PSE meat is produced when oxidative stress occurs, and because more water leaks from the surface of PSE pork, and because the contraction of muscle fibres compresses the water storage space, most of the light that hits the surface of the muscle is reflected back, so the meat appears pale (Daiwen et al. 2002). The above studies suggest that oxidation of myocyte ferrous haemoglobin and water exudation due to oxidative stress are the factors responsible for the change in meat colour.

3.3.3 Interaction of Muscle Oxidative Stress with Intracellular Ca^{2+}

Reactive oxygen species (ROS) can be produced by a variety of pathways within cells. It has been found that mitochondria are the organelles that produce ROS in the cell. During the electron transfer process of the mitochondrial respiratory chain, some electrons are leaked to oxygen molecules, resulting in the univalent reduction of oxygen molecules and the continuous production of free radicals. The regulation of intracytoplasmic Ca^{2+} concentration by ROS is related to the concentration of ROS, which does not work at low concentrations and can increase cytoplasmic Ca^{2+} concentration when a certain concentration is reached (Gen et al. 2001). These studies suggest that ROS and calcium ions simultaneously determine cellular physiological functions and that there may be an interactive relationship between the two.

It has been found that the amount of ROS production is related to the efficiency of cellular energy metabolism (Pérez-Campo et al. 1998). Ca^{2+} in mitochondria increases ROS production by stimulating ATP synthase, adenine nucleoside transporter enzymes, and also activates the triphosphate cycle to accelerate energy metabolism (Mildaziene et al. 1995). In addition, Ca^{2+} activates nitric oxide (NO) synthase to promote NO synthesis, which is oxidative in nature, and increases the amount of electron leakage from the respiratory chain, increasing ROS production (Cleeter et al. 1994). Membrane potential is altered and mitochondrial oxidative phosphorylation is uncoupled, resulting in high ROS production. These studies suggest that the concentration of calcium ions is one of the important factors influencing ROS production.

Under the influence of stress, ROS are not effectively cleared from the organism's organs and accumulate in the body, which may damage the organism's function and the lipid structure of cell membranes, resulting in the overflow of cell contents and apoptosis, while in muscle tissue may result in the production of poor-quality meat.

4 Measures to Alleviate Transport Stress and Improve Meat Quality

4.1 Breeding Stress-Resistant Varieties

Different animals and different species of animals of the same species have different tolerances to stress. For breeding animals, genetic selection to eliminate stress-resistant breeds can be effective in slowing down the level of response to stress. In pig breeding it has been found that the halothane gene causes the production of PSE meat in pigs, so genetic selection to eliminate the halothane gene can reduce PSE meat production. Pigs that are dominantly pure also produce PSE meat, suggesting that there are still other important genetic factors besides the halothane gene such as obesity genes, melanocortin receptor genes, Leptin genes, hormone-sensitive lipase genes, lipoprotein lipase genes and other candidate genes that influence the development of PSE meat (Pommier et al. 1993). The acid meat gene is also a master effector gene, and pigs carrying the RN-genes have higher than normal glycogen content in their muscles, resulting in excessive lactic acid production from glycolysis and rapid pH reduction after slaughter (Rosenvold and Houde 2001). Molecular genetic marker methods were used to isolate meat-harming genes and select new stress-resistant strains to improve the stress resistance of livestock and poultry and fundamentally improve the quality of pork after slaughter.

4.2 Improving Transport Conditions

Livestock are subjected to a variety of transport conditions during transport, such as loading and unloading tools, the condition of the vehicle, road conditions, as well as transport time and transport temperature, all of which can have a significant impact on the performance of livestock and meat quality (Hambrecht et al. 2005). Prolonged fasting and water deprivation can lead to metabolic disturbances and disruption of the acid–base balance, making animals susceptible to dehydration and acidosis. Noise during transport can alter the physiological and biochemical parameters of the animal's blood and hormone levels (Agnes et al. 1990). The high temperature and humidity in the transport carriage and the outside temperature, especially in summer, can cause increased mortality during transport (Drain et al. 2007). Herding, herd transfer and up-downloading can all cause nervous tension and induce stress in animals (Broom 2005). Therefore, the physiological characteristics of livestock need to be taken into account to improve mid-transport management and to screen for appropriate transport methods to minimise stressful stimuli. The main measures include choosing the time of transport in the hot summer, reducing transport density, improving ventilation, ensuring water supply, avoiding mixing herds to reduce the occurrence of fighting and using tools wisely when loading and downloading to avoid jerking and catching.

4.3 Nutrient Regulation Strategies

The non-enzymatic antioxidant system plays a complementary role, preventing reactive oxygen species from reacting with highly unsaturated phospholipids on subcellular membranes (microsomes, mitochondria) to produce free radicals; preventing the occurrence of lipid peroxidation chain reactions; protecting pork from loss of flavour, colour and nutritional value (Arai et al. 1999). The addition of certain antioxidant substances to animal diets has received increasing attention from researchers as a means of improving meat quality by reducing systemic stress, increasing antioxidant capacity, reducing the release of reactive oxygen species from cell membranes and protecting tissues from free radical attack (Simitzis et al. 2010; Wolfgang et al. 2009).

4.3.1 Nutritional Strategies for Modulating the HPA Axis Through Gut Health

The gastrointestinal tract of animals is colonised by a large number of bacteria, the ratio of which maintains a dynamic balance and acts as a biological barrier, the first barrier to pathogen invasion in the gastrointestinal tract. The gut nervous and immune systems influence the central nervous system through electrophysiological stimulation, cytokines and neuropeptides (Riether et al. 2008), and CRF are important for the transmission of messages from the gut to the central nervous system (Fekete and Zorrilla 2007). Probiotics are beneficial bacteria that live in the intestinal tract and act to balance the microecological environment in the gut and can colonise the gastrointestinal tract through exogenous addition (Mack et al. 1999). *Lactobacillus* and *Bifidobacterium* are important beneficial bacteria in the intestinal tract and were the first probiotics to be used to maintain intestinal flora balance, inhibit colonisation by harmful pathogenic microorganisms and enhance the barrier function of the intestine (Zareie et al. 2006). *Bifidobacterium* can regulate the expression of intestinal inflammatory cytokines by affecting tight junction proteins (Fasano and Nataro 2004), as well as affecting intestinal cell Ca^{2+}-dependent ion channels to enhance or diminish vagal excitability, which in turn affects the CRF and central nervous HPA axis (Wolfgang et al. 2009).

Plant extracts have antimicrobial properties both in vivo and in vitro, extending the shelf life of meat in vitro by killing harmful microorganisms and in vivo by selectively inhibiting a wide range of harmful bacteria (Skandamis et al. 2002; Djenane et al. 2003; Chouliara et al. 2007). Active packaging (AP) based on oregano extract reduces the number of total bacteria and *Pseudomonas* in beef and is effective against *Salmonella typhimurium* and *E. coli* O157:H7 (Oussalah et al. 2006). Skandamis and Nychas suggest that the reason why oregano extract inhibits the growth of microorganisms is mainly due to the fact that its components can affect the glucose and lactate consumption of microbial growth and metabolic processes (Skandamis and Nychas 2001). Therefore, the antibacterial and bactericidal properties of plant extracts may affect the number and homeostasis of intestinal microorganisms, which in turn affects the central nervous system.

4.3.2 Nutritional Strategies to Improve the Antioxidant Capacity of the Body

An increase in the body's antioxidant capacity is essential to protect the body and tissues from free radical attack. It has been found that magnesium has the ability to resist cellular lipid oxidation. In vivo and in vitro studies in rats have demonstrated that magnesium has the effect of reducing MDA, a product of cellular lipid peroxidation (Güther et al. 1992). It is possible that magnesium has many binding sites on the surface of the cell membrane and acts as a stabiliser of the lipid bilayer structure. Magnesium is a cofactor required for the synthesis of the antioxidant enzyme glutathione in the body and plays an important role in stabilising the redox state in the body (Shivakumar and Kumar 1997). Magnesium is an antagonist of calcium, maintaining mitochondrial integrity and reducing apoptosis due to intracellular calcium ion overload. It also competes for the calcium-binding site of troponin, inhibiting muscle contraction and reducing PSE occurrence (Heaton and Elie 1984).

Selenium is a component of GSH-Px, an important antioxidant enzyme in the body, and works synergistically with vitamin E to scavenge peroxides formed in cells. The addition of selenium to the diet reduces the MDA content in the body, increases the activity of GSH-Px, reduces the level of free radicals, enhances the antioxidant capacity of the body and effectively protects the quality of the meat during storage (Homma et al. 1993). Excessive additions of inorganic selenium can produce cytotoxicity, and the addition of organic selenium to broiler diets can more effectively reduce drip losses in chicken meat. Edens (1997) found that the combination of vitamin E and selenium in feed improved the ability of chicken meat to resist free radical attack during refrigeration, delayed meat oxidation and fatty acid decay, and effectively extended the shelf life of meat. Therefore, minerals and vitamins with antioxidant properties are now widely used in feed production as important antioxidant components.

4.3.3 Nutritional Strategies to Mitigate Oxidative Damage to Myocyte Membranes

Free radicals are a key factor in breaking down cell membranes and causing loss to occur in the body or tissues. For this reason, a number of substances with the ability to scavenge free production are also used. The molecular structure of vitamin E contains phenolic hydroxyl groups, which react with oxidation groups through electron transfer to produce stable tocopherol peroxide groups. Studies have confirmed that vitamin E is an important antioxidant in the body (Kamal-Eldin and Appelqvist 1996). Due to its lipid solubility, vitamin E can be embedded in the phospholipid layer of biofilms containing unsaturated fatty acids, scavenging free radical damage to biofilm unsaturated fatty acids and preventing lipid peroxidation from occurring, forming the first line of antioxidant defence for cell membranes. Vitamin E is oxidised by free radicals to tocopherol quinone or tocopherol hydroquinone, which is then metabolised to alpha-tocopheric acid or alpha-tocopherol lactone and excreted from the body. Long-term supplementation of vitamin E in feed can increase the cell membrane vitamin E (VE) content and enhance the oxidative stability of mitochondrial membrane lipids (Asghar et al. 1991; Buckley et al. 1996). It was found that VE could reduce the body's sensitivity to stress, reduce the loss of immunoglobulins and thus improve the body's immunity to transport stress. Vitamin C (VC) is a water-soluble antioxidant, found in the cytosol, and is the most important antioxidant in the cytosol, working together with the enzymatic antioxidant system to reduce damage to the body from oxidative stress by limiting the increase in reactive oxygen radicals (Stocker and Frei 1991). Vitamin C is also an electron donor and is oxidised by free radicals to dehydrovitamin C, which protects proteins and DNA, the large molecules within biological membranes, from oxidation by free radicals. Vitamin C and vitamin E work synergistically to maintain the antioxidant capacity of cells and the integrity of cell membrane function. Peeters showed that the addition of vitamin C to feed significantly improved the ability of farrowing pigs to resist transport stress and improved the quality of pork (Peeters and Geers 2006). Bai reported that vitamin C could scavenge free radicals and improve the antioxidant and repair capacity of nerve cells.

Certain plant extracts have a similar phenolic hydroxyl structure to VE and improve meat quality by retarding cell membrane lipid peroxidation (Chaves et al. 2008), which can inhibit meat rancidity and meat colour changes (Djenane et al. 2003). The addition of oregano and sage extracts to beef and pork (Fasseas et al. 2008) and the spraying of rosemary and VC solutions on the surface of the meat (Djenane et al. 2003) both reduced the oxidation of the meat during the chilling process. In addition, the addition of oregano, rosemary and sage extracts to the diet can also retard the formation of MDA in meat (Botsoglou et al. 2003; Simitzis et al. 2010). Numerous studies have shown that oregano extracts exert the strongest antioxidant effect at doses of 100 mg/kg for broilers (Botsoglou et al. 2003) and 200 mg/kg for turkeys (Botsoglou et al. 2003). Janz is the only report comparing the effects of oregano extract with other plant extracts (rosemary, garlic and ginger) in pigs (Janz et al. 2007). The results showed the lowest MDA values in the oregano plant extract group, probably because the antioxidants in the oregano plant extract were absorbed into the circulatory system and eventually reached and were retained in the muscle and other tissues.

Some researchers have reported that natural plant extracts can prevent loss of flesh colour by increasing redness and delaying the formation of myohaemoglobin (MMG). Djenane reported that spraying rosemary and VC on the surface of fresh steaks also reduced the production of MMG and increased the redness of the meat (Djenane et al. 2003). The above studies suggest that vitamin-based or natural plant extracts are good at scavenging the production of free radicals in muscle cells and have a good effect on improving meat quality, and may serve as an important nutritional regulator to alleviate the production of free radicals under stressful conditions and protect the high quality of meat.

5 Effects of Supplying Oregano Essential Oil on Pork Quality

5.1 Effects of Supplying Oregano Essential Oils on Meat Quality

Effects of supplying oregano essential oils (OEOs) on meat quality of finishing pigs under transport stress has been reported (Zou et al. 2016). The results of the electrical conductivity, pH value and Opto-star value (meat colour) measurements in the longissimus thoracis et lumborum (LTL) are presented in Table 1. Compared with the control group, the average pH at 45-min postmortem and Opto-star values at 24-h postmortem increased in the vit E, OEO or quercetin groups ($p < 0.05$). The pH values at 24-h postmortem of the vit E or quercetin groups increased ($p < 0.05$) compared with the control group, with no difference between the vit E and quercetin groups. Compared with pigs fed the control diet, the drip loss at 24-h postmortem from the LTL of the vit E group was no different, but those of the OEO or quercetin groups decreased ($p < 0.05$). No main effects ($p > 0.05$) were observed in $EC_{45\ min}$, $EC_{24\ h}$ and Opto-star$_{45\ min}$ of pig meat between the four diets.

Table 1 Effects of dietary vitamin E, oregano essential oil or quercetin supplementation on carcass characteristics and meat quality in transported pigs

Measures	Control	Vit E	Quercetin	OEO	SEM	p-value
Number of pigs	12	12	12	12		
Pre-slaughter BW (kg)	101.3	100.9	100.1	99.5	0.42	0.56
Hot carcass weights (kg)	67.9^{b}	68.6^{b}	71.0^{a,b}	73.5	0.71	0.04
Carcass straight length (cm)	100.0	99.9	99.9	99.8	0.49	0.87
Carcass slanting length (cm)	86.1	85.7	85.7	84.9	0.34	0.42
Back-fat thickness (mm)	18.2	17.3	17.8	18.1	0.44	0.90
Dressing out (%)	67.4^{b}	68.2^{b}	70.5^{a,b}	72.9	0.74	0.03
pHi (45 min)	6.12^{b}	6.33^{a}	6.42^{a}	6.46^{a}	0.04	< 0.01
pHu (24 h)	5.67^{b}	5.81^{a}	5.85^{a}	5.73^{a,b}	0.02	0.03
EC (45 min) mS cm^{-1}	4.65	4.64	4.63	4.66	0.02	0.98
EC (24 h) mS cm^{-1}	4.51	4.50	4.67	4.59	0.03	0.18
Opto-star (45 min)	74.3	76.9	76.2	76.2	0.51	0.30
Opto-star (24 h)	45.3^{b}	51.9^{a}	50.2^{a}	52.9^{a}	0.80	< 0.01
Drip loss (24 h, %)	1.70^{a}	1.40^{a,b}	1.09^{b}	1.10^{b}	0.08	0.01

BW, bodyweight; EC, electrical conductivity; OEO, oregano essential oil; SEM, standard error of mean; vit E, vitamin E. The pre-slaughter BW was used as a covariate for hot carcass weights and dressing out. ^{a,b}Letters within a row denote statistical differences between means
Source: Zou et al. (2016)

5.2 Meat Quality Parameters and Fatty Acid Profile of Intramuscular Fat

Effects of dietary treatments on the physical parameters, chemical parameters and fatty acid composition (%) of longissimus thoracis (LT) muscle in growing-finishing pigs have been reported (Cheng et al. 2017).

The meat quality parameters, pH values, incarnadine score (subjective LT colour scores), marbling score, colour indices (objective colour measurements), cooking loss and shear force are shown in Table 2. Compared with the reduced protein, amino-acid-supplemented treatment (normal protein treatment [NPD]), the OEO treatment showed a greater ($p < 0.05$) yellowness. No effects of dietary treatments were detected on incarnadine score, marbling scores, pH values, lightness, redness, cooking loss and shear force.

For chemical parameters, pigs fed with RPD or OEO had greater ($p < 0.05$) intramuscular fat (IMF) and lower ($p < 0.05$) crude protein in LT muscle compared with pigs fed with the NPD diet (Table 2). Dietary treatments did not affect muscle moisture and ash content.

The fatty acid profile of total fat in LT muscle is shown in Table 2. Pigs fed with RPD or OEO had lower ($p < 0.05$) percentage of the total SFAs in LT muscle compared with pigs fed with the NPD diet. In addition, the concentration of myristic acid (C14:0) in pigs fed with the RPD diet showed no significant difference

Table 2 Effects of dietary treatments on the meat quality parameters and fatty acid profile of intramuscular fat of longissimus thoracis (LT) muscle in growing-finishing pigs

Item	NPD	RPD	OEO	SEM	*p*-value
Physical parameters					
pH, 45 min	6.50	6.48	6.41	0.04	0.775
pH, 24 h	5.94	5.74	5.76	0.04	0.117
Incarnadine score (subjective LT colour scores)	3.58	3.50	3.58	0.06	0.799
Marbling score	2.00	2.25	2.33	0.14	0.631
Colour indices, 45 min					
Lightness	39.10	40.13	39.14	0.38	0.482
Redness	4.04	4.31	4.72	0.28	0.632
Yellowness	10.78^{b}	$11.51^{a,b}$	12.02^{a}	0.21	0.036
Cooking loss, %	24.80	25.22	25.14	0.41	0.915
Shear force, N/cm^2	43.00	41.23	42.96	1.59	0.899
Chemical parameters					
Moisture, %	74.43	74.46	74.67	0.20	0.874
Crude protein, %	22.88^{a}	22.06^{b}	22.02^{b}	0.14	0.012
Ash, %	1.91	1.85	2.04	0.08	0.607
Intramuscular fat, %	1.67^{b}	2.13^{a}	2.00^{a}	0.08	0.049
Fatty acid composition (%)					
C14:0	1.44^{a}	$1.19^{a,b}$	1.05^{b}	0.06	0.046
C16:0	25.28	24.38	24.20	0.50	0.663
C16:1n-9	3.20	3.11	2.71	0.11	0.155
C17:0	0.19	0.21	0.23	0.01	0.394
C18:0	12.17	11.48	11.99	0.23	0.468
C18:1n-9	38.14^{b}	40.62^{a}	38.67^{b}	0.41	0.019
C18:2n-6	11.34	10.48	11.71	0.37	0.417
C18:3n-3	0.26^{a}	0.10^{b}	0.27^{a}	0.03	0.016
C20:0	0.18	0.13	0.16	0.02	0.750
C20:2n-6	0.28	0.31	0.28	0.01	0.355
C20:3n-3	0.23	0.27	0.31	0.02	0.170
C20:4n-6	2.54	2.41	2.80	0.16	0.638
C22:5n-3	0.34	0.43	0.47	0.03	0.334
n-6 PUFAs[1]	14.44	13.13	14.79	0.47	0.334
n-3 PUFAs[2]	$0.83^{a,b}$	0.75^{b}	1.10^{a}	0.06	0.037
SFAs[3]	39.99^{a}	37.18^{b}	37.55^{b}	0.54	0.025
MUFAs[4]	41.60^{b}	44.54^{a}	41.67^{b}	0.51	0.015
PUFAs[5]	15.14	13.80	15.89	0.49	0.214

NPD, normal protein treatment; *OEO*, 250 mg/kg oregano essential oil treatment; *RPD*, reduced protein, amino-acid-supplemented treatment

Samples from six barrows were measured per treatment. The fatty acid results were presented as g/100 g fatty acids (wt %). Within a row, means with different superscript letters are significantly different ($p < 0.05$).

[1]n-6 polyunsaturated fatty acids (n-6 PUFAs) percentage is the sum of 18:2n-6, 20:2n-6 and 20:4n-6

[2]n-3 polyunsaturated fatty acids (n-3 PUFAs) percentage is the sum of 18:3n-3, 20:3n-3 and 22:5n-3

[3]Saturated fatty acids (SFAs) percentage is the sum of 14:0, 16:0, 17:0, 18:0 and 20:0

[4]Monounsaturated fatty acids (MUFAs) percentage was calculated as the sum of c16:1n-9 and c18:1n-9

[5]Polyunsaturated fatty acids (PUFAs) percentage was calculated as the sum of n-3 PUFAs and n-6 PUFAs

Source: Cheng et al. (2017)

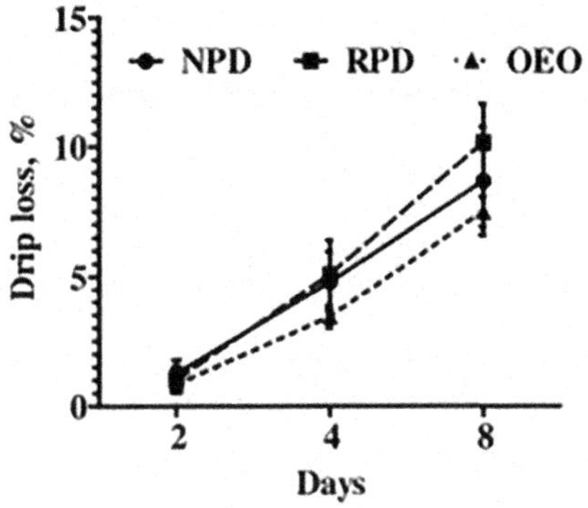

Fig. 1 Longissimus thoracis muscle drip loss in relation to storage time at 4 °C. *NPD*, normal protein treatment; *OEO*, 250 mg/kg oregano essential oil treatment; *RPD*, reduced protein, amino-acid-supplemented treatment. Samples from six barrows were measured per treatment. Effects of treatment, $p < 0.001$; time, $p < 0.01$; time $*$ treatment, $p = 0.112$. (Source: Cheng et al. 2017)

($p > 0.05$), whereas those of the OEO treatment significantly decreased ($p < 0.05$) compared with pigs fed with the NPD diet. Compared with the RPD treatment, the NPD or OEO treatment contained a lower percentage of oleic acid (C18:1n-9) and monounsaturated fatty acids (MUFAs) and a higher percentage of α-linolenic acid (C18:3n-3). A higher ($p < 0.05$) percentage of total n-3 PUFAs in the OEO treatment was detected compared with the RPD treatment.

The changes of drip loss of pork LT muscle in relation to dietary treatment (NPD, RPD and OEO) and storage time at 4 °C are shown in Fig. 1. Drip loss was significantly affected by storage time ($p < 0.01$) and dietary treatment ($p < 0.001$). No interaction between time and treatment was observed.

6 Conclusions

In the present study, pre-slaughter transport is one of the major sources of stress in pigs. OEO can act as an efficient dietary supplement to alleviate transport stress in finishing pigs, which can improve pork meat quality of finishing pigs. OEO was superior to vitamin E in increasing antioxidant enzyme activity, thereby reducing transportation-induced oxidative stress and improving meat quality. OEO supplementation to a reduced protein, amino-acid-supplemented diet can improve the content of intramuscular fat and change the fatty acid composition towards a higher percentage of n-3 PUFAs in *longissimus thoracis muscle*, which have a higher overall sensory acceptance score. In addition, long-term supplementation with OEO has the potential to prevent lipid oxidation, and to delay lipid oxidation, without affecting other meat quality parameters.

References

Aberle E, Reeves E, Judge M et al (1981) Palatability and muscle characteristics of cattle with controlled weight gain: time on a high energy diet. J Anim Sci 52:757–763

Agnes F, Sartorelli P, Abdi BH et al (1990) Effect of transport loading or noise on blood biochemical variables in calves. Am J Vet Res 51:1679–1681

Arai M, Imai H, Koumura T et al (1999) Mitochondrial phospholipid hydroperoxide glutathione peroxidase plays a major role in preventing oxidative injury to cells. J Biol Chem 274:4294–4933

Arthington JD, Eicher SD, Kunkle WE et al (2003) Effect of transportation and commingling on the cute-phase protein response, growth, and feed intake of newly weaned beef calves. J Anim Sci 81(5):1120–1125

Asghar A, Gray JI, Booren AM et al (1991) Effects of supranutritional dietary vitamin E levels on subcellular deposition of α-tocopherol in the muscle and on pork quality. Sci Food Agric 57:31–41

Blackshaw JK, Thomas FJ, Lee JA (1997) The effete of a fixed or free toy on the growth rate and aggressive behaviour of weaned Pigs and the influence of hierarchy on initial Investigation of the toys. Appl Anim Behav Sci 53:203–212

Boros M, Takaichi S, Hatanaka K et al (1995) Ischemic time-dependent microvascular changes and reperfusion injury in the rat small intestine. J Surg Res 59(2):311–320

Borovikova LV, Ivanova S, Zhang M et al (2000) Vagus nerve stimulation attenuates the systemic inflammatory response to endotoxin. Nature 405:458–462

Botsoglou N, Fletouris D, Florou-Paneri P et al (2003) Inhibition of lipid oxidation in long-term frozen stored chicken meat by dietary oregano essential oil and α-tocopheryl acetate supplementation. Food Res Int 36:207–213

Breckenridge DG, Germain M, Mathai JP et al (2003) Regulation of apoptosis by endoplasmic reticulum pathways. Oncogene 22:8608–8618

Broom DM (2005) The effects of land transport on animal welfare. Rev Sci Tech 24:683–691

Buckham Sporer KR, Burton JL, Earley B et al (2007) Transportation stress in young bulls alters expression of neutrophil genes important for the regulation of apoptosis, tissue remodeling, margination, and anti-bacterial function. Vet Immunol Immunopathol 118(1):19–29

Buckley DJ, Gray JI, Asghar A et al (1996) Effects of dietary antioxidants and oxidized oil on membranal lipid stability and pork product quality. J Food Sci 54(5):1193–1197

Cernak I, Savic VJ, Kotur J et al (2000) Characterization of plasma magnesium concentration and oxidative stress following graded traumatic brain injury in humans. J Neurotraum 17(1):53–68

Chameau P, Qin Y, Spijker S et al (2007) Glucocorticoids specifically enhance L-type calcium current amplitude and affect calcium channel subunit expression in the mouse hippocampus. J Neurophysiol 97:5–14

Chaves A, Stanford K, Gibson L et al (2008) Effects of carvacrol and cinnamaldehyde on intake, rumen fermentation, growth performance, and carcass characteristics of growing lambs. Anim Feed Sci Technol 145:396–408

Cheng C, Liu Z, Zhou Y (2017) Effect of oregano essential oil supplementation to a reduced-protein, amino acid-supplemented diet on meat quality, fatty acid composition, and oxidative stability of Longissimus thoracis muscle in growing-finishing pigs. Meat Sci 133:103–109

Chen-Quay SC, Eiting KT, Li AWA et al (2009) Identification of tight junction modulating lipids. J Pharm Sci 98:606–619

Chouliara E, Karatapanis A, Savvaidis I et al (2007) Combined effect of oregano essential oil and modified atmosphere packaging on shelf-life extension of fresh chicken breast meat, stored at 4 °C. Food Microbiol 24:607–617

Chrousos GP, Kino T (2007) Glucocorticoid action networks and complex psychiatric and/or somatic disorders. *Stress* 10:213–219

Cleeter MWJ, Cooper JM, Darley-Usmar VM et al (1994) Reversible inhibition of cytochrome c oxidase, the terminal enzyme of the mitochondrial respiratory chain, by nitric oxide: implications for neurodegenerative diseases. FEBS Lett 345(1):50–54

D'Souza DN, Pethick DW, Dunshea FR et al (2003) Nutritional manipulation increases intramuscular fat levels in the Longissimus muscle of female finisher pigs. Aust J Agr Res 54(8):745–749

Daiwen C, Keying Z, Zuyu H (2002) The mechanism for developing pork quality traits. J Sichuan Agric Univ 20(1):60–66

Davey CL, Gilbert K, Carse W (1976) Carcass electrical stimulation to prevent cold shortening toughness in beef. N Z J Agric Res 19:13–18

De Smet S, Pauwels H, De Bie S et al (1996) Effect of halothane genotype, breed, feed withdrawal, and lairage on pork quality of Belgian slaughter pigs. J Anim Sci 74:1854–1863

Del Barrio AS, Schrama JW, van Der Hel W et al (1993) Energy metabolism of growing pigs after transportation, regrouping, and exposure to new housing conditions as affected by feeding level. J Anim Sci 71:1754–1760

Djenane D, Sanchez-Escalante A, Beltran JA et al (2003) Extension of the shelf life of beef steaks packaged in a modified atmosphere by treatment with rosemary and displayed under UV-free lighting. Meat Sci 64:417–426

Dokmanovic M, Baltic MZ, Duric J et al (2015) Correlations among Stress Parameters, Meat and Carcass Quality Parameters in Pigs. Asian-Australas J Anim 28:435

Drain ME, Whiting TL, Rasali D et al (2007) Warm weather transport of broiler chickens in Manitoba. I. Farm management factors associated with death loss in transit to slaughter. Can Vet J 48:76–80

Edens FW (1997) Potential for organic selenium to replace selenite in poultry diets. Zootecnica Int 20(1):33–38

Eisenhofer G, Aneman A, Friberg P et al (1997) Substantial production of dopamine in the human gastrointestinal tract. J Clin Endocrinol Metab 82:3864–3871

Elenkov IJ, Wilder RL, Chrousos GP et al (2000) The sympathetic nerve: an integrative interface between two supersystems: the brain and the immune system. Pharmacol Rev 52(4):595–638

Fasano A, Nataro JP (2004) Intestinal epithelial tight junction as targets for enteric bacteria-derived toxins. Adv Drug Deliv Rev 56(6):795–807

Fasseas M, Mountzouris K, Tarantilis P et al (2008) Antioxidant activity in meat treated with oregano and sage essential oils. Food Chem 106:1188–1194

Faustman C, Cassens RG, Schaefer DM et al (1989) Improvement of pigment and lipid stability in Holstein steer beef by dietary supplementation with vitamin E. J Food Sci 54(4):858–862

Fazio E, Ferlazzo A (2003) Evaluation of stress during transport. Vet Res Commun 27:519–524

Fekete E, Zorrilla EP (2007) Physiology, pharmacology, and therapeutic relevance of urocortins in mammals: ancient CRF paralogs. Front Neuroendocrinol 28:1–27

Ferguson D, Warner R (2008) Have we underestimated the impact of pre-slaughter stress on meat quality in ruminants. Meat Sci 80:12–19

Forman H, Torres M (2003) Reactive oxygen species and cell signaling: respiratory burst in macrophage signaling. Am J Resp Crit Care 166:S4–S8

Freeman DA (2008) Multiple neuroendocrine pathways mediate seasonal immunity. Am J Physiol 294(2):R382–R383

Gay J, Kokkotou E, O'Brien M et al (2008) Corticotropin-releasing hormone deficiency is associated with reduced local inflammation in a mouse model of experimental colitis. Endocrinology 149(7):3403–3409

Geers R, Bleus E, Van Schie T et al (1994) Transport of pigs different with respect to the halothane gene: stress assessment. J Anim Sci 72(10):2552–2558

Gen W, Tani M, Takeshita J et al (2001) Mechanisms of Ca 2+ overload induced by extracellular H 2 O 2 in quiescent isolated rat cardiomyocytes. Basic Res Cardiol 96:623–629

Gispert M, Faucitano L, Oliver M et al (2000) A survey of pre-slaughter conditions, halothane gene frequency, and carcass and meat quality in five Spanish pig commercial abattoirs. Meat Sci 55: 97–106

Goerl K, Eilert S, Mandigo R et al (1995) Pork characteristics as affected by two populations of swine and six crude protein levels. J Anim Sci 73:3621–3626

Grammatopoulos DK, Chrousos GP (2002) Functional characteristics of CRH receptors and potential clinical application of CRH-receptor antagonists. Trends Endocrinol Met 13(10): 436–444

Guardia MD, Estany J, Balasch S et al (2005) Risk assessment of DFD meat due to pre-slaughter conditions in pigs. Meat Sci 70(4):709–716

Güther T, Vormann J, Hollriegl V et al (1992) Role of lipid peroxidation and vitamin E in magnesium deficiency. Magnes Bull 14:57–64

Guzik A, Matthews J, Kerr B et al (2006) Dietary tryptophan effects on plasma and salivary cortisol and meat quality in pigs. J Anim Sci 84:2251–2259

Haddad JJ, Saade NE, Safieh GB (2003) Cytokines and neuro-immune-endocrine interactions: a role for the hypothalamic-pituitary-adrenal revolving axis. Neuroimmunol 145(2):154

Hambrecht E, Eissen JJ, Newman DJ et al (2005) Negative effects of stress immediately before slaughter on pork quality are aggravated by suboptimal transport and lairage conditions. J Anim Sci 83(2):440–448

Heaton FW, Elie JP (1984) Metabolic activity of liver mitochondria from magnesium-deficient rats. Magnesium 3:21–28

Homma S, Sat M, Tanignchi T et al (1993) Pathological and biochemical findings in vitamin E and selenium deficiency in pigs. J Vet Med Jpn

Irie M, Sakimoto M (1992) Fat characteristics of pigs fed fish oil containing eicosapentaenoic and docosahexaenoic acids. J Anim Sci 70:470–477

Janz J, Morel P, Wilkinson B (2007) Preliminary investigation of the effects of low-level dietary inclusion of fragrant essential oils and oleoresins on pig performance and pork quality. Meat Sci 75(2):350–355

John CD, Buckingham JC (2003) Cytokines: regulation of the hypothalamo-pituitary-adrenocortical axis. Curr Opin Pharmacol 3(1):78–84

Kamal-Eldin A, Appelqvist L-Å (1996) The chemistry and antioxidant properties of tocopherols and tocotrienols. Lipids 31:671–701

Kannan G, Terrill T, Kouakou B et al (2000) Transportation of goats: effects on physiological stress responses and live weight loss. J Anim Sci 78:1450–1457

Kapoor A, Winn J (2006) Located hidden random fields: learning discriminative parts for object detection. Comput Vis ECCV 3953:302–315

Kietzman ML (1985) Sensory and perceptual behavioural markers of schizophrenia. Psychopharmacol Bull 21:518–522

Lambillotte C, Gilon P, Henquin JC (1997) Direct glucocorticoid inhibition of insulin secretion. An in vitro study of dexamethasone effects in mouse islets. J Clin Invest 99(3):414–423

Lambooij E (2000) Transport of pigs. In: Grandin T (ed) Livestock handling and transport, 2nd edn. CABI Publ, Wallingford, Oxford

Larzul C, Le Roy P, Gueblez R et al (1997) Effect of halothane genotype (*NN*, *Nn*, *nn*) on growth, carcass and meat quality traits of pigs slaughtered at 95 kg or 125 kg live weigh. J Anim Breed Genet 114:309–320

Lawrie RA (2006) Lawrie's meat science, 7th edn. Woodhead Publishing, Cambridge, pp 55–56

Lee WR, Shen SC, Lin HY et al (2002) Wogonin and fisetin induce apoptosis in human promyeloleukemic cells, accompanied by a decrease of reactive oxygen species, and activation of caspase 3 and Ca^{2+}-dependent endonuclease. Biochem Pharmacol 63(2):225–236

Lencesova L, O'Neill A, Resneck WG et al (2004) Plasma membrane-cytoskeleton-endoplasmic reticulum complexes in neurons and astrocytes. J Bio Chem 279:2885–2893

Le Roy P, Naveau J, Elsen JM et al (1990) Evidence for a new major gene influencing meat quality in pigs. Genet Res 55(1):33–40

Lightman SL (2008) The neuroendoerinology of stress: a never ending story. Neuroendocrinology 20(6):880–884

Mack DR, Michail S, Wei S et al (1999) Probiotics inhibit enteropathogenic E.coli adherence in vitro by inducing intestinal mucin gene expression. Am J Physiol 276:941–950

Marchiando AM, Graham WV, Turner JR (2010) Epithelial barriers in homeostasis and disease. Annu Rev Pathol 5:119–144

Mastorakos G, Weber JS, Magiakou MA et al (1994) Hypothalamic pituitary-adrenal axis activation and stimulation of systemic vasopressin secretion by recombinant interleukin-6 in humans: potential implications for the syndrome of inappropriate vasopressin secretion. Clin Endocrinol Metab 79(4):934–939

McLennan DH, Phillips MS (1992) Malignant hyperthermia. Science 256:789–794

Mildaziene V, Baniene R, Nauciene Z et al (1995) Calcium indirectly increases the control exerted by the adenine nucleotide translocator over 2-oxoglutarate oxidation in rat heart mitochondria. Arch Biochem Biophys 324(1):130–134

Monin G, Sellier P (1985) Pork of low technological quality with a normal rate of muscle pH fall in the immediate postmortem period: the case of the Hampshire breed. Meat Sci 13:49–63

Monnikes H, Tebbe J, Hildebrandt M et al (2001) Role of stress in functional gastrointestinal disorders. Digest Dis 19:201–211

Musch MW, Walsh-Reitz MM, Chang EB (2006) Roles of ZO-1, occludin, and actin in oxidant-induced barrier disruption. Am J Physiol Gastrointest Liver Physiol 290:G222–G231

Nakaya H, Tohse N, Kanno M (1987) Electrophysiological derangements induced by lipid peroxidation in cardiac tissue. Am J Phys 253(5):H1089–H1097

Nilsson UA, Schoenberg MH, Aneman A et al (1994) Free radicals and pathogenesis during ischemia and reperfusion of the cat small intestine. Gastroenterology 106(3):629–636

Odore R, D'angelo A, Badino P et al (2004) Road transportation affects blood hormone levels and lymphocyte glucocorticoid and β-adrenergic receptor concentrations in calves. Vet J 168:297–303

Onmaz A, Van Den Hoven R, Gunes V et al (2011) Oxidative stress in horses after a 12-hours transport period. Rev Med Vet 162:213–217

Oussalah M, Caillet S, Saucier L, Lacroix M (2006) Antimicrobial effects of alginate-based film containing essential oils for the preservation of whole beef muscle. J Food Prot 69:2364–2369

Passaquin AC, Lhote P, Rüegg UT (1998) Calcium influx inhibition by steroids and analogs in C2C12 skeletal muscle cells. Br J Pharmacol 124:1751–1759

Peeters E, Geers R (2006) Influence of provision of toys during transport and lairage on stress responses and meat quality of pigs. Anim Sci 82:591–595

Pérez-Campo R, Lopez-Torres M, Cadenas S et al (1998) The Rate of free radical production as a determinant of the rate of aging: evidence from the comparative approach. J Comp Physiol B 168(3):149–158

Perez MP, Palacio J, Santolaria MP et al (2002) Influence of lairage time on some welfare and meat quality parameters in pigs. Vet Res 33(3):239–250

Pisula A (1996) Current technical, technological and quality problems facing the Polish meat industry. Przemysl Spozywczy 50(1):5–7

Pommier SA, Houde A (1993) Effect of the genotype for malignant hyperthermia as determined by a restriction endonuclease assay on the quality characteristics of commercial pork loins. J Anim Sci 71(2):420–425

Radwan MA, El-Gendy KS, Gad AF (2010) Oxidative stress biomarkers in the digestive gland of Theba pisana exposed to heavy metals. Arch Environ Con Tox 58(3):828–835

Rempel W, Lu MY, Mickelson J et al (1995) The effect of skeletal muscle ryanodine receptor genotype on pig performance and carcass quality traits. Anim Sci 60:249–257

Riether C, Doenlen R, Pacheco-Lopez G et al (2008) Behavioural conditioning of immune functions: how the central nervous system controls peripheral immune responses by evoking associative learning processes. Rev Neurosci 19(1):1–17

Rivier C (1993) Effect of peripheral and central cytokines on the hypothalamo-pituitary-adrenal axis of the rat. Ann NY Acad Sci 697:97–105

Rosenvold K, Andersen HJ (2003) Factors of significance, for pork quality - a review. Meat Sci 64 (3):219–237

Rosenvold K, Patersen JS, Laerke HN et al (2001) Muscle glycogen stores and meat quality as affected by strategic finishing feeding of slaughter pigs. J Anim Sci 79:382–391

Rosmond R (2003) Stress induced disturbances of the HPA axis: a pathway to type 2 diabetes. Med Sci Monit 9(2):RA35–39

Rubenfeld GD, Caldwell E, Peabody E et al (2005) Incidence and outcomes of acute lung injury. N Engl J Med 353(16):1685–1693

Sapolsky RM, Romero LM, Munck AU (2000) How do glucocorticoids influence stress responses? Integrating permissive, suppressive, stimulatory, and preparative actions. Endocr Rev 21(1):55–89

Secchi A, Ortanderl JM, Schmidt W et al (2000) Effect of endotoxemia on hepatic portal and sinusoidal blood flow in rats. J Surg Res 89(1):26–30

Shivakumar K, Kumar BP (1997) Magnesium deficiency enhances oxidative stress and collagen synthesis in vivo in the aorta of rats. Int J Biochem Cell Biol 29:1273–1278

Shixuya K, Komori T, Fujiwara A et al (1997) The influence of restraint stress on the exposure of mRNAs for IL-6 and the IL-6 receptor in the hypothalamus and midbrain of the rat. Life Sci 61(10):135–140

Simitzis PE, Symeon GK, Charismiadou MA et al (2010) The effects of dietary oregano oil supplementation on pig meat characteristics. Meat Sci 84(4):670–676

Skandamis P, Nychas GJ (2001) Effect of oregano essential oil on microbiological and physico-chemical attributes of minced meat stored in air and modified atmospheres. J Appl Microbiol 91: 1011–1022

Skandamis P, Tsigarida E, Nychas GJE (2002) The effect of oregano essential oil on survival/death of Salmonella typhimurium in meat stored at 5 C under aerobic, VP/MAP conditions. Food Microbiol 19(1):97–103

Stocker R, Frei B (1991) Endogenous antioxidant defences in human blood plasma. Oxidative stress: oxidants and antioxidants. Springer, London, pp 213–242

Stratakis CA, Chrousos GP (1995) Neuroendocrinology and pathophysiology of the stress system. Ann NY Acad Sci 771:1–18

Sugino K, Dohi K, Yamada K, Kawasaki T (1987) The role of lipid peroxidation in endotoxin-induced hepatic damage and the protective effect of antioxidants. Surgery 101:746–752

Suzuki T (2013) Regulation of intestinal epithelial permeability by tight junctions. Cell Mol Life Sci 70:631–659

Suzuki T, Tanabe S, Hara H (2011) Kaempferol enhances intestinal barrier function through the cytoskeletal association and expression of tight junction proteins in Caco-2 cells. J Nutr 141 (1):87–94

Swaran JS (2011) Arsenic-induced oxidative stress and its reversibility. Free Radical Biol Med 51: 257–281

Tache Y (1999) Cyclic vomiting syndrome: the corticotropin-releasing-factor hypothesis. Digest Dis Sci 44:79S–86S

Tanida S, Mizoshita T, Mizushima T et al (2011) Involvement of oxidative stress and mucosal addressin cell adhesion molecule-1 (MAdCAM-1) in inflammatory bowel disease. J Clin Biochem Nutr 48:112–116

Turner JR (2009) Intestinal mucosal barrier function in health and disease. Nat Rev Immunol 9(11): 799–809

Vanitallie T (2002) Stress: a physiological, clinical, and public health perspective – foreword. Metabolism. 51(6):1–1

Warriss PD (2003) Optimal lairage times and conditions for slaughter pigs: a review. Vet Rec 153: 170–176

Warriss PD, Brown SN, Knowles TG et al (1998) The effect of stocking density in transit on the carcass quality and welfare of slaughter pigs: 2. Results from the analysis of blood and meat samples. Meat Sci 50:447–456

Wolfgang AK, MaoY K, Wang B et al (2009) Lactobacillus reuteri enhances, excitability of colonic AH neurons by inhibiting calcium-dependent potassium channel opening. J Cell Mol Med 13(8B):2261–2270

Yao J, Wang JY, Liu L et al (2010) Anti-oxidant effects of resveratrol on mice with DSS-induced ulcerative colitis. Arch Med Res 41:288–294

Yoshioka G, Imaeda N, Torimoto Y et al (2004) Influence of transport stress on serum cortisol and thyroid hormones in pigs with Halothane gene. Anim Sci J 75:451–456

Zareie M, Johnson-Henry K, Jury J et al (2006) Probiotics prevent bacterial translocation and Improve intestinal barrier function in rats following chronic psychological stress. Gut 55(11): 1553–1560

Zimmerman BJ, Granger DN (1994) Oxygen free radicals and the gastrointestinal tract: role in ischemia-reperfusion injury. Hepatogastroenterology 41(4):337–342

Zou Y, Xiang Q, Wang J (2016) Effects of oregano essential oil or quercetin supplementation on body weight loss, carcass characteristics, meat quality and antioxidant status in finishing pigs under transport stress. Livest Sci 192:33–38

Feed Additives as Dietary Tools to Support Gut Health of Pigs

Runlin Li, Jianming Wu, Xuedong Ding, and Jing Wang

Abstract

Both thymol and rosmarinic acid are polyphenolic compounds that have antibacterial, anti-inflammatory, and antioxidant functions, and can be rapidly absorbed and metabolized in the body, thereby exerting a variety of physiological functions. In this chapter, the research on and application of thymol and rosmarinic acid as feed additives in weaned pigs are reviewed, and the role of thymol and rosmarinic acid in improving animal growth performance, regulating the intestinal microbiota, and improving intestinal health is expounded, which provides a reference for further improving the application of thymol and rosmarinic acid as feed additives in livestock and poultry health breeding.

Keywords

Plant extracts · Thymol · Rosmarinic acid · Weaned piglets · Feed additives · Growth performance · Intestinal microbiota · Intestinal health

List of Abbreviations

DPPH	1,1-diphenyl-2-picryl-hydrazyl radical
E. coli	*Escherichia coli*
EO	Essential oil
GPX	Glutathione peroxidase
IL-1β	Interleukin 1β

R. Li · J. Wu · X. Ding · J. Wang (✉)
National Center for International Research on Animal Gut Nutrition, Jiangsu Key Laboratory of Gastrointestinal Nutrition and Animal Health, Laboratory of Gastrointestinal Microbiology, College of Animal Science and Technology, Nanjing Agricultural University, Nanjing, China
e-mail: 2019205031@njau.edu.cn; jwang8@njau.edu.cn

G. Arsenos, I. Giannenas (eds.), *Sustainable Use of Feed Additives in Livestock*,
https://doi.org/10.1007/978-3-031-42855-5_24

IL-6	Interleukin 6
LPS	Lipopolysaccharide
MIC	Minimum inhibitory concentration
NF-κB	Nuclear factor kappa-B
RA	Rosmarinic acid
ROS	Reactive oxygen species
SOD	Superoxide dismutase
TLR-4	Toll-like receptor 4
TNF-α	Tumor necrosis factor

1 Introduction

Feed additives are generally divided into essential substances (amino acids, major and trace elements, vitamins) and non-essential feed additives (e.g., antibiotics, enzymes, organic acids, probiotics and plant extracts) (Flachowsky and Schulz 1997). In recent years, plant extracts as a kind of important non-essential feed additives play a variety of prebiotic functions, such as modulating intestinal microbiota, improving animal immunity, and prolonging the shelf life of feed (Zeng et al. 2020). And the results of animal experiments show that they can be used as substitutes for antibiotics to promote the growth of piglets, enhance disease resistance, and improve pork quality (Silva et al. 2020; Li et al. 2012).

In the modern pig breeding system, the adoption of pre-weaning technology improves the reproductive efficiency of sows, shortens the slaughter cycle of pigs, and also makes the breeding process of sows more standardized and periodic, thus increasing the economic benefits of the entire pig industry. Although early weaning can help improve the reproductive efficiency of sows, due to the imperfect development of the gastrointestinal tract of early weaned piglets and the immature intestinal immune system, weaned piglets are very vulnerable to the invasion of pathogenic bacteria in the environment leading to the damaged intestinal barrier and immune imbalance, which in turn causes diarrhea, growth retardation, and even death in piglets (Lin et al. 2020). Yellow and white diarrhea and edema disease caused by enterotoxin-producing *Escherichia coli* during the nursery stage are the main causes of diarrhea in weaned piglets (Okello et al. 2015), which seriously restricts their survival rate. The use of antibiotics can inhibit the influence of pathogenic bacteria on the health of piglets and promote the growth of piglets (Hofer 2014), but a large number of antibiotic products will cause a series of hazards, affecting human health and undermining the stability of the living environment. China has banned the use of all growth-promoting drug feed additives from January 1, 2020. Plant extracts are widely used in livestock and poultry breeding as feed additives because of their green and pollution-free nature, and because they can regulate intestinal microbiota and its metabolites, improve body immunity, etc., thus playing a role in promoting intestinal health (Nour et al. 2017; Sood et al. 2009). Thymol is a monoterpene phenolic compound found mainly in the Lamiaceae family plants and has the effect of reducing inflammatory cell infiltration and intestinal permeability (Xiaona et al. 2019; Yeh et al. 2007). Rosmarinic acid (RA) is a water-soluble phenolic acid compound found in the Lamiaceae family plants, which can reduce tissue ulceration to a certain extent and has antioxidant and antibacterial effects (Formiga et al. 2020; Zdarilová et al. 2009). In this chapter, the role of thymol and rosmarinic acid as feed additives in improving animal growth performance, regulating the intestinal microbiota, and improving intestinal health is clarified through literature review, which provides a reference for further improving the application of thymol and rosmarinic acid as feed additives in livestock and poultry health breeding.

2 Thymol and Rosmarinic Acid

2.1 Source and Properties of Thymol and Rosmarinic Acid

Thymol is widely found in the Lamiaceae family plants (thyme, oregano, basil, and peppermint) and in the *Scropholariaceae*, Verbenaceae, Apiaceae, and Ranunculaceae, and belongs to the monoterpene phenol family. Thymol has a variety of pharmacological effects such as antibacterial, antioxidant, and anti-inflammatory (Salehi et al. 2018), and is widely used in cosmetics, food, and pharmaceutical industries. Rosmarinic acid is a water-soluble polyphenolic compound, widely found in the Lamiaceae and Comfrey families, in Cucurbitaceae, Apiaceae, and linden plants. Rosmarinic acid and its derivatives have antibacterial, anti-inflammatory, antioxidant, antitumor, and antidepressant functions, making it to have broad development prospects in the fields of pharmaceuticals, cosmetics, and food. Both thymol and rosmarinic acid can be prepared by biosynthesis, of which rosmarinic acid can be synthesized by *L*-phenylalanine and *L*-tyrosine (Sundaram et al. 2010), while thymol is mainly extracted from thyme oil and oregano oil treated with alkaline solutions such as sodium hydroxide (Lagouri et al. 2011). At present, thymol synthesis through catalyzing *m*-cresol with metal oxides has a high conversion rate and a low cost, and is expected to achieve an industrial production (Lazarevic et al. 2017).

2.2 Metabolism of Thymol and Rosmarinic Acid in the Body

Studies have shown that polyphenols can be degraded in the intestine and metabolized in the liver (Rechner et al. 2002). Seigo Baba et al. administered 50 mg/kg body weight (BW) of rosmarinic acid orally to rats and measured the concentration of relevant metabolites in plasma and urine at 0.5, 1, 2, 3, 5, 8, 12, and 18 hours after ingestion. It was found that the concentration of total rosmarinic acid (including free and conjugated forms) in the blood plasma reached its highest concentration after 0.5 hour of administration, the total methylated rosmarinic acid reached its concentration peak after 1 hour, and its metabolite conjugated form of mesavin reached its concentration peak after 8 hours. It shows that rosmarinic acid is rapidly absorbed by the intestine after oral administration, and most of rosmarinic acid is present in plasma in the form of binding or methylation. Within 8–18 hours of administration, free, glucuronylated, and methylated forms of rosmarinic acid were found in urine (Baba et al. 2004). Nakazawa and Ohsawa (1998) gave rats 200 mg/kg rosmarinic acid by gavage to explore its metabolic process, and rosmarinic acid metabolites were found in plasma and urine, but no rosmarinic acid metabolites were found in bile. Rosmarinic acid was mainly changed to *m*-hydroxyphenylpropionic acid, trans-*m*-coumaric acid, 3-*O*-sulfate, and other degraded forms that excreted from the urine. The metabolism of thymol in rats and humans was mainly based on liver and kidney metabolism (Ocel'ova et al. 2016), and its metabolites were excreted with urine. After oral administration of thymol, it can quickly enter the bloodstream and is almost completely absorbed in the small intestine. Thymol was detected in the blood after 1.35–1.39 hours, when an oral administration of 13.2 mg/kg BW of thymol was carried out in weaned piglets. A full digestive tract half-life of thymol is approximately 1.84–2.05 hours (Michiels et al. 2008). The metabolites of thymol are mainly excreted in the form of thymol glucuronide and thymol sulfate, but salts in the form of thymol hydroquinone sulfate appear in the human body (Kohlert et al. 2002; Austgulen et al. 1987). Studies have also found that thymol is not metabolized by small intestinal microorganisms after the ingestion by animals under in vitro mimicking in vivo conditions, but is partially degraded in the cecum (Michiels et al. 2008; Varel 2002). The above studies show that thymol mainly plays a role in the small intestine and can be quickly absorbed into the blood circulation, while rosmarinic acid can also be rapidly absorbed into the blood, both of which reach the highest concentration in the blood after about one hour of ingestion and are excreted through urine; so thymol and rosmarinic acid are conditionally able to be used as a combination for a better application.

3 The Biological Function of Thymol and Rosmarinic Acid

3.1 Broad-Spectrum Bacteriostatic Effect of Thymol and Rosmarinic Acid

During weaning piglets' breeding process, diseases such as diarrhea caused by pathogenic bacterial infections seriously affect the health of piglets, including enterotoxin-producing *E. coli*, *Salmonella*, *Shigella*, *Staphylococcus aureus*, and *Clostridia*, which cause diarrhea, slow growth, and even death in piglets (Fairbrother 2015; Vanbeersschreurs et al. 1992). Therefore, the added plant extracts with antibacterial activity can effectively prevent diseases caused by pathogenic bacterial infections during weaning piglet breeding. Thymol mainly damages the cell membrane lipid layer of pathogenic bacteria, such as *Escherichia coli* and *Listeria*, so that the cell membrane permeability is increased and intracellular substances are exuded, resulting in the death of bacteria (Trombetta et al. 2005). In addition, thymol can restrain telomerase activity by inhibiting the transcription of telomerase reverse transcriptase, thereby accelerating the rate of bacterial aging and death (Emad et al. 2013). Thymol has a broad-spectrum bacteriostatic effect, and studies showed that it significantly inhibited the growth of pathogenic bacteria such as *Escherichia coli*, *Clostridium perfringens*, and *Salmonella*, and has the strongest antibacterial effect on *E. coli*, with a minimum inhibitory concentration (MIC) of 187.5 μg/mL (El et al. 2011). Rosmarinic acid can significantly inhibit the growth of *Escherichia coli* and *Staphylococcus aureus*, which mainly affects bacterial cell metabolism, protein metabolism, and DNA replication by increasing the permeability of bacterial cell membranes to leak sugars and proteins (Sun et al. 2005; Ellis and Towers 1970). Studies have shown that 6% rosmarinic acid has an inhibitory effect on both *Escherichia coli* and *Staphylococcus aureus*, and has a stronger inhibitory effect on *Staphylococcus aureus* (Sun et al. 2005). In addition, the esterification of rosmarinic acid significantly inhibits the growth of *Bacillus cereus* (Zhu et al. 2018), and rosmarinic acid has a synergistic effect with vancomycin, ofloxacin, and amoxicillin in inhibiting *Staphylococcus aureus* (Ekambaram et al. 2016). The above studies have shown that both thymol and rosmarinic acid have broad-spectrum bacteriostatic effects, and there are differences in the inhibitory effect and bacteriostatic mode of these two plant extracts on pathogenic bacteria; so the combination of them may increase antibacterial effects of pathogenic bacteria.

3.2 Antioxidant Effects of Thymol and Rosmarinic Acid

The stress caused by early weaning can lead to redox imbalance in the intestines and livers of piglets, resulting in oxidative damage, which in turn causes diarrhea in piglets and increases the risk of organ damage in piglets. Both thymol and rosmarinic acid have good antioxidant functions, and, within a certain concentration range, the ability of thyme essential oil (EO) to scavenge 1,1-diphenyl-2-picryl-hydrazyl (DPPH) free radicals is proportional to the concentration and free radical scavenging rate of thyme essential oil under in vitro conditions (Beena et al. 2013). In isoprenaline-induced oxidative stress tests in mice, 7.5 mg/kg thymol reduced lipid peroxide production, increased plasma concentrations of reducing glutathione (GSH) and vitamin C, and enhanced the activity of antioxidant enzymes such as superoxide dismutase (SOD), glutathione peroxidase (GPX), and catalase in cardiomyocyte mitochondria (Nagoor Meeran et al. 2016; Meeran and Prince 2012). Rosmarinic acid relieved the oxidative stress response induced by lipopolysaccharide (LPS) in human gingival fibroblasts, reduced the concentration of intracellular reactive oxygen species (ROS), and increased the concentration of GSH (Zdařilová et al. 2009). More studies have also found that the antioxidant activity of rosmarinic acid is higher than that of flavonoids, which may be due to the fact that the hydrogen atoms of the ortho-hydroxyl groups on the A and B rings in the molecular structure of rosmarinic acid are easily lost, and for rosmarinic acid it is easy to form a semiquinone structure, or even a quinone structure (Cao et al. 2005). The above studies show that thymol and rosmarinic acid can improve the antioxidant function of animals, and thymol mainly plays a role by affecting the activity of antioxidant enzymes, while rosmarinic acid reduces the probability of peroxidation by reducing the production of ROS in the body; so the combination of them may better exert their antioxidant functions, which are conducive to improving the ability of weaned piglets to resist oxidative damage.

3.3 Anti-Inflammatory Effects of Thymol and Rosmarinic Acid

The infection of pathogenic bacteria and the loss of balance in redox reaction during the feeding process of weaned piglets cause inflammation in the body, which in turn leads to diarrhea and decreased resistance of piglets (Lauridsen 2019; Melin et al. 2004). Both thymol and rosmarinic acid inhibited the process of inflammation in the animal body and improved the body's immunity (Chang et al. 2022; Omonijo et al. 2019). The anti-inflammatory effect of thymol has been verified in animal models and cell tests, in that thymol inhibited the toll-like receptor 4 (TLR-4) signaling pathway induced by LPS to exert anti-inflammatory effects and alleviated the inflammatory response by reducing the concentrations of inflammatory factors TNF-α (tumor necrosis factor-alpha) and interleukin (IL)-6 in mouse mammary epithelial cells and blocking the phosphorylation pathway of p38MAPK (Dejie et al. 2014). Thymol reduced the secretion of inflammatory factors TNF-α, IL-6, and IL-1β in lung tissue of rats with lung injury by inhibiting the nuclear factor kappa-B (NF-κB) signaling pathway, and upregulated the expression of erythroid-derived nuclear factor Nrf2 (Lan et al. 2018). Rosmarinic acid also has anti-inflammatory effects. Sonia Marinho et al. constructed an experimental acute colitis model in mice, giving rats 5, 10, and 20 mg/kg/d of nanovesicle rosmarinic acid per day, and found that rosmarinic acid reduced Dextran Sulfate Sodium Salt (DSS)-induced colitis in mice, relieved inflammatory cell infiltration, and decreased the concentrations of TNF-α (tumor necrosis factor-alpha) and IL-1β (interleukin-1beta) and reduced inflammation and oxidative stress by modulating the Nrf2/HO-1 signaling pathway (Marinho et al. 2021). Further studies have found that rosmarinic acid could be used to treat mouse models of acute lung injury, exerting anti-inflammatory effects by modulating Extracellular regulated protein kinases/Mitogen-activated protein kinases (ERK/MAPK) signaling pathways and reducing the concentration of inflammatory factors TNF-α, IL-6, and IL-1β (Chu et al. 2012). Moreover, rosmarinic acid inhibited the activation of the NF-κB pathway by inhibiting the phosphorylation of IκB, thereby reducing the inflammatory response (Moon et al. 2010). The above results show that there are differences in the way that thymol and rosmarinic acid exert anti-inflammatory effects, but both of them can reduce the concentration of inflammatory factors to reduce the body's inflammatory response; so the combination of them may increase their anti-inflammatory effects by inhibiting different inflammatory pathways.

3.4 Other Effects of Thymol and Rosmarinic Acid

Plant extracts often have a variety of physiological effects, among which the anticancer effect is particularly pronounced. Studies have shown that thymol induced phospholipase-dependent Ca^{2+} release in human prostate cancer cells and entered the cytoplasm through other channels, increasing Ca^{2+} concentration and causing prostate cancer cell death (Jeng-Hsien et al. 2017). In addition, thymol exhibited other properties such as growth, pro-growth, genotoxicity, anthelmintic, antitumor, and vasorelaxant (Nordin et al. 2020; Aanyu et al. 2018). Rosmarinic acid affected the Warburg effect in the tumor microenvironment by inhibiting the IL-6/STAT3 and miR-155 pathways in cancer cells (Xu et al. 2016). Furthermore, rosmarinic acid attenuated the expression of mRNA of fatty acid synthase in ethanol-induced hepatocyte lines and reduced the content of glycine and succinic acid associated with carnitine synthesis, helping to improve ethanol-induced lipid synthesis in cells (Guo et al. 2020).

4 Application of Thymol and Rosmarinic Acid in Pig Production

4.1 Effects of Thymol and Rosmarinic Acid on Growth Performance

Thymol has been found to improve the performance of weaned piglets and fattening pigs in many studies, and the addition of 50 mg/kg thymol to the diet of weaned piglets increased the average daily weight gain of piglets, but this diet treatment had no significant effect on feed conversion and diarrhea rate (Jiang et al. 2017). However, the addition of 500 mg/kg of thymol to the diet not only increased the daily weight gain and feed conversion rate of weaned piglets, but also significantly reduced the diarrhea rate of weaned piglets (Van Noten et al. 2020). A higher dosage of thymol added to the diet of weaned piglets also promoted the growth performance, while adding 1% thymol to the diet of weaned piglets could also improve the daily feed intake and average daily gain of piglets, but had no significant effect on the feed conversion rate (Trevisi et al. 2007). This result suggests that the effect of adding 1% (10 g/kg) to the diet of weaned piglets on the growth performance is lower than that of adding 500 mg/kg group. There are few studies on the application of thymol in fattening pig breeding. Luo et al. (2020) showed that the addition of 0.01% thymol to fattening pig feed increased the daily weight gain of fattening pigs, but had no significant effect on feed conversion rate. The use of 200 mg/kg rosmarinic acid extract as a feed additive has been reported to significantly improve the growth performance of weaned piglets and significantly reduce the rate of diarrhea in weaned piglets (Yang et al. 2021; Cullen et al. 2005). The above studies have shown that both thymol and rosmarinic acid can promote the growth performance of pigs, especially weaned piglets, and reduce diarrhea in piglets (Table 1).

Table 1 Effects of thymol and rosmarinic acid on animal growth performance

Objects	Day/weight	Dosage	Concentration (%)	Average daily feed intake (g)	Average daily gain (g)	Feed conversion ratio	Diarrhea rate (%)	References
Piglets (Duroc × [large white × landrace])	24 d 8.63 ± 1.69 kg	50 mg/kg thymol	13.50		↑*		/	Jiang et al. (2017)
Piglets (Duroc × [large white × landrace])	28 d 6.5 ± 1.5 kg	500 mg/kg thymol	99.50	↑*	↑*	↑*	↓*	Van Noten et al. (2020)
Piglets (Duroc × [large white × landrace])	24 d 6.58 ± 0.58 kg	1%	99	↑*	↑*		/	Trevisi et al. (2007)
Finishing pigs (Duroc × landrace × Yorkshire)	150 d	0.01% thymol	99		↑*		/	Luo et al. (2020)
Piglets (Duroc × landrace × Yorkshire)	8.37 ± 1.58 kg	0.01% essential oils blend	18		↑*		↓*	Li et al. (2012)
Piglets (Duroc × landrace × Yorkshire)	6.59 ± 0.90 kg	50 mg carvacrol and 50 mg thymol/kg	99					Wei et al. (2017)
Piglets (Duroc × [large white × landrace])	23 d 6.65 ± 0.33 kg	100 mg/kg rosemary extract	/					Yang et al. (2021)
		200 mg/kg rosemary extract		↑*	↑*	↑*	↓*	
		400 mg/kg rosemary extract		↑*	↑*	↑*	↓*	
Male pigs (large white × [large white × landrace])	91 d 42.0 ± 1.2 kg	1 g/kg freeze-dried rosemary	/	↑*		↑*	/	Cullen et al. (2005)
		10 g/kg freeze-dried rosemary						

*Indicates a significant change

4.2 Effects of Thymol and Rosmarinic Acid on the Intestinal Microbiota

Both thymol and rosmarinic acid have a bacteriostatic effect, which affects the composition of the animal's intestinal microbiota. Studies showed that the addition of 500 mg/kg of thymol to the feed of weaned piglets reduced the gene copy numbers of *E. coli* in the intestines of piglets and significantly increased the gene copy numbers of *Lactobacilli* (Van Noten et al. 2020). Similarly, the addition of 0.01% essential oils blend (thymol 18%) to the diet of weaned piglets also significantly reduced the number of *Escherichia coli* in ileum and colon contents, and significantly increased the number of *Lactobacilli* (Li et al. 2012). And it was also found that 50 mg/kg essential oil addition can also increase the number of *Lactobacilli* in ileum contents, and reduce the number of *Escherichia coli* (Jiang et al. 2015). Wei et al. study found that 50 mg carvacrol and 50 mg thymol/kg could significantly increase the relative abundance of *Lactobacilli* in jejunum contents of weaned piglets, and significantly reduce the relative abundance of *Escherichia coli* and *Enterococcus* (Wei et al. 2017). These results indicate that the addition of thymol (18 mg/kg) at a lower level can also change the gene copy numbers of *Lactobacillus* and *Escherichia coli* in the intestinal tract of weaned piglets and play a prebiotic role in both the small intestine and the large intestine. The addition of different doses of rosemary extract such as 100, 200, and 400 mg/kg to weaned piglet diets had similar results, and the relative abundance of *E. coli* in the ileum of piglets was significantly reduced, but the addition of 100 mg/kg rosemary extract did not have a significant effect on the relative abundance of *Lactobacilli* and *Bifidobacteria*, and the addition of 200 mg/kg and 400 mg/kg of rosemary extracts significantly increased the relative abundance of *Lactobacilli* and *Bifidobacteria* (Yang et al. 2021). In addition, in vitro studies showed that the Chao1 index of microorganisms in the rumen juice fermentation system was significantly increased by adding 200 mg/L thymol to the system, which indicated that the richness of microorganisms in the fermentation system was improved. In addition, the relative abundance of *Lactobacilli* in the treatment group with 200 mg/L thymol was higher than that in the control group, and the concentration of isobutyric acid and valeric acid produced by fermentation in the treatment group was also significantly increased. However, when 400 mg/L of thymol was added to the rumen juice fermentation system in vitro, the Chao1 index of the microbiota and the abundance of *Bifidobacteria* were significantly lower than those of 200 mg/L of thymol treatment. The concentrations of short-chain fatty acids and ammonia nitrogen in the 400 mg/L thymol rumen juice fermentation system were significantly lower than those in the 200 mg/L thymol treatment, suggesting that excessive concentrations of thymol may adversely affect the microbiota of the digestive tract (Yu et al. 2020) (Table 2).

Table 2 Effects of thymol and rosmarinic acid on animal intestinal microbiota

Objects	Day/ weight	Dosage	Sample	*Lactobacillus*	*Escherichia coli*	*Bifidobacterium*	*Enterococcus*	References
Piglets (Duroc × [large white × landrace])	28 d 6.5 ± 1.5 kg	500 mg/kg thymol	Jejunum contents	↑*	↓*			Van Noten et al. (2020)
Piglets (Duroc × landrace × Yorkshire)	8.37 ± 1.58 kg	0.01%essential oils blend (thymol 18%)	Cecum and colon contents	↑*	↓*			Li et al. (2012)
Piglets (Duroc × landrace × Yorkshire)	6.59 ± 0.90 kg	50 mg carvacrol and 50 mg thymol/ kg	Jejunum contents	↑*	↓*		↓*	Wei et al. (2017)
Piglets (Stambo HBI × Dalland)	24 d 8.10 kg	50 mg/kg EO blend	Ileum contents	↑*	↓*			Jiang et al. (2015)
Piglets (Duroc × [large white × landrace])	23 d 6.65 ± 0.33 kg	100 mg/kg rosemary extract	Cecum contents		↓*			Yang et al. (2021)
		200 mg/kg rosemary extract	Cecum contents	↑*	↓*	↑*		
		400 mg/kg rosemary extract	Cecum contents	↑*	↓*	↑*		

*Indicates a significant change

4.3 Effects of Thymol and Rosmarinic Acid on Intestinal Health

Plant extracts have a variety of biological functions such as anti-inflammatory and antioxidant, and the addition of plant extracts to feed grains shows beneficial effects on animal intestinal health. Previous studies have shown that 0.01% essential oils blend (thymol 18%) added to the diet of weaning piglets can significantly improve the digestibility of dry matter and crude protein in the ileum of weaning piglets (Li et al. 2012), and adding 50 mg/kg essential oil in the diet of weaning piglets can improve the digestibility of lipid, crude protein, and neutral detergent fiber (NDF, internal factor fiber) in the feces of piglets (Jiang et al. 2015). Several studies have shown that thymol can significantly increase the villus height and the ratio of villus to crypt of jejunum in weaned piglets (Wei et al. 2017; Li et al. 2012), and the addition of 200 mg/kg and 400 mg/kg rosmarinic acid extracts can also significantly increase the villus height and the ratio of villus to crypt of jejunum and ileum in weaned piglets (Yang et al. 2021). Thymol can also improve the intestinal barrier function of weaned piglets; a study found that the dietary addition of 500 mg/kg thymol can significantly reduce intestinal permeability and the concentration of Fluorescein isothiocyanate-labeled dextran 4 (FD4) (Jiang et al. 2017). Thymol and rosmarinic acid both have anti-inflammatory effects, and studies have found that different concentrations of thymol and rosmarinic acid can alleviate intestinal inflammation in weaned piglets and reduce the expression of inflammatory factors (Luo et al. 2020; Van Noten et al. 2020; Wei et al. 2017; Jiang et al. 2015; Li et al. 2012). In addition, the cell experiment found that the treatment of small intestinal epithelial cells (IPEC-J2) with thymol and rosmarinic acid significantly increased the expression of small intestinal cell tight junction proteins and enhanced the intestinal tight junction functions (Pomothy et al. 2020). In alleviating intestinal inflammation, the addition of thymol and rosmarinic acid improved intestinal inflammatory response, increased the content of immunoglobulins in the blood, inhibited the expression of NF-κB pathway proteins, and reduced the expression of inflammatory factors in the intestinal mucosa (Cai et al. 2019; Hashemipour et al. 2013).

5 The Effect of Thymol and Rosmarinic Acid When Used in Combination with Other Substances

Both thymol and rosmarinic acid are polyphenols that contain a variety of functional groups, which give them a variety of biochemical properties, and when used in conjunction with some other substances, may increase their biological activity or increase their bioavailability. However, there have been few previous studies on this, and the existing results show that the combination of thymol, carvacrol, and acids such as benzoic acid can enhance their effect. The study by Diao et al. (2015) found that adding 1000 mg/kg benzoic acid + 100 mg/kg thymol, 1000 mg/kg benzoic acid + 200 mg/kg thymol, and 2000 mg/kg benzoic acid + 100 mg/kg thymol significantly improved the digestibility of crude protein and crude fat in weaned piglets, but the 2000 mg/kg benzoic acid + 100 mg/kg thymol treatment significantly improved the growth performance of piglets, significantly increased the villi height and villi ratio of height to crypt in the jejunum and ileum of piglets, significantly increased the activities of digestive enzymes such as lactase, lipase, and maltase in the jejunum and ileum of piglets, significantly reduced the relative abundance of *E. coli* in the ileum of piglets, and significantly increased the relative abundance of *Lactobacilli* and *Bifidobacteria*, which indicates that thymol has a better effect when combined with higher concentrations of organic acids. Wei et al. (2017) also found that 50 mg/kg thymol + 50 mg/kg carvacrol relieved weaning stress in piglets, significantly increased the villi height and villi ratio of height to crypt in the jejunum of weaned piglets, improved the antioxidant function of the jejunum, reduced the concentrations of TNF-α and IL-1β in the jejunal mucosa, significantly reduced the relative abundance of *E. coli* in the jejunum, and significantly increased the relative abundance of *Lactobacilli*. Studies on the combination of thymol and rosmarinic acid have not been reported, but the combination of oregano oil and rosemary suggests a synergistic effect. Jiang et al. (2015) study found that thymol combined with enzymes had a combination effect, and the results showed that the combined use of 50 mg/kg thymol and 100 mg/kg enzymes (xylanase and β-glucanase) significantly reduced the villus height/crypt depth in ileum, and significantly reduced the gene expression of IL-1β in ileum's mucosa; the numbers of *Escherichia coli* were significantly decreased and *Lactobacillus* were significantly increased in ileum of piglet. There was no report on the combination of rosmarinic acid with thymol, but, as important active ingredients in oregano oil and rosemary extract, studies have confirmed that oregano oil (2 mg /kg) and rosemary extract (2 mg /kg) can exert the same effect of enrofloxacin, reduce diarrhea of newborn piglets, and significantly reduce the counts of *Escherichia coli* and *Clostridium perfringens* in ileum (Rossi et al. 2015). The above research shows that thymol has a good composite effect, which can cause a combined effect with a variety of phenolic acids and enzymes, and the combination of thymol and rosmarinic acid may have a synergistic effect.

Thymol and rosmarinic acid have a variety of biological functions, and animal experiments have also shown that thymol and rosmarinic acid as feed additives have the effects of improving the growth performance, reducing diarrhea, modulating intestinal microbiota, and improving intestinal health.

Acknowledgments The authors thank the National Key R&D program of China (2017YFE0135200) for financial support.

Glossary

Feed additives Feed additive refers to a small amount or trace number of substances added in the process of feed production, processing, and use; the amount of feed additive is small but the effect is significant.

Weaned piglets The stage of piglets from delivery room to nursery has a great impact on the stress of weaned piglets.

Intestinal microbiota There are a large number of microorganisms in the intestines of animals, which help the host to complete a variety of physiological and biochemical functions.

Intestinal health Intestinal health is a very important part of the body. It can affect many factors such as weight, disease, immunity, and even mood.

Metabolism Molecules are absorbed by the body, hence a series of chemical reactions occur under the action of body enzymes, also known as biological transformation.

Broad-spectrum bacteriostatic Drugs that can inhibit or kill a variety of different kinds of bacteria and have a wide range of antibacterial effects are called broad-spectrum antibacterial drugs. They not only have antibacterial effects on Gram-positive and Gram-negative bacteria, but also have inhibitory effects on mycoplasma, chlamydia, rickettsia, etc.

Antioxidant effects Antioxidation is any substance that can effectively inhibit the oxidation reaction of free radicals in low concentration, and its mechanism of action can be direct action on free radicals or indirect consumption of substances that are easy to generate free radicals to prevent further reactions.

Inflammatory It is a kind of defense response of the body to stimulation, manifested as redness, swelling, heat, pain, and dysfunction. Inflammation can be infectious inflammation caused by infection or non-infectious inflammation not caused by infection.

Minimum inhibitory concentration The sensitivity of bacteria to drugs is determined by tube dilution method. Thus, minimum inhibitory concentration is the amount of drug contained in 1 mL of the maximum dilution tube that completely inhibits the growth of bacteria.

References

Aanyu M, Betancor MB, Monroig O (2018) Effects of dietary limonene and thymol on the growth and nutritional physiology of Nile tilapia (Oreochromis niloticus). Aquaculture 488:217–226

Austgulen LT, Solheim E, Scheline RR (1987) Metabolism in rats of p-cymene derivatives: carvacrol and thymol. Pharmacol Toxicol 61:98–102

Baba S, Osakabe N, Natsume M, Terao J (2004) Orally administered rosmarinic acid is present as the conjugated and/or methylated forms in plasma, and is degraded and metabolized to conjugated forms of caffeic acid, ferulic acid and m-coumaric acid. Life Sci 75:165–178

Beena, Kumar D, Rawat DS (2013) Synthesis and antioxidant activity of thymol and carvacrol based Schiff bases. Bioorg Med Chem Lett 23:641–645

Cai X, Yang F, Zhu L, Xia Y, Wu Q, Xue H, Lu Y (2019) Rosmarinic acid, the main effective constituent of Orthosiphon stamineus, inhibits intestinal epithelial apoptosis via regulation of the Nrf2 pathway in mice. Molecules 24.17

Cao H, Cheng WX, Li C, Pan XL, Li TH (2005) Dft study on the antioxidant activity of rosmarinic acid. J Mol Struct Theochem 719:177–183

Chang SY, Song MH, Lee JH, Oh HJ, Kim YJ, An JW, Go YB, Song DC, Cho HA, Cho SY, Kim DJ, Kim MS, Kim HB, Cho JH (2022) Phytogenic feed additives alleviate pathogenic Escherichia coli-induced intestinal damage through improving barrier integrity and inhibiting inflammation in weaned pigs. J Anim Sci Biotechnol 13:1

Chu X, Ci X, He J, Jiang L, Wei M, Cao Q, Guan M, Xie X, Deng X, He J (2012) Effects of a natural prolyl Oligopeptidase inhibitor, rosmarinic acid, on lipopolysaccharide-induced acute lung injury in mice. Molecules 17:3586–3598

Cullen S, Monahan F, O'doherty, J. (2005) The effect of dietary garlic and rosemary on grower-finisher pig performance and sensory characteristics of pork. J Anim Sci 83:29–29

Dejie L, Fengyang L, Yunhe F, Yongguo C, Xiaojing S, Tiancheng W, Wei W, Mengyao G, Ershun Z, Depeng L, Zhengtao Y, Naisheng Z (2014) Thymol inhibits LPS-stimulated inflammatory response via down-regulation of Nf-κB and Mapk signaling pathways in mouse mammary epithelial cells. Inflammation 37:214–222

Diao H, Zheng P, Yu B, He J, Mao XB, Yu J, Chen DW (2015) Effects of benzoic acid and thymol on growth performance and gut characteristics of weaned piglets. Asian Australas J Anim Sci 28:827–839

Ekambaram SP, Perumal SS, Balakrishnan A, Marappan N, Gajendran SS, Viswanathan V (2016) Antibacterial synergy between rosmarinic acid and antibiotics against methicillin-resistant Staphylococcus aureus. J Intercult Ethnopharmacol 5:358–363

El AS, Ibnsouda KS, Latrache H, Zineb G, Mouradi H, Remmal A (2011) Carvacrol and thymol components inhibiting Pseudomonas aeruginosa adherence and biofilm formation. Afr J Microbiol Res 5:3229

Ellis BE, Towers GHN (1970) Biogenesis of rosmarinic acid in Mentha. Biochem J 118:291

Emad D, Mansoor O, Akbar BA, Ashkan G, Smith ML (2013) Thymol antifungal mode of action involves telomerase inhibition. Med Mycol 51(8):826–834

Fairbrother JM (2015) E-coli diarrhea in weaned piglets etiology, clinical symptoms and risk factors. Tieraerztliche Umschau 70:87–89

Flachowsky G, Schulz E (1997) Feed supplements and their significance for performance and ecology. Archiv Fur Tierzucht 40:101–107

Formiga R, Balbino E, Júnior A, Vasconcelos RC, Batista LM (2020) p-cymene and Rosmarinic acid ameliorate TNBS-induced intestinal inflammation Upkeeping ZO-1 and MUC-2: role of antioxidant system and immunomodulation. Int J Mol Sci 21:5870

Guo C, Shangguan Y, Zhang M, Ruan Y, Xue G, Ma J, Yang J, Qiu L (2020) Rosmarinic acid alleviates ethanol-induced lipid accumulation by repressing fatty acid biosynthesis. Food Funct 11:2094–2106

Hashemipour H, Kermanshahi H, Golian A, Veldkamp T (2013) Effect of thymol and carvacrol feed supplementation on performance, antioxidant enzyme activities, fatty acid composition, digestive enzyme activities, and immune response in broiler chickens. Poult Sci 92:2059–2069

Hofer U (2014) Effects of in-feed antibiotics on pig microbiota. Nat Rev Microbiol 12:234–234

Jeng-Hsien Y, Chou C-T, Chen IS et al (2017) Effect of thymol on Ca^{2+} homeostasis and viability in PC3 human prostate cancer cells. Chin J Physiol

Jiang X, Awati A, Agazzi A, Vitari F, Ferrari A, Bento H, Crestani M, Domeneghini C, Bontempo V (2015) Effects of a blend of essential oils and an enzyme combination on nutrient digestibility, ileum histology and expression of inflammatory mediators in weaned piglets. Animal 9:417–426

Jiang XR, Li XL, Awati A, Bento H, Zhang HJ, Bontempo V (2017) Effect of an essential oils blend on growth performance, and selected parameters of oxidative stress and antioxidant defence of Escherichia coli challenged piglets. J Anim Feed Sci 26:38–43

Kohlert C, Schindler G, Marz RW, Abel G, Brinkhaus B, Derendorf H, Grafe EU, Veit M (2002) Systemic availability and pharmacokinetics of thymol in humans. J Clin Pharmacol 42:731–737

Lagouri V, Guldas M, Gurbuz O (2011) In vitro antioxidant/free radical scavenging and antibacterial properties of endemic oregano and thyme extracts from Greece. Food Sci Biotechnol 20:1487–1493

Lan Y, Guo H, Lu W, Xiao-Shu Z, Zhou L (2018) Protective effects of thymol on LPS-induced acute lung injury in mice. Microb Pathog 116:8

Lauridsen C (2019) From oxidative stress to inflammation: redox balance and immune system. Poult Sci 98:4240–4246

Lazarevic J, Kolarevic A, Dordevic A, Stojanovic G, Smelcerovic A, Ciuffreda P, Santaniello E (2017) Synthesis, antimicrobial activity and in silico studies on thymol esters. Acta Chim Slov 64:603–612

Li P, Piao X, Ru Y, Han X, Xue L, Zhang H (2012) Effects of adding essential oil to the diet of weaned pigs on performance, nutrient utilization, immune response and intestinal health. Asian Australas J Anim Sci 25:1617

Lin F, Li X, Wen J, Wang C, Peng Y, Feng J, Hu C (2020) Effects of coated sodium butyrate on performance, diarrhea, intestinal microflora and barrier function of pigs during the first 2-week post-weaning. Anim Feed Sci Technol 263:114464

Luo P, Luo L, Zhao WJ, Wang LS, Sun LJ, Wu HY, Li YX, Zhang RX, Shu G, Wang SB, Gao P, Zhu XT, Xi QY, Zhang YL, Wang LN, Jiang QY (2020) Dietary thymol supplementation promotes skeletal muscle fibre type switch in longissimus dorsi of finishing pigs. J Anim Physiol Anim Nutr 104:570–578

Marinho S, Illanes M, Vila-Román J, Motilva V, Talero E (2021) Anti-inflammatory effects of rosmarinic acid-loaded Nanovesicles in acute colitis through modulation of NLRP3 Inflammasome. Biomol Ther 11:162

Meeran MFN, Prince PSM (2012) Protective effects of thymol on altered plasma lipid peroxidation and nonenzymic antioxidants in isoproterenol-induced myocardial infarcted rats. J Biochem Mol Toxicol 26:368–373

Melin L, Mattsson S, Katouli M, Wallgren P (2004) Development of post-weaning diarrhoea in piglets. Relation to presence of Escherichia coli strains and rotavirus. J Vet Med B Infect Dis Vet Public Health 51:12–22

Michiels J, Missotten J, Dierick N, Fremaut D, Maene P, De Smet S (2008) In vitro degradation and in vivo passage kinetics of carvacrol, thymol, eugenol and trans-cinnamaldehyde along the gastrointestinal tract of piglets. J Sci Food Agric 88:2371–2381

Moon DO, Kim MO, Lee JD, Choi YH, Kim GY (2010) Rosmarinic acid sensitizes cell death through suppression of TNF-α-induced NF-κB activation and ROS generation in human leukemia U937 cells. Cancer Lett 288:183–191

Nagoor Meeran MF, Jagadeesh GS, Selvaraj P (2016) Thymol, a dietary monoterpene phenol abrogates mitochondrial dysfunction in β-adrenergic agonist induced myocardial infarcted rats by inhibiting oxidative stress. Chem Biol Interact 244:159

Nakazawa T, Ohsawa K (1998) Metabolism of rosmarinic acid in rats. Jnatprod 61:993–996

Nordin N, Othman SH, Rashid SA, Basha RK (2020) Effects of glycerol and thymol on physical, mechanical, and thermal properties of corn starch films. Food Hydrocoll 106:105884

Nour V, Trandafir I, Cosmulescu S (2017) Bioactive compounds, antioxidant activity and nutritional quality of different culinary aromatic herbs. Notulae Botanicae Horti Agrobotanici Cluj-Napoca 45:179–184

Ocel'ova V, Chizzola R, Pisarcikova J, Novak J, Ivanisinova O, Faix S, Placha I (2016) Effect of thyme essential oil supplementation on thymol content in blood plasma, liver, kidney and muscle in broiler chickens. Nat Prod Commun 11:1545–1550

Okello E, Moonens K, Erume J, De Greve H (2015) Enterotoxigenic Escherichia coli strains are highly prevalent in Ugandan piggeries but disease outbreaks are masked by antibiotic prophylaxis. Trop Anim Health Prod 47:117–122

Omonijo FA, Liu SX, Hui QR, Zhang H, Lahaye L, Bodin JC, Gong JS, Nyachoti M, Yang CB (2019) Thymol improves barrier function and attenuates inflammatory responses in porcine intestinal epithelial cells during lipopolysaccharide (LPS)-induced inflammation. J Agric Food Chem 67:615–624

Pomothy JM, Barna RF, Pászti E, Babiczky K, Gere EP (2020) Beneficial effects of rosmarinic acid on IPEC-J2 cells exposed to the combination of Deoxynivalenol and T-2 toxin. Mediat Inflamm 2020:1–10

Rechner AR, Kuhnle G, Bremner P, Hubbard GP, Moore KP, Rice-Evans CA (2002) The metabolic fate of dietary polyphenols in humans. Free Radic Biol Med 33:220–235

Rossi CA, Soares M, Luchese FC, Santurio JM (2015) Use of essential oils in the control of clinical signs of neonatal diarrhea in piglets born to sows with different parturition orders. Ciencia Animal Brasileira 16:93–102

Salehi B, Mishra AP, Shukla I, Sharifi-Rad M, Del Mar Contreras M, Segura-Carretero A, Fathi H, Nasrabadi NN, Kobarfard F, Sharifi-Rad J (2018) Thymol, thyme, and other plant sources: health and potential uses. Phytother Res 32:1688–1706

Silva CD, Martins CCS, Dias FTF, Sitanaka NY, Ferracioli LB, Moraes JE, Pizzolante CC, Budino FEL, Pereira R, Tizioto P, Paula VRC, Coutinho LL, Ruiz US (2020) The use of an alternative feed additive, containing benzoic acid, thymol, eugenol, and piperine, improved growth performance, nutrient and energy digestibility, and gut health in weaned piglets. J Anim Sci 98

Sood S, Arora B, Bansal S, Muthuraman A, Gill NS, Arora R, Bali M, Sharma PD (2009) Antioxidant, anti-inflammatory and analgesic potential of the Citrus decumana L. peel extract. Inflammopharmacology 17:267–274

Sun X, Wang J, Li H, Du G, Guo D (2005) A study on the antibacterial mechanism of rosmarinic acid. J Qingdao Univ 4:44–48

Sundaram S, Tripathi A, Gupta DK (2010) Metabolic modeling of rosmarinic acid biosynthetic pathway. Bioinformation 5:168–172

Trevisi P, Merialdi G, Mazzoni M, Casini L, Tittarelli C, De Filippi S, Minieri L, Lalatta-Costerbosa G, Bosi P (2007) Effect of dietary addition of thymol on growth, salivary and gastric function, immune response, and excretion of salmonella enterica serovar typhimurium, in weaning pigs challenged with this microbe strain. Ital J Anim Sci 6:374–376

Trombetta D, Castelli F, Sarpietro MG, Venuti V, Cristani M, Daniele C, Saija A, Mazzanti G, Bisignano G (2005) Mechanisms of antibacterial action of three monoterpenes. Antimicrob Agents Chemother 49:2474–2478

Van Noten N, Degroote J, Van Liefferinge E, Taminiau B, De Smet S, Desmet T, Michiels J (2020) Effects of thymol and thymol α-D-glucopyranoside on intestinal function and microbiota of weaned pigs. Animals 10:329

Vanbeersschreurs HMG, Vellenga L, Wensing T, Breukink HJ (1992) The pathogenesis of the post-weaning syndrome in weaned piglets - a review. Vet Q 14:29–34

Varel VH (2002) Carvacrol and thymol reduce swine waste odor and pathogens: stability of oils. Curr Microbiol 44:38–43

Wei H-K, Xue H-X, Zhou Z, Peng J (2017) A carvacrol–thymol blend decreased intestinal oxidative stress and influenced selected microbes without changing the messenger RNA levels of tight junction proteins in jejunal mucosa of weaning piglets. Animal 11:193–201

Xiaona A, Xiaotong S, Yonghao H, Xiaomei Y, Hongli C, Peng Z, Jianbo W (2019) Protective effect of oxytocin on LPS-induced acute lung injury in mice. Sci Rep 9:1

Xu Y, Han S, Lei K, Chang X, Wang K, Li Z, Liu J (2016) Anti-Warburg effect of rosmarinic acid via miR-155 in colorectal carcinoma cells. Eur J Cancer Prev 25:481

Yang M, Yin Y, Wang F, Bao X, Long L, Tan B, Yin Y, Chen J (2021) Effects of dietary rosemary extract supplementation on growth performance, nutrient digestibility, antioxidant capacity, intestinal morphology, and microbiota of weaning pigs. J Anim Sci 99:9

Yeh CC, Kao SJ, Lin CC, Wang SD, Liu CJ, Kao ST (2007) The immunomodulation of endotoxin-induced acute lung injury by hesperidin in vivo and in vitro. Life Sci 80:1821–1831

Yu J, Cai L, Zhang J, Yang A, Wang Y, Zhang L, Guan LL, Qi D (2020) Effects of thymol supplementation on goat rumen fermentation and rumen microbiota in vitro. Microorganisms 8:8

Zdarilová A, Svobodová A, Simánek V, Ulrichová J (2009) Prunella vulgaris extract and rosmarinic acid suppress lipopolysaccharide-induced alteration in human gingival fibroblasts. Toxicol In Vitro 23:386–392

Zeng ZK, Zhang S, Wang HL, Piao XS (2020) Essential oil and aromatic plants as feed additives in non-ruminant nutrition: a review. J Anim Sci Biotechnol 6:7

Zhu F, Wang J, Takano H, Xu Z, Tamura H (2018) Rosmarinic acid and its ester derivatives for enhancing antibacterial, α-glucosidase inhibitory, and lipid accumulation suppression activities. J Food Biochem 43(2):e12719

Feed Additives as Immune-Boosting Factors in Swine Health

Panagiotis D. Tassis

Abstract

Modern swine production systems globally are evolving, and the swine industry should adapt to new circumstances that include reduced antibiotics usage, reduced environmental impact, enhanced welfare requirements, increased sustainability under the scope of climate change, as well as exclusion of formerly used substances in pig feed, such as zinc oxide. The role of novel feed additives in health and productivity of pigs is gaining increased interest, as a factor that can improve response to the health challenges that pigs are facing in intensive units. The immune system is the major tool in combating diseases and disorders of the pigs and the vaccination scheme is the key component of a disease prevention program on farm. However, novel feed additives supporting or boosting the immune system, either at the organ/system level or at the total adaptive immune system level, are gradually becoming significant parts of swine health management programs on farm. Critical points of such immune support with feed additives are the intestinal immune response and intestinal homeostasis of the newly weaned pigs, as well as the proper immune preparation of sows, for improved passive immunity through colostrum in piglets.

Previous research data suggested the beneficial effects of basic feed ingredients and nutrients (amino acids, fatty acids, carbohydrates, etc.), as well as vitamins and mineral complexes, on the proper development and function of pigs' immune system. Recent advances provide evidence that eubiotics such as phytogenic feed additives and respective products of botanical origin, as well as other feed additives (probiotics, prebiotics, enzymes, etc.), could have anti-inflammatory, antioxidative, or antibacterial effects, or/and can support an

P. D. Tassis (✉)
Farm Animals Clinic, Thessaloniki, Greece

School of Veterinary Medicine, Aristotle University of Thessaloniki, Thessaloniki, Greece
e-mail: ptassis@vet.auth.gr

G. Arsenos, I. Giannenas (eds.), *Sustainable Use of Feed Additives in Livestock*,
https://doi.org/10.1007/978-3-031-42855-5_25

improved immune response after antigenic exposure. The present chapter will focus on recent data, developments, and modes of action of novel feed additives that can improve the immune response of pigs.

Keywords

Pigs · Weaning · Feed additives · Immune response · Immune system · Intestinal barrier · Phytogenics · Prebiotics · Probiotics · Antioxidants

1 The Immune System of the Pig and Intestinal Homeostasis

The immune system of the pig is the predominant defense mechanism against infectious and other agents. Basic aspects of the immune response are inflammation, and cellular and humoral response (Oswald et al. 2005). Briefly, after engagement of the immune system (e.g., after contact with an infectious agent) a first defense mechanism takes action. This mechanism includes the innate immune response with phagocytic cells and production of various cytokines, chemokines, and proteins that provide antimicrobial protection, recruit T-cells through the inflammatory process, and further activate the adaptive, or acquired, immune response. Part of the innate system are also the natural killer (NK) cells that present a dual function including an innate response to attack infectious-agent-infected cells and production of cytokines for assisting in the activation of acquired immunity (Gerner et al. 2009; Mair et al. 2014; Chase and Lunney 2019). Thus, NK cells and other parts of the innate immune system such as defensins (host defense peptides), complement system, toll-like receptors (TLRs), type I interferons (IFNs), tumor necrosis factor-α (TNF-α), interleukin (IL)-6, and IL-8 (pro-inflammatory cytokines) defend against pathogens, control infection, and activate the cascade of events of inflammation and adaptive immunity response (Chase and Lunney 2019). As previously demonstrated, toll-like receptors (TLRs) act as pattern-recognition receptors binding microbial ligands present in the lumen and determine the host-immune defense reaction, immune cell recruitment, and induction of mucosal inflammation, aiming at the preservation of intestinal homeostasis (Aderem and Ulevitch 2000).

The adaptive immune system utilizes B-cells, T-cells, cytokines, and antibodies, in order to provide pathogen-specific memory for protection from subsequent infections with the same pathogen. Lymphocyte populations (B- and T-lymphocytes) and antibody production reach optimal functional capacity two to three weeks after the first exposure to the pathogen. Immunoglobulins (Igs) show differences among each type, therefore IgG is the predominant Ig class (>80%) in the serum and colostrum of the pig, whereas IgM accounts for approximately 5–10% of total Igs in serum and colostrum. On the other hand, IgA is considered as mucosal Ig for swine. Mucosa epithelial cells secrete IgA, which assists to neutralize microbes in the lumen of respiratory and gastrointestinal (GIT) tracts (Kaetzel 2014). T-cell subsets express certain polarizing cytokines, the interleukins (ILs) or IFNs, as the immune response matures. Furthermore, cell-mediated immunity includes antigen-presenting cells such as macrophages and dendritic cells that phagocytize and process antigens and then present these antigenic fragments bound to swine leukocyte antigen (SLA) molecules as they contact T-lymphocytes, while also release cytokines (Chase and Lunney 2019). Taken together interactions with macrophages, T-cells and B-cells determine the balance between immune activation and tolerance induction, affecting the mucosal immune system in the GIT and consequently the overall health status of the animals (Bouwens and Savelkoul 2019).

It should be pointed out that piglets have limited ability in terms of T- and B-cell responses during the first few weeks of life, since it takes weeks after birth before B- and T-cell areas are formed in the bronchus-associated lymphoid tissue (BALT); thus they are heavily reliant on innate immunity given the maturity of this system at that time period (Lalles et al. 2007; Humphrey et al. 2019). Moreover, TLR-induced antiviral responses of plasmacytoid dendritic cells are reduced at birth in pigs but develop to a full response within weeks after birth (Jamin et al. 2006). Taken together, the innate defense mechanisms that neither require previous exposure to antigen nor have an immunological "memory" provide a first almost immediate response to the infectious agent and control infection, while at the same time assist in the activation of the adaptive immune system. The latter has immunological "memory" and will produce antibody- and cell-mediated immune responses (Chase and Lunney 2019). However, according to Netea and Van der Meer (2017), the adaptive immune response and the innate immune system can adapt to previous infections and develop memory, thus providing better protection against the same or heterologous infection.

Significant parts of the immune defense system in pigs are the mucosal epithelium (e.g., intestinal and respiratory tract), the microbiome (intestinal microbial ecosystem), and the lymphoid system consisting of the lymph nodes, lymphoid follicles, tonsils, thymus, and spleen (Rothkötter 2009; Wilson and Obradovic 2015). Thus, the first line of defense includes epithelial cells (epithelial layer is rapidly regenerating once every two to three days in pigs), bactericidal fatty acids, normal flora, and the mucus layer that is continuously produced, as well as cells with phagocytic abilities such as granular leukocytes (neutrophils, basophils, mast cells, and eosinophils), and mononuclear phagocytes (circulating blood monocytes and tissue macrophages), providing a mucosal immune system underneath the mucus-covered epithelial cell layer (Uni et al. 2001; Kelly and Mulder 2012; Williams et al. 2015). Furthermore, macrophage-mediated immunity, the activation of aryl hydrocarbon receptor (AHR), and the heat-shock protein (HSP)70 chaperone gene expression are well connected with the modulation of gut-associated immune response (Bouwens and Savelkoul 2019). The mucosal immune system prevents the uptake of pathogens and coordinates immune response, functioning as a defense tool with innate and adaptive cells that are accumulated in, or in transit between, various mucosa-associated lymphoid tissues (MALT) (Bouwens and Savelkoul 2019). The Peyer's patches, mesenteric lymph node, solitary follicles in the intestine, and the tonsils are points of immune response initiation and are included in MALT (Bailey 2009). Each mucosal tissue has its own associated lymphoid tissue resulting in

NALT (nasal cavity), BALT (bronchus/lower airways), GENALT (urogenital tract), and GALT (gut; GIT) (Bouwens and Savelkoul 2019).

An optimal intestinal microbiota prevents colonization of the intestinal epithelium by pathogens and penetration of the gut barrier, modulates the gut-associated lymphoid tissue (GALT) and systemic immunity, and influences gastrointestinal development (Broom 2015). The intestinal microbiota is a dynamic and complex environment and the interactions between the immune system and nutritional signaling at the intestinal level play a significant role in proper immune function. Immune reactions are generated from microbiota signals that direct responses with effector T-cells against pathogens or, in the case of commensals, induce a state of tolerance via modulation of regulatory T-cells (Tregs) and release of immunosuppressive cytokines such as interleukin-10 (IL-10) and transforming growth factor-β (TGF-β; Bouwens and Savelkoul 2019). According to Gresse et al. (2017), weaning age alterations include a reduction in obligate anaerobic bacteria (e.g., *Clostridia* and *Bacteroidia*) and an increase in facultative anaerobic bacteria (e.g., *Enterobacteriaceae*), which translate to decreased microbial diversity and a pro-inflammatory state within the intestine. Factors such as nutrition and management/husbandry approaches have a significant role in the microbiota alterations. Previous studies have presented facts that suggest differences among microbial communities and intestinal immune responses among piglets raised with the sow or fed with milk replacers (Lewis et al. 2012). Moreover, the age of weaning and the contact with solid feed affect the microbiota significantly (Bian et al. 2016). Taken together, various findings suggest that weaning is a predominant event in pigs' life that has multiple effects on gastrointestinal function and immune response, including activation of pathways related to inflammatory responses; alterations in hormonal activity; reduction in gastric motility; reduced villous height; reduction in nutrient, fluid, and electrolyte absorption; and increased permeability to antigens and toxins (Gresse et al. 2017).

2 Challenges for the Immune System and Gastrointestinal Eubiosis of Pigs

As described in the previous part, the intestinal immune system provides protection along the intestinal tract and balances the host response to microflora alterations and intestinal pathogens (Humphrey et al. 2019). However, there are plenty of challenges for the intestinal integrity and homeostasis during all stages of pig production. Diet alterations and stressful conditions, the transfer between units, and mixing of animals are benchmarks during these stages, which could affect proper intestinal function. On the other hand, interactions between the sows, embryos, and suckling piglets affect the intestinal microbiota development in early life and the intestinal immune response. As reported by Everaert et al. (2017), interventions through the maternal diet or direct to piglets in the pre-weaning period may have a beneficial effect on intestinal structural and functional aspects. Such modulations of the intestinal microbiota of sows can affect bacterial populations of the suckling piglets' GIT due to their contact with sow's feces.

The evaluation of the underlying mechanisms associated with intestinal function and immunity during the weaning transition is critical since successful weaning transition can result to an improved final body weight (BW) and reduced production days in total (Tokach et al. 1992). Efforts during that period include provision of highly digestible diets that will assist piglets' transition from milk to solid feed. Findings demonstrated by Main et al. (2004) suggested that an improvement of increasing weaning age by 1 day between days 12 and 21 would result in an extra 0.93 kg of day 42 BW and up to 0.47% reduction in wean-to-finish mortality. On the other hand, a transient anorexia at weaning leads to compromised intestinal barrier function and localized inflammatory response, as well as to an increase of reactive oxygen species (ROS) (Humphrey et al. 2019). The absence of feed stimuli in the intestinal tract at weaning affects immune cells' activity and balance of cytokines, chemokines, and other immune markers (e.g., IL-4, IL-10, and IFN-γ), thus decrease IgA content in intestinal and mucosal tissues and alters intestinal microbiota (Anastasilakis et al. 2013; Demehri et al. 2013). Therefore, it is crucial to establish a steady feed intake at weaning and a smooth transition to solid feed, in order to avoid inflammatory responses at the intestinal level or disruption of the intestinal barrier function.

Challenges from various pathogens such as enterotoxigenic *Escherichia coli* affect the nitrogen flow to the distal ileum, as well as the intestinal structure (e.g., villous height) and intestinal pH, with respective consequences on nutrient digestion and absorption processes (Heo et al. 2009, 2010). Effects of bacterial toxins on mineral and water intestinal absorption can contribute to the induction of post-weaning diarrhea in piglets (Sun and Kim 2017). Parasitic infections such as *Ascaris suum* alone, as well as their possible interactions with enteric bacterial pathogens such as *Lawsonia intracellularis*, can also contribute as disrupting factors to the intestinal function (Boes et al. 2010; Tassis et al. 2022). Viral challenges at the intestinal level such as *Porcine deltacoronavirus*, leading to intestinal microbiota alterations toward dysbiosis conditions (Li et al. 2020a), as well as particular mycotoxins affecting the intestinal barrier integrity and function (e.g., deoxynivalenol or fumonisins), are significant factors that affect intestinal health (Bracarense et al. 2012). Furthermore, reactive oxygen species (ROS), nitric oxide, hydrogen peroxide, or thiobarbituric acid-reactive substances are increased after weaning, due to weaning and weaning-associated intestinal inflammation (Zhu et al. 2012; Wei et al. 2017). Such substances have been associated with overgrowth of *E. coli* populations, whereas they could also result in reduction of natural antioxidants such as vitamin E, thus affect intestinal immune status (Kim et al. 2016; Wei et al. 2017).

Taken together, it is obvious that a variety of infectious and non-infectious challenges affect immune response of pigs during their life (Fig. 1). Furthermore, the antibiotics usage in intensive pig production is also a significant factor adding up to the above-mentioned challenges to intestinal eubiosis. Recent legal requirements for reduction of antibiotics usage and the ban of ZnO in pig feeds create a novel

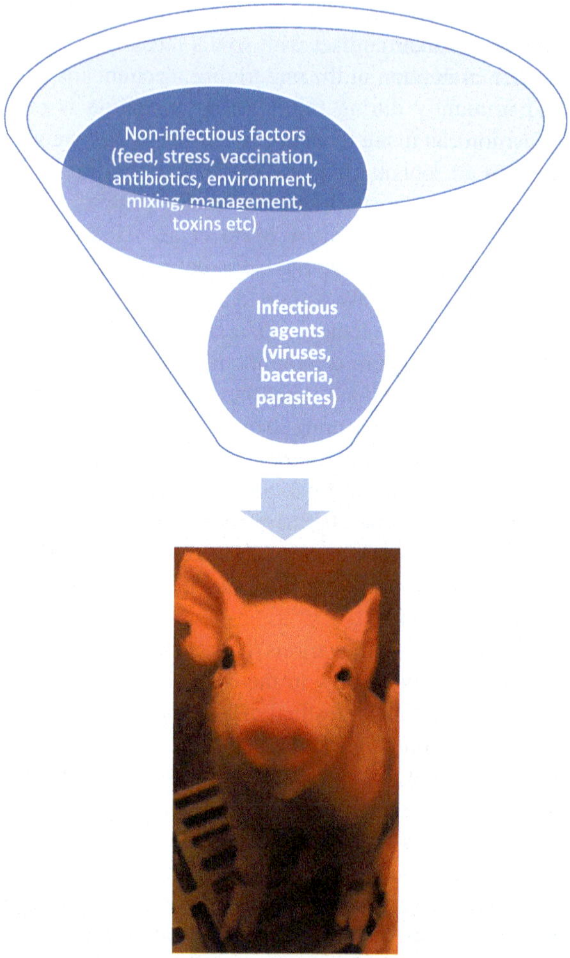

Fig. 1 Challenges for the immune system of the pig at weaning

environment for sustainable and welfare-friendly pig production. Therefore, the aim of optimizing gut health and immune responses is of major significance for the health and performance of the animals with reduced antibiotic usage.

3 Nutrition Interventions for the Improvement of Immune Response in Pigs

Nutritional interventions can affect the immune system at the intestinal as well as at the airways level in response to viral (e.g., porcine reproductive and respiratory syndrome virus) and bacterial (e.g., *Bordetella bronchiseptica*) infections, affecting pig health and performance (Opriessnig et al. 2011; Bouwens and Savelkoul 2019). Long-term overuse of antimicrobials in global pig production resulted in increased interest as regards the administration of alternative substances such as immunomodulators like cytokines, pharmaceuticals, microbial products, traditional medicinal plants, and nutraceuticals (Hardy et al. 2013; Bouwens and Savelkoul 2019).

According to the meta-analysis of Vanrolleghem et al. (2019), many studies have focused on potential dietary feed additives with antibacterial effects on weaned piglets, such as:

I. Antimicrobial peptides (small molecules [<10 kDa] with a broad-spectrum activity against bacteria, fungi, protozoa, and some viruses (Lai and Gallo 2009))
II. Chitosan (obtained from the shell water of industries processing crab, shrimp, and crawfish, with possible antimicrobial property (Singla and Chawla 2001))
III. Lysozyme (naturally occurring enzyme, with ability to cleave the glycosidic linkage of bacterial cell walls peptidoglycan (Ellison and Giehl 1991))
IV. Medium-chain fatty acids (MCFAs)/triglycerides (organic acids [OAs] with 6–12 carbon atoms acting as non-ionic surfactants, which are incorporated into the bacterial cell membrane (Desbois and Smith 2010))
V. Plant extracts and essential oils (with bacteriostatic and/or bactericidal and antioxidant effects (Franz et al. 2010))

The aforementioned analysis findings supported the possibility of antibiotics replacement with the above-mentioned feed additives, especially at the weaning period. In addition to those antibiotic alternatives, major interventions that could affect immune response and defense predominantly against intestinal pathogens, thus supporting eubiosis and proper intestinal function in pigs, will be demonstrated in the present chapter (Fig. 2).

4 Low-Protein Diets and Amino Acids Supplementation

Several studies have presented evidence between diets with increased protein levels and detrimental effects on the gut health of pigs, and a special relationship has been presented between feeding high levels of proteins and the incidence of post-weaning diarrhea in pigs, as reviewed by Rodrigues et al. (2022). On the other hand, it has been reported that diets with lower levels of proteins could improve gut health by suppressing the proliferation of pathogenic bacteria and increasing beneficial microbial populations (Wellock et al. 2008; Heo et al. 2009; Rist et al. 2013). Thus, since many years, a lower protein diet supplemented with crystalline amino acids (AAs) to meet requirements for essential AAs has been recommended, especially post-weaning, since through that intervention reduced undigested protein and harmful metabolites are available for the overgrowth of pathogenic bacteria populations in the gut lumen (Nyachoti et al. 2006). Nevertheless, it should be noted that additional factors, other than simply total dietary protein content, such as indigestible content,

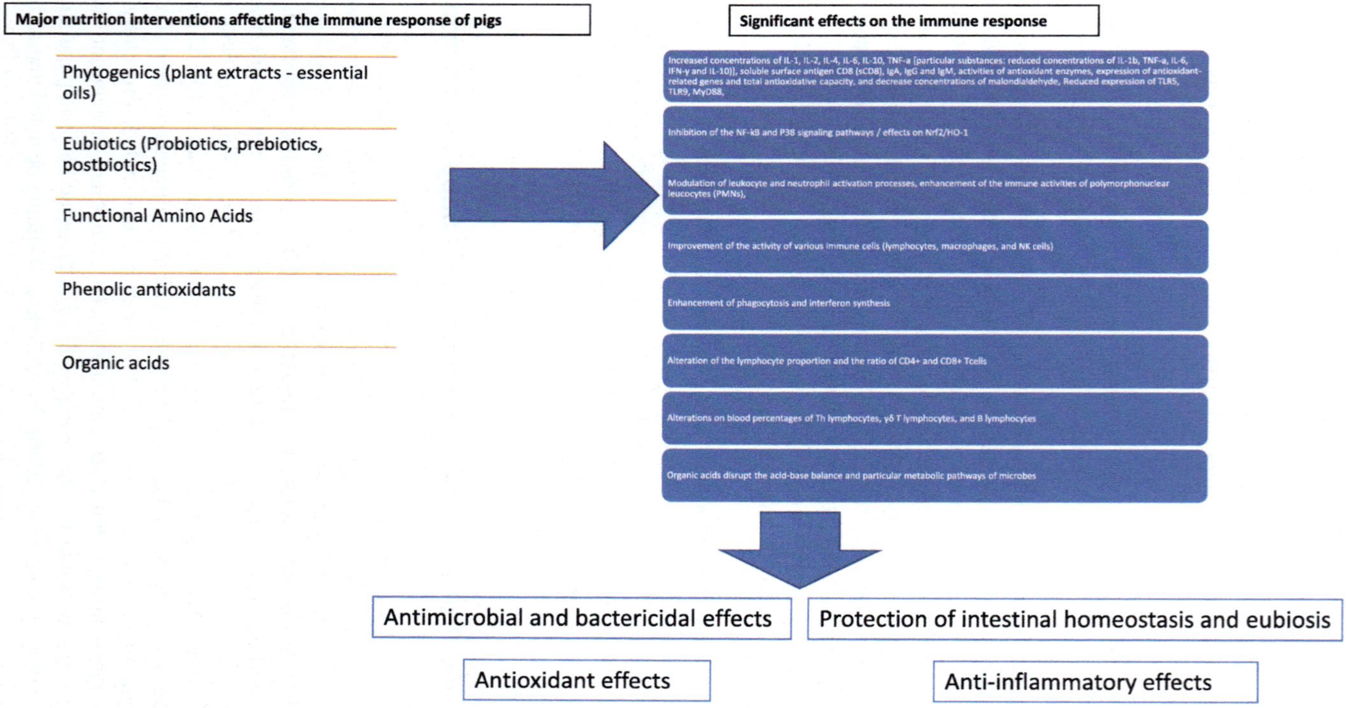

Fig. 2 Major groups of nutrition interventions and their mode of action that affects immune response of pigs

or protein type, could be also involved in the worsening of post-weaning diarrhea (Rodrigues et al. 2022).

A prioritization of AA utilization for the immune response at the expense of growth has been demonstrated (Reeds et al. 1994). It should be noted that a significant expenditure of AAs is present during inflammation, due to the need of AAs for endogenous antioxidants syntheses to cope with oxidative stress, as well as the necessity to support the activated immune system under such circumstances (Rodrigues et al. 2022). The circulating acute-phase proteins (e.g., C-reactive protein, serum amyloid A, haptoglobin, pig-major acute phase protein (MAP)) as well as the proliferation of immune cells (e.g., clonal lymphocyte and monocyte differentiation) and lymphoid tissue hyperplasia, along with the secretion of molecules such as cytokines and immunoglobulins by immune cells, are major points of AA expenditure during inflammation (Le Floc'h et al. 2004; Parra et al. 2006). Therefore, the AA needs in pigs are not stable and differ between physiological and inflammatory status. Evidence suggests a positive boost on immune response after AA supplementation, through the reduction of body protein loss and acceleration of recovery procedures (Le Floc'h et al. 2018). The "functional" roles of AAs beyond their role as constituents of lean gain are a major point of ongoing research.

As previously demonstrated, excess nitrogen available for fermentation in the distal ileum and colon can negatively affect the intestinal barrier function and immunity and contribute to post-weaning diarrhea at weaning (Kim et al. 2012). Tryptophan (Trp) has gained significant attention due to its metabolism to functional metabolites that possess immune regulatory properties (regulation of T-cell function and response) (Humphrey et al. 2019). Gao et al. (2018) reported that endogenous (kynurenine, serotonin, and melatonin) and bacterial (indole, indolic acid, skatole, and tryptamine) Trp metabolites can affect gut microbial composition and metabolism, immune response of the host, and host microbiome interaction. Findings from other studies support the beneficial outcome of increased Trp intake in piglets, through the improvement of intestinal microbiome diversity, decreased abundance of opportunistic bacteria, and increased mucosal IL-8 mRNA level and zonula occludens (ZO)-1 (Liang et al. 2018). Based on a study by Wang et al. (2010), an optimal ratio of Trp:Lys of 0.89% for weanling piglets tofto support intestinal barrier function was concluded. Other amino acids of importance include glutamine, glutamate, proline, aspartate, ornithine, and citrulline. As regards lysine in nursery pigs, an average total lysine:crude protein (CP) ratio of 6.8% is often suggested (total Lys: CP ratio should not exceed 7.1%) (Nemechek et al. 2011; Humphrey et al. 2019).

Glutamic acid (GLU) is a non-essential amino acid, present in the body, which is a precursor of protein synthesis associated with cellular metabolism and immune responses. According to Kyoung et al. (2021), supplementation of weaners' diet with GLU provides benefits in terms of immune response and intestinal microbiota and function, since increased villus height to crypt depth ratio, number of goblet cells, and ileal gene expression of claudin family and occludin, and decreased serum TNF-α and IL-6 and ileal gene expression of TNF-α, were observed. Moreover, increased relative composition of bacterial communities of genus *Prevotella* and *Anaerovibrio* and decreased populations of the *Clostridium* and *Terrisporobacter* genera were reported for the GLU-supplemented groups. Additionally, findings from the study of Koo et al. (2020) demonstrated an increased villus height and goblet cell density, along with increased expression of jejunal occludens and downregulation of IL-6 in pigs fed 115% standardized ileal digestible (SID) requirements of *L*-threonine.

5 Phenolic Antioxidants

Several agricultural by-products could be suggested as excellent sources of phenolic and antioxidant compounds that can be administered as functional ingredients in livestock feeding (Castrica et al. 2019). Up today, about 8000 phenolic compound structures have been identified (Vuolo et al. 2019), whereas the most widely tested are phenolic acid, flavonoids, tannins, avenanthramides, alkylresorcinols, oligomeric proanthocyanidins, and lignans (Dykes and Rooney 2007; Christaki et al. 2020) originating from plant tissues like grains, vegetables, fruits, trees, and their extract (Dykes and Rooney 2007; Jamwal et al. 2018; Rosa et al. 2019). Natural phenolic antioxidants include the following (Shahidi and Ambigaipalan 2015):

I. Phenolic acids (e.g., benzoic acid, ferulic acid, gallic acid, vanillin)
II. Flavonoids (flavonols, flavononols, flavones, flavanones, anthocyanidins, isoflavonoids)
III. Stilbenes (resveratrol)
IV. Coumarins
V. Lignans
VI. Tannins

Phenols have naturally antioxidant properties; thus, they are capable of protecting biomolecules (proteins, nucleic acids, polyunsaturated lipids, and sugars) from oxidative damage via free radical-mediated reactions (Heleno et al. 2015). Reactive oxygen radicals could disrupt nutrient absorption after affecting the intestinal mucosa, whereas antioxidants can neutralize reactive oxygen radicals and improve intestinal function (Valenzuela-Grijalva et al. 2017). The increase of reactive oxygen species (ROS) around weaning and disturbance of cellular antioxidant systems balance, as demonstrated by reduced superoxide dismutase (SOD) and glutathione peroxidase (GPX), are related to disruption of intestinal function (Humphrey et al. 2019). Substances with antioxidant properties could enhance immunocompetence or through their co-enzymatic activity could affect cell-to-cell communication, thereby modulating immune system reactions (Catoni et al. 2008).

Polyphenols and, in some cases, flavonoids have been shown to reduce the effect of the above-mentioned phenomena supporting immune and inflammatory cell functions (Shi et al. 2003; Xu et al. 2014). Carotenoids, vitamin C, and vitamin E have been also suggested as substances that can improve both specific and non-specific immune responses in several species (Catoni et al. 2008). Furthermore, in a study with piglets, dietary supplementation of a polyphenol mixture (from apples, grape seeds, green tea leaves, and olive leaves) resulted in reduced plasma malondialdehyde (Zhang et al. 2014), whereas supplementation of grape seed procyanidins as phenolic compounds increased resistance to weaning stress through the enhancement of glutathione peroxidase (GSH-Px), SOD, and catalase (CAT) genes expression (Fang et al. 2020).

Anti-inflammatory properties of phenolic compounds are based on the suppression of inflammatory prostaglandins and nitric oxide production (Valenzuela-Grijalva et al. 2017). Phenolic compounds support, through their mode of action, the production of immunoglobulins and secretion of cytokines, increase of phagocytosis by influencing mitogen-activated protein kinase (MAPK) and nuclear factor κB (NF-κB) signaling pathways (Artuso-Ponte et al. 2020), as well as release of IFN-γ (Christaki et al. 2020). Polyphenol-rich diets in piglets have shown to reduce the expression of different pro-inflammatory genes in duodenum, ileum, and colon (Fiesel et al. 2014), whereas reduced inflammatory mediators NF-κB and Nrf2 (nuclear factor erythroid 2–related factor 2) have been detected in duodenal mucosa of pigs fed with phenol-rich supplemented diets containing grape seed and grape pomace extract (Gessner et al. 2013). Findings from Coddens et al. (2017) on cranberry extracts (rich in proanthocyanin) supported its efficacy on the inhibition of $F4^+$ and $F18^+$ *E. coli* adhesion to the ileum in vitro. Grape by-products have been reported as a beneficial polyphenol source for pigs (Brenes et al. 2016), since the introduction of fermented grape pomace (48.5% dietary inclusion) to 20 days old piglets feed for 30 days (Kafantaris et al. 2018) resulted in the enhancement of the antioxidant defense system along with the increase of *Bifidobacterium* and lactic acid bacteria counts and the reduction of *Enterobacteriaceae* counts. Additionally, tea polyphenols have showed multiple effects on immune response in pigs, since they could influence the activities of T-lymphocyte, increase the ratio of $CD4^+/CD8^+$, and reduce the outcome of oxidative stress. Furthermore, they could improve cell-mediated immune response and the secretion of pro-inflammatory cytokines such as IFN-γ (Deng et al. 2010). They can also improve intestinal mucosal immunity via increasing the content of IL-2, IL-10 in jejunum and ileum and activate the Notch2 signaling pathway in small intestine (Dong et al. 2019).

Antimicrobial and bactericidal properties of phenolic compounds have been also reported. The latter is attributed to their hydroxyl (–OH) groups (Park et al. 2002), whereas antimicrobial effects are based on their structural and lipophilic properties that negatively affect the cellular membrane function of bacteria and can cause cell death (Mahfuz et al. 2021). Findings from trials with supplementation of pig diets with benzoic acid and thyme (Diao et al. 2015), or chestnut wood tannins and organic acids (Brus et al. 2013), or polyphenol-rich grape extract or hop (Fiesel et al. 2014), provided evidence of an improved outcome in terms of reduction of harmful bacteria and improvement of intestinal microbiota.

Significant research findings have been demonstrated from the evaluation of resveratrol (trans-3,5,4′-trihydroxystilbene), which is a stilbenoid, a type of natural polyphenol and aromatic phytoalexin found predominantly in grapes, berries (mulberries), and Japanese knotweed (Ahmed et al. 2013). As reviewed recently by Meng et al. (2023), dietary resveratrol has therapeutic effects on the oxidative stress and inflammation, as well as beneficial effects on growth and meat quality. Significant findings have been described as regards its capability to modulate immune response and inflammation processes. Such effects include stimulation of peripheral blood and splenic lymphocytes proliferation, and improved immune responses to vaccination against classical swine fever and foot-and-mouth disease, whereas it was also found to promote IgG production, regulate the release of IFN-γ, and downregulate the release of TNF-α (Fu et al. 2018). The capability of resveratrol to regulate various signaling pathways such as sirtuin 1, NF-κB, and Nrf enhances the expression of various antioxidant defensive enzymes such as heme oxygenase 1, catalase, GPX, and SOD, and induces glutathione levels responsible for maintaining the cellular redox balance (Truong et al. 2018).

The dietary supplementation of resveratrol alone or with essential oils (oregano, anise, orange peel, and chicory essential oils) in weaned piglets challenged with *E. Coli* and *Salmonella typhimurium* resulted in improved IgG content in the group fed resveratrol only, as well as reduced fecal *Salmonella* and *E. coli* counts in all treatment groups and increased fecal *Lactobacillus* spp. count in the group that received both phytogenic products. The aforementioned results were considered as strong indication of the potential of resveratrol to be used as an antibiotic alternative under the conditions described in that particular study (Ahmed et al. 2013).

Antiviral activity of resveratrol has been reported in various species, whereas its capability to modulate immune response has been reported in rotavirus (RV) and pseudorabies virus (PRV) studies with pigs (Cui et al. 2018; Zhao et al. 2018). A challenge study in piglets suggested that it can be considered as a possible RV infection control measure. It was able to reduce inflammation response by inhibiting the TNF-α production whereas the immune function in RV-infected piglets was maintained by enhancing the IFN-γ content and $CD4^+/CD8^+$ ratio (Cui et al. 2018). Zhao et al. (2018) suggested resveratrol as alternative control measure for PRV infection, since it showed an inhibitory effect on viral reproduction, alleviated PRV-induced inflammation, and enhanced animal immunity (increased the levels of serum TNF-α, IFN-α, IFN-γ, and IL-12). Furthermore, an in vitro study suggested that resveratrol could alleviate *E. coli* K88 infection-induced damage in the porcine intestinal epithelial cell by activating sirtuin 1 signaling pathway (Luo et al. 2022).

6 Eubiotics: Phytobiotics Probiotics, Prebiotics, Postbiotics

According to Wiemann (2013), eubiotics are feed additives that include direct acting gut flora modulators, probiotics, prebiotics, and immune modulators to stimulate a healthy microbiota. Probiotics as beneficial microbes in combination with prebiotics (indigestible dietary fiber/carbohydrate, e.g., inulin) provide health benefits to the animal, through several pathways such as normalization of the microbiota due to probiotics addition, or via products resulting from prebiotics anaerobic fermentation, or through their immunomodulatory role (Hardy et al. 2013). Postbiotics are metabolites and cell contents extracted from probiotics (Teame et al. 2020).

Probiotics, prebiotics, and postbiotics could act as anti-inflammatory factors (Cheng and Kim 2022) since they can stimulate TLR to inhibit NF-κB and activate an anti-inflammatory response (Suda et al. 2014; Poulsen et al. 2018). Selected *Bacillus* species (Taras et al. 2005), as well as *Lactic acid* bacteria (LAB), including *Enterococcus* species (e.g., *Enterococcus faecium*) and *Lactobacillus* species, have shown promising results as functional feed additives for improved immune responses in nursery pigs (Pessione 2012; Suda et al. 2014). Research findings have demonstrated that *Enterococcus faecium* could reduce newborns' mortality and post-weaning diarrhea when fed to sows (Taras et al. 2006) or decrease serum IgG (Broom et al. 2006) and chlamydial infection in newborn piglets from infected sows (Pollmann et al. 2005).

Effects of feeding mannan oligosaccharide (MOS) and *Lactobacillus mucosae* (LM) as prebiotic and probiotic sources in weanling pigs have been previously tested under *Escherichia coli* lipopolysaccharide (LPS) challenge conditions (Li et al. 2021a). Results demonstrated an increase of circulating but not secretory IgG antibodies in MOS-fed groups, as well as a mild increase in both secretory and circulating IgA concentrations in pigs fed LM. In another study by Yu et al. (2021), MOS supplementation in *E. coli*-challenged pigs resulted in reduced IL-1β concentration. Decreased pro-inflammatory cytokines were observed also in pigs fed levan-type fructan (Li and Kim 2013). The prebiotic lactulose has showed immune-boosting effects in pigs (Liu et al. 2018), as it was able to induce greater concentrations of serum IgM and IgA and improved immunity against *Salmonella typhimurium* (Naqid et al. 2015). Moreover, increased cell-mediated immune response, IL-1β gene expression, and serum levels of IL-1β, IL-2, and IL-6 were observed after supplementation of weaned pigs' diets with chitosan and galacto-mannan oligosaccharides (Yin et al. 2008).

Previous studies with *E. coli*-challenged pigs reported increased secretory IgA in animals receiving *Lactobacillus rhamnosus* (Zhang et al. 2010) or *Lactobacillus acidophilus* with increased concentration of IgA in the jejunum (Li et al. 2018). Other studies further supported *Lactobacillus* species' beneficial effects, such as their ability to alleviate gut inflammation, improve intestinal barrier function, and decrease pro-inflammatory cytokines (*L. rhamnosus*) according to Mao et al. (2020) and downregulate IL-1β (*L. fermentum*) based on Wang et al. (2019) findings.

A connection between modulation of sow's intestinal microbiota and suckling piglets' bacterial colonization of the GIT has been reported. Supplementations of probiotics such as *Enterococcus faecium* or *Bacillus subtilis* (Macha et al. 2004; Baker et al. 2013; Starke et al. 2013) and prebiotics such as inulin (Paßlack et al. 2015) in sows diets were able to induce such microbiota alterations. As regards immune response, enrichment of sows' diets with oligosaccharides such as short-chain fructooligosaccharides (scFOS), MOS, or a seaweed extract containing laminarin could be related with an increase of colostral immunity (IgA, IgG, or TGFß) (Czech et al. 2010; Leonard et al. 2012; Le Bourgot et al. 2014).

Beta-glucans, mannoprotein, and chitin are the main cell wall components of yeasts. Beta-glucans derived from yeast cell walls bind to the TLR2 and C-type lectin receptors (CLR) family and dectin-1 receptor on enterocytes and immune cells (monocyte-macrophage cell lineage and other antigen-presenting immunocompetent cells) (Akira et al. 2006; Li et al. 2019a). Through their mode of action, they result in the increase of pro-inflammatory cytokines and chemokines inducing antigen presentation and improvement of humoral and cellular immunity (Vetvicka et al. 2014). Supplementation of sows' gestation and lactation diets with yeasts (e.g., *Saccharomyces cerevisiae* fermentation product) improved growth performance of piglets (Kim et al. 2008, 2010; Shen et al. 2011) and improved beneficial microbiota status in the GIT (Lu et al. 2019).

Moreover, Li et al. (2006) evaluated whether supplementation of pig diet with β-glucan could affect immune response, and presented results of increased plasma IL-6, IL-10, and TNF-α, hours after LPS challenge. Ryan et al. (2012) provided evidence that glucan incorporation in pig feeds results in decreased Th-related cytokine production (reduction of the Th17 signature molecule IL-17a in the porcine colon), whereas increased IgA levels in serum were observed at lower glucan doses provided as yeast (*Saccharomyces cerevisiae*) cell wall extract (Sauerwein et al. 2007). Proportion of $CD4^+$ T-cell subpopulations has been found greater in mesenteric lymph nodes and Peyer's patches, as well as $CD8^+$ T-cells in peripheral blood in pigs fed glucan (Vetvicka et al. 2014). Moreover, stimulation of IL-2 and phagocytosis as well as suppression of TNF-α due to glucan administration in pigs' feed has been reported (Vetvicka and Oliveira 2014).

Fibers include a broad spectrum of oligosaccharides and starch resistant to proximal intestine hydrolysis, as well as non-starch polysaccharides such as pectin, cellulose, hemicellulose, β-glucans, and fructans (Rodrigues et al. 2022). Digestion process of dietary fiber in pigs includes their fermentation primarily in the colon producing gases and several physiologically active by-products. On the other hand, insoluble fiber increases diet bulkiness due to its metabolic inert characteristic (Jarrett and Ashworth 2018). It has been demonstrated that supplementation of a diet with 25% sugar beet pulp in pregnant gilts resulted in increased white blood cells, without affecting natural killer cell cytotoxicity, neutrophil chemotaxis and chemokinesis, mitogen-induced lymphocyte proliferation, and differential counts (McGlone and Fullwood 2001). Moreover, feeding diets rich in crude fibers (different roughage sources, i.e., straw, hay, clover grass silage, maize silage, or Jerusalem artichoke) during pregnancy, but not lactation, resulted in a decrease of C-reactive protein levels in colostrum, suggesting a possible reduction of inflammatory processes (Werner et al. 2014). In growing pigs, it was reported that pigs fed soluble fiber (sugar beet fiber) had reduced fecal egg counts following *Oesophagostomum dentatum* challenge (Petkeviius et al. 2003).

The positive effects of fermentable fibers on immune response and intestinal function are various: the improvement in colonic barrier function and immune/metabolism-related gene expression (Che et al. 2014), the maintenance of microbial community homeostasis, the improvement in microbiota diversity, proliferation of potentially beneficial microorganisms (Li et al. 2020b), as well as the attenuation of the release of inflammatory intermediates (Li et al. 2019b) are among those effects. Moreover, it has been demonstrated that through the promotion of the growth of lactic acid bacteria, the prebiotic inulin has an indirect beneficial immune effect, since it can affect the production of anti-inflammatory cytokines, mononuclear cells, and phagocytic macrophages (Grela et al. 2021). Additionally, it has been associated with the induction of immunoglobulins synthesis, in particular IgA (Macfarlane and Cummings 1999). Thus, it seems that inulin has a positive effect on the intestinal immune system, blood flow through the mucosa, and the activity of the local nervous system (Grela et al. 2021).

7 Phytogenics and Essential Oils

Phytogenics or plant secondary metabolites (Rodrigues et al. 2022) can be classified into:

I. Terpenes (e.g., carvacrol, thymol)
II. Phenolics (e.g., eugenol, resveratrol, quercetin, tannins), which are emphasized for their antioxidant capabilities in a previous part of this chapter
III. N-containing compounds
IV. S-containing compounds (e.g., alliin and allicin)

Novel technologies allow us to chemically synthesize some of the above-mentioned substances and these products can be called nature-identical compounds (Rossi et al. 2020). Such compounds (e.g., thymol and vanillin) can act synergically in combination with other feed additives in pigs' diets (Rodrigues et al. 2022). Essential oils are either terpenes or phenolics and are usually extracted from plants, whereas their antimicrobial, anti-inflammatory, and antioxidative properties are attributed mainly to their phenolic ring, or capacity to disturb microbial membranes and intracellular homeostasis (Omonijo et al. 2018).

Plants from the *Echinacea* family are known to modulate immune functions, stimulating the innate immune system, and increasing the resistance to infection (Bauer et al. 1999), whereas improved immune response after vaccination against *Erysipelothrix rhusiopathiae* was observed after inclusion of *Echinacea purpurea* into the diet of finishers (Maass et al. 2005). On the other hand, Taranu et al. (2012) provided further evidence that introduction of *Chlorella vulgaris* powder (eukaryotic freshwater green microalga), Na-alginate, inulin, and a mixture of essential oils into diets of weaned piglets resulted in increased IgG in the plasma, modulation of cytokine production, and mineral retention (increased liver concentrations of IL-1β, IL-8, TNF-α, IFN-γ, Cu, and Fe). A possible interaction of active molecules from the test products (polyphenols, vitamins, minerals, etc.) as additional ligands with Fc-gamma receptors for IgG (FcgammaRs) and their further influence on the immune system could be supported as an explanation for the observed increased IgG levels (Nimmerjahn and Ravetch 2010).

Plant extracts like cinnamon, thyme, oregano (Namkung et al. 2004), and saponin (Ilsley et al. 2005) were able to increase IgG concentration in pigs. On the other hand, studies from Ilsley et al. (2005) and Ariza-Nieto et al. (2011) suggested absence of IgG levels increase after dietary curcumin or oregano essential oil supplementation. As regards modes of action of the plant extracts on cellular immunity and cytokine production, a possible potentiation of the immune reaction through the increase of IFN-γ production, and the involvement of a Th1 rather than Th2 type of cellular immunity, as well as their anti-inflammatory properties (activation of the NF-κB pathway), have been reported (Taranu et al. 2012). Novel herbal feed additives (Guizhi Li-Zhong Tang extract granules) have been tested with encouraging results on alleviating or preventing pneumonia in weaned piglets, through the inhibition of angiotensin-converting enzyme 2 expression along with increased IgA and IgG, but reduced IgE levels (Lu et al. 2021). A greater antioxidant capacity and lower cytotoxicity of those herbal feed additives was based on findings of enhanced expression of antioxidant-related SOD2 and lower expression of oxidative-stress-related 3-nitrotyrosine (NT), inflammation-related TNF-α and NF-κB, and apoptosis-related caspase-3 in lung tissue (Wang et al. 2021).

Thymol and carvacrol, active components of plant essential oils, can increase the percentage of $CD4^+$, $CD8^+$, major histocompatibility complex (MHC class II, and non-T/non-B-cells in peripheral blood, and $CD4^+$, $CD8^+$ double-positive T-lymphocytes in peripheral blood and mesenteric lymph nodes in pigs (Walter and Bilkei 2004). Moreover, thymol enhances total IgA and IgM serum levels, and exhibits particular local anti-inflammatory properties, as demonstrated with the reduction of TNF-α mRNA in the stomach of post-weaned pigs (Trevisi et al. 2007). Additional evidence provided by Li et al. (2012) after the introduction of an essential oil product, which contained 18% thymol and cinnamaldehyde in weaned pigs' diets, suggested immune-modulating beneficial properties that could position such products as antibiotic replacements in pigs' diets. Results included reduced IL-6 concentration and increased TNF-α and total antioxidant capacity levels in plasma, as well as greater villus height to crypt depth ratio and reduced *E. coli* populations in cecum, colon, and rectum. Quite similarly, Nofrarías et al. (2006) supported the capability of essential oils to induce an improved immune response, as demonstrated in immune cell subsets of gut tissues and blood after the introduction of a plant extracts mixture with 5% carvacrol, 3% cinnamaldehyde, and 2% capsicum oleoresin (CAP) in weaned pigs' diets. Essential oil compounds from oregano, clove, and cinnamon were also tested by Halas et al. (2011) and results suggested enhancement of the non-specific immunocompetence of 28-days-old pigs.

As regards viral challenges, capsicum oleoresin (CAP), garlic botanical (GAR), or turmeric oleoresin (TUR) were tested in vivo under porcine reproductive and respiratory syndrome virus (PRRSV) challenge conditions (Liu et al. 2013). Findings suggested various effects on the immune response of animals fed the plant extracts. Feeding GAR increased B-cells and $CD8^+$ T-cells of PRRSV-infected pigs, suggesting an improvement of immune response. Anti-inflammatory effects of the extracts were suggested due to suppressed serum TNF-α and IL-1β production in PRRSV-challenged animals that received the test products. Quite similarly, Kim et al. (2020) provided evidence that the aforementioned substances (i.e., CAP, GAR, TUR) altered the expression of 46 genes (24 up, 22 down), 134 genes (59 up, 75 down), or 98 genes (55 up, 43 down) in alveolar macrophages of PRRSV-infected pigs. Supplementation of diets with TUR or GAR reduced the expression of genes associated with antigen processing and presentation, whereas introduction of CAP upregulated the expression of genes involved in those processes.

As reviewed by Li et al. (2021b), sugar cane extracts can enhance immune response in PRV-challenged pigs, through the increase of natural killer cytotoxicity, lymphocyte proliferation, phagocytosis by monocytes, and IFN-γ production of $CD4^+$ and γδ T-cells (Lo et al. 2006). Studies on porcine epidemic diarrhea virus (PEDV) suggested isoflavonoid (major component of puerarin from the Chinese herb Gegen) could regulate the interferon and NF-κB signaling pathways and provide antiviral and anti-inflammatory functions in PEDV-infected piglets (Wu et al. 2020), whereas tomatidine can inhibit the virus replication mainly by targeting 3-chymotrypsin-like (3CL) protease (Wang et al. 2020). Recent data on the devastating African swine fever virus (ASFV) suggested that the introduction of a formulation with three essential oils, i.e., *Eucalyptus globulus*, *Pinus sylvestris*, and *Lavandula latifolia*, can improve immune response resulting in enhanced IgG levels and reduced IgM levels and minimize ASFV transmission in pigs in vivo (Babikian et al. 2021).

Taken together, immune-boosting functions of phytogenic products could include the modulation of leukocyte and neutrophil activation processes of the innate immune system (Firmino et al. 2021). They improve the activity of various immune cells such as lymphocytes, macrophages, and NK cells, thus enhancing phagocytosis and IFN synthesis (Kuralkar and Kuralkar 2021). The enhancement of the immune activities of polymorphonuclear leucocytes (PMNs), alteration of the lymphocyte proportion and the ratio of $CD4^+$ and $CD8^+$ T-cells, downregulation of NF-κB (and p38 pathway) on peripheral blood mononuclear cells (PBMCs), which is responsible for gene transcription, thus of encoding many pro-inflammatory cytokines and chemokines, are among the major immune-modulating actions provoked by phytogenics (Oeckinghaus et al. 2011; Huang et al. 2012; Stelter et al. 2013; Cappelli et al. 2021). Furthermore, alterations on blood percentages of Th lymphocytes, γδ T-lymphocytes, and B-lymphocytes could be observed (Lo Verso et al. 2020). Depending on each essential oil mode of action, they could increase serum concentrations of IL-1, IL-2, IL-4, IL-6, TNF-α, soluble surface antigen CD8 (sCD8), immunoglobulins IgA, IgG, and IgM, the activities of antioxidant enzymes, and total antioxidative capacity, and decrease concentrations of malondialdehyde (e.g., water extract of *Artemisia ordosica*) (Xing et al. 2019). Other cases (e.g., anethole) could result in reduced expression of TLR5, TLR9, MyD88, IL-1β, TNF-α, IL-6, and IL-10 in the jejunum (Yi et al. 2021). Additional anti-inflammatory properties have been attributed to the inhibition of the NF-κB and P38 signaling pathways, which result in decrease of inflammatory cytokine expression (e.g., *Scutellaria baicalensis* extracts) (Huang et al. 2019). In addition to the NF-κB pathway, the antioxidative, anti-apoptotic, and anti-inflammatory effects of essential oils (e.g., *Tagetes erecta* flowers essential oils) may rely on Nrf2/HO-1 since Nrf2 has a significant role in protection of cells against oxidative damage as well as in cell survival (Shaw and Chattopadhyay 2020). The increased expression of antioxidant-related genes could also explain the antioxidant capabilities of essential oils as observed for cinnamaldehyde and thymol in pigs (Su et al. 2018).

8 Organic Acids (OAs)

Classification of OAs includes three categories based on the carbon chain:

I. Short-chain fatty acids (SCFAs; e.g., formic acid, acetic acid, propionic acid, butyric acid)
II. Medium-chain fatty acids (MCFAs; e.g., caproic acid, caprylic acid, capric acid, lauric acid)
III. Tricarboxylic acids (e.g., citric acid, fumaric acid, and malic acid) (Rodrigues et al. 2022)

OAs possess bacteriostatic and bactericidal actions, since they can diffuse across the bacterial cell membranes, release H^+ ions intracellularly, and disrupt the acid–base balance and particular metabolic pathways of microbes (Nguyen et al. 2020).

A combination of OA utilization may be significantly beneficial for pigs, since SCFAs have been proven efficacious against Gram (−) bacteria, including *E. coli* and *Salmonella* spp., whereas MCFAs were efficacious against Gram (+) bacteria, such as *C. perfringens* and *Streptococcus* spp. (Zentek et al. 2011; Gómez-García et al. 2019). Furthermore, particular immunomodulatory effects of OAs have been demonstrated in enterotoxic *Escherichia coli* (ETEC)-challenged pigs. Among those, decreased concentration of pro-inflammatory cytokines (IL-1β, IL-6, TNF-α, and IFN-γ in plasma) was observed at levels comparable to antibiotics supplementation (Ren et al. 2019). According to Jiménez et al. (2020), feed supplementation with an organic acid-based feed additive reduced the number of inflammatory cells in the jejunal and ileal lamina propria, which were elevated due to inoculation with an enterotoxic strain of E. coli (K88) in weaned pigs. Moreover, it has been demonstrated that conjugated linoleic acid fed in late gestation and (or) lactation to sows could induce positive immunomodulatory effects on colostrum, milk, and progeny serum IgG concentrations (Craig et al. 2019).

Immunomodulatory properties have been also reported for certain MCFAs and monoglycerides, such as the C12 monoglyceride (glycerol monolaurate, GML), which is known to affect T-cell lymphocytes, due to membrane interactions linked to cell signaling pathways (Jackman et al. 2020; Zhang et al. 2018). GML supplementation can also decrease cytokine production in vitro, thus inducing immunosuppressive effects that can be useful for anti-inflammatory applications (Zhang et al. 2016). Oral administration of GML could reduce intestinal inflammation in vivo (Zhang et al. 2018). Findings from an in vitro study demonstrated that caprylic (C8), capric (C10), and lauric (C12) acids could enhance immune response at a porcine intestinal cell line (Martínez-Vallespín et al. 2016).

9 Concluding Remarks

A significant number of feed additives that could improve the immune response of pigs and act as anti-inflammatory and antioxidant agents have been investigated and the list will continue to expand as novel research findings are presented. As observed from the above-mentioned studies, a combination of the beneficial feed additives is expected to result in improvement of the pig's response against pathogens and provide solutions as alternative to antibiotics substances. However, the addition of the additives should take into account the observed health challenges at the farm, as well as restrictions on the use of particular feed additives either alone or in combination with others. It is possible that the target of eubiosis and improved health status of pigs could be achieved or assisted through appropriate feed interventions with the administration of more "natural" substances, which enhance immune response in pigs.

References

Aderem A, Ulevitch RJ (2000) Toll-like receptors in the induction of the innate immune response. Nature 406(6797):782–787. https://doi.org/10.1038/35021228

Ahmed S, Hossain M, Kim G et al (2013) Effects of resveratrol and essential oils on growth performance, immunity, digestibility and fecal microbial shedding in challenged piglets. Anim Biosci 26(5):683–690. https://doi.org/10.5713/ajas.2012.12683

Akira S, Uematsu S, Takeuchi O (2006) Pathogen recognition and innate immunity. Cell 124:783–801. https://doi.org/10.1016/j.cell.2006.02.015

Anastasilakis CD, Ioannidis O, Gkiomisi AI et al (2013) Artificial nutrition and intestinal mucosal barrier functionality. Digestion 88:193–208

Ariza-Nieto C, Bandrick M, Baidoo SK et al (2011) Effect of dietary supplementation of oregano essential oils to sows on colostrum and milk composition, growth pattern and immune status of suckling pigs. J Anim Sci 89:1079–1089

Artuso-Ponte V, Pastor A, Andratsch M (2020) The effects of plant extracts on the immune system of livestock: the isoquinoline alkaloids model. In: Florou-Paneri P, Christaki E, Giannenas I (eds) Feed additives. Academic Press, London, pp 295–310

Babikian HY, Jha RK, Truong QL et al (2021) Novel formulation with essential oils as a potential agent to minimize African swine fever virus transmission in an in vivo trial in swine. Vet World 14(7):1853–1866. https://doi.org/10.14202/vetworld.2021.1853-1866

Bailey M (2009) The mucosal immune system: recent developments and future directions in the pig. Dev Comp Immunol 33:375–383

Baker AA, Davis E, Spencer JD et al (2013) The effect of a Bacillus-based direct-fed microbial supplemented to sows on the gastrointestinal microbiota of their neonatal piglets. J Anim Sci 91: 3390–3399. https://doi.org/10.2527/jas.2012-5821

Bauer R, Hoheisel O, Stuhlfauth I et al (1999) Extract of the Echinacea purpurea herb: an allopathic phytoimmunostimulant. Wien Med Wochenschr 149:85–189

Bian G, Ma S, Zhu Z et al (2016) Age, introduction of solid feed and weaning are more important determinants of gut bacterial succession in piglets than breed and nursing mother as revealed by a reciprocal cross-fostering model. Environ Microbiol 18:1566–1577

Boes J, Kanora A, Havn KT et al (2010) Effect of Ascaris suum infection on performance of fattening pigs. Vet Parasitol 172:269–276

Bouwens M, Savelkoul HFJ (2019) Animal nutrition and immunity in pigs and poultry. In: Hendriks WH, Verstegen MWA, Babinszky L (eds) Poultry and pig nutrition. Wageningen Academic Publishers, Wageningen, The Netherlands, pp 105–127

Bracarense APF, Lucioli J, Grenier B et al (2012) Chronic ingestion of deoxynivalenol and fumonisin, alone or in interaction, induces morphological and immunological changes in the intestine of piglets. Br J Nutr 107:1776–1786

Brenes A, Viveros A, Chamorro S et al (2016) Use of polyphenol-rich grape by-products in monogastric nutrition. A review. Anim Feed Sci Technol 211:1–17

Broom L (2015) Mycotoxins and the intestine. Anim Nutr 1:262–265
Broom LJ, Miller HM, Kerr KG et al (2006) Effects of zinc oxide and Enterococcus faecium SF68 dietary supplementation on the performance, intestinal microbiota and immune status of weaned piglets. Res Vet Sci 80:45–54
Brus M, Dolinsek J, Cencic A et al (2013) Effect of chestnut (Castanea sativa Mill.) wood tannins and organic acids on growth performance and faecal microbiota of pigs from 23 to 127 days of age. Bulg J Agri Sci 19:841–847
Cappelli K, Sabino M, Trabalza-Marinucci M et al (2021) Differential effects of dietary oregano essential oil on the inflammation related gene expression in peripheral blood mononuclear cells from outdoor and indoor reared pigs. Front Vet Sci 8:602811. https://doi.org/10.3389/fvets.2021.602811
Castrica M, Rebucci R, Giromini C et al (2019) Total phenolic content and antioxidant capacity of agri-food waste and byproducts. Ital J Anim Sci 18(1):336–341
Catoni C, Peters A, Schaefer HM (2008) Life history trade-offs are influenced by the diversity, availability and interactions of dietary antioxidants. Anim Behav 76:1107–1119
Chase C, Lunney JK (2019) The immune system. In: Zimmerman JJ, Karriker LA, Ramirez A, Schwartz KJ, Stevenson GW, Zhang J (eds) Disease of swine, 11th edn. Wiley-Blackwell, Hoboken, NJ, pp 264–291
Che L, Chen H, Yu B et al (2014) Long-term intake of pea fiber affects colonic barrier function, bacterial and transcriptional profile in pig model. Nutr Cancer 66:388–399
Cheng Y-C, Kim SW (2022) Use of microorganisms as nutritional and functional feedstuffs for nursery pigs and broilers. Animals 12:3141. https://doi.org/10.3390/ani12223141
Christaki E, Giannenas I, Bonos E et al (2020) Innovative uses of aromatic plants as natural supplements in nutrition. In: Florou-Paneri P, Christaki E, Giannenas I (eds) Feed additives. Academic Press, London, pp 19–34
Demehri FR, Barrett M, Ralls MW, Miyasaka EA, Feng Y, Teitelbaum DH (2013) Intestinal epithelial cell apoptosis and loss of barrier function in the setting of altered microbiota with enteral nutrient deprivation. Front Cell Infect Microbiol 3:105. https://doi.org/10.3389/fcimb.2013.00105
Desbois AP, Smith VJ (2010) Antibacterial free fatty acids: activities, mechanisms of action and biotechnological potential. Appl Microbiol Biotechnol 85:1629–1642. https://doi.org/10.1007/s00253-009-2355-3
Coddens A, Loos M, Vanrompay D et al (2017) Cranberry extract inhibits in vitro adhesion of F4 and F18+ Escherichia coli to pig intestinal epithelium and reduces in vivo excretion of pigs orally challenged with F18+ verotoxigenic E. coli. Vet Microbiol 202:64–71
Craig JR, Dunshea FR, Cottrell JJ et al (2019) Feeding conjugated linoleic acid without a combination of medium-chain fatty acids during late gestation and lactation improves pre-weaning survival rates of gilt and sow progeny. Animals 9(2):62. https://doi.org/10.3390/ani9020062
Cui Q, Fu Q, Zhao X et al (2018) Protective effects and immunomodulation on piglets infected with rotavirus following resveratrol supplementation. PLoS One 13:e0192692. https://doi.org/10.1371/journal.pone.0192692
Czech A, Grela ER, Mokrzycka A et al (2010) Efficacy of mannanoligosaccharides additive to sows diets on colostrum, blood immunoglobulin content and production parameters of piglets. Pol J Vet Sci 13:525–531
Deng Q, Xu J, Yu B et al (2010) Effect of dietary tea polyphenols on growth performance and cell-mediated immune response of post-weaning piglets under oxidative stress. Arc Anim Nutr 64(1):12–21
Diao H, Zheng P, Yu B et al (2015) Effects of benzoic acid and thymol on growth performance and gut characteristics of weaned piglets. Asian-Australas J Anim Sci 28:827–839
Dong L, Liu J, Zhong Z et al (2019) Dietary tea tree oil supplementation improves the intestinal mucosal immunity of weanling piglets. Anim Feed Sci Technol 255:114209

Dykes L, Rooney L (2007) Phenolic compound in Cereal grains and their health benefits. Cereal Foods World 52(3):105–111

Ellison RT, Giehl TJ (1991) Killing of gram-negative bacteria by lactoferrin and lysozyme. J Clin Invest 88:1080–1091

Everaert N, Van Cruchten S, Weström B et al (2017) A review on early gut maturation and colonization in pigs, including biological and dietary factors affecting gut homeostasis. Anim Feed Sci Technol 233:89–103. https://doi.org/10.1016/j.anifeedsci.2017.06.011

Fang L, Li M, Zhao L et al (2020) Dietary grape seed procyanidins suppressed weaning stress by improving antioxidant enzyme activity and mRNA expression in weanling piglets. J Anim Physiol Anim Nutr 104(4):1178–1185

Fiesel A, Gessner DK, Most E et al (2014) Effects of dietary polyphenol-rich plant products from grape or hop on pro-inflammatory gene expression in the intestine, nutrient digestibility and faecal microbiota of weaned pigs. BMC Vet Res 10:196

Firmino JP, Vallejos-Vidal E, Balebona MC et al (2021) Diet, immunity, and microbiota interactions: an integrative analysis of the intestine transcriptional response and microbiota modulation in Gilthead Seabream (Sparus Aurata) fed an essential oils-based functional diet. Front Immunol 12:625297. https://doi.org/10.3389/fimmu.2021.625297

Franz C, Baser KHC, Windisch W (2010) Essential oils and aromatic plants in animal feeding – a European perspective. A review. Flavour Fragr J 25:327–340

Fu Q, Cui Q, Yang Y, Zhao X, Song X, Wang G et al (2018) Effect of resveratrol dry suspension on immune function of piglets. Evid Based Complement Alternat Med 2018:595270

Gao J, Xu K, Liu H et al (2018) Impact of the gut microbiota on intestinal immunity mediated by tryptophan metabolism. Front Cell Infect Microbiol 8:13

Gerner W, Käser T, Saalmüller A (2009) Porcine T lymphocytes and NK cells: an update. Dev Comp Immunol 33:310–320. https://doi.org/10.1016/j.dci.2008.06.003

Gessner DK, Fiesel A, Most E et al (2013) Supplementation of a grape seed and grape marc meal extract decreases activities of the oxidative stress-responsive transcription factors NF-KB and Nrf2 in the Duodenal Mucosa of Pigs. Acta Vet Scand 55:18

Gómez-García M, Sol C, de Nova PJG et al (2019) Antimicrobial activity of a selection of organic acids, their salts and essential oils against swine enteropathogenic bacteria. Porc Health Manag 5:32

Grela ER, Świątkiewicz M, Florek M et al (2021) Effect of inulin source and a probiotic supplement in pig diets on carcass traits, meat quality and fatty acid composition in finishing pigs. Animals 11:2438. https://doi.org/10.3390/ani11082438

Gresse R, Chaucheyras-Durand F, Fleury MA et al (2017) Gut microbiota dysbiosis in postweaning piglets: understanding the keys to health. Trends Microbiol 25:851–873

Halas V, Nochta I, Pásti Z et al (2011) Cellular immune response of weaned pigs fed diet supplemented with an essential oil. Agr Consept Scient 2011:76

Hardy H, Harris J, Lyon E et al (2013) Probiotics, prebiotics and immunomodulation of gut mucosal defences: homeostasis and immunopathology. Nutrients 5:1869–1912

Heleno SA, Martins A, Queiroz MJRP et al (2015) Bioactivity of phenolic acids: metabolites versus parent compounds: a review. Food Chem 173:501–513

Heo JM, Kim JC, Hansen CF, Mullan BP et al (2009) Feeding a diet with decreased protein content reduces indices of protein fermentation and the incidence of postweaning diarrhea in weaned pigs challenged with an enterotoxigenic strain of Escherichia coli. J Anim Sci 87:2833–2843

Heo JM, Kim JC, Hansen CF et al (2010) Feeding a diet with a decreased protein content reduces both nitrogen content in the gastrointestinal tract and post-weaning diarrhoea but does not affect apparent nitrogen digestibility in weaner pigs challenged with an enterotoxigenic strain of Escherichia coli. Anim Feed Sci Technol 160:148–159

Huang CW, Lee TT, Shih YC et al (2012) Effects of dietary supplementation of Chinese Medicinal herbs on polymorphonuclear neutrophil immune activity and small intestinal morphology in weanling pigs. J Anim Physiol Anim Nutr (Berl) 96(2):285–294. https://doi.org/10.1111/j.1439-0396.2011.01151.x

Huang C, Wang Y, He X et al (2019) The involvement of NF-kb/P38 pathways in Scutellaria Baicalensis extracts attenuating of Escherichia Coli K88-induced acute intestinal injury in weaned piglets. Br J Nutr 122(2):152–161. https://doi.org/10.1017/s0007114519000928

Humphrey B, Zhao J, Faris R (2019) Link between intestinal immunity and practical approaches to swine nutrition. Animal 13:2736–2744. https://doi.org/10.1017/S1751731119001861

Ilsley SE, Miller HM, Kamel C (2005) Effects of dietary quillaja saponin and curcumin on the performance and immune status of weaned piglets. J Anim Sci 83:82–88

Jackman JA, Boyd RD, Elrod CC (2020) Medium-chain fatty acids and monoglycerides as feed additives for pig production: towards gut health improvement and feed pathogen mitigation. J Anim Sci Biotechnol 11:44. https://doi.org/10.1186/s40104-020-00446-1

Jamin A, Gorin S, Le Potier M-F et al (2006) Characterization of conventional and plasmacytoid dendritic cells in swine secondary lymphoid organs and blood. Vet Immunol Immunopathol 114:224–237

Jamwal K, Bhattacharya S, Puri S (2018) Plant growth regulator mediated consequences of secondary metabolites in medicinal plants. J Appl Res Med Aromat Plants 9:26–38

Jarrett S, Ashworth CJ (2018) The role of dietary fibre in pig production, with a particular emphasis on reproduction. J Anim Sci Biotechnol 9:59

Jiménez MJ, Berrios R, Stelzhammer S et al (2020) Ingestion of organic acids and cinnamaldehyde improves tissue homeostasis of piglets exposed to enterotoxic Escherichia coli (ETEC). J Anim Sci 98(2):skaa012

Kaetzel CS (2014) Cooperativity among secretory IgA, the polymeric immunoglobulin receptor, and the gut microbiota promotes host–microbial mutualism. Immunology Letters 162:10–21

Kafantaris I, Stagos D, Kotsampasi B, Hatzis A, Kypriotakis A, Gerasopoulos K, Makri S, Goutzourelas N, Mitsagga C, Giavasis I (2018) Grape pomace improves performance, antioxidant status, fecal microbiota and meat quality of piglets. Animal 12:246–255. https://doi.org/10.1017/S1751731117001604

Kelly D, Mulder IE (2012) Microbiome and immunological interactions. Nutr Rev 70(Suppl 1):S18–S30

Kim SW, Brandherm M, Freeland M et al (2008) Effects of yeast culture supplementation to gestation and lactation diets on growth of nursing piglets. Asian Australas J Anim Sci 21:1011–1014. https://doi.org/10.5713/ajas.2008.70438

Kim SW, Brandherm M, Newton B et al (2010) Effect of supplementing *Saccharomyces cerevisiae* fermentation product in sow diets on reproductive performance in a commercial environment. Can J Anim Sci 90:229–232. https://doi.org/10.4141/CJAS09100

Kim JC, Hansen CF, Mullan BP et al (2012) Nutrition and pathology of weaner pigs: nutritional strategies to support barrier function in the gastrointestinal tract. Anim Feed Sci Technol 173:3–16

Kim JC, Mullan BP, Black JL et al (2016) Acetylsalicylic acid supplementation improves protein utilization efficiency while vitamin E supplementation reduces markers of the inflammatory response in weaned pigs challenged with enterotoxigenic E. coli. J Anim Sci Biotechnol 7:58

Kim K, Ji P, Song M, Che TM, Bravo D, Pettigrew JE, Liu Y (2020) Dietary plant extracts modulate gene expression profiles in alveolar macrophages of pigs experimentally infected with porcine reproductive and respiratory syndrome virus. J Anim Sci Biotechnol 11:74. https://doi.org/10.1186/s40104-020-00475-w

Koo B, Choi J, Yang C et al (2020) Diet complexity and l-threonine supplementation: effects on growth performance, immune response, intestinal barrier function, and microbial metabolites in nursery pigs. J Anim Sci 98(5):skaa125. https://doi.org/10.1093/jas/skaa125

Kuralkar P, Kuralkar SV (2021) Role of herbal products in animal production – an updated review. J Ethnopharmacol 278:114246. https://doi.org/10.1016/j.jep.2021.114246

Kyoung H, Lee JJ, Cho JH et al (2021) Dietary glutamic acid modulates immune responses and gut health of weaned pigs. Animals 11:504. https://doi.org/10.3390/ani11020504

Lai Y, Gallo RL (2009) AMPed up immunity: how antimicrobial peptides have multiple roles in immune defense. Trends Immunol 30(3):131–141. https://doi.org/10.1016/j.it.2008.12.003

Lalles JP, Bosi P, Smidt H et al (2007) Nutritional management of gut health in pigs around weaning. Proc Nutr Soc 66:260–268

Le Bourgot C, Ferret-Bernard S, Le Normand L et al (2014) Maternal short-chain fructooligosaccharide supplementation influences intestinal immune system maturation in piglets. PLoS One 9:1–12. https://doi.org/10.1371/journal.pone.0107508

Le Floc'h N, Melchior D, Obled C (2004) Modifications of Protein and Amino Acid Metabolism during Inflammation and Immune. System Activation. Livest Prod Sci 87:37–45

Le Floc'h N, Wessels A, Corrent E, Wu G et al (2018) The relevance of functional amino acids to support the health of growing pigs. Anim Feed Sci Technol 245:104–116

Leonard SG, Sweeney T, Bahar B et al (2012) Effect of maternal seaweed extract supplementation on suckling piglet growth, humoral immunity, selected microflora, and immune response after an ex vivo lipopolysaccharide challenge. J Anim Sci 90:505–514. https://doi.org/10.2527/jas.2010-3243

Lewis MC, Inman CF, Patel D (2012) Direct experimental evidence that early-life farm environment influences regulation of immune responses. Pediatr Allergy Immunol 23:265–269

Li J, Kim IH (2013) Effects of levan-type fructan supplementation on growth performance, digestibility, blood profile, fecal microbiota, and immune responses after lipopolysaccharide challenge in growing pigs. J Anim Sci 91:5336–5343. https://doi.org/10.2527/jas.2013-6665

Li J, Li DF, Xing JJ et al (2006) Effects of β-glucan extracted from Saccharomyces cerevisiae on growth performance, and immunological and somatotropic responses of pigs challenged with Escherichia coli lipopolysaccharide1. J Anim Sci 84:2374–2238

Li P, Piao X, Ru Y et al (2012) Effects of adding essential oil to the diet of weaned pigs on performance, nutrient utilization, immune response and intestinal health. Asian Australas J Anim Sci 25:1617–1626. https://doi.org/10.5713/ajas.2012.12292

Li H-H, Jiang X-R, Wang W-J et al (2018) Effects of *Lactobacillus acidophilus* and zinc oxide on the growth performance, jejunal morphology and immune function of weaned piglet following an *Escherichia coli* K88 challenge. Italian J Anim Sci 17:114–120. https://doi.org/10.1080/1828051X.2017.1344573

Li T-H, Liu L, Hou Y-Y et al (2019a) C-type lectin receptor-mediated immune recognition and response of the microbiota in the gut. Gastroenterol Rep (Oxf) 7:312–321. https://doi.org/10.1093/gastro/goz028

Li Q, Burrough ER, Gabler NK et al (2019b) A soluble and highly fermentable dietary fiber with carbohydrases improved gut barrier integrity markers and growth performance in F18 ETEC challenged pigs. J Anim Sci 97:2139–2153

Li H, Li B, Liang Q et al (2020a) Porcine deltacoronavirus infection alters bacterial communities in the colon and feces of neonatal piglets. Microbiologyopen. https://doi.org/10.1002/mbo3.1036

Li Q, Peng X, Burrough ER et al (2020b) Dietary soluble and insoluble fiber with or without enzymes altered the intestinal microbiota in weaned pigs challenged with enterotoxigenic E. coli F18. Front Microbiol 11:1110

Li YS, San Andres JV, Trenhaile-Grannemann MD et al (2021a) Effects of mannan oligosaccharides and *Lactobacillus mucosae* on growth performance, immune response, and gut health of weanling pigs challenged with *Escherichia coli* lipopolysaccharides. J Anim Sci 99:skab286. https://doi.org/10.1093/jas/skab286

Li L, Sun X, Zhao D, Dai H (2021b) Pharmacological applications and action mechanisms of phytochemicals as alternatives to antibiotics in pig production. Front Immunol 12:798553. https://doi.org/10.3389/fimmu.2021.798553

Liang H, Dai Z, Liu N et al (2018) Dietary L-tryptophan modulates the structural and functional composition of the intestinal microbiome in weaned piglets. Front Microbiol 9:1736

Liu Y, Che TM, Song M et al (2013) Dietary plant extracts improve immune responses and growth efficiency of pigs experimentally infected with porcine reproductive and respiratory syndrome virus. J Anim Sci 91:5668–5679. https://doi.org/10.2527/jas.2013-6495

Liu Y, Espinosa CD, Abelilla JJ et al (2018 Jun) Non-antibiotic feed additives in diets for pigs: a review. Anim Nutr 4(2):113–125. https://doi.org/10.1016/j.aninu.2018.01.007

Lo DY, Chien MS, Yeh KS et al (2006) Effects of sugar cane extract on pseudorabies virus challenge of pigs. J Vet Med Sci 68(3):219–225. https://doi.org/10.1292/jvms.68.219

Lo Verso L, Talbot G, Morissette B, Guay F, Matte JJ, Farmer C et al (2020) The combination of nutraceuticals and functional feeds as additives modulates gut microbiota and blood markers associated with immune response and health in weanling piglets. J Anim Sci 98(8):skaa208. https://doi.org/10.1093/jas/skaa208

Lu H, Wilcock P, Adeola O et al (2019) Effect of live yeast supplementation to gestating sows and nursery piglets on postweaning growth performance and nutrient digestibility. J Anim Sci 97: 2534–2540. https://doi.org/10.1093/jas/skz150

Lu C-W, Wang S-E, Wu W-J et al (2021) Alternative antibiotic feed additives alleviate pneumonia with inhibiting ACE-2 expression in the respiratory system of piglets. Food Sci Nutr 9:1112–1120. https://doi.org/10.1002/fsn3.2089

Luo X, Wu S, Jia H et al (2022) Resveratrol alleviates enterotoxigenic Escherichia coli K88-induced damage by regulating SIRT-1 signaling in intestinal porcine epithelial cells. Food Funct 13:7346–7360. https://doi.org/10.1039/d1fo03854k

Maass N, Bauer J, Paulicks BR et al (2005) Efficiency of Echinacea purpurea on performance and immune status in pigs. J Anim Physiol Anim Nutr 89:244–252

Macfarlane GT, Cummings JH (1999) Probiotics and prebiotics: can regulating the activities of intestinal bacteria benefit health? BMJ 318:999–1003. https://doi.org/10.1136/bmj.318.7189.999

Macha M, Taras D, Vahjen W et al (2004) Specific enumeration of the probiotic strain Enterococcus faecium NCIMB 10415 in the intestinal tract and in faeces of piglets and sows. Arch Anim Nutr 58:443–452. https://doi.org/10.1080/00039420400020058

Mahfuz S, Shang Q, Piao X (2021) Phenolic compounds as natural feed additives in poultry and swine diets: a review. J Animal Sci Biotechnol 12:48. https://doi.org/10.1186/s40104-021-00565-3

Main RG, Dritz SS, Tokach MD et al (2004) Increasing weaning age improves pig performance in a multisite production system. J Anim Sci 82:1499–1507

Mair KH, Sedlak C, Käser T et al (2014) The porcine innate immune system: an update. Dev Comp Immunol 45:321–343. https://doi.org/10.1016/j.dci.2014.03.022

Mao J, Qi S, Cui Y et al (2020) Lactobacillus rhamnosus GG attenuates lipopolysaccharide-induced inflammation and barrier dysfunction by regulating MAPK/NF-κB signaling and modulating metabolome in the piglet intestine. J Nutr 150:1313–1323. https://doi.org/10.1093/jn/nxaa009

Martínez-Vallespín B, Vahjen W, Zentek J (2016) Effects of medium-chain fatty acids on the structure and immune response of IPEC-J2 cells. Cytotechnology 68(5):1925–1936. https://doi.org/10.1007/s10616-016-0003-1

McGlone JJ, Fullwood SD (2001) Behavior, reproduction, and immunity of crated pregnant gilts: effects of high dietary fiber and rearing environment. J Anim Sci 79:1466

Meng Q, Li J, Wang C, Shan A (2023) Biological function of resveratrol and its application in animal production: a review. J Anim Sci Biotechnol. 14(1):25. https://doi.org/10.1186/s40104-022-00822-z

Namkung H, Li J, Gong M, Yu H et al (2004) Impact of feeding blends of organic acids and herbal extracts on growth performance, gut microbiota and digestive function in newly weaned pigs. Can J Anim Sci 84:697–704

Naqid IA, Owen JP, Maddison BC et al (2015) Prebiotic and probiotic agents enhance antibody-based immune responses to Salmonella Typhimurium infection in pigs. Anim Feed Sci Technol 201:57–65

Nemechek JE, Tokach M, Britz SS et al. (2011) Effect of total lysine: crude protein ratio on growth performance of nursery pigs from 15 to 25 lb. In: Proceedings of the Swine Industry Day, Kansas State University 1056

Netea MG, van der Meer JWM (2017) Trained immunity: an ancient way of remembering. Cell Host Microbe 21:297–300

Nguyen DH, Seok WJ, Kim IH (2020) Organic acids mixture as a dietary additive for pigs—a review. Animals 10:952

Nimmerjahn F, Ravetch JV (2010) Antibody-mediated modulation of immune responses. Immunol Rev 236:265–275

Nofrarías M, Manzanilla EG, Pujols J et al (2006) Effects of spray-dried porcine plasma and plant extracts on intestinal morphology and on leukocyte cell subsets of weaned pigs1. J Anim Sci 84: 2735–2742

Nyachoti CM, Omogbenigun FO, Rademacher M et al (2006) Performance responses and indicators of gastrointestinal health in early-weaned pigs fed low-protein amino acid-supplemented diets. J Anim Sci 84:125–134

Oeckinghaus A, Hayden MS, Ghosh S (2011) Crosstalk in NF-kb signaling pathways. Nat Immunol 12(8):695–708. https://doi.org/10.1038/ni.2065

Omonijo FA, Ni L, Gong J et al (2018) Essential oils as alternatives to antibiotics in swine production. Anim Nutr 4:126–136

Opriessnig T, Giminez-Lirola LG, Halbur PG (2011) Polymicrobial respiratory disease in pigs. Anim Health Res Rev. 12(2):133–148

Oswald IP, Marin DE, Bouhet S et al (2005) Immunotoxicological risk of mycotoxins for domestic animals. Food Addit Contam 22:354–360. https://doi.org/10.1080/02652030500058320

Parra MD, Fuentes P, Tecles F et al (2006) Porcine acute phase protein concentrations in different diseases in field conditions. J Vet Med Ser B 53:488–493

Park S-Y, Bok S-H, Jeon S-M, Park YB, Lee S-J, Jeong T-S, Choi M-S (2002) Effect of rutin and tannic acid supplements on cholesterol metabolism in rats. Nutr Res 22(3):283–295. https://doi.org/10.1016/S0271-5317(01)00398-0

Paßlack N, Vahjen W, Zentek J (2015) Dietary inulin affects the intestinal microbiota in sows and their suckling piglets. BMC Vet Res 11:51. https://doi.org/10.1186/s12917-015-0351-7

Pessione E (2012) Lactic acid bacteria contribution to gut microbiota complexity: lights and shadows. Front Cell Inf Microbiol 2:86. https://doi.org/10.3389/fcimb.2012.00086

Petkeviius S, Bach Knudsen KE, Murrell KD et al (2003) The effect of inulin and sugar beet fibre on Oesophagostomum dentatum infection in pigs. Parasitology 127:61–68

Pollmann M, Nordhoff M, Pospischil A et al (2005) Effects of a probiotic strain of *Enterococcus faecium* on the rate of natural chlamydia infection in swine. Infect Immun 73:4346–4353. https://doi.org/10.1128/IAI.73.7.4346-4353.2005

Poulsen A-SR, de Jonge N, Nielsen JL et al (2018) Impact of Bacillus spp. spores and gentamicin on the gastrointestinal microbiota of suckling and newly weaned piglets. PLoS One 13: e0207382. https://doi.org/10.1371/journal.pone.0207382

Reeds PJ, Fjeld CR, Jahoor F (1994) Do the differences between the amino acid compositions of acute-phase and muscle proteins have a bearing on nitrogen loss in traumatic states? J Nutr 124: 906–910

Ren C, Zhou Q, Guan W et al (2019) Immune response of piglets receiving mixture of formic and propionic acid alone or with either capric acid or *Bacillus Licheniformis* after *Escherichia coli* challenge. Biomed Res Int 2019:1–9

Rist VTS, Weiss E, Eklund M et al (2013) Impact of dietary protein on microbiota composition and activity in the gastrointestinal tract of piglets in relation to gut health: a review. Animal 7:1067–1078

Rodrigues LA, Koo B, Nyachoti M et al (2022) Formulating diets for improved health status of pigs: current knowledge and perspectives. Animals 12:2877. https://doi.org/10.3390/ani12202877

Rosa LA, Moreno-Escamilla JO, Rodrigo-García J, Alvarez-Parrilla E (2019) Phenolic compounds. In: Yahia EM (ed) Postharvest physiology and biochemistry of fruits and vegetables. Woodhead Publishing, Duxford, pp 253–271

Rossi B, Toschi A, Piva A et al (2020) Single components of botanicals and nature-identical compounds as a non-antibiotic strategy to ameliorate health status and improve performance in poultry and pigs. Nutr Res Rev 33:218–234

Rothkötter HL (2009) Anatomical particularities of the porcine immune system—a physician's view. Dev Comp Immunol 33:267–272. https://doi.org/10.1016/j.dci.2008.06.016

Ryan MT, O'Shea CJ, Collins CB et al (2012) Effects of dietary supplementation with Laminaria hyperborea, Laminaria digitata, and Saccharomyces cerevisiae on the IL-17 pathway in the porcine colon. J Anim Sci 90:263–265

Sauerwein H, Schmitz S, Hiss S (2007) Effects of a dietary application of a yeast cell wall extract on innate and acquired immunity, on oxidative status and growth performance in weanling piglets and on the ileal epithelium in fattened pigs. J Anim Physiol Anim Nutr 91:369–380

Shahidi F, Ambigaipalan P (2015) Phenolics and polyphenolics in foods, beverages and spices: antioxidant activity and health effects – a review. J Funct Foods 18. https://doi.org/10.1016/j.jff.2015.06.018

Shaw P, Chattopadhyay A (2020) Nrf2-ARE signaling in cellular protection: mechanism of action and the regulatory mechanisms. J Cell Physiol 235(4):3119–3130. https://doi.org/10.1002/jcp.29219

Shen YB, Carroll JA, Yoon I et al (2011) Effects of supplementing Saccharomyces cerevisiae fermentation product in sow diets on performance of sows and nursing piglets. J Anim Sci 89:2462–2471. https://doi.org/10.2527/jas.2010-3642

Shi J, Yu J, Pohorly JE et al (2003) Polyphenolics in grape seeds-biochemistry and functionality. J Med Food 6:291–299

Singla AK, Chawla M (2001) Chitosan: some pharmaceutical and biological aspects—an update. J Pharm Pharmacol 53:1047–1067

Starke IC, Pieper R, Neumann K et al (2013) Individual responses of mother sows to a probiotic Enterococcus faecium strain lead to different microbiota composition in their offspring. Benef Microbes 4:345–356. https://doi.org/10.3920/BM2013.0021

Stelter K, FrahmJ PJ et al (2013) Effects of oregano on performance and immunmodulating factors in weaned piglets. Arch Anim Nutr 67(6):461–476. https://doi.org/10.1080/1745039x.2013.858897

Su G, Zhou X, Wang Y et al (2018) Effects of plant essential oil supplementation on growth performance, immune function and antioxidant activities in weaned pigs. Lipids Health Dis 17(1):139. https://doi.org/10.1186/s12944-018-0788-3

Suda Y, Villena J, Takahashi Y et al (2014) Immunobiotic Lactobacillus jensenii as immune-health promoting factor to improve growth performance and productivity in post-weaning pigs. BMC Immunol 15:24. https://doi.org/10.1186/1471-2172-15-24

Sun Y, Kim SW (2017) Intestinal challenge with enterotoxigenic Escherichia coli in pigs, and nutritional intervention to prevent postweaning diarrhea. Anim Nutr 3:322–330

Taranu I, Marin DE, Untea A et al (2012) Effect of dietary natural supplements on immune response and mineral bioavailability in piglets after weaning. Czech J Anim Sci 57:332–343. https://doi.org/10.17221/6008-CJAS

Taras D, Vahjen W, Macha M et al (2005) Response of performance characteristics and fecal consistency to long-lasting dietary supplementation with the probiotic strain *Bacillus cereus* var. toyoi to sows and piglets. Arch Anim Nutr 59:405–417. https://doi.org/10.1080/17450390500353168

Taras D, Vahjen W, Macha M et al (2006) Performance, diarrhea incidence, and occurrence of Escherichia coli virulence genes during long-term administration of a probiotic Enterococcus faecium strain to sows and piglets. J Anim Sci 84:608–617. https://doi.org/10.2527/2006.843608x

Tassis P, Symeonidou I, Gelasakis AI et al (2022) Serological assessment of Ascaris suum exposure in Greek Pig Farms and associated risk factors including Lawsonia intracellularis. Pathogens 11:959. https://doi.org/10.3390/pathogens11090959

Teame T, Wang A, Xie M et al (2020) Paraprobiotics and postbiotics of Probiotic Lactobacilli, their positive effects on the host and action mechanisms: a review. Front Nutr 7:570344. https://doi.org/10.3389/fnut.2020.570344

Tokach MD, Goodband RD, Nelssen JL et al (1992) Influence of weaning weight and growth during the first week postweaning on subsequent pig performance. Swine Day 1992:14–17

Trevisi P, Merialdi G, Mazzoni M et al (2007) Effect of dietary addition of thymol on growth, salivary and gastric function, immune response, and excretion of Salmonella enterica serovar Typhimurium, in weaning pigs challenged with this microbe strain. Ital J Anim Sci 6:374–376. https://doi.org/10.4081/ijas.2007.1s.374

Truong VL, Jun M, Jeong WS (2018) Role of resveratrol in regulation of cellular defense systems against oxidative stress. Biofactors 44(1):36–49. https://doi.org/10.1002/biof.1399

Uni Z, Gal-Garber O, Geyra A et al (2001) Changes in growth and function of chick small intestine epithelium due to early thermal conditioning. Poult Sci 80:438–445

Valenzuela-Grijalva NV, Pinelli-Saavedra A, Muhlia-Almazan A et al (2017) Dietary inclusion effects of phytochemicals as growth promoters in animal production. J Anim Sci Technol 59:8

Vanrolleghem W, Tanghe S, Verstringe S et al (2019) Potential dietary feed additives with antibacterial effects and their impact on performance of weaned piglets: a meta-analysis. Vet J 249:24–32. https://doi.org/10.1016/j.tvjl.2019.04.017

Vetvicka V, Vannucci L, Sima P (2014) The effects of β-glucan on pig growth and immunity. Open Biochem J 8:89–93. https://doi.org/10.2174/1874091X01408010089

Vetvicka V, Oliveira C (2014) β(1–3)(1–6)-D-glucans modulate immune status in pigs: potential importance for efficiency of commercial farming. Ann Transl Med 2(2):16

Vuolo MM, Lima VS, Maróstica Junior MR (2019) Phenolic compounds: structure, classification, and antioxidant power. In: Bioactive compounds. Woodhead Publishing, Duxford, pp 33–50

Walter BM, Bilkei G (2004) Immunostimulatory effect of dietary oregano etheric oils on lymphocytes from growth-retarded, low-weight growing-finishing pigs and productivity. Tijdschr Diergeneeskd 129:178–181

Wang W, Zeng X, Mao X et al (2010) Optimal dietary true ileal digestible threonine for supporting the mucosal barrier in small intestine of weanling pigs. J Nutr 140(5):981–986

Wang S, Yao B, Gao H et al (2019) Combined supplementation of Lactobacillus fermentum and Pediococcus acidilactici promoted growth performance, alleviated inflammation, and modulated intestinal microbiota in weaned pigs. BMC Vet Res 15:239. https://doi.org/10.1186/s12917-019-1991-9

Wang P, Bai J, Liu X, Wang M, Wang X, Jiang P (2020) Tomatidine inhibits porcine epidemic diarrhea virus replication by targeting 3CL protease. Vet Res 51(1):136. https://doi.org/10.1186/s13567-020-00865-y

Wang CH, Wu WJ, Su LY et al (2021 Sep 7) Chinese herbal medicine formula Guizhi Li-Zhong Tang as an alternative to antibiotic feed additives for preventing pneumonia in piglets through antioxidant, anti-inflammatory, and antiapoptotic protection. Evid Based Complement Alternat Med 2021:4978783. https://doi.org/10.1155/2021/4978783

Wei HK, Xue HX, Zhou ZX et al (2017) A carvacrol-thymol blend decreased intestinal oxidative stress and influenced selected microbes without changing the messenger RNA levels of tight junction proteins in jejunal mucosa of weaning piglets. Animal 211(2):193–201. https://doi.org/10.1017/S1751731116001397

Wellock IJ, Fortomaris PD, Houdijk JG et al (2008) Effects of dietary protein supply, weaning age and experimental enterotoxigenic Escherichia coli infection on newly weaned pigs: performance. Animal 2(6):834–842. https://doi.org/10.1017/S1751731108001559

Werner C, Schubbert A, Schrödl W et al (2014) Effects of feeding different roughage components to sows in gestation on bacteriological and immunological parameters in colostrum and immune response of piglets. Arch Anim Nutr 268(1):29–41. https://doi.org/10.1080/1745039X.2013.876184

Wiemann M (2013) Optimized piglet feeding without antibiotics - a European perspective. Pig Feed Quality Conference

Williams JM, Duckworth CA, Burkitt MD et al (2015) Epithelial cell shedding and barrier function: a matter of life and death at the small intestinal villus tip. Vet Pathol 52(3):445–455. https://doi.org/10.1177/0300985814559404

Wilson HL, Obradovic MR (2015) Evidence for a common mucosal immune system in the pig. Mol Immunol 66:22–34. https://doi.org/10.1016/j.molimm.2014.09.004

Wu M, Zhang Q, Yi D et al (2020) Quantitative proteomic analysis reveals antiviral and anti-inflammatory effects of Puerarin in piglets infected with porcine epidemic diarrhea virus. Front Immunol 11:169. https://doi.org/10.3389/fimmu.2020.00169

Xing Y, Wu Y, Mao C et al (2019) Water extract of Artemisia ordosica enhances antioxidant capability and immune response without affecting growth performance in weanling piglets. J Anim Physiol Anim Nutr (Berl) 103(6):1848–1856. https://doi.org/10.1111/jpn.13171

Xu J, Xu C, Chen X et al (2014) Regulation of an antioxidant blend on intestinal redox status and major microbiota in early weaned piglets. Nutrition 30:584–589. https://doi.org/10.1016/j.nut.2013.10.018

Yi Q, Liu J, Zhang Y et al (2021) Anethole attenuates enterotoxigenic Escherichia Coli-induced intestinal barrier disruption and intestinal inflammation via modification of TLR signaling and intestinal microbiota. Front Microbiol 12:647242. https://doi.org/10.3389/fmicb.2021.647242

Yin YL, Tang ZR, Sun ZH et al (2008) Effect of galacto-mannan-oligosaccharides or chitosan supplementation on cytoimmunity and humoral immunity in early-weaned piglets. Asian Australas J Anim Sci 21:723e31

Yu E, Chen D, Yu B et al (2021) Manno-oligosaccharide attenuates inflammation and intestinal epithelium injury in weaned pigs upon enterotoxigenic *Escherichia coli* K88 challenge. Br J Nutr 126:993–1002. https://doi.org/10.1017/S0007114520004948

Zentek J, Buchheit-Renko S, Ferrara F et al (2011) Nutritional and physiological role of medium-chain triglycerides and medium-chain fatty acids in piglets. Anim Health Res Rev. 12(1):83–93. https://doi.org/10.1017/S1466252311000089

Zhang L, Xu Y-Q, Liu H-Y et al (2010) Evaluation of Lactobacillus rhamnosus GG using an Escherichia coli K88 model of piglet diarrhoea: Effects on diarrhoea incidence, faecal microflora and immune responses. Vet Microbiol 141:142–148

Zhang HJ, Jiang XR, Mantovani G et al (2014) Modulation of plasma antioxidant activity in weaned piglets by plant polyphenols. Ital J Anim Sci 13:2

Zhang MS, Sandouk A, Houtman JC (2016) Glycerol Monolaurate (GML) inhibits human T cell signaling and function by disrupting lipid dynamics. Sci Rep 6:30225. https://doi.org/10.1038/srep30225

Zhang MS, Tran PM, Wolff AJ (2018) Glycerol monolaurate induces filopodia formation by disrupting the association between LAT and SLP-76 microclusters. Sci Signal 11(528):eaam9095. https://doi.org/10.1126/scisignal.aam9095

Zhao X, Tong W, Song X et al (2018) Antiviral effect of resveratrol in piglets infected with virulent pseudorabies virus. Viruses 10:457. https://doi.org/10.3390/v10090457

Zhu LH, Zhao KL, Chen XL, Xu JX (2012) Impact of weaning and an antioxidant blend on intestinal barrier function and antioxidant status in pigs. J Anim Sci 90(8):2581–2589. https://doi.org/10.2527/jas.2012-4444

Probiotics as Feed Additives in Monogastric Animals

Soumela Savvidou, George K. Symeon, and Vassilios Dotas

Abstract

Persistent demand of animal and food industries for both high production performance and reduction in production cost, immense regulatory pressure for safe food supply systems and increasing consumer awareness of healthy animals and animal products regarding the antibiotic resistance arising from excessive administration of antibiotics as growth promoters led to the investigation for safer alternative biotechnological advancements and consequently to the usage of probiotics as feed and food additives in livestock.

Among numerus definitions, probiotics have more accurately been defined as "live micro-organisms which when administered in adequate amounts confer a health benefit on the host" (FAO/WHO 2001). Their main mode of action focuses on the maintenance of the microbial equilibrium in the gastrointestinal tract, the increase of beneficial microbial population, and the competitive exclusion of potentially pathogenic bacteria as well as the generation of antimicrobial substances and toxin reduction. Moreover, it includes several biological mechanisms, such as the alteration of metabolism by increasing digestive enzyme activity, the decreasing of bacterial enzyme activity and ammonia production, the alteration in gene expression in pathogenic microorganisms as well as the immunomodulation and histological alterations of the gastrointestinal tract.

Recent research results have shown that the supplementation of monogastric (poultry and swine) diets with probiotics significantly enhances their health and immunity, increases nutritional digestibility and growth performance, and improves the animal products' quantity and quality. Nevertheless, there is still

S. Savvidou (✉) · G. K. Symeon
Research Institute of Animal Science, ELGO-DEMETER, Paralimni Gannitsa, Greece

V. Dotas
Department of Animal Production, Faculty of Agriculture, Aristotle University of Thessaloniki, Thessaloniki, Greece

G. Arsenos, I. Giannenas (eds.), *Sustainable Use of Feed Additives in Livestock*,
https://doi.org/10.1007/978-3-031-42855-5_26

an ongoing discussion among scientists with respect to the selected microorganism, the suitable doses, as well as the administration methods and techniques for obtaining the optimal results.

Keywords

Feed additives · Probiotics · Monogastric animals · Poultry nutrition · Pig nutrition

1 Introduction

As the antibiotic over-use and misuse in animal production contributed to the emerging threat of antibiotic resistance, the World Health Organization was led to announce a warning of a post-antibiotic era arrival. Nowadays, more than ever, animal nutrition scientists have been challenged by the prohibition of antibiotics used as growth promoters and impelled to studying innovating alternatives, such as probiotics.

Among numerous definitions, probiotics have recently defined as "live microorganisms which when administered in adequate amounts confer a health benefit on the host" (FAO/WHO 2001). Their main mode of action focuses on the maintenance of the microbial equilibrium in the gastrointestinal tract, the increase of beneficial microbial population and the competitive exclusion of potentially pathogenic bacteria as well as the generation of antimicrobial substances and toxin reduction.

Recent research results have shown that the supplementation of monogastric diets with probiotics significantly enhances their health and immunity, increases nutritional digestibility and growth performance, and improves the animal products' quantity and quality. Nevertheless, there is still an ongoing discussion among scientists with respect to the selected microorganism as well as the administration methods and techniques for obtaining the optimal results.

This chapter will review the scientific literature related to the use of probiotics as feed additives in monogastric animal feeding (poultry and pigs and rabbits). It will provide an overview of the current situation, limitations and prospects for probiotic formulations applied to monogastric animals and will focus on their mode of action and their effects on animal well-being, production performance, and the quality of animal products.

2 Definitions

The potential role of intestinal bacteria in the maintenance of health and consequently longevity of humans consuming milk products fermented by lactic acid bacteria was first reported by Metchnikoff (1907). Later in (1965), the term "probiotic" was introduced by Lilly and Stillwell and referred to unknown growth promoting substances produced by microorganisms. The term "probiotic" consists of two words of the Greek language, "pro" and "bios" meaning "for" and "life," respectively. It is actually the opposite of the term "antibiotic" which means "against life" (Ghadban 2002). Parker (1974), included broader group of organisms and defined probiotics as microorganisms or substances that contribute to intestinal microbial balance. Fuller (1989) narrowed the definition of "probiotics" by Parker (1974), excluded antibiotic preparations, and he refined "probiotic" as a live microbial feed supplement, with a favorable effect on the host animal by improving its intestinal microbial balance. Fuller's (1989) definition was restricted to feed supplements, animals, and their intestinal tract, but later Havenaar and Huis In′t Veld (1992) extended it and redefined "probiotic" as "a mono or mixed culture of living microorganisms which applied to animal or man, beneficially affects the host by improving the properties of the indigenous microflora". In 2005, Curbelo et al. indicated that the majority of researchers considered probiotics as feed additives, formed by live microbes that beneficially affect the health of the host. In recent years, however, due to an increase in interest and research into probiotics, this definition has been broadened to include microbial products as well as fermentation products. Therefore, probiotics have been defined as "concoctions of known viable microbes in adequate numbers that modify the host microflora to confer health benefits" (Callaway et al. 2008).

In 2001, FAO/WHO proposed a widely accepted definition of probiotics, that of the joint Food and Agriculture Organization of the United Nations (FAO) and World Health Organization (WHO) Working Group, as "live micro-organisms that, when administered in adequate amounts, confer a health benefit on the host," which is also approved by the International Scientific Association for Probiotics and Prebiotics (Hill et al. 2014). The most recent definition of probiotics suggested by FAO (2016) though is "live microorganisms that confer health benefits on the host once ingested in the correct amounts."

3 Classification

A probiotic preparation for livestock consists mainly of one species of a microorganism or a mixture of species, such as lactic acid bacteria (LAB), yeasts, *Bifidobacterium*, *Enterococcus*, and *Bacillus* (Callaway et al. 2008).

According to FAO (2016), probiotic microorganisms are categorized as follows:

1. Bacterial (species of Lactobacillus, Bifidobacterium, Bacillus, Enterococcus) vs Non-bacterial probiotics (Aspergillus oryzae, Candida pintolopesii, Saccharomyces boulardii, Saccharomyces cerevisiae).
2. Spore-forming (Bacillus subtilis, Bacillus amyloliquefaciens) vs Non-spore-forming probiotics (Lactobacillus and Bifidobacterium).
3. Multi-species or multi-strain probiotics (PoultryStar ME, which contains Enterococcus faecium, Lactobacillus reuteri, Lactobacillus salivarius and Pediococcus acidilactici; PrimaLac, which contains Lactobacillus spp., E. faecium, and Bifidobacterium thermophilum; Microguard, which contains various species of

Lactobacillus, Bacillus, Streptococcus, Bifidobacterium, and Saccharomyces) vs Single-species or single-strain probiotics (Species of Lactobacillus, Bifidobacterium, Bacillus, Aspergillus, Streptococcus, Saccharomyces, Pediococcus, Enterococcus, etc., and commercial products such as Bro-biofair—Saccharomyces cerevisiae and Anta Pro EF—E. faecium).
4. Allochthonous probiotics (not present in the GIT of animals, like yeasts) vs Autochthonous probiotics (natural indigenous microorganisms of the gastrointestinal tract (e.g., Lactobacillus and Bifidobacterium).

For a microorganism to be selected as a probiotic, it must possess certain attributes like the tolerance to gastrointestinal conditions, the ability to adhere to the gastrointestinal mucosa, and the competitive exclusion of pathogens (Gadde et al. 2017). Moreover, from a technical point of view, the microorganism should be able to maintain viability and desirable characteristics through manufacturing, transportation, storage, and application processes (Bajagai et al. 2016).

4 Mechanisms of Probiotic Action

The mode of action of probiotics seems to be multifactorial and their efficacy is characterized by strain specificity. Probiotics exert their beneficial effects through diverse mechanisms either in the gastrointestinal lumen or the wall of the gastrointestinal tract. Figure 1 summarizes schematically the basic mechanisms of probiotic action (adapted by Anee et al. 2021).

Probiotic mode of action includes several biological mechanisms, such as the modification of the microbial population of the gastrointestinal tract and promotion of the favorable gastrointestinal microflora, the alteration of metabolism by increasing digestive enzyme activity, the decreasing bacterial enzyme activity and ammonia production, the alteration in gene expression in pathogenic microorganisms, immunomodulation and histological alterations of the gastrointestinal tract.

5 Modification of the Microbial Population of the GIT and Promotion of the Favorable Gastrointestinal Microflora

Probiotics can affect the microbial population dynamics in the gastrointestinal tract, by maintaining mainly the beneficial bacteria, such as *Lactobacillus* and *Bifidobacteria* and suppressing viable numbers of specific harmful ones, such as Escherichia coli and Clostridium spp. (Mountzouris et al. 2007; An et al. 2008).

The decrease of pathogenic microorganisms in the gastrointestinal tract may be attributable to mechanisms such as antagonistic activity, competitive exclusion, adhesion to epithelial cells, autoaggregation and coaggregation with pathogens, competition for essential nutrients.

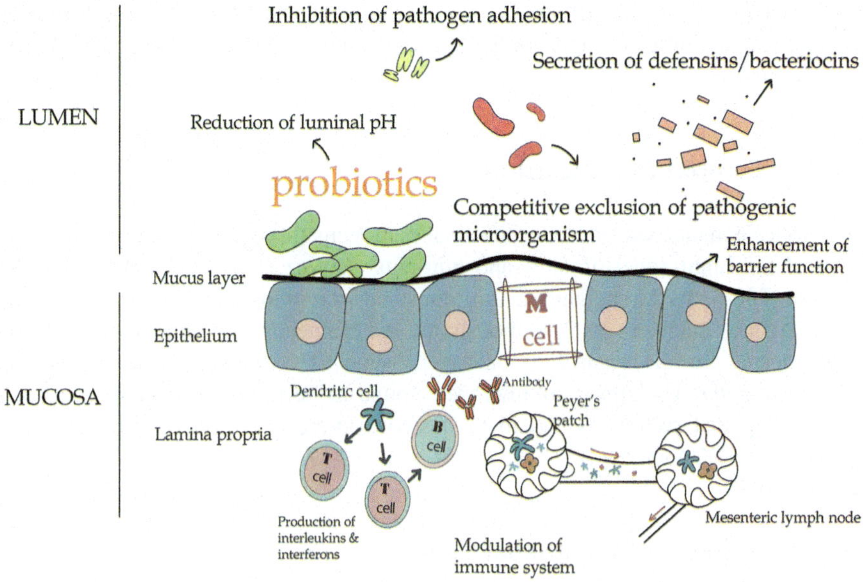

Fig. 1 Mechanism of probiotic action (adapted by Anee et al. 2021)

5.1 Antagonistic Activity

Probiotic bacteria are able to produce antibacterial substances such as defensives/ bacteriocins (acidolin, bulgaricin, nisin, reuterin), bacteriocin-like substances (Mahajan and Sahoo 1998), short chain volatile fatty acids (lactic, propionic, butyric, and acetic), and hydrogen peroxides (Fuller 1989; Jin et al. 1997; Mahajan and Sahoo 1998; Ghadban 2002; Curbelo et al. 2005), which have been reported to be inhibitory to pathogenic microorganisms (Chaucheyras-Durand and Durand 2010). Plaza-Diaz et al. (2019) suggest that such probiotic metabolic elements can regulate various metabolic pathways in cells (apoptosis, cell proliferation, inflammation, and differentiation). *Lact. acidophilus* has been reported to produce a large number of bacteriocins with an inhibitory effect on pathogenic bacteria such as *Salmonella*, Coliforms, and *Campylobacter* (Tahara and Kanatani 1997; Curbelo et al. 2005). Additionally, hydrogen peroxide due to its intense oxidizing effect on the bacterial cells or through the destruction of basic molecular structures of nucleic acid and cell proteins manages to prevent the growth of pathogens (Jin et al. 1996). The bactericidal effect of organic acids (lactic, acetic, butyric and propionic), which are major end products of the metabolism of lactic acid bacteria such as *Lactobacillus*, *Enterococcus,* and *Bifidobacterium* strains, is attributed to their ability to decrease intestinal pH (Jin et al. 1997) in the gastrointestinal tract and thus making it less favorable for pathogen growth. Though, the EU Council Regulations 429/2008 (Paragaraph 2.2) prohibits the use of microorganisms as additives or as production strains that are capable of producing antibiotic substances that are relevant as antibiotics in humans and animal, for example reuterin (European Patent EP0357673).

5.2 Competitive Exclusion

Probiotics enhance resistance to intestinal pathogens through competitive colonization of intestinal adhesion sites and nutrients (Parvez et al. 2006; Boirivant and Strober 2007), by strengthening this intestinal communication system (Mookiah et al. 2014). Ghadban (2002) demonstrated the efficiency of competitive exclusion mechanisms for the control of pathogens such as *Salmonella, Esch. coli,* and *Campylobacter*. Undefined cultures were found to be more effective than defined cultures against pathogens and in vivo studies were more variable than in vitro (Stavric and D'aoust 1993).

5.3 Adhesion to Epithelial Cells

Ability to compete pathogens for same receptors and adhere to the gut epithelial cells, is an ambiguous requirement for the potential probiotic strains in order to exert their probiotic effect in the gastrointestinal tract, with a lot of reporting that it is a desirable prerequisite researchers (Neeser et al. 2000; Lorca et al. 2002) and less suggesting that is not essential (Bezkorovainy 2001). The presence of adhesion-promoting proteins on the surface of *Lact. fermentum* and *Lact. brevis* (Lorca et al. 2002 and Rojas et al. 2002) or lectinlike protein structures on the surface of *Lact. animalis*, *Lact. fermentum* and particularly *Lact. fermentum spp.* (Gusils et al. 2002) suggests mechanism via which probiotic bacteria engage their potential binding sites in the gut, inhibiting pathogenic microorganisms such as *E. coli* O157:H7, *Salmonella*, etc., to attach to the epithelium (Bernet et al. 1994; Hudault et al. 1997; Johnson-Henry et al. 2007).

5.4 Autoaggregation and Coaggregation with Pathogens

Autoaggregation or "clumping" of lactic acid bacteria may also be necessary for adhesion to intestinal epithelial cells and with the addition of their potential coaggregation ability with enteropathogens, may form a barrier that prevents colonization or even more, entirely exclude enteropathogens from the gastrointestinal tract (Kos et al. 2003; Ghadban 2002).

5.5 Competition for Essential Nutrients

According to Mahajan and Sahoo (1998) glucose and amino acids and other nutrients that might be otherwise utilized by pathogens for their proliferation are preferably accessible and consequently are used by lactic acid bacteria in the gut. This hypothetical competition between gut bacteria and pathogens for nutrients as a competitive exclusion defense mechanism requires much more evidence to be confirmed though.

6 Alteration of Metabolism by Increasing Digestive Enzyme Activity, Decreasing Bacterial Enzyme Activity and Ammonia Production

Improved feed intake and feed conversion of monogastric animals provided with probiotics have been reported extensively, both for poultry (Bai et al. 2017; Gao et al. 2022; Palamidi et al. 2016; He et al. 2019; Zaghari et al. 2020) and pigs (Konstantinov et al. 2008; Yirga 2015). Probiotics may alter the digestive enzyme activity in the gastrointestinal tract of poultry and pigs either by producing the enzymes or by altering the microbial population and consequently enzyme production, and as digestibility of nutrients is inextricably linked with the enzyme activity in the intestine affects positively the metabolism of dietary carbohydrates, protein, lipids and minerals and the synthesis of vitamins and thence feed conversion.

Probiotics such as *Aspergillus oryzae* (Lee et al. 2006) and *Lact. casei* (Yeo and Kim 1997) have also found to significantly ($P < 0.05$) decrease ammonia production and urease activity, when fed to poultry, which may be linked with an improvement of animal health and enhanced growth (Jin et al. 1997), as ammonia produced by ureolysis and urea's activity in the intestinal mucosa can be toxic to the cell wall.

7 Alteration in Gene Expression in Pathogenic Microorganisms

Medellin-Peña et al. (2007) suggested that fermentation products from *L. acidophilus* La-5 significantly reduced secretion of autoinducer-2, a chemical signal, by human enterohemorrhagic *E. coli* serotype O157:H7, which resulted in the in vitro suppression of the virulence gene (LEE—locus of enterocyte effacement) expression. The disruption of quorum sensing and finally the prevention by E. coli serotype O157:H7 colonization in the gastrointestinal tract was reported.

8 Immunomodulation

The immunostimulatory effect of probiotics was reported in several studies. Probiotics exert an essential role in either stimulating or suppressing the immune system.

On the one hand, probiotics are reported to enhance disease resistance and reduce metabolic stress and mortality (Ezema 2013; Casas and Dobrogosz 2000). A suggested mechanism might be by increasing expression of toll-like receptors (TLRs), which causes the production of cytokines such as tumor necrosis factor-α (TNF-α), interleukin-4 (IL-4), and interferon-γ (IFN-γ) (Ashraf and Shah 2014). Anee et al. (2021) suggest that probiotics can regulate both the innate immunity and adaptive immunity by T- and B-cell stimulation through the dendritic cell in the lamina propria, surface epithelium, and Peyer's patches in the mucosa. Another proposed method might be increasing serum immunoglobulin levels (IgA and IgM) (Zhang and Kim 2014). A probiotic comprising by *L. fermentum* and *S. cerevisiae* increased the intestinal T-cell immune system (CD3+, CD4+ and CD8+ T-lymphocytes) in the gastrointestinal tract of broiler chickens. Probiotic *B. cereus* var. toyoi likewise increased significantly the intraepithelial population of CD8+ T cells in intestine of piglets (Scharek et al. 2007). Asgari et al. (2016) reported increased amounts of CD8+, CD4+, TCR1 + T cell in the gastrointestinal tract as well as in the peripheral blood system of poultry, when provided with supplementation with 1×10^9 CFU/kg of *Lactobacillus acidophilus*.

On the other hand, some studies have demonstrated immunosuppressive activity of probiotics in the host, for example Siepert et al. (2014) reported that E. faecium NCIMB 10415 delayed early immune response to antigens in post-weaning piglets. Dose-dependency and specificity of certain probiotic bacteria, or the gastrointestinal microbial profile of the host might provide justification for the variable results (Borchers et al. 2009).

9 Histological Alterations of the Gastrointestinal Tract

Histological alteration of the gastrointestinal tract and hence intestinal mucosa development as a mode of action of probiotics has been reported as well. Broilers fed *Bacillus subtilis*-based probiotic showed significantly ($P < 0.01$) greater villus height in jejunum and ileum and greater crypt depths in the duodenum, jejunum, and ileum, compared with control, according to Pelicano et al. (2005). Afsharmanesh and Sadaghi (2014) reported enhanced villus height and villus-to-crypt ratio of intestinal mucus layer, as well as increased absorption of nutrients due to a greater surface area, when *Bacillus subtilis* was provided to poultry. Wang et al. (2018) reached in similar conclusions and also reported improved intestinal homeostasis when probiotics were provided to weaned piglets.

There is a diversity in the proposed mechanisms of action of probiotics reported by the scientific community, either associated with an improvement of the animal performance or the inhibition of enteric pathogenic microorganisms. Although several probiotics microorganisms appeared to have similar mode(s) of action, closely related ones appear to have different modes of action, while at the same time a specific strain might function through multiple mechanisms. The exact mechanisms of certain probiotics need to be studied on a case-by-case basis. The application of molecular methods and DNA sequencing used to study microbial ecology will greatly facilitate our understanding of the way probiotics work.

10 Probiotic Application in Different Monogastric Livestock Production Systems

10.1 Probiotics in Poultry Nutrition

10.1.1 Introduction

The use of probiotics in the poultry industry rose when, in 2006, the European Union decided to ban the use of antibiotics as growth promoters (AGP) in farm animals' diets, in an attempt to reduce the emergence of microbial resistance to antibiotics (Turnidge 2004). This marked a turning point for the poultry industry since, till then, the use of growth promoters had improved growth of broilers by 4–8% and feed utilization by 2–5%, while it had also reduced the incidence of subclinical intestinal infections (Ajuwon 2016). Nevertheless, the use of probiotics has produced some very promising results, making them, among other feed additives, significant candidates for the replacement of AGPs (Lutful Kabir 2009).

The most commonly used probiotic microorganisms in poultry nutrition (Table 1) are bacteria belonging to the genera *Bifidobacterium, Bacillus, Enterococcus, Lactococcus, Lactobacillus,* and *Streptococcus* (Jha et al. 2020), but certain yeasts and fungal probiotics are also used; e.g., *Aspergillus oryzae, Candida pintolopesii, Saccharomyces, and Saccharomyces cerevisiae* (Bajagai et al. 2016).

Table 1 The most widely used microorganisms as probiotics in poultry production

Genus	Species	References
Bacillus	*B. subtilis*	Ramlucken et al. (2020); Zaghari et al. (2020); Zhu et al. (2020)
	B. licheniformis	Balamuralikrishnan et al. (2017); Flores et al. (2016); He et al. (2019)
Lactobacillus	*L. acidophilus*	de Souza et al. (2018); Pourakbari et al. (2016); Tayeri et al. (2018)
	L. plantarum	Gao et al. (2022); Meyer et al. (2020)
Bifidobacterium	*B. bifidum*	Stęczny and Kokoszynski (2019); Zhang et al. (2021)
	B. animalis	Palamidi et al. (2016)
Enterococcus	*E. faecium*	Park et al. (2016); Peralta-Sánchez et al. (2019); Z. F. Zhang and Kim (2014)
Candida	*C. utilis*	Chang et al. (2020)
Saccharomyces	*S. cerevisiae*	He et al. (2019); Stęczny and Kokoszynski (2019)

The alleged benefits of the use of probiotics in poultry production, i.e., increased growth rates, higher laying performance, and improved gut histomorphology, have led to a constant increase of their use and the respective market was estimated at 80 million USD in 2018. In fact, the probiotics market in poultry feed is projected to reach 125 million USD by 2025, accomplishing an annual growth rate of 7.7% (Jha et al. 2020). For the scopes of this review, the benefits of the use of probiotics in poultry production will be evaluated for each production direction separately, e.g., meat and egg production.

10.2 Probiotics in Poultry Meat Production

10.2.1 Growth Performance and Carcass Traits

There have been many reports in the late years which confirm that the dietary supplementation of broilers with probiotics has a beneficial effect on growth performance. Those results have been obtained either by using probiotics of the *Bacillus* strain (Ramlucken et al. 2020; Zaghari et al. 2020), or probiotics of the *Lactobacillus* strain (Wang et al. 2017; Gao et al. 2022). Moreover, the similar positive results have also been obtained by the use of mixtures of the above-mentioned strains, with or without the addition of fungi species (He et al. 2019; Chang et al. 2020), as well as by using commercial mixtures like Protexin (Pourakbari et al. 2016; Tayeri et al. 2018) or PoultryStar (Palamidi et al. 2016).

In terms of which growth parameter is most influenced, body weight and body weight gain are most prominently elevated by the use of probiotics in poultry diets (Zhang and Kim 2014; Zhang et al. 2021). Feed intake, on the other hand, is not commonly affected by the dietary inclusion of probiotics (Balamuralikrishnan et al. 2017; Tayeri et al. 2018) but there have been reports where feed intake was also increased (Zhang et al. 2021; Gao et al. 2022). When both increase in body weight and decrease in feed intake are reported simultaneously, it is concluded that the feed conversion ratio declines, which might be very promising since it affects directly the economic output of the production (Bai et al. 2017; Gao et al. 2022). The mode of action of probiotics for the production of such results is the apparent improvement of the digestibility of the feed nutrients, especially crude protein, dry matter, and organic matter (Palamidi et al. 2016; He et al. 2019; Zaghari et al. 2020).

There have been works, though, where the use of probiotics did not affect significantly the growth performance of broilers. Fathi et al. (2017), as well as Flores et al. (2016) who used *Bacillus* strains in different concentrations did not observe any effects and the same applies for works that used the *Lactobacillus* strains (Gheorghe et al. 2018; Meyer et al. 2020) or mixtures of bacteria with the addition or not of fungi (de Souza et al. 2018; Al-Khalaifa et al. 2019). Nevertheless, the majority of papers reports the beneficial effects of the probiotic use, more eminently in recent works like the one from Sjofjan et al. (2021a). The authors in this paper performed a meta-analysis of 49 papers that used *Lactobacillus spp.* as probiotics in the years 2008–2020 and concluded that body weight and body weight gain are positively, and feed conversion ratio and mortality negatively affected by the dietary inclusion of the bacteria. Therefore, the variation observed in the efficacy or not of the probiotics in the literature could be attributed to differences in the probiotic used (one strain or a mixture), the dosage as well as to the different management conditions of each flock.

The results in the literature are much more homogeneous when the carcass traits are concerned. Most authors conclude that the use of probiotics in poultry feeding has none or a very minimum impact on the carcass traits. For example, Wang et al. (2016), who used a probiotic based on Bacillus strains, reported that the proportional weight of neither of the commercial parts (breast, leg, wing) or the abdominal fat pad has been affected. Comparable results are also reported by Pourakbari et al. (2016), Pournazari et al. (2017) as well as Zhang et al. (2021). On the other hand, Wang et al. (2017)) reported that the use of *Lactobacillus johnsonii BS15* in the concentration of 1×10^6 cloning-forming units [cfu]/g as fed reduced that proportion of the abdominal fat pad in the carcass, but this is not a common finding. The same overall image is also reported for the weights of the edible internal organs of broilers, where the weight is not affected by the dietary inclusion of probiotics (Olnood et al. 2015; Balamuralikrishnan et al. 2017; Tayeri et al. 2018; Gao et al. 2022).

10.2.2 Meat Quality

Meat quality is a critical attribute since it determines to a great extent the possibilities for its storage or further processing and affects the consumer acceptance of the product. Therefore, a successful feed additive should improve meat quality indices or, at the very least, not affect them negatively. On the other hand, most meat quality parameters are intercorrelated, and slight changes on one of them often affect the whole product.

With respect to meat quality, the effect of probiotics seems to be minor. Most researchers agree that meat pH, either 45 min or 24 h after slaughter, is not affected by the dietary inclusion of probiotics (Kim et al. 2016; Balamuralikrishnan et al. 2017) and the same applies also for meat color (Pelicano et al. 2003; Stęczny and Kokoszynski 2019), irrespective of the type of probiotic. Nevertheless, Bai et al. (2017) reported some minor changes on meat color (L, a*, b*) after the inclusion of a probiotic based on the *Bacillus* strain, but it is questionable if these changes could be detected by the consumer. Likewise, Ivanovic et al. (2012) reported some minor changes on meat pH after using a *Streptococcus*-based probiotic.

As far as the water holding capacity of meat is concerned, the results from the literature are controversial. Some authors found that the dietary supplementation of broilers with probiotics does not affect drip or cook loss (Pelicano et al. 2003; Kim et al. 2016; Stęczny and Kokoszynski 2019) while others reported an improvement on drip loss (Bai et al. 2017; Balamuralikrishnan et al. 2017) or even a negative effect (Kim et al. 2016). The same applies also for the shear values of meat (Alfaig et al. 2013-no effect; Bai et al. 2017-positive effect; Stęczny and Kokoszynski 2019-negative effect). As with pH and color though, it is questionable if any effects that exist could be identified by the consumers and at what extent.

Another important aspect of meat quality is lipid composition and oxidation of meat. Popova (2017) reviewed papers where the supplementation of poultry feed with probiotics reduced the saturated and increased the mono and poly unsaturated fatty acids in poultry meat, which is a very positive result. The increase in the mono- and polyunsaturated fatty acids in poultry meat indicates that the meat is more susceptible to oxidation however, interestingly, the use of *Aspergillus awamori* decreased significantly the content of TBARS in broiler breast (Saleh et al. 2011, 2012). Other researchers also reported that the use of probiotics reduced the production of malondialdehyde in broilers' muscle tissues (Hossain et al. 2012a, b; Bobko et al. 2015). The above findings could be very promising, but due to the many parameters involved (type of probiotic, dosage, diets, management conditions, etc.), further confirmation from the literature is needed if they are to be taken into account, since there are also contradictory results, e.g. Hossain et al. (Hossain et al. 2012b), who reported a reduction in the n-6 PUFA in both breast and thigh.

10.2.3 Health Status

In terms of overall health, it seems to be a clear benefit for the broilers receiving dietary probiotics. In most cases, there were an improvement of homeostasis and health of gut ecosystem (de Souza et al. 2018; Ramlucken et al. 2020; Zhu et al. 2020; Gao et al. 2022) and a reduction in the presence of *E. coli* and *Salmonella* (Zhang and Kim 2014; Pourakbari et al. 2016; Gheorghe et al. 2018; Zhang et al. 2021). The general immune system of broilers is also positively affected by the dietary inclusion of probiotics (Palamidi et al. 2016; Fathi et al. 2017; Meyer et al. 2020; Sjofjan et al. 2021b). In terms of blood constituents, the probiotics seem to reduce the concentrations of blood cholesterol, triglycerides, and LDL (Wang et al. 2016; Zhu et al. 2020) and increase the concentration of HDL (Pourakbari et al. 2016; Tayeri et al. 2018). There have been works, however, that did not report any beneficial effects of probiotics on the heath profile of broilers (Balamuralikrishnan et al. 2017; Pournazari et al. 2017; Zaghari et al. 2020), but these differences could be attributed to the numerous variable factors among studies (type of probiotic, dosage, diets, etc.).

10.3 Probiotics in Egg Production

There has been extensive research on the effects of the dietary inclusion of probiotics in hen's diets and, similarly with broilers, there are inconsistent results. It is well established that the addition of probiotics improves the digestibility of dry matter, crude protein, and energy (Park et al. 2016) and has a beneficial effect on the gut morphology (Xiang et al. 2019; Obianwuna et al. 2022). In most cases, feed intake is not affected (Fathi et al. 2018; Sjofjan et al. 2021a; Ray et al. 2022) but there have also been reports where feed intake was reduced after the probiotic dietary supplementation (Lokapirnasari et al. 2019; Xiang et al. 2019).

In terms of productivity, most authors agree that egg production is enhanced by the use of probiotics (Abdelqader et al. 2013; Mazanko et al. 2018; Peralta-Sánchez et al. 2019) but the same doesn't apply for egg weight, which is not affected (Mikulski et al. 2012; Chung et al. 2015; Carvalho et al. 2022). Nevertheless, the combination of increased productivity and unaffected feed intake produces an improvement of feed conversion ratio (Panda et al. 2008; Xiang et al. 2019; Sjofjan et al. 2021a), resulting thus in better productivity of the farm.

With respect to egg quality, the effects of probiotic dietary supplementation are moderate. The most prominent effect is an increase of eggshell weight and thickness (Panda et al. 2008; Abdelqader et al. 2013; Sjofjan et al. 2021a) which in turn results in higher eggshell strength (Fathi et al. 2018; Obianwuna et al. 2022). Yolk weight is not affected by the probiotic inclusion (Mohebbifar et al. 2013; Aalaei et al. 2018; Carvalho et al. 2022) and the same applies also for yolk color (Chung et al. 2015; Aalaei et al. 2018) and Haugh units (Mikulski et al. 2012; Gnanadesigan et al. 2014).

Despite the minimum effects on egg quality, there is a significant beneficial effect of probiotic dietary supplementation on the overall hen health status. Most authors report a reduction of plasma cholesterol concentration (Zhang and Kim 2014; Sjofjan et al. 2021a) as well as an elevation of the immune system of hens (Panda et al. 2008; Obianwuna et al. 2022). Moreover, the presence of *E. coli* in hens' feces is reduced after the inclusion of probiotics in the diets (Aalaei et al. 2018; Upadhaya et al. 2019). Overall, the combination of increased productivity and better health mark the probiotics as very favorable supplements for commercial egg production.

10.4 Probiotics in Pig Nutrition

10.4.1 Introduction

It is estimated that 109.8 million tons of pork are produced worldwide, which accounts for about 33% of the global meat production (FAO 2020). Pig farming for commercial purposes is primarily focused on increasing the proportion of carcass meat in growing-finishing pigs (Cameron 1990). It has been documented that stress following weaning adversely affects piglets, resulting in major economic losses for pig farmers (Dowarah et al. 2017). The symptoms after weaning can be devastating, including diarrhea, weight loss, and death in extreme cases (Kreuzer-Redmer et al. 2016). For the above reasons antibiotics have been used as standard practice by modern pig production methods for a long time. Table 2 presents microorganisms commonly used as probiotics in swine nutrition. The use of antibiotics as feed additives in swine production has been reported to promote growth, feed conversion efficiency, and reduced morbidity and mortality, particularly in piglets (Cromwell 2002). The use of probiotics has been shown to have a significant impact on the stimulation of the immune system in animals (Kiczorowska et al. 2017), improving health status and production indices in fatteners, and reducing post-weaning diarrhea incidences and mortality rates in piglets (Yan and Kim 2013; Zimmermann et al. 2016).

Table 2 Microorganisms commonly used as probiotics in swine nutrition

Genus	Species	Reference
Lactobacillus	*L. plantarum*	Suo et al. (2012); Barszcz et al. (2016); Shin et al. (2019); Rybarczyk et al. (2020)
	L. casei	Barszcz et al. (2016); Rybarczyk et al. (2020)
	L. acidophilus	Joysowal et al. (2018)
Bacillus	*B. subtilis*	Meng et al. (2010); Balasubramanian et al. (2017)
	B. licheniformis	Balasubramanian et al. (2017)
Bifidobacterium	*B. lactis*	Merrifield et al. (2013)
Enterococcus	*E. faecium*	Kreuzer-Redmer et al. (2016)
Saccharomyces	*S. cerevisiae*	Suo et al. (2012); Barszcz et al. (2016); Rybarczyk et al. (2020)

10.5 Growth Performance and Carcass Traits

The use of probiotics is beneficial to animal growth, due to improvements in digestion, absorption, and utilization of nutrients (Konstantinov et al. 2008; Yirga 2015). The results of studies conducted on the efficacy of probiotics in pigs have been inconclusive. In the majority of studies, supplementation with LAB and yeast (S. cerevisiae) results in improved growth in weaned piglets, mainly by preventing diarrhea (Suo et al. 2012). LAB administration, on the other hand, proved to have no effect on improving growth efficiency in older pigs because they already have a balanced population of microorganisms (Shon et al. 2005). Some studies have, however, demonstrated that probiotics may be beneficial to the health and growth performance of grown pigs (Meng et al. 2010; Tufarelli et al. 2017).

An experiment conducted by Dowarah et al. (2018) with 36 weaning piglets found no significant differences between the control group and those receiving probiotics with their diets in terms of pre-slaughter weight, carcass weight, dressing percentage, carcass length, total back-fat thickness, and fat free index. Similarly, Anna et al. (2005) reported that back-fat thickness and carcass percentage did not change between probiotic-treated and non-treated animals. According to Barszcz et al. (2016), the weight gain of piglets receiving probiotics (0.5 g/kg diet) was similar to that of those not receiving any probiotics from the day of weaning through to the end of the experiment.

On the other hand, Kumar et al. (2009) reported significant improvement ($P < 0.01$) in carcass weight, dressing percentage, back-fat thickness, and meat percentage when a commercial probiotic (5 g/pig/day) was given. In the study conducted by Balasubramanian et al. (2017), 150 pigs were allocated to three diets, each supplemented with 0, 0.1, or 0.2 g/kg of multi-species probiotic (MSP). Inclusion of dietary MSP resulted in an increase ($P < 0.05$) in body weight, average daily gain, and gain/feed ratio, with no effect on average daily feed intake throughout the overall experiment and on apparent total tract digestibility of dry matter and nitrogen at week 16. Joysowal et al. (2018) conducted a trial with 27 crossbred grower pigs (3 groups with 3 replicates) that were supplemented with a basal diet either without probiotics or with a dairy-based or swine-based probiotic and concluded that supplementation of both probiotics significantly improved ($P < 0.05$) final body weight, average daily gain (g/d), and feed conversion ratio in pigs.

10.6 Meat Quality

The effects of different treatments with probiotics affect not only pig health directly, but also the properties of the meat produced. This is of rising importance to consumers who are concerned about the quality and safety of their food. Some studies have shown that probiotics improve meat quality (Alexopoulos et al. 2004; Česlovas et al. 2005), while others have found the opposite (Quadros et al. 2001). These discrepancies could be attributed to differences in the animals and probiotics used in their research.

Tian et al. (2021) concluded that long-term dietary probiotic *Lactobacillus reuteri* 1 enhanced meat quality compared to antibiotics by decreasing drip loss and shear force and increasing inosinic acid and glutamic acid, factors that may improve flavor and modify muscle fiber characteristics, in 144 crossbred male piglets randomly assigned to control, antibiotic, and probiotic treatments. Dowarah et al. (2018) reported that meat from pigs fed with swine-origin probiotics displayed improved water holding capacity and TBARS as compared to control meat. It was further found that ether extract, pH, and some sensory attributes (juiciness and appearance) were also improved ($P < 0.05$) with the probiotic, suggesting that the specific probiotic could contribute to improving pork carcass quality and physicochemical properties without negatively affecting blood metabolites. A dietary MSP supplementation led to an increase ($P < 0.05$) in sensory evaluation of color and back-fat thickness at week 16 in pigs as determined by Balamuralikrishnan et al. (2017). Furthermore, the tendency for increased firmness and reduced cooking and drip loss ($P < 0.1$) was observed. MSP inclusion in the diet has the potential to improve meat quality in pigs, as indicated by these findings. As a result of probiotic supplementation, Meng et al. (2010) found that pigs fed with probiotic-supplemented diets had significantly greater scores on meat color, marbling, and redness ($P < 0.01$).

Conversely, Rybarczyk et al. (2020) conducted an experiment with supplementation of two dosages (0.3 and 0.5%) of probiotic preparation and concluded that meat from pigs that were given probiotics had a higher electrical conductivity, a greater cooking loss, and a lower degree of juiciness of the longissimus lumborum muscle when compared to pigs that did not receive any probiotics. Additionally, the 0.5% supplemented group scored lower on color assessment of cooked meat when compared to the 0.3% and control groups, along with the highest redness (a*) and yellowness (b*) scores.

Clearly, probiotics have several beneficial effects on the sensory evaluation of meat. However, further research is needed to determine the effects of supplementation on all meat quality traits in growing and finishing pigs.

10.7 Health Status

The effects of dietary probiotics on animal performance have been firmly established by studies that show that they produce antibacterial substances in the intestine of animals, which compete with harmful gut microbes and stimulate the immune system (Hossain et al. 2015; Lan et al. 2017). Bacteria must have the ability to tolerate acidic conditions in the stomach and bile acids in the small intestine in order to function as a probiotic in the gastrointestinal tract (Erkkilä and Petaja 2000).

There has been considerable research showing that probiotics supplementation can improve growth performance, feed conversion efficiency, intestinal microbiota modulation, nutrient utilization, gut health, and immune system regulation in swine diets (Iournals 2011; Merrifield et al. 2013; Dowarah et al. 2018). A number of anti-infectious properties of probiotics have also been demonstrated, including the reduction of colonization and shedding of *Salmonella*, and the reduction of post-weaning Colibacillosis caused by Enterotoxigenic *Escherichia coli* (ETEC) (Fairbrother et al. 2005; Callaway et al. 2008). It has been recognized that probiotics have a variety of mechanisms of action through which they affect their hosts, such as manipulating gut microbial populations, competing for adhesion sites on the mucosa, strengthening the gut epithelial barrier function, and regulating immune responses (Corcionivoschi et al. 2010; Brown 2011).

Based on a study conducted by Balasubramanian et al. (2017), the inclusion of MSP in growing-finishing pig diets increased the number of fecal Lactobacilli and decreased the number of *Escherichia coli* ($P < 0.05$). Compared to pigs fed the control diet, pigs fed the MSP diet had a lower diarrhea score ($P < 0.05$). There may have been an improvement in the utilization of nutrients and a reduction in diarrhea scores as a result of optimization of the microbiota balance in the gut. The probiotic properties of two strains of *Lactobacillus* species were studied by Zhang et al. (2019) in an experiment involving 120 crossbred (Landrace x Large White) growing pigs. Both strains showed tolerance to acid and heat, as well as antimicrobial properties in vitro. Furthermore, the two Lactobacillus treatments elevated fecal Lactobacillus populations at the end of the experiment. In addition, both *Escherichia coli* and *Staphylococcus aureus* populations in feces decreased in comparison with the control. In a similar study, Shin et al. (2019) reported that *Lactobacillus plantarum* JDFM LP11 can increase lactic acid bacteria populations in pig feces and contributing to the development of villi in the small intestine. *L. plantarum* JDFM LP11 exhibited a higher diversity and richness in the microbial community and a decreased expression of the gut inflammation-related genes in weaned piglets, promoting intestinal development. Probiotics have also been demonstrated to increase metabolic pathways related to branched chain amino acids biosynthesis and butyrate metabolism, which indicates their beneficial effects on gut health.

11 Conclusion

Considering the threat to human and animal health, as well as the environment, efforts were made to restrict the use of antibiotics in animal feed due to the emergence of microbes resistant to antibiotics (Gonzalez Ronquillo and Angeles Hernandez 2017; Liu et al. 2019). A ban was eventually imposed by the European Union in 2006 on the use of antimicrobial growth promoters in livestock for subtherapeutic purposes.

The use of probiotics has generally been shown to be an excellent alternative to antibiotics for improving animal health and biological performance, as well as promoting economic sustainability of production. However, supplementation protocols should be standardized based on the dosage, onset, and duration of treatment for each probiotic strain. In order to augment the literature related to using probiotics as alternatives to antibiotics, more field studies should be conducted in natural and commercial farm settings.

References

Aalaei M et al (2018) Comparison of single- and multi-strain probiotics effects on broiler breeder performance, egg production, egg quality and hatchability. Br Poult Sci 59(5):531–538. https://doi.org/10.1080/00071668.2018.1496400

Abdelqader A, Irshaid R, Al-Fataftah A-R (2013) Effects of dietary probiotic inclusion on performance, eggshell quality, cecal microflora composition, and tibia traits of laying hens in the late phase of production. Trop Anim Health Prod 45(4):1017–1024. https://doi.org/10.1007/s11250-012-0326-7

Afsharmanesh M, Sadaghi B (2014) Effects of dietary alternatives (probiotic, green tea powder, and Kombucha tea) as antimicrobial growth promoters on growth, ileal nutrient digestibility, blood parameters, and immune response of broiler chickens. Comp Clin Pathol 23(3):717–724

Ajuwon KM (2016) Toward a better understanding of mechanisms of probiotics and prebiotics action in poultry species1. J Appl Poult Res 25(2):277–283. https://doi.org/10.3382/japr/pfv074

Alexopoulos C, Georgoulakis IE, Tzivara A, Kyriakis CS, Govaris A, Kyriakis SC (2004) Field evaluation of the effect of a probiotic-containing *Bacillus licheniformis* and *Bacillus subtilis* spores on the health status, performance, and carcass quality of grower and finisher pigs. J. Vet Med A Physiol Pathol Clin Med 51:306–312

Alfaig et al (2013) Effect of probiotics and thyme essential oil on the texture of cooked chicken breast meat. Acta Sci Pol Technol Aliment 12(4):379–384

Al-Khalaifa H et al (2019) Effect of dietary probiotics and prebiotics on the performance of broiler chickens. Poult Sci 98(10):4465–4479. https://doi.org/10.3382/ps/pez282

An B, Cho B, You S, Paik H, Chang H, Kim S, Yun C, Kang C (2008) Growth performance and antibody response of broiler chicks fed yeast derived β-glucan and single-strain probiotics. Asian Australas J Anim Sci 21(7):1027–1032

Anee IJ, Alam S, Begum RA et al (2021) The role of probiotics on animal health and nutrition. J Basic Appl Zool 82:52

Anna R, Wiecek J, Dziuba M (2005) Effect of feed additive on results of fattening and selected slaughter and quality traits of pork meat of pigs with different genotypes. Czech J Anim Sci 12: 561–567

Asgari F, Madjd Z, Falak R, Bahar MA, Nasrabadi MH, Raiani M, Shekarabi M (2016) Probiotic feeding affects T cell populations in blood and lymphoid organs in chickens. Benefic Microbes 7(5):669–675

Ashraf R, Shah NP (2014) Immune system stimulation by probiotic microorganisms. Crit Rev Food Sci Nutr 54(7):938–956

Bai K et al (2017) Supplemental effects of probiotic Bacillus subtilis fmbJ on growth performance, antioxidant capacity, and meat quality of broiler chickens. Poult Sci 96(1):74–82. https://doi.org/10.3382/ps/pew246

Bajagai YS et al. (2016) 'Probiotics in animal nutrition: production, impact and regulation', FAO Animal Production and Health Paper (FAO) eng no. 179 [Preprint]. Available at: http://www.fao.org/3/a-i5933e.pdf. Accessed 24 Oct 2022

Balamuralikrishnan B, Lee SI, Kim IH (2017) Dietary inclusion of different multi-strain complex probiotics; effects on performance in broilers. Br Poult Sci 58(1):83–86. https://doi.org/10.1080/00071668.2016.1257112

Balasubramanian B, Lee SI, Kim IH (2017) Inclusion of dietary multi-species probiotic on growth performance, nutrient digestibility, meat quality traits, faecal microbiota and diarrhoea score in growing–finishing pigs. Ital J Anim Sci 17(1):100–106

Barszcz M, Taciak M, Skomial J (2016) The effects of inulin, dried Jerusalem artichoke tuber and a multispecies probiotic preparation on microbiota ecology and immune status of the large intestine in young pigs. Arch Anim Nutr 70(4):278–292. https://doi.org/10.1080/1745039X.2016.1184368

Bernet M-F, Brassart D, Neeser J, Servin A (1994) Lactobacillus acidophilus LA 1 binds to cultured human intestinal cell lines and inhibits cell attachment and cell invasion by enterovirulent bacteria. Gut 35(4):483–489

Bezkorovainy A (2001) Probiotics: determinants of survival and growth in the gut. Am J Clin Nutr 73:399S–405S

Bobko M et al (2015) Lipid oxidation in chicken meat after application of bee pollen extract, propolis extract and probiotic in their diets. Potr S J Food Sci 9(1):342–346. https://doi.org/10.5219/495

Boirivant M, Strober W (2007) The mechanism of action of probiotics. Curr Opin Gastroenterol 23: 679–692

Borchers AT, Selmi C, Meyers FJ, Keen CL, Gershwin ME (2009) Probiotics and immunity. J Gastroenterol 44(1):26–46

Brown M (2011) Modes of action of probiotics: recent developments. J Anim Vet Adv 10(14): 1895–1900

Callaway, T.R., Edrington, T.S., Anderson, R.C., Harvey, R.B., Genovese, K.J., . . ., Nisbet, D.J., 2008. Probiotics, prebiotics and competitive exclusion for prophylaxis against bacterial disease. Anim Health Res Rev 9, 217–225

Cameron ND (1990) Genetic and phenotypic parameters for carcass traits, meat and eating quality traits in pigs. Livest Prod Sci 26:119–135

Carvalho C et al (2022) Effects of dietary probiotic supplementation on egg quality during storage. Poultry 1:180–192. https://doi.org/10.3390/poultry1030016

Casas IA, Dobrogosz WJ (2000) Validation of the probiotic concept: Lactobacillus reuteri confers broad-spectrum protection against disease in humans and animals. Microb Ecol Health Dis 12(4):247–285

Česlovas J, Vigilijus J, Almantas Š (2005) The effect of probiotics and phytobiotics on meat properties and quality in pigs. Vet Zootech 29:80–84

Chang J et al (2020) Compound probiotics alleviating aflatoxin B1 and zearalenone toxic effects on broiler production performance and gut microbiota. Ecotoxicol Environ Saf 194:110420. https://doi.org/10.1016/j.ecoenv.2020.110420

Chaucheyras-Durand F, Durand H (2010) Probiotics in animal nutrition and health. Benef Microbes 1:3–9

Chung S, Lee J, Kong C (2015) Effects of multi strain probiotics on egg production and quality in laying hens fed diets containing food waste product. Int J Poult Sci 14:19–22. https://doi.org/10.3923/ijps.2015.19.22

Corcionivoschi N, Drinceanu D, Pop IM, Stack D, Ştef L, Julean C et al (2010) The effect of probiotics on animal health. Scientific Papers Anim Sci Biotechnol 43(1):35–41
Cromwell GL (2002) Why and how antibiotics are used in swine production. Anim Biotechnol 13: 7–27
Curbelo YG, García Y, López A, Boucourt R (2005) Probiotics: An alternative to improve animal performance. Cuba J Agric Sci 39:123–134
de Souza LFA et al (2018) Probiotics on performance, intestinal morphology and carcass characteristics of broiler chickens raised with lower or higher environmental challenge. Austral J Vet Sci 50(1):35–41. https://doi.org/10.4067/S0719-81322018000100107
Dowarah R, Verma AK, Agarwal N (2017) The use of Lactobacillus as an alternative of antibiotic growth promoters in pigs: a review. Anim Nutr 3(1):1–6. https://doi.org/10.1016/j.aninu.2016.11.002
Dowarah R, Verma AK, Agarwal N, Singh P (2018) Efficacy of species-specific probiotic Pediococcus acidilactici FT28 on blood biochemical profile, carcass traits and physicochemical properties of meat in fattening pigs. Res Vet Sci 117:60–64. https://doi.org/10.1016/j.rvsc.2017.11.011
Erkkilä S, Petaja E (2000) Screening of commercial meat starter cultures at low pH and in the presence of bile salts for potential probiotic use. Meat Sci 55(3):297–300. https://doi.org/10.1016/S0309-1740(99)00156-4
Ezema C (2013) Probiotics in animal production: a review. J Vet Med Anim Health 5(11):308–316
Fairbrother JM, Nadeau E, Gyles CL (2005) Escherichia coli in postweaning diarrhea in pigs: an update on bacterial types, pathogenesis, and prevention strategies. Anim Health Res Rev 6(1): 17–39
FAO (2016) Probiotics in animal nutrition – Production, impact, and regulation: by Bajagai, Y.S., Klieve, A.V. Dart, P.J. & Bryden, W.L. Ed. P.S. Harinder. FAO Animal Production and Health Paper No. 179. FAO, Rome
FAO (2020) Food and Agriculture Organization of the United Nations. FAOSTAT Statistical Database. FAO, Rome
FAO/WHO (2001). Report of a Joint of FAO/ WHO Expert Consultation on Evaluation of Health and Nutritional Properties of Probiotics in Food Including Powder Milk with Live Lactic Acid Bacteria, (October 2001)
Fathi M et al (2018) Effects of dietary probiotic (Bacillus subtilis) supplementation on productive performance, immune response and egg quality characteristics in laying hens under high ambient temperature. Ital J Anim Sci 17(3):804–814. https://doi.org/10.1080/1828051X.2018.1425104
Fathi MM et al (2017) Influence of probiotic supplementation on immune response in broilers raised under hot climate. Br Poult Sci 58(5):512–516. https://doi.org/10.1080/00071668.2017.1332405
Flores C et al (2016) Direct-fed microbial and its combination with xylanase, amylase, and protease enzymes in comparison with AGPs on broiler growth performance and foot-pad lesion development. J Appl Poult Res 25(3):328–337. https://doi.org/10.3382/japr/pfw016
Fuller R (1989) Probiotics in man and animals. J Appl Bacteriol 66:365–378
Gadde U et al (2017) Alternatives to antibiotics for maximizing growth performance and feed efficiency in poultry: a review. Anim Health Res Rev 18(1):26–45. https://doi.org/10.1017/S1466252316000207
Gao J et al (2022) Effects of novel microecologics combined with traditional Chinese medicine and probiotics on growth performance and health of broilers. Poult Sci 101(2):101412. https://doi.org/10.1016/j.psj.2021.101412
Ghadban GS (2002) Probiotics in broiler production - a review. Archiv Fur Geflugelkunde 66:49–58
Gheorghe A et al (2018) Effect of dietary supplementation with probiotic mixture based on Lactobacillus strains on performance, gastrointestinal development and ileal microflora in broilers. J Biotechnol 280:S41. https://doi.org/10.1016/j.jbiotec.2018.06.129

Gnanadesigan M et al (2014) Quality evaluation of egg composition and productivity of layers in EM (effective microorganisms) treatments: a field report. Egypt J Basic Appl Sci 1(3):161–166. https://doi.org/10.1016/j.ejbas.2014.06.003

Gonzalez Ronquillo M, Angeles Hernandez JC (2017) Antibiotic and synthetic growth promoters in animal diets: review of impact and analytical methods. Food Control 72:255–267

Gusils C, Bujazha M, Gonzalez S (2002) Preliminary studies of design a probiotic for use in swine feed. Interciencia 27:409–413

Havenaar R, Huis In't Veld MJH (1992) Probiotics: a general view. In: Lactic acid bacteria in health and disease. Elsevier Applied Science Publishers, Amsterdam

He T et al (2019) Effects of probiotics as antibiotics substitutes on growth performance, serum biochemical parameters, intestinal morphology, and barrier function of broilers. Animals 9(11): 985. https://doi.org/10.3390/ani9110985

Hill C, Guarner F, Reid G, Gibson GR, Merenstein DJ, Pot B et al (2014) Expert consensus document: the International Scientific Association for Probiotics and Prebiotics consensus statement on the scope and appropriate use of the term probiotic. Nat Rev Gastroenterol Hepatol 11:506

Hossain MM, Begum M, Kim IH (2015) Effect of *Bacillus subtilis*, *Clostridium butyricum* and *Lactobacillus acidophilus* endospores on growth performance, nutrient digestibility, meat quality, relative organ weight, microbial shedding and excreta noxious gas emission in broilers. Vet Med Czech 60:77–86

Hossain ME et al (2012a) Evaluation of probiotic strains for development of fermented Alisma canaliculatum and their effects on broiler chickens. Poult Sci 91(12):3121–3131. https://doi.org/10.3382/ps.2012-02333

Hossain ME et al (2012b) Growth performance, meat yield, oxidative stability, and fatty acid composition of meat from broilers fed diets supplemented with a medicinal plant and probiotics. Asian Australas J Anim Sci 25(8):1159–1168. https://doi.org/10.5713/ajas.2012.12090

Hudault S, Liévin V, Bernet-Camard M-F, Servin AL (1997) Antagonistic activity exerted in vitro and in vivo by *Lactobacillus casei* (strain GG) against *Salmonella typhimurium* C5 infection. Appl Environ Microbiol 63(2):513–518

Iournals M (2011) Probiotics as a dietary additive for pigs: a review. J Anim Vet Adv 10(16):2

Ivanovic S et al (2012) Influence of probiotics on quality of chicken meat. Afr J Agric Res 7(14): 2191–2196. https://doi.org/10.5897/AJAR11.1693

Jha R et al (2020) Probiotics (direct-fed microbials) in poultry nutrition and their effects on nutrient utilization, growth and laying performance, and gut health: a systematic review. Animals 10(10):1863. https://doi.org/10.3390/ani10101863

Jin LZ, Ho YW, Abdullah N, Jalaludin S (1997) Probiotics in poultry: modes of action. Worlds Poult Sci J 53:351–368

Jin LZ, Ho YW, Abdullah N, Ali MA, Jalaludin S (1996) Antagonistic effects of intestinal Lactobacillus isolates on pathogens of chicken. Lett Appl Microbiol 23:67–71

Johnson-Henry KC, Hagen KE, Gordonpour M, Tompkins TA, Sherman PM (2007) Surface-layer protein extracts from *Lactobacillus helveticus* inhibit enterohaemorrhagic *Escherichia coli* O157: H7 adhesion to epithelial cells. Cell Microbiol 9(2):356–367

Joysowal M, Saikia B, Dowarah R, Tamuly S, Kalita D, Choudhury K (2018) Effect of probiotic Pediococcus acidilactici FT28 on growth performance, nutrient digestibility, health status, meat quality, and intestinal morphology in growing pigs. Vet World 11:1669–1676. https://doi.org/10.14202/vetworld.2018.1669-1676

Kiczorowska B, Samolińska W, Arm A-Y, Kiczorowski P, Winiarska-Mieczan A (2017) The natural feed additives as immunostimulants in monogastric animal nutrition–a review. Ann Anim Sci 17:605–625

Kim HW et al (2016) Effects of probiotics feeding on meat quality of chicken breast during postmortem storage. Poult Sci 95(6):1457–1464. https://doi.org/10.3382/ps/pew055

Kim KH, Ingale SL, Kim JS, Lee SH, Lee JH, Kwon IK, Chae BJ (2014) Bacteriophage and probiotics both enhance the performance of growing pigs but bacteriophage are more effective. Anim Feed Sci Technol 196:88–95. https://doi.org/10.1016/j.anifeedsci.2014.06.012

Konstantinov SR, Smidt H, Akkermans ADL, Casini L, Trevisi P, Mazzoni M, De Filippi S, Bosi P, de Vos W (2008) Feeding of *Lactobacillus sobrius* reduces Escherichia coli F4 levels in the gut and promotes growth of infected piglets. FEMS Microbiol Ecol 66:599–607. https://doi.org/10.1111/j.1574-6941.2008.00517.x

Kos B, Suskovic J, Vukovic S, Simpraga M, Frece J, Matosic S (2003) Adhesion and aggregation ability of probiotic strain *Lactobacillus acidophilus* M92. J Appl Microbiol 94:981–987

Kreuzer-Redmer S, Bekurtz JC, Arends D, Bortfeldt R, Kutz-Lohroff B, Sharbati S et al (2016) Feeding of *Enterococcus faecium* NCIMB 10415 leads to intestinal miRNA-423-5p-induced regulation of immune-relevant genes. Appl Environ Microbiol 82(8):2263–2269. https://doi.org/10.1128/AEM.04044-15

Kumar SG, Gnanaraj APT, Sivakumar T, Karthickeyan SMK, Murugan M (2009) Effect of probiotic supplementation on the carcass traits and sensory qualities of swill fed pork. Tamilnadu J Vet Anim Sci 5(4):157–160

Lan R, Tran H, Kim IH (2017) Effects of probiotic supplementation in different nutrient density diets on growth performance, nutrient digestibility, blood profiles, fecal microflora and noxious gas emission in weaning pig. J Sci Food Agric 97:1335–1341

Lee KW, Lee SK, Lee BD, Gungdong Y, Daejeon SK (2006) Aspergillus oryzae as probiotic in poultry - a review. Int J Poult Sci 5:01–03

Lilly DM, Stillwell RH (1965) Probiotics: growth-promoting factors produced by microorganisms. Science 147:747–748

Liu Z, Klumper U, Shi L, Ye L, Li M (2019) From pig breeding environment to subsequently produced pork: comparative analysis of antibiotic resistance genes and bacterial community composition. Front Microbiol 10:43

Lokapirnasari WP et al (2019) Potency of probiotics Bifidobacterium spp. and *Lactobacillus casei* to improve growth performance and business analysis in organic laying hens. Veterinary World 12(6):860–867. https://doi.org/10.14202/vetworld.2019.860-867

Lorca G, Torino MI, de Valdez GF, Ljungh A (2002) Lactobacilli express cell surface proteins which mediate binding of immobilized collagen and fibronectin. FEMS Microbiol Lett 206:31–37

Lutful Kabir SM (2009) The role of probiotics in the poultry industry. Int J Mol Sci 10(8): 3531–3546. https://doi.org/10.3390/ijms10083531

Mahajan P, Sahoo J (1998) Probiotics feeding to improve poultry meat quality - a critical appraisal. Indian Food Ind 17:333–344

Mazanko MS et al (2018) ‘Bacillus probiotic supplementations improve laying performance, egg quality, hatching of laying hens, and sperm quality of roosters. Probiotics Antimicrob Proteins 10(2):367–373. https://doi.org/10.1007/s12602-017-9369-4

Medellin-Peña MJ, Wang H, Johnson R, Anand S, Griffiths MW (2007) Probiotics affect virulence-related gene expression in Escherichia coli O157: H7. Appl Environ Microbiol 73(13): 4259–4267

Meng QW, Yan L, Ao X, Zhou TX, Wang JP, Lee JH, Kim IH (2010) Influence of probiotics in different energy and nutrient density diets on growth performance, nutrient digestibility, meat quality, and blood characteristics in growing-finishing pigs. J Anim Sci 88:3320–3326. https://doi.org/10.2527/jas.2009-2308

Merrifield CA, Lewis MC, Claus SP, Pearce JT, Cloarec O, Duncker S, Heinzmann SS, Dumas ME, Kochhar S, Rezzi S, Mercenier A, Nicholson JK, Bailey M, Holmes E (2013 Jun) Weaning diet induces sustained metabolic phenotype shift in the pig and influences host response to Bifidobacterium lactis NCC2818. Gut 62(6):842–851. https://doi.org/10.1136/gutjnl-2011-301656

Metchnikoff E (1907) Lactic acid as inhibiting intestinal putrefaction. In: Chalmers Mitchell P (ed) The prolongation of life: optimistic studies. William Heinemann, London

Meyer MM, Fries-Craft KA, Bobeck EA (2020) Composition and inclusion of probiotics in broiler diets alter intestinal permeability and spleen immune cell profiles without negatively affecting performance. J Anim Sci 98(1):skz383. https://doi.org/10.1093/jas/skz383

Mikulski D et al (2012) Effects of dietary probiotic (Pediococcus acidilactici) supplementation on performance, nutrient digestibility, egg traits, egg yolk cholesterol, and fatty acid profile in laying hens. Poult Sci 91(10):2691–2700. https://doi.org/10.3382/ps.2012-02370

Mohebbifar A et al (2013) Effects of commercial prebiotic and probiotics of diet on performance of laying hens, egg traits and some blood parameters. Annu Res Rev Biol:921–934

Mookiah S, Sieo CC, Ramasamy K, Abdullah N, Ho YW (2014) Effects of dietary prebiotics, probiotic and synbiotics on performance, caecal bacterial populations and caecal fermentation concentrations of broiler chickens. Journal of the Science of Food and Agriculture 94(2): 341–348

Mountzouris KC, Tsirtsikos P, Kalamara E, Nitsch S, Schatzmayr G, Fegeros K (2007) Evaluation of the efficacy of a probiotic containing Lactobacillus, Bifidobacterium, Enterococcus, and Pediococcus strains in promoting broiler performance and modulating cecal microflora composition and metabolic activities. Poult Sci 86:309–317

Neeser JR, Granato D, Rouvet M, Servin A, Teneberg S, Karlsson EA (2000) Lactobacillus johnsonii Lal shares carbohydrate-binding specificities with several enteropathogenic bacteria. Glycobiology 10:1193–1199

Obianwuna UE et al (2022) Enhancing egg production and quality by the supplementation of probiotic strains (Clostridium and Brevibacillus) via improved amino acid digestibility, intestinal health, immune response, and antioxidant activity. Front Microbiol 13:987241. https://doi.org/10.3389/fmicb.2022.987241

Olnood CG et al (2015) Novel probiotics: their effects on growth performance, gut development, microbial community and activity of broiler chickens. Anim Nutr 1(3):184–191. https://doi.org/10.1016/j.aninu.2015.07.003

Palamidi I et al (2016) Probiotic form effects on growth performance, digestive function, and immune related biomarkers in broilers. Poult Sci 95(7):1598–1608. https://doi.org/10.3382/ps/pew052

Panda AK et al (2008) Effect of probiotic (Lactobacillus sporogenes) feeding on egg production and quality, yolk cholesterol and humoral immune response of white Leghorn layer breeders. J Sci Food Agric 88(1):43–47. https://doi.org/10.1002/jsfa.2921

Park JW et al (2016) Effect of dietary supplementation with a probiotic (*Enterococcus faecium*) on production performance, excreta microflora, ammonia emission, and nutrient utilization in ISA brown laying hens. Poult Sci 95(12):2829–2835. https://doi.org/10.3382/ps/pew241

Parker RB (1974) Probiotics, the other half of the antibiotic story. Anim Nutr Health 29:4–8

Parvez S, Malik KA, Ah Kang S, Kim HY (2006) Probiotics and their fermented food products are beneficial for health. J Appl Microbiol 100:1171–1185

Pelicano ERL et al (2003) Effect of different probiotics on broiler carcass and meat quality. Braz J Poult Sci 5:207–214. https://doi.org/10.1590/S1516-635X2003000300009

Pelicano ERL, Souza P, Souza HBA, Figueiredo DF, Boiago MM, Carvalho SR, Bordon VF (2005) Intestinal mucosa development in broiler chickens fed natural growth promoters. Braz J Poult Sci 7:221–229

Peralta-Sánchez, J.M. et al. (2019) Egg production in poultry farming is improved by probiotic bacteria. Front Microbiol, 10. https://www.frontiersin.org/articles/10.3389/fmicb.2019.01042. Accessed 29 Nov 2022

Plaza-Diaz J, Ruiz-Ojeda FJ, Gil-Campos M, Gil A (2019) Mechanisms of action of probiotics. Adv Nutr 10:S49–S66

Popova T (2017) Effect of probiotics in poultry for improving meat quality. Curr Opin Food Sci 14: 72–77. https://doi.org/10.1016/j.cofs.2017.01.008

Pourakbari M et al (2016) Probiotic level effects on growth performance, carcass traits, blood parameters, cecal microbiota, and immune response of broilers. An Acad Bras Cienc 88:1011–1021. https://doi.org/10.1590/0001-3765201620150071

Pournazari M et al (2017) Prebiotics, probiotics and thyme (*Thymus vulgaris*) for broilers: performance, carcass traits and blood variables. Revista Colombiana de Ciencias Pecuarias 30(1): 3–10. https://doi.org/10.17533/udea.rccp.v30n1a01

Quadros ARB, Kiefer C, Ribeiro NLC, Zink LA (2001) Características qualitativas da carne de suínos alimentados com rações contendo ou não probióticos. Reunião Anual da Sociedade Brasileira de Zootecnia, Piracicaba. Anais, Piracicaba, Brazil, pp 794–795

Ramlucken U et al (2020) A novel *Bacillus* based multi-strain probiotic improves growth performance and intestinal properties of *Clostridium perfringens* challenged broilers. Poult Sci 99(1): 331–341. https://doi.org/10.3382/ps/pez496

Ray BC et al (2022) Comparative effects of feeding single- and multi-strain probiotics to commercial layers on the productive performance and egg quality indices. J Appl Poult Res 31(3): 100257. https://doi.org/10.1016/j.japr.2022.100257

Rojas M, Ascencio F, Conway PL (2002) Purification and characterization of a surface protein from *Lactobacillus fermentum* 104R that binds to porcine small intestinal mucus and gastric mucin. Appl Environ Microbiol 68:2330–2336

Rybarczyk A, Bogusławska-Wąs E, Łupkowska A (2020) Effect of EM® probiotic on gut microbiota, growth performance, carcass and meat quality of pigs. Livest Sci 241:104206

Saleh AA et al (2011) Effects of feeding *Aspergillus awamori* and *Aspergillus niger* on growth performance and meat quality in broiler chickens. J Poult Sci 48(3):201–206. https://doi.org/10.2141/jpsa.011019

Saleh AA et al (2012) The modification of the muscle fatty acid profile by dietary supplementation with *Aspergillus awamori* in broiler chickens. Br J Nutr 108(9):1596–1602. https://doi.org/10.1017/S0007114511007069

Scharek L, Altherr B, Tölke C, Schmidt M (2007) Influence of the probiotic *Bacillus cereus var. toyoi* on the intestinal immunity of piglets. Vet Immunol Immunopathol 120(3):136–147

Shin D, Chang SY, Bogere P, Won K, Choi JY, Choi YJ, Lee HK, Hur J, Park BY, Kim Y, Heo J (2019) Beneficial roles of probiotics on the modulation of gut microbiota and immune response in pigs. PLoS One 14(8):e0220843. https://doi.org/10.1371/journal.pone.0220843

Shon KS, Hong JW, Kwon OS, Min BJ, Lee WB, Kim IH, Park YH, Lee IS (2005) Effects of Lactobacillus reuteri-based direct-fed microbial supplementation for growing-finishing pigs. Asian Australas J Anim Sci 18:370–374. https://doi.org/10.5713/ajas.2005.370

Siepert B, Reinhardt N, Kreuzer S, Bondzio A, Twardziok S, Brockmann G, Nöckler K, Szabó I, Janczyk P, Pieper R (2014) Enterococcus faecium NCIMB 10415 supplementation affects intestinal immune-associated gene expression in post-weaning piglets. Vet Immunol Immunopathol 157(1):65–77

Sjofjan O et al (2021a) The effects of probiotics on the performance, egg quality and blood parameters of laying hens: a meta-analysis. J Anim Feed Sci 30(1):11–18. https://doi.org/10.22358/jafs/133432/2021

Sjofjan O et al (2021b) The effects of lactic acid bacteria and yeasts as probiotics on the growth performance, relative organ weight, blood parameters, and immune responses of broiler: a meta-analysis. F1000Research 10:183. https://doi.org/10.12688/f1000research.51219.3

Stavric S, D'aoust JY (1993) Undefined and defined bacterial preparations for the competitive-exclusion of Salmonella in poultry - a review. J Food Prot 56:173–180

Stęczny K, Kokoszynski D (2019) Effects of probiotics and sex on physicochemical, sensory and microstructural characteristics of broiler chicken meat. Ital J Anim Sci 18(1):1385–1393. https://doi.org/10.1080/1828051X.2019.1667269

Suo C, Yin Y, Wang X, Lou X, Song D, Wang X, Gu Q (2012) Effects of *Lactobacillus plantarum* ZJ316 on pig growth and pork quality. BMC Vet Res 8:89–101. https://doi.org/10.1186/1746-6148-8-89

Tahara T, Kanatani K (1997) Isolation and partial amino acid sequence of bacteriocins produced by *Lactobacillus acidophilus*. Biosci Biotechnol Biochem 61:884–886

Tayeri V et al (2018) A comparison of the effects of antibiotics, probiotics, synbiotics and prebiotics on the performance and carcass characteristics of broilers. Vet Res Commun 42(3):195–207. https://doi.org/10.1007/s11259-018-9724-2

Tian Z, Cui Y, Lu H, Wang G, Ma X (2021) Effect of long-term dietary probiotic *Lactobacillus reuteri* 1 or antibiotics on meat quality, muscular amino acids and fatty acids in pigs. Meat Sci 171:108234. https://doi.org/10.1016/j.meatsci.2020.108234

Tufarelli V, Crovace AM, Rossi G, Laudadio V (2017) Effect of a dietary probiotic blend on performance, blood characteristics, meat quality and faecal microbial shedding in growing-finishing pigs. S Afr Anim Sci 47:875–882. https://doi.org/10.4314/sajas.v47i6.15

Turnidge J (2004) Antibiotic use in animals—prejudices, perceptions and realities. J Antimicrob Chemother 53(1):26–27. https://doi.org/10.1093/jac/dkg493

Upadhaya SD, Rudeaux F, Kim IH (2019) Efficacy of dietary *Bacillus subtilis* and *Bacillus licheniformis* supplementation continuously in pullet and lay period on egg production, excreta microflora, and egg quality of Hyline-Brown birds. Poult Sci 98(10):4722–4728. https://doi.org/10.3382/ps/pez184

Wang H et al (2017) Live probiotic *Lactobacillus johnsonii* BS15 promotes growth performance and lowers fat deposition by improving lipid metabolism, intestinal development, and gut microflora in broilers. Front Microbiol 8. https://www.frontiersin.org/articles/10.3389/fmicb.2017.01073. Accessed: 21 Nov 2022

Wang X et al (2016) Effects of prebiotics, probiotics, and their combination on growth performance, small intestine morphology, and resident Lactobacillus of male broilers1. Poult Sci 95(6):1332–1340. https://doi.org/10.3382/ps/pew030

Wang Y, Xie Q, Sun S, Huang B, Zhang Y, Xu Y et al (2018) Probiotics-fermented Massa Medicata Fermentata ameliorates weaning stress in piglets related to improving intestinal homeostasis. Appl Microbiol Biotechnol 102(24):10713–10727

Xiang Q et al (2019) Effects of different probiotics on laying performance, egg quality, oxidative status, and gut health in laying hens. Animals 9(12):1110. https://doi.org/10.3390/ani9121110

Yan L, Kim IH (2013) Effect of probiotics supplementation in diets with different nutrient densities on growth performance, nutrient digestibility, blood characteristics, faecal microbial population and faecal noxious gas content in growing pigs. J Appl Anim Res 41:23–28

Yeo J, Kim KI (1997) Effect of feeding diets containing an antibiotic, a probiotic, or yucca extract on growth and intestinal urease activity in broiler chicks. Poult Sci 76:381–385

Yirga H (2015) The use of probiotics in animal nutrition. J Prob Health 3:1–10. https://doi.org/10.4172/2329-8901.1000132

Zaghari M, Sarani P, Hajati H (2020) Comparison of two probiotic preparations on growth performance, intestinal microbiota, nutrient digestibility and cytokine gene expression in broiler chickens. J Appl Anim Res 48(1):166–175. https://doi.org/10.1080/09712119.2020.1754218

Zhang DY, Ji HF, Wang SX, Liu H, Wang J, Wang YM (2019) *In vitro* characterisation of two *Lactobacillus* strains and evaluation of their suitability as probiotics for growing-finishing pigs. Anim Prod Sci 59:1537–1545

Zhang L et al (2021) Supplementation of probiotics in water beneficial growth performance, carcass traits, immune function, and antioxidant capacity in broiler chickens. Open Life Sci 16(1): 311–322. https://doi.org/10.1515/biol-2021-0031

Zhang ZF, Kim IH (2014) Effects of multistrain probiotics on growth performance, apparent ileal nutrient digestibility, blood characteristics, cecal microbial shedding, and excreta odor contents in broilers. Poult Sci 93(2):364–370. https://doi.org/10.3382/ps.2013-03314

Zhu C et al (2020) Effect of heat-inactivated compound probiotics on growth performance, plasma biochemical indices, and cecal microbiome in yellow-feathered broilers. Front Microbiol 11. Available at: https://www.frontiersin.org/articles/10.3389/fmicb.2020.585623. Accessed 21 Nov 2022

Zimmermann JA, Fusari ML, Rossler E, Blajman JE, Romero-Scharpen A, Astesana DM, Olivero CR, Berisvil AP, Signorini ML, Zbruna MV et al (2016) Effects of probiotics in swines growth performance: a meta-analysis of randomised controlled trials. Anim Feed Sci Technol 219:280–293

Feed Additives for Insect Production

Christos I. Rumbos and Christos G. Athanassiou

Abstract

Insect production has attracted a lot of scientific and business interest during the last decade. Insects are a well-promising nutrient source for food and feed, whereas they have a great exploitation potential in industrial and technological applications, e.g., cosmetics, pharmaceuticals, or bioplastics. As in most traditional livestock production systems, the optimization of insect production is calling for diets of high nutritional value to generate large quantities of nutritious insect biomass. Therefore, several studies have focused on the supplementation of insect diets with feed additives in order to increase larval growth and performance. In this context, several mineral elements, amino acids, vitamins, or probiotics have been evaluated as feeding substrate supplements to improve the diet properties and correct nutrient deficiencies and subsequently enhance insect development, in terms of total insect biomass production, growth rate, feed conversion efficiency, or even insect health. Apart from affecting the growth and development, insect diet greatly impacts also the nutritional composition of the insects produced, thus a considerable amount of research has been directed to the evaluation of the diet effect on insect composition. Several recent studies have focused on how the dietary enrichment with specific feed additives can alter the composition of the insects, conferring specific nutritional traits to the insects produced. For instance, the enrichment of insect feeding substrate with small amounts of specific lipids, carbohydrates, or other micronutrients has resulted in the production of insects with functional properties and desired characteristics, e.g., improved amino acid and fatty acid profile or increased content of specific nutrients. Based on the above, the aim of this chapter is to review the use of feed

C. I. Rumbos · C. G. Athanassiou (✉)
Laboratory of Entomology and Agricultural Zoology, Department of Agriculture, Crop Production and Rural Environment, University of Thessaly, Volos, Greece
e-mail: athanassiou@uth.gr

G. Arsenos, I. Giannenas (eds.), *Sustainable Use of Feed Additives in Livestock*,
https://doi.org/10.1007/978-3-031-42855-5_27

additives in insect production in order to promote insect growth and attribute specific nutritional properties to the produced insects.

Keywords

Insect composition · Feed quality · Probiotics · Metals · Growth enhancers

1 Introduction

The human population expands and is expected to reach approximately 10 billion by 2050 (UN 2022). Along with the global population increase, the demand for animal protein is also expected to rise as the annual meat consumption will further increase to reach 52 kg per capita by 2050 (FAO 2009). Driven by the population growth and the increased demand for livestock products, the animal production is expected to further grow to meet the augmented protein needs (FAO 2018; OECD/FAO 2022). Concerns have been raised though that the production of traditional feedstocks may not suffice to cover the nutritional needs of the farmed animals in the near future. Moreover, the production of several feedstocks, such as the soybean meal, is associated with serious environmental and social impacts, e.g., deforestation, biodiversity loss, farmers displacement, etc. (Herrero et al. 2015; Song et al. 2021). In this context, the livestock sector is in search of alternative, more sustainable nutrient sources as feedstocks for the produced animals.

Among the alternatives proposed, insects have several advantages which render their use as animal feed ingredients truly promising. Insects are highly nutritious (Van Huis et al. 2021) and very efficient feed converters (Van Huis 2013), they can be fed on low economic value side-streams and by-products being aligned with circular economy strategies (Varelas 2019; Van Peer et al. 2021), whereas their production has low environmental footprint, i.e., low GHG emissions, reduced water and land needs, etc. (Van Huis and Oonincx 2017). Due to these advantages, there are currently several applications for insects as animal feed (Hawkey et al. 2021). For instance, in July 2017 EU authorized the use of insects in aquafeeds (Commission Regulation 2017/893; EU 2017), whereas more recently, in September 2021, the use of insects as ingredients of poultry and swine feeds was also approved (Commission Regulation 2021/1372; EU 2021a).

2 Insect Feedstocks

A broad spectrum of organic substrates can be utilized as feedstocks for insect production, depending on the legislation in the different regions, the availability and seasonality of the substrates, and of course their market price (Pinotti and Ottoboni 2022). For instance, in the European Union, due to the feed ban regulation that prohibits the use of animal proteins in farmed animal feeds, insects can be fed, with few exceptions, mainly with substrates of vegetal origin, whereas materials containing meat and fish, e.g., former foodstuffs, kitchen and catering waste, are banned (EU 2001, 2011). In contrast, in African countries with less strict legislative framework, a wider range of organic wastes of different origin, such as municipal biowastes, etc., can be used for insect production (Münke-Svendsen et al. 2017; Vernooij et al. 2019). The insect feedstocks have been recently extensively reviewed by other researchers (Van Peer et al. 2021; Pinotti and Ottoboni 2022), therefore, they will not be the focus of this chapter. In the present chapter we will emphasize on the materials that have been tested and used as insect feed additives, in order to improve the quality of the feed and subsequently to enhance the animals' performance and health.

3 Feed Additives

Feed additives are substances, microorganisms, enzymes, or other preparations that are in purpose supplemented to the bulk feed material to improve the characteristics of the feed itself (e.g., flavor, smell, etc.) and enhance its nutritional quality in order to meet the certain nutritional requirements of the farmed animals (McCartney 2018). In this way, feed additives aim to favorably affect animal production, performance or welfare, e.g., improving the digestibility of the feed. Moreover, feed additives may be used to enhance the quality of food products that derive from the farmed animals or even alleviate the environmental consequences of animal production (McCartney 2018). In EU, according to the Feed Additives Regulation (EC Regulation 1831/2003), feed additives have to be authorized prior to being put on the market following a scientific evaluation carried out by EFSA (EU 2003). To the best of the authors' knowledge, no feed additives have been officially registered for use in insect production. However, there are several research works reporting the ability of specific substances and materials to act as feed additives, i.e., to promote insect growth and performance and improve insect composition. Therefore, in the present chapter we will summarize recent published data on the use of specific compounds or materials as feed additives in insect feeds.

3.1 Feed Additives to Manipulate Insect Composition

It has been extensively shown that the diet can affect insect composition (Oonincx and Finke 2021). Several studies have investigated the possibility to alter the composition and improve the nutritional quality of insects by the addition of functional additives to the insect diets. For instance, Oonincx et al. (2020) fed larvae of the lesser mealworm, *Alphitobius diaperinus* (Panzer) (Coleoptera: Tenebrionidae), and the black soldier fly, *Hermetia illucens* (L.) (Diptera: Stratiomyidae), as well as nymphs of the house cricket, *Acheta domesticus* (L.) (Orthoptera: Gryllidae), with diets enriched with flaxseed oil, which is known to be rich in omega-3 fatty acids, at rates between 1 and 4%. Based on their results, the enrichment of the diet with flaxseed oil as feed additive increased the omega-3 polyunsaturated fatty acids (PUFA) concentrations particularly that of alpha-linolenic acid and decreased the omega-6/omega-3 ratios to levels close to the ones considered optimal for human health. A similar effect was shown for larvae of the yellow mealworm, *Tenebrio molitor* L. (Coleoptera: Tenebrionidae), grown on four basal diets supplemented with 10% linseed flour (Francardi et al. 2017). The addition of linseed resulted in a higher PUFA content for all enriched diets. Particularly, in these diets there was a 20- to 30-fold increase of the α-linolenic acid content compared to the basal diet used and a subsequent decrease of the omega-6/omega-3 ratio. Along the same lines, when a standard diet consisting of wheat bran and soybean meal was supplemented with 7.5% of sunflower, linseed, or fish oil, the omega-3 level of the PUFAs of *T. molitor* larvae, mainly the alpha-linolenic acid and the eicosapentaenoic acid, was positively correlated with the respective level in the feeding substrates (Rossi et al. 2022). However, in the same study, the addition of the edible oils resulted in all cases in slower larval development compared to the control, indicating that the dietary influence on the fatty acid profile of *T. molitor* larvae may also affect larval performance and should be always taken into consideration in mass-rearing scenarios. The ability to change the fatty acid composition profiles of edible insects by manipulating the diet has also been shown by other studies. For instance, Lawal et al. (2021) supplemented a basal diet with various levels of seed meals (flax seed, chia seed, hemp seed, and rapeseed) and observed higher levels of PUFA and monounsaturated fatty acids (MUFA) for *T. molitor* and *H. illucens* larvae fed the seed meal supplemented diets compared to the control. The authors suggested that the concentration of beneficial fatty acids, particularly omega-3 PUFA, and subsequently the omega-6 to omega-3 ratio can be improved by feeding.

Feed additives are also used for the fortification of edible insects with specific elements. In this context, Keil et al. (2020) spiked wheat bran with $ZnSO_4$ at different concentrations (84–3577.5 mg/kg bran) in order to fortify the larvae of *T. molitor* with Zn, which is an essential trace element for human and animal nutrition. Based on their results, Zn-enrichment of the *T. molitor* diet resulted in higher Zn content of the larvae and lower Cd content, and proposed the use of Zn-enriched feeding substrates as a means to decrease the bioaccumulation of the toxic heavy metal Cd. Similarly, Dong et al. (2021) fed *T. molitor* larvae with sodium selenite and reported a significant increase of the total selenium content of larvae, the organic form of which accounting for over 97%, and suggested that the selenium-enriched larvae could be used as a dietary selenium supplementation source. In this framework, insects enriched with nutrients could be used for specific food and feed applications for the targeted delivery of nutrients to humans, as well as farmed animals.

Natural feed additives have also been tested with success for insect rearing. Brai et al. (2022) added distillery by-products, e.g., fresh grape pomace obtained after wine production, exhausted grape marcs, grape skin pulp, grape seeds and winery waste sludge, as feed supplements (10%) to a standard diet fed to *T. molitor* larvae. All feed supplements were found to have high content of antioxidant compounds, in particular phenols, flavonols, flavonoids, and condensed tannins, and subsequently the phenol content, antioxidant activity, and radical scavenging activity of larvae fed the diets supplemented with the distillery by-products was in most cases increased compared to larvae fed the standard diet. Apart from the increase of the antioxidant activity, in the case of the winery waste sludge, larval growth was also significantly enhanced compared to control. Essential oils of aromatic plants are also considered to be a natural source of feed additives that are widely used due to their known antimicrobial activity (Stevanović et al. 2018). The post-distillation residues of aromatic plants though, which are produced at large volumes during essential oil production, are also considered as a source of valuable compounds, e.g., phenols (De Elguea-Culebras et al. 2022), and could be potentially used in insect production. Recently, Andreadis et al. (2022) investigated the inclusion of essential oils distillation residues of medicinal and aromatic plants, i.e., lavender (*Lavandula angustifolia* Mill.), rosemary (*Salvia rosmarinus* Spenn.), Greek oregano (*Origanum vulgare* L. subsp. *hirtum*), and olive-cake (*Olea europaea* L. subsp. *europaea*), as feed additives at 10 and 20% ratio in agri-by-products-based diets. In that study, the high total phenolic content and the increased antioxidant activity of the tested substrates had a respective impact on the *T. molitor* larvae composition, namely larvae fed the diets with the essential oils post-distillation residues had in most cases a higher total flavonoid content.

The inclusion of the brown algae *Ascophyllum nodosum* (L.) Le Jolis in the substrate of *H. illucens* larvae up to a level of 50% resulted in larvae with high eicosapentaenoic acid (EPA), iodine, and vitamin E content, without negatively affecting larval growth and survival (Liland et al. 2017). However, this is not always the case. The addition of baker's yeast, *Saccharomyces cerevisiae* Meyen ex. E. C. Hansen, which is known to produce the unsaturated oleic acid, to the feeding substrate of *H. illucens* larvae, did not significantly affect their fatty acid composition (Opatovsky et al. 2021). Therefore, it becomes evident that the effect of the different feed additives on the growth and composition of edible insects is case- and species-specific and should be exhaustingly evaluated.

3.2 Feed Additives to Enhance Insect Growth

One of the major roles of feed additives is to serve as growth promoters. For example, Martynov and Brygadyrenko (2017) tested the effect of 18 food additives on the growth of *T. molitor* larvae and reported that three of them, namely sodium glutamate, sodium cyclamate, and sodium benzoate, significantly increased the larval weight. Similarly, supplementation of wheat bran with milled whole carob pods at 25% inclusion level enhanced the survival growth rate of *T. molitor* larvae resulting in shorter development times compared to the wheat bran that served as control (Antonopoulou et al. 2022). For the same species, enhanced larval growth was reported following the enrichment of *T. molitor* diet with winery and distillery residues (Brai et al. 2022). For a not so extensively studied species, the Mediterranean field cricket, *Gryllus bimaculatus* (L.) (Orthoptera: Gryllidae), the incorporation of Jew weeds, *Commelina sinensis* (L.) Kuntze, into its diet led to the increase of insect body weight, as well as to improved microbial quality (Ng'ang'a et al. 2020).

However, the addition of feed additives does not always lead to improved larval growth. The supplementation of coconut endosperm and soybean curd residue-based diets with a mixture of macro- and micro-minerals, vitamins, and amino acids did not have a positive effect on the growth performance and waste reduction efficiency of *H. illucens* larvae; however, it resulted into an almost three-fold increase of ash content of both produced larvae and prepupae (Pliantiangtam et al. 2021). Thus, it is suggested that an increase in the mineral content of the feeding substrate could be reflected to a respective rise of these minerals in *H. illucens* larvae and prepupae, as it has been shown previously for *A. domesticus* and *T. molitor* (Anderson 2000). Similarly, when crystalline lysine was added to different percentages (0.5–3.0%) to a basal diet mimicking household and restaurant food waste, larval growth, development time, and nutritional value of *H. illucens* larvae were not improved due to the lysine addition (Koethe et al. 2022). In contrast, when high amounts of the amino acid were provided to the larvae, larval size, survival and prepupation rates were reduced. In a similar study with *H. illucens* larvae, Cohn et al. (2022) supplemented a chicken feed-based diet with ten carbohydrates at a 20% inclusion level, larval survival, bioconversion, waste reduction, and conversion efficiency were in several cases reduced. The authors used these results to understand the preferences of *H. illucens* larvae in dietary carbohydrates, and subsequently to design suitable waste-based diets that will favor larval growth.

3.3 Probiotics as Insect Feed Additives

According to the most widely adopted definition, probiotics are non-pathogenic "live microorganisms which when administered in adequate amounts confer a health benefit on the host" (FAO/WHO 2001; Hill et al. 2014). These microorganisms are added in the animals' diet as supplements in order to improve the animal growth and performance and promote their health. Probiotics have already been extensively investigated as growth promoters for poultry (Abd El-Hack et al. 2020), aquaculture (Kuebutornye et al. 2019), ruminants (Nalla et al. 2022), and other livestock animals. The addition of probiotics in insect diets has also been proposed as a means to enhance insect performance. Using the red flour beetle, *Tribolium castaneum* (Herbst) (Coleoptera: Tenebrionidae), as a model species, Grau et al. (2017a) showed the protective effect of an *Enterococcus mundtii* Collins et al. isolate for *T. castaneum* larvae against *Bacillus thuringiensis* infection and suggested that this approach could also be adopted for edible insects, e.g., *T. molitor* (Grau et al. 2017b). However, current information about the use of probiotics as insect dietary supplement is still limited compared to other livestock animals. A recent review summarized all available data on the use of probiotics for insect mass-rearing (Savio et al. 2022). Although from another perspective, that of silk production, several researchers have studied the impact of various probiotic bacteria (Lactobacillaceae, Staphylococcaceae, etc.), yeasts (*Saccharomyces* spp.), and fungi (*Trichoderma* sp.)

on the performance of the silkworm, *Bombyx mori* (L.) (Lepidoptera: Bombycidae), as indicated by survival, larval growth, improved cocoon characteristics, etc. (Savio et al. 2022 and references therein). For instance, when *B. mori* larvae were supplemented with *Lactobacillus casei* (Orla-Jensen) Zheng et al., they could better tolerate *Nosema bombycis* Näegeli infection resulting in higher larval and cocoon weight and improved cocooning and pupation ratio compared to the control (Suraporn and Terenius 2021). Apart from its significance for silk production, *B. mori* has been recently authorized as ingredient of animal feed in the EU (EU Regulation 2021/1925; EU 2021b), therefore the already acquired knowledge on the use of probiotics as dietary supplement for *B. mori* reared for silk production could also be of significance for its mass-rearing as food and feed.

Regarding *T. molitor*, the individual inclusion of three probiotic bacteria, i.e., *Bacillus subtilis* (Ehrenberg) Kohn, *Bacillus toyonensis* Jimenez et al. and *Enterococcus faecalis* (Andrewes and Horder) Schleifer and Kilpper-Balz, into the wet feed had in most of the cases a positive result on larval growth, in terms of larval weight and length, as well as on development time and nutritional value (Rizou et al. 2022). Specifically, probiotic treatment with *E. faecalis* resulted in significantly higher larval weight and length compared to the control, whereas larvae provided with carrots inoculated with each one of the bacterial strains tested had significantly higher protein levels than the control. In a similar study with a mixture of probiotic bacteria, *T. molitor* larval growth and weight gain were also positively affected when larvae were provided with a mixed culture of *Bifidobacterium bifidum* (Tissier) Orla-Jensen, *Clostridium butyricum* Prazmowski, *Bacillus licheniformis* (Weigmann) Chester and *B. subtilis* on a solid-state medium (Zhong et al. 2017). Lecocq et al. (2022) isolated a strain of *Pediococcus pentosaceus* Mees from the gut of *T. molitor* larvae and showed that this strain was able to inhibit the growth of selected insect pathogens, e.g., *Bacillus*, *Serratia*, and *Pseudomonas* spp. in vitro. Moreover, when a powdered formulation of live or autoclaved lactic acid bacteria was provided to *T. molitor* larvae in *in vivo* bioassays, larval survival and growth rate were significantly enhanced with both dietary supplements compared to the control.

Several studies have also investigated the effect of probiotics inclusion in the diets of *H. illucens*. Kooienga et al. (2020) supplemented the diet of *H. illucens* larvae with either two oleaginous bacterial species, e.g. *Arthrobacter* AK19 and *Rhodococcus rhodochrous* 21,198, or *Bifidobacterium breve* Reuter, a well-studied human probiotic. The first hypothesis tested, i.e., that the dietary inclusion of the two high lipid bacteria can boost larval growth, was verified by their results. Surprisingly, the *B. breve* supplementation negatively affected larval growth and resulted in smaller and less active larvae, indicating that the probiotic effect is species-specific. In a similar study with *R. rhodochrous*, the supplementation of the probiotics resulted in higher growth rate, improved feed conversion ratio, and subsequently larger *H. illucens* larvae (Franks et al. 2021). When *H. illucens* larvae were fed with poultry manure inoculated individually with four *B. subtilis* strains isolated from *H. illucens* larvae or their feed, increased larval weight has been reported for the treated larvae compared to control (Yu et al. 2011). For another dipteran species, the housefly, *Musca domestica* L. (Diptera: Muscidae), which has been already authorized for feed use in the EU, the dietary inclusion of the intestinal symbiotic bacterium *Enterobacter hormaechei* O'Hara et al. increased larval and pupal weight, as well as the larval growth rate, and this effect was mainly attributed to changes in the housefly larvae intestinal flora (Zhang et al. 2021).

4 Conclusions

Edible insects have recently gained soaring interest as an alternative source of nutrients for the food and feed industry. However, as far as insect nutrition is concerned, there is still room for optimization in many aspects. The inclusion of feed additives, which is a common practice and well-studied and documented in conventional livestock farming, e.g., poultry and swine production, as well as aquaculture, is still a new field of investigation for the insect sector. In the present chapter, we aimed to summarize the results of relevant studies on the effect of dietary supplementation with functional ingredients on insect performance. Several materials, compounds, or substrates could serve as feed additives, in order to enhance insect growth and improve the nutritional quality of the insects; some of them have already been successfully tested as analyzed above. However, further research is warranted to design not only species-specific, but also application-specific feed additives that could promote insect growth and provide edible insects with desirable functional traits.

References

Abd El-Hack ME, El-Saadony MT, Shafi ME, Qattan SYA, Batiha GE, Khafaga AF, Abdel-Moneim A-ME, Alagawany M (2020) Probiotics in poultry feed: a comprehensive review. J Anim Physiol Anim Nutr 104:1835–1850. https://doi.org/10.1111/jpn.13454

Anderson SJ (2000) Increasing calcium levels in cultured insects. Zoo Biol 19:1–9

Andreadis SS, Panteli N, Mastoraki M, Rizou E, Stefanou V, Tzentilasvili S, Sarrou E, Chatzifotis S, Krigas N, Antonopoulou E (2022) Towards functional insect feeds: agri-food by-products enriched with post-distillation residues of medicinal aromatic plants in *Tenebrio molitor* (Coleoptera: Tenebrionidae) breeding. Antioxidants 11:68. https://doi.org/10.3390/antiox11010068

Antonopoulou E, Panteli N, Feidantsis K, Mastoraki M, Koutsogeorgiou EI, Grivaki E, Papagrigoriou T, Christias SP, Chatzifotis S, Lazari D, Andreadis SS, Krigas N (2022) Carob (*Ceratonia siliqua*) as functional feed is beneficial in yellow mealworm (*Tenebrio molitor*) rearing: evidence from growth, antioxidant status and cellular responses. Antioxidants 11:1840. https://doi.org/10.3390/antiox11091840

Brai A, Vagaggini C, Pasqualini C, Poggialini F, Tarchi F, Francardi V, Dreassi E (2022) Use of distillery by-products as *Tenebrio molitor* mealworm feed supplement. J Insects Food Feed, In press 9:611

Cohn Z, Latty T, Abbas A (2022) Understanding dietary carbohydrates in black soldier fly larvae treatment of organic waste in the circular economy. Waste Manag 137:9–19. https://doi.org/10.1016/j.wasman.2021.10.013

De Elguea-Culebras GO, Bravo EM, Sánchez-Vioque R (2022) Potential sources and methodologies for the recovery of phenolic compounds from distillation residues of

Mediterranean aromatic plants. An approach to the valuation of by-products of the essential oil market – a review. Ind Crop Prod 175:114261. https://doi.org/10.1016/j.indcrop.2021.114261
Dong Z, Lin Y, Wu H, Zhang M (2021) Selenium accumulation in protein fractions of *Tenebrio molitor* larvae and the antioxidant and immunoregulatory activity of protein hydrolysates. Food Chem 334:127475. https://doi.org/10.1016/j.foodchem.2020.127475
EU (2001) Regulation (EC) No 999/2001 of the European Parliament and of the Council of 22 May 2001 laying down rules for the prevention, control and eradication of certain transmissible spongiform encephalopathies. Off J Eur Union:L 147
EU (2003) Regulation (EC) No 1831/2003 of the European Parliament and of the Council of 22 September 2003 on additives for use in animal nutrition. Off J Eur Union:L 268/29
EU (2011) Regulation (EU) No. 142/2011 of 25 February 2011 implementing regulation (EC) no 1069/2009 of the European Parliament and of the Council laying down health rules as regards animal by-products and derived products not intended for human consumption and implementing Council Directive 97/78/EC as regards certain samples and items exempt from veterinary checks at the border under that Directive, EU 142/2011, Brussels, Belgium
EU (2017) Commission Regulation (EU) 2017/893 of 24 May 2017 amending Annexes I and IV to Regulation (EC) No 999/2001 of the European Parliament and of the Council and Annexes X, XIV and XV to Commission Regulation (EU) No 142/2011 as regards the provisions on processed animal protein. Available online: https://eur-lex.europa.eu/legal-content/EN/TXT/PDF/?uri=CELEX:32017R0893&from=EN. Accessed 11 Oct 2022
EU (2021a) Commission Regulation (EU) 2021/1372 of 17 August 2021 amending Annex IV to Regulation (EC) No 999/2001 of the European Parliament and of the Council as regards the prohibition to feed non-ruminant farmed animals, other than fur animals, with protein derived from animals. Off J Eur Union L 295:1–17. Available at: https://eur-lex.europa.eu/legal-content/EN/TXT/HTML/?uri=OJ:L:2021:295:FULL&from=EN. Accessed 11 Oct 2022
EU (2021b) Commission Regulation (EU) 2021/1925 of 5 November 2021 amending certain Annexes to Regulation (EU) No 142/2011 as regards the requirements for placing on the market of certain insect products and the adaptation of a containment method. Available at: https://eur-lex.europa.eu/legal-content/EN/TXT/PDF/?uri=CELEX:32021R1925&from=EN. Accessed 11 Oct 2022
FAO (2009) How to feed the world in 2050. Food and Agriculture Organization. Available online: https://www.fao.org/fileadmin/templates/wsfs/docs/expert_paper/How_to_Feed_the_World_in_2050.pdf
FAO, 2018. The future of food and agriculture – alternative pathways to 2050. Summary version. Rome. 60 pp. Licence: CC BY-NC-SA 3.0 IGO. Available online: https://www.fao.org/3/CA1553EN/ca1553en.pdf
FAO/WHO (2001) Food and Agricultural Organization of the United Nations and World Health Organization. Health and nutritional properties of probiotics in food including powder milk with live lactic acid bacteria. Report of a Joint FAO/WHO Expert Consultation, 1–4 October 2001, Córdoba, Argentina
Francardi V, Cito A, Fusi S, Botta M, Dreassi E (2017) Linseed to increase n-3 fatty acids in *Tenebrio molitor* (Coleoptera Tenebrionidae). Redia 100:73–76. https://doi.org/10.19263/REDIA-100.17.08
Franks K, Kooienga E, Sanders M, Pendarvis K, Yang F, Tomberlin JK, Jordan HR (2021) The effect of *Rhodococcus rhodochrous* supplementation on black soldier fly (Diptera: Stratiomyidae) development, nutrition, and waste conversion. J Insects Food Feed 7:397–408. https://doi.org/10.3920/JIFF2020.0033
Grau T, Vilcinskas A, Joop G (2017a) Probiotic *Enterococcus mundtii* isolate protects the model insect *Tribolium castaneum* against *Bacillus thuringiensis*. Front Microbiol 8:1261. https://doi.org/10.3389/fmicb.2017.01261
Grau T, Vilcinskas A, Joop G (2017b) Sustainable farming of the mealworm *Tenebrio molitor* for the production of food and feed. Z Naturforsch 72:337–349. https://doi.org/10.1515/znc-2017-0033

Hawkey KJ, Lopez-Viso C, Brameld JM, Parr T, Salter AM (2021) Insects: A potential source of protein and other nutrients for feed and food. Annu Rev Anim Biosci 9:8.1–8.22. https://doi.org/10.1146/annurev-animal-021419-083930

Herrero M, Wirsenius S, Henderson B, Rigolot C, Thornton P, Havlík P, De Boer I, Gerber PJ (2015) Livestock and the environment: what have we learned in the past decade? Annu Rev Environ Resour 40:177–202. https://doi.org/10.1146/annurev-environ-031113-093503

Hill C, Guarner F, Reid G, Gibson GR, Merenstein DJ, Pot B, Morelli L, Canani RB, Flint HJ, Salminen S et al (2014) The international scientific association for probiotics and prebiotics consensus statement on the scope and appropriate use of the term probiotic. Nat Rev Gastroenterol Hepatol 11:506–514. https://doi.org/10.1038/nrgastro.2014.66

Keil C, Maares M, Kröncke N, Benning R, Haase H (2020) Dietary zinc enrichment reduces the cadmium burden of mealworm beetle (*Tenebrio molitor*) larvae. Sci Rep 10:20033. https://doi.org/10.1038/s41598-020-77079-x

Koethe M, Taubert J, Vervuert I (2022) Impact of lysine supplementation on growth and development of *Hermetia illucens* larvae. J Insects Food Feed 8:35–44. https://doi.org/10.3920/JIFF2020.0059

Kooienga EM, Baugher C, Currin M, Tomberlin JK, Jordan HR (2020) Effects of bacterial supplementation on black soldier fly growth and development at benchtop and industrial scale. Front Microbiol 11:587979. https://doi.org/10.3389/fmicb.2020.587979

Kuebutornye FKA, Abarike ED, Lu Y (2019) A review on the application of *Bacillus* as probiotics in aquaculture. Fish Shellfish Immunol 87:820–828. https://doi.org/10.1016/j.fsi.2019.02.010

Lawal KG, Kavle RR, Akanbi TO, Mirosa M, Agyei D (2021) Enrichment in specific fatty acids profile of *Tenebrio molitor* and *Hermetia illucens* larvae through feeding. Future Foods 3: 100016. https://doi.org/10.1016/j.fufo.2021.100016

Lecocq A, Natsopolou ME, Berggreen IE, Eilenberg J, Heckmann LHL, Nielsen HV, Stensvold CR, Jensen AB (2022) Probiotic properties of an indigenous *Pediococcus pentosaceus* strain on *Tenebrio molitor* larval growth and survival. J Insects Food Feed 7:975–986. https://doi.org/10.3920/JIFF2020.0156

Liland NS, Biancarosa I, Araujo P, Biemans D, Bruckner CG, Waagbø R, Torstensen BE, Lock E-J (2017) Modulation of nutrient composition of black soldier fly (*Hermetia illucens*) larvae by feeding seaweed-enriched media. PLoS One 12:e0183188. https://doi.org/10.1371/journal.pone.0183188

Martynov VO, Brygadyrenko VV (2017) The influence of synthetic food additives and surfactants on the body weight of larvae of *Tenebrio molitor* (Coleoptera, Tenebrionidae). Biosyst Divers 25:236–242. https://doi.org/10.15421/011736

McCartney E (2018) Focus on European feed additives: complexity, change and challenges ahead. Regul Rapp 15:16–19

Münke-Svendsen C, Halloran A, Oloo J, Magara H, Nyakeri E, Roos N (2017) Technical brief #2: insect production systems for food and feed in Kenya. GREEiNSECT. https://doi.org/10.13140/RG.2.2.24053.99040

Nalla K, Manda NK, Dhillon HS, Kanade SR, Rokana N, Hess M, Puniya AK (2022) Impact of probiotics on dairy production efficiency. Front Microbiol 13:805963. https://doi.org/10.3389/fmicb.2022.805963

Ng'ang'a J, Imathiu S, Fombong F, Borremans A, Van Campenhout L, Broeck JV, Kinyuru J (2020) Can farm weeds improve the growth and microbiological quality of crickets (*Gryllus Bimaculatus*)? J Insects Food Feed 6:199–209. https://doi.org/10.3920/JIFF2019.0051

OECD/FAO (2022) OECD-FAO Agricultural Outlook 2022-2031. OECD Publishing, Paris. https://doi.org/10.1787/f1b0b29c-en

Oonincx DGAB, Finke MD (2021) Nutritional value of insects and ways to manipulate their composition. J Insects Food Feed 7:639–659. https://doi.org/10.3920/JIFF2020.0050

Oonincx DGAB, Laurent S, Veenenbos ME, van Loon JJA (2020) Dietary enrichment of edible insects with omega 3 fatty acids. Insect Science 27:500–509. https://doi.org/10.1111/1744-7917.12669

Opatovsky I, Vitenberg T, Jonas-Levi A, Gutman R (2021) Does consumption of baker's yeast (*Saccharomyces cerevisiae*) by black soldier fly (Diptera: Stratiomyidae) larvae affect their fatty acid composition? J Insect Sci 21:1–5. https://doi.org/10.1093/jisesa/ieab031

Pinotti L, Ottoboni M (2022) Substrate as insect feed for bio-mass production. J Insects Food Feed 7(5):585–596. In press. https://doi.org/10.3920/JIFF2020.01101

Pliantiangtam N, Chundang P, Kovitvadhi A (2021) Growth performance, waste reduction efficiency and nutritional composition of black soldier fly (*Hermetia illucens*) larvae and prepupae reared on coconut endosperm and soybean curd residue with or without supplementation. Insects 12:682. https://doi.org/10.3390/insects12080682

Rizou E, Kalogiouri N, Bisba M, Papadimitriou A, Kyrila G, Lazou A, Andreadis S, Hatzikamari M, Mourtzinos I, Touraki M (2022) Amelioration of growth, nutritional value, and microbial load of *Tenebrio molitor* (Coleoptera: Tenebrionidae) through probiotic supplemented feed. Eur Food Res Technol 248:727–739. https://doi.org/10.1007/s00217-021-03925-5

Rossi G, Mattioli S, Rondoni G, Dal Bosco A, Servili M, Castellini C, Conti E (2022) Characterisation of fatty acid profiles of *Tenebrio molitor* larvae reared on diets enriched with edible oils. J Insects Food Feed 8:901–912. https://doi.org/10.3920/JIFF2021.0164

Savio C, Mugo-Kamiri L, Upfold JK (2022) Bugs in bugs: the role of probiotics and prebiotics in maintenance of health in mass-reared insects. Insects 13:376. https://doi.org/10.3390/insects13040376

Song XP, Hansen MC, Potapov P, Adusei B, Pickering J, Adami M, Lima A, Zalles V, Stehman SV, Di Bella CM, Conde MC, Copati EJ, Fernandes LB, Hernandez-Serna A, Jantz SM, Pickens AH, Turubanova S, Tyukavina A (2021) Massive soybean expansion in South America since 2000 and implications for conservation. Nat Sustain 4:784–792. https://doi.org/10.1038/s41893-021-00729-z

Stevanović ZD, Bošnjak-Neumüller J, Pajić-Lijaković I, Raj J, Vasiljević M (2018) Essential oils as feed additives—future perspectives. Molecules 23:1717. https://doi.org/10.3390/molecules23071717

Suraporn S, Terenius O (2021) Supplementation of *Lactobacillus casei* reduces the mortality of *Bombyx mori* larvae challenged by *Nosema bombycis*. BMC Res Notes 14:398. https://doi.org/10.1186/s13104-021-05807-1

United Nations (UN) Department of Economic and Social Affairs, Population Division 2022. World population prospects 2022: summary of results. UN DESA/POP/2022/TR/NO. 3

Van Huis A (2013) Potential of insects as food and feed in assuring food security. Annu Rev Entomol 58:563–583. https://doi.org/10.1146/annurev-ento-120811-153704

Van Huis A, Oonincx DGAB (2017) The environmental sustainability of insects as food and feed: a review. Agron Sustain Dev 37:43. https://doi.org/10.1007/s13593-017-0452-8

Van Huis A, Rumpold B, Maya C, Roos N (2021) Nutritional qualities and enhancement of edible insects. Annu Rev Nutr 41:551–576. https://doi.org/10.1146/annurev-nutr-041520-010856

Van Peer M, Frooninckx L, Coudron C, Berrens S, Álvarez C, Deruytter D, Verheyen G, Van Miert S (2021) Valorisation potential of using organic side streams as feed for *Tenebrio molitor*, *Acheta domesticus* and *Locusta migratoria*. Insects 12:796. https://doi.org/10.3390/insects12090796

Varelas V (2019) Food wastes as a potential new source for edible insect mass production for food and feed: a review. Fermentation 5:81. https://doi.org/10.3390/fermentation5030081

Vernooij A, Veldkamp T, Ndambi A (2019) Insects for Africa: developing business opportunities for insects in animal feed in Eastern Africa. Wageningen Livestock Research, Wageningen

Yu G, Cheng P, Chen Y, Li Y, Yang Z, Chen Y, Tomberlin JK (2011) Inoculating poultry manure with companion bacteria influences growth and development of black soldier fly (*Diptera*: *Stratiomyidae*) larvae. Environ Entomol 40:30–35. https://doi.org/10.1603/en10126

Zhang Q, Wang S, Zhang X, Zhang K, Liu W, Zhang R, Zhang Z (2021) *Enterobacter hormaechei* in the intestines of housefly larvae promotes host growth by inhibiting harmful intestinal bacteria. Parasit Vectors 14:598. https://doi.org/10.1186/s13071-021-05053-1

Zhong J, Zhang F, Peng Y, Ji Z, Li H, Li S, Zhang X, Shi Q, Zhang J (2017) Mixed culture of probiotics on a solid-state medium: an efficient method to produce an affordable probiotic feed additive. Biotechnol Bioprocess Eng 22:758–766. https://doi.org/10.1007/s12257-017-0038-y

Feed Additives in Aquaculture

Ángel Hernández-Contreras, Andressa Teles,
Joan Sebastián Salas-Leiva, Elena Chaves-Pozo,
and Dariel Tovar-Ramírez

Abstract

Aquaculture is an important sector within animal production worldwide. Despite its rapid growth in recent decades, it has encountered certain challenges in the availability of raw materials for feed production, the appearance of epidemic outbreaks, and the overall environmental sustainability of the sector. One emerging strategy for addressing these challenges is the development and use of feed additives. These additives are products, not raw materials or premixes, intentionally added to feed to achieve an effect on the feed itself, on animals, on food products obtained from animals, or on the environment.

The research carried out on this subject in recent decades has allowed the development of a large number of additives that can currently be used in aquaculture, although their use and regulation are more advanced in terrestrial farmed species. Among the most used additives, there are pro- and prebiotics, enzymes, vitamins, amino acids, fatty acids, minerals, and other functional compounds obtained from plants, bacteria, micro and macroalgae, fungus, or other animals (i.e., insects or crustaceans).

Á. Hernández-Contreras (✉) · D. Tovar-Ramírez
Laboratory of Comparative Physiology and Functional Genomics, Northwest Biological Research Center (CIBNOR), La Paz, BCS, Mexico

A. Teles
Department of Fisheries Engineering, Autonomous University of Baja California Sur (UABCS), La Paz, BCS, Mexico

J. S. Salas-Leiva
Department of Environment and Energy, CONACYT-Center for Research in Advanced Materials, Chihuahua, Chih, Mexico

E. Chaves-Pozo
Oceanographic Center of Murcia, Spanish Institute of Oceanography, Spanish National Research Council (IEO-CSIC), Puerto de Mazarrón, Murcia, Spain

G. Arsenos, I. Giannenas (eds.), *Sustainable Use of Feed Additives in Livestock*,
https://doi.org/10.1007/978-3-031-42855-5_28

In accordance with the safety demands of consumers and with the strategies developed worldwide to reduce the climate and environmental footprint of food production, the current trend is toward the development of increasingly sustainable feed additives, of natural origin and not toxic. Its effects on aquaculture species are being studied through physiology, immunology, and different –omics using cutting-edge techniques. In order to analyze all these advances in feed additives, this chapter gathers the main research in this regard, emphasizing natural additives or those that favor sustainability. In addition, recent changes in legislation concerning development and use of feed additives are analyzed, as well as future perspectives on their use as key tools to increase the sustainability of the food sector.

Keywords

Aquaculture · Functional additives · Feed additives · Sustainability · Growth promotion · Health · Immunostimulation

1 Introduction

One of the main FAO messages is that aquaculture has great potential to feed and nourish the world's growing population, but growth must be sustainable (*The State of World Fisheries and Aquaculture 2022* 2022). Global aquaculture reached the record of 122.6 million tons (Mt) in 2020 and continues growing (another 14% has been estimated by 2030) while fisheries production has remained stable since the 90s. This growth has often occurred at the expense of the environment, but current European Union strategies as "Blue Transformation" and "Farm-to-Fork" can turn the tide through environmental-friendly policies in aquaculture and technological innovations.

Despite the rapid growth, and partly because of it, aquaculture faces new challenges that threaten environmental-friendly and sustainable growth, as the availability of raw materials, biosecurity and disease control, food quality and security, and environmental sustainability. These challenges are still being addressed from food science, biology, genetics and agricultural engineering with green and sustainable solutions gaining more and more prominence. Feed additives are among these types of solutions, especially those of natural origin and from a sustainable source. Feed additives have been defined by the European Food Safety Authority (EFSA) as "products used in animal nutrition to achieve an effect on the feed itself, the animals, food products obtained from the animals consuming the feed additive or the environment" (https://www.efsa.europa.eu/en/topics/topic/feed-additives). Plenty of examples are available of how feed additives can address the aquaculture challenges previously mentioned and how successful they have been. Innovative feed additives to improve digestibility and nutrient bioavailability or probiotics to increase animal resistance to pathogen challenges are examples exposed within the priority areas for the blue transformation of Asian aquaculture (*The State of World Fisheries and Aquaculture 2022* 2022).

One of the main challenges facing aquaculture is the sustainability of aquafeeds, because it depends on raw materials, such as fish oil and fish meal, ingredients that are increasingly scarce and call into question the environmental sustainability of the aquafeed sector. Currently, the percentage of fish meal in aquaculture feed has been greatly reduced (Ytrestøyl et al. 2015), and commercial diets free of it already exist for the main cultured species. However, its substitution with plant protein-rich ingredients has negative effects as reduced ingestion, low nutrient and energy digestibility, and a poorer physical and organoleptic quality of fillets, which, on the other hand, can be counteracted by feed additives (Zarantoniello et al. 2022). In addition, the lack of certain nutrients found in fish meal or fish oil can be supplied with the addition of extracts from micro and macroalgae (Bélanger et al. 2021). Other challenges of crucial importance in an intensified industry as finfish aquaculture are animal well-being, health, and disease control that are perhaps the aspects that have been more widely addressed in the design and evaluation of feed additives where an abundant scientific bibliography in this regard exists. These additives are often used as immunostimulants to improve the general health status of aquatic animals that prevent or hold back infectious outbreaks. Moreover, some of these substances have been found to possess anti-stress or anti-choleretic properties, for example, improving well-being of animals. All these properties—along with the current trend toward natural additives—can increase the consumer confidence of farmed fish (Dawood et al. 2018). Finally, interest is growing in additives that are capable of reducing the production of organic waste that can harm the environment. In aquaculture, this aspect is usually approached from two perspectives: the control of waste production from unconsumed or chewed feed (technological additives) and the control of the waste production by the animal at a physiological level (Alcaraz et al. 2021; Ballester-Moltó et al. 2016; Bureau and Hua 2010). All the variety of additives with different functions related to aquaculture production sustainability, jointly with the current trend toward the use of exclusively natural additives, may be the key to developing a safety, sustainable, and environmentally friendly aquaculture whose growth does not imply a deterioration of ecosystems or human health.

The European legislation Regulation (EC) No 1831/2003 classifies feed additives into the following categories: Technological, Sensorial, Nutritional, Zootechnical, Coccidiostat, and histomonostat additives. Most of them are described in this chapter but from a different point of view: their sustainable sources and contribution to aquaculture sustainability, paying special attention to those that represent original and innovative solutions to the challenges facing aquaculture today.

2 Feed Additives in Aquaculture

2.1 New Approaches in Probiotic Bacteria

Bacteria are present in culture systems and can be beneficial or malignant (pathogenic) to cultured organisms. In the first case, they are bacteria that have been studied and used as probiotics in aquaculture, given that they can promote the intestinal microbial balance, inhibit the growth of pathogens, and improve the immune response and digestion through enzymatic activities of the host, among others (Saravanan et al. 2021). Several reports of favorable effects exist on growth, immune response, and resistance to diseases of various species of fish and crustaceans, among others because of the effect of probiotic bacteria (i.e., *Bacillus* spp., *Lactobacillus* sp., *Pediococcus*, *Enterococcus*, *Lactococcus*, *Aeromonas*, *Alteromonas*, *Arthrobacter*, *Bifidobacterium*, *Brochothrix*, *Clostridium*, *Kocuria*, *Microbacterium*, *Micrococcus*, *Paenibacillus*, *Phaeobacter*, *Pseudoalteromonas*, *Pseudomonas*, Rhodococcus, Rhodopseudomonas, Rhodosporidium, Roseobacter, Shewanella and Vibrio) (Ringø et al. 2022; Saravanan et al. 2021; Van Doan et al. 2021; Yilmaz et al. 2022). The study of probiotic bacteria in aquaculture began more than 30 years ago with the use of Toyocerin (Bacillus toyoi) (Kozasa 1986) and has a very broad state of development in knowledge. However, most studies have focused on the use of "pure" bacterial monocultures (one probiotic bacterial population) and little is known about the use of bacterial consortia (two or more probiotic bacterial populations) (Puvanasundram et al. 2021).

Probiotic bacterial consortia (PBC) can be a natural or artificial association of two or more strains (multi-strain) or microbial populations of different species (multi-species), acting together as a community in a complex system where all of them benefit from the activities of others (Festa et al. 2017). Although recently terms that can be adjusted to such a definition have been used, such as multi-strain probiotics (MSP) (Melo-Bolívar et al. 2021; Puvanasundram et al. 2021) or a combination of probiotic strains (Barman et al. 2017), PCBs are not simple mixtures of probiotic bacteria, since knowledge of the ecological relationships/social interactions between strains (commensalism, amensalism, neutralism, cooperation, competition, and predation) in PBC is ideal for achieving the desired purpose(s). (Kong et al. 2018; Pascual et al. 2018). In a current literature review, we observed that the most tested strains for probiotic consortia in shrimp and farmed fish are the genera Bacillus, Lactobacillus, and Lactococcus (Table 1).

Probiotic bacterial consortia and a single-strain probiotics (SPB) have the ability to modulate the gut microbiota through (i) colonization of the intestinal mucosa; (ii) competitive adherence and exclusion; and (iii) production of beneficial substances and signals to the host, leading to modulation and regulation of allergic responses by the immune system, reduction of virulence through manipulation of quorum sensing, among other mechanisms (El-Saadony et al. 2021; Kwoji et al. 2021; McFarland 2021; Ningtyas et al. 2020; Vargas-Albores et al. 2021). However, PBC have been reported to be more effective and broad spectrum than an SPB (Chapman et al. 2011; Ningtyas et al. 2020; Wang et al. 2019), although it remains

Table 1 Effects of diets supplemented with probiotic mixtures on shrimp and fish performance

Probiotic mixtures	Aquaculture species (host)	Dosage	Duration	Synergy	Major findings	Reference
Lactobacillus plantarum SGLAB01 + *Lactococcus lactis* SGLAB02	*Litopenaeus vannamei*	2–4×10^8 CFU/g diet	16 d	Yes	Microbial activity against *Vibrio parahaemolyticus*, Modulation of proPO system, hemolymph PO activity, Increase in the relative mRNA expression of LvproPO1 and LvproPO2 and resistance to *V. parahaemolyticus* infection	Chomwong et al. (2018)
Lb pentosus BD6 + Lac. fermentum LW2 + *Bacillus subtilis* E20 + *Saccharomyces cerevisiae* P13	*L. vannamei*	10^7 -10^9 CFU/kg diet	56	Yes	Diets with BD6 and E20 significantly improved shrimp growth and health status. The probiotic mixture had higher survival after injection with the *Vibrio alginolyticus* and an increased PO activity, respiratory burst, and lysozyme activity of hemocytes	Wang et al. (2019)
B. subtilis L10 and G1	*L. vannamei*	10^5 y 10^8 CFU/g diet	56	NA	At the highest concentration, an increase of SGR, digestive enzyme activity, survival %. Probiotic mix up-regulated proPO and eroxinectin (PE) and other genes-related immune response	Zokaeifar et al. (2012)
B. licheniformis LS-1 + *B. flexus* LD-1	*L. vannamei*	LS-1+LD-1 (1.0×10^9 CFU/g diet LS-1 + 1.0×10^9 CFU/g diet LD-1)	21	No	Mix probiotics improved survival rate (S%) after *V. harveyi* challenge	Cai et al. (2019)

(continued)

Table 1 (continued)

Probiotic mixtures	Aquaculture species (host)	Dosage	Duration	Synergy	Major findings	Reference
Bacillus sp. JL47 + Bacillus sp. JL1	*Penaeus monodon*	4 g wet bacterial weight kg feed	30	No	No improvement was observed in the parameters evaluated with respect to the control (AW, length, S%)	Laranja et al. (2014)
B. subtilis + *B. clausii* + *B. cereus* + *B. megaterium* + *B. pumilus* + *B. polymyxa*	*L. vannamei*	ND	120	NA	Multi-strain Bacillus species improved growth (weight and length) and S% of shrimp compared to control	Rajasulochana and Gummadi (2022)
B. subtilis + *B. licheniformis*	*Oncorhynchus mykiss*	$2x10^9$ CFU/kg diet	56	No	Improved AWG and specific growth rate SGR of fed fish, and increased SOD and lysozyme activities relative to the control diet. In the *Aeromonas salmonicida* challenge test, S% was higher than that of the control group.	Park et al. (2017)
B. amyloliquefaciens 54A + *B. pumilus* 47B	*Pangasianodon hypophthalmus*	$1x10^8$, $3x10^8$ and $5x10^8$ CFU/g diet	90	NA	At the highest concentration, an increase was observed in AWG, phagocytic activity, respiratory burst, lysozyme activity, and S %.	Truong Thy et al. (2017)

B. licheniformis + B subtilis	*Lates calcarifer*	1×10^3, 1×10^6 , and 1×10^9 CFU/g diet	56	NA	The best results were obtained with 1×10^6; at this concentration, the growth, protein level, digestive enzyme activity and hematological parameters improved with respect to the control (without probiotic).	Adorian et al. (2019)

CFU colony forming units, *NA* not available data, *ProPO* prophenoloxidase, *PO* phenoloxidase, *SGR* specific growth rate, *AWG* Absolute weight gain, *SOD* superoxide dismutase, *S%* cumulative survival rate controversial (Korada et al. 2018; McFarland 2021) mainly because in several studies the physiological effects of each PBC strain/species are not evaluated separately. Therefore, it is not possible to know if synergistic effects will take place (Table 1). In this sense, the synergy between the strains/species that constitute PBC results in the enhancement of probiotic mechanisms and additive metabolic effects (Kwoji et al. 2021). Some researchers recommend the use of PBC when they are polyculture-type farming systems since in this way they could cover a range of hosts (Ghosh et al. 2021). It is important to mention that in single-strain bacterial probiotics, as well as in PBC, the kind of strain(s), the concentration (i.e., colony forming units, CFU/ml), and handling (i.e., administration) are determining factors for success in aquaculture (El-Saadony et al. 2021).

2.2 Probiotic Yeasts

Although much less studied than other probiotics (Staniszewski and Kordowska-Wiater 2021), yeasts are gaining interest in aquaculture because of certain advantages they present, such as the production of a wide range of bioactive substances (glucans, glutathione, toxins, enzymes, phytase, nucleotides, and vitamins) or the possibility of being used for bioremediation of water quality (Ceseña et al. 2021). Yeasts are unicellular fungus with budding or fission reproduction widely distributed in several natural environments, such as soil, freshwater, and seawater, which are part of the normal microbiota of fish and other cultured organisms. Probiotic yeast has been identified and isolated from both wild and farmed individuals, and their role in health and nutrition has been addressed in the literature: the effects of yeast administration can be observed either alive, their cell components or after processing as a feed ingredient (Navarrete et al. 2014). Regarding advantages as probiotics, yeast can be a hundred times larger than bacterial cells, which may explain the fact that the introduction of a low yeast population (10^4 CFU/g) through feed can induce beneficial effects in the host (Navarrete et al. 2014). In addition, yeast releases very interesting bioactive compounds, such as polyamines that enhance cell proliferation and differentiation or extracellular enzymes that help in the digestive process (Ceseña et al. 2021).

Probably the most studied yeast as a probiotics in aquaculture is *Saccharomyces cerevisiae*, in part because of its high availability as a by-product of the brewing industry. However, other species, which are halotolerant and naturally present in marine organisms, are currently more studied because they can be administered alive and grow inside the intestine of the animal producing bioactive compounds. *Debaryomyces hansenii* in finfish aquaculture or several species from the genera *Debaryomyces, Candida, Phaffia, Rhodosporidium* and *Yarrowia* in shrimp have been tested for several years (Ceseña et al. 2021; Navarrete et al. 2014).

In fish, such as the olive flounder *Paralichthys olivaceus*, rainbow trout *Oncorhynchus mykiss*, grouper *Epinephelus coioides*, Nile tilapia *Oreochromis niloticus,* and channel catfish *Ictalurus punctatus*, diet supplementation with *S. cerevisiae* has shown to be an effective tool. Growth promotion and improved survival against challenges with pathogens or up-regulation of cellular and humoral immune responses are some of the main effects of this yeast with optimal doses ranging from 10^6 to 10^8 CFU/kg. (Abdel-Tawwab et al. 2008; Chiu et al. 2010; Harikrishnan et al. 2011; Lara-Flores et al. 2003; P. Li and Gatlin 2005; Tukmechi et al. 2011). In addition, derivatives of this yeast as mannan oligosaccharide (MOS) have also improved growth and the immune response in several species of farmed fish (Torrecillas et al. 2014), which is the suggested mode of action to reinforce the intestinal integrity and functionality, together with the stimulation of the innate immune system. Several studies have also demonstrated the effectiveness of live *S. cerevisiae* or its bioactive compounds on immunostimulant and growth promotion in shrimp and other crustaceans (Ceseña et al. 2021).

Relative to yeast species naturally present in marine organisms, *D. hansenii* has been one of the most studied. Its ability to colonize the intestine of rainbow trout and turbot was demonstrated (Andlid et al. 1995) some decades ago. After that, reduced spinal malformation and increased survival and gastrointestinal enzymatic activity were observed in European seabass *Dicentrarchus labrax* larvae fed with *D. hansenii* compared to the *S. cerevisiae* group (Tovar et al. 2002). Additionally, enhanced growth and oxidative status during larval stages was demonstrated in the same fish species (Tovar-Ramírez et al. 2010). Several studies regarding the probiotic effects of *D. hansenii* have been carried out in fish in previous years, revealing properties as growth promotion, humoral and cellular immune stimulation, and improved resistance to pathogens as *Amyloodinium ocellatum* and *Aeromonas hydrophila* (Angulo et al. 2020). In a recent study, a new approach to functional feed design was studied, combining probiotics and phytogenics (*D. hansenii* with oregano essential oil) (Hernández-Contreras et al. 2021). These authors observed an up-regulation of TNF-α expression and increase of humoral immune response on the skin mucus of Longfin yellowtail *Seriola rivoliana* juveniles fed *D. hansenii*. However, an antagonistic effect was found with the combination of both additives, probably due to reduced yeast growth and adhesion in fish gut (Hernández-Contreras et al. 2021). The probiotic effects of *D. hansenii* on shrimp have been evaluated in several studies, often in comparison with *S. cerevisiae* and showing a greater effect on the immune status and resistance to common diseases as the white spot syndrome. Those differences have been attributed to physicochemical characteristics of *D. hansenii* glucans, such as higher molecular weight and lower level of branching than those of *S. cerevisiae* (Angulo et al. 2020).

The need for sustainable production of juvenile fish for the industry makes an eco-friendly strategy necessary, and the use of probiotics to promote larval quality enhancement is a way to reach it. Recently, a holistic study (Teles et al. 2022) revealed that live yeast (*D. hansenii*) administration promoted higher growth and survival, a precocious maturation of the digestive tract, a higher degree of bone mineralization, and a reduction in skeletal deformities, demonstrating their probiotic effect at the whole organism.

Certainly, yeast can be used in an ample range of cultured organisms (favoring those belonging to fish and crustacean) not only as probiotics, but also as single cell protein in aquafeeds or for eco-friendly ammonia removal from aquaculture ponds (Yun et al. 2021).

2.3 Prebiotics and Synbiotics

Intensive animal husbandry made extensive use of industrialized feed, and antimicrobial substances were added to conventional feed with the aim of promoting growth. However, at present, the interest of consumers for safe products without drugs and the need for a sustainable aquaculture have encouraged researchers to seek alternatives to carry out studies and develop technologies that allow profitability and safety in crops. In this sense, the use of functional additives, such as prebiotics and synbiotics, is an environmentally friendly health strategy to improve animal nutrition (Carnevali et al. 2017; Dawood et al. 2020; Márquez-Couturier 2015).

The current prebiotics definition is "a selectively fermented ingredient that results in specific changes in the composition and/or activity of the gastrointestinal microbiota, thus conferring benefit(s) upon host health" (Gibson et al. 2017). Currently established prebiotics are non-digestible carbohydrates (NDC), composed by simple carbohydrate polymers with three or more monomeric units. These polymers are neither digested nor absorbed in the intestine and have importance for gut health since they are fermented in the intestine by the intestinal microbiota. Health-promoting compounds are subsequently produced enhancing growth performance, feed utilization, immunity, disease resistance, and fish survival against pathogens and environmental stress (Figueroa-Lozano et al. 2020; Ganguly et al. 2013; Gibson et al. 2017; Mugwanya et al. 2022). It is worth to mention that not all fibers can be classified as prebiotic. In contrast, most prebiotics can be classified as dietary fibers, which to fit at the concept of prebiotics have to meet three criteria: (1) be resistant to gastric acidity and hydrolysis by enzymes and gastrointestinal absorption; (2) be fermented by intestinal microbiota; and (3) selectively stimulate growth and/or activity of intestinal bacteria associated with health and well-being (Gibson et al. 2017; Guerreiro et al. 2018; Holscher 2017). In terms of their production, the prebiotics are obtained by extraction from plants; for example, inulin is extracted from chicory by enzymatic or chemical hydrolysis of plant polysaccharides or by enzymatic synthesis using disaccharides or other substrates, as done for Galactooligosaccharide (GOS) and Fructooligosaccharides (FOS) (Cunningham et al. 2021).

Several studies were performed on the use of prebiotics in aquaculture from the class of oligosaccharides, such as FOS, Mannan oligosaccharide (MOS), GOS, Arabinoxylan-oligosaccharide (AXOS), Xylo-oligosaccharide (XOS), and some from the class of polysaccharides, such as Inulin and β-glucan (W.-L. Chen et al. 2022; Y. Li et al. 2021; Munir et al. 2018; Poolsawat et al. 2020; Sepúlveda-Quiroz et al. 2021). These prebiotics were reported to promote growth performance, increase feed utilization efficiency, enhance immune system, and stimulate disease resistance of aquaculture species (Wee et al. 2022).

Some organic acids are principal end products of non-digestible carbohydrate fermentation by host microorganisms (Gibson et al. 2017). Prebiotics are not digested by the host itself but are fermented by bacteria present in the host gut. For example, *Bifidobacterium* and *Lactobacillus* lead to the production of short-chain fatty acids (SCFAs), namely acetic, propionic, and butyric acids cause a pH drop in the gut and lead to colonic and systemic health effects, which might be absorbed by the host and used as energy sources (Guerreiro et al. 2018). Moreover, prebiotics regulate local cytokine and antibodies to increase intestinal SCFA production and improve the binding capability of SCFAs to G-protein coupled receptors on leukocytes and carbohydrate receptors on intestinal epithelium (Butt et al. 2021).

To improve the positive effects of prebiotics and probiotics in the host organisms, these additives can be added together. The term synbiotics denotes a dietary supplement that includes both prebiotics and probiotics in a synergistic relationship (Butt et al. 2021). Nonetheless, the current definition for synbiotics by The International Scientific Association for Probiotics and Prebiotics (ISAPP) is "a mixture comprising live microorganisms and substrate(s) selectively utilized by the host microorganisms that confers a health benefit on the host" (Swanson et al. 2020).

The composition of synbiotics depends mainly on the mechanism of action of prebiotics in the host species. However, prebiotics with low degree of polymerization is better suited for the development of synbiotics than the ones with higher degree of polymerization, since they hydrolyze the grown beneficial strains and produce primary and secondary metabolites that may have beneficial effects on the host (Butt et al. 2021). Different bacterial strains, such as *Bacillus* sp. *Lactobacillus* sp., *Pediococcus* sp., are broadly used combined with prebiotics as synbiotics in aquaculture (Huynh et al. 2018; Khosravi-Katuli et al. 2021; Sewaka et al. 2019; Torrecillas et al. 2018; Yao et al. 2021). The use of synbiotics has been previously proposed to treat fish dysbiosis and a potential strategy to move toward more sustainable land-based aquaculture. Since its administration has been proven to be more effective than the use of probiotics alone, a better survival of probiotics has been observed when they are used in combination with prebiotics and vice versa (Butt et al. 2021; Vargas-Albores et al. 2021).

Increasing evidence exists that the microbial community of the intestinal tract provides both nutritional benefits and protection against pathogens in aquatic organisms. Therefore, manipulation of the gut microbiota is an important mechanism to achieve increased feed efficiency, growth, and health of fish and shellfish in aquaculture (Sutriana et al. 2021).

2.4 Phytogenic or Phytobiotic Compounds

The use of phytogenic compounds as feed additives has been extensively studied, most of them coming from aromatic plants and their essential oils (Citarasu 2010). Phytogenics are a heterogeneous group of feed additives originating from plants that have a wide range of activities including antimicrobial, anti-helminthic, antioxidant, growth enhancer, and immune modulator (Karásková et al. 2015). Aromatic plants are the source of many of these active compounds, of which the vast majority are fat-soluble and integrated into essential oils. These oils contain mostly terpenes and other molecules derived from them, such as terpenoids, aldehydes, ketones, acids, phenols, lactones, ethers, and esters (Tongnuanchan and Benjakul 2014). The use of phytogenic additives in aquaculture has shown promising results and is supported by a large number of scientific studies (Reverter et al. 2014; Sutili et al. 2018). In addition, these additives have economic and marketing advantages, since they address the current demands of society in terms of food safety and the production of animals in a way that respects the environment and their welfare.

One of the most widely used phytogenic compounds in aquaculture is allicin and its derivatives, present in garlic and onion. Garlic has been used for decades as antiparasitic in fish with good results (Boxaspen and Holm 1991; Peña et al. 1988). Subsequently, its use in aquaculture as an extract, essential oil, or fresh product has been studied. For example, an improvement in general aspects of *O. niloticus* physiology has been observed, not only in growth, hematology, and apparent health (Diab et al. 2008; Metwally 2009; Shalaby et al. 2006) but also in specific mechanisms of the innate immune system (Ndong and Fall 2011). Moreover, in Asian sea bass, an important growth promotion effect was reported for garlic, as well as positive effects on immunological and survival aspects (Talpur and Ikhwanuddin 2012). This growth-promoting effect was also observed in juvenile specimens of *D. labrax* with garlic and onion, of which the optimal doses were 30 g/kg for garlic and 10 g/kg for onion. Regarding salmonids, studies with this compound in Atlantic salmon *Salmo salar* have not been found. However, in *O. mykiss*, garlic (Nya and Austin 2009, 2011) or allicin, its bioactive compound (Nya et al. 2010) has demonstrated a protective effect against *A. hydrophila*, in addition to improvements in specific mechanisms of the innate immune system; also noteworthy is that harmful effects have been observed at high doses (2%) of stabilized allicin extract in rainbow trout (Breyer et al. 2015). Garlic has also been occasionally tested in crustaceans and shown to be effective in *Macrobrachium rosenbergii* (Poongodi et al. 2012) and *Litopenaeus vannamei* (Javadzadeh et al. 2012).

Other compounds from aromatic plants widely studied and known to have growth -promotion and immunoregulatory properties are the monoterpene phenols, thymol and carvacrol. These compounds are not only present in aromatic plants of the Lamiaceae family like thyme, oregano, or satureja but also found in other families, such as Chenopodiaceae, Plantaginaceae, Umbelliferae, Verbenaceae, etc. (Kirimer et al. 1995). They also have been used as technological additives to improve feed oxidative stability and fillet quality (Hernández et al. 2014a, b, 2015). The first complete study to achieve the use of these additives in fish feed was carried out by Zheng et al. (2009)) adding thymol, carvacrol, the combination of both and an essential oil of oregano *Origanum heracleoticum* to the diets of *I. punctatus*. These authors observed an improvement in the activity of enzymes related to the immune system, such as lysozyme, superoxide dismutase, and catalase besides a better resistance to *A. hydrophila* infection and increased growth with essential oil of oregano, which was the most effective in general. Subsequently, isolated compounds, as carvacrol (E. Yilmaz et al. 2015) or complete essential oils such as those from oregano (Diler et al. 2017; Hernández-Contreras et al. 2021), rosemary *Rosmarinus officinalis,* Shirazi thyme *Zataria multiflora,* dill *Anethum graveolens,* and blue gum *Eucalyptus globulus* have been studied (Roomiani et al. 2013), finding beneficial effects in all the cases related to antibacterial, immunological, or health status and eventually in growth performance. In the last decade, most research on the effects of extracts and essential oils from aromatic plants has been company-driven studies carried out with commercial additives based on a combination of different plant-based products and even probiotics (Table 2), which implies a new tendency in fish additives research.

2.5 Microalgae and Their Derivatives

In the last decades, microalgae have raised interest since they easily grow in land structure, and huge biomass is feasible to be reached in a well-controlled culture. This issue, together with the fact that they are a source of compounds with multiple properties, has led to several biotechnological applications from food derivative for human and animals to energy production, including cosmetic industries amongst others (Balasubramaniam et al. 2021). Regarding aquaculture, microalgae have traditionally been used as live feeds; however, multiple different applications of microalgae and their derivatives have recently been raised.

Microalgae are a wide diverse group of living organism; until now, more than 10,000 species of freshwater and marine microalgae have been identified (Charoonnart et al. 2018). They are the basis of aquatic food chains and adapted to multiple habitats (Camacho et al. 2019). Thus, they are the natural source of food for different animals with aquaculture importance, such as zooplankton, bivalves, larval shrimp, or fish fry. Several microalgal species cultures had been developed accordingly with their profile of nutrients and pigments for different aims in aquaculture (Hemaiswarya et al. 2011). In consequence, a small group of microalgal species (*Chorella, Tetraselmis, Isochrysis, Diacronema, Phaeodactylum, Chaetoceros, Nannochloropsis, Skeletonema* or *Thalassiosira*) has been characterized in terms of nutrient compositions and pigments (reviewed in Hemaiswarya et al. 2011). Apart from the well-known nutrients, such as proteins, polysaccharides, polyunsaturated fatty acids, pigments, vitamins, and minerals, they also produce oligosaccharides and a wide range of secondary metabolic compounds, many of which have biological functions (reviewed in (Cannell 1993; F. Chen et al. 2021)). The potential use of all these compounds in different biotechnological applications has contributed to an increase in bioprospecting different microalgal species from different seas (Grubišić et al. 2022; Guzmán et al. 2001). The ability of microalgae to produce bioactive compounds together with their microorganism growth characteristics allows the control of secondary metabolite production, which depends on temperature, salinity, osmotic pressure, light changes, or nutrient availability (Camacho et al. 2019) and favors their biotechnological use. Interestingly, those secondary metabolites are also secreted, which is one of the reasons for growth enhancement and the survival observed in the green- and pseudo-green-water techniques normally

Table 2 Effects of commercial additives based on a combination of different plant-based products on shrimp and fish performance and immunity

Commercial additive/ *composition*	Aquaculture species	Dosage	Duration	Major findings	Reference
AROTEC-G®, TECNOVIT-FARMFAES, S. L., Spain/*garlic, carvacrol, and thymol*	*Sparus aurata*	0.5%	65 days	Transcriptional regulation of factors of the skin innate immunity and stimulation of the secretory pathway. Decrease of bacterial growth capacity in mucus. Lower glucose, lactate, and cortisol levels	Firmino et al. (2021)
HERB; Delacon, Austria / *garlic and labiatae herbal extracts*	*Seriola dumerili*	0.02%	70 days	Plasmatic cortisol reduction after stress challenge, higher skin mucus protease and lysozyme activities. Up-regulation of genes related to skin immune response: muc-2, piscidin, tnf-α, il1-β, hep, c3, cd8 and casp3.	Fernández-Montero et al. (2021)
PHYTO, Delacon, Austria / *garlic and labiatae essential oils*	*Dicentrarchus labrax*	0.02%	63 days	Reduction of circulating plasma cortisol concentration during confinement challenge. Reinforced fish capacity of stress response via protection of head kidney leucocytes from stress-related apoptotic processes.	Serradell et al. (2020)

(continued)

Table 2 (continued)

Commercial additive/ *composition*	Aquaculture species	Dosage	Duration	Major findings	Reference
				Increased fish serum lysozyme.	
Next Enhance® 150, Novus International, Inc., Missouri, USA/*thymol and carvacrol (1:1)*	*Sparus aurata*	0, 0.005%, 0.01%, 0.02%, 0.03%	63 days	Improved feed gain ratio and reduced feed intake in a dose-dependent manner. Induction of an anti-inflammatory and antiproliferative transcriptomic profile. Most changes in the transcriptome of posterior intestine. Higher number of mucosal foldings.	Pérez-Sánchez et al. (2015)
Biomin® P.E.P. 1000, Biomin GmbH, Herzogenburg, Austria / *Blend of EOs rich in carvacrol*	*Oncorhynchus mykiss*	0.1%	56 days	Improved feed efficiency, lower total anaerobe counts, less lipid oxidation during cold storage of fish fillet. Higher lysozyme and catalase activity and total complement concentrations.	Giannenas et al. (2012)
Biomin® P.E.P. MGE 1000, Biomin GmbH, Herzogenburg, Austria / *Blend of EOs rich in thymol*	*O. mykiss*	0.1%	56 days	Improved feed efficiency, lower lactobacillus total anaerobe counts in gut, less lipid oxidation during cold storage of fish fillet. Higher lysozyme and catalase activity and total complement concentrations.	Giannenas et al. (2012)

(continued)

Table 2 (continued)

Commercial additive/ *composition*	Aquaculture species	Dosage	Duration	Major findings	Reference
				Reduction of serum nitric oxide levels.	
Digestarom® PEP MGE 150, Biomin GmbH, Herzogenburg, Austria / *Blend of anise, citrus, and oregano EOs*	*D. labrax*	0.02%	60 days	Improved performance and nutrient utilization, increased protein and energy retention. Compensation of intestinal changes produced by low fish meal diet. Increase in lysozyme activity,	Gonçalves et al. (2019)
Digestarom® PEP MGE, Biomin GmbH, Herzogenburg, Austria/ *carvacrol, thymol, anethol, and limonene*	*Ictalurus punctatus*	0.02%	42 days	No effect on growth performance. Higher survival and up-regulation of mannose binding lectin gene expression after *Edwardsiella ictaluri* challenge.	Peterson et al. (2015)

used in larval rearing of several fish species (Papandroulakis et al. 2001; Planas and Cunha 1999). In fact, recent studies have demonstrated that the presence of several microalgal species in the larval rearing water decreases the horizontal transmission of betanodavirus (Hernandez-Corona et al. 2002; Vázquez-Salgado et al. 2020).

Other culture technology where the microalgae are a key element is the bio-floc technology (BFT), in which bacteria and microalgae are cultured within the shrimp or fish ponds for helping in water quality control since they are able to grow on animal excreta compounds (De Schryver et al. 2008). However, BFT is not suitable for all aquaculture species since bio-floc culture species have to tolerate high levels of solids, poor water quality, and be able to digest microbial protein to take advantage of bio-floc as a food resource (reviewed in Hargreaves 2013).

The use of microalgae as an aquaculture feed ingredient has focused on a high interest in the last decades (Chauton et al. 2015; Shah et al. 2018) not only because they represent a more complete, sustainable, and cheaper alternative for replacing fish meal and fish oil in aquafeeds (F. Chen et al. 2021) but also to their cell wall components and secondary metabolites that are known to have biological functions, such as antimicrobial, antitumoral, antiviral, anti-inflammatory, anticoagulant, antioxidant, or even hormonal activities (Amaro et al. 2011; Cakmak et al. 2014; Cannell 1993). However, the cell wall of microalgae hampers the digestibility of the feed. In fact, the direct dietary administration of microalgae enhances the inflammatory process in the fish gut (Cerezuela et al. 2012; Reyes-Becerril et al. 2013), although they can improve the immunoglobulin serum levels (Zhang et al. 2014). In that sense and to avoid the main disadvantage, several strategies for increasing microalgae digestibility have been studied (Teuling et al. 2019; Tibbetts 2018), although a simple, economical, and cost-effective method has not been reached yet. Anyway, benefits on growth, welfare, or immune activities of marine fish feed with aquafeed containing different levels of microalgae have been described (Ayala et al. 2020; Galafat et al. 2022; Sáez et al. 2022; Villar-Navarro et al. 2022).

Focusing on the antimicrobial properties of microalgae, they could also be used for inhibition of biofilm formation or bacterial load control in culture systems (Charoonnart et al. 2018). In fact, certain molecules of microalgae interfere with bacterial cell-to-cell communication (quorum sensing) that plays a key role in bacterial virulence (Natrah et al. 2011).

Genetic engineering has focused on microalgae because they represent a simpler system for genetic manipulation than plant model systems. In the last decade, several strategies have been developed to introduce new genes into microalga cells (Hallmann 2016; Spicer and Molnar 2018). However, no more than 20 different microalga species have been genetically modified (Spicer and Molnar 2018). The changes cope with inhibition or activation of specific genes that impact the metabolic output including oil, protein, pigments, or cell wall components or other factors related to microalgae growth for making the genetically modified strain more suitable for massive production (Spicer and Molnar 2018). It is worthy to note that genetic engineering of microalgae is still a very young strategy, and transgenic microalgae might be negatively affected in cellular metabolism, which might represent a significant threat to the ecosystem if their manipulation would not cope with the right constraints measures (reviewed in Hallmann 2016; Pulz and Gross 2004). However, genetically engineered microalgae can be used to increase the industrial production of essential nutrients, such as polyunsaturated fatty acids, pigments, or other functional molecules for aquafeed formulation (reviewed in Osmond and Colombo 2019; Sprague et al. 2017), so their potential should not be underestimated.

2.6 Additives from Seaweeds

Seaweeds or macroalgae have been part of the animal feed in coastal areas from ancient times (Morais et al. 2020), although their main uses have been in human food, medicine, or as a fertilizer in agriculture. Seaweeds have low amount of lipids, but most of them are polyunsaturated fatty acids (PUFA) with high percentage of 20:5 n-3 eicosapentaenoic acid (EPA) and 20:4 n-5 arachidonic acid (ARA), especially in brown and red algae (Mišurcová 2011). These characteristics make them very attractive to be used as aquafeed ingredient or to extract PUFAs as an additive. Their protein content can be up to 50% in dry weight in some Rhodophyta species and vary depending on the alga genus, species, season, and environmental factors (Wan et al. 2019). This protein has an excellent amino acid profile comparable to the other conventional protein sources (Thiviya et al. 2022). They contain high levels of minerals, thanks to their capacity to absorb inorganic substances from seawater with concentration factors for some metals 10 to 20 times more than land plants (Mišurcová et al. 2011). In addition to all, macroalgae can be a great source of bioactive compounds. They produce a variety of secondary metabolites, commonly attributed to their challenging habitat and physical stressors, such as free amino acids, peptides, lectins, phycobiliproteins, carotenoids, phenolics, vitamins, and functional carbohydrates (Wan et al. 2019). Thanks to these bioactive components, their extracts have antihypertensive, antidiabetic, antioxidant, anti-inflammatory, antitumoral, antiviral, antimicrobial, and many other beneficial functional properties (Dagnino-Leone et al. 2022; Safavi et al. 2019; Wan et al. 2019).

Among the most extracted and used functional compounds from seaweeds are sulfated polysaccharides (alginate, fucoidan, carrageenan, laminarin, ulvan, galactan, agar). They have antiviral, antibacterial, anti-inflammatory, anthelmintic, antioxidant, immunomodulatory, and anticoagulant activities in terms of physiological activities (Safavi et al. 2019). However, they are also used as technological additives, since they have gelling properties and can be used in the production of microencapsulated, semi-moist, or extruded aquafeed (Borgogna et al. 2011; Rodríguez-Miranda et al. 2012). Alginate is the most studied functional compound in brown seaweed, which has proven its immunostimulant activity in cultured crustaceans, fish, and mollusks. The improvement of the immune response and disease resistance may be attributed to the presence of β-glucan that stimulates the nonspecific immune response through direct activation of the macrophages (Van Doan et al. 2019). Alginate has also been studied as a technological additive to improve aquafeed stability and reduce nutrient leaching in extruded fish feed (Rodríguez-Miranda et al. 2012) or in semi-moist diets for mollusks (Knauer et al. 1993). Fucoidan is also in the spotlight because it has proven to improve growth, intestinal health, antioxidant capacity, and immune responses of several finfish and shellfish species (Abdel-Latif et al. 2022). It has also been used to counteract the adverse effects of low fishmeal diets, enhancing growth, immunity, hematological indices, and resistance to oxidative stress in red sea bream *Pagrus major* with a 0.3–0.4% dose (Sony et al. 2020). As growth promoter, doses of 1% fucoidan improve growth performance in Asian seabass juveniles and 3% doses do in striped catfish, but its mechanism of action is still unknown (Abdel-Latif et al. 2022). In shellfish (*L. vannamei, M. rosenbergii, Penaeus monodon*) dozens of studies have demonstrated a growth promoter effect, apart from its immunostimulatory effect and improved resistance to infections. This improved growth has often been attributed to improved digestion, assimilation, and gut health; in fact, some studies have found improved enzyme activities and changes in histomorphology and bacterial communities (Cui et al. 2020; Ikeda-Ohtsubo et al. 2020; Mahgoub et al. 2020).

Carotenoids have several functional properties as growth promoter, antioxidant, free radical scavenger, singlet oxygen quencher, anti-inflammatory activity, antitumor activity, and immunostimulant activity (Wan et al. 2019). Despite all these physiological and immunological properties, carotenoids are currently of greater interest as technological additives. Some of them as astaxanthin and fucoxanthin have a high demand in aquafeed production, since they can substitute synthetic colorants that are commonly used in the aquaculture industry, for example, in organically certified salmon production. Other fish or shellfish species have been objective of studies that demonstrate the effectiveness of carotenoids in skin coloration enhancement (Rajauria 2015). Vitamins are other of the many forms of antioxidants present in seaweed, which make them an attractive source for sustainable feed additives for aquaculture. Both water- and fat-soluble vitamins are present including vitamins A, B, C, and E, with relatively high levels compared to land vegetables. In addition, seaweeds are one of the few vegetable sources of vitamin B12 (MacArtain et al. 2007).

Phycobiliproteins are one of the major pigments produced by algae, and they are light harvesting proteins found in seaweeds (Rhodophytes). They have multiple biological activities, including antioxidant, antibacterial, and antitumor (Dagnino-Leone et al. 2022), thus they have recently been studied as feed additives in fish. Phycocyanin (50 mg/kg) increased growth performance, feed utilization, and survival rate of *O. niloticus*, apart from increasing digestive enzyme secretion and immune response and reducing oxidative stress; all of these in a greater magnitude than beta carotene (Hassaan et al. 2021). Phycoerythrin and allophycocyanin have not yet been studied at this level as an additive in fish, although similar properties are attributed to them.

2.7 Additives from Mushrooms

Mushrooms are rich sources of prebiotic compounds as hemicellulose, chitin, α- and β-glucans, mannans, xylans, and galactans (Mohan et al. 2022). Some of them are also rich in phenolic compounds as melanins and lanostane-type triterpenoids, which are responsible for the antioxidant, antitumor, antiviral, anticancer, antimicrobial, and immunomodulatory effects on those species (Wold et al. 2020). Although the use of mushrooms as an ingredient in fish diets (as a substitute for fishmeal) has been studied by several authors, this chapter focuses on their use as a feed additive, looking at especially sustainable solutions. The species that have attracted the attention of researchers have been those that are useful in traditional medicine or have demonstrated functional properties in mammals, as described below.

Ganoderma lucidum is a basidiomycete fungus with a limited native distribution in Europe and some parts of China. The genus Ganoderma has been used in traditional Asian medicine for 2000 years and is currently cultivated for commercial use, whose name *reishi* in Japanese has become popular. It contains various bioactive compounds that have been studied, including polysaccharides as ganoderan A and B, triterpenes or bioactive proteins as ganodermin. In aquaculture, polysaccharides, extracts, and powders of *G. lucidum* have been tested in different species of crustaceans and fish, improving several innate immune parameters and, in most cases, growth performance (Mohan et al. 2022).

The genus *Pleurotus sp.* includes several edible mushroom species, ranking third in the commercial production of edible mushrooms in the world, which have been reported to have antitumor, antiviral, immunostimulants, antioxidant, antimicrobial, antimitogenic, and antiproliferative activities (Sartori et al. 2015). Based on these known properties, many studies have been carried out on different fish species (Mohan et al. 2022) and recently in white leg shrimp (Prabawati et al. 2022). These studies have shown the improvement of digestive enzymes activity in carp and a general immunostimulants effect in many fish species. It is also noticeable that enhancement in growth performance is achieved with powder, extract, or mushroom stalk meal of *Pleurotus sp.* but not with polysaccharides.

Other mushroom highly commercialized for culinary and medicinal use is *Lentinula edodes*. It is called *Shitake* and is very common in Asian food. *L. edodes* is rich in bioactive compounds with demonstrated effects against hypertension, hyperlipidemia, cardiovascular irregularities, depressed immunity, hepatic disorders, and cancer (Van Doan et al. 2019). Lentinan, a polysaccharide extracted from *L. edodes*, has shown immunomodulation properties in cultured fish (Djordjevic et al. 2009; Ren et al. 2019) while extract or whole mycelium also improved growth performance (Chitsaz et al. 2018; Pascual et al. 2018).

Coriolus versicolor is a well-known mushroom used in Asian traditional medicine to treat cancer and immune deficiency-related diseases (Van Doan et al. 2019). Its main properties are antioxidant, antibacterial, antitumor, immunomodulatory, and anticancer. The few reports on its activity in cultured fish shows a general immunomodulatory activity and, exceptionally, an improvement in growth performance in *P. olivaceus* (Mohan et al. 2022).

Other medicinal mushrooms as *Cordyceps sinensis, Schizophyllum commune,* or *Inonotus obliquus* are being tested as fish feed additives in a lesser extent, showing also beneficial effects on immunological parameters, growth, or stress, among others.

Given the large commercial production of some of these edible mushrooms, sustainable solutions are currently being sought as by-products of this industry, which is where feed additives for aquafeed can play an important role as scientific studies show in this regard. In the last decade, several authors have valued the functional properties of mushroom stalk waste and mushroom spent substrate as feed additives, since it is a sustainable solution for the use of by-products of mushroom cultivation through its incorporation into fish feed. The promising results of research on this kind of feed additive are gathered in Table 3.

Table 3 Effects of diets supplemented with mushroom cultivation by-products on fish performance and immunity

Mushroom cultivation by-product	Aquaculture species	Dosage	Duration	Major findings	Reference
Agaricus blazei spent mushroom substrate	*Oreochromis niloticus*	0, 1, 5%	7/14 days	In vitro promotion of the phagocytic activity of macrophages and tilapia head kidney cells. Higher expression of four cytokine genes. *S. agalactiae* growth inhibition in disc diffusion test. Higher survival rate after challenge with *S. agalactiae.* Protective effects were sustained at day 7 after the feed was switched. (1% diet).	Lee et al. (2022)
Cordyceps militaris spent mushroom substrate	*O. niloticus*	0, 0.5, 1, 2, 4 %	60 days	Increased skin mucus lysozyme and peroxidase activities and serum immune parameters. Higher resistance to *S. agalactiae* and better SGR, WG, FW, and FCR (10 g kg^{-1} diet).	Van Doan et al. (2017)
Pleurotus pulmonarius stalk waste E-HWE	*O. niloticus*	0, 1%	30 days	Improved feed conversion ratio, protein efficiency ratio, and hepatosomatic index but no effect on growth. After PAMPs challenge on fish: 1. Increased phagocytic activity and respiratory burst activity of blood leukocytes. 2. Increased serum hemolytic and	Ching et al. (2022)

(continued)

Table 3 (continued)

Mushroom cultivation by-product	Aquaculture species	Dosage	Duration	Major findings	Reference
				bactericidal activity. 3. Up-regulation of immune-related genes.	
Pleurotus pulmonarius stalk waste HWE	*O. niloticus*	0, 0.5, 1 %	30 days	Hot water extraction process optimization: higher nutrient and β-glucan content after 5 hours of extraction. Better fish growth and higher antioxidant enzyme activities (5 g kg^{-1}).	Ahmed et al. (2017)
Raw polysaccharide from mushroom stalk waste	*O. niloticus*	0, 0.5, 1%	30 days	Higher SGR, post stress improved SOD and CAT activity (5 g kg^{-1}). Better WG and HSI (10 g kg^{-1}).	Ahmed et al. (2017)
Mushroom stalk meal	*Clarias gariepinus*	0, 2.5, 5, 7.5,10 %	56 days	No differences in growth performance but higher protein efficiency ratio (5 g x $100g^{-1}$ diet). Favorable gut microflora and gut histomorphology changes in fish fed supplemented diets.	Adejonwo et al. (2020)

SGR specific growth rate, *WG* weight gain, *FW* final weight; *FCR* feed conversion rate, *PAMP* pathogen-associated molecular pattern, *HWE* hot water extract, *E-HWE* ethanol-pretreated hot water extract

3 Current Regulation, Future Perspectives, and Research Gaps

Europe has taken the lead in recent times through the Farm-to-Fork strategy (European Commission 2020) to accelerate the transition toward sustainable food systems. The main goals of this strategy are “to reduce the environmental and climate footprint of the EU food system and strengthen its resilience, ensure food security in the face of climate change and biodiversity loss and lead a global transition towards competitive sustainability from farm to fork and tapping into new opportunities." In this framework, natural or organic feed additives can be a key tool in this transition as they achieve the mentioned goals that are generally safe and sustainable. They can reduce antimicrobial resistance, improve animal welfare, and mitigate the effects of climate change and environmental degradation (Angelis and Carreño 2021).

Although the increase and diversification of natural additives in aquaculture feed is palpable, the growth of this industry could largely depend on the success of organic aquaculture. The fostering of this type of aquaculture by the EU is expressly contemplated in the Farm-to-Fork strategy and already has a specific regulation since 2007 (Council Regulation (EC) No. 834/2007), which was improved by Regulation (EU) 2018/848. This regulation establishes priorities in the design of fish feed, such as animal health and welfare, high product quality, and low environmental impact, many of which can be addressed through feed additives. In addition, in the treatment of diseases, it specifically establishes that the use of allopathic veterinary medicinal products are only be carried out where the use of phytotherapeutic, homeopathic, and other products is inappropriate and establishes the following order of preference when a health problem arises: (i) substances from plants, animals, or minerals in a homeopathic dilution; (ii) plants and their extracts not having anesthetic effects; and (iii) substances such as trace elements, metals, natural immunostimulants, or authorized probiotics.

The placing on the European Union market, labeling and using feed additives is regulated by Regulation (EC) 1831/2003 of the European Parliament and of the Council of 22 September 2003 on additives for use in animal nutrition. As established in Regulation (EC) 1831/2003, the commission created the European Union Register of Feed Additives, which is updated regularly and contains all the additives with current authorization. Many additives described in this chapter are represented in this register of nearly 1000 compounds, especially probiotic bacteria and phytogenic (https://food.ec.europa.eu/safety/animal-feed/feed-additives/eu-registered). Some newer compounds as mushroom extracts have not been registered for authorization yet, although it may not be long. Similarly, in the USA, the FDA is the organization responsible for the authorization for the sale and use of food additives and they maintain an updated register where current authorizations can be reviewed (https://www.fda.gov/food/ food-additives-petitions/food-additive-status-list). In this case, mentioned register contains both food additives for humans and feed additives for animals. Generally, food additives of natural origin such as those discussed in this chapter are classified as "Generally Recognized as Safe" (GRAS) and have very light restrictions. Such restrictions include, for example, not exceed ing the amount reasonably required to accomplish its intended effect in food by reducing to the extent reasonably possible substances coming from manufacturing, processing or packaging, or preparing and handling it as a food ingredient (https://www.ecfr.gov/current/title-21/chapter-I/subchapter-B/part-182). In the case of probiotics, they are not even considered within such register, since they can be regulated as foods, food supplements, cosmetics, or drugs. Their regulation is complex and depends largely on the claims made toward the product (Degnan 2008).

The diversity of legislation to be taken into account when using feed additives and the procedures for obtaining new authorizations, it is feared that it could be an impediment to the development of the natural feed additives industry within the framework of organic production. To avoid this problem, the European Commission (EC) is carrying out an evaluation of the current EU rules on feed additives, with which they foresee to make it easier to produce sustainable and innovative additives, speeding up the authorization process without compromising health & food safety (https://ec.europa.eu/info/law/better-regulation/have-your-say/initiatives/12624-Animal-nutrition-modernisation-of-EU-rules-on-feed-additives_en). The preliminary results of the evaluation suggest some potential areas for improvement as efficacy criteria more focused on animal well-being or the environment, more flexibility to promote technical and scientific innovation, and the dependence on imports due to the high cost of authorization and production.

In addition to a better regulation and more agile authorization process for the use and manufacture of feed additives, research gaps still need to be filled. To improve the design of natural feed additives, knowing the mechanism of action of separate bioactive compounds is essential, which has been addressed in common additives such as garlic but needs more research in others, such as those derived from seaweeds or mushrooms, commonly used as extracts or powders. Furthermore, given that the industry is committed to the combined use of several functional compounds, it would be interesting to study the potential interactions among or between them when they have the possibility of being used together (Hernández-Contreras et al. 2021). Nevertheless, more studies focused on optimal dosage and duration of treatment are needed, since they can be essential aspects in the effectiveness and profitability of the product.

Some aspects with less interest for researchers in decades but very important for the industry are the technological aspects, such as the physical quality of the pellet or stability and interactions of the additives during feed conservation. A new trend in this sense is appearing in fish feed research as it is closely related to profitability but also to sustainability of the activity through the reduction of waste production (Alcaraz et al. 2021; Hernández et al. 2014b).

Moreover, related to sustainability, additives from the food industry by-products are a promising research field that can boost natural and sustainable feed additives for the industry to get the objectives of EU strategies as Green Deal or Farm-to-Fork. Finally, a multidisciplinary approach in feed additives for aquafeeds achieving technological, nutritional, and immunological aspects is needed to foster the development of multipurpose, safe, and sustainable additives.

References

Abdel-Latif HMR, Dawood MAO, Alagawany M, Faggio C, Nowosad J, Kucharczyk D (2022) Health benefits and potential applications of fucoidan (FCD) extracted from brown seaweeds in aquaculture: an updated review. Fish Shellfish Immunol 122:115–130. https://doi.org/10.1016/j.fsi.2022.01.039

Abdel-Tawwab M, Abdel-Rahman AM, Ismael NEM (2008) Evaluation of commercial live bakers' yeast, *Saccharomyces cerevisiae* as a growth and immunity promoter for Fry Nile tilapia, *Oreochromis niloticus* (L.) challenged in situ with *Aeromonas hydrophila*. Aquaculture *280*(1):185–189. https://doi.org/10.1016/j.aquaculture.2008.03.055

Adejonwo OA, Omitoyin BO, Ajani EK, Ogunwole OA, Omitoyin SA (2020) Growth, gut morphology and microflora of African Catfish Fed Mushroom () stalk meal supplemented diets. Croat J Fish *78*(2):79–90. https://doi.org/10.2478/cjf-2020-0008

Adorian TJ, Jamali H, Farsani HG, Darvishi P, Hasanpour S, Bagheri T, Roozbehfar R (2019) Effects of probiotic bacteria bacillus on growth performance, digestive enzyme activity, and hematological parameters of Asian Sea Bass, *Lates calcarifer* (Bloch). Probiotics Antimicrob Proteins *11*(1):248–255. https://doi.org/10.1007/s12602-018-9393-z

Ahmed M, Abdullah N, Shuib AS, Abdul Razak S (2017) Influence of raw polysaccharide extract from mushroom stalk waste on growth and pH perturbation induced-stress in Nile tilapia, *Oreochromis niloticus*. *Aquaculture 468*:60–70. https://doi.org/10.1016/j.aquaculture.2016.09.043

Alcaraz R, Hernández-Contreras A, Iglesias P, Hernández MD (2021) Effect of the inclusion of microalgae on the physical properties of extruded feed for gilthead seabream (*Sparus aurata* L.). Algal Res *53*:102167. https://doi.org/10.1016/j.algal.2020.102167

Amaro HM, Guedes AC, Malcata FX (2011) Antimicrobial activities of microalgae: An invited review. In: Mendez-Vilas A (ed) Science against microbial pathogens: communicating current research and technological advances. Formatex Research Center, Badajoz

Andlid T, Juárez R-V, Gustafsson L (1995) Yeast colonizing the intestine of rainbow trout (*Salmo gairdneri*) and turbot (*Scophtalmus maximus*). Microb Ecol *30*(3):321–334. https://doi.org/10.1007/BF00171938

Angelis FD, Carreño I (2021) Achieving sustainability of the EU food chain with feed additives as a key tool: The European Commission intends to modernise the legal framework to foster innovation. Eur J Risk Regul *12*(4):866–870. https://doi.org/10.1017/err.2021.10

Angulo M, Reyes-Becerril M, Medina-Córdova N, Tovar-Ramírez D, Angulo C (2020) Probiotic and nutritional effects of *Debaryomyces hansenii* on animals. Appl Microbiol Biotechnol 104(18):7689–7699. https://doi.org/10.1007/s00253-020-10780-z

Ayala MD, Galián C, Fernández V, Chaves-Pozo E, García De La Serrana D, Sáez MI, Galafaz Díaz A, Alarcón FJ, Martínez TF, Arizcun M (2020) Influence of low dietary inclusion of the microalga *Nannochloropsis gaditana* (Lubián 1982) on performance, fish morphology, and muscle growth in juvenile gilthead seabream (*Sparus aurata*). Animals 10(12) Art. 12. https://doi.org/10.3390/ani10122270

Balasubramaniam V, Gunasegavan RD-N, Mustar S, Lee JC, Mohd Noh MF (2021) Isolation of industrial important bioactive compounds from Microalgae. Molecules *26*(4) Art. 4. https://doi.org/10.3390/molecules26040943

Ballester-Moltó M, Sanchez-Jerez P, García-García B, García-García J, Cerezo-Valverde J, Aguado-Giménez F (2016) Controlling feed losses by chewing in gilthead sea bream (*Sparus aurata*) ongrowing may improve the fish farming environmental sustainability. Aquaculture *464*:111–116. https://doi.org/10.1016/j.aquaculture.2016.06.018

Barman P, Raut S, Sen SK, Shaikh U, Bandyopadhyay P, Mohapatra PKD (2017) Effect of a three-component bacterial consortium in white shrimp farming for growth, survival and water quality management. Acta Biol Szeged *61*(1):35–44

Bélanger A, Sarker PK, Bureau DP, Chouinard Y, Vandenberg GW (2021) Apparent digestibility of macronutrients and fatty acids from Microalgae (*Schizochytrium* sp.) fed to rainbow trout (*Oncorhynchus mykiss*): a potential candidate for fish oil substitution. Animals *11*(2):Art. 2. https://doi.org/10.3390/ani11020456

Borgogna M, Bellich B, Cesàro A (2011) Marine polysaccharides in microencapsulation and application to aquaculture: "from sea to sea". Mar Drugs *9*(12):2572–2604. https://doi.org/10.3390/md9122572

Boxaspen K, Holm JC (1991) A new treatment against sea lice. Aquac Environ ProcAquac Europe 91:36–37

Breyer KE, Getchell RG, Cornwell ER, Wooster GA, Ketola HG, Bowser PR (2015) Efficacy of an Extract from Garlic, *Allium sativum*, Against Infection with the Furunculosis Bacterium, *Aeromonas salmonicida*, in Rainbow Trout, *Oncorhynchus mykiss*. *J World Aquacult Soc 46*(3):273–282. https://doi.org/10.1111/jwas.12195

Bureau DP, Hua K (2010) Towards effective nutritional management of waste outputs in aquaculture, with particular reference to salmonid aquaculture operations. *Aquac Res 41*(5):777–792. https://doi.org/10.1111/j.1365-2109.2009.02431.x

Butt UD, Lin N, Akhter N, Siddiqui T, Li S, Wu B (2021) Overview of the latest developments in the role of probiotics, prebiotics and synbiotics in shrimp aquaculture. Fish Shellfish Immunol *114*:263–281. https://doi.org/10.1016/j.fsi.2021.05.003

Cai Y, Yuan W, Wang S, Guo W, Li A, Wu Y, Chen X, Ren Z, Zhou Y (2019) In vitro screening of putative probiotics and their dual beneficial effects: To white shrimp (*Litopenaeus vannamei*) postlarvae and to the rearing water. *Aquaculture 498*:61–71. https://doi.org/10.1016/j.aquaculture.2018.08.024

Cakmak YS, Kaya M, Asan-Ozusaglam M (2014) Biochemical composition and bioactivity screening of various extracts from *Dunaliella salina*, a green microalga. *EXCLI J 13*:679–690

Camacho F, Macedo A, Malcata F (2019) Potential industrial applications and commercialization of Microalgae in the functional food and feed industries: a short review. Mar Drugs *17*(6):312. https://doi.org/10.3390/md17060312

Cannell RJP (1993) Algae as a source of biologically active products. *Pestic Sci 39*(2):147–153. https://doi.org/10.1002/ps.2780390208

Carnevali O, Maradonna F, Gioacchini G (2017) Integrated control of fish metabolism, wellbeing and reproduction: The role of probiotic. *Aquaculture 472*:144–155. https://doi.org/10.1016/j.aquaculture.2016.03.037

Cerezuela R, Guardiola FA, Meseguer J, Esteban MÁ (2012) Enrichment of gilthead seabream (*Sparus aurata* L.) diet with microalgae: effects on the immune system. *Fish Physiol Biochem 38*(6):1729–1739. https://doi.org/10.1007/s10695-012-9670-9

Ceseña CE, Vega-Villasante F, Aguirre-Guzman G, Luna-Gonzalez A, Campa-Cordova A (2021) Update on the use of yeast in shrimp aquaculture: a mini review. Int Aquat Res *13*(1):1–16. https://doi.org/10.22034/iar.2021.1904524.1066

Chapman CMC, Gibson GR, Rowland I (2011) Health benefits of probiotics: are mixtures more effective than single strains? *Eur J Nutr 50*(1):1–17. https://doi.org/10.1007/s00394-010-0166-z

Charoonnart P, Purton S, Saksmerprome V (2018) Applications of microalgal biotechnology for disease control in aquaculture. *Biology 7*(2):24. https://doi.org/10.3390/biology7020024

Chauton MS, Reitan KI, Norsker NH, Tveterås R, Kleivdal HT (2015) A techno-economic analysis of industrial production of marine microalgae as a source of EPA and DHA-rich raw material for aquafeed: research challenges and possibilities. *Aquaculture 436*:95–103. https://doi.org/10.1016/j.aquaculture.2014.10.038

Chen F, Leng Y, Lu Q, Zhou W (2021) The application of microalgae biomass and bio-products as aquafeed for aquaculture. *Algal Res 60*:102541. https://doi.org/10.1016/j.algal.2021.102541

Chen W-L, Ge Y-P, Sun M, He C-F, Zhang L, Liu W-B, Li H-X, Li X-F (2022) Insights into the correlations between prebiotics and carbohydrate metabolism in fish: Administration of xylooligosaccharides in *Megalobrama amblycephala* offered a carbohydrate-enriched diet. *Aquaculture 561*:738684. https://doi.org/10.1016/j.aquaculture.2022.738684

Ching JJ, Shuib AS, Abdullah N, Abdul Majid N, Mohd Taufek N, Sutra J, Azmai MNA (2022) Effects of *Pleurotus pulmonarius* stalk waste extract-supplemented feed on the immune responses of red hybrid tilapia *Oreochromis* sp. following challenge with pathogen-associated molecular patterns. *Aquac Res* 53:1–19. https://doi.org/10.1111/are.16050

Chitsaz H, Akrami R, Ahmadi Z (2018) Effects of mushroom (*Lentinula edodes*) extract on growth performance, immune response and hemato-biochemical parameters of great sturgeon juvenile (*Huso huso* Linnaeus, 1754). Sustain Aquac Health Manag J *4*(1):29–48. https://doi.org/10.29252/ijaah.4.1.29

Chiu C-H, Cheng C-H, Gua W-R, Guu Y-K, Cheng W (2010) Dietary administration of the probiotic, *Saccharomyces cerevisiae* P13, enhanced the growth, innate immune responses, and disease resistance of the grouper, *Epinephelus coioides*. Fish Shellfish Immunol *29*(6): 1053–1059. https://doi.org/10.1016/j.fsi.2010.08.019

Chomwong S, Charoensapsri W, Amparyup P, Tassanakajon A (2018) Two host gut-derived lactic acid bacteria activate the proPO system and increase resistance to an AHPND-causing strain of *Vibrio parahaemolyticus* in the shrimp *Litopenaeus vannamei*. Dev Comp Immunol *89*:54–65. https://doi.org/10.1016/j.dci.2018.08.002

Citarasu T (2010) Herbal biomedicines: a new opportunity for aquaculture industry. Aquacult Int 18:403–414. https://doi.org/10.1007/s10499-009-9253-7

Council Regulation (EC) No 834/2007 of 28 June 2007 on organic production and labelling of organic products and repealing regulation (EEC) No 2092/91. OJ L 189, 20.7.2007, p 1–23.

Cui H, Wang Z, Liu J, Wang Y, Wang Z, Fu J, Wan Z, Li R, Li Q, Helen Fitton J, Liu Y, Zhang M (2020) Effects of a highly purified fucoidan from *Undaria pinnatifida* on growth performance and intestine health status of gibel carp *Carassius auratus gibelio*. *Aquac Nutr 26*(1):47–59. https://doi.org/10.1111/anu.12966

Cunningham M, Vinderola G, Charalampopoulos D, Lebeer S, Sanders ME, Grimaldi R (2021) Applying probiotics and prebiotics in new delivery formats – is the clinical evidence transferable? *Trends Food Sci Technol 112*:495–506. https://doi.org/10.1016/j.tifs.2021.04.009

Dagnino-Leone J, Figueroa CP, Castañeda ML, Youlton AD, Vallejos-Almirall A, Agurto-Muñoz A, Pavón Pérez J, Agurto-Muñoz C (2022) Phycobiliproteins: structural aspects, functional characteristics, and biotechnological perspectives. Comput Struct Biotechnol J *20*:1506–1527. https://doi.org/10.1016/j.csbj.2022.02.016

Dawood MAO, Koshio S, Esteban MÁ (2018) Beneficial roles of feed additives as immunostimulants in aquaculture: a review. *Rev Aquac 10*(4):950–974. https://doi.org/10.1111/raq.12209

Dawood MAO, Eweedah NM, Moustafa EM, Shahin MG (2020) Synbiotic effects of *Aspergillus oryzae* and β-glucan on growth and oxidative and immune responses of Nile Tilapia, *Oreochromis niloticus*. Probiotics Antimicrob Proteins *12*(1):172–183. https://doi.org/10.1007/s12602-018-9513-9

De Schryver P, Crab R, Defoirdt T, Boon N, Verstraete W (2008) The basics of bio-flocs technology: the added value for aquaculture. *Aquaculture 277*(3):125–137. https://doi.org/10.1016/j.aquaculture.2008.02.019

Degnan FH (2008) The US food and drug administration and probiotics: regulatory categorization. *Clin Infect Dis 46*(Supplement_2):S133–S136. https://doi.org/10.1086/523324

Diab A, Aly S, John G, Abde-Hadi Y, Fathi M (2008) Effect of garlic, black seed and Biogen as immunostimulants on the growth and survival of Nile tilapia, *Oreochromis niloticus* (Teleostei: Cichlidae), and their response to artificial infection with *Pseudomonas fluorescens*. *Afr J Aquat Sci 33*:63–68. https://doi.org/10.2989/AJAS.2007.33.1.7.391

Diler O, Gormez O, Diler I, Metin S (2017) Effect of oregano (*Origanum onites* L.) essential oil on growth, lysozyme and antioxidant activity and resistance against *Lactococcus garvieae* in rainbow trout, *Oncorhynchus mykiss* (Walbaum). *Aquac Nutr 23*(4):844–851. https://doi.org/10.1111/anu.12451

Djordjevic B, Škugor S, Jørgensen SM, Øverland M, Mydland LT, Krasnov A (2009) Modulation of splenic immune responses to bacterial lipopolysaccharide in rainbow trout (*Oncorhynchus mykiss*) fed lentinan, a beta-glucan from mushroom *Lentinula edodes*. Fish Shellfish Immunol *26*(2):201–209. https://doi.org/10.1016/j.fsi.2008.10.012

El-Saadony MT, Alagawany M, Patra AK, Kar I, Tiwari R, Dawood MAO, Dhama K, Abdel-Latif HMR (2021) The functionality of probiotics in aquaculture: an overview. Fish Shellfish Immunol *117*:36–52. https://doi.org/10.1016/j.fsi.2021.07.007

European Commission (2020) Farm to fork strategy: for a fair, healthy and environmentally-friendly food system. Communication from the Commission to the European Parliament, the Council, the European Economic and Social Committee and the Committee of the Regions. Brussels, Belgium. https://eur-lex.europa.eu/resource.html?uri=cellar:ea0f9f73-9ab2-11ea-9d2d-01aa75ed71a1.0001.02/DOC_1&format=PDF

FAO (2022) The state of world fisheries and aquaculture 2022. Towards blue transformation. FAO, Rome. https://doi.org/10.4060/cc0461en

Fernández-Montero Á, Torrecillas S, Acosta F, Kalinowski T, Bravo J, Sweetman J, Roo J, Makol A, Docando J, Carvalho M, Izquierdo MS, Montero D (2021) Improving greater amberjack (*Seriola dumerili*) defenses against monogenean parasite *Neobenedenia girellae* infection through functional dietary additives. *Aquaculture* 534:736317. https://doi.org/10.1016/j.aquaculture.2020.736317

Festa S, Coppotelli BM, Madueño L, Loviso CL, Macchi M, Tauil RMN, Valacco MP, Morelli IS (2017) Assigning ecological roles to the populations belonging to a phenanthrene-degrading bacterial consortium using omic approaches. PLoS One *12*(9):e0184505. https://doi.org/10.1371/journal.pone.0184505

Figueroa-Lozano S, Ren C, Yin H, Pham H, van Leeuwen S, Dijkhuizen L, de Vos P (2020) The impact of oligosaccharide content, glycosidic linkages and lactose content of galacto-oligosaccharides (GOS) on the expression of mucus-related genes in goblet cells. Food Funct *11*(4):3506–3515. https://doi.org/10.1039/D0FO00064G

Firmino JP, Fernández-Alacid L, Vallejos-Vidal E, Salomón R, Sanahuja I, Tort L, Ibarz A, Reyes-López FE, Gisbert E (2021) Carvacrol, thymol, and garlic essential oil promote skin innate immunity in gilthead seabream (*Sparus aurata*) through the multifactorial modulation of the secretory pathway and enhancement of mucus protective capacity. Front Immunol 12:633621. https://www.frontiersin.org/articles/10.3389/fimmu.2021.633621

Galafat A, Vizcaíno AJ, Sáez MI, Martínez TF, Arizcun M, Chaves-Pozo E, Alarcón FJ (2022) Assessment of dietary inclusion of crude or hydrolysed *Arthrospira platensis* biomass in starter diets for gilthead seabream (*Sparus aurata*). *Aquaculture 548*:737680. https://doi.org/10.1016/j.aquaculture.2021.737680

Ganguly S, Dora KC, Sarkar S, Chowdhury S (2013) Supplementation of prebiotics in fish feed: a review. Rev Fish Biol Fish *23*(2):195–199. https://doi.org/10.1007/s11160-012-9291-5

Ghosh K, Mukherjee A, Dutta D, Banerjee S, Breines EM, Hareide E, Ringø E (2021) Endosymbiotic pathogen-inhibitory gut bacteria in three Indian Major Carps under polyculture system: a step toward making a probiotics consortium. Aquac Fish *6*(2):192–204. https://doi.org/10.1016/j.aaf.2020.03.009

Giannenas I, Triantafillou E, Stavrakakis S, Margaroni M, Mavridis S, Steiner T, Karagouni E (2012) Assessment of dietary supplementation with carvacrol or thymol containing feed additives on performance, intestinal microbiota and antioxidant status of rainbow trout (*Oncorhynchus mykiss*). *Aquaculture* 350–353:26–32. https://doi.org/10.1016/j.aquaculture.2012.04.027

Gibson GR, Hutkins R, Sanders ME, Prescott SL, Reimer RA, Salminen SJ, Scott K, Stanton C, Swanson KS, Cani PD, Verbeke K, Reid G (2017) Expert consensus document: The International Scientific Association for Probiotics and Prebiotics (ISAPP) consensus statement on the definition and scope of prebiotics. *Nat Rev Gastroenterol Hepatol 14*:491–502. https://doi.org/10.1038/nrgastro.2017.75

Gonçalves RA, Serradeiro R, Machado M, Costas B, Hunger C, Dias J (2019) Interactive effects of dietary fishmeal level and plant essential oils supplementation on European sea bass, *Dicentrarchus labrax*: growth performance, nutrient utilization, and immunological response. *J World Aquacult Soc 50*(6):1078–1092. https://doi.org/10.1111/jwas.12616

Grubišić M, Šantek B, Zorić Z, Čošić Z, Vrana I, Gašparović B, Čož-Rakovac R, Ivančić Šantek M (2022) Bioprospecting of Microalgae isolated from the Adriatic Sea: characterization of biomass, pigment, lipid and fatty acid composition, and antioxidant and antimicrobial activity. *Molecules 27*(4):1248. https://doi.org/10.3390/molecules27041248

Guerreiro I, Serra CR, Pousão-Ferreira P, Oliva-Teles A, Enes P (2018) Prebiotics effect on growth performance, hepatic intermediary metabolism, gut microbiota and digestive enzymes of white sea bream (*Diplodus sargus*). *Aquac Nutr 24*(1):153–163. https://doi.org/10.1111/anu.12543

Guzmán S, Gato A, Calleja JM (2001) Antiinflammatory, analgesic and free radical scavenging activities of the marine microalgae *Chlorella stigmatophora* and *Phaeodactylum tricornutum*. *Phytother Res 15*(3):224–230. https://doi.org/10.1002/ptr.715

Hallmann A (2016) Algae biotechnology – green cell-factories on the rise. Curr Biotechnol *4*(4): 389–415. https://doi.org/10.2174/2211550105666151107001338

Hargreaves JA (2013) Biofloc production systems for aquaculture, vol 4503. Southern Regional Aquaculture Center, Stoneville, p 12

Harikrishnan R, Kim M-C, Kim J-S, Balasundaram C, Heo M-S (2011) Immunomodulatory effect of probiotics enriched diets on *Uronema marinum* infected olive flounder. Fish Shellfish Immunol *30*(3):964–971. https://doi.org/10.1016/j.fsi.2011.01.030

Hassaan MS, Mohammady EY, Soaudy MR, Sabae SA, Mahmoud AMA, El-Haroun ER (2021) Comparative study on the effect of dietary β-carotene and phycocyanin extracted from *Spirulina platensis* on immune-oxidative stress biomarkers, genes expression and intestinal enzymes, serum biochemical in Nile tilapia, *Oreochromis niloticus*. Fish Shellfish Immunol *108*:63–72. https://doi.org/10.1016/j.fsi.2020.11.012

Hemaiswarya S, Raja R, Ravi Kumar R, Ganesan V, Anbazhagan C (2011) Microalgae: a sustainable feed source for aquaculture. *World J Microbiol Biotechnol 27*(8):1737–1746. https://doi.org/10.1007/s11274-010-0632-z

Hernández A, García García B, Jordán MJ, Hernández MD (2014a) Improved conservation of gilthead seabream (*Sparus aurata*) in ice storage. The influence of doses of rosemary extract added to feed. *Aquaculture* 426–427:31–40. https://doi.org/10.1016/j.aquaculture.2014.01.018

Hernández A, García García B, Jordán MJ, Hernández MD (2014b) Natural antioxidants in extruded fish feed: Protection at different storage temperatures. *Anim Feed Sci Technol 195*:112–119. https://doi.org/10.1016/j.anifeedsci.2014.06.003

Hernández A, García García B, Jordán M, j., & Hernández, M. d. (2015) Study of the dose of thyme essential oil in feed to prolong the shelf life of gilthead seabream (*Sparus aurata*). *Aquac Nutr 21*(5):740–749. https://doi.org/10.1111/anu.12196

Hernández-Contreras Á, Tovar-Ramírez D, Reyes-Becerril M (2021) Modulatory effect of *Debaryomyces hansenii* and oregano essential oil on the humoral immunity of skin mucus in Longfin yellowtail *Seriola rivoliana*. *Aquac Res 52*(2):749–762. https://doi.org/10.1111/are.14931

Hernandez-Corona A, Nieves I, Meckes M, Chamorro G, Barron B (2002) Antiviral activity of *Spirulina maxima* against herpes simplex virus type 2. *Antivir Res 56*(3):279–285. https://doi.org/10.1016/S0166-3542(02)00132-8

Holscher HD (2017) Dietary fiber and prebiotics and the gastrointestinal microbiota. *Gut Microbes 8*(2):172–184. https://doi.org/10.1080/19490976.2017.1290756

Huynh T-G, Cheng A-C, Chi C-C, Chiu K-H, Liu C-H (2018) A synbiotic improves the immunity of white shrimp, *Litopenaeus vannamei*: metabolomic analysis reveal compelling evidence. Fish Shellfish Immunol *79*:284–293. https://doi.org/10.1016/j.fsi.2018.05.031

Ikeda-Ohtsubo W, López Nadal A, Zaccaria E, Iha M, Kitazawa H, Kleerebezem M, Brugman S (2020) Intestinal Microbiota and Immune Modulation in Zebrafish by Fucoidan From Okinawa Mozuku (*Cladosiphon okamuranus*). Front Nutr *7*. https://www.frontiersin.org/articles/10.3389/fnut.2020.00067

Javadzadeh M, Salarzadeh AR, Yahyavi M, Hafezieh M, Darvishpour H (2012) Effect of garlic extract on growth and survival rate in Litopenaeus vannamei post larvae. Iran J Fish Sci 21 (1):39–46

Karásková K, Suchý P, Straková E (2015) Current use of phytogenic feed additives in animal nutrition: a review. Czech J Anim Sci *60*(12):521–530

Khosravi-Katuli K, Mohammadi Y, Ranjbaran M, Ghanaatian H, Khazaali A, Paknejad H, Santander J (2021) Effects of mannan oligosaccharide and synbiotic supplementation on growth performance and immune response of Gilthead Sea Bream (*Sparus aurata*) before and after thermal stress. Aquac Res *52*(8):3745–3756. https://doi.org/10.1111/are.15220

Kirimer N, Başer KHC, Tümen G (1995) Carvacrol-rich plants in Turkey. *Chem Nat Compd 31*(1): 37–41. https://doi.org/10.1007/BF01167568

Knauer J, Britz PJ, Hecht T (1993) The effect of seven binding agents on 24-hour water stability of an artificial weaning diet for the South African abalone, *Haliotis midae* (Haliotidae, Gastropoda). *Aquaculture 115*(3):327–334. https://doi.org/10.1016/0044-8486(93)90146-P

Kong W, Meldgin DR, Collins JJ, Lu T (2018) Designing microbial consortia with defined social interactions. *Nat Chem Biol 14*(8):821–829. https://doi.org/10.1038/s41589-018-0091-7

Korada SK, Yarla NS, Mishra V, Daim MA, Sharma B, Ashraf GM, Reggi R, Palmery M, Peluso I, Kamal MA (2018) Single probiotic versus multiple probiotics—a debate on current scenario for alleviating health benefits. *Curr Pharm Des 24*(35):4150–4153

Kozasa M (1986) Toyocerin (*Bacillus toyoi*) as growth promotor for animal feeding. *Microbiol Aliment Nutrition 4*:121–135

Kwoji ID, Aiyegoro OA, Okpeku M, Adeleke MA (2021) Multi-strain probiotics: synergy among isolates enhances biological activities. *Biology 10*(4):Art. 4. https://doi.org/10.3390/biology10040322

Lara-Flores M, Olvera-Novoa MA, Guzmán-Méndez BE, López-Madrid W (2003) Use of the bacteria *Streptococcus faecium* and *Lactobacillus acidophilus*, and the yeast *Saccharomyces cerevisiae* as growth promoters in Nile tilapia (*Oreochromis niloticus*). *Aquaculture 216*(1): 193–201. https://doi.org/10.1016/S0044-8486(02)00277-6

Laranja JLQ, Ludevese-Pascual GL, Amar EC, Sorgeloos P, Bossier P, De Schryver P (2014) Poly-β-hydroxybutyrate (PHB) accumulating *Bacillus* spp. Improve the survival, growth and robustness of *Penaeus monodon* (Fabricius, 1798) postlarvae. *Vet Microbiol 173*(3):310–317. https://doi.org/10.1016/j.vetmic.2014.08.011

Lee P-T, Wu Y-S, Tseng C-C, Lu J-Y, Lee M-C (2022) Dietary *Agaricus blazei* spent substrate improves disease resistance of Nile Tilapia (*Oreochromis niloticus*) against *Streptococcus agalactiae* in vivo. *J Mar Sci Eng* 10(1) Art. 1. https://doi.org/10.3390/jmse10010100

Li P, Gatlin DM (2005) Evaluation of the prebiotic GroBiotic®-A and brewers yeast as dietary supplements for sub-adult hybrid striped bass (*Morone chrysops*×*M. saxatilis*) challenged in situ with *Mycobacterium marinum. Aquaculture 248*(1):197–205. https://doi.org/10.1016/j.aquaculture.2005.03.005

Li Y, Yuan W, Zhang Y, Liu H, Dai X (2021) Single or combined effects of dietary arabinoxylan-oligosaccharide and inulin on growth performance, gut microbiota, and immune response in Pacific white shrimp *Litopenaeus vannamei. J Oceanol Limnol 39*(2):741–754. https://doi.org/10.1007/s00343-020-9083-z

MacArtain P, Gill CIR, Brooks M, Campbell R, Rowland IR (2007) Nutritional value of edible seaweeds. *Nutr Rev 65*(12):535–543. https://doi.org/10.1111/j.1753-4887.2007.tb00278.x

Mahgoub HA, El-Adl MAM, Ghanem HM, Martyniuk CJ (2020) The effect of fucoidan or potassium permanganate on growth performance, intestinal pathology, and antioxidant status in Nile tilapia (*Oreochromis niloticus*). *Fish Physiol Biochem 46*(6):2109–2131. https://doi.org/10.1007/s10695-020-00858-w

Márquez-Couturier G (2015) Estado de arte de la biología y cultivo de pejelagarto (*Atractosteus tropicus*). *Agro Productividad 8*(3) https://www.revista-agroproductividad.org/index.php/agroproductividad/article/view/660

McFarland LV (2021) Efficacy of single-strain probiotics versus multi-strain mixtures: systematic review of strain and disease specificity. *Dig Dis Sci 66*(3):694–704. https://doi.org/10.1007/s10620-020-06244-z

Melo-Bolívar JF, Ruiz Pardo RY, Hume ME, Villamil Díaz LM (2021) Multistrain probiotics use in main commercially cultured freshwater fish: a systematic review of evidence. *Rev Aquac 13*(4): 1758–1780. https://doi.org/10.1111/raq.12543

Metwally M (2009) Effects of garlic (*Allium sativum*) on some antioxidant activities in Tilapia Nilotica (*Oreochromis niloticus*). *World J Fish Marine Sci 1*(1):56–64

Mišurcová L (2011) Chemical composition of seaweeds. In: *Handbook of marine macroalgae*. Wiley, Chichester, pp 171–192. https://doi.org/10.1002/9781119977087.ch7

Mišurcová L, Machů L, Orsavová J (2011) Chapter 29—Seaweed Minerals as Nutraceuticals. In: Kim S-K (ed) *Advances in Food and Nutrition Research*, vol 64. Academic Press, pp 371–390. https://doi.org/10.1016/B978-0-12-387669-0.00029-6

Mohan K, Karthick Rajan D, Muralisankar T, Ramu Ganesan A, Marimuthu K, Sathishkumar P (2022) The potential role of medicinal mushrooms as prebiotics in aquaculture: a review. *Rev Aquac 14*(3):1300–1332. https://doi.org/10.1111/raq.12651

Morais T, Inácio A, Coutinho T, Ministro M, Cotas J, Pereira L, Bahcevandziev K (2020) Seaweed potential in the animal feed: a review. J Mar Sci Eng *8*(8) Art. 8. https://doi.org/10.3390/jmse8080559

Mugwanya M, Dawood MAO, Kimera F, Sewilam H (2022) Updating the role of probiotics, prebiotics, and synbiotics for Tilapia aquaculture as leading candidates for food sustainability: a review. Probiotics Antimicrob Proteins *14*(1):130–157. https://doi.org/10.1007/s12602-021-09852-x

Munir MB, Hashim R, Nor SAM, Marsh TL (2018) Effect of dietary prebiotics and probiotics on snakehead (*Channa striata*) health: haematology and disease resistance parameters against *Aeromonas hydrophila*. Fish Shellfish Immunol *75*:99–108. https://doi.org/10.1016/j.fsi.2018.02.005

Natrah FMI, Kenmegne MM, Wiyoto W, Sorgeloos P, Bossier P, Defoirdt T (2011) Effects of micro-algae commonly used in aquaculture on acyl-homoserine lactone quorum sensing. *Aquaculture 317*(1-4):53–57. https://doi.org/10.1016/j.aquaculture.2011.04.038

Navarrete P, Tovar-Ramírez D, Navarrete P, Tovar-Ramírez D (2014) Use of yeasts as probiotics in fish aquaculture. Sustainable aquaculture techniques. IntechOpen. https://doi.org/10.5772/57196

Ndong D, Fall J (2011) The effect of garlic (*Allium sativum*) on growth and immune responses of hybrid tilapia (*Oreochromis niloticus* x *Oreochromis aureus*). J Clin Immunol Immunopathol Res *3*(1):1–9. https://doi.org/10.5897/JCIIR.9000010

Ningtyas MD, Isdiantoni, Ida E, Koentjoro MP, Prasetyo EN (2020) Microbial consortium synergism for promising freshwater culture probiotic. 2215:070009. https://doi.org/10.1063/5.0000813

Nya EJ, Austin B (2009) Use of garlic, *Allium sativum*, to control *Aeromonas hydrophila* infection in rainbow trout, *Oncorhynchus mykiss* (Walbaum). *J Fish Dis 32*(11):963–970. https://doi.org/10.1111/j.1365-2761.2009.01100.x

Nya EJ, Austin B (2011) Development of immunity in rainbow trout (*Oncorhynchus mykiss*, Walbaum) to *Aeromonas hydrophila* after the dietary application of garlic. *Fish Shellfish Immunol 30*(3):845–850. https://doi.org/10.1016/j.fsi.2011.01.008

Nya EJ, Dawood Z, Austin B (2010) The garlic component, allicin, prevents disease caused by *Aeromonas hydrophila* in rainbow trout, *Oncorhynchus mykiss* (Walbaum). *J Fish Dis 33*(4): 293–300. https://doi.org/10.1111/j.1365-2761.2009.01121.x

Osmond ATY, Colombo SM (2019) The future of genetic engineering to provide essential dietary nutrients and improve growth performance in aquaculture: Advantages and challenges. *J World Aquacult Soc 50*(3):490–509. https://doi.org/10.1111/jwas.12595

Papandroulakis N, Divanach P, Anastasiadis P, Kentouri M (2001) The pseudo-green water technique for intensive rearing of sea bream (*Sparus aurata*) larvae. *Aquac Int 9*(3):205–216. https://doi.org/10.1023/A:1016813623122

Park Y, Lee S, Hong J, Kim D, Moniruzzaman M, Bai SC (2017) Use of probiotics to enhance growth, stimulate immunity and confer disease resistance to *Aeromonas salmonicida* in rainbow trout (*Oncorhynchus mykiss*). *Aquac Res 48*(6):2672–2682. https://doi.org/10.1111/are.13099

Pascual MM, Hualde JP, Bianchi VA, Moreno P, Castro JM, Luquet CM (2018) Wheat grains fermented by fungal mycelia (*Pleurotus ostreatus* or *Lentinus edodes*) as alternative feed ingredients for juvenile rainbow trout (*Oncorhynchus mykiss*). *Aquac Int 26*(6):1339–1352. https://doi.org/10.1007/s10499-018-0286-7

Peña N, Auró A, Sumano H (1988) A comparative trial of garijc, its extract and ammonium-potassium tartrate as antheijviintics in carp. *J Ethnopharmacol 24*(2):199–203. https://doi.org/10.1016/0378-8741(88)90152-3

Pérez-Sánchez J, Benedito-Palos L, Estensoro I, Petropoulos Y, Calduch-Giner JA, Browdy CL, Sitjà-Bobadilla A (2015) Effects of dietary NEXT ENHANCE®150 on growth performance and expression of immune and intestinal integrity related genes in gilthead sea bream (*Sparus aurata* L.). *Fish Shellfish Immunol 44*(1):117–128. https://doi.org/10.1016/j.fsi.2015.01.039

Peterson BC, Peatman E, Ourth DD, Waldbieser GC (2015) Effects of a phytogenic feed additive on growth performance, susceptibility of channel catfish to *Edwardsiella ictaluri* and levels of mannose binding lectin. *Fish Shellfish Immunol 44*(1):21–25. https://doi.org/10.1016/j.fsi.2015.01.027

Planas M, Cunha I (1999) Larviculture of marine fish: Problems and perspectives. *Aquaculture 177*(1-4):171–190. https://doi.org/10.1016/S0044-8486(99)00079-4

Poolsawat L, Li X, Yang H, Yang P, Kabir Chowdhury MA, Yusuf A, Leng X (2020) The potentials of fructooligosaccharide on growth, feed utilization, immune and antioxidant parameters, microbial community and disease resistance of tilapia (*Oreochromis niloticus* × *O. aureus*). *Aquac Res 51*(11):4430–4442. https://doi.org/10.1111/are.14786

Poongodi R, Saravana Bhavan P, Muralisankar T, Radhakrishnan S (2012) Growth promoting potential of garlic, ginger, turmeric and fenugreek on the freshwater prawn *Macrobrachium rosenbergii*. *Int J Pharm Bio Sci 3*(4):914–926

Prabawati E, Hu S-Y, Chiu S-T, Balantyne R, Risjani Y, Liu C-H (2022) A synbiotic containing prebiotic prepared from a by-product of king oyster mushroom, *Pleurotus eryngii* and probiotic, *Lactobacillus plantarum* incorporated in diet to improve the growth performance and health status of white shrimp, *Litopenaeus vannamei*. *Fish Shellfish Immunol 120*:155–165. https://doi.org/10.1016/j.fsi.2021.11.031

Pulz O, Gross W (2004) Valuable products from biotechnology of microalgae. *Appl Microbiol Biotechnol 65*(6):635–648. https://doi.org/10.1007/s00253-004-1647-x

Puvanasundram P, Chong CM, Sabri S, Yusoff MS, Karim M (2021) Multi-strain probiotics: Functions, effectiveness and formulations for aquaculture applications. *Aquac Rep 21*:100905. https://doi.org/10.1016/j.aqrep.2021.100905

Rajasulochana P, Gummadi SN (2022) A probiotic based product using multi-strain Bacillus species and predictive models for shrimp growth following probiotic intervention. *Aquaculture 551*:737869. https://doi.org/10.1016/j.aquaculture.2021.737869

Rajauria G (2015) Chapter 15 - Seaweeds: a sustainable feed source for livestock and aquaculture. In: Tiwari BK, Troy DJ (eds) *Seaweed sustainability*. Academic Press, London, pp 389–420. https://doi.org/10.1016/B978-0-12-418697-2.00015-5

Regulation (EC) No 1831/2003 of the European Parliament and of the Council of 22 September 2003 on additives for use in animal nutrition. OJ L 268, 18.10.2003, pp 29–43

Regulation (EU) 2018/848 of the European Parliament and of the Council of 30 May 2018 on organic production and labelling of organic products and repealing Council Regulation (EC) No 834/2007. OJ L 150, 14.6.2018, pp 1–92

Ren G, Xu L, Lu T, Zhang Y, Wang Y, Yin J (2019) Protective effects of lentinan on lipopolysaccharide induced inflammatory response in intestine of juvenile taimen (*Hucho taimen*, Pallas). *Int J Biol Macromol 121*:317–325. https://doi.org/10.1016/j.ijbiomac.2018.09.121

Reverter M, Bontemps N, Lecchini D, Banaigs B, Sasal P (2014) Use of plant extracts in fish aquaculture as an alternative to chemotherapy: current status and future perspectives. *Aquaculture 433*:50–61. https://doi.org/10.1016/j.aquaculture.2014.05.048

Reyes-Becerril M, Guardiola F, Rojas M, Ascencio-Valle F, Esteban MÁ (2013) Dietary administration of microalgae *Navicula* sp. affects immune status and gene expression of gilthead seabream (*Sparus aurata*). *Fish Shellfish Immunol 35*(3):883–889. https://doi.org/10.1016/j.fsi.2013.06.026

Ringø E, Li X, Van Doan H, Ghosh K (2022) Interesting probiotic bacteria other than the more widely used lactic acid bacteria and Bacilli in finfish. *Front Mar Sci 9*. https://doi.org/10.3389/fmars.2022.848037

Rodríguez-Miranda J, Delgado E, Hernández-Santos B, Medrano Roldan H, Aguilar-Palazuelos E, Navarro-Cortez C, Gomez-Aldapa C, Castro-Rosas J (2012) Effect of sodium alginate on functional properties of extruded feed for fish for human consumption. *J Anim Sci Adv 2*:608–615

Roomiani L, Soltani M, Akhondzadeh Basti A, Mahmoodi A, Taheri A, Taheri Mirghaed A, Yadollahi F (2013) Evaluation of the chemical composition and in vitro antimicrobial activity of *Rosmarinus officinalis*, *Zataria multiflora*, *Anethum graveolens* and *Eucalyptus globulus* against *Streptococcus iniae* the cause of zoonotic disease in farmed fish. *Iran J Fish Sci 12*:702–716. https://aquadocs.org/handle/1834/11663

Sáez MI, Galafat A, Vizcaíno AJ, Chaves-Pozo E, Ayala MD, Arizcun M, Alarcón FJ, Suárez MD, Martínez TF (2022) Evaluation of *Nannochloropsis gaditana* raw and hydrolysed biomass at low inclusion level as dietary functional additive for gilthead seabream (*Sparus aurata*) juveniles. *Aquaculture 556*:738288. https://doi.org/10.1016/j.aquaculture.2022.738288

Safavi SV, Kenari AA, Tabarsa M, Esmaeili M (2019) Effect of sulfated polysaccharides extracted from marine macroalgae (*Ulva intestinalis* and *Gracilariopsis persica*) on growth performance, fatty acid profile, and immune response of rainbow trout (*Oncorhynchus mykiss*). *J Appl Phycol 31*(6):4021–4035. https://doi.org/10.1007/s10811-019-01902-w

Saravanan K, Sivaramakrishnan T, Praveenraj J, Kiruba-Sankar R, Haridas H, Kumar S, Varghese B (2021) Effects of single and multi-strain probiotics on the growth, hemato-immunological, enzymatic activity, gut morphology and disease resistance in Rohu. *Labeo rohita Aquaculture 540*:736749. https://doi.org/10.1016/j.aquaculture.2021.736749

Sartori SB, Ferreira LFR, Messias TG, Souza G, Pompeu GB, Monteiro RTR (2015) Pleurotus biomass production on vinasse and its potential use for aquaculture feed. *Mycology 6*(1):28–34. https://doi.org/10.1080/21501203.2014.988769

Sepúlveda-Quiroz CA, Peña-Marín ES, Pérez-Morales A, Martínez-García R, Alvarez-Villagomez CS, Maytorena-Verdugo CI, Camarillo-Coop S, Vissio PG, Pérez Sirkin D, Tovar-Ramírez D, Galaviz M, Álvarez-González CA (2021) Fructooligosaccharide supplementation in diets for tropical gar (*Atractosteus tropicus*) juvenile: effects on morphophysiology and intestinal barrier function. *Aquac Res 52*(1):37–50. https://doi.org/10.1111/are.14867

Serradell A, Torrecillas S, Makol A, Valdenegro V, Fernández-Montero A, Acosta F, Izquierdo MS, Montero D (2020) Prebiotics and phytogenics functional additives in low fish meal and fish oil based diets for European sea bass (*Dicentrarchus labrax*): effects on stress and immune responses. Fish Shellfish Immunol *100*:219–229. https://doi.org/10.1016/j.fsi.2020.03.016

Sewaka M, Trullas C, Chotiko A, Rodkhum C, Chansue N, Boonanuntanasarn S, Pirarat N (2019) Efficacy of synbiotic Jerusalem artichoke and *Lactobacillus rhamnosus* GG-supplemented diets on growth performance, serum biochemical parameters, intestinal morphology, immune parameters and protection against *Aeromonas veronii* in juvenile red tilapia (*Oreochromis* spp.). Fish Shellfish Immunol *86*:260–268. https://doi.org/10.1016/j.fsi.2018.11.026

Shah MR, Lutzu GA, Alam A, Sarker P, Kabir Chowdhury MA, Parsaeimehr A, Liang Y, Daroch M (2018) Microalgae in aquafeeds for a sustainable aquaculture industry. *J Appl Phycol 30*(1):197–213. https://doi.org/10.1007/s10811-017-1234-z

Shalaby AM, Khattab YA, Abdel Rahman AM (2006) Effects of Garlic (*Alliumsativum*) and chloramphenicol on growth performance, physiological parameters and survival of Nile tilapia

(*Oreochromis niloticus*). J Venom Anim Toxins Incl Trop Dis *12*(2):172–201. https://doi.org/10.1590/S1678-91992006000200003

Sony NM, Hossain MS, Ishikawa M, Koshio S, Yokoyama S (2020) Efficacy of mozuku fucoidan in alternative protein-based diet to improve growth, health performance, and stress resistance of juvenile red sea bream, *Pagrus major. Fish Physiol Biochem 46*(6):2437–2455. https://doi.org/10.1007/s10695-020-00881-x

Spicer A, Molnar A (2018) Gene editing of microalgae: scientific progress and regulatory challenges in Europe. *Biology 7*(1):21. https://doi.org/10.3390/biology7010021

Sprague M, Betancor MB, Tocher DR (2017) Microbial and genetically engineered oils as replacements for fish oil in aquaculture feeds. *Biotechnol Lett 39*(11):1599–1609. https://doi.org/10.1007/s10529-017-2402-6

Staniszewski A, Kordowska-Wiater M (2021) Probiotic and potentially probiotic yeasts—characteristics and food application. *Foods 10*(6):1306. https://doi.org/10.3390/foods10061306

Sutili FJ, Gatlin DM III, Heinzmann BM, Baldisserotto B (2018) Plant essential oils as fish diet additives: benefits on fish health and stability in feed. *Rev Aquac 10*(3):716–726. https://doi.org/10.1111/raq.12197

Sutriana A, Akter MN, Hashim R, Nor SAM (2021) Effectiveness of single and combined use of selected dietary probiotic and prebiotics on growth and intestinal conditions of striped catfish (*Pangasianodon hypophthalmus*, Sauvage, 1978) juvenile. *Aquac Int 29*(6):2769–2791. https://doi.org/10.1007/s10499-021-00777-4

Swanson KS, Gibson GR, Hutkins R, Reimer RA, Reid G, Verbeke K, Scott KP, Holscher HD, Azad MB, Delzenne NM, Sanders ME (2020) The International Scientific Association for Probiotics and Prebiotics (ISAPP) consensus statement on the definition and scope of synbiotics. *Nat Rev Gastroenterol Hepatol* 17(11) Art. 11. https://doi.org/10.1038/s41575-020-0344-2

Talpur AD, Ikhwanuddin M (2012) Dietary effects of garlic (*Allium sativum*) on haemato-immunological parameters, survival, growth, and disease resistance against *Vibrio harveyi* infection in Asian sea bass, *Lates calcarifer* (Bloch). *Aquaculture 364–365*:6–12. https://doi.org/10.1016/j.aquaculture.2012.07.035

Teles A, Alvarez-González CA, Llera-Herrera R, Gisbert E, Salas-Leiva J, del Carmen Rodríguez-Jaramillo M, Tovar-Ramírez D (2022) *Debaryomyces hansenii* CBS 8339 promotes larval development in *Seriola rivoliana. Aquaculture 560*:738587. https://doi.org/10.1016/j.aquaculture.2022.738587

Teuling E, Wierenga PA, Agboola JO, Gruppen H, Schrama JW (2019) Cell wall disruption increases bioavailability of *Nannochloropsis gaditana* nutrients for juvenile Nile tilapia (*Oreochromis niloticus*). *Aquaculture 499*:269–282. https://doi.org/10.1016/j.aquaculture.2018.09.047

Thiviya P, Gamage A, Gama-Arachchige NS, Merah O, Madhujith T (2022) Seaweeds as a source of functional proteins. *Phycology 2*(2) Art. 2. https://doi.org/10.3390/phycology2020012

Tibbetts SM (2018) The potential for 'next-generation', microalgae-based feed ingredients for salmonid aquaculture in context of the blue revolution. In: Jacob-Lopes E, Zepka LQ, Queiroz MI (eds) Microalgal biotechnology. InTech. https://doi.org/10.5772/intechopen.73551

Tongnuanchan P, Benjakul S (2014) Essential oils: extraction, bioactivities, and their uses for food preservation. *J Food Sci 79*(7):R1231–R1249. https://doi.org/10.1111/1750-3841.12492

Torrecillas S, Montero D, Izquierdo M (2014) Improved health and growth of fish fed mannan oligosaccharides: Potential mode of action. Fish Shellfish Immunol *36*(2):525–544. https://doi.org/10.1016/j.fsi.2013.12.029

Torrecillas S, Rivero-Ramírez F, Izquierdo MS, Caballero MJ, Makol A, Suarez-Bregua P, Fernández-Montero A, Rotllant J, Montero D (2018) Feeding European sea bass (*Dicentrarchus labrax*) juveniles with a functional synbiotic additive (mannan oligosaccharides and *Pediococcus acidilactici*): An effective tool to reduce low fishmeal and fish oil gut health effects? Fish Shellfish Immunol *81*:10–20. https://doi.org/10.1016/j.fsi.2018.07.007

Tovar D, Zambonino J, Cahu C, Gatesoupe FJ, Vázquez-Juárez R, Lésel R (2002) Effect of live yeast incorporation in compound diet on digestive enzyme activity in sea bass (*Dicentrarchus labrax*) larvae. *Aquaculture 204*(1):113–123. https://doi.org/10.1016/S0044-8486(01)00650-0

Tovar-Ramírez D, Mazurais D, Gatesoupe JF, Quazuguel P, Cahu CL, Zambonino-Infante JL (2010) Dietary probiotic live yeast modulates antioxidant enzyme activities and gene expression of sea bass (*Dicentrarchus labrax*) larvae. *Aquaculture 300*(1):142–147. https://doi.org/10.1016/j.aquaculture.2009.12.015

Truong Thy HT, Tri NN, Quy OM, Fotedar R, Kannika K, Unajak S, Areechon N (2017) Effects of the dietary supplementation of mixed probiotic spores of *Bacillus amyloliquefaciens* 54A, and *Bacillus pumilus* 47B on growth, innate immunity and stress responses of striped catfish (*Pangasianodon hypophthalmus*). Fish Shellfish Immunol *60*:391–399. https://doi.org/10.1016/j.fsi.2016.11.016

Tukmechi A, Rahmati Andani HR, Manaffar R, Sheikhzadeh N (2011) Dietary administration of beta-mercapto-ethanol treated *Saccharomyces cerevisiae* enhanced the growth, innate immune response and disease resistance of the rainbow trout, *Oncorhynchus mykiss*. Fish Shellfish Immunol *30*(3):923–928. https://doi.org/10.1016/j.fsi.2011.01.016

Van Doan H, Hoseinifar SH, Tapingkae W, Chitmanat C, Mekchay S (2017) Effects of *Cordyceps militaris* spent mushroom substrate on mucosal and serum immune parameters, disease resistance and growth performance of Nile tilapia, (*Oreochromis niloticus*). Fish Shellfish Immunol *67*:78–85. https://doi.org/10.1016/j.fsi.2017.05.062

Van Doan H, Hoseinifar SH, Esteban MÁ, Dadar M, Thu TTN (2019) Chapter 2 - Mushrooms, seaweed, and their derivatives as functional feed additives for aquaculture: an updated view. In: Atta-ur-Rahman (ed) *Studies in natural products chemistry*, vol 62. Elsevier, Amsterdam, pp 41–90. https://doi.org/10.1016/B978-0-444-64185-4.00002-2

Van Doan H, Soltani M, Ringø E (2021) In vitro antagonistic effect and in vivo protective efficacy of Gram-positive probiotics versus Gram-negative bacterial pathogens in finfish and shellfish. *Aquaculture 540*:736581. https://doi.org/10.1016/j.aquaculture.2021.736581

Vargas-Albores F, Martínez-Córdova LR, Hernández-Mendoza A, Cicala F, Lago-Lestón A, Martínez-Porchas M (2021) Therapeutic modulation of fish gut microbiota, a feasible strategy for aquaculture? *Aquaculture 544*:737050. https://doi.org/10.1016/j.aquaculture.2021.737050

Vázquez-Salgado L, Olveira JG, Dopazo CP, Bandín I (2020) Role of rotifer (*Brachionus plicatilis*) and *Artemia* (*Artemia salina*) nauplii in the horizontal transmission of a natural nervous necrosis virus (NNV) reassortant strain to Senegalese sole (*Solea senegalensis*) larvae. *Vet Q 40*(1):205–214. https://doi.org/10.1080/01652176.2020.1810357

Villar-Navarro E, Ruiz J, Garrido-Pérez C, Perales JA (2022) Microalgae biotechnology for simultaneous water treatment and feed ingredient production in aquaculture. *J Water Process Eng 49*:103115. https://doi.org/10.1016/j.jwpe.2022.103115

Wan AHL, Davies SJ, Soler-Vila A, Fitzgerald R, Johnson MP (2019) Macroalgae as a sustainable aquafeed ingredient. *Rev Aquac 11*(3):458–492. https://doi.org/10.1111/raq.12241

Wang Y-C, Hu S-Y, Chiu C-S, Liu C-H (2019) Multiple-strain probiotics appear to be more effective in improving the growth performance and health status of white shrimp, *Litopenaeus vannamei*, than single probiotic strains. Fish Shellfish Immunol *84*:1050–1058. https://doi.org/10.1016/j.fsi.2018.11.017

Wee W, Abdul Hamid NK, Mat K, Khalif RIAR, Rusli ND, Rahman MM, Kabir MA, Wei LS (2022) The effects of mixed prebiotics in aquaculture: a review. *Aquac Fish*. https://doi.org/10.1016/j.aaf.2022.02.005

Wold CW, Gerwick WH, Wangensteen H, Inngjerdingen KT (2020) Bioactive triterpenoids and water-soluble melanin from *Inonotus obliquus* (Chaga) with immunomodulatory activity. *J Funct Foods 71*:104025. https://doi.org/10.1016/j.jff.2020.104025

Yao W, Li X, Zhang C, Wang J, Cai Y, Leng X (2021) Effects of dietary synbiotics supplementation methods on growth, intestinal health, non-specific immunity and disease resistance of Pacific white shrimp, *Litopenaeus vannamei*. Fish Shellfish Immunol *112*:46–55. https://doi.org/10.1016/j.fsi.2021.02.011

Yilmaz E, Ergün S, Ilmaz S (2015) Influence of carvacrol on the growth performance, hematological, non-specific immune and serum biochemistry parameters in rainbow trout (*Oncorhynchus mykiss*). *Food Nutr Sci 06*(05):523–531. https://doi.org/10.4236/fns.2015.65054

Yilmaz S, Yilmaz E, Dawood MAO, Ringø E, Ahmadifar E, Abdel-Latif HMR (2022) Probiotics, prebiotics, and synbiotics used to control vibriosis in fish: a review. *Aquaculture 547*:737514. https://doi.org/10.1016/j.aquaculture.2021.737514

Ytrestøyl T, Aas TS, Åsgård T (2015) Utilisation of feed resources in production of Atlantic salmon (*Salmo salar*) in Norway. *Aquaculture 448*:365–374. https://doi.org/10.1016/j.aquaculture.2015.06.023

Yun L, Wang W, Li Y, Xie M, Chen T, Hu C, Luo P, Li D (2021) Potential application values of a marine red yeast, *Rhodosporidiums sphaerocarpum* YLY01, in aquaculture and tail water treatment assessed by the removal of ammonia nitrogen, the inhibition to *Vibrio* spp., and nutrient composition. PLoS One *16*(2):e0246841. https://doi.org/10.1371/journal.pone.0246841

Zarantoniello M, Pulido Rodriguez LF, Randazzo B, Cardinaletti G, Giorgini E, Belloni A, Secci G, Faccenda F, Pulcini D, Parisi G, Capoccioni F, Tibaldi E, Olivotto I (2022) Conventional feed additives or red claw crayfish meal and dried microbial biomass as feed supplement in fish meal-free diets for rainbow trout (*Oncorhynchus mykiss*): possible ameliorative effects on growth and gut health status. *Aquaculture 554*:738137. https://doi.org/10.1016/j.aquaculture.2022.738137

Zhang Q, Qiu M, Xu W, Gao Z, Shao R, Qi Z (2014) Effects of dietary administration of Chlorella on the immune status of Gibel Carp, *Carassius Auratus Gibelio*. *Ital J Anim Sci 13*(3):3168. https://doi.org/10.4081/ijas.2014.3168

Zheng ZL, Tan JYW, Liu HY, Zhou XH, Xiang X, Wang KY (2009) Evaluation of oregano essential oil (*Origanum heracleoticum* L.) on growth, antioxidant effect and resistance against *Aeromonas hydrophila* in channel catfish (*Ictalurus punctatus*). *Aquaculture 292*(3-4):214–218. https://doi.org/10.1016/j.aquaculture.2009.04.025

Zokaeifar H, Balcázar JL, Saad CR, Kamarudin MS, Sijam K, Arshad A, Nejat N (2012) Effects of *Bacillus subtilis* on the growth performance, digestive enzymes, immune gene expression and disease resistance of white shrimp, *Litopenaeus vannamei*. Fish Shellfish Immunol *33*(4):683–689. https://doi.org/10.1016/j.fsi.2012.05.027

Certain Major Diseases Having Nutritional Disorders in their Etiology and Economic Importance in Horses

C. Ayvazoğlu and P. A. Demir

Abstract

Nutrition is the foundation of health and well-being that goes hand in hand with horse breeding. Improper nutrition plays an important role in the development of various diseases in horses. Changes in feed type and quality can vary greatly and predispose horses to a range of problems. The rations to be applied to horses in breeding enterprises are the main element of animal welfare and business profitability. This chapter aims to contribute positively to the existing literature by evaluating some important diseases (equine metabolic syndrome, laminitis, paralytic myoglobinuria, and colic) in horses with malnutrition in their etiology.

Keywords

Colic · Equine · Equine metabolic syndrome · Laminitis · Nutritional disorders

1 Introduction

Horses were used in agriculture and transportation services and played important roles in wars. Having provided great services to humanity throughout history, horses have lost their former importance with the increase in mechanization over time. Today, horses are raised for sports and racing purposes in parallel with the economic power of the society (Çelik et al. 2015).

C. Ayvazoğlu (✉)
Nihat Delibalta Göle Vocational High School, Ardahan University, Ardahan, Turkey

P. A. Demir
Faculty of Veterinary Medicine, Department of Animal Health Economics and Management, Kırıkkale University, Kars, Turkey

G. Arsenos, I. Giannenas (eds.), *Sustainable Use of Feed Additives in Livestock*,
https://doi.org/10.1007/978-3-031-42855-5_29

Nutrition is the foundation of health and welfare, which goes hand in hand with horse breeding. Without proper nutrition, all of the horse's body systems are compromised, thus negatively affecting the horse's health and welfare. In this case, speed, endurance, growth, fertility, and body shape, which are very important for horses, cannot be achieved. In order to raise a healthy horse, horses should be fed with adequate and balanced rations in terms of essential amino acids, fats, vitamins, and minerals in accordance with their development periods from fertilized eggs until they reach adult weight (Noble 2023).

Improper nutrition in horses plays an important role in the development of various diseases. Changes in feed type and quality can vary greatly and predispose horses to a range of problems. Modern nutritional practices applied in horse breeding today are often associated with problems of the gastrointestinal tract. For example, sudden changes in diet have been identified as the most important risk factor for the development of abdominal pain (colic) in horses (Archer and Proudman 2006). The risk appears to increase up to 14 days after a dietary change (Cohen et al. 1999). Furthermore, continuous access to pasture is considered protective and there is a lower incidence of colic in this population than in horses kept predominantly in stables (Hudson et al. 2001). The sudden and typically incidental consumption of large amounts of starch, usually grain, has long been identified as a cause of colic, abdominal distension, diarrhea, toxemia, and laminitis. Indeed, the presence of diseases caused by feeding errors can cause significant economic losses. In addition to mortality and yield losses, high treatment costs, lost time, and labor bring high costs to the business (Köseman and Şeker 2016).

The rations to be applied to horses in breeding enterprises are the basic element of animal welfare and business profitability. In this section, it is aimed to contribute positively to the existing literature knowledge by evaluating some important diseases (equine metabolic syndrome, laminitis, paralytic myoglobinuria, and colic) in horses with malnutrition in their etiology.

1.1 Equine Metabolic Syndrome

Equine metabolic syndrome refers to a constellation of clinical abnormalities associated with an increased risk of laminitis (Frank et al. 2010). Although the presence of laminitis has historically or currently been recognized as a characteristic feature of equine metabolic syndrome (EMS) (Treiber et al. 2006), our understanding of the pathophysiology of EMS and the risk factors associated with laminitis is still in its infancy (McCue et al. 2015). It has also been reported that metabolic syndrome (MetS) in humans is similar to EMS and that obesity and insulin resistance are factors shared by both syndromes (Johnson et al. 2007).

The diagnosis of EMS requires physical examination and laboratory analysis such as body condition score (BCS) or crest of neck measurement in horses between 5–15 years of age. Radiographic evaluation of the feet is also important for diagnosis (McCue et al. 2015). The European College of Equine Internal Medicine (ECEIM) reported that ungulates affected with EMS should include symptoms of regional

Fig. 1 Cresty neck scores (Carter et al. 2009)

adiposity or obesity, insulin resistance, and susceptibility to laminitis (Durham et al. 2019).

Laboratory diagnosis of EMS is based on measurement of blood insulin and glucose concentrations after a 6-hour fast (Frank 2009). Horses with EMS tend to have a resting blood glucose concentration that is usually at the upper end of the normal range rather than hyperglycemic (Secombe and Lester 2012). A serum insulin concentration greater than 20 μU/mL is consistent with insulin resistance (Frank et al. 2010). Conditions such as stress and pain can affect test specificity. Laboratory tests should only be performed when the animal is acclimatized to the environment and free of other diseases such as lameness. It has also been reported that the sensitivity of laboratory tests is affected in the early stages of the disease or when pancreatic insufficiency occurs with type 2 diabetes mellitus (Frank 2009; Frank et al. 2010).

Adipose tissue is hormonally active and produces adipokines and adipocytokines (Rasouli and Kern 2008). Studies have identified more than 100 different adipokines (Hutley and Prins 2005; Secombe and Lester 2012). Inappropriate secretion of adipokines leads to the pathophysiological consequences of obesity (Hutley and Prins 2005). Specific adipokines involved in the EMS include leptin, adiponectin, and resistin. Adipocytokines released by adipose tissue or from macrophages within fat are proinflammatory and lead to a state of chronic, low-level inflammation (Rasouli and Kern 2008). A cresty neck scoring system (cresty neck score) has been developed to help differentiate horses that develop regional obesity rather than generalized obesity. The scoring is between 0–5, with scores of 3 and above frequently reported in horses with EMS. Accordingly, 0: No palpable crest; 5: Extremely large crest area (Carter et al. 2009; Fig. 1).

The most common clinical signs in EMS are laminitis and obesity. Studies have reported that physiologic hyperinsulinemia can cause laminitis in horses (Asplin et al. 2007). The effect of high insulin levels on the sensitive laminated tissues of the hoof has not yet been fully explained (Frank et al. 2010; Secombe and Lester 2012).

The mainstay of EMS treatment is ration modification and increased physical activity. Pharmacological intervention can be used to enhance the response to ration changes, but is unlikely to result in a successful response when used alone. Horses should be fed low amounts of non-structural carbohydrates to reduce the insulinemic response to meals.

Weight loss is essential for successful treatment of EMS, but horses should not be completely starved. Starvation can worsen insulin resistance or cause pathological mobilization of fat stores (Secombe and Lester 2012). Horses should be fed hay at a rate of 1.5% of body weight daily for the first 30 days of treatment and then this rate should be reduced to 1% daily. In addition, horses exhibiting insulin resistance and regional adiposity rather than obesity are more difficult to treat (Frank et al. 2010).

Exercise has been reported to be effective in improving insulin sensitivity in people with insulin resistance (Goodyear and Kahn 1998; Crandall et al. 2008). Similar results were reported in horses with insulin resistance (Pratt et al. 2006). Although the amount of exercise required to improve insulin sensitivity in humans has been determined, this is not the case in horses (Bajpeyi et al. 2009). However, Frank et al. (2010) reported that 2–3 exercises of 20–30 minutes per week are beneficial in horses without laminitis.

Pharmacological treatment of EMS includes levothyroxine sodium, insulin sensitizing drugs such as metformin, or mineral supplements such as magnesium/ chromium. In one study, levothyroxine was found to cause weight loss and improve insulin sensitivity in horses in the short term (Frank et al. 2005).

The earlier the EMS disease is diagnosed, the lower the economic loss can be. However, in interviews with veterinary specialists, it was reported that the total cost of diagnosis and treatment in Turkey is approximately $123.56/horse (min:83-max:178), including an average of $48.80/horse for the diagnosis of the disease and $65.25/horse for hoof care and medication costs per horse in cases of EMS laminitis. However, it can be said that this cost is even higher when additional maintenance costs and economic loss due to poor performance are taken into account.

1.2 Laminitis (Pododermatitis aseptica diffusa-Furbür)

Laminitis is a musculoskeletal disease that causes inflammation, degeneration, and necrosis of the dermal and epidermal laminae of the nail wall (Hunt 1993; Linford 2002). Laminitis is a local manifestation of a general metabolic disorder, such as digestive disorders and diseases causing endotoxemia, and often occurs as a reflection of diseases that produce gram-negative bacteria (Linford 2002; Stashak 2002). If preventive measures are not taken, laminitis is often seen after cases such as colonic torsion, enteritis, pleurapneumonia, metritis and excessive grain feed consumption (Hunt 1993; Linford 2002). In addition, past hoof trauma, irregular overweight, pasture grazing, excessive grain feeding, housing/working on hard surfaces and long walks predispose horses to laminitis (de Laat et al. 2019). Furthermore, infectious diseases such as pneumonia, metritis, retention of plecenta, kidney diseases, castration, Cushing's (hyperadrenocortism), hypothyroidism, intensive use of glucocorticoid type drugs and irregularities in the immune system can also cause laminitis (Stashak 2002). In addition, it has been reported that horses that have had laminitis are more likely to get laminitis again than other horses (Linford 2002; Stashak 2002; Johnson et al. 2004).

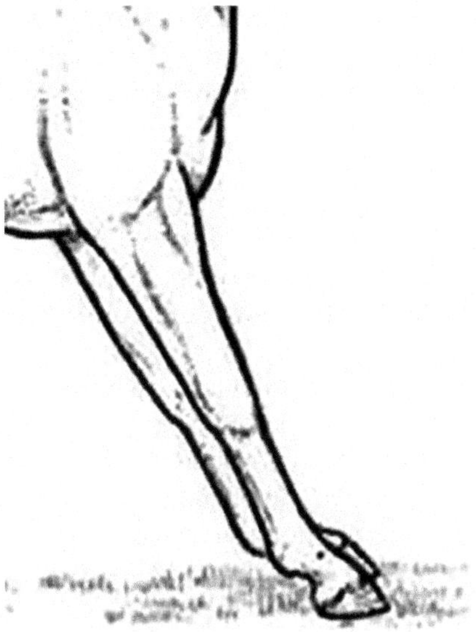

Fig. 2 Position of the forefoot in laminitis

Fever in the hoof, increased pulse rate in the foot arteries, pain in the feet, reluctance to move and deterioration in the general condition of the animal help in the diagnosis of the disease. Although laminitis can be confused with tetanus, azoturia, gastric/bladder rupture, and colic, it is differentiated from these diseases by local pain on palpation of the feet (Linford 2002; Stashak 2002). Clinically, the pulsation of the arteries in the leg is high. Serum infiltration occurs between the nail and the soft layer. In the later stages, lamina enlargement and separation may be observed (Auer 2006). In general, the anterior legs are more severely affected than the posterior legs (Fig. 2). In cases where all hooves are affected, the horse is reluctant to walk and tends to lie down (Auer 2006; Sundra et al. 2022). It can be acute or chronic depending on the diet (de Laat et al. 2019).

The important thing in the treatment is to regulate the ration and then medical treatment, soft litter and hoof treatment should be applied (Auer 2006; Parsons et al. 2007). Ration energy and carbohydrate levels should be adjusted according to the horse's age and lifestyle. Hoof treatment should be corrected by cutting and filing according to the anatomical shape and appropriate shoeing should be done. The aim here is to reduce the pressure on the damaged lamellae by minimizing the tensile forces that cause rotation of the distal phalanx (Harris et al. 2006; Auer 2006).

Laminitis is one of the leading diseases that cause significant economic loss as well as pain and welfare loss in horses. In a study conducted to determine the economic loss caused by laminitis, it was reported that the total loss ranged between \$678 million and \$1 billion and that 66% of this value was due to loss of use of horses, 29% to veterinary services, medication, and additional care costs, and 5% to deaths of horses (Seitzinger et al. 2000).

In a study conducted in the USA, it was reported that 13% of equine facilities had one or more horses affected by laminitis annually (Kane et al. 2000). Eades et al.

(2002) reported that the cost of diagnosis and treatment for laminitis is approximately $eight million per year and if 75% of horses affected by this disease are euthanized, the loss of these horses is $five million per year.

1.3 Paralytic Myoglobinuria/Azoturia

Paralytic myoglobinuria (PM) in horses is a metabolic muscle disease observed in horses of all age groups that are suddenly exercised after feeding with feeds high in carbohydrates, ration, and rest (Baran and Alkan 2014). PM is also known as azoturia and Monday's disease because it is expressed in different ways physiopathologically. This is due to the fact that the disease cannot be explained by a single physiopathological event (Demirtaş et al. 2019).

Excessive carbohydrate consumption, local hypoxia, metabolic changes, reproduction, electrolyte imbalance, vitamin E and selenium deficiency are the main etiological causes of the disease (Bilal 2003). PM causes destruction especially in type II muscle fibers (Başoğlu and Sevinç 2004). In muscles, glucose is stored as glycogen and glycogen is first aerobically and then anaerobically broken down as a result of exercise. Pyruvates released as a result of anaerobic breakdown of glucose cannot enter the aerobic breakdown cycle due to lack of oxygen and are reduced to pyruvic acid, causing excessive production of lactic acid. Lactic acid causes necrosis in muscle cells and myoglobinemia occurs when myoglobin enters the blood (Baran and Alkan 2014). Myoglobinuria also occurs when myoglobin is excreted from the body through urine. Increased creatine kinase (CK) and aspartate aminotransferase (AST) enzyme levels have been reported in clinically healthy horses after racing and exercise (Nollet and Deprez 2005). It has been reported that muscle injuries commonly occur following repetitive exercise and subclinical muscle injuries are the primary underlying cause of the disease (Bilal 2003).

Moderate PM can be confused with laminitis. However, the typical animal does not want to move even if it is forced or takes small steps. Sick animals are completely immobilized and lie down (Demirtaş et al. 2019). Pain and stiffness are felt on palpation of the affected muscles (Harris and Rivero 2017). It shows visible clinical signs such as intense sweating, stiffness in muscles, increased respiration and pulse rate, and pain (EL-Deeb and El-Bahr 2014). Urine is dark brown in color and is characteristic for this disease. The muscle groups most affected by PM are those with large volumes such as m. semitendinosus and m. semimembranaceus. Although clinical symptoms vary according to the intensity of exercise, they are much more severe in the acute period. The occurrence of musculoskeletal disorders as a result of intense exercise is considered important anamnestic information. A match between anamnesis and clinical symptoms is sufficient to suspect the disease (Demirtaş et al. 2019).

Treatment of paralytic myoglobinuria may vary according to the degree of clinical signs (Başoğlu and Sevinç 2004). The first task is to rest the horse and the animals should not be allowed to work or walk. Sodium bicarbonate should be given I.V. and orally as soon as possible to eliminate the acidosis that occurs in horses with

PM. Vitamins B1, E, selenium, and calcium gluconate can be given. In addition, the animal should be relieved with cortisone or tranquilizer injections. Dimethylglycine (DMG), Methylsulfonylmethane (MSM), Acepromazine can be used to prevent paralytic myoglobinuria (Bilal 2003). In addition, reducing the carbohydrate ratio in the ration is an important prophylactic measure in horses with PM. Feeding rations with high protein levels instead of a carbohydrate-rich ration reduces the risk of developing PM disease. This is because muscle atrophy and cachexia are caused by decreased utilization of amino acids, which are necessary for muscle energy. For this reason, high protein level helps to maintain and recycle muscle volume in horses with PM (Bilal 2003; Başoğlu and Sevinç 2004).

Paralytic myoglobinuria is a metabolic disease caused by malnutrition and the economic loss varies according to the severity of the disease. In line with the interviews with specialized veterinarians, it was reported that the average cost of veterinary services for the diagnosis and treatment of the disease in Turkey is $38.77/horse, and the cost of serum, vitamin, and mineral injections and cortisone treatment is $46.45/horse, for a total of $85.22/horse. However, other costs related to the disease (additional maintenance costs and loss of performance) were not taken into account.

1.4 Colic

Colic is a common problem in equine clinics (Onmaz and Van Den Hoven 2012) and its incidence in a normal population ranges from 10% to 36% each year (Uhlinger 1990; Tinker et al. 1997). In another study, it was reported that 10 out of every 100 horses followed up within a year showed signs of colic (Tinker et al. 1997). Equine colic is a term used to describe pain of abdominal origin (Abutarbush et al. 2005; Curtis et al. 2019; Dybkjær et al. 2022) and is reported as the leading cause of death in horses (Egenvall et al. 2006; Scantlebury et al. 2011).

Colic in horses can be categorized as gastrointestinal and non-gastrointestinal (Hillyer et al. 2002; Onmaz and Van Den Hoven 2012). Non-gastrointestinal pain cases usually originate from the kidney, biliary tract, uterus, nervous system, and urinary system (Hillyer et al. 2002; Kızıl 2007). Gastrointestinal pain cases originate from gastrointestinal problems and are characterized by gastrointestinal ulcer, peritoneal pain, ischemia, and mesenteric ischemia (Onmaz and Van Den Hoven 2012).

The symptoms vary according to the cause and severity of the pain. Very few horses exhibit all symptoms at once. Although there are many different causes of pain, it is stated that the real cause is motility disorder in the gastrointestinal system (Onmaz and Van Den Hoven 2012). Horses are sensitive to pain due to the anatomical structure of the gastrointestinal system (Kızıl 2007). Poor care conditions, abdominal operations, climatic conditions, sudden environmental changes, malnutrition, excessive physical activity, bacterial, viral, or parasitic infections create predisposition for labor (Onmaz and Van Den Hoven 2012). In addition, it has been reported that horses younger than 2 years of age and older than

10 years of age have a low risk of developing labor, but this risk is higher in middle-aged horses (Tinker et al. 1997).

It is stated that anamnesis, physical examinations, rectal examination, abdominocentesis (punctate examination), abdominal chirurgy, blood examinations, biochemical analyses, ultrasonography-radiography, and fecal examination findings should be taken into consideration in order to make a definitive diagnosis and effective treatment in a horse with pain (Furr et al. 1995). In mild pain, animals are observed to dig the ground with their forelimbs, sweating, restlessness, frequently looking at the abdomen and keeping their hind legs tense in the lying position (Singer and Smith 2002; Onmaz and Van Den Hoven 2012). In cases of severe pain, the clinical picture is characterized by abdominal distension, muscle tremors or straining, severe sweating, dog sitting, hypothermia or hyperthermia, increased heart and respiratory frequency, decreased or absent bowel sounds (Reeves et al. 1996; Singer and Smith 2002).

The treatment of pain is done in two categories: medical and surgical treatment. Although it has been reported that as high as 93% of labor pain cases can be managed without the need for surgical intervention (Proudman 1992), it is reported that medication or surgery should be decided as early as possible (Van der Linden et al. 2003). If the pain is caused by intestinal compression, parasites, or inflammation, medical treatment is usually applied, while surgical intervention is required in cases of obstruction or torsion. Some pain conditions that cannot be corrected with medical treatment require surgical intervention. Surgical interventions are sometimes applied to remove gastrointestinal or secum contents and gas in the stomach and intestines, and sometimes in cases of intestinal obstruction and displacement (Colic 2008; Onmaz and Van Den Hoven 2012).

In the treatment of pain, pain relief is the first priority. Metamizole 50% (Novalgin®) can be used to relieve pain (Huskamp et al. 1999). Wintzer (1999) reported that Atropine can be administered at a dose of 0.02–0.1 g S.C. in intestinal transit disorders (Obstipatio caeci). Besides, Keller (1978) reported that butylscopolamine-metamizole combination (Busc. comp.®) should be used at a dose of 20–25 ml/500 kg body weight (VA) in all cases of labor. Among the non-steroidal anti-inflammatory drugs (NSAIDs), flunixin-meglumine (Finadyne®) is one of the most commonly used drugs in labor (Johnson 2003). In addition to these parenteral applications in the treatment of labor, laxative salts such as sodium sulfate (50 g sodium sulfate/1 L water) and mineral oils such as paraffin liquid (2–3 L) can be given orally through a nasal catheter. It is also important to administer 40–80 L of solutions such as Ringerlactate or Ringeracetate every 24 hours as supportive fluid therapy in cases of obstruction (Wintzer 1999).

Colic, which is one of the most common diseases in horse breeding, is one of the most common diseases that cause both horse loss and economic loss among equine diseases. In a study conducted in Turkey, the average incidence rate of colic cases was reported as 16.36%. In the study, it was determined that the average veterinarian and medication cost of colic cases for a horse was $215.60/horse, while the euthanasia application rate in colicky animals was 11.80% and the euthanasia cost was approximately $212/horse per animal.

In the interviews with veterinarians, it was also stated that surgical intervention is generally not preferred by horse owners due to the low recovery rate in the treatment of colicky horses, the occurrence of significant complications after the operation and the high cost of the operation (min: $ 620; max: $ 4500). As a matter of fact, it was reported that the rate of recurrence of colic within one year after treatment was 48.2% and the rate of development of laminitis after colic was 11.9%. For this reason, depending on the severity of colic horses, the value of horses decreases due to the difficulty of returning to its former state after treatment and/or operation. In this context, it can be said that the loss will increase even more when the loss of performance value is taken into account when calculating the economic loss due to colic.

In another study, it was reported that the cost of veterinary services, medication, and additional care in colic cases was $35 million (30%), the loss due to death was $115 million (66%), and the loss of utilization was $four million (4%) (Seitzinger et al. 2000).

As a result, due to the high economic return in the equine sector and the high economic value of horses used in every field of equestrian sports, it is necessary to establish a mechanism for disease prevention and control before the disease occurs. Because in addition to the productivity and performance losses that may occur due to a disease that may occur in horses used in races, high treatment costs, time and labor costs, and deaths can cause high costs to the business and the sector. In this context, feeding horses with appropriate rations under the control of a veterinarian is one of the most important measures against diseases.

References

Abutarbush SM, Carmalt JL, Shoemaker RW (2005) Causes of gastrointestinal colic in horses in western Canada: 604 cases (1992 to 2002). Can Vet J 46:800–805

Archer DC, Proudman CJ (2006) Epidemiological clues to preventing colic. Vet J 172(1):29–39

Asplin KE, Sillence MN, Pollitt CC et al (2007) Induction of laminitis by prolonged hyperinsulinaemia in clinically normal ponies. Vet J 174(3):530–535

Auer JA (2006) Musculoskeletal system. In: Fathman L (ed) Equine surgery, 3rd edn. Missouri, Saunders Elsevier, pp 1213–1217

Bajpeyi S, Tanner CJ, Slentz CA et al (2009) Effect of exercise intensity and volume on persistence of insulin sensitivity during training cessation. J Appl Physiol 106(4):1079–1085

Baran MS, Alkan O (2014) Tayların beslenmesi ve beslenme hastalıkları. Uludağ Univ Vet Fak Derg 33(1–2):63–69

Başoğlu A, Sevinç M (2004) Evcil hayvanlarda metabolik ve endokrin hastalıklar. Pozitif matbaacılık, Konya, pp 213–224

Bilal T (2003) Tek tırnaklıların iç hastalıkları ve beslenmesi. İstanbul Üniversitesi basım ve yayınevi müdürlüğü, İstanbul, pp 473–481

Carter RA, Geor RJ, Staniar WB et al (2009) Apparent adiposity assessed by standardised scoring systems and morphometric measurements in horses and ponies. Vet J 179(2):204–210

Çelik Ş, Coşkun F, Yılmaz O (2015) Türk Alaca Atlarda Yaş Grubuna Göre Vücut Ölçülerinin Farklı Ortogonal Karşılaştırma Yöntemleriyle İncelenmesi. ÇOMÜ Ziraat Fak derg 3(1):81–87

Cohen N, Gibbs P, Woods A (1999) Dietary and other management factors associated with equine colic. J Am Vet Med Assoc 45:96–98

Colic A (2008) An age-old problem. CEH Horse Report 26(1):2–11
Crandall JP, Knowler WC, Kahn SE et al (2008) The prevention of type 2 diabetes. Nat Clin Pract Endocrinol Metab 4(7):382–393
Curtis L, Burford JH, England GC et al (2019) Risk factors for acute abdominal pain (colic) in the adult horse: a scoping review of risk factors, and a systematic review of the effect of management-related changes. PLoS One 14(7):e0219307
de Laat MA, Reiche DB, Sillence MN et al (2019) Incidence and risk factors for recurrence of endocrinopathic laminitis in horses. J Vet Intern Med 33(3):1473–1482
Demirtaş R, Şahinduran Ş, Akar A (2019) Atlarda Paralitik Myoglobinüri/Azotüri. International Young Researchers Student Congress 28–30 November 2019 Burdur, Turkey, p 101–106
Durham AE, Frank N, McGowan CM et al (2019) ECEIM consensus statement on equine metabolic syndrome. J Vet Intern Med 33(2):335–349
Dybkjær E, Steffensen KF, Honoré ML et al (2022) Short-term survival rates of 1,397 horses referred for colic from 2010 to 2018. Acta Vet Scand 64(1):1–9
Eades SC, Holm EMS, Moore RM (2002) A review of the pathophysiology and treatment of acute laminitis: pathophysiologic and therapeutic implications of endothelin-1. AAEP Proceedings 48:353–361
Egenvall A, Penell JC, Bonnett BN et al (2006) Mortality of Swedish horses with complete life insurance between 1997 and 2000: variations with sex, age, breed and diagnosis. Vet Rec 158(12):397–406. https://doi.org/10.1136/vr.158.12.397. PMID: 16565338
EL-Deeb WM, El-Bahr SM (2014) Selected biochemical indicators of equine rhabdomyolysis in Arabian horses: acute phase proteins and trace elements. J Equine Vet Sci 34(4):484–488
Frank N (2009) Equine metabolic syndrome. J Equine Vet Sci 29(5):259–267
Frank N, Sommardahl CS, Eiler H et al (2005) Effects of oral administration of levothyroxine sodium on concentrations of plasma lipids, concentration and composition of very-low-density lipoproteins, and glucose dynamics in healthy adult mares. Am J Vet Res 66(6):1032–1038
Frank N, Geor RJ, Bailey SR et al (2010) Equine metabolic syndrome. J Vet Intern Med 24(3): 467–475
Furr MO, Lessard P, Iı NAW (1995) Development of a colic severity score for predicting the outcome of equine colic. Vet Surg 24(2):97–101
Goodyear LJ, Kahn BB (1998) Exercise, glucose transport, and insulin sensitivity. Annu Rev Med 49:235
Harris PA, Rivero JLL (2017) Nutritional considerations for equine rhabdomyolysis syndrome. Equine Vet Educ 29(8):459–465
Harris P, Bailey SR, Elliott J et al (2006) Countermeasures for pasture-associated laminitis in ponies and horses. J Nutr 136(7):2114–2121
Hillyer MH, Taylor FG, Proudman CJ et al (2002) Case control study to identify risk factors for simple colonic obstruction and distension colic in horses. Equine Vet J 34(5):455–463. https://doi.org/10.2746/042516402776117746
Hudson JM, Cohen ND, Gibbs PG et al (2001) Feeding practices associated with colic in horses. J Am Vet Med Assoc 219(10):1419–1425
Hunt RJ (1993) A retrospective evaluation of laminitis in horses. Equine Vet J 25(1):61–64
Huskamp B, Kopf N, Scheidemann W et al (1999) Magen-Darm-Trakt. Handbuch Pferde Praxis. Ferdinand Enke Verlag, Stuttgart, pp 411–502
Hutley L, Prins JB (2005) Fat as an endocrine organ: relationship to the metabolic syndrome. Am J Med Sci 330(6):280–289
Johnson A (2003) Ergebnisse von 1431 stationär behandelten Kolikpatienten, unter besonderer Berücksichtigung der 285 chirurgisch versorgten Pferde, in den Jahren 1990 bis 1997 in der Klinik für Pferde, allgemeine Chirurgie und Radiologie der Freien Universität Berlin, Doctoral dissertation
Johnson PJ, Messer NT, Slight SH et al (2004) Endocrinopathic laminitis in the horse. Clin Tech Equine Pract 3(1):45–56

Johnson JL, Slentz CA, Houmard JA et al (2007) Exercise training amount and intensity effects on metabolic syndrome (from studies of a targeted risk reduction intervention through defined exercise). Am J Cardiol 100(12):1759–1766

Kane AJ, Traub-Dargatz J, Losinger WC et al (2000) The occurrence and causes of lameness and laminitis in the US horse population. Proc Am Assoc Equine Pract 46:277–280

Keller H (1978) Diagnose, Therapie und Prognose bei der konservativen Behandlung der Kolik des. Pferdes, Tierarztliche Umschau

Kızıl Ö (2007) Atlarda sancı ve sancılı atlarda muayene parametreleri. FU Vet Fak Derg 21(6): 285–290

Köseman A, Şeker İ (2016) Atçılık İşletmelerinde Biyogüvenlik ve Önemi. Bahri Dağdaş Hay Araş Derg 5(1):33–39

Linford RL (2002) Laminitis (Founder). In: Stover SM, Bradford PS (eds) Large animal internal medicine, 3rd edn. Mosby, Missouri, pp 1116–1124

McCue ME, Geor GJ, Schultz N (2015) Equine metabolic syndrome: a complex disease influenced by genetics and the environment. J Equine Vet Sci 35(5):367–375

Noble GK (2023) Horse husbandry–nutrition, management and welfare. Anim 13(1):169

Nollet H, Deprez P (2005) Hereditary skeletal muscle diseases in the horse: a review. Vet Q 27(2): 65–75

Onmaz A, Van Den Hoven R (2012) Atlarda Sancı: Cerrahi veya İlaç ile Sağaltım Ölçütleri. Erciyes Üniv Vet Fak Derg 9(2):123–131

Parsons CS, Orsini JA, Krafty R et al (2007) Risk factors for development of acute laminitis in horses during hospitalization: 73 cases (1997–2004). J Am Vet Med Assoc 230(6):885–889

Pratt SE, Geor RJ, McCutcheon LJ (2006) Effects of dietary energy source and physical conditioning on insulin sensitivity and glucose tolerance in Standardbred horses. Equine Vet J 38(S36): 579–584

Proudman CJ (1992) A two year, prospective survey of equine colic in general practice. Equine Vet J 24(2):90–93

Rasouli N, Kern PA (2008) Adipocytokines and the metabolic complications of obesity. J Clin Endocrinol Metab 93(11):64–73

Reeves MJ, Salman MD, Smith G (1996) Risk factors for equine acute abdominal disease (colic): results from a multi-center case-control study. Prev Vet Med 26(3–4):285–301

Scantlebury CE, Archer DC, Proudman CJ et al (2011) Recurrent colic in the horse: incidence and risk factors for recurrence in the general practice population. Equine Vet J Suppl 39:81–88. https://doi.org/10.1111/j.2042-3306.2011.00383.x

Secombe CJ, Lester GD (2012) The role of diet in the prevention and management of several equine diseases. Anim Feed Sci Technol 173(1–2):86–101

Seitzinger AH, Traub-Dargatz JL, Kane AJ et al (2000) A comparison of the economic costs of equine lameness, colic, and equine protozoal myeloencephalitis (EPM). In: Proceedings of the 9th international symposium on veterinary epidemiology and economics, p 1–3

Singer ER, Smith MA (2002) Examination of the horse with colic: is it medical or surgical? Equine Vet Educ 14(2):87–96

Stashak TS (2002) The foot. In: Troy D (ed) Adam's lameness in horses, 5th edn. Lippincott WilliEMS&Wilkins, Philadelphia, pp 645–663

Sundra T, Kelty E, Rendle D (2022) Preliminary observations on the use of ertugliflozin in the management of hyperinsulinaemia and laminitis in 51 horses: a case series. Equine Vet Educ 00: 1–10

Tinker MK, White NA, Lessard P et al (1997) Prospective study of equine colic risk factors. Equine Vet J 29(6):454–458

Treiber KH, Kronfeld DS, Hess TM et al (2006) Evaluation of genetic and metabolic predispositions and nutritional risk factors for pasture-associated laminitis in ponies. J Am Vet Med Assoc 228(10):1538–1545

Uhlinger C (1990) Effects of three anthelmintic schedules on the incidence of colic in horses. Equine Vet J 22(4):251–254
Van der Linden MA, Laffont CM, van Oldruitenborgh-Oosterbaan MMS (2003) Prognosis in equine medical and surgical colic. J Vet Intern Med 17(3):343–348
Wintzer H J (1999) Krankheiten des Pferdes, herausgegeben. 3. Aufl. Parey Buchverlag, Berlin

Feed Additives in Rabbit Nutrition to Ensure Sustainability

Takwa Belhassen

Abstract

Consumers nowadays are interested in natural, healthy foods made from animals, and they often even want to pay more for them. Also, the European Union's restriction on the non-therapeutic use of antibiotic growth promoters in animal nutrition has stoked interest in using natural alternatives as dietary components in animal nutrition. This ban is intended to limit the development of antibiotic-resistant bacteria. The discovery of antibiotic substitutes to preserve or enhance animal health and performance has been the subject of much investigations. Prebiotics, probiotics, organic acids, bacteriocins, and phytobiotics are examples of natural feed additives that can meet these needs and fulfill the growing demand for natural products among customers. The most researched and advanced compounds that are often employed as dietary antibiotic alternatives in rabbit nutrition are the topic of this chapter.

Keywords

Probiotics · Prebiotics · Rabbits · Phytobiotics · Enzymes · Organic acids

1 Introduction

The weaning period is one of the most crucial times in rabbit breeding since the kits are under a lot of stress from being separated from their mothers and their owns' habitat and milk is substituted with solid feed. Significant changes occur, in fact, in the intestine with a progressive installation of a digestive flora. The kit is unable to handle this stress well, which causes raised cortisol levels, greater susceptibility to

T. Belhassen (✉)
Department of Animal Production, National Institute of Agronomy of Tunis, University of Carthage, Tunis, Tunisia

G. Arsenos, I. Giannenas (eds.), *Sustainable Use of Feed Additives in Livestock*,
https://doi.org/10.1007/978-3-031-42855-5_30

illness, and increased digestive disturbances. These latter cause substantial economic losses (stunting and mortality) (Dalle Zotte et al. 2016).

In order to prevent these digestive disturbances, antibiotics have been used in animal nutrition. Unfortunately, widespread use of antibiotics over a long period of time has increased bacterial drug resistance, raising concerns about food residues that might be harmful to human health (Chen et al. 2017). However, the use of antibiotics as growth promoters in animal feed has been banned due to the problem of antibiotic resistance (European Union Commission 2005).

To deal with this problem, researchers and feed companies have increased their efforts to develop safer and more natural feed additives, improving both the intestinal health and productivity of rabbits in the meantime. Probiotics, prebiotics, enzymes, phytobiotics, and organic acids are the different categories of these natural additives (Gibson and Roberfroid 1995; Guarner and Schaafsma 1998; García-Ruiz et al. 2006; Choct 2006).

These alternatives benefit the animal's performance and wellbeing, especially since they regulate the intestinal flora, which is crucial for preserving the host's digestive health (Tuohy et al. 2005). A healthy gut microbiota acts as an immune system booster and an efficient barrier against infections colonizing the digestive tract.

1.1 Probiotics

Metchnikoff's research, which was included in the book "Optimistic Essays," where he hypothesized that consuming live lactic acid bacteria lengthens life by reducing the number of bacteria in the digestive system, is where the idea for probiotics originated. Probiotics have been defined differently over time based on advancements in science and technology, but it is generally accepted that a probiotic is a live microbial feed supplement which beneficially affects the host animal by improving its intestinal microbial balance (Falcão-e-Cunha et al. 2007).

In general, probiotic properties are often assessed in vitro by examining their antibacterial potential, capacity to stick to the host's intestinal mucin, and tolerance to the gastrointestinal environment. Some probiotic properties manifested in vivo, however, are more challenging to assess. Examples include their capacity to promote gut immune system development and control intestinal innate immunological and inflammatory homeostasis (Mancini and Paci 2021).

Numerous reviews (Ziermer and Gibson 1998; Ouwehand et al. 1999; Simon et al. 2003) have proposed a variety of potential mechanisms of action for probiotics, including a decrease in metabolic reactions that produce toxic substances, stimulation of host enzymes, production of vitamins or antimicrobial substances, competition for adhesion to epithelial cells, an increase in resistance to colonization, and stimulation of the host immune system (Fig. 1). Several probiotics have been utilized to assist the digestive system in rabbits, increase feed conversion and digestibility, and enhance animal health. The effects of several probiotic microorganisms in rabbit's performances and health are summarized in Table 1.

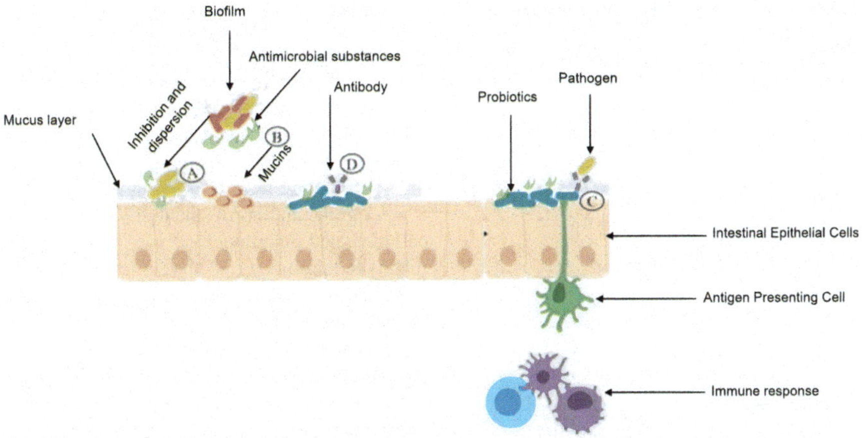

Fig. 1 Modes of action of probiotics. (A) Competitive exclusion of pathogenic microorganisms. (B) Production of antimicrobial substances. (C) Increased adhesion to the intestinal mucosa and improvement of the epithelial barrier. (D) Stimulation of the immune system (Silvaa et al. 2020)

Probiotics may contribute to the weight increase of rabbits as well as the animals' ability to digest the nutrients in their feed and successfully convert them to body mass. The results of the various studies on the effects of probiotics on reproductive and growth performances in rabbits show that the responses to probiotics probably depend on the nature of the probiotic (the strain, the dose, and means of administration), the individual animal (breeds, sex, ages), type of basal diets, and the environmental conditions (rearing climate).

1.2 Prebiotics

A prebiotic is described as a selectively fermented food component that modifies the gastrointestinal microbiota's composition and/or activity in a specific way to promote the health of the host (Fig. 2) (Ducatelle et al. 2015).

Prebiotics can be produced via transglycosylation processes, partial acid or enzymatic hydrolysis of polysaccharides, or direct extraction from natural sources (plants, yeasts, milk) (Oku 1996). Fructo-oligosaccharides (FOS), galacto-oligosaccharides (GOS), trans-galacto-oligosaccharides (TOS), mannan-oligosaccharides (MOS), and xilo-oligosaccharides are now the primary commercial oligosaccharides (XOS).

Oligosaccharides are intended to selectively activate the beneficial microorganisms already present in the gut, whereas probiotics are intended to introduce beneficial microbes there. In comparison with probiotics, they offer two distinct advantages: one technological, as there are no significant issues with the thermal processing of the feed and the acid conditions of the stomach, and one

Table 1 Effect of some probiotics in rabbit's performances

Strain	Dose	Results	Trial
Bacillus subtilis	(200 and 400) g/t feed	• A significant improvement in percentage of dressed carcass, mid and hind parts was recorded for rabbits fed a diet containing 400 g probiotic/t feed • A significant improvement in cell-mediated immunity for rabbits fed probiotic • Final weight and body weight gain were not significantly affected by dietary probiotic levels	Fathi et al. (2017)
AmPhi-Bact[a]	(150, 300, 450) mg/kg diet	• Positive effect on FCR and a higher digestibility of crude protein. • An increased weight gain in the final part of the trial in the rabbits fed with higher concentrations of probiotics • A negative effect on the carcass yield percentage	Abdel-Wareth et al. (2021)
Dry live yeast (RUMI YEAST Saccharomyces cerevisiae Sc 47-NeoviaFrance)	0.5, 1.0, 1.5, and 2.0%	• Any statistical differences to the control in the FCR and digestibility coefficient	Tag El Din (2019)
Saccharomyces cerevisiae	0.12 g/kg diet	• A lack of variations in FCR, final body weight, and average daily weight gain	Emmanuel et al. (2019)
Saccharomyces cerevisiae (Actisaf Sc 47, S.I. LESAFFRE, France)	1 g/kg of feed	• The fertility rate of rabbits does and the viability rate of kits at birth have significantly improved	Belhassen et al. (2016)
Enterococcus faecium EF9a bacteriocin-producing strain	Ringer solution with *Enterococcus faecium* EF9a 1.0 x10^9 CFU/mL, in a dose 500 µL/animal/day into drinking water	• A significant decrease in the coliforms, coagulase-positive staphylococci, pseudomonads, and coagulase-negative staphylococci in the rabbit feces. • An antimicrobial effects in the cecum against coliforms, coagulase-negative staphylococci and pseudomonads. • Any modification in jejunal morphometry or morphological parameters in the ileum	Pogány Simonová et al. (2020)

(continued)

Table 1 (continued)

Strain	Dose	Results	Trial
Lactobacillus plantarum	*Lactobacillus plantarum* 1x _10^6 CFU/g (0.25 and 0.5) g/kg of feed	• No significant effect of probiotics on total protein, immunoglobulins, and IgG. • An improvement in goblet cells and crypts in the base of the tissue or surface and production of mucus.	El-Shafei et al. (2019)

[a]American Pharmaceutical Innovations Company®, containing a mix of lactic acid bacteria culture, *Lactobacillus acidophilus*, *Lactobacillus plantarum*, *Bifidobacterium bifidum*, *Bacillus subtilis* fermentation extract, and *Aspergillus niger* fermentation extract

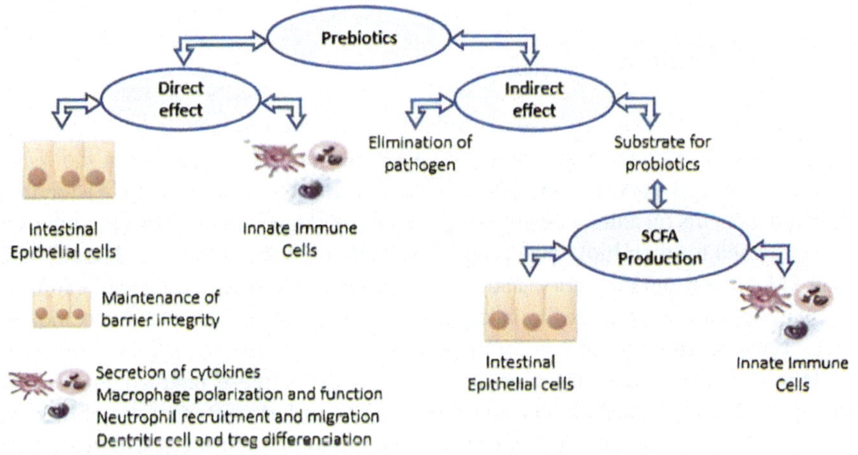

Fig. 2 Effect of prebiotics on the innate immune response and health of the host (Pujari and Banerjee 2021)

safety-related, as they do not introduce alien microbial species into the gut. If encouraged, beneficial microorganisms will be better able to compete with harmful ones (Falcão-e-Cunha et al. 2007). The majority of published investigations on the subject of prebiotics' mechanisms of action have been conducted on laboratory animals and in vitro. Improvements in daily weight gain, (FCR), and/or health status have all been seen in farm animals; however, the benefits have been shown to differ depending on the oligosaccharide and the usage circumstances.

Prebiotics have already been studied in rabbits. Most recent research has focused on their impacts on gut morphology. The majority of papers published so far address their effects on cecal microbiota and/or production performances. Abdel-Aziz et al. (2022) found that fructooligosaccharide supplementation enhanced growth and carcass traits by improving the hematobiochemical parameters and antioxidant status and reducing cecal pathogenic bacteria in two rabbit breeds. Moreover, the inclusion of Inmunair17.5® (Propionibacterium acnes and coli lipopolysaccharides) as a prebiotic to the water consumed by NZW fattening rabbits improved body weight (BW) at marketing, body weight gain (BWG), and FCR (Mousa et al. 2018). In addition, Iser et al. (2016) indicate that Agave fourcroydes powder, as a prebiotic supplement, had beneficial effects on increasing the growth performance and serum concentration of IgG, as well as improving the gut morphology without affecting the hematology parameters in broiler rabbits. On the contrary, Lebas (1996) did not get any effect on growth rate and FCR, whereas Gidenne (1995) got a significant negative effect of GOS on morbidity and mortality.

It should not be overlooked that rabbit diets naturally contain a variety of fibrous feedstuffs, some of which contain large levels of oligosaccharides. Choosing the feedstuffs having the most desired oligosaccharides at each stage of the rabbit's life would be a potential substitute for commercial prebiotics.

1.3 Phytobiotics

Research on plant extracts, essential oils (EO), and by-products of plant origin as alternatives to synthetic vitamins in livestock nutrition is on the rise in the recent decades due to a growing interest in natural plant feed additives (PFA) for their potential benefits on animal health and productivity (Ognik et al. 2016). Phytobiotics are renowned for their biological properties, including their antibacterial, antifungal, antioxidant, anti-inflammatory, and antiviral effects. These properties are exhibited by many plant components like alkaloids, terpenes, flavonoids, and glucosinolates (Fig. 3) (Koné et al. 2018).

However, each plant has a unique combination of these phytochemicals, and therefore their biological effects are expected to be different. Regardless of the field of application, factors linked to plant species, growth circumstances, harvest time, and plant chemotype affect the bioactivity of phytobiotics (Figueiredo et al. 2008). Phytobiotics's biological activities result in better production in poultry, pigs, ruminant and aquaculture animals as well as feed palatability, digestive processes, and intestinal microbiota structure.

In fact, phytobiotics have a variety of uses in animal production, including sensory phytogenic additives, technological additives to improve feed quality and safety, and additives promoting animal health and welfare by acting as immunomodulators, antioxidants, digestive stimulants, and substances that can improve the performance and quality of animal products (Karásková et al. 2015).

In rabbits, phytobiotics has been shown to significantly affect an animal's feed intake, growth, and productivity in several investigations. Effects of plant extracts may result from improved feed consumption, which boosts growth. In addition, Cardinali et al. (2015) reported that the BW and ADG of rabbits were increased by an average of 2.87 and 5.1%, respectively, after 50 days of dietary supplementation with oregano extract (OE, at 2.0% of the diet) or oregano (Origanum vulgare) and rosemary extract (Rosmarinus officinalis), both at a concentration of 0.1%. The findings imply that OO supplementation may be more beneficial than VitE for enhancing growth and several indices of productive performance. Thymol and

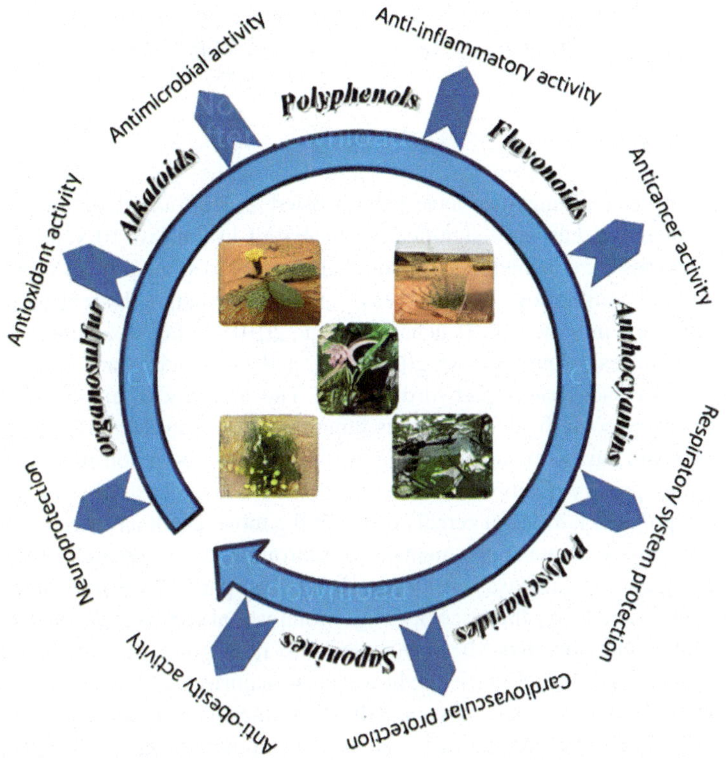

Fig. 3 The most important biological activities of phytobiotics (Hemmami et al. 2023)

carvacrol (polyphenols) are the primary components of OO, while p- cymene and γ-terpinene, their predecessors, are the two most common monoterpenes. The ability of polyphenols to exert antioxidant effects within the gastrointestinal tract, where they come into direct contact with the cells without being absorbed and metabolized (Surai et al. 2004) with a potential protective action on the gut mucosal cell membranes, could be the explanation for these findings. In other study, Abou-Kassema et al. (2021) discovered that adding different levels of a herbal mixture (30%, 0.50% and 0.70%) to the diet of growing NZW rabbits had a significant impact on their growth and carcass traits as well as a favorable impact on the population of lactic acid bacteria and a decrease in the number of pathogenic microorganisms.

It has been demonstrated that a number of phytobiotics have an impact on the metabolic and hematological parameters, as well as the health, fertility, and oxidative state in rabbits. Hashem et al. (2017) reported that 150 or 300 mg propolis/kg and 150 or 300 mg/kg moringa roots added to the diet of rabbits for five weeks did not affect hematological parameters which were in the physiological ranges and were comparable to those in the control animals. On the other hand, they observed that adding propolis and moringa roots to one's diet enhanced the blood reducing capacity (TAC). Nevertheless, these improvements in TAC were accompanied by significantly smaller decreases in LDL and MDA (lipid peroxidation indicator).

1.4 Enzymes

Despite the fact that enzymes have been utilized as feed additives for more than 50 years, their use has only significantly increased in the last 10 years (Tuoying 2005). Enzymes are considered as a natural alternative to some growth promoters as hormones and antibiotics. They increase animal production by enhancing their endogenous enzymes' activity (Cachaldora et al. 2010). In fact, exogenous enzymes such as cellulases, α-amylases, β-glucanases, and β-xylanases are utilized in the process of polysaccharides prehydrolyzation. The majority of enzymatic extracts used in industry come from filamentous fungus and bacteria (Mounsey 2006).

Recent years have seen an increase in the use of enzymes in non-ruminant animal diets, notably poultry (Campbell and Bedford 1992). The digestibility of starch and non-starch polysaccharides in cereals is one of the most recent advancements in this field. Exogenous enzyme supplements have a strong body of research behind them. These enzymes can increase nutritional absorption, partly hydrolyze non-soluble protein (NSP), and lower the viscosity of gastrointestinal contents (Attia et al. 2012).

Exogenous enzymes have been included into rabbit diets in several studies to increase the availability of nutrients; however in majority of these experiments, the rabbits showed less responsiveness, and different impacts were seen on their performance. García-Ruiz et al. (2006) who tested supplementation of protease and xylanase in 35-day-old rabbits for 5 weeks reported a positive effect on ileal apparent digestibility and fattening mortality, whereas growth traits were unaffected. Similarly, Eiben et al. (2004) found that the use of cellulase improved the FCR and mortality of rabbits weaned at 23 days of age, but had no effect on ADG. Likewise, Oloruntola (2018) recorded a significantly higher total BW and total WG in rabbits fed 0.35 g/kg of multi-enzyme.

1.5 Organic Acids

Organic acids have a long history in the food and feed sectors, where they are frequently used as preservatives due to their antifungal action. All organic carboxylic acids with the generic structural formula R-COOH are regarded as organic acids (including fatty acids and amino acids) (Papatsiros and Christodoulopoulos 2011). Organic acids such as lactic, citric, formic, and fumaric and their salts are routinely included in diets for animals in order to replace antibiotics as growth promoters. When added to non-ruminant diets, organic acids have been shown to enhance growth performance (such as enhanced palatability, feed efficiency, mineral absorption, and phytate-P utilization), to reduce the incidence of diarrhea, feed costs and time to market, subsequently, it enhances economic return (Liu et al. 2018). Recent

Table 2 Some results of trials with organic acids in rabbits

Trial	Organic acids	Results
Abdel-Kader and Emam (2020)	0.5 and 0.25% citric acid 0.5 and 0.25% lactic acid	• Any significant differences for the carcass characteristics and blood parameters • Great effect on the growth characteristics and feeding efficiency
Sherif (2018)	Organic acids mixture (Fordex)[a] (1.0 g/kg)	• Positive effects on live body weight, daily weight gain, and feed conversion of growing rabbits • No significant effects on blood parameters or carcass traits
Romero et al. (2011)	0.4% and 0.2% formic and citric acids	• Improvements in weight gain in the 56–77 period • A trophic effect on the jejunal mucosa at 56 d
Dorra et al. (2013)	0.4% formic and propionic acids	• No positive effects on either growth performance or economic efficiency of rabbit production

[a]propionic acid (99.5%) 192.50 g; formic acid (99.5%) 92.35 g; ammonium propionate (99%) 9.30 g; ammonium formate (99%) 55.70 g; ascorbic acid (98%) 80 g; silicon dioxide 50 g and calcium carbonate up to 1 kg

studies have highlighted additional advantages of organic acids that might result from an immediate stimulation of gastrointestinal mucosa growth.

The most prominent mechanisms through which dietary organic acids may produce beneficial effects on health status and growth performance are the following: reduction of gastric pH, increase of proteolytic enzymes activity, reduced survival of pathogens through the stomach, balancing the microbial population, stimulating the activity of digestive enzymes, depolarization of the bacterial membrane and promotion of beneficial bacterial growth (Diebold and Eidelsburger 2006; Tung and Pettigrew 2006).

Young rabbit's small intestine mucosa represents an important region of defense against antigenic aggressions and plays a significant function in the digestion and absorption of nutrients (Gallois et al. 2005). The use of organic acids seems intriguing, despite the fact that there are few and sometimes contradicting scientific studies on how they affect rabbit microbiota populations, mucosal immunity, and growth performance (Falcão-e-Cunha et al. 2007). Table 2 shows the effect of several organic acids on rabbit performance. Considerable variations in the results of their response due to possible dietary and other factors such as: inclusion rate and dose of supplemented acids, intrinsic acid activity, hygiene and welfare standards, age of animals, type/composition of diets and their acid–base or buffering capacity and feed palatability.

2 Conclusion

According to some research, alternatives to antibiotics, previously mentioned, can play a variety of functions in rabbit farming, from biological control against pathogenic bacteria to growth promoters, or active substances to boost the amount and quality of the finished product. The ability of these alternatives to interact with the host and with all the microbiota present in the various regions or organs of the GIT, altering the overall production process, is often attributed to the favorable benefits observed in rabbits fed the experimental diets. The study of the many research papers reveals that feed additives have various mechanisms of action and that a wide range of factors might affect the responses. This kind of research is particularly crucial for minimizing the therapeutic use of antibiotics by preventing disease in animals through diet.

It is important to pay attention to the proper dosage, feed additives activity, and timing of administration. By boosting digestive efficiency, directly improving nutrient digestibility, and activating immunological mechanisms, the composition of the microbiota may be optimized to maximize production profitability. In order to ensure that these alternatives to antibiotics will reach the target site of action in large numbers, future research must also concentrate on the administration methods (microencapsulation, cell immobilization, and continuous fermentation).

References

Abdel-Aziz AH, Abo-Ghanima MM, Alsanie WF, Gaber A, Alsenosy A, Easa AA, Moawed SA, Raza SA, Elfadadwy A et al (2022) Fructooligosaccharide supplementation boosts growth performance, antioxidant status and cecal microbiota differently in two rabbit breeds. Animals 12:1528

Abdel-Kader IA, Emam AM (2020) Testing two levels of lactic and citric acids as growth promoters in rabbits diets. Egypt J Nutr Feeds 23(3):435–444

Abdel-Wareth AAA, Elkhateeb FSO, Ismail ZSH, Ghazalah AA, Lohakare J (2021) Combined effects of fenugreek seeds and probiotics on growth performance, nutrient digestibility, carcass criteria, and serum hormones in growing rabbits. Livest Sci 251:104616

Abou-Kassema D, Mahrosea KH, El-Samahyb RA, Shafic ME, El-Saadonyd MT, Abd El-Hackb ME, Emame M, El-Sharnoubyf M, Tahagand AM, Ashour EA (2021) Influences of dietary herbal blend and feed restriction on growth, carcass characteristics and gut microbiota of growing rabbits. Ital J Anim Sci 20(1):896–910. https://doi.org/10.1080/1828051X.2021.1926348

Attia AK, Sohair YS, Abd El-hamid S, Zaki A, El-Sawy M (2012) Effects of exogenous multi-enzyme feed additive (Kemzyme) on the activities of certain digestive enzymes and intestinal morphology in growing rabbits. J Agric Sci 4(3):35–44

Belhassen T, Simon E, Potel A, Auclair E, Bergaoui R (2016) Effect of diet supplementation with live yeast (Saccharomyces cerevisiae) on performance of rabbit does and their progenies. World Rabbit Sci 2016(24):77–82. https://doi.org/10.4995/wrs.2016.3917

Cachaldora P, Nicodemus N, Garcia J, Carabano R, De Blas JC (2010) Efficacy of Amylofeed® in growing rabbit diets. World Rabbit Sci 12(1):23–31. https://doi.org/10.4995/wrs.2004.583

Campbell GL, Bedford MR (1992) Enzyme applications for monogastric feeds: a review. Can J Anim Sci 72:449. https://doi.org/10.4141/cjas92-058

Cardinali R, Cullere M, Dal Bosco A, Mugnai C, Ruggeri S, Mattioli S, Castellini C, Trabalza Marinucci M, Dalle Zotte A (2015) Oregano, rosemary and vitamin E dietary supplementation in growing rabbits: effect on growth performance, carcass traits, bone development and meat chemical composition. Livest Sci 175:83–89

Chen H, Yang W, Wang C (2017) The review on the function of intestinal flora and the regulatory effects of probiotics on the intestinal health of rabbits. Adv Biol Sci Res. 2nd international conference on biological sciences and technology 6:49–55. https://doi.org/10.2991/bst-17.2018.8

Choct M (2006) Enzymes for the feed industry: past, present and future. World Poultry Sci J 62:5–15

Dalle Zotte A, Celia C, Szendrő Z (2016) Herbs and spices inclusion as feedstuff or additive in growing rabbit diets and as additive in rabbit meat: a review. Livest Sci 189:82. https://doi.org/10.1016/j.livsci.2016.04.024

Diebold G, Eidelsburger U (2006) Acidification of diets as an alternative to antibiotic growth promoters. In: Barug D, de Jong J, Kies AK, Verstegen MWA (eds) Antimicrobial growth promoters. Wageningen Academic Publishers, Wageningen, pp 311–327

Dorra TMI, Ismail FSA, Sherif KE, Rabie MH (2013) Growth performance of fattening rabbits as affected by stocking density and added dietary organic acids. J Anim and Poult Prod 4(5): 249–262

Ducatelle R, Eeckhaut V, Haesebrouck F, Van Immerseel F (2015) A review on prebiotics and probiotics for the control of dysbiosis: present status and future perspectives. Animal 9(1): 43–48. © The Animal Consortium 2014. https://doi.org/10.1017/S1751731114002584

Eiben CS, Mézes M, Zijártó N, Kustos K, Gódor-Surmann K, Erdélyi M (2004) Dose-dependent effect of cellulose supplementation on performance of early-weaned rabbit. In: Proc. 8th world rabbit congress, Puebla,México, pp 799–804

El-Shafei AA, Younis TM, Al-Gamal MA, Hesham AM (2019) Impact of probiotic (Lactobacillus planterium L) supplementation on productive and physiological performance of growing rabbits under Egyptian conditions. Egypt J Rabbit Sci 29:125–148

Emmanuel D, Amaka A, Okezie E, Sunday U, Ethelbert O (2019) Epididymal sperm characteristics, testicular morphometric traits and growth parameters of rabbit bucks fed dietary saccharomyces cerevisiae and/or zinc oxide. Braz J Poult Sci 21. https://doi.org/10.1590/1806-9061-2018-0803

European Union Commission (2005) Ban on antibiotics as growth promoters in animal feed enters into effect. Regulation 1831/2003/ec on additives for use in animal nutrition, replacing directive 700/524//333c on additives in feedstuffs, Brussels, 22 December

Falcão-e-Cunha L, Castro-Solla L, Maertens L, Marounek M, Pinheiro V, Freire J, Mourão JL (2007) Alternatives to antibiotic growth promoters in rabbit feeding: a review. World Rabbit Sci 15:127–140

Fathi M, Abdelsalam M, Al-Homidan I, Ebeid T, El-Zarei M, Abou-Emera O (2017) Effect of probiotic supplementation and genotype on growth performance, carcass traits, hematological parameters and immunity of growing rabbits under hot environmental conditions. Anim Sci 88(10):1644–1655

Figueiredo AC, Barroso JG, Pedro LG, Scheffer JJC (2008) Factors affecting secondary metabolite production in plants: volatile components and essential oils. Flavour Fragr J 23:213–226

Gallois M, Gidenne T, Forthun-Lamothe L, Le Huerou-Luron I, Lallès JP (2005) An early stimulation of solid feed intake slightly influences the morphological gut maturation in the rabbit. Reprod Nutr Dev 45:109–122

García-Ruiz AI, García-Palomares J, García-Rebollar P, Chamorro S, Carabaño R, De Blas C (2006) Effect of protein source and enzyme supplementation on ileal protein digestibility and fattening performance in rabbits. Span J Agric Res 4:297–303

Gibson GR, Roberfroid MB (1995) Dietary modulation of the human colonic microbiota. Introducing the concept of prebiotics. J Nutr 125:1401–1412

Gidenne T (1995) Effect of fibre level reduction and glucooligosaccharide addition on the growth performance and caecal fermentation in the growing rabbit. Anim Feed Sci Techn 56:253–263

Guarner F, Schaafsma GJ (1998) Probiotics. Int J Food Microbiol 39:237–238

Hashem NM, Abd El-Hady AM, Hassan OA (2017) Inclusion of phytogenic feed additives comparable to vitamin E in diet of growing rabbits: effects on metabolism and growth. Ann Agric Sci 62:161–167

Hemmami H, Ben Seghir B, Zeghoud S, Ben Amor I, Kouadri I, Rebiai A, Zaater A, Messaoudi M, Benchikha N, Sawicka B, Atanassova M (2023) Desert endemic plants in Algeria: A review on traditional uses, phytochemistry, polyphenolic compounds and pharmacological activities. Molecules 28:1834

Iser M, Martínez Y, Ni H, Jiang H, Valdivié Navarro M, Wu X, Al-Dhabi NA, Rosales M, Duraipandiyan V, Fang J (2016) The effects of *Agave fourcroydes* powder as a dietary supplement on growth performance, gut morphology, concentration of IgG, and hematology parameters in broiler rabbits. Biomed Res Int 7:1

Karásková K, Suchý P, Straková E (2015) Current use of phytogenic feed additives in animal nutrition: A review. Czech J Anim Sci 60:521–530

Koné AP, Desjardins Y, Gosselin A, Cinq-Mars D, Guay F, Saucier L (2018) Plant extracts and essential oil product as feed additives to control rabbit meat microbial quality. Meat Sci 150: 111–121. https://doi.org/10.1016/j.meatsci.2018.12.013

Lebas F (1996) Effects of fructo-oligo-saccharides origin on rabbit's growth performance in 2 seasons. In Proc.: 6th world rabbit congress. Toulouse, France, pp 211–215

Liu L, Xu Y, Xu X (2018) Effect of supplementation with two combinations of alternative to antimicrobials by stages on cecal fermentation in rabbits. Czech J Anim Sci 63:419–427

Mancini S, Paci G (2021) Probiotics in rabbit farming: growth performance, health status, and meat quality. Animals 11:3388

Mounsey SP (2006) Handbook of feed additives, 24th edn. DMS, Chesterfield

Mousa YI, Abdel-Monem U, Bazid A (2018) Effect of spirulina and prebiotic (Inmunair 17.5®) on New-Zealand white rabbits performance. Zagazig J Agric Res 97:3622–3634

Ognik K, Cholewinska E, Sembratowicz I, Grela E, Czech A (2016) The potential of using plant antioxidants to stimulate antioxidant mechanisms in poultry. Worlds Poult Sci J 72:291–298

Oku T (1996) Oligosaccharides with beneficial health effects: a Japanese perspective. Nutr Rev 54(11):s59–s66

Oloruntola OD (2018) Influence of enzyme supplementation on rabbits fed rumen liquor with poultry waste fermented cassava peels based diets. Anim Res Int 15(1):2950–2964

Ouwehand AC, Kirjavainen PV, Shortt C, Salminen S (1999) Probiotics: mechanisms and established effects. Intern Dairy J 9:43–53

Papatsiros VG, Christodoulopoulos G (2011) The use of organic acids in rabbit farming. Online J Anim Feed Res 1(6):434–438

Pogány Simonová M, Lauková A, Chrastinová L, Plachá I, Szabóová R, Kandriˇcáková A, Žitnan R, Chrenková M, Ondruška L, Bónai A et al (2020) Beneficial effects of enterococcus Faecium EF9a Administration in Rabbit Diet. World Rabbit Sci 28:169

Pujari R, Banerjee G (2021) Impact of prebiotics on immune response: from the bench to the clinic. Immunol Cell Biol 99:255–273

Romero C, Rebollar PG, Dal Bosco A, Castellini C, Cardinali R (2011) Dietary effect of short-chain organic acids on growth performance and development of intestinal lymphoid tissues in fattening restricted rabbits. World Rabbit Sci 19:133–142

Sherif KS (2018) Effect of dietary additives on growth performance, carcass traits and some blood constituents of rabbits. J Agric Sci 10(1):139

Silvaa DR, Sardi JCO, Pitanguic NC, Roquee SM, Da Silva ACB, Rosalena PL (2020) Probiotics as an alternative antimicrobial therapy: current reality and future directions. J Funct Foods 73: 104080

Simon O, Vilfried V, Scharek L (2003) Micro-organisms as feed additives – probiotics. In Proc.: 9th intern. symp. on digestive physiology in pigs, Banff, AB, Canada, pp 295–318

Surai KP, Surai PF, Speake BK, Sparks NHC (2004) Antioxidant-prooxidant balance in the intestine: food for thought 2 Antioxidants. Curr Top Nutraceutical Res 2:27–46

Tag El Din N (2019) Effect of dry live yeast inclusion in fattening rabbit diets on productive performance and digestibility of nutrients. Egypt J Rabbit Sci 29:219–233

Tung CM, Pettigrew JE (2006) Critical review of acidifiers. National Pork Board. Available at: http://www.pork.org/Documents/PorkScience/

Tuohy KM, Rouzaud GCM, Brück WM, Gibson GR (2005) Modulation of the human gut microflora towards improved health using prebiotics – assessment of efficacy. Curr Pharm Des 11:75–90

Tuoying AO (2005) Exogenous enzymes and organic acids in the nutrition of broiler chicks: effects on growth performance and in vitro and in vivo digestion. Ph.D.Thesis. The Graduate School University of Kentucky

Ziermer CJ, Gibson G (1998) An overview of probiotics, prebiotics and symbiotics in the functional food concept: perspectives and futures strategies. Dairy J 8:473–479

Part V

Analytical Methods, Future Outlook and Perspectives

Modern Applications in the Determination of Food and Feed Additives

Christos Papaneophytou

Abstract

The constant growth of the human population has led to increasing demand for food of animal and plant origin. Additionally, consumers demand foods of animal origin with high safety standards and quality obtained via animal-friendly and environmental practices. To this end, food industries have employed various natural food- and feed-additives to increase the quality of foods and improve the health of animals, respectively. Food additives are added to food to improve eating quality, preserve flavor, enhance its texture and appearance, or for other technological functions, for example, to facilitate industrial processing and improve its organoleptic properties. However, using chemical compounds as food additives and preservatives is a sensitive issue because of health concerns. Moreover, feed additives have been used in animal agriculture to improve animal health and welfare. In previous years, medicinal products, including antibiotics, have been widely used as feed additives to prevent and/or treat infectious diseases and as growth promoters. However, the overuse of antibiotics resulted in the development of drug-resistant bacteria, posing a threat to consumers' health and negatively affecting the environment. Therefore, using both food and feed additives requires safety policies and regulations because some compounds have been implicated in the development of diseases, making regulation and control vital for food safety and public health. Agriculture and food industries must follow national and international quality standards that establish the conditions under which food and feed additives may be used. Therefore, industries and authorities must monitor the safety of food- and feed-additives while their use must be closely controlled. This chapter discusses the most current

C. Papaneophytou (✉)
Department of Life Sciences, School of Life and Health Sciences, University of Nicosia, Nicosia, Cyprus
e-mail: papaneophytou.c@unic.ac.cy

G. Arsenos, I. Giannenas (eds.), *Sustainable Use of Feed Additives in Livestock*,
https://doi.org/10.1007/978-3-031-42855-5_31

analytical methods used in detecting and quantifying food and feed additives, as well as their advantages and disadvantages.

Keywords

Food additives · Feed additives · Analysis · Regulations · HPLC · GC

1 Introduction

It is estimated that by 2050 the world's population will exceed 9 billion (Markowiak and Śliżewska 2018). The increase in human population has created a demand to provide more animal- and plant-origin food. At the same time, it has also enhanced food production, which must be cheaper and meet high-quality standards (Carocho et al. 2014). Moreover, globally, consumers increasingly demand animal products with high nutritional value, safety standards, and sensory quality, which must be obtained using environmental- and animal-friendly methods (Minchin et al. 2010; Prache et al. 2020). In addition to the increase in the human population, lifestyle changes promoted significant modifications in the composition of foods (Martins et al. 2019). Thus, food scientists are looking for solutions to allow food production intensification, reduction of production costs, and compliance with high standards of safety and quality and for both people and the environment (Markowiak and Śliżewska 2018). To this end, a variety of natural and synthetic food additives (i.e., natural or synthetic chemicals added to food) have been employed to preserve flavor, improve the quality or appearance of a food, improve eating characteristics or appeal to the consumer, or for other technological functions(Martins et al. 2019).

Today, humanity depends on food additives. The term food additive (additional substance) means "*any substance the intended use of which results or may reasonably be expected to result, directly or indirectly, in its becoming a component or otherwise affecting the characteristics of any food*" (Hoadley 2011). Overall, more than 2500 permitted compounds are used as food additives worldwide to maintain specific properties, including taste, smell, and texture, and/or to extend the shelf-life of the food [see Partridge et al. (2019) and references cited therein]. A food additive may be a natural product, e.g., beeswax, a chemical identical to one found in nature, e.g., citric acid, or an entirely new compound, e.g., saccharin (Emerton and Choi 2008). In the European Union (EU), all food additives are labeled with the letter "E" (i.e., Europe) and a specific number (e.g., E901 stands for beeswax). The "E" nomenclature was extended to the Codex Alimentarius Commission to identify food additives worldwide quickly (https://www.foodchemicalscodex.org/). In the EU, food additives are divided into 26 functional groups, depending on their function in food (Wiley and Yen Nee 2020): Preservatives, sweeteners, colorants, antioxidants, carriers, acidity regulators, acids, anticaking agents, antifoaming agents, flavor enhancers, bulking agents, emulsifiers, emulsifying salts, firming agents, foaming agents, gelling agents, glazing agents, humectants, modified starches, packaging gases, fuels, raising agents, sequestrants, stabilizers, thickeners, and flour treatment agents. More than 3000 food additives are used in the United States (US), which are distributed into six groups: Preservatives, nutritional additives, coloring agents, flavoring agents, texturizing agents, and miscellaneous agents [reviewed in Carocho et al. (2014)]. A list of the chemical additives used in food industries based on their function purposes is available at https://food.ec.europa.eu/safety/food-improvement-agents/additives/database_en. The US Food and Drug Administration (FDA) has also developed a food and color additives database, which is available at https://www.fda.gov/food/food-ingredients-packaging/food-additive-listings. The use of food additives must be in compliance with the i) specific legislation of each country and ii) food safety policy proposed by the Codex Alimentarius (https://www.fao.org/gsfaonline/docs/CXS_192e.pdf). However, several additives were banned throughout the years, some at a global level and others in specific countries (Carocho et al. 2014). For example, potassium bromate (E924) is widely used as a flour improver in the US, but its use has been banned in the EU.

The increase in the human population has led to the rise of animal agriculture due to a higher demand for animal-derived products (such as milk, eggs, and meat) (Bradford 1999; Vranken et al. 2014). However, farming seriously affects animal health and welfare (Silbergeld 2019) due to chronic stress and infectious and inflammatory diseases, which affect animal health and performance (Blokhuis et al. 1998). Thus, various feed additives have been employed to improve animal performance and well-being. For example, amino acids and vitamins provide essential nutrients and micro-components (Karau and Grayson 2014). Other additives, such as coccidiostats, have been employed to prevent invasive diseases (e.g., coccidiosis), and antioxidants are used to reduce the oxidation processes of feed components. Furthermore, other additives, such as carotenoids, enhance the dietary value and quality of food products of animal origin (e.g., egg yolk coloration) (Rubaj et al. 2012). Antibiotics have also been used as feed additives, while the employment of synthetic non-steroidal anti-inflammatory drugs and antibiotics is a common practice in farms for disease treatment. However, the use of antibiotics in animal feed as growth promoters appears to promote the emergence of antibiotic-resistant strains (Markowiak and Śliżewska 2018). Moreover, using chemical reagents as food and feed additives and preservatives is a sensitive issue because of health concerns (Mursalat et al. 2013). In several countries, chemical preservatives and food additives are regulated and monitored by various acts, rules, and government authorities (Amit et al. 2017). Official feed control is implemented to monitor the fair and safe use of feed additives in animal nutrition. Moreover, controlling feed production improves the quality and safety of animal products for consumers. Food industries must follow both national and international quality standards and regulations on how to use food additives, while food safety is warranted by strict quality control. Notably, in recent years, intense market pressure has been applied on livestock to become clean, green, and ethical (CGE). The CGE approach involves the reduction of synthetic drugs, the reduction of the impact of animal industries on the environment, and the improvement of animal welfare (Martin and Ferasyi 2016). Furthermore, in 2006, the European Union banned the employment of antibiotic growth promoters in animal nutrition to fight the emergence of antibiotic-resistant bacteria (Parliament 2003).

To this end, regulatory agencies and industries require rigorous quality control in identifying the class and quantities of food and feed additives to avoid any illegal additives. This chapter describes the analytical methods routinely employed to analyze these compounds. Each technique's main advantages and disadvantages are discussed—perspectives of chemical analysis in the food industry and the analysis of additives in animal feed are also provided.

2 General Concepts on Chemical Analysis of Food and Feed Additives

Food additives should be used with a strict food safety policy because some compounds have been implicated in the development of diseases, including diabetes, obesity, and metabolic disorders (Martins et al. 2019). Food additives are widely used to improve the organoleptic characteristics of food, increase its shelf-life or facilitate industrial processing. Additives in the food industry are not used without controversy; their use is monitored by specific regulations and controls to maintain food safety and public health (Palacios-Jordan et al. 2022). There are various methods for determining and quantifying the food additives; the choice depends on the nature of the product (i.e., food) and expected levels of the additive(s).

Likewise, feed additives must be used with strict regulations. Several feed additives, such as antibiotics and coccidiostats, have been used in animal agriculture to control the colonization of pathogens in animals' guts. Some organic acids, e.g., propionic and formic acids, have been added to broiler birds' feed to eliminate feed-borne pathogens from their intestines and inhibit the shedding of these microorganisms. However, their administration is essential for livestock health and productivity, and the widespread use of these substances represents a risk for consumers (Sanz et al. 2015). Recent reports have highlighted the presence of drug residues in animal-derived products, such as milk, eggs, and meat (Rana et al. 2019). Animal feed and feed additives must follow strict criteria without increasing the cost of animal breeding. In the past, medicinal products, including antibiotics, had been widely used to modify the dietary microbiota and facilitate animal growth and productivity. However, the uncontrolled and overuse of antibiotics and other medicinal products led to the development of drug-resistant bacterial strains, posing a public health issue worldwide and a negative effect on the environment (Markowiak and Śliżewska 2018). Animal feed is an essential part of the food supply chain and is vital to the efficient and profitable production of quality and safe food (Malomo and Ihegwuagu 2017). Therefore, feed safety is essential to food safety, and stakeholders interested in producing safe foods must provide animal feeds of high quality. Research evidence highlights the risks associated with consuming contaminated feeds. Several epidemics detectable to animal feeds in different countries have made the demands for safe feed even more severe (Alkhalaf et al. 2010).

Fig. 1 A schematic diagram of the food and feed additives analysis workflow showing the four key steps followed and indicating the consideration point(s) (marked with warning signs) during each stage. Homogenous food or animal feed samples should be obtained and stored at conditions that minimize degradation (step 1). Subsequently, samples undergo a pretreatment step to eliminate the factors affecting the downstream applications (step 2). The selection of the method for analyzing and quantifying additives in food or feed is based on the number of samples, the available instrumentation, and whether the technique has been validated based on specific regulations (step 3). Various methods are employed to interpret the results (step 4) and quantify food and feed additives based on the analytical technique

Based on the above, food and feed additives should be continuously monitored and controlled. Over the past years, several analytical methods have been developed to analyze food and feed additives. Regardless of the analytical process and instrumentation, determining and quantifying food and feed additives involves common steps (Fig. 1), i.e., i) sample collection, ii) pretreatment, iii) analysis, and iv) interpretation of the results (Mitra and Brukh 2003).

These main steps followed during the analysis of food and feed additives are discussed further in the following paragraphs.

Step 1: Sample Collection This step must assure the representativeness of the food/feed sample, and the sampling process depends on the sample size (Nielsen 2014). In brief, a sample would be identical to the food/feed from which it has been obtained. Importantly, food or feed samples must be collected without contamination and handled adequately for the analytical results to be reliable. Samples must be preserved to prevent deterioration if the analysis is not carried out immediately. Containers must be dry, sterile, and unbreakable. Codex code has also set specific guidelines for sample collection (https://www.fao.org/uploads/media/Codex_2004_sampling_CAC_GL_50.pdf).

Step 2: Pretreatment The pretreatment of samples is usually carried out because some substances, such as fats, lipids, oils, proteins, etc., may interfere with downstream processes (chemical analysis). Microwave-assisted extraction, which enhances the recovery yield of food- or feed-additives and minimizes extraction duration and volume of solvents, has been successfully used as a pretreatment step in liquid chromatography (LC) and infrared (IR) spectrometric analysis (Nescatelli et al. 2017). Other techniques, including centrifugation, simple solvent extraction, Soxhlet extraction, supercritical fluid extraction, pressurized-fluid extraction distillation, and more straightforward methods such as mixing, homogenization, and dilution, have been used as a pretreatment step in food analysis (Raina-Fulton 2021). The pretreatment method must be simple, fast, cheap, and environmentally

Table 1 Frequency of analytical methods used in the detection of food and feed additives

Method	Food additives	Feed additives
Spectrometric		
UV-Vis[a]	Moderate	Moderate
IR[b]	Moderate	Moderate
NIR[c]/MIR[d]	Moderate	High
FTIR[e]	High	High
Raman	High	High
NMR[f]	Low/Moderate	Low
Chromatographic		
HPLC[g]	High	High
GC[h]	High	High
Electroanalytic		
Voltammetry	Moderate	Low

[a]*UV-Vis* Ultraviolet-Visible, [b]*IR* Infrared, [c]*NIR* Near infrared, [d]*MIR* Mid-infrared spectroscopy, [e]*FTIR* Fourier transform infrared, [f]*NMR* Nuclear magnetic resonance, [g]*HPLC* High-performance liquid chromatography, [h]*GC* Gas chromatography

friendly (Wrolstad et al. 2005). Modern pretreatment techniques have been reviewed elsewhere (Buldini et al. 2002; Smith 2003) and will not be discussed in this chapter.

Step 3: Analysis The choice of the analytical technique depends on many factors, including the sample's nature (gas, liquid, solid) expected concentration of the additive, the reagents and the equipment available, the number of samples, and notably, the presence and levels of substances, that may interfere with the method (Martins et al. 2019). Official methods for determining food and feed additives have been developed by AOAC International (2019) and Food Chemical Codex (https://www.foodchemicalscodex.org/).

Overall, an ideal method for the characterization and quantification of additives in foods and animal feeds should be characterized by (Martins et al. 2019; Tunick 2005):

I. Ease of identification and quantification.
II. The number of additives that can be determined in a single experiment which is a prerequisite for automation and high throughput.
III. Short analytical time and low cost per analysis.
IV. Ability to store and retrieve data and results.

Therefore, it is essential to employ analytical techniques that provide high selectivity and sensitivity reliability, speed, and operational security. The methods should also be environmentally friendly and cheap (Wrolstad et al. 2005).

Traditional and modern analytical methods are usually employed to analyze food and feed additives. The methods are spectroscopic, chromatographic, and electroanalytical (Table 1). Various titrimetric, gravimetric, volumetric, physical, and biological methods have also been developed to analyze food and feed additives

(Martins et al. 2019; Malomo and Ihegwuagu 2017). However, these methods are beyond the scope of this chapter and will not be discussed. The analytical techniques routinely used for analyzing food and feed additives are discussed in the following paragraphs.

Step 4: Interpretation of the Results The raw data in food/feed analysis are analyzed and validated using statistical methods to ensure specific legislation's repeatability, reliability, robustness, and adequacy. All analytical techniques should be validated, while their performance and detection limits must be determined before use. The analytical validation parameters include accuracy, precision, sensitivity, quantification limits, and robustness (Wrolstad et al. 2005). Other techniques require more sophisticated methods of analysis of results (discussed further below).

3 Analytical Methods for the Determination of Food Additives

In the following paragraphs, the basic principles of each of the analytical techniques employed in food analysis are presented. The advantages and limitations of each method are discussed, while some examples where analytical methods are routinely used for monitoring food additives are provided.

3.1 Spectrometric Methods

Spectrometric techniques offer a rapid and full screenshot of foods' composition and properties by determining specific biomolecules, including organic acids, amino acids, minerals, sugars, minerals, polyphenols, volatile compounds, additives, etc. (Karabagias 2020). The spectrometric techniques study the interaction of different frequency components of electromagnetic radiation (EMR) with matter (Akash and Rehman 2020). Thus, spectroscopic methods are widely used to detect food additives that absorb or emit light (radiation) in a specific region of the electromagnetic spectrum. The main spectroscopic methods are based on two fundamental concepts, i.e., vibrational and spin-based spectroscopy. The most common spectrometric methods used for the analysis of food use ultraviolet (UV), visible (Vis), infrared (IR), radio (nuclear magnetic resonance, NMR), and microwave (electron spin resonance, ESR) regions of the spectrum (Table 2). Spectrometric techniques, including Fourier transform infrared spectroscopy (i.e., mid-infrared; MIR and near-infrared; NIR) and Raman spectroscopy (Table 2), have also successfully monitored food quality (Hassoun et al. 2020).

Spectroscopic techniques are sensitive, have reasonable precision and accuracy, robustness, and a high linear range, and permit automation that decreases the analysis time and minimizes systematic errors. Despite their numerous advantages, some spectroscopic techniques require skilled labor due to the increased risk associated with their operation in the sample preparation steps, have poor selectivity,

Table 2 An overview of the most common techniques used in the analysis of food additives

Technique	Wavelength (nm)	It measures	Advantages	Disadvantages
UV-Vis[a]	100–390 (UV) 400–750 (Vis)	The amount of UV or Vis light absorbed by a sample	Low-cost, fast, suitable for color determination	High background noise, requirement of standard curves
NIR[b]	750–2500	Vibrational overtones of biochemical bonds	High sensitivity	Low specificity
MIR[c]	2500–25,000	Vibrations of common functional groups	High sensitivity	Water interference
FTIR[d]	2500–25,000	Full spectrum	High sensitivity, fast, and easy	Expensive equipment, water interference
Raman	750–1060	Vibrational transitions	Quantitative results	Low sensitivity
NMR [e](MHz)	Low field: 10–50 High field: >300	Nuclei orientation into a magnetic field	High accuracy, determination of precise structures	Expensive equipment

[a]*UV-Vis* Ultraviolet-Visible, [b]*NIR* Near infrared, [c]*MIR* Mid-infrared spectroscopy, [d]*FTIR* Fourier transform infrared, [e]*NMR* Nuclear magnetic resonance

and require sample preparation steps to minimize the interferences (Lohumi et al. 2015).

3.1.1 UV-Vis Spectroscopy

UV–Vis spectrophotometry is the most commonly used detection method of food additives because of its simplicity, high efficiency, low cost, and good reliability. Food industries use UV/Vis spectroscopy (200–800 nm) to determine the chemical composition, textural parameters, and quality-related of several food samples, including processed foods, dairy products, meat, oils, fresh and processed fruits, and vegetables, beverages, wine, spices, and flavors. Using the Beers-Lambert law, UV–Vis spectroscopy can give high accuracy and precision (Haque et al. 2021). However, UV–Vis spectroscopy's application in food analysis can be limited due to low-concentration target additives in foods and high detection limits. Furthermore, using food samples with complex matrices in UV–Vis spectrophotometry is challenging due to the low concentration of some additives in food samples. Hence, a sample preparation step is commonly used before an instrumental analysis to reduce interference from the sample matrix and enrich the analyte (Liu et al. 2016). Sample preparation for UV–Vis spectroscopy is usually non-destructive, low cost, fast, and straightforward (Haque et al. 2021).

The visual spectrum (400 to 800 nm) facilitates analysis of the samples' color and any changes (see Hassoun et al. 2020) and references cited therein); thus, UV-Vis

Table 3 Example of analysis of food additives using UV-Vis spectroscopy

Additive	Category	Sample	Reference
Indigotin & Ponceau-4R	Color	Powdered drinks, sweets, and jelly beans	Altınöz and Toptan (2003)
Curcumin	Color	Ginger and curry powder	Liu et al. (2016)
Carotenoids	Color	Carrot	Lawaetz et al. (2016)
Nitrite	Preservatives	Meats	Noor et al. (2016)
Nitrate and nitrite	Preservatives	Meats, vegetables, dairy products, vegetables, fruits	Bahadoran et al. (2016)

spectroscopy is routinely used to monitor color additives in food. In addition to smell and taste, color is probably the first parameter determining food quality, affecting consumers' acceptance and selection (Silva et al. 2022). Notably, the choice of analytical method depends not only upon whether the color is artificial or natural but also on its production method and, thus, on its physicochemical properties (Scotter 2015). For example, Altınöz and Toptan (2003) developed a simple UV-spectrophotometry-based method for determining Indigotin and Ponceau-4R in food samples. This straightforward method was accurate, precise, reproducible, and easily applicable for analyzing Ponceau-4R, and Indigotin in food samples. Importantly the results obtained with the UV method were comparable ($p > 0.05$) with data from HPLC methods. In general, synthetic indigotin dyes, synthetic azo dye, and Ponceau-4R are colorants in foods including ice cream, drinks, and sweets. The analytical control of these dyes is essential in the food industry because they are toxic and have been implicated in cancer development (Altınöz and Toptan 2003). UV-VIS spectroscopy has been used to determine other color additives and preservatives, as shown in Table 3.

3.1.2 Infrared Spectroscopy

Infrared (IR) radiation is the region of the electromagnetic spectrum between the visible and the microwave wavelengths. IR spectroscopy is one of the most commonly used analytic techniques in the food industry to assess food quality. The range of wavelengths for near-infrared (NIR) and mid-infrared (MIR) is 750–2500 nm (13,400 to 4000 cm^{-1}), 2500–25,000 nm (4000 to 400 cm^{-1}), respectively. Gaseous, liquid, or solid samples can absorb IR radiation at specific wavelengths resulting in a "fingerprint" or spectrum (Cozzolino 2011). Therefore, several molecular groups absorb IR light at specific wavelengths in an infrared spectrum, with the fingerprint region identifying compounds (Davies 2004). NIR spectra of foods include broad bands arising from overlapping absorptions corresponding mainly to overtones and combinations of vibrational modes involving C–H, O–H, and N–H chemical bonds (Huang et al. 2008). This long-standing technique has been expanded upon in recent years. The main steps in the NIR technique include the pretreatment of spectral data, the partition of sample sets, the extraction of character bands, and the establishment of the model. However, the collected information is sometimes affected by external interference, which impacts the accuracy of the prediction model. Therefore, it is necessary to preprocess spectra data before finalizing the model.

IR provides quantitative and qualitative information quickly, cost-effectively, and non-destructive. Importantly it does not require polluting solvents and can be carried by a novice analyst (Capuano and van Ruth 2016). The absorption of radiation in the NIR (800 to 2500 nm), which used detectors approximately 100 times more sensitive than other IR detectors, is helpful for qualitative analysis, such as identifying specific functional groups in substances in a fast and straightforward way. In NIR spectroscopy, multivariate statistical techniques and calibration steps are employed to quantify food additives in various foods. However, the high initial cost of the instrumentation and the necessity of the particular calibrations for each product measured are the main drawbacks of NIR spectroscopy in the food industries (Martins et al. 2019).

On the other hand, MID-IR (2500 to 25,000 nm) is used in quantitative research, and it has attracted significant interest mainly due to technical and instrumental advantages compared with NIR (Pizarro et al. 2013). In contrast to NIR, which provides a unique "fingerprint" in sample peaks and peak intensities, the spectra generated by MIR are characterized by high quality, selectivity, and sensitivity (Castritius et al. 2012).

NIR has been widely used to analyze food additives in processed meat products, biscuits, muesli bars, ready-to-eat breakfast products, butter, and margarine [see Martins et al. (2019) and references cited therein]. Recently, Zhang et al. (2017) used a modern IR-based spectrometric method (Terahertz radiation) to determine glutamic acid, a flavor enhancer, in cereals. Terahertz radiation ranges from 0.1 to 10 THz (1 THz $= 10^{12}$ Hz, 0.03 mm to 3 mm in wavelength) and lies between the microwave and infrared portions of the electromagnetic spectrum. In addition to additives, IR spectroscopy has been used to detect harmful substances in food, including melamine and the "Sudan red duck egg" (Qi et al. 2022). IR-based methods have also been used to monitor changes in food quality by applying natural additives, including preservatives (Hassoun et al. 2020). In their study, Che et al. (2017) proposed a straightforward/noninvasive NIR-based method for determining azodicarbonamide (ADA) concentration in wheat flour. Notably, ADA has been widely used as a flour gluten fortifier; however, its use has been banned or limited in several countries as it has been implicated in the development of diseases (Lopez-Gonzalez et al. 2019). IR-based spectrometric methods have been developed for the detection of harmful substances. Yazici et al. (2020) established a method for rapid, non-destructive detection of pesticide residues in strawberries based on near-infrared spectroscopy. Guo et al. (2016a) reported a quantitative approach to determine the levels of Dextrin in Danshen (*Salvia miltiorrhiza*) granules using mid-infrared spectroscopy (MIR) coupled with chemometrics.

In addition to IR spectroscopy, Raman spectroscopy and Fourier transform infrared (FTIR) spectroscopy have been widely used for the analysis of food samples [reviewed in Li-Chan (1996)]. The principles of Raman and FTIR spectroscopy are described in the following paragraphs.

3.1.2.1 Fourier Transform Infrared Spectroscopy

FTIR is widely used for the quantitative and qualitative analysis of organic and inorganic molecules, including food additives (Rodriguez-Saona and Allendorf 2011; Baker et al. 2014). FTIR collects spectrochemical information on the absorption intensities for each wavenumber of the MIR region of the spectrum. The IR bands provide vibrational details to identify the molecular components and their respective structures. Therefore, the spectra produce a characteristic molecular fingerprint to screen and scan samples in various segments. The fingerprint spectrum for biological samples (1800 to 900 cm^{-1}), or the "bio-fingerprint" region, provides information about the key biomolecules, including proteins, carbohydrates, and nucleic acids (Baker et al. 2014). The modern FTIR spectrometers do not split light into individual wavelengths for measurements; every interferogram has information from each wavelength of the light being measured (Duraipandian et al. 2013). Advances in FTIR instrumentation combined with powerful multivariate data analysis methods make this technique ideal for large volume, rapid screening, and characterization of minor food components at low concentrations. FTIR spectrometers are used on production lines to determine fat, protein, and moisture concentrations. Notably, Luna et al. (2015) combined FT-MIR and chemometrics to discriminate transgenic from non-transgenic soybean oil.

3.1.2.2 Raman Spectroscopy

In 1928 Sir Chandrasekhara Venkata Raman introduced and explained the effects of light changing its wavelength when it passes through a transparent object (Rodríguez Vera et al. 2022). Like FTIR, Raman spectroscopy obtains a sample's vibrational spectrochemical information (molecular polarizability changes). The Raman bio-fingerprint region (2000 and 500 cm^{-1}) comprises mainly molecules' stretching vibrations. Raman scattering is based on changes in the polarizability of functional groups; thus, nonpolar groups produce intense responses.

Overall, IR-based techniques (Raman, NIR, MIR) are the primary fingerprinting techniques that collect signals from all chemicals in a sample. Raman and IR spectroscopies have been extensively used to monitor food composition due to their simple instrumentation, increased accessibility, and rapid and non-destructive features (Jawaid et al. 2013; Rodrigues Júnior et al. 2016). However, because of the low sensitivity, these two spectroscopic techniques are not ideal for determining chemicals at low concentrations in food samples. Raman spectroscopy is now being used to define and quantify food additives. For example, Raman techniques have been successfully employed for the vibrational characterization of the food additive carmoisine (E122) (Snehalatha et al. 2009). Howes et al. (2019) used Raman spectroscopy to examine the effects of several antimicrobial and antioxidant additives, such as ascorbic acid (E300), sodium ascorbate (E301), sodium citrate tribasic (E331), and on meat drip from defrosted yellowfin tuna fish loins. Table 4 summarizes some applications of IR-based techniques in analyzing food additives.

Table 4 Examples of analysis of food additives using IR-based spectrometric methods

Additive	Category	Sample	Method	Reference
Glutamic acid	Flavor enhancer	Cereals	IR[a]	Zhang et al. (2017)
Sodium alginate (E401)	Thickener, stabilizer, and gelling agent	Seaweed	IR	Fertah et al. (2017)
Azodicarbonamide	Flour quality improver	Flour	NIR[b]	Che et al. (2017)
Mineral oil	Glazing agent	Oil	FTIR[c]	Polansky et al. (2017)
Benzoyl peroxide	Bleaching agent	Flour	FTIR	Guo et al. (2016b)
Dextrin	Stabilizer, emulsifier	Danshen	FT-MIR[d]	Guo et al. (2016a)
Guar gum	Thickener, stabilizer, and emulsifier	Various gums	Raman	Srivastava et al. (2016)
Aspartame, cyclamate, Saccharin, acesulfame-K	Sweeteners	Powder tabletop sweeteners	FT-Raman[e]	Duarte et al. (2017)
β-carotene	Color	Oil/water emulsions	Raman	Wan Mohamad et al. (2017)

[a]*IR* infrared, [b]*NIR* near-infrared, [c]*FTIR* Fourier transform infrared, [d]*FT-MIR* Fourier transform infrared, [e]*FT-Raman* Fourier transform Raman

3.1.3 Nuclear Magnetic Resonance and Electron Spin Resonance

Nuclear magnetic resonance (NMR) is based on nuclei spin with a magnetic momentum. Nuclei (^{1}H, ^{13}C, ^{23}Na, and ^{31}P) are excited at energy states created by an external magnetic field; i.e., electromagnetic pulse trains excite and change the spin direction of the nuclei. The nuclei return to their fundamental spin state after turning off the pulses. The time the nucleic needs to return to the ground state is characteristic of the molecule and its environment (Hatzakis 2019). NMR methods are classified according to the external magnetic field strength: i) Low-field NMR (10–50 MHz), which is used for the analysis of foods and determines food characteristics (Ezeanaka et al. 2019), and ii) High-field NMR (>300 MHz) widely used in pharmaceutical and organic chemistry applications. Notably, high-field NMR has been used in the food industry for the structural analysis of complex biomolecules, such as carbohydrates, lipids, and proteins (Belton 1993).

Although IR-based techniques have been extensively used to analyze food additives, these techniques measure the vibration of functional groups in chemicals, and the variety of functional groups in chemicals (including food additives) is limited. On the contrary, NMR spectroscopy identifies different compounds because even slight differences in the electron cloud of nuclei can be precisely determined by NMR spectra (Hu et al. 2017).

Table 5 Examples of analysis of food additives using NMR

Additive	Category	Sample	Reference
Polydimethylsiloxanes (E900)	Antifoaming agent	Various types of food	Palacios-Jordan et al. (2022)
Acesulfame potassium	Sweetener	Various types of food	Ohtsuki et al. (2015)
Alginate and carrageenan	Antifoaming agent, gelling agent, glazing agent, bulking agent, foaming agent, stabilizer and thickener, and emulsifier	Seaweeds	Youssouf et al. (2017)
Sodium alginate (E401)	Thickener, stabilizer, and gelling agent	Seaweeds	Fertah et al. (2017)

A significant advantage of NMR spectroscopy over IR-based techniques is that it measures intact biomaterials and foods with little or no sample preparation. NMR is a straightforward method for analyzing food additives (Table 5) because it can determine molecular structure relatively quickly, is a non-destructive analysis, and can quantify multiple compounds (Palacios-Jordan et al. 2022). However, its low sensitivity limits the determination of trace compounds in complex samples, including foods. Nevertheless, in recent years, NMR-based methods have been employed to detect food additives. NMR is a quantitative tool in food analysis with excellent repeatability and reproducibility (Dais et al. 2007). NMR-based methods have been successfully used to determine silicone-based antifoaming agents (polydimethylsiloxanes -PDMS or E900) (Palacios-Jordan et al. 2022). Ohtsuki et al. (2015) reported an NMR-based method that couples quantitative ^{1}H NMR with solvent extraction for quantifying acesulfame potassium, an artificial sweetener, in various food samples.

In contrast to other analytical techniques, e.g., HPLC, the NMR method does not require an authentic analyte standard. Although GC remains the standard method for analyzing fish oils, ^{1}H NMR spectroscopy has been successfully used to determine fish oil quality and composition(Giese et al. 2018). NMR-based methods have also been employed to determine illegal additives in various food additives. Hu et al. (2017) used ^{1}H NMR spectroscopy to determine Sudan dye I, a carcinogenic dye (used industrial dye for coloring clothes and wax) that is added illegally into spices in paprika powder. Notably, Sudan I has been categorized as a third-group genotoxic carcinogen by the International Agency for Research on Cancer. All countries have banned it as a food dye (Ertaş et al. 2007).

3.2 Chromatographic Methods

Chromatographic techniques facilitate the separation, identification, and quantification of organic and inorganic compounds using the distribution of samples between a mobile phase (gas, liquid, or supercritical fluid) and a stationary phase (solid or liquid absorbed on a solid matrix). Based on the physicochemical principles used to separate the component of the mixture, chromatographic techniques are categorized into i) adsorption chromatography, which uses a solid stationary phase on which the sample components from the mobile phase are adsorbed, ii) partition chromatography, which uses a liquid conjugated on a solid support as a stationary phase, iii) ion-exchange chromatography (IEC) in which an ion-exchange solid matrix is used as the stationary phase (Coskun 2016), iv) exclusion chromatography which separates molecules based on their size. These separation methods are employed, in gas chromatography (GC) and liquid chromatography (LC), which are the two primary chromatography techniques used to analyze food additives (Martins et al. 2019).

3.2.1 High-Performance Liquid Chromatography

High-performance liquid chromatography (HPLC) with UV detection is widely employed for the simultaneous detection and quantitation of food additives [see De Girolamo et al. (2022) and references cited therein]. In normal-phase HPLC, the stationary phase is a polar adsorbent, and the mobile phase is a nonpolar solvent. In reversed-phase HPLC, a nonpolar stationary phase and a polar mobile phase are used. Reverse-phase HPLC is more popular. A functionalized organic resin is the stationary phase in ion-exchange HPLC. HPLC is the most popular food analysis technique, followed by GC, and has been successfully used to analyze food additives (Nie and Nie 2019). For example, HPLC is frequently used with mass spectrometry (LC-MS) to quantify color additives and preservatives in foods (Bajkacz and Kycia-Słocka 2020). Notably, several food additives are unstable at high temperatures and polar and, therefore, cannot be analyzed by gas chromatography and mass spectrometry (Bajkacz and Kycia-Słocka 2020).

However, chromatographic data have some difficulties, such as background effects, noise, overlapping peaks, and elution time shifts, which significantly affect obtained results (Ghafghazi et al. 2017). To this end, chemometric algorithms, which are based on collecting multidimensional chromatographic data without separating the analyte peaks, have been used in the quantitative analysis by HPLC (Mortera et al. 2018). Overall, chemometrics methods have been introduced in LC with diode-array detection (DAD), where spectra are recorded at several retention times for each sample. The most popular multiway analysis methods are the generalized rank annihilation method (GRAM), parallel factor analysis (PARAFAC), and multivariate curve resolution alternating least square (MCR-ALS) [reviewed in Escandar and Olivieri (2019)]. Other methodologies use latent variables for processing second-order data, such as principal component regression (PCR) and partial least squares (PLS) (Dinç et al. 2017).

Nevertheless, HPLC-based methods have been successfully used to detect and quantify various food additives. Recently Riman et al. (2019) described a novel electrochemical detector for HPLC using pencil graphite as a disposable working electrode to detect tocopherols. The electrochemical sensor, in combination with HPLC, exhibited nanomolar concentrations of tocopherols. Aksu Dönmez et al.

(2020) determined sodium benzoate, potassium sorbate, Ponceau-4R, and carmoisine in a beverage using a novel fast HPLC-diode array detection (HPLC-DAD) analytical approach. The average recoveries for all target analytes were 98.27–101.37%, with average relative prediction errors of less than 3%. Ma et al. (2012) reported a method that combines HPLC with a photodiode array detector (PDA) for the simultaneous determination of 20 synthetic food additives at low concentrations (0.005–0.150 μg/mL), including seven food colorants, three sweeteners, nine artificial preservatives, and caffeine. In their study, Ertaş et al. (2007) developed a rapid HPLC with a UV–VIS detector to separate and determine Sudan dyes (I, II, III, and IV) and Para Red in red chili peppers. Interestingly, all dyes were separated in less than 9 min while the limit of ditection (LOD) and limit of quantitation (LOQ) ranged from 1.2 to 5.4 μg/kg and 4 to 18 μg/kg, respectively, depending on the dye. Gao et al. (2013) detected simultaneously 30 synthetic food additives, including preservatives, artificial colorants, and sweeteners in soft drinks, by combining SPE with HPLC/electrospray ionization-MS/MS. All additives were efficiently separated in a single run within 18 min and with LODs ranging from 0.01 to 20 mg/kg, which are acceptable. Further validation of the methods revealed recoveries ranging from 80.8 to 106.4%. Petigara Harp et al. (2013) proposed a new approach for the determination of seven color additives, namely, FD&C Blue No. 1, FD&C Blue No. 2, FD&C Green No. 3, FD&C Red No. 3, FD&C Red No. 40, FD&C Yellow No. 5, and FD&C Yellow No. 6 in food products. The color additives are extracted from a product using one of two procedures developed for various product types, isolated from the noncolored components, and analyzed by liquid chromatography with photodiode array detection. Depending on the color additive, and the food sample, the LOD and LOQ of the method ranged from 13 to 410 μg/kg and from 43 to 2370 μg/kg, respectively, depending on the dye. HPLC-based methods have been successfully also used for the detection of preservatives (e.g., benzoate, sorbate) in beverages (Petanovska-Ilievska et al. 2017), cheese and yogurts (Zamani Mazdeh et al. 2017), color additives and preservatives in juices and energy drink (Aşçı et al. 2016), sweeteners in jams, yogurts, and other food (Bartholomees et al. 2016), color additives in intact formulas, and breakfast cereal (Lee et al. 2016), as well as in other types of foods [for a review on the topic, see Martins et al. (2019), Wu et al. (2021), and Di Stefano et al. (2012)].

In addition to the traditional HPLC, Fast-LC is a significant trend in modern food analysis because there is a continuously increasing demand for the quality control of large samples in the minimum possible time (Fekete et al. 2014a). In the past years, ultra-high-performance (pressure) liquid chromatography (UHPLC) has been intruded, enhancing the separation and throughput capacities of LC (Fekete et al. 2014b). UHPLC uses narrow-bore columns packed with microscopic particles (<2 μm) and mobile phase delivery systems operating at high back pressures. In contrast to the conventional HPLC, where the maximum back pressure ranges from 35 to 40 MPa, in UHPLC, back pressures may reach up to 103.5 MPa (Nováková et al. 2006). Furthermore, the development of detectors has improved, and all of them can be coupled to UHPLC. Common detectors are ultraviolet (UV), refractive index (RI), evaporative light-scattering detector (ELSD), UV-diode-array detector

(UV-DAD), corona aerosol-discharge (CAD), fluorescence detector (FD), and mass spectrometry (MS). Recently, Han et al. (2016) developed an LC–ES-MS/MS method to determine four paraben-type preservatives in plant and animals seafood.

3.2.2 Gas Chromatography

Gas chromatography (GC) is one of the most widely used techniques for analyzing (semi)volatile compounds. The high separation capacity of GC combined with a wide range of detectors makes this technique vital in determining various components in such complex matrices as food crops and products (Hajslova and Cajka 2008). GC mainly combines a mass spectrometer (MS), a powerful detector, and various multivariate data processing tools in the last few decades have been increasingly applied (Cordero et al. 2020).

GC is widely used in the food industry because it provides high selectivity, sensitivity, and resolution, good accuracy and precision, and a wide dynamic concentration range. In GC, the carrier gas (mobile phase), which contains the analytes (sample), flows continuously to the column so that they can be separated and eluted. Subsequently, the carrier gas and sample pass through a detector that measures the sample's quantity. An electrical signal is generated, giving rise to a chromatogram for sample qualitative/quantitative data collection and analysis (Feng et al. 2019). The detector signals are collected and displayed by a computer as a function of time. The area under the peak is proportional to the concentration. A calibration curve is used to determine the amount of substance (Wrolstad et al. 2005).

Furthermore, various detectors, such as MS , flame ionization, thermal conductivity, photoionization, electron capture, flame photometric, and electrolytic conductivity, are available and can be connected to GC. The detector's choice depends on the sample's nature and the desired sensitivity or selectivity in the analysis. Santos and Schug (2017) developed a vacuum UV spectrophotometer detector that measures the gas-phase absorption at 120–240 nm. All chemical compounds present unique absorption spectra at this range, permitting the determination of co-eluting food additives.

However, despite the apparent advantages of GC over other analytical techniques, most (>80%) known compounds are not adequately volatile or stable to be separated by GC, as this method requires volatile and stable samples at operational temperatures ranging from 50 to 300 °C. Furthermore, analysis of food additives using GC involves laborious preparation steps, including liquid-liquid or solid-phase, headspace analysis, and distillation. In addition, GC instrumentation is expensive, while the maintenance cost is relatively high. The analyst should have a strong background and expertise in the technique, limiting its use to small food industries (Martins et al. 2019).

Nevertheless, GC has been widely used for the determination of food additives. In their study, Zheng et al. (2016) employed GC-MS to determine organic acids, sugars, aroma components, and carotenoids in grapefruit pulps. Cacho et al. (2016) optimized a GC-MS method using microvial insert large volume injection to detect three synthetic phenolic antioxidants, namely butylated hydroxytoluene

(BHT), butylated hydroxyanisole (BHA), and tert-butyl hydroquinone (TBHQ), in edible vegetable oils. The matrix-matched calibration method was used to quantify additives, while carvacrol was used as a standard. Depending on the additive, the LODs of the method ranged from 0.08 to 0.10 ng/g. In another study, Tang and Duan (2017) optimized an extraction method using porous ionic liquid polymer to analyze organic acids (acidity regulators), namely fumaric acid, lactic acid, and malic acid in grape wine, using GC-MS. The detection limits of the method were lower than 0.07 ng/ml. The LODs ranged from 0.07 to 1.15 ng/ml, indicating good sensitivity for the organic acids. LOQs (0.40–3.85 ng/ml) also presented an excellent quantitative analysis capability.

Furthermore, Fuller et al. (1964) developed an effective tool for discriminating relevant food flavor components by combining GC with olfactometry. The potential applications of gas chromatography-olfactometry (GC-O) in food flavor dairy products (milk and cheese), coffee, meat, and fruits have been reviewed elsewhere (d'Acampora Zellner et al. 2008).

3.3 Electroanalytical Techniques

The requirements for exact and fast quantification of many vital compounds by food analysis led to the rapid development of various electrochemical techniques (Brainina et al. 2000). Among the electrochemical techniques applied in food analysis, the principal ones are voltammetric and polarographic techniques, while the former is the most widely used. Electroanalytical methods require cheap instrumentation while they provide high sensitivity with wide linear concentration ranges for both inorganic and organic compounds, rapid analysis times (in seconds), and simultaneous determination of several analytes.

In voltammetric techniques, the potential is altered to facilitate an electron transfer reaction. The resulting current is recorded, producing a voltammogram that is used to identify (peak location) and quantification (peak weight) of the target analyte; however, this method can detect only electroactive (reducible or oxidizable) compounds (Mendoza et al. 2015). Furthermore, various sensors have been developed, including graphite, carbon paste, thick film carbon/graphite, and thin film metallic electrodes modified in-situ or beforehand, hanging mercury drop electrodes and mercury-coated glassy carbon electrodes (Herzog et al. 2008). A subdivision of voltammetry is stripping voltammetry (SV), which has received much attention as it is the most sensitive electroanalytical technique (Alghamdi 2010). In SV-based methods, the analyte of interest is accumulated on a working electrode by controlled potential electrolysis. Following a short pause step, a stripping step is applied, which involves the dissolution of the deposit when a linear ramp is applied to the electrode resulting in the production of a detectable current at the electrode surface following the oxidation or reduction of the analyte at a characteristic potential. A peak shape current–potential voltammogram is obtained during the stripping step [see Alghamdi (2010) and references cited therein]. The main advantage of SV is its low detection limits (10^{-10}–10^{-12} mol/L). At the same time, it offers multielement and speciation

Table 6 Examples of analysis of food additives using voltammetric methods

Additive	Category	Sample	Electrode	Reference
Mannitol	Sweetener, stabilizer, anticaking agent	Sugarcane vinasse	AuNP[a]/ RGO-GCE[b]	Beluomini et al. (2017)
Monosodium glutamate	Flavor enhancer	Beef OXO cube	MB[c]-Screen printed carbon	Hughes et al. (2015)
tert-butylhydroquinone butylated hydroxyanisole	Antioxidant	Oil, margarine, and mayonnaise	GCE-MIP[d] MWCNT[e]	dos Santos et al. (2016)
Indigo carmine	Color	Chocolate, jelly power	CPE/NBE	Arvand et al. (2017)
Ascorbic acid	Acidity regulator	Orange juice	Gold	Kuss and Compton (2017)
Butylated hydroxyanisole propyl gallate	Antioxidant	Oil, potato chips, and cookies	Au-Pt nanotubes	Wu et al. (2016)

[a]*AuNP* gold nanoparticles, [b]*RGO-GCE* reduced graphene oxide-glassy carbon electrode, [c]*MB* Meldola's Blue, [d]*MIP* molecularly imprinted polymers, [e]*MWCNT* multi-walled carbon nanotubes

capabilities that are suitable for automated online and in situ measurements and utilizes low-cost instrumentation. The two versions of SV are anodic stripping voltammetry (ASV) and cathodic stripping voltammetry (CSV). In the former method, a cathodic preconcentration step is undertaken, followed by a potential scan toward more positive potentials and the anodic (oxidative) currents are determined. In the latter method, an anodic preconcentration step is undertaken, which is followed by a potential scan toward more negative potentials and the cathodic (reductive) currents are determined (Achterberg et al. 2018).

Voltammetric techniques are also categorized based on how the potential is imposed on the electrode. For example, in square wave voltammetry (SWV) and differential pulse voltammetry (DPV), the responses are based on the superior elimination of the capacitive/background current, exhibiting a sensitivity comparable to chromatographic techniques (Spinola Machado and Cincotto 2022). Over the last years, a variety of voltammetry methods and electrodes have been developed for the determination and qualification of food additives such as preservatives (Schneider et al. 2014), color additives (Li et al. 2016), flavor enhancers (Dorozhko et al. 2015), and others as summarized in Table 6.

4 Analytical Methods for the Determination of Feed Additives

In contrast to food additives used to improve the quality, taste, and smell of the food, feed additives are mainly used to supplement the animals' needs for nutrients (e.g., amino acids) and essential compounds (e.g., vitamins). Other additives prevent invasive diseases (e.g., coccidiostats), reduce oxidation of feed's components (antioxidants), enhance the dietary value of feeds, or improve the quality of food products of animal origin (Rubaj et al. 2012). Feed additives currently authorized within the EU are summarized in the feed additive register (https://food.ec.europa.eu/safety/animal-feed_en) and represent an entirely different group of products, including pure substances, preparations, and microorganisms. Five classes of additives are identified: technological, organoleptic, nutritional, zoo-technical, coccidiostats, and histomonostatic. The main categories of feed additives examined in this chapter are illustrated in Table 7.

Specific attention should be given to the feed sampling to determine additives. The sampling procedure is a multistage process involving sampling and sample preparation, and the sample should represent the whole material. As feed additives have various characteristics and physicochemical properties, the EURL-for feed additives (EURL-FA; https://joint-research-centre.ec.europa.eu/eurl-fa-eurl-feed-additives_en) suggests the storage of the samples at three different temperature

Table 7 Examples of feed additives

Additive	Purpose	Example(s)
Antibiotics[a]	• Treatment of infection • Improvement of growth	Bacitracin, colistin, virginiamycin
Coccidiostats	• Elimination of Coccidia parasites	Amprolium, Roxarsone, Sulfadiazine
Natural extracts	• Alternative to antibiotics • Promote livestock performance	Thymol, Oregano essential oils
Vitamins	• Regulators of key metabolic pathways	Vitamin D, vitamin E
Minerals	• Components of various biomolecules (e.g., enzyme co-factors) • Essential elements of the poultry immune system	Zinc, manganese, and selenium
Antioxidants	• Prevent deterioration caused by oxidation • Improve feed quality	Ethoxyquin, BHA[b], BHT[c]
Color additives	• Improve the acceptance of the product by customers	Carotenoids
Enzymes	• Improve animal performance by hydrolyzing feed substrates that are only partially degraded by the animal's endogenous enzymes	Proteases Xylanases

[a]The use of antibiotics as growth promoters in animal farms has been banned as of 2003
[b]*BHA* Butylated Hydroxyanisole
[c]*BHT* Butylated hydroxytoluene

conditions, i.e., RT (~25 °C), fridge (4 °C), and freezer (−18 °C). A certificate of analysis and information about the product's shelf-life is also required for transporting the samples. The EURL-FA repository contains about 2100 samples, corresponding to 700 feed additives (Omar et al. 2017).

All feed evaluation techniques aim to identify the degree to which individual feedstuffs contribute to the animal's nutritional requirements (Mould 2003). Although various analytical methods have been developed for determining and quantifying additives in animal feeds, the methods of choice are spectrophotometric and chromatographic methods. For example, spectrometric techniques have also been employed to identify feed additives. Since the ban on growth-promoting antibiotics throughout the EU, essential oils have been used increasingly as feed additives. Spectrometric methods were developed to ensure a reliable authenticity assessment of those oils (Greule et al. 2008). Spectrophotometric methods have also been used to determine the levels of water-soluble proteins, including B_1, B_2, and B_6.

HPLC-based methods have been widely used in detecting and quantifying various feed additives, including antioxidants, amino vitamins, feed colorants, and acids, in animal feeds and premixes. HPLC methods have also been employed for detecting coccidiostats in feed preparations, premixes, and blends. Although IEC is the method of choice for determining amino acids in animal feeding, HPLC methods are becoming increasingly popular as they make the analyses shorter in time. As mentioned before, the main advantages of HPLC include a plethora of stationary phases, high sensitivity, and ease of identifying and quantifying multiple components in several sample types with a high resolution.

Furthermore, the EURL-FA recommended that all feed additive samples be analyzed using two complementary vibrational spectroscopic techniques, i.e., NIR and Raman spectroscopy. These techniques can characterize the feed additives in a non-destructive manner without time-consuming pretreatment steps (Ambrose et al. 2016; Graham et al. 2012). The evaluation of spectra often requires multivariate statistics, which is a handy tool when differentiating compounds considering all the information in the spectrum simultaneously (Cozzolino 2014). Omar et al. (2017) used an IR-based method to analyze feed additive samples from products containing (1) selenium (Se), (2) zinc (Zn) and manganese (Mn), (3) vitamins, (4) essential oils such as thyme and oregano. Chemometrics methods were used to analyze the samples, especially as spectra differentiation by visual inspection was challenging (Omar et al. 2017). In their study, Owens et al. (2009) used NIR to monitor additives in wheat for broiler chickens. This study highlighted that NIR is a valuable tool in accurately and rapidly determining wheat chemical parameters and nutritional value and could be highly beneficial to poultry and wheat industries.

Laboratories performing official feed analyses must be accredited according to the ISO/IEC 17025:2005 standard, and the analytical methods used must be validated (Borràs et al. 2011). Optimization of analytical methodologies is essential in developing multi-residue methods because different additives exhibit different physicochemical characteristics. The following paragraphs discuss the main analytical techniques employed for analyzing the main categories of additives in animal feeds.

4.1 Antibiotics

Antibiotics in feed have been used in animal production in Europe since 1953. Antibiotics have been widely used as animal growth promoters (AGPs) to improve the health and well-being of animals. However, the misuse and overuse of nontherapeutic antibiotics in animals have contributed to the development of antimicrobial resistance strains (ABR) (Paulson et al. 2015). To this end, regulation EC 1831/2003 (available at https://www.legislation.gov.uk/eur/2003/1831) banned all antibiotics, other than coccidiostats and histomonostats, as feed additives. Therefore, monitoring the presence of antibiotics in animal feed is of particular interest. Although various methods have been developed for detecting antibiotics in animal feed, such as column and thin layer chromatography (TLC), these methods lack specificity and sensitivity. Thus, novel methods for detecting antibiotic contamination in animal feed are sorely needed. To this end, various LC-based methods have been developed over the past years. For example, Macy and Loh (1983) reported a fast and simple HPLC method for determining monensin, a polyether antibiotic, in feed premixes. Cronly et al. (2010) developed and validated a rapid multi-class method to confirm virginiamycin and 13 prohibited medicinal additives in pig and poultry compound feed with HPLC-MS/MS. In their study, Van Poucke et al. (2005) used LC-MS to detect spiramycin, olaquindox, bacitracin, tylosin, and virginiamycin in animal feeds. Sin et al. (2004) employed a sensitive LC-ESI-MS method for determining lincomycin and virginiamycin M_1 in swine muscle, liver, and kidney. De Alwis and Heller (2010)) used LC-ion trap tandem MS to detect virginiamycin M_1 and another 12 antibiotics residues in distillers grains. Tao et al. (2018) reported a quantitative LC-MS/MS based on matrix solid-phase dispersion for simultaneous determination of various antibiotics in animal feeds, including bacitracin A, bacitracin B, colistin A, colistin B, and virginiamycin M1. These additives have been extensively used as growth promoters to prevent/control diseases in farm animals (Smith and Weinberg 1962).

Another example is the detection of Zn-Bacitracin (Zn-BC), one of the most common antibiotics used as additives in animal feeds due to its growth-promoting effects. However, the use of Zn-BC in animal feed has been banned since 1999 (Phillips 1999). For the detection of Zn-BC, several microbiological methods have been developed [see Capitan-Vallvey et al. (2002) and references cited therein]. As investigating and determining the presence of banned substances in feedstuffs is essential, several chemical methods have been reported recently to determine BC in animal feeds.

Capitan-Vallvey et al. (2002) reported a sensitive and selective method for determining Zn-BC in adulterated animal feed by reversed phase ion-pair high-performance liquid chromatography and post-column derivatization with *o*-phthalaldehyde before fluorescence detection. The same group has proposed a method to determine Zn-BC in feedstuffs by HPLC with ultraviolet detection at levels from 5.0 to 200.0 mg kg^{-1} (Capitán-Vallvey et al. 2001). Webster (2020) developed an HPLC method to determine bacitracin methylene disalicylate, another stable form of BC also used as an additive in animal diets, in feedstuffs at 50.0 and

250.0 mg kg^{-1}. Hormazábal and Yndestad (2000) reported a rapid assay for determining Zn-BC in chicken and swine feed by LC-MS. For the analysis of colistin (polymyxin E) in animal feed, Morales-Muñoz and de Castro (2005) used a dynamic ultrasound-assisted extraction (UAE) method, an online pre-column derivatization/HPLC coupled with a fluorescence detector. Kaufmann and Widmer (2013) described an HPLC-MS/MS method to determine residual polymyxin B_1, polymyxin B_2, bacitracin, colistin A, and colistin B in muscle, liver, kidney, egg, and milk. Govaerts et al. (2002) investigated the major polymyxins of the B and E series by direct infusion in time of flight (TOF) and ion trap devices. Sniegocki et al. (2014) determined metabolites of carbadox (quinoxaline-2-carboxylic, desoxycarbadox) and olaquindox (3-methylquinoxaline-2-carboxylic acid) residues in pig muscle tissue using LC-MS/MS. Following a one-step preparation of samples by extraction and clean-up, they underwent LC separation using a C8 column with a gradient system consisting of isopropanol/water/acetic acid and methanol as the mobile phase. The decision limit of the method varied from 1.04 to 2.11 μg/kg. This method was validated based on the European Commission Decision 2002/657/EC.

4.2 Coccidiostats

Coccidiostats are often used to prevent poultry infections from the genus *Eimeria* that cause *coccidiosis*. Maximum levels of coccidiostats in food and feed have been determined by the European Union (EU) [reviewed in Martins et al. (2022)]. Several methods have been reported for determining coccidiostat residues in eggs, chicken, and feed, including LC coupled with UV or fluorescence detection systems and LC-MS [see Shao et al. (2009) and references cited therein]. Polyether antibiotics used to control coccidiosis exhibit low UV absorbance; thus, a time-consuming derivation is required when a UV or fluorescence detection system is used. Therefore, LC-MS is the preferred method for the analysis of coccidiostats.

Shao et al. (2009) developed a rapid LC–MS–MS method for multi-class determination of 14 coccidiostat residues in eggs and chicken. Bienenmann-Ploum et al. (2012) reported a multiplex flow cytometric immunoassay for detecting six coccidiostats (i.e., salinomycin, lasalocid, narasin, monensin, maduramicin, and nigericin residue) in feed and eggs. The method was subsequently validated according to the EURL guidelines (Bienenmann-Ploum et al. 2013). In the study by Olejnik et al. (2011), the samples were analyzed using a multi-residue LC-MS/MS method, validated according to the 2002/657/EC, enabling the determination of 12 coccidiostats. In a recent study, González-Rubio et al. (2020) used supramolecular solvents (SUPRAS) during the extraction step and dispersive SPE, to determine several coccidiostats, including monensin, narasin, lasalocid, salinomycin, maturemycin, and semduramicin, in swine fat, chicken muscle, bovine muscle eggs, and milk. Barreto et al. (2017) used a straightforward multi-residue method (LC–QqLIT-MS/MS) to detect 14 coccidiostats in chicken muscle samples and eggs. An extraction step was followed by a cleaning step at low temperatures before the analysis. In their study, Matus and Boison (2016) developed and validated a sensitive multi-residue LC-MS/MS to determine 17 anticoccidials in poultry meat. Before the analysis, an extraction and clean-up step was performed using a solvent mixture with salts, followed by centrifugation, dilution, and filtration. Buiarelli et al. (2017) developed a sensitive and straightforward LC-MS/MS method for simultaneously detecting seven analytes (five additives and two metabolites) in eggs. Yoshikawa et al. (2017) proposed an LC-MS method to determine 37 compounds, including the coccidiostats lasalocid, maduramicin, diclazuril, and nicarbazin, in processed chicken samples. Extraction of additives was carried out with ethyl acetate followed by acetonitrile. Dasenaki and Thomaidis (2019) used hydrophilic interaction LC with tandem MS detection (HILIC-MS/MS) to determine 16 coccidiostats in chicken muscle and eggs. The method exhibited high sensitivity (LOD <0.6 mg/kg for all analytes). Rusko et al. (2019) proposed a fast and selective method by combining LC and an Orbitrap high-resolution mass spectrometer as a detector (LC-Orbitrap-HRMS) to determine 17 coccidiostats in birds and eggs. Extraction of analytes was carried out using acetonitrile, followed by purification at 0 °C, preconcentration, reconstitution, and filtration.

4.3 Natural Extracts and Essential Oils

As mentioned above, the EU has banned antibiotic use in animal farms. However, antibiotic removal has led to animal performance problems and a rise in the incidence of certain animal diseases, e.g., necrotic enteritis. Essential oils (EOs) and natural extracts have been considered an alternative to antibiotics in animal feed, possibly substituting antibiotics. The results of some studies demonstrated the efficacy of EOs as feed additives in animal breeding (Cheng et al. 2018). EOs have been shown as promising substitutes for antibiotics in calves (Santos et al. 2015). Furthermore, specific essential oil blends and probiotics used as dietary supplementation have been shown to promote broiler growth (Giannenas et al. 2014a; Oviedo-Rondón et al. 2010) and in combination with benzoic acid have improved gut integrity and intestinal microbiota in turkey poults (Giannenas et al. 2014b). The potential of EOs as an alternative to antibiotics has been recently reviewed (Horky et al. 2019). However, the toxicity of EOs remains inconclusive; thus, the presence and levels of EOs in feeds should be monitored. Analysis of natural extracts in feeds is usually carried out using GC (Ching et al. 2012) and LC coupled to different detectors (Cantalapiedra et al. 2014), usually MS, but UV photometers (Ghiasvand et al. 2015), fluorimetric (Viñas et al. 2006), and electrochemical detectors (Cantalapiedra et al. 2014) have also been reported for liquid chromatography. Jiménez-Salcedo and Tena (2017) used a pressurized liquid extraction method and GC-MS to determine cinnamaldehyde, carvacrol, and thymol in feedstuff additives. Truzzi et al. (2021) reported the optimization and validation of an HPLC-based process with a UV detector to analyze hesperidin and carvacrol in animal feed.

4.4 Vitamins

Vitamins are low-molecular-weight organic chemicals participating in biochemical reactions and metabolism of the animal body. Vitamins are the most widely used additives in animal feeds. The official methods of the Association of Official Analytical Chemists (AOAC 2019) for vitamins A, D, and E are separate assays with detection based on colorimetry. The complexity and time-consuming assays make the AOAC methods challenging to use for routine analysis. Qian and Sheng (1998) reported a straightforward analytical method using one-step extraction and simultaneous determination of vitamins A, D, and E and pro-vitamin D_2 in animal feeds by HPLC. The detection limit for the four vitamins was 10 ng/g of the feed sample. HPLC-based methods for determining nicotinic acid and nicotinamide vitamins B_1, B_2, and B_6, in mineral preparations and mixtures and vitamin B_1 in feed mixtures and premixes were published. Cohen and Lapointe (1979) described the separation and quantification of Vitamin D3 from different feeds and premixes using HPLC. A liquid partition clean-up step was initially carried out, followed by chromatography on activated Silica gel 60. Subsequently, samples were analyzed on an HPLC using a LiChrosorb NH_2 column in only 15 min. A variable wavelength UV detector was used and set at 264 nm. The HPLC methods for the determination of both soluble and insoluble vitamins have been previously reviewed (Rubaj et al. 2012).

4.5 Minerals

Minerals are essential components of all cells, including soft tissue, nerves, blood, hormones, muscles, bones, and teeth (Bhagwat et al. 2021). Specific minerals are essential components of biocatalysts (enzymes) that catalyze biochemical reactions, including metabolism, energy production, muscle contraction, nerve-impulse transmission, and cell permeability (Bhagwat et al. 2021). Among the nutrients in the chicken diet, the microminerals have received particular attention as they exert essential functions in the organism. Significantly, the immune function of all poultry species depends on trace minerals. These compounds can be provided in inorganic (traditional) and organic (chelate) forms (Aksu et al. 2012). However, the overuse of inorganic minerals in poultry farms has led to severe environmental contamination because they have lower inclusion rates and reduced excretion (Bao and Choct 2009).

Furthermore, copper sulfate and zinc oxide, which are used as sources of copper and zinc used in the poultry diet, are obtained from the steel industry. Thus, they contain high amounts of contaminants (e.g., fluorine and cadmium), which are transferred into the poultry diet (Lu et al. 2007). Recent evidence has highlighted the use of minerals in organic form to replace inorganic minerals (Ao and Pierce 2013). The National Research Council (NRC) (Council NR 1994) has recommended specific nutrient requirements for poultry, including the levels of minerals such as Zinc (Zn), Manganese (Mn), Copper (Cu), Iron (Fe), Selenium (Se), and Iodine (I).

Thus, analytical methods for determining both inorganic and organic minerals and contaminants in poultry feed are required to monitor their levels. Wang et al. (2010) developed a novel high-performance liquid chromatography coupled to an inductively coupled plasma mass spectrometer (HPLC-ICP-MS) method for determining organic arsenic feed additives and speciation of anionic arsenic in animal feed.

4.6 Antioxidants

Production animals are often exposed to several oxidative stress conditions. Antioxidants are natural or synthetic substances that may prevent or delay some types of cell damage, and they are mainly added to diets containing high concentrations of lipids (Wang et al. 2022). Antioxidants should also be added in feeds susceptible to oxidation due to unfavorable storage conditions. Common antioxidants used in animal feed to protect them against lipid peroxidation are vitamin C, citric acid, ethoxyquin (EQ, 6-ethoxy-1,2-dihydro-2,2,4-trimethylquinoline) (Błaszczyk et al. 2013), butylated hydroxyanisole (BHA), and butylated hydroxytoluene (BHT) (Babich 1982). However, vitamin C and citric acid are too expensive to be used as antioxidants in most piglet diets. Natural antioxidants are frequently marketed, but their effectiveness remains inconclusive.

Some harmful effects in animals and people occupationally exposed to antioxidants were reported in the 1980s, resulting in new studies to reevaluate their toxicity. Thus, the maximum amount of antioxidants in feed mixtures for animals used for food production should be monitored and controlled (Wang et al. 2022). The presence and levels of antioxidants in animal feed in routinely monitored with HPLC methods [reviewed in Rubaj et al. (2012)].

4.7 Color Additives

Carotenoids are included in the feed as part of the nutrients that support animal health and product quality. Carotenoids have various biological effects on animals which are classified into functions, actions, and associations. The most common carotenoids used in animal nutrition are astaxanthin, capsanthin, β-carotene, β-apo-8′-carotenoic acid ethyl ester, lutein, zeaxanthin, and canthaxanthin. Carotenoids have been additives used in poultry feed for pigmentation of egg yolk, skin (and fat), legs, beak, comb, and feathers (Langi et al. 2018).

Like other additives, the presence and levels of carotenoids and other color additives in animal feeds should be monitored and controlled. Vincent et al. (2021) developed a reversed-phase HPLC coupled to an optical detector for the detection and quantification of or the determination of the ten carotenoids authorized by European legislation as feed additives in fish and poultry feed, namely astaxanthin (AXN), canthaxanthin (CXN), adonirubin (ADR), astaxanthin dimethyl succinate (AXN DMDS), capsanthin (CSN), beta-carotene (BCAR), the ethyl ester of beta-apo-8′-carotenoic acid (BACARE), citranaxanthin (CIXN), lutein (LUT), and zeaxanthin (ZEA). Notably, these additives are used as colorants under the category "sensory additives" (European Union 2003) to adapt the color of egg yolk and fish flesh to the consumer's preferences. The relative standard deviations for repeatability (RSDr) varied from 2.2 to 16.2%, with a mean value of 6%. In comparison, the relative standard deviations for reproducibility (RSDR) ranged from 6.8 to 39%, with a mean value of 21%.

4.8 Enzymes

Feed enzymes improve animal performance by hydrolyzing feed substrates that are only partially or not broken down by the animal's endogenous enzymes, especially in the case of young animals whose gastrointestinal tract (GIT) is relatively immature and unable to produce adequate amounts of endogenous enzymes (Ravindran 2013). For example, monogastric animals can only partially degrade and utilize non-starch polysaccharides (NSP), such as xylans, β-glucans, and cellulose. Thus, adding enzyme preparations containing xylanases, β-glucanases, and cellulases to the feed reduces the antinutritional effect of NSP (Kiarie et al. 2013). Furthermore, exogenous enzymes have been considered as alternatives to antibiotic growth promoters (Cheng et al. 2014). Enzymes used in animal nutrition (feed enzymes) are considered feed additives and thus regulated by Regulation 1831/2003 on feed additives (https://www.legislation.gov.uk/eur/2003/1831/contents).

Since enzyme users must know the enzyme product's activity to formulate their diets rationally, it is necessary to assay enzymes in the feed. Various classic biochemical methods are available to analyze enzymes in feed (Sabatier and Fish 1996). For example, the NSP-hydrolyzing enzymes degrade the polysaccharide chains into more and increasingly smaller fragments. Some approaches for determining NSP-hydrolyzing activity monitor the decreasing degree of polymerization of the substrate by measuring the decrease in the assay solution's viscosity. Other methods measure the increase in reducing equivalents in the assay solution using an oxidizing reagent. The cleavage of the glycosidic bond forms a hemiacetal group that is oxidized. Other methods utilize dye-labeled substrates. The NSP-hydrolyzing enzymes progressively produce water-soluble fragments, increasing the absorbance of the supernatant molecules linked randomly to the building blocks of the model substrates (König et al. 2002). Over the past years, alternative methods have been developed to detect feed enzyme activity. König et al. (2002) described a straightforward and accurate approach for determining xylanase, beta-glucanase, and cellulase in animal feeds. The method is based on the measurement of reducing moieties released by the enzymes from arabinoxylan, beta-glucan, or carboxymethylcellulose (CMC) and is independent of enzyme standards. Wang et al. (1999) reported an enzyme-linked sorbent assay (ELSA) for determining beta-glucanase and xylanase activity by coating the wells of a micro-well plate with biotinylated beta-glucan and biotinylated arabinoxylan, respectively. The activity of the bound indicator enzyme, alkaline phosphatase, is proportionally related to the beta-glucanase activity in the sample. ELSA is a straightforward method which can be employed for the routine analyses assay of a high number samples (~ 200 per analyst/day) with good precision and high sensitivity.

5 Analytical Trends in Determining Additives in Foods and Feeds

The primary goals in analyzing food and feed additives, i.e., to achieve better accuracy, lower detection limits, and higher selectivity with more accessible, faster, and low-cost methods using more robust, highly versatile, and smaller instruments, have not changed the past years. Furthermore, lower detection limits and higher selectivity goals have developed into actual trends. Overall, many techniques today provide greater sample throughput with more ease due to automation. Still, they are rarely cheaper (Lehotay and Hajšlová 2002).

Several novel methods have been developed to analyze food and feed additives. For example, Chen et al. (2021) reported a selective and sensitive sensor using Ag-NPs as a colorimetric probe to detect azodicarbonamide, a whitening agent of cereal flour. Well-dispersed Ag-NPs aggregated following reaction with glutathione via Ag-SH covalent bonding and electrostatic attraction. This reaction causes changes in the color of the mixture from bright yellow to red. In this work, the approximate concentration of azodicarbonamide was estimated by the naked eye, with a visual detection at 0.33 μM. This method exhibited good linearity from 0.33 μM to 1.7 μM. In addition, Han et al. (2017) reported a rapid and sensitive detection of sodium saccharin in soft drinks by silver nanorod array SERS substrates. Sodium saccharin was detected with an excellent LOD and a broad linear range. A one-step liquid–liquid extraction procedure was applied to detect sodium saccharin from soft drinks. However, these methods should be validated for routine food and feed additives analysis. The critical requirement for this validation is to prevent false-negative results; therefore, the β-error at the level of interest should be lower than or equal to 5%.

6 Conclusions

This chapter highlights the importance of balancing cost, speed, and analytical quality for determining and quantifying food and feed additives. Various analytical methods have been developed to assess food and feed additives. The choice of the method depends on several factors, including the nature of the sample and additive and the legislation at national and international levels. The main drawback of the current state-of-the-art food/feed analysis technologies is the necessity of a pretreatment step such as filtration, dilution, pH adjustment, extraction, heating, etc. Thus, more advanced food and feed analysis tools are needed that can be used without expert knowledge or bulky/expensive instrumentation.

References

Achterberg EP, Gledhill M, Zhu K (2018) Voltammetry-Cathodic stripping. In: Reedijk J (ed) Reference module in chemistry, molecular sciences and chemical engineering. Elsevier, Cambridge, MA

Akash MSH, Rehman K (2020) Introduction to spectrophotometric techniques. In: Akash MSH, Rehman K (eds) Essentials of pharmaceutical analysis. Springer Nature Singapore, Singapore, pp 19–27

Aksu Dönmez Ö, Dinç-Zor Ş, Aşçı B et al (2020) Quantitative analysis of food additives in a beverage using high performance liquid chromatography and diode array detection coupled with chemometrics. J AOAC Int 103(3):779–783. https://doi.org/10.1093/jaocint/qsz009

Aksu D, Aksu T, Önel E (2012) Does Inclusion at low levels of Organically complexed minerals versus inorganic forms create a weakness in performance or antioxidant defense system in broiler diets? Int J Poult Sci 11:666–672. https://doi.org/10.3923/ijps.2012.666.672

Alghamdi AH (2010) Applications of stripping voltammetric techniques in food analysis. Arab J Chem 3(1):1–7. https://doi.org/10.1016/j.arabjc.2009.12.001

Alkhalaf NA, Osman AK, Salama KA (2010) Monitoring of aflatoxins and heavy metals in some poultry feeds. Afr J Food Sci 4(4):192–199. https://doi.org/10.5897/AJFS.9000193

Altınöz S, Toptan S (2003) Simultaneous determination of Indigotin and Ponceau-4R in food samples by using Vierordt's method, ratio spectra first order derivative and derivative UV spectrophotometry. J Food Compost Anal 16(4):517–530. https://doi.org/10.1016/S0889-1575(03)00022-X

Ambrose A, Lohumi S, Lee W-H et al (2016) Comparative nondestructive measurement of corn seed viability using Fourier transform near-infrared (FT-NIR) and Raman spectroscopy. Sens Actuators B Chem 224:500–506. https://doi.org/10.1016/j.snb.2015.10.082

Amit SK, Uddin MM, Rahman R et al (2017) A review on mechanisms and commercial aspects of food preservation and processing. Agric Food Secur 6(1):51. https://doi.org/10.1186/s40066-017-0130-8

Ao T, Pierce J (2013) The replacement of inorganic mineral salts with mineral proteinates in poultry diets. Worlds Poult Sci J 69(1):5–16. https://doi.org/10.1017/S0043933913000019

AOAC (2019) Official methods of analysis, 21st edn. AOAC, Washington

Arvand M, Saberi M, Ardaki MS et al (2017) Mediated electrochemical method for the determination of indigo carmine levels in food products. Talanta 173:60–68. https://doi.org/10.1016/j.talanta.2017.05.062

Aşçı B, Dinç Zor Ş, Aksu DÖ (2016) Development and validation of HPLC method for the simultaneous determination of five food additives and caffeine in soft drinks. Int J Anal Chem 2016:2879406. https://doi.org/10.1155/2016/2879406

Babich H (1982) Butylated hydroxytoluene (BHT): a review. Environ Res 29(1):1–29. https://doi.org/10.1016/0013-9351(82)90002-0

Bahadoran Z, Mirmiran P, Jeddi S et al (2016) Nitrate and nitrite content of vegetables, fruits, grains, legumes, dairy products, meats and processed meats. J Food Compos Anal 51:93–105. https://doi.org/10.1016/j.jfca.2016.06.006

Bajkacz S, Kycia-Słocka E (2020) Liquid chromatography in food analysis. In: Pico Y (ed) Chemical analysis of food, 2nd edn. Academic Press, Washington, DC, pp 391–455

Baker MJ, Trevisan J, Bassan P et al (2014) Using Fourier transform IR spectroscopy to analyze biological materials. Nat Protoc 9(8):1771–1791. https://doi.org/10.1038/nprot.2014.110

Bao Y, Choct M (2009) Trace mineral nutrition for broiler chickens and prospects of application of organically complexed trace minerals: a review. Anim Prod Sci 49(4):269–282. https://doi.org/10.1071/EA08204

Barreto F, Ribeiro C, Hoff RB et al (2017) A simple and high-throughput method for determination and confirmation of 14 coccidiostats in poultry muscle and eggs using liquid chromatography - quadrupole linear ion trap - tandem mass spectrometry (HPLC-QqLIT-MS/MS): validation according to European Union 2002/657/EC. Talanta 168:43–51. https://doi.org/10.1016/j.talanta.2017.02.007

Bartholomees U, Struyf T, Lauwers O et al (2016) Validation of an HPLC method for direct measurement of steviol equivalents in foods. Food Chem 190:270–275. https://doi.org/10.1016/j.foodchem.2015.05.102

Belton PS (1993) New methods for monitoring changes in proteins. Food Rev Int 9(4):551–573. https://doi.org/10.1080/87559129309540978

Beluomini MA, da Silva JL, Sedenho GC et al (2017) D-mannitol sensor based on molecularly imprinted polymer on electrode modified with reduced graphene oxide decorated with gold nanoparticles. Talanta 165:231–239. https://doi.org/10.1016/j.talanta.2016.12.040

Bhagwat VG, Balamurugan E, Rangesh P (2021) Cocktail of chelated minerals and phytogenic feed additives in the poultry industry: a review. Vet World 14(2):364–371. https://doi.org/10.14202/vetworld.2021.364-371

Bienenmann-Ploum ME, Huet AC, Campbell K et al (2012) Development of a five-plex flow cytometric immunoassay for the simultaneous detection of six coccidiostats in feed and eggs. Anal Bioanal Chem 404(5):1361–1373. https://doi.org/10.1007/s00216-012-6214-1

Bienenmann-Ploum ME, Vincent U, Campbell K et al (2013) Single-laboratory validation of a multiplex flow cytometric immunoassay for the simultaneous detection of coccidiostats in eggs and feed. Anal Bioanal Chem 405(29):9571–9577. https://doi.org/10.1007/s00216-013-7362-7

Błaszczyk A, Augustyniak A, Skolimowski J (2013) Ethoxyquin: an antioxidant used in animal feed. Int J Food Sci 2013:585931. https://doi.org/10.1155/2013/585931

Blokhuis HJ, Hopster H, Geverink NA et al (1998) Studies of stress in farm animals. Comp Haematol Int 8(2):94–101. https://doi.org/10.1007/BF02642498

Borràs S, Companyó R, Granados M et al (2011) Analysis of antimicrobial agents in animal feed. Trends Analyt Chem 30(7):1042–1064. https://doi.org/10.1016/j.trac.2011.02.012

Bradford GE (1999) Contributions of animal agriculture to meeting global human food demand. Livest Prod Sci 59(2):95–112. https://doi.org/10.1016/S0301-6226(99)00019-6

Brainina KZ, Malakhova NA, Stojko NY (2000) Stripping voltammetry in environmental and food analysis. Fresenius J Anal Chem 368(4):307–325. https://doi.org/10.1007/s002160000525

Buiarelli F, Di Filippo P, Riccardi C et al (2017) Liquid chromatography tandem mass spectrometry snalysis of synthetic coccidiostats in eggs. Separations 4(2):15

Buldini PL, Ricci L, Sharma JL (2002) Recent applications of sample preparation techniques in food analysis. J Chromatogr A 975(1):47–70. https://doi.org/10.1016/S0021-9673(02)01335-3

Cacho JI, Campillo N, Viñas P et al (2016) Determination of synthetic phenolic antioxidants in edible oils using microvial insert large volume injection gas-chromatography. Food Chem 200: 249–254. https://doi.org/10.1016/j.foodchem.2016.01.026

Cantalapiedra A, Gismera MJ, Sevilla MT et al (2014) Sensitive and selective determination of phenolic compounds from aromatic plants using an electrochemical detection coupled with HPLC method. Phytochem Anal 25(3):247–254. https://doi.org/10.1002/pca.2500

Capitán-Vallvey LF, Navas N, Titos A et al (2001) Determination of the antibiotic zinc bacitracin in animal food by high-performance liquid chromatography with ultraviolet detection. Chromatographia 54(1–2):15–20. https://doi.org/10.1007/BF02491826

Capitan-Vallvey LF, Titos A, Checa R et al (2002) High-performance liquid chromatography determination of Zn-bacitracin in animal feed by post-column derivatization and fluorescence detection. J Chromatogr A 943(2):227–234. https://doi.org/10.1016/S0021-9673(01)01442-X

Capuano E, van Ruth SM (2016) Infrared spectroscopy: applications. In: Caballero B, Finglas PM, Toldrá F (eds) Encyclopedia of food and health. Academic Press, Washington, DC, pp 424–431

Carocho M, Barreiro MF, Morales P et al (2014) Adding molecules to food, pros and cons: a review on synthetic and natural food additives. Compr Rev Food Sci Food Saf 13(4):377–399. https://doi.org/10.1111/1541-4337.12065

Castritius S, Geier M, Jochims G et al (2012) Rapid determination of the attenuation limit of beer using middle-infrared (MIR) spectroscopy and a multivariate model. J Agric Food Chem 60(25):6341–6348. https://doi.org/10.1021/jf3006382

Che W, Sun L, Zhang Q et al (2017) Application of visible/near-infrared spectroscopy in the prediction of azodicarbonamide in wheat flour. J Food Sci 82(10):2516–2525. https://doi.org/10.1111/1750-3841.13859

Chen F, Liu L, Zhang W et al (2021) Visual determination of azodicarbonamide in flour by label-free silver nanoparticle colorimetry. Food Chem 337:127990. https://doi.org/10.1016/j.foodchem.2020.127990

Cheng G, Hao H, Xie S et al (2014) Antibiotic alternatives: the substitution of antibiotics in animal husbandry? Front Microbiol 5:217. https://doi.org/10.3389/fmicb.2014.00217

Cheng C, Xia M, Zhang X et al (2018) Supplementing oregano essential oil in a reduced-protein diet improves growth performance and nutrient digestibility by modulating intestinal bacteria, intestinal morphology, and antioxidative capacity of growing-finishing pigs. Animals (Basel) 8(9):159. https://doi.org/10.3390/ani8090159

Ching J, Soh WL, Tan CH et al (2012) Identification of active compounds from medicinal plant extracts using gas chromatography-mass spectrometry and multivariate data analysis. J Sep Sci 35(1):53–59. https://doi.org/10.1002/jssc.201100705

Cohen H, Lapointe M (1979) Quantitative analysis of vitamin D3 in a feed using normal phase high pressure liquid chromatography. J Chromatogr Sci 17(9):510–513. https://doi.org/10.1093/chromsci/17.9.510

Cordero C, Cagliero C, Liberto E et al (2020) Gas chromatography. In: Pico Y (ed) Chemical Analysis of Food, 2nd edn. Academic Press, Washington, DC, pp 457–498

Coskun O (2016) Separation techniques: chromatography. North Clin Istanb 3(2):156–160. https://doi.org/10.14744/nci.2016.32757

Council NR (1994) Nutrient requirements of poultry: Ninth revised edition. The National Academies Press, Washington, DC

Cozzolino D (2011) Infrared methods for high throughput screening of metabolites: food and medical applications. Comb Chem High Throughput Screen 14(2):125–131. https://doi.org/10.2174/138620711794474105

Cozzolino D (2014) An overview of the use of infrared spectroscopy and chemometrics in authenticity and traceability of cereals. Food Res Int 60:262–265. https://doi.org/10.1016/j.foodres.2013.08.034

Cronly M, Behan P, Foley B et al (2010) Development and validation of a rapid multi-class method for the confirmation of fourteen prohibited medicinal additives in pig and poultry compound feed by liquid chromatography–tandem mass spectrometry. J Pharm Biomed Anal 53(4):929–938. https://doi.org/10.1016/j.jpba.2010.06.027

d'Acampora Zellner B, Dugo P, Dugo G et al (2008) Gas chromatography-olfactometry in food flavour analysis. J Chromatogr A 1186(1–2):123–143. https://doi.org/10.1016/j.chroma.2007.09.006

Dais P, Spyros A, Christophoridou S et al (2007) Comparison of analytical methodologies based on 1 h and 31p NMR spectroscopy with conventional methods of analysis for the determination of some olive oil constituents. J Agric Food Chem 55(3):577–584. https://doi.org/10.1021/jf061601y

Dasenaki ME, Thomaidis NS (2019) Multi-residue methodology for the determination of 16 coccidiostats in animal tissues and eggs by hydrophilic interaction liquid chromatography - Tandem mass spectrometry. Food Chem 275:668–680. https://doi.org/10.1016/j.foodchem.2018.09.138

Davies T (2004) Book review: infrared spectroscopy: fundamentals and applications. NIR News 15(6):12–12. https://doi.org/10.1255/nirn.794

De Alwis H, Heller DN (2010) Multiclass, multiresidue method for the detection of antibiotic residues in distillers grains by liquid chromatography and ion trap tandem mass spectrometry. J Chromatogr A 1217(18):3076–3084. https://doi.org/10.1016/j.chroma.2010.02.081

De Girolamo A, Lippolis V, Pascale M (2022) Overview of recent liquid chromatography mass spectrometry-based methods for natural toxins detection in food products. Toxins 14(5):328. https://doi.org/10.3390/toxins14050328

Di Stefano V, Avellone G, Bongiorno D et al (2012) Applications of liquid chromatography–mass spectrometry for food analysis. J Chromatogr A 1259:74–85. https://doi.org/10.1016/j.chroma.2012.04.023

Dinç E, Ertekin ZC, Büker E (2017) Multiway analysis methods applied to the fluorescence excitation-emission dataset for the simultaneous quantification of valsartan and amlodipine in tablets. Spectrochim Acta A Mol Biomol Spectrosc 184:255–261. https://doi.org/10.1016/j.saa.2017.04.081

Dorozhko EV, Korotkova EI, Shabaeva AA et al (2015) Electrochemical determination of L-glutamate on a carbon-containing electrode modified with gold by voltammetry. Procedia Chem 15:365–370. https://doi.org/10.1016/j.proche.2015.10.058

dos Santos ME, de Oliveira FM, Scheel GL et al (2016) Synthesis of surface molecularly imprinted poly(methacrylic acid-hemin) on carbon nanotubes for the voltammetric simultaneous determination of antioxidants from lipid matrices and biodiesel. Electrochim Acta 212:322–332. https://doi.org/10.1016/j.electacta.2016.06.174

Duarte LM, Paschoal D, Izumi CMS et al (2017) Simultaneous determination of aspartame, cyclamate, saccharin and acesulfame-K in powder tabletop sweeteners by FT-Raman spectroscopy associated with the multivariate calibration: PLS, iPLS and siPLS models were compared. Food Res Int 99:106–114. https://doi.org/10.1016/j.foodres.2017.05.006

Duraipandian S, Zheng W, Ng J et al (2013) Near-infrared-excited confocal Raman spectroscopy advances in vivo diagnosis of cervical precancer. Biomedical Optics 18(6):067007–067007. https://doi.org/10.1117/1.jbo.18.6.067007

Emerton V, Choi E (2008) Food additives and why they are used. In: Emerton V, Choi E (eds) Essential guide to food additives. The Royal Society of Chemistry, London, pp 1–21

Ertaş E, Özer H, Alasalvar C (2007) A rapid HPLC method for determination of Sudan dyes and Para Red in red chilli pepper. Food Chem 105(2):756–760. https://doi.org/10.1016/j.foodchem.2007.01.010

Escandar GM, Olivieri AC (2019) Multi-way chromatographic calibration-A review. J Chromatogr A 1587:2–13. https://doi.org/10.1016/j.chroma.2019.01.012

Ezeanaka MC, Nsor-Atindana J, Zhang M (2019) Online low-field nuclear magnetic resonance (LF-NMR) and magnetic resonance imaging (MRI) for food quality optimization in food processing. Food Bioproc Tech 12(9):1435–1451. https://doi.org/10.1007/s11947-019-02296-w

Fekete S, Schappler J, Veuthey J-L et al (2014a) Current and future trends in UHPLC. Trends Analyt Chem 63:2–13. https://doi.org/10.1016/j.trac.2014.08.007

Fekete S, Kohler I, Rudaz S et al (2014b) Importance of instrumentation for fast liquid chromatography in pharmaceutical analysis. J Pharm Biomed Anal 87:105–119. https://doi.org/10.1016/j.jpba.2013.03.012

Feng T, Sun M, Song S et al (2019) Gas chromatography for food quality evaluation. In: Zhong J, Wang X (eds) Evaluation technologies for food quality. Woodhead Publishing, Oxford, pp 219–265

Fertah M, Belfkira A, Em D et al (2017) Extraction and characterization of sodium alginate from Moroccan Laminaria digitata brown seaweed. Arab J Chem 10:S3707–S3714. https://doi.org/10.1016/j.arabjc.2014.05.003

Fuller GH, Steltenkamp R, Tisserand GA (1964) The gas chromatograph with human sensor: perfumer model. Ann N Y Acad Sci 116(2):711–724. https://doi.org/10.1111/j.1749-6632.1964.tb45106.x

Gao H, Yang M, Wang M et al (2013) Determination of 30 synthetic food additives in soft drinks by HPLC/electrospray ionization-tandem mass spectrometry. J AOAC Int 96(1):110–115. https://doi.org/10.5740/jaoacint.12-046

Ghafghazi S, Moini Zanjani T, Vosough M et al (2017) Interference-free determination of carbamazepine in human serum using high performance liquid chromatography: a comprehensive rresearch with three-way calibration methods. Iran J Pharm Res 16(1):120–131

Ghiasvand A, Dowlatshah S, Nouraei N et al (2015) A solid-phase microextraction platinized stainless steel fiber coated with a multiwalled carbon nanotube-polyaniline nanocomposite film for the extraction of thymol and carvacrol in medicinal plants and honey. J Chromatogr A 1406: 87–93. https://doi.org/10.1016/j.chroma.2015.06.052
Giannenas I, Papaneophytou C, Tsalie E et al (2014a) The effects of benzoic acid and essential oil compounds in combination with protease on the performance of chickens. J Animal Feed Sci 23: 73–81. https://doi.org/10.22358/jafs/65719/2014
Giannenas I, Papaneophytou CP, Tsalie E et al (2014b) Dietary supplementation of benzoic acid and essential oil compounds affects buffering capacity of the feeds, performance of turkey poults and their antioxidant status, pH in the digestive tract, intestinal microbiota and morphology. Asian-Australas J Anim Sci 27(2):225–236. https://doi.org/10.5713/ajas.2013.13376
Giese E, Winkelmann O, Rohn S et al (2018) Determining quality parameters of fish oils by means of 1H nuclear magnetic resonance, mid-infrared, and near-infrared spectroscopy in combination with multivariate statistics. Food Res Int 106:116–128. https://doi.org/10.1016/j.foodres.2017.12.041
González-Rubio S, García-Gómez D, Ballesteros-Gómez A et al (2020) A new sample treatment strategy based on simultaneous supramolecular solvent and dispersive solid-phase extraction for the determination of ionophore coccidiostats in all legislated foodstuffs. Food Chem 326: 126987. https://doi.org/10.1016/j.foodchem.2020.126987
Govaerts C, Orwa J, Van Schepdael A et al (2002) Liquid chromatography-ion trap tandem mass spectrometry for the characterization of polypeptide antibiotics of the colistin series in commercial samples. J Chromatogr A 976(1–2):65–78. https://doi.org/10.1016/s0021-9673(02)00375-8
Graham SF, Haughey SA, Ervin RM et al (2012) The application of near-infrared (NIR) and Raman spectroscopy to detect adulteration of oil used in animal feed production. Food Chem 132(3): 1614–1619. https://doi.org/10.1016/j.foodchem.2011.11.136
Greule M, Hänsel C, Bauermann U et al (2008) Feed additives: authenticity assessment using multicomponent−/multielement-isotope ratio mass spectrometry. Eur Food Res Technol 227(3):767–776. https://doi.org/10.1007/s00217-007-0785-7
Guo T, Feng W-H, Liu X-Q et al (2016a) Fourier transform mid-infrared spectroscopy (FT-MIR) combined with chemometrics for quantitative analysis of dextrin in Danshen (Salvia miltiorrhiza) granule. J Pharm Biomed Anal 123:16–23. https://doi.org/10.1016/j.jpba.2015.11.021
Guo X-X, Hu W, Liu Y et al (2016b) Rapid determination and chemical change tracking of benzoyl peroxide in wheat flour by multi-step IR macro-fingerprinting. Spectrochim Acta Mol Biomol Spectrosc 154:123–129. https://doi.org/10.1016/j.saa.2015.10.017
Hajslova J, Cajka T (2008) Gas chromatography in food analysis. In: Otles S (ed) Handbook of food analysis instruments. CRC Press, Boca Raton, FL, pp 119–144
Han C, Xia B, Chen X et al (2016) Determination of four paraben-type preservatives and three benzophenone-type ultraviolet light filters in seafoods by LC-QqLIT-MS/MS. Food Chem 194: 1199–1207. https://doi.org/10.1016/j.foodchem.2015.08.093
Han C, Yao Y, Wang W et al (2017) Rapid and sensitive detection of sodium saccharin in soft drinks by silver nanorod array SERS substrates. Sens Actuators B Chem 251:272–279. https://doi.org/10.1016/j.snb.2017.05.051
Haque F, Bubli SY, Khan MS (2021) UV–Vis spectroscopy for food analysis. In: KhanM MS, Rahman S (eds) Techniques to measure food safety and quality: microbial, chemical, and sensory. Springer International Publishing, Cham, pp 169–193
Hassoun A, Carpena M, Prieto MA et al (2020) Use of spectroscopic techniques to monitor changes in food quality during application of natural preservatives: a review. Antioxidants 9(9). https://doi.org/10.3390/antiox9090882
Hatzakis E (2019) Nuclear magnetic resonance (NMR) spectroscopy in food science: a comprehensive review. Compr Rev Food Sci Food Saf 18(1):189–220. https://doi.org/10.1111/1541-4337.12408

Herzog G, Kam V, Berduque A et al (2008) Detection of food additives by voltammetry at the liquid–liquid interface. J Agric Food Chem 56(12):4304–4310. https://doi.org/10.1021/jf7035966

Hoadley JE (2011) U.S. regulation of functional foods. In: Saarela M (ed) Functional foods, 2nd edn. Woodhead Publishing, Witney/Oxford, pp 41–63

Horky P, Skalickova S, Smerkova K et al (2019) Essential oils as a feed additives: pharmacokinetics and potential toxicity in monogastric animals. Animals (Basel) 9(6). https://doi.org/10.3390/ani9060352

Hormazábal V, Yndestad M (2000) Rapid assay for the determination of zinc bacitracin in feed by liquid chromatography-mass spectrometry. J Liq Chromatogr Relat Technol 23(7):1083–1088. https://doi.org/10.1081/JLC-100101509

Howes BD, Milazzo L, Droghetti E et al (2019) Addition of sodium ascorbate to extend the shelf-life of tuna meat fish: a risk or a benefit for consumers? J Inorg Biochem 200:110813. https://doi.org/10.1016/j.jinorgbio.2019.110813

Hu Y, Wang S, Wang S et al (2017) Application of nuclear magnetic resonance spectroscopy in food adulteration determination: the example of Sudan dye I in paprika powder. Sci Rep 7(1): 2637. https://doi.org/10.1038/s41598-017-02921-8

Huang H, Yu H, Xu H et al (2008) Near infrared spectroscopy for on/in-line monitoring of quality in foods and beverages: a review. J Food Eng 87(3):303–313. https://doi.org/10.1016/j.jfoodeng.2007.12.022

Hughes G, Pemberton RM, Fielden PR et al (2015) Development of a novel reagentless, screen-printed amperometric biosensor based on glutamate dehydrogenase and NAD+, integrated with multi-walled carbon nanotubes for the determination of glutamate in food and clinical applications. Sens Actuators B Chem 216:614–621. https://doi.org/10.1016/j.snb.2015.04.066

International AOAC (2019) Official methods of analysis of AOAC international, 19th edn. International AOAC, Gaithersburg, MD

Jawaid S, Talpur FN, Sherazi ST et al (2013) Rapid detection of melamine adulteration in dairy milk by SB-ATR-Fourier transform infrared spectroscopy. Food Chem 141(3):3066–3071. https://doi.org/10.1016/j.foodchem.2013.05.106

Jiménez-Salcedo M, Tena MT (2017) Determination of cinnamaldehyde, carvacrol and thymol in feedstuff additives by pressurized liquid extraction followed by gas chromatography–mass spectrometry. J Chromatogr A 1487:14–21. https://doi.org/10.1016/j.chroma.2017.01.042

Karabagias IK (2020) Advances of spectrometric techniques in food analysis and food authentication implemented with chemometrics. Foods 9(11):1550

Karau A, Grayson I (2014) Amino acids in human and animal nutrition. Adv Biochem Eng Biotechnol 143:189–228. https://doi.org/10.1007/10_2014_269

Kaufmann A, Widmer M (2013) Quantitative analysis of polypeptide antibiotic residues in a variety of food matrices by liquid chromatography coupled to tandem mass spectrometry. Anal Chim Acta 797:81–88. https://doi.org/10.1016/j.aca.2013.08.032

Kiarie E, Romero LF, Nyachoti CM (2013) The role of added feed enzymes in promoting gut health in swine and poultry. Nutr Res Rev. 26(1):71–88. https://doi.org/10.1017/s0954422413000048

König J, Grasser R, Pikor H et al (2002) Determination of xylanase, beta-glucanase, and cellulase activity. Anal Bioanal Chem 374(1):80–87. https://doi.org/10.1007/s00216-002-1379-7

Kuss S, Compton RG (2017) Electrocatalytic detection of ascorbic acid using N,N,N′,N′-tetramethyl-para-phenylene-diamine (TMPD) mediated oxidation at unmodified gold electrodes; reaction mechanism and analytical application. Electrochim Acta 242:19–24. https://doi.org/10.1016/j.electacta.2017.05.003

Langi P, Kiokias S, Varzakas T et al (2018) Carotenoids: from plants to food and feed industries. Methods Mol Biol 1852:57–71. https://doi.org/10.1007/978-1-4939-8742-9_3

Lawaetz AJ, Christensen SMU, Clausen SK et al (2016) Fast, cross cultivar determination of total carotenoids in intact carrot tissue by Raman spectroscopy and partial least squares calibration. Food Chem 204:7–13. https://doi.org/10.1016/j.foodchem.2016.02.107

Lee J, Song Y-S, Sim H-J et al (2016) Isotope dilution-liquid chromatography/mass spectrometric method for the determination of riboflavin content in multivitamin tablets and infant formula. J Food Compos Anal 50:49–54. https://doi.org/10.1016/j.jfca.2016.05.008

Lehotay SJ, Hajšlová J (2002) Application of gas chromatography in food analysis. Trends Analyt Chem 21(9):686–697. https://doi.org/10.1016/S0165-9936(02)00805-1

Li J, Wang X, Duan H et al (2016) Based on magnetic graphene oxide highly sensitive and selective imprinted sensor for determination of sunset yellow. Talanta 147:169–176. https://doi.org/10.1016/j.talanta.2015.09.056

Li-Chan ECY (1996) The applications of Raman spectroscopy in food science. Trends Food Sci Technol 7(11):361–370. https://doi.org/10.1016/S0924-2244(96)10037-6

Liu X, Zhu L, Gao X et al (2016) Magnetic molecularly imprinted polymers for spectrophotometric quantification of curcumin in food. Food Chem 202:309–315. https://doi.org/10.1016/j.foodchem.2016.02.015

Lohumi S, Lee S, Lee H et al (2015) A review of vibrational spectroscopic techniques for the detection of food authenticity and adulteration. Trends Food Sci Technol 46(1):85–98. https://doi.org/10.1016/j.tifs.2015.08.003

Lopez-Gonzalez E, Salmazo LO, Lopez-Gil A et al (2019) Study of the effect of different electron irradiation doses on the decomposition temperature of azodicarbonamide. Polym Eng Sci 59(4):791–798. https://doi.org/10.1002/pen.25007

Lu L, Luo X, Ji C et al (2007) Effect of manganese supplementation and source on carcass traits, meat quality, and lipid oxidation in broilers. J Anim Sci 85:812–822. https://doi.org/10.2527/jas.2006-229

Luna AS, da Silva AP, Pinho JSA et al (2015) A novel approach to discriminate transgenic from non-transgenic soybean oil using FT-MIR and chemometrics. Food Res Int 67:206–211. https://doi.org/10.1016/j.foodres.2014.11.011

Ma K, Yang YN, Jiang XX et al (2012) Simultaneous determination of 20 food additives by high performance liquid chromatography with photo-diode array detector. Chin Chem Lett 23(4):492–495. https://doi.org/10.1016/j.cclet.2011.12.018

Macy TD, Loh A (1983) High pressure liquid chromatographic determination of Monensin in feed premixes. J Assoc Off Anal Chem 66(2):284–286

Malomo GA, Ihegwuagu NE (2017) Some aspects of animal feed sampling and analysis. In: Stauffer M (ed) Ideas and applications toward sample preparation for food and beverage analysis. IntechOpen, London

Markowiak P, Śliżewska K (2018) The role of probiotics, prebiotics and synbiotics in animal nutrition. Gut Pathog 10(1):21. https://doi.org/10.1186/s13099-018-0250-0

Martin G, Ferasyi T (2016) Clean, green, ethical (CGE) management: what research do we really need? Int J Trop Vet Biomed Res 1:1–9. https://doi.org/10.21157/ijtvbr.v1i1.5066

Martins FCOL, Sentanin MA, De Souza D (2019) Analytical methods in food additives determination: compounds with functional applications. Food Chem 272:732–750. https://doi.org/10.1016/j.foodchem.2018.08.060

Martins RR, Silva LJG, Pereira A et al (2022) Coccidiostats and poultry: a comprehensive review and current legislation. Foods 11(18). https://doi.org/10.3390/foods11182738

Matus JL, Boison JO (2016) A multi-residue method for 17 anticoccidial drugs and ractopamine in animal tissues by liquid chromatography-tandem mass spectrometry and time-of-flight mass spectrometry. Drug Test Anal 8(5–6):465–476. https://doi.org/10.1002/dta.2019

Mendoza S, Bustos E, Manríquez J et al (2015) Voltammetric techniques. In: Escarpa A, González MC, López MÁ (eds) Agricultural and food electroanalysis. Wiley, Maitland, FL, pp 21–48

Minchin W, Buckley F, Kenny DA et al (2010) An evaluation of over-wintering feeding strategies prior to finishing at pasture for cull dairy cows on live animal performance, carcass and meat quality characteristics. Meat Sci 85(3):385–393. https://doi.org/10.1016/j.meatsci.2010.02.005

Mitra S, Brukh R (2003) Sample preparation: an analytical perspective. In: Mitra S (ed) Sample preparation techniques in analytical chemistry. Wiley, Maitland, FL, pp 1–36

Morales-Muñoz S, de Castro MD (2005) Dynamic ultrasound-assisted extraction of colistin from feeds with on-line pre-column derivatization and liquid chromatography-fluorimetric detection. J Chromatogr A 1066(1–2):1–7. https://doi.org/10.1016/j.chroma.2005.01.010
Mortera P, Zuljan FA, Magni C et al (2018) Multivariate analysis of organic acids in fermented food from reversed-phase high-performance liquid chromatography data. Talanta 178:15–23. https://doi.org/10.1016/j.talanta.2017.09.005
Mould FL (2003) Predicting feed quality—chemical analysis and in vitro evaluation. Field Crops Res 84(1):31–44. https://doi.org/10.1016/S0378-4290(03)00139-4
Mursalat M, Rony AH, Rahman A et al (2013) A critical analysis of artificial fruit ripening: scientific, legislative and socio-economic aspects. ChE Thoughts 3(1):1–7
Nescatelli R, Carradori S, Marini F et al (2017) Geographical characterization by MAE-HPLC and NIR methodologies and carbonic anhydrase inhibition of Saffron components. Food Chem 221: 855–863. https://doi.org/10.1016/j.foodchem.2016.11.086
Nie Q, Nie S (2019) High-performance liquid chromatography for food quality evaluation. In: Zhong J, Wang X (eds) Evaluation technologies for food quality. Woodhead Publishing, Witney, Oxford, pp 267–299
Nielsen SS (2014) Food analysis. Springer, Berlin
Noor NSM, Tan LL, Heng LY et al (2016) Acrylic microspheres-based optosensor for visual detection of nitrite. Food Chem 207:132–138. https://doi.org/10.1016/j.foodchem.2016.03.088
Nováková L, Matysová L, Solich P (2006) Advantages of application of UPLC in pharmaceutical analysis. Talanta 68(3):908–918. https://doi.org/10.1016/j.talanta.2005.06.035
Ohtsuki T, Sato K, Abe Y et al (2015) Quantification of acesulfame potassium in processed foods by quantitative 1H NMR. Talanta 131:712–718. https://doi.org/10.1016/j.talanta.2014.08.002
Olejnik M, Szprengier-Juszkiewicz T, Jedziniak P et al (2011) Residue control of coccidiostats in food of animal origin in Poland during 2007–2010. Food Addit Contam Part B Surveill 4(4): 259–267. https://doi.org/10.1080/19393210.2011.637238
Omar J, Slowikowski B, Boix A et al (2017) Spectroscopy applied to feed additives of the European Union Reference Laboratory: a valuable tool for traceability. Food Addit Contam Part A Chem Anal Control Expo Risk Assess 34(8):1272–1284. https://doi.org/10.1080/19440049.2017.1303196
Oviedo-Rondón EO, Hume ME, Barbosaa NA et al (2010) Ileal and caecal microbial populations in broilers given specific essential oil blends and probiotics in two consecutive grow-outs. Avian Biol Res 3(4):157–169. https://doi.org/10.3184/175815511X12919853724050
Owens B, McCann MEE, McCracken KJ et al (2009) Prediction of wheat chemical and physical characteristics and nutritive value by near-infrared reflectance spectroscopy. Br Poult Sci 50(1): 103–122. https://doi.org/10.1080/00071660802635347
Palacios-Jordan H, Jané-Brunet A, Jané-Brunet E et al (2022) Considerations on the analysis of E-900 food additive: an NMR perspective. Foods 11(3):297
Parliament E (2003) Regulation (EC) No 1831/2003 on the European Parliament and of the Council of 22 September 2003. On additives for use in animal nutrition. Off J Eur Commun 268:29–43
Partridge D, Lloyd KA, Rhodes JM et al (2019) Food additives: assessing the impact of exposure to permitted emulsifiers on bowel and metabolic health – introducing the FADiets study. Nutr Bull 44(4):329–349. https://doi.org/10.1111/nbu.12408
Paulson JA, Zaoutis TE, The Council On Environmental H et al (2015) Nontherapeutic use of antimicrobial agents in animal agriculture: implications for pediatrics. Pediatrics 136(6):e1670–e1677. https://doi.org/10.1542/peds.2015-3630
Petanovska-Ilievska B, Velkoska-Markovska L, Jankulovska MS (2017) Development of reverse-phase high-performance liquid chromatography method for simultaneous determination of sodium benzoate and potassium sorbate in beverages. Acta Chromatogr 29(3):345–358. https://doi.org/10.1556/1326.2017.29.3.05
Petigara Harp B, Miranda-Bermudez E, Barrows JN (2013) Determination of seven certified color additives in food products using liquid chromatography. J Agric Food Chem 61(15):3726–3736. https://doi.org/10.1021/jf400029y

Phillips I (1999) The use of bacitracin as a growth promoter in animals produces no risk to human health. J Antimicrob Chemother 44(6):725–728. https://doi.org/10.1093/jac/44.6.725

Pizarro C, Esteban-Díez I, Rodríguez-Tecedor S et al (2013) Determination of the peroxide value in extra virgin olive oils through the application of the stepwise orthogonalisation of predictors to mid-infrared spectra. Food Control 34(1):158–167. https://doi.org/10.1016/j.foodcont.2013.03.025

Polansky R, Prosr P, Vik R et al (2017) Comparison of the mineral oil lifetime estimates obtained by differential scanning calorimetry, infrared spectroscopy, and dielectric dissipation factor measurements. Thermochim Acta 647:86–93. https://doi.org/10.1016/j.tca.2016.12.002

Prache S, Martin B, Coppa M (2020) Review: authentication of grass-fed meat and dairy products from cattle and sheep. Animal 14(4):854–863. https://doi.org/10.1017/S1751731119002568

Qi W, Tian Y, Lu D et al (2022) Research progress of applying infrared spectroscopy technology for detection of toxic and harmful substances in food. Foods 11(7). https://doi.org/10.3390/foods11070930

Qian H, Sheng M (1998) Simultaneous determination of fat-soluble vitamins A, D and E and pro-vitamin D2 in animal feeds by one-step extraction and high-performance liquid chromatography analysis. J Chromatogr A 825(2):127–133. https://doi.org/10.1016/s0021-9673(98)00733-x

Raina-Fulton R (2021) Sample preparation methods. In: Galanakis CM (ed) Innovative food analysis. Academic Press, Washington, DC, pp 85–98

Rana MS, Lee SY, Kang HJ et al (2019) Reducing veterinary drug residues in animal products: a review. Food Sci Anim Resour 39(5):687–703. https://doi.org/10.5851/kosfa.2019.e65

Ravindran V (2013) Feed enzymes: the science, practice, and metabolic realities. J Appl Poult Res 22(3):628–636. https://doi.org/10.3382/japr.2013-00739

Riman D, Prodromidis MI, Jirovsky D et al (2019) Low-cost pencil graphite-based electrochemical detector for HPLC with near-coulometric efficiency. Sens Actuators B Chem 296:126618. https://doi.org/10.1016/j.snb.2019.05.095

Rodrigues Júnior PH, de Sá OK, de Almeida CE et al (2016) FT-Raman and chemometric tools for rapid determination of quality parameters in milk powder: classification of samples for the presence of lactose and fraud detection by addition of maltodextrin. Food Chem 196:584–588. https://doi.org/10.1016/j.foodchem.2015.09.055

Rodríguez Vera A, Rosales López MJ, Ortiz Ortega E et al (2022) Coupled Characterization Techniques. In: Ortiz Ortega E, Hosseinian H, Aguilar Meza IB et al (eds) Material characterization techniques and applications. Springer Singapore, Singapore, pp 269–305

Rodriguez-Saona LE, Allendorf ME (2011) Use of FTIR for rapid authentication and detection of adulteration of food. Annu Rev Food Sci Technol 2:467–483. https://doi.org/10.1146/annurev-food-022510-133,750

Rubaj J, Korol W, Bielecka G (2012) Using high performance liquid chromatography (HPLC) for analyzing feed additives. In: de Azevedo Calderon L (ed) Chromatography - the most versatile method of chemical analysis. IntechOpen, London

Rusko J, Jansons M, Pugajeva I et al (2019) Development and optimization of confirmatory liquid chromatography-Orbitrap mass spectrometry method for the determination of 17 anticoccidials in poultry and eggs. J Pharm Biomed Anal 164:402–412. https://doi.org/10.1016/j.jpba.2018.10.056

Sabatier AM, Fish NM (1996) Method of analysis for feed enzymes: methodological problems? J Appl Poult Res 5(4):408–413. https://doi.org/10.1093/japr/5.4.408

Santos IC, Schug KA (2017) Recent advances and applications of gas chromatography vacuum ultraviolet spectroscopy. J Sep Sci 40(1):138–151. https://doi.org/10.1002/jssc.201601023

Santos FH, De Paula MR, Lezier D et al (2015) Essential oils for dairy calves: effects on performance, scours, rumen fermentation and intestinal fauna. Animal 9(6):958–965. https://doi.org/10.1017/s175173111500018x

Sanz D, Razquin P, Condon S et al (2015) Incidence of antimicrobial residues in meat using a broad spectrum screening strategy. EJNFS 5(3):156–165. https://doi.org/10.9734/EJNFS/2015/13795

Schneider M, Türke A, Fischer W-J et al (2014) Determination of the wine preservative sulphur dioxide with cyclic voltammetry using inkjet printed electrodes. Food Chem 159:428–432. https://doi.org/10.1016/j.foodchem.2014.03.049

Scotter MJ (2015) Methods of analysis for food colour additive quality and safety assessment. In: Scotter MJ (ed) Colour additives for foods and beverages. Woodhead Publishing, Oxford, pp 131–188

Shao B, Wu X, Zhang J et al (2009) Development of a rapid LC–MS–MS method for multi-class determination of 14 coccidiostat residues in eggs and chicken. Chromatographia 69(9): 1083–1088. https://doi.org/10.1365/s10337-009-1009-z

Silbergeld EK (2019) One health and the agricultural transition in food animal production. Glob Transit 1:83–92. https://doi.org/10.1016/j.glt.2019.01.003

Silva MM, Reboredo FH, Lidon FC (2022) Food colour additives: a synoptical overview on their chemical properties, applications in food products, and health side effects. Foods 11(3):379. https://doi.org/10.3390/foods11030379

Sin DWM, Ho C, Wong YC et al (2004) Simultaneous determination of lincomycin and virginiamycin M1 in swine muscle, liver and kidney by liquid chromatography–electrospray ionization tandem mass spectrometry. Anal Chim Acta 517(1):39–45. https://doi.org/10.1016/j.aca.2004.04.037

Smith RM (2003) Before the injection--modern methods of sample preparation for separation techniques. J Chromatogr A 1000(1–2):3–27. https://doi.org/10.1016/s0021-9673(03)00511-9

Smith JL, Weinberg ED (1962) Mechanisms of antibacterial action of bacitracin. J Gen Microbiol 28:559–569. https://doi.org/10.1099/00221287-28-3-559

Snehalatha M, Ravikumar C, Hubert Joe I et al (2009) Spectroscopic analysis and DFT calculations of a food additive carmoisine. Spectrochim Acta A Mol Biomol Spectrosc 72(3):654–662. https://doi.org/10.1016/j.saa.2008.11.017

Sniegocki T, Gbylik-Sikorska M, Posyniak A et al (2014) Determination of carbadox and olaquindox metabolites in swine muscle by liquid chromatography/mass spectrometry. J Chromatogr B 944:25–29. https://doi.org/10.1016/j.jchromb.2013.09.039

Spinola Machado SA, Cincotto FH (2022) Electrochemical methods applied for bioanalysis: differential pulse voltammetry and square wave voltammetry. In: Kubota LT, da Silva JAF, Sena MM et al (eds) Tools and trends in bioanalytical chemistry. Springer International Publishing, Cham, pp 273–282

Srivastava HK, Wolfgang S, Rodriguez JD (2016) Rapid screening of guar gum using portable Raman spectral identification methods. J Pharm Biomed Anal 118:387–392. https://doi.org/10.1016/j.jpba.2015.11.013

Tang Z, Duan Y (2017) Fabrication of porous ionic liquid polymer as solid-phase microextraction coating for analysis of organic acids by gas chromatography – mass spectrometry. Talanta 172: 45–52. https://doi.org/10.1016/j.talanta.2017.05.032

Tao Y, Xie S, Zhu Y et al (2018) Analysis of major components of Bacitracin, Colistin and Virginiamycin in feed using matrix solid-phase dispersion extraction by liquid chromatography-electrospray Ionization Tandem Mass Spectrometry. J Chromatogr Sci 56(3): 285–291. https://doi.org/10.1093/chromsci/bmx096

Truzzi E, Benvenuti S, Bertelli D et al (2021) Optimization and validation of a high-performance liquid chromatography method for the analysis of hesperidin and carvacrol for veterinary use. AJAC 12:459–470. https://doi.org/10.4236/ajac.2021.1212029

Tunick MH (2005) Selection of techniques used in food analysis. In: Otles S (ed) Methods of analysis of food components and additives. CRC Press, Boca Raton, FL

Van Poucke C, Dumoulin F, Van Peteghem C (2005) Detection of banned antibacterial growth promoters in animal feed by liquid chromatography–tandem mass spectrometry: optimisation of the extraction solvent by experimental design. Anal Chim Acta 529(1):211–220. https://doi.org/10.1016/j.aca.2004.08.076

Viñas P, Soler-Romera MJ, Hernández-Córdoba M (2006) Liquid chromatographic determination of phenol, thymol and carvacrol in honey using fluorimetric detection. Talanta 69(5): 1063–1067. https://doi.org/10.1016/j.talanta.2005.12.030

Vincent U, Serano F, von Holst C (2021) Validation of a multi-analyte HPLC method for the determination of carotenoids used as feed additives in fish and poultry feed: results of an interlaboratory study. Food Addit Contam Part A Chem Anal Control Expo Risk Assess 38(3):396–408. https://doi.org/10.1080/19440049.2020.1869325

Vranken L, Avermaete T, Petalios D et al (2014) Curbing global meat consumption: emerging evidence of a second nutrition transition. Environ Sci Policy 39:95–106. https://doi.org/10.1016/j.envsci.2014.02.009

Wan Mohamad WAF, Buckow R, Augustin M et al (2017) In situ quantification of β-carotene partitioning in oil-in-water emulsions by confocal Raman microscopy. Food Chem 233:197–203. https://doi.org/10.1016/j.foodchem.2017.04.086

Wang G, Marquardt RR, Xiao H et al (1999) Development of a 96-well enzyme-linked solid-phase assay for beta-glucanase and xylanase. J Agric Food Chem 47(3):1262–1267. https://doi.org/10.1021/jf980702w

Wang P, Zhao G, Tian J et al (2010) High-performance liquid chromatography−Inductively coupled plasma mass spectrometry based method for the determination of organic arsenic feed additives and speciation of anionic arsenics in animal feed. J Agric Food Chem 58(9): 5263–5270. https://doi.org/10.1021/jf1001205

Wang J, Si W, Du Z et al (2022) Antioxidants in animal feed. Antioxidants (Basel) 11(9). https://doi.org/10.3390/antiox11091760

Webster GK (2020) Liquid chromatographic analysis of bacitracin methylene disalicylate in feed. J AOAC Int 80(4):732–736. https://doi.org/10.1093/jaoac/80.4.732

Wiley D, Yen Nee CN (2020) Food ingredients. In: Gibson M (ed) Food and society. Academic Press, Washington, DC, pp 377–391

Wrolstad RE, Acree TE, Decker EA et al (2005) Handbook of food analytical chemistry, volume 1: water, proteins, enzymes, lipids, and carbohydrates. Wiley, New Jersey

Wu L, Yin W, Tang K et al (2016) Enzymatic biosensor of horseradish peroxidase immobilized on Au-Pt nanotube/Au-graphene for the simultaneous determination of antioxidants. Anal Chim Acta 933:89–96. https://doi.org/10.1016/j.aca.2016.06.020

Wu L, Zhang C, Long Y et al (2021) Food additives: from functions to analytical methods. Crit Rev Food Sci Nutr 62:1–21. https://doi.org/10.1080/10408398.2021.1929823

Yazici A, Tiryaki GY, Ayvaz H (2020) Determination of pesticide residual levels in strawberry (Fragaria) by near-infrared spectroscopy. J Sci Food Agric 100(5):1980–1989. https://doi.org/10.1002/jsfa.10211

Youssouf L, Lallemand L, Giraud P et al (2017) Ultrasound-assisted extraction and structural characterization by NMR of alginates and carrageenans from seaweeds. Carbohydr Polym 166: 55–63. https://doi.org/10.1016/j.carbpol.2017.01.041

Zamani Mazdeh F, Sasanfar S, Chalipour A et al (2017) Simultaneous determination of preservatives in dairy products by HPLC and chemometric analysis. Int J Anal Chem 2017: 3084359. https://doi.org/10.1155/2017/3084359

Zhang X, Lu S, Liao Y et al (2017) Simultaneous determination of amino acid mixtures in cereal by using terahertz time domain spectroscopy and chemometrics. Chemom Intell Lab Syst 164:8–15. https://doi.org/10.1016/j.chemolab.2017.03.001

Zheng H, Zhang Q, Quan J et al (2016) Determination of sugars, organic acids, aroma components, and carotenoids in grapefruit pulps. Food Chem 205:112–121. https://doi.org/10.1016/j.foodchem.2016.03.007

Antimicrobials as Feed and Food Additives, their Technological Properties and Perspectives

Nikola Puvača and Vincenzo Tufarelli

Abstract

As awareness and concerns about antibiotic residues in animal products have increased among consumers, the search for alternatives to antibiotic growth promoters has become more critical. In recent decades, efforts have been renewed to identify natural additives for use in the food industry as biopreservatives for fermented foods and in animal feeding as natural growth promoter additives. Essential oils are natural antimicrobials that have been highlighted for their numerous beneficial effects. When essential oils or medicinal plants with multiple modes of action are combined, their preservative properties are enhanced and the risk of antimicrobial resistance is reduced. Due to their ability to modulate gut microbiota, which is strongly related to animal and human wellbeing, essential oils are increasingly recognized as health modulators in addition to their benefits as food preservatives. As antibacterials, antivirals, antifungals, antioxidants, digestive stimulants, immunomodulators, hypolipidemic agents, and heat stress alleviators, essential oils have been utilized in animal feeding for centuries. It has been demonstrated that essential oils can act as a dietary antibiotic replacement without causing side effects. Additionally, essential oils play a role in the preservation of feed in which they are incorporated, as well as the meat that is obtained from animals fed with them. Essential oils not only reduce oxidative damage to the animal in vivo, but they also exert an anti-oxidative influence on feed that is incorporated with them. Researchers exploited essential oils as feed

N. Puvača (✉)
Department of Engineering Management in Biotechnology, Faculty of Economics and Engineering Management, University Business Academy in Novi Sad, Novi Sad, Serbia
e-mail: nikola.puvaca@fimek.edu.rs

V. Tufarelli
Department of Precision and Regenerative Medicine and Jonian Area, Section of Veterinary Science and Animal Production, University of Bari Aldo Moro, Valenzano, Italy

G. Arsenos, I. Giannenas (eds.), *Sustainable Use of Feed Additives in Livestock*,
https://doi.org/10.1007/978-3-031-42855-5_32

additives after obtaining favorable results as an alternative to antibiotics. The expanding research horizons in essential oils should bring an end to the extensive use of antibiotics as feed additives. Having in mind that the future is now, the role of essential oils as natural growth promoters and immunity enhancers in animal feeding is huge, but still more investigations are needed for the development of this part of the industry. Due to antimicrobial functions in animal and human health, this chapter reviews recent advancements in essential oil research that will contribute to the development of functional foods and feeds.

Keywords

Antibiotics · Essential oils · AMR · Animal nutrition · Antibiotic residues

1 Introduction

In commercial poultry production, antibiotic use has been practiced for decades as a growth promoter in the daily diet. Antibiotics are increasingly used in animal production, both as a disease preventative and as growth enhancers. As a result of opportunistic pathogens inhabiting the intestinal tract and the microbial load in the animal's environment, the growth rate may be slowed down. During the treatment of animals with antibiotics, pathogenic bacteria within their gut will produce toxins that will harm the animals' livelihoods or production ability (Puvača et al. 2020b). It is most likely that antibiotics used in this manner will end up accumulating in the tissues, causing antibiotic resistance in humans via the food chain that leads ultimately to therapeutic failure (Ljubojević et al. 2017).

Antibiotics are no longer used as feed additives in many countries (Dawood et al. 2018). As such, finding an alternative to antibiotics is of utmost importance. Hence, the World Health Organization has encouraged the development of alternatives to antibiotics to control enteric diseases. There have been many attempts by scientists to find antibiotic alternatives (Czaplewski et al. 2016; Joerger 2003; Roca et al. 2015; Zorzet 2014).

Plant parts are used to extract essential oils, which are oily mixtures of aromatic compounds. Steam distillation is the most commonly used method for commercially producing essential oils, along with expression, fermentation, and extraction. Essential oils are classified as "generally recognized as safe" (GRAS), as supported by the Flavor and Extract Manufacturers Association (FEMA) and the Food and Drug Administration (FDA) (Nieto 2017). As a growth and health promoter, essential oils can enhance productivity (Brenes and Roura 2010). Since essential oils are versatile, they can be used in animal production to promote growth with terpenes and phenylpropenes as the main constituents of essential oils.

Each essential oil has its characteristic odor. Depending on the ambient temperature, they exist as liquids or volatiles. Benzene, ether, acetone, and other organic solvents readily dissolve essential oils (Castilho et al. 2012). There is a difference between the specific gravity of the clove and cinnamon oils and that of the essential

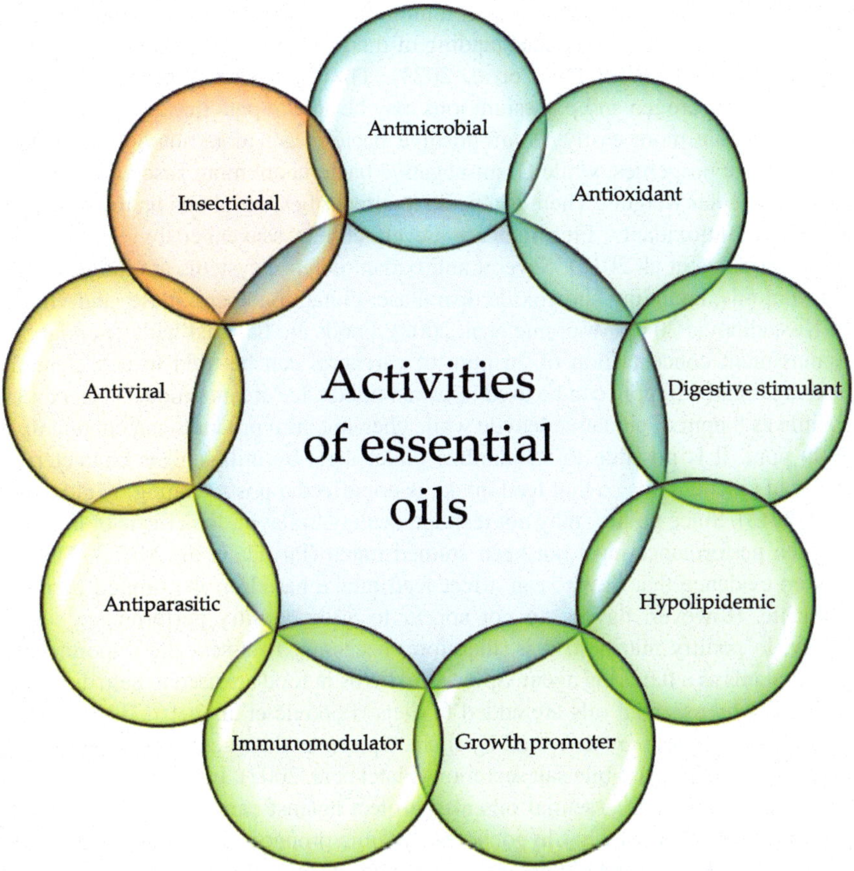

Fig. 1 Main activities of essential oils (Puvača et al. 2013; Puvača et al. 2022a)

oils, with most being between 0.8 and 1.17. In Fig. 1, we summarize the activities of essential oils.

2 Properties of Essential Oils as Feed and Food Antimicrobials

Essential oils have been exploited since ancient times for their antimicrobial properties, and researchers are interested in using them as an alternative to antibiotics because of their beneficial properties (Chouhan et al. 2017). Antimicrobial properties have been demonstrated in vitro for pure compounds. *Clostridium perfringens* and *Bacteroides fragilis* were strongly inhibited by cinnamaldehyde, while *Bifidobacterium longum* and *Lactobacillus acidophilus* were moderately inhibited. It is possible to balance the microbial population in the poultry intestine by exploiting this property of selective inhibition of intestinal pathogenic bacteria. There is currently no clear understanding of the mechanism of antimicrobial action (Hyldgaard et al. 2012; Pelić et al. 2023). The main site of action is the cell membrane. Hydrogen and potassium ions may be able to pass through the cytoplasmic membrane more easily. Gram-positive bacteria are more susceptible to their hydrophobic properties, while Gram-negative bacteria are more susceptible to their small molecular weights. There is a positive effect of essential oils on the biological system as antioxidants. Free radicals are effectively scavenged by essential oils (Bhavaniramya et al. 2019). Several antioxidant defense systems are influenced by essential oils, including superoxide dismutase, glutathione peroxidase, and vitamin E (Kostadinović 2023; Mourente et al. 2002). Foods are flavored with essential oils. A maximum concentration of 26 ppm of carvacrol can be used in non-alcoholic beverages and 120 ppm can be used in baked goods. Ice cream products can contain as little as 8 ppm of cinnamaldehyde while chewing gum products can contain up to 4900 ppm. It is possible to standardize piglet diets by utilizing the characteristic flavor of essential oils so that feed intake is not affected post-weaning (Middelkoop et al. 2018). Since poultry may not respond acutely to flavor, the effects of flavor on chicken performance have not been studied much (Puvača et al. 2015a). There is limited evidence that flavors can affect feed intake based on the limited literature available. However, flavors do not appear to affect poultry performance significantly. In poultry nutrition, it is, therefore, necessary to assess the importance of essential oils as a flavoring agent. Spices and herbs in food are used to help digestion, which is why essential oils are added to diets (Popović et al. 2016). The enzymes trypsin and amylase are activated by them. Spices or their active ingredients have been shown to increase bile salt secretion (Platel et al. 2002). In addition to lowering cholesterol, herbs and essential oils also protect against cancer through their anticancer properties. Garlic has hypocholesterolemic properties due to its inhibition of hepatic 3-hydroxy-3-methylglutaryl coenzyme A reductase, a key enzyme that regulates cholesterol synthesis (Puvača et al. 2015b).

3 Essential Oils' Antibacterial Properties and Mode of Action

It has been well-documented that essential oils and their components have antimicrobial properties, but their mechanisms of action have not been thoroughly examined. Fig. 2 shows possible antimicrobial mechanism of essential oil (Wang et al. 2020). Essential oils contain a variety of chemical compounds, making it most likely that their antibacterial properties are not due to a single mechanism, but respond to multiple targets within the body (Dosoky and Setzer 2018; Tsai et al. 2013). A high percentage of phenolic compounds such as carvacrol, eugenol, and thymol can be found in essential oils that have the strongest antibacterial properties against foodborne pathogens (Abbaszadeh et al. 2014).

Antibacterial activity and the precise mode of action of essential oils are affected by the chemical structure of their constituents. Carvacrol and thymol, which contain hydroxyl groups, have been proven to be important phenolic compounds (Bassolé

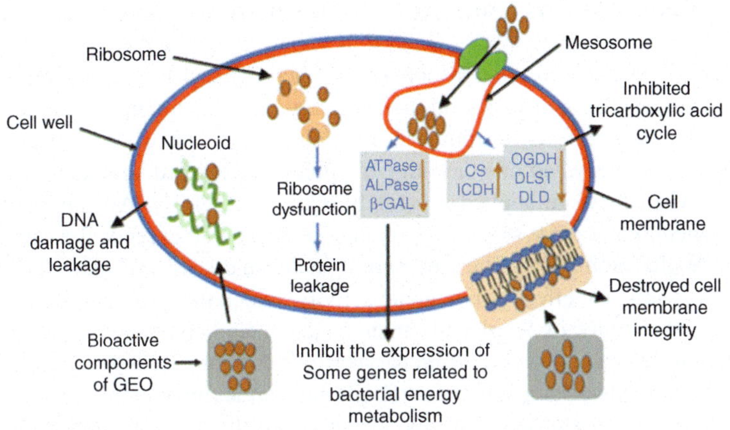

Fig. 2 Possible antimicrobial mechanism of essential oil (Wang et al. 2020)

and Juliani 2012). There does not seem to be a strong influence of the relative position of the hydroxyl group on the phenolic ring on thymol's antibacterial activity; it appears to be comparable to carvacrol in its ability to kill *B. cereus*, *Staphylococcus aureus,* and *Pseudomonas aeruginosa*. There was, however, a difference between the effects of carvacrol and thymol against Gram-positive and Gram-negative bacteria. Carvacrol displays significantly greater activity than menthol, which demonstrates the importance of the phenolic ring. It has been found that the type of alkyl group influencing the activity of essential oils is alkenyl rather than alkyl for the non-phenolic components (Raei et al. 2017). There is greater activity in limonene than in p-cymene, for example.

Cell proteins embedded in the cytoplasm seem to also be affected by components of essential oils. The cytoplasmic membrane is known to contain enzymes like ATPase and to be surrounded by lipid molecules (Nazzaro et al. 2013). The action of cyclic hydrocarbons on these has been suggested by two possible mechanisms. Lipid–protein interactions can be affected by lipophilic hydrocarbon molecules accumulating in the lipid bilayer; else, lipophilic compounds can directly interact with hydrophobic parts of proteins. Certain yeasts can grow pseudo mycelia when some essential oils are applied. The synthesis of structural components or energy regulation appears to be controlled by essential oils (Mantzourani et al. 2022). *Enterobacter aerogenes* is inhibited by cinnamon oil and its components (Ooi et al. 2006), while it was thought that protein binding was the mechanism of action. A study using milk with different levels of protein also indicated that essential oil components might act on proteins (Burt 2004).

4 Essential Oils Constitute Synergism and Antagonism

There is an expected correlation between oil activity and the configuration of its constituents, their proportions, and their interactions. A combined effect equals the sum of individual effects, resulting in an additive effect. It is possible to observe antagonistic effects when one or both compounds are applied simultaneously compared to when they are applied separately (Cox et al. 2001). A synergistic effect occurs when two or more substances interact in a way that exceeds the sum of their actions (Williamson 2001). The minor components in essential oils have been found to have a synergistic effect or potentiate activity, which may account for the greater antibacterial activity of whole oils than mixed major components. When tested against *S. aureus* and *P. aeruginosa*, carvacrol, and thymol, two structurally similar major components of oregano essential oil, gave an additive effect (Lambert et al. 2001). *B. cereus* vegetative cells respond synergistically to carvacrol and p-cymene. Carvacrol and p-cymene appear to have different effects on bacterial cell membrane swelling, as well. When used together with carvacrol, p-cymene likely facilitates more efficient transport of the drug into the cell, leading to a synergistic effect (Delgado et al. 2004).

5 Essential Oils in Poultry Production

The use of essential oils in poultry diets is regarded as a growth promoter. There are considerable variations in animal trial results, however. By introducing essential oils to chick diets, some studies indicate that feed intake was unchanged or slightly reduced (Lika et al. 2021; Mohamed et al. 2020; Puvača et al. 2020a). The lower feed consumption could be explained by the irritating smell of essential oils, which makes birds dislike the taste of their diet. Additionally, some studies found broilers' daily feed intake was numerically decreased by supplementing their diets with thyme, star anise, and origanum leaves, and the essential oils within them (Reyer et al. 2017). In addition, the inclusion of essential oils in graded amounts significantly reduced the feed intake of broiler chickens (Amad et al. 2011). A limited amount of information was available regarding the feed preferences of poultry, in contrast to that of pigs. The bird is more tolerant to moderate levels of essential oils than the pig, and poultry might be less sensitive to flavor than pigs. A major observation was an improvement in weight gain and feed conversion ratio rather than feed intake (Zhai et al. 2018). Boosting digestible enzyme secretion and stabilizing the gut microbiome are two powerful mechanisms that improve feed utilization and reduce growth-depressing digestive and metabolic disorders. Many studies have shown that essential oils increase the secretion of digestive enzymes from the pancreas and mucosa of the intestines (Jang et al. 2004, 2007). Growth performance wasn't improved, even though nutrient digestibility increased. A poor hygienic environment might have contributed to essential oils' positive effects on broiler growth performance, and the environment in which these trials were conducted is not adequately described. Many in vitro studies have examined essential oils for their antimicrobial properties, which have shown high antimicrobial activity against pathogenic bacteria like *E. coli* and *S. typhimurium*, which are both potential risk factors for enteric infections. Thymol, eugenol, and carvacrol have been demonstrated to have high antimicrobial activity (Abbaszadeh et al. 2014). The optimization of their formulation, therefore, requires the unraveling of the synergistic mechanism. Studies have demonstrated clear growth inhibition of *C. perfringens* and *E. coli* in the hindgut when essential oils were used both individually and in combination (Puvača et al. 2022a). These treatments improved intestinal lesions and weight loss compared to challenged control groups. It is well known that antibacterial activity originates from their hydrophobic behavior, which disrupts membrane permeability and cell homeostasis, causing cellular death or an influx of other substances. Gram-positive bacteria are generally resistant to essential oil action due to the presence of hydrophilic components in the outer membrane (Chao et al. 2000), while Gram-negative bacteria, however, are more susceptible (Tavares et al. 2020).

6 Essential Oils in Pig Production

There are many factors affecting voluntary feed intake, including housing, social environment, and dietary characteristics, one of which could become the dominant determinant under certain circumstances. Usually, essential oils have a pungent smell, so they might attract pigs to the feed more and thereby arouse their interest in exploring more (Franz et al. 2010). An ambiguous effect was observed on pig feed intake by essential oils. Supplementation with essential oils resulted in changes in feed intake ranging from −3% to 19%. Generally, growth-promoting additives improve the growth rate of animals, which may explain the improvement in feed intake (Yan et al. 2010). Pigs' feeding preferences and growth performance are not consistent when influenced by essential oils. For this reason, it is important to take care when interpreting the results of different types of animal trials, and more detailed research is needed to reconcile the incongruent results (Omonijo et al. 2018). Spices digestive stimulant action is primarily mediated by bile and enzyme secretion, which in part explains the increased apparent digestibility of fats and nutrients (Puvača 2018).

The passage of digesta and the interaction between feed and endogenous enzymes in the gut can also be influenced by essential oils by relaxing and contracting the gut (Zhai et al. 2018). It has been shown that tea tree oil has spasmolytic properties and strong spasmogenic effects (Ali et al. 2015). Following oral administration, piglets absorbed nearly completely carvacrol, thymol, eugenol, and cinnamaldehyde in their stomachs and small intestines within two hours (Michiels et al. 2008). Therefore, essential oils need to be protected so that they are not absorbed in the foregut, which would compromise their antimicrobial properties. The gut microflora could be selectively influenced by some essential oils. As a green preservative, essential oils could also be used to prevent pathogen contamination in food (Singh et al. 2021). Total coliform populations were reduced in Tuscan sausages treated with bay leaf essential oils (da Silveira et al. 2014). Essential oils must be used at ten to hundred-fold higher concentrations in food to achieve the same antimicrobial effects in vitro.

7 Essential Oils in Meat Hygiene and Food Preservation

During livestock rearing, the fatty acid content of meat and the oxidation of lipids can be affected by supplementing the diet with essential oils (Cheng et al. 2017). Plant compounds with antioxidant properties were found to enhance meat quality during storage (Al Jumayi et al. 2022). Meat and meat products can also be infused with essential oils. As antimicrobial and antioxidant additives, essential oils prevent the spoilage of meat and can be used as natural meat preservatives (Yu et al. 2021). The addition of essential oils to animal-based products, such as raw or thermally processed meat, can improve their quality and microbiological safety (Cutter 2006). In addition to improving meat products' sensory attributes and overall acceptability, essential oils serve as natural taste enhancers. Nevertheless, essential oils must be dosed carefully, since at high concentrations they can cause toxicity or impart unfavorable tastes (Mucha and Witkowska 2021).

In several studies, researchers have examined the effects of essential oils as food additives. A variety of proteins were analyzed, including those from meat and milk products. Some essential oils tested against pathogenic bacteria demonstrated strong antimicrobial activity (Puvača et al. 2021). Nevertheless, fermented milk products resulted in lower counts of beneficial microbiota as a consequence of their antibacterial effect. Bacteria and pathogens that cause food spoilage can be inhibited most effectively by some essential oils. Numerous studies have shown that essential oils can effectively prevent lipid oxidation in food products (Keshvari et al. 2013; Shahidi and Zhong 2010). Increasing the palatability of meat is one of the most popular uses for oregano essential oil, which has antioxidant properties. After five and ten days of storage, the addition of sage essential oil decreased Enterococcus counts and reduced lipid oxidation in pork (Botsoglou et al. 2004; Vital et al. 2016). Raw and thermally processed minced pork and beef, stored for 12 days at a temperature of 4 °C, had significantly lower lipid autoxidation when 3% sage essential oil was added. When meat is thermally processed, the sage essential oil is more effective as an antioxidant than when it is raw (Jaworska et al. 2021).

As an alternative to conventional food packaging, edible films contain essential oils that can be consumed with food (Vapa Tankosić et al. 2022). Through their ability to inhibit biological, biochemical, and physicochemical changes inside the food matrix, they preserve the quality of food products and prolong their shelf lives. It is becoming increasingly popular to use edible films in place of plastic packaging as an environmentally friendly alternative (Gheorghita Puscaselu et al. 2020).

Oxygen, light, temperature, and pH are physicochemical factors that affect essential oils (Dima and Dima 2015). When oxygen and light are present, unsaturated compound oxidation occurs, causing free radical formation (Turek and Stintzing 2013). The poor solubility of essential oils also limits their use. In addition to encapsulation, the controlled release of essential oils is one of the new processing

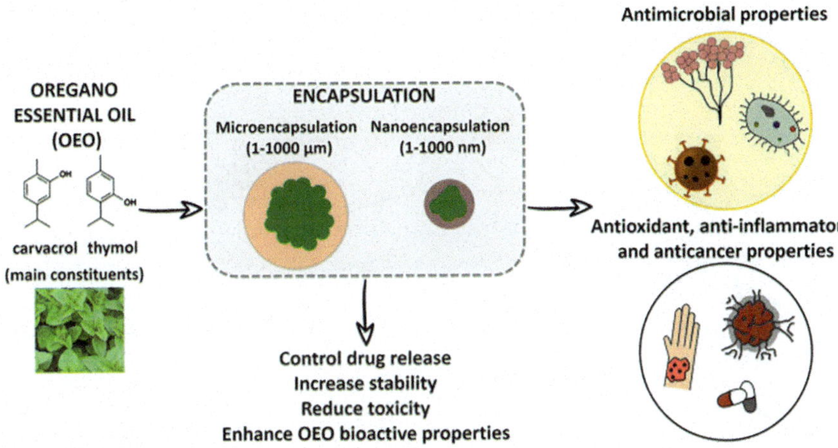

Fig. 3 Essential oil encapsulation (Pontes-Quero et al. 2021)

techniques, essential oils can be encapsulated and released slowly for a controlled effect (Fig. 3).

In dairy and meat products, polymer microcapsules and nanocapsules are encapsulated with essential oils. By slowly releasing essential oil components, microcapsules preserve flavor and extend shelf life. By encapsulating essential oils in zein nanoparticles, essential oils can be dispersed in water (Wu et al. 2012), which greatly enhances their potential to be used for food preservation and bacteria control (Alinaqi et al. 2021).

A healthy diet and lifestyle have become consumers' top priorities. Food choices have become more critical as consumers pay more attention to what is on their plates, becoming aware of different possibilities of using essential oils in animal production (Fig. 4) (Mucha and Witkowska 2021).

Food manufacturers of animal-based products should strive to meet consumer expectations regarding food safety, sustainability, convenience, and the absence of veterinary drug residues (Puvača et al. 2022b) in products raised organically (Milić and Tomaš Simin 2023). Consumers who believe animals have sentient beings may have moral dilemmas when it comes to eating meat (Lin-Schilstra and Fischer 2020). Hence, improving the sustainability of the feed production chain is a challenge and a great opportunity for the industry.

8 Conclusions and Future Perspectives

Arguably the most interesting area of application for essential oils is the inhibition of growth and reduction in numbers of the more serious food-borne pathogens such as *Salmonella* spp., *E. coli,* and *L. monocytogenes*. The delay of spoilage and improvement of organoleptic qualities in vacuum-packed meat or fish may also be interesting from a commercial point of view. Given their organoleptic properties, essential oils

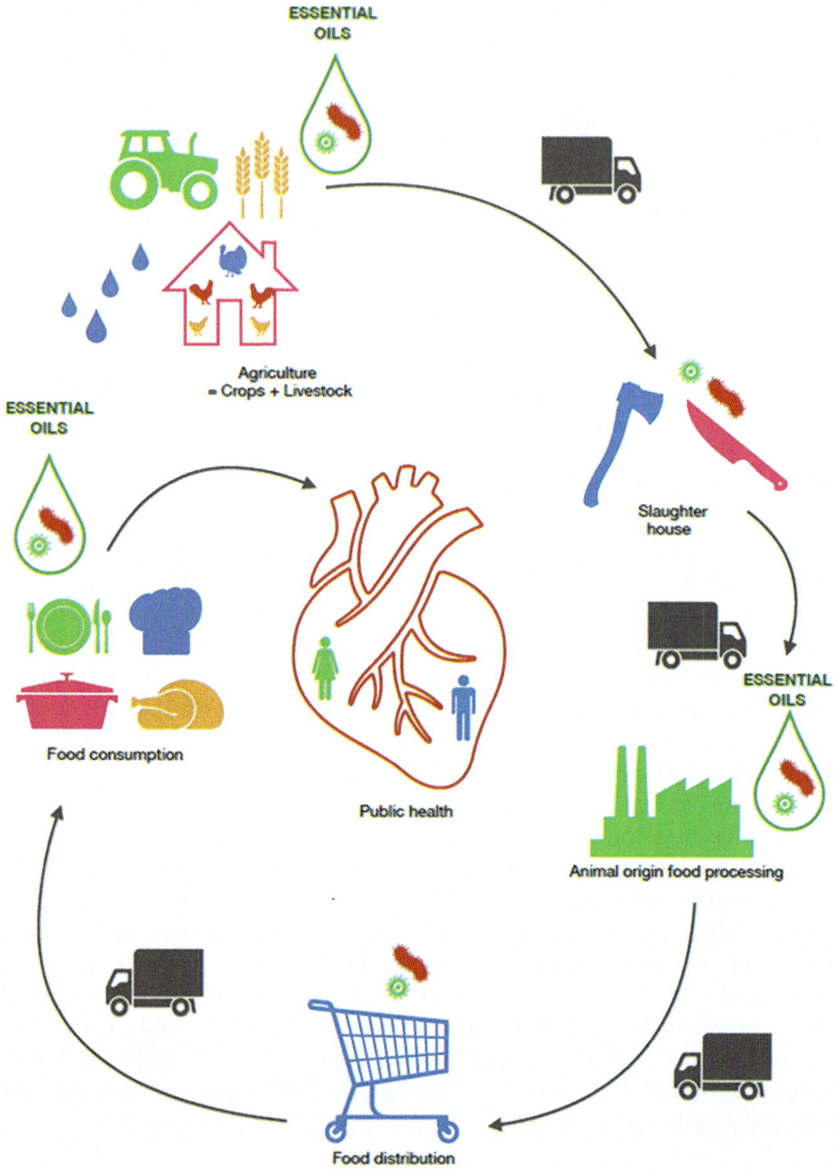

Fig. 4 Possibilities of using essential oils in animal production (Mucha and Witkowska 2021)

could most readily be incorporated into manufactured foods that are traditionally associated with herbs or with spices. It may be possible to use essential oils in foods not previously associated with a herby or spicy flavor if the presence of one or more synergists can produce the desired antibacterial effect at a concentration that does not produce undesirable changes in the flavor or aroma. The use of essential oils in consumer goods is expected to increase in the future due to the rise of green consumerism, which stimulates the use and development of products derived from plants. This applies to the food and cosmetic sectors but also medicinal products. If essential oils were to be required in much greater volumes than at present, bioengineering of their synthesis in plants could provide greater yields. International standardization of the composition of commercially available essential oils would be essential for reliable applications.

Further on, research is necessary to identify the effect of essential oils on proteins and phospholipids embedded in the cytoplasmic membrane. It would be useful for technological applications if these mechanisms were further elucidated. It is particularly relevant to study antibacterial activity against stationary bacterial cells. *B. cereus* has been shown to become less sensitive to carvacrol after growing in the presence of nonlethal concentrations of essential oils in foods. Furthermore, the extent to which bacteria can adapt to the presence of essential oils in foods will also need to be evaluated further. The decrease in sensitivity was achieved by changing the fatty acid and phospholipid head-group composition in the membrane, which reduced the fluidity and passive permeability of the cell membrane. Also, research is needed on the interaction between essential oils, their constituents, and other food ingredients and additives. When clove or oregano oils come into contact with iron, they can become darkly pigmented; this may limit their uses. Essential oils could be combined to maximize their antibacterial activity and minimize the concentration requirements for achieving a particular antibacterial effect through synergistic effects. Antagonism between essential oil and food ingredients is undesirable and research is needed so it can be avoided in practical applications. The stability of essential oils during food processing will also need to be studied. The heat stability of cinnamaldehyde has been investigated; it was found to decompose to benzalde hyde at temperatures approaching 60 °C when heated alone. When combined with eugenol or cinnamon leaf oil, however, cinnamaldehyde was stable even after 30 min at 200 °C. Possible secondary or indirect consequences of the use of essential oils would need to be explored. L. monocytogenes has been shown to become more tolerant of mild heat (56 °C) after being stressed by the presence of ethanol, hydrogen peroxide, or low pH. In the past, there has been little standardization of test methods for testing antibacterials for use in food. This is a field where a selection of standard methods would accelerate the study of promising antibacterial components and their synergistic or antagonistic action with each other and with food ingredients.

Additionally, the environment plays a crucial role in animal production besides essential oils. It would be beneficial to conduct more research on the interactions between the environment and essential oils. To interpret experimental results and variations correctly, a detailed description of hygienic conditions is necessary. Research on the anticoccidial and immunomodulatory properties of essential oils in poultry requires further exploration. To track the active compounds and their metabolites, it is necessary to investigate the metabolic fates of essential oils in animals. Identifying practical applications for essential oils requires research on their interactions with other feed additives. Essential oils have not been extensively studied for their olfactory effects. There might be a significant link between essential oils and improved digestibility of energy and nutrients due to their direct physiologic effects on relaxation, contraction, resistance, and peristalsis in the gut. Because essential oils are easily absorbed, it is imperative to experiment with different methods of delivering them to the gastrointestinal tract. To determine which essential oils modulate the immune system specifically, different challenge models should be used in the perspectives.

References

Abbaszadeh S, Sharifzadeh A, Shokri H, Khosravi AR, Abbaszadeh A (2014) Antifungal efficacy of thymol, carvacrol, eugenol and menthol as alternative agents to control the growth of food-relevant fungi. J Mycol Méd 24(2):e51–e56. https://doi.org/10.1016/j.mycmed.2014.01.063

Al Jumayi HA, Allam AY, El-Beltagy AE-D, Algarni EH, Mahmoud SF, El Halim Kandil AA (2022) Bioactive compound, antioxidant, and radical scavenging activity of some plant aqueous extracts for enhancing shelf life of cold-stored rabbit meat. Antioxidants 11(6) Article 6. https://doi.org/10.3390/antiox11061056

Ali B, Al-Wabel NA, Shams S, Ahamad A, Khan SA, Anwar F (2015) Essential oils used in aromatherapy: a systemic review. Asian Pac J Trop Biomed 5(8):601–611. https://doi.org/10.1016/j.apjtb.2015.05.007

Alinaqi Z, Khezri A, Rezaeinia H (2021) Sustained release modeling of clove essential oil from the structure of starch-based bio-nanocomposite film reinforced by electrosprayed zein nanoparticles. Int J Biol Macromol 173:193–202. https://doi.org/10.1016/j.ijbiomac.2021.01.118

Amad AA, Männer K, Wendler KR, Neumann K, Zentek J (2011) Effects of a phytogenic feed additive on growth performance and ileal nutrient digestibility in broiler chickens. Poult Sci 90(12):2811–2816. https://doi.org/10.3382/ps.2011-01515

Bassolé IHN, Juliani HR (2012) Essential oils in combination and their antimicrobial properties. Molecules 17(4):3989–4006. https://doi.org/10.3390/molecules17043989

Bhavaniramya S, Vishnupriya S, Al-Aboody MS, Vijayakumar R, Baskaran D (2019) Role of essential oils in food safety: antimicrobial and antioxidant applications. Grain Oil Sci Technol 2(2):49–55. https://doi.org/10.1016/j.gaost.2019.03.001

Botsoglou NA, Florou-Paneri P, Christaki E, Giannenas I, Spais AB (2004) Performance of rabbits and oxidative stability of muscle tissues as affected by dietary supplementation with oregano essential oil. Arch Anim Nutr 58(3):209–218. https://doi.org/10.1080/00039420410001701404

Brenes A, Roura E (2010) Essential oils in poultry nutrition: Main effects and modes of action. Anim Feed Sci Technol 158(1–2):1–14. https://doi.org/10.1016/j.anifeedsci.2010.03.007

Burt S (2004) Essential oils: their antibacterial properties and potential applications in foods—a review. Int J Food Microbiol 94(3):223–253. https://doi.org/10.1016/j.ijfoodmicro.2004.03.022

Castilho PC, Savluchinske-Feio S, Weinhold TS, Gouveia SC (2012) Evaluation of the antimicrobial and antioxidant activities of essential oils, extracts and their main components from oregano from Madeira Island, Portugal. *Food Control* 23(2):552–558. https://doi.org/10.1016/j.foodcont.2011.08.031

Chao SC, Young DG, Oberg CJ (2000) Screening for inhibitory activity of essential oils on selected bacteria, fungi and viruses. J Essent Oil Res 12(5):639–649. https://doi.org/10.1080/10412905.2000.9712177

Cheng C, Liu Z, Zhou Y, Wei H, Zhang X, Xia M, Deng Z, Zou Y, Jiang S, Peng J (2017) Effect of oregano essential oil supplementation to a reduced-protein, amino acid-supplemented diet on

meat quality, fatty acid composition, and oxidative stability of Longissimus thoracis muscle in growing-finishing pigs. Meat Sci 133:103–109. https://doi.org/10.1016/j.meatsci.2017.06.011
Chouhan S, Sharma K, Guleria S (2017) Antimicrobial activity of some essential oils—present status and future perspectives. Medicines 4(3) Article 3. https://doi.org/10.3390/medicines4030058
Cox SD, Mann CM, Markham JL (2001) Interactions between components of the essential oil of melaleuca alternifolia. J Appl Microbiol 91(3):492–497. https://doi.org/10.1046/j.1365-2672.2001.01406.x
Cutter CN (2006) Opportunities for bio-based packaging technologies to improve the quality and safety of fresh and further processed muscle foods. Meat Sci 74(1):131–142. https://doi.org/10.1016/j.meatsci.2006.04.023
Czaplewski L, Bax R, Clokie M, Dawson M, Fairhead H, Fischetti VA, Foster S, Gilmore BF, Hancock REW, Harper D, Henderson IR, Hilpert K, Jones BV, Kadioglu A, Knowles D, Ólafsdóttir S, Payne D, Projan S, Shaunak S et al (2016) Alternatives to antibiotics—A pipeline portfolio review. Lancet Infect Dis 16(2):239–251. https://doi.org/10.1016/S1473-3099(15)00466-1
da Silveira SM, Luciano FB, Fronza N, Cunha A, Scheuermann GN, Vieira CRW (2014) Chemical composition and antibacterial activity of Laurus nobilis essential oil towards foodborne pathogens and its application in fresh Tuscan sausage stored at 7 °C. LWT Food Sci Technol 59(1):86–93. https://doi.org/10.1016/j.lwt.2014.05.032
Dawood MAO, Koshio S, Esteban MÁ (2018) Beneficial roles of feed additives as immunostimulants in aquaculture: A review. Rev Aquac 10(4):950–974. https://doi.org/10.1111/raq.12209
Delgado B, Fernández PS, Palop A, Periago PM (2004) Effect of thymol and cymene on Bacillus cereus vegetative cells evaluated through the use of frequency distributions. Food Microbiol 21(3):327–334. https://doi.org/10.1016/S0740-0020(03)00075-3
Dima C, Dima S (2015) Essential oils in foods: extraction, stabilization, and toxicity. Curr Opin Food Sci 5:29–35. https://doi.org/10.1016/j.cofs.2015.07.003
Dosoky NS, Setzer WN (2018) Chemical composition and biological activities of essential oils of curcuma species. Nutrients 10, 10(9) Article 9. https://doi.org/10.3390/nu10091196
Franz C, Baser K, Windisch W (2010) Essential oils and aromatic plants in animal feeding – a European perspective: a review. Flavour Fragr J 25(5):327–340. https://doi.org/10.1002/ffj.1967
Gheorghita Puscaselu R, Gutt G, Amariei S (2020) The use of edible films based on sodium alginate in meat product packaging: an eco-friendly alternative to conventional plastic materials. Coatings 10(2) Article 2. https://doi.org/10.3390/coatings10020166
Hyldgaard M, Mygind T, Meyer R (2012) Essential oils in food preservation: mode of action, synergies, and interactions with food matrix components. Front Microbiol 3. https://www.frontiersin.org/articles/10.3389/fmicb.2012.00012
Jang IS, Ko YH, Yang HY, Ha JS, Kim JY, Kang SY, Yoo DH, Nam DS, Kim DH, Lee CY (2004) Influence of essential oil components on growth performance and the functional activity of the pancreas and small intestine in broiler chickens. Asian Australas J Anim Sci 17(3):394–400. https://doi.org/10.5713/ajas.2004.394
Jang IS, Ko YH, Kang SY, Lee CY (2007) Effect of a commercial essential oil on growth performance, digestive enzyme activity and intestinal microflora population in broiler chickens. Anim Feed Sci Technol 134(3):304–315. https://doi.org/10.1016/j.anifeedsci.2006.06.009
Jaworska D, Rosiak E, Kostyra E, Jaszczyk K, Wroniszewska M, Przybylski W (2021) Effect of herbal addition on the microbiological, oxidative stability and sensory quality of minced poultry meat. Foods 10(7) Article 7. https://doi.org/10.3390/foods10071537
Joerger R (2003) Alternatives to antibiotics: Bacteriocins, antimicrobial peptides and bacteriophages. Poult Sci 82(4):640–647. https://doi.org/10.1093/ps/82.4.640

Keshvari M, Asgary S, Jafarian-dehkordi A, Najafi S, Ghoreyshi-Yazdi SM (2013) Preventive effect of cinnamon essential oil on lipid oxidation of vegetable oil. ARYA Atheroscler 9(5): 280–286

Kostadinović L (2023) Influence of wormwood seeds on enzymatic and non–enzymatic activity in blood of broilers with coccidiosis. J Agron Technol Eng Manag 6(1):857–865. https://doi.org/10.55817/IHHY9641

Lambert RJW, Skandamis PN, Coote PJ, Nychas G-JE (2001) A study of the minimum inhibitory concentration and mode of action of oregano essential oil, thymol and carvacrol. J Appl Microbiol 91(3):453–462. https://doi.org/10.1046/j.1365-2672.2001.01428.x

Lika E, Kostić M, Vještica S, Milojević I, Puvača N (2021) Honeybee and plant products as natural antimicrobials in enhancement of poultry health and production. Sustainability 13(15):8467. https://doi.org/10.3390/su13158467

Lin-Schilstra L, Fischer ARH (2020) Consumer moral dilemma in the choice of animal-friendly meat products. Sustainability *12*(12) Article 12. https://doi.org/10.3390/su12124844

Ljubojević D, Pelić M, Puvača N, Milanov D (2017) Resistance to tetracycline in Escherichia coli isolates from poultry meat: epidemiology, policy and perspective. Worlds Poult Sci J 73(2): 409–417. https://doi.org/10.1017/S0043933917000216

Mantzourani I, Daoutidou M, Dasenaki M, Nikolaou A, Alexopoulos A, Terpou A, Thomaidis N, Plessas S (2022) Plant extract and essential oil application against food-borne pathogens in raw pork meat. Foods 11(6) Article 6. https://doi.org/10.3390/foods11060861

Michiels J, Missotten J, Dierick N, Fremaut D, Maene P, De Smet S (2008) In vitro degradation and in vivo passage kinetics of carvacrol, thymol, eugenol and trans-cinnamaldehyde along the gastrointestinal tract of piglets. J Sci Food Agric 88(13):2371–2381. https://doi.org/10.1002/jsfa.3358

Middelkoop A, Choudhury R, Gerrits WJJ, Kemp B, Kleerebezem M, Bolhuis JE (2018) Dietary diversity affects feeding behaviour of suckling piglets. Appl Anim Behav Sci 205:151–158. https://doi.org/10.1016/j.applanim.2018.05.006

Milić D, Tomaš Simin M (2023) Economics of organic carrot production through case study report. J Agron Technol Eng Manag 6(1):873–884. https://doi.org/10.55817/UQRF7578

Mohamed EA, Tulcan C, Alexa E, Morar D, Dumitrescu E, Muselin F, Radulov I, Puvača N, Cristina RT (2020) Sea buckthorn and grape extract might be helpful and sustainable phyto-resources as associated hypolipidemic agents—preliminary study. Sustainability 12(21):9297. https://doi.org/10.3390/su12219297

Mourente G, Diaz-Salvago E, Bell JG, Tocher DR (2002) Increased activities of hepatic antioxidant defence enzymes in juvenile gilthead sea bream (*Sparus aurata* L.) fed dietary oxidised oil: attenuation by dietary vitamin E. Aquaculture 214(1):343–361. https://doi.org/10.1016/S0044-8486(02)00064-9

Mucha W, Witkowska D (2021) The applicability of essential oils in different stages of production of animal-based foods. Molecules *26*, 26(13) Article 13. https://doi.org/10.3390/molecules26133798

Nazzaro F, Fratianni F, De Martino L, Coppola R, De Feo V (2013) Effect of essential oils on pathogenic bacteria. Pharmaceuticals 6(12):1451, Article 12. https://doi.org/10.3390/ph6121451

Nieto G (2017) Biological activities of three essential oils of the Lamiaceae Family. Medicines 4(3) Article 3. https://doi.org/10.3390/medicines4030063

Omonijo FA, Ni L, Gong J, Wang Q, Lahaye L, Yang C (2018) Essential oils as alternatives to antibiotics in swine production. Anim Nutr 4(2):126–136. https://doi.org/10.1016/j.aninu.2017.09.001

Ooi LSM, Li Y, Kam S-L, Wang H, Wong EYL, Ooi VEC (2006) Antimicrobial activities of cinnamon oil and Cinnamaldehyde from the Chinese medicinal herb Cinnamomum cassia Blume. Am J Chin Med 34(03):511–522. https://doi.org/10.1142/S0192415X06004041

Pelić M, Puvača N, Kartalović B, Živkov Baloš M, Novakov N, Ljubojević Pelić D (2023) Antibiotics and sulfonamides in water, sediment, and fish in an integrated production system. J Agron Technol Eng Manag 6(1):851–856. https://doi.org/10.55817/YVRR1215

Platel K, Rao A, Saraswathi G, Srinivasan K (2002) Digestive stimulant action of three Indian spice mixes in experimental rats. Food Nahrung 46(6):394–398. https://doi.org/10.1002/1521-3803(20021101)46:6<394::AID-FOOD394>3.0.CO;2-D

Pontes-Quero GM, Esteban-Rubio S, Pérez Cano J, Aguilar MR, Vázquez-Lasa B (2021) Oregano essential oil micro- and Nanoencapsulation with bioactive properties for biotechnological and biomedical applications. Front Bioeng Biotechnol 9:703684. https://doi.org/10.3389/fbioe.2021.703684

Popović S, Puvača N, Kostadinović L, Džinić N, Bošnjak J, Vasiljević M, Đuragić O (2016) Effects of dietary essential oils on productive performance, blood lipid profile, enzyme activity and immunological response of broiler chickens. Eur Poult Sci 80. https://doi.org/10.1399/eps.2016.146.CORR

Puvača N (2018) Honeybee and medicinal plants products in poultry postantibiotic era production. J Agron Technol Eng Manag 1(1):8–17

Puvača N, Stanaćev V, Glamočić D, Lević J, Perić L, Stanaćev V, Milić D (2013) Beneficial effects of phytoadditives in broiler nutrition. Worlds Poult Sci J 69(1):27–34. https://doi.org/10.1017/S0043933913000032

Puvača N, Ljubojević D, Kostadinović L, Lević J, Nikolova N, Miščević B, Könyves T, Lukač D, Popović S (2015a) Spices and herbs in broilers nutrition: hot red pepper (*Capsicum annuum* L.) and its mode of action. Worlds Poult Sci J 71(4):683–688. https://doi.org/10.1017/S004393391500241X

Puvača N, Ljubojević D, Kostadinović L, Lukač D, Lević J, Popović S, Đuragić O (2015b) Spices and herbs in broilers nutrition: effects of garlic (*Allium sativum* L.) on broiler chicken production. Worlds Poult Sci J 71(3):533–538. https://doi.org/10.1017/S0043933915002214

Puvača N, Lika E, Cocoli S, Shtylla Kika T, Bursić V, Vuković G, Tomaš Simin M, Petrović A, Cara M (2020a) Use of tea tree essential oil (*Melaleuca alternifolia*) in laying Hen's nutrition on performance and egg fatty acid profile as a promising sustainable organic agricultural tool. Sustainability 12(8):3420. https://doi.org/10.3390/su12083420

Puvača N, Lika E, Tufarelli V, Bursić V, Ljubojević Pelić D, Nikolova N, Petrović A, Prodanović R, Vuković G, Lević J, Giannenas I (2020b) Influence of different tetracycline antimicrobial therapy of mycoplasma (*Mycoplasma synoviae*) in laying hens compared to tea tree essential oil on table egg quality and antibiotics residues. Foods 9(5):612. https://doi.org/10.3390/foods9050612

Puvača N, Milenković J, Galonja Coghill T, Bursić V, Petrović A, Tanasković S, Pelić M, Ljubojević Pelić D, Miljković T (2021) Antimicrobial activity of selected essential oils against selected pathogenic bacteria: in vitro study. Antibiotics 10(5):546. https://doi.org/10.3390/antibiotics10050546

Puvača N, Tufarelli V, Giannenas I (2022a) Essential oils in broiler chicken production, immunity and meat quality: review of Thymus vulgaris, Origanum vulgare, and Rosmarinus officinalis. Agriculture 12(6):874. https://doi.org/10.3390/agriculture12060874

Puvača N, Vapa Tankosić J, Ignjatijević S, Carić M, Prodanović R (2022b) Antimicrobial resistance in the environment: review of the selected resistance drivers and public health concerns. J Agron Technol Eng Manag 5(5):793–802. https://doi.org/10.55817/CSCQ3326

Raei P, Pourlak T, Memar MY, Alizadeh N, Aghamali M, Zeinalzadeh E, Asgharzadeh M, Kafil HS (2017) Thymol and carvacrol strongly inhibit biofilm formation and growth of carbapenemase-producing gram negative bacilli. Cell Mol Biol 63(5) Article 5. https://doi.org/10.14715/cmb/2017.63.5.20

Reyer H, Zentek J, Männer K, Youssef IMI, Aumiller T, Weghuber J, Wimmers K, Mueller AS (2017) Possible molecular mechanisms by which an essential oil blend from Star Anise, Rosemary, Thyme, and Oregano and Saponins increase the performance and Ileal protein digestibility of growing broilers. J Agric Food Chem 65(32):6821–6830. https://doi.org/10.1021/acs.jafc.7b01925

Roca I, Akova M, Baquero F, Carlet J, Cavaleri M, Coenen S, Cohen J, Findlay D, Gyssens I, Heure OE, Kahlmeter G, Kruse H, Laxminarayan R, Liébana E, López-Cerero L, MacGowan A, Martins M, Rodríguez-Baño J, Rolain J-M et al (2015) The global threat of antimicrobial resistance: science for intervention. New Microbes New Infect 6:22–29. https://doi.org/10.1016/j.nmni.2015.02.007

Shahidi F, Zhong Y (2010) Lipid oxidation and improving the oxidative stability. Chem Soc Rev 39(11):4067–4079. https://doi.org/10.1039/B922183M

Singh BK, Tiwari S, Dubey NK (2021) Essential oils and their nanoformulations as green preservatives to boost food safety against mycotoxin contamination of food commodities: a review. J Sci Food Agric 101(12):4879–4890. https://doi.org/10.1002/jsfa.11255

Tavares TD, Antunes JC, Padrão J, Ribeiro AI, Zille A, Amorim MTP, Ferreira F, Felgueiras HP (2020) Activity of specialized biomolecules against gram-positive and gram-negative bacteria. Antibiotics 9(6) Article 6. https://doi.org/10.3390/antibiotics9060314

Tsai ML, Wu CT, Lin TF, Lin WC, Huang YC, Yang CH (2013) Chemical composition and biological properties of essential oils of two mint species. Trop J Pharm Res 12(4):Article 4. https://doi.org/10.4314/tjpr.v12i4.20

Turek C, Stintzing FC (2013) Stability of essential oils: a review: stability of essential oils. Compr Rev Food Sci Food Saf 12(1):40–53. https://doi.org/10.1111/1541-4337.12006

Vapa Tankosić J, Puvača N, Giannenas I, Tufarelli V, Ignjatijević S (2022) Food safety policy in the European Union. J Agron Technol Eng Manag 5(2):712–717. https://doi.org/10.55817/EMRK6646

Vital ACP, Guerrero A, de Monteschio JO, Valero MV, Carvalho CB, de Filho BA, Madrona GS, do Prado IN (2016) Effect of edible and active coating (with Rosemary and Oregano essential oils) on beef characteristics and consumer acceptability. PLoS One 11(8):e0160535. https://doi.org/10.1371/journal.pone.0160535

Wang X, Shen Y, Thakur K, Han J, Zhang J-G, Hu F, Wei Z-J (2020) Antibacterial activity and mechanism of ginger essential oil against *Escherichia coli* and *Staphylococcus aureus*. Molecules 25(17) Article 17. https://doi.org/10.3390/molecules25173955

Williamson EM (2001) Synergy and other interactions in phytomedicines. Phytomedicine 8(5): 401–409. https://doi.org/10.1078/0944-7113-00060

Wu Y, Luo Y, Wang Q (2012) Antioxidant and antimicrobial properties of essential oils encapsulated in zein nanoparticles prepared by liquid–liquid dispersion method. LWT Food Sci Technol 48(2):283–290. https://doi.org/10.1016/j.lwt.2012.03.027

Yan L, Wang JP, Kim HJ, Meng QW, Ao X, Hong SM, Kim IH (2010) Influence of essential oil supplementation and diets with different nutrient densities on growth performance, nutrient digestibility, blood characteristics, meat quality and fecal noxious gas content in grower–finisher pigs. Livest Sci 128(1):115–122. https://doi.org/10.1016/j.livsci.2009.11.008

Yu HH, Chin Y-W, Paik H-D (2021) Application of natural preservatives for meat and meat products against food-borne pathogens and spoilage bacteria: a review. Foods 10(10) Article 10. https://doi.org/10.3390/foods10102418

Zhai H, Liu H, Wang S, Wu J, Kluenter A-M (2018) Potential of essential oils for poultry and pigs. Anim Nutr 4(2):179–186. https://doi.org/10.1016/j.aninu.2018.01.005

Zorzet A (2014) Overcoming scientific and structural bottlenecks in antibacterial discovery and development. Ups J Med Sci 119(2):170–175. https://doi.org/10.3109/03009734.2014.897277

Feed Additives and Future Perspectives

Alireza Seidavi, Babak Hosseintabar-Ghasemabad, Ivan Fedorovich Gorlov, Marina Ivanovna Slozhenkina, Alexander Anatolyevich Mosolov, Marianna Oteri, Biagina Chiofalo, and Luigi Liotta

Abstract

From a strategic perspective, food security is one of the most important pillars of national security for human societies. Introduction of effective new feed additives in livestock and poultry nutrition can be a key to sustainable food security by increasing animal performance and improving animal health. Continuous research in this area is necessary because certain feed additives can also combat pathogens and ameliorate the use of antibiotics. Future studies on feed additives should be aimed at determining their optimal levels to maximize the goals of improved production efficiency and conversion ratio and consequently the quality and quantity of livestock products. By improving bioavailability, digestibility, and effectiveness of additives, we will be able to use them more accurately in animal diets without any worry about their side effects. Synthetic additives with new formulations along with bioactive compounds derived from plants can help attain this goal. Some of the known compounds include phenolic compounds, essential oils, terpenoids, phytosterols, polycosanols, and tocopherols, which can improve feed conversion ratio and production performance just like synthetic additives. The use of these compounds can also lead to the production of high-quality organic products. Therefore, understanding the mechanism of action of bioactive compounds in the body physiology and biochemical status of domestic

A. Seidavi (✉)
Department of Animal Science, Rasht Branch, Islamic Azad University, Rasht, Iran

B. Hosseintabar-Ghasemabad
Department of Animal Science, Faculty of Agriculture, University of Tabriz, Tabriz, Iran

I. F. Gorlov · M. I. Slozhenkina · A. A. Mosolov
Volga Region Research Institute of Manufacture and Processing of Meat-and-Milk Production, Volgograd, Russia

M. Oteri · B. Chiofalo · L. Liotta
Department of Veterinary Sciences, University of Messina, Messina, Italy

G. Arsenos, I. Giannenas (eds.), *Sustainable Use of Feed Additives in Livestock*,
https://doi.org/10.1007/978-3-031-42855-5_33

animals can be one of the goals of future studies. In this chapter, we provide information on the selection of additives by presenting the latest studies on the use of well-known and lesser-known additives as well as the simultaneous use of several additives in animal nutrition by discussing their mechanisms of action and beneficial physiological and biochemical effects. We also list a series of new plant supplements containing valuable bioactive compounds to encourage researchers to study the use of these new plants in various forms in livestock and poultry nutrition.

Keywords

Animal health · Animal performance · Bioactive compounds · Food security · Human demand · New plant supplements · Synthetic additives

1 Introduction

The main drivers of the global market for animal feed additives include increasing demand for animal products, increasing awareness of the quality and safety of animal products, capability for mass production, and the necessity of preventing the spread of various types of stresses and diseases in livestock (Pandey et al. 2019). Therefore, it is absolutely important to design a targeted strategy to draw successful health-oriented and economic perspectives. "Zero hunger" and "responsible consumption and production" include the second and twelfth sustainable development goals in order to achieve food security, both of which are considered in the animal feed industry (Pandey et al. 2019). Poultry strains used in the industry have been subjected to intensive genetic selection for rapid growth and increased feed efficiency along with high metabolic rate (Klasing 2007; Seidavi et al. 2014). However, the biggest challenge for the poultry producers and specialists is to achieve balanced rations at a minimum price (Pandey et al. 2019; Alagawany and Abd El-Hack 2020). Thus, appropriate feed additives can play an important role in achieving the important and strategic goals of health and economy.

2 Microencapsulations and Nanotechnology of Feed Additives and Future Perspective

Despite the versatile uses of plants and their components as poultry feed supplements, it seems that their beneficial aspects face certain limitations. The sensitivity of bioactive components of plants to peroxidation and oxidation damage, hydrophobicity and volatile nature of the compounds are some of the important limitations of the direct use of these resources. In addition, some reports show that the supplementation of plant extracts in high doses reduces palatability and feed intake (Pliego et al. 2022).

Recently, the application of microencapsulation and nanoencapsulations as a solution of utilizing the potential of plants in poultry nutrition has received increased scientific attention. Microencapsulation is a technique in which small particles or droplets are surrounded by a coating wall or are embedded in a homogeneous or heterogeneous matrix to form small capsules. In general, microcapsules are tiny particles between 1 and 1000 mm that contain a bioactive agent surrounded by a natural or synthetic polymer membrane (Calvo et al. 2011; Pliego et al. 2022). Reports show that microencapsulated organic acids and essential oils of medicinal plants, alone or mixed, as feed additives in broiler chickens improve the function and intestinal microflora, reduce intestinal pathogenic microbes, and strengthen health and improve safety (Scherer et al. 2014; Pliego et al. 2022). Nano materials are essentially different from their larger materials due to differences in surface effects and quantum effects and spatial arrangement allows them to have new and element-specific properties, such as creating a magnetic environment in metals in the form of nanoparticles (Singh 2016; Abd El-Ghany 2019).

Nanoparticles can generally be categorized into inorganic, organic, emulsions, dispersions, and nano clays. The use of inorganic and organic nanoparticles in poultry feed additives is a precedent (Singh 2016; Anwar et al. 2019). Micelles are organic nanoparticles that can assemble by a thermodynamic process called self-assembly. It has been reported that Nano-sized Self-assembled Liquid Structures (NSLS) particles act as carriers for compounds that are more easily absorbed into the bloodstream from the intestine and increase their bioavailability. However, nano-technology allows the feed additive to increase the parameters related to the health of chickens and also help to improve the bioavailability of nutrients with the optimal physiological function of the body (Singh 2016; Swain et al. 2021).

Heydarian et al. (2020) also investigated the effects of dietary inclusion of encapsulated thyme essential oil (TEO), oregano essential oil (OEO) and probiotics on growth performance, immune response and intestinal morphology of broilers. They showed that dietary inclusion of TEO + OEO feed in an encapsulated form along with probiotics increases humoral immunity in broiler chickens.

Ahmadi et al. (2020) studied the optimal level of nano-selenium supplement on broiler chickens with regard to performance, economic parameters, plasma composition, and safety. The results showed that Nano-S supplementation had a positive effect on body weight gain (BWG) and feed conversion ratio (FCR), without negatively affecting bird health status. Increase in productivity and reduction of the environmental hazards related to the additive source as an effect of nanotechnology appear as a very promising finding. In the food industry, the use of liposomal nanovesicles for the encapsulation and delivery of nutrients and functional components such as proteins, enzymes, flavors, and antimicrobial compounds has precedent (Wen et al. 2006). In another study, the effects of nano-encapsulated extracts (0.02, 0.025 and 0.05% w/w) of aloe vera, dill and nettle root on performance, carcass traits, and immunoglobulin serum concentration (IgM and IgY) in broiler chickens were investigated and the results indicated that nanoencapsulation of plant extracts can improve the growth performance of broiler chickens (Meimandipour et al. 2017).

However, the challenges of using nano nutrients include changes in metabolism, toxicity, and environmental effects, which should be minimized. To accomplish this, more studies and efficient research by researchers and experts in the future are needed. Entering nanotechnology due to the change of molecular nature requires high precision and sensitivity in order to prevent its possible economic, social, ethical, and legal consequences (Singh 2016). It is hoped that using this technology in the future of animal nutrition (feed additive) can help to improve the bioavailability of nutrients and production performance and meet the goals of the poultry industry (Ireland 2008; Radha et al. 2014; Singh 2016; Rani et al. 2018).

3 Bacteriophage Additives and Future Perspective

Bacteriophages are viruses that infect and destroy bacteria. Due to the fact that bacteriophages do not cause contamination of plant and animal cells, they can be used instead of antibiotics in the prevention and treatment of diseases in humans and animals, including poultry. Therefore, this type of additive can be suggested as a successful prospect due to its special properties (Evran et al. 2022). There are specific bacteriophages for different bacteria. After entering the bacterial cell, bacteriophages multiply and disrupt the normal activities of the bacteria, thereby facilitating their death. The mutual effects of the combination of bacteriophage and antibiotic therapy have increased their therapeutic aspect (Huff et al. 2006; Żbikowska et al. 2020; Evran et al. 2022).

At the same time, due to the combined use of bacteriophages and therapeutic antibiotics, lower levels of antibiotics can be used. Bacteriophage can be used to treat and prevent Klebillus in poultry and this product can be an effective alternative to antibiotics. Bacteriophages were jointly discovered by Twort (1915) and De Hérelle (1917) in the early twentieth century. Bacteriophages are special viruses and their structure consists of a protein layer that surrounds nucleic acid (nucleic acid can be in the form of DNA or RNA). Bacteriophages are divided into two main groups, benign and malignant, whose life cycles are different. Malignant bacteriophages destroy bacteria in a multi-step process. That is, they attach themselves to the surface of the bacteria by finding specific binding sites (receptors). Then they inject their nucleic acid into the bacteria. In the next step, viruses multiply inside bacteria. The release of an average of 50 to 200 cells causes the bacteria to break down and finally the bacteria is destroyed. Benign bacteriophages do not reproduce so quickly in bacteria, but exist as prophages along with plant bacteria and reproduce with them and turn the bacteria into a lysogenic strain. When lysogenic bacteria are under stress, the prophages become active and cause the virus to multiply and eventually kill the bacteria. In general, malignant bacteria provide the highest probability of destruction of bacteria in different situations (Huff et al. 2006). As soon as bacteriophages were discovered, the tendency to use them as a tool for the prevention and treatment of bacterial infections was proposed and started. It is considered that with further studies and information updating by experts, the use of this additive will be developed without any worries and perspective a successful measure of this technique and technology was envisioned (Huff et al. 2006; Lang 2006; Żbikowska et al. 2020; Islam et al. 2021; Olson et al. 2021; Han et al. 2022).

4 Phytogenic Additives and Future Perspective

Plant feed additives (PFAs) are classified as sensory and flavor compounds according to the EU law (EC 1831/2003) (Regulation 2003; Shehata et al. 2022). It is estimated that among the 100,000 different compounds of natural origin, approximately 80,000 are derived from plants (Upadhaya and Kim 2017; Pirgozliev et al. 2019).

However, Pandey et al. (2019) stated in their review report that the main reasons for using phytogenics as additives by researchers include antimicrobial effect, APG replacement strategy, growth promotion, anti-inflammatory, good past experience, better FCR, higher feed intake, used with AGPs and others, respectively.

Phytogenic substances improve the performance and health of animals through: (1) improved digestibility, (2) antimicrobial activities, (3) anti-inflammatory and antioxidant effects, (4) stabilization of intestinal microbiota, (5) improvement of animal traits, and (6) reduction of environmental emissions (Shehata et al. 2022). Medicinal plants have a huge impact on the poultry industry by improving the performance and productivity of birds (Pliego et al. 2022).

All assumptions related to the use of plant-based bioactive compounds have always been considered and supported by the "REPLACE" project in the framework of European Union programs (Alloui et al. 2014). Phytogenic products are claimed to positively affect palatability and gut function, thus promoting feed utilization and increasing production performance (Windisch et al. 2008).

4.1 Phytogenics Additives Based on Cultivated Plants

In the field of agricultural products, a cultivated plant is widespread and its consumption market is safe and risk free. If specialized research related to phytogenic plants reaches a point where the producers of the poultry industry are convinced that these natural additives can be consumed with safety, while their health-enhancing effects are evident, a positive perspective of their use is expected. Therefore, by continuously studying the effects of phytogenic additives in poultry nutrition, the desired plants can be cultivated and produced in a practical way to achieve the strategic goals of additives and the successful perspective of the poultry industry. Because in the future, many of these plants can be cultivated based on production and mechanized principles and the final price can be greatly reduced.

With the optimal and economical production of raw materials of plant origin in bulk and mass, the poultry industry can be a stable market and customer for phytogenic additives and the producers of medicinal plants and also develop phytogenic products with the support of governments and organizations and specialized centers that can put them in their agenda and priority. Therefore, it is recommended that researchers with numerous and extensive studies related to phytogenics can prepare the ground for the development of valuable and health-giving medicinal plants in poultry nutrition and industry.

Windisch et al. (2008) introduced the classification of different herbal compounds based on their origin and processing as follows: Plants (flowering, non-woody, and non-resistant plants), Spices are herbs with a strong smell or taste, essential oils (volatile lipophilic compounds) and oleoresins (extracts from non-aqueous solvents). It should be noted that the active compounds of a plant species might vary depending on different geographical regions and weather conditions. These changes are the results of genetic and environmental interaction and the manifestation of biodiversity in the same plant species (Zhang et al. 2011). Many medicinal plants and sources of phytogenics have been cultivated traditionally and industrially in many countries for many years and there is technical knowledge to produce them, such as tea, alfalfa, spices, etc. But if it is used in poultry feeding as an additive, the products of this sector can be developed and expanded quickly. Other medicinal plants, with a limited cultivation or that they are obtained from nature in a semi-domestic and wild form, they will be mentioned in the next parts of the chapter.

In their review report, Seidavi et al. (2020) examined the use of green tea (*Camellia sinensis* L) in poultry nutrition and reported that supplementing diets with green tea increased feed intake and final body weight. It improves FCR and can be used as a growth stimulant instead of antibiotics. In addition, green tea improved immunity against coccidiosis, Newcastle disease, reduced plasma low-density lipoprotein, and reduced serum cholesterol. It improved the quality of animal products by reducing meat peroxidation, reducing abdominal fat, improving meat quality, and increasing the chemical composition of meat. Similarly, egg yolk cholesterol decreased, egg yolk color improved, and yolk alpha-tocopherol increased. In quails, green tea improved intestinal weight, decreased plasma low-density lipoprotein, egg yolk cholesterol, and improved feed intake and weight gain (Seidavi et al. 2020). Saraei et al. (2016) showed that the use of green tea powder (15 grams per kilogram) significantly reduced plasma low-density lipoprotein (LDL) without affecting high-density lipoprotein (HDL). As a result, the ratio of LDL to HDL improved. In addition, the use of green tea improved the antioxidant status due to the increase of plasma uric acid. In general, green tea can be supplemented in the diet of broilers, laying hens and quails with numerous positive functional and health-giving results (Seidavi et al. 2020) (Fig. 1).

Several commercial phytogenic products obtained from cultivated plants include Anise seeds (*Pimpinella anisum*), Caraway seeds (*Carum carvi*), Cinnamon bark (*Cinnamomum verum*), Chamomile flowers (*Matricaria recutita*), Citrus peel (*Citrus* spp.), Clove buds (*Syzygium aromaticum*), Fennel seeds (*Foeniculum vulgare*), Garlic bulbs (*Allium sativum*), Ginger rhizome (*Zingiber officinale*), Melissa leaves (*Melissa officinalis*), Onion bulbs (*Allium cepa*), Oregano leaves (*Origanum vulgare*), Peppermint leaves (*Mentha piperita*), Rosemary leaves (*Rosmarinus officinalis*), Sage leaves (*Salvia officinalis*), Thyme leaves (*Thymus vulgaris*), and Valerian root/rhizome (*Valeriana officinalis*). These plants have a series of valuable

Fig. 1 Cultivated tea (*Camellia sinensis* L) gardens in Iran and Gilan province–GH. Hosseintabar an agricultural expert and manager of the Darvash Giah Khazar Medicinal Herbs Complex in 2022

bioactive compounds that can have a promising effect on the health and performance of poultry (Máthé 2009; Shehata et al. 2022).

Alfalfa (*Medicago sativa* L.) is also a well-known plant for animal nutrition, which is widely cultivated in many countries. But due to the existence of various vitamins, minerals, and proteins and valuable and balanced active compounds, it is known as the king of medicinal plants. It has also been proven that this plant is an important source of various minerals and vitamins, flavonoids, phenolic acid, xanthophylls, xantophylls, carotenoids, and other nutrients. Also, other studies have found that alfalfa or *Medicago sativa* contains saponins, flavonoids, and phytoestrogens (Pliego et al. 2022).

The active compounds of this plant can increase the secretion of neutral steroids and bile acids into feces. It has anti-fungal properties and can even reduce cholesterol. In research, the nutritional effects of low-fiber alfalfa (15% by weight on ration weight) on production and egg characteristics were investigated. The results showed that supplementation of low-fiber alfalfa meal in the diet of laying hens can positively affect the yolk quality, without having a negative effect on production traits (Laudadio et al. 2014). The effect of polysavone, a natural extract from alfalfa, on the deposition of abdominal fat and immunity of broiler chickens was determined and the results showed that it reduced the weight of abdominal fat at 5 and 6 weeks of age and the relative weight of thymus and spleen and bursa weight improved compared to the control group. In addition, the proliferation of T and B lymphocytes in the poly-sawn group was significantly higher than the control group. When the birds were 4 and 5 weeks old, polysavone led to a significant increase in serum antibody titer and inhibition of anti-Newcastle disease virus hemagglutination. In total, these results showed that polysavone reduced abdominal fat deposition and strengthened immunity without adversely affecting the performance of broiler chickens (Dong et al. 2007). However, many phytogenic resources are obtained from plants that have been widely cultivated in many regions of the world for many years, and according to the needs of the poultry industry, these phytogenic resources can be used as an alternative to antibiotics. Because of these plants, while having phytogenic properties, it is possible to mass- produce and access the raw material economically (Pliego et al. 2022).

4.2 Phytogenic Additives Based on Essential Oils

Many studies show that the use of essential oils (EOs) can provide superior effects compared to the direct consumption of a series of plants. Therefore, a series of plants from which it is possible to economically extract essential oils can play an important role as a feed additive in the future. EOs are a mixture of aromatic oily liquids obtained from plant materials such as flowers, buds, seeds, leaves, branches, tree bark, herbs, wood, fruits, and roots (Craig 1999; Stevanović et al. 2018; Abdelli et al. 2021). EOs are classified into two categories: Terpenes and phenylpropenes (Cooke et al. 1998; Gopi et al. 2014). Based on the 5-carbon isoprene unit (building block), terpenes are divided into mono, sesqui, and diterpenes, where the number of isoprene units is 2, 3, and 4, respectively, while phenylpropenes are composed of 6-carbon aromatic ring with 3 rings (Gopi et al. 2014; Abdelli et al. 2021; Youssef et al. 2021). To date, more than 1000 monoterpenes and 3000 sesquiterpenes have been identified (Cooke et al. 1998; Gopi et al. 2014). Terpenes and phenylpropenes are synthesized by mevalonic and shichemic routes, respectively. Mevalonic acid (six carbons) is formed by the condensation of three acetate units by HMG-CoA reductase that is converted into 5-carbon isopentenyl pyrophosphate (IPP) and dimethylallyl pyrophosphate (DMAPP). Then, IPP and DMAPP are combined in a molar ratio of 1:1 to produce 10-carbon geranyl pyrophosphate (GPP), a precursor of monoterpenes. The conversion of IPP to GPP produces the 15-carbon sesquiterpene compound, farnesyl pyrophosphate (FPP) (Seigler 1998; Gopi et al. 2014). Thymol and carvacrol are derived from GPP and are classified as monoterpenoids or isoprenoids. β-Ionone is derived from FPP and is therefore classified as a sesquiterpene or isoprenoid. Eugenol, trans cinnamaldehyde, safrole, as well as capsaicin and piperine can be mentioned among the important compounds of phenylpropanoids (Gopi et al. 2014; Hundal et al. 2019). It has been found that after entering the bacteria, the essential oils block the cytosolic proteins and the release of the essential compounds inside the cells, along with the destruction of the bacteria. Essential oils and compounds derived from them act on the cytoplasm and change the shape of the cell wall by breaking down the membrane (Delaquis et al. 2002; Singh and Yadav 2020). Essential oils destroy pathogenic bacteria by leaking cell contents, damaging the cytoplasmic membrane, and reducing the proton force and destroying membrane proteins (Nazzaro et al. 2013; Singh and Yadav 2020).

Essential oils (eucalyptus, rosemary, lavender, millefolia) and other plants (pine, cloves, and myrrh) are also suggested and used as anti-inflammatory agents in mixed formulations (Miguel 2010). According to Yitbarek (2015), phytogenic essential oils have the ability to disrupt the cytoplasmic membrane of pathogens. It has also been found that plants and their essential oils play a role in lowering cholesterol and by doing so protect them against cancer (Craig 1999). The hypocholesterolemic effect of lemon balm oil has been proven due to the inhibition of hepatic 3-hydroxy-3-methylglutaryl coenzyme A (HMG-CoA) reductase activity (key regulatory enzyme in cholesterol synthesis (Qureshi et al. 1988; Chung et al. 2010; Karimi et al. 2010). Case et al. (1995) reported that a 5% inhibition of HMC-CoA reductase activity would lead to a 2% reduction in serum cholesterol. The hypolipidemic and immune-modulating properties of essential oils have always been of interest among poultry manufacturers and if these compounds are extracted, it is possible to take advantage of its beneficial effects in poultry feed additives and expand the use of medicinal plants and provide a successful perspective for phytogenics (Gopi et al. 2014).

4.3 Phytogenic Additives Based on Neglected and Underutilized Species (NUS) Plants

According to the World Health Organization (WHO) report in 2018, a number of plants are considered as Neglected and Underutilized Species (NUS) in the world, and their cultivation can lead to improved food security, increased production, and economic growth of communities and enhance coping mechanisms against climate change for mankind in the future (WHO 2018).

Considering the importance and strategic characteristics of the plants defined in the NUS system, phytogenic additives can be selected and extracted from these sources.

Neglected and underutilized species (NUS) constitute a large pool of plant genetic resources, including cultivated, semi-domesticated, or wild species cultivated using low-cost inputs (WHO 2018; Hosseintabar-Ghasemabad et al. 2022).

NUS plants play a key role in creating employment for communities and due to their ecological and environmental compatibility, they can be considered as low-risk options. Especially with the extreme consequences of climate change, these plants are considered as a safety net for producers and their promotion, production, and development will be welcomed by governments and organizations (WHO 2018; Janmohammadi et al. 2022).

For example, in the Andes region of South America, the International University with the support of the International Fund for Agricultural Development (IFAD) placed the development of three types of NUS products on the agenda, including pseudo-cereals (such as quinoa and amaranth); partial millet (such as finger millet, small millet, and barnyard millet); and medicinal and aromatic plants (such as argel, caper, oregano, and mint) (WHO 2018).

Due to the strategic features of NUS plants, which possess valuable nutritional and medicinal properties and have a good genetic contribution to the vast nature of the world, they can be used as a safe productive and economic network in production and health stable, therefore, the selection of plant feed additives from plants of the NUS system can be a bright prospect for the future in poultry nutrition (Baumhardt et al. 2009; Asfaw and Lipper 2016; Lipper et al. 2017; WHO 2018; FAO 2020).

Some studies on phytogenic additives in poultry nutrition were based on plants of the NUS system. However, considering the wide range and variety of NUS system plants, the plants proposed in this report will be introduced as a perspective based on the characteristics of bioactive compounds and their possible effects in the poultry industry.

Mint family is one of the most important suggested plants. *Labiatae* and *Lamiaceae* family plants have a great chance to be used as phytogenic additives in poultry nutrition.

Brenes and Roura (2010) reported that a wide range of plants and their extracts have potential antioxidant functions, and especially those from the *Labiatae* plant family such as rosemary, oregano, and thyme. Ruberto et al. (2002) as well stated that plants of the *Labiatae* family (such as mints) have attracted a lot of attention. Their antioxidant activities are due to the presence of phenolic terpenes (Cuppett and Hall 1998). *Mentha pulegium* L. or Pennyroyal is a medicinal plant from the *Lamiaceae* family. The most important components of Pennyroyal essential oil are trans-caryophyllene, eucalyptol, germacrene-D, and viridiflorol (Faithful et al. 2017). The effects of pennyroyal powder or Pennyroyal (1 to 3% by weight) as an alternative to antibiotics for broilers showed that the use of 2% by weight in the diet had a positive effect on performance and carcass characteristics (Goodarzi and Nanekarani 2014). In another study, the effects of Pennyroyal levels (0, 0.25, or 0.5% w/w) on growth performance and bacterial count in the jejunum of broiler chickens were investigated. Supplementing the diet with Pennyroyal reduced the number of *E. coli* and increased the number of lactic acid bacteria in the jejunum, and subsequently, an improvement of the feed conversion ratio was observed (Erhan et al. 2012). The authors investigated the effects of different levels (0.5% to 2% by weight) of the extract of this plant on the immune response, so that the reduction of Newcastle disease was noticeable after feeding this plant (Nobakht et al. 2011). The supplement of pennyroyal extract at different levels (0, 32.5, 65, and 130 mg/kg extract) in the diets of laying hens improved feed conversion ratio, egg production, egg weight, and shell strength (Aydin and Bolukbasi 2020). In another report, Pennyroyal essential oil at different levels (200, 300 and 400 ppm/kg diet) significantly improved the feed conversion ratio in Japanese quail. In addition, serum triglyceride level decreased in groups that received experimental diets containing different levels of essential oil (Dehghani et al. 2018). The findings indicate that the essential oils of the mentioned plant can be used as an ideal substitute for antibiotic growth parameters without negatively affecting quail health (Dehghani et al. 2018; Pliego et al. 2022). In another report, consuming different levels (0.5–1.5% w/w) of pennyroyal extract led to a significant improvement in performance, egg traits, blood biochemical (triglyceride, albumin, total protein, and uric acid content) and immune (heterophil and lymphocyte) parameters of laying hens (Paymard et al. 2013). It has been determined that different doses of Pennyroyal powder and extract can improve egg traits, blood biochemical parameters, and immunity of laying hens (Arjomandi et al. 2011; Paymard et al. 2013; Pliego et al. 2022). In addition, other plant species with antioxidant properties such as thyme and oregano contain large amounts of monoterpenes, thymol, and carvacrol (Ruberto et al. 2002; Alloui et al. 2014). However, the valuable antioxidant properties of *Labiatae* plants in animal nutrition are very promising and encouraging (Cuppett and Hall 1998). The effects of phenolic compounds from plants of the *Labiatae* family also led to the stability of animal products in broilers, which has been shown by several researchers (Botsoglou et al. 2003; Basmacioglu et al. 2004; Giannenas et al. 2005). However, the use of PFAs as an antioxidant is important not only for the health of poultry, but also for the oxidative stability of their products (meat). For example, supplementation of turkey diet with 200 mg/kg of oregano extracts significantly reduced lipid peroxidation during refrigerated storage of fresh and cooked meat (Botsoglou et al. 2003).

Thymus vulgaris L. contains 1–2.5% essential oil consisting of monoterpenes, mainly thymol and its phenol isomer carvacrol. The phenols contain caffeic acid and p-cymene-2,3-diol, and some biphenylic and flavonoid compounds, including flavonoid glycosides and flavonoid aglycones, which have various beneficial effects in animals (Fachini-Queiroz et al. 2012; Hosseinzadeh et al. 2015; Yalcin et al. 2020).

Placha et al. (2013) also reported that thyme oil may strengthen the intestinal barrier. Dietary supplementation with 0.5 g/kg dry matter thyme oil may improve intestinal integrity. Yalcin et al. (2020) showed the antioxidant effects and reduction of blood lipids by supplementing thyme in the nutrition of laying hens and further improving the humoral immune response without negative effects on egg performance and quality characteristics. Yoshino et al. (2006) reported that oregano extract exhibited iron-reducing activity, although its potency was approximately one-fifth that of ascorbic acid. Administration of oregano extract prevented the gastritis of mice caused by cold restraint stress. It seems that the antioxidant activities of oregano extract contribute to its preventive effects against inflammatory diseases (Shehata et al. 2022).

Rosemary (*Rosmarinus officinalis*) is a plant member of the *Labiatae* family with needle-like leaves that is widely cultivated in many regions with a Mediterranean climate (Abd El-Hack et al. 2020). Rosemary flowers may be pink, purple, white, or blue. Rosemary is accepted as one of the valuable plant species with the highest antioxidant activity (Peng et al. 2005). The analysis of the active compounds of rosemary showed that it has several types of antioxidants including flavonoids such as carnosol, carnosic and rosmarinic acid, and volatile oils (Okamura et al. 1994; Angelini et al. 2003). The main bioactive compounds of rosemary are rosmarinic acid, caffeic acid, betulinic acid, ursolic acid, and camphor in which carnosol and carnosic acid are the strongest antioxidants (Nakatani 2000; Jordán et al. 2014). It is interesting to note that the antioxidant activity of carnosic acid is three times higher than that of carnosol and seven times higher than the synthetic antioxidant supplement butylated hydroxyanisole and butylated hydroxytoluene (Richheimer et al. 1996).

Lavandula stoechas L. lavender is also a perennial plant from the *Lamiaceae* family, with a pleasant smell, whose cultivation has many fans because of its various properties. The essential oil of this plant contains compounds such as linalyl acetate linalool, butyric acid, propionic acid, valeric acid, free linalool, and grambol (Barazandeh 2002). Lavender essential oil (LEO) feed supplement increased superoxide dismutase activity in serum and liver and glutathione peroxidase activity in serum, but decreased malondialdehyde concentration in serum of broilers (Barbarestani et al. 2020). In another study, Salarmoini et al. (2019) investigated the effects of lavender (*Lavandula angustifolia*) extract as a substitute for flavophospholipol on performance, carcass characteristics, antioxidant status, meat quality, intestinal morphology, and intestinal microflora of broilers. Coliform counts in the digestive tract of birds fed with 400 ppm of lavender extract decreased significantly. In addition, by supplementing the basal diet with 300 or 400 ppm lavender extract, level of thiobarbituric acids (TBARS) after cooking the meat was significantly reduced. In general, the results of this experiment showed that lavender extract, especially at the level of 400 ppm, can be used as a suitable substitute for antibiotic growth stimulants in broiler diets.

Another plant mentioned in the NUS system is amaranth, which has been mentioned as a "Hidden treasure" due to its biological value and favorable protein quality according to the FAO report in 2018 (Li and Siddique 2018; WHO 2018). Amaranth is considered a pseudocereal, but due to its superior bioactive compounds, it can be used as an additive in ration formulation. For example, amaranth oil contains at a level of more than 5% a special compound called squalene (2, 6, 10, 15, 19, 23-hexamethyl-2, 6, 10, 14, 18, 22-tetracozahexane), which is actually a valuable 3-carbon isoprenoid that plays a role in cholesterol biosynthesis. Its values are between 584–684 mg/kg (He et al. 2002; Cai et al. 2003; Ryan et al. 2007). Squalene can play an important antioxidant role in animal cells by scavenging individual O and protecting against Sister Chromatid Exchange (SCE) by H_2O_2 in V79 cells (O'Sullivan et al. 2002; Janmohammadi et al. 2023). Squalene leads to the stimulation of macrophages and evidence shows that biological membranes are preserved by squalene during phagocytosis and in fact, these membranes are protected against oxidative stress (Ronco and De Stéfani 2013). It has been proven that two precursors of squalene, including geranyl and farnesyl, play a significant role in the synthesis and secretion of cytokines (Ronco and De Stéfani 2013). However, due to its abundance in squalene and vitamin E complex, amaranth can have a good option in the future and perspective of poultry nutrition as an additive (Hosseintabar-Ghasemabad et al. 2022).

Among other prominent plants of the *Lamiaceae* family, we can suggest the valuable *Perilla frutescens* plant. Perilla is a medicinal, aromatic, functional food and feed annual ornamental plant that belongs to the mint family. The perilla plant contains both valuable leaves and an excellent edible seed product. So far, 271 bioactive compounds have been identified in the organs of the perilla plant, including phenolic acids, flavonoids, essential oils, triterpenes, carotenoids, phytosterols, fatty acids, tocopherols, and policosanols and in fact, perilla is considered a rich source of active medicinal compounds (Ahmed 2018; Hou et al. 2022). In addition to its solvent extracts, it contains individual compounds (rosmarinic acid, perillaldehyde, luteolin, apigenin, tormentic acid, and isogumactone) that have attracted the attention of researchers due to its medicinal properties. Tian et al. (2014) were able to identify 119 compounds of perilla essential oil from 11 different regions and showed that their dominant compounds are 2-acetylfuran (maximum 82.17%), perillaldehyde (maximum 53.41%), caryophyllene (max. 38.34%), laurolene (max. 40.6%), 2-hexanofuran (max. 33.03%), 2-butylamine (max. 22.22%), asarone (max. 11.85%), farnesene (max. 9.25%), caryophyllene (max. 9.16%) %, and (Z, E)-farnesene (max. 7.14%).

Perilla leaves are a rich source of carotenoids so that compared to products rich in carotene (carrots, spinach) and rich in lutein (spinach, broccoli, lettuce), the content of perilla carotenoids is up to 5 times higher (Müller-Waldeck et al. 2010). In addition, this plant contains rare triterpenes such as tormentic acid, oleanolic acid, and ursolic acid (Chen et al. 2003). Also, some phytosterol compounds including campestrol, stigmasterol, −sitosterol, −amyrin, oxalic acid, triglycerol have been found in perilla seeds. Perilla seed contains 30–40% oil and its oil contains 54–64% omega-3 in the form of alpha-linolenic acid (ALA). It was shown that the high content of beta-sitosterol is definitely related to the content of linolenic acid. In addition, the content of omega-6 (linoleic acid) and omega-9 (oleic acid) in perilla oil is reported to be about 14 and 9.9%, respectively. Perilla seeds contain various polyphenols and flavones, including rosmarinic acid, luteolin, chrysoeriol, quercetin, catechin, apigenin, and shikonin (Ahmed 2018).

Perilla seed oil also contains important phenolic compounds including rosmarinic acid, luteolin, chrysoeriol, quercetin, catechin, and apigenin. Perilla seeds also contain the polyphenol luteolin, and research shows that this compound has the strongest antimicrobial properties and effects in the world compared to other known phenolic compounds (Yamamoto and Ogawa 2002; Ahmed 2018; Chen et al. 2022).

Other studies show that the antibacterial activity of perilla leaf essential oil on Gram-negative bacteria indicates the effectiveness of this essential oil in inhibiting the growth of the tested bacteria. Cui et al. (2019) showed that supplementing broiler diets using perilla oil at a level of 1% led to the production of meat enriched with omega-3 with a higher content of ALA and also, docosahexaenoic acid (DHA) and in total the ratio of omega-6 to omega-3 was significantly improved compared to the control group. However, all groups of chickens consuming perilla oil as an additive showed a decrease in lipid-related indices and an improvement in the fatty acid profile of breast meat. There was no adverse effect on the antioxidant status and meat quality, and when perilla oil was mixed with anise oil and ginger oil, synergistic effects were shown and functional improvement was significantly increased. However, studies on the use of perilla (seeds and leaves) in poultry nutrition are limited, but it is hoped that with the economic and mass production of products extracted from the perilla plant (leaves and seeds), the valuable properties of this amazing plant can be used as a feed additive in the industry poultry.

4.4 Phytogenic Additives Based on Good Agricultural and Collection Practices (GACP) Plants

The production of many well-known medicinal and food plants, as well as NUS plants, due to their importance in the field of food and medicine, in many countries of the world, there is on the agenda based on agricultural programs, mass cultivation and mechanization.

The possibility of agricultural development of some plants in countries is faced with climatic, economic, and social obstacles and limitations, therefore, their preparation is based on WHO recommendations for GACP system plants (WHO 2006). Plants of the GACP system refer to those plants that have valuable nutritional and medicinal properties, but it is possible to obtain them from the wild nature with proper scientific and technical methods. Plants of the GACP system refer to plants that have valuable nutritional and medicinal properties and are obtained from the wild with appropriate scientific and technical methods. In fact, the operation of preparing this plant is carried out by observing the basic methods of harvesting and post-harvest operations, with the priority of preserving the environment and not harming the genetic resources of the plant in its habitat, and the products obtained in the field of food and medicine are used (Pierce and Kathe 2004; WHO 2006; Andre et al. 2011; Hagels et al. 2013; Rastogi and Shukla 2013).

The range of these plants is very wide and the diversity of their cover and abundance is different in different lands. If the plant is properly known in terms of its scientific name, distribution power, climatic compatibility, and knowledge of active compounds, it can be used as a valuable source of phytogenic additive and its production can be developed in poultry feeding and a successful prospect is envisioned. In this part, a number of plants that are mentioned in the form of GACP system are also introduced and scientific documents related to their role as additives in poultry nutrition will be mentioned.

One of the prominent and valuable plants in the GACP system is the *Artemisia annua*, which can be considered a very important additive due to its valuable bioactive compounds (WHO 2006). The medicinal value of *Artemisia annua* became famous due to the presence of the active compound artemisinin in relation to the fight against malaria, but later in the field of nutrition, it was able to show other useful and significant effects (Delazar et al. 2007; Nahed et al. 2022).

Cherian et al. (2013) reported that feeding broilers with *Artemisia annua* led to a significant decrease in the amount of TBARS in breast and thigh meat, which is due to the antioxidant properties of polyphenol compounds and vitamin E of *Artemisia annua*. In a report by Baghban-Kanani et al. (2019), it was shown that *Artemisia annua* at the levels of 5 and 7.5% led to improvement of feed conversion ratio, yolk color index, shell thickness, and subsequently reduced egg cholesterol in laying hens. Also, consumption of Artemisia led to a decrease in the atherogenic index of plasma and a significant improvement in the blood antioxidant status. Khalaji et al. (2011) reported that the inclusion of *Artemisia annua* leaves can have a positive effect on intestinal health and growth performance of broiler chickens. Previous studies showed that *Artemisia annua* extract increases GSH-Px activity and also decreases MDA production in the liver of rats (Ryu et al. 1998). Wan et al. (2016) reported that *Artemisia annua* increases GSH-Px activity and decreases MDA levels in the serum and liver of poultry and can improve the antioxidant status of their blood. *Artemisia annua* contains phenolic compounds and flavonoids with superior antioxidant activity (Brisibe et al. 2009; Gouveia and Castilho 2013). The antioxidant compounds of *Artemisia annua*, such as vitamin E and other phenolic compounds, are able to turn this plant into a natural herbal additive that could be recommended in poultry diets (Cherian et al. 2013). The same researchers stated that the main active compounds of *Artemisia annua* included essential oils (such as camphene, beta-camphene), phenolics, sesquiterpenoids, flavonoids, amino acids, minerals, and vitamins. In addition, in broilers, the use of *Artemisia annua* dietary supplement as an anti-parasitic drug was quite efficient and effective against coccidiosis (Brisibe et al. 2008; Dragan et al. 2010; Nahed et al. 2022). Feeding broilers under heat stress with *Artemisia annua* led to a decrease in serum levels of corticosterone, MDA, aspartate aminotransferase (AST), and alanine aminotransferase (ALT) activity, while the serum levels of superoxide dismutase (SOD), triiodothyronine (T3), and triiodothyronine/thyroxine (T3/T4) increased compared to broilers under heat stress (Wan et al. 2017). Therefore, *Artemisia annua* can be used as a promising phytogenic additive to reduce heat stress in poultry production and as a natural and valuable antioxidant (Wan et al. 2017; Wan et al. 2018).

Among other valuable plants that have the ability to grow in a great extent of heat stress conditions and have a favorable distribution in many regions is the milk thistle (*Silybum marianum*) plant. In some countries, its production is based on agricultural systems, but in many countries, this plant is prepared based on the GACP system.

Zaker-Esteghamati et al. (2020) in a review report on the effect of *Silybum marianum* and its derivatives stated that the use of this plant and its compounds as a natural feed supplement could be quite valuable for improving performance and health of broilers. Khazaei et al. (2022) in their review report regarding the mechanism of action of silymarin extracted from thistle plant (*Silybum marianum*) stated that its seeds contain about 70–80% flavonolignans of silymarin and about 20–30%, they are oxidized polymer and polyphenol compounds (such as tannins). Consumption (0.5% and 1%) of milk thistle powder in Japanese quail diet improved performance and parameters of blood and antioxidant health. The active ingredient silymarin is completely effective in regenerating and regulating the level of liver enzymes and can prevent cell growth and DNA synthesis of cancer cells (Feshanghchi et al. 2022).

Astragalus is one of the other valuable plants that has a favorable and very compact distribution in the mountainous heights, and its aerial parts and herb can play a great role in improving the performance and health of poultry as a prominent phytogenic additive. Also, *Astragalus* is an important immune regulatory plant that increases the proliferation of B cells and antibody production and even strengthens the spleen and has protective effects against toxins in animal models. Having three strong and extraordinary compounds including saponins, various flavonoids, and polysaccharides has led to the creation of a huge potential for *Astragalus* to develop pharmaceuticals and food additives to improve the health of the host (Qiao et al. 2018; Qiao and Kyselov 2021; Liang et al. 2021; Liao et al. 2021).

Farag and Alagawany (2019) stated that *Astragalus membranaceus* (AM) is a member of the *Leguminosae* family and plays a very important role in regulating poultry immunity.

Xi et al. (2014) investigated the effects of ultrafine powder from stem and leaves of *Astragalus* (SLASUP) on the immune system of hyaline brown chickens. They showed that 1.5% feeding could improve the immune function by increasing the antibody titer, IL-2, and IFN-γ content and it was found that the stem and leaf part of the powder can be clinically effective and useful for strengthening the immune response and preventing poultry farm infections. Farag and Alagawany (2019) by feeding *Astragalus membranaceus* (AM) in the diet of broilers at levels of 0, 100, 200 and 300 mg/kg showed that all groups fed with *Astragalus* increased the weight of immune organs and IgG levels and improved liver and kidney function and blood antioxidant status. Several studies have shown that the oral administration of APS extracted from the plant is highly effective as a potentiator of poultry vaccine and the immune function of poultry significantly increased. It has been found that *Astragalus* polysaccharides (APS) have promising biological activities such as immune system strengthening, antioxidant, anti-viral, antimicrobial, and anti-parasitic (Wu 2018; Shan et al. 2019; Awadallah 2019; Qiao and Kyselov 2020). Jiang et al. (2010) believe that probiotics when prescribed (APS) can have a synergistic and positive effect on the balance of immunity and the improvement of intestinal microbiota, and they hope for the discovery and effectiveness of new prebiotics in future research. Horsetail plant (*Equisetum arvense*) is one of the other valuable and abundant plants with wide distribution in many countries and as a rich source of minerals, including silica (70% ash) which can be a valuable additive along with other phytogenic resources in the format of the GACP system is presented and obtained.

Tufarelli et al. (2021) investigated the effects of horsetail (*Equisetum arvense*) and spirulina (*Spirulina platensis*) nutritional supplementation in laying hens. They showed that the groups fed with 0.5% of horsetail along with 2% of spirulina improved the thickness and strength of the shell and the color of the yolk of the produced eggs. It also reduced the cholesterol in the egg yolk as well as the serum concentration of aspartate aminotransferase (AST) and alanine aminotransferase (ALT), while total superoxide dismutase (TSOD) and antioxidant capacity (TAC) were significantly increased. In general, the findings indicated that the combination of horsetail plant and spirulina algae can be used as a valuable additive by improving the physical parameters of eggs and blood characteristics and the oxidative status of birds without any negative effect on production performance.

Among other valuable plants with wide distribution and growth power, the licorice plant (*Glycyrrhiza glabra L*) is from the *Leguminosae* family, which has valuable nutritional and medicinal properties and antibacterial, antioxidant, antiseptic, antispasmodic, anti-inflammatory, and other properties. Its anti-viral, anti-hepatic, and anti-fungal effects have been identified in numerous reports (Seidavi et al. 2021). Licorice radix has many compounds such as different sugars (up to 18%), flavonoids, sterols, amino acids, gum and starch, essential oils, and saponins.

Its main saponin is glycyrrhizic acid or glycyrrhizin, which is 50 times sweeter than sucrose. It is used in pharmaceutical and food industries. In addition to the above-mentioned compounds, licorice root contains glycyrrhetinic acid, tannic acid, resins, volatile oils, flavonoid compounds such as liquiritigenin, as well as coumarin compounds such as herniarin (Li et al. 2000; Al-Snafi 2018). Aoki et al. (2007) reported that the use of licorice essential oil in the feeding of rats causes a significant reduction of fat in the ventricular cavity. At first, it was thought that the reduction of intra-abdominal fat was due to decrease in feed intake, absorption of lipids, biosynthesis of fatty acids, and increase in oxidation of these acids. However, Tominaga et al. (2006) observed no change in feed intake and this finding was confirmed by Nowakowska (2007). Also, the results of the reports showed that the blood triglyceride level did not also change, which indicates a lack of change in the absorption of dietary fat. Therefore, the decrease in abdominal fat of chickens is probably due to the decrease in the synthesis of fatty acids and the increase in the oxidation of these acids. Fuhrman et al. (2002) showed that consumption of licorice extract in hypercholesterolemic patients reduced total cholesterol, LDL, cholesterol, and triglycerides. Licorice extract probably affects the metabolism of arachidonic acid through the flavonoids hispaglabridin B, hispaglabridin A, and formononetin present in it and causes a decrease in the production of free radicals. These compounds also have anti-platelet, anti-inflammatory, and antioxidant properties (Somjen et al. 2004). Antioxidant compounds isolated from licorice root include isoprenyl chalcone, glabridin, and formononetin. Glabridin constitutes the major part of the extract (Tominaga et al. 2006). Glabridin prevents the oxidation of LDL cholesterol by reducing the activity of NADPH oxidase (Rosenblat et al. 1999). The reduction of oxidation of LDL cholesterol when consuming licorice extract may be due to reasons such as binding of its compounds to LDL cholesterol, elimination of free radicals, and protection of antioxidants related to cholesterol oxidation such as carotenoids (Fuhrman et al. 2002). Hussain et al. (2017) reported that the level of 300 mg/kg methanolic extract of licorice root led to an increase in humoral immunity of broiler chickens infected with coccidiosis. Also, lower levels of methanol extract of licorice radix (100 and 200 mg/kg) led to improved performance compared to the negative control group (infected with coccidiosis) and close to the positive control group (vitamin E).

4.5 Phytogenic Additives Based on New Plants

Although bioactive compounds of many plants have not been extensively studied in the poultry model, many in vitro and in vivo evaluations have been performed in mice and humans, indicating that these compounds can have potential beneficial effects in strengthening immunity and improving the health and performance of poultry, without the need for antibiotic stimulants (Hasted et al. 2021).

There are a number of plants that, due to their valuable bioactive compounds, can be promising as phytogenic additives and antibiotic substitutes in nutrition and poultry industry for future research.

These plants, due to their high distribution in nature and their superior ecological adaptability in a number of climates, can be provided in the format and recommendations of the GACP system. But they can be widely applied and domesticated in the cultivation system in a controllable and manageable manner, on a large scale and with the views of the NUS system, so that they can meet the needs of the poultry industry in relation to alternative additives for the future will be responsive.

For example, we can take *Sambucus ebulus* as an example, whose fruit is known as a valuable source of antioxidants, immune stimulants, anti-influenza, and an alternative to antibiotics and a strong anti-viral and has a great role in reducing various diseases. On the other hand, the high growth power, the significant abundance of the mentioned source, and the distribution and availability of this plant in many countries are significant and valuable points of this food source, which can be added to poultry nutrition as an additive show its valuable place in the future of this industry (Shokrzadeh and Saravi 2010; Karimi et al. 2014). Tasinov et al. (2021) reported that among the phytochemical compounds identified in the aqueous extract of the fruit of the plant, 15 amino acids, 10 organic acids, 36 sugar acids and alcohols, 25 mono, di, and trisaccharides respectively. 13 cases of fatty acids (saturated and unsaturated) and their esters and 38 cases of phenolic compounds have been identified. However, the high diversity of bioactive and unique compounds in the fruit product of this plant has made it a preventive food additive against oxidative stress, strengthening the anti-inflammatory and antiseptic properties as well as significant power in eliminating free radicals, but the use of this source in poultry nutrition is very limited and requires more studies (Tasinov et al. 2021).

Considering the diverse perspectives of using phytogenics as feed additives, new plants containing valuable bioactive compounds can be introduced so that the field of further research studies by researchers can be put on the agenda.

The proposed plants as phytogenic additives have not yet been widely studied in relation to poultry nutrition, or they have been limited, and the need for more and more detailed studies is important.

Among other valuable plants, we can mention the *Stevia rebaudiana* plant, which despite the availability of studies on the effects of consumption in poultry, its practical aspects still need more research. Stevia leaves contain a mixture of 8 glycosides derived from the tetracyclic diterpene steviol. Stevioside, as the most abundant sweetening compound, contains 5 to 10% of the dry leaf weight. This leaf is 300 times sweeter than sugar (sucrose), which is due to the presence of Ent Kaurene Diterpene Glycosides, which is soluble in water (Ramesh et al. 2006). Other glycosides include ribodiozide A (2 to 4% of dry weight), ribodiozide B and C (up to 2%), ribodiozide D and E (0.4 to 0.7%), as well as dulcoside A and steviolbioside, which have the strength. Their sweetness is 250 to 300, 350 to 450, 300 to 350, 50 to 120, 200 to 300, 250 to 300, 50 to 120, 100 to 125 times compared to sugar, respectively (Ramesh et al. 2006; Chatsudthipong and Muanprasat 2009). Stevia also contains protein, fiber, carbohydrates, phosphorus, iron, calcium, sodium, potassium, magnesium, zinc, vitamin A and C. Its other effective compounds include sterols, flavonoids, triterpene, monoterpene, sesquiterpene, tannin, and chlorophyll (Gardana et al. 2010). Atteh et al. (2008) investigated the effects of stevia leaves and stevioside on the performance and blood parameters of broiler chickens in the form of experimental treatments including 1) a positive control group in the form of a basal diet without antibiotics but containing enzymes, 2) a negative control group the basal diet without antibiotics and enzymes, 3) two percent stevia leaves in the negative control diet, and 4) 130 ppm stevioside in the negative control group showed that birds fed with stevia leaves and stevioside had more feed consumption than the positive control group. Weight gain was the highest in the positive control group and the stevioside receiving group in the starter period. The best FCR was in the positive control group and the stevioside receiving group compared to other experimental groups. Also, chickens fed with stevia leaves and stevioside decreased the concentration of short chain fatty acids (SCFA) in the cecum and a decrease in glucose, triglyceride, and triiodothyronine (T_3) was observed in the blood.

Daneshyar et al. (2012) reported that after the injection of human serum albumin to broiler chickens fed with stevioside, they observed an increase in body weight and a decrease in alpha-glycoprotein and reported that the inclusion of stevioside in the diet of broilers can respond reduce primary inflammation after innate immune response stimulation.

Takahashi et al. (2001) conducted an experiment to determine whether *Stevia rebaudiana* stem extract in the diet of broiler chickens can prevent calculus erosion and wound caused by histamine. The experimental treatments included control group, control diet with 0.4% histamine and control diet with 0.4% histamine +0.2% stevia stem extract (based on dry matter) for 14 days. Dietary histamine caused growth reduction and caused erosion and wound. The feeding of stevia extract in the diet partially prevents the reduction of growth and wound caused by histamine. These researchers reported that stevia stem extract contains active compounds that reduce histamine activity, and the antioxidant power of stevia stem is much higher than its leaf and even green tea extract.

It is necessary for researchers to make additional efforts to replace antibiotic and synthetic additives by choosing new valuable plants and resources and dynamically update information on phytogenic additives, in order to provide a promising perspective for the future of additives (WHO 2006; Stevanović et al. 2018; WHO 2018; FAO 2020; Abdelli et al. 2021).

In Table 1, a checklist of plants that contain valuable bioactive compounds is presented. These resources have the potential to be used as valuable phytogenic resources and in the form of the GACP and NUS system and that they can answer part of the needs of the poultry industry and strengthen the perspective of feed additives in the way of improving health, environment, performance, and economy (Figs. 2, 3, 4, 5, and 6).

Table 1 A list of new and valuable proposed plants containing bioactive compounds to be used as a phytogenic feed additive

Row	Latin name	Row	Latin name	Row	Latin name
1	*Saturia* spp	40	*Lithospermum officinale*	79	*Conium maculatum*
2	*Sesbania* spp	41	*Anthemis cotula*	80	*Hibiscus* spp
3	*Verbascum* spp	42	*Petasites hybridus*	81	*Lotus* spp
4	*Xanthium strumarium*	43	*Ephedra* spp	82	*Moringa oleifera*
5	*Echinophora* spp	44	*Capparis spinosa*	83	*Suaeda physophora*
6	*Sida* spp	45	*Abutilon theophrasti*	84	*Caccinia strigosa*
7	*Tanacetum parthenium*	46	*Salvia glutinosa*	85	*Lindelofia kandavanesis*
8	*Crataegus* spp	47	*Anchusa italica*	86	*Lithospermum officinale*
9	Aesculus hippocastanum	48	*Juniperus communis*	87	*Nonnea longifolia*
10	*Traxacum* spp	49	*Plantago lanceolata*	88	*Rindera regia*
11	Melissa officinalis	50	*Malva sylvestris*	89	*Callicephalus nitens*
12	*Rubus idaeus*	51	*Calendula officinalis*	90	*Centaurea elbursensis*
13	Vitex agnus castus	52	*Heracleum persicum*	91	*Cousinia erinacea*
14	Phytolacca americana	53	*Diospyros piscatorial*	92	*Cousinia hypochionea*
15	*Suaeda physophora*	54	*Cymbaria mongolica*	93	*Cousinia pinarocephala*
16	*Sophora* spp	55	*Verbena litoralis*	94	*Iranecio elbursensis*
17	Centella asiatica	56	*Nepeta cataria*	95	*Jurinella frigida*
18	*Persicaria* spp	57	*Tribulus terrestris*	96	*Leontodon kotschyi*
19	*Ammi visnaga*	58	*Bidens tripartite*	97	*Onopordum acanthium*
20	*Origanum vulgar*	59	*Rhamnus frangula*	98	*Outrey acarduiformis*
21	Cynoglossum creticum	60	*Marrubium globosum*	99	*Rhagadiolus stellatus*
22	*Symphytum asperum*	61	*Scutellaria pinnatifida*	100	*Rhaponticum pulchrum*
23	Artemisia absinthium	62	*Taxus baccata*	101	*Scorzonera* spp
24	*Centaurea behen*	63	*Viscum album*	102	*Tragopogon longirostris*
25	*Conyza canadensis*	64	*Cirsium* spp	103	*Camelina rumelica*
26	Calamintha officinalis	65	*Helichrysum globiferum*	104	*Turritis glabra*
27	*Lamium album*	66	*Carex* spp	105	*Ecballium elaterium*
28	*Mentha aquatica*	67	*Impatiens Noli tangere*	106	*Cyperus esculentus*
29	*Phlomis herba venti*	68	*Prosopis farcta*	107	*Datisca cannabina*
30	*Prunella vulgaris*	69		108	*Tamus communis*

(continued)

Table 1 (continued)

Row	Latin name	Row	Latin name	Row	Latin name
			Hypericum perforatum		
31	*Stachys byzantina*	70	*Nasturtium*	109	*Aruncus vulgaris*
32	*Teucrium polium*	71	*Lepidium sativum*	110	*Agriophyllum squarrosum*
33	*Laurus nobilis*	72	*Ziziphus* spp	111	*Physalis alkekengi*
34	*Laser trilobum*	73	*Myrtus communis*	112	*Physalis divaricata*
35	*Lemna* spp	74	*Trapa natans*	113	*Typha latifolia*
36	*Oenothera biennis*	75	*Eclipta prostrate*	114	*Sorghum halepense*
37	*Salix* spp	76	*Smilax excelsa*	115	*Senecio vulgaris*
38	*Cnicus benedictus*	77	*Inula helenium*	116	*Tussilago farfara*
39	*Sumbucus ebulus*	78	*Stevia rebaudiana*	117	*Salicornia* spp.

Fig. 2 *Sambucus ebulus* plant in Iran country and Gilan province in 2021

Fig. 3 *Phlomis herba venti* plant in Iran country and Gilan province in 2022

5 Conclusion

The future perspective of feed additives in poultry nutrition indicates the potential of this field in the development of the poultry industry in order to improve performance and health of the flock. On the other hand, reducing the use of antibiotics and producing organic poultry products is completely aligned with the consumer demand for healthy food in the perspective of health-oriented nutrition. The strategy of using microcapsule and nanotechnology in order to significantly improve the bioavailability of active and effective feed additive compounds in the body shows the promising state of this knowledge in order to serve the poultry industry. The innovative point of view of using bacteriophage additive is a special proposal in order to take advantage of the potential of microorganisms in feed additives. The amazing, diverse, and complex potential of phytogenic effects in poultry nutrition as a feed additive in the form of using essential oils, cultivated plants, NUS plants and GACP plants shows that phytogenic sources and its natural derivatives can be a permanent and sustainable strategy for applied research in the future so that all the potentials of these resources can be used correctly and at the service of the poultry industry. However, the selection of these sources with the aim of the least pressure on the environment, in harmony with climate change and global warming, helping to reduce the

Fig. 4 *Stevia rebaudiana* farms in Iran country and Mazandaran province—GH. Hosseintabar an agricultural expert and manager of the Darvash Giah Khazar Medicinal Herbs Complex in 2016

consumption and side effects of antibiotics and helping to increase the production of healthy products for the future perspective can be planned and suggested. A checklist of new plants based on the mentioned reasons that high growth power and new bioactive compounds were proposed in the end of the chapter report that can be a transformation and an innovative idea for experts and researchers of different countries to use the potential of the vegetation power of their land for the optimal use of these resources be in feed additives.

Fig. 5 *Salicornia* plant that has grown in Iran country and Semnan province in 2022 under drought stress conditions

Fig. 6 *Suaeda physophora* plants habitat in Iran country and Semnan province—GH. Hosseintabar an agricultural expert and manager of the Darvash Giah Khazar Medicinal Herbs Complex in 2022

Funding The research was carried out with the help of the Russian Science Foundation grant No. 21-16-00025 (section 4.1) and 22-16-00041 (section 4.2), State Scientific Institution NIIMMP. Also, this study was supported by the Italian Ministry of Economic Development (MISE) and by European Union–PON Business and Competitiveness 2014–2020 FESR, Axis I—Innovation, Action 1.1.3., Thematic Area—National Strategy of Intelligent Specialization "AGRIFOOD," CUP B41B20000280005, project codex n. F/200078/03/X45, Title of the project "Feed valorization of biological waste from poultry production, industrial thermal waste and CO2-(BIO = C = O) "; Scientific Responsible: Biagina Chiofalo. Financial support by Rasht Branch, Islamic Azad University, Rasht, Iran, grant number 17-16-1-575 is gratefully acknowledged.

References

Abd El-Ghany WA (2019) Nanotechnology and its considerations in poultry field: an overview. J Hellenic Vet Med Soc 70(3):1611–1616

Abd El-Hack ME, Abdelnour SA, Taha AE, Khafaga AF, Arif M, Ayasan T, Abdel-Daim MM (2020) Herbs as thermoregulatory agents in poultry: an overview. Sci Total Environ 703:134399

Abdelli N, Solà-Oriol D, Pérez JF (2021) Phytogenic feed additives in poultry: achievements, prospective and challenges. Animals 11(12):3471

Ahmadi M, Poorghasemi M, Seidavi A, Hatzigiannakis E, Milis C (2020) An optimum level of nano-selenium supplementation of a broiler diet according to the performance, economical parameters, plasma constituents and immunity. J Elem 25(3):1187–1198

Ahmed HM (2018) Ethnomedicinal, phytochemical and pharmacological investigations of *Perilla frutescens* (L) Britt. Molecules 24(1):102

Alagawany M, Abd El-Hack ME (eds) (2020) Natural feed additives used in the poultry industry. Bentham Science Publishers, Singapore

Alloui MN, Agabou A, Alloui N (2014) Application of herbs and phytogenic feed additives in poultry production: a review. Glob J Anim Sci 2(3):234–243

Al-Snafi AE (2018) *Glycyrrhiza glabra*: a phytochemical and pharmacological review. IOSR J Pharm 8(6):1–17

Andre P, Su HB, Gao WY (2011) Ensuring the compliance of raw herbal materials stemmed from China with European Good Agricultural and Collection Practice (GACP). Chin Herb Med 3: 251–256

Angelini LG, Carpanese G, Cioni PL, Morelli I, Macchia M, Flamini G (2003) Essential oils from Mediterranean *Lamiaceae* as weed germination inhibitors. J Agric Food Chem 51(21): 6158–6164

Anwar MI, Awais MM, Akhtar M, Navid MT, Muhammad F (2019) Nutritional and immunological effects of nano-particles in commercial poultry birds. Worlds Poult Sci J 75(2):261–272

Aoki F, Honda S, Kishida H, Kitano M, Arai N, Tanaka H, Mae T (2007) Suppression by licorice flavonoids of abdominal fat accumulation and body weight gain in high-fat diet-induced obese C57BL/6J mice. Biosci Biotech Bioch 71(1):206–214

Arjomandi M, Nobakht A, Pishchang J, Mehmannavaz Y, Chekaniazar S (2011) Evaluation the effects of using of probiotic, and pennyroyal (*Mentha pulegium* L) medicinal plant on perormance of laying hens. J Appl Environ Biol Sci 1:164–167

Asfaw S, Lipper L (2016) Managing climate risk using climate-smart agriculture. FAO, Rome

Atteh JO, Onagbesan OM, Tona K, Decuypere E, Geuns JMC, Buyse J (2008) Evaluation of supplementary stevia (Stevia rebaudiana, bertoni) leaves and stevioside in broiler diets: effects on feed intake, nutrient metabolism, blood parameters and growth performance. J Anim Physiol Anim Nutr 92(6):640–649

Awadallah MRF (2019) The role of *Astragalus membranaceus* as immunomodulator in poultry. Worlds Poult Sci J 75:43

Aydin A, Bolukbasi SC (2020) Effect of supplementation of hen diet with pennyroyal extract (*Mentha pulegium*) on performance, egg quality and yolk TBARS values. Pak J Zool 52(3): 1045

Baghban-Kanani P, Hosseintabar-Ghasemabad B, Azimi-Youvalari S, Seidavi A, Ragni M, Laudadio V, Tufarelli V (2019) Effects of using *Artemisia annua* leaves, probiotic blend, and organic acids on performance, egg quality, blood biochemistry, and antioxidant status of laying hens. J Poult Sci 56(2):120–127

Barazandeh MM (2002) Essential oil composition of *Lavandula latifolia* Medik from Iran. J Essent Oil Res 14(2):103–104

Barbarestani SY, Jazi V, Mohebodini H, Ashayerizadeh A, Shabani A, Toghyani M (2020) Effects of dietary lavender essential oil on growth performance, intestinal function, and antioxidant status of broiler chickens. Livest Sci 233:103958

Basmacioglu H, Tokusoglu Ö, Ergül M (2004) The effect of oregano and rosemary essential oils or alpha-tocopheryl acetate on performance and lipid oxidation of meat enriched with n-3 PUFA's in broilers. S Afr J Anim Sci 34(3):197

Baumhardt F, Lasage R, Suarez P, Chadza C (2009) Farmers become filmmakers: climate change adaptation in Malawi. Participatory Learning and Action 60(1):129–138

Botsoglou NA, Grigoropoulou SH, Botsoglou E, Govaris A, Papageorgiou G (2003) The effects of dietary oregano essential oil and α-tocopheryl acetate on lipid oxidation in raw and cooked Turkey during refrigerated storage. Meat Sci 65(3):1193–1200

Brenes A, Roura E (2010) Essential oils in poultry nutrition: Main effects and modes of action. Anim Feed Sci Technol 158(1–2):1–14

Brisibe EA, Umoren UE, Owai PU, Brisibe F (2008) Dietary inclusion of dried *Artemisia annua* leaves for management of coccidiosis and growth enhancement in chickens. Afr J Biotechnol 7(22)

Brisibe EA, Umoren UE, Brisibe F, Magalhäes PM, Ferreira JF, Luthria D, Prior RL (2009) Nutritional characterisation and antioxidant capacity of different tissues of *Artemisia annua* L. Food Chem 115(4):1240–1246

Cai Y, Corke H, Wu HX (2003) Amaranth. In: Wrigley CW, Corke H, Walker CE (eds) Encyclopedia of grain science. Elsevier Academic Press, Amsterdam

Calvo P, Castaño ÁL, Hernández MT, González-Gómez D (2011) Effects of microcapsule constitution on the quality of microencapsulated walnut oil. Eur J Lipid Sci Technol 113(10): 1273–1280

Case GL, He L, Mo H, Elson CE (1995) Induction of geranyl pyrophosphate pyrophosphatase activity by cholesterol-suppressive isoprenoids. Lipids 30(4):357–359

Chatsudthipong V, Muanprasat C (2009) Stevioside and related compounds: therapeutic benefits beyond sweetness. Pharmacol Ther 121(1):41–54

Chen JH, Xia ZH, Tan RX (2003) High-performance liquid chromatographic analysis of bioactive triterpenes in *Perilla frutescens*. J Pharm Biomed Anal 32(6):1175–1179

Chen Z, Wu K, Zhu W, Wang Y, Su C, Yi F (2022) Chemical compositions and bioactivities of essential oil from perilla leaf (*Perillae folium*) obtained by ultrasonic-assisted hydro-distillation with natural deep eutectic solvents. Food Chem 375:131834

Cherian G, Orr A, Burke IC, Pan W (2013) Feeding *Artemisia annua* alters digesta pH and muscle lipid oxidation products in broiler chickens. Poult Sci 92(4):1085–1090

Chung MJ, Cho SY, Bhuiyan MJH, Kim KH, Lee SJ (2010) Anti-diabetic effects of lemon balm (*Melissa officinalis*) essential oil on glucose-and lipid-regulating enzymes in type 2 diabetic mice. Br J Nutr 104(2):180–188

Cooke CJ, Nanjee MN, Dewey P, Cooper JA, Miller GJ, Miller NE (1998) Plant monoterpenes do not raise plasma high-density-lipoprotein concentrations in humans. Am J Clin Nutr 68(5): 1042–1045

Craig WJ (1999) Health-promoting properties of common herbs. Am J Clin Nutr 70(3 Suppl):491S–499S. https://doi.org/10.1093/ajcn/70.3.491s

Cui X, Gou Z, Fan Q, Li L, Lin X, Wang Y, Jiang Z (2019) Effects of dietary perilla seed oil supplementation on lipid metabolism, meat quality, and fatty acid profiles in yellow-feathered chickens. Poult Sci 98(11):5714–5723

Cuppett SL, Hall CA (1998) Antioxidant activity of the *Labiatae*. Adv Food Nutr Res 42:245–272

Daneshyar M, Geuns JMC, Willemsen H, Ansari Z, Darras VM, Buyse JG, Everaert N (2012) Evaluation of dietary stevioside supplementation on anti-human serum albumin immunoglobulin G, Alpha-1-glycoprotein, body weight and thyroid hormones in broiler chickens. J Anim Physiol Anim Nutr 96(4):627–633

Dehghani N, Afsharmanesh M, Salarmoini M, Ebrahimnejad H, Bitaraf A (2018) Effect of pennyroyal, savory and thyme essential oils on Japanese quail physiology. Heliyon 4(10): e00881

Delaquis PJ, Stanich K, Girard B, Mazza G (2002) Antimicrobial activity of individual and mixed fractions of dill, cilantro, coriander and eucalyptus essential oils. Int J Food Microbiol 74(1–2): 101–109

Delazar A, Naseri M, Nahar L, Moghadam SB, Esnaashari S, Nazemiyeh H, Sarker SD (2007) GC-MS analysis and antioxidant activities of essential oils of two cultivated *Artemisia* species. Chem Nat Compd 43(1):112–114

Dong XF, Gao WW, Tong JM, Jia HQ, Sa RN, Zhang Q (2007) Effect of polysavone (alfalfa extract) on abdominal fat deposition and immunity in broiler chickens. Poult Sci 86(9): 1955–1959

Dragan L, Titilincu A, Dan I, Dunca I, Dragan M, Mircean V (2010) Effects of *Artemisia annua* and *Pimpinella anisum* on *Eimeria tenella* (*Phylum Apicomplexa*) low infection in chickens. Sci Parasitol 11(2):77–82

Erhan MK, Bölükbaşı ŞC, Ürüşan H (2012) Biological activities of pennyroyal (*Mentha pulegium* L) in broilers. Livest Sci 146(2–3):189–192

Evran S, Tayyarcan EK, Acar-Soykut E, Boyaci IH (2022) Applications of bacteriophage cocktails to reduce *Salmonella* contamination in poultry farms. Food Environ Virol 14(1):1–9

Fachini-Queiroz FC, Kummer R, Estevao-Silva CF, Carvalho MDDB, Cunha JM, Grespan R, Cuman RKN (2012) Effects of thymol and carvacrol, constituents of *Thymus vulgaris* L essential oil, on the inflammatory response. Evid Based Complement Alternat Med 2012:1

Faithful M, Mohsenpour M, Meghani H, Vatankhah A (2017) The impact of environmental factors on yield and chemical compositions of essential oil of water mint, *Mentha aquatica* L from different habitats of Mazandaran province. J Poult Res 30:432–444

FAO (UN Food and Agriculture Organization) (2020) Migration, agriculture and climate change: reducing vulnerabilities and enhancing resilience. FAO, Rome

Farag MR, Alagawany M (2019) The role of *Astragalus membranaceus* as immunomodulator in poultry. Worlds Poult Sci J 75(1):43–54

Feshanghchi M, Baghban-Kanani P, Kashefi-Motlagh B, Adib F, Azimi-Youvalari S, Hosseintabar-Ghasemabad B, Tufarelli V (2022) Milk thistle (*Silybum marianum*), marine algae (*Spirulina platensis*) and toxin binder powders in the diets of broiler chickens exposed to aflatoxin-b1: growth performance, humoral immune response and cecal microbiota. Agriculture 12(6):805

Fuhrman B, Volkova N, Kaplan M, Presser D, Attias J, Hayek T, Aviram M (2002) Antiatherosclerosis effects of licorice extract supplementation on hypercholesterolemic patients: increased resistance of LDL to atherogenic modifications, reduced plasma lipid levels, and decreased systolic blood pressure. Nutrition 18(3):268–273

Gardana C, Scaglianti M, Simonetti P (2010) Evaluation of steviol and its glycosides in Stevia rebaudiana leaves and commercial sweetener by ultra-high-performance liquid chromatography-mass spectrometry. J Chromatogr A 1217(9):1463–1470

Giannenas IA, Florou-Paneri P, Botsoglou NA, Christaki E, Spais AB (2005) Effect of supplementing feed with oregano and/or alpha-tocopheryl acetate on growth of broiler chickens and oxidative stability of meat. J Anim Feed Sci 14(3):521

Goodarzi M, Nanekarani S (2014) Effects of feeding *Mentha pulegium* L as an alternative to antibiotics on performance of broilers. APCBEE Procedia 8:53–58

Gopi M, Karthik K, Manjunathachar HV, Tamilmahan P, Kesavan M, Dashprakash M, Purushothaman MR (2014) Essential oils as a feed additive in poultry nutrition. Adv Anim Vet Sci 2(1):1–7

Gouveia SC, Castilho PC (2013) *Artemisia annua* L: essential oil and acetone extract composition and antioxidant capacity. Ind Crop Prod 45:170–181

Hagels H, Tegtmeier M, Strube J, Both S (2013) GACP regulation for herbal raw materials. Planta Med 79(13):WS22

Han S, Byun KH, Mizan MFR, Kang I, Ha SD (2022) Bacteriophage and their lysins: a new era of biocontrol for inactivation of pathogenic bacteria in poultry processing and production: a review. Food Control 137:108976

Hasted TL, Sharif S, Boerlin P, Diarra MS (2021) Immunostimulatory potential of fruits and their extracts in poultry. Front Immunol 12:1634

He HP, Cai Y, Sun M, Corke H (2002) Extraction and purification of squalene from *Amaranthus* grain. J Agric Food Chem 50(2):368–372

Heydarian M, Ebrahimnezhad Y, Meimandipour A, Hosseini SA, Banabazi MH (2020) Effects of dietary inclusion of the encapsulated thyme and oregano essential oils mixture and probiotic on growth performance, immune response and intestinal morphology of broiler chickens. Poult Sci J 8(1):17–25

Hosseintabar-Ghasemabad B, Janmohammadi H, Hosseinkhani A, Amirdahri S, Baghban-Kanani-P, Gorlov IF, Slozhenkina MI, Mosolov AA, Ramirez LS, Seidavi A (2022) Effects of using processed amaranth grain with and without enzyme on performance, egg quality, antioxidant status and lipid profile of blood and yolk cholesterol in laying hens. Animals 12(22):3123

Hosseinzadeh S, Jafarikukhdan A, Hosseini A, Armand R (2015) The application of medicinal plants in traditional and modern medicine: a review of *Thymus vulgaris*. Int J Clin Med 6(09): 635

Hou T, Netala VR, Zhang H, Xing Y, Li H, Zhang Z (2022) *Perilla frutescens*: a rich source of pharmacological active compounds. Molecules 27(11):3578

Huff WE, Huff GR, Rath NC, Balog M, Donoghue AM (2006) Bacteriophage: a safe and natural alternative to AGPs. In: Barug D (ed) Antimicrobial growth promoters: where do we go from here, vol 978. Wageningen Academic Publishers, Wageningen, pp 341–360

Hundal JS, Wadhwa M, Bakshi MPS (2019) Herbal feed additives containing essential oil: 1 impact on the nutritional worth of complete feed in vitro. Trop Anim Health Prod 51(7):1909–1917

Hussain K, Iqbal Z, Zahid Abbas R, Kasib Khan M, Kashif Saleemi M (2017) Immunomodulatory activity *of Glycyrrhiza glabra* extract against mixed *Eimeria* infection in chickens. Int J Agric Biol 19(4):928

Ireland F (2008) The relevance for food safety of applications of nanotechnology in the food and feed industries. Food Safety Authority of Ireland, Dublin, p 88

Islam MR, Martinez-Soto CE, Lin JT, Khursigara CM, Barbut S, Anany H (2021) A systematic review from basics to omics on bacteriophage applications in poultry production and processing. Crit Rev Food Sci Nutr 63:1–33

Janmohammadi H, Hosseintabar-Ghasemabad B, Amirdahri S, Gorlov IF, Vladimirovna KE, Slozhenkina MI, Bilal RM, Seidavi A, Phillips CJC (2022, 1810) The energy value for broiler chickens of heat-treated and untreated amaranth grain, with and without enzyme addition. Agriculture 12(11)

Janmohammadi H, Hosseintabar-Ghasemabad B, Oliyai M, Alijani S, Gorlov IF, Slozhenkina MI, Mosolov AA, Ramirez LS, Seidavi A, Laudadio V, Tufarelli V, Ragni M (2023) Effect of dietary Amaranth (Amaranthus hybridus chlorostachys) supplemented with enzyme blend on egg quality, serum biochemistry and antioxidant status in laying hens. Antioxidants 12(2):456

Jiang J, Wu C, Gao H, Song J, Li H (2010) Effects of *Astragalus* polysaccharides on immunologic function of erythrocyte in chickens infected with infectious bursa disease virus. Vaccine 28(34): 5614–5616

Jordán MJ, Castillo J, Bañón S, Martínez-Conesa C, Sotomayor JA (2014) Relevance of the carnosic acid/carnosol ratio for the level of rosemary diterpene transfer and for improving lamb meat antioxidant status. Food Chem 151:212–218

Karimi I, Hayatgheybi H, Razmjo M, Yousefi M, Dadyan A, Hadipour M (2010) Anti-hyperlipidaemic effects of an essential oil of *Melissa officinalis* L in cholesterol-fed rabbits. J Appl Biol Sci 4(1):17–22

Karimi S, Mohammadi A, Dadras H (2014) The effect of Echinacea purpurea and Sambucus nigra L. on H9N2 avian influenza virus in infected chicken embryo. Veterinarski Arhiv 84(2): 153–165

Khalaji S, Zaghari M, Hatami KH, Hedari-Dastjerdi S, Lotfi L, Nazarian H (2011) Black cumin seeds, Artemisia leaves (*Artemisia sieberi*), and *Camellia* L plant extract as phytogenic products in broiler diets and their effects on performance, blood constituents, immunity, and cecal microbial population. Poult Sci 90(11):2500–2510

Khazaei R, Seidavi A, Bouyeh M (2022) A review on the mechanisms of the effect of silymarin in milk thistle (*Silybum marianum*) on some laboratory animals. Vet Med Sci 8(1):289–301

Klasing KC (2007) Nutrition and the immune system. Br Poult Sci 48(5):525–537

Lang LH (2006) FDA approves use of bacteriophages to be added to meat and poultry products. Gastroenterology 131(5):1370

Laudadio V, Ceci E, Lastella NMB, Introna M, Tufarelli V (2014) Low-fiber alfalfa (*Medicago sativa* L) meal in the laying hen diet: effects on productive traits and egg quality. Poult Sci 93(7): 1868–1874

Li X, Siddique KH (2018) Future smart food. Rediscovering hidden treasures of neglected and underutilized species for Zero Hunger in Asia. Food and Agriculture Organization of the United Nations, Bangkok

Li W, Asada Y, Yoshikawa T (2000) Flavonoid constituents from *Glycyrrhizaglabra* hairy root cultures. Phytochemistry 55(5):447–456

Liang W, Li H, Zhou H, Wang M, Zhao X, Sun X, Zhang X (2021) Effects of *Taraxacum* and *Astragalus* extracts combined with probiotic *Bacillus subtilis* and *Lactobacillus* on *Escherichia coli* infected broiler chickens. Poult Sci 100(4):101007

Liao L, Li J, Li J, Huang Y, Wu Y (2021) Effects of *Astragalus* polysaccharides on intestinal morphology and intestinal immune cells of Muscovy ducklings infected with Muscovy duck reovirus. Poult Sci 100(1):64–72

Lipper L, McCarthy N, Zilberman D, Asfaw S, Branca G (2017) Climate smart agriculture: building resilience to climate change. Springer Nature, Cham, p 630

Máthé A (2009) Essential oils-biochemistry, production and utilization. In: Steiner T (ed) Phytogenic in animal nutrition: natural concepts to optimize gut health and performance, vol 1. 5M Publishing Ltd, Nottingham, pp 1–18

Meimandipour A, Nouri Emamzadeh A, Soleimani A (2017) Effects of nano encapsulated *Aloe vera*, dill and nettle root extract as feed antibiotic substitutes in broiler chickens. Arch Anim Breed 60(1):1–7

Miguel MG (2010) Antioxidant and anti-inflammatory activities of essential oils: a short review. Molecules 15(12):9252–9287

Müller-Waldeck F, Sitzmann J, Schnitzler WH, Graßmann J (2010). Determination of toxic perilla ketone, secondary plant metabolites and antioxidative capacity in five Perilla frutescens L. varieties. Food Chem Toxicol 48(1):264–270

Nahed A, Abd El-Hack ME, Albaqami NM, Khafaga AF, Taha AE, Swelum AA, Elbestawy AR (2022) Phytochemical control of poultry coccidiosis: a review. Poult Sci 101(1):101542

Nakatani N (2000) Phenolic antioxidants from herbs and spices. Biofactors 13(1–4):141–146

Nazzaro F, Fratianni F, De Martino L, Coppola R, De Feo V (2013) Effect of essential oils on pathogenic bacteria. Pharmaceuticals 6(12):1451–1474

Nobakht A, Norani J, Safamehr A (2011) The effects of different amounts of *Mentha pulegium* L (pennyroyal) on performance, carcass traits, hematological and blood biochemical parameters of broilers. J Med Plant Res 5(16):3763–3768

Nowakowska Z (2007) A review of anti-infective and anti-inflammatory chalcones. Eur J Med Chem 42(2):125–137

O'sullivan L, Woods JA, O'Brien NM (2002) Squalene but not n-3 fatty acids protect against hydrogen peroxide-induced sister chromatid exchanges in Chinese hamster V79 cells. Nutr Res 22(7):847–857

Okamura N, Haraguchi H, Hashimoto K, Yagi A (1994) Flavonoids in *Rosmarinusofficinalis* leaves. Phytochemistry 37(5):1463–1466

Olson EG, Micciche AC, Rothrock MJ Jr, Yang Y, Ricke SC (2021) Application of bacteriophages to limit campylobacter in poultry production. Front Microbiol 12:458721

Pandey AK, Kumar P, Saxena MJ (2019) Feed additives in animal health. In: Nutraceuticals in veterinary medicine. Springer, Cham, pp 345–362

Paymard J, Nobakht A, Mazlum F, Moghaddam M (2013) The effects of different levels of dried aerial parts powder and extract of pennyroyal (*Mentha pulegium*) medicinal plant on performance, egg quality, blood biochemical and immunity parameters of laying hens. Iran J Appl Anim Sci 3(3):589–594

Peng Y, Yuan J, Liu F, Ye J (2005) Determination of active components in rosemary by capillary electrophoresis with electrochemical detection. J Pharm Biomed Anal 39(3–4):431–437

Pierce A, Kathe W (2004) Practices (GACP) for medicinal plants. Medicinal Plant Conserv 56:1

Pirgozliev V, Rose SP, Ivanova S (2019) Feed additives in poultry nutrition. Bulgarian J Agr Sci 25(1):8–11

Placha I, Chrastinova L, Laukova A, Cobanova K, Takacova J, Strompfova V, Faix S (2013) Effect of thyme oil on small intestine integrity and antioxidant status, phagocytic activity and gastrointestinal microbiota in rabbits. Acta VeterinariaHungarica 61(2):197–208

Pliego AB, Tavakoli M, Khusro A, Seidavi A, Elghandour MM, Salem AZ, Rene Rivas-Caceres R (2022) Beneficial and adverse effects of medicinal plants as feed supplements in poultry nutrition: a review. Anim Biotechnol 33(2):369–391

Qiao Y, Kyselov O (2020) The immune regulation mechanism of *Astragalus* polysaccharide and its application in poultry industry

Qiao Y, Kyselov O (2021) Using *Astragalus* polysaccharide herbal supplement to improve disease resistance in poultry

Qiao H, Zhang L, Shi H, Song Y, Bian C (2018) *Astragalus* affects fecal microbial composition of young hens as determined by 16S rRNA sequencing AMB. Express 8(1):1–10

Qureshi AA, Mangels WR, Din ZZ, Elson CE (1988) Inhibition of hepatic mevalonate biosynthesis by the monoterpene, d-limonene. J Agric Food Chem 36(6):1220–1224

Radha K, Thomas A, Sathian CT (2014) Application of nano technology in dairy industry: prospects and challenges-a review. Indian J Dairy Sci 67(5):367–374

Ramesh K, Singh V, Megeji NW (2006) Cultivation of stevia (*Stevia rebaudiana*): a comprehensive review. Adv Agron 89:137–177

Rani A, Singh R, Singh C, Singh M, Lakhera K (2018) Role of nanotechnology in food sector. Biosci Res Bull 34(1):33–35

Rastogi MK, Shukla SK (2013) Challenges and impact of disinvestment on Indian economy. Int J Manag Bus Stud 3(4):44–53

Regulation OJEU (2003) No 1831/2003 of the European Parliament and Council of 22 September 2003 on additives for use in animal nutrition. Official J Eur Commun 268:29–43

Richheimer SL, Bernart MW, King GA, Kent MC, Beiley DT (1996) Antioxidant activity of lipid-soluble phenolic diterpenes from rosemary. J Am Oil Chem Soc 73(4):507–514

Ronco AL, De Stéfani E (2013) Squalene: a multi-task link in the crossroads of cancer and aging. Funct Foods Health Disease 3(12):462–476

Rosenblat M, Belinky P, Vaya J, Levy R, Hayek T, Coleman R, Aviram M (1999) Macrophage enrichment with the isoflavone glabridin inhibits NADPH oxidase-induced cell-mediated oxidation of low-density lipoprotein: a possible role for protein kinase C. J Biol Chem 274(20): 13790–13799

Ruberto G, Baratta MT, Sari M, Kaâbeche M (2002) Chemical composition and antioxidant activity of essential oils from *Algerian Origanum glandulosum*. Desf Flavour Fragr J 17(4):251–254

Ryan E, Galvin K, O'Connor TP, Maguire AR, O'Brien NM (2007) Phytosterol, squalene, tocopherol content and fatty acid profile of selected seeds, grains, and legumes. Plant Foods Hum Nutr 62(3):85–91

Ryu BK, Ahn BO, Oh TY, Kim SH, Kim WB, Lee EB (1998) Studies on protective effect of DA-9601, *Artemisia asiatica* extract, on acetaminophen-and CCI4-induced liver damage in rats. Arch Pharm Res 21(5):508–513

Salarmoini M, Salajegheh A, Salajegheh MH, Afsharmanesh M (2019) The effect of lavender (*Lavandula angustifolia*) extract in comparison to antibiotic on growth performance, intestinal morphology, ileal microflora, antioxidant status and meat quality of broilers. Iran J Appl Anim Sci 9(4):717–725

Saraei MA, Seidavi A, Dadashbeiki M, Edens FW (2016) Response of plasma constituents and body measurement in broiler chickens fed fish oil and green tea powder. Arch Med Vet 48(1): 61–68

Scherer R, Junior SB, De Albuquerque R, Godoy HT (2014) Microencapsulated eucalyptol and eugenol as growth promoters in broilers. Braz J Food Res 5:26–32

Seidavi A, Asadpour L, Dadashbeiki M, Payan-Carreira R (2014) Effects of dietary fish oil and green tea powder supplementation on broiler chickens' immunity. Acta Sci Vet 42(1):1–13

Seidavi A, Belali M, Elghandour MM, Adegbeye MJ, Salem AZM (2020) Potential impacts of dietary inclusion of green tea (*Camellia sinensis* L) in poultry feeding: a review. Agrofor Syst 94(4):1161–1170

Seidavi A, Tavakoli M, Slozhenkina M, Gorlov I, Hashem NM, Asroosh F, Swelum AA (2021) The use of some plant-derived products as effective alternatives to antibiotic growth promoters in organic poultry production: a review. Environ Sci Pollut Res 28(35):47856–47868

Seigler DS (1998) Phenylpropanoids. In: Seigler DS (ed) Plant secondary metabolism. Springer, Boston, MA, pp 106–129

Shan C, Sun B, Dalloul RA, Zhai Z, Sun P, Li M, Luan W (2019) Effect of the oral administration of astragalus polysaccharides on jejunum mucosal immunity in chickens vaccinated against Newcastle disease. Microb Pathog 135:103621

Shehata AA, Yalçın S, Latorre JD, Basiouni S, Attia YA, Abd El-Wahab A, Tellez-Isaias G (2022) Probiotics, prebiotics, and phytogenic substances for optimizing gut health in poultry. Microorganisms 10(2):395

Shokrzadeh M, Saravi SS (2010) The chemistry, pharmacology and clinical properties of *Sambucus ebulus*: a review. J Med Plant Res 4(2):095–103

Singh P (2016) Use of nano feed additives in livestock feeding. Int J Livest Res 6(1):1–14

Singh J, Yadav AN (eds) (2020) Natural bioactive products in sustainable agriculture. Springer Nature, Singapore

Somjen D, Knoll E, Vaya J, Stern N, Tamir S (2004) Estrogen-like activity of licorice root constituents: glabridin and glabrene, in vascular tissues in vitro and in vivo. J Steroid Biochem Mol Biol 91(3):147–155

Stevanović ZD, Bošnjak-Neumüller J, Pajić-Lijaković I, Raj J, Vasiljević M (2018) Essential oils as feed additives: future perspectives. Molecules 23(7):1717

Swain PS, Prusty S, Rao SBN, Rajendran D, Patra AK (2021) Essential nano minerals and other nanomaterials in poultry nutrition and production. In: Patra AK (ed) Advances in poultry nutrition research. IntechOpen, London

Takahashi K, Akiba Y, Nakano T, Yamaguchi T, Sato M, Sato N (2001) Effect of dietary stevia (*Stevia rebaudiana*) extract on gizzard erosion and Ulceration induced by dietary histamine in broiler chicks. J Poult Sci 38(2):181–184

Tasinov O, Dincheva I, Badjakov I, Kiselova-Kaneva Y, Galunska B, Nogueiras R, Ivanova D (2021) Phytochemical composition, anti-inflammatory and ER stress-reducing potential of Sambucus ebulus L. fruit extract. Plants 10(11):2446

Tian J, Zeng X, Zhang S, Wang Y, Zhang P, Lü A, Peng X (2014) Regional variation in components and antioxidant and antifungal activities of *Perilla frutescens* essential oils in China. Ind Crop Prod 59:69–79

Tominaga Y, Mae T, Kitano M, Sakamoto Y, Ikematsu H, Nakagawa K (2006) Licorice flavonoid oil effects body weight loss by reduction of body fat mass in overweight subjects. J Health Sci 52(6):672–683

Tufarelli V, Baghban-Kanani P, Azimi-Youvalari S, Hosseintabar-Ghasemabad B, Slozhenkina M, Gorlov I, Laudadio V (2021) Effects of horsetail (*Equisetum arvense*) and spirulina (*Spirulina platensis*) dietary supplementation on laying hens productivity and oxidative status. Animals 11(2):335

Upadhaya SD, Kim IH (2017) Efficacy of phytogenic feed additive on performance, production and health status of monogastric animals–a review. Ann Anim Sci 17(4):929–948

Wan XL, Niu Y, Zheng XC, Huang Q, Su WP, Zhang JF, Wang T (2016) Antioxidant capacities of *Artemisia annua* L leaves and enzymatically treated *Artemisia annua* L in vitro and in broilers. Anim Feed Sci Technol 221:27–34

Wan X, Jiang L, Zhong H, Lu Y, Zhang L, Wang T (2017) Effects of enzymatically treated *Artemisia annua* L on growth performance and some blood parameters of broilers exposed to heat stress. Anim Sci J 88(8):1239–1246

Wan X, Ahmad H, Zhang L, Wang Z, Wang T (2018) Dietary enzymatically treated *Artemisia annua* L improves meat quality, antioxidant capacity and energy status of breast muscle in heat-stressed broilers. J Sci Food Agric 98(10):3715–3721

Wen HW, DeCory TR, Borejsza-Wysocki W, Durst RA (2006) Investigation of Neutr Avidin-tagged liposomal nanovesicles as universal detection reagents for bioanalytical assays. Talanta 68(4):1264–1272

WHO (World Health Organization) (2006) WHO monograph on good agricultural and collection practices (GACP) for *Artemisia annua* L. World Health Organization, Geneva

WHO (World Health Organization) (2018) The state of food security and nutrition in the world 2018: building climate resilience for food security and nutrition. Food Agriculture Org, Geneva

Windisch W, Schedle K, Plitzner C, Kroismayr A (2008) Use of phytogenic products as feed additives for swine and poultry. J Anim Sci 86(suppl_14):E140–E148

Wu S (2018) Effect of dietary *Astragalus membranaceus* polysaccharide on the growth performance and immunity of juvenile broilers. Poult Sci 97(10):3489–3493

Xi N, Kang J, Hao L, Li R, Bao Y, Shi W (2014) Effects of ultrafine powder of the stem and leaf of Astragalus on immunity in chickens. Ital J Anim Sci 13(1):3022

Yalcin S, Eser H, Onbaşilar İ, Yalcin S (2020) Effects of dried thyme (*Thymus vulgaris* L) leaves on performance, some egg quality traits and immunity in laying hens. Ank Univ Vet Fak Derg 67(3):303

Yamamoto H, Ogawa T (2002) Antimicrobial activity of perilla seed polyphenols against oral pathogenic bacteria. Biosci Biotechnol Biochem 66(4):921–924

Yitbarek MB (2015) Phytogenics as feed additives in poultry production: a review. Int J Extensive Res 3:49–60

Yoshino K, Higashi N, Koga K (2006) Antioxidant and anti-inflammatory activities of oregano extract. J Health Sci 52(2):169–173

Youssef IM, Männer K, Zentek J (2021) Effect of essential oils or saponins alone or in combination on productive performance, intestinal morphology and digestive enzymes' activity of broiler chickens. J Anim Physiol Anim Nutr 105(1):99–107

Zaker-Esteghamati H, Seidavi A, Bouyeh M (2020) A review on the effect of Silybum marianum and its derivatives on broilers under healthy and aflatoxicosis conditions: part 1: performance, carcass and meat characteristics, and intestinal microflora. Worlds Poult Sci J 76(2):318–327

Żbikowska K, Michalczuk M, Dolka B (2020) The use of bacteriophages in the poultry industry. Animals 10(5):872

Zhang X, Zhou T, Guo L, Huang L, Jiang W, Yang Z, Ma C (2011) Volatile oil contents correlate with geographical distribution patterns of the miao ethnic herb fructus cinnamomic. Shengtai Xuebao/Acta Ecologica Sinica 31(18):5299–5306

Index

G. Arsenos, I. Giannenas (eds.), *Sustainable Use of Feed Additives in Livestock*,
https://doi.org/10.1007/978-3-031-42855-5

The manufacturer's authorised representative in the EU is Springer Nature Customer Service Centre GmbH, Europaplatz 3, 69115 Heidelberg, Germany. If you have any concerns regarding our products, please contact ProductSafety@springernature.com

Printed and bound by CPI Group (UK) Ltd, Croydon, CR0 4YY
01/10/2026
02225866-0020